MW01627236

Sensors, Actuators, and their Interfaces

Sensors, Actuators, and their Interfaces

A Multidisciplinary Introduction

Nathan Ida
University of Akron

Edison, NJ
scitechpub.com

Published by SciTech Publishing, an imprint of the IET.
www.scitechpub.com
www.theiet.org

Editor: Dudley R. Kay
Cover Design: Brent Beckley

10 9 8 7 6 5 4 3 2 1

ISBN 978-1-61353-006-1 (hardback)
ISBN 978-1-61353-195-2 (PDF)

Typeset in India by MPS Limited
Printed in the USA by Sheridan Books, Inc.
Printed in the UK by Hobbs The Printers Ltd

Contents

Preface with Publisher's Acknowledgements

The book in front of you introduces the subjects of sensing and actuation. At first it would seem that nothing could be easier; we may think we know what a sensor is, and certainly we know what an actuator is. But do we know it so well that it actually escapes us? There are literally *thousands* of devices all around us that qualify in either category. In Chapter 1 there is an example that lists many of the sensors and actuators just in a car. The count is approximately 200, and this is merely a partial list! The approach adopted here is to view all devices as belonging to three categories: sensors, actuators, and processors (interfaces). Sensors are the devices that provide *input* to systems and actuators are those devices that serve as *outputs*. In between, linking, interfacing, processing, and driving are the processors. In other words, the view advocated in this text is one of general sensing and actuation. In that sense, a switch on the wall is a sensor (a force sensor), and the light bulb that turns on as a result is an actuator (it does something). In between there is a "processor" – the wiring harness or, in case a dimmer is used, an actual electronic circuit – that interprets the input data and does something with it. In this case it may be no more than a wiring harness, but in other cases it may be a microprocessor or an entire system of computers.

CHALLENGES

The process of sensing and actuation permeates the whole spectrum of science and engineering. The principles involved in sensing and actuation derive from all corners of our knowledge, sometimes from corners obscure to all but the most specialized experts. And the principles are mixed. It is not unusual for a sensor to span two or more disciplines. Take, for example, an infrared sensor. It may be built in different ways, but one method is to measure the temperature rise produced by the infrared radiation. Thus, a number of semiconductor thermocouples are built, and their temperature relative to a reference temperature is measured. Not a particularly complex sensor but, to fully understand it, one would have to resort to theories of heat transfer, optics, and semiconductors, at the very least. In addition, one must at least consider the electronics needed to make it work and the interfacing with a controller, such as a microprocessor. It would therefore be difficult, nay, impossible, to cover all principles and all theories in any detail. Necessarily then, we must limit ourselves to a pragmatic approach with "high-level" detail and, at times, to limited explanations. It is not reasonable to assume that one can be proficient in all areas of science and, fortunately, for successful application of sensors and actuators, this "high-level" approach is sufficient. That is, we will often view a device as a "black box" with its inputs and outputs and operate on these rather than concern ourselves with the physics and the detailed operation of the "black box." Nevertheless, ignoring the box entirely is detrimental: the user must understand in sufficient detail the principles involved, the materials used, and the construction of sensors and actuators. The textbook is sufficiently detailed to allow reasonable understanding of principles.

To bridge the divide between the theory of sensors and actuators and their application and to gain insight into design of sensing and actuation, we note that most sensors have electrical output, whereas most actuators have electrical input. Virtually all interfacing issues in sensors and actuators are electrical in nature. This means that to understand and use these devices, and certainly to interface them and integrate them into a system, requires elements of electrical engineering. Conversely, the sensed quantities pertain to all aspects of engineering, and electrical engineers will find that mechanical, biological, and chemical engineering issues must be considered alongside electrical engineering issues. This multidisciplinary textbook is intended for all engineers and all those interested in sensing and actuation. Each discipline will find components in it that are familiar and others that need to be learned. This is, in fact, the lot of the present-day engineer, an engineer who must either assimilate various disciplines or work in teams to accomplish tasks across disciplines. However, not all sensors or actuators are "electrical." Some have nothing to do with electricity. A meat thermometer will sense the temperature (sensor) and display it (actuator) without any electrical signal being involved. The expansion of a bi-metal piece activates a dial against a spring, so the whole process is mechanical. Similarly, a vacuum motor in a car can open an air conditioning vent by entirely mechanical means.

MULTIDISCIPLINARY APPROACH

In each chapter the student will find a number of examples taken from varied areas and adapted to emphasize the issues discussed. Many examples are based on actual experiments, some are based on simulations, and some deal with theoretical issues. At the end of each chapter there is a set of problems, further expanding on the content of the chapter and exploring details and applications related to the subject matter. An effort has been made to make the examples and problems realistic, applicable, and relevant whenever possible, while still keeping each problem focused and self-contained. Because of the uniqueness of the subject, that is, the multidisciplinary focus, the student is faced with using units that he or she may not be familiar with. To mitigate this, a section on units is included in Chapter 1. Some chapters containing unfamiliar units also contain a section that defines these units and the conversions between them. SI units are used throughout as a rule but, on occasion, common units (such as the PSI or the electron-volt) are also defined and used because of their widespread use.

ORGANIZATION

The textbook is divided into three main parts. The first two chapters serve as an introduction and expose the general properties and issues involved in sensing and actuation. The second part includes seven chapters, each dealing with a class of sensors. These are grouped by broad area of detection. For example, those sensors and actuators based on acoustic waves – from audio microphones to surface acoustic waves and including ultrasonic devices – are grouped together. Similarly, sensors and actuators based on temperature and heat are grouped together. This grouping scheme does not indicate any exclusivity. An optical sensor may well use thermocouples to sense, but it is classified as an optical sensor and discussed in conjunction with optical sensors. Similarly, a radiation sensor may use a semiconducting junction for sensing, but its function is to sense radiation, and therefore it is classified and discussed as such. The third part of the textbook includes the last two chapters that deal exclusively with interfacing and circuits that are needed for interfacing. Emphasis is given to the microprocessor as a general-purpose controller. In addition, a chapter on micro-electro-mechanical devices (MEMS) and smart sensors is included and placed between the second and third parts of the text.

The text includes 12 chapters. Chapter 1 is an introduction. Following a short historical perspective, we define the various terms including sensors, transducers, and actuators. Then, the issue of classification of sensors is introduced followed by a short discussion of sensing and actuation strategies and the general requirements for interfacing.

Chapter 2 discusses the performance characteristics of sensors and actuators. We discuss the transfer function, span, sensitivity and sensitivity analysis, errors, nonlinearities, as well as frequency response, accuracy, and other properties including issues of reliability, response, dynamic range, and hysteresis. At this point the discussion is general, although the examples given rely on actual sensors and actuators.

Chapters 3 through 9 look at classes of devices, starting with temperature sensors and thermal actuators in Chapter 3. The starting point is thermoresistive sensors, including metal resistance temperature detectors (RTDs), silicon resistive sensors, and thermistors. These are followed by thermoelectric sensors and actuators. We discuss metal junction and semiconductor thermocouples, as well as Peltier cells, both as sensors and as actuators. PN junction temperature sensor and thermo-mechanical devices are introduced, as are thermal actuators. One interesting aspect of temperature sensors is that in many common applications the sensor and actuator are one and the same, although this duality is not limited to thermal devices. The whole class of bimetal sensors is of this type with applications in thermostats, thermometers, and in MEMS, a topic that will be expanded upon in Chapter 10.

The important issue of optical sensing is the subject of Chapter 4. Thermal and quantum-based sensors are discussed, first through the photoconducting effect and then through silicon-based sensors that include photodiodes, transistors, and photovoltaic sensors. Photoelectric cells, photomultipliers, and CCD sensors are a second important class, followed by thermal-based optical sensors that include thermopiles, infrared sensors, pyroelectric sensors, and bolometers. Although one seldom thinks of optical actuators, these do exist, and they conclude the chapter.

In Chapter 5 we address electric and magnetic sensors and actuators. Naturally, a large number of devices fall into this class, and, as a consequence, the chapter is rather extensive. It starts with electric and capacitive devices, followed by magnetic devices. We discuss here a variety of sensors and actuators including position, proximity, and displacement sensors, as well as magnetometers, velocity, and flow sensors. The principles involved, including the Hall effect and magnetostrictive effect, are discussed side by side with more common effects. A rather extensive discussion of motors and solenoids covers many of the principles of magnetic actuation, but capacitive actuators are discussed as well.

Chapter 6 is dedicated to mechanical sensors and actuators. The classical strain gauge is featured as a generic device for sensing of forces and the related quantities of strain and stress. But it is also used in accelerometers, load cells, and pressure sensors. Accelerometers, force sensors, pressure sensors, and inertial sensors take the bulk of the chapter. Mechanical actuators are exemplified by the bourdon tube, bellows, and vacuum motors.

Chapter 7 discusses acoustic sensors and actuators. By acoustic, we mean sensors and actuators based on elastic, sound-like waves. These include microphones, hydrophones based on magnetic, capacitive, and piezoelectric principles, the classical loudspeaker, ultrasonic sensors and actuators, piezoelectric actuators, and surface acoustic wave (SAW) devices. Thus, although acoustics may imply sound waves, the frequency range here is from near zero to many GHz.

Chemical sensors and actuators are among the most common, ubiquitous, and, unfortunately, least understood devices by most engineers. For this reason, Chapter 8 discusses these in some detail. A fairly large section of the existing chemical sensors, including electrochemical and potentiometric sensors, thermochemical, optical, and mass sensors, is given. Chemical actuation is not neglected; it is much more prevalent than normally thought. Actuators include catalytic conversion, electroplating, cathodic protection, and others.

Chapter 9 introduces radiation sensors. Aside from classical ionization sensors, we take a wider view that includes nonionizing, microwave radiation as well. Here we look at reflection, transmission, and resonant sensors. Since any antenna can radiate power, it can serve as an actuator to affect specific tasks such as cauterization during surgery, low-level treatment for cancer or hypothermia, and microwave cooking and heating.

The subject of Chapter 10 is micro-electro-mechanical sensors and actuators, or MEMS, as well as smart sensors. It is somewhat different than the previous chapters in that it discusses methods of sensor production in addition to classes of sensors. However, the importance of these sensors and actuators justifies their introduction and the deviation in the usual method of presentation. Some of the methods of production are first given, followed by a number of common classes of sensors and actuators including inertial and electrostatic sensors and actuators, optical switches, valves, and others. In the context of smart sensors, we emphasize issues associated with wireless transmission, modulation, encoding, and sensors networks.

Chapters 11 and 12 form a unit on interfacing. Many of the common circuits applicable to interfacing are described in Chapter 11. These start with the operational amplifier and its many applications, followed by power amplifiers and pulse width modulation circuits for use with actuators. The A/D and D/A in their various forms, including voltage to frequency and frequency to voltage converters, follow these before we get into discussion of bridge circuits and data transmission methods. A section on excitation circuits deals with linear and switching power supplies, current and voltage references, and oscillators. The chapter ends with a discussion on noise and interference. Chapter 12 introduces the microprocessor and its role in interfacing sensors and actuators. Although the emphasis is on 8-bit microprocessors, the issues addressed are general and pertain to all microprocessors. In this last chapter we deal with the architecture, memory, and peripherals of the microprocessor, the general requirements for interfacing, and properties of signals, resolution, and errors.

LIMITATIONS

The discussion in this book focuses on sensors and actuators as independent components. There will be little discussion about systems; rather, we will talk on the sensor/actuator level, the lowest level at which these devices are useful as building blocks for the engineer and a bit lower than that into the operation and physical principles. For example, magnetic resonance imaging (MRI) is an extremely useful system for medical diagnostics and for chemical analysis relying on sensing the precession of molecules (normally hydrogen) in the body or in a solution. But the system is so very complicated and its operation so intrinsically linked with this complexity that the principle, that of precession, cannot really be utilized on a low level. Discussion of a system of this type requires discussion of ancillary issues including superconductivity, generation of uniform, high magnetic fields, interaction between DC and pulsed, high-frequency magnetic fields, and the atomic level issues of excitation and precession. All of these are interesting and important, but they are beyond the scope of this text. A different example is radar. A ubiquitous system but one that again requires many additional components to operate and to be useful although on the low level is not different than a flashlight and our eye, the flashlight sending a beam (an actuator), and the eye receiving a reflection (the sensor). We will discuss sensors based on the principle of reflection of electromagnetic waves that is identical to radar, without the need to discuss the ancillary issues of how radar is made to work.

RESOURCES

Instructors who adopt this book should check the publisher's website page for this book to see the latest instructor resources: solutions to problems, PowerPoint slides, student projects, etc. They and more will be developed as the book finds acceptance for a growing number of courses. To inquire about availability, send an e-mail to marketing@scitechpub.com.

CONCLUSION

This book has been in the works over a number of years with considerable feedback from students in electrical, mechanical, civil, chemical, and biomedical engineering, both undergraduate and graduate. The subject was taught each summer, either as a traditional course or online. Much of the text, very fittingly, was written on the TGV (Train a Grande Vitese) during a twice daily, 230 km commute between Paris and Lille, France, traveling at over 300 km/h during fall 2009, summer 2010 and summer 2011. I have made use of a variety of sources, but much of the material, including all examples, problems, circuits, and photographs arose from my own and my students' work on sensors and actuators. When experimental data is indicated, it means that the experiment has actually been carried out and the data collected specifically for, either the purpose of the given example or something very close. Simulation is an important issue in all aspects of engineering, and the present subject is no exception. For this reason, some examples and problems, especially in Chapter 11, rely on or assume simulated configurations. In examples and problems I tried to be as practical as possible without unnecessarily complicating the issues. In some cases, simplifications had to be resorted to. Nevertheless, many of the examples and problems can serve as starting points for more complex developments and, indeed, for implementation in the laboratory or extended projects.

Nathan Ida
August 2013

PUBLISHER'S ACKNOWLEDGEMENTS

Authors work very hard and long to complete textbooks, and the best textbooks result from rewriting in order to improve them. In the process of iterations, an author's best friend can be the "other pair of eyes" brought to the process by peer reviewers who want little more than to see a good, clear textbook result for them and their students. The publisher gratefully acknowledges the valuable comments of this book's manuscript by the following selfless reviewers:

Prof. Fred Lacy – Southern University, Los Angeles
Dr. Randy J. Jost – Ball Aerospace and Technology Systems and Adjunct Faculty, Colorado School of Mines
Prof. Todd J. Kaiser – Montana State University
Prof. Yinchao Chen – University of South Carolina
Prof. Ronald A. Coutu, Jr. – Air Force Institute of Technology, Ohio
Prof. Shawn Addington – Virginia Military Institute
Mr. Craig G. Rieger – ICIS Distinctive Signature Lead
Prof. Kostas S. Tsakalis – Arizona State University
Prof. Jianjian Song – Rose-Hulman Institute of Technology, Indiana

CHAPTER 1

Introduction

The senses

The five senses—vision, hearing, smell, taste, and touch—are universally recognized as the means by which humans and most animals perceive their universe. They do so through optical sensing (vision), acoustic sensing (hearing), chemical sensing (smell and taste), and mechanical (or tactile) sensing (touch). But humans, animals, and even lower level organisms rely on many other sensors as well as on actuators. Most organisms can sense heat and estimate temperature, can sense pain, and can locate a sensation on and in the body. Any stimulus on the body can be precisely located. Touching of a single hair on the body of an animal is immediately located exactly through the kinesthetic sense. If an organ is affected, the brain knows exactly where it occurred. Organisms can sense pressure and have a mechanism for balance (the inner ear in humans). Some animals such as bats can echolocate using ultrasound, while others, including humans, make use of binaural hearing to locate sounds. Still others, such as sharks and fish (as well rays and the platypus), sense variations in electric fields for location and hunting. Birds and some other animals can detect magnetic fields and use these for orientation and navigation. Pressure is one of the main mechanisms fish use to detect motion and prey in the water, and vibration sensing is critical to a spider's ability to hunt. Bees use polarized light to orient themselves, as do some species of fish. And these represent only a small selection of the sensing mechanisms used by organisms.

Organisms also have a variety of actuators to interact with their environment. In humans, the hand is an exquisite mechanical actuator capable of a surprising range of motion, but it is also a tactile sensor. The feet, as well as many muscles, allow interaction with the environment. But here as well there are other mechanisms that can be used to affect actuation. A human can use its mouth to blow away dust and can close and open its eyelids, a cat can unsheathe its claws, and a chameleon can move each eye independently and shoot its tongue to catch a fly. Other actuators allow for voice communication (vocal chords in humans) or the stunning of prey (ultrasound in dolphins, electrical shock in eels), direct mechanical impact used by some species of shrimp, and many other specialized functions.

With respect to the sensory and actuation diversity in organisms we are still very far behind and our mimicking of natural sensors and actuators is still in its infancy. It has taken the better part of 40 years to develop a working artificial heart, whereas seemingly simple organs, such as the esophagus, have not yet been developed. Where are we in comparison with the nose of a dog?

1.1 | INTRODUCTION

It would be a cliché to say that sensors are important or that they are in widespread use. And this is not because it is so often stated, but rather because it is an understatement. In fact, they are in such widespread use that we take them for granted, just as we take for granted computers or cars.

Yet, whereas most people will acknowledge their existence, sensors, and to a smaller extent actuators, are not as visible as other devices. The main reason is that they are usually integrated in larger systems and the operation of the sensor or actuator is normally not directly observable nor is it usually self-contained. That is, a sensor or actuator can rarely operate on its own, but rather it is part of a larger system that may include many sensors, actuators, and processing elements, as well as auxiliary components such as power supplies and drive mechanisms, among others. For this reason, most people only come in contact with sensors and actuators indirectly. A few examples may be useful here to demonstrate these statements.

A car may contain dozens of sensors and actuators derived from various disciplines but we almost never come into direct contact with any of them. The engine temperature sensor (often more than one) exists, but how many drivers know where it is, how it is connected, and exactly what it measures? As a matter of fact, it does not actually measure engine temperature, but rather coolant temperature in the engine. Air bags in cars save lives and reduce injury by deploying before the driver's head hits the steering wheel. This is done by one or more accelerometers (acceleration sensors) that detect the deceleration of a vehicle involved in an accident, activating an explosive charge (actuator) that fills the bag with a gas. In between the sensors and the actuator there is a processor that decides whether an accident has occurred based on set sensing parameters. Ask the driver and he or she would probably not know that accelerometers are involved and most likely will not know where these sensors are physically located. It may come as a surprise to many to find out that actuation of the airbag is through an explosion using substances that an explosives expert would find familiar. Another example is the catalytic converter in the car. This is a unique chemical actuator whose purpose is to convert otherwise noxious gases into more benign substances through the use of a number of sensors, and by doing so reducing pollution. Most people do not know where this device is or what it does and are ignorant as to its operation.

Similarly, when we change the temperature setting at home, an actuator is activated to operate the furnace (or heater) or air conditioner, and somewhere there is a sensor (or a number of sensors) to turn off the furnace or air conditioner at the set temperature. Even if we have an idea of where these sensors/actuators are, it is usually vague. Moreover, we know very little of what type of sensors/actuators are involved and even less about how they operate or how they are connected, the types of signals they use, etc. In most homes in the United States there is at least one thermostat that regulates heating and/or cooling in the house. The homeowner would be hard pressed to know where the temperature sensor is (if indeed a separate temperature sensor is used) or if a "classical" thermostat is used instead. It would probably come as a surprise to many to find out that a fairly primitive mercury switch combined with a bimetal sensor/actuator is used in many low-cost thermostats.

And how many of us have ever given much thought to the pop-up lid on almost any jar or can of food we consume? Yet it is there to detect if the jar or can is properly sealed and hence to detect possible spoiled food. It is, in effect, a pressure sensor, and perhaps one of the most common sensor we come in contact with.

All in all, it is estimated that in daily life a person comes into contact with a few hundred sensors and actuators in the home, transport, work, and entertainment, although as a rule most people are scarcely aware of them.

1.2 | A SHORT HISTORICAL NOTE

We tend to think of sensors and actuators as a product of the information age and of the rapid development of electronics associated with it. Indeed, as far as the sheer number and variety of sensors available and their sophistication, this perception is justified. However, sensors existed before electronics, before the transistor, before the vacuum tube, and even before electricity. We shall see in **Chapter 3** that some of the most common temperature sensors we use today, thermocouples, have been in use since 1826 when Antoine Cesar Becquerel used them for the first time to measure temperature. The Peltier effect, which allowed heating and refrigeration in space starting in the early 1960s and then became a fixture in portable coolers and heaters, was discovered in 1834 by Charles Athanase Peltier. The device has been used as a thermoelectric generator since at least the 1890s and later was further developed for cooling and heating purposes. Resistive temperature sensing based on the change in conductivity of metals has been in use since 1871 when William Siemens proposed its use with a platinum wire. The modern equivalent is the thermoresistive sensor (or resistive temperature detector [RTD]) and most of these use platinum wires. And there are more. The photoelectric sensor has been in widespread use since the early 1930s, as were others, including photoemissive sensors. Actuators based on thermal expansion have existed since the mid-1880s, and actuators based on electric motors have been used since the invention of the electric motor in 1824 by Michael Faraday. Air speed on modern aircraft and in wind tunnels is measured using a Pitot tube, invented in 1732 by Henri Pitot to measure water flow in rivers. And then there is the compass, a sensor critical to the development of the modern world that existed in Europe since at least 1100 CE and in China since 2400 BCE.

It should be noted that electronics had its beginning with the invention of the electronic vacuum tube. Invented in 1904 by John A. Fleming, it remained in its simple form (a diode) until 1906 when Lee De Forest developed the audion—the first electronic amplifier—a step that started the electronic age, and with it the development and use of newer sensors that were electronic in nature or that required electronics for their operation.

But before there were sensors in the way we know them today there were others that could be called "primitive" or "natural" sensors, and of course, actuators. Our five senses and the senses of animals are yet to be challenged by modern sensors in sensitivity and sophistication. The acuity of a dog's sense of smell searching for a lost person, explosives in an airport, or in some cases cancer in the body of a person, or that of a pig searching for truffles in the French or Italian countryside are yet to find their match. Similarly, a shark can detect traces of blood at distances of 30 km or more. No tactile sensor can even come close to our own skin, in which we can locate the touching of a single hair. The dexterity of the human hand is the subject of much imitation in robotics and mechatronics, but no existing tactile sensors comes close. The binaural positioning capabilities of animals is legendary. A fox can locate a mouse under thick snow by hearing alone, pouncing directly on it through the snow. Other senses are even more amazing. An elephant can both generate infrasound for long-distance communication and "hear" infrasound through its legs, which then propagate the vibration to its inner ear through its bone structure. The use of ultrasound by bats to locate insects in flight and to avoid obstacles constitutes a refined system of ultrasound generation (actuation) and sensing of an incredible resolution. Dolphins are not far behind in these capabilities, using ultrasound not only to detect (sensor) but also to communicate and to stun prey (actuator). There are also indications that animals can detect impeding earthquakes or storms. It is likely that they use their highly acute senses to detect precursors such as electric fields, variations in magnetic fields, and minute tremors to which we, and sometimes our instruments, are oblivious. Some of these exceptional capabilities are due in part to highly developed processors—the brains of these animals.

But primitive sensing is not limited to the five senses. Traveling through Eastern Europe one may still encounter the use of fish as sensors for water quality. In many a well, one can find a small fish or two, typically trout. They serve two purposes. First, the fish eat insects that have fallen into the well, helping to keep it clean. But more importantly, being sensitive to water quality, the fish will die, or at least show signs of distress, when the water is no longer safe to drink. A dead fish in a well is a clear indication not to drink the water. The same, seemingly primitive method is in use in the United States in some municipal water treatment facilities. After all the chemical tests are done, and the water is properly treated, the final "test" is a minnow left in the water overnight. If it is alive the next morning, the water is "safe." One can buy commercial water quality testing systems in which small fish are placed in the water to be tested and their breathing pattern is monitored electronically for signs of stress (primarily changes in breathing rate). This is then correlated to water quality. In earlier times, going back to at least tenth-century France, salamanders were kept in water sources for exactly the same purpose and for the same reason—their sensitivity to any change in water quality (the concern was mainly poisoning of water sources, apparently a common occurrence at the time). The canary in the coal mine is another example. It turns out that the canary is quite sensitive to methane and other noxious gases. Methane or carbon monoxide accumulation is indicated when the bird stops singing. Higher concentrations will kill it. These are clear indications of the need to evacuate the mine before an explosion occurs. And it is rather curious that canaries were used for the purpose as late as 1986. Other animals, most notably cats, were used in a similar manner. Miners also noticed that methane and carbon monoxide changed the color and intensity of their gas-burning lanterns and have used these as "sensors" for the presence of methane and carbon monoxide. We will see that the modern equivalent sensors are used in a similar but more controlled manner.

Plants have also been harnessed in our quest to improve our environment. For ages, wine producers have been relying on the simple rose bush to detect fungi that attack and can devastate their grape vines. The rose, it turns out, is much more sensitive to fungi and hence shows signs of its presence much earlier than the grape vine. To this day one can see beautiful roses growing at the edges of vineyards, serving to detect the fungus and acting as a warning to the vintner. Of course, the roses also add a welcome splash of color.

1.3 | DEFINITIONS

Sensors and actuators are unique devices. First, they come in a wide range of types that sometimes defy classification. Also, the operating principles of sensors and actuators span the whole spectrum of physical laws. They are used in all engineering disciplines and for almost all conceivable applications. It is therefore not surprising that various definitions of sensors and actuators may be found and, more importantly, that all of these definitions are more or less correct and more or less useful. For example, sensor, transducer, probe, gauge, detector, pickup, receptor, perceptron, transmitter, and transponder are often used interchangeably and sometimes incorrectly. In particular, there seems to be confusion between the terms transducer and sensor, in spite of the fact that these are very distinct terms. Similarly, the terms actuator, driver, and operating element are often used interchangeably. And often also, actuators will be called by their function or by their primary use (motor, valve, solenoid, etc.) rather than using the term actuator. Then there is the "babel" of units that again, because of the various disciplines involved, includes almost all possible combinations of units, at times with little regard to standards.

To add another dimension of uncertainty as to what a sensor or actuator is, it should be noted that sometimes the boundary between the two is blurred. There are sensors that double as actuators and devices that perform both functions. For example, a bimetallic switch is a temperature sensor that activates a switch or creates a direct contact (cooking thermometer, thermostat). Since it performs both the functions of a sensor and an actuator, it is difficult to decide what it is and the only proper definition would be to call it a sensor-actuator. In some cases even the quantity sensed is not obvious. An example is the common fuse. One might say that it senses current and disconnects the circuit. But in fact it is not the current that is the direct cause of fusing, but heat generated by the current. Therefore one might also say that the fuse senses temperature. In either case it is clearly a sensor-actuator whose stated function is current sensing.

We will try to properly define these and other useful terms before continuing, and then stick to these definitions to avoid confusion. A proper, useful definition that encompasses the array of devices we need to deal with is not easily found. Nevertheless, we will start with the dictionary, both to see what has been defined and to demonstrate the inadequacy of these definitions.

SENSOR

1. A device that responds to a physical stimulus and transmits a resulting impulse. (*Webster's New Collegiate Dictionary*, 1998)

   Problem: What is an impulse? Does every sensor "transmit" an impulse?

2. A device, such as a photoelectric cell, that receives and responds to a signal or stimulus. (*American Heritage Dictionary*, 3rd ed., 1996)

   Problem: The definition uses an example (photoelectric cell) that may not be representative of all sensors. What does "receives" mean?

3. A device that responds to a physical stimulus (as heat, light, sound, pressure, magnetism, or a particular motion) and transmits a resulting impulse (as for measurement or operating a control). (*Webster's New World Dictionary*, 3rd ed., 1999)

   Problem: What is "impulse" and why "as for measurement or operating a control"?

TRANSDUCER

1. A device that is actuated by power from one system and supplies power usually in another form to a second system. (*Webster's New Collegiate Dictionary*, 1998)

   Problem: Why "power" and is a transducer an actual physical device?

2. A substance or device, such as a piezoelectric crystal, that converts input energy of one form into output energy of another (from: trans-ducere—to transfer, to lead) (*American Heritage Dictionary*, 3rd ed., 1996)

   Problem: What is meant by "substance" and "input energy"? Is the example of the piezoelectric crystal appropriate and representative?

3. A device that is actuated by power from one system and supplies power usually in another form to a second system (a loudspeaker is a transducer that transforms electrical signals to sound energy). (*Webster's New World Dictionary*, 3rd ed., 1990)

   Problem: Is the loudspeaker a transducer or does transduction occur in the loudspeaker as part of its function?

Note: Some use transducer as a term that covers both sensors and actuators.

ACTUATOR

1. A mechanism for moving or controlling something indirectly instead of by hand. (*Webster's New Collegiate Dictionary*, 1998)

   Problem: Does it require specifically a motion? Does that mean that a direct control such as in a thermostat does not qualify as actuation?

2. One that activates, especially a device responsible for actuating a mechanical device such as one connected to a computer by a sensor link. (*American Heritage Dictionary*, 3rd ed., 1996)

   Problem: Does an actuator have to be a mechanical device (see first definition)? An example is given, but is it appropriate as a definition?

3. One that actuates; a mechanical device for moving or controlling something. (*Webster's New World Dictionary*, 3rd ed., 1990)

   Problem: Does it have to be a mechanical device? Does it have to move or control something?

These definitions (and there are others) show what the problem is: one can easily take the definition of "transducer" to mean both a sensor or an actuator and the definitions are not broad enough to represent the wide variety of sensors and actuators in existence. For example, a loudspeaker is clearly an actuator—it converts electrical energy into acoustic energy. But one can connect the same loudspeaker as an input device and use it as a microphone. Now the same device is a sensor—it senses pressure (stimulus) but it is also a transducer (the conversion of energy is from acoustic to electrical). And this duality is not limited to loudspeakers—many actuators can operate as sensors or actuators (other than, perhaps, the power levels involved—an actuator usually needs to supply more power than a sensor can generate or needs to operate and therefore a microphone is physically much smaller than a loudspeaker). So we are back to the original question: what is a sensor, what is an actuator, and what is a transducer? To add to this, some writers have taken the position that a transducer is more than a sensor; it includes a "sensing element" and an "energy conversion element" as well as auxiliary elements such as filters, signal conditioning, perhaps a power source, etc. Others have taken the exact opposite view, whereby a transducer is part of a sensor. Others simply assume that they are one and the same—a transducer is just another name for a sensor. And these views are only a small selection. What is it then? Who is right? The answer is, of course, that everybody is right simply because all of the above are true under given conditions and stemmed from the complexity and variety of sensors, actuators, and transducers and the physical laws involved as well as the construction of the devices.

To better understand these issues, consider again the loudspeaker and microphone, but now let's specify what we are talking about. First, let's look at a magnetic loudspeaker (there are other types). If we use it as a microphone, the motion of the loudspeaker cone moves a coil in a magnetic field and this generates a voltage across the coil. When connected in a circuit, a measurable current appears in the circuit. This is a passive sensor—it generates power and does not require an external power source to sense. Thus our statement that energy is converted (transducer) is correct. In fact, in principle, we could connect two loudspeakers as in **Figure 1.1**. Speaking into loudspeaker 1 results in sound being generated by loudspeaker 2 (a direct connection between a sensor and an actuator). Transduction from sound pressure to electrical voltage occurs in one loudspeaker, while transduction from electrical current to pressure waves occurs in the other loudspeaker, and the process is reversible. This is the same idea we used as children to communicate using two tin cans and a string (**Figure 1.2**). Here transduction is from sound waves to vibrations in the string and vice versa.

Usually a direct connection between sensors and actuators is not possible and we need to use a processing element—in this case an amplifier—as in **Figure 1.3**. This is typically the way sensors and actuators operate and interact.

Power and Transduction

Consider now a (simplified) telephone link that includes a carbon microphone and a loudspeaker. The carbon microphone (to be discussed in **Chapter 7**) operates on the principle of changes in resistance: acoustic power moves a membrane, which in turn

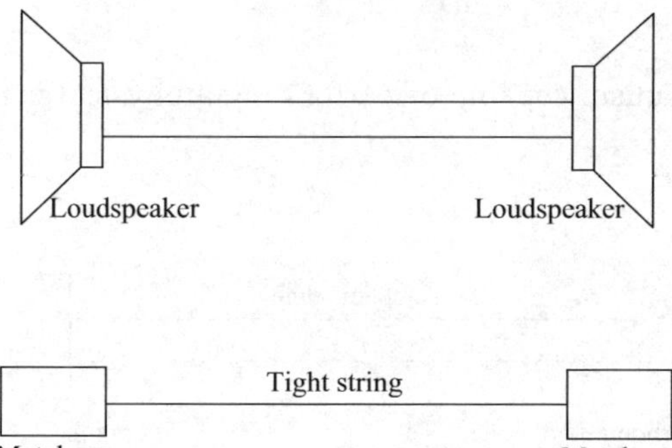

FIGURE 1.1 ■ Two loudspeakers used to demonstrate the ideas of sensing, actuation, and transduction.

Tight string

Metal can

Metal can

FIGURE 1.2 ■ Another sensor and actuator with transduction at each end.

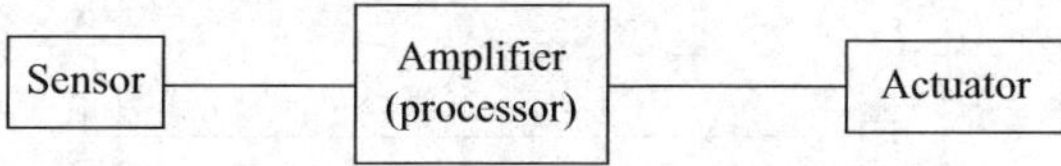

FIGURE 1.3 ■ The three elements of a sensor-actuator system. The amplifier is the "processor" or "controller" in the system.

presses on carbon particles. This changes the resistance between the two electrodes of the microphone. Suppose we again connect the microphone directly to a loudspeaker, as in **Figure 1.4**. No communication can occur since the microphone does not convert power—acoustic power is converted into changes in resistance, but not into any form of usable power, and to communicate we need power. Interchanging between the loudspeaker and microphone is of no use either. The loudspeaker does generate power, but the microphone cannot convert this power into acoustical power. Now the microphone is not a transducer, but it clearly is a sensor (active sensor). To make the system in **Figure 1.4** work, we need to add a power source as in **Figure 1.5**. Now the changes in resistance result in changes in current in the circuit that then result in changes in the position of the loudspeaker's cone and these changes result in variations in air pressure (sound waves). In this system we can view the microphone and the battery as a transducer (lending credence to the idea that the sensor is part of the transducer) or we can view them as a sensor (lending credence to the idea that the transducer is part of the sensor). We could, and perhaps should, keep them separate and view the microphone as a sensor, the sensor plus the battery as the transducer, and the loudspeaker as the actuator. By doing so we can avoid some difficulties. Specifically, in this case, since the microphone cannot serve both as a sensor and an actuator, by viewing them as separate functional elements, one is not tempted to automatically assume functional duality between them while, at the same time, not excluding it either. On the other hand, we will also have to be flexible, as in the case of the transducer—sometimes the transducer will be clearly identifiable as a separate element from the sensor, sometimes it will include the sensor.

Following this rather long introduction, the definitions we will use are as follows:

Sensor

A device that responds to a physical stimulus.

Transducer

A device or mechanism that converts power of one form into power of another form.

Actuator

A device or mechanism capable of performing a physical action or effect.

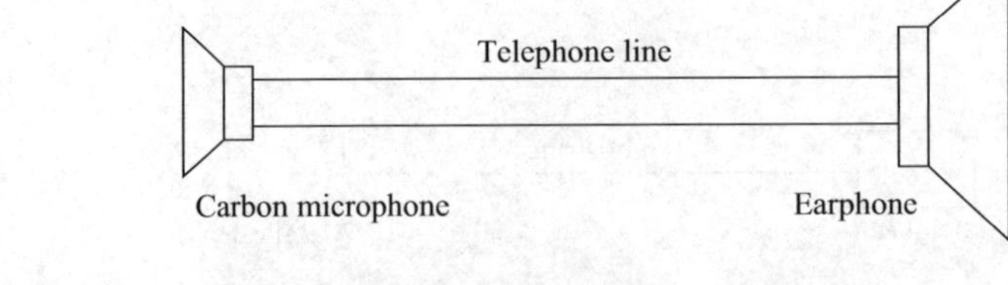

FIGURE 1.4 ■ A telephone link that cannot work. The microphone now is an active sensor and requires power for transduction.

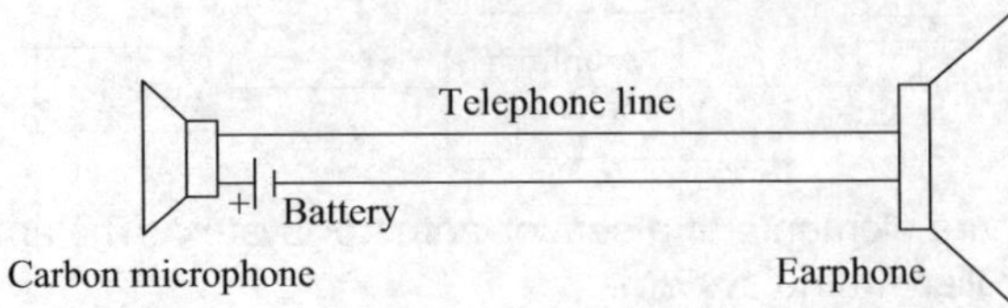

FIGURE 1.5 ■ A "proper" telephone link based on an active (carbon) microphone.

These are very general definitions and encompass all (or nearly all) available devices. Even the term "device" should be understood in the broadest sense. For example, a paper strip imbued with a glucose-sensitive substance used to test for sugar in the blood is a "device." At times we may have to narrow these definitions. For example, it is common to assume that most sensors have an electrical output or that (most) actuators perform some type of motion or involve the exertion of force. We shall make no such assumptions here, but shall do so often in subsequent chapters. In some cases the output of a sensor will indeed be electrical, but in others it can be mechanical. Similarly the physical action of an actuator may not involve force at all, such as in the case of a lightbulb used as the output in a system or a display used to monitor status. In fact, an actuator may perform a chemical action, such as the conversion of carbon monoxide (CO) into carbon dioxide (CO_2) in the catalytic converter of a car.

A more general definition is for a sensor to be the input to a system, whereas an actuator is an output. In this view, one that we will adopt to a large extent in this text, sensors of various types and complexity serve as inputs to systems, whereas actuators serve as outputs. In between there is a processor that accepts the inputs, processes the data, and acts through the actuators connected to the output of the system. In general, one may say that the processor interfaces between the sensors and actuators. This is shown in **Figure 1.6**. The sensors and actuators can be of a very general nature. A switch on the front of a washing machine is a sensor and a light-emitting diode (LED) showing that the washing machine is in operation is an actuator. It is not absolutely necessary that an actuator physically produces motion or force, but rather that it acts as an output to the system producing an effect. Many actuators are in fact mechanical and based on the use of motors. However, even a motor needs to be understood in its broadest sense, as these may be electrical (direct current [DC], alternating current [AC], continuous, stepping, linear, etc.), pneumatic, or even a micromachined electric motor. And some motors can easily serve as sensors. Indeed, a small DC motor used to sense wind speed operates as a generator and is viewed as a sensor, whereas the same motor used to run a fan is considered an actuator.

The processor or controller itself may be trivially simple or terribly complicated depending on needs. It may be as simple as a direct connection or it may be an amplifier, a set of resistors, a filter, a microprocessor, or a distributed system of computers. In extreme cases the processor is not necessary at all. In these cases the sensor acts as well as an actuator. Bimetal thermometers and thermostats are typical examples, in that the expansion of metals is a measure of temperature and that same expansion can be viewed directly on a scale or it can operate a switch.

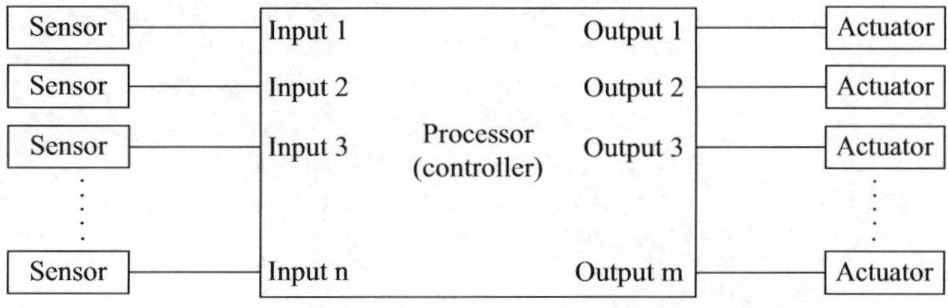

FIGURE 1.6 ■ A generic system with sensors as inputs and actuators as outputs. The processor or controller interfaces between the sensors and actuators.

EXAMPLE 1.1 Sensors and actuators in the car

A modern car contains dozens of sensors and actuators. All are connected to a processor (often called an electronic control unit [ECU]) as inputs and outputs as shown in **Figure 1.6** (sometimes multiple control units may be used, each dedicated to a set of related functions). Some of the "sensors" are switches or relays used to detect conditions (e.g., that the air conditioning is on or off, that the transmission is in gear, that the doors are closed, and many others), whereas others are true sensors. Most of the actuators are solenoids, valves, or motors, but some are indicators such as the low oil pressure lamp or an "open door" buzzer. Not all cars have the same sensors and actuators depending on the make and model. Most of the sensors and actuators in a car are monitored by an onboard diagnostics (OBD) system that gives the driver, the mechanic, and regulators an indication of the condition of the systems in the car. A partial list of sensors and actuators monitored by the OBD system is given below. In addition to those listed, many sensors are "hidden" within other components. For example, the cruise control system uses pressure sensors to maintain speed and the voltage regulator uses current and voltage sensors to keep the voltage constant, but these are not monitored directly. Similarly there are many other actuators not monitored by the OBD system, including motors and valves within other systems such as those used to open and close windows, doors, sunroofs, etc. It should also be noted that many of these sensors are "smart sensors," often containing their own microprocessors.

Sensors

Crankshaft position (CKP) sensor
Camshaft position (CMP) sensor (two)
Heated oxygen sensor (HO_2S) (two or four)
Mass air flow (MAF) sensor
Manifold absolute pressure (MAP) sensor
Intake air temperature (IAT) sensor
Engine coolant temperature (ECT) sensor
Engine oil pressure sensor
Throttle position (TP) sensor (one to four)
Fuel composition sensor (for alternative fuels)
Fuel temperature sensor (one or two)
Fuel rail pressure sensor
Engine oil temperature sensor
Turbocharger boost sensor (one or two)
Rough road sensor
Knock sensor (KS) (one or two)
Exhaust gas recirculation sensor (one or two)
Fuel tank pressure sensor
Evaporative emission control pressure sensor
Fuel level sensor (one or two)
Purge flow sensor
Exhaust pressure sensor
Vehicle speed sensor (VSS)
Cooling fan speed sensor
Transmission fluid temperature (TFT) sensor
A/C refrigerant pressure sensor
Rear vertical sensor
Horizontal sensor
A/C low-side temperature sensor
A/C evaporator temperature sensor
A/C high-side temperature sensor
A/C refrigerant overpressure
Left A/C discharge sensor
Right A/C discharge sensor
Power steering pressure (PSP) switch
Transmission range sensor
Input/turbine speed sensor
Output speed sensor
Secondary vacuum sensor
Alternative fuel gas mass sensor
Accelerator pedal position sensor (two)
Barometric pressure sensor
Cruise servo position sensor
Brake boost vacuum (BBV) sensor
Wheel speed sensor (one on each wheel)
Steering handwheel speed sensor
Left heater discharge sensor
Right heater discharge sensor
Mirror horizontal position sensor
Mirror vertical position sensor
Driver recline sensor
Driver lumbar horizontal sensor
Driver lumbar vertical sensor
Driver belt tower vertical sensor
Recline sensor
Right rear position sensor

Front vertical sensor
Lumbar forward/aft sensor
Lumbar up/down sensor
Left front mirror vertical position sensor
Right front mirror vertical position sensor
Driver front vertical sensor
Driver rear vertical sensor
Driver seat assembly horizontal sensor
Twilight photocell
Seat back heater sensor
Telescope position sensor
Tilt position sensor
Security system sensor
Automatic headlamp leveling device (AHLD) front axle sensor
AHLD rear axle sensor
Window position sensor
Evaporative emission (EVAP) system leak detector
Left front position sensor
Right front position sensor
Left rear position sensor
Right rear position sensor
Level control position sensor
Tire pressure monitor (TPM) system sensor (four)
Vehicle stability enhancement system (VSES) sensor
Yaw rate sensor
Lateral accelerometer sensor
Steering sensor
Brake fluid pressure sensor
Left front/driver side impact sensor (SIS)
Electronic front end sensor (one or two)
Outside air temperature sensor
Ambient air temperature sensor
Passenger compartment temperature sensor (one or two)
Output air temperature sensor (one or two)
Solar load sensor (one or two)
Right-hand panel discharge temperature sensor
Rear discharge temperature sensor
Discrete sensor
Evaporator inlet temperature sensor
Left-hand sun load sensor
GPS antennas, satellite antennas, radio antennas, ultrasound and accelerometers for theft prevention, etc.

Actuators

Turbocharger wastegate solenoid (two)
Exhaust gas recirculation (EGR) solenoid
Secondary air injection (AIR) solenoid
Secondary air injection switching valve (two)
Secondary air injection (AIR) pump
EVAP purge solenoid valve
Evaporative emission (EVAP) vent solenoid
Intake manifold tuning (IMT) valve solenoid
TCC enable solenoid
Torque converter clutch
Shift solenoid A
1–2 shift solenoid valve
Shift solenoid B
2–3 shift solenoid valve
Shift solenoid C
Shift solenoid D
Shift solenoid E
3–2 shift solenoid
Shift/timing solenoid
1–4 upshift (skip shift) solenoid
Line pressure control (PC) solenoid
Shift pressure control (PC) solenoid
Shift solenoid (SS) 3
Shift solenoid (SS) 4
Shift solenoid (SS) 5
Intake resonance switchover solenoid
Reverse inhibit solenoid
Pressure control (PC) solenoid
A/T solenoid
Torque converter clutch (TCC)/shift solenoid
Brake band apply solenoid
Intake manifold runner control (IMRC) solenoid
Left front ABS solenoid (two)
Right front ABS solenoid (two)
Left rear ABS solenoid (two)
Right rear ABS solenoid (two)
Left TCS solenoid (two)
Right TCS solenoid (two)
Steering assist control solenoid
Left front solenoid
Right front solenoid
Left rear solenoid
Right rear solenoid
Exhaust solenoid valve
Secondary air injection switching valve (two)
Evaporative emission system purge control valve
Exhaust pressure control valve
Intake plenum switchover valve
Exhaust gas recirculation system valve 1
Exhaust gas recirculation system valve 3
Throttle valve
Electronic brake control module (EBCM) control valve

Fuel solenoid
Cruise vent solenoid
Cruise vacuum solenoid
Right front inlet valve solenoid
Right front outlet valve solenoid
Left rear inlet valve solenoid
Left rear outlet valve solenoid
Right rear inlet valve solenoid
Right rear outlet valve solenoid
Left front TCS master cylinder isolation valve
Left front TCS prime valve
Right front TCS master cylinder isolation valve
Right front TCS prime valve
Exhaust solenoid valve short to ground (GND)
Throttle actuator control (TAC) motor
Pump motor
Mirror motor (one on each side)
Tilt/telescope motor
Level control exhaust valve
Left front inlet valve solenoid
Left front outlet valve solenoid
Front washer motor
Rear washer motor
Front wiper relay
Rear wiper relay
HVAC actuator
Coolant thermostat
Injectors (air, fuel) (one per cylinder)
Window motors
Electric door motors
Cooling fans in engine
Cooling/heating fans in compartment
Starter motor
Alternator
Catalytic converter

1.4 | CLASSIFICATION OF SENSORS AND ACTUATORS

Sensors and actuators may be classified in any number of ways. Classification can be based on the physical laws governing their operation, on their application, or on some other convenient distinction between them. There is no single method of classification general enough to include all types and therefore various classifications are used for various purposes. However, certain distinctions between classes of sensors and actuators can be useful. Starting with sensors, we can distinguish between **active** and **passive** sensors. An active sensor is a sensor that requires an external power source. Active sensors are also called parametric sensors because of the dependence of their output on changes in sensor properties (parameters). Simple examples are sensors like strain gauges (resistance changes as a function of strain), thermistors (resistance changes as a function of temperature), capacitive or inductive proximity sensors (capacitance or inductance is a function of position), and others. In all of these, the sensing function is a change in the device properties, but they can only be used after a source is connected so that an electric signal can be modulated by the respective property change. In contrast, passive sensors operate by changing one or more of their own properties to generate an electric signal. Passive sensors are sensors that do not require external power sources. These are also called **self-generating** sensors. Examples are thermoelectric sensors, solar cells, magnetic microphones, piezoelectric sensors, and many others.

Note: Some sources define active and passive sensors in exactly the opposite way.

Another distinction that can be made is between **contact** and **noncontact** sensors, a distinction that may be important in certain applications. For example, strain gauges are contact sensors, but a proximity sensor is not. However, it should be understood that the same sensor may sometimes be used in either mode (e.g., a thermistor measuring the temperature of an engine is a contact sensor, but when measuring ambient temperature in the car it is not). Sometimes we even have a choice as to how a sensor may be mounted. Other sensors can only be used in one mode. For example, a Geiger tube cannot be a contact sensor since radiation must penetrate into the tube from the outside.

Sensors are sometimes classified as **absolute** or **relative**. An absolute sensor reacts to a stimulus in reference to an absolute scale. An example is the thermistor. Its output is absolute. That is, its resistance relates to the absolute temperature. Similarly the capacitance proximity sensor is an absolute sensor—its capacitance variations are due to the physical distance to the sensing position. A relative sensor's output depends on a relative scale. For example, the output of a thermocouple depends on the temperature difference between two junctions. The sensed (measured) quantity is the temperature difference rather than absolute temperature. Another example is the pressure sensor. All pressure sensors are relative sensors. When the reference pressure is vacuum, the sensor is said to be absolute, although the very idea of vacuum is relative. A relative pressure sensor senses the pressure difference between two pressures such as, say, that in the intake manifold of an internal combustion engine and atmospheric pressure.

Most classification schemes use one or more of the "descriptors" associated with sensing. Sensors may be classified by the application, by the physical phenomena used, by the detection method, by sensor specifications, and many others. Some of the possible classifications are shown in **Table 1.1**, but it should be borne in mind that ad hoc classifications are common. For example, one may classify sensors as low or high temperature, low or high frequency, low or high accuracy, etc., when specific applications are considered. It is also common to classify sensors by materials used. Thus one can talk of semiconductor (silicon) sensors, biological sensors, and the like. Sometimes even the physical size is used as a method of classification (miniature sensors, microsensors, nanosensors, etc.). Many of these qualifications are relative and depend on the application area. A "miniature" sensor in an automobile may not be on the same scale as a miniature sensor in a laptop computer or cell phone. For example, the airbag deploying system uses accelerometers, as do cell phones to orient the display when the phone is flipped. Likely the two sensors will be on a vastly different scale in size.

The classification of actuators is somewhat different in that actuators are understood (in most cases) to generate motion, apply force (i.e., a motor in the general sense), or generate an effect. Thus some of the classification schemes rely on motion descriptors while others rely on physical laws used for activation. Therefore, in addition to the classifications in **Table 1.1**, which apply to actuators as well as to sensors, there are others, as can be seen in **Table 1.2**.

One of the main difficulties in discussing sensors is that there are so many different sensors, sensing a myriad of quantities, using various principles and physical laws that it is rather difficult to discuss them in a logical way. Often these various issues are so intertwined that some sensors even defy classification.

The approach to sensing taken in this text is to look at sensors and actuators based on the broad area of detection or actuation. This has the advantage that in a particular class of sensors, only one or a few related physical principles are used, simplifying understanding of the theory behind sensing and actuation. Thus we will discuss temperature sensors, optical sensors, magnetic sensors, chemical sensors, and so on. Each of these classes of sensors is based on a few principles at most, sometimes on a

TABLE 1.1 ■ Classifications of sensors

By area of detection	By measured output	By physical effects and laws	By specifications	By area of application	Other classifications
Electric	Resistive	Electrostrictive	Accuracy	Consumer products	Power
Magnetic	Capacitive	Electroresistive	Sensitivity	Military applications	Interfaces
Electromagnetic	Inductive	Electrochemical	Stability	Infrastructure	Structure
Acoustic	Current	Electro-optic	Response time	Energy	
Chemical	Voltage	Magnetoelectric	Hysteresis	Heat/thermal	
Optical	Resonant	Magnetocaloric	Frequency response	Manufacturing	
Thermal	Optical	Magnetostrictive	Input (stimulus) range	Transportation	
Temperature	Mechanical	Magnetoresistive	Resolution	Automotive	
Mechanical		Photoelectric	Linearity	Avionics	
Radiation		Photoelastic	Hardness	Marine	
Biological		Photomagnetic	Cost	Space	
		Photoconductive	Size	Scientific	
		Thermomagnetic	Weight		
		Thermoelastic	Construction materials		
		Thermo-optic	Operating temperature		
		Thermoelectric			

TABLE 1.2 ■ Additional classification methods for actuators

By type of motion	By power
Linear	Low-power actuators
Rotary	High-power actuators
One axis	Micropower actuators
Two axes	Miniature actuators
Three axes	Microactuators
	MEMs actuators
	Nanoactuators

single principle. However, in each class the same principles are used for a variety of physical sensing and actuating quantities. For example, an optical sensor may be used to measure light intensity, but it may also be used to measure temperature. Conversely, a temperature sensor may be used to measure light intensity, pressure, temperature, or air speed. Similarly, when we talk about magnetic sensors, the principles may be applied to sense position, distance, temperature, or pressure.

EXAMPLE 1.2 Classification of the pop-up lid used on food jars

The pop-up lid on food jars detects pressure loss in the jar. When the lid pops-up, it creates a visual indication of pressure loss inside the jar. It is therefore a sensor-actuator.

Classification as a sensor
Area of detection: mechanical sensor
Measurand: pressure
Application: consumer products
Specifications: low cost
Type: passive (it requires no power to operate)

Classification as an actuator
Area of detection: mechanical actuator
Application: consumer products
Specifications: low cost
Type: linear
Power: low

Other classifying terms may be used. For example, we might say it is a visual actuator, or that it is an embedded sensor-actuator (i.e., embedded in the production or integral with the lid as opposed to a separate, attached sensor). One can say as well that this is not a pressure sensor, but rather a "spoilage" or even biological sensor and it simply uses pressure as an indication of spoilage.

EXAMPLE 1.3 **Classification of an oxygen sensor**

Oxygen sensors are common in vehicles. All vehicles that use catalytic converters must incorporate these sensors.

Broad area of detection: chemical (or electrochemical)
Measured output: voltage
Physical law: electrochemical
Specifications: high temperature
Area of application: automotive
Power: none (the sensor is a passive sensor—it does not require external power to operate)

Thus the oxygen sensor is a high-temperature, passive, electrochemical sensor used in automotive applications whose output is a voltage and senses oxygen concentration in the exhaust flow of vehicles.

1.5 | GENERAL REQUIREMENTS FOR INTERFACING

Sensors and actuators almost never operate by themselves. They are more often parts of more complex systems and function within these larger systems. It is indeed a rare occurrence when the specifications of sensors or actuators match the needs of the system. Therefore most sensors and actuators need to be interfaced with the system in which they operate. A simple, yet very general configuration is shown in **Figure 1.7**. Here a sensor is connected to a "processor" to sense a physical property, say, temperature. An actuator, also connected to the processor, reacts to the sensed temperature in some way, such as by displaying the temperature, closing a valve, turning on a fan at a predetermined temperature, or any of a number of other possible functions. The processor is viewed here as some sort of a controller that may be a microprocessor or a simpler circuit that implements the needs of the system.

To make this example more concrete, suppose the sensor is a thermocouple and the actuator is a motor whose speed is proportional to temperature (operating a fan to cool a computer processor). As we shall see later, a thermocouple is a passive sensor, so it does not require a power source to operate. However, its output is very low—on the order of 10–50 μV/°C. The motor operates at 12 V DC, whereas the controller, which we will take to be a small microprocessor, operates at 5 V DC. Apart from the fact that we must provide power to operate the processor and the controller, as well as to program the processor, we must also provide interfacing circuits between the sensor and the microprocessor and between the microprocessor and the actuator. A possible implementation is shown in **Figure 1.8**. Here, the thermocouple is placed in contact with the computer

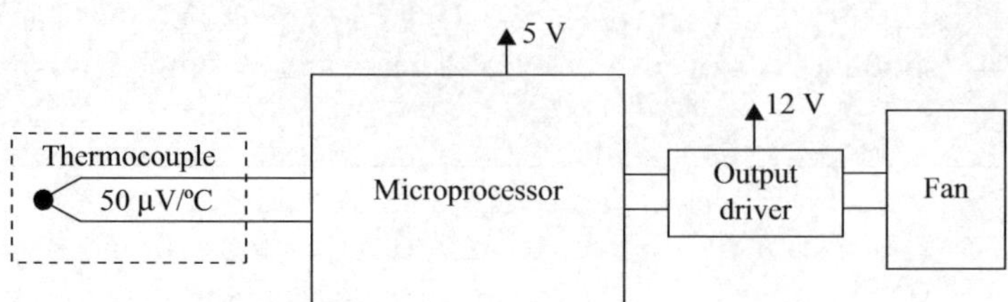

FIGURE 1.7 ■ A system comprising a sensor, a processor, and an actuator.

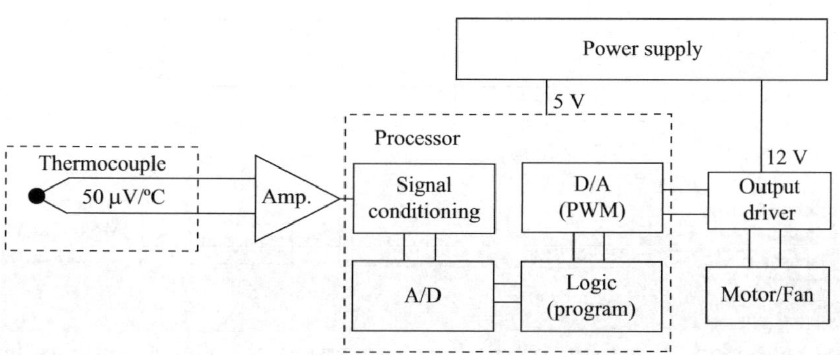

FIGURE 1.8 ■ A complete system for sensing of temperature and activation of a fan to cool a device.

processor or its heat sink to sense its temperature. Because the thermocouple measures temperature difference, a reference temperature T_0 must be available (say, the ambient temperature). The signal from the thermocouple is amplified to a more convenient signal so that the range is between zero and 5 V (5 V represents the highest input voltage of the processor and therefore should correspond to the highest temperature the system is expected to sense). This signal is an analog signal. It must be converted to a digital signal before the microprocessor can act on it. This is the function of the analog-to-digital converter (A/D or ADC), which may be internal or external to the microprocessor. The amplifier and the ADC may be viewed as forming a transducer. The microprocessor responds to this input by supplying an output signal that is proportional to the temperature. Since it is a digital device, it supplies a digital signal that must be converted back into an analog signal. The digital-to-analog converter (D/A or DAC) does that. A power amplifier is provided to indicate that a small power signal must somehow operate a higher power motor. In practice, this function may be achieved by different means, but the configuration here demonstrates the principle. The DAC together with the power amplifier form a transducer.

In addition, there may be other requirements that influence the design. For example, we may need to isolate the actuator from the microprocessor for safety or functional reasons. This is particularly true if the actuator operates at grid voltages (usually 120–480 V AC).

The need for interfacing and the method of interfacing a sensor or actuator should be taken into account in the design since it may influence the choice of sensors, actuators, and the processor. For example, temperature sensors with digital outputs are available, and if used instead of the thermocouple, the system is simplified considerably. On the other hand, a digital sensor is not, by itself, necessarily the best approach in all cases. Similarly we may choose to use a 5 V motor instead of a 12 V motor to simplify power management, if such a choice is practical. The choice of processor is in itself influenced by the sensor and actuator. Some microprocessors include ADCs internally and some have proportional outputs ideally suited to drive power devices (pulse width modulation [PWM] modules), allowing the elimination of the DAC and power amplifier and replacing these with a single transistor. **Figure 1.9** shows a different implementation using some of these alternative options and a semiconductor temperature sensor. This is much simpler because many of the requirements of interfacing have been integrated into the sensor and the microprocessor. Of course, there are consequences to any alternative design. The configuration in **Figure 1.9** can operate at up to about 100°C (the upper temperature limit of semiconducting temperature sensors is less than 125°C), whereas thermocouples can operate at well over 2000°C.

Interfacing of any device depends on the specifications of the device and the requirements of the system to which the device is interfaced, but in almost all cases this

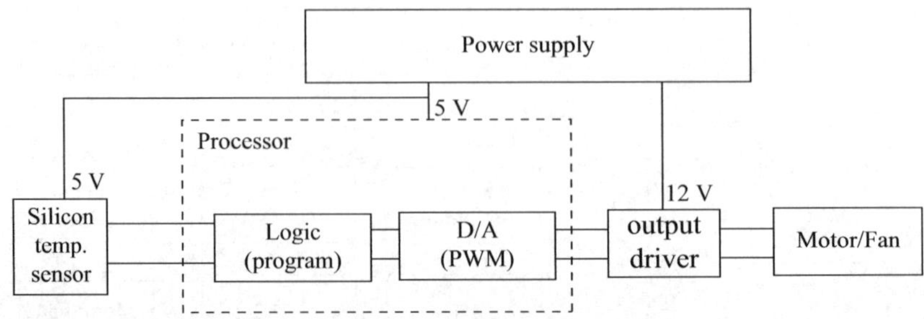

FIGURE 1.9 ■ An alternative design to the temperature controller in **Figures 1.7** and **1.8**.

involves conversion of one sort or another. Conversion of voltages, currents, and impedances are very common, but sometimes interfacing may involve conversions in other parameters, such as frequency. These conversions take place in the "transduction" section of the system and may involve multiple steps. While it may look simple enough in principle, the actual implementation in an interfacing circuit may be very complex. For example, a piezoelectric sensor may generate a few hundred volts that will instantly destroy a microprocessor. On the other hand, the impedance of the sensor is practically infinite, whereas the input impedance of the microprocessor may be much lower, severely loading the sensor and, at best, influencing its properties (sensitivity, output, linearity, etc.) or, at worst, rendering it useless. Thus we need to both reduce the voltage from a few hundred volts to about 5 V and match the impedance of the sensor to that of the microprocessor. Other sensors have totally different properties and requirements. A magnetic sensor usually includes a coil that has very low impedance, so now we have exactly the opposite problem.

For all of these reasons, interfacing circuits vary from one application to another and cover the whole spectrum of electronic circuits. Many of these will be discussed in **Chapters 11** and **12**.

1.6 UNITS

The system of units adopted throughout this book is the Systeme Internationale (SI) system. However, in the literature on sensors and in practical design there is a tendency to use mixed units. This is a result of the multidisciplinary aspects of sensors and the fact that these units have evolved over time in different engineering areas. Because of this, whenever non-SI units are used, they are usually mentioned together with the appropriate SI units. This is particularly noticeable when we discuss pressure sensors, where pounds per square inch (psi) is the common unit in the United States, whereas the SI unit is the pascal (Pa). Similarly we will occasionally mention units such as bar, rem, curie, electron-volt, and the like, all non-SI units, but we will keep these occurrences to the minimum necessary. A short discussion of relevant units and conversion between them is supplied at the beginning of the chapters in which they are relevant.

1.6.1 Base SI Units

The SI units are defined by the International Committee for Weights and Measures and includes seven base units as shown in **Table 1.3**. The base units are defined as follows:

Length. The **meter** (m) is the distance traveled by light in a vacuum during a time interval equal to 1/299,792,458 s.

TABLE 1.3 ■ The base SI units

Physical quantity	Unit	Symbol
Length	meter	m
Mass	kilogram	kg
Time	second	s
Electric current	ampere	A
Temperature	kelvin	K
Luminous intensity	candela	cd
Amount of substance	mole	mol

Mass. The **kilogram** (kg) is the prototype kilogram, a body made of a platinum-iridium compound and preserved in a vault in Sevres, France.

Time. The **second** (s) is the duration of 9,192,631,770 periods of the radiation corresponding to the transition between the two hyperfine levels of the ground state of the cesium-133 atom.

Electric current. The **ampere** (A) is the constant current that, if maintained in two straight conductors of infinite length and of negligible circular cross section, placed 1 m apart in a vacuum, produces between the conductors a force of 2×10^{-7} newtons per meter (N/m).

Temperature. The **kelvin** (K) unit of thermodynamic temperature is 1/273.16 of the thermodynamic temperature of the triple point of water (the temperature and pressure at which ice, water, and water vapor are in thermodynamic equilibrium). The triple point of water is 273.16 K at 611.73 Pa.

Luminous intensity. The **candela** (cd) is the luminous intensity in a given direction of a source that emits monochromatic radiation of frequency 540×10^{12} Hz and has a radiation intensity in that direction of 1/683 watts per steradian (W/sr) (see **Section 1.6.3** below).

Amount of substance. The **mole** (mol) is the amount of substance of a system that contains as many elementary entities as there are atoms in 0.012 kg of carbon-12. (The entities may be atoms, molecules, ions, electrons, or any other particles.) The accepted number of entities (i.e., molecules) is known as Avogadro's number and equals approximately 6.02214×10^{23}.

1.6.2 Derived Units

Most other metric units in common use are derived from the base units. We will discuss some of these in the following chapters as it becomes necessary, but it is useful to note here that these units have been defined for convenience based on some physical law, even though they can be expressed directly in the base units. For example, the unit of force is the newton (N). This is derived from Newton's law of force as $F = ma$. The unit of mass is the kilogram and the unit of acceleration is meters per second squared (m/s^2). Thus the newton is in fact kilogram meters per second squared (kg·m/s^2).

$$\text{N} = (mass \cdot acceleration) = \left(\frac{\text{kg} \cdot \text{m}}{\text{s}^2}\right)$$

Similarly the unit of electric potential is the volt (V). The derived unit starts from the definition of electric field intensity in terms of force F and charge q (Coulomb's law):

$E = F/q$, whose units are newtons/coulomb (N/C). The coulomb is the unit of charge and has units of ampere seconds (A·s). Now, 1 N/C = 1 V/m, a result that can be seen directly from Faraday's law. Therefore,

$$\mathrm{V} = \left(\frac{\mathrm{N \cdot m}}{\mathrm{C}}\right) = \left(\frac{\mathrm{N \cdot m}}{\mathrm{A \cdot s}}\right) = \left(\frac{\mathrm{kg \cdot m^2}}{\mathrm{A \cdot s^3}}\right).$$

This also shows the value of derived units: it is hard to imagine the volt in terms of kilograms, meters, amps, and seconds, in addition to the cumbersome nature of the expression.

Therefore derived units are common and useful, but they can all be related, if need be, to the base units.

EXAMPLE 1.4 **The unit of capacitance, the farad (F)**

The farad is derived from the relation between charge and voltage: $C = Q/V$. Since Q has units of charge (coulombs [C]) it can be written in units of ampere seconds (A·s). The unit of voltage is $\mathrm{kg \cdot m^2/A \cdot s^3}$ and we have

$$\mathrm{F} = C/V = \left(\frac{\mathrm{A \cdot s}}{\mathrm{kg \cdot m^2/A \cdot s^3}}\right) = \left(\frac{\mathrm{A^2 \cdot s^4}}{\mathrm{kg \cdot m^2}}\right).$$

EXAMPLE 1.5 **The unit of energy, the joule (J)**

Energy is force integrated over distance. Therefore it has units of newton meters. The newton is calculated from $F = ma$. The newton is then $\mathrm{kg \cdot m/s^2}$. Therefore the joule is

$$\mathrm{J} = (\mathrm{N \cdot m}) = \left(\frac{\mathrm{kg \cdot m^2}}{\mathrm{s^2}}\right).$$

1.6.3 Supplementary Units

The system of units also includes so-called derived nondimensional units, also termed "supplementary units." These are the unit for the plane angle, the radian (rad), and the unit for the solid angle, the steradian (sr). The radian is defined as the planar angle at the center of a circle of radius R subtended by an arc of length R. The steradian is defined as the solid angle at the center of a sphere of radius R subtended by a section of its surface, whose area equals R^2.

1.6.4 Customary Units

In addition to the SI units there are many other units, some current, some obsolete. These are usually referred to as "customary units." They include commonly used units such as the calorie (cal) or the kilowatt-hour (kWh) and less common units (except in the United

States) such as foot, mile, gallon, psi (pounds per square inch), and many others. Some units are associated almost exclusively with particular disciplines. The units may be SI, metric (current or obsolete), or customary. These have been defined for convenience and, as with any other unit, they represent a basic quantity that is meaningful in that discipline. For example, in astronomy one finds the astronomical unit (AU), which is equal to the average distance between earth and the sun (1 AU = 149,597,870.7 km). In physics, the angstrom (Å) represents atomic dimensions (1 Å = 0.1 nm). Similarly utilitarian units are the electron volt (eV) for energy (1 eV = 1.602×10^{-19} J), the atmosphere for pressure (1 atm = 101,325 N/m^2), ppm (parts per million) for chemical quantities, and the sievert (1 sv = 1 J/kg) for dose equivalents in radiation exposure. Although we will stick almost exclusively to SI units, it is important to remember that should the need arise to use customary units, conversion values to SI units can be substituted as necessary. Additional derived and supplementary units will be introduced and discussed in the chapters where they occur.

A concise source for units, conversion tables, and definitions can be found in Wildi [1].

EXAMPLE 1.6 Conversion of customary units

Pounds per square inch (psi) is commonly used in the United States as a measure of pressure.

a. Convert psi into metric units.
b. Convert psi into base units.

Solution: The pound (lb) and inch (in.) are converted as follows:

1 lb = 0.45359237 kg (mass)

Since psi is pressure, or force/area, we must convert the pound to newtons by multiplying by the gravitational acceleration, g = 9.80665 m/s^2:

1 lbf = $0.45359237 \times 9.80665$ = 4.4822161526 N.

The inch is converted as follows:

1 inch = 0.0254 m.

a. Thus psi becomes

$$\text{psi} = \frac{\text{lbf}}{\text{inch}^2} = \frac{4.4822161526}{(0.0254)^2} = 6894.76\ \text{N/m}^2.$$

The unit N/m^2 is a derived unit called a pascal (Pa). Therefore we can also write

psi = 6894.76 Pa.

b. Since the newton is kg·m/s^2, we can write

$$\text{psi} = 6894.76\ \frac{\text{kg}}{\text{m.s}^2}.$$

EXAMPLE 1.7 Molecular mass and mass of a molecule

The mole was defined above, but it can also be explained as follows: "the mass of a mole of an element is equal to the atomic mass of the element in grams" and by extension "the mass of a mole of a molecule is equal to the atomic mass of the molecule in grams." This is different than the mass of the atom (or molecule). To see these differences, calculate the mass of a mole of iron oxide (Fe_2O_3) and the mass of a molecule of iron oxide.

Solution: A mole of any substance contains 6.0221×10^{22} entities, in this case molecules of iron oxide. Using the periodic table (see inside back cover) we first calculate the mass of a mole as follows:

A mole of Fe_2O_3 is made of 2 moles of iron (Fe) and 3 moles of oxygen (O). Looking up the atomic mass we get:

- For iron, the atomic mass is 55.847 g/mol. Thus 1 mole has a mass of 55.847 g.
- For oxygen, the atomic mass is 15.999 g/mol. Thus 1 mole has mass of 15.999 g.

The molecular mass of iron oxide is therefore

$$M_{\mathrm{mass}} = 2 \times 55.847 + 3 \times 15.999 = 159.691 \text{ g/mol}.$$

The mass of a single molecule of iron oxide is therefore

$$M_{\mathrm{molecule}} = \frac{159.691}{6.02214 \times 10^{23}} = 2.6517 \times 10^{-22} \text{ g}.$$

1.6.5 Prefixes

In conjunction with units, the SI also defines the proper prefixes that provide standard notation of very small or very large units. The prefixes allow one to express large and small numbers in a compact and universal fashion and are summarized in **Table 1.4**. Again, this is mostly a convenience, but since their use is common, it is important to use the proper notation to avoid mistakes and confusion. Some of the prefixes are commonly used, others are rare, and still others are used in specialized areas. Atto, femto, peta, and exa are rarely used, whereas prefixes such as deca, deci, and hecto are more commonly used with liquids. They can be used with any quantity, but in practice they are not. One can say 100 hHz (meaning 10,000 Hz), but that would be very unusual. On the other hand, 100 hl (meaning 10,000 liters) is appropriate and is commonly used, for example, in the wine and dairy industries.

1.6.6 Other Units and Measures

1.6.6.1 Units of Information

There are a few other measures that are in common use in designating specific quantities, some are old and some very new. Just as in the past, it became convenient to define a quantity like a "dozen" or a "gross" (1 gross = 12 dozen or 144), some new quantities became convenient with the advent of the digital age. Since digital systems

TABLE 1.4 ▪ The common prefixes used in conjunction with the SI system of units

Prefix	Symbol	Multiplier	Examples	Notes
yocto	y	10^{-24}		
zepto	z	10^{-21}		
atto	a	10^{-18}		
femto	f	10^{-15}	fs (femtosecond)	Optics, chemistry
pico	p	10^{-12}	pF (picofarad)	Electronics, optics
nano	n	10^{-9}	nH (nanohenry)	Electronics, materials,
micro	μ	10^{-6}	μm (micrometer)	Electronics, distances, weights
milli	m	10^{-3}	mm (millimeter)	Distances, chemistry, weights
centi	c	10^{-2}	cl (centiliter)	Fluids, distances
deci	d	10^{-1}	dg (decigram)	Fluids, distances, weights
deca	da	10^{1}	dag (decagram)	Fluids, distances, weights
hecto	h	10^{2}	hl (hectoliter)	Fluids, surfaces
kilo	k	10^{3}	kg (kilogram)	Fluids, distances, weights
mega	M	10^{6}	MHz (megahertz)	Electronics
giga	G	10^{9}	GW (gigawatt)	Electronics, power
tera	T	10^{12}	Tb (terabit)	Optics, electronics
peta	P	10^{15}	PHz (petahertz)	Optics
exa	E	10^{18}		
zetta	Z	10^{21}		
yotta	Y	10^{24}		

use base 2, base 8, or base 16 counting and mathematics, the decimal system is not particularly convenient as a measure. Therefore special prefixes have been devised for digital systems. The basic unit of information is the bit (a 0 or a 1). Bits are grouped into bytes, where 1 byte contains 8 bits, sometimes also called a "word." A kilobyte (kbyte or kb) is 2^{10} bytes or 1024 bytes or 8192 bits. Similarly a megabyte (mb) is 2^{20} (or 1024^2) bytes or 1,048,576 bytes (or 8,388,608 bits). Although these prefixes are confusing enough, their common usage is even more confusing, as it is common to mix digital and decimal prefixes. As an example, it is common to rate a storage device or memory board as containing, say, 100 Gb. The digital prefix should mean that the device contains 2^{30} or 1024^3 bytes, or approximately 107.4×10^9 bytes. Rather, the device contains 100×10^9 bytes. In digital notation, the device actually contains only 91.13 Gb.

1.6.6.2 The Decibel (dB) and Its Use

There are instances in which the use of the common prefixes is inconvenient at the very least. In particular, when a physical quantity spans a very large range of numbers, it is difficult to properly grasp the magnitude of the quantity. Often, too, a quantity only has meaning with respect to a reference value. Take, for example, the human eye. It can see in luminance from about 10^{-6} cd/m^2 to 10^6 cd/m^2. This is a vast range of luminance and the natural reference value is the lowest luminance the eye can detect. Another example is the Richter magnitude scale, used for earthquake "strength" identification (displacement or energy). The scale for earthquakes is understood to be from 0 to 10, but in fact it is open-ended and covers vast values from 0 to (in principle) infinity.

The use of normal scientific notation for such vast scales is inconvenient and is not particularly telling for a number of reasons. Using again the example of our eyes response to light, it is not linear, but rather logarithmic. That is, for an object to appear twice as bright, the illumination needs to be about 10 times higher. The same applies to sound and to many other quantities. Another example is in the amplification of signals. In some cases one may need low amplification or none at all. In others one may need very high amplification, such as in amplifying the signal from a microphone. In still other cases, one may need to attenuate the signals rather than amplify them. In such instances, the quantities in question are described as ratios on a logarithmic scale using the notation of decibel (dB). The basic ideas in the use of the decibel are as follows:

1. Given a quantity, divide it by the reference value for that quantity. That may be a "natural" value, such as the threshold of vision or that of hearing, or it may be a constant, agreed upon value such as 1 or 10^{-6}.
2. Take the base 10 logarithm of the ratio.
3. If the quantities involved are power related (power, power density, energy, etc.), multiply by 10:

$$p = 10\log_{10}\frac{P}{P_0} \quad [\text{dB}].$$

4. If the quantities involved are field quantities (voltage, current, force, pressure, etc.), multiply by 20:

$$v = 20\log_{10}\frac{V}{V_0} \quad [\text{dB}].$$

For example, in the case of vision, the reference value is 10^{-6} cd/m^2. A luminance of 10^{-6} cd/m^2 is therefore 0 dB. A luminance of 10^3 cd/m^2 is $10\log_{10}(10^3/10^{-6}) = 90$ dB. One can say that the human eye has a span of 120 dB (between 10^{-6} cd/m^2 to 10^6 cd/m^2).

When dealing with quantities of a specific range, the reference value can be selected to accommodate that range. For example, if one wishes to describe quantities that are typically in milliwatts (mW), the reference value is taken as 1 mW and power values are indicated in decibel milliwatts (dBm). Similarly, if one needs to deal with voltages in the microvolt (μV) range, the reference value is taken as 1 μV and the result is given in decibel microvolts (dBμV). For example, a power sensor may be said to operate in the range −30 dBm and 20 dBm. This means it can detect power from 0.001 mW and 100 mW (see **Example 1.8**). The use of a specific reference value simply places the 0 dB point at that value. As an example on the dBm scale, 0 dBm means 1 mW. On the normal scale, 0 dB means 1 W. It is therefore extremely important to indicate the scale used else confusion may occur. There are many different scales, each clearly denoted to make sure the reference value is known.

As indicated above, the use of the dB scale has certain practical advantages, accounting for its widespread use. The most important are

- A very large range reduces to a short, easily comprehensible scale: a change in power ratio of 10 corresponds to 10 dB, a change in field ratio of 10 corresponds to 20 dB.
- The logarithmic scale means that a product of ratios becomes a sum in decibels.
- In many cases, such as acoustics, the dB scale is closer to the way devices (such as loudspeakers) produce output and organs (such as the eye or the ear) perceive physical quantities such as light, power, or pressure.

EXAMPLE 1.8 Use of decibels

A power sensor for detection of cellular phone transmissions is rated for an input power range of −32 dBm to 18 dBm. Calculate the range and span of the sensor in terms of power.

Solution: The fact that the range is given in dBm means that the reference value is 1 mW. Starting with the low range value, we write

$$p = 10\log_{10}\frac{P}{1\text{ mW}} = -32\text{ dBm}.$$

Dividing both sides by 10 gives

$$\log_{10}\frac{P}{1\text{ mW}} = -3.2.$$

Now we can write

$$\frac{P}{1\text{ mW}} = 10^{-3.2} \rightarrow P = 10^{-3.2} = 0.00063\text{ mW}.$$

For the upper range, we have

$$p = 10\log_{10}\frac{P}{1\text{ mW}} = 20 \rightarrow \log_{10}\frac{P}{1\text{ mW}} = 2 \rightarrow P = 10^2 = 100\text{ mW}.$$

The range is therefore 0.00063 mW to 100 mW for a span of 99.99937 mW. Note that the span in decibels is 50 dBm.

EXAMPLE 1.9 Voltage amplification and dB

An audio amplifier is used to amplify the signal from a microphone. The peak voltage produced by the microphone is 10 μV and the amplifier is required to produce a peak output voltage of 1 V as input to a power amplifier. Calculate the amplification of the amplifier in dB.

Solution: The amplification of the amplifier is the ratio of the output voltage and input voltage:

$$a = \frac{V_{out}}{V_{in}} = \frac{1\text{ V}}{10\ \mu\text{V}} = \frac{1}{10 \times 10^{-6}} = 10^5.$$

Since this is a ratio of voltages, we write

$$a = 20\log_{10}10^5 = 100\text{ dB}.$$

The amplification is said to be 100 dB rather than saying amplification is 100,000.

REFERENCES

[1] T. Wildi, "Units and Conversion Charts," IEEE Press, New York, NY, 1991.

1.7 | PROBLEMS

Sensors and actuators—general

1.1 **Sensors and actuators in the home.** List the sensors and actuators one expects to find in an average home.

1.2 **Sensors and actuators in an appliance.** List the sensors and actuators in a washing machine. These include functional devices needed for the proper execution of the tasks and safety devices to protect the user and to protect the machine from damage.

1.3 **Identification of transducers.** The mercury thermometer is a sensor-actuator that senses temperature and indicates it on a scale based on the expansion of mercury with temperature. Identify the sensing, transduction, and actuation functions.

1.4 **Identification of sensors and actuators.** An ultrasonic sensor is often called an "ultrasonic transducer." Identify the transduction process when the device is used as a sensor and when it is used as an actuator. In your opinion, can it be a passive sensor or must it be an active sensor. Give your reasons either way.

1.5 **Passive and active sensors.** A passive sensor is one that does not require external power, whereas an active sensor requires external power. Identify which of the following sensors are active and which are passive.

a. Alcohol thermometer
b. Thermostat in a car
c. pH meter in an aquarium
d. Pressure sensor in the lid of a jar
e. Microphone in a cell phone
f. Water level sensor in a dishwasher
g. Temperature sensor in a refrigerator
h. Acceleration sensor in a car

1.6 **Identification of sensing functions.** Consider a flying insect such as a butterfly. What are the necessary sensors it must possess for it to survive as a species. List the sensing mechanisms and their purpose in survival.

1.7 **Sensing of the environment.** Sensing of the environment is an essential activity in efforts to protect the planet, its resources, and its many life forms. Consider the monitoring of a river. What properties should be monitored to ensure the health of the river?

1.8 **Classification of sensors.** An outdoor thermometer uses a bimetal strip to sense and display temperature. It is based on the fact that different metals expand at different rates in response to temperature, hence the strip bends at a rate that depends on the temperature. This property is used to move a dial that displays the temperature.

a. Is this device a sensor, an actuator or both? Explain.

b. Is the device active or passive?

c. What is (are) the transduction mechanism(s) that allow the device to function? Explain based on the information given.

1.9 Identification of sensing and transduction. In ancient times the divining rod was an accepted means for detecting water underground (the "method" consisted of a person holding a forked twig or two rods, one in each hand, and when water was detected, the twig/rods "moved" or the person experienced a distinct sensation). Suppose for a moment that this is a valid sensing mechanism. If so, what is the sensor and what is the transducer?

Classification of sensors and actuators

1.10 Classification of a sensor. A mass airflow meter in a car engine uses a hot wire in the airflow and measures the power or the current needed to keep the temperature of the wire constant at a temperature well above ambient. As the mass flow increases, the wire cools and its resistance decreases. The power needed to restore its temperature is measured. This is then correlated with the mass air flow. Classify the sensor based on the information given here.

1.11 Classification of an actuator. A small DC motor is used to drive a variable speed fan to cool the processor in a computer. Classify the motor based on the information given here.

1.12 Classification of a sensor. The oxygen sensor in a vehicle senses the oxygen concentration in the exhaust system. It is made of two electrodes with a solid electrolyte between them, requires no external power, and produces a potential as output. List all possible classifications of this sensor.

1.13 Classification of a sensor/actuator. Consider a fuse for electrical equipment. It is designed to disconnect at 2 A after a delay of 100 ms. The operation of the fuse is based on heating of a thin wire. When the temperature reaches the melting point, the wire fuses. Classify the device in all possible ways.

1.14 Classification of a thermometer. Classify the outdoor thermometer described in **Problem 1.8** as a sensor and as an actuator.

1.15 Classification of a temperature sensor. The thermocouple is a device based on the fact that the junction of two metals produces a potential difference across the junction proportional to the temperature of the junction (the Seebeck effect). Classify the thermocouple based on the information given.

Units

1.16 Derived units. Show that the unit of electrical resistance, the ohm (Ω), in base SI units is $kg{\cdot}m^2/A^2{\cdot}s^3$.

1.17 Derived units. Show that the unit of magnetic flux density, the tesla (T), in base SI units is $kg/A{\cdot}s^2$. Use the fact that magnetic force on a moving charge is the product of charge, velocity, and magnetic flux density: $F_m = qv \times B$, where q is charge, v is velocity, and B is flux density.

1.18 **Derived units.** Show that the unit of inductance, the henry (H), which is the ratio of magnetic flux and current, in base SI units is $kg \cdot m^2/A^2 \cdot s^2$. Magnetic flux is the integral of magnetic flux density over area and the magnetic flux density was defined in **Problem 1.17**.

1.19 **Derivation/conversion of units.** Write the unit of torque (N·m) in base SI units.

1.20 **Derivation/conversion of units.** Show that the unit of power, the watt (W), in base SI units is $kg \cdot m^2/s^3$.

1.21 **Conversion of units.** Although some customary units are not metric, the need sometimes arises to convert to and from these units. As an example, convert the derived unit for torque from units of newton meters (N·m) to the customary unit pound-force·foot (lbf·ft). The pound (lb) is a unit of mass equal to 0.45359237 kg, whereas the pound-force (lbf) is a unit of force, $F = mg$, where m is mass (in kilograms). The foot is equal to 12 inches or $12 \times 0.0254 = 0.3048$ m.

1.22 **Moles and mass.** Calculate the number of moles of water if the water is weighed on an accurate scale and weighs 35 grams-force.

1.23 **Mass and molecular mass.** Urea is an organic compound and has the chemical formula $(NH_2)_2CO$. Calculate the mass of one molecule and the molecular mass of the compound.

1.24 **Use of units in chemical reactions.** A gasoline engine uses 8 liters (L) of gasoline per 100 km. The combustion reaction is

$$2(C_8H_{18}) + 25(O_2) \rightarrow 16(CO_2) + 18(H_2O),$$

where the formula for gasoline is C_8H_{18}. The by-products are carbon dioxide (CO_2) and water (H_2O). Calculate the amount of carbon dioxide produced, in g/km, assuming complete combustion. The density of gasoline is 740 kg/m^3.

1.25 **Digital data units.** A storage device in a computer is rated at 1.5 Tb.

a. How many bytes and how many bits of data can be stored on the device based on the customary commercial notation?

b. How many bytes and how many bits of data should the device contain based on the digital notation?

1.26 **Digital data units.** A 256 Mb memory chip is formed in silicon using basic 8-bit structures. How many individual bits does the chip contain?

1.27 **Power loss in optical fibers and the use of dB.** An optical fiber is rated as having a loss of 4 dB/km. Given an input light power density of 10 mW/mm^2, what is the light power density at the end of a fiber 6 km long?

1.28 **Use of dB in a multistage voltage amplifier.** When high amplification is required, it is not usually possible to do so with a single amplification stage, for a number of reasons. For example, a simple transistor amplifier is typically limited to an amplification of about 100 or less. More complex amplifiers, such as the operational amplifier (discussed in **Chapter 11**) are also limited to reasonably low amplification because of bandwidth requirements. Therefore multistage amplifiers are a necessity whereby the output of one stage is fed as the input to the next stage. Suppose a high-frequency amplifier is required with an amplification (often called gain) of 120 dB. Assuming that individual amplifiers with amplification between 1 and 50 are available, what is the minimum number of amplifiers needed and what are their amplifications? Is the solution unique?

1.29 **Acoustic pressure and dB.** The human ear responds to pressures between 2×10^{-5} Pa (the threshold of hearing) and 200 Pa (the threshold of pain) (1 Pa = 1 N/m). Beyond 200 Pa hearing can be permanently impaired.

a. Calculate the span of the human ear in dB.

b. A jet engine at a short distance produces a sound pressure of 5000 Pa. An operator must wear hearing protection. What must be the minimum attenuation of the hearing protector in dB?

1.30 **The Richter magnitude scale.** The Richter magnitude scale is commonly used to characterize earthquakes and their relative levels and it can be viewed as a particular form of the logarithmic scale. The Richter scale is defined as follows:

$$R_m = \log_{10} \frac{A}{A_0},$$

where A is the maximum excursion of the seismograph (i.e., the amplitude of the seismograph) and A_0 is a reference amplitude calculated based on the distance of the seismograph from the epicenter of the earthquake. Consider a magnitude 8.0 earthquake and a magnitude 9.4 earthquake.

a. What is the ratio between the actual strengths of the two earthquakes in terms of amplitudes (often called shaking amplitude)?

b. Energy released in an earthquake relates to the amplitude as $A^{3/2}$, where A is the amplitude. What is the ratio in energy released in the two earthquakes?

CHAPTER 2

Performance Characteristics of Sensors and Actuators

Humans, sensing, and actuation

Beyond the natural senses and actuation in living organisms, sensing and actuation is almost exclusively a human activity whose ultimate purpose is to improve our lives and our interactions with the universe. Sensors and actuators are ubiquitous in our lives, whether we are aware of them or not. But beyond industrial sensors, those that produce many of the products we use, keep our transportation moving, and watch over our safety, there are two types of sensors and actuators that merit separate attention. The first class of devices includes those used to improve and sustain our health. From artificial limbs and organs to implantable devices, robot-assisted surgery, medical tests, and the manipulation of tissue and cells, this class of sensors and actuators is an important part of our health system and, indeed, life. They include systems such as X-ray imaging, magnetic resonance imaging (MRI), computed tomography (CT or CAT) scans, ultrasound scanning, and robotic surgery systems. Still others, of a perplexing variety, are used to test for every conceivable substance and condition in the body.

The second class of devices expands our knowledge of the universe around us and, hopefully, allows us to better understand the universe, our place in it, and ultimately to live in harmony with it. Sensing of the environment not only benefits us, but contributes to the environment itself and all organisms in it. Off the planet, sensors allow us to protect ourselves from radiation, the effects of solar flares, and maybe even to avoid catastrophic collisions with meteorites, but perhaps most of all, they satisfy our curiosity.

2.1 | INTRODUCTION

Aside from the functionality of a sensor or actuator, that is, its basic function, the performance characteristics of the device or system are the most important issues the engineer is faced with. If we need to sense temperature, then of course a temperature sensor is needed. But what kind of sensor, what temperature range can it sense? How "accurate" does it need to be? Is it important that it be a linear measurement, and how critical is the repeatability of the sensor? Does it need to respond quickly or can we use a slow responding sensor? Similarly, if we use a motor to position a writing head in a printer or to machine a metal piece, what performance characteristics are important in its selection? These questions and others will be addressed here. The basic properties of sensors and actuators, that is, their performance characteristics, will be defined with a view to their interface to controllers.

The characteristics of a device start with its transfer function, that is, the relation between its input and output. This includes many other properties, such as span (or range), frequency response, accuracy, repeatability, sensitivity, linearity, reliability, and resolution, among others. Of course, not all are equally important in all sensors and actuators, and often the choice of properties will depend on the application. And it is important to have the application in mind when selecting a device, since the very best performing sensor or actuator may not be the best choice for all applications.

The properties of sensors and actuators are usually given by the manufacturer and engineers can usually rely on these data. There are instances, however, in which one might wish to use a device outside its stated range or improve on one of its properties (say, linearity), or even use it for an unintended use (e.g., use a microphone as a dynamic pressure sensor or as a vibration sensor). In these cases, the engineer will need to evaluate the characteristics or, at the very least, derive its calibration curve rather than relying on the manufacturer's calibration curve. Sometimes, too, the available data may be lacking certain information, again necessitating an evaluation. In cases such as these, the engineer needs to understand what affects these properties and what can be done to control them.

2.2 | INPUT AND OUTPUT CHARACTERISTICS

Before we can properly define input and output characteristics, it is best to first define the input and output of sensors and actuators. For a sensor, the input is the stimulus or the measured quantity (measurand). The output may be any number of quantities, including voltage, current, charge, frequency, phase, or a mechanical quantity such as displacement. For an actuator, the input is usually electric (voltage or current) and the output may be electrical or mechanical (displacement, force, a dial gauge, a light indication, a display, etc.). But one should keep in mind that the input and output may be more general. They may be mechanical or even chemical. We can describe both types of devices by a transfer function that relates input and output regardless of what these quantities are. In addition, we must take into account input and output properties such as impedance, temperature, and environmental conditions in order to provide proper operating conditions for the device.

2.2.1 Transfer Function

Also called the transfer characteristic function, the input/output characteristic function or response of a device is a relationship between the output and input of the device, usually defined by some kind of mathematical equation and a descriptive curve or graphical representation in a given range of inputs and outputs. The function may be linear or nonlinear, single valued or multivalued, and may at times be very complex. It may represent a one-dimensional relation (between a single input and a single output) or may be multidimensional (between multiple inputs and one output). In simple terms, it defines the response of a sensor or actuator to a given input or set of inputs and is one of the main parameters used in design. With the exception of linear transfer functions, it is usually difficult to describe the transfer function mathematically, although we can indicate it at least symbolically as

$$S = f(x), \tag{2.1}$$

where x is the input (stimulus in sensors or, say, current to an actuator) and S is the output. The dependence of the output S on x indicates that this function can be (and often is) nonlinear.

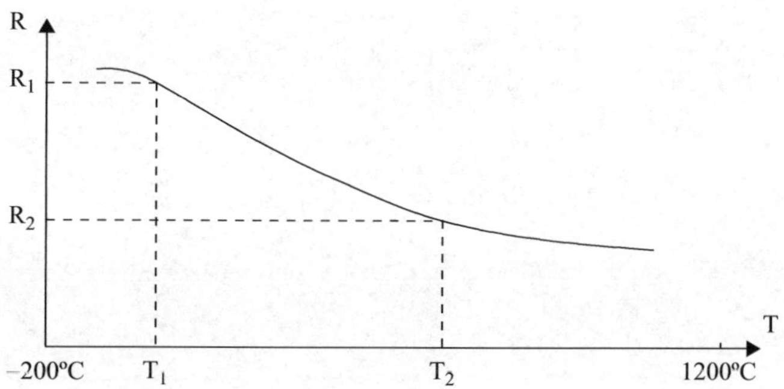

FIGURE 2.1 ■ Resistance-temperature relationship in a hypothetical temperature sensor.

Often the transfer function will be given graphically and will be limited to a range of inputs and outputs. **Figure 2.1** shows the input-output relationship for a hypothetical temperature sensor. The range between T_1 and T_2 is approximately linear and may be described by the following transfer function:

$$aT + b = R, \tag{2.2}$$

where R is the resistance of the sensor (output) and T is the temperature it senses (input) in the range $T_1 < T < T_2$.

However, the ranges below T_1 and above T_2 are nonlinear and require much more complex transfer functions, which may actually be found experimentally or may be polynomials derived by curve fitting. In many cases the sensor is restricted to operate in the linear range, in which case the graphical representation is sufficient. The curve in **Figure 2.1** contains additional data about the sensor, such as saturation, sensitivity, and range, which will be discussed in subsequent sections. It is rather rare that a transfer function is available other than in the form of a general curve or some statement as to its shape (linear, quadratic, etc.). Often it is necessary to derive it experimentally through a calibration process. There are exceptions however. Thermocouple transfer functions are available as high-order polynomials giving the transfer functions in very accurate form, as can be seen in **Example 2.1**.

EXAMPLE 2.1 Transfer function of a thermocouple

The output (voltage) of a thermocouple (temperature sensor) for a given temperature is given by a polynomial that can range from a 3rd order to a 12th order polynomial depending on the type of thermocouple. The output of a particular type of thermocouple is given by the following relation in the range 0°C–1820°C:

$$\begin{aligned} V = (&-2.4674601620 \times 10^{-1} \times T + 5.9102111169 \times 10^{-3} \times T^2 \\ &- 1.4307123430 \times 10^{-6} \times T^3 + 2.1509149750 \times 10^{-9} \times T^4 \\ &- 3.1757800720 \times 10^{-12} \times T^5 + 2.4010367459 \times 10^{-15} \times T^6 \\ &- 9.0928148159 \times 10^{-19} \times T^7 + 1.3299505137 \times 10^{-22} \times T^8) \times 10^{-3}\ \text{mV}. \end{aligned}$$

This is a rather involved transfer function (most sensors will have a much simpler response) and is nonlinear. The main purpose of the elaborate function is to provide very accurate representation over the range of the sensor (in this case 0°C–1820°C). The transfer function is shown in **Figure 2.2**.

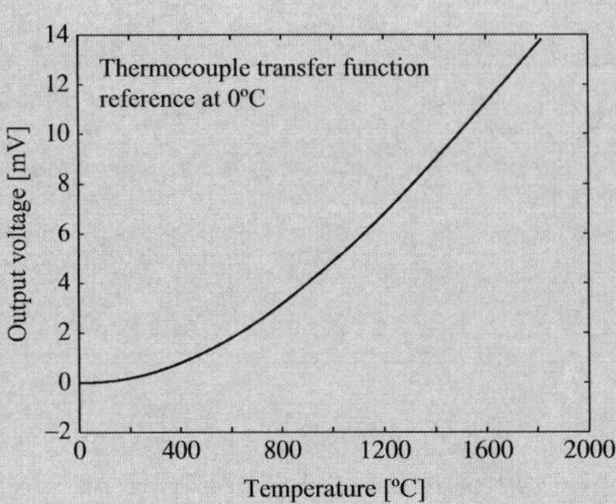

FIGURE 2.2 ■ Transfer function of the thermocouple in the range 0°C–1820°C. The output is shown in volts (V), although the polynomial gives it in millivolts (mV).

EXAMPLE 2.2 Experimental evaluation of the transfer function of a sensor

A force sensor is connected in a circuit that produces a digital output in the form of a train of pulses. The frequency of these pulses is the output of the sensor (actually of the system that includes the sensor and the circuit that converts the output to frequency—we will later call this a smart sensor). The plotted measurements are shown in **Figure 2.3**. The input force range is 0–7.5 N, for which the output produces a frequency between 25.98 kHz and 39.35 kHz. The range between 1 N and about 5 N may be useful as a "linear" range provided that the error incurred by doing so is acceptable. Below 1 N and above 6 N the output is not usable because of reduced response (saturation).

Note: With appropriate circuitry the response of a sensor (or actuator) may be linearized to a large degree if that is deemed useful. Also, it should be noted that the nonlinearity of the curve as well as its saturation may be due to the sensor itself, the electronics in the circuit, or both.

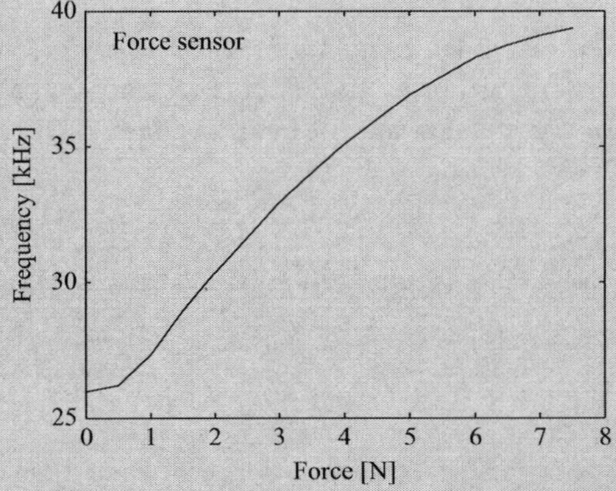

FIGURE 2.3 ■ Experimental evaluation of the transfer function of a force sensor.

Another type of input-output characteristic function, sometimes called a transfer function, is the frequency response of a device. It may be called a transfer function because it gives the output response to input over a frequency range. It is sometimes called a frequency transfer function. We will simply call it a frequency response, and it will be discussed separately in **Section 2.2.7**.

The inputs and outputs of sensors or actuators may be further characterized by the type of signal they provide or require. The output signal of a sensor may be voltage or current or it may be frequency, phase, or any other measurable quantity. The output of actuators is usually mechanical, manifest in motion or force, but it may be in other forms such as light or electromagnetic waves (antennas are actuators when they transmit or sensors when they receive), or it may be chemical. Within these characteristics, some devices operate at very low or very high levels. For example, the output of a thermocouple is typically 10–50 μV/°C, whereas a piezoelectric sensor can produce 300 V or more in response to motion. A magnetic actuator may require, say, 20 A at 12 V, whereas an electrostatic actuator may operate at 500 V (or higher) at a very low current.

2.2.2 Impedance and Impedance Matching

All devices have an internal impedance that may be real or complex. Although we can view both sensors and actuators as two-port devices, we will only be concerned here with the output impedance of sensors and the input impedance of actuators, since these are the properties necessary for interfacing and are readily measured. The range of impedances of devices is large and the importance of impedance matching cannot be overstated; more often than not, failure to properly match a device means failure of the sensing strategy, or in the case of actuators, it may mean physical destruction of the actuator, its drivers, or worse.

The input impedance of a device is defined as "the ratio of the rated voltage and the resulting current through the input port of the device with the output port open" (no load). The output impedance is defined as "the ratio of the rated output voltage and the short circuit current of the port" (i.e., the current when the output is shorted).

The reason these properties matter is that they affect the operation of the device. To understand this, consider first the output impedance of a hypothetical strain gauge (a strain sensor) that has an output resistance of 500 Ω at no strain and 750 Ω at a higher strain. The strain gauge is an active sensor so we must connect it to a source as in **Figure 2.4a**. As the strain increases, the sensor's resistance increases. The strain is measured by measuring this change in resistance in terms of the change in voltage on the sensor. At no strain, this voltage is 2.5 V (corresponding to 500 Ω) and at the measured strain it is 3 V (corresponding to 750 Ω). Suppose now we connect this sensor to a processor, which has an input impedance of 500 Ω as well. As soon as the connection is made, the voltage across the sensor goes down to 1.666 V at no strain and rises to 1.875 V at the measured strain (see **Figure 2.4b** and **2.4c**). Two things should be noted here. First is the reduction in the output of the sensor from 2.5 V to 1.666 V. We refer to this as **loading** of the sensor by the input impedance of the processor. Second, and more importantly, whereas the output at no load rose by 0.5 V, when connected it rose by only 0.209 V. This change may be viewed as a reduction in sensitivity (see **Section 2.2.5**), and unless special measures are taken, it may result in an erroneous measurement of strain. In this case, the obvious solution is for the input impedance of the processor to be

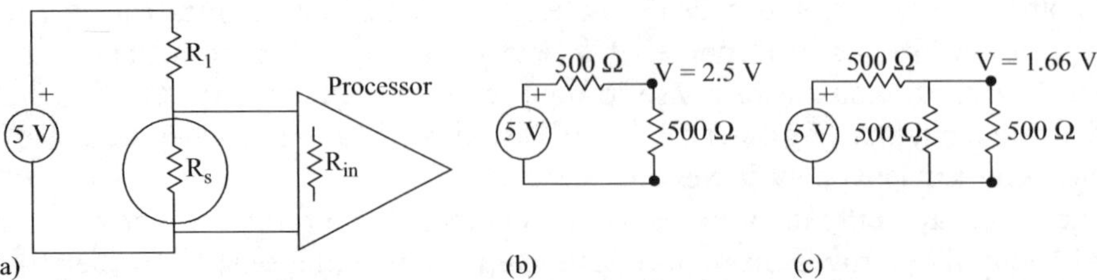

FIGURE 2.4 ■ A strain sensor connected to a processor. (a) The sensor and connections. (b) Equivalent circuit of a sensor alone. (c) Equivalent circuit of a sensor and processor.

as high as possible (ideally infinite) or an impedance matching circuit can be connected between the sensor and the processor. This matching circuit must have very high input impedance and low output impedance. We shall see later that such circuits are available and commonly used.

On the other hand, if the output of a sensor is current, it must be connected to as small an external impedance as possible to avoid changes in the sensor's current, or again a matching circuit with low input and high output impedance must be used. The same exact considerations apply to actuators.

EXAMPLE 2.3 Force sensor

A force sensor is used in an electronic scale to weigh items ranging from 1 gf to 1000 gf. The sensor's resistance changes linearly from 1 MΩ to 1 kΩ as the force changes from 1 gf to 1000 gf (9.80665 mN to 9.80665 N). To measure the resistance, the sensor is connected to a constant current source of 10 μA and the voltage across the sensor is used as the measured quantity. The voltage is measured with a voltmeter with internal impedance of 10 MΩ. The configuration is shown in **Figure 2.5a**. Calculate the error produced by the connection of the voltmeter. What will the actual readings be for the range of forces given?

Solution: The voltage on the sensor changes from 10 V for a mass of 1 g to 0.01 V for a mass of 1000 g before the voltmeter is connected. When the voltmeter is connected, the net resistance is lower and the voltage across the sensor is lower as well. The voltmeter adds its impedance in parallel with the sensor's as shown in **Figure 2.5b**. The net resistances at the ends of the range are at 1 g,

$$R(1\text{ g}) = \frac{R_s R_v}{R_s + R_v} = \frac{10^6 \times 10^7}{10^6 + 10^7} = \frac{10}{11} \times 10^6 = 0.909090 \times 10^6\ \Omega,$$

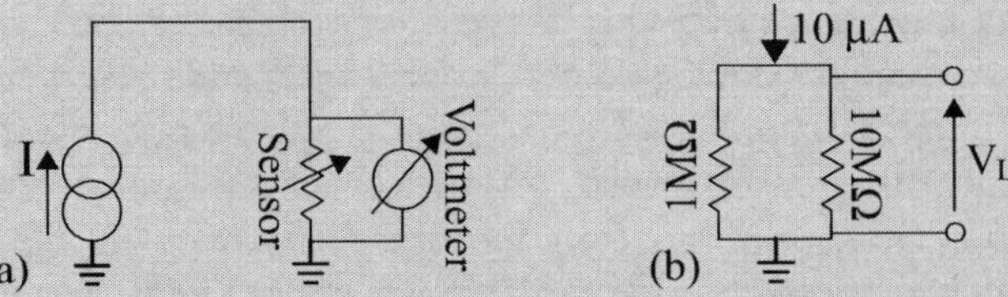

FIGURE 2.5 ■ Loading of a sensor by the measuring instrument. (a) The circuit. (b) The equivalent circuit.

and at 1000 g,

$$R(1000\text{ g}) = \frac{R_s R_v}{R_s + R_v} = \frac{10^3 \times 10^7}{10^3 + 10^7} = \frac{1}{1.001} \times 10^3 = 0.99900 \times 10^3\ \Omega.$$

The measured voltages are

$$V(1\text{ g}) = I_s R(1\text{ g}) = 10 \times 10^{-6} \times 0.909090 \times 10^6 = 9.09\text{ V}$$

and

$$V(1000\text{ g}) = I_s R(1000\text{ g}) = 10 \times 10^{-6} \times 0.99900 \times 10^3 = 0.00999\text{ V}.$$

Clearly the error at 1 g is 9.1%, whereas at 1000 g the error is 0.1%. The actual readings are 0.909 g and 999 g. The error at 1000 g is small and acceptable but at 1 g it may be too high. If higher accuracy is needed, the impedance of the voltmeter must be higher still. This can again be done through electronic circuits, as we shall see in **Chapter 11**.

In some cases, especially in actuators, the output is power rather than voltage or current. In such cases we are usually interested in transferring maximum power from the processor to the actuated medium (say, e.g., from an amplifier to air through a loudspeaker). Maximum power transfer is achieved through conjugate matching, which means simply that given an output impedance of a processor, $R + jX$, the input impedance of the actuator must be equal to $R - jX$. In the case of resistances (real impedances), conjugate matching means that the output resistance of the processor and the input resistance of the actuator must be the same. A very simple example of this may be found in audio amplifiers; an amplifier will transfer maximum power to an 8 Ω speaker if the amplifier's output equals 8 Ω (however, see **Example 2.5**). Although not common, there are sensors that also operate in this mode, in which case conjugate matching applies to them as well.

EXAMPLE 2.4 Impedance matching in actuators

A voice coil actuator (an actuator that operates on the principle of the loudspeaker; we shall discuss these in **Chapter 5**) is pulse driven by an amplifier. The amplifier provides an amplitude $V_s = 12$ V. The internal impedance of the amplifier is $R_s = 4\ \Omega$.

a. Calculate the power transferred to an impedance-matched actuator.
b. Show that the power transmitted to an actuator with lower or higher impedance is lower than that for the matched actuator in (a).
c. What is the power supplied to a 4 Ω actuator if the internal impedance of the amplifier is 0.5 Ω and supplies the same voltage (12 V)?

Solution: Because the actuator is pulse driven, the power is considered instantaneous, but since the voltage is constant through the duration of the pulse, we will calculate the power as if it were a DC source (i.e., the power during the ON portion of the pulse).

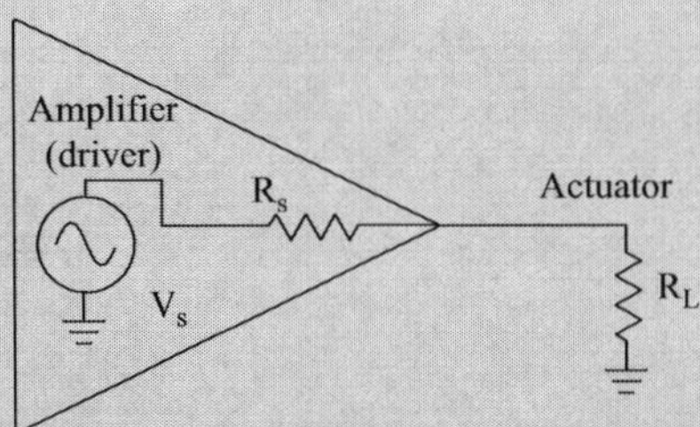

FIGURE 2.6 ■ The concept of matching the load impedance to the internal impedance of a processor.

a. The equivalent circuit for the matched condition is shown in **Figure 2.6**. The actuator's resistance is $R_L = 4\ \Omega$ and the power supplied to the actuator is

$$P_L = \frac{V_L^2}{R_L} = \left(\frac{V_s}{R_s + R_L}R_L\right)^2 \frac{1}{R_L} = \left(\frac{12}{4+4}4\right)^2 \frac{1}{4} = 9\ \text{W}.$$

Note that this is exactly half the power supplied by the source, the other half being dissipated on the internal resistance of the source in the form of heat.

b. With a lower or higher actuator impedance we have, for $R_L = 2\ \Omega$ for example,

$$P_L = \frac{V_L^2}{R_L} = \left(\frac{V_s}{R_s + R_L}R_L\right)^2 \frac{1}{R_L} = \left(\frac{12}{4+2}2\right)^2 \frac{1}{2} = 8\ \text{W}.$$

Similarly, for a higher actuator impedance, say for $R_L = 6\ \Omega$,

$$P_L = \frac{V_L{}^2}{R_L} = \left(\frac{V_s}{R_s + R_L}R_L\right)^2 \frac{1}{R_L} = \left(\frac{12}{4+6}6\right)^2 \frac{1}{6} = 8.64\ \text{W}.$$

Clearly, maximum power is transferred for matched impedances.

Note: The exact condition for matching is that of $Z_L = Z_s^*$, that is, if Z_s is complex, $Z_s = R_s + jX_s$, then a matched load has impedance $Z_L = R_L - jX_s$. In the case shown here, $X_s = 0$; hence the matching condition is $R_L = R_s$.

c.

$$P_L = \frac{V_L^2}{R_L} = \left(\frac{V_s}{R_s + R_L}R_L\right)^2 \frac{1}{R_L} = \left(\frac{12}{0.5+4}4\right)^2 \frac{1}{4} = 28.44\ \text{W}.$$

Note that as the internal impedance of the amplifier approaches zero, the power delivered to the load approaches 36 W and the power dissipated on the internal impedance of the amplifier approaches zero. This may seem to contradict the maximum power transfer condition. Note, however, that if the load were equal to the internal impedance (0.5 W in this case), the power delivered to the load would have been 288 W.

There are also sensors and actuators that operate at high frequencies. Impedance matching to these devices is a much more complex issue and will be discussed briefly in **Chapter 9**. It suffices to say here that the general requirement is that the connection of a sensor or actuator produces no reflection of voltage or current. In such cases the impedance of the sensor or actuator must equal the input impedance of the processor. This requirement does not guarantee maximum power transfer, only nonreflection.

2.2.3 Range, Span, Input and Output Full Scale, Resolution, and Dynamic Range

The ***range*** of a sensor refers to the lower and upper limit operating values of the stimulus, that is, the minimum and maximum input for which a valid output is obtained. Typically we would say that the range of the sensor is between the minimum and maximum values. For example, a temperature sensor may operate between −45°C and +110°C. These are the range points.

The ***span*** of a sensor is the arithmetic difference between the highest and lowest values of the stimulus that can be sensed within acceptable errors (i.e., the difference between the range values). This may also be called the ***input full scale*** (IFS) of the sensor. The ***output full scale*** (OFS) is the difference between the upper and lower ranges of the output of the sensor corresponding to the span of the sensor. For example, a sensor measures temperature between −30°C and + 80°C and produces an output between 2.5 V and 1.2 V. The span (IFS) is 80°C − (−30°C) = 110°C and the OFS is 2.5 V − 1.2 V = 1.3 V. The range of the sensor is between −30°C and +80°C. The IFS and OFS apply equally to actuators.

The range and span of a sensor or actuator express essentially the same information in slightly different ways and therefore they are often used interchangeably.

The **resolution** of a sensor is the minimum increment in stimulus to which it can respond. It is the magnitude of the input change that results in the smallest discernible output. For example, a sensor may be said to have a resolution of 0.01°C, meaning that an increment in temperature of 0.01°C produces a readily measurable output. In common usage this is sometimes, erroneously, referred to as sensitivity. Resolution and sensitivity are two very distinct properties (**sensitivity** is the ratio of change in output to the change in input and will be discussed separately). Resolution of analog devices is said to be infinitesimal, that is, their response is continuous and hence the resolution depends on our ability to discern the change. Often the resolution is defined by the noise level, since for a signal to be discernible it must be larger than the noise level. The resolution of a sensor or actuator is often defined by the instrument or processor used to measure the output. For example, a sensor producing, say, an output of 0–10 V for a temperature range of 0°C–100°C must be connected to an instrument to monitor that voltage. If this instrument is analog (an analog voltmeter, for example), the resolution may be, say, 0.01 V (1000 graduations on the voltmeter's scale) or 0.001 V (10,000 graduations). Even then, when reading the voltmeter, one may interpolate between the graduations, extending the resolution. If, however, the voltmeter is digital and it has an increment of 0.01 V, then this is the resolution of the instrument, and by extension, that of the sensing system made of the sensor and the instrument.

Resolution may be specified in the units of the stimulus (e.g., 0.5°C for a temperature sensor, 1 mT for a magnetic field sensor, 0.1 mm for a proximity sensor, etc.) or may be specified as a percentage of the span (0.1%, for example).

The resolution of an actuator is the minimum increment in its output that it can provide. For example, a DC motor is capable of infinitesimal resolution, whereas a stepper motor may have 200 steps/revolution for a resolution is 1.8°.

In digital systems, resolution may be specified in bits (such as N-bit resolution) or in some other means of expressing the idea of resolution. In an analog-to-digital (A/D) converter, the resolution means the number of discrete steps the converter can convert. For example, a 12-bit resolution means the device can resolve $2^{12} = 4096$ steps. If the converter digitizes a 5 V input, each step is $5/4096 = 1.22 \times 10^{-3}$ V. On the analog side, the resolution may be described as 1.22 mV, but on the digital side it is described as 12-bit resolution. In digital cameras and in display monitors the resolution is typically given as the total number of pixels. Thus a digital camera may be said to have a resolution of a number (x) of megapixels.

EXAMPLE 2.5 Resolution of a system

Suppose a signal is digitized and measured with a two-digit digital voltmeter capable of measuring up to 1 V (after proper amplification). The possible measurement is between 0 and 0.99 V with a resolution of 0.01 V or 1%. In this case the resolution is limited by the voltmeter (the actuator in this system), whereas the signal is continuous, and given a "better" voltmeter, we may well be able to resolve it further. We shall see in **Chapter 11** that A/D converters are capable of much higher resolution than the one shown here.

EXAMPLE 2.6 Resolution of anolog and digital sensors

A digital pressure sensor has a range between 100 kPa (approximately 1 atm) and 1 MPa (approximately 10 atm). The sensor is an analog sensor and produces an output voltage that varies between 1 V at the lowest range and 1.8 V at the highest range. The digital output display shows the pressure directly using a 3½-digit panel meter (a 3½-digit panel meter has three digits that can display 0 to 9 and one digit that can display 0 and 1). What is the resolution of the sensor itself (analog) and what is the resolution of the digital sensor, assuming that the display is autoranging, that is, it is capable of placing the digital point automatically?

Solution: The analog sensor has infinitesimal resolution and its output changes continuously between 1 V and 1.8 V. That is, for a change in pressure of 9×10^6 Pa, the voltage changes by 0.8 V or 88.89 nV/Pa. Of course, the pascal is a very small unit, so we might say as well that the output changes by 88.89 μV/kPa. That is, if we were to assume that an output of 100 nV can be read reliably, the resolution of the sensor is $100/88.89 = 1.125$ Pa. On the other hand, if the output can only read, say, 10 μV, the resolution is $10{,}000/88.89 = 112.5$ Pa. The practical resolution is limited only by our ability to measure voltage and by any noise that may exist.

The digital panel meter will display from 0.100 to 10.00 MPa. In the range 0.100 to 9.999 MPa, the resolution is clearly 0.001 MPa, or 1 kPa. At 10.00 MPa the resolution decreases to 0.01 MPa, or 10 kPa. Note that the limitation is imposed by the display itself.

The **dynamic range of a device** (sensor or actuator) is the ratio of the span of the device and the minimum discernible quantity the device is capable of (resolution). Typically the lowest discernible value is taken as the noise floor, that is, the level at which the signal is "lost" in the noise. The use of dynamic range is particularly useful in devices with large spans and for that reason is usually expressed in decibels. Since the ratio represents either powerlike (power, power density) or voltagelike quantities (voltage, current, force, fields, etc.), the dynamic range is written as follows:

For voltage-like quantities:

$$\text{Dynamic range} = 20 \log|\text{span}/\text{lower measurable quantity}|. \tag{2.3}$$

For power-like quantities:

$$\text{Dynamic range} = 10 \log|\text{span}/\text{lower measurable quantity}|. \tag{2.4}$$

Suppose we look at a 4-digit digital voltmeter capable of measuring between 0 and 20 V. The total span is 19.99 V and the resolution (smallest increment) is 0.01 V. The dynamic range is thus

$$\text{Dynamic range} = 20 \log(19.99/0.01) = 20 \times 3.3 = 66 \text{ dB}.$$

On the other hand, a 4-digit digital wattmeter measuring up to 20 W in increments of 0.01 W will have a dynamic range of

$$\text{Dynamic range} = 10 \log(19.99/0.01) = 10 \times 3.3 = 33 \text{ dB}.$$

Span, IFS, and OFS are usually measured in the respective quantity at the input and output of the device (pressure and voltage in the example above), but in some cases, where the dynamic range is very large, these may also be given in decibels. The same consideration may be applied to actuators, but whereas dynamic range is often used for sensors, the inputs and outputs of actuators are usually defined by the range, especially when motion is involved.

EXAMPLE 2.7 Dynamic range of a temperature sensor

A silicon temperature sensor has a range between 0°C and 90°C. The accuracy is defined in the data sheet as ±0.5°C. Calculate the dynamic range of the sensor.

The resolution is not given, so we will take the accuracy as the minimum measurable quantity. In general, these need not be the same. Since the minimum resolution is 0.5°, the dynamic range is

$$\textit{Dynamic range} = 20\log_{10}\left(\frac{90}{0.5}\right) = 45.1 \text{ dB}.$$

EXAMPLE 2.8 Dynamic range of a loudspeaker

A loudspeaker is rated at 6 W and requires a minimum power of 0.001 W to overcome internal friction. What is its dynamic range? Clearly, any change smaller than 1 mW will not change the position of the speaker's cone and hence no change in output will be produced. Thus 1 mW is the resolution of the speaker and the dynamic range of the speaker is

$$\textit{Dynamic range} = 10\log_{10}\left(\frac{6}{0.001}\right) = 37.78 \text{ dB}.$$

In digital sensors and actuators, the signal levels change in increments of bits. In general, one can define a dynamic range based on the digital representation or on the equivalent analog signal. In an N-digit device, the ratio between the highest and lowest representable level is $2^N/1 = 2^N$. Therefore the dynamic range may be written as

$$\textit{Dynamic range} = 20\log_{10}\left(2^N\right) = 20N\log_{10}(2) = 6.0206N \quad [\text{dB}]. \tag{2.5}$$

EXAMPLE 2.9 Dynamic range of an A/D converter

A 16-bit analog to digital [A/D] converter (to be discussed in **Chapter 11**) is used to convert an analog music recording into digital format so it can be stored digitally and played back (by converting it back to analog form). The amplitude varies between -6 V and $+6$ V.

a. Calculate the smallest signal increment that can be used.

b. Calculate the dynamic range of the A/D conversion.

Solution:

a. A 16-bit A/D converter can represent $2^{16} = 65{,}536$ levels of the signal. The smallest signal increment in the case discussed here is

$$\Delta V = \frac{12}{2^{16}} = 1.831 \times 10^{-4}\ \text{V}.$$

That is, the signal is represented in increments of 0.1831 mV.

b. The dynamic range of the A/D converter is

$$\textit{Dynamic range} = 6.0206N = 96.33\ \text{dB}.$$

2.2.4 Accuracy, Errors, and Repeatability

The errors involved in sensing and actuating define the accuracy of the device. These may stem from various sources, but they all represent deviations of the device output from the ideal. Inaccuracies in the output (i.e., in the transfer function) stem from a variety of sources, including materials, construction tolerances, aging, operational errors, calibration errors, matching (impedance) or loading errors, and many others. The definition of error is rather simple and can be represented as the difference between the measured and actual value. In practice, it may be represented in a number of ways.

1. The most obvious is as a difference: $e = |V - V_0|$, where V_0 represents the actual (correct) value and V is that measured by the device. Often the error is given as $\pm$e.
2. A second way is to represent this as a percentage of IFS (span), $e = (\Delta t/(t_{max} - t_{min})) \times 100$, where t_{max} and t_{min} are the maximum and minimum values at which the device is designed to operate (range values).
3. A third method is to specify the error in terms of the output signal expected rather than the stimulus. Again, it may simply be the difference between values or it may be represented as a percentage of OFS.

EXAMPLE 2.10 **Errors in sensing**

A thermistor is used to measure temperatures between −30°C and + 80°C and produces an output voltage between 2.8 V and 1.5 V. The ideal transfer function is shown in **Figure 2.7** (solid line). Because of errors, the accuracy in sensing is ±0.5°C. The errors may be specified as follows:

a. In terms of the input as ±0.5°C.
b. As a percentage of the input range:

$$e = [0.5/(80 + 30)] \times 100 = 0.454\%.$$

c. In terms of the output range. This may be taken off the curve as the difference shown or may be calculated by first evaluating the transfer function and its maximum and minimum limits, as shown in **Figure 2.7**. This gives an error of ±0.059 V. It may also be given in percentage of OFS as $\pm[0.059/(2.8 - 1.5)] \times 100 = 4.54\%$. Note that these are not the same and the appearance of accuracy will depend on how it is expressed. In most cases, the error given in terms of the measurand or percent of IFS is the best measure of the sensor's accuracy.

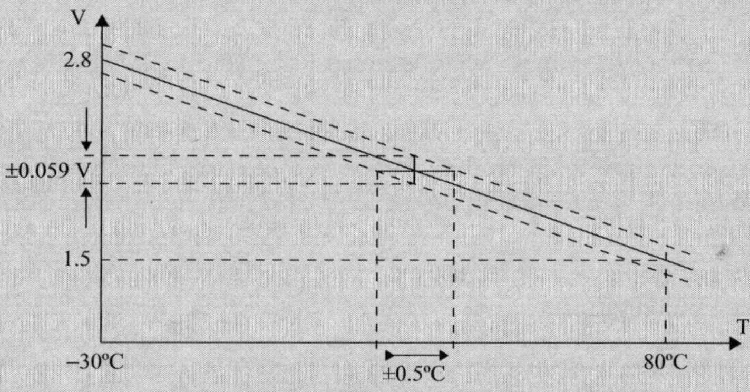

FIGURE 2.7 ■ Transfer function and error limits.

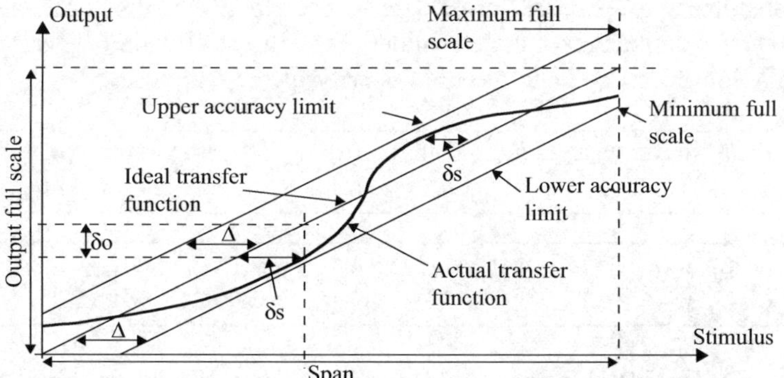

FIGURE 2.8 ■ Error and accuracy limits in nonlinear transfer functions.

In general, the transfer function is nonlinear and the error varies throughout the range of the sensor, as illustrated in **Figure 2.8**. When this happens, either the maximum error or an average error are taken as representative of the error of the device. **Figure 2.8** shows the error limits or accuracy limits as parallel curves that limit the actual transfer

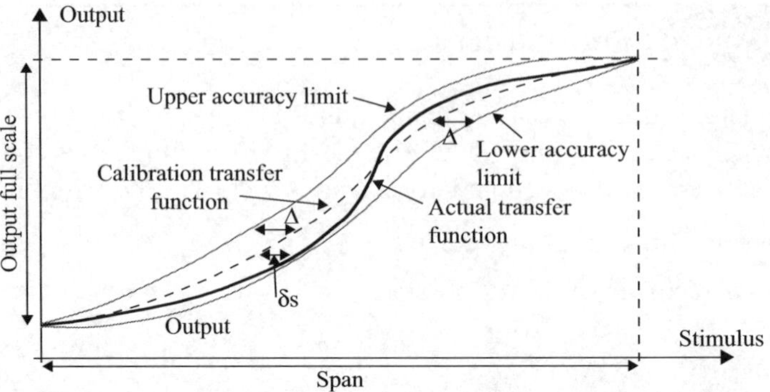

FIGURE 2.9 ■ Error in nonlinear transfer functions may be taken around the calibration transfer function rather than the ideal transfer function.

function. The accuracy limits are parallel to the ideal transfer function and do not have to be straight lines (see **Figure 2.9**). In some cases, the sensor may be calibrated during production or during installation. In such cases, the error may be taken around the calibration curve rather than around the ideal transfer function, as shown in **Figure 2.9**. In effect, rather than using the ideal transfer function, the actual transfer function (calibration curve) is used. This has the effect of being more realistic for the specific device, but does not apply to other devices of the same type—each must be calibrated separately.

So far errors have been assumed to be static. That is, they do not depend on time. However, errors can also be dynamic or time dependent, but the calculation and meaning of errors is the same as for static errors.

Some errors are random, whereas others are constant or **systemic**. If various samples of a device exhibit different errors in a particular parameter, or if a particular device exhibits different error values each time it is operated, these errors are said to be random. If the errors are constant, they are said to be systemic. A device may have both systemic and random errors.

EXAMPLE 2.11 Nonlinearity errors

In a capacitive accelerometer the distance between two plates is related to the force due to acceleration, $F = ma$, where m is the mass of the moving plate and a is the acceleration (see **Figure 2.10a**). A spring generates the restoring force. The relation between the force and the distance between the plates (hence capacitance) is determined experimentally and is given in the table below. Establish the maximum error due to the nonlinearity of the response.

d [mm]	0.52	0.5	0.48	0.46	0.44	0.42	0.4	0.38	0.36	0.34	0.32	0.3	0.28	0.26
F [μN]	0	6	9	13	17	21	25	28	31	35	39	43	46	49

d [mm]	0.24	0.22	0.2	0.18	0.16	0.14	0.12	0.1	0.08	0.06	0.04	0.02	0.012	0
F [μN]	52	55	58	61	64	67	70	73	76	79	82	84	85	86

Solution: The transfer function and the upper and lower accuracy limits are plotted in **Figure 2.10b**. The maximum error is 8.0 μN and occurs at the very beginning and end of the curve (i.e., at $d = 0$ and at $d = 0.520$ mm. If we disregard these as the extremes, at which the sensor is likely to be less accurate, the maximum error is 7.6 μN and occurs at $d = 0.22$ mm.

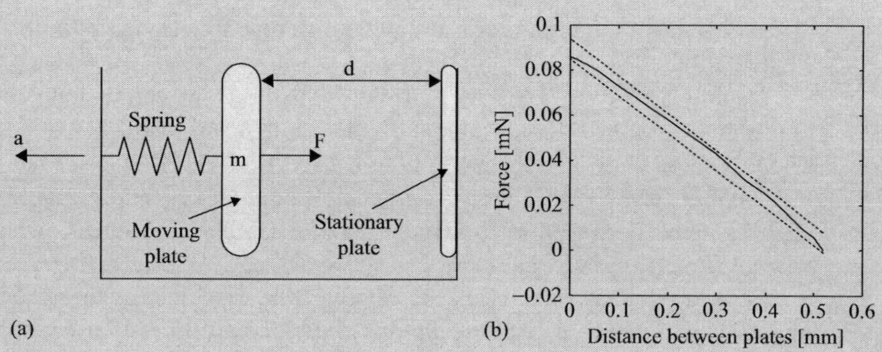

FIGURE 2.10 ■ (a) Capacitive accelerometer. (b) Transfer function of the accelerometer in (a) showing the upper and lower accuracy limits.

The error in sensed distance is 0.04 mm. These errors are derived from the accuracy limits shown since the ideal transfer function is linear ($F = ma = kx$, where k is the spring constant and x is the displacement). The errors may equally be given in percentage of full scale. The error in displacement is $(0.04/0.52) \times 100 = 7.69\%$, whereas the error in force is $(7.6/86) \times 100 = 8.84\%$.

Note: Typically the error would be taken from the ideal transfer function to the accuracy limit. Here, however, we look at the error due to nonlinearity, hence the error is taken between the two accuracy limits that bound the transfer function. In effect, the error is taken as if the ideal transfer function were the lower accuracy limit.

Repeatability, sometimes called reproducibility, of sensors and actuators is an important design characteristic and simply indicates the failure of the sensor or actuator to represent the same value (i.e., stimulus for sensors or output for actuators) under identical conditions when measured at different times. It is usually associated with calibration and is viewed as an error. It is given as the maximum difference between two readings (either two calibration readings or two measurement readings) taken at different times under identical input conditions. Usually the error will be given as a percentage of IFS.

2.2.5 Sensitivity and Sensitivity Analysis

The sensitivity of a sensor or actuator is defined as the change in output for a given change in input, usually a unit change in input. Clearly, sensitivity represents the slope of the transfer function. We can write

$$s = \frac{dS}{dx} = \frac{d}{dx}(f(x)). \tag{2.6}$$

For the linear transfer function in **Equation (2.2)**, where the output is resistance (R) and the input is temperature (T), we have

$$s = \frac{dR}{dT} = \frac{d}{dT}(aT + b) = a \quad [\Omega/^\circ\mathrm{C}]. \tag{2.7}$$

Note in particular the units: in this case, since the output is in ohms, and the stimulus is in degrees Celsius, the sensitivity is given in ohms per degree Celsius ($\Omega/^\circ$C).

Sensitivity may be constant throughout the span (linear transfer function), it may be different in different regions, or it may be different at every point in the span (such as in **Figure 2.8**).

Usually sensitivity is associated with sensors. However, as long as a transfer function can be defined for an actuator, the same ideas can be extended to actuators. Therefore it is quite appropriate to define a sensitivity of, say, a speaker as dP/dI, where P is the pressure (output) the speaker generates per unit current I (input) into the speaker, or the sensitivity of a linear positioner as dl/dV, where l is linear distance and V is the voltage (input) to the positioner.

Sensitivity analysis is usually a difficult task, especially when, in addition to the stimulus, there is noise. In addition, the sensitivity of a sensor is often a combined function of sensitivities of various components of the sensor, including the transduction section or sections (if multiple transduction steps are involved). Further, the sensor may be rather complex with multiple transduction steps, each one with its own sensitivity, sources of noise, and other parameters that come into play, such as nonlinearities, accuracy, and others, some of which may be known, but many of which may not be known or may only be known approximately. Nevertheless, sensitivity analysis is an important step in the design process, especially when complex sensors are used, since, in addition to providing information on the output range of signals one can expect, it also provides information on noise and errors. Sensitivity analysis may also provide clues as to how the effects of noise and errors can be minimized by proper choice of sensors, their connections, and other steps that can be taken to improve performance (amplifiers, feedback, etc.).

EXAMPLE 2.12 Sensitivity of a thermocouple

Consider the thermocouple discussed in **Example 2.1**. The transfer function is given as

$$V = (-2.4674601620 \times 10^{-1} \times T + 5.9102111169 \times 10^{-3} \times T^2 - 1.4307123430 \times 10^{-6} \times T^3 + 2.1509149750 \times 10^{-9} \times T^4 - 3.1757800720 \times 10^{-12} \times T^5 + 2.4010367459 \times 10^{-15} \times T^6 - 9.0928148159 \times 10^{-19} \times T^7 + 1.3299505137 \times 10^{-22} \times T^8) \times 10^{-3}\text{mV}.$$

The sensitivity of the sensor can be found by direct differentiation:

$$s = \frac{dV}{dT} = (-2.4674601620 \times 10^{-1} + 1.182042223 \times 10^{-2} \times T - 4.292137029 \times 10^{-6} \times T^2 + 8.6036599 \times 10^{-9} \times T^3 - 1.587890036 \times 10^{-11} \times T^4 + 1.406220476 \times 10^{-14} \times T^5 - 6.3649703711 \times 10^{-18} \times T^6 + 1.06396041 \times 10^{-21} \times T^7) \times 10^{-3}\text{mV}/^\circ\text{C}.$$

However, this may not be as useful as it looks. Indeed, one can obtain the sensitivity at any temperature by simply substituting the temperature in this relation. But suppose we need to use the sensor to measure temperature between 0°C and 150°C. It may be more useful to obtain a single "average" value for sensitivity by first passing a linear best fit between the points on the transfer function (see **Appendix A**). Once that is done, the sensitivity is the slope of the transfer function. The steps are as follows:

1. We first obtain the output voltage using the transfer function at a number of points (the more points, the better).
2. **Equation (A.15)** is then used to obtain a linear best fit of the form $V = aT + b$.
3. The sensitivity of this linearized transfer function is a.

By generating the values of V between $T = 0$°C and $T = 150$°C and using **Equation (A.15)** or **Equation (2.19)** the linear transfer function is

$$V = aT + b = 6.15540978 \times 10^{-4}T - 0.02122939 \text{ mV}.$$

This is plotted together with the exact curve (eighth-order polynomial above) in **Figure 2.11**. The sensitivity now becomes

$$s = \frac{dV}{dt} = a = 6.15540978 \times 10^{-4} \text{ mV/°C}.$$

This is a low sensitivity of only 6.1554 μV/°C, but is not out of line in thermocouples.

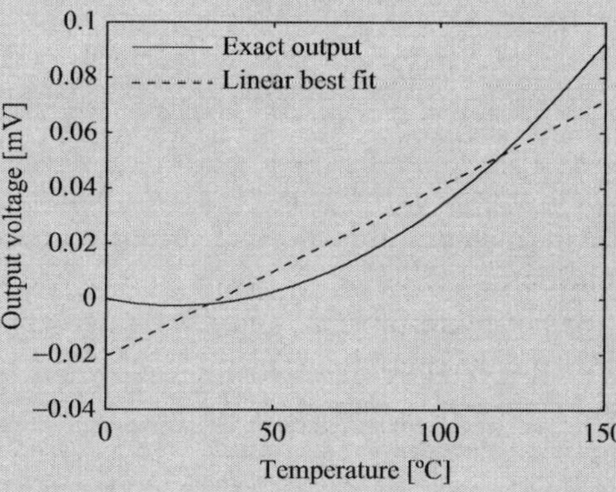

FIGURE 2.11 ■ Transfer function of a thermocouple and the linear best fit approximation.

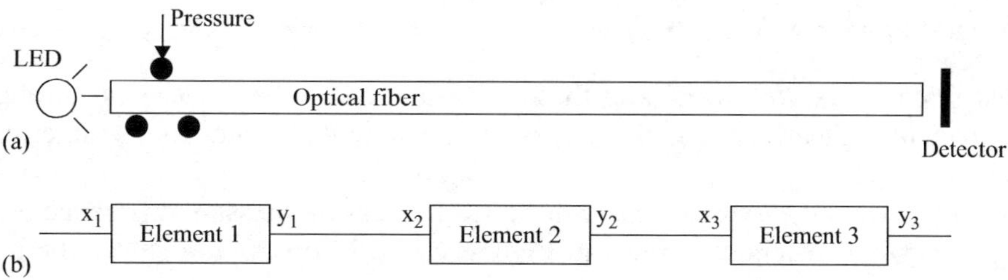

FIGURE 2.12 ■ A sensing system made of a source, optical fiber pressure sensor, and processor. (a) The sensor. (b) Equivalent configuration showing the transducing elements.

To understand some of the issues involved in sensitivity analysis, consider a sensor with three conversion steps in series. An example of a sensor of this type is the optical fiber pressure sensor shown schematically in **Figure 2.12a**. The operation is as follows: The optical fiber transmits light generated by a laser or LED to a detector and the phase of this signal is calibrated to read pressure. When pressure is applied on the optical fiber, it is tensioned (i.e., its length changes slightly). This means the light travels a longer distance in the fiber and its phase at the detector will be larger. This is a complex sensor that includes three transduction steps. First, an electric signal is converted into light and is coupled into the fiber. Then pressure is converted to displacement and, in the detector, light is converted into an electric signal. Each one of these transduction steps has its own errors, its own transfer function, and its own sensitivity.

The three transducers are connected in series and their errors are additive. The sensitivity of each element is

$$s_1 = \frac{dy_1}{dx_1}, \quad s_2 = \frac{dy_2}{dx_2}, \quad s_3 = \frac{dy_3}{dx_3}, \tag{2.8}$$

where y_i is the output of transducer i and x_i is its input. Suppose first that there are no errors in the system. Then we can write

$$S = s_1 s_2 s_3 = \frac{dy_1}{dx_1}\frac{dy_2}{dx_2}\frac{dy_3}{dx_3}. \tag{2.9}$$

But clearly $x_2 = y_1$ (the output of transducer 1 is the input to transducer 2) and $x_3 = y_2$. With these we get

$$S = s_1 s_2 s_3 = \frac{dy_3}{dx_1}. \tag{2.10}$$

This is both simple and logical. The internal conversion steps are not seen in **Equation (2.10)**, meaning that we only need to take into account the sensor as a unit with input and output.

If errors or noise are present, and assuming each transducer element has different errors, we can write the output of transducer element 1 as $y_1 = y_1^0 + \Delta y_1$, where y_1^0 is the output without errors. Assuming that we know the sensitivities of the elements, we can write the output of element 2 as

$$y_2 = s_2\left(y_1^0 + \Delta y_1\right) + \Delta y_2 = y_2^0 + s_2\Delta y_1 + \Delta y_2, \tag{2.11}$$

where $y_2^0 = s_1 y_1^0$ is the output of element 2 without errors and Δy_2 is the errors introduced by element 2. Now this becomes the input to element 3 and we can write

$$y_3 = s_3\left(y_2^0 + s_2\Delta y_1 + \Delta y_2\right) + \Delta y_3 = y_3^0 + s_2 s_3\Delta y_1 + s_3\Delta y_2 + \Delta y_3. \tag{2.12}$$

The last three terms are errors, and these are summed as they propagate through the series elements. Clearly, then, the errors in the output depend on the intermediate transduction steps.

Consider now a differential sensor designed to measure pressure difference between two locations in a system, as shown in **Figure 2.13a**. **Figure 2.13b** shows the transfer functions, inputs, and outputs. Assuming first that there are no errors and that each transducer has a different transfer function, the sensitivity of each sensor is

$$s_1 = \frac{dy_1}{dx_1}, \quad s_2 = \frac{dy_2}{dx_2}. \tag{2.13}$$

The outputs of the two sensors are $y_1 = s_1 x_1$ and $y_2 = s_2 x_2$ and the overall output is

$$y = y_1 - y_2 = s_1 x_1 - s_2 x_2. \tag{2.14}$$

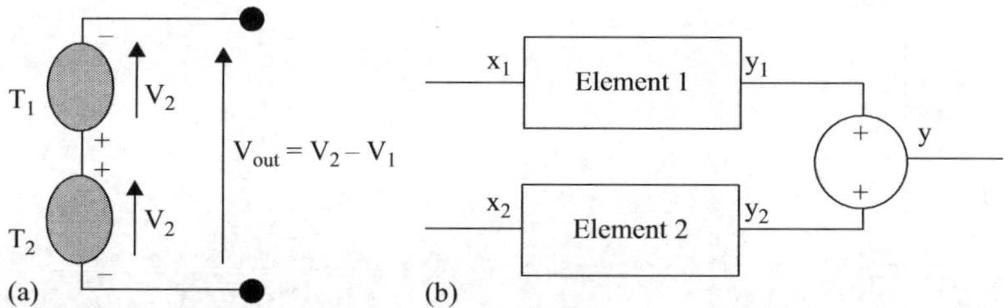

FIGURE 2.13 ■ A differential sensor. (a) The two sensors measure temperatures at different locations connected in opposition. (b) Equivalent configuration showing the transducer elements.

If the two sensors are identical, so that $s_1 = s_2 = s$, we get

$$y = s(x_1 - x_2), \tag{2.15}$$

and the sensitivity of the differential sensor can be calculated as

$$s = \frac{d(y_1 - y_2)}{d(x_1 - x_2)}. \tag{2.16}$$

If the two sensors are identical (assumed), the errors produced by them will be identical (or nearly so). Hence, in taking the difference between the two outputs, the errors cancel out and the differential sensor is virtually error free. Noise, which is one source of errors, cancels as well, as long as it is common to both sensors (common-mode noise). In practice, total cancellation does not occur because of mismatching between the two sensors and other effects, such as the fact that they may be installed at different locations and hence experience different conditions.

EXAMPLE 2.13 Sensitivity to noise

The response of a pressure sensor is determined experimentally and given in the table below. A noise in the form of pressure variations of 330 Pa exists due to local atmospheric changes. Calculate the output due to noise and the error in output produced by this noise.

Pressure [kPa]	100	120	140	160	180	200	220	240	260	280	300	320	340	360	380	400
Voltage [V]	1.15	1.38	1.6	1.86	2.1	2.35	2.6	2.89	3.08	3.32	3.59	3.82	4.05	4.29	4.54	4.78

To calculate the output due to noise we must first find the sensitivity of the sensor. Since the output is experimental, we first need to pass a best fit curve through the data. This is necessary since the sensor output is not perfectly linear, as can be seen by plotting the data (**Figure 2.14**). The linear least squares process in **Appendix A** (see **Equation (A.15)**) gives

$$V = aP + b = 0.0122P - 0.0783 \text{ V},$$

where P is pressure in kilopascals. Thus the sensitivity is

$$s = \frac{dV}{dP} = a = 0.0122 \text{ V/kPa}.$$

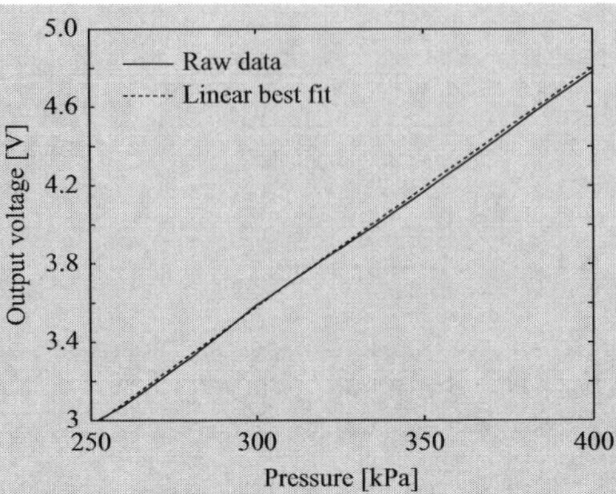

FIGURE 2.14 ■ Plot of sensor output as a function of pressure: raw data and linear best fit.

The output of the sensor at any pressure is that due to pressure and that due to noise (see **Equation (2.11)** or **(2.12)**). Because the curve is linear, we take a convenient point, say, $P = 200$ kPa. The total pressure is 200.33 kPa and the output is

$$V = 0.0122 \times 200.33 - 0.0783 = 2.365726 \text{ V}.$$

Without noise the output is

$$V = 0.0122 \times 200 - 0.0783 = 2.3617 \text{ V}.$$

The output noise is 0.004 V and the error is

$$error = \frac{2.365726 - 2.3617}{2.3617} \times 100 = 0.17\%.$$

EXAMPLE 2.14 Nondestructive testing of materials using a differential inductive probe

One of the reasons to use differential sensors is the cancellation of common-mode effects, including those due to temperature variations and noise, as well as many others. Differential sensors also remove the mean value of the output, leaving only the changes in the output. This is very convenient for further processing of the signal, including amplification, and in some cases it is critical to the operation of the sensor.

As an example, consider a differential eddy current probe used for testing of metals for defects (such as cracks in tubes or in flat surfaces such as the skin of an airplane or a component in an engine). A probe is made of two coils (the sensors), each 1 mm in diameter and separated 2.5 mm apart. Each coil has an inductance L that depends on what is present in the vicinity of the coil. The probe is used to sense a defect in the surface of a steel item by moving it across the material to detect any flaw that may exist. As the leading coil approaches the defect, its inductance goes down, whereas the trailing coil has a higher inductance. In turn, as the leading coil moves past the defect, its inductance increases back to the original value, whereas the trailing coil's inductance goes down. The difference in inductance is the differential output of the probe. **Figure 2.15a** shows the measured inductance of the two coils as they move across the flaw. Note that the behavior of the coils is identical, but the changes occur 2.5 mm apart, as expected.

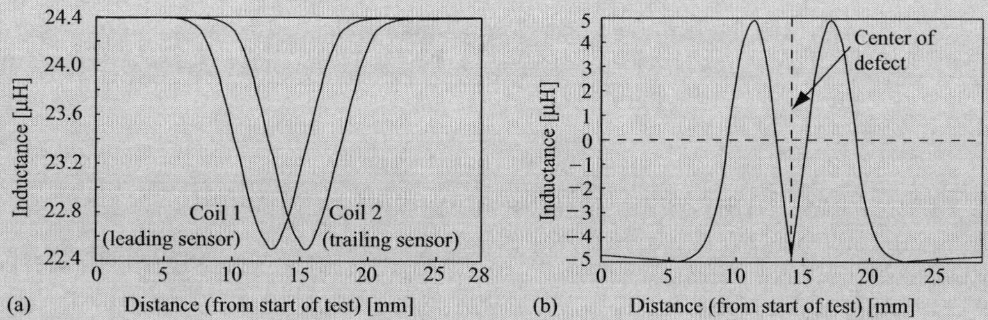

FIGURE 2.15 ■ Eddy current testing of flaws in steel. (a) The inductances of two sensors separated 2.5 mm apart. (b) Differential inductance of the two sensors due to a small flaw.

Taking the differences between the inductances of the two coils at each position as the probe moves leads to the result in **Figure 2.15b**. This differential inductance shows two important aspects. First, the inductance changes around zero—the large inductance (about 24.4 μH) has been removed and all we are left with is the change in inductance due to the defect. This is necessarily so since the inductance due to the solid metal is common to both sensors. Second, the signal is zero when the probe is centered above the flaw—that is, at that point both coils see identical conditions, and since these conditions are common to both, they cause identical changes in the inductance of the coils, cancelling them. This allows identification of the exact position of the defect, an important aspect of testing since the flaw may not be visible with the naked eye or it may be under the surface or under a coating.

It should be noted that the sensitivity has not changed—the change in output per change in input stays the same for each of the two sensors.

There are other ways of connecting sensors. For example, a thermopile ("a pile of thermocouples") is a sensor made of n thermocouples electrically connected in series to increase the electrical output, whereas the input (temperature) to all thermocouples is the same (they are said to be connected in parallel thermally), as shown in **Figure 2.16**. The output now becomes

$$y = y_1 + y_2 + y_3 + \cdots + y_n = (s_1 + s_2 + s_3 + \cdots + s_n) = nsx, \tag{2.17}$$

where it was assumed all thermocouples have identical transfer functions (and therefore sensitivities). The overall sensitivity is

$$S = ns, \tag{2.18}$$

and hence the value of a thermopile.

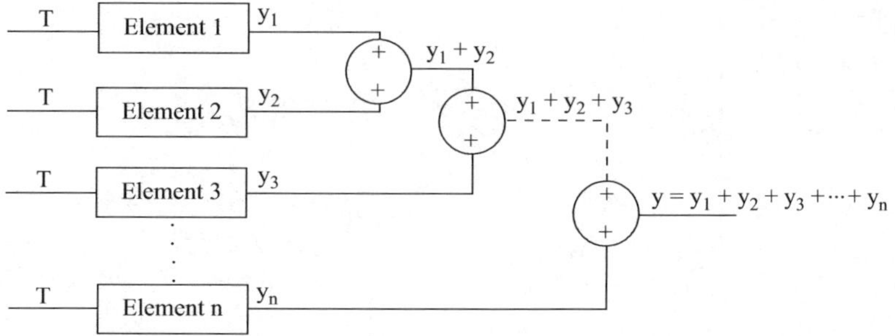

FIGURE 2.16 ■ A sensing system made of n thermocouples connected in series electrically. The output is the sum of the outputs of the individual sensors.

Noise and errors are additive since the outputs are in series. Because the inputs are in parallel thermally, and because the thermocouples are essentially identical, the errors are also identical (or nearly so) and therefore the output error is n times the error (or noise) of a single thermocouple.

2.2.6 Hysteresis, Nonlinearity, and Saturation

Hysteresis (literally lag) is the deviation of the sensor's output at any given point when approached from two different directions (see **Figure 2.17**). Specifically, this means that the output at a given value of stimulus when it increases and when it decreases is different. For example, if temperature is measured, at a rated temperature of 50°C, the output might be 4.95 V when the temperature increases, but 5.05 V when the temperature decreases. This is an error of ±0.5% (for an OFS of 10 V in this idealized example). The sources of hysteresis are either mechanical (friction, slack in moving members), electrical (such as due to magnetic hysteresis in ferromagnetic materials), or due to circuit elements with inherent hysteresis. Hysteresis is also present in actuators and, in the case of motion, is more common than in sensors. There it may manifest itself as positioning errors. Also, hysteresis may be introduced artificially for specific purposes.

Nonlinearity may be either a property of a sensor (see, e.g., **Figure 2.1**) or an error due to deviation of a device's ideal, linear transfer function. A nonlinear transfer function is a property of the device and, as such, is neither good nor bad. One simply has to design with it or around it. However, a nonlinearity error is a quantity that influences the accuracy of the device. It must be known to the designer, must be taken into account, and possibly minimized. If the transfer function is nonlinear, the maximum deviation from linearity across the span is stated as the nonlinearity of the device. However, this measure of linearity is not always possible or desirable. Therefore there are various valid ways of defining the nonlinearity of a sensor or actuator. If the transfer function is close to linear, an approximate line may be drawn and used as the reference linear function. Sometimes this is done simply by connecting the end points (range values) of the transfer function (line 1 in **Figure 2.18**). Another method is to draw a least squares line through the actual curve (line 2 in **Figure 2.18**), usually by first selecting a reasonable number of points on the curve and then, given the selected (or measured) pairs of input and output values (x_i, y_i), draw the line $y = a + bx$ by calculating the slope a and the

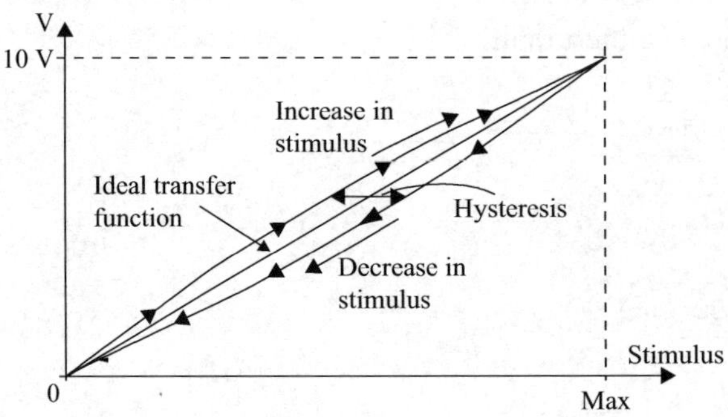

FIGURE 2.17 ■ Hysteresis in a sensor.

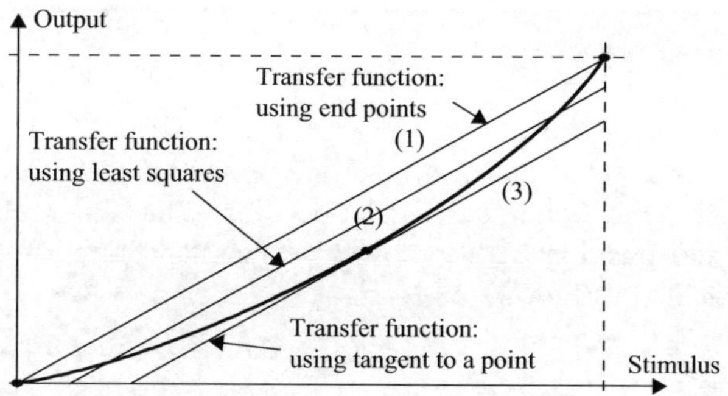

FIGURE 2.18 ▪ Linear approximations of nonlinear transfer functions.

axis intercept b as shown in detail in **Appendix A**. The linear best fit to the data (see **Equation (A.15)**) is

$$a = \frac{n\sum_{i=1}^{n} x_i y_i - \left\{\sum_{i=1}^{n} x_i\right\}\left\{\sum_{i=1}^{n} y_i\right\}}{n\sum_{i=1}^{n} x_i^2 - \left\{\sum_{i=1}^{n} x_i\right\}^2}, \quad b = \frac{\left\{\sum_{i=1}^{n} y_i\right\}\left\{\sum_{i=1}^{n} x_i^2\right\} - \left\{\sum_{i=1}^{n} x_i\right\}\left\{\sum_{i=1}^{n} x_i y_i\right\}}{n\sum_{i=1}^{n} x_i^2 - \left\{\sum_{i=1}^{n} x_i\right\}^2}. \quad (2.19)$$

This provides a "best" fit (in the least squares sense) through the points of the transfer function and can be used for purposes of measuring the nonlinearity of the actual sensor or actuator. If this method is used, nonlinearity is the maximum deviation from this line.

There are many variations on both of these methods. In some cases a sensor is only expected to operate in a small section of its span. In this case, either method may be applied for that portion of the span. Another method that is sometimes employed is to take a midpoint in this reduced range and draw a tangent to the transfer function through the selected point and use this tangent as the "linear" transfer function for purposes of defining nonlinearity (line 3 in **Figure 2.18**). Needless to say, each method results in different values for nonlinearity and, while these are valid, it is important for the user to know the exact method used.

It should also be noted that in spite of the foregoing, nonlinearity is not necessarily a "bad" thing or something that needs correction. In fact, there are instances in which a nonlinear response is superior to a linear response and there are sensors and actuators that are intentionally and carefully designed to be nonlinear. An example is the common potentiometer used as a volume control, especially in audio systems. Notwithstanding the fact that most current volume control systems tend to be digital and many are linear, our hearing is not linear—in fact, it is logarithmic. This allows the ear to respond to minute pressure changes (as low as 10^{-5} Pa) as well as high pressures (high power sound)—as high as 60 Pa. Normally the range is given between 0 and 130 dB. To accommodate this natural response, potentiometers for volume control were also designed as logarithmic to conform with our ears' response. Even some digital potentiometers are logarithmic. The following example discusses these issues a bit more, but the important point is that this nonlinear response has been designed on purpose to fit a particular need and in this case a nonlinear response is superior.

EXAMPLE 2.15 Rotary logarithmic potentiometer

A 100 kΩ rotary logarithmic potentiometer turns from zero resistance to 100 kΩ in 300° of the slider position. **Figure 2.19a** shows schematically how the potentiometer functions and **Figure 2.19b** shows the resistance as a function of the angular position of the slider relative to the starting point. The resistance at any position of the moving, center tap of the potentiometer (sometimes called a wiper) is calculated as follows:

$$R = 100{,}000\left[1 - \frac{1}{K}\log_{10}\left(\frac{\alpha_{max} - \alpha + \alpha_{min}}{\alpha_{min}}\right)\right] \quad [\Omega],$$

where the normalization factor K is

$$K = \log_{10}\left(\frac{\alpha_{max} + \alpha_{min}}{\alpha_{min}}\right)$$

and $\alpha_{max} = 300°$ and $\alpha_{min} = 10°$ represent the maximum and minimum slider angular positions between which a resistance is measurable and α is the slider angular position. This formula ensures the resistance is zero at $\alpha = 0$ and 100 kΩ at $\alpha = 300°$. Note that at, for example, $\alpha = 150°$ the resistance is only 19.75 kΩ, whereas at $\alpha = 225°$ it rises to 38 kΩ.

Notes:

1. There are also antilogarithmic potentiometers in which the resistance increases quickly initially and then levels off.
2. Some logarithmic potentiometers are in fact exponential rather than logarithmic. They are called logarithmic because their response is linear on a log scale.

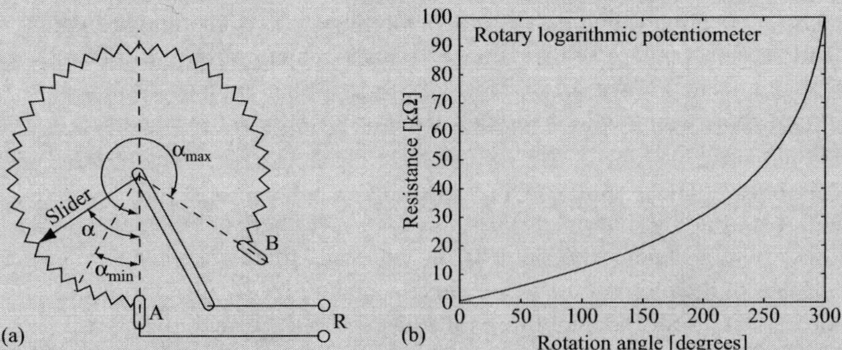

FIGURE 2.19 ■ (a) Schematic structure of the rotary logarithmic potentiometer. (b) The resistance between the tap and point A as a function of the angle α.

Finally, we mention that sometimes one has a choice as to what quantity to use, and a proper choice can make all the difference. In **Chapter 5** we will talk about resistive force sensors, that is, sensors whose resistance changes with the force applied. Naturally in a sensor of this type the resistance is measured, but it so happens that the resistance is highly nonlinear with respect to the applied force. If instead of resistance one measures the conductance (the reciprocal of resistance), the transfer function is perfectly linear. Although the difference between the two representations of the measurand (force) may be dramatic, usage is simple in both cases. If we apply a current source and measure the voltage across the sensor, we obtain a nonlinear response to force. If, on the other hand, we apply a constant voltage source and measure the current through the sensor, the curve is linear and the response of the sensor to force is linear. These issues are shown in **Example 2.16** for an experimentally evaluated resistive force sensor. The choice of selecting the response is not always available, but when it is, it can simplify interfacing considerably.

EXAMPLE 2.16 Linear and nonlinear transfer functions in the same sensor

The response of a resistive force sensor is found experimentally by measuring its resistance as a function of applied force as follows:

Force [N]	0	44.5	89	133	178	222	267	311	356	400	445	489	534
Resistance [Ω]	910	397	254	187	148	122	108	91	80	72	65	60	55

The plot of resistance as a function of force is shown in **Figure 2.20a** and is, as expected, highly nonlinear. The reciprocal plot, that of conductance as a function of force, is shown in **Figure 2.22b** and, within the limitations of measurements, is linear. If a linear response is deemed desirable, one should simply apply a constant voltage source across the sensor and measure its current directly rather than measuring resistance or voltage.

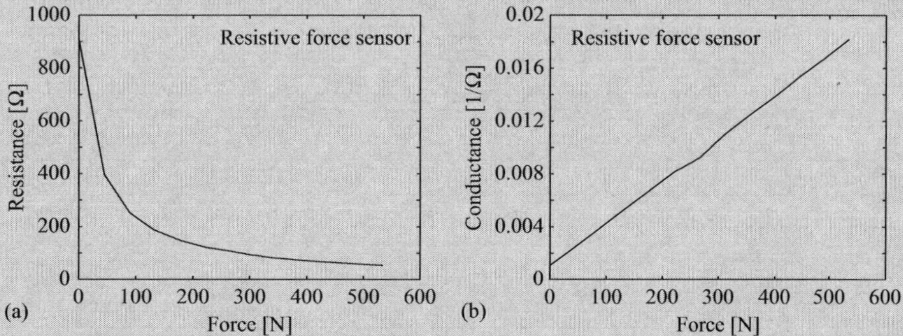

FIGURE 2.20 ■ The transfer function of a resistive force sensor. (a) Plot of the sensor's resistance versus force. (b) Plot of the sensor's conductance versus force.

Saturation refers to the behavior of sensors or actuators when they no longer respond to the input or, more likely, their response is reduced. This usually occurs at or near the ends of their span and indicates that the output is no longer a function of the input or, more likely, is a highly reduced function of the input. In **Figure 2.1** the sensor exhibits saturation at points below T_1 and above T_2, as seen by the flattening of the curves. In both sensors and actuators, one should avoid saturation for two reasons. First, sensing is inaccurate at best and the sensitivity, and often the response, is reduced. Second, saturation may, in some cases, damage the device. In particular, in actuators this may mean that any additional power supplied does not produce an increase in the output power of the device (i.e., sensitivity is reduced), leading to internal heating and possible damage.

2.2.7 Frequency Response, Response Time, and Bandwidth

Frequency response (also called the **frequency transfer function**) of a device indicates the ability of the device to respond to a harmonic (sinusoidal) input. Typically the frequency response shows the output (as a magnitude of the output or gain) of a device as a function of the frequency at the input, as shown in **Figure 2.21**. Sometimes the phase of the output is also given (the pair amplitude gain-phase responses is known as the Bode diagram). The frequency response is important in that it indicates the range of frequencies of the stimulus for which the output is adequate (i.e., does not deteriorate or increase the error due to the inability of the device to operate at a frequency or range of frequencies). For sensors and actuators that are required to operate over a range of frequencies, the frequency response provides three important design parameters. One is the **bandwidth** of the device. This is the frequency range between the two pre-agreed-upon points A and B in **Figure 2.21**. These points are almost always taken as the half-power points (at which power is half that of the flat region). The gain (magnitude) at the half-power points is $1/\sqrt{2}$ of the gain in the flat region, or 70.7%. Often the frequency response is given in decibels, in which case the half-power points are points at which the gain is 3 dB down ($10 \times \log 0.5 = -3$ dB or $20 \times \log(\sqrt{(2)}/2) = -3$ dB). The second parameter that is sometimes used is the **useful frequency range** or **flat frequency range** (or **static range**), which, as its name implies, is that portion of the bandwidth that is flat. However, most devices do not have a frequency response anywhere near the ideal response shown in **Figure 2.21**. Therefore the useful frequency range may be

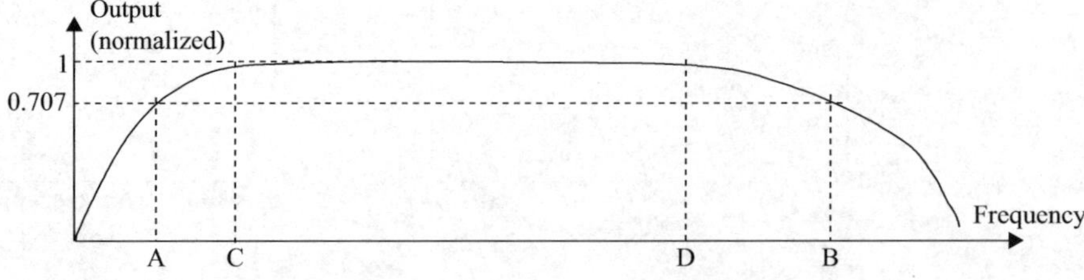

FIGURE 2.21 ■ Frequency response of a device showing the half-power points.

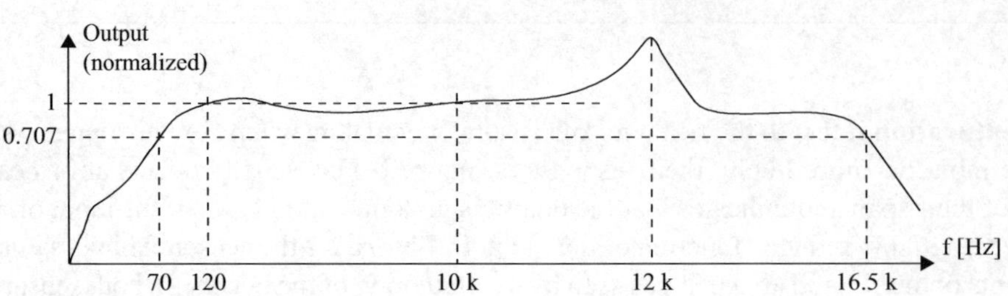

FIGURE 2.22 ■ Definition of bandwidth for a device with nonflat frequency response (frequency axis not to scale).

considerably smaller than the bandwidth (depending on how flat it needs to be for the application) or it may be a compromise between flatness and width. An example is shown in **Figure 2.22**, where the frequency response of a hypothetical loudspeaker is shown. The bandwidth extends between 70 Hz and 16.5 kHz and most often is indicated as 16.5 kHz–70 Hz = 16,430 Hz (the difference between the half-power points). The response at 12 kHz is said to be resonant (maximum in this case, but can also be minimum). The flat region is not entirely flat, but we could reasonably take the range between 120 Hz and 10 kHz as "flat." In this region the speaker will reproduce the input most faithfully. The half-power points on the frequency response curve are viewed as **cutoff frequencies**, essentially indicating that beyond these points the device is "useless." Of course, these are arbitrary points to a large extent and the device can operate beyond these points, but with a reduced response. In some cases the lower cutoff frequency does not exist, indicating that the device responds down to DC.

Related to frequency response is the **response time** (or delay time) of the device, which indicates the time needed for the output to reach steady state (or a given percentage of steady state) for a step change in input. This is often specified for slow-responding devices such as temperature sensors or thermal actuators. Typically the response time will be given as the time needed to reach 90% of steady-state output upon exposure to a unit step change in input. The response time of the device is due to the inertia of the device (mechanical, thermal, electrical). For example, in a temperature sensor, it may simply be due to the time needed for the sensor's body to reach the temperature it is trying to measure (thermal time constant) or may be due to the electrical time constants inherent in the device due to capacitances and inductances, or, most likely, due to both. Clearly this means that the higher the response time, the less the sensor can follow rapid changes in the stimulus, and consequently the narrower its frequency response (bandwidth). Response time is an important design parameter that the engineer must take into account. As a rule, since response time is mostly related to mechanical time constants, and these are in general related to physical dimensions, smaller sensors tend to have shorter response times, whereas bulky sensors tend to respond slower (longer response times). Response time is most often specified with devices that respond slowly. Fast-acting devices will be specified in terms of frequency response.

EXAMPLE 2.17 Frequency response of a magnetic sensor

The frequency response of a magnetic sensor used to detect flaws in conducting structures is shown in **Figure 2.23**. This sensor is called an eddy current sensor (really a rather simple coil) and is a common sensor in testing of tubes for internal or external flaws. The frequency response is rather narrow, indicating that the sensor is resonant, in this case with a center frequency of about 290 kHz. Nevertheless, the resonance is not very sharp, indicating a lossy resonant circuit. In a device of this type one tries to operate at a frequency around the resonant frequency. The output of the sensor is typically a voltage when fed with a constant current source or a current when fed with a constant voltage source (i.e., either the voltage across the sensor or the current in the sensor is measured, depending on the way the sensor is fed). The amplitude and phase of the output are then related to the size, type, and location of flaws. (Eddy current sensors will be discussed in **Chapter 5**).

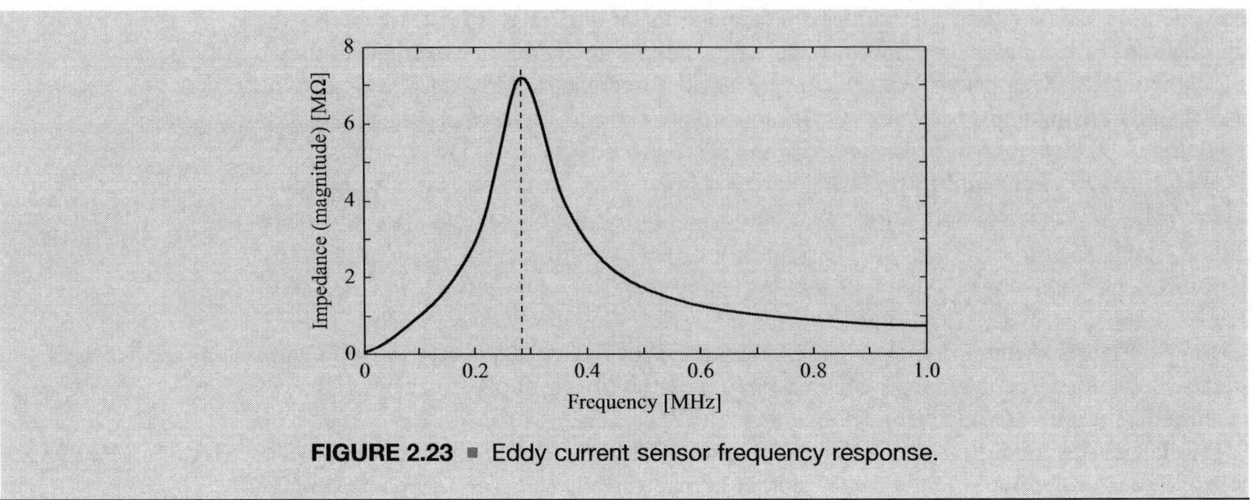

FIGURE 2.23 ■ Eddy current sensor frequency response.

2.2.8 Calibration

Calibration is the experimental determination of the transfer function of a sensor or actuator. Typically calibration of a device will be needed when the transfer function is not known or, more likely, when the device must be operated at tolerances below those specified by the manufacturer. Since tolerances indicate the maximum (sometimes typical) deviations of the device's transfer function from the ideal, if the device needs to operate at lower tolerances, we must specify the exact transfer function for the specific device. For example, we may wish to use a thermistor with a 5% tolerance on a full scale from 0°C to 100°C (i.e., the variations in measurements are ±5°C) to measure temperature with an accuracy of, say, ±0.5°C. The only way this can be done is by first establishing the transfer function of the sensor. For best results, this must be done for each device. There are two ways this can be accomplished.

One method assumes that the equation of the transfer function is known, in which case the constants in the equation must be determined experimentally. Suppose that the thermistor above has a linear transfer function between the range points given as $R = aT + b$, where T is the measured temperature and R is the resistance of the sensor, with a and b constants. To establish the transfer function for the thermistor we must specify the constants a and b, and for this we require two measurements at two different temperatures, T_1 and T_2. Then we can write

$$V_1 = aT_1 + b, \quad V_2 = aT_2 + b. \tag{2.20}$$

These are solved for a and b as

$$a = \frac{V_1 - V_2}{T_1 - T_2}, \quad b = V_1 - \frac{V_1 - V_2}{T_1 - T_2} T_1 = V_1 - aT_1. \tag{2.21}$$

This relation can now be supplied to the processor and from it the measured temperature is deduced through the transfer function $R = aT + b$.

If the transfer function is described by a more complex function, more measurements may be needed, but in all cases the constants in the relation must be determined.

For example, suppose that an actuator's output force is given as $F = aV + bV^2 + cV^3 + d$. The constants a, b, c, and d need to be determined and this will require four measurements and solution of a system of four equations.

In the calibration process one should be careful and choose measurement points within the span of the device and, especially for nonlinear transfer functions, space the measurement points more or less equally within the span. For linear functions, any two points will do, but even here they should not be too close to each other.

The second method is to assume no knowledge of the transfer function and determine it by a series of experiments. Typically a number of measurements will be needed measuring T_i and reading the resulting resistance R_i. These represent points of the transfer function. The points are plotted and a curve (best fit curve) passed through the points. It may turn out that the points define a (more or less) linear curve, in which case the linear least squares method described in **Appendix A (Equation (A.15))** may be used. Otherwise a polynomial fit through the points may be needed (see **Equation (A.24)** in **Appendix A**). Alternatively, the points may be supplied to the processor (especially if this is a microprocessor) as a table and the processor can then be programmed to retrieve these values and perhaps to interpolate between them for stimuli that fall between two measurements. This linear piecewise approximation may be quite sufficient, especially if the calibration uses a reasonably large number of points.

Calibration is a critical step in the use of sensors or actuators and should be undertaken with the utmost care. Measurements must be meticulous, instruments as accurate as possible, and conditions as close as possible to those under which the sensor or actuator will operate. One should also establish the errors in calibration or, at the very least, have a good estimate of the errors.

2.2.9 Excitation

Excitation refers to the electrical supply required for operation of a sensor or actuator. It may specify the range of voltages under which the device should operate (say, 2–12 V), the range of current, power dissipation, maximum excitation as a function of temperature, and sometimes frequency. Together with other specifications (such as mechanical properties and electromagnetic compatibility [EMC] limits), it defines the normal operating conditions of the sensor. Failure to follow rated values may result in erroneous outputs or premature failure of the device.

2.2.10 Deadband

Deadband is the lack of response or insensitivity of a device over a specific range of the input. In this range, which may be small, the output remains constant. A device should not operate in this range unless this insensitivity is acceptable. For example, an actuator that is not responding to inputs in a small range around zero may be acceptable, but one that "freezes" over the normal range may not be.

2.2.11 Reliability

Reliability is a statistical measure of the quality of a device that indicates the ability of the device to perform its stated function, under normal operating conditions, without failure, for a stated period of time or number of cycles. Reliability may be specified in hours or years of operation or as number of cycles or number of failures in a sample. Electronic components including sensors and actuators are rated in a number of ways.

The **failure rate** is the number of components that fail per given time period, typically per hour. A more common method of specifying the reliability of devices is in **mean time between failures** (MTBF). MTBF is the reciprocal of the failure rate: MTBF = 1/(failure rate). Another common term used to specify reliability is the **failure in time** (FIT) value. This measure gives the number of failures in 10^9 device-hours of operation. The device-hour figure can be made of any number of devices and hours as long as the product is 10^9 (say, 10^6 devices tested for 1000 hours). It can also be done with fewer device-hours and scaled to the required value. For example, one may test 1000 devices for 1000 hours and multiply the result by 1000.

Reliability data are usually provided by manufacturers and are based on accelerated lifetime testing conducted by the manufacturer. Although specification sheets do not usually provide much data about reliability or the methods used to obtain reliability data, most manufacturers will supply these data upon request and some may also have certification data based on standards of testing, when applicable.

It should be noted that reliability is heavily influenced by the operating conditions of devices. Elevated temperatures, higher voltages and currents, as well as environmental conditions (such as humidity) reduce reliability, sometimes dramatically. Any condition exceeding rated values must be taken into account and the reliability data derated accordingly. In some cases these data are available from manufacturers or professional organizations dedicated to issues of reliability. Calculators are available allowing the user to estimate reliability.

EXAMPLE 2.18 Failure rate

To test a component, 1000 identical components are tested for 750 hours and 8 of them fail during the test. The failure rate is

$$FR = \frac{8}{750 \times 1{,}000} = 1.067 \times 10^{-5}.$$

That is, the failure rate is 1.067×10^{-5} failures/h. The mean time between failures is MTBF = 93,750 hours. One can also estimate the FIT rate. Since the device-hour figure is $750 \times 1000 = 750{,}000$ device-hours, the FIT rate (number of failures for 10^9 device-hours) is

$$FIT = \frac{8}{750 \times 1{,}000} \times 10^9 = 10{,}666.$$

This (fictitious) component has extremely low reliability. Typical values for FIT rates are 2–5 and MTBF is usually measured in billions of hours.

2.3 | PROBLEMS

The transfer function

2.1 **Error in the simplified transfer function.** In **Example 2.1**, suppose that the transfer function is simplified to a third-order function by neglecting all terms except for the first three. Calculate the largest error expected over the range between 0°C and 1800°C.

2.2 **Transfer function of a position sensor.** The transfer function of a small position sensor is evaluated experimentally. The sensor is made of a very small magnet and the position with respect to the centerline (see **Figure 2.24**) is sensed by the horizontal, restoring force on the magnet. The magnet is held at a fixed distance, h, from the iron plate. The measurements are given in the table below.

a. Find the linear transfer function that best fits these data.

b. Find a transfer function in the form of a second-order polynomial ($y = a + bf + cf^2$), where y is the displacement and f is the restoring force by evaluating the constants a, b, and c.

c. Plot the original data together with the transfer functions in (a) and (b) and discuss the errors in the choice of approximation.

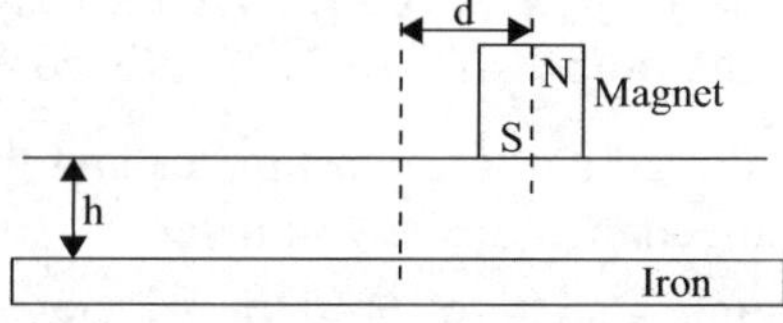

FIGURE 2.24 ■ A simple position sensor.

Displacement, d [mm]	0	0.08	0.16	0.24	0.32	0.4	0.48	0.52
Force [mN]	0	0.578	1.147	1.677	2.187	2.648	3.089	3.295

2.3 **Analytic form of transfer function.** In certain cases the transfer function is available as an analytic expression. One common transfer function used for resistance temperature sensors (to be discussed in **Chapter 3**) is the Callendar–Van Duzen equation. It gives the resistance of the sensor at a temperature T as

$$R(T) = R_0\left(1 + AT + BT^2 - 100CT^3 + CT^4\right) \quad [\Omega],$$

where the constants A, B, and C are determined by direct measurement of resistance for the specific material used in the sensor and R_0 is the temperature of the sensor at 0°C. Typical temperatures used for calibration are the oxygen point (−182.962°C; the equilibrium between liquid oxygen and its vapor), the triple point of water (0.01°C; the point of equilibrium temperature between ice, liquid water, and water vapor), the steam point (100°C; the equilibrium point between water and vapor), the zinc point (419.58°C; the equilibrium point between solid and liquid zinc), the silver point (961.93°C), and the gold point (1064.43°C), as well as others. Consider a platinum resistance sensor with a nominal resistance of 25 Ω at 0°C. To calibrate the sensor its resistance is measured at the oxygen point as 6.2 Ω, at the steam point as 35.6 Ω, and at the zinc point as 66.1 Ω. Calculate the coefficients A, B, and C and plot the transfer function between −200°C and 600°C.

Impedance matching

2.4 **Loading effects on actuators.** Suppose an 8 Ω loudspeaker is driven from an amplifier under matched conditions, meaning that the amplifier's output impedance equals 8 Ω. Under these conditions the amplifier transfers maximum power to the loudspeaker. The user decides to connect a second, identical speaker in parallel to the first to better distribute sound in the room. The amplifier supplies an output voltage $V = 48$ V.

a. Show that the total power on the two speakers is lower than that supplied to a single speaker.

b. If one desires to maintain the same power, what must be the impedance of the speakers assuming they are still connected in parallel to the amplifier?

2.5 **Loading effects.** The ratings of a piezoelectric sensor indicate that at a certain stimulus the sensor provides an output of 150 V at no load. The short-circuit current of the sensor is measured and found to be 10 μA (this is done by shorting the output through a current meter). To measure the output voltage of the sensor, a voltmeter with an internal impedance of 10 MΩ is connected across the sensor.

a. Calculate the actual reading of the voltmeter and the error in reading the stimulus due to the impedance of the voltmeter.

b. What must be the impedance of the voltmeter to reduce the error in reading below 1%?

2.6 **Impedance matching effects.** A pressure sensor produces an output varying between 0.1 and 0.5 V as the pressure sensed varies from 100 to 500 kPa. To read the pressure the sensor is connected to an amplifier with an amplification of 10 so that the output varies between 1 and 5 V for ease of interpretation of the sensed pressure. If the sensor has an internal impedance of 1 kΩ and the input impedance of the amplifier is 100 kΩ, what is the voltage range at the output of the amplifier?

2.7 **Power output of an electrical motor.** The torque of a DC motor is linear with the speed of the motor and given in **Figure 2.25**. Find the power transfer function of

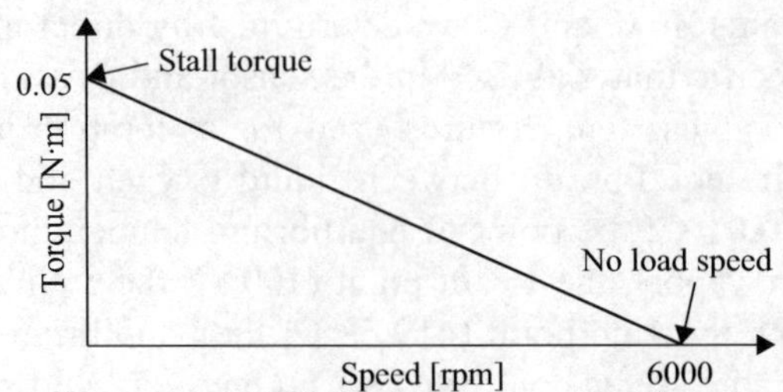

FIGURE 2.25 ■ Speed–torque curve of a DC motor.

the motor, that is, the relation between speed and power. Show that maximum mechanical power is obtained at half the no-load speed and/or half the stall torque.

Note: Power is the product of torque and angular velocity.

2.8 Power transfer to an actuator and matching. A power amplifier may be modeled as an ideal voltage source with a series internal impedance whereas a load is modeled as an impedance, both shown in **Figure 2.26**. The internal impedance of the amplifier is $Z_{in} = 8 + j2\ \Omega$ at frequency f.

a. Calculate and plot the power supplied to the load if $V_0 = 48$ V for a resistive load varying between 0 and 20 Ω.

b. Calculate and plot the power supplied to the load if $V_0 = 48$ V for a load varying between $Z_L = 8 + j0\ \Omega$ to $Z_L = 8 + j20\ \Omega$.

c. Show that the maximum power to the load is transferred if $Z_L = Z_{in}{}^* = 8 - j2\ \Omega$.

d. How do you explain the result in (c) from a physical point of view?

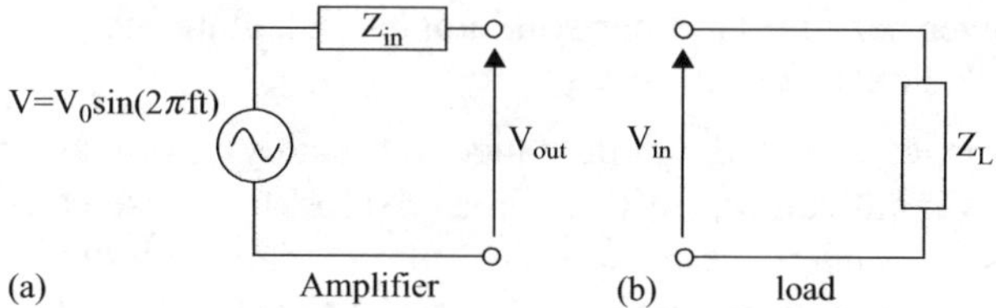

FIGURE 2.26 ■ Power transfer and matching. (a) Model of the amplifier output. (b) The load.

2.9 Matching of frequency-dependent impedances. An actuator is given with a resistance of 8 Ω at 100 Hz. Its inductance is measured and found to be 1 mH. The actuator is used at frequencies ranging from 10 to 2000 Hz. In attempting to match the actuator to a driving source the following are proposed:

1. Use a driver with an output resistance equal to 8 Ω and a source with peak voltage of 12 V.
2. Use a driver with an output impedance of $8 + j0.006f\ \Omega$, where f is the frequency [Hz] and a source with peak voltage of 12 V.
3. Use a driver with an output impedance of $8 - j0.006f\ \Omega$, where f is the frequency [Hz] and a source with peak voltage of 12 V.

a. Calculate and plot the output power supplied to the actuator for the three proposed drivers. Which one is better and why?

b. What are the requirements of an ideal driver for the given load?

Range, span, input and output full scale, and dynamic range

2.10 The human ear is a uniquely sensitive sensor. Its range is given either in pressure or in power per unit area. The ear can sense pressures as low as 2×10^{-5} Pa (on the order of a billionth of the atmospheric pressure) and can still function properly at levels of 20 Pa at the high end (the threshold of pain). Alternatively, the range can be specified from 10^{-12} W/m^2 to 10 W/m^2. Calculate the dynamic range for pressure and power.

2.11 **The human eye.** The human eye is sensitive from roughly 10^{-6} cd/m^2 (dark night, rod-dominated vision, essentially monochromatic) to about 10^6 cd/m^2 (bright sunlight). Calculate the dynamic range of the eye.

2.12 A loudspeaker is rated at 10 W, that is, it can produce 10 W of acoustic power. Since it is an analog actuator, the minimum range point is not well defined but there is a minimum power necessary to overcome friction. We will assume here that it is 10 mW. What is the dynamic range of the loudspeaker?

Note: A loudspeaker's dynamic range is measured in various ways, some of them intended to boost marketability rather than to describe the physical properties of the loudspeaker.

2.13 A digital frequency meter needs to measure the frequency of a microwave sensor whose frequency varies between 10 MHz and 10 GHz in increments of 100 Hz. What is the dynamic range of the frequency meter?

2.14 A liquid crystal display is said to have a contrast ratio of 3000:1, giving the ratio between the highest and lowest luminance it can display. Luminance is a measure of power per unit area per solid angle. Calculate the dynamic range of the display in decibels.

Note: In displays, the dynamic range is usually given as the contrast ratio, whereas in digital cameras it is given as *f*-stops, but these can always be written in decibels if necessary. For example, a digital camera with a contrast of 1024:1 is said to have a dynamic range of 2^{10} or 10 *f*-stops.

2.15 Digital signal processors, especially those handling audio and video data, must have large dynamic ranges. If the dynamic range of the A/D conversion of a signal processor must be at least 89 dB, what is the number of bits required of the D/A converter?

Sensitivity, accuracy, errors, and repeatability

2.16 **Linear approximation of nonlinear transfer function.** The response of a temperature sensor is given as

$$R(T) = R_0 e^{\beta\left(\frac{1}{T}-\frac{1}{T_0}\right)} \quad [\Omega],$$

where R_0 is the resistance of the sensor at temperature T_0 and β is a constant that depends on the material of the sensor. $T_0 = 20°\text{C}$. Temperatures T and T_0 are in K. Given: $R(T) = 1000\ \Omega$ at 25°C and 3000 Ω at 0°C. The sensor is intended for use between −45°C and 120°C.

a. Evaluate β for this sensor and plot the sensor transfer function for the intended span.

b. Approximate the transfer function as a straight line connecting the end points and calculate the maximum error expected as a percentage of full scale.

c. Approximate the transfer function as a linear least squares approximation and calculate the maximum error expected as a percentage of full scale.

Hysteresis

2.17 Hysteresis in a torque sensor. A torque sensor is calibrated by applying static torque to it (i.e., a certain torque is applied, the sensor response is measured, and then the torque is increased or decreased to measure another point on the curve). The following data are obtained. The first set is obtained by increasing torque, the second by decreasing it.

Applied torque [N·m]	2.3	3.14	4	4.84	5.69	6.54	7.39	8.25	9.09	9.52	10.37	10.79
Sensed torque [N·m]	2.51	2.99	3.54	4.12	4.71	5.29	5.87	6.4	6.89	7.1	7.49	7.62

Applied torque [N·m]	10.79	10.37	9.52	9.09	8.25	7.39	6.54	5.69	4.84	4	3.14	2.3
Sensed torque [N·m]	7.68	7.54	7.22	7.05	6.68	6.26	5.8	5.29	4.71	4.09	3.37	2.54

a. Plot the transfer function of the torque sensor using a second-order least squares approximation.

b. Calculate the maximum error due to hysteresis as a percentage of full scale.

2.18 The Schmitt trigger. Hysteresis is not necessarily a negative property. The use of hysteresis in electronic circuits can accomplish specific purposes. One of these is the Schmitt trigger. It is an electronic circuit whose output voltage (V_{out}) changes according to the input voltage (V_{in}) as follows:

$$\text{if } V_{in} \geq 0.5V_0,\ V_{out} = 0$$

$$\text{if } V_{in} \leq 0.45V_0,\ V_{out} = V$$

a. Draw the transfer function of the device for $0 \leq V_{in} \leq V_0$ with $V_0 = 5$ V.

b. Suppose the input is a sinusoidal voltage $V_{in} = 5\sin 2\pi f$, where t is time and $f = 1000$ Hz is the frequency of the signal. Plot the input and the output as a function of time. What is the purpose of the Schmitt trigger in this case?

2.19 Mechanical hysteresis. Springs are often used in sensors, particularly in sensing forces, taking advantage of the fact that under an applied force the spring contracts (or expands) based on the formula $F = -kx$ (Hooke's law), where F is the applied force, x is the spring contraction or expansion, and k is a constant called the spring constant. The negative sign indicates that compression reduces the length of the spring. However, the spring constant is force dependent and has slightly different variations under compression and expansion. Consider the spring in **Figure 2.27a** and its force–displacement curve given in **Figure 2.27b**.

a. Explain the meaning of the curve and, in particular, that of the horizontal lines on which additional force does not change the displacement.

b. What is the maximum error in displacement as a percentage of full scale due to hysteresis?

c. What is the maximum error in force as a percentage of full scale due to hysteresis?

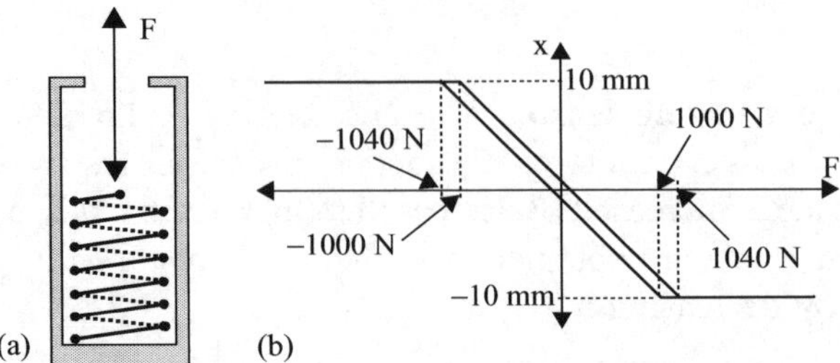

FIGURE 2.27 ■ Mechanical hysteresis in a spring. (a) Spring and applied force. (b) Response.

2.20 Hysteresis in thermostats. Hysteresis is often intentionally built into sensors and actuators. A simple example is the common thermostat. These devices are designed to switch off at a certain temperature and then switch back on at a certain temperature. The switch-on and switch-off temperatures must be different, otherwise the thermostat's status is undetermined and it will switch on and off rapidly when the set temperatures are reached. The hysteresis can be mechanical, thermal, or electronically imposed. Consider a thermostat intended to control temperature in a home. The set temperature is 18°C but the thermostat has a ±5% hysteresis.

a. For a temperature between 15°C and 24°C, determine the transfer function of the thermostat if it is used to turn on a heater to keep the room warm.
b. For the same range as in (a), determine the transfer function of the thermostat if it is to turn on an air conditioner to keep the room cool.

In both cases, show when the thermostat switches on and off.

Deadband

2.21 Deadband due to linkage slack. A linear rotary potentiometer of nominal value 100 kΩ is used to sample the voltage across a sensor (such as a microphone). The shaft has a rotational slack of 5°, that is, if one rotates it in one direction to a certain point and then rotates it in the opposite direction, the shaft must be rotated 5° before the slider responds. If the full-scale rotation is 310°, calculate the error due to the slack in the linkage in terms of resistance and as a percentage of full scale.

Reliability

2.22 Reliability. An electronic component's data sheet shows a mean time between failure (MTBF) of 4.5×10^8 hours when tested at 20°C. At 80°C the MTBF decreases to 62,000 hours. Calculate the failure rate (FR) and the failure in time (FIT) value for the component at the two temperatures.

2.23 Reliability. In testing a pressure sensor for reliability, 1000 sensors are tested for 850 hours. If six sensors fail during that period, what is the MTBF of the sensor?

CHAPTER 3

Temperature Sensors and Thermal Actuators

The human body and heat

A person consuming 2000 kcal/day, assuming all calories are expended in 24 hours, produces $2000 \times 1000 \times 4.184 = 8.368 \times 10^6$ J of energy. That translates to an average power of $8.368 \times 10^6/24/3600 = 96.85$ W. That's an average power of about 100 W, with lower values during sleep and higher values while active. During exertion such as exercise, a person can produce in excess of 1.5 kW. But the real story is that the body both expends energy and requires the resulting heat to regulate the body temperature and to supply its energy needs. The body requires a fairly narrow range of temperatures. The normal body temperature for most individuals is 37°C, but it can fluctuate somewhat, with women having an average temperature about 0.5°C lower than men. At temperatures above 37°C the body experiences fever, an elevated temperature due to failure of the body to regulate its temperature (typically because of sickness). Hyperthermia is an elevated body temperature that is not a fever, but rather is a reaction to external heat or to drugs or stimulants. Above 41.5°C the body enters into a state called hyperpyrexia, a dangerous state that can lead to death or serious side effects. Lower body temperature, hypothermia, is equally dangerous. Defined as a body core temperature below 35°C, it typically occurs due to exposure to extreme cold or extended immersion in cold water, but can also be due to trauma. Heat regulation in the body is controlled by the hypothalamus in the brain and can be accomplished using a number of methods, including sweating, increasing or decreasing heart rate, shivering, and constriction of blood vessels.

3.1 | INTRODUCTION

Temperature sensors are the oldest sensors in use (excluding the magnetic compass), dating back to the very beginning of the scientific age. Early thermometers were introduced in the early 1600s. Around the middle of the 1600s, the need for standards of temperature measurement were voiced by Robert Boyle and others. Shortly after, about 1700, some temperature scales were already in use, devised by Lorenzo Magalotti, Carlo Renaldini, Isaac Newton, and Daniel Fahrenheit. By 1742, all temperature scales, including the Celsius scale (devised in 1742 by Andres Celsius), but excluding the Kelvin scale, were established. Following the work of Leonard Carnot on engines and heat, Lord Kelvin proposed the absolute scale bearing his name in 1848 and established its relation with the Celsius scale. The temperature scales were further developed and improved until the establishment of the International Practical Temperature Scale in 1927, followed by further revisions to improve accuracy.

The classical thermometer is clearly a sensor, but in its common configuration it does not provide an output signal (but see **Section 3.2**). Its output is read directly. The establishment of temperature sensing was not far behind the development of thermometers. The Seebeck effect was discovered in 1821 by Thomas Johann Seebeck and within a few years it became an accepted method of temperature measurement when Antoine Cesar Becquerel used it in 1826 to build the first modern temperature sensor—the **thermocouple**. The Seebeck effect is the basis of thermocouples and thermopiles, the workhorses of industrial temperature sensing. A second effect related to the Seebeck effect was discovered in 1834 by Charles Athanase Peltier. Named the Peltier effect, it is used for sensing, but more often it is used for cooling or heating as well as for thermoelectric power generation (TEG). Whereas the Seebeck effect has been used from very early on for sensing, applications based on the Peltier effect had to wait for the discovery of semiconductors, in which the effect is much larger than in conductors. These thermoelectric effects are discussed in **Section 3.3** in detail, as well as in **Chapter 4** in conjunction with optical sensing, especially in the infrared region.

Following the observation by Humphry Davey in 1821 that the conductivity of metals decreases with temperature, William Siemens described in 1871 a method for measurement of temperature based on the resistance–temperature relation in platinum. This method has become the basis of **thermoresistive sensors** on which resistance temperature detectors (RTDs) are based. Its extension to semiconductors gave birth to **thermistors** (thermal resistors) as well as to a variety of other useful sensors including semiconductor temperature sensors.

Temperature sensing is both well established and in very widespread use. Many of the sensors available are deceptively simple in construction and can be extremely accurate. On the other hand, they sometimes require special instrumentation and considerable care to achieve this accuracy. A good example is the thermocouple. Arguably the most widespread of temperature sensors, especially at higher temperatures, it is also a very delicate instrument whose output signal is tiny and requires special techniques for measurement, special connectors, elimination of noise, and calibration, as well as a reference temperature. Yet, in its basic construction it is the "mere welding of any two dissimilar conductors" to form a junction. Others seem to be even simpler. A length of copper wire (or any other metal) connected to an ohmmeter can make an instant thermoresistive sensor of surprisingly good quality. Further adding to the widespread use and availability of temperature sensors is the fact that additional physical attributes may be measured indirectly through measurement of temperature. Examples are the use of temperature sensors to measure air velocity or fluid flow (what is measured is the cooling effect of moving air or that of a fluid with reference to a constant temperature) and radiation intensity, especially in the microwave and infrared spectra, by measuring the temperature increase due to absorption of radiation energy.

It should be noted that there are important thermal actuators as well. The fact that metals as well as gases expand with temperature allows these to be used for actuation. In many cases, sensing and actuation can be achieved directly. For example, a column of water, alcohol, or mercury, or a volume of gas will expand to indicate temperature directly or indirectly and, at the same time, this expansion can be used to actuate a dial or a switch. Direct reading cooking thermometers and thermostats are of this type. Another simple example is the bimetal switch commonly used in direction indicators in cars, where the sensing element activates a switch directly without the need of an intermediate controller.

3.1.1 Units of Temperature, Thermal Conductivity, Heat, and Heat Capacity

The SI unit of absolute temperature is the kelvin (K). It is based on the absolute zero. The common unit in day-to-day work is the degree Celsius (°C). The two are the same except for the reference zero temperature. The Kelvin scale's zero is the absolute zero, whereas the Celsius scale is based on the triple point of water. Thus 0°C = 273.15 K and one says that the absolute zero temperature is 0 K or −273.15°C. Conversions between the three common temperature scales are as follows:

From °C to K: N [°C] = (N + 273.15) [K]
From °C to °F: P [°C] = ($P \times 1.8 + 32$) [°F]
From K to °C: M [K] = ($M - 273.15$) [°C]
From °F to °C: Q [°F] = ($Q - 32$)/1.8 [°C]
From K to °F: S [K] = ($S - 273.15$) × 1.8 + 32 [°F]
From °F to K: U [°F] = ($U - 32$)/1.8 + 273.15 [K]

Heat is a form of energy. Therefore its SI derived unit is the joule (J). The joule is a small unit, so units of megajoules (MJ), gigajoules (GJ), and even terajoules (TJ) are used, as are smaller units all the way down to nanojoules (nJ) and even attojoules (aJ). Although there is a whole list of units of energy, some metric and some customary, we mention here only a few of the more common ones. A joule equals a watt-second (1 J = 1 W·s) or a newton meter (N·m), but perhaps in more common use is the kilowatt-hour (kWh) (1 kWh = 3.6 MJ). Another commonly used unit of energy, especially in relation to heat energy is the calorie (cal). The calorie is a thermochemical unit equal to 0.239 J. It should be noted here that in the United States the term calorie usually refers to 1000 calories or the kilocalorie (kcal), that is, what is typically called a calorie in the United States is in fact 1000 cal or 239 J. Neither the calorie nor the watt-hour are SI units. Both are considered obsolete units and their use is discouraged.

Thermal conductivity, denoted as k or λ, is measured in watts per meter per kelvin (W/m/K) and is a measure of the ability of materials to conduct heat. **Heat capacity**, denoted as C, refers to the amount of heat necessary to change the temperature of a substance by a given amount. The SI unit for heat capacity is the joule per Kelvin (J/K). In conjunction with chemical sensors, the **molar heat capacity** is often employed and has units of joules per mole per Kelvin (J/mol/K). Other useful quantities are the **specific heat capacity** (J/kg/K), the amount of heat required to increase the temperature of 1 kg of a substance by 1 K, and **volumetric heat capacity** (J/m^3/K), the amount of heat needed to increase the temperature of 1 m^3 of a substance by 1 K. Often, too, the units of specific heat capacity are given in joules per gram per kelvin (J/g/K) and those of volumetric heat capacity as joules per cubic centimeter per kelvin (J/cm^3/K).

3.2 | THERMORESISTIVE SENSORS: THERMISTORS, RESISTANCE TEMPERATURE SENSORS, AND SILICON RESISTIVE SENSORS

The bulk of thermoresistive sensors may be divided into two basic types: **resistance temperature detectors (RTDs)** and **thermistors** (the name is a concatenation of the words thermal and resistor). RTDs have come to indicate thermoresistive sensors based on solid conductors, usually in the form of metal wires or films. In these devices the resistance of the sensor increases with temperature—that is, the materials used have positive temperature coefficients (PTCs) of resistance. Silicon-based RTDs have also been developed and have the distinct advantage of being much smaller than conductor-based RTDs, as well as exhibiting higher resistances and higher temperature coefficients. Thermistors are semiconductor-based devices and usually have a negative temperature coefficient (NTC) of resistance, but PTC thermistors also exist.

3.2.1 Resistance Temperature Detectors

The earlier sensors of this type were made of an appropriate metal such as platinum, nickel, or copper, depending on the application, temperature range, and often the cost. All RTDs are based on the change in resistance due to the temperature coefficient of resistance (TCR) of the metal being used. The resistance of a conductor of length L with constant cross-sectional area S and conductivity σ (**Figure 3.1**) is

$$R = \frac{L}{\sigma S} \quad [\Omega]. \tag{3.1}$$

The conductivity of the material itself is temperature dependent and given as

$$\sigma = \frac{\sigma_0}{1 + \alpha\,[T - T_0]} \quad [\mathrm{S/m}], \tag{3.2}$$

where α is the temperature coefficient of resistance of the conductor, T is the temperature, and σ_0 is the conductivity of the conductor at the reference temperature T_0. T_0 is usually given at 20°C, but may be given at other temperatures as necessary. The resistance of the conductor as a function of temperature is therefore

$$R(T) = \frac{L}{\sigma_0 S}(1 + \alpha[T - T_0]) \quad [\Omega] \tag{3.3}$$

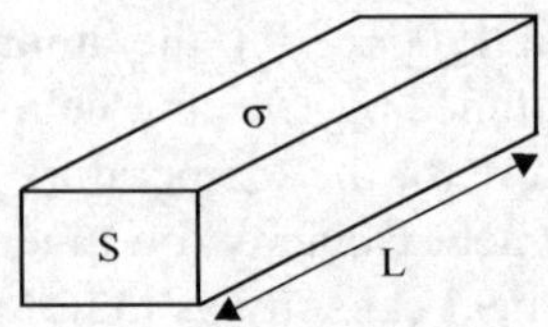

FIGURE 3.1 ■ Geometry used for calculation of resistance of a conductor of length L and uniform cross-sectional area S.

or

$$R(T) = R_0(1 + \alpha[T - T_0]) \quad [\Omega], \tag{3.4}$$

where R_0 is the resistance at the reference temperature T_0. In most cases the temperatures T and T_0 are given in degrees Celsius, but they can be specified in kelvin as well, provided both are given on the same scale.

Although this relation is linear, the coefficient α is usually quite small and the conductivity σ_0 is large. For example, for copper, $\sigma_0 = 5.8 \times 10^7$ S/m and $\alpha = 0.0039$/°C (the coefficient can also be indicated as Ω/Ω/°C) at $T_0 = 20$°C. Taking a wire with a cross-sectional area S of 0.1 mm^2 and length L of 1 m gives a change in resistance of 6.61×10^{-5} Ω/°C and a base resistance of 0.017 Ω at 20°C. This is a change of 0.38%. Thus, for the sensor to be practical, the conductor must be long and thin and/or conductivity must be low. A large temperature coefficient is also useful in that the change in resistance is large and hence processing of the signal obtained is easier. The TCRs and conductivities for a number of useful materials are given in **Table 3.1**.

Equation (3.3) is useful in establishing the physics of RTDs and gives useful insight into how resistance changes with temperature. However, in practice, one typically does not know the properties of the material used in producing the sensor. The typical data available are the nominal resistance of the RTD, usually given at 0°C, the range

TABLE 3.1 ■ Conductivities and temperature coefficients of resistance for selected materials (at 20°C unless otherwise indicated)

Material	Conductivity σ [S/m]	Temperature coefficient of resistance [per °C]
Copper (Cu)	5.8×10^7	0.0039
Carbon (C)	3.0×10^5	−0.0005
Constantan (60% Cu, 40% Ni)	2.0×10^6	0.00001
Chromium (Cr)	5.6×10^6	0.0059
Germanium (Ge)	2.2	−0.05
Gold (Au)	4.1×10^7	0.0034
Iron (Fe)	1.0×10^7	0.0065
Mercury (Hg)	1.0×10^6	0.00089
Nichrome (NiCr)	1.0×10^6	0.0004
Nickel (Ni)	1.15×10^7	0.00672
Platinum (Pl)[2]	9.4×10^6	0.003926 (at 0°C)
Silicon (Si) (pure)	4.35×10^{-6}	−0.07
Silver (Ag)	6.1×10^7	0.0016
Titanium (Ti)	1.8×10^6	0.042
Tungsten (W)	1.8×10^7	0.0056
Zinc (Zn)	1.76×10^7	0.0059
Aluminum (Al)	3.6×10^7	0.0043

Notes:

1. Instead of conductivity, σ [S/m], some sources list resistivity, ρ, measured in ohm-meters ($\rho = 1/\sigma$ [Ωm]).
2. Platinum is a particularly important material and there are different grades with different TCRs in use. The TCR is often given at 0°C. The most common TCRs at 0°C are 0.00385 (European curve), 0.003926 (American curve), and 0.00375 (common in thin-film sensors). The TCR of pure platinum at 0°C is 0.003926. Some alloys in use have TCRs of 0.003916 and 0.003902 (at 0°C). Other grades can be made by alloying pure platinum with materials such as rhodium.
3. The TCR of a material changes with temperature (see **Problem 3.9**). For example, the TCR of pure platinum at 20°C is 0.003729/°C.

EXAMPLE 3.1 **Wire-spool sensor**

A spool of magnet wire (copper wire insulated with a thin layer of polyurethane) contains 500 m of wire with a diameter of 0.2 mm. It is proposed to use the spool as a temperature sensor to sense the temperature in a freezer. The proposed range is between −45°C and +10°C. A milliammeter is used to display the temperature by connecting the sensor directly to a 1.5 V battery and measuring the current through it.

a. Calculate the resistance of the sensor and the corresponding currents at the minimum and maximum temperatures.

b. Calculate the maximum power the sensor dissipates.

Solution: The resistance of a length of wire, disregarding temperature is

$$R = \frac{l}{\sigma s},$$

where l is the length of the wire, S is its cross-sectional area, and σ is its conductivity.

The conductivity is temperature dependent. For copper, **Table 3.1** gives the conductivity as $\sigma_0 = 5.8 \times 10^7$ S/m at 20°C. Thus the resistance is written as a function of temperature using **Equation (3.3)**:

$$R(T) = \frac{l}{\sigma_0 S}(1 + \alpha[T - 20°]).$$

The TCR, α, for copper is given in **Table 3.1**. At −45°C,

$$R(-45°) = \frac{500}{5.8 \times 10^7 \times \pi \times (0.0001)^2}(1 + 0.0039[-45 - 20°]) = 204.84\ \Omega.$$

At +10°C,

$$R(+10°) = \frac{500}{5.8 \times 10^7 \times \pi \times (0.0001)^2}(1 + 0.0039[10 - 20°]) = 263.7\ \Omega.$$

The resistance changes from 204.84 Ω at −45°C to 263.7 Ω at +10°C. The currents are

$$I(-45°) = \frac{1.5}{204.84} = 7.323 \text{ mA}$$

and

$$I(+10°) = \frac{1.5}{263.7} = 5.688 \text{ mA}.$$

The current is linear with temperature and amounts to a sensitivity of 29.72 μA/°C. It is not a particularly large current, but even the simplest digital multimeter should be able to measure a change of 10 μA, or a temperature change of about 0.3°C. With a better microammeter, resolution down to 1 μA, or 0.03°C is possible.

The power dissipated is

$$P(+10^\circ) = I^2R = \left(5.688 \times 10^{-3}\right)^2 \times 263.7 = 8.53 \text{ mW}$$

and

$$P(-45^\circ) = I^2R = \left(7.323 \times 10^{-3}\right)^2 \times 204.84 = 10.98 \text{ mW}.$$

The power is low, an important property in temperature sensors since, as we shall see shortly, the power dissipated in the sensor can lead to errors due to self-heating. Note also the absolute simplicity of this sensor.

EXAMPLE 3.2 Wire RTD resistance and sensitivity

A wire-wound RTD sensor is made of pure platinum wire, 0.01 mm in diameter, to have a resistance of 25 Ω at 0°C. Assume here that the TCR is constant with temperature.

a. Find the necessary length for the wire.
b. Find the resistance of the RTD at 100°C.
c. Find the sensitivity of the sensor in ohms/degree Celsius [Ω/°C].

Solution:

a. The resistance is written as a function of temperature using **Equation (3.3)**:

$$R(T) = \frac{l}{\sigma_0 S}(1 + \alpha[T - 20^\circ]).$$

The TCR, α, for platinum is given in **Table 3.1** at 0°C but the conductivity is given at 20°C. The resistance of the RTD at 0°C is

$$25 = \frac{l}{9.4 \times 10^6 \times \pi \times \left(0.05 \times 10^{-3}\right)^2}(1 + 0.003926[0 - 20]) = 12.48154l \quad [\Omega].$$

This gives

$$l = \frac{25}{12.48154} = 2.003 \text{ m}.$$

The sensor requires 2 m of platinum wire.

b. At 100°C,

$$R(100^\circ\text{C}) = \frac{2.003}{9.4 \times 10^6 \times \pi \times \left(0.05 \times 10^{-3}\right)^2}(1 + 0.003926[100 - 20]) = 35.652 \ \Omega.$$

The resistance changes from 25 Ω at 0°C to 35.652 Ω at +100°C.

c. Sensitivity is calculated at an arbitrary temperature by calculating the resistance at that temperature then increasing the temperature by 1°C, calculating the resistance, and subtracting the former from the latter. At a temperature T we have

$$R(T) = \frac{l}{\sigma_0 S}(1 + \alpha[T - 20°]).$$

At a temperature $T+1$,

$$R(T+1) = \frac{l}{\sigma_0 S}(1 + \alpha[(T+1) - 20°]).$$

The difference in resistance between the two is

$$R(T+1) - R(T) = \frac{l}{\sigma_0 S}(1 + \alpha[T + 1 - 20°]) - \frac{l}{\sigma_0 S}(1 + \alpha[T - 20°]) = \frac{l\alpha}{\sigma_0 S}.$$

Note that this expression is the slope of the function $R(T)$. We get

$$\Delta R = \frac{l\alpha}{\sigma_0 S} = \frac{2.003 \times 0.003926}{9.4 \times 10^6 \times \pi \times (0.05 \times 10^{-3})^2} = 0.1065\ \Omega.$$

The sensitivity is therefore 0.1065 Ω/°C.

Check: Since the resistance is linear with temperature, the sensitivity is the same everywhere and thus we can write the resistance at 100°C as

$$R(100°\text{C}) = R(0°\text{C}) + 100 \times \Delta R = 25 + 100 \times 0.1065 = 35.65\ \Omega.$$

The small difference is due to truncation of the numbers during evaluation.

(say between −200°C and +600°C) and additional data on performance such as self-heat, accuracy, and the like. The transfer function can be obtained in one of two ways. First, the manufacturers conform to existing standards that specify the coefficient α that a sensor uses. For example, standard EN 60751, dealing with platinum RTDs, specifies $\alpha = 0.00385$ (this is sometimes called the "European curve"). Other values are 0.003926 ("American curve"), 0.003916, and 0.003902, among others, and relate to grades of platinum (see the notes to **Table 3.1**). This value is part of the sensor's specifications. This allows one to establish an approximate transfer function as follows:

$$R(T) = R(0)[1 + \alpha T] \quad [\Omega], \tag{3.5}$$

where $R(0)$ is the resistance at 0°C and T is the temperature at which the resistance is sought. This is an approximate value because α is itself temperature dependent.

To improve on this, the resistance as a function of temperature is calculated based on relations established from the actual measurements and given, again, by the same standards. This is given as follows:

For $T \geq 0$°C:

$$R(T) = R(0)[1 + aT + bT^2] \quad [\Omega]. \tag{3.6}$$

The coefficients are calculated for each material based on fixed temperatures (see **Problem 3.6**). For example, for platinum (standard EN 60751, $\alpha = 0.00385$), the coefficients are

$$a = 3.9083 \times 10^{-3},\ b = -5.775 \times 10^{-7}.$$

For $T < 0°C$:

$$R(T) = R(0)[1 + aT + bT^2 + c(T - 100)T^3] \quad [\Omega]. \tag{3.7}$$

The coefficients are (again from standard EN 60751, $\alpha = 0.00385$):

$$a = 3.9083 \times 10^{-3},\ b = -5.775 \times 10^{-7},\ c = -4.183 \times 10^{-12}.$$

These relations are known as the Callendar–Van Dusen equations or polynomials. Instead of using the polynomials one can use design tables that list the values of resistance at various temperatures. For other values of α, the coefficients are different, but are specified by the standards or can be calculated based on accurate measurements. It should be noted that for small sensing spans close to the nominal temperature, the temperature curve is nearly linear and **Equation (3.5)** is sufficiently accurate. **Equations (3.6)** and **(3.7)** are only needed for larger spans or if sensing is done at low or high temperatures (see **Example 3.3**).

EXAMPLE 3.3 RTD representation and accuracy

An RTD with nominal resistance of 100 Ω at 0°C is specified for the range −200°C to +600°C. The engineer has the option of using the approximate transfer function in **Equation (3.5)** or the exact transfer function in **Equations (3.6)** and **(3.7)**. Assume $\alpha = 0.00385$/°C.

a. Calculate the error incurred by using the approximate transfer function at the extremes of the range.

b. What are the errors if the range used is −50°C to +100°C?

Solution:

a. From **Equation (3.5)**:

At 600°C,

$$R(600°C) = R(0)[1 + \alpha T] = 100\,[1 + 0.00385 \times 600] = 331\ \Omega.$$

At −200°C,

$$R(-200°C) = 100[1 - 0.00385 \times 200] = 23\ \Omega.$$

From **Equation (3.6)**:

$$\begin{aligned} R(600°C) &= R(0)[1 + aT + bT^2] \\ &= 100\,[1 + 3.9083 \times 10^{-3} \times 600 - 5.775 \times 10^{-7} \times 600^2] = 313.708\ \Omega. \end{aligned}$$

From **Equation (3.7)**:

$$\begin{aligned} &100[1 + 3.9083 \times 10^{-3} \times (-200) - 5.775 \times 10^{-7} \times 200^2 \\ &- 4.183 \times 10^{-12} \times (-200)^3] = 18.52\ \Omega. \end{aligned}$$

The resistance calculated with the approximate formula is higher by 5.51% at 600°C and higher by 24.19% at −200°C. These deviations are not acceptable and therefore one cannot use the approximate formula for the whole range—the use of the Callendar–Van Dusen relations is essential.

b. From **Equation (3.5)**:

At 100°C,

$$R(100°C) = R(0)[1 + \alpha T] = 100\,[1 + 0.00385 \times 100] = 138.5\ \Omega.$$

At −50°C,

$$R(-50°C) = 100[1 - 0.00385 \times 50] = 80.75\ \Omega.$$

From **Equation (3.6)**:

$$R(100°C) = 100\,[1 + 3.9083 \times 10^{-3} \times 100 - 5.775 \times 10^{-7} \times 100^2] = 138.5055\ \Omega.$$

From **Equation (3.7)**:

$$100[1 + 3.9083 \times 10^{-3} \times (-50) - 5.775 \times 10^{-7} \times 50^2 - 4.183 \times 10^{-12} \times (-50)^3] = 80.3063\ \Omega.$$

The resistance calculated with the approximate formula is only 0.0397% lower at 100°C and lower by 0.552% at −50°C. These deviations may be quite acceptable and the approximate formula can be used.

In designing RTDs, one has to be careful to minimize the effect of tension or strain on the wires. The reason for this is that tensioning a conductor changes its length and cross-sectional area (constant volume), which has exactly the same effect on resistance as a change in temperature. An increase in strain on the conductor increases the resistance of the conductor. This effect will be explored in detail in **Chapter 6** when we talk about strain gauges. There we shall see that the opposite problem occurs—changes in temperature cause errors in strain readings and these errors must be compensated for.

A characteristic property of wire RTDs is their relatively low resistance. High resistances would require very long wires or excessively thin wires. Another consideration is cost. High resistance RTDs require more material, and since most RTDs are based on platinum, material costs can be significant. For wire sensors, a satisfactory resistance is from a few ohms to a few tens of ohms, but thin film sensors with higher resistances can be made. In wire sensors, the wire is made of a fairly thin uniform wire wound in a small diameter coil (typically, but not always) and the coil is then supported on a suitable support such as mica or glass. If the total length of the wire is small, this support is not necessary and the wire coil, or sometimes just a length of wire, may be free standing or threaded around pegs to keep it in place. Depending on the intended use, the wire and support may be enclosed in an evacuated glass (typically Pyrex) tube with connecting wires going through the tube or in a highly conductive metal (sometime stainless steel) to allow better heat transfer to the sensing wire and therefore faster response of the sensor, or in a ceramic enclosure for higher temperature applications. They may also be flat for surface applications, encircling, and other forms.

When precision sensors are needed, platinum or platinum alloy is the first choice because of its excellent mechanical and thermal properties. In particular, platinum is chemically stable even at elevated temperatures, it resists oxidation, can be made into thin wires of high chemical purity, resists corrosion, and can withstand severe environmental conditions. For these reasons it can be used at temperatures up to about 850°C and down to below −250°C. On the other hand, platinum is very sensitive to strain and to chemical contaminants, and because its conductivity is high, the wire length needed is long (a few meters, depending on the required resistance). As a consequence, the resulting sensor is physically large and not suitable for sensing where temperature gradients are high.

For less demanding applications, both in terms of stability and temperature, nickel, copper, and other conducting materials offer less expensive alternatives at reduced performance. Nickel can be used from about −100°C to about 500°C, but its *R-T* curve is not as linear as that of other materials. Copper has excellent linearity, but can only be used from about −100°C to about 300°C at best. At higher temperatures, tungsten is often a good choice.

Thin film sensors are produced by depositing a thin layer of a suitable material, such as platinum or one of its alloys, on a thermally stable, electrically nonconducting ceramic that must also be a good heat conductor. The thin film can then be etched to form a long strip (typically in a meander fashion) and the sensor is potted in epoxy or glass to protect it. The final package is typically small (a few millimeters), but can vary in size depending on the application and required resistance. Typical resistance is 100 Ω, but much higher resistances, upwards of 2000 Ω are available. Thin film sensors are typically small and relatively inexpensive and are often the choice in modern sensors, especially when the very high precision of platinum wire sensors is not needed. **Figure 3.2** shows schematic constructions of wire and thin film RTDs.

RTDs typically come in relatively low resistances, especially in the case of wire RTDs. Because of this, an important issue in precision sensing is the resistance of the lead wires, which, necessarily, are made of other materials compatible with external circuits (copper, tinned copper, etc.). The resistance of the lead wires also changes with temperature, and these effects can add to errors in the sensing circuit since these resistances are not negligible (except in some high-resistance thin-film RTDs). Because of this, some commercial sensors come in two-, three-, or four-wire configurations as shown in **Figure 3.3**. The purpose of these configurations is to facilitate compensation for the lead wires. We shall see how this is done in **Chapter 11**, but it should be noted here that the two-wire sensor's leads cannot be compensated, whereas the three- and four-wire sensors allow compensation of the lead resistance and should be used when high precision is essential. To understand why these configuration are important, it is

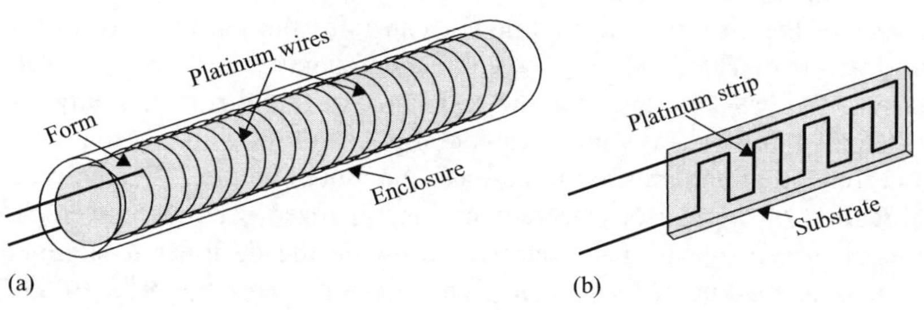

FIGURE 3.2 ■ Schematic construction of RTDs. (a) Wire-wound RTD. (b) Thin film RTDs.

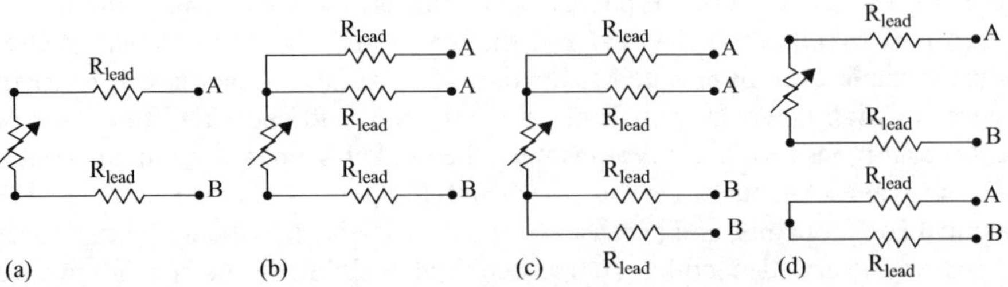

FIGURE 3.3 ■ RTD connection styles. (a) Two-wire (uncompensated). (b) Three-wire. (c) Four-wire. (d) Two-wire with compensation loop. The three- and four-wire styles allow compensation for temperature variations and resistance of the lead wires.

sufficient to note that if the resistance in *A-A* in **Figure 3.3b** is subtracted from the total resistance (measured between *A-B*), one obtains the resistance of the RTD regardless of the resistance of the lead wires. Similar compensation can be achieved with the configuration in **Figure 3.3c** and **3.3d**, but we will see in **Chapter 11** that other methods are often more effective and easier to implement.

Thermoresistive sensors must be calibrated for operation in the range of temperatures for which they were designed. Calibration procedures and calibration temperatures are specified in standards.

The accuracy of thermoresistive sensors can vary considerably depending on materials, temperature range, construction, and methods of measurement. Typical accuracies are on the order of ±0.01°C to ±0.05°C. Higher- and lower-accuracy sensors are available.

The stability of RTDs is measured in degrees Celsius per year (°C/year) and is on the order of 0.05°C/year or less for platinum sensors. Other materials have poorer stability.

3.2.1.1 Self-Heat of RTDs

We shall discuss the connection of sensors in measuring circuits in **Chapter 11**. At this point, however, it is important to mention the fact that many temperature sensors, including thermoresistive sensors, are very much subject to errors due to increases in their own temperature produced by the heat generated in them by the current used to measure their resistance. This is of course a problem with any active sensor, but it is particularly acute in temperature sensors. The rise in temperature may be understood qualitatively from the fact that the higher the current in the sensor, the larger the output signal available. This is particularly important for wire sensors whose resistances are small. On the other hand, power dissipated in the conductor is proportional to the square of the current, and this power can raise the temperature of the sensor, introducing an error. The power can be calculated quantitatively as $P_d = I^2R$, where I is the current (DC or RMS) and R is the resistance of the sensor. In many sensors the power dissipated follows a much more complicated relation so that the relation between current and temperature increase can be quite complex. Typically, as part of the specification of the sensor, the temperature increase per unit power (°C/mW in most cases) is given by the manufacturer, allowing the designer to compensate for these errors in the reading of the sensor. Typical errors are on the order of 0.01°C/mW to 0.2°C/mW, depending on the sensor and on environmental factors such as the cooling conditions (moving air or standing air, contact with heat sinks, in stationary or moving fluids, etc.).

EXAMPLE 3.4 Self-heat of RTDs

Consider the self-heat of an RTD operating in the range −200°C to +850°C that has a nominal resistance of 100 Ω at 0°C and a temperature coefficient of resistance of 0.00385/°C. Its self-heat is provided in its data sheet as 0.08°C/mW in air (typically this value is given at a low airspeed of 1 m/s). Calculate the maximum error expected due to self-heat if

a. The resistance is measured by applying a constant voltage of 0.1 V across the sensor.

b. The resistance is measured by applying a constant current of 1 mA through the sensor.

Note: Both of these measurements provide a nominal current of 1 mA at 0°C.

Solution: First we need to calculate the resistances at the extremes of the span using **Equations (3.6)** and **(3.7)**. These provide the following values:

$$\begin{aligned} R(-200^\circ\mathrm{C}) &= R(0)[1 + aT + bT^2 + c(T - 100)T^3] \\ &= 100[1 + 3.9083 \times 10^{-3} \times (-200) - 5.775 \times 10^{-7} \times 200^2 \\ &\quad -4.183 \times 10^{-12} \times (-200)^3] \\ &= 18.52\ \Omega \end{aligned}$$

and

$$\begin{aligned} R(850^\circ\mathrm{C}) &= R(0)[1 + aT + bT^2] \\ &= 100[1 + 3.9083 \times 10^{-3} \times 850 - 5.775 \times 10^{-7} \times 850^2] \\ &= 390.48\ \Omega. \end{aligned}$$

a. For a constant voltage source, we write the power dissipated as follows:

At −200°C and 850°C,

$$P(-200^\circ\mathrm{C}) = \frac{V^2}{R} = \frac{0.01}{18.52} = 0.54\ \mathrm{mW}$$

and

$$P(850^\circ\mathrm{C}) = \frac{V^2}{R} = \frac{0.01}{390.48} = 0.0256\ \mathrm{mW}.$$

The maximum error occurs at −200°C and equals

$$\text{error at } -200^\circ\mathrm{C} = \frac{0.54}{0.08} = 6.75^\circ\mathrm{C}$$

and

$$\text{error at } 850^\circ\mathrm{C} = \frac{0.0256}{0.08} = 0.32^\circ\mathrm{C}.$$

At the high end of the span the error is only 0.32°C.

b. With a current source, we write

$$P(-200°C) = I^2R = (1 \times 10^{-3})^2 \times 18.52 = 0.0185 \text{ mW}$$

and

$$P(850°C) = I^2R = (1 \times 10^{-3})^2 \times 390.48 = 0.39 \text{ mW}.$$

The maximum error occurs at 850°C and equals

$$\text{error at } 850°C = \frac{0.39}{0.08} = 4.875°C.$$

At the low end of the range the error is only 0.23°C:

$$\text{error at } -200°C = \frac{0.0185}{0.08} = 0.23°C.$$

Both methods are used and in both the errors vary with temperature. To reduce the error the current can be reduced, but it cannot be too small or difficulties in measurements as well as noise may be encountered.

3.2.1.2 Response Time

The response of most temperature sensors is slow, especially if they are physically large. Typically these are on the order of a few seconds (90% of steady state) and are given in the specification data published by manufacturers. It can range from as little as 0.1 s in water to 100 s in air. It also changes with flowing or standing water or air and these data are usually available. Wire RTDs have the slowest response because of their physical size. Typical specifications are for 50% and 90% of steady-state response in moving air and flowing water (see **Example 3.5**). The response time may be specified for other response levels and under other conditions as necessary. The response time is measured by applying a step change in temperature, ΔT, and measuring the time it takes the sensor to reach a certain temperature (usually one measures the sensor's resistance to deduce its temperature). For example, 50% of steady state means the sensor has reached a temperature equal to its initial value before the step change has been applied plus 50% of the step. The time needed to reach this temperature is the 50% response time.

EXAMPLE 3.5 Specification of response time in RTDs

The response time for the sensor in **Example 3.4** is evaluated experimentally in moving air and in flowing water as follows: For the measurement in air, the RTD is placed in a stream of air at an ambient temperature of 24°C moving at approximately 1 m/s. At a time $t = 0$, a heater is turned on, heating the air to 50°C. For the flowing water test, the RTD is placed in a pipe and allowed to settle at the ambient temperature of 24°C. A stream of cold water at 15°C moving at approximately 0.4 m/s is turned on at $t = 0$. The data obtained are shown in the tables below. The resistance of the sensor was measured and its temperature calculated from these data.

RTD in moving air

Time [s]	0	1	2	3	4	5	6	7	8	9	10
Temperature [°C]	24	25	26.4	28.6	31.6	35	38.3	40.5	42.1	43.5	44.4
Time [s]	11	12	13	14	15	16	17	18	19	20	
Temperature [°C]	45.6	46	46.6	47.1	47.5	47.7	48	48.2	48.5	48.8	

RTD in moving water

Time [s]	0	.05	.1	.15	.2	.25	.3	.35	.4	.45	.5
Temperature [°C]	24	23.1	21.7	20.3	18.8	17.6	16.8	15.9	15.6	15.3	15.2
Time [s]	.55	.6	.65	.7	.75	.8	.85	.9	.95	1.0	
Temperature [°C]	15.1	15	15	15	15	15	15	15	15	15	

a. Estimate the 50% and 90% response time in air.
b. Estimate the 50% and 90% response time in water. A 50% and 90% response means that the sensor has reached 50% or 90% of the final reading expected.

Solution: In many cases the data will be given in plots, but in this case we have tabulated data allowing direct calculation of the response time.

a. 50% of steady state for the measurement in air means $24+(50-24)\times 0.5=37°C$. 90% of steady state means $24+(50-24)\times 0.9=47.4°C$.

The response time may be estimated from the tables above (using linear interpolation between tabulated values):

- The 50% response time is approximately 5.5 s.
- The 90% response time is approximately 14.75 s.

b. In water, the step is negative, equal to $-9°C$. Therefore 50% of steady state means $24+(15-24)\times 0.5=19.5°C$ and 90% of steady state means $24+(15-24)\times 0.9=15.9°C$. Using the table above and interpolating between values, we get

- the 50% response time is approximately 0.23 s.
- the 90% response time is approximately 0.35 s.

3.2.2 Silicon Resistive Sensors

The conductivity of semiconductors is best explained in terms of quantum effects. We will discuss the quantum effects in semiconductors in the next chapter in conjunction with the photoconductive effect, but for the purpose of this discussion it is useful to point out some of the thermal effects that affect conductivity. To do so we use the classical model of valence and conduction electrons and known relations in semiconductors. Valence electrons may be viewed as those bound to atoms and hence not free to move. Conduction electrons are free to move and affect the current through the semiconductor. In a pure semiconductor, most electrons are valence electrons and are said to be in the valence band. For an electron to move into the conduction band it must acquire additional energy, and in moving into the conduction band it leaves behind a hole (positively charged particle). This energy is called the band gap energy and is material dependent. In the case we are dealing with here, the additional energy comes from heat, but of course it can come from radiation (light, nuclear, electromagnetic). Based on this description, the higher the temperature, the higher the number of electrons (and holes) available and hence the higher the current that can flow through the device (i.e., the lower its resistance). As temperature increases, the resistance decreases and hence pure semiconductors such as silicon typically have negative temperature coefficient (NTC) characteristics. Of course, the behavior of the material is much more complex than that given here and must take into account many parameters such as carrier mobility (which in itself may be temperature dependent) and the purity level of the semiconductor.

Semiconductors are rarely used as pure (intrinsic) materials. More often, impurities are introduced into the intrinsic material in a process called doping. Specifically, when doping silicon with an *n*-type impurity such as arsenic or antimony, the reverse effect is observed below a certain temperature. Typically for *n*-type silicon, a positive temperature coefficient (PTC) is observed below about 200°C. Above that temperature the properties of intrinsic (pure) silicon predominate and the behavior is NTC. The explanation is that at these higher temperatures the energy is so high as to spontaneously generate carriers (move them into the conduction band) regardless of doping. Of course, for practical sensing applications using silicon semiconductor devices, the range of interest is below 200°C.

The conductivity of semiconductors is given by the following relation:

$$\sigma = e(n\mu_e + p\mu_h) \quad [\mathrm{S/m}], \tag{3.8}$$

where e is the charge of the electron (1.61×10^{-19} C), n and p are the concentrations of the electrons and holes in the material (in units of particles/cm^3), and μ_e and μ_p are the mobilities of the electrons and holes, respectively (typically given in units of cm^2/V/s). This relation clearly indicates that conductivity depends on the type of material, since both concentrations and mobilities are material dependent. They are also temperature dependent. In an intrinsic material, the concentration of electrons and holes is equal ($n = p$), but of course in doped materials they are not. However, the concentrations are related through the **mass-action law**:

$$np = n_i^2, \tag{3.9}$$

where n_i is the intrinsic concentration. As the concentration of one type of carrier increases through doping, the other decreases proportionally. In the limit, one type of carrier dominates and the material becomes an *n*-type or *p*-type material. In that case the conductivity of the semiconductor is

$$\sigma = en_d\mu_d \quad [\mathrm{S/m}], \tag{3.10}$$

where n_d is the concentration of the dopant and μ_d is its mobility.

The relations for conductivity apply to all semiconductors and can be used to calculate conductivity based on the properties of the semiconductor. In particular, the concentrations of carriers are temperature dependent and that relation is nonlinear. Nevertheless, the conductivity variation with temperature is a useful measure of temperature. A class of temperature sensors based on silicon in which the nonlinearity is

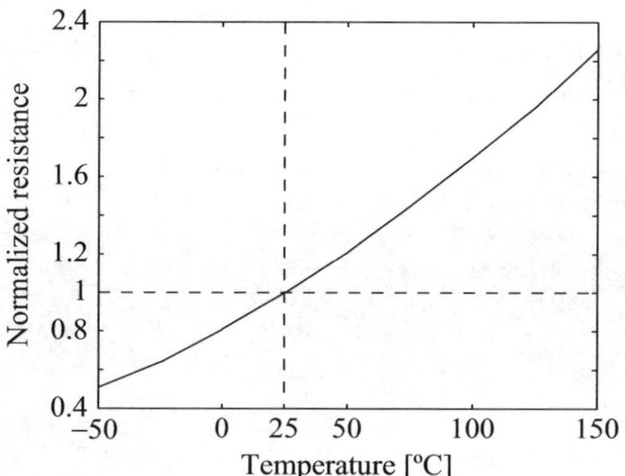

FIGURE 3.4 ■ Normalized resistance (at 25°C) versus temperature for a silicon resistive sensor. The nominal resistance of 1 kΩ at 25°C is indicated by the intersection of the dashed lines.

relatively mild exists. These are called **silicon resistive sensors**. The reduction in nonlinearity is based on construction and the proper selection of dopants, allowing for a useful sensor with sensitivity much higher than that of metal-based sensors, but, as expected, with a much narrower range of temperatures.

Silicon resistive sensors are somewhat nonlinear and offer sensitivities on the order of 0.5–0.7%/°C. They can operate in a limited range of temperatures like most semiconductor devices based on silicon (between −55°C to +150°C). Physically these sensors are very small, made of a piece of silicon with two electrodes deposited on it and encapsulated, usually in epoxy or glass. Typical resistances are on the order of 1 kΩ, specified at a temperature in the span of the device (typically 25°C). Because of the problem of self-heat, the current through these sensors must be kept to a minimum. As a whole, these devices are simple and inexpensive, but their accuracy is limited, with most devices exhibiting errors between 1% and 3%. The normalized resistance of a silicon resistive sensor is shown in **Figure 3.4**.

The transfer function of silicon resistive sensors is given by the manufacturer as a table, as a plot (similar to the one shown in **Figure 3.4**), or as a polynomial. These specification are for individual sensors or for families of sensors and depend on construction and materials. The resistance may be written in general using the Callendar–Van Dusen equation:

$$R(T) = R(0)[1 + a(T - T_{ref}) + b(T - T_{ref})^2 + HOT] \quad [\Omega], \tag{3.11}$$

where HOT (higher-order term) is a correction term that may be added to accommodate specific curves, especially at high temperatures. The coefficients are derived from the response of the specific sensor being evaluated (see **Example 3.6**).

EXAMPLE 3.6 Silicon resistive sensor

A silicon resistive sensor is described by the first two terms in **Equation (3.11)** with coefficient $a = 7.635 \times 10^{-3}$, $b = 1.731 \times 10^{-5}$ with a reference resistance of 1 kΩ at 25°C. The sensor is to be used for temperature sensing in the range 0°C–75°C. Calculate the maximum deviation of resistance from linearity, where the linear response is given by **Equation (3.4)** with a temperature coefficient of 0.013/°C.

Solution: It is best to calculate and plot the response to see the behavior. The response is plotted in **Figure 3.5**. Clearly the maximum error in output (resistance) is at the lowest temperature.

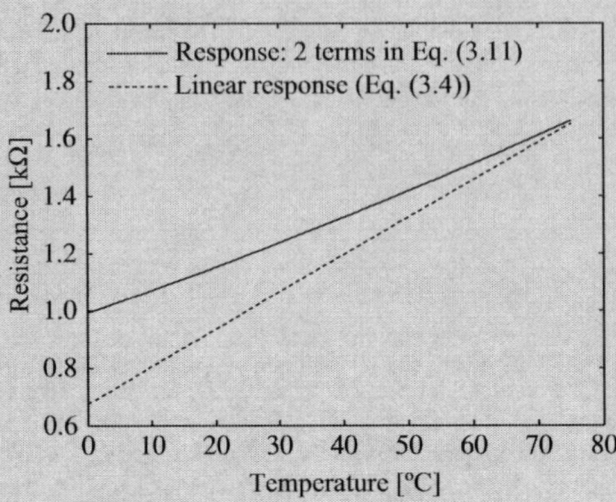

FIGURE 3.5 ■ Response of a silicon resistive sensor showing the deviation between the approximate linear formula and the second-order formula.

At 0°C using **Equation (3.4)**:

$$R(0°\text{C}) = R_0(1 + \alpha[T - T_0]) = 1000(1 + 0.013[0 - 25]) = 675.0\ \Omega.$$

Using **Equation (3.11)**:

$$\begin{aligned} R(0°\text{C}) &= R_0[1 + a(T - T_{ref}) + b(T - T_{ref})^2] \\ &= 1000\left(1 + 7.635 \times 10^{-3}[0 - 25] + 1.731 \times 10^{-5}[0 - 25]^2\right) \\ &= 992.4\ \Omega. \end{aligned}$$

The difference is 317.4 Ω. This is 31.98% and clearly the linear formula cannot be used for practical implementation. In application of a sensor of this type, the resistance of the sensor is measured and the temperature is calculated from **Equation (3.11)**, perhaps through the use of a lookup table stored in a microprocessor or direct evaluation of the expressions.

3.2.3 Thermistors

Thermistors (**therm**al res**istors**) came into existence together with other semiconductor devices and have been used for temperature sensing starting in the 1960s. As their name implies, these are thermal resistors, made of semiconducting metal oxides that have high temperature coefficients. Most metal oxide semiconductors have NTCs and their resistance at reference temperatures (typically 25°C) can be rather high. A simple model of a thermistor is the following:

$$R(T) = R_0 e^{\beta(1/T - 1/T_0)} = R_0 e^{-\beta/T_0} e^{\beta/T} \quad [\Omega], \tag{3.12}$$

where R_0 [Ω] is the resistance of the thermistor at the reference temperature T_0, β [K] is the **material constant** and is specific for the particular material used in a device, $R(T)$ is the resistance of the thermistor and T is the temperature sensed [K]. This relation is clearly nonlinear and is only approximate. The inverse relation to **Equation (3.12)** is also useful, particularly in evaluating temperature from measured resistance, and is often used in sensing:

$$T = \frac{\beta}{\ln(R(T)/R_0 e^{-\beta/T_0})} \quad [\text{K}]. \tag{3.13}$$

The model in **Equation (3.12)** can be improved by using the **Steinhart–Hart** equation, which gives the resistance as

$$R(T) = e^{\left(x-\frac{y}{2}\right)^{1/3} - \left(x+\frac{y}{2}\right)^{1/3}} \quad [\text{in } \Omega], \quad y = \frac{a - 1/T}{c}, \quad x = \sqrt{\left(\frac{b}{3c}\right)^3 + \frac{y^2}{4}}. \tag{3.14}$$

The constants a, b, and c are evaluated from three known points on the thermistor response. The inverse relation is often used as well:

$$T = \frac{1}{a + b\ln(R) + c\ln^3(R)} \quad [\text{K}]. \tag{3.15}$$

Equations (3.12) and **(3.14)** establish approximate transfer functions for the thermistor suitable for many applications.

Because the variation between devices can be large, it is often necessary to establish the transfer function through calibration, as discussed in **Chapter 2**. In most cases **Equation (3.15)** is used to evaluate the coefficients a, b, and c and only then does one use **Equation (3.14)** to write the resistance as a function of temperature. The coefficients are often evaluated by manufacturers of thermistors and are available in tables for use in calibration. Similarly, if the simplified form of the transfer function is used, one starts with **Equation (3.13)**, from which the term β may be easily evaluated.

Thermistors are available in higher accuracy models through the processes of trimming the device. The method of production itself can also affect the transfer function of the device. Thermistors are produced by a number of methods. Bead thermistors (**Figure 3.6a**) are essentially a small volume of the metal oxide with two conductors (platinum alloy for high-quality thermistors, copper or copper alloy for inexpensive devices) attached through a heat process. The bead is then coated, usually with glass or epoxy. Another method is to produce chips with surface electrodes (**Figure 3.6b**) that are then connected to leads and the device encapsulated. Chips can be easily trimmed for specific resistance values. A third method of production is to deposit the semiconductor on a substrate following standard semiconductor production methods (see **Figure 3.7b**). These are particularly useful in integrated devices and in complex sensors such as radiation sensors. While the method of production is important, perhaps more important is the encapsulation, since that is the main means of ensuring long-term stability. This encapsulation is one of the main differences between various thermistors.

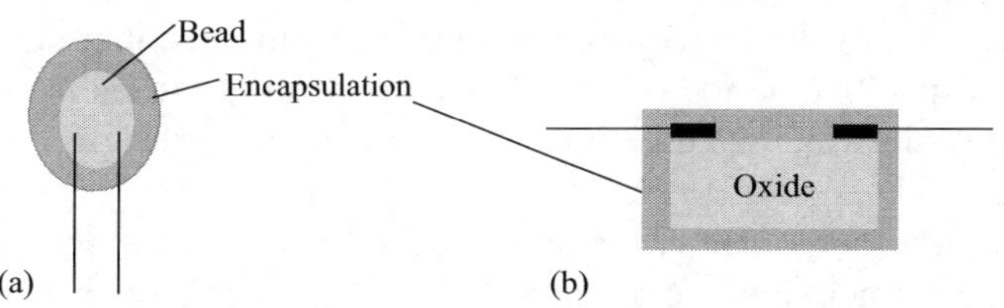

FIGURE 3.6 ■ (a) Construction of a bead thermistor. (b) Construction of a chip thermistor.

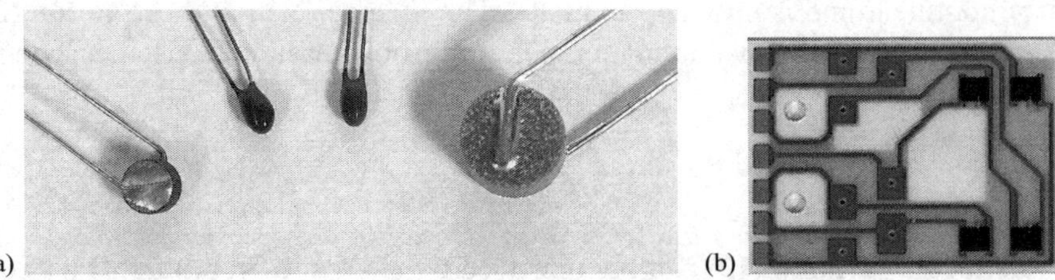

FIGURE 3.7 ■ (a) Two common types of thermistors. (b) Deposited thermistors on a ceramic substrate (the thermistors are the four dark, rectangular areas on the right).

Good thermistors are glass encapsulated, while epoxy potting is common for less expensive devices. Sometimes stainless steel jackets are added for protection in harsh environments. **Figure 3.7a** shows a picture of two epoxy-encapsulated bead thermistors and two chip thermistors. The size of the device is also dictated by the production method (with bead thermistors being the smallest) and this dictates the thermal response of the device. Typically the thermal response of thermistors is relatively short, primarily because of their small physical size.

Although most thermistors are NTC devices, PTC devices can also be made from special materials. These are usually based on barium titanate ($BaTiO_3$) or strontium titanate ($SrTiO_3$) with the addition of doping agents that make them semiconducting. These materials have very high resistances and a highly nonlinear transfer function. Nevertheless, in a small useful range they exhibit a mildly nonlinear curve with a PTC of resistance. Unlike NTC thermistors, PTC thermistors have a steep curve in the useful range (large changes in resistance) and are therefore more sensitive than NTCs in that range. Overall, PTC thermistors are not as common as NTC thermistors, but they have one advantage that is common to all PTC devices (including wire-wound sensors). If connected to a voltage source, as the temperature increases, the current decreases, and therefore they cannot overheat due to self-heating. This is an intrinsic protection mechanism that can be very useful in high-temperature applications. In contrast, NTC thermistors can overheat under the same conditions.

Thermistors exhibit errors due to self-heating similar to those of RTDs. Typical values are between 0.01°C/mW in water and 1°C/mW in air. However, since thermistors are available in a wide range of resistances that can reach a few megaohms with high sensitivity, currents through thermistors are typically very low and self-heating is not usually a problem. On the other hand, some thermistors can be very small, increasing the effect. There are instances in which a thermistor is deliberately heated by passing a current through it, taking advantage of its self-heating properties. An example of this will be discussed in **Chapter 6**. Self-heating is specified by the manufacturer in a manner analogous to that given for RTDs.

The long-term stability of thermistors has been an issue in the past since all thermistors exhibit changes in resistance with aging, especially immediately after production. For this reason they are aged prior to shipment by maintaining them at an elevated temperature for a specified period of time. Good thermistors exhibit negligible drift after the aging process, allowing accurate measurements on the order of 0.25°C with excellent repeatability.

The temperature range of thermistors is higher than that of silicon RTDs and can exceed 1500°C and reach down to about −270°C. The thermistor is often the sensor of

EXAMPLE 3.7 **NTC thermistor**

A thermistor has a nominal resistance of 10 kΩ at 25°C. To evaluate the thermistor, the resistance at 0°C is measured as 29.49 kΩ. Calculate and plot the resistance of the thermistor between −50°C and +50°C.

Solution: Using **Equation (3.12)** we must first evaluate the coefficient β. This is done as follows:

$$R(0°\text{C}) = 29{,}490 = 10{,}000e^{\beta(1/(273.15)-1/(273.15+25))} \quad [\Omega].$$

To calculate β, we write

$$\ln\left(\frac{29{,}490}{10{,}000}\right) = \beta(1/(273.15) - 1/(298.15))$$

or

$$\beta = \ln\left(\frac{29{,}490}{10{,}000}\right)\left(\frac{1}{(1/(273.15) - 1/(298.15))}\right) = 3.523 \times 10^3 \text{ K}.$$

Now the expression for resistance is

$$R(T) = 10{,}000e^{3.523\times10^3\times(1/T-1/(298.15))} \quad [\Omega].$$

The resistance at −50°C is 530,580 Ω and at + 50°C is 4008 Ω:

$$R(T = 223.15 \text{ K}) = 10{,}000e^{3.523\times10^3\times(1/223.15-1/(298.15))} = 530{,}580\ \Omega$$

$$R(T = 323.15 \text{ K}) = 10{,}000e^{3.523\times10^3\times(1/323.15-1/(298.15))} = 4008\ \Omega.$$

Note the nonlinear behavior of the thermistor over this wide range (**Figure 3.8**). Also, because this is an approximation, the narrower the span, the better the approximation, provided that β is calculated (or given) at a temperature within the span. Clearly this also assumes that β is independent of temperature.

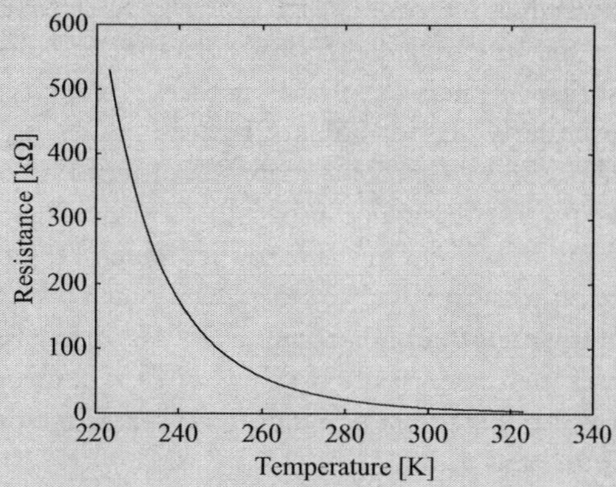

FIGURE 3.8 ■ Response of a thermistor between −50°C (223.15 K) and 50°C (323.15 K).

choice in many consumer products because of the low cost, small size, and simple interfacing needed to make it work.

3.3 | THERMOELECTRIC SENSORS

As indicated in the introduction, thermoelectric sensors are among the oldest sensors, some of the most useful and most commonly used, and have been in use for well over 150 years. And yet, at first sight, this seems curious since the signals produced by thermoelectric sensors are small and difficult to measure and are plagued by noise problems. Perhaps the main reason for their success, particularly in the early years, is the fact that these are passive sensors—they generate electric emfs (voltages) directly and hence all one needs to do is measure the voltage. In the early years, in the absence of amplifiers and controllers, one could still measure the emf, though small, and get an accurate reading of the temperature. Also, they can be produced by anyone with minimum skill. They have other properties that have ensured transcendence into the modern era. In addition to being well developed, simple, rugged, and inexpensive, thermoelectric sensors can operate on almost the entire practically useful range of temperatures from near absolute zero to about 2700°C. No other sensor technology (other than perhaps infrared temperature sensors) can match even a fraction of this range.

There is really only one type of thermoelectric sensor, often called a **thermocouple**. However, there are variations in nomenclature and in construction. Thermocouple usually refers to a junction made of two dissimilar conductors. A number of these junctions connected in series is referred to as a **thermopile**. Semiconductor thermocouples and thermopiles have similar functions but also serve in the reverse function, to generate heat or to cool, and can therefore be used as actuators. These devices are usually called **thermoelectric generators** (TEGs), or sometimes **Peltier cells,** indicating their actuation use, but they can be used as sensors.

Thermocouples are based on the Seebeck effect, which in turn is the sum of two other effects—the Peltier effect and the Thomson effect. These two effects and the resulting Seebeck effect can be described as follows:

The **Peltier effect** is heat generated or absorbed at the junction of two dissimilar materials when an emf exists across the junction due to current in the junction. The effect occurs either by connecting an external emf across the junction or it may be generated by the junction itself, depending on the mode of operation. In either case a current must flow through the junction. This effect has found applications in cooling and heating, particularly in portable refrigerators and in cooling electronic components. Discovered in 1834 by Charles Athanase Peltier, it was developed into its current state in the 1960s as part of the space program. The devices in existence have benefited considerably from developments in semiconductors, and particularly high-temperature semiconducting materials.

The **Thomson effect**, discovered in 1892 by William Thomson (Lord Kelvin), functions such that a current-carrying wire, if unevenly heated along its length, will either absorb or radiate heat depending on the direction of current in the wire (from hot to cold or from cold to hot).

The **Seebeck effect** is an emf produced across the junction between two dissimilar conducting materials. If both ends of the two conductors are connected and a temperature difference is maintained between the two junctions, a thermoelectric current will flow through the closed circuit (**Figure 3.9a**). Alternatively, if the circuit is opened (**Figure 3.9b**), an emf will appear across the open circuit. It is this exact emf that is measured in a thermocouple sensor. The effect was discovered in 1821 by Thomas Johann Seebeck.

In the following simplified analysis we assume that the two junctions in **Figure 3.9b** are at different temperatures, T_1 and T_2, and the conductors are homogeneous. We can then define the Seebeck emf across each of the conductors, *a* and *b*, as

$$emf_a = \alpha_a(T_2 - T_1) \quad \text{and} \quad emf_b = \alpha_b(T_2 - T_1). \tag{3.16}$$

In these relations α_A and α_B are the absolute Seebeck coefficients given in microvolts per degree Celsius (μV/°C) and are properties of the materials involved. The thermoelectric emf generated by a thermocouple made of two wires, A and B, is therefore

$$emf_T = emf_a - emf_b = (\alpha_a - \alpha_b)(T_2 - T_1) = \alpha_{ab}(T_2 - T_1). \tag{3.17}$$

The term α_{ab} is the relative Seebeck coefficient of the material combination *a* and *b* (**Table 3.3**). These coefficients are available for various material combinations and indicate the sensitivity of the thermocouple. Some are listed in **Table 3.3**. Other relative Seebeck coefficients may be obtained from the absolute coefficients in **Table 3.2** and similar tables for other materials by subtracting the absolute coefficients one from the other.

The Seebeck coefficients are rather small—on the order of a few microvolts per degree to a few millivolts per degree for the largest coefficients. This means that in many cases the output from the thermocouple will have to be amplified before it can be

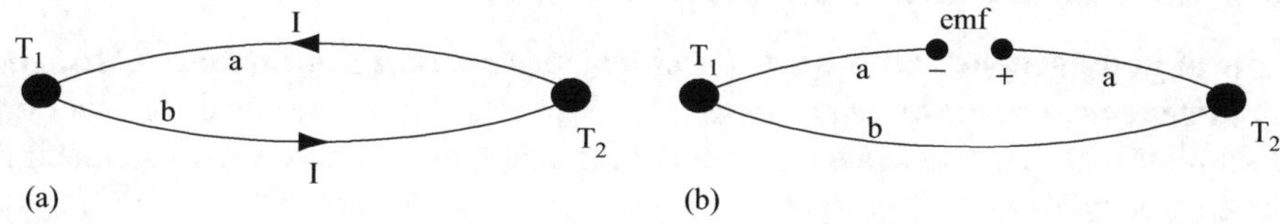

FIGURE 3.9 ■ (a) A thermoelectric current flows in a circuit comprised of two junctions at different temperatures. (b) An emf is developed across the open circuit.

TABLE 3.2 ■ Absolute Seebeck coefficients for selected elements (thermoelectric series)

Material	α [μV/K]
p-Silicon	100–1000
Antimony (Sb)	32
Iron (Fe)	13.4
Gold (Au)	0.1
Copper (Cu)	0
Silver (Ag)	−0.2
Aluminum (Al)	−3.2
Platinum (Pt)	−5.9
Cobalt (Co)	−20.1
Nickel (Ni)	−20.4
Bismuth (Sb)	−72.8
n-Silicon	−100 to −1000

TABLE 3.3 ■ Relative Seebeck coefficients for some material combinations

Materials	Relative Seebeck coefficient at 25°C [μV/°C]	Relative Seebeck coefficient at 0°C [μV/°C]
Copper/constantan	40.9	38.7
Iron/constantan	51.7	50.4
Chromel/alumel	40.6	39.4
Chromel/constantan	60.9	58.7
Platinum (10%)/rhodium-platinum	6.0	7.3
Platinum (13%)/rhodium-platinum	6.0	5.3
Silver/palladium	10	
Constantan/tungsten	42.1	
Silicon/aluminum	446	
Carbon/silicon carbide	170	

used in practical applications. This also implies that special care must be taken in connecting to thermocouples to avoid noisy signals and errors due to, for example, induced emfs from to external sources. Direct measurement of the output, which was the main method of using these sensors in the past, is still possible if used strictly for temperature measurement and no further processing of the output is needed. More often, however, the signal will be used to take some action (turn on or off a furnace, detect a pilot flame before turning on the gas, etc.), and that implies at least some signal conditioning and a controller to affect the action.

The operation of thermocouples is based on three laws—the thermoelectric laws—that summarize the discussion above. These are

1. **Law of homogeneous circuit:** ***A thermoelectric current cannot be established in a homogeneous circuit by heat alone***. This law establishes the need for junctions of dissimilar materials since a single conductor is not sufficient to establish an emf and hence a current.
2. **Law of intermediate materials:** ***The algebraic sum of the thermoelectric forces (emfs) in a circuit composed of any number and combination of dissimilar materials is zero if all junctions are at the same temperature***. This establishes the fact that additional materials may be connected in the thermoelectric circuit without affecting the output of the circuit as long as any junctions added to the circuit are kept at the same temperature. Also, the law indicates that voltages are additive so multiple junctions may be connected in series to increase the output (thermopiles).
3. **Law of intermediate temperatures:** ***If two junctions at temperatures T_1 and T_2 produce Seebeck voltage V_2 and temperatures T_2 and T_3 produce Seebeck voltage V_1, then temperatures T_1 and T_3 produce Seebeck voltage $V_3 = V_1 + V_2$***. This law establishes methods of calibration for thermocouples.

Note: Some sources list five laws, looking at more detailed behavior. The three laws listed here are inclusive and describe all observed effects.

Based on the principles described above, thermocouples are usually used in pairs (but there are exceptions and variations) so that one junction is at the sensing temperature while the second is at a reference temperature, usually a lower temperature, but it can also be a higher temperature. This is shown in **Figure 3.10**, where the voltmeter

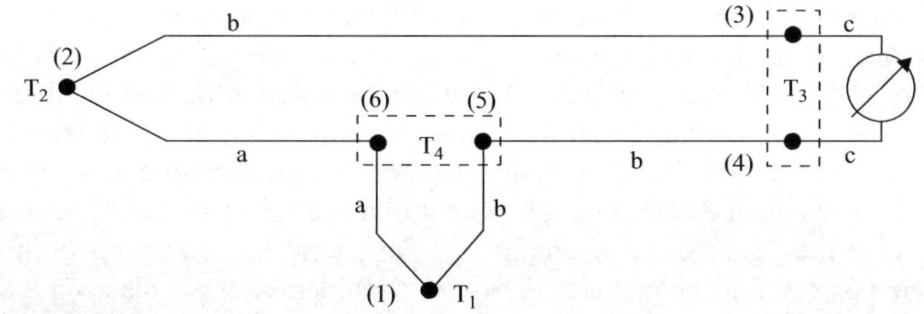

FIGURE 3.10 ■ A measuring thermocouple (hot junction 2) and reference thermocouple (cold junction 1) and additional junctions introduced by connections.

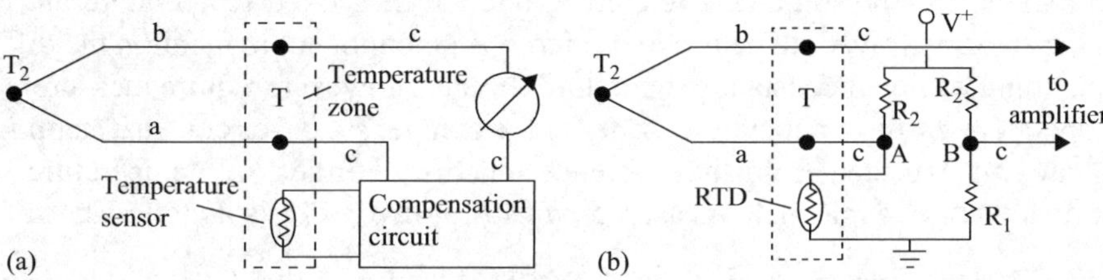

FIGURE 3.11 ■ (a) Connection of a thermocouple through a uniform temperature zone (T) and a compensation circuit to replace the reference junction. (b) The compensation circuit.

represents the device to which the sensor is connected (usually an amplifier). Any connection in the circuit between dissimilar materials adds an emf due to that junction. However, any pair of junctions at identical temperatures may be added without changing the output. In **Figure 3.10** the output is produced by junctions (1) and (2) for the following reason: junctions (3) and (4) are identical (one between material b and c and one between material c and b) and their temperature is the same. Thus no net emf due to this pair is produced. Junctions (5) and (6) also produce zero emf since they are junctions between identical materials that produce zero emf at any temperature. Note that each connection (to the reference junction and to the measuring instrument) necessarily adds two junctions. This then indicates the strategy in sensing: any junction that is not sensed or is not a reference junction must either be between identical materials or must come in pairs and both junctions in the pair must be maintained at the same temperature. In addition, it is a good precaution to use unbroken wires leading from the sensor to the reference junction or to the measuring instrument. If splicing is necessary to extend the length, identical wires must be used to avoid additional emfs.

Connection of thermocouples can be done in many ways, each with their own advantages. One of the most common connections is shown in **Figure 3.11a**. The two junctions (between material b and c and between material a and c) are placed in a so-called **uniform temperature zone** or **isothermal zone**. This can be a small junction box or may in fact be just two junctions in very close proximity—as long as they can be guaranteed to be at identical temperatures. In this case there is no cold junction, but a compensation circuit is added to ensure that the potentials on the junction b–c and a–c, together with the compensation potential, "simulate" the effect of the cold junction. The compensation circuit guarantees that the output is zero at the reference temperature (usually 0°C).

If a reference junction as in **Figure 3.10** is used, it is critical that the temperature is known and constant. In such cases the temperature of the reference junction may be measured separately by a sensor (usually an RTD, sometimes a thermistor, but in any case not a thermocouple) and the reading used to compensate for any change in the temperature of the reference junction. The reference junction may be held at the temperature of a water–ice mixture that guarantees a temperature of 0°C (some variations from this can occur if the water is contaminated or if the atmospheric pressure changes). The temperature of the ice bath should be monitored even if no specific compensation is incorporated in the circuits. An alternative to the ice–water mixture is boiling water, with the same precautions being taken. These two temperatures are also commonly used for calibration of thermocouples. In normal operation, use of an ice–water mixture or boiling water is inconvenient at best. In many applications, the method in **Figure 3.11a** is used, which does not require a reference junction or a fixed temperature and hence avoids the errors involved in the use of a reference junction, not to mention the difficulty of maintaining a known constant temperature. It does, however, require measurement of the temperature in the temperature zone and a compensation circuit that supplies the equivalent emf expected from the reference junction. Further, the temperature sensor cannot be a thermocouple. The measured emf in **Figure 3.11a** is as follows:

$$emf = \alpha_{\rm ba} T_2 - [\alpha_{\rm bc} + \alpha_{\rm ca}]T + emf_{\rm comp} \quad [\mu{\rm V}], \tag{3.18}$$

where $\alpha_{\rm ba}$, $\alpha_{\rm bc}$, and $\alpha_{\rm ca}$ are the relative Seebeck coefficients. Of the terms in **Equation (3.18)**, $\alpha_{\rm ba}T_2$ is of interest and the term $[\alpha_{\rm bc}+\alpha_{\rm ca}]T$ may be viewed as part of the reference emf, measured at the temperature T. This term may be written in terms of the absolute Seebeck coefficients of the three materials as follows:

$$[\alpha_{bc} + \alpha_{ca}]T = [(\alpha_b - \alpha_b) + (\alpha_c - \alpha_a)]T = (\alpha_b - \alpha_a)T = \alpha_{ba}T. \tag{3.19}$$

Equation (3.18) now becomes

$$emf = \alpha_{\rm ba} T_2 - \alpha_{\rm ba} T + emf_{\rm comp} \quad [\mu{\rm V}]. \tag{3.20}$$

The compensation term $emf_{\rm comp}$ is added to ensure that T_2 is correctly measured. The purpose of the compensation circuit is to cancel the term $\alpha_{\rm ba}T$. That is, the emf supplied by the compensation circuit must be

$$emf_{\rm comp} = \alpha_{ba} T \quad [{\rm mV}], \tag{3.21}$$

where T is the temperature of the temperature zone. Under these conditions the measured emf is $emf = \alpha_{\rm ba}T_2$ and is entirely due to T_2. It should also be noted that the coefficient $\alpha_{\rm ba}$ is the relative Seebeck coefficient of the sensing junction, so that compensation is based on the sensitivity of the sensing junction alone.

To understand the method of compensation, consider **Figure 3.11b**. The cold junction is replaced with a potential difference $V_{\rm BA}$. The resistance R_1 is selected to equal the resistance of the RTD at 0°C, denoted as R_T. The resistors R_2 are equal and are selected to produce a potential difference per degree Celsius as required by the type of thermocouple being used. Typically the RTD is a platinum RTD with a resistance around 100 Ω and the reference potential V^+ is regulated at an arbitrary level, but typically between 5 V and 12 V. The potential at point A is

$$V_A = \frac{V^+}{R_1 + R_2} R_1. \tag{3.22}$$

The potential at point B is temperature dependent as follows:

$$V_B = \frac{V^+}{R_2 + R_T(1 + \alpha T)} R_T(1 + \alpha T), \quad (3.23)$$

where α is the resistance temperature coefficient of the RTD in use and R_T is its resistance at 0°C. The potential that replaces the cold junction is

$$emf_{comp} = V_{BA} = \frac{V^+}{R_2 + R_T(1 + \alpha T)} R_T(1 + \alpha T) - \frac{V^+}{R_1 + R_2} R_1. \quad (3.24)$$

This relation allows immediate calculation of the resistance R_2 since, for any given type of thermocouple, V_{BA} is known for any temperature T. **Example 3.8a** shows how the actual calculation of resistance is done.

The method of compensation discussed above does have a slight shortcoming: the range of temperatures of the temperature zone cannot be too far from the reference temperature of the RTD. The reason for this is that the compensation circuit is designed to produce zero emf at the reference temperature of the RTD (as can be verified from the equations above and from **Figure 3.11b**). As long as the temperature range is relatively small, the method is very accurate (see **Example 3.8**).

EXAMPLE 3.8 Cold junction compensation of a K-type thermocouple

Consider the cold junction compensation of a chromel-alumel thermocouple using a platinum RTD, as shown in **Figure 3.11b**. The RTD has a resistance of 100 Ω at 0°C and has a TCR coefficient of 0.00385. The relative Seebeck coefficient (sensitivity) for the K-type thermocouple at 0°C is 39.4 μV/°C (see **Table 3.3**).

a. Given a regulated voltage source of 10 V, calculate the resistance R_2 required for this type of thermocouple.

b. Calculate the error in temperature measurement at 45°C if the temperature zone is at $T = 27$°C and explain the source of this error.

Solution: We use **Equation (3.24)** directly. However, since the Seebeck coefficient is given per degree Celsius, the temperature T can be taken as any value above (or below) 0°C. We will take it as 1°C for convenience. With $R_1 = 100\ \Omega$, we have

a.

$$39.4 \times 10^{-6} = \frac{10}{R_2 + 100(1 + 0.00385 \times 1)} 100(1 + 0.00385 \times 1) - \frac{10}{100 + R_2} 100$$

or

$$39.4 \times 10^{-6} = \frac{1003.85}{R_2 + 100.385} - \frac{1000}{100 + R_2}.$$

Cross-multiplying and separating R_2 we get

$$R_2^2 - 97515.351 R_2 + 10038.5 = 0.$$

Solving this equation gives

$$R_2 = 97{,}515.3\ \Omega.$$

We will take the value, 97,500 Ω, as a resistance that can be commercially made.

b. At the zone temperature of 27°C, the circuit above provides an emf (again from **Equation (3.24)**):

$$emf_{\text{comp}} = \frac{10}{97{,}500 + 100(1 + 0.00385 \times 27)} 100(1 + 0.00385 \times 27) - \frac{10}{97{,}500 + 100} 100$$

$$= 1.063857\ \text{mV}.$$

The term $\alpha_{\text{ba}}T$ (see **Equation (3.19)**) is

$$\alpha_{\text{ba}}T = 39.4 \times 10^{-6} \times 27 = 1.0638 \times 10^{-3}\ \text{V}.$$

The emf of the circuit including the compensation is therefore (from **Equation (3.20)**)

$$emf = 39.4 \times 10^{-6} \times 45 - 1.0638 \times 10^{-3} + 1.063857 \times 10^{-3} = 1.773057\ \text{mV}.$$

This value corresponds to a temperature T_2:

$$T_2 = \frac{1.773057 \times 10^{-3}}{39.4 \times 10^{-6}} = 45.00145°\text{C}.$$

The error is minute—only 0.003%.

The main source of error is the selection of the resistor R_2, since we have chosen a resistance that can be made commercially. A more exact value would reduce the error (the resistor may be replaced with an adjustable resistor or potentiometer). There are other sources of error, the most important of which is the nonlinear transfer function of the thermocouple (we shall discuss this next). In practice we should also expect the resistors themselves to have some tolerance as well as some temperature dependence, adding to errors. Overall, however, this method is very accurate and commonly used in thermocouple sensing.

3.3.1 Practical Considerations

Some of the properties of thermocouples have been discussed above. The choice of materials used to make the junctions is an important consideration that affects the output emf, temperature range, and resistance of the thermocouple. To aid in the selection of thermocouples and thermocouple materials, the thermocouple reference tables have been established and are supplied by standards organizations. There are three basic tables available. The first is called the thermoelectric series table, shown in **Table 3.4** for selected materials. Each material in this table is thermoelectrically negative with respect to all materials above it and positive with respect to all materials below it. This also indicates that the farther from each other a pair is, the larger the emf output that will be produced.

TABLE 3.4 ▪ The thermoelectric series: selected elements and alloys at selected temperatures

100°C	500°C	900°C
Antimony	Chromel	Chromel
Chromel	Copper	Silver
Iron	Silver	Gold
Nichrome	Gold	Iron
Copper	Iron	90% platinum, 10% rhodium
Silver	90% platinum, 10% rhodium	Platinum
90% platinum, 10% rhodium	Platinum	Cobalt
Platinum	Cobalt	Alumel
Cobalt	Alumel	Nickel
Alumel	Nickel	Constantan
Nickel	Constantan	
Constantan		

TABLE 3.5 ▪ Seebeck coefficients with respect to platinum 67

	Thermoelement type—Seebeck coefficient [μV/°C]					
Temperature [°C]	JP	JN	TP	TN, EN	KP, EP	KN
0	**17.9**	**32.5**	5.9	32.9	25.8	13.6
100	17.2	37.2	9.4	37.4	30.1	11.2
200	14.6	40.9	11.9	41.3	32.8	7.2
300	11.7	43.7	14.3	43.8	34.1	7.3
400	9.7	45.4	16.3	45.5	34.5	7.7
500	9.6	46.4		46.6	34.3	8.3
600	11.7	46.8		46.9	33.7	8.8
700	15.4	46.9		46.8	33.0	8.8
800				46.3	32.2	8.8
900				45.3	31.4	8.5
1000				44.2	30.8	8.2

The second standard table lists the Seebeck coefficients of various materials with reference to platinum 67 and of various common thermocouple types, as shown in **Tables 3.5** and **3.6**. In these tables the first material in each type (E, J, K, R, S, and T) is positive and the second is negative. In **Table 3.5** the Seebeck emf is given for the base elements of thermocouples with respect to platinum 67. For example, J-type thermocouples use iron and constantan. Thus column JP lists the Seebeck emf for iron with respect to platinum 67, whereas JN lists the emfs for constantan with respect to platinum 67. Adding the two together gives the corresponding value for the J-type thermocouple in **Table 3.6**. Thus, for example, taking the JP and JN values at 0°C in **Table 3.5** and adding them, $17.9 + 32.5 = 50.4$ μV/°C, gives the entry in the J column at 0°C in **Table 3.6**. Note also that these tables list limits on the high- or low-temperature use of elements and thermocouples and that the Seebeck coefficients vary with temperature. This means that the output of thermocouples cannot be linear, as we shall see shortly.

The third table, called the thermoelectric reference table, gives the thermoelectric emf produced by the thermocouple (in effect, this is the transfer function) for each type

TABLE 3.6 ■ Seebeck coefficients for various types of thermocouples

	Thermocouple type—Seebeck coefficient [μV/°C]					
Temperature [°C]	**E**	**J**	**K**	**R**	**S**	**T**
−200	25.1	21.9	15.3			15.7
−100	45.2	41.1	30.5			28.4
0	58.7	**50.4**	39.4	5.3	5.4	38.7
100	67.5	54.3	41.4	7.5	7.3	46.8
200	74.0	55.5	40.0	8.8	8.5	53.1
300	77.9	55.4	41.4	9.7	9.1	58.1
400	80.0	55.1	42.2	10.4	9.6	61.8
500	80.9	56.0	42.6	10.9	9.9	
600	80.7	58.5	42.5	11.3	10.2	
700	79.8	62.2	41.9	11.8	10.5	
800	78.4		41.0	12.3	10.9	
900	76.7		40.0	12.8	11.2	
1000	74.9		38.9	13.2	11.5	

of thermocouple as an *n*th-order polynomial in a range of temperatures. In fact, the tables give the coefficients of the polynomials. The standard tables provide the emf with a reference junction at 0°C. These tables ensure accurate representation of the thermocouple's output and can be used by the controller to accurately represent the temperature sensed by the thermocouple. There are in fact two tables. One provides the thermocouple output (with reference to zero temperature), whereas the second provides the temperature corresponding to an output emf. As an example of how these tables represent the transfer function consider **Table 3.7**, which shows the table entry

TABLE 3.7 ■ Standard thermoelectric reference table (transfer function) for type E thermocouples (chromel-constantan) with reference junction at 0°C.

Polynomial: $emf = \sum_{i=0}^{n} c_i T^i$ [μV]

Temperature range [°C]	−270°C to 0°C	0°C to 1000°C
C_0	0.0	0.0
C_1	$5.8665508708 \times 10^{1}$	$5.8665508708 \times 10^{1}$
C_2	$4.5410977124 \times 10^{-2}$	$4.5032275582 \times 10^{-2}$
C_3	$-7.7998048686 \times 10^{-4}$	$2.8908407212 \times 10^{-5}$
C_4	$-2.5800160843 \times 10^{-5}$	$-3.3056896652 \times 10^{-7}$
C_5	$-5.9452583057 \times 10^{-7}$	$6.5024403270 \times 10^{-10}$
C_6	$-9.3214058667 \times 10^{-9}$	$-1.9197495504 \times 10^{-13}$
C_7	$-1.0287605534 \times 10^{-10}$	$-1.2536600497 \times 10^{-15}$
C_8	$-8.0370123621 \times 10^{-13}$	$2.1489217569 \times 10^{-18}$
C_9	$-4.3979497391 \times 10^{-15}$	$-1.4388041782 \times 10^{-21}$
C_{10}	$-1.6414776355 \times 10^{-17}$	$3.5960899481 \times 10^{-25}$
C_{11}	$-3.9673619516 \times 10^{-20}$	
C_{12}	$-5.5827328721 \times 10^{-22}$	
C_{13}	$-3.4657842013 \times 10^{-26}$	

In explicit polynomial form we have:

In the range −270°C to 0°C,

$$emf = 5.8665508708 \times 10^{1} T^{1} + 4.5410977124 \times 10^{-2} T^{2} - 7.7998048686 \times 10^{-4} T^{3} - 2.5800160843 \times 10^{-5} T^{4} - 5.9452583057 \times 10^{-7} T^{5} - 9.3214058667 \times 10^{-9} T^{6} - 1.0287605534 \times 10^{-10} T^{7} - 8.0370123621 \times 10^{-13} T^{8} - 4.3979497391 \times 10^{-15} T^{9} - 1.6414776355 \times 10^{-17} T^{10} - 3.9673619516 \times 10^{-20} T^{11} - 5.5827328721 \times 10^{-22} T^{12} - 3.4657842013 \times 10^{-26} T^{13} \quad [\mu V].$$

In the range 0°C to 1000°C,

$$emf = 5.8665508708 \times 10^{1} T^{1} + 4.5032275582 \times 10^{-2} T^{2} + 2.8908407212 \times 10^{-5} T^{3} - 3.3056896652 \times 10^{-7} T^{4} + 6.5024403270 \times 10^{-10} T^{5} - 1.9197495504 \times 10^{-13} T^{6} - 1.2536600497 \times 10^{-15} T^{7} + 2.1489217569 \times 10^{-18} T^{8} - 1.4388041782 \times 10^{-21} T^{9} + 3.5960899481 \times 10^{-25} T^{10} \quad [\mu V].$$

(coefficients of the polynomial) for type E thermocouples, and **Table 3.8**, which shows the coefficients of the inverse polynomial, that is, the coefficients of the polynomial that provides the temperature given the emf. **Table 3.8** also shows the accuracy expected in the various temperature ranges. Note that the temperature is given in degrees Celsius and emf is given in microvolts.

The polynomials are considered to be exact. Truncation of the polynomials should be avoided since any truncation may cause large errors.

The thermoelectric reference tables for the most common types of thermocouples are given in **Appendix B**.

TABLE 3.8 ▪ Coefficients of the inverse polynomials, type E thermocouples

$$T = \sum_{i=0}^{n} c_i E^i \quad [°C]$$

Temperature range [°C]	−200°C to 0°C	0°C to 1000°C
Voltage range [μV]	$E = -8825$ to 0	$E = 0$ to 76,373
C_0	0.0	0.0
C_1	1.6977288×10^{-2}	1.7057035×10^{-2}
C_2	$-4.3514970 \times 10^{-7}$	$-2.3301759 \times 10^{-7}$
C_3	$-1.5859697 \times 10^{-10}$	$6.5435585 \times 10^{-12}$
C_4	$-9.2502871 \times 10^{-14}$	$-7.3562749 \times 10^{-17}$
C_5	$-2.6084314 \times 10^{-17}$	$-1.7896001 \times 10^{-21}$
C_6	$-4.1360199 \times 10^{-21}$	$8.4036165 \times 10^{-26}$
C_7	$-3.4034030 \times 10^{-25}$	$-1.3735879 \times 10^{-30}$
C_8	$-1.1564890 \times 10^{-21}$	$1.0629823 \times 10^{-35}$
C_9		$-3.2447087 \times 10^{-41}$
Error range	0.04°C to −0.01°C	0.02°C to −0.02°C

In explicit form we get (E denotes the thermocouple emf in μV):

In the range −270°C to 0°C,

$$T = 1.6977288 \times 10^{-2}E^1 - 4.3514970 \times 10^{-7}E^2 - 1.5859697 \times 10^{-10}E^3 - 9.2502871 \times 10^{-14}E^4 - 2.6084314 \times 10^{-17}E^5 - 4.1360199 \times 10^{-21}E^6 - 3.4034030 \times 10^{-25}E^7 - 1.1564890 \times 10^{-21}E^8 \quad [°C].$$

In the range 0°C to 1000°C,

$$T = 1.7057035 \times 10^{-2}E^1 - 2.3301759 \times 10^{-7}E^2 + 6.5435585 \times 10^{-12}E^3 - 7.3562749 \times 10^{-17}E^4 - 1.7896001 \times 10^{-21}E^5 + 8.4036165 \times 10^{-26}E^6 - 1.3735879 \times 10^{-30}E^7 + 1.0629823 \times 10^{-35}E^8 - 3.2447087 \times 10^{-41}E^9 \quad [°C].$$

EXAMPLE 3.9 Thermoelectric reference tables

A chromel-constantan thermocouple is intended for use in a steam generator, normally operating at 350°C. To measure the temperature it is suggested to use a reference junction at 100°C (boiling water) since that is easier to maintain in the steam plant. In addition, to simplify interfacing, it is suggested to use only the first three terms in the polynomial for the reference emf.

a. Calculate the thermoelectric emf produced by the thermocouple at the nominal temperature (350°C).
b. Calculate the error incurred by using only the first three terms in the polynomial.

Solution:

a. The polynomial in **Table 3.7** gives the output for the chromel-constantan thermocouple (E-type) with a 0°C reference junction. We calculate the output using the first three terms and subtract from it the emf of the reference junction, also calculated with the three-terms polynomial.

The emf of the thermocouple using three terms in the polynomial is

$$emf(0) = (5.8695857799 \times 10 \times T + 4.3110945462 \times 10^{-2} \times T^2 + 5.7220358202 \times 10^{-5} \times T^3) \times 10^3,$$

where $T = 350$°C. This gives 28.278 mV.

Using the same relation, the reference emf at 100°C is 6.3579 mV. Thus the measured emf is

$$emf = 28.278 - 6.3579 = 21.9201 \text{ mV}.$$

b. Using the full polynomial in **Table 3.8**, we get

$$emf(0) = 24.9614 \text{ mV}.$$

The reference emf (at 100°C, also calculated with the ninth-order polynomial) is

$$emf_{ref} = 6.3171 \text{ mV}.$$

The emf is therefore

$$emf = 24.9614 - 6.3171 = 18.6443 \text{ mV}.$$

This produces an error of

$$error = \frac{21.9201 - 18.6443}{18.6443} \times 100 = 17.6\%.$$

Note that the error is due to the incomplete polynomial and has nothing to do with the reference temperature. Nevertheless, using a zero reference temperature would produce a larger output with a lower error.

Using the results in (a) and (b) for the 0°C reference values, the error would be

$$error = \frac{28.278 - 24.9614}{24.9614} \times 100 = 13.3\%.$$

TABLE 3.9 ■ Common thermocouple types and some of their properties

Materials	Sensitivity [μV/°C] at 25°C	Standard type designation	Recommended temperature range [°C]	Notes
Copper/constantan	40.9	T	0 to 400 (−270 to 400)	Cu/60% Cu 40% Ni
Iron/constantan	51,7	J	0 to 760 (−210 to 1200)	Fe/60% Cu 40% Ni
Chromel/alumel	40.6	K	−200 to 1300 (−270 to 1372)	90% Ni 10% Cr/95% Ni 2% Al 2% Mg 1% Si
Chromel/constantan	60.9	E	−200 to 900 (−270 to 1000)	90% Ni 10% Cr/60% Cu 40% Ni
Platinum (10%)/ rhodium-platinum	6.0	S	0 to 1450 (−50 to 1760)	Pt/90% Pt 10% Rh
Platinum (13%)/ rhodium-platinum	6.0	R	0 to 1600 (−50 to 1760)	Pt/87% Pt 13% Rh
Silver/palladium	10		200 to 600	
Constantan/tungsten	42.1		0 to 800	
Silicon/aluminum	446		−40 to 150	
Carbon/silicon carbide	170		0 to 2000	
Platinum (30%)/ rhodium-platinum	6.0	B	0 to 1820	Pt/70% Pt 30% Rh
Nickel/chromium-silicon alloy		N	(−270 to 1260)	
Tungsten 5%-rhenium/ tungsten 26%-rhenium		C	0 to 2320	
Nickel-18% molybdenum/ nickel-0.8% cobalt		M	−270 to 1000	
Chromel-gold/iron	15		1.2 to 300	

The common thermocouple types are shown in **Table 3.9**, which lists their basic range and transfer functions together with some additional properties. There are many other thermocouples available commercially and still more that can be made. Two chromel/alumel thermocouples (K-type) with exposed junctions are shown in **Figure 3.12**.

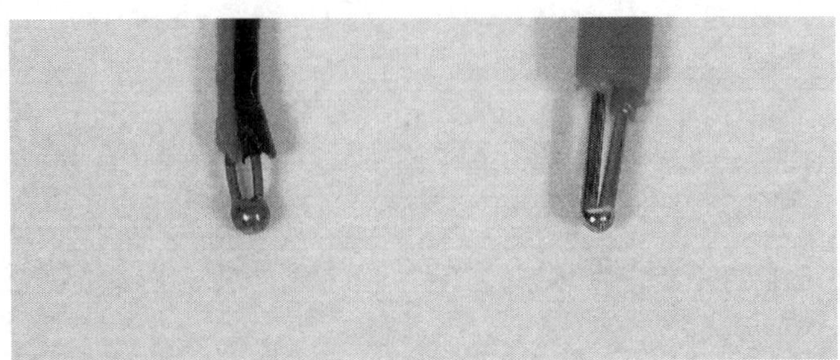

FIGURE 3.12 ■ Chromel-alumel (K-type) thermocouples showing the junction.

Note: The temperature ranges shown are recommended. Nominal ranges are shown in parentheses and are higher than the recommended ranges. The sensitivity of a thermocouple is the relative Seebeck coefficient of the combination of two materials used for the thermocouple (see **Table 3.3**).

EXAMPLE 3.10 Errors in the use of thermocouples

Thermocouples must be handled carefully and connections must be properly done or significant errors will occur in the measured output. To understand this, consider a type K thermocouple (chromel-alumel) used to measure the temperature of glass in a furnace in a glass-blowing studio. The temperature needed for proper blowing is 900°C. The thermoelectric voltage is measured using the configuration in **Figure 3.13a** with a chromel-alumel reference junction at 0°C. In the connection process, the wires for the reference junction have been inadvertently inverted and now the configuration is as in **Figure 3.13b**. The junction box is at the ambient temperature of 30°C.

a. Calculate the error in the measured voltage due to the error in connection.

b. What is the temperature that the measuring instrument will show?

Solution:

a. The two connections in the temperature zone in **Figure 3.13b** are in fact two K-type thermocouples with polarity opposing the polarity of the measuring thermocouple as indicated in the figure. This has the net effect of reducing the output emf and hence showing a lower temperature.

To calculate the emf, we use the polynomial for the K-type thermocouple in **Appendix B, Section B.2**. The emf of the sensing junction is

$$\begin{aligned}E = &-1.7600413686 \times 10^1 + 3.8921204975 \times 10^1 \times 900 + 1.8558770032 \\ &\times 10^{-2} \times 900^2 - 9.9457592874 \times 10^{-5} \times 900^3 + 3.1840945719 \times 10^{-7} \\ &\times 900^4 - 5.6072844889 \times 10^{-10} \times 900^5 + 5.6075059059 \times 10^{-13} \\ &\times 900^6 - 3.2020720003 \times 10^{-16} \times 900^7 + 9.7151147152 \times 10^{-20} \times 900^8 \\ &- 1.2104721275 \times 10^{-23} \times 900^9 + 1.185976 \times 10^2 e^{-1.183432\times10^{-4}(900-126.9686)^2} \\ = &\ 37{,}325.915\ \mu\text{V}.\end{aligned}$$

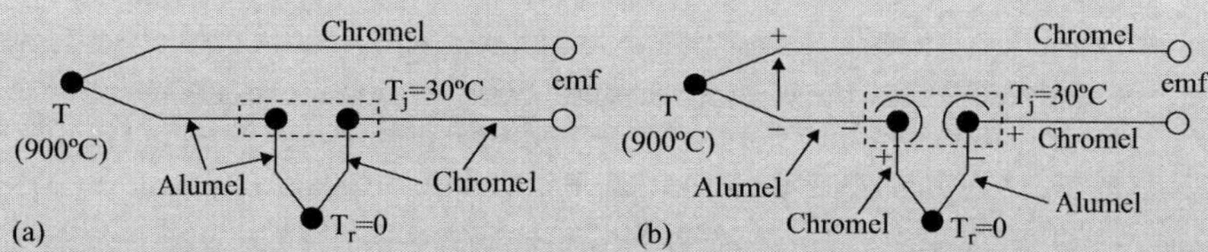

FIGURE 3.13 ■ (a) A properly connected reference junction in a chromel-alumel thermocouple. (b) The reference junction with inverted connections.

The emf of each of the two connections in the temperature zone is

$$\begin{aligned} E = & -1.7600413686 \times 10^1 + 3.8921204975 \times 10^1 \times 30 + 1.8558770032 \\ & \times 10^{-2} \times 30^2 - 9.9457592874 \times 10^{-5} \times 30^3 + 3.1840945719 \times 10^{-7} \\ & \times 30^4 - 5.6072844889 \times 10^{-10} \times 30^5 + 5.6075059059 \times 10^{-13} \times 30^6 \\ & - 3.2020720003 \times 10^{-16} \times 30^7 + 9.7151147152 \times 10^{-20} \times 30^8 \\ & - 1.2104721275 \times 10^{-23} \times 30^9 + 1.185976 \times 10^2 e^{-1.183432 \times 10^{-4}(30-126.9686)^2} \\ = & \ 1203.275\ \mu\text{V}. \end{aligned}$$

The net emf at the instrument is

$$emf = emf(900) - 2 \times emf(30) = 37.3259 - 2 \times 1.2033 = 34.9193 \text{ mV}.$$

The error is the difference between the correct value and the actual reading, or

$$error = 37.3259 - 34.9193 = 2.4066 \text{ mV}.$$

This error is due to the two reversed connections, each contributing 1.2033 mV.

b. To find the temperature that corresponds to this reading we use the inverse polynomial and substitute $E = 34{,}919.3\ \mu\text{V}$:

$$\begin{aligned} T = & -1.318058 \times 10^2 + 4.830222 \times 10^{-2} \times 34919.3 - 1.646031 \times 10^{-6} \\ & \times 34919.3^2 + 5.464731 \times 10^{-11} \times 34919.3^3 - 9.650715 \times 10^{-16} \times 34919.3^4 \\ & + 8.802193 \times 10^{-21} \times 34919.3^5 - 3.110810 \times 10^{-26} \times 34919.3^6 \\ = & \ 839.97^\circ\text{C}. \end{aligned}$$

This represents an error of 6.67% in the temperature reading.

3.3.2 Semiconductor Thermocouples

As can be seen from **Table 3.2**, semiconductors such as p and n silicon (p-doped and n-doped silicon) have absolute Seebeck coefficients that are orders of magnitude higher than those of conductors. The advantage of using semiconductors lies primarily in the large emf that develops at the junction of an n or p semiconductor and a metal (typically aluminum) or at the junction between an n and a p material. In addition, these junctions can be produced by standard semiconductor fabrication techniques, adding to their widespread use in integrated electronics. They suffer from one major shortcoming—the range of temperatures at which they are useful is limited. Silicon in general cannot operate below −55°C or above about 150°C. However, there are semiconductors such as bismuth telluride (Bi_2Te_3) that extend the range to about 225°C, and newer materials can reach about 800°C. Most semiconductor thermocouples are used either in thermopiles for sensing or in thermopiles designed for cooling and heating (Peltier cells). The latter are viewed here as actuators since their main purpose is generation of power or cooling/heating. Peltier cells can be used as sensors since their output is directly proportional to the temperature gradient across the cells. As far as their properties and use are concerned, they are similar to other semiconductor thermocouples.

3.3.3 Thermopiles and Thermoelectric Generators

A thermopile is an arrangement of a number of thermocouples so that their emfs are connected in series. The purpose of this arrangement is to provide much higher outputs than are possible with single junctions. An arrangement of this type is shown in **Figure 3.14**. Note that whereas the electrical outputs are in series, the thermal inputs are in parallel (all cold junctions are at one temperature and all hot junctions are at a second temperature). If the output of a single junction pair is emf_1, and there are n pairs in the thermopile, the output of the thermopile is $n \times emf_1$. The use of thermopiles dates back to the end of the last century and they are commonly used today for a variety of applications. In particular, semiconductor thermocouples are easily produced and integrated with electronics to form the basis for advanced integrated sensors. We will discuss these again in the following chapter where thermopiles are used in infrared sensors. Metal-based thermopiles are used both as sensors and as electricity generators. A very common thermopile sensor used in gas furnaces to detect the pilot flame has a nominal output of 750 mV (0.75 V) and uses a few dozen thermocouples to operate at temperatures up to about 800°C (see **Example 3.11**). Other thermopile assemblies are used in gas-fired generators for the purpose of generating electricity for small, remote installations.

Semiconductor thermopiles made of crystalline semiconductor material such as bismuth telluride (Bi_2Te_3) are being used in Peltier cells for cooling and heating in dual-purpose refrigerators/heaters, primarily for outdoor use and transportation of medical materials. These can also be used as sensors and can have output voltages of a few volts. In this type of semiconductor thermopile, the base semiconductor is doped to make a junction between an n- and p-type material and processed to yield oriented polycrystalline semiconductors with anisotropic thermoelectric properties. Because the junctions are small, hundreds of pairs may be built into a single device to produce outputs on the order of 20 V or more.

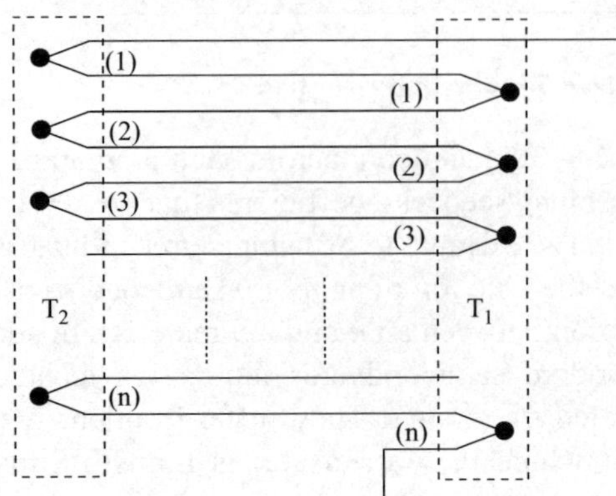

FIGURE 3.14 ■ Principle of a thermopile.

(a)

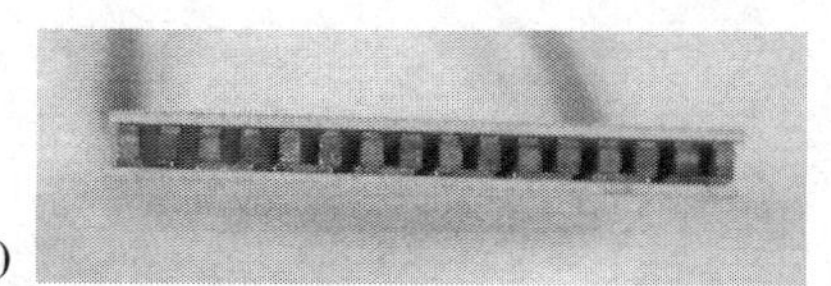
(b)

FIGURE 3.15 ■ (a) Various Peltier cells. (b) Detail of the construction of a Peltier cell.

Figure 3.15a shows a number of thermoelectric devices (Peltier cells) designed primarily for cooling electronics components such as computer processors. **Figure 3.15b** shows the internal construction where the cold junctions are connected thermally to one ceramic plate and the hot junctions are connected to the opposite plate. The junctions are connected in series in rows. The number of junction pairs can be quite high—typically 31, 63, 127, 255, etc. (the odd number allows space for connection of the lead wires in a matrix of junctions which is usually $n \times n$). A typical cooling cell operating at 12 V (or generating 12 V nominal) will contain 127 junctions. Reversing the current in a cooling device will produce heating.

EXAMPLE 3.11 Thermoelectric furnace pilot sensor

A thermopile is needed to sense the presence of a pilot light in a gas furnace to ensure that the gas valve is not opened in the absence of the pilot light. The thermopile needs to provide thermoelectric voltage (emf) of 750 mV for a temperature difference of 650°C. The cold junction is supplied by the body of the furnace, which is at 30°C.

a. What are the options for thermocouples one can use for this purpose? Select an appropriate thermocouple.

b. For the selection in (a), how many thermocouples are needed?

c. Can one use a Peltier cell for this purpose?

Solution:

a. Many of the thermocouple types can be used with the exception of type T, semiconductor thermocouples, and some like the constantan-tungsten thermocouples. A K-type (chromel-alumel), a J-type (iron-constantan), or an E-type (chromel-constantan) should work well. We will select the E-type thermocouple to build the thermopile because it produces higher emfs and thus fewer thermocouples will be needed.

b. The emf of an individual thermocouple is calculated using the coefficients in **Table 3.7** with a reference temperature of 30°C as follows:

At 650°C,

$$\begin{aligned} emf &= 5.8665508708 \times 10^{1} \times 650 + 4.5032275582 \times 10^{-2} \times 650^{2} + 2.8908407212 \\ &\quad \times 10^{-5} \times 650^{3} - 3.3056896652 \times 10^{-7} \times 650^{4} + 6.5024403270 \times 10^{-10} \\ &\quad \times 650^{5} - 1.9197495504 \times 10^{-13} \times 650^{6} - 1.2536600497 \times 10^{-15} \times 650^{7} \\ &\quad + 2.1489217569 \times 10^{-18} \times 650^{8} - 1.4388041782 \times 10^{-21} \times 650^{9} \\ &\quad + 3.5960899481 \times 10^{-25} \times 650^{10} \\ &= 49{,}225.67\ \mu\text{V}. \end{aligned}$$

At 30°C,

$$\begin{aligned} emf &= 5.8665508708 \times 10^{1} \times 30 + 4.5032275582 \times 10^{-2} \times 30^{2} + 2.8908407212 \\ &\quad \times 10^{-5} \times 30^{3} - 3.3056896652 \times 10^{-7} \times 30^{4} + 6.5024403270 \times 10^{-10} \\ &\quad \times 30^{5} - 1.9197495504 \times 10^{-13} \times 30^{6} - 1.2536600497 \times 10^{-15} \times 30^{7} \\ &\quad + 2.1489217569 \times 10^{-18} \times 30^{8} - 1.4388041782 \times 10^{-21} \times 30^{9} \\ &\quad + 3.5960899481 \times 10^{-25} \times 30^{10} \\ &= 1801.022\ \mu\text{V}. \end{aligned}$$

The emf at 650°C with respect to zero is 49.225 mV. The emf at 30°C with respect to zero is 1.801 mV. Therefore the emf at 650°C with reference to 30°C is 49.225−1.801 = 47.424 mV.

For a 750 mV output one needs

$$n = \frac{750}{47.424} = 15.8 \rightarrow n = 16.$$

c. No and perhaps yes. Most Peltier cells are based on low-temperature semiconductors and thus cannot be used directly. However, there are high-temperature Peltier cells that may be used, but even in the absence of these, one can place the Peltier cell with the cold junction on the furnace body and provide a metal structure to conduct heat from the pilot light to the hot surface of the Peltier cell while ensuring that the temperature on the hot surface does not exceed about 80°C (most Peltier cells operate at a temperature difference below 50°C between the hot and cold surfaces). The advantage of the Peltier cell is its physical size for a given thermoelectric voltage and, of course, the fact that it can generate higher emfs than metal thermopiles.

3.4 | *p-n* JUNCTION TEMPERATURE SENSORS

Returning now to semiconductors, suppose that an intrinsic semiconductor is doped so that part of it is of *p*-type while the other is of *n*-type, as shown in **Figure 3.16a**. By so doing a *p-n* junction is created. This is usually indicated as shown in **Figure 3.16b** and is known as a diode. The direction of the arrow shows the direction of flow of current (holes). Electrons flow in the opposite direction. The diode is said to conduct when forward biased, as shown in **Figure 3.16c**. When reverse biased (**Figure 3.16d**), the diode does not conduct. The current–voltage (I-V) characteristics of a *p-n* junction are shown in **Figure 3.17**.

When a *p-n* junction is forward biased, the current through the diode is temperature dependent. This current can be measured and used to indicate temperature. Alternatively, the voltage across the diode can be measured (almost always a preferable approach) and its dependence on temperature used as the sensor's output. This type of sensor is called a *p-n* junction temperature sensor or bandgap temperature sensor. It is particularly useful because it can be easily integrated in microcircuits and is rather linear in output. Needless to say, any diode or a junction in a transistor can be used for this purpose.

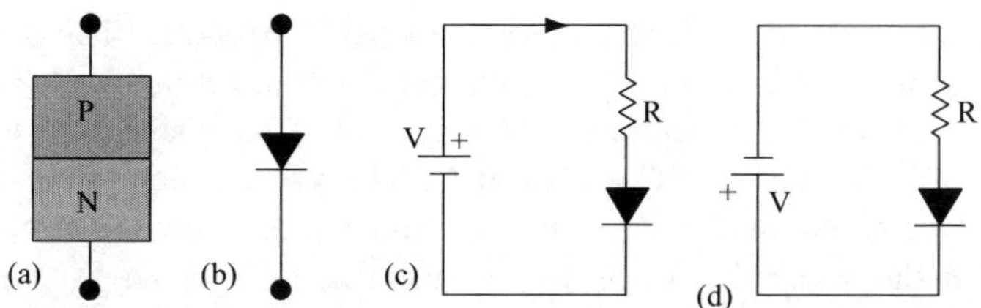

FIGURE 3.16 ■ (a) Schematic of a *p-n* junction. (b) The symbol of the junction as a diode. (c) Forward biasing of the diode. (d) Reverse biasing of the diode.

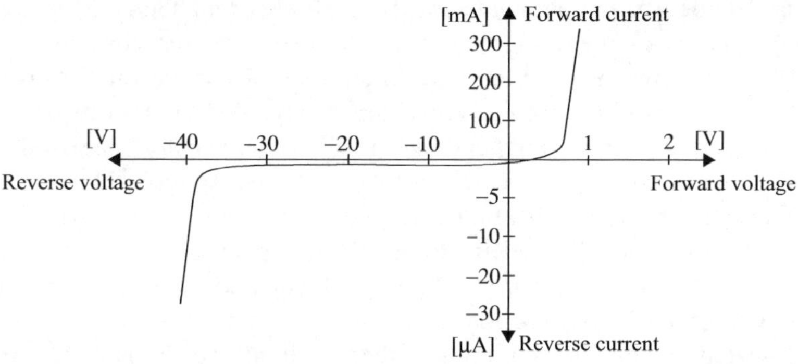

FIGURE 3.17 ■ The *I-V* characteristics of a silicon diode.

Assuming that the junction is forward biased, the *I-V* characteristic is described by the following relation:

$$I = I_0\left(e^{qV/nkT} - 1\right) \quad [\text{A}], \tag{3.25}$$

where I_0 is the saturation current (a small current, on the order of a few nanoamps, dominated by temperature effects), q is the charge of the electron, k is Boltzmann's constant, and T is the absolute temperature (K). n is a constant between 1 and 2 depending on a number of properties including the materials involved and may be viewed as a property of the device. In junction temperature sensors, the current I is reasonably large compared with I_0, so that the term -1 can be neglected. Also, $n = 2$ for this type of sensor, so a good approximation to the forward current in a diode is

$$I \approx I_0 e^{qV/2kT} \quad [\text{A}]. \tag{3.26}$$

Even if the terms affecting the *p-n* junction characteristics are not known, a junction can always be calibrated to measure temperature in a given range. The relation between current and temperature is nonlinear, as can be seen in **Equation (3.26)**. Usually the voltage across the diode is both easier to sense and more linear. The latter is given as

$$V_f = \frac{E_g}{q} - \frac{2kT}{q}\ln\left(\frac{C}{I}\right) \quad [\text{V}], \tag{3.27}$$

where E_g is the bandgap energy (in joules) of the material (to be discussed more fully in **Chapter 4**; see **Table 4.3** for specific values), C is a temperature-independent constant for the diode, and I is the current through the junction. If the current is constant, then the voltage is a linear function of temperature with negative slope. The slope (dV/dT) is clearly current dependent and varies with the semiconductor material. For silicon it is between 1.0 and 2.5 mV/°C depending on the current. This is shown in **Figure 3.18** for silicon diodes in the range −50°C to +150°C. The voltage across the diode at room temperature is approximately 0.7 V for silicon diodes (the larger the current through the diode, the higher the forward voltage drop across the diode at any given temperature). **Equation (3.27)** may be used to design a sensor based on almost any available diode or transistor. In general the published values for the bandgap energy for silicon can be used and the constant C determined by measuring the forward voltage drop at a given temperature and current through the diode.

In using a diode as a temperature sensor, a stable current source is needed. In most practical applications a voltage source with a relatively large resistor may be used, as in **Figure 3.19**, to bias the junction and establish a small current on the order of 100–200 μA. Because V_f is not constant with temperature, this method of biasing is only sufficient for general purpose applications or when the sensing span is small. Sensitivities are between 1 and 10 mV/°C with a thermal time response of less than 1 second. Like any thermal sensor, self-heating must be taken into account when connecting power to junction sensors. The self-heating effects are similar to those of thermistors and are on the order of 0.1–1 mW/°C. More sophisticated methods of biasing will be discussed in **Chapter 11** when we take up the issue of current sources. Junction sensors can be fabricated on a silicon chip together with all accompanying components, including current regulation, and may be quite complicated. The sensitivity of these junction sensors is usually improved to about 10 mV/°C and may be calibrated to produce output proportional to temperature on the Celsius, Fahrenheit, or Kelvin scales. The device is usually connected to a constant voltage source (say 5 V) and produces an output voltage directly proportional to the selected scale, with excellent linearity and

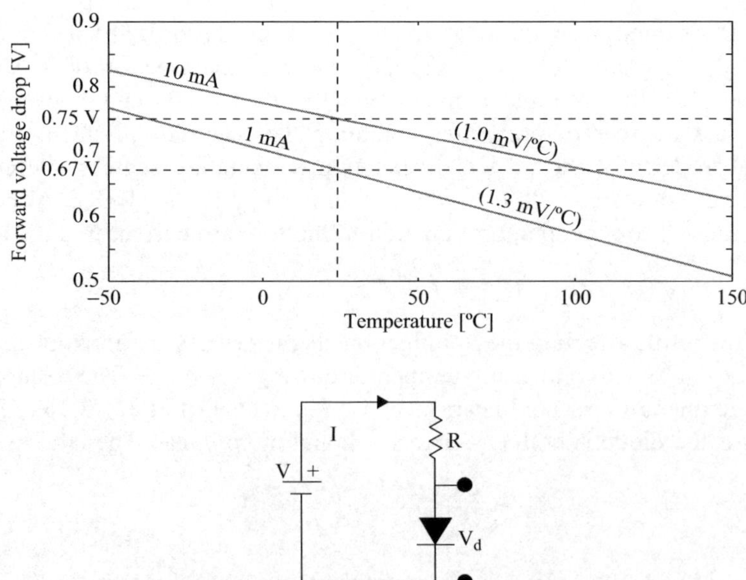

FIGURE 3.18 ■ Potential drop on a forward-biased *p-n* junction versus temperature (1N4148 silicon switching diode, evaluated experimentally).

FIGURE 3.19 ■ Forward-biased *p-n* junction with a rudimentary current *source*. *R* must be large to produce a low current and small variations with variations in *V*.

FIGURE 3.20 ■ Junction temperature sensors.

accuracy typically around ±0.1°C. The range of temperatures that can be measured with these sensors is rather small and cannot exceed the operating range of the base material. Typically a silicon sensor can operate between −55°C and about 150°C, although they can be designed for somewhat wider ranges, whereas some devices are rated for narrower ranges (usually at lower cost). **Figure 3.20** shows three junction temperature sensors in various packages.

EXAMPLE 3.12 Silicon diode as a temperature sensor

A silicon diode is proposed for use as a temperature sensor in a vehicle to sense ambient temperature between −45°C and +45°C. To determine its response, the diode is forward biased with a 1 mA current and its forward voltage drop is measured at 0°C as 0.712 V. The bandgap energy of silicon is 1.11 eV. Calculate

a. The output expected for the span needed.

b. The sensitivity of the sensor.

c. The error in measuring temperature if the self-heating specification for the diode in still air is 220°C/W.

Solution: We first calculate the constant C in **Equation (3.27)** from the known voltage drop at 0°C followed by the forward voltage drop at the range points of the span.

a. At 0°C we have

$$V_f = \frac{E_g}{q} - \frac{2kT}{q}\ln\left(\frac{C}{I}\right) = 0.712$$

$$= \frac{1.11 \times 1.602 \times 10^{-19}}{1.602 \times 10^{-19}} - \frac{2 \times 1.38 \times 10^{-23} \times 273.15}{1.602 \times 10^{-19}}\ln\left(\frac{C}{10^{-3}}\right).$$

Note that 1 eV = 1.602×10^{-19} J and that the temperature T is in degrees kelvin. Thus

$$0.712 = 1.11 - 0.04706\ln\left(\frac{C}{10^{-3}}\right) \rightarrow \ln(10^3 C) = \frac{0.712 - 1.11}{-0.04706} = 8.457.$$

This gives

$$e^{8.457} = 10^3\mathrm{C} \rightarrow \mathrm{C} = \frac{e^{8.457}}{10^3} = 4.7.$$

Now the forward voltage at −45°C and +45°C are

$$\begin{aligned} V_f(-45°\mathrm{C}) &= \frac{E_g}{q} - \frac{2kT}{q}\ln\left(\frac{\mathrm{C}}{I}\right) \\ &= \frac{1.11 \times 1.602 \times 10^{-19}}{1.602 \times 10^{-19}} - \frac{2 \times 1.38 \times 10^{-23} \times (273.15 - 45)}{1.602 \times 10^{-19}}\ln(4.7 \times 10^3) \\ &= 0.77765 \text{ V} \end{aligned}$$

and

$$\begin{aligned} V_f(+45°\mathrm{C}) &= \frac{E_g}{q} - \frac{2kT}{q}\ln\left(\frac{\mathrm{C}}{I}\right) = 1.11 - \frac{2 \times 1.38 \times 10^{-23} \times (273.15 + 45)}{1.602 \times 10^{-19}}\ln(4.7 \times 10^3) \\ &= 0.64654 \text{ V}. \end{aligned}$$

The forward voltage drop varies between 0.77765 V at −45°C and 0.64654 V at +45°C.

b. Since **Equation (3.27)** is linear with temperature, the sensitivity of the device is the difference between the two range points divided by the difference in temperature:

$$s = \frac{0.64654 - 0.77765}{90} = 1.457 \text{ mV/°C}.$$

Comparing this to **Figure 3.18**, it is clear that the diode selected here is somewhat less sensitive than the one described in **Figure 3.18**.

c. The self-heating effect causes an increase in temperature of 220°C/W or 0.22°C/mW. Since the current through the diode is 1 mA, the power dissipated is,

at −45°C,

$$P(-45°\mathrm{C}) = 0.77765 \times 10^{-3} = 0.778 \text{ mW}.$$

The increase in temperature is $0.778 \times 0.22 = 0.171$°C and the forward voltage decreases by $0.171 \times 1.457 = 0.249$ mV. This represents an error of 0.38% in the temperature reading.

At 45°C,

$$P(+45°\mathrm{C}) = 0.64654 \times 10^{-3} = 0.647 \text{ mW}.$$

The increase in temperature is $0.647 \times 0.22 = 0.142$°C and the forward voltage decreases by $0.142 \times 1.457 = 0.207$ mV. The error in temperature reading is 0.32%.

These errors are small, but not necessarily negligible.

3.5 | OTHER TEMPERATURE SENSORS

Almost all physical quantities and phenomena that can be measured are temperature dependent and therefore, in principle, a sensor can be designed around almost any of these. For example, the speed of light and/or its phase in an optical fiber, the speed of sound in air or a fluid, the frequency of vibration of a piezoelectric membrane, the length of a piece of metal, the volume of a gas, and so on, are temperature dependent. Rather than discuss all possible sensors of this type, we briefly discuss here a few representative sensors.

3.5.1 Optical and Acoustical Sensors

Optical temperature sensors are of two basic types. One type is the noncontact sensors that measure the infrared radiation of a source. With proper calibration the temperature of the source can be sensed and accurately measured. We shall discuss infrared radiation sensors in the **Chapter 4.** There are, however, many other temperature sensors based on the optical properties of materials. For example, the index of refraction of silicon is temperature dependent and the speed of light through a medium is inversely proportional to the index of refraction. By comparing the phase of a beam propagating through a silicon fiber that is exposed to heat with the phase of a beam in a reference fiber, the phase can be used as a measure of temperature. This type of sensor is an interferometric sensor and can be extremely sensitive, especially if the fiber is long.

Acoustic sensors can act in a similar way, but because the speed of sound is low, it can be measured directly from the time of flight of an acoustic signal through a known distance. Typically a sensor of this type includes a source that generates an acoustic signal, such as a loudspeaker or ultrasonic transmitter (a device very similar to a loudspeaker, but much smaller and operating at higher frequencies). A tube filled with a gas or a fluid is exposed to the temperature to be measured and a microphone or a second ultrasonic device (a receiver) is placed at the other end. This is shown schematically in **Figure 3.21**. A signal is transmitted and the delay between the time of its transmission and the time of arrival at the receiver is calculated. The length of the tube divided by the time difference (time of flight) gives the speed of sound. This can then be calibrated to read the temperature. For example, if the tube is air filled, the relation between temperature and the speed of sound in air is

$$v_s = 331.5\sqrt{\frac{T}{273.15}} \quad [\mathrm{m/s}], \tag{3.28}$$

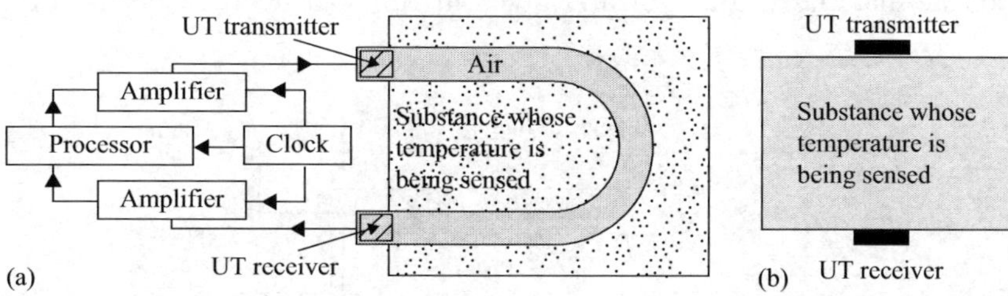

FIGURE 3.21 ■ Acoustic temperature sensing. (a) Sound travels through a fluid-filled channel. (b) Sound travels through the working fluid itself.

where T is the absolute temperature (K) and the speed at 273.15 (0°C) is 331.5 m/s. In some installations, the tube may be eliminated and the substance whose temperature is being sensed may serve instead (**Figure 3.21b**).

The speed of sound in water is also temperature dependent, and this dependency can be used to measure temperature or the temperature may be used to compensate for changes in the speed of sound due to temperature. In seawater the speed of sound depends on depth as well as salinity. Disregarding salinity and depth (i.e., regular water at the surface), a simplified relation is

$$v_s = a + bT + cT^2 + dT^3 \quad [\text{in m/s}], a = 1449, b = 4.591, \\ c = -5.304 \times 10^{-2}, \quad d = 2.374 \times 10^{-4}, \tag{3.29}$$

where $a = 1449$ m/s is the speed of sound in water at 0°C and T is given in degrees Celsius. The other terms may be viewed as corrections to the first term. In seawater and at other depths additional correction terms are needed.

3.5.2 Thermomechanical Sensors and Actuators

An important and common class of temperature sensors and actuators is the so-called thermomechanical sensors and actuators. The general idea is that the temperature being sensed changes a physical property such as length, pressure, volume, etc. These properties then serve to measure the temperature and often perform actuation. For this reason, and since it is often difficult to distinguish between sensing and actuation, these devices will be discussed together. Common examples are the change in length of a metal or the volume of a gas. Another is the glass thermometer in which the height of a capillary column of fluid (mercury or alcohol) indicates the temperature through expansion of the fluid. Many of these types of sensors also feature a direct reading of temperature without the need for an intermediary processing stage and typically require no external power.

A simple example of a sensor based on the expansion of gas (or an expandable fluid such as alcohol) is shown in **Figure 3.22**. The volume of the gas, and therefore the position of the piston is directly proportional to the temperature being measured.

The volume expands based on the volume expansion coefficient of the medium, β. The change in volume due to the change in temperature is

$$\Delta V = \beta V \Delta T \quad [\text{m}^3], \tag{3.30}$$

where V is the volume and ΔT is the change in temperature. The coefficient β is a property of the material and is typically given in 10^{-6} per degree Celsius at a specific

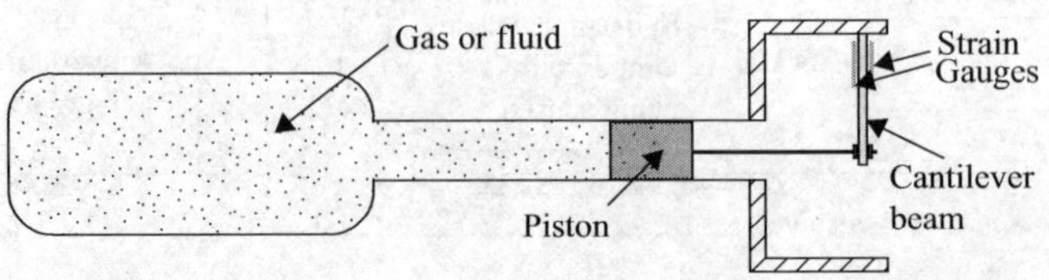

FIGURE 3.22 ■ Temperature sensing based on expansion of gases (or liquids). The piston may be replaced with a diaphragm.

temperature (usually 20°C). The volume of a material at a temperature T is then calculated as

$$V = V_0[1 + \beta(T - T_0)] \quad [\text{m}^3], \tag{3.31}$$

where V_0 is the volume at the reference temperature T_0 (usually 20°C).

In many solids and fluids (isotropic materials), the relation between the volume coefficient of expansion, β, and the linear coefficient of expansion, α, is

$$\alpha = \frac{\beta}{3}. \tag{3.32}$$

If necessary, one can also define a coefficient of surface expansion as

$$\gamma = \frac{2\beta}{3}. \tag{3.33}$$

In the case of gases, the situation is somewhat different as the expansion depends on the conditions under which the gas expands and, of course, only a volume expansion is physically meaningful. For an ideal gas, under isobaric expansion (i.e., the gas pressure remains unchanged), the coefficient of volume expansion is

$$\beta = \frac{1}{T}, \tag{3.34}$$

where T is the absolute temperature.

The expansion of gasses can also be deduced from the ideal gas law, stated as follows:

$$PV = nRT, \tag{3.35}$$

where P is pressure in pascals, V is volume in cubic meters, T is temperature in degrees kelvin, n is the amount of gas in moles, and R is the gas constant, equal to 8.314 J/K/mol or 0.08206 L/atm/K/mol. This relation gives the state of the gas and may be used to calculate pressure under constant volume conditions or volume under constant pressure conditions as a function of temperature. One can then use either **Equation (3.31)** with the coefficient in **Equation (3.34)** or start with the ideal gas law in **Equation (3.35)** to calculate the volume of the gas. The choice depends on the conditions (see **Problem 3.29**).

Coefficients of linear and volume expansion are given in **Table 3.10** for a number of metals, fluids, and other materials. It should be noted that some materials have low coefficients, whereas others, particularly fluids, have large coefficients. Clearly the use of fluids like ethanol or water is favored because the resulting expansion is larger, leading to a more sensitive sensor/actuator. Of course, gases expand much more and thus a gas-filled sensor will have higher sensitivity and faster response time, but, on the other hand, the sensor will be nonlinear.

The position of the piston (**Figure 3.22**) is sensed in any of a number of ways. It may drive a potentiometer and the resistance is then a measure of temperature. Alternatively, it may have a mirror connected to it that tilts with the increase in pressure and a light beam is then deflected accordingly, or one can measure the strain in a diaphragm to get the temperature. Or it may even drive a needle directly to indicate temperature on a scale. In the configuration shown in **Figure 3.22**, the temperature is sensed using strain gauges that measure the strain in the cantilever beam (we shall discuss strain gauges in

TABLE 3.10 ■ Coefficients of linear and volume expansion for some materials given per degree Celsius at 20°C

Material	Coefficient of linear expansion (α), $\times 10^{-6}$/°C	Coefficient of volume expansion (β), $\times 10^{-6}$/°C
Aluminum	23.0	69.0
Chromium	30.0	90.0
Copper	16.6	49.8
Gold	14.2	42.6
Iron	12.0	36
Nickel	11.8	35.4
Platinum	9.0	27.0
Phosphor-bronze	9.3	27.9
Silver	19.0	57.0
Titanium	6.5	19.5
Tungsten	4.5	13.5
Zinc	35	105
Quartz	0.59	1.77
Rubber	77	231
Mercury	61	182
Water	69	207
Ethanol	250	750
Wax	16,000–66,000	50,000–200,000

Chapter 6). The piston may be replaced with an appropriate diaphragm. A very sensitive implementation of this sensor is shown in **Figure 3.23**. This sensor is called a Golay cell (sometimes also called a thermopneumatic sensor). The trapped gas (or liquid) expands the diaphragm and the position of the light beam indicates the temperature. Although gases or liquids may be used, gases have lower heat capacities (require less energy to raise their temperature) and hence have better response times.

We shall see in **Chapter 4** that this device can be used to measure infrared radiation as well. It should be obvious that the expansion and the resulting motion of the piston can also be used for actuation or, as is the case in alcohol and mercury thermometers, for direct indication of temperature (a sensor-actuator). However, most actuators based on expansion use metals and will be discussed below.

The diameter must be 0.309 mm. This is in fact a capillary tube. Because it is so thin, the alcohol is dyed (typically red or blue) and the tube is fitted with a cylindrical lens along its visible surface to facilitate reading of the temperature.

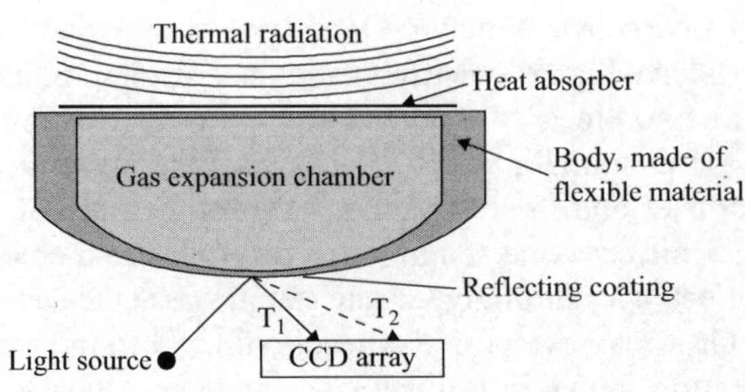

FIGURE 3.23 ■ The Golay cell is a thermopneumatic sensor based on the expansion of gasses. The charge-coupled device (CCD) array is a light sensor that will be discussed in **Chapter 4**.

EXAMPLE 3.13 **The alcohol thermometer**

A medical thermometer is to be manufactured using a thin glass tube, as shown in **Figure 3.24**, with a range of 34°C–43°C. The volume of the bulb (which serves as a reservoir) is 1 cm^3. To properly read the temperature, the graduations are to be 1 cm/°C (so that a change of 0.1°C raises the alcohol level by 1 mm). Assuming the glass does not expand (i.e., its coefficient of volume expansion is negligible), calculate the inner diameter of the glass tube needed to produce the thermometer.

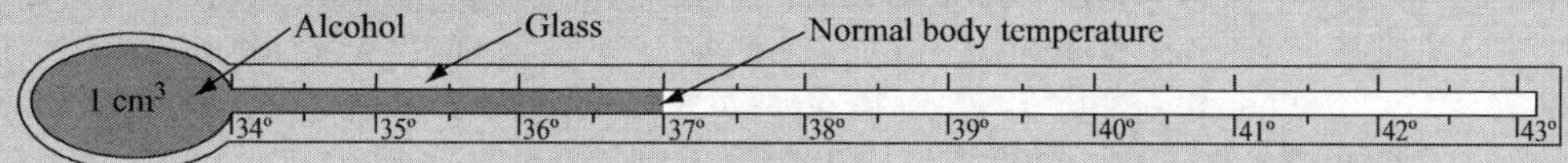

FIGURE 3.24 ■ An alcohol thermometer.

Solution: We can use **Equation (3.30)** to calculate the change in volume for a change in temperature of 9°C (43°C–34°C). This change in volume is the volume of the thin tube between the two extreme graduations (9 cm long):

$$\Delta V = \beta V \Delta T = 750 \times 10^{-6} \times 1 \times 9 = 6750 \times 10^{-6}\ \text{cm}^3,$$

where we have used the coefficient of volume expansion of ethanol (we assume, implicitly, that the coefficient is the same as at 20°C). Taking the inner diameter of the thin tube as d, we have

$$\pi \frac{d^2}{4} \times L = \Delta V \rightarrow d = \sqrt{\frac{4\Delta V}{\pi L}}\quad [\text{cm}],$$

where L is the length of the tube (9 cm in our case). Thus

$$d = \sqrt{\frac{4\Delta V}{\pi L}} = \sqrt{\frac{4 \times 6750 \times 10^{-6}}{\pi \times 9}} = 3.09 \times 10^{-2}\ \text{cm}.$$

Fluid thermometers are not as common as they used to be, but they can still be found, especially as outdoor thermometers.

Some of the oldest temperature sensors are based on the thermal expansion of metals with temperature and constitute direct conversion of temperature into displacement. These are commonly used as direct reading sensors because this mechanical expansion can be used as an actuator as well to move a dial or some other type of indicator. A conductor made in the form of a bar or wire of length l will experience an elongation with an increase in temperature. If the length is l_1 at T_1, then at $T_2 > T_1$, $l_2 > l_1$ as follows:

$$l_2 = l_1[1 + \alpha(T_2 - T_1)]\quad [\text{m}]. \tag{3.36}$$

If $T_2 < T_1$, the bar will contract. The change in length can then be measured to represent the temperature being measured. The coefficient α is called the coefficient of linear expansion of the metal (see **Table 3.10** for the coefficients for some materials). Although the coefficient of expansion is small, it is nevertheless measurable and with proper care becomes a useful method of sensing. There are two basic methods that can be

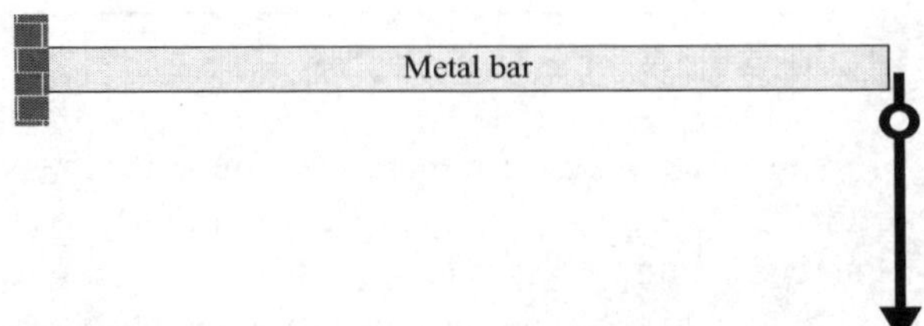

FIGURE 3.25 ■ A simple direct indication by an expanding linear bar. The dial is pushed to indicate temperature.

used. One is shown in **Figure 3.25**. It is a simple bar pushing against an indicator. An increase in temperature will move the arrow to read on a dial. An alternative way is for the bar to rotate a potentiometer or to press against a pressure gauge, in which case the electric signal may be used to indicate temperature or to connect to a processor. Although a sound principle, this is difficult to use because the expansion is small (see **Table 3.10**) and because of hysteresis and mechanical slack. However, this method is often used in microelectromechanical systems (MEMS) where expansions of a few micrometers are sufficient to affect the necessary actuation. We will discuss thermal actuation in MEMS in **Chapter 10** (but see **Example 3.14**). One glaring exception is wax (especially paraffin), with its large coefficient. Waxes of various compositions are used in direct actuation in thermostats, especially in vehicles (see **Problems 3.32** and **3.33**).

EXAMPLE 3.14 Linear thermal microactuator

An actuator is built as a thin chromium rectangular wire as shown in **Figure 3.26a**. A current passes through the wire to heat it up. The free end serves as the actuator (i.e., it can serve to close a switch or to tilt a mirror for an optical switch; we shall encounter these applications in **Chapter 10**). If the temperature of the chromium wire can vary from 25°C to 125°C, what is the largest change in length of the actuator? The dimensions in the figure are given at 20°C.

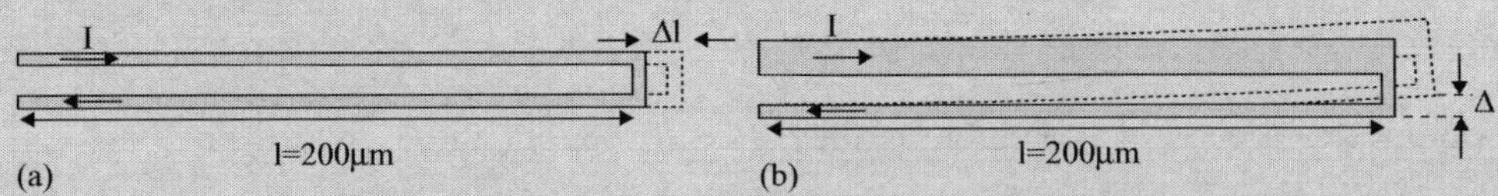

FIGURE 3.26 ■ Microthermal actuator. (a) Linear motion/actuation. (b) Angular actuation.

Solution: By direct application of **Equation (3.36)** we get

$$l(25°\text{C}) = 200[1 + 30 \times 10^{-6} \times (25 - 20)] = 200.03\ \mu\text{m}$$

and

$$l(125°\text{C}) = 200[1 + 30 \times 10^{-6} \times (125 - 20)] = 200.63\ \mu\text{m}.$$

The actuator changes in length by 0.6 μm. This may seem small, but it is both measurable and linear, and in the context of microactuators is sufficient for many applications. Thermal actuators are some of the simplest and commonly utilized microdevices. The main problem plaguing thermal actuators on the macroscopic level, that of slow response time as well as the power needed, are not relevant in microdevices. Their small size makes them sufficiently responsive and the power needed is small as well.

Note also that if one leg of the structure, say, the upper, is made thicker, it will heat to a lower temperature since it can dissipate more heat, and the whole frame will bend upwards with an increase in current. The position of the tip of the frame can then be controlled by the current in the frame (**Figure 3.26b**).

A useful implementation of the expansion of metals is the bimetal bar shown in **Figure 3.27a**. Two metals with different expansion coefficients are bonded together. Suppose that the layer on top has a higher coefficient than the layer on the bottom. When the temperature is raised, the top layer will expand more and thus the tip will move downward (**Figure 3.27b**). If the temperature is lowered, the tip will move upward. This may be used to move a dial or strain a gauge to measure the motion. It may also be used to close or open a switch. The latter is still used in direction indicators in cars, where a bimetal sensor is used to sense the current through the indicating bulb by passing the current through the bimetal material, which forms a switch as shown in **Figure 3.29a**. The bimetal bar heats up and bends downward, disconnecting the switch. The lamp goes off and the bimetal element cools down, moving upward and reconnecting the lamp. As is the case with most thermomechanical devices, the bimetal sensor is really a sensor-actuator.

The bimetal principle can also be used for direct dial indication, as shown in **Figure 3.27c**. Here a long strip of bimetal material is bent into a coil. The change in

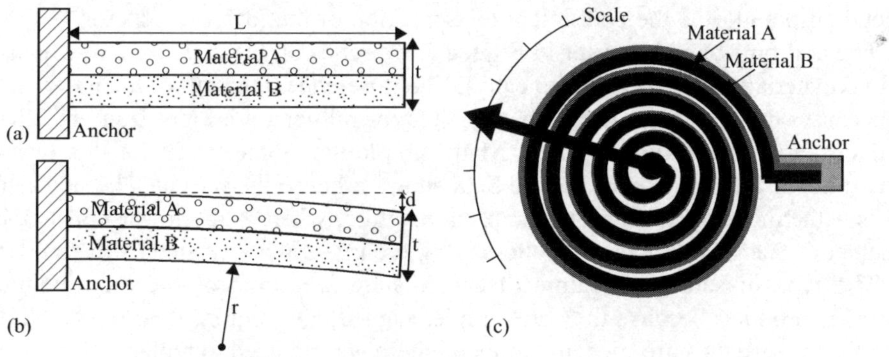

FIGURE 3.27 ■ Bimetal sensor. (a) Basic construction. (b) Displacement with temperature. (c) Coil bimetal temperature sensor.

FIGURE 3.28 ■ Bimetal thermometer and thermostat. The upper bimetal coil is the thermometer, the lower is the thermostat. The thermostat coil activates the mercury switch at the front of the picture.

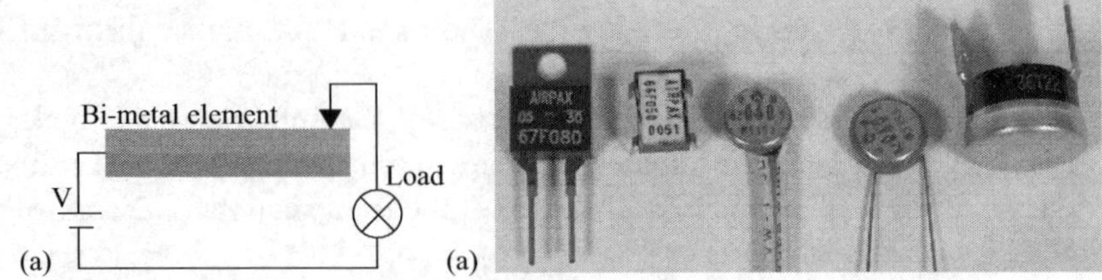

FIGURE 3.29 ■ (a) Schematic of a bimetal switch as used for directional flashers in automobiles. (b) Small bimetal thermostats.

temperature causes a much larger change in the length of the strip and the motion of the strip rotates the dial in proportion to temperature.

The displacement of the free-moving end of a bimetal strip can be calculated from approximate expressions. In **Figure 3.27b**, the displacement of the free-moving tip is given by

$$d = r\left[1 - \cos\left(\frac{180L}{\pi r}\right)\right] \quad [\text{m}], \tag{3.37}$$

where

$$r = \frac{2t}{3(\alpha_u - \alpha_l)(T_2 - T_1)} \quad [\text{m}]. \tag{3.38}$$

T_1 is the reference temperature at which the bimetal bar is flat, T_2 is the sensed temperature, t is the thickness of the bar, L is the length of the bar, and r is the radius of curvature or warping. α_u is the coefficient of expansion of the upper conductor in the bimetal strip and α_l is the coefficient of expansion of the lower conductor.

The coil-type bimetal sensor in **Figure 3.27c** relies on the difference in expansion of the two materials to turn a dial (typically). The difference in length of the inner and outer strips causes the coil to rotate, and because the overall length is significant, the change is relatively large (see **Example 3.15**). Many simple thermometers are of this type (especially outdoor thermometers). **Figure 3.28** shows a house thermostat. The upper bimetal coil is a thermometer to indicate room temperature, whereas the lower coil is the thermostat coil. The glass bulb at the bottom is a mercury switch activated by the bimetal coil.

The type of sensors/actuators discussed here are some of the most common in consumer products because they are simple, rugged, and require no power, and at least partially because they are tried and true and there was no need to replace them. They can be found in kitchen thermometers (meat thermometers), appliances, thermostats, and outdoor thermometers.

As indicated earlier, these devices are equally useful as actuators since they convert heat into displacement. In selected cases this actuation is a natural choice, as in the example of the turn indicator switch, thermostats, and simple thermomechanical thermometers.

Whereas the bimetal principle exploits the linear expansion of metals, the volume expansion of gases and fluids can be used for sensing and for actuation, as has been exemplified by the Golay cell and the alcohol thermometer. There are also some solid materials that expand or contract significantly when a change of phase occurs. For example, water, when it freezes, expands by about 10%. One of the more interesting materials in this category is paraffin wax. When melting, its volume expands by anywhere from 5% to 20% (see **Table 3.10**) depending on the composition of the wax. More significantly from the point of view of actuation is the fact that different

EXAMPLE 3.15 **Bimetal coil thermometer**

An outdoor bimetal thermometer is made in the form of a coil, as shown in **Figure 3.27c**. The coil has six full turns, with the inner turn having a radius of 10 mm and the outer turn a radius of 30 mm when the thermometer is at 20°C. The bimetal strip is 0.5 mm thick and is made of chromium (outer strip) and nickel (inner strip) to resist corrosion. The thermometer is intended to operate between −45°C and +60°C. Estimate the change in angle the needle makes as the temperature changes from −45°C to +60°C.

Solution: It is not possible to calculate the change in angle exactly since this depends on additional parameters. However, we may assume that if the strip were straight, each of the conductors must expand according to **Equation (3.36)**. The difference in the expansion coefficients of the two metals forces them to coil. Therefore we will calculate the change in length of the outer strip, and that change is what causes the needle to move (the lower expansion coefficient of the inner, nickel strip is what causes curling of the strip). To be able to approximate the angle of the needle, we first calculate the length of the strip using an average radius. Then we calculate its length at −45°C and at +60°C. The difference between the two moves the strip along the inner loop.

The average radius of the coil is

$$R_{avg} = \frac{30 + 10}{2} = 20 \text{ mm}.$$

The length of the strip at the nominal temperature (20°C) is

$$L = 6(2\pi R_{avg}) = 6 \times 2\pi \times 20 \times 10^{-3} = 0.754 \text{ m}.$$

Now, at the temperature extremes we have

$$L(-45°\text{C}) = 0.754[1 + 30 \times 10^{-6} \times (-45 - 20)] = 0.75253 \text{ m}$$

and

$$L(60°\text{C}) = 0.754[1 + 30 \times 10^{-6} \times (60 - 20)] = 0.7549 \text{ m}.$$

The difference between the two is $\Delta L = 0.7549 – 0.75253 = 0.002375$ m, or 2.375 mm.

To estimate the angle the needle moves, we argue that this expansion is small and therefore the inner loop remains of the same radius. If its circumference is $2\pi r$, then $\Delta\alpha = (\Delta L/2\pi r) \times 360°$:

$$\Delta\alpha = \frac{\Delta L}{2\pi r_{inner}} \times 360 = \frac{0.002375 \times 360}{2 \times \pi \times 0.01} = 13.6°.$$

This means that the scale shown schematically in **Figure 3.27c** covers a 13.6° section.

compositions can be made to melt at specific temperatures. Furthermore, the transition is gradual, first the wax becomes soft (and expands slightly) and then melts, expanding much more. The opposite occurs when it solidifies. This introduces an inherent hysteresis in the process, which is often useful. All of these properties are exploited in thermostats for car engines. The thermostat (a car thermostat is shown in **Figure 3.30**) is essentially a cylinder and a piston, with the cylinder filled by a solid wax pellet. During operation, the piston is pushed by the melting wax to open the cooling water

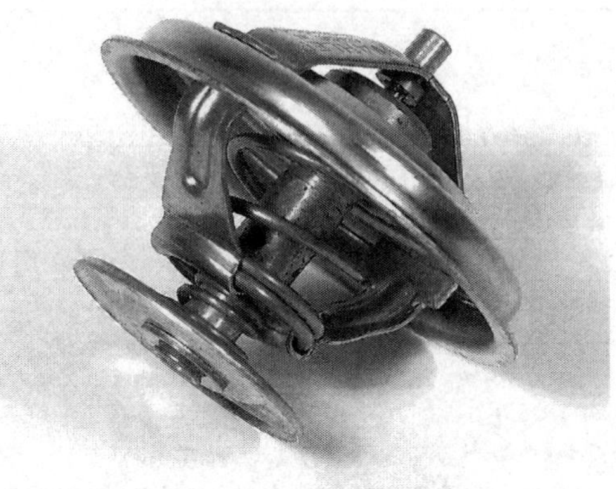

FIGURE 3.30 ■ A car thermostat.

line when the engine has reached the preset temperature. The thermostat not only allows cooling, but ensures quick warming of the engine by keeping the water line closed until the proper temperature has been reached. Then it opens gradually as the temperature rises or closes gradually as the temperature drops to keep the engine coolant temperature within a narrow range of temperatures and by so doing ensures maximum engine efficiency. Thermostats come in a variety of shapes, sizes, and set temperatures. The temperature is set during production by the choice of the wax composition. Wax pellets for temperatures from about 20°C to 175°C exist for a variety of applications, not all of them automotive. See also **Problems 3.32** and **3.33**.

3.6 | PROBLEMS

Units of temperature and heat

3.1 Convert the absolute temperature (0 K) to degrees Celsius and to degrees Fahrenheit.

3.2 The calorie (cal) is a unit of energy equal to 0.239 J. How many electron volts does it represent? The electron-volt (eV) is the energy needed to move an electron (charge equal to 1.602×10^{-19} C) across a potential difference equal to 1 V.

Resistance temperature detectors

3.3 **Simple RTD.** An RTD can be built relatively easily. Consider a copper RTD made of magnet wire (copper wire insulated with a polymer). The wire is 0.1 mm thick and the nominal resistance required is 120 Ω at 20°C. Neglect the thickness of the insulating polymer.

a. How long must the wire be?

b. Assuming we wish to wind the copper wire into a single spiral winding 6 mm in diameter so that it can be enclosed in a stainless steel tube, what is the minimum length of the RTD?

c. Calculate the range of resistance of the RTD for use between −45°C and 120°C.

3.4 Self-heat in RTDs and errors in sensing. A platinum wire RTD enclosed in a ceramic body is designed to operate between −200°C and +600°C. Its nominal resistance at 0°C is 100 Ω and its TCR is 0.00385/°C (European curve). The sensor has a self-heat of 0.07°C/mW. The sensor is fed from a constant voltage source of 6 V through a fixed 100 Ω resistor and the voltage across the sensor is measured directly. Calculate the error in temperature sensed in the range 0°C–100°C. Plot the error as a function of temperature. What is the maximum error and at what temperature does it occur? Explain.

3.5 Temperature sensing in a light bulb. Incandescent lightbulbs use a tungsten wire as the light-producing filament by heating it to a temperature at which it is bright enough to produce light. The temperature of the wire can be estimated directly from the power rating and the resistance of the wire when it is cold. Given a 120 V, 100 W lightbulb with a resistance of 22 Ω at room temperature (20°C):

a. Calculate the temperature of the filament when the lightbulb is lit.

b. What are the possible sources of error in this type of indirect sensing? Explain.

3.6 Accurate representation of resistance of RTDs. The Callendar–Van Dusen polynomials (**Equations (3.6)** and **(3.7)**) can be used either with published data from common RTD materials or the coefficients of the polynomial can be determined from measurements. Suppose one decides to introduce a new line of RTDs made of nichrome (a nickel-chromium alloy) for the range between −200°C and +900°C. To evaluate the behavior of the new type of sensors, one must determine the constants a, b, and c in **Equations (3.6)** and **(3.7)**. There are common calibration points that guarantee exact known temperatures at which the resistance is measured. The common calibration points are the oxygen point (−182.962°C, equilibrium between liquid oxygen and its vapor), the triple point of water (0.01°C, the point of equilibrium temperature between ice, liquid water, and water vapor), the steam point (100°C, equilibrium point between water and its vapor), the zinc point (419.58°C, equilibrium point between solid and liquid zinc), the silver point (961.93°C), and the gold point (1064.43°C), as well as others. By selecting the appropriate temperature points and measuring the resistance at those points one can determine the coefficients. The resistance measurements for the new type of RTD are as follows: $R = 9$ Ω at the oxygen point, $R = 38$ Ω at 0°C, $R = 52$ Ω at the steam point, and $R = 98$ Ω at the zinc point. The TCR for the nichrome alloy used here is 0.004/°C.

a. Find the coefficients of the polynomials.

b. Find the resistance at −150°C and at 800°C. Compare the results with those obtained using **Equation (3.5)**. What are the errors?

3.7 Effect of temperature gradient on accuracy. Wire RTDs are often relatively long sensors and the temperature gradient on the sensor itself may be of concern in certain applications or in dynamic situations where the temperature changes quickly. To understand the effect, consider a platinum wire RTD of length 10 cm and a nominal resistance of 120 Ω at 20°C.

a. Calculate the temperature reading at 80°C if one end is at that temperature while the other is 1°C lower and the distribution within the sensor is linear.

b. Calculate the temperature reading as in (a) but for a parabolic distribution in which the temperature of the center of the sensor is at 79.25°C.

3.8 **Indirect temperature sensing: pyrometric temperature sensor.** A relatively old and well-established method of sensing high temperatures, especially that of molten metals and molten glass, is the use of color comparison. The premise is that if the color of the molten material and that of a control heated filament are the same, their temperatures must also be the same. With proper selection of comparison filaments the method can be very accurate, and it is an entirely noncontact method of sensing. A sensor of this type uses a lamp, heated through a variable resistor, and the voltage and current are read as shown in **Figure 3.31**.

a. Given a reading of V and I for a color match and given the resistance of the filament as R_0 at 20°C, calculate the temperature sensed. Take the temperature coefficient of resistance as α.

b. In an actual sensor, the filament is made of tungsten and has a resistance of 1.2 Ω at 20°C. In a particular application, the voltage across the lamp is measured as 4.85 V with a current of 500 mA. What is the temperature being sensed?

c. Discuss the possible errors involved in this type of measurement.

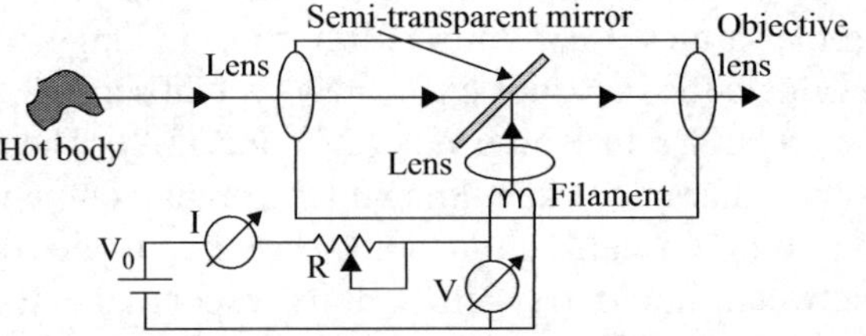

FIGURE 3.31 ■ The pyrometric temperature sensor.

3.9 **TCR and its dependence on temperature.** The temperature coefficient of resistance is not constant but depends on the temperature at which it is given or evaluated. Nevertheless, the formula in **Equation (3.4)** is correct at any temperature, no matter what the temperature T_0 is, as long as α is measured (or given) at T_0. Suppose α is measured at 0°C and is equal to $\alpha = 0.00385$/°C (for platinum).

a. Calculate the coefficient α at 50°C.

b. Generalize the result in (a) as follows: given α_0 at T_0, what is α_1 at T_1?

Silicon resistive sensors

3.10 **Semiconducting resistive sensor.** A semiconducting resistive sensor is made as a simple rectangular bar 2 mm × 0.1 mm in cross section and 4 mm long. The intrinsic carrier concentration at 20°C is 1.5×10^{10}/cm^3 and the mobilities of electrons and holes are 1350 cm^2/V/s and 450 cm^2/V/s, respectively. The TCR for the particular device being used here is −0.009/°C and is assumed to be unaffected by doping. Treat the sensor as a simple resistor.

a. If intrinsic material is used, calculate the resistance of the sensor at 50°C.

b. Now the material is heavily doped with an n-type dopant at a concentration of $10^{15}/m^3$. Calculate the resistance of the sensor at 50°C.

c. What is the resistance of the sensor if instead it is doped with a p-type dopant at the same concentration as in (b)?

3.11 Silicon resistive sensors and their transfer functions. A silicon resistive sensor has a nominal resistance of 2000 Ω at 25°C. To calculate its transfer function its resistance is measured at 0°C and 90°C and found to be 1600 Ω and 3200 Ω, respectively. Assuming the resistance is given by a second-order Callendar–Van Dusen equation, calculate the coefficients of the equation and plot the transfer function between 0°C and 100°C.

Thermistors

3.12 Thermistor transfer function. The transfer function of an NTC thermistor is best approximated using the Steinhart–Hart model in **Equation (3.14)** or **Equation (3.15)**. To evaluate the constants, a thermistor's resistance is measured experimentally at three temperatures, giving the following results: $R = 1.625$ kΩ at 0°C, $R = 938$ Ω at 25°C, and $R = 154$ Ω at 80°C.

a. Calculate the transfer function using the Steinhart–Hart model.

b. Using the resistance at 25°C as the reference temperature, calculate the transfer function using the simplified model in **Equation (3.12)**.

c. Plot the two transfer functions in the range 0°C–100°C and discuss the differences between them.

3.13 Thermistor transfer function. A new type of thermistor rated at 100 kΩ at 20°C is used to sense temperatures between −80°C and +100°C. It is expected that the transfer function is a second-order polynomial. To evaluate its transfer function, the resistance of the thermistor is measured at −60°C as 320 kΩ and at +80°C as 20 kΩ.

a. Find and plot the transfer function for the required span using a second-order polynomial.

b. Calculate the resistance expected at 0°C.

3.14 Thermistor simplified transfer function. The transfer function of a thermistor over the range 0°C–120°C is required. The thermistor is rated at 10 kΩ at 20°C. Three measurements are taken, at 0°C, 60°C, and 120°C, resulting in 24 kΩ, 2.2 kΩ, and 420 Ω, respectively. The simple exponential model in **Equation (3.12)** is used to derive the model. However, since the model only has one variable, β, one can choose any of the three temperatures to derive the transfer function.

a. Derive the transfer function using, in turn, the three measurements. Compare the values of β obtained.

b. Calculate the errors at the three points for the three transfer functions.

c. Plot the three transfer functions and compare them. Discuss the differences and the "proper" choice of a temperature in deriving the simplified model.

3.15 Self-heat of a thermistor. A thermistor with a nominal resistance of 15 kΩ at 25°C carries a current of 5 mA. At an ambient temperature of 30°C (measured with a thermocouple), the resistance of the thermistor is 12.5 kΩ. The current is now removed and the resistance of the thermistor drops to 12.35 kΩ. Calculate the error due to self-heat of the thermistor in degrees Celsius per milliwatt (°C/mW).

Thermoelectric sensors

3.16 **Improper junction temperatures.** A K-type thermocouple measures temperature T_1 and has reference T_r as shown in **Figure 3.32**. Calculate the reading of the voltmeter under the following conditions:

a. $T_1 = 100°C$, $T_r = 0°C$, and the junctions x–x' and y–y' are each in their own temperature zones. (c = chromel, a = alumel).

b. $T_1 = 100°C$, $T_r = 0°C$, and junctions y–y' are in a temperature zone. The junctions x–x' are not in a temperature zone with a temperature difference of 5°C (x is at the higher temperature).

c. Which reading is correct and what is the error in temperature using the incorrect reading?

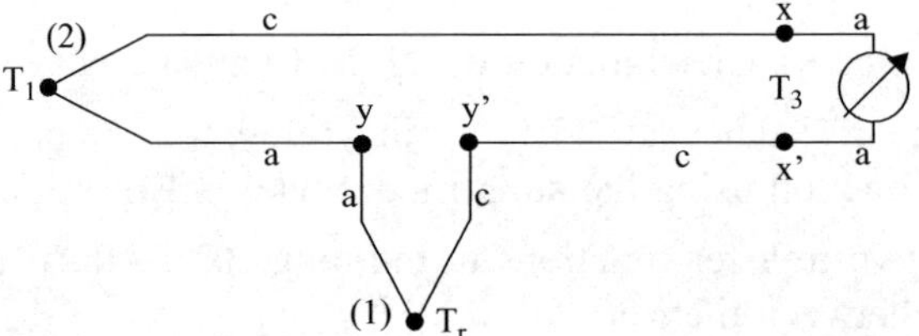

FIGURE 3.32 ■ Connection of a K-type thermocouple.

3.17 **Extension of thermocouple wires.** A K-type thermocouple measures temperature T_1 and has reference T_r as shown in **Figure 3.33**. Calculate the reading of the voltmeter under the following conditions:

a. $T_1 = 100°C$, $T_r = 0°C$, and the extension section is absent. The y–y' junctions are in a temperature zone. (c = chromel, a = alumel). The extension section is not present (i.e., x and w are the same point and x' and y' are the same point).

b. $T_1 = 100°C$, $T_r = 0°C$, and the extension section is present with the alumel wire on top and the chromel wire on the bottom. The junctions x–x', y–y', and z–z' are each in their own temperature zones, all three at 25°C. The junction w is at 20°C. Calculate the error in the reading of temperature T_1.

c. In an attempt to reduce the error, the extension is flipped so that now the chromel wire is on top and the alumel wire is on the bottom. Does that resolve the issue?

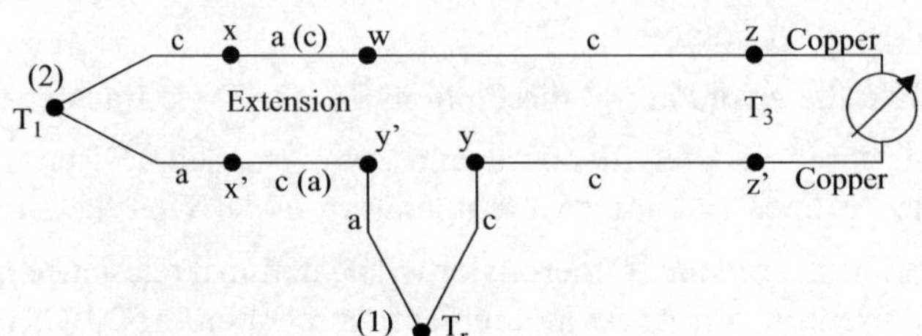

FIGURE 3.33 ■ Extension of thermocouple wires.

3.18 Reference junction with measured temperature. The configuration in **Figure 3.34** is used in a temperature-sensing system. The sensing and reference junctions are both K-type thermocouples. The sensing junction measures the temperature of molten glass at 950°C, whereas the reference thermocouple's temperature is measured as 54°C. The two connections marked as A and B are also within the same temperature zone as the reference junction. The temperature zone T_1 contains the connections to the measuring instrument.

a. Calculate the emf measured by the voltmeter using **Table 3.6**.

b. Show how that same emf can be obtained from the thermoelectric reference table for the K-type thermocouple by taking into account the emf of the reference junction and discuss the differences between the two methods.

c. Show that if the reference junction is held at 0°C the output is identical to that predicted by the thermoelectric reference table for temperature T.

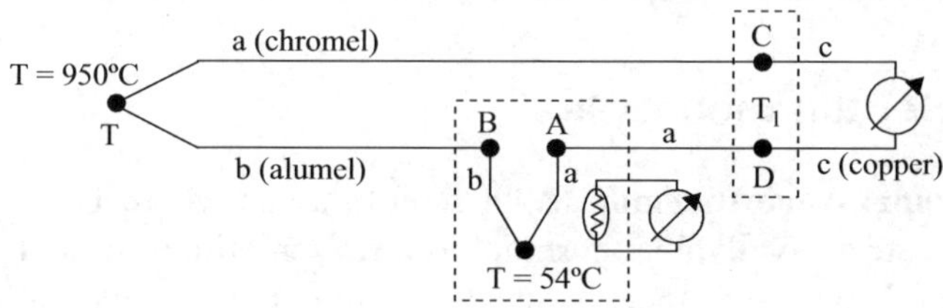

FIGURE 3.34 ■ Temperature sensing of the cold junction.

3.19 Thermoelectric reference emf and temperature for a type T thermocouple. The reference emf and reference temperature for copper-constantan (type T thermocouple) in the range 0°C–400°C are given as follows:

$$E = 3.8748106364 \times 10^1 T^1 + 3.3292227880 \times 10^{-2} T^2 + 2.0618243404 \times 10^{-4} T^3 - 2.1882256846 \times 10^{-6} T^4 + 1.0996880928 \times 10^{-8} T^5 - 3.0815758772 \times 10^{-11} T^6 + 4.5479135290 \times 10^{-14} T^7 - 2.7512901673 \times 10^{-17} T^8 \ \mu\text{V}$$

$$T_{90} = 2.5949192 \times 10^{-2} E^1 - 7.602961 \times 10^{-7} E^2 + 4.637791 \times 10^{-11} E^3 - 2.165394 \times 10^{-15} E^4 + 6.048144 \times 10^{-20} E^5 - 7.293422 \times 10^{-25} E^6.$$

a. Calculate the emf expected at 200°C using the first term (first-order approximation of the transfer function), first two terms (second-order approximation of the transfer function), first three terms, and so on until the complete eighth-order polynomial is used. Use a reference temperature of 0°C.

b. Calculate the error incurred in using reduced-order approximations compared to the exact value using all eight terms of the approximation. Plot the error as a function of the number of terms. What are your conclusions from these results.

c. Take the value found in (a) for the eighth-order approximation and calculate the reference temperature corresponding to the emf using one term in the expression for T (first-order approximation), first two terms (second-order approximation), first three terms, and so on up to six terms. Compare the results to the nominal temperature (200°C). What are the errors in the calculated temperatures with the various approximations? What are your conclusions from these calculations?

3.20 Cold junction compensation of an E-type thermocouple. Consider the cold junction compensation of a chromel-constantan thermocouple using a platinum RTD as shown in **Figure 3.11**. The RTD has a resistance of 120 Ω at 0°C and a TCR coefficient of 0.00385/°C. The relative Seebeck coefficient (sensitivity) for the E-type thermocouple at 0°C is 58.7 μV/°C (see **Table 3.3**).

a. Given a regulated voltage source of 5 V, calculate the resistance R_2 required for this type of thermocouple.

b. Calculate the error in temperature measurement at 45°C if the temperature zone is at $T = 25$°C. Use **Equation (3.20)** to calculate the emf for the T-type thermocouple.

c. Suppose the same configuration is used to sense a temperature of 400°C. What is the error? Discuss the sources of the error. Use the inverse polynomial for the T-type thermocouple in **Appendix B, Section B.4**, to calculate the emf due to $T_2 = 1{,}200$°C.

Semiconductor thermocouples

3.21 High-temperature thermopile. In remote areas where fuel, such as natural gas, is readily available, thermopiles are sometimes used to supply small amounts of power for specific needs, such as communication equipment and cathode protection of pipelines, among others. Consider the following example: A thermopile is needed to supply 12 V DC for emergency refrigeration using a thermopile. To do so, the hot junctions are heated to 450°C. The cold junctions are thermally connected to a conducting structure with fins, cooled by air, and expected to fluctuate between 80°C and 120°C depending on air temperature and wind speed. Because of the high temperatures involved, Peltier cells are not practical and it is proposed to use J-type thermocouples for the purpose.

a. Calculate the number of junctions needed to ensure a minimum output of 12 V.

b. What is the range of the output voltage?

3.22 Automotive thermogenerator. One of the more interesting attempts at using Peltier cells is in the replacement of alternators in trucks and, in the process, to recover some of the power loss through heat in the exhaust. The device is in the form of a cylindrical arrangement of junctions placed over the exhaust pipe, with the inner, hot junctions kept at the temperature of the exhaust pipe. The outer, cold junctions are kept at a temperature differential by a set of cooling fins on the outer surface of the cylindrical structure (or by circulating radiator coolant). Assuming that a minimal temperature difference of 60°C can be maintained between the hot and cold junctions, calculate the number of junctions needed to supply a minimum voltage of 27 V for trucks that operate at 24 V. The material used for construction of the junctions is carbon/silicon carbide because of its temperature range and high sensitivity (see **Table 3.9**).

Note: Prototypes capable of supplying about 5 kW have been built. However, there are some problems with this type of device. They are relatively expensive because of the need for high-temperature materials and can only supply power after the exhaust pipe has reached its normal operating temperature.

p-n junction temperature sensors

3.23 Errors in *p-n* junction sensor. There are two main errors that one should be aware of when using *p-n* junction sensors. One is the self-heat of the junction, the other is introduced by variations in the current through the diode. Consider a germanium diode with a known forward voltage drop of 0.35 V at a current of 5 mA and an ambient temperature of 25°C. The self-heat of the sensor is given in the device data sheet as 1.3 mW/°C.

a. Calculate the sensitivity of the sensor and the expected voltage reading at 50°C neglecting effects of self-heat.

b. Suppose that the current in the diode varies by ±10% due to variations in the power supply. Calculate the error in the measured temperature due to this variation as a percentage. Evaluate it at the values given above ($V_f = 0.35$ V, $T = 25$°C, $I = 5$ mA).

c. Calculate the error in the measured temperature due to self-heat at a junction current of 5 mA and an ambient temperature of 50°C.

d. Discuss these errors, their relative importance, and means of reducing them.

Optical and acoustical sensors

3.24 Acoustic temperature sensing in seawater. To sense the average water temperature close to the surface, an ultrasound transmitter and receiver are set at a distance 1 m apart and the time of flight of the ultrasound wave is measured using a microprocessor, as shown in **Figure 3.35**. The time Δt gives a direct indication of the temperature since the speed of sound in seawater is temperature dependent.

a. Calculate the sensitivity of the sensor.

b. Calculate and plot the measured time of flight as a function of temperature for the expected range of seawater temperatures between 0°C and 26°C.

Note: A sensor of this type is probably not something one would build, but if ultrasound measurements are being used for some other purpose the temperature can then be deduced as well.

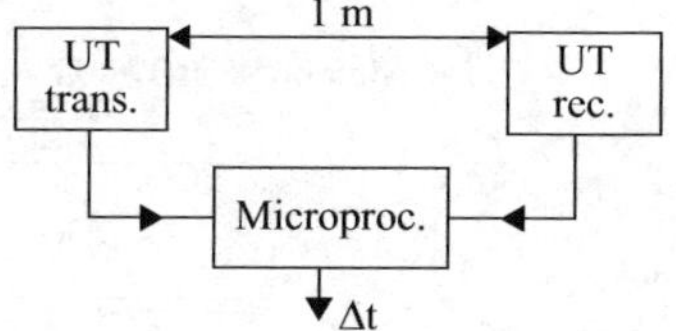

FIGURE 3.35 ■ Ultrasonic water temperature sensor.

3.25 **Ultrasonic autofocusing and errors due to temperature variations.** An ultrasonic sensor is used as a range meter for a camera to autofocus the lens. The method works based on measuring the time of flight of the ultrasound beam to the subject and back to the camera. Suppose the autofocus system is calibrated at 20°C.

a. Calculate the error in reading the distance as a percentage due to changes in the temperature of the air.

b. What is the actual distance measured at −20°C and at +45°C if the subject is 3 m away from the camera?

Thermomechanical sensors and actuators

3.26 **The mercury thermometer.** A mercury thermometer is made of glass as shown in **Figure 3.24**. The thermometer is intended for laboratory use with a scale of 0.5°C/mm and a range between 0°C and 120°C. If the thin tube is 0.2 mm in diameter, what is the volume of mercury necessary?

3.27 **Gas temperature sensor.** A gas temperature sensor is built in the form of a small container and a piston as in **Figure 3.22**. The piston is 3 mm in diameter. If the total volume of gas at 20°C is 1 cm^3, calculate the sensitivity of the sensors in millimeters per degree Celsius (mm/°C). Assume the pressure of the gas remains unchanged.

3.28 **Fluid-filled Golay cell.** A Golay cell is built as a cylindrical container with a flexible membrane, as shown in **Figure 3.36**. Its radius is $a = 30$ mm and its height is $h = 10$ mm. The membrane is stretched between the rim of the cylinder and a rigid disk of radius $b = 10$ mm. When the cell expands due to heating, the rigid disk lifts, stretching the membrane, and in so doing it creates a cone above the cylinder as shown in **Figure 3.36b**. The cell is filled with alcohol and the output is read optically as follows: A small mirror is attached to the membrane and a laser beam is reflected off the mirror. The cell in **Figure 3.36a** is shown at 0°C with the surface of the membrane perfectly flat. The reflected laser beam is read (using an optical sensor) on a scale at a distance of 60 mm from the mirror.

a. Calculate the sensitivity of the sensor for small changes in temperatures close to 0°C. *Note*: The reflection angle of the light equals the incidence angle, where the angles are measured with respect to the normal to the mirror at the

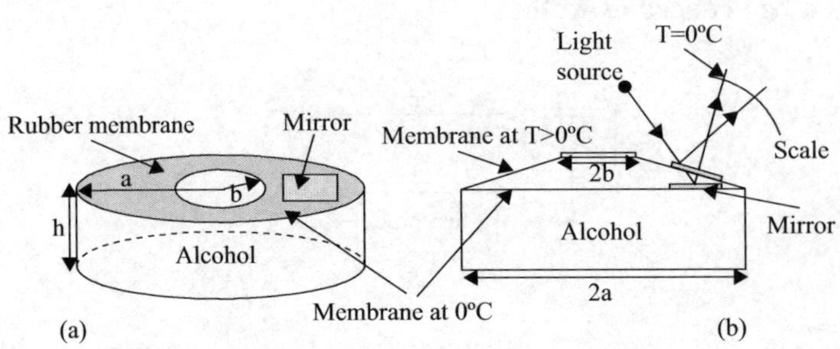

FIGURE 3.36 ■ A fluid-filled Golay cell. a. At 0°C. b. At T>0°C.

point of incidence. Since the sensed quantity is temperature (input) and the output is a linear length on the scale, sensitivity is given in millimeters per degree Celsius (mm/°C).

b. Assuming that the sensor is capable of distinguishing a beam separation of 0.1 mm on the scale, calculate the resolution of the Golay cell for small variations in temperature around 0°C.

3.29 Piston-type Golay sensor. A temperature sensor is made in the form of a glass container filled with air and a piston, as shown in **Figure 3.37**. The total volume of the gas at 0°C is 10 cm^3 and the piston's location indicates the temperature on a scale marked on the cylinder. Assume ideal gas behavior and no friction due to the piston (i.e., the internal pressure equals the external pressure).

a. If the external pressure is constant and equal to 1 atm (1013.25 mbar or 101,325 Pa), calculate the sensitivity of the sensor in degrees Celsius per millimeter displacement of the piston.

b. What is the maximum error in the reading if the external pressure changes from a low pressure of 950 mbar (95,000 Pa) to a high pressure of 1100 mbar (110,000 Pa) while the internal pressure stays at 1013.25 mbar (101,325 Pa).

c. What are the conclusions from (a) and (b)?

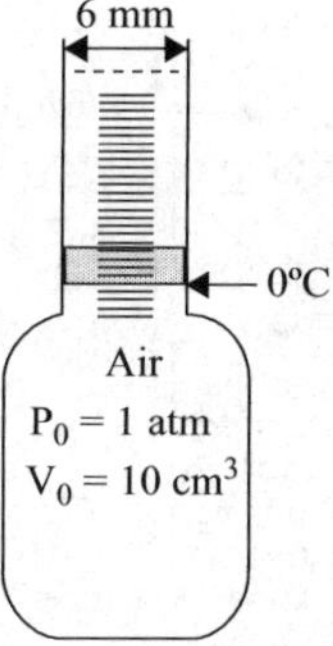

FIGURE 3.37 ■ Piston-type Golay sensor.

3.30 Bimetal thermostat. A thermostat is built as a bimetal bar and a snap switch as shown in **Figure 3.38** and used to control temperature in a small chamber. A snap switch operates by pressing against a strip spring, which when pushed beyond a certain point, snaps to a position that opens (or closes, depending on the type of switch) the contacts. When released, the contacts close (or open). The travel necessary is short, often less than 1 mm. In the case discussed here, the switch travel required is $d = 0.5$ mm. The bimetal bar is $t = 1$ mm thick made of iron (on the bottom) and copper (on the top).

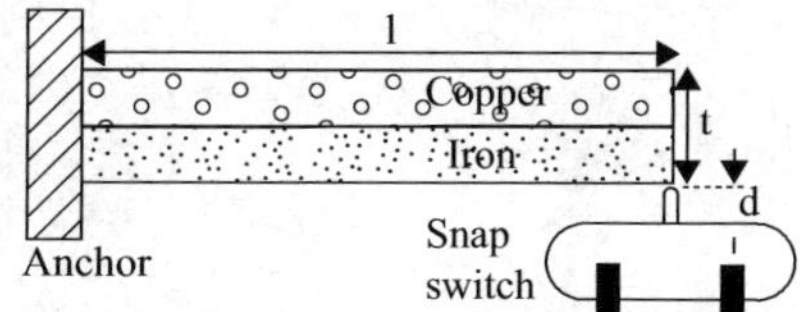

FIGURE 3.38 ■ Bimetal switch.

a. Assuming that at room temperature of 20°C the bar just touches the switch actuator and the thermostat must open the switch at 350°C, what is the minimum bar length needed.

b. If the distance *l* can be adjusted to a minimum of 25 mm, what is the highest temperature to which the thermostat can be set?

3.31 **Coil bimetal thermometer.** A coil bimetal thermometer is designed to operate in a range of 0°C–300°C. Assuming the inner diameter of the coil is 10 mm, calculate the length of the strip required to produce a 30° circular scale. The strip is made of copper (outer metal) and iron (inner metal). Assume that as the strip expands, all its coils maintain their diameter. The dial is moved directly by the inner coil.

3.32 **The car thermostat: principle.** A thermostat for use in a car engine is designed to open at 110°C. It does so through the use of a solid wax pellet in a cylinder that melts at a given temperature and as it does, expands considerably. The volume of the wax used expands by 5% as it melts. The configuration employed is a straight cylinder of diameter $d = 15$ mm with a piston connected directly to a disk that blocks the flow of water to the engine (**Figure 3.39**). The disk must move a distance $a = 6$ mm to fully open. Calculate the volume of the pellet necessary and the dimensions of the cylinder to accomplish this.

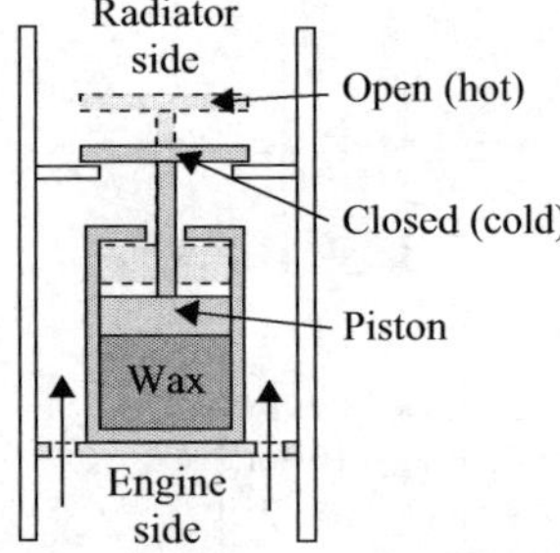

FIGURE 3.39 ■ The car thermostat: principle.

3.33 **The car thermostat: a practical design.** A practical configuration for a car thermostat is shown in **Figure 3.40**. Here the piston has been reduced in diameter to 3 mm, whereas the diameter of the cylinder remains the same as in **Problem 3.32**. In practical designs the cylinder moves down against a spring, whereas the piston itself is stationary. Calculate the volume necessary and the length of the actuating cylinder.

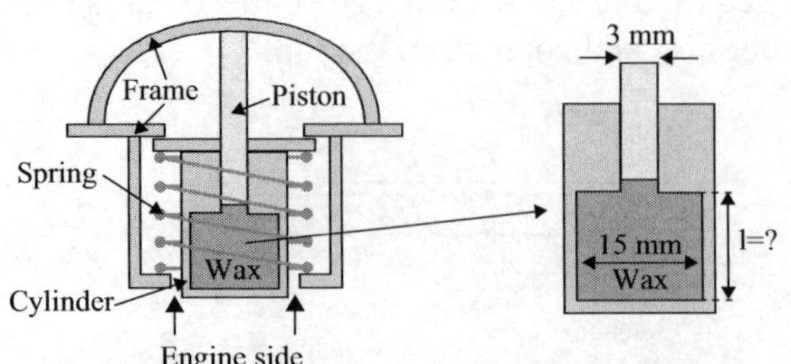

FIGURE 3.40 ■ A modified car thermostat.

CHAPTER 4

Optical Sensors and Actuators

The eye

The human eye, like that of other vertebrates, is a marvelous, complex sensor allowing us to perceive the world around us in minute detail and true colors. In fact, the eye is akin to a video camera. It consists of a system of lenses (the cornea and crystalline lens), an aperture (iris and pupil), an image plane (retina), and a lens cover (eye lids). In humans and animals that prey, the eyes point forward to create binocular vision with excellent depth perception. Many prey animals have side-facing eyes to increase their field of view, but the vision is monocular and lacks perception of depth. The eyelids, in addition to protecting the eye, also keep it clean and moist by distributing tears as well as lubricants (the conjunctiva) and protect it from dust and foreign objects in conjunction with the eyelashes. The front dome of the eye is made of the cornea, a clear, fixed lens. This is a unique organ, as it has no blood vessels and is nourished by tears and the fluid inside the eye sphere. Behind it is the iris, which controls the amount of light that enters the eye. On the periphery of the iris there is a series of slits that allow fluid to pass out from the eye sphere. This passes nutrients to the front of the eye and relieves the pressure in the eye (when this is not perfectly regulated one has glaucoma, a condition that can affect the retina and eventually can cause blindness). Behind is the crystalline lens, an adjustable lens that allows the eye to focus on objects as close as about 10 cm and as far as infinity. The lens is controlled by the ciliary muscle. When this muscle loses some function, the ability of the lens to focus is impaired, leading to the need for corrective action (glasses or surgery). The lens itself can cloud over time (cataracts), a condition that requires replacement of the lens. At the back of the eye lies the optical sensor proper—the retina. It is made of two types of cells: cone cells that perceive color and rod or cylindrical cells that are responsible for low-light (night) vision. The cone cells are divided into three types, sensitive to red, green, and blue light, with a total of about 6 million cells, most of them in the center of the retina (the macula). Rod cells are distributed mostly on the peripheral parts of the retina and are responsible for low-light vision. They do not perceive color but are as much as 500 times more sensitive than cone cells. There are also many more rod cells—as many as 120 million of them. The retina is connected to the visual cortex in the brain through the optical nerve. Although the lens of the eye is adjustable, the size of the optical ball also plays a role in vision. Individuals with larger eyeballs are nearsighted, those with smaller eyeballs are farsighted.

The sensitivity of the human eye ranges from roughly 10^{-6} cd/m^2 (dark night, rod-dominated vision, essentially monochromatic) to about 10^6 cd/m^2 (bright sunlight, cone-dominated vision, full color). This is a vast dynamic range (120 dB). The spectral sensitivity of the eye is divided into four partially overlapping zones. Blue cones are sensitive between about 370 nm and 530 nm, with peak sensitivity at 437 nm; green cones between 450 and 640 nm, with peak sensitivity at 533 nm; and red cones between 480 and 700 nm, with peak sensitivity at 564 nm. Rods are sensitive between about 400 and 650 nm, with peak at 498 nm. This peak is in the blue-green range. For this reason, low-light vision tends to be dark green.

It should also be noted that the human-type eye, a structure shared by many animals, is not the only type of eye. There are some 10 different structures ranging from simple light-sensitive cells that allow the organism to detect light but not to create images, to compound eyes made of thousands of simple, individual "eyes" particularly suited to detect motion but can only create "pixilated" images.

4.1 | INTRODUCTION

Optics is the science of light and light is an electromagnetic radiation that manifests itself either as an electromagnetic wave or as photons (particles with quanta of energy). Before continuing it is worth mentioning that the term light refers specifically to the visible spectrum of electromagnetic radiation as perceived by the human eye (see **Figure 4.1**), but because both below and above this spectrum the behavior of radiation is similar, the term light is normally extended to include a much wider spectrum that includes infrared (IR) radiation (below the frequency of visible light or "below red") and ultraviolet (UV) radiation (above the visible range or "above violet"). Even the nomenclature has been modified and we sometimes say IR light or UV light. These terms are incorrect but are in widespread use. The range that is properly called light is defined by the response of the human eye between 430 THz and 750 THz (1 THz) = 10^{12} Hz). In characterizing light it is more common to use wavelength, defined as the distance in meters the light wave propagates in one cycle or $\lambda = c/f$, where c is the speed of light and f is its frequency. The range of wavelengths in the visible light region is between 700 nm (deep red) and 400 nm (violet). However, the ranges of IR and UV radiation are not as well defined and, as can be seen in **Figure 4.1**, the lower range of IR radiation overlaps the higher range of microwave radiation (sometimes this upper range is called millimeter wave radiation), whereas the upper reaches of UV radiation reach into the X-ray spectrum. For the purpose of this discussion, the IR range is between 1 mm and 700 nm and the UV range is between 400 nm and 1 nm. What unifies this wide range for the purpose of this chapter is the fact that the principles of sensing are similar and based on essentially the same effects. It should also be pointed out that the term radiation here means electromagnetic radiation.

Optical sensors are those sensors that detect electromagnetic radiation in what is generally understood as the broad optical range—from far IR to UV. The sensing methods may rely on direct methods of transduction from light to electrical quantities such as in photovoltaic or photoconducting sensors or indirect methods such as

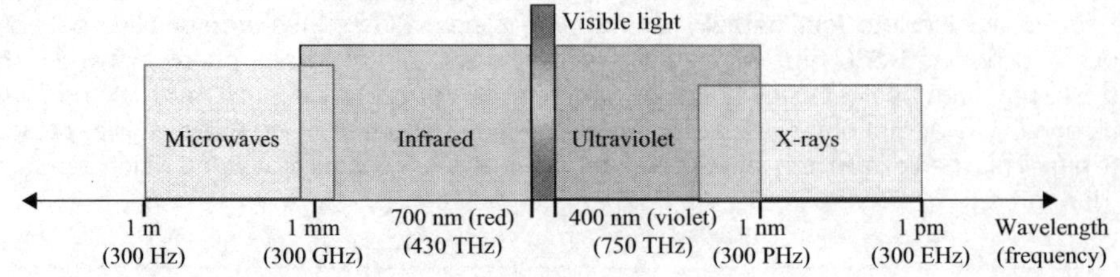

FIGURE 4.1 ■ Spectrum of infrared, visible, and ultraviolet radiation.

conversion first into temperature variation and then into electrical quantities such as in passive IR (PIR) sensors and bolometers.

There is a third method of sensing related to optics—sensors based on light propagation and its effects (reflection, transmission refraction), which will not be discussed here because the optical aspect is usually not the sensing mechanism, but rather an intermediate transduction mechanism. Nevertheless, the physics will be mentioned briefly for completeness.

4.2 | OPTICAL UNITS

The units used in optics seem to be more obscure than most. Thus it is useful to discuss these at this point. First, the SI units provide for a measure of **luminous intensity**, the candela (cd) (see **Section 1.6.1**). The candela is defined as the luminous intensity, in a given direction, of a source that emits monochromatic radiation of frequency 540×10^{12} Hz and that has a radiation intensity of 1/683 W/sr. In short, the candela is a measure of radiation intensity.

Other units are often used. The **lumen** (luminous flux) is a candela steradian (cd·sr) and is a measure of power. The **lux** (a measure of illumination) is a candela steradian per square meter (cd·sr/m^2) and is therefore a power density. These are summarized in **Table 4.1**.

TABLE 4.1 ■ Optical quantities and their units

Quantity	Name	Unit	Derived unit	Comments
Luminous intensity	Candela	[cd]	[W/sr]	Power radiated per steradian
Luminous flux	Lumen	[cd·sr]	[W]	Power radiated
Illuminance	Lux	[cd·sr/m^2]	[W/m^2]	Power density
Luminance	Candela per meter square	[cd/m]^2	[W/sr·m^2]	Density of luminous intensity

EXAMPLE 4.1 Conversion of optical units

A point source emits uniformly in all directions in space (e.g., the sun may be considered a point source when viewed from Earth). Given a total power radiated of 100 W, calculate the source's luminous intensity and illuminance at a distance of 10 m from the source.

Solution: Since there are 4π solid unit angles in a sphere, the luminous intensity is

$$\text{luminous intensity} = \frac{100}{4\pi} = 7.958 \text{ W/sr}.$$

Since a candela is 1/683 W/sr, the luminous intensity is

$$\text{luminous intensity} = \left(\frac{100}{4\pi}\right) \times 683 = 7.958 \times 683 = 5435.14 \text{ cd}.$$

Although the units of illuminance are cd·sr/m^2, it is best to start from the power radiated. That power is spread over the sphere of radius $R = 10$ m, so we get

$$\text{Illuminance} = \frac{100}{4\pi R^2} = \frac{7.958}{10^2} = 0.0796 \text{ lux.}$$

It should be noted that whereas the luminous intensity is a fixed value that only depends on the source, illuminance depends as well on the distance from the source.

Note: A uniformly radiating source is called an isotropic source.

4.3 | MATERIALS

The sensors/actuators discussed in **Chapter 3** and those that will follow take advantage of many physical principles. But, in addition, they take advantage of specific material properties, either of elements, alloys, or in other forms available, including synthetic and naturally occurring salts, oxides, and others. As we will discuss some of these, especially in conjunction with semiconducting materials, it is perhaps useful to bear in mind the periodic table (see the inside back cover). Many of the properties of materials are not specific to a single element, but rather belong to a group (often a column in the table of elements), and one can expect that if an element in a specific column is used for a given purpose, other elements from the same column may have similar properties and be equally useful. For example, if potassium (alkali column I) is useful in the production of cathodes for photoelectric cells, then lithium, sodium, rubidium, and cesium should also be useful. But there are clear limits. Hydrogen and francium, which are also in the same column, are not useful. The first because it is a gas, the second because it is radioactive. Similarly, if gallium-arsenide (GaAs) makes a useful semiconductor, so should indium-antimonide (InSb), and so on. We already saw some of these principles in discussing thermocouples. The elements in the VIII column—nickel, palladium, and platinum are used for various types of thermocouples together with elements from the IB and IIB columns. We shall refer to the periodic table often, but will also refer to many simple or complex compounds with specific properties that have been found to be useful in sensors and actuators. Here we will be concerned primarily with semiconductors, but other materials will become important in subsequent chapters.

4.4 | EFFECTS OF OPTICAL RADIATION

4.4.1 Thermal Effects

The interaction of light (radiation) with matter results in absorption of energy in two distinct ways. One is thermal and is usually viewed as absorption of electromagnetic waves. The other is a quantum effect. The thermal effect is based on electromagnetic energy absorbed by the medium and converted into heat through the increased motion of atoms. This heat is sensed and translated into a measure of the incident radiation. Here we will not go beyond the understanding that by raising the temperature of a material, its electrons gain kinetic energy and may be released given sufficient energy and, of course, that this interaction can be used for sensing.

4.4.2 Quantum Effects

4.4.2.1 The Photoelectric Effect

The second effect is a quantum effect and is governed by photons, the particle-like manifestation of radiation. In this representation of light, and in general radiation, energy travels in bundles (photons) whose energy is given by Plank's equation:

$$e = hf, \tag{4.1}$$

where $h = 6.6262 \times 10^{-34}$ J/s or 4.1357×10^{-15} eV, which is Planck's constant, and f is the frequency in hertz. e is the photon energy and is clearly frequency dependent. The higher the frequency (the shorter the wavelength), the higher the photon energy. In the quantum mode, energy is imparted to materials by elastic collision of photons and electrons. The electrons acquire energy and this energy allows the electron to release itself from the surface of the material by overcoming the **work function** of the substance. Any excess energy imparts the electron kinetic energy. This theory was first postulated by Albert Einstein in his photon theory, which he used to explain the photoelectric effect in 1905 (and for which he received the Nobel Prize). This is expressed as

$$hf = e_0 + k, \tag{4.2}$$

where e_0 is the work function and is the energy required to leave the surface of the material (see **Table 4.2**). The work function is a given constant for each material. $k = mv^2/2$ represents the maximum kinetic energy the electron may have outside the material. That is, the maximum velocity electrons can have outside the material is $v = \sqrt{2k/m}$, where m is the mass of the electron.

A photon with energy higher than the work function will, in principle, release an electron and impart a kinetic energy according to **Equation (4.2)**. But does in fact each

TABLE 4.2 ■ Work functions for selected materials

Material	Work function [eV]
Aluminum	3.38
Bismuth	4.17
Cadmium	4.0
Cobalt	4.21
Copper	4.46
Germanium	4.5
Gold	4.46
Iron	4.4
Nickel	4.96
Platinum	5.56
Potassium	1.6
Silicon	4.2
Silver	4.44
Tungsten	4.38
Zinc	3.78

Note: 1 eV = 1.602×10^{-19} J.

photon release an electron? That depends on the quantum efficiency of the process. Quantum efficiency is the ratio of the number of electrons released (N_e) to number of photons absorbed (N_{ph}):

$$\eta = \frac{N_e}{N_{ph}}. \tag{4.3}$$

Typical values are around 10%–20%. This simply means that not all photons release electrons.

Clearly, for electrons to be released, the photon energy must be higher than the work function of the material. Since this energy depends on frequency alone, the frequency at which the photon energy equals the work function is called a **cutoff frequency**. Below it quantum effects do not exist (except for tunneling effects) and only thermal effects are observed. Above it, thermal and quantum effects are present. For this reason, low-frequency radiation (IR in particular) can only give rise to thermal effects, whereas at high frequencies (UV radiation and above) the quantum effect dominates.

This then describes the photoelectric effect, which is the basis for a number of sensing methods, as we shall discuss next. In all of these methods, surface electrons are released.

EXAMPLE 4.2 Minimum wavelength for photoelectric emission

Consider a photoelectric device intended for light detection.

a. Assuming it is made of a potassium-coated surface, what is the lowest wavelength that the device can detect?

b. What is the kinetic energy of an emitted electron under red light radiation at a wavelength of 620 nm?

Solution:

a. The photon energy is given in **Equation (4.2)**. With the photon energy equal to the work function we have

$$hf = e_0 \rightarrow f = \frac{e_0}{h} \quad [\text{Hz}].$$

Since photons travel at the speed of light, the frequency may be written as

$$f = \frac{c}{\lambda} \quad [\text{Hz}],$$

where c is the speed of light and λ is the wavelength. The longest wavelength detectable is

$$\lambda = \frac{ch}{e_0} = \frac{3 \times 10^8 \times 4.1357 \times 10^{-15}}{1.6} = 7.7544 \times 10^{-7} \text{ m}.$$

This is 775.44 nm. From **Figure 4.1**, this is in the very near IR region.

b. At 620 nm, the frequency is c/λ and from **Equation (4.2)** the kinetic energy is

$$k = hf - e_0 = 4.1357 \times 10^{-15} \times \frac{3 \times 10^8}{620 \times 10^{-9}} - 1.6 = 0.4 \qquad [\text{eV}]$$

This kinetic energy is rather low because the red light is close to the longest wavelength to which the photoelectric device responds.

4.4.2.2 Quantum Effects: The Photoconducting Effect

Many modern sensors are based on quantum effects in the solid state, and particularly in semiconductors. Although some electrons may still leave the surface based on the photoelectric effect, when a semiconductor material is subjected to photons they can transfer the energy to electrons. If this energy is sufficiently high, the electrons become mobile, resulting in an increase in the conductivity of the material and, as a result, in an increase in current through the material. This current or its effects become a measure of the radiation intensity (visible light, UV radiation, and, to a lesser extent, IR radiation) that strikes the material. The model for this effect is shown in **Figure 4.2a**. Electrons are normally in the valence band—they are bound to lattice sites within the crystal (i.e., bound to the atoms that make up the crystal) and have specific densities and momentum. Valent electrons are those that are bound to an individual atom. Covalent electrons are also bound, but are shared between two neighboring atoms in the crystal. An electron can only move into the conduction band if its energy is larger than the energy gap (bandgap energy, W_{bg}) specific to the material and if the momentum of the site in the conducting band is the same as the momentum of the electron in the valence band (law of conservation of momentum). This energy may be supplied thermally, but here we are interested in energy absorbed from photons. If the radiation is of sufficiently high frequency (sufficiently energetic photons), valence or covalence electrons may be released from their sites and moved across the bandgap into the conduction band (**Figure 4.2a**).

There are two mechanisms for this transition to occur. In direct bandgap materials, the momentum in the top of the valence band and in the bottom of the conduction band are the same and an electron can transit without the need for a change in momentum, provided it acquires sufficient energy from the photon interaction. In indirect bandgap materials, the electron must interact with the crystal lattice to either gain or lose momentum before it can occupy a site in the conduction band. This process is characterized by a lattice vibration called a phonon and is a less efficient process than that in direct bandgap materials. When in the conduction band, electrons are mobile and free to move as a current. When electrons leave their sites, they leave behind a "hole," which is simply a positive charge carrier. This hole may be taken by a neighboring electron with little additional energy (unlike the original electron released by the photon; see **Figure 4.2b**) and therefore, in semiconductors, the current is due to the net concentrations of electrons and holes. The release of electrons is manifested as a change in the

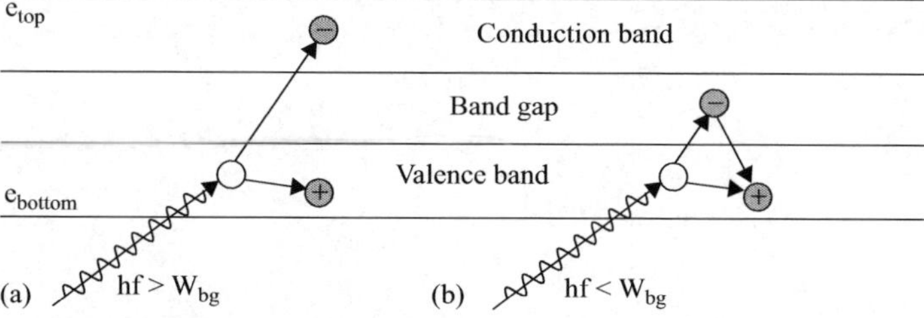

FIGURE 4.2 ■ A model of the photoconductive effect. (a) The photon energy is sufficiently high to move an electron across the bandgap, leaving behind a hole. (b) The photon energy is too low, resulting in recombination of the electron and the hole.

concentration of electrons in the conduction band and of holes in the valence band. The conductivity of the medium is due to the concentrations of both carriers and their mobilities:

$$\sigma = e(\mu_e n + \mu_p p) \quad [\mathrm{S/m}], \tag{4.4}$$

where σ is conductivity, μ_e and μ_p are the mobilities (in m^2/V·s or, often, cm^2/V·s) of electrons and holes, respectively, and n and p are the concentrations (particles/m^3 or particles/cm^3) of electrons and holes. This change in conductivity or the resulting change in current is then the basic measure of the radiation intensity in photoconducting sensors.

The effect just described is called the **photoconducting effect** and is most common in semiconductors because the bandgaps are relatively small. It exists in insulators as well, but there the bandgaps are very high and therefore it is difficult to release electrons except at very high energies. In conductors the valence and conduction bands overlap (there is no bandgap). Most electrons are free to move, indicating that photons will have minimal or no effect on the conductivity of the medium. Therefore semiconductors are the obvious choice for sensors based on the photoconducting effect, whereas conductors will most often be used in sensors based on the photoelectric effect.

From **Table 4.3**, it is clear that some semiconductors are better suited for low-frequency radiation whereas others are better at high-frequency radiation. The lower the bandgap, the more effective the semiconductor will be at detection at low frequencies (long wavelength, hence lower photon energies). The longest wavelength specified for the material is called the **maximum useful wavelength**, above which the effect is negligible. For example, InSb (indium antimony) has a maximum wavelength of 5.5 μm,

TABLE 4.3 ■ Bandgap energies, longest wavelength, and working temperatures for selected semiconductors

Material	Band gap [eV]	Longest wavelength λ_{max} [μm]	Working temperature [K]
ZnS	3.6	0.35	300
CdS	2.41	0.52	300
CdSe	1.8	0.69	300
CdTe	1.5	0.83	300
Si	1.2	1.2	300
GaAs	1.42	0.874	300
Ge	0.67	1.8	300
PbS	0.37	3.35	
InAs	0.35	3.5	77
PbTe	0.3	4.13	
PbSe	0.27	4.58	
InSb	0.18	6.5	77
Ge:Cu		30	18
Hg/CdTe		8–14	77
Pb/SnTe		8–14	77
InP	1.35	0.95	300
GaP	2.26	0.55	300

Note: Properties of semiconductors vary with doping and other impurities. The values shown should be viewed as representative only.

making it useful in the near IR range. Its bandgap is very low, which also makes it very sensitive. However, that also means that electrons can be easily released by thermal sources, and in fact, the material may be totally useless for sensing at room temperatures (300 K), because at that temperature most electrons will be in the conduction band and these available conduction carriers serve as a thermal background noise for the photon-generated carriers. For this reason, it is often necessary to cool these long-wavelength sensors to make them useful by reducing the thermal noise. The third column in **Table 4.3** shows the (highest) working temperature of the material.

4.4.2.3 Spectral Sensitivity

Each semiconducting material has a range of the spectrum in which it is sensitive, given as a function of frequency or wavelength. The upper range (longest wavelength or minimum energy) is defined by the bandgap (about 1200 nm in **Figure 4.3**). Above the bandgap, the response of a material (i.e., the concentration of conduction electrons in the conduction band due to photon interaction) increases steadily to a maximum and then decreases, as shown schematically in **Figure 4.3**. The reason for the increase and decrease in response is that electron density and momentum are highest in the middle of the valence band and taper off to zero at its boundaries. Because of the law of conservation of momentum, an electron can only transit into the conduction band to a site of like momentum, and the probability of this first increases with an increase in energy (decrease in wavelength), but after most of the electrons in the middle of the valence band have been displaced, the probability of electron transiting to sites of ever-increasing momentum decreases until, at an electron energy equal to the difference between the top of the conduction band and the bottom of the valence band ($e_{top}-e_{bottom}$ in **Figure 4.2**), this probability goes to zero. In **Figure 4.3** that occurs at approximately 650 nm.

4.4.2.4 Tunneling Effect

Another important quantum effect is the **tunneling effect** in semiconductor devices. A simple explanation of this curious effect is that although carriers may not have sufficient energy to go "over" the gap, they can tunnel "through" the gap. Although this explanation is shaky at best, the tunneling effect is real, is a direct consequence of quantum mechanics, and is fully predicted by the Schrödinger equation. The tunneling effect explains behavior on the microscopic level that cannot be explained through classical physics but which, nevertheless, manifests itself on the macroscopic level. Semiconductor devices based on this effect, particularly tunnel diodes, are common, and the effect is used extensively in optical sensors.

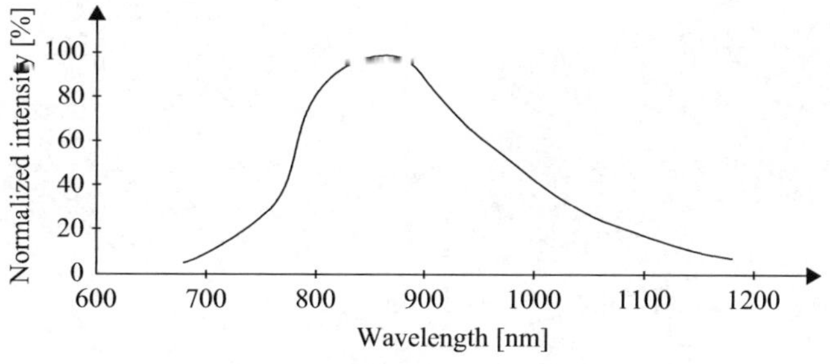

FIGURE 4.3 ■ Spectral sensitivity of a semiconductor.

4.5 | QUANTUM-BASED OPTICAL SENSORS

Optical sensors are divided into two broad classes: **quantum sensors** (or detectors) and **thermal sensors** (or detectors). (Optical sensors are most often called detectors.) A quantum optical sensor is any sensor based on the quantum effects described above and include photoelectric and photoconductive sensors as well as photodiodes and phototransistors (variations of the photoconductive sensor). Thermal optical sensors are mostly encountered in the IR region (and particularly in the low IR region) and come in many variations, including **passive infrared (PIR)** sensors, **active far infrared (AFIR)** sensor bolometers, and others, as we shall see shortly.

4.5.1 Photoconducting Sensors

Photoconducting sensors, or as they are sometimes called, **photoresistive sensors** or **photoresistive cells**, are possibly the simplest optical sensors. They are made from a semiconducting material connected to two conducting electrodes and are exposed to light through a transparent window. A schematic view of the sensor is shown in **Figure 4.4a**. Materials used for these sensors are cadmium sulfide (CdS), cadmium selenide (CdSe), lead sulfide (PbS), indium antimonide (InSb), and others, depending on the range of wavelengths for which the sensor is designed to operate and other requisite properties of the sensor. Of these, CdS is the most common material.

In terms of construction, the electrodes are typically set on top of the photoconductive layer, which in turn is placed on top of a substrate layer. The electrodes may be very simple (**Figure 4.4a**) or may resemble a meandering or comblike shape (**Figure 4.4b**), depending on requirements. In either case, the area exposed between the electrodes is the sensitive area. **Figure 4.5** shows a few sensors of various sizes and construction. The photoconductor is an active sensor that must be connected to a source. The current through or the voltage on the sensors is taken as the output, but what changes with light intensity is the conductivity of the semiconductor and hence its resistance.

The conductivity of the device, given in **Equation (4.4)**, results from the charge of electrons e, the mobilities of electrons and holes (μ_e and μ_p), and the concentrations of electrons n and holes p from whatever source. In the absence of light, the material exhibits what is called dark conductivity, which in turn results in a dark current. Depending on the construction and materials, the resistance of the device may be very high (a few megaohms) or may be in the range of a few kilo-ohms. When the sensor is

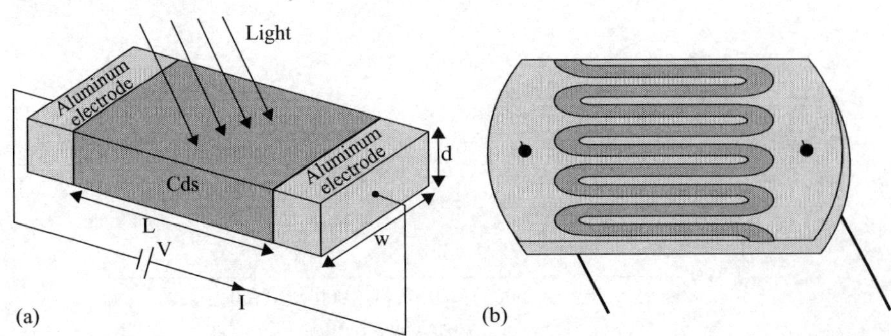

FIGURE 4.4 ■ Structure of a photoconductive sensor. (a) Simple electrodes. (b) Schematic of a sensor showing the connections.

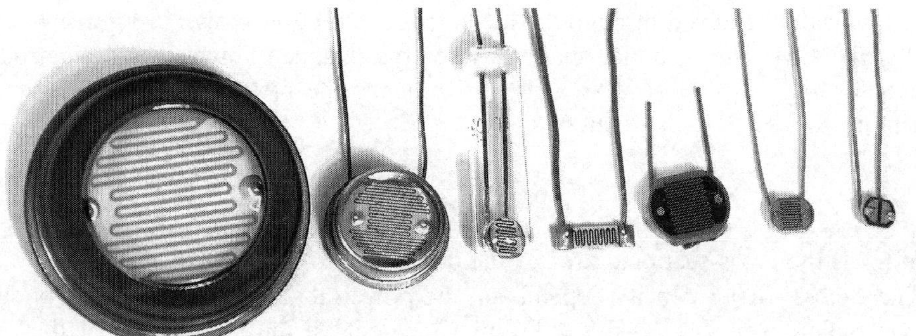

FIGURE 4.5 ■ Examples of photoconductive sensors. The sensor on the right has simple electrodes. The others have comblike electrodes.

illuminated, its conductivity changes (the conductivity increases and hence the resistance decreases) depending on the change in carrier concentrations (excess carrier concentrations). This change in conductivity is

$$\Delta\sigma = e(\mu_e \Delta n + \mu_p \Delta p) \quad [\mathrm{S/m}], \tag{4.5}$$

where Δn and Δp are the excess carrier concentrations generated by the radiation (light). The carriers are generated by the radiation at a certain generation rate (the number of electrons or holes per second per unit volume), but they also recombine at a set recombination rate. The generation and recombination rates depend on a variety of properties, including the absorption coefficient of the material, dimensions, incident power density (of the radiation) wavelength, and the carrier lifetime (the lifetime of carriers is the time it takes for excess carriers to decay—recombine). Both generation and recombination exist simultaneously, and under a given illumination a steady state is obtained when these are equal. Under this condition, the change in conductivity may be written as

$$\Delta\sigma = ef(\mu_n \tau_n + \mu_p \tau_p) \quad [\mathrm{S/m}], \tag{4.6}$$

where τ_n and τ_p are the lifetimes of electrons and holes, respectively, and f is the number of carriers generated per second per unit volume. These properties are material dependent and are generally known, although they are temperature as well as concentration dependent. Although carrier generation is in pairs, if the preexisting carrier density of one type dominates, the excess carrier density of the second type will be negligible with respect to the density of the dominant carrier. If electrons dominate, the material is said to be an *n*-type semiconductor, whereas if the dominant carriers are holes, the semiconductor is said to be *p*-type. In each of these, the concentration of the opposite type is negligible and the change in conductivity is due to the dominant carrier.

An important property of a photoresistor is its sensitivity to radiation (sometimes called its efficiency). Sensitivity, also called gain, is given as

$$G = \frac{V}{L}(\mu_n \tau_n + \mu_p \tau_p) \quad [\mathrm{V/V}], \tag{4.7}$$

where L is the length of the sensor (distance between electrodes) and V is the voltage across the sensor. Note that the units in **Equation (4.7)** are volts per volt, hence this is a dimensionless quantity. Sensitivity gives the ratio of carriers generated per photon of the input radiation. To increase sensitivity, one should select materials with high carrier lifetimes, but one must also keep the length of the photoresistor as small as possible. The latter is typically achieved through the meander construction shown in **Figure 4.4b** (see also **Figure 4.5**). The meander shape ensures the distance between two electrodes is reduced for a given exposure area. It also reduces the resistance of the sensor that, referring to **Figure 4.4a**, is given by

$$R = \frac{L}{\sigma wd} \quad [\Omega], \tag{4.8}$$

where wd is the cross-sectional area of the device and σ is its conductivity.

The excess carrier density depends on the power absorbed by the photoconductor. Given a radiation power density P [W/m^2] incident on the top surface of the photoconductor in **Figure 4.4a**, and assuming that a fraction T of this power penetrates into the photoconductor (the rest is reflected off the surface), the power entering the device is $PTS = PTwL$ [W]. This is, by definition, the energy per unit time absorbed in the device. Since the photons have an energy hf, we can write the total number of excess carrier pairs released per unit time as

$$\Delta N = \eta \frac{PTwL}{hf} \quad [\text{carriers/s}], \tag{4.9}$$

where η is the quantum efficiency of the material (a known, given property, dependent on the material used). The latter indicates how efficient the material is at converting photon energy into carriers and clearly indicates that not all photons participate in the process. Assuming carrier generation is uniform throughout the volume of the photoconductor (an assumption only valid for thin photoconductors), we can calculate the rate at which carriers are generated per unit volume per second:

$$\Delta n = \eta \frac{PTwL}{hfwLd} = \eta \frac{PT}{hfd} \quad [\text{carriers/m}^3\text{s}]. \tag{4.10}$$

As mentioned before, the recombination rate influences the net excess carrier density. The carrier density (concentration) is then obtained by multiplying the rate of generation by the lifetime of the carriers, τ:

$$\Delta n = \eta \frac{PT\tau}{hfd} \quad [\text{carriers/m}^3]. \tag{4.11}$$

Both majority and minority excess carriers generated by light have the same densities.

Some of the various terms in **Equation (4.11)** are not necessarily constant with concentrations and some may only be estimates as well as being temperature dependent. However, the equation shows the link between light intensity and excess carrier concentration and hence the dependence of conductivity on light intensity.

Other parameters to consider are the response time of the sensor, its dark resistance (which depends on doping), the range of resistance for the span of the sensor, and the spectral response of the sensor (i.e., the portion of the spectrum in which the sensor is usable). These properties depend on the semiconductor used as well as on the manufacturing processes used to produce the sensor.

Noise in photoconducting sensors is another important factor. Much of the noise is thermally induced and becomes worse at longer wavelengths. Hence many IR sensors must be cooled for proper operation. Another source of noise is the fluctuations in the rates of generation and recombination of the carriers. This noise is particularly important at shorter wavelengths.

From a sensor production point of view, photoresistive sensors are made either as a single crystal semiconductor, by deposition of the material on a substrate, or by sintering (essentially an amorphous semiconductor made of compressed, powdered material sintered at high temperatures to form the photoconductive layer). Usually sensors made by deposition are the least expensive, whereas single crystal sensors are the most expensive, but with better properties. A particular method may be chosen based on requirements. For example, large surface area sensors may need to be made by sintering because large single crystals are both difficult to make and more expensive.

EXAMPLE 4.3 Properties of a photoresistor

Properties of many semiconductors are determined experimentally primarily because of the variability of constituents of the semiconductors and the difficulties of obtaining reliable data. Nevertheless, especially for cadmium sulfide (CdS) sensors, some reliable data are available. To see what the properties of a photoresistor are, consider a simple CdS structure as in **Figure 4.4a** of length 4 mm, width 1 mm, and thickness 0.1 mm. The mobility of electrons in CdS is approximately 210 $cm^2/V{\cdot}s$ and that of holes is 20 $cm^2/V{\cdot}s$. The dark concentration of carriers is approximately 10^{16} carriers/cm^3 (for both electrons and holes). At a light density of 1 W/m^2 the carrier density increases by 11%:

a. Calculate the conductivity of the material and the resistance of the sensor under dark conditions and under the given illumination.
b. Assuming a rate of carrier generation due to light of 10^{15} carriers/cm^3, estimate the sensitivity of the sensor to radiation at a wavelength of 300 nm.

Solution:
a. The conductivity is calculated directly from **Equation (4.4)**:

$$\sigma = e(\mu_e n + \mu_p p) = 1.602 \times 10^{-19} \times (210 \times 10^{16} + 20 \times 10^{16}) = 0.36846\ [\text{S/cm}].$$

Because of the units of mobility and carrier density, the result is in siemens per centimeter. By multiplying this by 100 (1 m = 100 cm), we get $\sigma = 36.85$ S/m.

Under light conditions, the carrier density increases by a factor of 1.11 and we get:

$$\sigma = e(\mu_e n + \mu_p p) = 1.602 \times 10^{-19} \times (210 \times 10^{16} \times 1.11 + 20 \times 10^{16} \times 1.11) = 0.409\ [\text{S/cm}].$$

The conductivity under light conditions is $\sigma = 40.9$ S/m.

The resistance is found from **Equation (4.9)**:

$$R = \frac{L}{\sigma WH} = \frac{0.004}{36.85 \times 0.001 \times 0.0001} = 1085.5\ [\Omega].$$

$$R = \frac{L}{\sigma WH} = \frac{0.004}{40.9 \times 0.001 \times 0.0001} = 978.0\ [\Omega].$$

Note that the resistance is directly proportional to the increase in carrier density but the increase in carrier density is not linear with illumination. For this reason the resistance decreases rather quickly initially but then levels off, since at high illumination levels there are fewer and fewer carriers available to be released into the conduction band.

b. The sensitivity of the sensor could be calculated from **Equation (4.7)** directly if we had information on the lifetimes of electrons and holes. In their absence we write from **Eqs. (4.5) and (4.6)**:

$$e(\mu_e \Delta n + \mu_p \Delta p) = ef(\mu_n \tau_n + \mu_p \tau_p) \rightarrow (\mu_n \tau_n + \mu_p \tau_p) = \frac{(\mu_e \Delta n + \mu_p \Delta p)}{f}.$$

Thus we can rewrite **Equation (4.7)** as

$$G = \frac{\mathrm{V}}{L^2}\left(\frac{\mu_e \Delta n + \mu_p \Delta p}{f}\right) = \frac{\mathrm{V}}{(0.004)^2}\left(\frac{210 \times 10^{-4} \times 1.0 \times 10^{15} + 20 \times 10^{-4} \times 1.0 \times 10^{15}}{10^{15}}\right)$$

$$= 1437/\mathrm{V}.$$

Note that we have converted the units of mobility to $m^2/V{\cdot}s$ but the units of carrier density do not need to be converted, as they appear in the numerator and denominator similarly, the frequency was writtern as $f = c/\lambda$, with c the speed of light and λ the given wavelength. This gives a sensitivity of 1437/volt, that is, for every 1 V potential difference between the electrodes, a photon generates 1437 carriers. This is a very large sensitivity, typical of CdS sensors.

The most common materials for inexpensive sensors are CdS and CdSe. These offer high sensitivities (on the order of 10^3–10^4), but at a reduced response time, typically about 50 ms. Construction is by deposition and electrodes are then deposited to create the typical comblike shape seen in **Figures 4.4b** and **4.5**, which provides a short distance between the electrodes and a large sensing area. CdS and CdSe can also be sintered. The spectral response of these sensors covers the visible range, although CdS tends to respond better at shorter wavelengths (violet) while CdSe responds better at longer wavelengths (red). Materials can be combined to tailor specific responses. The use of PbS, which is typically deposited as a thin film, shifts the response into the IR region (1000–3500 nm) and improves response to less than 200 μs, but as is typical of IR sensors, at an increase in thermal noise and hence the need for cooling. Examples of single crystal sensors are those made from indium antimonide (InSb). A sensor of this type can operate down to about 7000 nm and can have a response time of less than 50 ns but must be cooled to operate at the longer wavelengths, typically to 77 K (by liquid nitrogen). For specialized application in the IR region, and especially in the far IR, mercury cadmium telluride (HgCdTe) and germanium boronide (GeB) materials may be used. These, especially GeB, can extend operation down to about 0.1 mm if cooled to 4 K (by liquid helium).

In general, cooling of a sensor made of any material extends its spectral response into longer wavelengths, but often slows its response. On the other hand, it increases sensitivity and reduces thermal noise. Many of the far IR applications are military or space applications. These specialized sensors must be made of single crystals and must be housed in a package that is compatible with the low temperature requirements.

4.5.2 Photodiodes

If the junction of a semiconducting diode is exposed to light radiation, the generation of excess carriers due to photons adds to the existing charges in the conduction band

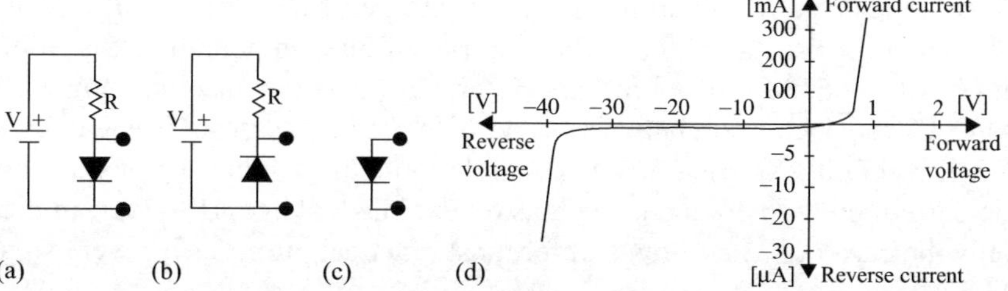

FIGURE 4.6 ■ The semiconducting (*p-n*) junction. (a) Forward biased. (b) Reverse biased. (c) Unbiased. (d) The *I-V* characteristics of the junction.

exactly in the same fashion as for a pure semiconductor. The diode itself may be reverse biased (**Figure 4.6a**), forward biased (**Figure 4.6b**), or unbiased (**Figure 4.6c**). **Figure 4.6d** shows the current–voltage (*I-V*) characteristics of the diode. Of the three configurations in **Figure 4.6**, the forward-biased mode is not useful as a photosensor because in this mode the normal current (not due to photons) is large in comparison to the current generated by photons. In the reverse-biased mode, the diode carries a minute current (i.e., a "dark" current) and the increase in current due to photons is large in comparison. In this mode the diode operates in a manner similar to the photoconducting sensor and is therefore called the **photoconducting mode** of the diode. If the diode is not biased it operates as a sensor in the **photovoltaic mode** (**Figure 4.6c**).

The equivalent circuit of a diode in the photoconductive mode (**Figure 4.6b**) is shown in **Figure 4.7a**. In addition to the current that would exist in the ideal diode (I_d) there is also a leakage current (I_0) defined by the "dark" resistance R_0 and a current through the capacitance (I_c) of the junction. The series resistance R_s is due to conductors connecting the diode. The photons release electrons from the valence band either on the *p* or *n* side of the junction. These electrons and the resulting holes flow toward the respective polarities (electrons toward the positive pole, holes toward the negative pole) generating a current, which in the absence of a bias current in the diode constitutes the only current (the diode is reverse biased). In practice there will be a small leakage current, shown in the equivalent circuit as I_0. The attraction of electrons by the positive pole will tend to accelerate them and in the process they can collide with other electrons and release them across the bandgap, especially if the reverse voltage across the diode is high. This is called an **avalanche effect** and results in multiplication of the carriers available. Sensors that operate in this mode are called **photomultiplier sensors**.

In any diode the current in the forward-biased mode is

$$I_d = I_0\left(e^{eV_d/nKT} - 1\right) \quad [\text{A}] \qquad (4.12)$$

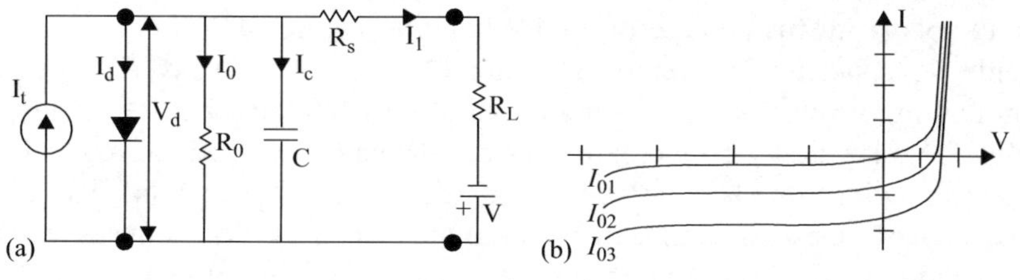

FIGURE 4.7 ■ A photodiode connected in the photoconductive mode (reverse biased). (a) Equivalent circuit. (b) *I-V* characteristics.

where I_0 is the leakage (dark) current, e is the charge of the electron, V_d is the voltage across the junction, also called the voltage barrier or built-in potential, k is Boltzmann's constant ($k = 1.3806488 \times 10^{-23}$ m^{2}kg/s^2/K or J/K), T is the absolute temperature [K], and n is an efficiency constant between 1 and 2. In practical cases it is equal to 1 (but see **Section 3.4**, where it was equal to 2). This relation clearly indicates the dependence of the diode current on temperature as well as on the bias voltage. However, in the reverse mode, only the current I_0 may flow and for most practical purposes it is very small, often negligible.

The current produced by photons is given as

$$I_p = \frac{\eta PAe}{hf} \quad [\text{A}], \tag{4.13}$$

where P is the radiation (light) power density [W/m^2], f is the frequency, and h is Planck's constant. All other terms are constants associated with the diode or the semiconductor used. η is the quantum absorption efficiency, A is the exposed area of the diode (ηPA is the power absorbed by the junction). The total current available external to the diode is (using $n = 1$)

$$I_l = I_d - I_p = I_0\left(e^{eV_d/KT} - 1\right) - \frac{\eta PAe}{hf} \quad [\text{A}] \tag{4.14}$$

This is the current measured for a photodiode sensor under forward or backward biasing conditions (depending on the voltage V_d across the junction). Under reverse-bias conditions, the first term may be neglected altogether since I_0 is small (on the order of 10 nA) and V_d is negative. As a first approximation, especially at low temperatures, one obtains a simple relation for the photodiode:

$$I_l \approx \frac{\eta PAe}{hf} \quad [\text{A}]. \tag{4.15}$$

When measured, this current gives a direct reading of the power absorbed by the diode, but as can be seen, it is not constant since the relation depends on frequency and the power absorbed itself is frequency dependent except for monochromatic radiation. As the input power increases, the characteristic curve of the diode changes, as shown in **Figure 4.7b**, resulting in an increase in the reverse current, as expected.

Any diode can serve as a photodiode, provided that the *n* region, *p* region, or *p-n* junction are exposed to radiation. However, specific changes in materials and construction have been made to common diodes to improve one or more of their photoconducting properties (usually the dark resistance and response time). Taking as an example the planar diffusion type of diode shown in **Figure 4.8**, it consists of *p* and *n* layers and two contacts. The region immediately below the *p* layer is the so-called depletion region, which is characterized by an almost total absence of carriers. This is essentially a regular diode. To increase dark resistance (lower dark current), the *p* layer may be covered with a thin layer of silicon dioxide (SiO_2) (**Figure 4.8a**). The addition of an intrinsic layer of the semiconductor between the *p* and *n* layers produces the so-called PIN photodiode, which, because of the high resistance of the intrinsic layer, has lower

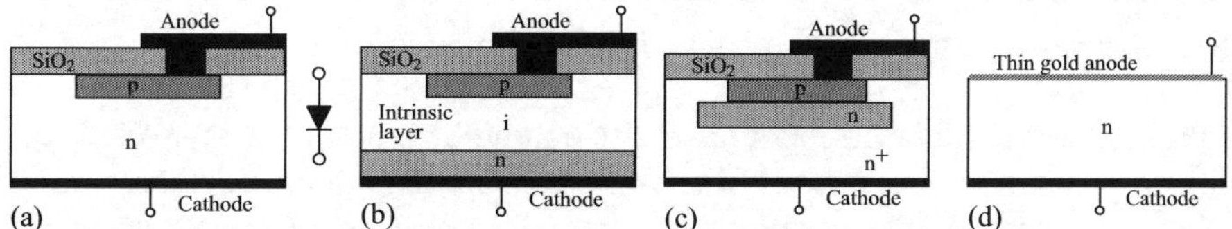

FIGURE 4.8 ▪ Various structures of photodiodes. (a) Common planar structure. (b) PIN diode. (c) pnn^+ structure. (d) Schottky diode.

dark current and lower junction capacitance (and hence better time response) (**Figure 4.8b**). The exact opposite is true in the *pnn*+ construction in which a thin, highly conductive layer is placed at the bottom of the diode. This reduces the resistance of the diode and improves low-wavelength sensitivity (**Figure 4.8c**). Another way of altering the response of a diode is through the use of a Schottky junction. In this diode, the junction is formed by use of a thin layer of sputtered conducting material (gold) on an *n* layer (the Schottky junction is a metal–semiconductor junction) (**Figure 4.8d**). This produces a diode with a very thin outer layer (metal) above the *n* layer, improving its long wavelength (IR) response. As mentioned before, a diode with high reverse bias may operate in avalanche mode, increasing the current and providing a gain or amplification (**photomultiplier diode**). The main requirement needed to obtain avalanche is the establishment of a high reverse electric field across the junction (on the order of 10^7 V/m or higher) to provide sufficient acceleration of electrons. In addition, low noise is essential. Avalanche photodiodes are available for high-sensitivity, low light level applications.

Photodiodes are available in various packages, including surface mount, plastic, and small can packages. **Figure 4.9a** shows one type of diode used as a detector for reflected laser light in CD players. They are also available in linear arrays of photodiodes, such as in **Figure 4.9b**, which shows a (512 element) linear array used as the sensor for a scanner. They are available for IR as well as for the visible range and some extend the range into the UV and even the X-ray range. Many photodiodes have a simple lens to increase the power density at the junction.

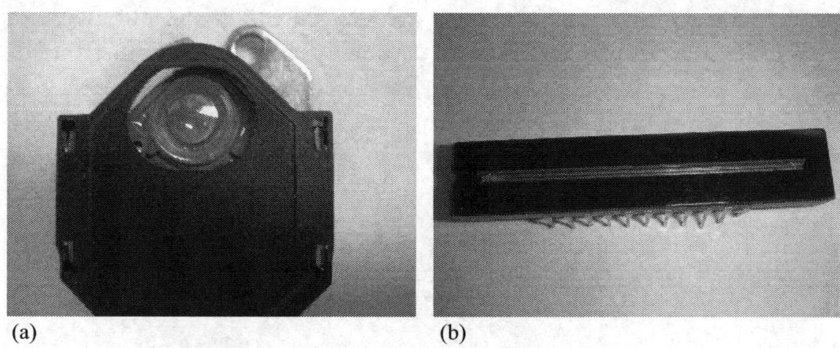

FIGURE 4.9 ▪ (a) A photodiode used as a sensor in a CD player shown installed in its holder. (b) A photodiode linear array (512 photodiodes) in a single integrated circuit used as the sensor element in a scanner. The top cover is glass and light is allowed in through the transparent slit.

EXAMPLE 4.4 Photodiode as a detector for fiber-optic communication

A digital communication link uses a red laser operating at 800 nm with an output of 10 mW. The optical link is 16 km long and is made of an optical fiber with an attenuation of 2.4 dB/km. At the receiving end of the link, a photodiode detects the pulses and the output is measured across a 1 MΩ resistor, as shown in **Figure 4.10**. Assuming the transmitter transmits a series of pulses and there are no losses on either end, that is, all power produced by the laser enters the optical fiber and all power at the diode is absorbed by the diode, calculate the amplitude of the received pulses. Assume that the dark current (leakage current) of the diode is 10 nA and the system operates at 25°C. How does the amplitude change if the temperature rises to 50°C?

Solution: We calculate the diode current using **Equation (4.12)** followed by the photon current using **Equation (4.13)**, or alternatively we can calculate the total current using **Equation (4.14)**. However, to do so we need the radiation power density P. That is calculated using the incident power density and the attenuation along the optical fiber as follows:

The laser produces 10 mW, but to calculate the power entering the diode we need to take into account the losses. To do so we first calculate the input power, P, in decibels:

$$P = 10 \times 10^{-3}\ \text{W} \rightarrow P = 10\log_{10}(10 \times 10^{-3}) = -20\ \text{dB}.$$

The total attenuation along the line is $2.4 \times 16 = 38.4$ dB. Therefore the power in decibels at the end of the line is

$$P = -20 - 38.4 = -58.4\ \text{dB}.$$

Now we convert this back to power. To do so we write

$$10\log_{10} P_0 = -58.4 \rightarrow \log_{10} P_0 = -5.84 \rightarrow P_0 = 10^{-6.84} = 1.445 \times 10^{-6}\ \text{W}.$$

Now we can use **Equation (4.14)**, but we first note that the term ηPA is the total power received by the diode. That, according to the problem statement, equals P_0. Thus we write

$$I_l = I_0\left(e^{eV_d/kT} - 1\right) - \frac{P_0 e}{hf} = 10 \times 10^{-9}\left(e^{1.61\times10^{-19}\times(-12)/1.3806488\times10^{-23}\times298} - 1\right)$$
$$- \frac{1.445 \times 10^{-6} \times 1.61 \times 10^{-19}}{6.6262 \times 10^{-34} \times 3.75 \times 10^{14}}$$
$$= -10 \times 10^{-9} - 936.3 \times 10^{-9}\ \text{A}$$

Clearly the current due to temperature is negligible. Now we can calculate the output voltage across the resistor. When the laser beam is off, the current through the photodiode is 10 nA and the output voltage is

$$V_0 = 10 \times 10^{-9} \times 1 \times 10^{6} = 10\ \text{mV}.$$

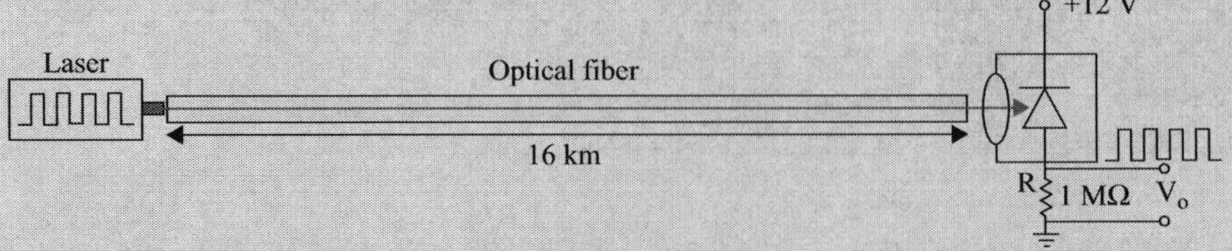

FIGURE 4.10 ■ An optical fiber communication link with the source (laser) and a photodiode used as a detector.

When the light beam is on, the current increases to 946.3 nA and the output voltage is

$$V_0 = 946.3 \times 10^{-9} \times 1 \times 10^6 = 946.3 \text{ mV}.$$

That is, the pulses will result in a voltage that changes from 10 mV for level "0" to 0.946 V for level "1" of the pulse. This may not be sufficient for interfacing and may require amplification.

At 50°C the solution is essentially the same since the component of the current that depends on temperature is negligible in this case. However, this is not a general conclusion.

4.5.3 Photovoltaic Diodes

Photodiodes may also operate in photovoltaic mode as shown in **Figure 4.11**. In this mode the diode is viewed as a generator and requires no biasing. The best known structure for photovoltaic diodes is the solar cell, which is a photodiode with a particularly large exposed area. All photodiodes can operate in this mode, but as a rule, the larger the surface area, the larger the junction capacitance. This capacitance is the main reason for the reduced time response of photovoltaic cells. In most other respects photodiodes operating in photovoltaic mode have the same properties as photodiodes operating in photoconducting mode. There are differences as well. For example, the avalanche effect cannot exist in this mode since there is no bias. **Figure 4.11** shows the equivalent circuit for a photovoltaic cell and **Figure 4.12** shows two small photovoltaic (solar) cells.

Although typically used in photovoltaic arrays for solar power generation as well as smaller arrays used to power small appliances (such as calculators), the photovoltaic diode also makes an exceedingly simple light sensor that needs little more than a voltmeter to measure light power density or light intensity.

Although the photovoltaic diode operates without a bias, under normal operation a voltage develops across the junction and the total current is described by **Equation (4.14)**, where the first term is the normal diode current and the second is the photocurrent. There are two important points in the operation of the diode. The first is the short-circuit current. If the diode is short-circuited, the voltage across the diode is zero and the only current that may exist is the photocurrent. Thus

$$I_{sc} = -I_p = -\frac{\eta PAe}{hf} \quad [\text{A}]. \tag{4.16}$$

The second term is the open circuit voltage, characterized by the fact that the normal diode current equals the photocurrent. The open circuit voltage, V_{oc}, can then be evaluated from this balance:

$$I_0\left(e^{eV_{oc}/kT} - 1\right) = -\frac{\eta PAe}{hf} = I_p. \tag{4.17}$$

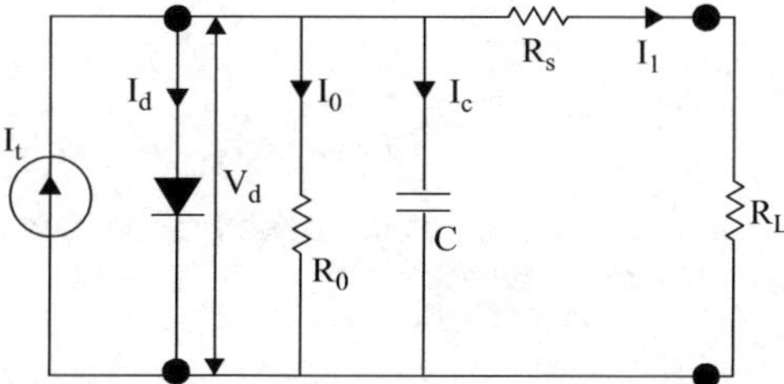

FIGURE 4.11 ■ Photodiode connected in the photovoltaic mode—equivalent circuit. The diode is unbiased.

FIGURE 4.12 ■ A particular type of photodiode, the photovoltaic cell or solar cell. Two types are shown. The one on the right is used in a calculator.

Clearly V_{oc} is equal to the built-in potential or potential barrier, which depends on the material, doping, and, through carrier concentration, temperature. The efficiency term in **Equations (4.16)** and **(4.17)** is an overall efficiency of the cell and is a product of the quantum absorption coefficient and the conversion coefficient of the cell.

EXAMPLE 4.5 Properties of a solar cell at low light

To establish the transfer function of a solar cell one must measure the input power or power density and the output voltage or power supplied by the cell. In particular, at low light, the conversion efficiency of the cell is low and the transfer function is very much dependent on the load (R_L in **Figure 4.11**). A solar cell, 11 cm × 14 cm in area, is connected to a 1 kΩ load, exposed to light from an artificial source, and the output voltage is measured. The voltage versus input power density is shown in **Figure 4.13a**. The curve is characteristic, indicating that the cell tends toward a saturation output at some (higher) power density. This particular solar cell only supplies a few milliwatts because of the low illuminance level.

One of the important parameters of solar cells is its conversion efficiency. Defined as the output power divided by the input power, it is usually given in percentages. As with all other

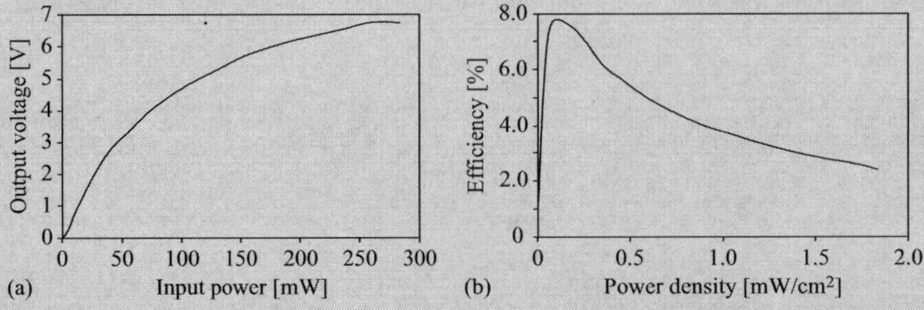

FIGURE 4.13 ■ Characteristics of a solar cell at low light. (a) Output voltage versus input power density. (b) Power conversion efficiency versus input power density.

characteristics, it depends on load and operating point (input power). The efficiency of the cell used here is calculated as follows:

$$e = \frac{P_{out}}{P_{in}} \times 100\% = \frac{V^2/R_L}{P_d \times S} \times 100\%,$$

where P_d is the input power density [mW/cm^2], S is the area of the cell [cm^2] (in this case 11 cm × 14 cm = 154 cm^2), R_L is the load [kΩ], and V is the output voltage [V]. The efficiency is plotted in **Figure 4.13b**. The efficiency reaches a maximum of 7.78% at 0.174 mW/cm^2 and declines beyond that, indicating the same tendency toward saturation. It should be noted again that the maximum efficiency point depends on the illuminance level and the load. Good solar cells will have efficiencies between 15% and 30%.

EXAMPLE 4.6 Optical position sensor

A position sensor is made as follows: A triangular slit is cut into an opaque material that covers a solar cell to create a triangular area of exposure (**Figure 4.14**). This makes the stationary part of the sensor. A moving part is placed above the stationary part and includes two items. One is a stationary source of light, and between the source and the stationary slit is a thin rectangular opening that allows light to go through that opening only. The opening is t m wide and the source supplies an illuminance of I lux. The position of the slit is the output of the sensor. Since the larger h (distance) is, the larger the light power on the cell, the larger the voltage of the solar cell. Assuming the voltage output to be linear with the power incident on the solar cell following the relation $V = kP$, where P is the incident power and k a constant of the cell, find a relation between the measured voltage of the cell and the position of the slit, h.

Solution: As h increases so does the amount of light reaching the cell because the width of the illuminated slit increases. Illuminance is measured in lux, which has units of watts per square meter. Therefore the power reaching the cell is proportional to the area of the triangular slit covered by the rectangle of width t. We calculate this area, multiply by the illuminance, and find the power P. It should be recalled, however, that 1 lux is (1/683) W/m^2. Then of course the output voltage is immediately available.

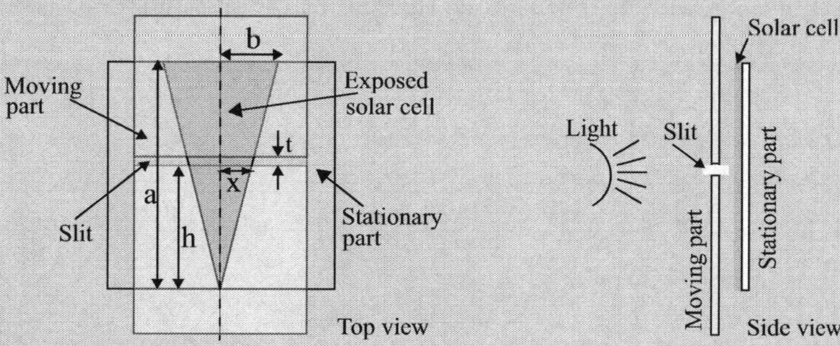

FIGURE 4.14 ■ An optical position sensor. The slit in the top layer indicates the position by exposing a strip width that depends on position.

Assuming the slit is at a height h as shown, the area exposed is

$$S = \frac{(2x' + 2x)t}{2} = (x' + x)t \quad [\text{m}^2].$$

To calculate x and x' we note the following:

$$\frac{b}{a} = \frac{x}{h} = \frac{x'}{h+t}.$$

Therefore

$$x = \frac{b}{a}h, x' = \frac{b}{a}(h+t) \quad [\text{m}].$$

We have

$$S = \left(\frac{b}{a}(h+t) + \frac{b}{a}h\right)t = \frac{bt}{a}(2h+t) \quad [\text{m}^2].$$

The power reaching the cell is

$$P = SI = \frac{btI}{a}(2h+t) \quad [\text{W}],$$

where I is the illuminance [W/m^2]. The output voltage is

$$V = kP = k\frac{btI}{a}(2h+t) \quad [\text{V}].$$

Since the output of the sensor is voltage and the measurand is the distance h, we write this in a more convenient way:

$$V = 2k\frac{btI}{a}h + k\frac{bt^2I}{a} \quad [\text{V}].$$

A linear relation is obtained as expected. Note also that the sensitivity can be increased by increasing b, t, and/or I.

4.5.4 Phototransistors

As an extension of the discussion on photodiodes, the phototransistor can be viewed as two diodes connected back to back, as shown in **Figure 4.15** for an *npn* transistor. With the bias shown, the upper diode (the collector–base junction) is reverse biased while the

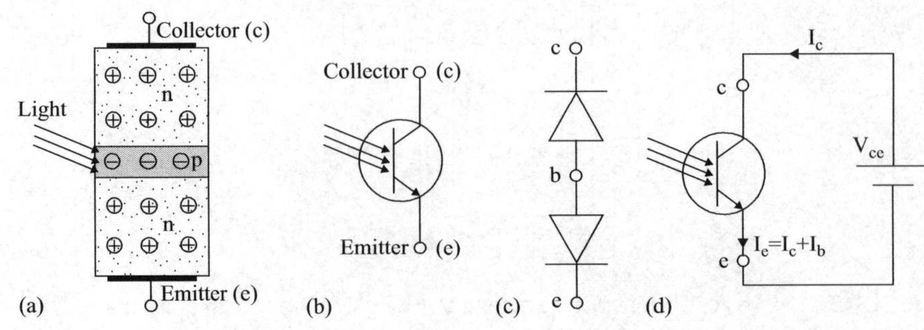

FIGURE 4.15 ■ An *npn* phototransistor. (a) Schematic structure and junctions. (b) The circuit schematic. (c) The two junctions form diodes as shown. (d) The biasing of a phototransistor.

lower (base–emitter) junction is forward biased. In a regular transistor, a current I_b injected into the base is amplified using the following simple relation:

$$I_c = \beta I_b \quad [\mathrm{A}], \tag{4.18}$$

where I_c is the collector current and β is the amplification or gain of the transistor, which depends on a variety of factors, including bias, construction, materials used, doping, etc. The emitter current I_e is

$$I_e = I_b(\beta + 1) \quad [\mathrm{A}]. \tag{4.19}$$

The relations above apply to any transistor. What is unique in a phototransistor is the means of generating the base current. When a transistor is made into a phototransistor, its base connection is usually eliminated and a provision is made for the radiation to reach the collector–base junction. The device operates as a regular transistor with its base current supplied by the photon interaction with the collector–base junction (which is reverse biased). The transistor described here is also called a bipolar junction transistor (BJT). This name distinguishes it from other types of transistors, some of which we will encounter later.

Under dark conditions, the collector current is small and is almost entirely due to leakage currents, designated here as I_0. This causes a dark current in the collector and emitter as

$$I_c = I_0\beta, I_e = I_0(\beta + 1) \quad [\mathrm{A}]. \tag{4.20}$$

When the junction is illuminated, the diode current is the current due to photons obtained in **Equation (4.13)**:

$$I_b = I_p = \frac{\eta PAe}{hf} \quad [\mathrm{A}]. \tag{4.21}$$

The collector and emitter currents are then

$$I_c = I_p\beta = \beta\frac{\eta PAe}{hf}, I_e = I_p(\beta + 1) = (\beta + 1)\frac{\eta PAe}{hf} \quad [\mathrm{A}], \tag{4.22}$$

where the leakage current was neglected in the final relations, as was done for the photodiode. Clearly then the operation of the phototransistor is identical to that of the photodiode except for the amplification, β, provided by the transistor structure. Since β, for even the simplest transistors, is on the order of 100 (and can be much higher), and the amplification is linear in most of the operation range (see **Figure 4.16**), the phototransistor is a very useful device and is commonly used for detection and sensing. The high amplification allows phototransistors to operate at low illumination levels. On the

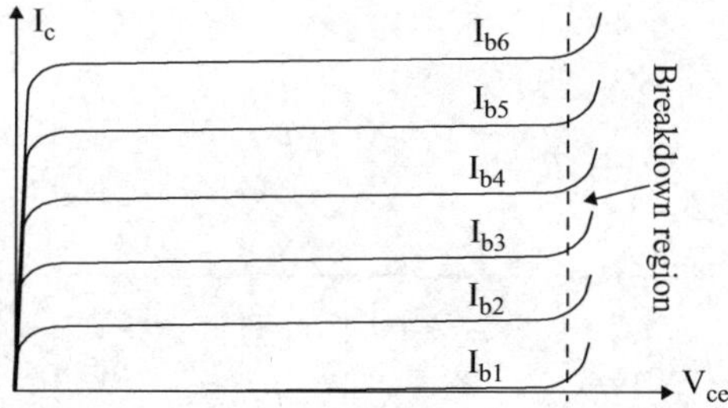

FIGURE 4.16 ■ The *I-V* characteristics of a transistor as a function of base current. In a phototransistor the base current is supplied by photon interaction.

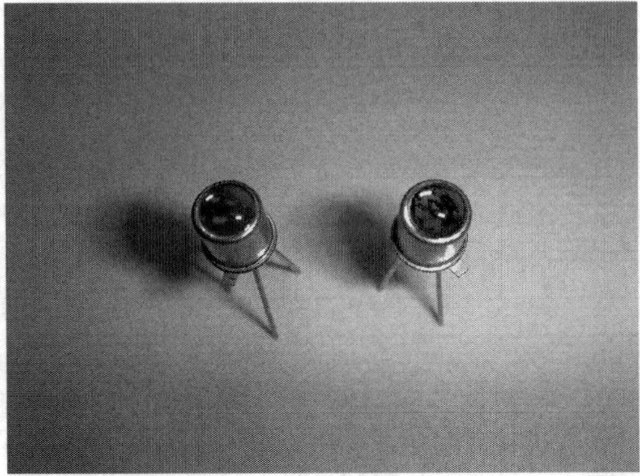

FIGURE 4.17 ■ Phototransistors equipped with lenses.

other hand, thermal noise can be a bigger problem, again because of the amplification. In particular, the base–emitter junction behaves as a regular diode as far as current through it. The latter is given in **Equation (4.12)**, where again, I_0 is the dark current. Although this current is small, the fact that the diode is forward biased, and due to the amplification of the transistor, the effects of temperature are significant.

In many cases a simple lens is also provided to concentrate the light on the junction, which for transistors is very small. A phototransistor equipped with a lens is shown in **Figure 4.17**.

Photoconducting sensors, photodiodes, and phototransistors can sense and measure directly the radiation power they absorb. However, they can easily be used to sense any other quantity or effect that can be made to generate or alter radiation in the range in which the sensor is sensitive. As such, they can be employed to sense position, distance, and dimensions, temperature, and color variations, in counting events, for quality control, and much more.

EXAMPLE 4.7 Sensitivity of a phototransistor

Figure 4.16 shows the *I-V* characteristics of a transistor as a function of base current. However, in a phototransistor, the base current is not measurable. Rather, the current is a function of light power density on the junction. The following is an experimental evaluation of the current in a phototransistor as a function of incident light power density. The table below shows selected values, but all values measured are plotted in **Figure 4.18**. Since the curve is linear for light densities between 0 and about 400 μW/cm^2, the sensitivity of the sensor can be written using any two columns in the table. Taking the first column and the eighth column, we find its sensitivity in the linear range as

$$S = \frac{0.13 - 0.00182}{152 - 2} = \frac{0.12818}{150} = 0.8545\ \mu\text{A}/\mu\text{W}/\text{cm}^2.$$

Light density [μW/cm^2]	2	9.57	20.7	46.2	60.4	83.9	113	152	343	409
Current [μA]	.00182	.00864	.0182	.0409	.0532	.0732	.0978	.13	.28	.324

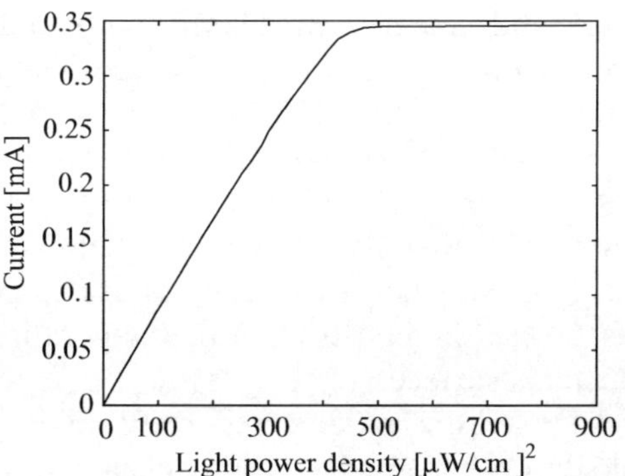

FIGURE 4.18 ■ Collector current in a phototransistor as a function of input light power density. Note the saturation above 400 $\mu W/cm^2$.

4.6 | PHOTOELECTRIC SENSORS

Photoelectric sensors, including photomultipliers, are based on the photoelectric effect (also called the photoemissive effect). As described in **Section 4.4.2**, the photoelectric effect relies on photons with energy *hf* impinging on the surface of a material. The radiation is absorbed by giving this energy to electrons and these are emitted from the surface, provided the energy of the photon is higher than the work function of the material. One can say that the collision between the photon and electron releases the electron if the energy exchanged is sufficiently high. This effect has been applied directly to the development of photoelectric sensors (sometimes called photoelectric cells). In fact, this type of optical sensor is one of the oldest optical sensors available.

4.6.1 The Photoelectric Sensor

The principle of the photoelectric sensor is shown in **Figure 4.19**. The photocathode is made of a material with relatively low work function to allow efficient emission of electrons. These electrons are then accelerated toward the photoanode because of the potential difference between the anode and cathode. The current in the circuit is then proportional to radiation intensity. The number of emitted electrons per photon is the quantum efficiency of the sensor and depends to a large extent on the material used for the photocathode (its work function). Many metals may be used for this purpose, but for the most part their efficiencies are low. More often, cesium-based materials are used because they have low work functions and fairly wide spectrum responses down to about 1000 nm (well into the IR region). Their response extends into the UV region as well. In older devices, highly resistant cathodes were made of a metal such

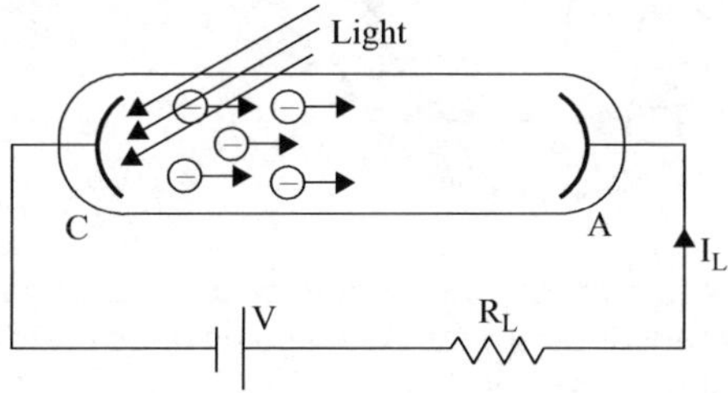

FIGURE 4.19 ■ Photoelectric sensor and biasing circuit.

as tantalum or chromium and coated with alkali compounds (lithium, potassium, sodium, or cesium, or more often, a combination of these; see the periodic table). This provides the low work function necessary. The electrodes are housed in an evacuated tube or in a tube with a noble gas (argon) at low pressure. The presence of gas increases the gain of the sensor (defined as the number of electrons emitted per incoming photon) by internal collisions between emitted electrons and the atoms of the gas through ionization of the gas. Newer devices use so-called negative electron affinity (NEA) surfaces. These are constructed by evaporation of cesium or cesium oxide onto a semiconductor's surface.

Classical photoelectric sensors require relatively high voltages for operation (sometimes a few hundred volts) to supply useful sensing currents. Negative electron afirity devices operate at much lower potentials.

4.6.2 Photomultipliers

Photomultipliers are a development of the classical photoelectric sensor. Whereas in a photoelectric sensor the current is low (the number of electrons emitted is small), photomultipliers, as their names imply, multiply the available current, resulting in sensors that are considerably more sensitive than the simple photoelectric cell. The construction is shown schematically in **Figure 4.20**. It consists of an evacuated tube (or a low-pressure gas-filled tube) made of metal, glass, or metal-coated glass with a window for the incoming radiation. The photocathode and photoanode of the basic photoelectric cell are maintained, but now there is a sequence of intermediate electrodes, as shown in **Figure 4.20a**. The intermediate electrodes are called **dynodes** are made of materials with low work functions, such as beryllium copper (BeCu), and are placed at potential differences with respect to preceding dynodes, as shown in **Figure 4.20b**. The operation is as follows: The incident radiation impinges on the cathode and releases a number of electrons, say, n. These are accelerated toward the first dynode by the potential difference, V_1. These electrons now have sufficient energy to release, say, n_1 electrons for each impinging electron. The number of electrons emitted from the first dynode is $n \times n_1$. These are again accelerated toward the second dynode, and so on, until they finally reach the photoanode. The multiplication effect at each dynode results in a very large number of electrons reaching the photoanode for each photon impinging on the cathode. Assuming there are k dynodes (10–14 is not unusual) and n is the average number of electrons emitted per dynode (secondary electrons), the gain may be written as

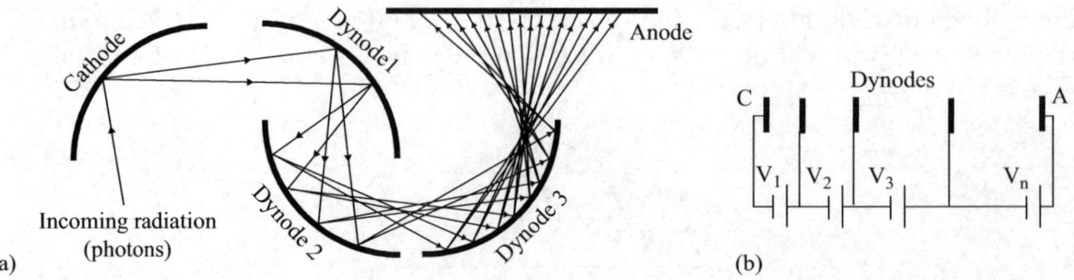

FIGURE 4.20 ▪ (a) Basic structure of a photomultiplier. (b) Biasing of the dynodes and photoanode. The typical potential difference between an anode and a cathode is about 600 V, about 60–100 V between each two dynodes.

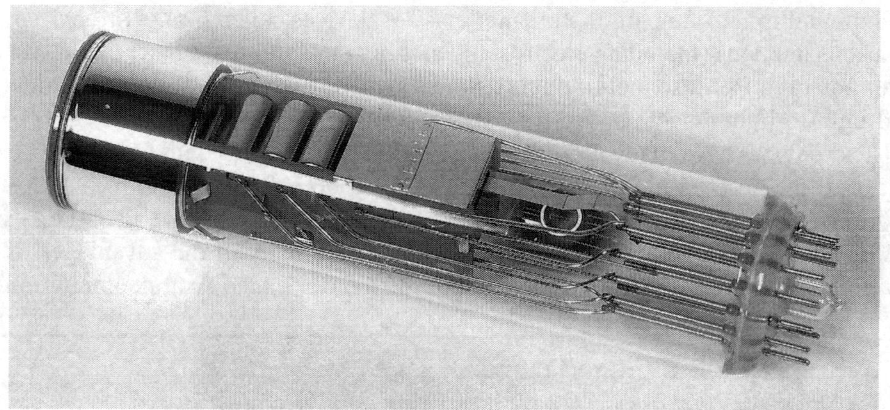

FIGURE 4.21 ■ A photomultiplier. The light enters through the top surface. The dynodes are the curved surfaces on top.

$$G = n^k. \tag{4.23}$$

This gain is the current amplification of the photomultiplier and depends on the construction, the number of dynodes, and the accelerating interelectrode voltages. Clearly, additional considerations must be employed for maximum performance. First, electrons must be "forced" to transit between electrodes at about the same time to avoid distortions in the signal. To do so the dynodes are often shaped as curved surfaces that also guide the electrons toward the next dynode. Additional grids and slats are added for the same purpose, to decrease transit time and improve quality of the signal, especially when the photomultiplier is used for imaging.

As with all sensors of this type, there are sources of noise, but because of the multiplying effect, noise is particularly important in photomultipliers. Of these, the dark current due to thermal emission, which is both potential and temperature dependent, is the most critical. The dark current in a photomultiplier is given as

$$I_0 = aAT^2 e^{-E_0/kT}, \tag{4.24}$$

where a is a constant depending on the cathode material, generally around 0.5, A is a universal constant equal to 120.173 A/cm^2, T is the absolute temperature, E_0 is the work function of the cathode material [e·V], and k is Boltzmann's constant. In photomultipliers this current is small because the cathode is cold and under these conditions the thermal emission is low. Nevertheless, a dark current between 1 and 100 nA is present because of the high gain of the photomultiplier. In addition, the shot noise due to fluctuations in the current of discrete electrons and multiplication noise due to the statistical spread of electrons limits the sensitivity of the device. A major concern with photomultipliers is their susceptibility to magnetic fields. Since magnetic fields apply a force on moving electrons, they can force electrons out of their normal paths reducing their gain and, more critically, distorting the signal.

Nevertheless, with proper construction (including possible cooling of the sensor to reduce dark current) an exceedingly sensitive device can be made. These sensors are therefore used for very low light applications, such as in night vision systems. For example, a photomultiplier sensor may be placed at the focal point of a telescope to view extremely faint objects in space.

Photomultipliers are part of a broader class of devices called **image intensifiers** that use various methods (including electrostatic and magnetic lenses) to increase the current due to radiation. Because their output is sometimes the image itself, they are sometimes called light-to-light detectors. **Figure 4.21** shows a small photomultiplier.

These devices have many disadvantages, including problems with noise, as discussed above, size, the need for high voltages (in excess of 2000 V for some), as well as cost. For these reasons, except for some applications in night vision, they have been largely displaced by charge-coupled devices (CCDs), which have many of the advantages of photomultipliers while eliminating most of the problems associated with photomultipliers.

EXAMPLE 4.8 Thermionic noise in a photomultiplier

A photomultiplier with 10 dynodes has a cathode coated with potassium to increase sensitivity. Calculate the thermally produced dark current at the cathode and at the anode at 25°C, assuming that each incoming photon is energetic enough to release six electrons and that each accelerated electron releases six electrons.

Solution: The work function of potassium is 1.6 eV (see **Table 4.2**). Room temperature is $273.15+25=298.15$°K. With the Boltzmann constant, $k=8.62\times10^{-5}$ eV/°K, we get

$$I_0 = aAT^2e^{-E_0/kT} = 0.5\times120.173\times10^4\times(298.15)^2e^{-1.6/8.62\times10^{-5}\times298.15} = 4.9\times10^{-17}\text{ A}.$$

This is a mere 4.9×10^{-8} nA. Since each accelerated electron releases six electrons, the gain of the photomultiplier is

$$G = n^k = 6^{10} = 6.05\times10^7.$$

The current at the anode due to thermionic emission is

$$I_a = 4.9\times10^{-17}\times6.05\times10^7 = 2.96\times10^{-9}\text{ A}.$$

This is just under 3 nA.

It is because of these very low dark currents that the photomultiplier is so very useful and has survived into the age of semiconductors.

4.7 | COUPLED CHARGE (CCD) SENSORS AND DETECTORS

Charge coupled devices (CCD) are typically made of a conducting substrate on which a *p*- or *n*-type semiconductor layer is deposited. Above it lies a thin insulating layer made of silicon dioxide to insulate the silicon from a transparent conducting layer above it, as shown in **Figure 4.22a**. This structure is called a metal oxide semiconductor (MOS) and is a simple and inexpensive structure. The conductor (also called a gate) and the substrate form a capacitor. The gate is biased positively with respect to the substrate (for an *n*-type semiconductor). This bias causes a depletion region in the semiconductor and, together with the silicon dioxide layer, makes this structure a very high resistance device. When optical radiation impinges on the device it penetrates through the gate and

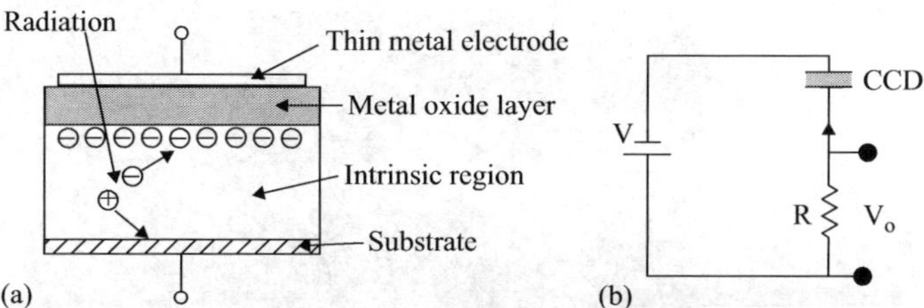

FIGURE 4.22 ▪ The basic CCD cell. (a) In forward-biased mode, electrons accumulate below the MOS layer. (b) In reversed-biased mode the charge is sensed by discharging it through an external load.

oxide layer to release electrons in the depletion layer. The charge density released is proportional to the incident radiation intensity. These charges are attracted toward the gate but cannot flow through the oxide layer and are trapped there. There are a number of methods to measure the charge (and hence the radiation intensity that produced it). In its simplest form, one can reverse bias the MOS device to discharge the electrons through a resistor, as shown in **Figure 4.22b**. The current through the resistor is a direct measure of the light intensity on the device. However, the main value of CCDs is in building one-dimensional (linear array) or two-dimensional arrays of MOS devices for the purpose of imaging. In such cases it is not possible to use the method in **Figure 4.22b** directly. Rather, the basic method is to move the charges of each cell to the next in a kind of "musical chairs" sequence by manipulating gate voltages. In this method the transfer is one cell per step and the current in the resistor for each step corresponds to a particular cell. This is shown in **Figure 4.23a** for one row in the two-dimensional array. At the end of this scan, the array can be scanned again to read a new image. The scan for a two-dimensional array is shown schematically in **Figure 4.23b**. The data are moved vertically one row at a time, that is, all cells move their data one row lower, whereas the lowest row moves it into a shift resister. The scan stops and the shift register is moved to the right to obtain the signal for one row (similar to that in **Figure 4.23a**). Then the next row is shifted until the whole array has been scanned. In practice, each cell is equipped with three electrodes, each covering one-third of the cell and the time step described above is made of three pulses or phases. All first electrodes in a row are connected to each other, all second electrodes in a row form a

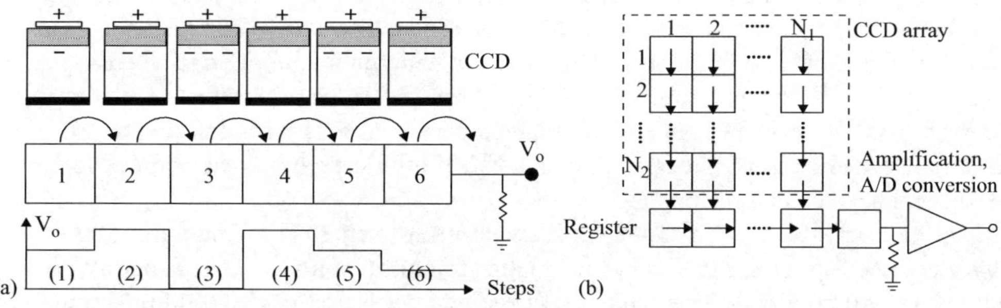

FIGURE 4.23 ▪ Method of sensing the charge in a CCD array. (a) The charge is moved in steps to the edge (by manipulating the gate voltages) and discharged through a resistor. (b) Two-dimensional scan of an $N_1 \times N_2$ image.

FIGURE 4.24 ■ A CCD used as the imaging element in a small camera.

second phase, and all third electrodes in a row form the third phase. The phases are powered in sequence, moving all charge in each row downward one-third of the cell. After three pulses the charge of each row is transferred to the row below. The signal obtained is typically amplified and digitized and used to produce the image signal, which can then be displayed on a display array such as a TV screen or a liquid crystal display. Of course, there are many variations of this basic process. For example, to sense color, filters may be used to separate colors into their basic components (red-green-blue [RGB] is one method). Each color is sensed separately and forms part of the signal. Thus a color CCD will contain four cells per "pixel," one reacting to red, two reacting to green (our eyes are most sensitive to green), and one reacting to blue. In some higher quality imaging systems, each color is sensed on a separate array, but a single array and filters arrangement is more economical.

CCD devices are the core of electronic cameras and video recorders but are also used in scanners (where linear arrays are typically used). They are also used for very low light applications by cooling the CCDs to low temperatures. Under these conditions their sensitivity is much higher, primarily due to reduced thermal noise. In this mode, CCDs have successfully displaced photomultipliers. An integrated CCD used in a video camera (500 lines, 625 pixels per line with four sensors [three colors] per pixel) is shown in **Figure 4.24**.

EXAMPLE 4.9 Some considerations in CCD imaging

Coupled charge devices are very common in still and video cameras, as they are inexpensive compared with other imaging devices and can be produced on a chip, making them ideal for miniaturization. The resolution, usually defined as number of pixels, can be very high while maintaining a small surface area, something that allows the use of small lenses with minimal motion for focusing and zooming. In the extreme, the imaging area can be quite large, with many millions of pixels. However, because even a simple camera contains an imaging sensor of a few million pixels, the transfer of images is not a mere formal process. Rather it is a limiting process that defines, for example, how quickly an image can be recorded.

Consider a digital camera imager with 12 megapixels in a 4:3 format. The image has 3000 columns with 4000 rows (see **Figure 4.23b** for the schematic structure). With three steps per row, the transfer of 3000 rows will take 9000 steps. Each row is first transferred into a 4000-position register that must be shifted out one cell at a time to generate the row signal. Assuming that each operation takes the same amount of time, there are $3000 \times 3 \times 4000 = 36 \times 10^6$ steps to be performed before the image is retrieved from the CCD. Assuming, arbitrarily that a step can be performed in 50 ns, it will take a minimum of 1.8 s to do so.

This means of course that a camera cannot record video in that resolution. Digital cameras that record video do so in a reduced resolution format, typically in video graphics array (VGA) or high-definition (HD) formats. For example, in VGA format the camera only records 640×480 pixels/frame. With the three steps needed, the transfer takes $640 \times 3 \times 480 = 921{,}400$ steps. At the same 50 ns step, this takes 46 ms and allows 21 frames/s, sufficient for good quality video.

Of course, many methods can and are being used to improve performance, but these simple and practical considerations give an idea of the issues involved. It should be noted as well that much higher clock speeds are not very practical, as the power consumption increases linearly with frequency, something that must also be taken into considerations in small, battery-operated cameras.

There are cameras and imaging systems that may include hundreds of megapixels. In these systems extraction of the image may take considerable time, but the quality and resolution are superior.

4.8 | THERMAL-BASED OPTICAL SENSORS

The thermal effects of radiation, that is, conversion of radiation into heat, are most pronounced at lower frequencies (longer wavelengths) and are therefore most useful in the IR portion of the spectrum. In effect, what is measured is the temperature associated with radiation. The sensors based on these principles carry different names—some traditional, some descriptive. Early sensors were known as pyroelectric sensors (from the Greek *πυρ*, "fire"). Bolometers (from the Greek *bolé*, "ray," so bolometer can be loosely translated as a ray meter) are also thermal radiation sensors that may take various forms, but all include an absorbing element and a temperature sensor in one form or another. Some are essentially thermistors and can be used in the whole range of radiation, including microwave and millimeter wave measurements. Bolometers date back to 1878 and were originally intended to sense low light levels from space. Other names like passive infrared (PIR) and active far infrared (AFIR) not only are more descriptive, but also broader, and encompass many types of sensors.

In effect, almost any temperature sensor may be used to measure radiation as long as a mechanism can be found to transform radiation into heat. Since most methods of temperature sensing were described in **Chapter 3**, we will discuss here the specific arrangements used to sense radiation and will view the temperature-sensing elements used in conjunction with various thermal radiation sensors as given and known.

In general, thermal radiation sensors are divided into two classes: PIR and AFIR. In a passive sensor, radiation is absorbed and converted to heat. The temperature increase is measured by a sensing element to yield an indication of the radiative power. In an active sensor, the device is heated from a power source and the variations in this power due to radiation (e.g., the current needed to keep the temperature of the device constant) give an indication of the radiation.

4.8.1 Passive IR Sensors

A PIR sensor has two basic components: an absorption section that converts radiation into heat and a proper temperature sensor that converts heat into an electrical signal. Without going into the questions of heat transfer and heat capacities (these were discussed briefly in **Chapter 3**), the absorption section of the sensor must be able to both absorb as much of the incoming radiated power at the sensor's surface as possible while at the same time quickly respond to changes in radiated power density. Typically the absorber is made of a metal of good heat conductivity (gold is a common choice in high-quality sensors) that is often blackened to increase absorption. The volume of the absorber is kept small to improve response (quick heating/cooling) to changes in radiation and hence keep the response time reasonable. Typically the absorber and the sensor are encapsulated or placed in a gas-filled or evacuated hermetic chamber to avoid variations in sensing signals due to the cooling effects of air motion. The absorber is located behind a transparent (to IR radiation) window, often made of silicon, but other materials may be used (germanium, zinc selenide, etc.). The choice of the sensor materials and structure dictates to a large extent the sensitivity, spectral response, and physical construction of the device.

4.8.1.1 Thermopile PIR

In this type of device, sensing is done by a thermopile. A thermopile is made of a number of thermocouples connected in series electrically but in parallel thermally (i.e., they are exposed to identical thermal conditions). Based on the thermoelectric effect, a thermocouple generates a small potential across a junction made of two different materials. Any two materials can be used, but some material combinations produce higher potential differences (see **Section 3.3**). Thermocouples can only measure temperature differences, hence the thermopile is made of alternating junctions, as shown in **Figure 4.25**. All "cold" junctions are held at a known (measured) lower temperature, while all "hot" junctions are held at the sensing temperature. In practical construction the cold junctions are placed on a relatively large frame that has a high thermal capacity and hence the temperature will fluctuate slowly while the hot junctions are in contact with the absorber, which is small and has a low heat capacity (**Figure 4.25**). In addition, the frame may be cooled, or a reference sensor may be used on the frame so that the temperature difference can be properly monitored and related to the radiated power

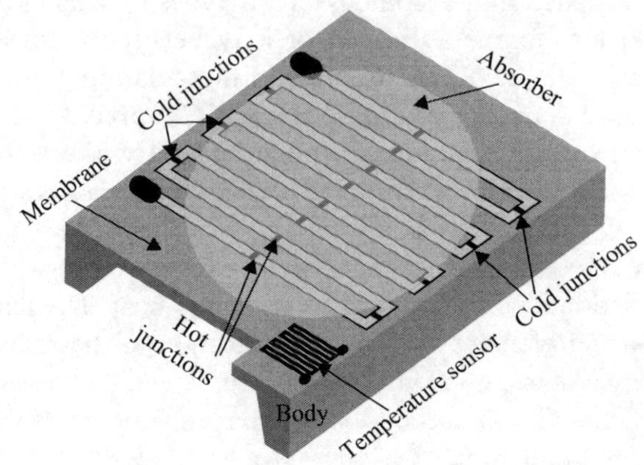

FIGURE 4.25 ■ The structure of a PIR sensor showing the thermopile used to sense temperature (under an IR absorber). A temperature sensor monitors the temperature of the cold junctions.

density at the sensor. Although any pair of materials may be used, in most PIRs crystalline or polycrystalline silicon and aluminum are used because silicon has a very high thermoelectric coefficient and is compatible with other components of the sensor, whereas aluminum has a low temperature coefficient and can be easily deposited on silicon surfaces. Other materials used (mostly in the past) are bismuth and antimony. The output of the thermocouple is the difference between the Seebeck coefficients of silicon and aluminum (see **Section 3.3**).

PIR sensors are used to sense mostly near-IR radiation but within this range they are quite common. When cooled they can be used further down into the far-IR radiation. One of the most common applications of PIR sensors is in motion detection (in which the transient temperature caused by motion is detected). However, for this purpose the pyroelectric sensors in the following section are often used because they are both simpler and less expensive than the structure described above.

EXAMPLE 4.10 Thermopile sensor for IR detection

An IR sensor is used to detect hot spots in forests to prevent fires. The sensor is made as in **Figure 4.25**, with 64 pairs of junctions using a silicon/aluminum junction (see **Tables 3.3** and **3.9**). The silicon/aluminum thermocouple has a sensitivity $S = 446$ mV/°C. The absorber is made of a thin gold foil, 10 μm thick and 2 cm^2 in area, coated black to increase absorption. Gold has a density $\rho = 19.25$ g/cm^3 and a heat capacity $C = 0.129$ J/g/°C (i.e., the absorber needs to absorb 0.129 J/g for its temperature to rise 1°C). The absorber is not ideal and its conversion efficiency is only 85% (i.e., 85% of the incoming heat is absorbed). We will denote efficiency as $e = 0.85$. The window of the sensor is $A = 2$ cm^2 and we will assume that it takes the sensor $t = 200$ ms to reach thermal steady state (i.e., for the temperature of the absorber to stabilize to a constant value at a given radiation power density). This temperature is due to the absorbed heat and heat loss. If a temperature difference between the hot and cold junction of 0.1°C can be reliably measured in the sensor, calculate the sensitivity of the sensor assuming the input is the power density of the IR radiation.

Solution: Because the input is a power density (i.e., measured in W/m^2 or in lux) and the temperature increase is due to heat (energy), the heat capacity is therefore the product of power density and time. Given a power density P_{in} [W·m^2], the power received by the sensor is the product of power density and the area of the absorber. With an 85% efficiency, the heat absorbed over 200 ms is

$$w = P_{in}tAe = P_{in} \times 0.2 \times 2 \times 10^{-4} \times 0.85 = 3.4 \times 10^{-5} P_{in} \quad [\text{J}].$$

To find the temperature increase of the absorber due to this heat, we divide it by the heat capacity of the foil, denoted as C_f. This is simply the specific heat multiplied by the mass of the absorber. The latter is

$$m = 10 \times 10^{-6} \times 2 \times 10^{-4} \times 19.25 = 3.85 \times 10^{-5} \text{g}.$$

The specific heat for the absorber is therefore

$$C_f = C_m = 0.129 \times 3.85 \times 10^{-5} = 4.96665 \times 10^{-6} \quad \left[\frac{\text{J}}{\text{K}}\right]$$

We obtain the temperature increase in the absorber by dividing the incoming heat by C_f:

$$T = \frac{w}{C_f} = \frac{P_{in}tAe}{C_f} = \frac{3.4 \times 10^{-5}}{4.96665 \times 10^{-6}} P_{in} = 6.846 P_{in} \ \text{[K]}.$$

Since the lowest measurable change in temperature is 0.1°C, we get the power density needed to raise the temperature by that amount:

$$P_{in} = \frac{0.1}{6.846} = 1.46 \times 10^{-2} \ \text{W/m}^2.$$

This is 14.6 mW/m^2.

The sensitivity, by definition (assuming, of course, a linear transfer function), is the output divided by input. We have the input power density. Now we need to calculate the output of the thermocouple for that same temperature difference of 0.1°C. Since we have a thermopile with 64 pairs of junctions and a sensitivity of 446 mV/°C, the output for a temperature difference of 0.1°C is

$$V_{out} = 446 \times 64 \times 0.1 = 2854.4 \ \text{mV}.$$

Thus the sensitivity of the sensor is

$$S = \frac{V_{out}}{P_{in}} = \frac{2854.4}{1.46 \times 10^{-2}} = 1.955 \times 10^5 \ \text{mV/W/m}^2.$$

In practical terms this means that, for example, the sensor will produce a 1.955 V output at an input power of 10^{-5} W/m^2. This kind of sensitivity is sufficient for most low-power sensing ranging from astronomy applications to highly sensitive motion sensing.

4.8.1.2 Pyroelectric Sensors

The pyroelectric effect is an electric charge generated in response to heat flow through the body of a crystal. The charge is proportional to the change in temperature and hence the effect may be viewed as heat flow sensing rather than temperature sensing. However, in the context of this section, our interest is in the measurement of radiation and thus pyroelectric sensors are best viewed as sensing changes in radiation. For this reason they have found applications in motion sensing in which the background temperature is not important—only that due to the motion of a "warm" source is sensed. Pyroelectricity was formally named in 1824 by David Brewster, but its existence in tourmaline crystals was described in 1717 by Louis Lemery. It is interesting to note that the effect is described in the writing of Theophrastus in 314 BC as the attraction of bits of straw and ash to tourmaline when the latter was heated. The attraction is due to the charge generated by the heat. As early as the end of the nineteenth century, pyroelectric sensors were made of Rochelle salt (potassium sodium tartrate [$KHC_4H_4O_6$]). Currently there are many other materials used for this purpose, including barium titanate oxide ($BaTiO_3$), lead titanate oxide ($PbTiO_3$), as well as lead zirconium titanate (PZT) materials ($PbZrO_3$), polyvinyl fluoride (PVF), and polyvinylidene fluoride (PVDF). When a pyroelectric material is exposed to a temperature change ΔT, a charge ΔQ is generated as

$$\Delta Q = P_Q A \Delta T \ \ \text{[C]}, \tag{4.25}$$

where A is the area of the sensor and P_Q is the pyroelectric charge coefficient defined as

$$P_Q = \frac{dP_s}{dT} \quad [\mathrm{C/m^2 \cdot K}] \tag{4.26}$$

and P_s is the spontaneous polarization [$\mathrm{C/m^2}$] of the material. Spontaneous polarization is a property of the material related to its electric permittivity.

A change in potential ΔV develops across the sensor as

$$\Delta V = P_V h \Delta T \quad [\mathrm{V}], \tag{4.27}$$

where h is the thickness of the crystal and P_V is its pyroelectric voltage coefficient,

$$P_V = \frac{dE}{dT} \quad [\mathrm{V/m^2 \cdot K}], \tag{4.28}$$

and E is the electric field across the sensor. The two coefficients (voltage and charge coefficients; see **Table 4.4**) are related as follows:

$$\frac{P_Q}{P_V} = \frac{dP_s}{dE} = \varepsilon_0 \varepsilon_r \quad [\mathrm{F/m}]. \tag{4.29}$$

By definition, the sensor's capacitance is

$$C = \frac{\Delta Q}{\Delta V} = \varepsilon_0 \varepsilon_r \frac{A}{h} \quad [\mathrm{F}]. \tag{4.30}$$

Hence one can write the change in voltage across the sensor as

$$\Delta V = P_Q \frac{h}{\varepsilon_0 \varepsilon_r} \Delta T \quad [\mathrm{V}]. \tag{4.31}$$

Clearly this change in voltage is linearly proportional to the change in temperature. It should be noted again that our main interest here is not in measuring the change in temperature, but rather the change in radiation that causes this change in temperature. Also to be noted is that all sensors must operate below their Curie temperature (at the Curie temperature their polarization vanishes). **Table 4.4** shows these properties for some materials used for pyroelectric sensors.

The structure of a pyroelectric sensor is quite simple. It consists of a thin crystal of a pyroelectric material between two electrodes, as shown in **Figure 4.26a**. Some sensors use a dual element, as in **Figure 4.26b**. The second element can be used as a reference by, for example, shielding it from radiation, and it is often used to compensate for

TABLE 4.4 ■ Pyroelectric materials and some of their properties

Material	P_Q [C/m^2/°K]	P_V [V/m/°K]	ε_r	Curie temperature [°C]
TGS (single crystal)	3.5×10^{-4}	1.3×10^{6}	30	49
$LiTaO_3$ (single crystal)	2.0×10^{-4}	0.5×10^{6}	45	618
$BaTiO_3$ (ceramic)	4.0×10^{-4}	0.05×10^{6}	1000	120
PZT (ceramic)	4.2×10^{-4}	0.03×10^{6}	1600	340
PVDF (polymer)	0.4×10^{-4}	0.4×10^{6}	12	205
$PbTiO_3$ (polycrystalline)	2.3×10^{-4}	0.13×10^{6}	200	470

TGS = TriGlycine Sulfate.

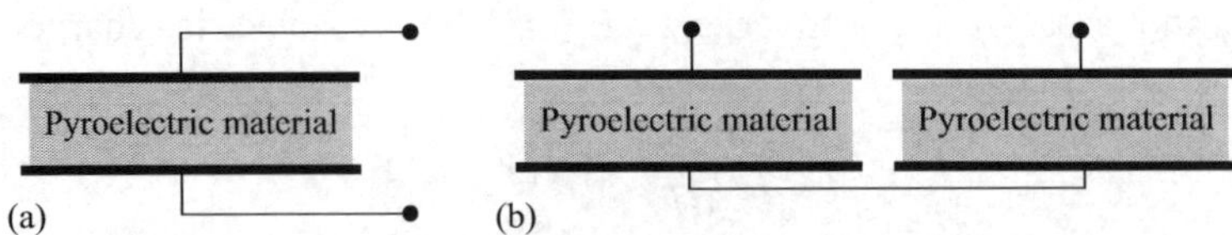

FIGURE 4.26 ■ The basic structure of a pyroelectric sensor. (a) Single element. (b) Dual element in series connected in a differential mode.

common mode effects such as vibrations or very rapid thermal changes, which can cause false effects. In the figure shown here the two elements are connected in series, but they may also be connected in parallel.

The most common materials in pyroelectric sensors are triglycine sulfate (TGS) and lithium tantalite crystals, but ceramic materials and, more recently, polymeric materials are also commonly used.

In applications of motion detection, especially of the human body (sometimes of animals), the change in temperature of IR radiation (between 4 and 20 μm) causes a change in the voltage across the sensor, which is then used to activate a switch or some other type of indicator.

An important property of all pyroelectric sensors is the decay time, during which the charge on the electrodes diffuses. This is on the order of 1–2 s because of the very high resistance of the materials, but it also depends on the external connection of the device. This response time is very important in the ability of the sensors to detect slow motion.

Figure 4.27 shows a dual IR sensor used for motion detection. This device includes a differential amplifier, operates at 3–10 V, and has a field of view of 138° horizontally (wide dimension of the window) and 125° vertically. The device has an optical bandwidth (sensitivity region) between 7 and 14 μm (in the near IR region).

FIGURE 4.27 ■ A PIR motion detector sensor. This is a dual sensor. Note the metal package and the window (3 mm × 4 mm).

EXAMPLE 4.11 Motion sensor

A motion sensor based on a PZT ceramic is used to turn on lights in a room as a person enters the room. The sensor is made of two conducting plates with a PZT chip (8 mm wide × 10 mm long × 0.1 mm thick) between them, forming a capacitor. One plate is exposed to the motion, whereas the other is connected to the body of the sensor and held at its temperature. As the person enters the room, the person's body temperature causes the exposed plate's temperature to temporarily rise by 0.01°C because of the IR radiation produced by the body. This temperature dissipates and eventually both plates will reach the same steady-state temperature. For this reason the sensor can detect motion but not presence. Calculate the charge produced on the plate and the potential difference across the sensor due to the rise in temperature.

Solution: The charge produced by the rise in temperature can be calculated from **Equation (4.25)** and the potential difference can be calculated from **Equation (4.31)**. However, we will start with **Equation (4.31)**, calculating the potential and then the charge from the relation between charge and capacitance (**Equation (4.30)**).

The potential across the plates is

$$\Delta V = P_Q \frac{h}{\varepsilon_0 \varepsilon_r} \Delta T = 4.2 \times 10^{-4} \times \frac{0.1 \times 10^{-3}}{1600 \times 8.854 \times 10^{-12}} \times 0.01 = 0.0296 \text{ V}.$$

This is a small voltage, but because the reference voltage (i.e., the output in the absence of motion) is zero, the small output voltage is easily measurable.

The charge produced by the change in temperature depends on the capacitance. The latter is

$$C = \frac{\varepsilon_0 \varepsilon_r A}{h} = \frac{1600 \times 8.854 \times 10^{-12} \times 0.008 \times 0.01}{0.1 \times 10^{-3}} = 1.1333 \times 10^{-8} \text{ F}.$$

The charge produced is

$$\Delta Q = C \Delta V = 1.1333 \times 10^{-8} \times 0.0296 = 3.355 \times 10^{-10} \text{ C}.$$

Clearly, to produce a useful output (i.e., to turn on a relay or an electronic switch) the output must be amplified. For example, if the output requires (typically) 5 V, then amplifying the voltage by about 170 will produce the required output. We will discuss these issues in **Chapter 11**, where we will see that in fact the amplification can be done using a charge amplifier. The main reason for that approach, rather than performing a classical voltage amplification, is that the impedance of pyroelectric sensors is very high, whereas the input impedance of conventional voltage amplifiers is much lower. The charge amplifier, which has a very high input impedance, is better suited for this application.

4.8.1.3 Bolometers

Bolometers are very simple radiation average power sensors useful over the whole spectrum of electromagnetic radiation, but they are most commonly used in microwave and far-IR ranges. They consist of any temperature-measuring device, but usually a small RTD or a thermistor is used. The operation is as follows: The radiation is absorbed by the device directly, causing a change in its temperature. This temperature increase is proportional to the radiated power density at the location of sensing. This change causes a change in the resistance of the sensing element that is then related to the power or power density at the location being sensed. Although there are many variations of this basic device, they all operate on essentially the same principle. However, since the temperature increase due to radiation is the measured quantity, it is important that the background temperature (i.e., air) be taken into account. This can be done by separate measurements or by a second bolometer that is shielded from the radiation (usually in a metal enclosure in the case of microwaves). The sensitivity of a bolometer to radiation can be written as follows:

$$\beta = \frac{\alpha\varepsilon_s}{2}\sqrt{\frac{Z_T R_0 \Delta T}{(1+\alpha_0\Delta T)[1+(\omega\tau)^2]}}, \tag{4.32}$$

where $\alpha = (dR/dT)/R$ is the temperature coefficient of resistance (TCR) of the bolometer, ε_s is its surface emissivity, Z_T is the thermal resistance of the bolometer, R_0 is its resistance at the background temperature, ω is the frequency, τ is the thermal time constant, and ΔT is the increase in temperature. Clearly then the ideal bolometer should have a large resistance at background temperatures and high thermal resistance. On the other hand, they must be small physically, which favors low thermal resistances.

In terms of construction, bolometers are fabricated as very small thermistors or RTDs, usually as individual components or as integrated devices. In all cases it is important to insulate the sensing element from the structure supporting it so that its thermal impedance is high. This can be done by simple suspension of the sensor by its wires or, as is sometimes done, by suspension of the sensor over a silicon groove.

Bolometers are some of the oldest devices used for this purpose and have been used for many applications in the microwave region, including mapping of antenna radiation patterns, detection of IR radiation, testing of microwave devices, and much more.

4.9 | ACTIVE FAR INFRARED (AFIR) SENSORS

In its simplest form an active infrared (AFIR) sensor can be thought of as a power source that heats the sensing element to a temperature above ambient and keeps its temperature constant. When used to sense radiation, additional heat is provided to the sensors through this radiation. The power necessary to keep the temperature constant is now reduced and the difference in power is a measure of the radiation power. Of course, in practice the process is more complicated. Assuming that the temperature of the sensing element is constant, the AFIR sensor can be viewed as being time independent. Under these conditions the power supplied to the sensor through an electric circuit that heats it up to a constant temperature T_s is

$$P = P_L + \Phi \quad [\mathrm{W}], \tag{4.33}$$

where $P = V^2/R$ is the heat supplied by a resistive heater (V is the voltage across the heating element and R its resistance), P_L is the power loss through conduction through the body of the sensor, and Φ is the radiation power being sensed.

The power loss is

$$P_L = \alpha_s(T_s - T_a) \quad [\mathrm{W}], \tag{4.34}$$

where α_s is a loss coefficient or thermal conductivity (which depends on materials and construction), T_s is the sensor's temperature, and T_a is the ambient temperature. Given the power supplied as $P = V^2/R$, a sensor with surface area A, and a total emissivity ε, the sensed temperature, T_m, is

$$T_m = \sqrt[4]{T_s^4 - \frac{1}{A\sigma\varepsilon}\left[\frac{V^2}{R} - \alpha_s(T_s - T_a)\right]}, \quad (4.35)$$

where σ is the electric conductivity of the sensor medium.

This relation gives the temperature as a function of voltage across the heating element. By measuring this voltage, a reading of the radiation power is obtained.

Although AFIR devices are much more complex than simple PIRs, including bolometers, they have the advantage of a much higher sensitivity and an independence from thermal noise that other IR sensors do not possess. Hence AFIR devices are used for low-contrast radiation measurements where PIRs are not suitable.

4.10 | OPTICAL ACTUATORS

When discussing optical devices it is not immediately clear what an optical actuator might be because of our tendency to think of actuators as devices that perform some kind of motion. However, based on our definition of actuators, in fact there are many optical actuators and they are quite common. The use of a laser beam to perform eye surgery, to machine a material, or to record data in a magneto-optical hard drive are some examples. Others are transmission of data on an optical fiber, transmission of a command using an IR remote control device, use of an LED or laser to illuminate a CD to read the data, scanning of an IPC code in a supermarket, or even turning on a light in a room. In **Chapter 10** we will discuss additional examples such as optical switching.

Optical actuation can be low or high power. In an optical link such as in optical fiber communication or in an optical isolator (to be discussed in **Chapter 12**), the transmitting element (the actuator) is a low-power LED in the IR or visible range (**Figure 4.28**). The power produced by the LED may only be a few milliwatts. On the other hand, industrial lasers, such as carbon dioxide (CO_2) lasers (in which CO_2 gas is excited and produces a beam in the IR region around 100 μm), can produce hundreds of kilowatts of useful power for a variety of industrial processing purposes, including machining, surface treatment, and welding. In between are lasers, mostly CO_2 lasers of moderate power (a few watts to a few hundred watts) used for medical applications, including surgery, skin ablation, and suturing. Other applications are for range finding, particularly in the military, and for speed detection and measurement. Laser actuation is also used for the production of electronic components, for trimming of devices, and even for recording of data on CD-ROMs, where they actually scribe the surface with the pattern representing the data.

A good example of an optical actuator of the type described here is the magneto-optical recording of data, a common method used for high-density data recording. The principle is shown schematically in **Figure 4.29** and is based on two principles. First, a laser beam, focused to a small point, can heat the surface of the disk to a high temperature in a few nanoseconds. Second, when a ferromagnetic material such as iron or its oxides is heated above a certain temperature (about 650°C), the material loses

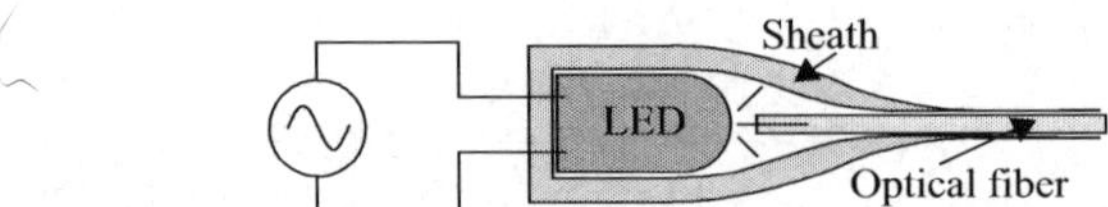

FIGURE 4.28 ■ An optical link. The fluctuations in the light intensity of the LED represent the data transmitted along the optical fiber.

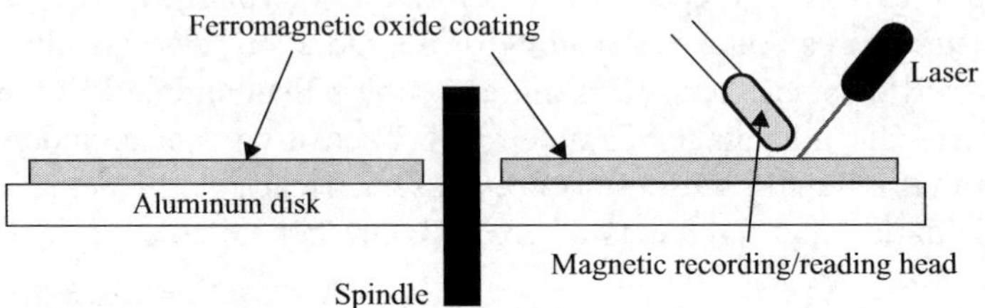

FIGURE 4.29 ■ Magneto-optical recording. The laser beam heats the recording medium above the Curie temperature and the magnetic recording head applies the magnetic field needed to record the data.

its magnetic properties. This temperature is called the **Curie temperature** and is characteristic of the particular material used as the recording medium (mostly Fe_2O_3). When cooled, the material becomes magnetized with the field supplied by the recording head. To record data, the laser is turned on to heat the point above the Curie temperature and the datum that needs to be recorded at that point is supplied in the form of a low-intensity magnetic field by a magnetic recording head. The beam is then switched off and the spot cools below the Curie temperature in the presence of the magnetic field, retaining the data permanently. Erasure of data is done by heating the spot and cooling it off without a magnetic field. The data are read using the magnetic recording head alone. The advantage of this method is that the data density is much higher than purely magnetic recording, which requires larger magnetic fields that in turn extend over larger surfaces and hence is only practical at lower data densities.

4.11 | PROBLEMS

Optical units

4.1 **Optical quantities.** An isotropic light source (a source that radiates uniformly in all directions in space) produces a luminance of 0.1 cd/m^2 at a distance of 10 m. What is the power of the source?

4.2 **Optical sensitivity.** Many optical instruments such as video cameras are rated in terms of sensitivity in lux, especially to indicate low-light sensitivity. One may encounter specifications such as "sensitivity – 0.01 lux." What is the sensitivity in terms of power density?

4.3 **Electron volts and joules.** The Planck constant is given as 6.6261×10^{-34} J/s or as 4.1357×10^{-15} eV/s. Show that the two quantities are identical but expressed in different units.

The photoelectric effect

4.4 **Photon energy and electron kinetic energy.** A photoelectric device intended for UV sensing is made of a platinum cathode. Calculate the range of kinetic energies of the electrons emitted by UV light between 400 nm and 1 pm, assuming that a photon emits a single electron.

4.5 **Work function and the photoelectric effect.** The work function for copper is 4.46 eV.

a. Calculate the lowest photon frequency that can emit an electron from copper.

b. What is the wavelength of the photons and in what range of optical radiation does that occur?

4.6 **Electron density in photoelectric sensor.** A photoelectric sensor has a cathode in the form of a disk of radius $a = 2$ cm coated with an alkali compound that has a work function $e_0 = 1.2$ eV and a quantum efficiency of 15%. The sensor is exposed to sunlight. Calculate the average number of electrons emitted per second assuming a power density of 1200 W/m^2 and uniform distribution of power over the spectrum between red (700 nm) and violet (400 nm).

4.7 **Work function, kinetic energy, and current in a photoelectric sensor.** A photoelectric sensor with unknown cathode material is subjected to experimental evaluation. The current in the sensor (see **Figure 4.19**) is measured while the wavelength of the radiation is recorded. Emission is observed starting with red light at 820 nm. The power density of the incoming radiation is kept constant at 5 mW/cm^2.

a. Calculate the work function of the cathode.

b. If the photoelectric sensor is now illuminated with the same power density, but with a blue light at a wavelength of 480 nm, what is the kinetic energy of the electrons released?

c. Assume each photon releases one electron. Calculate the current in the sensor if the cathode has an area of 2 cm^2.

The photoconducting effect and photoconducting sensors

4.8 **Bandgap energy and spectral response.** A semiconductor optical sensor is required to respond down to 1400 nm for use as a near-IR sensor. What is the range of bandgap energies of the semiconductor that can be used for this purpose?

4.9 **Germanium silicon and gallium arsenide photoconducting sensors.** The intrinsic concentrations and mobilities for germanium, silicon, and gallium arsenide are as follows:

	Germanium (Ge)	Silicon (Si)	Gallium Arsenide (GaAs)
Intrinsic concentration [per cm^3]	2.8×10^{13}	1.0×10^{10}	2.0×10^{6}
Mobility of electrons, μ_e [cm^2/V/s]	3900	1400	8800
Mobility of holes, μ_p [cm^2/V/s]	1900	450	400

Compare the dark resistance of an identical sensor for the three materials made as a rectangular bar of length 2 mm, width 0.2 mm, and thickness 0.1 mm to be used as photoconductors. The resistance calculated here is the nominal resistance of the sensor.

4.10 Gallium arsenide photoconductive sensor. A gallium arsenide (GaAs) photoconductive sensor is made as a small, rectangular chip 2 mm long, 1 mm wide, and 0.1 mm thick (see **Figure 4.4a** for the construction). A red light of intensity 10 mW/cm^2 and wavelength 680 nm is incident perpendicularly on the top surface. The semiconductor is *n*-type with an electron concentration of 1.1×10^{19} electrons/m^3. The mobilities of electrons and holes in GaAs are 8400 $cm^2/V{\cdot}s$ and 400 $cm^2/V{\cdot}s$, respectively. Assuming all incident power on the top surface of the device is absorbed and a quantum efficiency of 0.38, calculate

a. The "dark" resistance of the sensor.

b. The resistance of the sensor when light shines on it. The recombination time of electrons is approximately 10 μs.

4.11 Improved gallium arsenide photoconductive sensor. In an attempt to improve the sensor described in **Problem 4.10**, the meander shape in **Figure 4.4b** is adopted, keeping the total exposed area the same (2 mm^2) but reducing the length between the electrodes to 0.5 mm.

a. Calculate the "dark" resistance of the sensor.

b. Calculate the resistance of the sensor when light shines on it.

c. Compare the results with those for the rectangular sensor and comment on the sensitivity of the two devices.

4.12 Intrinsic silicon optical sensor. Suppose a photoconductive sensor is made of intrinsic silicon with structure and dimensions as shown in **Figure 4.30**. The intrinsic concentration is 1.5×10^{10} carriers/cm^3 and mobilities of electrons and holes are 1350 $cm^2/V{\cdot}s$ and 450 $cm^2/V{\cdot}s$, respectively. The carrier lifetime for electrons and holes depends on the concentration and changes with illumination, but for simplicity we will assume these are constant at 10 μs. Assume also that 50% of the incident power is absorbed by the silicon and 45% quantum efficiency.

a. Find the sensitivity of the device to input power density at a given wavelength in general terms.

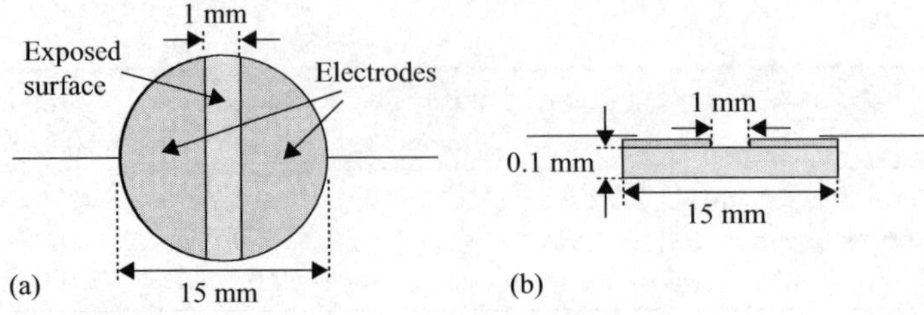

FIGURE 4.30 ■ A photoconducting sensor.

b. What is the sensitivity at 1 mW/cm^2 at a wavelength of 480 nm?

c. What is the cutoff wavelength, that is, the wavelength beyond which the sensor cannot be used?

Photodiodes

4.13 Photodiode in photoconductive mode. A photodiode is connected in reverse mode with a small reverse voltage to ensure low reverse current. The leakage current is 40 nA and the sensor operates at 20°C. The junction has an area of 500 μm^2 and operates from a 3 V source, as shown in **Figure 4.31**. The resistor is 240 Ω. Calculate the voltage across R in the dark and when illuminated with a red laser beam (800 nm) with power density of 5 mW/cm^2. Assume a quantum efficiency of 50%.

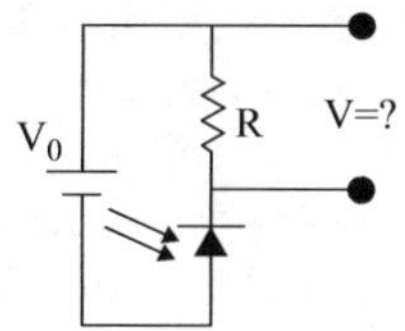

FIGURE 4.31 ▪ A photodiode in photoconductive mode.

4.14 Photodiode in forward bias. A photodiode is forward biased as shown in **Figure 4.32** so that the voltage across the diode is 0.2 V. The dark current through the diode is 10 nA.

a. What must be the voltage V_0 to obtain this bias at 20°C in the dark?

b. When illuminated with a certain power density, and with the voltage V_0 found in (a) applied, the bias changes to 0.18 V. Calculate the total power absorbed by the diode at 800 nm.

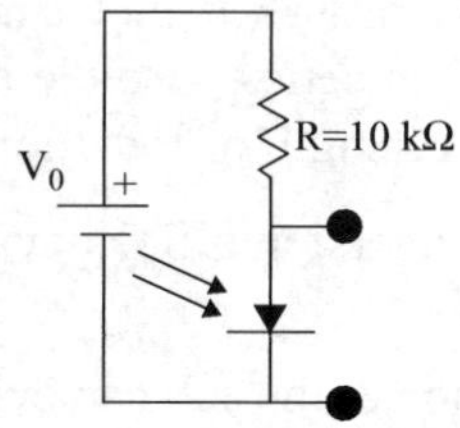

FIGURE 4.32 ▪ Forward-biased photodiode.

4.15 Dusk/dawn light switch. Many lighting systems, including street lightning, turn on and off automatically based on light intensity. To do so, it is proposed to use a photodiode in the configuration shown in **Figure 4.33**. The diode has negligible dark current, an exposed area of 1 mm^2, and a quantum efficiency of 35%. The electronic switch is designed so that to turn on the lights the voltage across R must be 8 V or less and to turn off the lights the voltage must be 12 V or greater. On a normal sunny day, the power density available at ground level is 1200 W/m^2.

a. Calculate the resistance R so that the lights turn on when available light intensity (in the evening) is 10% (or less) of the normal daylight. Assume radiation at an average wavelength of 550 nm.

b. At what light intensity will the lights turn off in the morning?

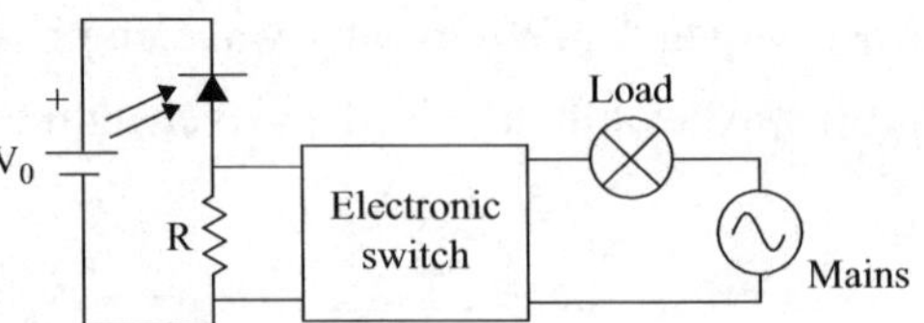

FIGURE 4.33 ■ A light-activated switch.

c. Repeat (a) and (b) if it is known that in the evening the average light wavelength tends to be more red, with an average wavelength of 580 nm, and in the morning it tends to be more blue, with an average wavelength of 520 nm.

Photovoltaic diodes

4.16 **Solar cells as actuators: power generation.** The use of solar cells for power generation is common in small-scale installations and for use in stand-alone equipment such as remote sensors and monitoring stations. To get some idea of what is involved, consider a solar cell panel with an overall power conversion efficiency of 30% (i.e., 30% of the power available at the surface of the solar cell is converted into electrical power). The panel is 80 cm × 100 cm in area and the maximum solar power density at the location is 1200 W/m^2. The panel is divided into 40 equal-size cells and the cells are connected in series. Assume that the cells are equally responsive over the entire visible spectrum (400–700 nm), have a quantum efficiency of 50%, and the internal resistance of the 40 cells in series is 10 Ω. The leakage current of each cell is 50 nA. Calculate the maximum power the solar cell can deliver and indicate what the conditions must be for that to happen.

4.17 **Overall efficiency of solar cells.** A solar panel supplies a current of 0.8 A into a 10 Ω load when exposed to the sun under optimal conditions (i.e., the sun's radiation is vertical, power density is maximal) at a location where the sun's radiation intensity is 1400 W/m^2. The panel is made of cells, each 10 cm × 10 cm, and the cells are connected in series. Use the average wavelength in the visible range of 550 nm as the wavelength of radiation.

a. Calculate the overall conversion efficiency of the solar cells under the stated conditions.

b. What is the maximum power the cell can deliver into a 10 Ω load if its overall efficiency can be increased to 30%? Assume the internal resistance is 10 Ω.

Phototransistors

4.18 **Phototransistor as a detector.** Consider again **Example 4.4**, but now the photodiode is replaced with a phototransistor biased as shown in **Figure 4.34**. The

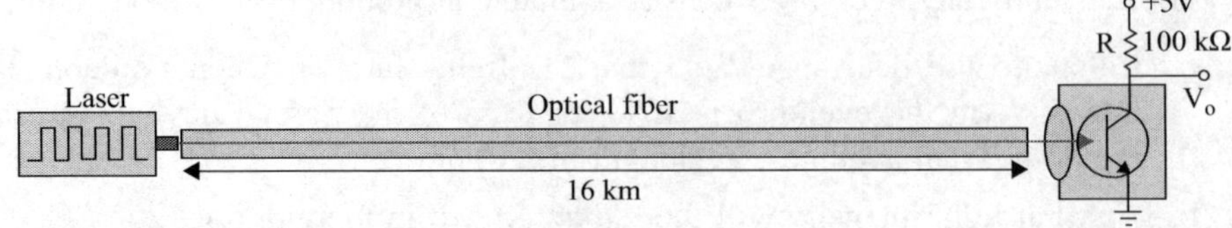

FIGURE 4.34 ■ Phototransistor as a detector in an optical link.

phototransistor has a gain of 50. For a given input pulse train, show the output in relation to the input and calculate the voltage levels expected. Assume that all power that reaches the phototransistor is absorbed in the base–emitter junction.

4.19 Phototransistor and saturation current. The phototransistor in **Figure 4.35** operates in its linear range. At an input power density of 1 mW/cm^2, the voltage measured between the collector and emitter is $V_{ce} = 10.5$ V. Calculate the span of the phototransistor as a light intensity sensor. That is, calculate the minimum light intensity (for which the current through the transistor is zero) and maximum power density (at which the current through the transistor saturates). The saturation V_{ce} voltage is 0.1 V. Neglect the effect of the dark current.

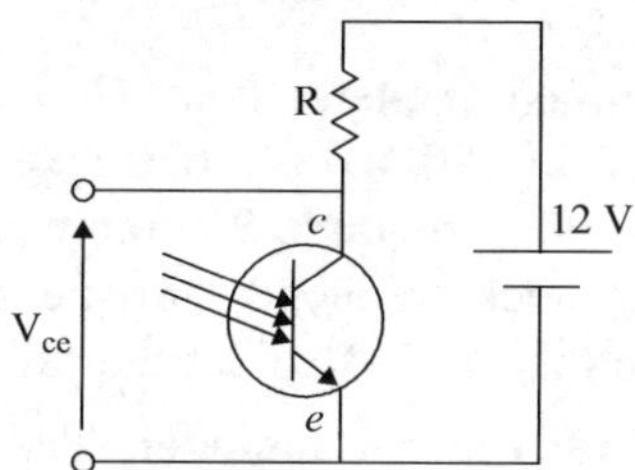

FIGURE 4.35 ■ A phototransistor and its operation.

4.20 Temperature effects on phototransistors. Consider again the configuration in **Figure 4.35**, where $R = 1$ kΩ and the amplification (gain) of the transistor is 100. The collector–emitter voltage at a light intensity of 1 mW/cm^2 is equal to 8 V at 20°C. If the light is removed, the collector–emitter voltage climbs to 11.8 V. Calculate the collector–emitter voltage if the temperature rises to 50°C. Assume that V_{BE} does not change with temperature or with light intensity and the dark current is 10 nA. Discuss the result.

Note: V_{be} decreases with temperature at a rate of 1.0–2.0 mV/°C (see **Section 3.4**), but we will neglect this change here.

Photoelectric sensors and photomultipliers

4.21 Current and electron velocity in a photomultiplier. To get some idea of the processes occurring in a photomultiplier, consider the following simplified configuration. A circular cathode and a circular anode each of radius $a = 20$ mm are separated $d = 40$ mm apart and connected to a potential difference $V = 100$ V as shown in **Figure 4.36**. The cathode is made of a potassium-based compound with a work function $e_0 = 1.6$ eV and a quantum efficiency $\eta = 18\%$.

a. If blue light at a wavelength of 475 nm and intensity of 100 mW/cm^2 impinges on the cathode, calculate the current expected in the device.

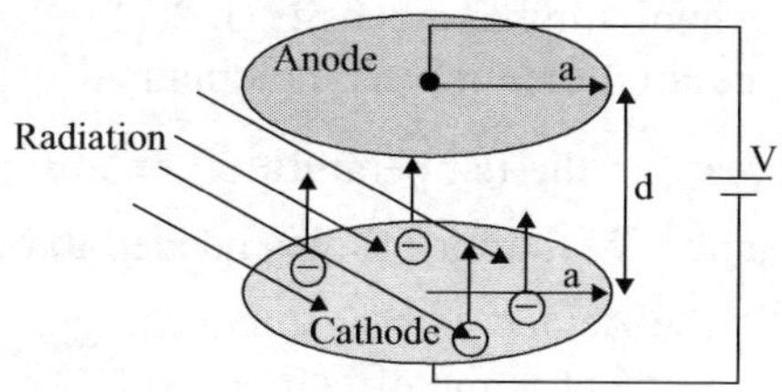

FIGURE 4.36 ■ A basic photomultiplier.

b. Calculate the velocity of the electrons when they reach the anode, given the mass of the electron $m_e = 9.1094 \times 10^{-31}$ kg.

4.22 Limit sensitivity of a photomultiplier in the visible range. Given a photomultiplier, its highest sensitivity occurs at the shorter wavelengths. Assume that the wavelength is 400 nm (violet light). If the cathode has a work function $e_0 =$ 1.2 eV and quantum efficiency of 20%, calculate the lowest illuminance the photomultiplier can discern if detection requires emission of at least 10 electrons/mm^2 of the cathode area per unit time.

CCD sensors and detectors

4.23 Digital video camera image transfer from CCD. A digital color video camera requires a picture format of 680 pixels $\times$ 620 pixels for display on a TV screen. Assuming that smooth video requires 25 frames/s (PAL or SECAM formats), calculate the minimum clock frequency for the stepping process necessary to retrieve the image from the CCD. Neglect the time needed to create the image.

4.24 CCD sensor for HD video image transfer. The CCD sensor in an HD video recorder is arranged as 1080 lines with 1920 pixels/line and produces 60 frames/s. Calculate the minimum clock frequency needed to transfer the image assuming the basic process in **Figure 4.23b** is used (there are other methods and modifications that can be used).

Thermopile PIR sensors

4.25 Thermopile PIR sensor. An IR sensor designed to operate at high temperatures is made as in **Figure 4.25** with 32 pairs of carbon/silicon carbide junctions (see **Chapter 3, Table 3.4**). The carbon/silicon carbide thermocouple has a sensitivity $s = 170$ mV/°C. The absorber is made of thin tungsten foil, 10 μm thick and 2 cm^2 in area, coated black to increase absorption. Tungsten has a density of 19.25 g/cm^3 (same as gold) and a heat capacity $C = 24.27$ J/mol/K. The absorber's conversion efficiency is 80%. The window of the sensor is $A = 5$ cm^2 and it takes the sensor $t = 300$ ms to reach thermal steady state, that is, for the temperature of the absorber to stabilize to a constant value at a given radiation power density. If a temperature difference between the hot and cold junction of 0.5°C can be reliably measured in the sensor, calculate the sensitivity of the sensor.

4.26 Thermopile IR sensor. An IR sensor is required to develop an output of 5 V for an IR radiation of 20 mW/cm^2. For increased sensitivity, aluminum/silicon thermocouples are used because they have an output of 446 mV/°C. An aluminum absorber of area 4 cm^2 and thickness 20 μm is used with an absorption efficiency of 80%. Aluminum has a density of 2.712 g/cm^3 and a heat capacity of 0.897 J/g/°C. The required output must be obtained within 100 ms.

a. What is the increase in the temperature of the absorber?

b. Calculate the number of thermocouples needed to obtain the required output.

c. If the output of the device is measured using a digital voltmeter with a resolution of 0.1 V, what is the effective resolution of the sensor?

Pyroelectric sensors

4.27 **Pyroelectric motion sensor.** A PZT motion sensor is required to detect the motion of a body within its range. The sensor is built as a dual element (see **Figure 4.26**) to reduce the influence of temperature changes that are not due to the motion of a body.

a. Calculate the sensitivity if each element is 0.1 mm thick. One element is exposed to the IR source, the other is shielded from it.

b. Show how temperature compensation for common-source heat (heat sources that affect both element equally) is accomplished.

4.28 **Time constant in motion sensors.** A pyroelectric sensor is made of a small barium titanate oxide ($BaTiO_3$) chip, 10 mm × 10 mm in area and 0.2 mm thick, sandwiched between two metal electrodes. In addition to the properties given in **Table 4.4**, barium titanate oxide has a conductivity of 2.5×10^{-9} S/m.

a. Calculate the time constant of charge decay after heat has been removed.

b. The sensor is used as a motion sensor to automatically trigger a camera for wildlife photography at night. A cat runs by the sensor, generating a temperature differential of 0.1°C, triggering the camera. If at least half the charge generated across the plates must discharge before the sensor can retrigger the camera, how much time is needed before the next event can be detected?

c. To discharge the sensor faster, one can connect a resistor across the sensor. If the sensor must be ready to trigger within 250 ms, what must be the resistance across the sensor? What is the side effect of connecting a smaller resistor across the sensor other than a quicker retrigger time?

Optical actuators

4.29 **Coupling of power in an optical link.** Optical links are very common in data communication (see **Figure 4.10** and **Problem 4.20**). However, linking optical power to optical fibers can be a very low efficiency affair if not done properly. Consider two ways of coupling the power from an LED to an optical fiber, shown in **Figure 4.37**. In **Figure 4.37a**, the optical fiber is simply held in front of the LED, whereas in **Figure 4.37b** an intervening light guide is used.

a. If the LED radiates 20 mW uniformly over a 5° cone and the optical fiber has a diameter of 130 μm, calculate the power coupled to the optical fiber using the method in **Figure 4.37a**. Neglect any reflections that may occur at the

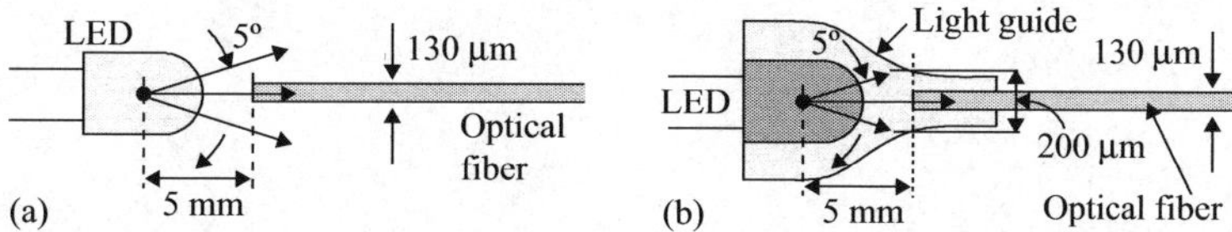

FIGURE 4.37 ■ Coupling light to optical fibers. (a) Direct coupling. (b) Use of a light guide to increase coupled power.

interface between the optical fiber and air. The distance between the LED source and surface of the fiber is 5 mm.

b. How much power is coupled in **Figure 4.37b**? Assume that all power follows the light guide with uniform power density across the light guide cross section and none can escape through its outer surface. The light guide has circular cross section as does the fiber.

4.30 **Magneto-optical recording.** In a hard disk storage device, writing data is done by magneto-optical means. Writing is done by heating the location where the data are written to the Curie temperature of the storage medium and applying the magnetic field representing the data to that spot while it cools below the Curie temperature (see **Figure 4.29**). The data are written on iron oxide (Fe_2O_3) coated on a conducting disk. Assume that 80% of the writing time is needed to heat the spot. The laser beam is 1 μm in diameter and supplies a power of 50 mW. The storage medium is 100 nm thick and has a heat capacity of 23.5 J/mol/K and a density of 5.242 g/cm^3. The curie temperature is 725°C.

a. Calculate the maximum writing data rate of the drive at an ambient temperature of 30°C.

b. Discuss possible ways the data rate can be increased.

c. Discuss effects that will reduce the data rate in practical applications.

CHAPTER 5

Electric and Magnetic Sensors and Actuators

Torpedoes, sharks, eels, pigeons, magnetic bacteria, and the platypus

Electric and magnetic fields are too important and too common to be neglected by nature in its grand design. Many animals and organisms have found ways to take advantage of these fundamental forces for sensing and actuation. The electric field in particular is used for both sensing and actuation. Almost all rays and sharks can sense electric fields produced by prey, as can some catfish, eels, and the platypus. Electric fields are sensed through use of special gelatinous pores that form electroreceptors called ampulae of Lorenzini. Sensing can be passive or active. Sharks and rays use passive sensing; prey is located by sensing weak electric fields produced by the muscles and nerves in the prey. Some animals, such as the electric fish, can generate electric fields for the purpose of active electrolocation of prey. The same basic sensory system is used by young sharks for protection by freezing in place when electrolocation fields are detected. But perhaps the best known example of electrolocation is the platypus, which uses electroreceptors in its bill to hunt by night. Actuation is just as common and is used primarily to stun prey, and also for protection. The torpedo or electric ray (genus *Torpedinidae*) is one of some 70 species of rays that can produce voltages and apply them in a manner similar to a battery. The voltage is produced in a pair of electric organs made of plates connected to a nervous system that controls them. In rays, these biological batteries are connected in parallel to produce low-voltage, high-current sources. The range is between 8 V to more than 200 V with currents that can reach a few amperes. Another example is the electric eel (*Electrophorus electricus*). Since it lives in freshwater, which is less conductive than seawater, it has its plates in series to produce higher voltages (up to 600 V at perhaps 1 A, in short pulses).

Magnetic fields are equally important in sensing. It is now well established that many birds can sense the terrestrial magnetic field and use this ability for navigation. Pigeons are known to have a biocompass made of magnetite particles in the upper tissue of their beaks and use that for magnetolocation. Traces of magnetite can be found even in the human brain, suggesting that perhaps we also had this ability in the distant past. Even bacteria have found a use for magnetite—to help them move along the lines of the terrestrial magnetic field. The bacterium *Magnetospirilllum magneticum* uses magnetite particles to orient itself along lines of the magnetic field to allow them to reach environments rich in oxygen.

5.1 INTRODUCTION

The class of electric and magnetic sensors and actuators is the broadest by far of all other classes, both in numbers and types and the variety within each type. Perhaps this should come as no surprise since in a majority of cases a sensor exploits the electrical properties of materials and, with few exceptions, the requisite output is electrical. In fact, we could argue that even sensors that were not placed in this category belong here as well. Thermocouples exploit electrical effects in conductors and semiconductors—an electrical phenomenon. Optical sensors are either based on wave propagation, which is an electromagnetic phenomenon, or on quanta, which are measured through electronic interaction with the atomic structure of the sensor. It would take little to argue that this is an electric phenomenon. In terms of actuation, most actuators are either electrical or, more commonly, magnetic. This is particularly true of actuators that need to provide considerable power. However, for the sake of simplicity and to follow the basic idea of limiting the number of principles involved in each class of sensors, we will limit ourselves here to the following types of sensors and actuators:

1. Sensors and actuators based on electric and electrostatic principles, including capacitive sensors (proximity, distance, level, material properties, humidity, and other quantities such as force, acceleration, and pressure) and related electric field sensors and actuators. This class of sensors includes microelectromechanical (MEMs) devices, but these will be discussed separately in **Chapter 10**.
2. Sensors based on direct measurement of resistance. There are many sensors that belong in this category, including both AC and DC sensing of current and voltage, position and level sensing, and many others.
3. Magnetic sensors and actuators based on the static and low-frequency time-dependent magnetic field. The variety here is large. It includes motors and valves for actuation, magnetic field sensors (hall element sensors, inductive sensors for position, displacement, proximity, etc.), and a variety of others, including magnetostrictive and magnetoresistive sensors and actuators.

A fourth group of electromagnetic sensors, based on radiation effects of the electromagnetic field will be discussed in **Chapter 9**.

One can classify the sensors and actuators discussed here as electric, magnetic, and resistive. Often we will simply call them electromagnetic devices, a term that encompasses all of them (as well as those discussed in **Chapter 9**).

All electromagnetic sensors and actuators are based on electromagnetic fields and their interaction with physical media. It turns out that the electric and magnetic fields in media are influenced by or influence a large number of properties. For this reason, electromagnetic sensors for almost any imaginable quantity or effect either exist or can be designed.

Before continuing it is perhaps useful to bear in mind the following definitions:

An electric field is the force per unit charge that exists in the presence of charges or charged bodies. An electric field may be static when charges do not move or move at constant velocity or may be time dependent if charges accelerate and/or decelerate.

Moving charges in conducting media or in space cause currents and currents produce magnetic fields. Magnetic fields are either static, when currents are constant (DC), or time dependent, when currents vary in time.

When currents vary in time, both an electric and a related magnetic field are established. This is called an electromagnetic field. Strictly speaking, an electromagnetic field implies that both an electric and a magnetic field exist. However, it is not entirely wrong to call all electric and magnetic fields by that name since, for example, an electrostatic field may be viewed as a time-independent electromagnetic field with zero magnetic field. Although the properties of the various fields are different, they are all described by Maxwell's equations.

Electromagnetic actuators are based on one of the two basic forces: electric force (best understood as the attraction between opposite polarity charges or repulsion between like polarity charges) and magnetic force. The latter is the attraction of conductors carrying currents in the same direction or repulsion of current carrying conductors with currents in opposite directions. Perhaps the best-known manifestation of this force is between two permanent magnets.

5.2 | UNITS

The basic SI electric unit is the ampere (A), as was discussed in **Section 1.6**. But the subject of electricity and magnetism includes a fairly large number of derived units based on the laws and relations in electromagnetics. The current itself is sometimes specified as a density either as current per unit area (ampere/meter^2 [A/m^2]) or in some cases as a current per unit length (ampere/meter [A/m]). These are current densities. Aside from the ampere as a unit of electric current, probably the most common unit is the volt, defined as energy per unit charge. Charge itself has units of coulombs (C) or ampere-seconds (A·s), but charge is often encountered as a density: coulomb/meter (C/m), coulomb/square meter (C/m^2), or coulomb/cubic meter (C/m^3). The derived unit for voltage is indicated as the volt (V) or as joules/coulomb (J/C), newton-meter/coulomb (N·m/C), or, as was shown in **Section 1.6**, the purely SI form is kg·m^2/A·s^3. The product of voltage and current is power, measured in watts (W), and although power can be described in other units (such as force × velocity, i.e., N·m/s or J/s), in electrical engineering it is rare to use any other unit except the watt or the ampere-volt (A·V). Work and energy are normally measured in joules (J), but in measuring and describing electrical energy it is customary to use the basic unit of watt × time, such as the watt-hour (W·h) or one of its multipliers such as kW·h. Energy density (per unit volume) is used to signify energy storage capacity in joules/cubic meter (J/m^3).

The ratio between voltage and current (V/A) is **resistance** (Ohm's law), so the unit of resistance is the ohm (Ω). The **conductivity** of a medium is one of the three fundamental electric properties of any medium, the other two being the **electric permittivity** and **magnetic permeability**. Conductivity is easier to understand by first defining its reciprocal, called **resistivity**, which is the measure of resistance of the material, given in ohm-meters (Ω·m). The conductivity, which is a measure of how well the medium conducts current, has units of 1/(Ω·m) and is the reciprocal of resistivity. The unit 1/Ω is known as the siemens (S). Hence the unit of conductivity is the siemens/meter (S/m). The ratio between charge and voltage is called **capacitance** and is indicated as the farad (F) (see **Example 1.4**). In addition, one encounters the **electric field intensity** in units of volts/meter (V/m) or newtons/coulomb (N/C). Permittivity is measured in farads/meter (F/m), a unit that can be easily derived from Coulomb's law (to be discussed later in this chapter). One can also define an **electric flux density** as the product of the electric field intensity and permittivity with units of coulombs/square meter (C/m^2) and, by integration of the **electric flux** density over an area, an electric flux with units of coulombs (C).

The magnetic field is typically given in terms of the **magnetic flux density** (sometimes called induction) or the **magnetic field intensity**. The derived unit for the magnetic field intensity is amperes/meter (A/m), derived from Ampere's law. The flux density is measured in units of tesla (T). The tesla is in fact force per unit length per unit current (N/A/m or kg/A·s^2). A commonly used nonmetric unit of magnetic flux density is the gauss (g), (1 T = 10,000 g). Integration of the magnetic flux density over an area produces a **magnetic flux** with units of tesla square meter (T·m^2), designated as the weber (Wb). Therefore the magnetic flux density can also be indicated as weber/square meter (Wb/m^2), explicitly showing the fact that it is a density. The ratio of magnetic flux and current is **inductance**, whose unit is webers/ampere (Wb/A) or tesla square meter/ampere (T·m^2/A), also known as the henry (H).

The ratio between magnetic flux density and magnetic field intensity is the magnetic permeability of the medium. Its units are clearly T·m/A. However, the quantity T·m^2/A is designated as the henry and hence permeability has units of henrys/meter (H/m). Another magnetic quantity of limited use is the **magnetic reluctance** or **reluctivity**, which may be best understood as a kind of "magnetic conductivity" with units of 1/henry (1/H). In addition to the above, one should also recall that the frequency of a signal is indicated in cycles/second (cycles/s) or hertz (Hz). Related to frequency is the angular frequency (also called angular velocity) in radians/second (rad/s).

There are other quantities that are sometimes used, such as the phase of a signal (degrees or radians), attenuation and phase constants, power density, and more, but these are best defined in the context of their use (some of these quantities will be introduced in **Chapter 9**). Some of the quantities described above (current density, electric and magnetic field intensities, and electric and magnetic flux densities) are vectors that have both a magnitude and direction, the rest are scalars and only posses a magnitude. Power can be represented as a scalar or a vector, but we will view it here strictly as a scalar. **Table 5.1** summarizes the quantities and units of many of the electric and magnetic quantities.

5.3 | THE ELECTRIC FIELD: CAPACITIVE SENSORS AND ACTUATORS

Electric field sensors and actuators are those that operate on the physical principles defining the electric field and its effects. The primary type of device is capacitive. The discussion here is in terms of capacitances because this affords a simple circuit approach, but it can equally well be done in terms of the electric field intensity. There are some sensors, such as charge sensors, that are better explained in terms of the electric field, but on the whole, discussion of capacitance and its use in sensing and actuation covers most aspects necessary for a thorough understanding of the principles involved without the need to study the intricacies of the electric field behavior.

All capacitive sensors are based on the change in capacitance due to the stimulus either directly or indirectly. First, it should be noted that capacitance, by definition, is the ratio between charge and voltage on a device:

$$C = \frac{Q}{V} \quad [\mathrm{C/V}]. \tag{5.1}$$

Capacitance is measured in coulombs/volt (C/V). This unit is called the farad (F). Because voltage is only properly defined as the difference in potentials between two

TABLE 5.1 ■ Electric and magnetic quantities and their units

	Unit	Notes	Symbol
Current	ampere (A)	(SI unit)	A
Current density	ampere/meter^2 (A/m^2), ampere/meter (A/m) (see text)	vector	J
Voltage	volt (V)	also emf	V
Charge	coulomb (C), (A·s)		Q, q
Charge density	coulombs/meter (C/m), coulomb/square meter (C/m^2), coulomb/cubic meter (C/m^3)		ρ_l, ρ_s, ρ_v
Permittivity	farad/meter (F/m)		ε
Electric field intensity	volt/meter (V/m) or newton/coulomb (N/C)	vector	E
Electric flux density	coulomb/square meter (C/m^2)	vector	D
Electric flux	coulomb (C)		Φ or Φ_e
Power	watt (W), (A·V)		P
Energy	watt-hour (W·h), joule (J)		W
Energy density	joule/cubic meter [(J/m^3)		w
Resistance	ohm (Ω)		R
Resistivity	ohm-meter (Ω·m)		ρ
Conductivity	siemens/meter (S/m) or 1/ohm-meter (1/Ω·m)	$\sigma = 1/\rho$	σ
Capacitance	farad (F)		C
Magnetic flux density	tesla (T), sometimes gauss (g), weber/square meter (Wb/m^2)	vector	B
Magnetic field intensity	ampere/meter (A/m)	vector	H
Magnetic flux	tesla-square meter (T·m^2) or weber (Wb)		Φ or Φ_m
Inductance	weber/ampere (Wb/A) or henry (H)		L
Magnetic permeability	henry/meter (H/m)		μ
Reluctance	1/henry (1/H)		R
Frequency	cycles/second (cycle/s) or hertz (Hz)		f
Angular frequency	radians/second (rad/s)	$\omega = 2\pi f$	ω

points, the capacitance is only defined for two conducting bodies across which the potential difference is connected. This is shown schematically in **Figure 5.1**. Body B is charged by the battery to a positive charge Q and body A to an equal but negative charge $-Q$. Any two conducting bodies, regardless of size and distance between them, have a capacitance. The capacitance of a single conducting body can also be defined in terms of the charge of that body and potential difference with respect to infinity as a special case of a two-body system. When a potential difference is connected across them, the bodies are charged and the relation in **Equation (5.1)** is satisfied. Nevertheless, capacitance is independent of voltage or charge—it only depends on physical dimensions and materials properties. To understand the principles, we start with the

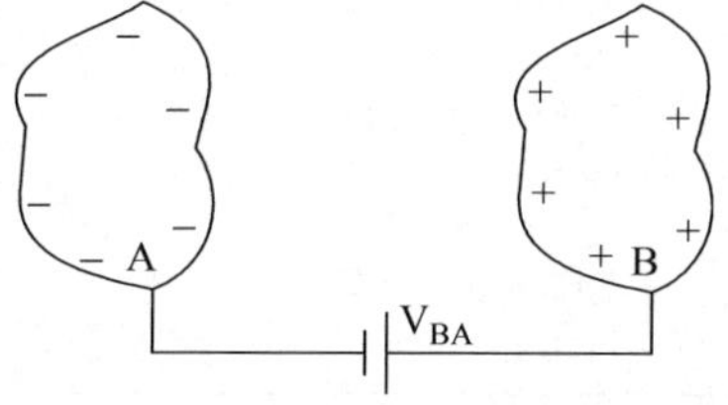

FIGURE 5.1 ■ The concept and definition of capacitance.

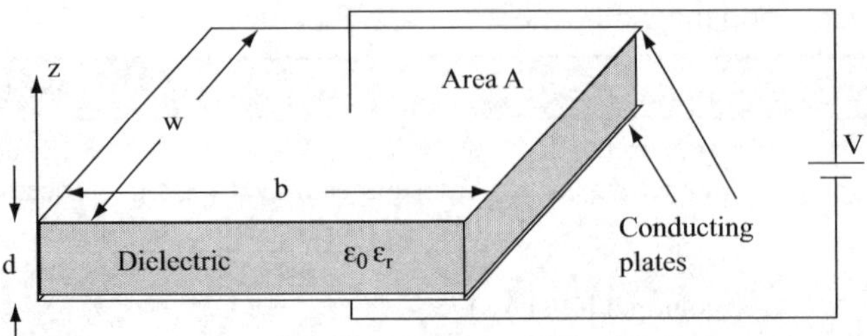

FIGURE 5.2 ■ The parallel plate capacitor connected to a DC source.

capacitance between two parallel plates, shown in **Figure 5.2**. Assuming first that the distance d between the two plates is small, the capacitance of the device is

$$C = \frac{\varepsilon_0 \varepsilon_r S}{d} \quad [\mathrm{F}], \tag{5.2}$$

where ε_0 is the permittivity of a vacuum, ε_r is the relative permittivity (dielectric constant) of the medium between the plates, S is the area of the plates, and d is the distance between the plates. ε_0 is a constant equal to 8.854×10^{-12} F/m, whereas ε_r is the ratio between the permittivity of the medium and that of free space (ε_0) and hence is dimensionless. Permittivity is measurable and available as part of the electrical properties of materials. Although not strictly necessary for the definition, it is usually assumed that the material between the plates of the capacitor is a nonconducting medium—a dielectric. The relative permittivities of some common dielectrics are listed in **Table 5.2**.

Any of the quantities in **Equation (5.2)** affect the capacitance and changes in these can be sensed. This allows for a wide range of stimuli, including displacement and anything that can cause displacement (pressure, force), proximity, permittivity (e.g., in moisture sensors), and a myriad of others. However, **Equation (5.2)** describes a very specific device and was obtained by assuming that the electric field intensity between the two plates does not leak (fringes) outside the space between the plates. This was done to obtain a simple expression. In the more general case, when d is not small, or if the plates are arranged in a different configuration (see **Figure 5.3**), we cannot calculate the capacitance directly, but we can still write the following:

$$C = \alpha[\varepsilon_0, \varepsilon_r, S, 1/d]. \tag{5.3}$$

TABLE 5.2 ■ Relative permittivities of various materials

Material	ε_r	Material	ε_r	Material	ε_r
Quartz	3.8–5	Paper	3.0	Silica	3.8
Gallium arsenide	13	Bakelite	5.0	Quartz	3.8
Nylon	3.1	Glass	6.0 (4–7)	Snow	3.8
Paraffin	3.2	Mica	6.0	Soil (dry)	2.8
Perspex	2.6	Water (distilled)	81	Wood (dry)	1.5–4
Polystyrene foam	1.05	Polyethylene	2.2	Silicon	11.8
Teflon	2.0	Polyvinyl chloride	6.1	Ethyl alcohol	25
Barium strontium titanate	10,000.0	Germanium	16	Amber	2.7
Air	1.0006	Glycerin	50	Plexiglas	3.4
Rubber	3.0	Nylon	3.5	Aluminum oxide	8.8

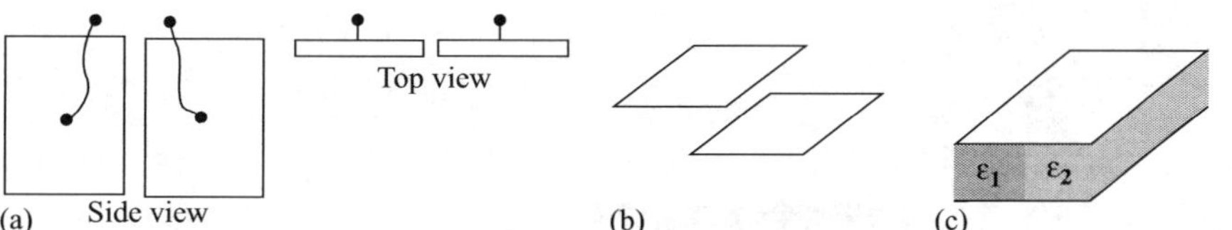

FIGURE 5.3 ■ Various configurations of parallel plate capacitors.

That is, capacitance is proportional to permittivity and the area of the conductors (plates) and inversely proportional to the distance between conductors. The parallel plate capacitor is only one possible device. As long as two conductors are involved, there will be a definable capacitance between them. **Figure 5.4** shows some other useful capacitive arrangements often exploited for sensing. Many capacitive sensors will be encountered in **Chapters 6–10**, but we will discuss here the sensing of position, proximity, displacement, and fluid level, as well as a number of capacitive actuators.

5.3.1 Capacitive Position, Proximity, and Displacement Sensors

Returning to **Equation (5.2)**, position and displacement can be used to change the capacitance of a device in three fundamental ways:

1. By allowing one conductor in a two-conductor capacitor (usually a plate) to move relative to the other. A number of configurations are shown in **Figure 5.5**. In **Figure 5.5a**, the sensor is made of a single plate while the second plate is a conductor relative to which the distance (proximity) is sensed. While this is a valid method, it is not a sensor one can obtain ready-made, but rather one has to build it, and this implies that proximity can only be sensed with respect to a conducting surface. A schematic position sensor of this type is shown in **Figure 5.6**. One plate

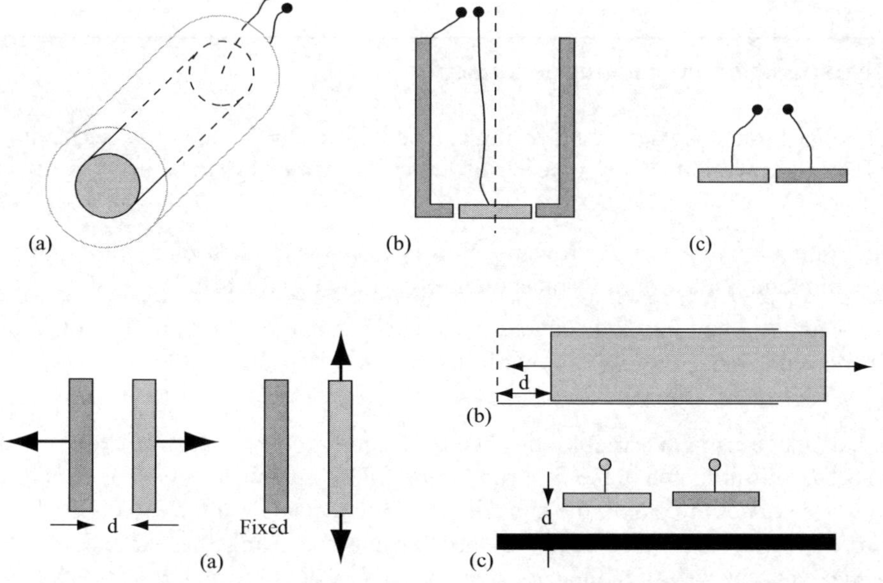

FIGURE 5.4 ■ Various types of capacitive sensor arrangements.

FIGURE 5.5 ■ Arrangements of capacitive sensors for position, proximity, and displacement sensing. (a) One plate is usually fixed and the measurand is the change in distance *d* or surface area *S* of the capacitor. (b) Change in permittivity. (c) Change in distance.

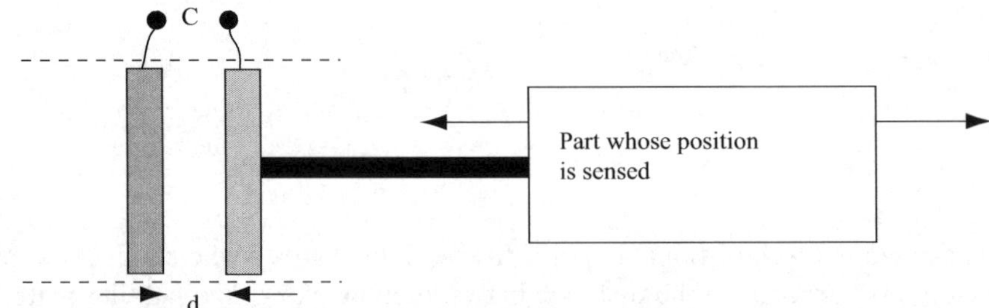

FIGURE 5.6 ■ A schematic capacitive position sensor.

is fixed while the other is pushed by the moving device. The position of the moving device causes a change in position of one of the plates, and this changes the capacitance. The capacitance is inversely proportional to distance, and as long as the distances sensed are small, the output is linear.

2. Alternatively, the plates remain fixed but the dielectric moves in or out, as in **Figure 5.5b**. This is a practical situation for some applications. For example, the dielectric may be connected to a float that then senses the fluid level, or it may be pushed by a device to sense the end of travel or position. The advantage of this device is that it is quite linear, and the range of motion is rather large since it can approach the width of the capacitor.
3. Another configuration is obtained by keeping the plates fixed, as in **Figure 5.5c**, and sensing the distance to a surface. This is a practical arrangement since the sensor is self-contained and requires no external electrical connections or physical arrangements to sense distance or position. However, the relation between capacitance and distance is nonlinear and distance is limited because the electric field does not extend very far.

EXAMPLE 5.1 Small capacitive displacement sensor

A small sensor capable of accurate displacement sensing can be built from two small plates, as in **Figure 5.5a**. The plates can move either toward each other or slide sideways. The sensors discussed here are as follows:

a. The two plates are 4 mm × 4 mm and move toward or away from each other with a minimum displacement of 0.1 mm and a maximum displacement of 1 mm (**Figure 5.7a**).

b. The two plates are 4 mm × 4 mm and are separated a fixed distance of 0.1 mm. They slide sideways with a displacement range of 0–2 mm (**Figure 5.7b**).

Solution:

a. The capacitance is calculated using a variable distance d (0.1 mm $< d <$ 1 mm) in **Equation (5.2)** based on **Figure 5.7a**. This is shown in the first row of the following table. Capacitance is in picofarads (pF). Because the plates are small, the capacitance is also small and the effects of the edges are likely to introduce errors in calculation of the capacitance using the parallel plate capacitor formula. To see what this effect is, the capacitance is also calculated numerically using a method called the method of moments, which allows exact computation of the capacitance and does not require the approximate capacitance formula in **Equation (5.2)**. The second row shows

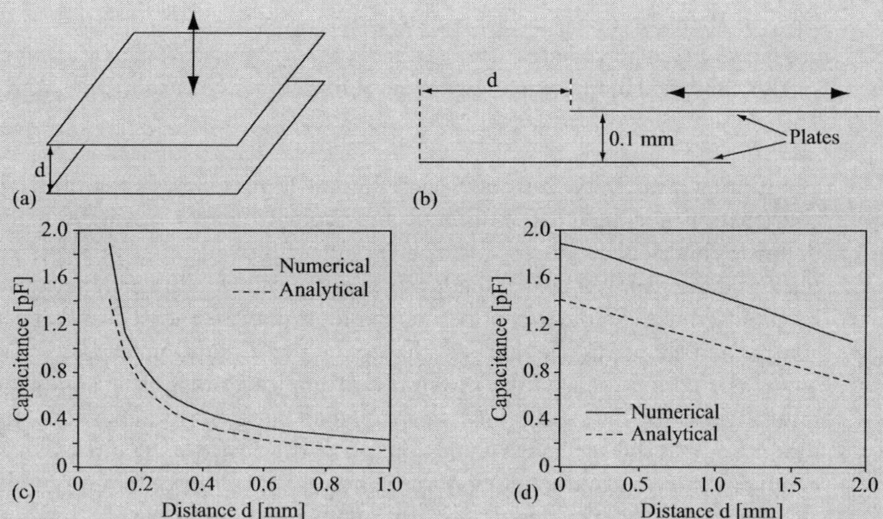

FIGURE 5.7 ■ Capacitive displacement sensors. (a) Lower plate fixed, upper plate moves up and down to indicate position. (b) Lower plate fixed, upper plate moves sideways to indicate position. (c) Capacitance as a function of displacement (analytical and numerical) for the sensor in (a). (d) Capacitance as a function of displacement for the sensor in (b).

the results obtained using the method of moments (the same can be done experimentally using a capacitance meter).

d (mm)	0.1	0.15	0.2	0.25	0.3	0.35	0.4	0.45	0.5	0.6	0.7	0.8	0.9	1.0
C (pF)	1.89	1.15	0.862	0.7	0.595	0.520	0.465	0.422	0.388	0.337	0.3	0.273	0.251	0.234
C (pF)	1.42	0.944	.708	0.567	0.472	0.405	0.354	0.315	0.283	0.236	0.202	0.177	0.157	0.142

b. The capacitance is calculated using a variable horizontal offset of the upper plate x, (0.0 mm $< x <$ 2 mm) in **Equation (5.2)**. This is shown in the first row of the following table. Capacitance is in picofarads. The second row shows the results obtained using the method of moments.

x (mm)	0.0	0.2	0.4	0.6	0.8	1.0	1.2	1.4	1.6	1.8	2.0
C (pF)	1.89	1.83	1.75	1.67	1.58	1.49	1.41	1.32	1.23	1.15	1.06
C (pF)	1.42	1.35	1.27	1.2	1.13	1.06	0.992	0.921	0.850	0.779	0.708

Clearly, the analytical results using **Equation (5.2)** and the numerical (or experimental) results seem to be very different, as can be expected from the small plates (the larger the plates and the smaller the distance between them, the closer the analytical and numerical results become). However, a plot of the two results reveals additional information. The plots of the data in the two tables are shown in **Figures 5.7c** and **5.7d**. The first thing to note is that the two sets of results in each figure are essentially the same in shape, but shifted with respect to each other. The second piece of information is that the lateral moving sensor is more linear but less sensitive (the change in capacitance is smaller). Both can be used successfully once properly calibrated, but it should be noted again that for the sensors shown here, the numerical (or experimental) data are appropriate to use whereas the analytical calculation gives the general behavior but not useful, accurate data.

In most proximity sensors, the method in **Figure 5.5c** is the most practical, but the construction is somewhat different. One typical type of sensor is made as follows: a hollow cylindrical conductor forms one plate of the sensor, as in **Figure 5.8**. The second plate of the sensor is a disk at the lower opening of the cylinder. The whole structure may be enclosed with an outer conducting shield or may be encased in a dielectric enclosure. The capacitance of the device is C_0 based on dimensions, materials, and structure. When any material is present in the proximity of the lower disk, it changes the effective permittivity seen by the sensor and its capacitance increases to indicate the distance between the sensor and the surface. The advantage of this sensor is that it can sense distances to conducting or nonconducting bodies of any shape, but the output is not linear. Rather, the smaller the sensed distance d, the larger the sensitivity of the sensor. The dimensions of the sensor have a large influence on its span and sensitivity. Large-diameter sensors will have a larger span but relatively low sensitivity, whereas smaller diameter sensors will have a smaller span with greater sensitivity. **Figure 5.9** shows some capacitive proximity sensors (of different physical sizes and sensing distances) that may be used to sense conducting surfaces and/or switch on at a preset distance.

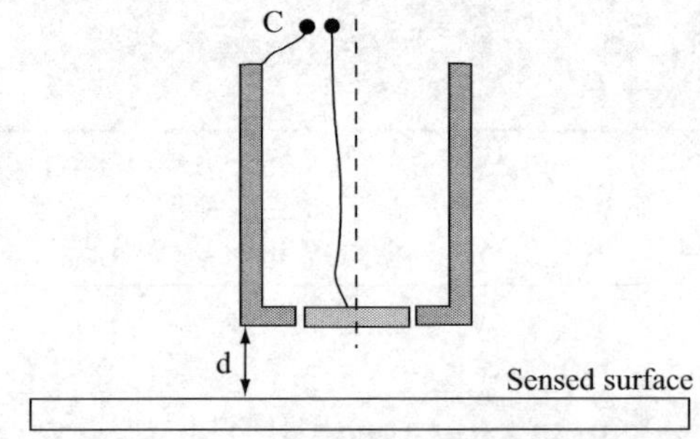

FIGURE 5.8 ■ A practical proximity sensor arrangement.

FIGURE 5.9 ■ Capacitive proximity sensors.

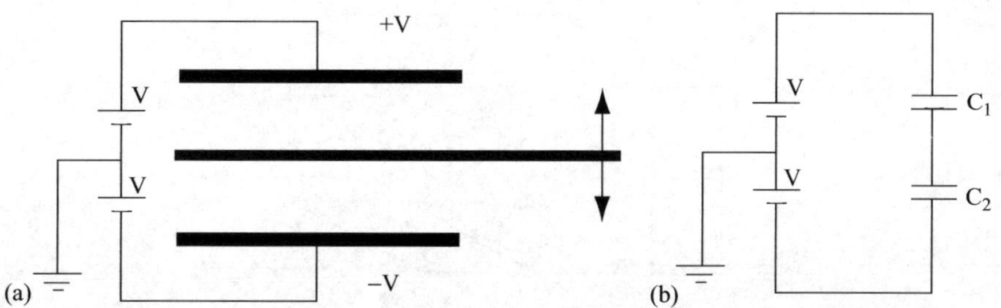

FIGURE 5.10 ■ Position sensor arrangement with improved linearity. (a) Sensor. (b) Equivalent circuit.

Capacitive position and proximity sensors may be made in other ways. One example is the sensor shown in **Figure 5.10**. The sensor has two fixed plates and one moving plate. When the plate is midway, its potential is zero with respect to ground since $C_1 = C_2$. As the plate moves up, its potential becomes positive (C_1 increases, C_2 decrease). When it moves down it is negative (C_2 increases, C_1 decreases). A sensor of this type tends to be more linear than the previous sensors, but the distance between the fixed plates must be small and consequently the motion must also be small or the capacitances will be very small and difficult to measure. Other capacitive sensors sense rotary motion by rotating one plate with respect to the other. Still others are cylindrical or made in any convenient shape such as a comb shape (see **Figure 5.11**).

5.3.2 Capacitive Fluid Level Sensors

Fluid level may be sensed by any of the position or proximity sensors discussed in the previous paragraph by sensing the position of the fluid surface either directly or through a float that then can change the capacitance of a linear or rotary capacitor. One of the simplest, direct methods for fluid level sensing is to allow the fluid to fill the space between the two conducting surfaces that make up the capacitor. For example, the capacitance of the parallel plate capacitor is linearly proportional to the permittivity between the two plates. Therefore the larger the amount of water between the plates, the larger the capacitance and therefore the capacitance is a measure of fluid level between the plates. **Figure 5.12** shows a parallel plate capacitor used as a fluid level sensor. The part of the plates under the surface of the water has a capacitance C_f:

$$C_f = \frac{\varepsilon_f h w}{d} \quad [\mathrm{F}], \tag{5.4}$$

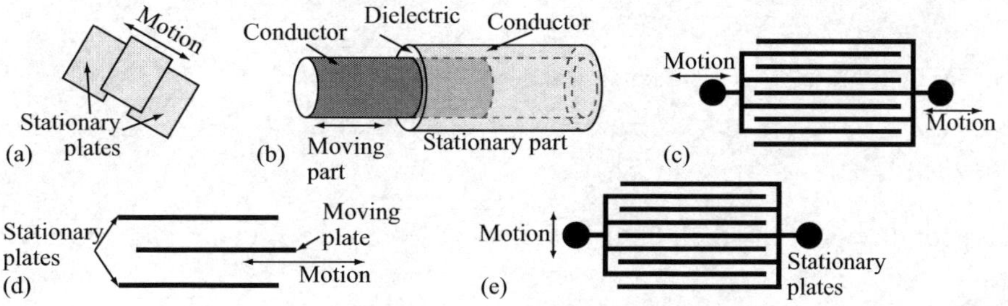

FIGURE 5.11 ■ Various arrangements for linear displacement capacitive sensors.

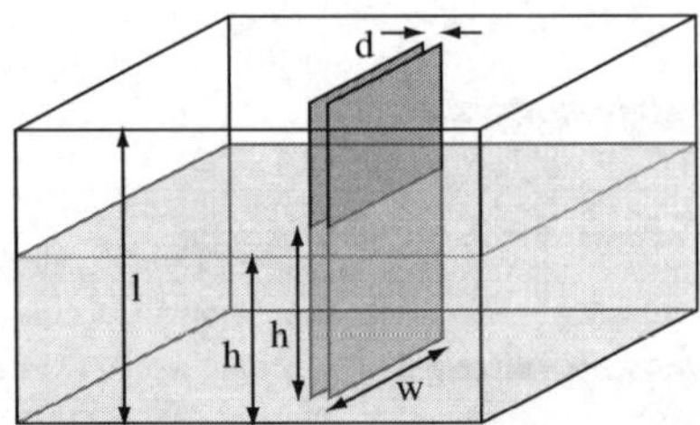

FIGURE 5.12 ■ The principle of a capacitive fluid level sensor. The fluid must be nonconducting.

where ε_f is the permittivity of the fluid, h is the height of the fluid, w is the width of the plates, and d is the distance between them. The part of the capacitor above the fluid has a capacitance C_0:

$$C_0 = \frac{\varepsilon_0(l-h)w}{d} \quad [\text{F}], \tag{5.5}$$

where l is the total height of the capacitor. The total capacitance of the sensor is the sum of the two:

$$C = C_f + C_0 = \frac{\varepsilon_f hw}{d} + \frac{\varepsilon_0(l-h)w}{d} = h\left[\frac{(\varepsilon_f - \varepsilon_0)w}{d}\right] - \frac{\varepsilon_0 lw}{d} \quad [\text{F}]. \tag{5.6}$$

Clearly this relation is linear and varies from the minimum capacitance $C_{\min} = \varepsilon_0 lw/d$ (for $h = 0$) to a maximum capacitance $C_{\max} = \varepsilon_f lw/d$ (for $h = l$). The sensitivity of the sensor may be calculated as dC/dh and is clearly linear.

Although a parallel plate capacitor can be used for this purpose, the expressions above are approximate (they neglect the effect of the edges by assuming the field is not affected by the finite size of the plates). In reality there will be a slight nonlinearity due to these effects and this nonlinearity depends on the distance between the plates. Also, it should be noted that the method is only practical with nonconducting fluids (oils, fuels, freshwater). For slightly conducting fluids, the plates must be coated with an insulating medium.

A more common implementation of this simple, rugged sensor is shown in **Example 5.2**.

EXAMPLE 5.2 Capacitive fuel gauge

A fuel tank gauge is made as shown in **Figure 5.13a**. A long capacitor is made from two coaxial tubes immersed in the fuel so that the fuel fills the space between them up to the fluid level. The empty cylinders (empty fuel tank) establish a capacitance C_0. The capacitance of a coaxial capacitor of length d, inner radius a, and outer radius b is given by

$$C_0 = \frac{2\pi\varepsilon_0 d}{\ln(b/a)} \quad [\text{F}].$$

If the fluid fills the capacitor to a height h, the capacitance of the device is

$$C_f = \frac{2\pi\varepsilon_0}{\ln(b/a)}(h\varepsilon_r + d - h) = \frac{2\pi\varepsilon_0}{\ln(b/a)}(\varepsilon_r - 1)h + \frac{2\pi\varepsilon_0}{\ln(b/a)}d \quad [\text{F}],$$

where ε_r is the relative permittivity of the fuel. Clearly then the capacitance is linear with respect to h from $h = 0$ to $h = d$.

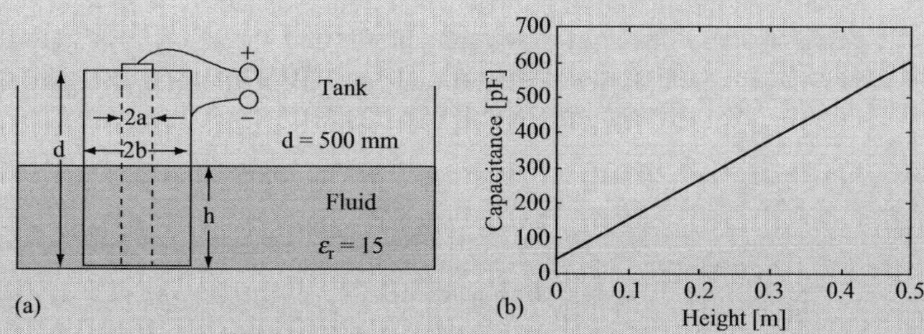

FIGURE 5.13 ■ A fluid level sensor (or fuel gauge) with improved transfer function. (a) The sensor. (b) The transfer function for the values given. For the best performance, the distance between the inner and outer conductor should be small to reduce edge effects and hence possible nonlinearities at very low and very high levels due to fringing at the edges.

The sensitivity of the sensor is

$$s = \frac{dC_f}{dh} = \frac{2\pi\varepsilon_0}{\ln(b/a)}(\varepsilon_r - 1) \quad [\text{F/m}].$$

The sensitivity is governed by the permittivity of the fluid and the dimensions of the two tubes.

Figure 5.13b shows the calculated transfer function of the fuel gauge. In practice the transfer function is slightly nonlinear for very low levels and very high levels of the fluid because of the capacitor edge effects.

Capacitive fuel gauges of this type are often used in diesel fuel tanks on ships and in aircraft fuel tanks. The idea can be used for any fluid that is nonconductive, such as oil or even water.

Capacitive sensors are some of the simplest, most rugged sensors that can be made and are useful in many applications beyond those described here (many more will be encountered in the following chapters). However, with few exceptions the capacitances are small and changes in capacitance even smaller. Therefore they require special methods of transduction. Often, instead of measuring DC voltages, the sensor is made as part of an oscillator whose frequency depends on capacitance and the frequency is measured, usually digitally. Others use an AC source and rather than sensing capacitance, the impedance or phase of the circuit is sensed. We shall discuss some of these issues in **Chapter 11**.

5.3.3 Capacitive Actuators

Capacitive actuation is exceedingly simple and may be understood from **Figure 5.1**. When a potential is connected across the two conductors, they acquire opposite sign charges. These charges attract each other based on Coulomb's law and this force tends to pull the two conductors closer together. Coulomb's law defines the force between the charge and the electric field intensity. Given a charge Q [C] in an electric field intensity $\mathbf{E}$ [V/m], a force $\mathbf{F}$ is exerted on the charge by the electric field as

$$\mathbf{F} = Q\mathbf{E} \quad [\text{N}]. \tag{5.7}$$

The boldface notation indicates that both the electric field intensity and the force are vectors, that is, that they have magnitude and direction in space. In a parallel plate capacitor, the magnitude of the total electric field intensity (i.e., the sum of the electric field produced by the upper and lower charged plates) is given as

$$E = \frac{V}{d} \quad [\text{V/m}], \tag{5.8}$$

where the direction is perpendicular to the plates and the electric field intensity points from the positively charged plate to the negatively charged plate (tending to close the gap between them; that is, they are attracted to each other since the charges on the two plates have opposite signs). Thus mechanical motion of the conductors is possible. In the particular case of the parallel plate capacitor in **Figure 5.2**, the force is found by substituting Q in **Equation (5.7)** with **Equation (5.1)** and E from **Equation (5.8)**, except that the electric field intensity in **Equation (5.8)** must be divided by 2, that is, the force produced by the lower plate on the upper plate is the electric field intensity of the lower plate at the location of the upper plate multiplied by the charge of the upper plate. The capacitance of the parallel plate capacitor is given in **Equation (5.2)**. Putting all these together, the force is

$$F = \frac{CV^2}{2d} = \frac{\varepsilon_0 \varepsilon_r S V^2}{2d^2} \quad [\text{N}]. \tag{5.9}$$

As before, if we cannot assume parallel plates with a small distance d between them, the equation will not be exact, but we can still expect the general relationships to hold—that is, force will be proportional to S, ε, and V^2 and inversely proportional to d^2.

If there is a force, we can also define an energy based on the fact that force is the rate of change of energy over distance:

$$F = \frac{dW}{dl} \quad [\text{N}]. \tag{5.10}$$

Therefore energy is

$$W = \int_0^d \mathbf{F}.d\mathbf{l} = \frac{CV^2}{2} = \frac{\varepsilon_0 \varepsilon_r S V^2}{2d} \quad [\text{J}]. \tag{5.11}$$

Note that this can also be written as

$$W = \frac{\varepsilon (Sd)}{2} \left(\frac{V^2}{d^2} \right) = \frac{\varepsilon E^2}{2} v \quad [\text{J}], \tag{5.12}$$

where v is the volume of the space between the capacitor's plates, ε is the permittivity of the medium, and $E = V/d$ is the electric field intensity between the plates. The quantity $\varepsilon E^2/2$ has units of [J/m^3] and is therefore the energy density in the capacitor.

Given a fixed conductor, a second conductor, if connected to a potential difference with respect to the first, will move with respect to the fixed conductor. This is shown schematically in **Figure 5.14**. This motion may be used for positioning or, as is the case here, as an electrostatic loudspeaker.

However, inspection of **Equation (5.9)** shows that the forces that can be achieved are rather small since ε_0 is very small. To make this device perform useful work, one of two things must be accomplished: either make the distance very small (very limited

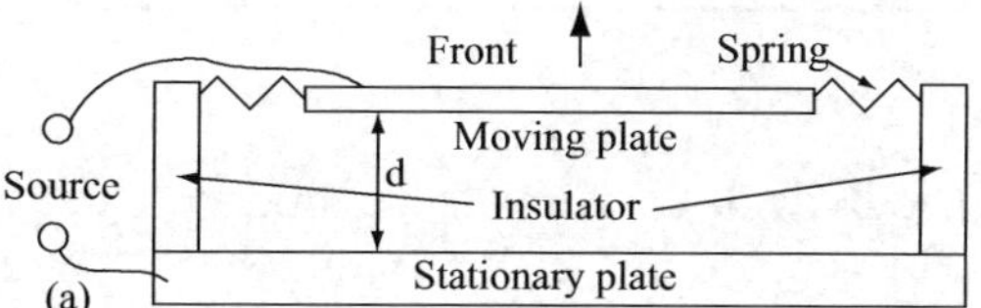

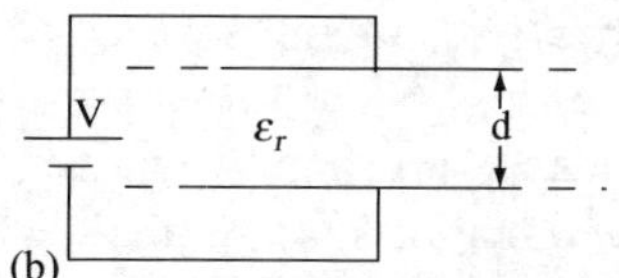

FIGURE 5.14 ■ (a) Schematic structure of a capacitive actuator and (b) its equivalent circuit.

motion) or increase the voltage to high values. In microelectromechanical (MEM) actuators, the distances are naturally small, but the voltage is also small (otherwise electric breakdown may occur). In electrostatic loudspeakers and headphones the displacement must be large so a high voltage (up to a few thousand volts) must be used.

But the obvious vertical attraction between two plates of a parallel plate capacitor is not the only mechanism for electrostatic actuation. Consider the asymmetric configuration of the two plates of a parallel plate capacitor in **Figure 5.15a**. The two plates attract each other, but now the force has a vertical and a horizontal component. The plates not only will attract and try to close the distance between them, but also to close the horizontal separation. This basic method may also be used to affect rotational motion using the device in **Figure 5.15b**. Here the distance between the plates is fixed, but the plates can rotate with respect to each other and will tend to do so until they are positioned exactly overlapping each other. With the addition of a spring to restore the initial position, this can be made into a very accurate positioner (see **Example 5.4**). The forces involved can be calculated from the virtual displacement principle, as follows: Using again the plates in **Figure 5.15a**, suppose we allow the upper plate to move a "virtual" distance dx to the left. The volume between the plates changes by a quantity $dv = wddx$, where w is the width of the plate and d is the distance between them. The change in energy due to this motion is dW and necessarily this must be equal to Fdx. With the energy density defined as $\varepsilon E^2/2$, we write

$$Fdx = dW = \frac{\varepsilon E^2}{2} wddx \tag{5.13}$$

and the magnitude of the lateral force is

$$F = \frac{\varepsilon E^2}{2} wd \quad [\text{N}]. \tag{5.14}$$

The direction of this force is such as to force the plates to the center with respect to each other since that is a situation of minimum (zero) lateral force.

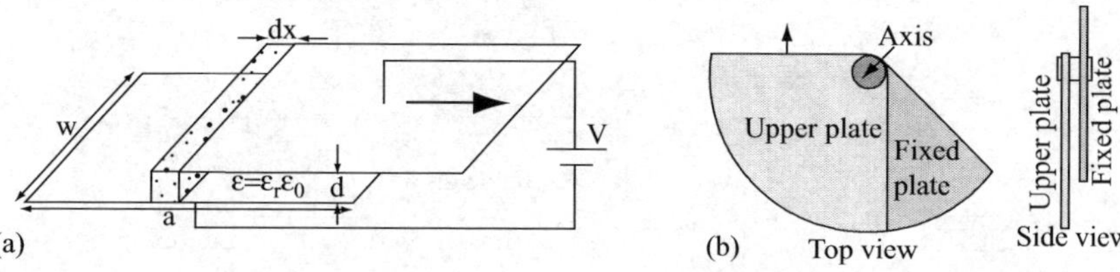

FIGURE 5.15 ■ Capacitive actuators. (a) Linear actuator. The lower plate, separated a distance d from the upper plate, is fixed. The upper plate can move relative to the lower plate. (b) Rotary actuator. The upper plate rotates under the influence of charges on the plates.

EXAMPLE 5.3 Electrostatic air pump

A small air pump is made as shown in **Figure 5.16a**. The body of the pump is a cylindrical cavity of radius $a = 50$ mm and height $h = 1$ mm. The top part is made of a disk of radius $b = 40$ mm suspended on a rubber diaphragm to allow the plate to move up and down. An insulating stop ring, 0.1 mm thick, prevents the moving disk from shorting to the body during compression. Two flap valves are used to affect pumping. The lower valve prevents air from leaking during the compression stroke, whereas the side valve closes during the suction stroke so air enters through the lower flap. Assume the rubber diaphragm is sufficiently flexible to offer no resistance so that the moving plate can move down until it encounters the plastic ring on the bottom plate. Assuming as well that the moving and stationary plates form a parallel plate capacitor and that the mass of the moving plate can be neglected, calculate

a. The volume of air the pump can move per minute if the source is sinusoidal at 60 Hz.
b. The maximum pressure at the outlet port if the peak voltage is 200 V.

Solution: The moving and stationary plates attract each other during each half cycle of the AC signal since during the positive half cycle the moving plate is positive with respect to the stationary plate, whereas during the negative half cycle the moving plate is negative with respect to the stationary body. Therefore the pump will effectuate 120 strokes/s. In each stroke the downward motion of the plate expels a volume of air equal to the change in volume of the chamber (**Figure 5.16b**). The maximum pressure occurs throughout the chamber when the plate rests against the plastic ring and is equal to the attraction force divided by the area of the moving plate.

a. The volume expelled per stroke is the volume of the chamber in **Figure 5.16b**. We write for this volume

$$\Delta v = \pi a^2 h - \pi a^2 c - \pi(a^2 - b^2)\frac{h-c}{2}$$

$$= \pi \times 50^2 \times 1 - \pi \times 50^2 \times 0.1 - \pi(50^2 - 40^2)\frac{1-0.1}{2}$$

$$= \pi(2500 - 250 - 405) = 5796.24 \text{ mm}^3$$

Since there are 120 strokes/s, or 120×60 strokes/min, the volume of air expelled per minute is

$$v = 120 \times 60\Delta v = 120 \times 60 \times 5796.24 = 4.17 \times 10^7 \text{ mm}^3.$$

The maximum volume that the pump can move is 0.0417 m^3/min or 41.7 L/min.

b. The force between the moving plate and the body of the pump is given in **Equation (5.9)**:

$$F = \frac{\varepsilon_0 S V^2}{2d^2} \quad [\text{N}]$$

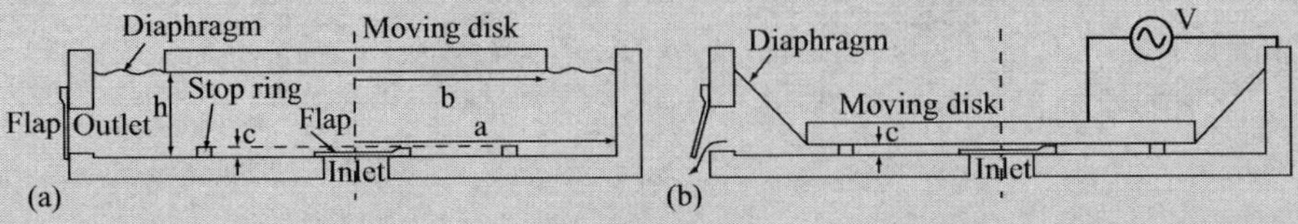

FIGURE 5.16 ■ (a) Principle of an electrostatic pump before application of the potential. (b) The electrostatic pump at the end of the compression stroke.

where ε_0 is the permittivity of air, S is the area of the moving plate, and d is the distance between the moving plate and the bottom surface. This varies from a minimum when $d = h = 1$ mm and a maximum when $d = c = 0.1$ mm. The maximum force is therefore

$$F = \frac{\varepsilon_0 S V^2}{2d^2} = \frac{8.854 \times 10^{-12} \times \pi \times 0.04^2 \times 200^2}{2 \times 0.0001^2} = 0.089 \text{ N}.$$

The pressure is

$$P = \frac{F}{S} = \frac{0.089}{\pi \times 0.04^2} = 17.7 \text{ N/m}^2.$$

This is a mere 17.7 Pa, a very low pressure indicating that the forces involved are very small, something that is typical in electrostatic devices. In fact, because of friction and tension in the diaphragm and flaps, this pressure may be even lower. Nevertheless, very small electrostatic forces are useful in microdevices, as we shall see in **Chapter 10**, where the dimensions and distances are much smaller than in this example.

EXAMPLE 5.4 Rotary capacitive actuator

Consider the rotary capacitive actuator in **Figure 5.15b**. The actuator is made of two half-circle plates of radius $a = 5$ cm and separated a distance $d = 0.5$ mm with a sheet of plastic of permittivity $\varepsilon = 4\varepsilon_0$ to separate the plates. Given the dimensions and permittivity of air and assuming the distance between the plates to be small enough to justify the use of formulas for the parallel plate capacitor, calculate (neglect any friction between the moving plate and plastic)

a. The force that the moving plate exerts as a function of the applied voltage.

b. The torque the moving plate can supply as a function of the applied voltage.

Solution:

a. Using the process in **Equations (5.13)** and **(5.14)** and referring to **Figure 5.17**, we calculate the force as follows: Suppose the lower plate rotates an angle $d\theta$. The change in volume is

$$dv = \frac{(ad\theta)ad}{2} \quad [\text{m}^3],$$

where the quantity $ad\theta$ is the arc length shown and the dotted surface area is taken as a triangle. The change in energy due to this (virtual) motion is

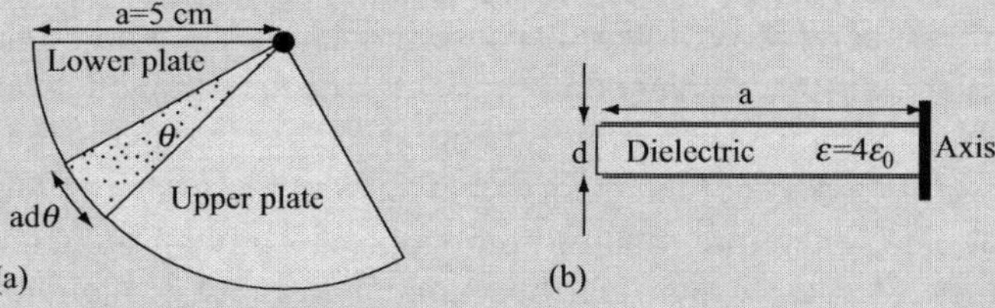

FIGURE 5.17 ■ The rotary capacitive actuator. (a) Top view. (b) Side view.

$$dW = \varepsilon\frac{E^2}{2}dv = \varepsilon\frac{V^2}{4d^2}a^2 dd\theta = \frac{\varepsilon V^2 a^2}{4d}d\theta \quad [\mathrm{J}].$$

By definition, force is the rate of change in energy:

$$F = \frac{dW}{d\theta} = \frac{\varepsilon V^2 a^2}{4d} \quad [\mathrm{N}].$$

The force acts at the center of gravity of the disk and depends on the applied voltage, permittivity between the plates, distance between the plates, and the radius of the plate.

Numerically we get

$$F = \frac{4 \times 8.845 \times 10^{-12} \times (0.05)^2}{4 \times 0.0005}V^2 = 44.27 \times 10^{-12}V^2 \ [\mathrm{N}].$$

b. The torque is the product of force and radial distance. Since the force acts at the center of gravity we must calculate its location or, at the very least, estimate it. In this case the result can be found in tables. The center of gravity is found at a radial distance of $4a\sqrt{2}/3\pi$ from the axis on the line dividing the quarter-circle plate. The torque is therefore

$$T = Fl = \frac{\varepsilon V^2 a^2}{4d}\frac{4a\sqrt{2}}{3\pi} = \frac{\varepsilon V^2 a^3\sqrt{2}}{3\pi d} \quad [\mathrm{N \cdot m}].$$

Or numerically,

$$T = \frac{\varepsilon V^2 a^3\sqrt{2}}{3\pi d} = \frac{4 \times 8.845 \times 10^{-12} \times (0.05)^3\sqrt{2}}{3 \times \pi \times 0.0005}V^2 = 1.33 \times 10^{-12}V^2 \ [\mathrm{N \cdot m}].$$

As expected, the force and torque are very small but increase quickly with the applied voltage. Although this actuator cannot be expected to be of use in regular applications, it is practical in MEMS devices, to be discussed in **Chapter 10**.

5.4 | MAGNETIC FIELDS: SENSORS AND ACTUATORS

Magnetic sensors and actuators are those governed by the magnetic field (more specifically, by the magnetic flux density, **B**) and its effects. The magnetic flux density is also called magnetic induction. Therefore these sensors tend to be inductive sensors. We should be careful, however, because induction has other connotations that will become clear later on. As we have done with electric sensors and actuators, we will try to rely as much as possible on simple properties of the magnetic field without getting into the nitty-gritty aspects of the theory (which would require a full understanding of Maxwell's equations). Therefore much of the work will rely on inductance, magnetic circuits, and magnetic forces, which can be explained, at least qualitatively, without resorting to Maxwell's equations. Needless to say, because of that, some quantities will be approximate and some merely qualitative.

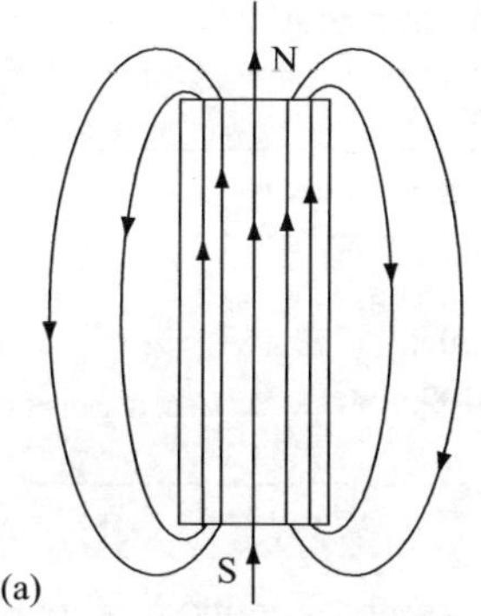

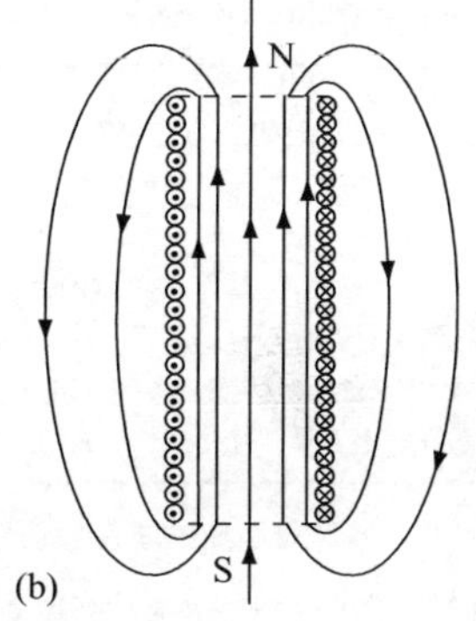

FIGURE 5.18 (a) Permanent magnet. (b) Coil with field equivalent to that of the magnet in (a).

To start with, a magnetic field may be understood by resorting to a permanent magnet. The magnet exerts a force on another magnet through space. We say that a "field" exists around the magnet through which it interacts. This force field is in fact the magnetic field (**Figure 5.18a**). The same can be observed by driving a current through a coil (**Figure 5.18b**). Since the two fields in **Figure 5.18** are identical, their sources must be identical and we conclude that all magnetic fields are generated by currents. In the case of the permanent magnet, the currents are atomic currents produced by electrons. A magnet will attract or repel another magnet—this gives us the first observable interaction in the magnetic field—but it will also attract a piece of iron. On the other hand, it will not attract a piece of copper. The conclusion is that there are different types of materials in terms of their magnetic properties. These properties are governed by the permeability of the material, μ [H/m]. The "strength" of the magnetic field is usually given by the magnetic flux density, **B** [T], or the magnetic field intensity, **H** [A/m]. The relation between the two fields is

$$\mathbf{B} = \mu_0 \mu_r \mathbf{H} \quad [\mathrm{T}], \tag{5.15}$$

where $\mu_0 = 4\pi \times 10^{-7}$ H/m is the permeability of vacuum and μ_r is the relative permeability of the medium in which the relation holds. μ_r is the ratio between the permeability of the medium and that of vacuum and hence is a dimensionless quantity associated with each material in nature. **Tables 5.3a–5.3d** list the permeabilities of some useful materials and classify them based on their relative permeabilities. If the relative permeability is <1, the materials are called diamagnetic, if >1, paramagnetic. Materials with $\mu_r >> 1$ are called ferromagnetic (ironlike) and are often the most useful materials

TABLE 5.3a ■ Permeability of various diamagnetic and paramagnetic materials

Material	Relative permeability	Material	Relative permeability
Silver	0.999974	Air	1.00000036
Water	0.9999991	Aluminum	1.000021
Copper	0.999991	Palladium	1.0008
Mercury	0.999968	Platinum	1.00029
Lead	0.999983	Tungsten	1.000068
Gold	0.999998	Magnesium	1.00000693
Graphite (Carbon)	0.999956	Manganese	1.000125
Hydrogen	0.999999998	Oxygen	1.0000019

TABLE 5.3b ■ Permeability of various ferromagnetic materials

Material	μ_r	Material	μ_r
Cobalt	250	Permalloy (78.5% Ni)	100,000
Nickel	600	Fe_3O_4 (magnetite)	100
Iron	6000	Ferrites	5000
Supermalloy (5% Mo, 79% Ni)	10^7	Mumetal (75% Ni, 5% Cu, 2% Cr)	100,000
Steel (0.9% C)	100	Permendur	5000
Silicon iron (4% Si)	7000		

TABLE 5.3c ■ Permeability of various soft magnetic materials

Material	Relative permeability (maximum) μ_r
Iron (0.2% impure)	9000
Pure iron (0.05% impure)	2×10^5
Silicon iron (3% Si)	55,000
Permalloy	10^6
Supermalloy (5% Mo, 79% Ni)	10^7
Permendur	5000
Nickel	600

TABLE 5.3d ■ Permeability of various hard magnetic materials

Material	μ_r
Alnico (aluminum-nickel-cobalt)	3–5
Ferrite (barium-iron)	1.1
Sm-Co (samarium-cobalt)	1.05
Ne-Fe-B (neodymium-iron-boron)	1.05

when working with magnetic fields. There are other types of magnetic materials (ferrites, magnetic powders, magnetic fluids, magnetic glasses, etc.) that we will encounter, and will describe them as necessary. Soft magnetic materials are those in which magnetization is reversible (i.e., they do not become permanent magnets following application of an external magnetic field), whereas hard magnetic materials are materials that retain magnetization and therefore are often used for production of permanent magnets.

Magnetic materials, especially ferromagnetic materials, have two related and important properties in addition to those discussed above. One is magnetic hysteresis and the other is nonlinearity of its magnetization curve. Magnetic hysteresis is shown in **Figure 5.19**. It indicates that as the magnetization changes (as shown by the change in magnetic field intensity), the curve traces different paths as the magnetization increases and as it decreases. The area of the magnetization curve is associated with losses. From a sensing point of view it is important to understand that the narrower the curve, the easier it is to reverse magnetization. This indicates that these materials are appropriate as magnetic cores in structures such as electric motors or transformers, especially those operating on alternating currents. Wide magnetization curves mean that the magnetization is not easily reversed, and these are usually materials used in permanent magnets.

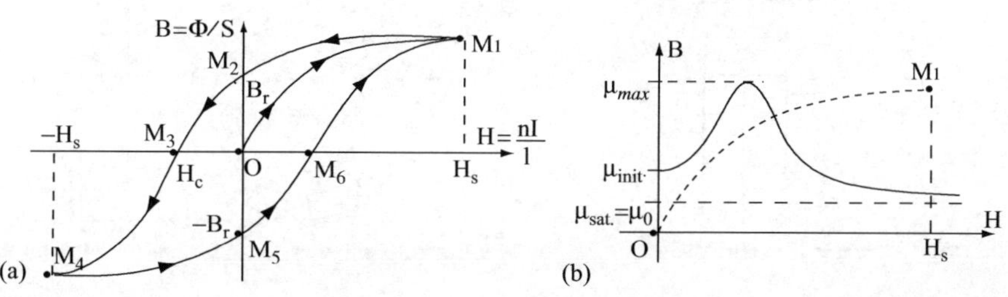

FIGURE 5.19 ■ Hysteresis (magnetization) curve and the resulting permeability for a ferromagnetic material.

Permeability is the slope of the magnetization curve (**Figure 5.19b**). Since this slope changes from point to point, permeability is nonlinear.

An important relation is that between current and magnetic flux density. Consider a very long straight wire (infinitely long) carrying a current I and placed in a medium of permeability $\mu = \mu_0\mu_r$. The magnitude of the magnetic flux density is

$$B = \mu_0\mu_r \frac{I}{2\pi r} \quad [\mathrm{T}], \tag{5.16}$$

where r is the distance from the wire to the location where the field is calculated (**Figure 5.20a**). The magnetic field has a direction as shown. If we place a system of coordinates (cylindrical in this case), the field can be described as a vector:

$$\mathbf{B} = \hat{\boldsymbol{\phi}}\mu_0\mu_r \frac{I}{2\pi r} \quad [\mathrm{T}]. \tag{5.17}$$

The important point to observe is that the field is perpendicular to the current. The relation between current and field is given by the right-hand rule shown in **Figure 5.20b**.

In more practical configurations, the wire may not be very long or it may be wound in a coil, but, nevertheless, the basic relations hold: the larger the current and/or the permeability, or the shorter the distance between the current and the location where the magnetic field is needed, the larger the magnetic flux density. However, **Equation (5.16)** is only correct for long, thin wires. In other configurations, the flux density may be quite different. In a long solenoid, with n turns per unit length (see **Figure 5.21a**), the magnetic flux density in the solenoid is constant, and with the geometry and system of coordinates shown is

$$\mathbf{B} = \hat{\mathbf{z}}\mu_0\mu_r nI \quad [\mathrm{T}]. \tag{5.18}$$

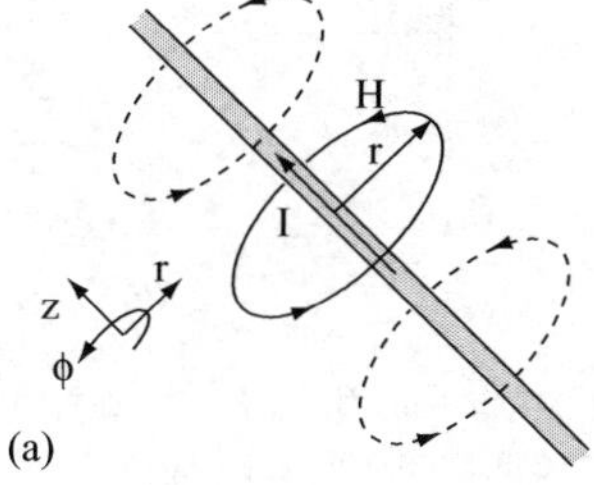

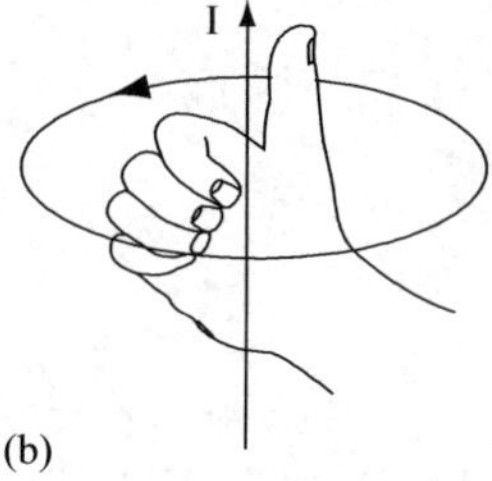

FIGURE 5.20 ■ (a) Magnetic field of a long, straight current-carrying conductor. (b) The relation between the direction of current and field (right-hand rule).

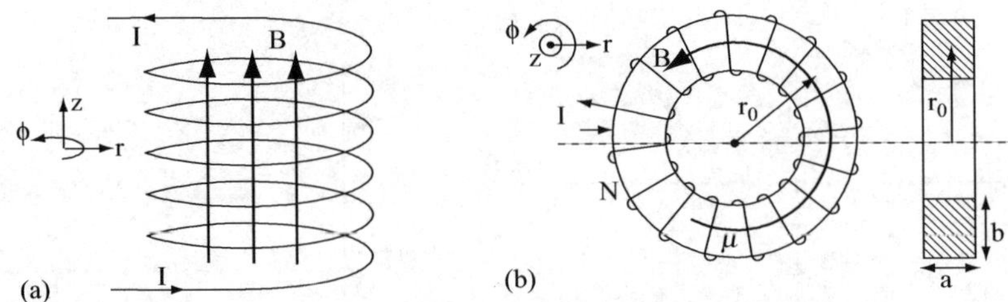

FIGURE 5.21 ■ (a) The long solenoid and its field. (b) The toroidal coil and its field.

The magnetic flux density outside the solenoid is zero. Similarly, the magnetic flux density in a toroidal coil of average radius r_0, cross-sectional area S, with N turns uniformly wound on the torus (**Figure 5.21b**) is

$$\mathbf{B} = \hat{\mathbf{z}}\frac{\mu_0\mu_r nI}{2\pi r_0} \quad [\mathrm{T}]. \tag{5.19}$$

The flux density outside the toroidal coil is zero. In other configurations the magnetic flux density is more complex and may not be calculable analytically.

If the flux density is integrated over an area, we obtain the flux of the magnetic field over that area:

$$\Phi = \int_S \mathbf{B}\cdot d\mathbf{s} \quad [\mathrm{Wb}]. \tag{5.20}$$

Of course, if $\mathbf{B}$ is constant over an area S and at an angle θ to the surface, the flux is $\Phi = BS\cos\theta$, as indicated by the scalar product in **Equation (5.20)**.

The force in a magnetic field is based on the fact that a charge moving at a velocity $\mathbf{v}$ in a magnetic field $\mathbf{B}$ experiences a force, called the **Lorentz force**, given as

$$\mathbf{F} = q\mathbf{v}\times\mathbf{B} \quad [\mathrm{N}], \tag{5.21}$$

where the force is perpendicular to the direction of both $\mathbf{v}$ and $\mathbf{B}$. The magnitude of the Lorentz force may be written as

$$F = qvB\sin\theta_{vB} \quad [\mathrm{N}], \tag{5.22}$$

where θ_{vB} is the angle between the direction of motion of the charge q and the direction of $\mathbf{B}$ as shown in **Figure 5.22a**. In most sensing applications charges are not moving in space (in some cases they do), but rather in conductors. In these important cases the force may be recast to act on the current rather than on charges. The starting relation is

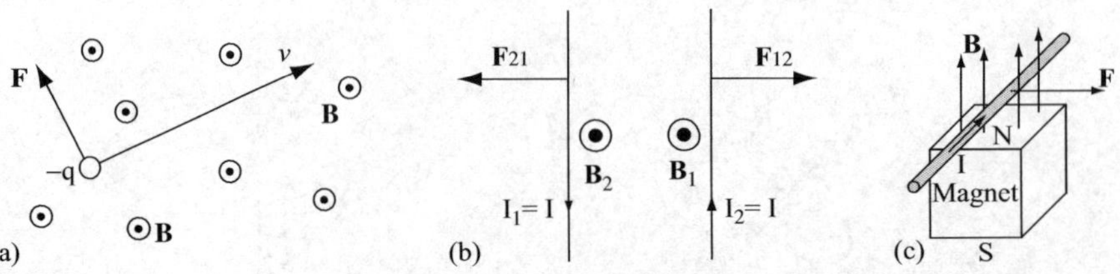

FIGURE 5.22 ■ (a) The relation between force and magnetic field for a moving charge. (b) Forces exerted by oppositely flowing currents on each other. (c) Force exerted by a magnet on a current-carrying wire.

Equation (5.21) and the current density in a volume containing n charges per unit volume is

$$\mathbf{J} = nq\mathbf{v} \quad [\mathrm{A/m^2}]. \tag{5.23}$$

That is, the force in **Equation (5.21)** can be written as a force per unit volume,

$$\mathbf{f} = \mathbf{J} \times \mathbf{B} \quad [\mathrm{N/m^3}], \tag{5.24}$$

or it may be integrated over a volume in which the current density flows to calculate the total force on the current:

$$\mathbf{F} = \int_v \mathbf{J} \times \mathbf{B}dv \quad [\mathrm{N}]. \tag{5.25}$$

As previously, the magnitude of the force density may be written as $f = JB\sin\theta$, where θ is the angle between the flux density and the current density.

The force between two wires carrying currents in opposite directions is shown in **Figure 5.22b**. The forces the wires exert on each other are in opposite directions and tend to separate the wires. These forces are evaluated from **Equations (5.17)** and **(5.25)**. If the currents were in the same direction (or the magnetic field reversed) the wires would attract. For long parallel wires, the force for a length L of the wire is

$$F = BIL \quad [\mathrm{N}]. \tag{5.26}$$

For other configurations the relation is much more complicated, but, in general, force is proportional to B, I, and L. A single wire carrying a current will be attracted or repelled by a permanent magnet, as shown in **Figure 5.22c**. These principles are the basis for magnetic actuation, and unlike electric field actuators, the forces can be very large since B, I, and L can be controlled and can be quite large. The relations above, and some we will use below, are very simple, but they suffice to help us understand the behavior of magnetic devices, at least qualitatively, and to explain how sensors and actuators operate.

5.4.1 Inductive Sensors

Inductance is a property of a magnetic device in the manner that capacitance is a property of an electric device. Inductance is defined as the ratio of magnetic flux and current:

$$L = \frac{\Phi}{I} \frac{\text{webber}}{\text{ampere}} \text{ or [henry]}. \tag{5.27}$$

The unit of inductance is the henry (H). Inductance is independent of current since Φ is linearly dependent on current (see **Equations (5.17)** and **(5.20)**). All magnetic devices have an inductance, but inductance is most often associated with electromagnets—in which the flux is produced by a current through conductors, usually in the form of coils. We define two types of inductance:

1. Self inductance: the ratio of the flux produced by a circuit (a conductor or a coil) in itself and the current that produces it. That is, the flux in **Equation (5.20)** is the flux through the device itself. Usually denoted as L_{ii}.

2. Mutual inductance: the ratio of the flux produced by circuit i in circuit j and the current in circuit i that produced it. Usually denoted as M_{ij}.

The concept of self-inductance is shown in **Figure 5.23a**, that of mutual inductance in **Figure 5.23b**. Thus any circuit (conductor, coil of conductors) has a self-inductance. A mutual inductance exists between any two circuits as long as there is a magnetic field (flux) that couples the two when a current passes through either circuit. This coupling can be large (tightly coupled circuits) or small (loosely coupled circuits).

Whereas the measurement of inductance is relatively easy, the calculation of inductance is not and depends on the geometry and its details. Nevertheless, exact or approximate formulas for inductance of various coils exist. For example, for a long circular coil of radius r and with n turns/m, the self-inductance can be approximated as

$$L = \mu n^2 \pi r^2 \quad [\text{H/m}]. \tag{5.28}$$

The self-inductance of toroidal coils can also be calculated relatively easily. Other approximate formulas for short coils as well as for inductance of straight wires are available.

It should also be recalled that the relation between voltage on an inductor and the current through it is strictly an AC relation given as

$$V = L\frac{dI}{dt} \quad [\text{V}], \tag{5.29}$$

where $I(t)$ is the current in the inductor and L is its total inductance. This voltage is called back-emf because its polarity opposes the polarity of the source that provides the current.

If a loop or coil that contains N turns is placed in the field produced by a second coil, the induced voltage (often called induced electromotive force or emf) is

$$emf = -N\frac{d\Phi}{dt} \quad [\text{V}], \tag{5.30}$$

where the negative sign indicates a phase difference between the current producing the flux and the voltage induced in the coil. For example, using **Figure 5.23b**, the voltage induced in coil 2 due to the field produced by coil 1 would be $-N_2 d\Phi_{12}/dt$, where N_2 is the number of turns in the coil on the right and Φ_{12} is the flux produced by the coil on the left in the coil on the right. Clearly **Equation (5.30)** is general and its usefulness depends on our ability to evaluate fluxes. In some cases that is relatively easy, but in others it is not.

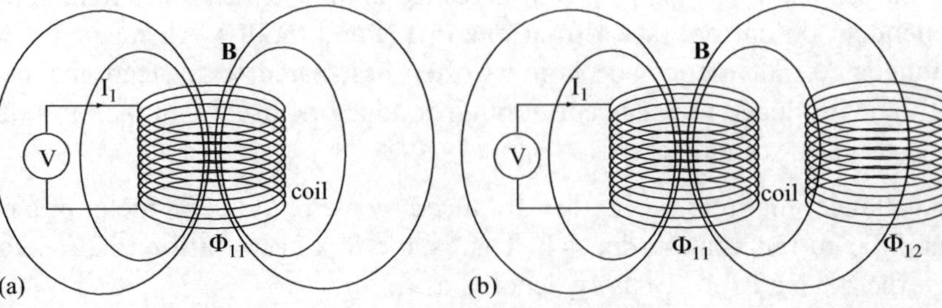

FIGURE 5.23 ■ The concept of inductance. (a) Self-inductance. (b) Self- and mutual inductance. Mutual inductance is due to the flux linking the two coils.

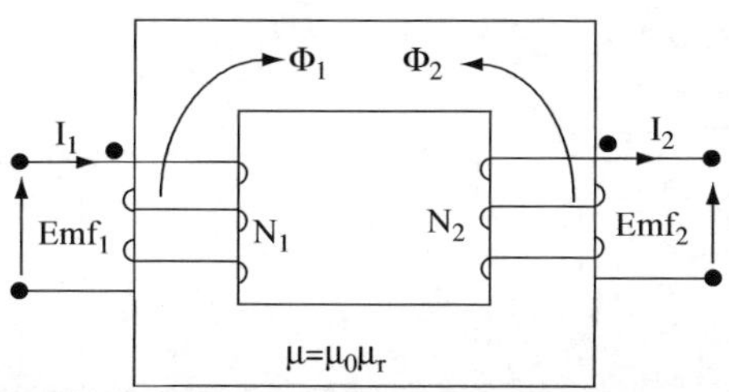

FIGURE 5.24 ■ The transformer.

The concepts of self- and mutual inductance are also associated with the principle of the transformer. In a transformer, which will contain two or more coils, an AC voltage applied to one circuit (or coil) produces a voltage in any other circuit that couples to the driving coil, as shown in **Figure 5.24**. The coils, with N_1 and N_2 turns, respectively, produce fluxes when a current exists through them. All flux produced by coil 1 couples with coil 2 through the magnetic circuit made of a ferromagnetic material (iron, for example). The voltages and currents relate as follows:

$$V_2 = \frac{N_2}{N_1} V_1 = \frac{1}{a} V_1, \; I_2 = \frac{N_1}{N_2} I_1 = a I_1, \tag{5.31}$$

where $a = N_1/N_2$ is the transformer ratio. If not all the flux produced by one coil links the other, the ratios above must be multiplied by a coupling coefficient that indicates how tightly the two coils are coupled. Under this condition the device is a loosely coupled transformer and occurs when the core is made of low-permeability materials such as air or when the core is not closed.

Most inductive sensors rely on self-inductance, mutual inductance, or transformer concepts to operate. It should be remembered, however, that these are active elements and require connection to a source to produce an output. In the process, the magnetic field described above is produced and the sensor can be said to respond to changes in this magnetic field. The output is sometimes given in terms of the magnetic flux density, as are sensitivity and error, but more often in terms of the output voltage the sensor produces. Similar considerations apply to actuators.

The most common types of stimuli sensed by inductive sensors are position (proximity), displacement, and material composition. These will be described next, taking advantage of the concepts of inductance and magnetic circuits. There are also sensors that use inductance and induction indirectly and some of these will be encountered later in this chapter and in later chapters.

5.4.1.1 Inductive Proximity Sensors

An inductive proximity sensor contains, at the very least, a coil (inductor) that, when a current passes through it, generates a magnetic field, as shown in **Figure 5.25a**. The coil has an inductance, which depends on the dimensions of the coil, number of turns, and materials around it. The current and the diameter of the coil define the extent to which the field projects away from the coil and therefore the range and span of the sensor. As the sensor gets closer to the sensed surface (**Figure 5.25b**), the inductance of the coil increases if the sensed surface is ferromagnetic (it does not change or changes very little if the surface is not ferromagnetic, but we shall see shortly that the same can be achieved in nonferromagnetic materials, provided they are conducting and the field is alternating). It is then sufficient to use an "inductance meter" and a calibration curve (transfer

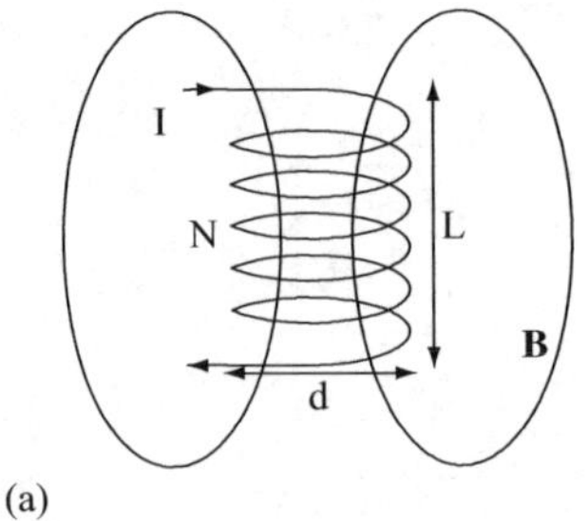

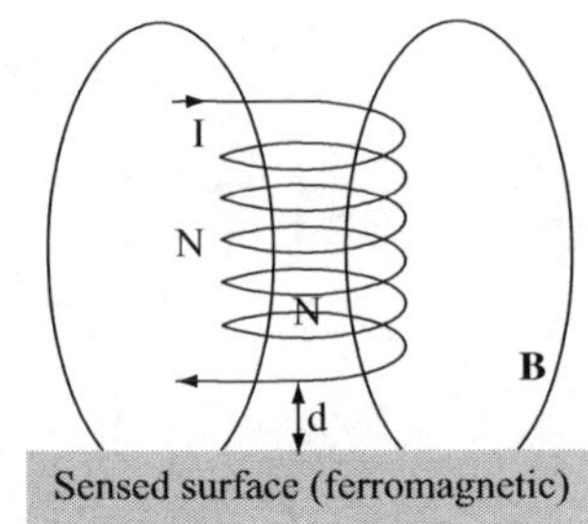

FIGURE 5.25 ■ The basic inductive proximity sensor. (a) Coil in air (no sensed surface). (b) The field produced by the coil interacts with the sensed surface, changing its inductance.

function) to devise a sensor. An inductance meter is usually made of an AC current source and a voltmeter or an AC bridge. By measuring the voltage across the inductor, the impedance can be evaluated, and since $Z = R + j\omega L$ (R is the ohmic resistance and $\omega = 2\pi f$ is the angular frequency), the inductance L is immediately available as a measure of the position of the coil or proximity to the surface being sensed. We assumed here that R is constant, but even if it is not, the impedance in air (nothing being sensed) is known and this can be used for calibration.

The advantage of the device in **Figure 5.25** is its simplicity, but it is a highly nonlinear sensor, and in practical sensors a few things are done to improve its performance. First, a ferromagnetic core is added to increase the inductance of the sensor. Most often this is made of a ferrite material (a powdered magnetic material such as iron oxide [Fe_2O_3] or other oxides in a binding substance and sintered into the shape needed). Ferrites have the advantage of being high-resistance materials (low conductivity). In addition, a shield may be placed around the sensor to prevent sensitivity to objects on the side of the sensor or at its back (**Figure 5.26a,b**). The net effect of the shield is to project the field in front of the sensor and hence increase both the field (inductance) and the span of the sensor. In other sensors there are two coils, one serving as a reference and the other as a sensor (**Figure 5.26b**). The reference coil's inductance remains constant and the two are balanced. When a surface is sensed, the sensing coil's inductance increases and the imbalance between the coils serves as a measure of distance (differential sensor). Other sensors, like the one is **Figure 5.26c**, may employ a closed magnetic circuit, which tends to concentrate the magnetic field in the gaps, and usually do not require shielding since the magnetic field is constrained within the ferromagnetic material making up the core. In all cases, exact calculation of the fields and the sensor response is only possible using numerical tools. In many cases, however, these can be determined rather easily by experiment.

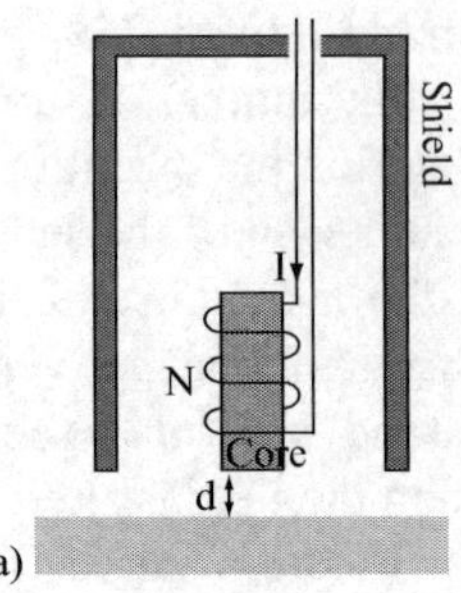

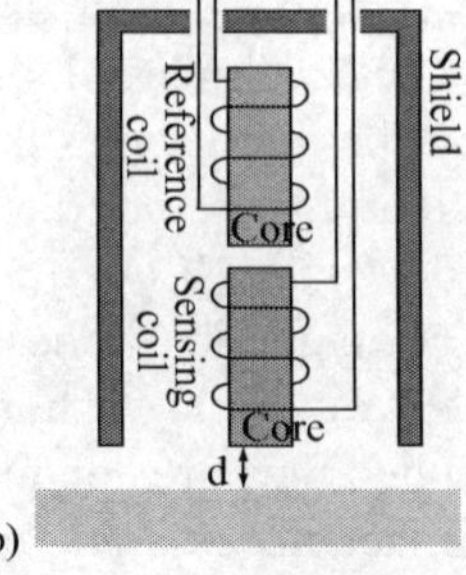

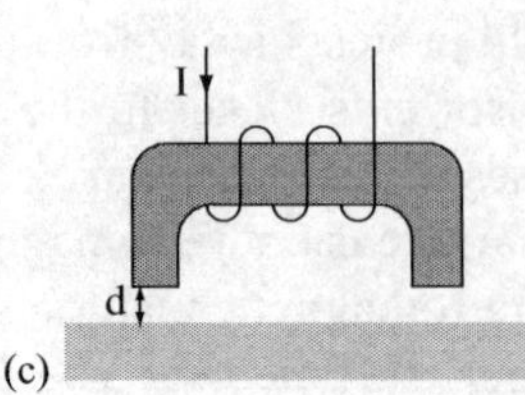

FIGURE 5.26 ■ Practical proximity sensors. (a) Shielded sensor. (b) Shielded sensor with reference coil. (c) Sensor employing a magnetic circuit to concentrate the field in a small gap.

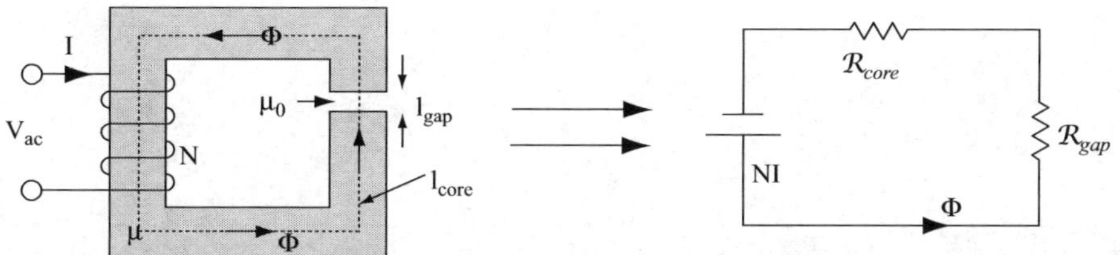

FIGURE 5.27 ■ The concept of a magnetic circuit and its equivalent as an electric circuit with modified quantities.

In certain cases, especially in transformers and in closed magnetic circuits such as the sensor shown in **Figure 5.26c**, an approximate method can be used to calculate parameters such as inductance, magnetic flux, or induced voltage based on the concept of the ***magnetic circuit***. In this method, the magnetic flux is viewed as a "current," the term NI (i.e., the product of the current) and the number of turns in a coil as a "voltage," and a term called reluctance replaces the resistance in a regular circuit. The basic concept is shown in **Figure 5.27**. The reluctance of a member of the magnetic circuit of length l_m [m], permeability μ [H/m], and cross-sectional area S [m^2] is

$$\mathcal{R}_m = \frac{l_m}{\mu S} \quad [1/\mathrm{H}]. \tag{5.32}$$

With these preliminaries, the flux in the circuit is

$$\Phi = \frac{\sum NI}{\sum \mathcal{R}_m} \quad [\mathrm{Wb}]. \tag{5.33}$$

In the case shown in **Figure 5.27**, there are two reluctances, one due to the path in the core (l_{core}) and one due to the path in the gap (l_{gap}).

Other quantities such as induced voltages in coils, forces, and fields can be calculated from these relations.

However, the equivalent circuit reveals the requirements as well. The flux must "flow" in the closed circuit just like an electric current must flow in a closed circuit. This can be achieved, at least approximately, if the permeability of the core is high. When the permeability is low, such as in the gap, the length of the gap must be short to prevent flux from "spreading" out and invalidating the assumption of the circuit. Nevertheless, the method is useful for approximate calculations. Magnetic circuits can be used with DC or AC sources.

EXAMPLE 5.5 Paint thickness sensor

A paint thickness sensor is built as in **Figure 5.26c** with a magnetic field sensor (a sensor that can quantify the magnetic flux density—a Hall element—to be discussed in **Section 5.4.2**) embedded into one of the surfaces as shown in **Figure 5.28**. The sensor's core is made of silicon steel (a steel often employed in electromagnetic devices because it has high permeability and low conductivity). The coil includes $N = 600$ turns and is supplied with a DC current $I = 0.1$ A. The core of the sensor has a cross-sectional area $S = 1$ cm^2 and an average magnetic path length $l_c = 5$ cm. The sensor can only be used reliably to test for paint thickness on ferromagnetic materials such as steel, cast iron, nickel, and their alloys. It can also sense the thickness of plating layers such as zinc or copper on steel. The purpose here is to establish a transfer function and sensitivity for the

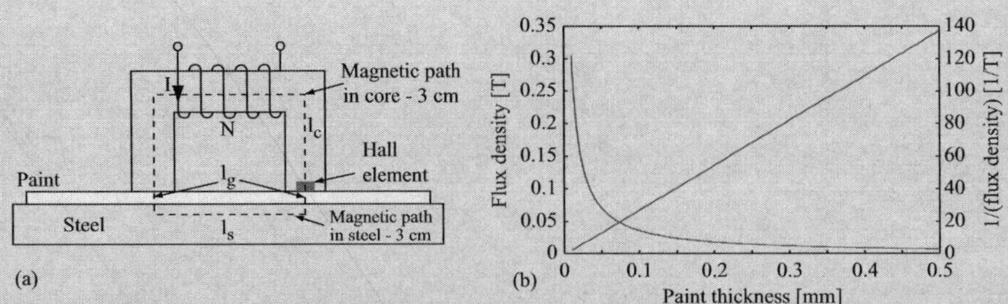

FIGURE 5.28 ▪ Paint thickness sensor. (a) Geometry and dimensions. (b) Flux density (B) measured as a function of paint thickness and the reciprocal of flux density (1/*B*) as a function of paint thickness.

sensor for paint thicknesses from 0.01 mm to 0.5 mm. The measurand is the paint thickness and the measured quantity is the magnetic flux density as measured by the field sensor. Silicon steel has a relative permeability of 5000, whereas steel has a relative permeability of 1000. Paint is nonmagnetic with relative permeability of 1.

Solution: We start by calculating the reluctances of the core, the two gaps, and an estimate of the reluctance in the steel material. Then we calculate the flux and finally the flux density as a function of gap thickness.

The reluctances of the core and the gap are

$$\mathcal{R}_c = \frac{l_c}{\mu_c S} = \frac{5 \times 10^{-2}}{5000 \times 4\pi \times 10^{-7} \times 10^{-4}} = 7.957 \times 10^4/\text{H}$$

and

$$\mathcal{R}_g = \frac{l_g}{\mu_c S} = \frac{l_g}{4\pi \times 10^{-7} \times 10^{-4}} = 7.957 \times 10^9 l_g/\text{H}.$$

Note that the reluctance of the gap is much higher than that of the core because of the low permeability of air.

In the steel we have the length of the path, but not the cross-sectional area. As a first approximation we will assume the same cross-sectional area as the core. In practice this unknown quantity will be part of the calibration of the sensor. With this assumption we have, for steel,

$$\mathcal{R}_s = \frac{l_s}{\mu_c S} = \frac{0.03}{1000 \times 4\pi \times 10^{-7} \times 10^{-4}} = 2.387 \times 10^5/\text{H}.$$

Now we can calculate the flux in the core, gap, and steel:

$$\Phi = \frac{NI}{\mathcal{R}_c + 2\mathcal{R}_g + \mathcal{R}_s} = \frac{600 \times 0.1}{7.957 \times 10^4 + 2 \times 7.957 \times 10^9 l_g + 2.387 \times 10^5}$$
$$= \frac{60}{3.183 \times 10^5 + 2 \times 7.957 \times 10^9 l_g} \quad [\text{Wb}].$$

Note that because the reluctance of the gap is at least four orders of magnitude larger than that of steel or of the core, the reluctances of steel and the core can be neglected and we obtain the approximate flux above. The flux density is the flux divided by the cross-sectional area:

$$B = \frac{\Phi}{S} = \frac{1}{1 \times 10^{-4}} \left(\frac{60}{3.183 \times 10^5 + 2 \times 7.957 \times 10^9 l_g} \right) = \frac{6}{3.183 + 2 \times 7.957 \times 10^4 l_g} \quad [\text{T}].$$

We could have neglected the first term in the denominator as small, but that would not be the case for very low values of l_g. The transfer function can now be established by entering values of the gap from $l_g = 0.01$ mm to 0.5 mm. The transfer function is shown in **Figure 5.28b**. The curve is highly nonlinear, but nevertheless, the relation between flux density and paint thickness is distinct and usable. In an instrument one can invert the result, that is, as a postprocessing step one can calculate the quantity $1/B$ and plot it against the paint thickness, also shown in **Figure 5.28b**. The result is a linear curve and a much easier to read output. The output units can be calibrated directly in terms of paint thickness.

5.4.1.2 Eddy Current Proximity Sensors

Inductive proximity sensors are sensitive to the presence (proximity) of nonconducting ferromagnetic materials or to any conducting media. Nonconductors in general do not affect inductive proximity sensors. Many inductive proximity sensors are of the eddy current type. The name eddy current comes from the fundamental property of AC magnetic fields to induce currents in conducting media (ferromagnetic or not). This is shown schematically in **Figure 5.29**. There are two related phenomena at work. The currents produced in the conductor, called eddy currents because they flow in closed loops, cause a field that opposes the original field that produces them (Lenz's law). This field reduces the net flux through the sensor's coil. Second, the currents flowing in the conductor being sensed dissipate power. The sensing coil is now forced to supply more power than it would otherwise supply and hence, given a constant current, its effective resistance increases. This change in impedance from $Z = R + j\omega L$ to $Z' = R' + j\omega L'$ is easily sensed either in absolute terms or as a change in the phase of the measured voltage (given a constant current). An AC magnetic field penetrating into a conducting medium is attenuated exponentially from the surface inward (and so are the eddy currents and other related quantities):

$$B = B_0 e^{-d/\delta} \quad [\text{T}], \text{or } J = J_0 e^{-d/\delta} \quad [\text{A/m}^2], \tag{5.34}$$

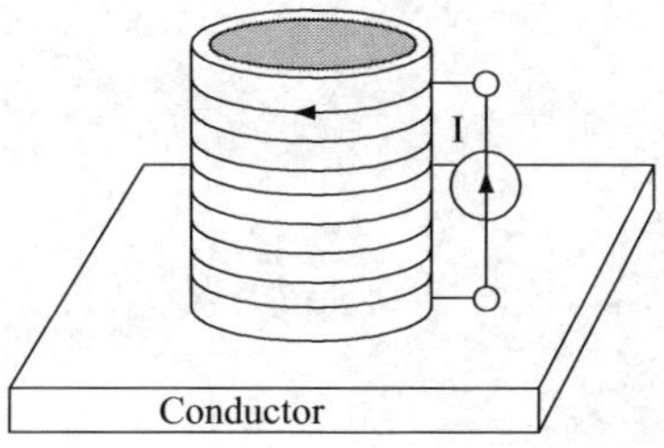

FIGURE 5.29 ■ An eddy current sensor. The AC in the coil induces eddy currents in the conducting plate.

where B_0 and J_0 are the flux density and the eddy current density at the surface, d is the depth in the medium, and δ is the skin depth. Skin depth is defined as the depth at which the field (or current density) is attenuated to $1/e$ of its value at the surface and for planar conductors is given as

$$\delta = \frac{1}{\sqrt{\pi f \mu \sigma}} \quad [\mathrm{m}], \tag{5.35}$$

where f is the frequency of the field. Clearly then, penetration depends on frequency, conductivity, and permeability. The main implication here is that the sensed conductor must be thick enough compared to skin depth. Alternatively, operation at higher frequencies may be needed to reduce the skin depth.

Figure 5.30 shows a number of inductive proximity sensors used for industrial control. A proximity sensor (either capacitive or inductive) can be used to sense distance. However, except for very short distances, their transfer function is much too nonlinear and their span too small to be effective for this purpose. For this reason proximity sensors are often used as switches to provide a clear indication when a certain, preset distance is reached. As with capacitive sensors, inductive sensors can produce an electric output, such as voltage, based on the change in their impedance, but often the inductor is part of an oscillator (LC oscillator is the most common, in which case $f = 1/2\pi\sqrt{LC}$) and the frequency of the sensor is then used as the output.

Figure 5.31 shows two eddy current sensors used for nondestructive testing of materials. The sensor on top is an absolute EC sensor (or probe) since it contains a single coil and measures the absolute change in impedance due to the presence of small flaws in conducting materials. The bottom sensor contains two small coils separated a short distance apart and is used as a differential sensor (see **Example 5.6**). The sensor on top operates at 100 kHz and its coil is embedded in a dielectric. The bottom sensor operates at 400 kHz and its coils are embedded in ferrite. **Figure 5.32** shows two differential eddy current probes as used for flaw detection inside tubes. The top sensor is 19 mm in diameter and is designed to operate at 100 kHz for the detection of flaws in stainless steel tubes used in nuclear reactor steam generators. The coil on the bottom is used for the detection of cracks and flaws in air conditioning tubing (8 mm) and operates at 200 kHz. The output is the difference between the outputs of the two coils.

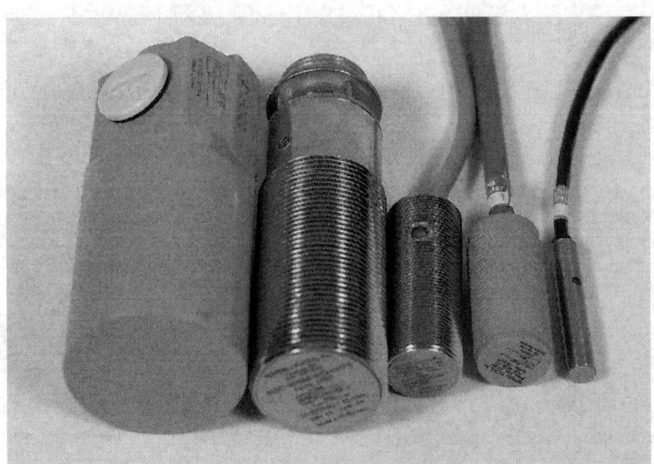

FIGURE 5.30 ■ Inductive proximity sensors.

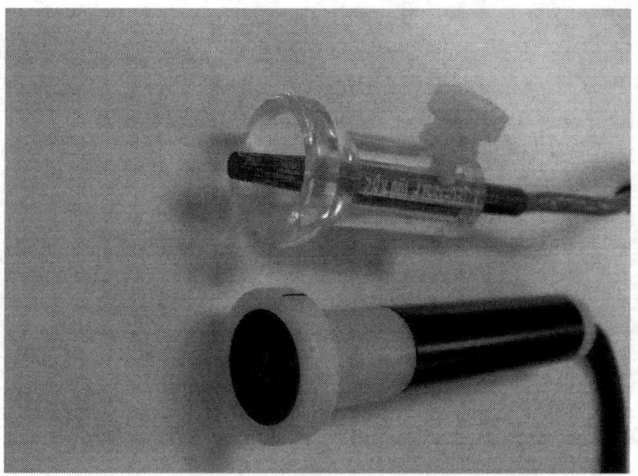

FIGURE 5.31 ■ Eddy current sensors. Top: 100 kHz absolute sensor embedded in a dielectric medium. Bottom: 400 kHz differential sensor embedded in ferrite.

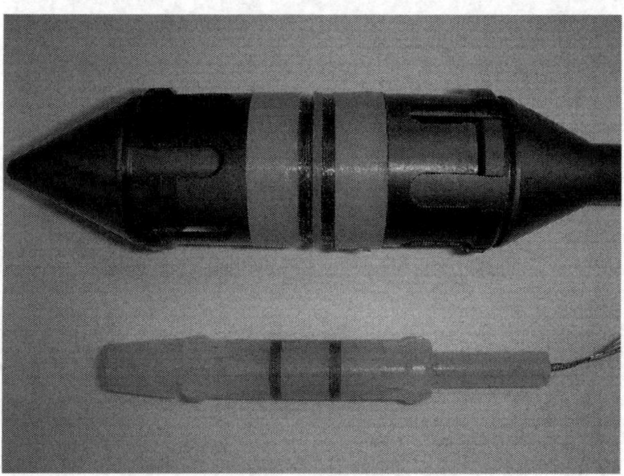

FIGURE 5.32 ■ Eddy current sensors (differential) for the detection of flaws in tubing. Top: 19 mm, 100 kHz sensor for stainless steel tubing in nuclear power plant steam generators. Bottom: 8 mm, 200 kHz sensor for air conditioning tubing.

EXAMPLE 5.6 Eddy current testing for flaws

The idea of position sensing can be used for nondestructive testing of materials to help in the detection of cracks, holes, and subsurface anomalies in conducting media. Consider the configuration in **Figure 5.33a**, where a thick aluminum conductor has a 2.4 mm hole drilled in it to some depth, representing a flaw. Two small coils, each 1 mm in diameter and separated 3 mm apart, are placed against the aluminum surface and slid to the right in small increments (the bottom probe in **Figure 5.31** was used for these measurements). The inductance of each coil is measured and the difference between the two inductances is used as an indication. This measurement is differential and is particularly useful when the environment is noisy.

Figure 5.33b shows a plot of the inductance versus position of the center of the probe, starting about 18 mm from the center of the hole and moving 18 mm past the hole. Since the two probes are identical, their inductance is identical, and as long as they see identical conditions, the output is zero. Far from the hole, or when the probe is centered over the hole, the output is zero. Anywhere else, one coil will have a higher or a lower inductance than the other and hence a variable output. Because the difference is taken between the leading coil and the trailing coil, and since the inductance in the vicinity of the flaw is lower, the curve first dips (negative difference) then rises to a positive difference.

In actual tests, the coils are fed with a constant AC and the potential across the two coils is measured. The potential is complex, but will vary in the same manner as the inductance.

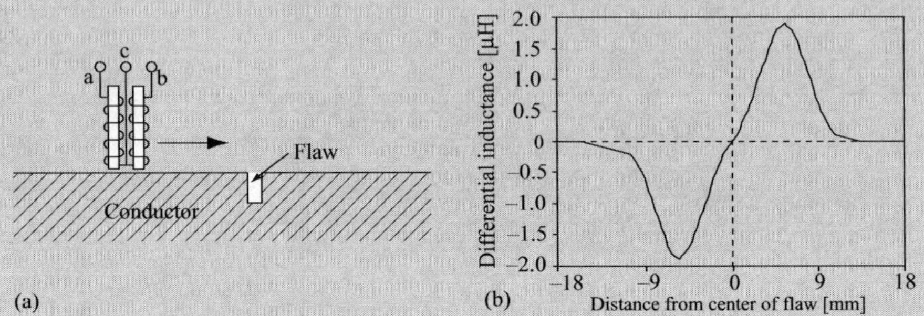

FIGURE 5.33 ■ Differential probe, eddy current nondestructive testing. (a) The probe and geometry. (b) The output given as the difference in inductance between the leading and trailing coils.

5.4.1.3 *Position and Displacement Sensing: Variable Inductance Sensors*

Position and displacement are usually understood as measuring the exact distance from a point or the travel of a point relative to another. This requires accurate measurements and possibly linear transfer functions of the sensors involved. One approach to this task is through the use of variable inductance sensors, sometimes called variable reluctance sensors. Magnetic reluctance is the equivalent magnetic term to electric resistance and is defined as (see also **Equation (5.32)**):

$$\mathcal{R} = \frac{L}{\mu S} \quad [1/\mathrm{H}]. \tag{5.36}$$

The reluctance is smaller the shorter the magnetic path, the larger its cross-sectional area, and the larger its permeability. Reluctance is then related to inductance through permeability, and reducing reluctance increases inductance and vice versa. Typically the reluctance of a coil can be changed by adding a gap in the magnetic path and changing the effective length of this gap.

Thus one of the simplest methods of changing the inductance of a coil is to provide it with a movable core as shown in **Figure 5.34**. In this sensor, the further the movable core moves in, the smaller the reluctance of the magnetic path and the larger

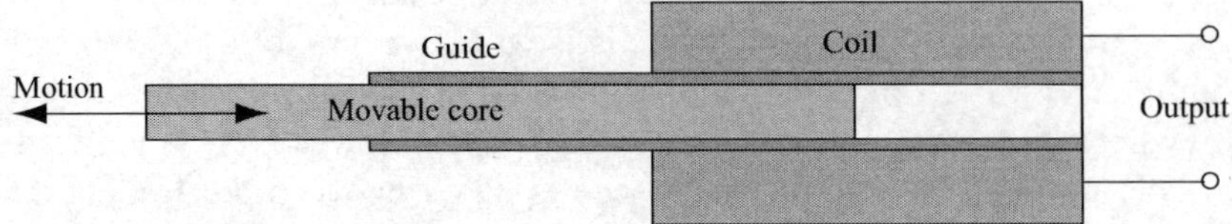

FIGURE 5.34 ■ Inductive sensor with a movable core.

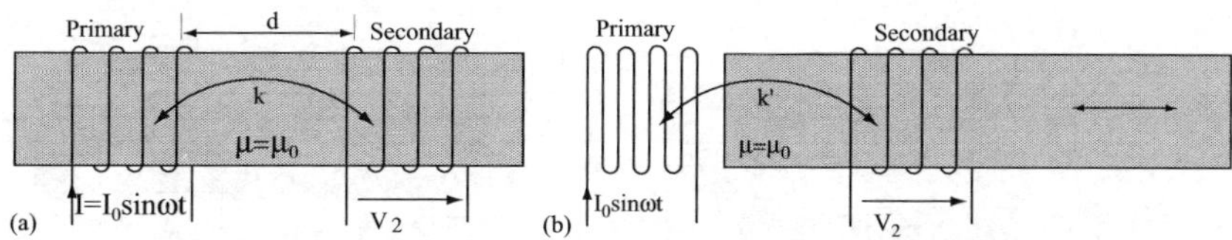

FIGURE 5.35 ■ Principle of the LVDT. (a) Changing the distance between the coils. (b) Moving the core between two fixed coils. In either case, the coupling coefficient k changes.

the inductance. If the core is made of a ferromagnetic material, the inductance increases, whereas a conducting material reduces the inductance (see explanation for eddy current proximity sensors). This type of sensor is called a ***linear variable inductance sensor***. Linear here means that the motion is linear, not necessarily that the transfer function is linear. By sensing the inductance, a measure of the position of the core is available. The same configuration may be used to measure force, pressure, or anything else that can produce linear displacement.

A better displacement sensor is a sensor based on the idea of the transformer. This is based on one of two related principles: either the distance between two coils of a transformer is varied (the coupling between the coils changes) or the coupling coefficient between the two coils is varied by physically moving the core while the two coils are fixed. Both principles are shown in **Figure 5.35**. A variation of the second of these is the **linear variable differential transformer** (LVDT), which will be discussed shortly. To understand the principles, first consider **Figure 5.35a**. Assuming a constant AC voltage V_{ref} is connected across the primary coil, the induced voltage in the secondary coil is the output voltage:

$$V_{out} = k\frac{N_2}{N_1}V_{ref} \quad [\mathrm{V}]. \tag{5.37}$$

k is a coupling coefficient that depends on the distance between the coils as well as the medium between them and any other material that may be present in the vicinity, such as a shield, enclosure, etc. Given a calibration curve, the output voltage, which can be measured directly, is a measure of the distance between the coils. In **Figure 5.35b**, the same relation holds except that now the moving core changes the coupling between the coils, thus changing the output voltage across the second coil.

Another type of position sensor may be made by connecting the two coils in **Figure 5.35a** in series and measuring the inductance of the two coils. The latter is $L_{11} + L_{22} + 2L_{12}$, where L_{12} is the mutual inductance between them, given as $L_{12} = k\sqrt{L_{11}L_{22}}$. The coupling coefficient k depends on the distance between the two coils, and by measuring the total inductance, one gets a measure of the position of one coil with respect to the second. This arrangement is called a ***coil-displacement sensor***.

The LVDT sensor is shown in **Figure 5.36**. This is similar to the variable inductance sensor, but now there are two coils in the output circuit whose voltages subtract. When the core is symmetrical about the coils (**Figure 5.36a**), the output of the sensor is zero. If the core moves to the left, the voltage on coil 2 decreases (**Figure 5.36b**) while the voltage on coil 1 remains the same, since only the coupling between the reference (primary) coil and coil 2 changes. The total voltage now increases and its phase is, say, positive. When the core moves to the right, the opposite happens, but the phase is

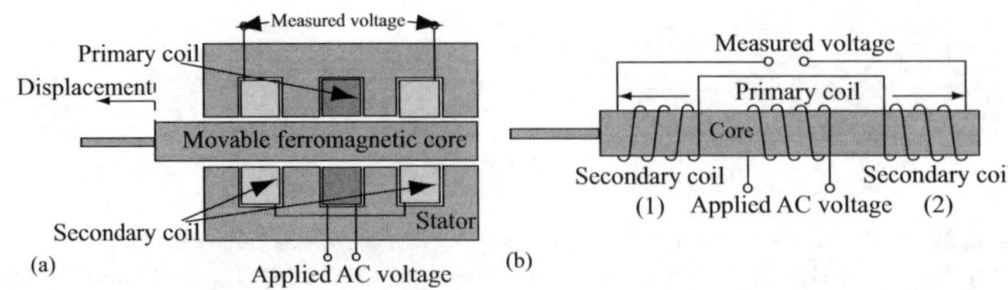

FIGURE 5.36 ■ The moving core LVDT. (a) Construction. (b) Principle.

opposite (negative in this case). This change in phase is used to detect the direction of motion, whereas the output voltage is a measure of distance the core travels from the zero output position. These devices are very sensitive and useful, and in a relatively small range of motion, the output is linear. The reference coil is driven with a stable sinusoidal source at a constant frequency and the core is ferromagnetic. The whole sensor is enclosed and shielded so that no field extends outside it and hence cannot be influenced by outside fields. The core slides in and out, and that motion is often used for accurate measurements of displacement for applications in industrial control and machine tools.

LVDT sensors are extremely rugged and come in various dimensions to suit many needs (some as small as 10 mm long). In most practical applications, the voltage output is measured (amplification is usually not needed), whereas the phase is detected with a zero-crossing phase detector (a comparator; see **Chapter 11**). The frequency of the source must be high enough with respect to the frequency of motion of the core (a figure of 10 times higher is common) to avoid errors in the output voltage due to slow response of the LVDT. The operation of LVDTs can be from AC sources or DC sources (with an internal oscillator providing the sinusoidal voltage). Typical working voltages are up to about 25 V, whereas output is usually below 5 V. Resolution can be very high, with a linear range about 10%–20% of the length of the coil assembly. Although the stated function of LVDTs is position sensing, anything that can be related to position can also be sensed. One can use an LVDT to sense fluid level, pressure, acceleration, and many other functions.

A variation of the LVDT is the ***rotary variable differential transformer*** (RVDT), intended for angular displacement and rotary position sensing. In all respects it is identical in operation to the LVDT device, but the rotary motion imposes certain restrictions on its construction. The RVDT is shown schematically in **Figure 5.37** and includes a ferromagnetic core that couples with the secondary coils based on angular position. The moving core is shaped to obtain linear output over the useful range of the

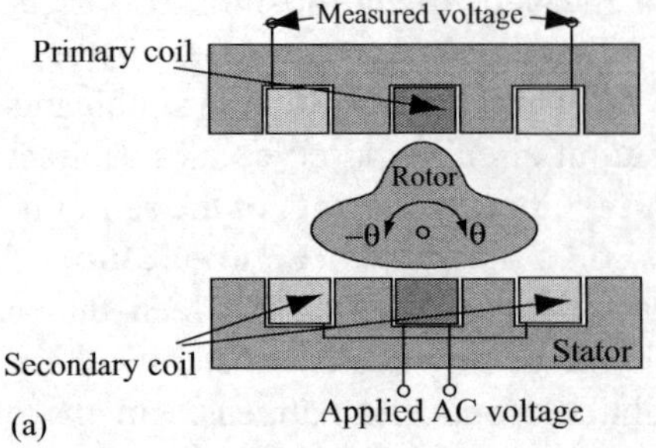

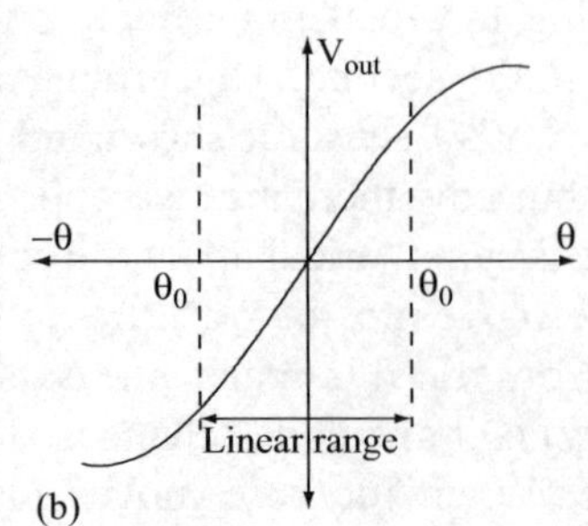

FIGURE 5.37 ■ Schematic view of an RVDT. (a) Construction. (b) Output.

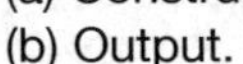

sensor. Other arrangements are possible as well. The span is given in angles and can be up to ±40°. Beyond that the output becomes nonlinear and less useful.

5.4.2 Hall Effect Sensors

The Hall effect was discovered in 1879 by Edward H. Hall. The effect exists in all conducting materials, but is particularly pronounced and useful in semiconductors. To understand the principle, consider a block of conducting medium through which a current of electrons is flowing caused by an external source, as shown in **Figure 5.38**. A magnetic flux density B is established across the conductor, making an angle θ with the direction of the current (in **Figure 5.38**, $\theta = 90°$). The electrons flow at a velocity v, and according to **Equation (5.22)**, a force perpendicular to both the current and field is established on the flowing electrons. Since the force is related to the electric field intensity as $F = qE$, we can write the electric field intensity in the conductor as

$$E_H = \frac{F}{q} = vB \sin\theta \quad [\mathrm{V/m}]. \tag{5.38}$$

The index H indicates this is the Hall electric field and it is perpendicular to the direction of current flow. To rewrite this in terms of current flowing in the element, we note that the current density may be written as $J = nqv$ [A/m^2], where nq is the charge density (n is number of electrons per cubic meter and q is the charge of the electron). v is the average velocity of electrons. Therefore the Hall electric field intensity is

$$E_H = \frac{nqvB \sin\theta}{nq} = \frac{JB \sin\theta}{nq} \quad [\mathrm{V/m}]. \tag{5.39}$$

The current density is the current I divided by the cross-sectional area perpendicular to the flow of current or $J = I/Ld$:

$$E_H = \frac{IB \sin\theta}{nqLd} \quad [\mathrm{V/m}]. \tag{5.40}$$

The force pulls the electrons toward the front surface of the conductor and therefore a voltage develops between the back (positive) and front (negative) surfaces. The potential

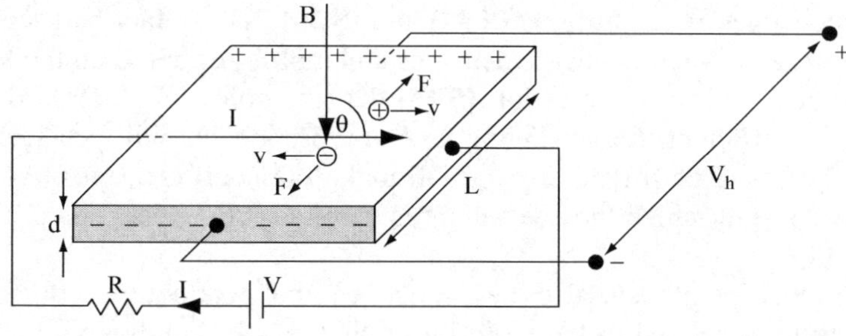

FIGURE 5.38 ■ The Hall element. The two opposite electrodes on the far and near surfaces measure the Hall voltage while a current flows across horizontally. The component of the magnetic flux density perpendicular to the element (shown here as B) is the sensed quantity.

difference is the integral of the electric field intensity E along the path of length L. Because Hall elements are typically small, we may assume E is constant along L and write

$$V_H = EL = \frac{IB \sin \theta}{qnd} \quad [\text{V}]. \tag{5.41}$$

This voltage is the ***Hall voltage***. In particular, in measurements, the angle θ is typically 90° and the Hall voltage is given by

$$V_H = \frac{IB}{qnd} \quad [\text{V}], \tag{5.42}$$

where d is the thickness of the hall plate, n is the carrier density [charges/m^3], and q is the charge of the electron [C].

It should be noted that if the current changes direction or the magnetic field changes direction, the polarity of the Hall voltage flips. Thus the Hall effect sensor is polarity dependent, a property that may be used to good advantage to measure the direction of a field or the direction of motion if the sensor is properly set up.

The term $1/qn$ [m^3/C] (or [m^3/A·s]) is material dependent and is called the **Hall coefficient** (K_H):

$$K_H = \frac{1}{qn} \quad [\text{m}^3/\text{A·s}]. \tag{5.43}$$

Strictly speaking, K_H in conductors is negative since q is the charge of the electron. The **Hall voltage** is usually represented as

$$V_{\text{out}} = K_H \frac{IB}{d} \quad [\text{V}]. \tag{5.44}$$

The relations above apply to all conductors. In semiconductors the Hall coefficient depends on both hole and electron mobilities and concentration as follows:

$$K_H = \frac{p\mu_h^2 - n\mu_e^2}{q(p\mu_h + n\mu_e)^2} \quad [\text{m}^3/\text{A·s}], \tag{5.45}$$

where p and n are the hole and electron densities, respectively, μ_h and μ_e are the hole and electron mobilities, respectively, and q is the charge of the electron. The net effect of this dependency is a large coefficient, so much so that all practical Hall sensors are based on semiconductors. **Equations (5.44)** and **(5.45)** may in fact be used to measure properties of materials such as charge densities and mobilities based on the Hall voltage. It should also be noted from **Equation (5.45)** that the hole and electron densities will affect the Hall coefficient. Large doping with n-type dopants will produce a negative coefficient, whereas large p-type doping will make the coefficient positive. There is a certain doping level at which the coefficient is zero, as can be calculated directly from **Equation (5.45)**.

The Hall coefficient can also be related to the conductivity of the medium. Since conductivity is related to the mobility of charges, for conductivity in conductors we have

$$\sigma = nq\mu_e \quad [\text{S/m}]. \tag{5.46}$$

In semiconductors the conductivity depends on the mobility of both electrons and holes:

$$\sigma = qn\mu_e + qp\mu_h \quad [\text{S/m}]. \tag{5.47}$$

Therefore the Hall coefficient in conductors can be written as

$$K_H = \frac{\mu_e}{\sigma} \quad [\text{m}^3/\text{A}\cdot\text{s}]. \tag{5.48}$$

In semiconductors we have

$$K_H = \frac{q\left(p\mu_h^2 - n\mu_e^2\right)}{\sigma^2} \quad [\text{m}^3/\text{A}\cdot\text{s}]. \tag{5.49}$$

In principle, the lower the conductivity, the higher the Hall coefficient. However, this is only true to a point. As the conductivity decreases, the resistance of the device increases and the current in the device decreases, reducing the Hall voltage (see **Equation (5.44)**).

It should also be recalled here that in doped semiconductors, the product of the electron and hole concentrations is related to the intrinsic concentration through the mass action law:

$$np = n_i^2. \tag{5.50}$$

An intrinsic material is that in which $n_i = n = p$.

These relations clearly indicate that the Hall effect can be used to measure conductivity or, alternatively, that any quantity that affects the conductivity of the medium also affects the Hall voltage. For example, a semiconducting Hall element exposed to light will read in error due to the change in conductivity of the semiconductor due to the photoconducting effect (see **Section 4.4.2.2**).

Hall coefficients vary from material to material and are particularly large in semiconductors. For example, in silicon it is on the order of -0.02 m^3/A·s, but it depends on doping and temperature, among other parameters. The most important aspect of this sensor is that it is linear with respect to the field for a given current and dimensions. However, the Hall coefficient is temperature dependent, and this must be compensated for if accurate sensing is needed. Because the Hall voltage in most materials is rather small—on the order of 50 mV/T—and considering the fact that most fields sensed are smaller than 1 T, the Hall voltage must in almost all cases be amplified. As an example, the earth's magnetic field is only about 50 μT, so the output of a Hall sensor in the terrestrial magnetic field is on the order of 25 μV. Nevertheless, these are easily measurable quantities and Hall sensors are among the most commonly used sensors for magnetic fields because they are simple, linear, can be integrated within semiconductor devices, and are inexpensive. They are available in various forms, sizes, sensitivities, and in arrays. The errors involved in measurement are mostly due to temperature variations, but the size of the Hall plate, if large, also introduces averaging errors due to its integration effect. Some of these effects can be compensated for by appropriate circuitry or compensating sensors. In terms of fabrication, a typical sensor is a thin rectangular wafer made of p- or n-doped semiconductor (InAs and InSb are the most commonly used materials because of their larger carrier densities, and hence larger Hall coefficients, but silicon may also be used with reduced sensitivity). The sensor is usually identified by two

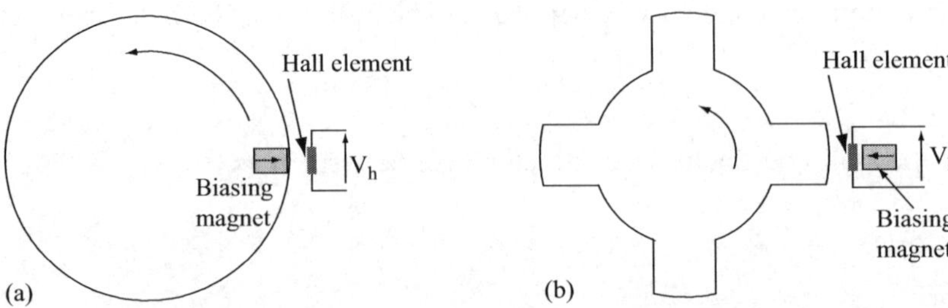

FIGURE 5.39 ■ (a) Sensing rotation of a shaft. The small magnet (arrow shows the direction of the field) induces a voltage pulse V_h every time it passes past the Hall element. (b) The Hall element senses the position of the rotating poles in a 4-cylinder engine to fire the appropriate cylinder in sequence and at the correct time.

transverse resistances: the control resistance through which the control current flows and the output resistance across which the Hall voltage develops.

In practical applications, the current is usually kept constant so that the output voltage is directly proportional to the field. The sensor may be used to measure the flux density (provided proper compensation can be incorporated) or may simply be used as a detector or operate a switch. The latter is very common in the sensing of rotation, which, in itself, may be used to measure a variety of effects (shaft position, frequency of rotation [rpm], differential position, torque, etc.). An example is shown in **Figure 5.39a**, where the rotation of a shaft is sensed. An emf is induced in the Hall element every time the small magnet passes by indicating the rotation of the shaft. Many variations of this basic configuration can be envisioned, including, for example, the measurement of angular displacement. Hall elements are integral to many electrical motors in disk drives, CD-ROM drives, and many other applications in which the rotation is sensed and controlled through these sensors.

Hall elements are also fabricated in pairs, separated a small distance apart, for the expressed purpose of sensing the gradient in the field rather than the field itself. This is particularly useful in position and presence sensing in which the edge of a ferromagnetic medium, such as gears in transmissions or in electronic ignition systems, is sensed. Some sensors come with their own biasing magnet to generate the magnetic field and may have either analog or digital output with onboard electronics. In these devices, changes in the magnetic field due to the presence of a ferromagnetic material are sensed. An example of this type of sensor used for position sensing is shown in **Figure 5.39b**. This particular configuration is common in electronic ignition systems (a 4-cylinder application is shown), where a pulse is produced by the Hall element every time one of the metal poles passes by. The material of the poles must be ferromagnetic (iron). This configuration, in many variants and for both linear and angular applications, is one of the simplest methods of sensing position.

The Hall sensor can be used for other applications. One example is the direct sensing of electric power. In an application of this type (**Figure 5.40**), the power, which is the product of current and voltage, is measured as follows: The voltage is connected to a coil that generates a magnetic field across the Hall element. The current, which now can be variable, is the control current in the sensor. The Hall voltage is proportional to power and, if properly calibrated, will measure power directly.

Hall element sensors are usually considered to be DC devices. Nevertheless, they can be easily used to sense alternating fields at relatively low frequencies. The specification sheet for Hall elements gives their response and the maximum useful frequency.

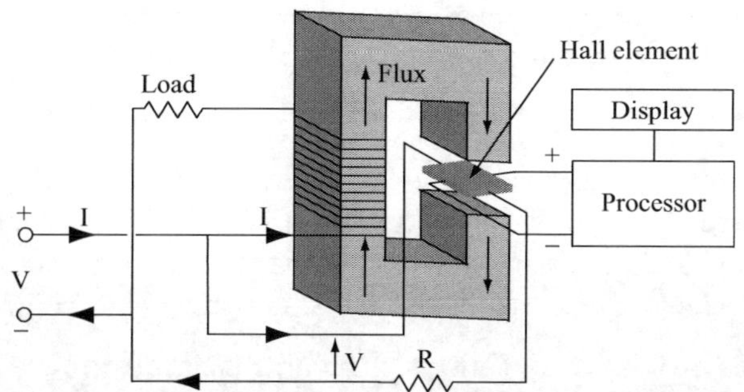

FIGURE 5.40 ■ Direct sensing of power with a single Hall element.

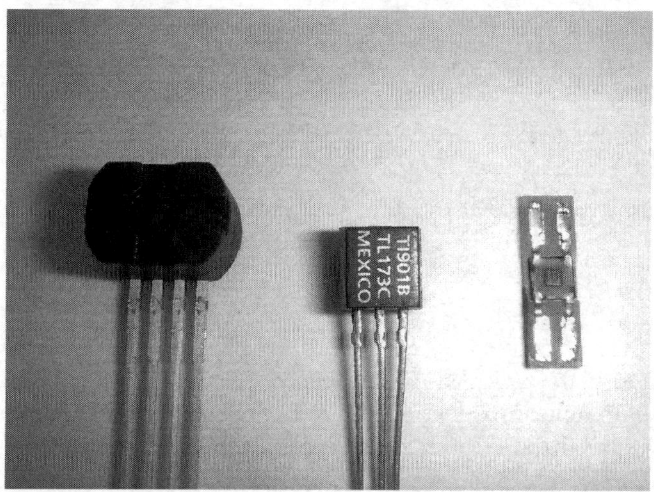

FIGURE 5.41 ■ Various Hall elements. Left: A dual Hall element with biasing magnet and digital output. Middle: An analog Hall sensor. Right: A Hall element chip, mounted on a small fiberglass piece. The chip is 1 mm × 1 mm.

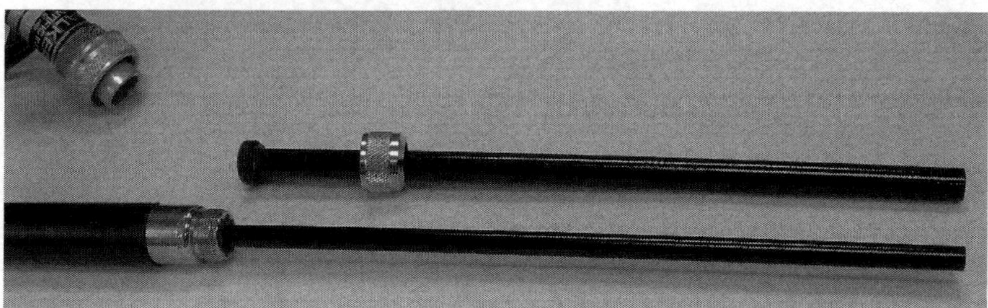

FIGURE 5.42 ■ A Hall element probe. The tip of the probe (bottom, right) contains three Hall element sensors at right angles to measure the three components of the magnetic field.

Figures 5.41 and **5.42** show a number of Hall elements/sensors, including a probe for three-axis measurement of fields down to 10 μT. The latter has three sensors, placed at right angles, and fed to three separate amplifiers.

Finally, it should be reemphasized that the only quantity that Hall sensors measure directly is the magnetic flux density, but they can be made to sense a whole range of quantities by judicious use of the sensors in conjunction with mechanical and electrical arrangements, the examples in **Figures 5.39** and **5.40** being representative examples.

EXAMPLE 5.7 Measurement of magnetic flux density and magnetic flux using a Hall element

The Hall element's primary function is sensing of magnetic fields, but it can also sense any quantity related to the magnetic field. The geometry in **Figure 5.43** uses a silicon element with a Hall coefficient of -10^{-2} m^3/A·s. The dimensions of the Hall element are $a = 2$ mm, $b = 2$ mm, and its thickness is $c = 0.1$ mm. Calculate the response of the Hall element:

a. For magnetic flux densities from 0 to 2 T. This is the range normally found in electric machines. What is the minimum field measurable if a digital voltmeter with a resolution of 2 mV is used to measure the Hall voltage?

b. For flux from 0 to 10 μWb.

Solution: The magnetic flux density is measured directly as indicated in **Equation (5.44)**. The flux is not measured directly, but rather, since $\Phi = B \times S$, we still measure the flux density B and translate the measurement into flux. Thus

a. The relation between flux density and the Hall voltage is

$$V_{\text{out}} = K_H \frac{I_H B}{d} = 0.01 \times \frac{5 \times 10^{-3}}{0.1 \times 10^{-3}} B = 0.5B \text{ V}.$$

This is a linear transfer function varying from 0 to 1 V for a flux density varying from 0 to 2 T. The sensitivity of the device is clearly 0.5 V/T. A 1 mV corresponds to 1 T/500 = 0.002 T. Thus a 2 mV voltmeter will measure a minimum flux density of 4 mT. This is not particularly sensitive (i.e., 4 mT = 4000 μT is much higher than the terrestrial magnetic flux density of 60 μT), but it is useful for higher fields.

b. To measure flux, we recall that flux is flux density integrated over area. Since the area of the sensor is small, $S = 4 \times 10^{-6}$ m^2, we may assume the flux density to be constant over the area and simply multiply the flux density by area:

$$\Phi = BS \quad [\text{Wb}].$$

But since we sense the flux density, the Hall voltage sensed will be

$$V_{out} = K_H \frac{I_H B}{d} = K_H \frac{I_H BS}{Sd} = K_H \frac{I_H \Phi}{Sd} = 0.01 \times \frac{5 \times 10^{-3}}{0.1 \times 10^{-3} \times 4 \times 10^{-6}} \Phi = 1.25 \times 10^5 \Phi \text{ V}.$$

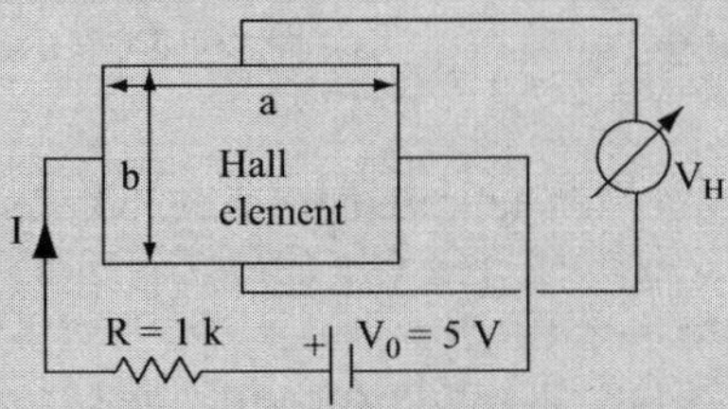

FIGURE 5.43 ■ Biasing a Hall element. The magnetic flux density is perpendicular to the Hall element plate.

The sensitivity is 1.25×10^5 V/Wb. For the range between 0 and 10 μWb, the output voltage will vary from 0 to $1.25 \times 10^5 \times 10 \times 10^{-6} = 1.25$ V. The minimum measurable flux using the same voltmeter is $10/625 = 0.016$ μWb.

EXAMPLE 5.8 Rotation speed of an engine

The rotation speed of an engine needs to be sensed for the purpose of speed regulation. To do so, a Hall element is used as a sensor using the configuration in **Figure 5.44a**. Two symmetric bumps or protrusions are added to the shaft (two bumps are needed to keep its mass balanced). The gap between the Hall element and the shaft varies from 1 mm when one of the bumps is aligned with the Hall element to 2 mm when it is not. Calculate the minimum and maximum reading of the Hall element if it is biased using the circuit in **Figure 5.44b**, which has a Hall coefficient of 0.01 m^3/A·s and is 0.1 mm thick. Assume the permittivity of the shaft and the iron ring are very high and the permeability of the Hall element is the same as air, equal to μ_0. The coil contains 200 turns and is supplied with 0.1 A to produce a magnetic flux density in the gap.

Solution: Because the permeability of the iron ring is large, we can neglect its reluctance, meaning that the flux in the gap only depends on the gap length. From **Equation (5.36)** we write for the gap:

$$\mathcal{R}_g = \frac{l_g}{\mu_0 S},$$

where S is the cross-sectional area of the gap. We do not have that quantity, but it is not necessary since we calculate the flux and then divide by the area S to find the magnetic flux density. The flux in the gap is

$$\Phi_g = \frac{NI}{\mathcal{R}_g} = \frac{NI\mu_0 S}{l_g}.$$

The flux density thus becomes

$$B_g = \frac{\Phi_g}{S} = \frac{NI\mu_0}{l_g}.$$

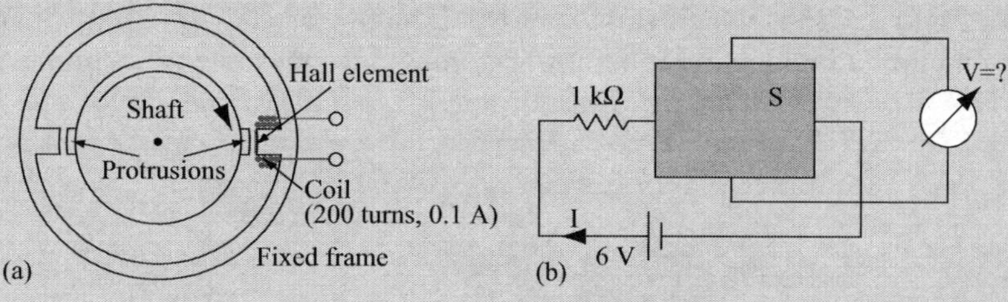

FIGURE 5.44 ■ An engine rotation speed sensor. (a) Geometry including the magnetizing coil and Hall element. (b) Electrical connections for the Hall element.

Now from **Equation (5.44)** we write

$$V_{out} = K_H \frac{I_H B}{d} = K_H \frac{I_H}{d} \frac{N I \mu_0}{l_g} \quad [\text{V}],$$

where I_H is the bias current in the Hall element (6 mA in this case) and I is the current in the coil (0.1 A). For the 1 mm gap we get (there are two identical gaps, one on each side):

$$V_{\max} = 0.01 \times \frac{6 \times 10^{-3}}{0.1 \times 10^{-3}} \times \frac{200 \times 0.1 \times 4\pi \times 10^{-7}}{2 \times 1 \times 10^{-3}} = 0.0075 \text{ V}.$$

For the 2 mm gap we get

$$V_{\min} = 0.01 \times \frac{6 \times 10^{-3}}{0.1 \times 10^{-3}} \times \frac{200 \times 0.1 \times 4\pi \times 10^{-7}}{2 \times 2 \times 10^{-3}} = 0.00375 \text{ V}.$$

The output voltage varies from 3.75 mV when the Hall element is not in the gap to 7.5 mV when it is. The output is a signal approximating a sinusoidal signal varying from a peak of 7.5 mV to a valley of 3.75 mV at a frequency twice the rate of rotation of the shaft. For example, an engine shaft rotating at 4000 rpm produces a frequency of $(4000/60) \times 2 = 133.33$ Hz. Measurement of this frequency (possibly after amplifying the signal and digitization) and proper calibration of the reading produces the necessary data.

5.5 | MAGNETOHYDRODYNAMIC (MHD) SENSORS AND ACTUATORS

The force relations in **Equations (5.21)** and **(5.25)** may be used in a number of ways for both sensing and actuation in addition to those discussed above. Because the magnetic forces involved act on charges, and therefore on currents, these are responsible for most methods of magnetic actuation and, as we have seen from the Hall effect, for sensing as well. One particularly interesting aspect of the magnetic force is the ability to create forces in moving media such as plasmas, charged gases, and liquids, and also in solid conductors. The phenomenon has been dubbed magnetohydrodynamics because of its use in moving charged gases, fluids, and molten metals, but the principle is fundamentally the same as used in electric motors and generators. The principles involved are demonstrated in **Figure 5.45a**, which shows an MHD generator (i.e., a sensor), and in **Figure 5.45b**, which shows an MHD pump or actuator. Both involve a channel containing a conducting medium. The medium may be any medium, provided it contains

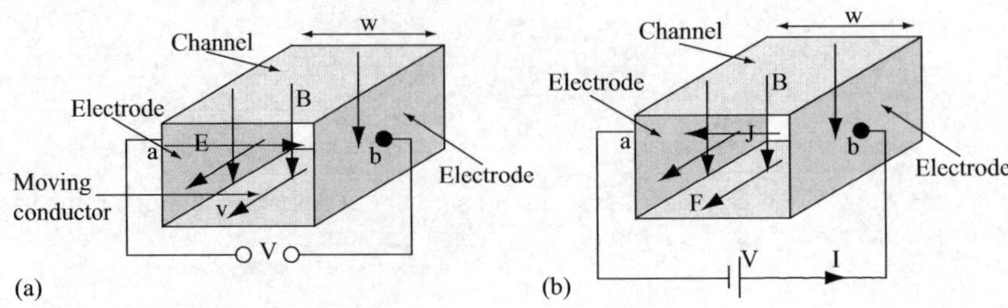

FIGURE 5.45 ■ (a) An MHD generator (sensor). (b) An MHD pump or actuator.

charges on which the magnetic field acts. The charges may be free charges in a conductor or a charged plasma.

5.5.1 MHD Generator or Sensor

The medium in the channel in **Figure 5.45a** moves at a velocity v, as shown. The magnetic field acts on the moving charges based on **Equation (5.21)**:

$$\mathbf{F} = q\mathbf{v} \times \mathbf{B} = q\mathbf{E} \quad [\mathrm{N}], \tag{5.51}$$

where the fact that the force on a charge may be written in terms of the electric field intensity that produces that force has been added. This means that the moving charged medium produces an electric field intensity, as shown in the figure by the horizontal arrow. Consequently, a potential difference is generated across the two electrodes. The latter is

$$V_{ba} = -\int_a^b \mathbf{E} \times d\mathbf{l} = -\int_a^b \mathbf{v} \times \mathbf{B} \times d\mathbf{l} = vBw \quad [\mathrm{V}]. \tag{5.52}$$

This is the basic principle of MHD generation and that of sensing: the velocity of the charged medium can be sensed by measuring the MHD voltage generated across the electrodes. However, for the potential to develop, there must be charges that can be separated, or on a macroscopic level, we may say that the medium must have a nonzero conductivity.

5.5.2 MHD Pump or Actuator

The generation process can be reversed by passing a current through the channel, perpendicular to the magnetic field, as shown in **Figure 5.45b**. The current density in the channel creates a force on the conducting medium, forcing it out of the channel based on **Equation (5.25)**:

$$\mathbf{F} = \int_v \mathbf{J} \times \mathbf{B} dv \quad [\mathrm{N}]. \tag{5.53}$$

This force may be used to pump any conducting liquid (such as molten metals or seawater), a conducting gas (plasma), or a solid conductor.

The main appeal of MHD actuators is that the system is trivially simple and requires no moving elements. It can, under the right conditions, generate enormous forces (and therefore accelerations). On the other hand, apart from some applications in pumping molten metals, it is an inefficient method of both generation and actuation. Nevertheless, it has found applications in sensing, where the efficiency of the generation process is not important, and in a number of actuators, including pumping, particle acceleration, and acceleration of solid objects in so-called rail guns, devices proposed for military and space applications.

An MHD flow sensor is shown in **Figure 5.46**. The two coils, one on top and one on the bottom, generate a magnetic flux density between the top and bottom surfaces (this can be done equally well with two permanent magnets). The fluid, such as water, must have free ions (primarily Na^+ and Cl^-) for the system to work. Fortunately most fluids, including water, have sufficient dissolved salts to provide ions for MHD sensing. The output is directly proportional to the velocity of the fluid and the magnetic flux density

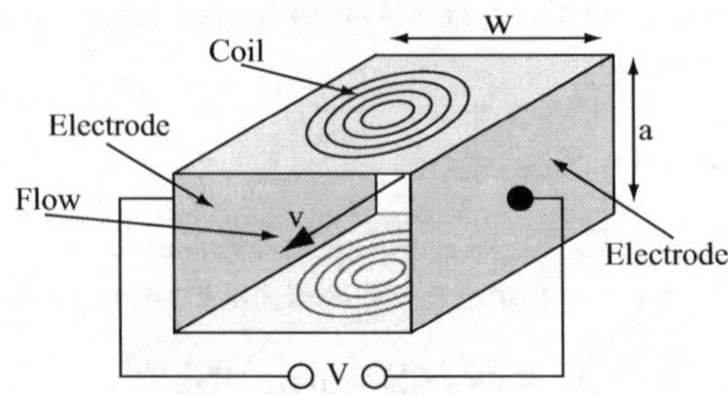

FIGURE 5.46 ■ An MHD flow rate sensor.

applied. Assuming that the flux density is constant throughout the width of the channel, the voltage measured is proportional to the fluid velocity:

$$V = Bwv \quad [\text{V}]. \tag{5.54}$$

The sensor can also sense flow rate (volume/second) as

$$Q = wav = \frac{aV}{B} \quad [\text{m}^3/\text{s}]. \tag{5.55}$$

The same configuration can be used, for example, to sense the velocity of a boat relative to water, especially in seawater.

The basic actuation method in **Figure 5.45b** can be used in a number of ways. **Figure 5.47** shows a pump for molten metals (aluminum, magnesium, sodium, etc.). The molten metal flows in a conduit equipped with two side electrodes, as shown. The force on the molten conductor is given by **Equation (5.53)**. In this case, the magnetic flux density, $\mathbf{B}_0$, is generated by a current in the coil, I_0. The volume of integration is that part of the volume of the pump in which the current I, produced by a separate source, and magnetic flux density interact, or approximately abd. Therefore, assuming that the current is uniform in this section and the power supply generates a current I, the current density is approximately I/bd, whereas the coil and armature generate a flux density B_0 and the force can be estimated as

$$F = B_0 \frac{I}{bd} abd = B_0 Ia \quad [\text{N}]. \tag{5.56}$$

The reason this force is only an approximation is that the flux density and the current density are not necessarily constant within the volume shown, but nevertheless, the result is a good approximation, especially in highly conducting media such as molten metals. The pump shown here has a significant advantage over other methods in that there is no mechanical interaction with the molten metal. However, it should also be

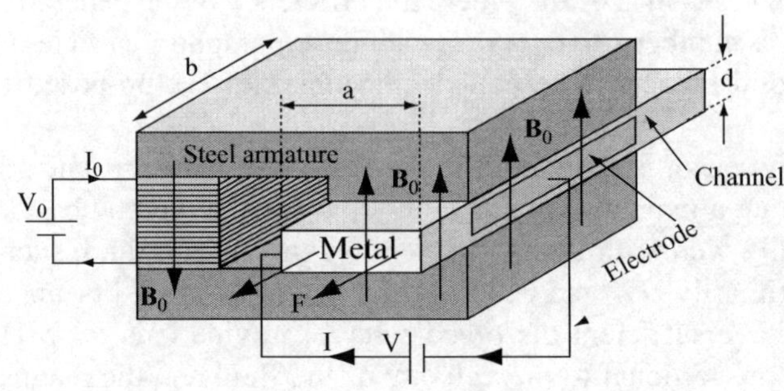

FIGURE 5.47 ■ An MHD pump for molten metals. The metal is contained within an enclosed channel.

remembered that the device, especially the coil, must be properly cooled and that it consumes a significant amount of power.

It should be noted that the force in **Equation (5.56)** is the same force that would apply on a current-carrying wire of length a, carrying a current I, in a magnetic field B_0, as was found previously in **Equation (5.26)**. This means that the same forces act in electric machines, but the designation MHD helps in separating applications that look very different.

EXAMPLE 5.9 Electromagnetic propulsion in seawater

A simple method of propulsion of vessels in seawater has been proposed based on MHD pumping. The dimensions of an actuator for this purpose are shown in **Figure 5.48**. The magnetic field is generated using a permanent magnet, producing a constant magnetic flux density $B = 0.6$ T. The conductivity of seawater is 4 S/m.

a. With the dimensions shown, calculate the power needed to generate 10 tons of force (10^5 N) to propel the vessel in seawater.

b. As a comparison, suppose that the same device is used to pump molten sodium (at 98°C) to cool a nuclear reactor. What is the power needed to produce the same force? The conductivity of molten sodium is 2.4×10^7 S/m.

Solution:

a. The force is given in **Equation (5.56)**.

$$F = B_0 I a \quad [\mathrm{N}].$$

Since B_0 is known, we need to calculate the necessary current. For 10 tons or 100,000 N we get

$$I = \frac{F}{Ba} = \frac{100000}{0.5 \times 0.8} = 250,000 \ \mathrm{A}.$$

To calculate power we need to calculate the resistance of the channel. The latter is

$$R = \frac{a}{\sigma bc} = \frac{0.5}{4 \times 0.25 \times 1} = 1 \ \Omega.$$

The power required is 62.5 GW.

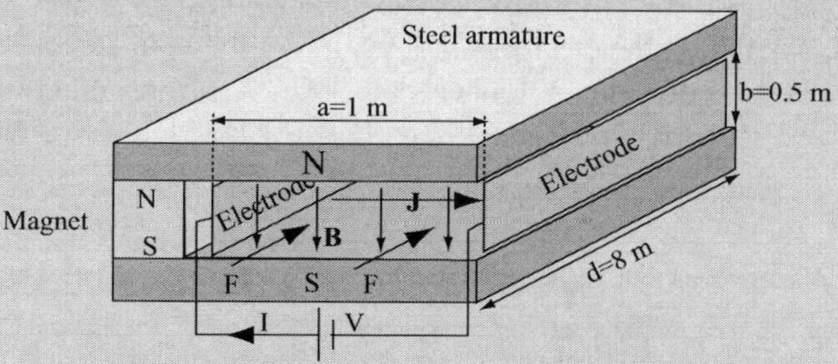

FIGURE 5.48 ■ MHD propulsion.

Clearly this is not a practical device. A 62.5 GW power source that generates 10 tons of thrust is not a viable system. The voltage source would have to supply 250 kV DC, something that is not trivial to generate even if one neglects the power needs. However, the method is valid under the right conditions, as part (b) shows.

b. What changes here is the resistance of the channel:

$$R = \frac{a}{\sigma bc} = \frac{0.5}{2.4 \times 10^7 \times 0.25 \times 1} = 8.33 \times 10^{-8}\ \Omega.$$

Therefore the power required is

$$P = I^2R = \frac{a}{\sigma bc} = (250,000)^2 \times 8.33 \times 10^{-8} = 5.208 \times 10^3\ \text{W},$$

or just over 5.2 kW. The voltage of the source is only 20 mV. Low-voltage, high-current sources are easier to build than high-voltage, high-current sources, although the case here is somewhat extreme. Nevertheless, one can use a transformer to reduce an AC voltage to a required level and then rectify it or one can produce this low-voltage, high-current source with special (homopolar) generators.

5.6 | MAGNETORESISTANCE AND MAGNETORESISTIVE SENSORS

Magnetoresistance is the effect the magnetic field has on the electrical resistance of a conductor or semiconductor. There are two mechanisms through which the resistance of a medium changes in the presence of a magnetic field. The first is due to the fact that electrons are attracted or repelled by magnetic fields, as was discussed above in conjunction with the Hall effect. The second mechanism exists in certain materials in which the direction of internal magnetization due to flow of a current changes with the application of an external magnetic field. This manifests itself in changes in the electrical resistance of the medium.

Magnetoresistors based on the first of these mechanisms are very similar to the Hall elements in that the same basic structure of **Figure 5.38** is used, but without the Hall voltage electrodes. The effect exists in all materials but is very pronounced in semiconductors. This is shown in **Figure 5.49a**. The electrons are affected by the magnetic field, as in the Hall element, and because of the magnetic force on them, they will flow in an arc, as shown in **Figure 5.49b**. The larger the magnetic flux density, the larger the force exerted on the electrons and the larger the radius of the arc. This effectively forces the electrons to take a longer path, which means that the resistance to their flow increases (exactly the same as if the effective length of the plate is larger). Thus a relationship between the magnetic field and current is established. This relation is

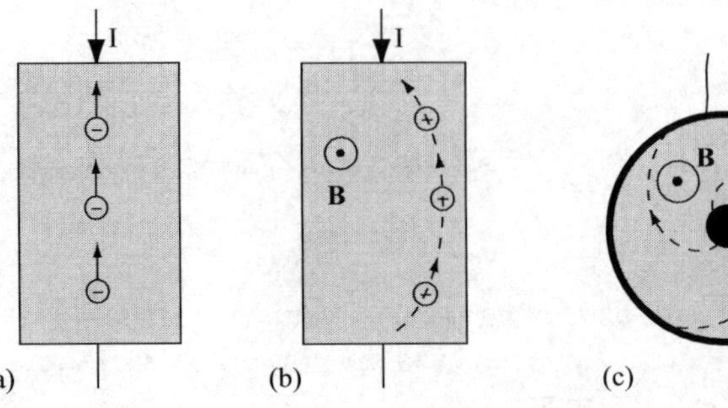

FIGURE 5.49 ■ Magnetoresistance in a semiconductor. (a) No magnetic field. (b) A magnetic field alters the flow path of the carriers. (c) The Corbino disk magnetoresistor.

proportional to B^2 for most configurations and is dependent on carrier mobility in the material used (usually a semiconductor). However, the exact relationship is rather complicated and depends on the geometry of the device. Therefore we will simply assume here that the following holds:

$$\frac{\Delta R}{R_0} = kB^2, \tag{5.57}$$

where k may be viewed as a calibration function. A particularly useful configuration for magnetoresistor is shown in **Figure 5.49c**. This is called the Corbino disk and has one electrode at the center of the disk and a second electrode on the perimeter. By so arranging the electrodes, the sensitivity of the device increases because of the long spiral paths electrons take in flowing from one electrode to the other.

Magnetoresistors are used in a manner similar to Hall elements, but their use is simpler since one does not need to establish a control current. Rather, the measurement of resistance is all that is necessary. The device is a two-terminal device built from the same types of materials as Hall elements (InAs and InSb in most cases). Magnetoresistors are often used where Hall elements cannot be used. One important application is in magnetoresistive read heads where the magnetic field corresponding to recorded data is sensed.

The second type of magnetoresistive sensor, which is even more sensitive than the basic element discussed above, is based on the property of some materials to change their resistance in the presence of a magnetic field when a current flows through them. Unlike the sensors discussed above, these are metals with highly anisotropic properties and the effect is due to the change in their magnetization direction due to the presence of a magnetic field. The effect is called **anisotropic magnetoresistance** (AMR), discovered in 1857 by William Thomson (Lord Kelvin).

One of the structures most commonly used in commercial magnetoresistive sensors is the following: A magnetoresistive material is exposed to the magnetic field to be sensed. A current passes through the magnetoresistive material at the same time. This is shown schematically in **Figure 5.50**. The magnetic field is applied perpendicular to the current. The sample has an internal magnetization vector parallel to the flow of current. When the magnetic field is applied, the internal magnetization changes direction by an angle α, as shown in **Figure 5.50**. The resistance of the sample becomes

$$R = R_0 + \Delta R_0 \cos^2\alpha, \tag{5.58}$$

where R_0 is the resistance without application of the magnetic field and ΔR_0 is the change in resistance expected from the material used. Both of these are properties of the material and the construction (for R_0). The angle α is proportional to the applied field and is material dependent. Some anisotropic magnetoresistive materials and their properties are shown in **Table 5.4**.

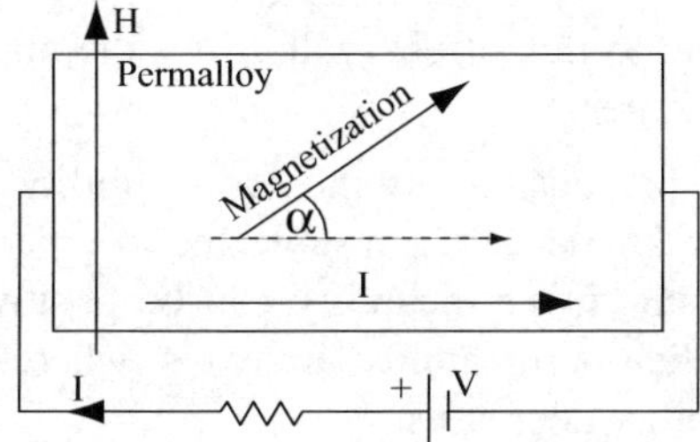

FIGURE 5.50 ■ Principle of operation of an AMR sensor.

TABLE 5.4 ■ Some AMR materials and some of their properties

Material	Resistivity $\rho = (1/\sigma) \times 10^{-8}$ Ωm	$\Delta\rho/\rho$ %
Fe19Ni81	22	2.2
Fe14Ni86	15	3
Ni50Co50	24	2.2
Ni70Co30	26	3.2
Co72Fe8B20	86	0.07

Note: The numerical values represent percentages (i.e., Fe19Ni81 means the material contains 19% iron and 81% nickel).

Magnetoresistive sensors usually come in a bridge configuration with four elements in the bridge. This allows adjustment of drift and increases the output from the sensor. As a whole, AMR sensors are very sensitive and can operate at low magnetic fields in a variety of applications, including electronic compasses and magnetic reading heads. Some materials exhibit enhanced magnetoresistive properties. These are often termed giant magnetoresistive (GMR) materials and their use results in GMR sensors with enhanced sensitivities and applications to very low field sensing.

5.7 | MAGNETOSTRICTIVE SENSORS AND ACTUATORS

The magnetostrictive effect is the contraction or expansion of a material under the influence of the magnetic field and the inverse effects of changes in magnetization due to stress in ferromagnetic materials due to the motion of the magnetic walls within the material. This bidirectional effect between the magnetic and mechanical states of a magnetostrictive material is a transduction capability that is used for both actuation and sensing. The effect is an inherent property of some materials. Most materials do not exhibit the effect, but others are strongly magnetostrictive. The effect was first observed in 1842 by James Prescott Joule. There are two effects and their reciprocals as follows:

Effects

1. The **Joule effect** is the change in length of a magnetostrictive sample due to an applied magnetization. This is the most common of the magnetostrictive effects and is quantified by the **magnetostrictive coefficient**, λ, defined as "the fractional change in length as the magnetization of the material increases from zero to its saturation value." Whereas the definition may be obscure to most people, its effects are common: the high-pitched sound emitted by a conventional TV or the humming of a transformer are due to this effect, caused by the magnetization and demagnetization of transformer cores.
2. The reciprocal effect, the change in the susceptibility (i.e., the permeability of the material changes) of a material when subjected to a mechanical stress, is called the **Villari effect**. The change in permeability can be positive or negative. A positive Villari effect is seen when permeability increases, whereas a negative Villari effect means that the permeability decreases.

Reciprocals

1. Twisting of a magnetostrictive sample. When an axial magnetic field is applied to the sample and a current passes through the magnetostrictive sample itself, the interaction between the two causes the twisting effect. This is known as the **Wiedemann effect**, and together with its inverse is used in torque magnetostrictive sensors.
2. The reciprocal effect, that of creation of an axial magnetic field by a magnetostrictive material when subjected to a torque, is known as the **Matteucci effect**.

The magnetostrictive effect is exhibited by the transitional metals, including iron, cobalt, and nickel, and some of their alloys. The magnetostrictive coefficients of some magnetostrictive materials are shown in **Table 5.5**. There are currently materials that exhibit what is called "giant magnetostriction," in which the magnetostrictive coefficient exceeds 1000 μL/L, such as the various metglas (metallic glass) materials and Terfenol-D. These materials are quickly becoming the materials of choice for magnetostrictive sensors and actuators.

The magnetostrictive coefficient is given for the saturation magnetization for each material and hence represents the largest expansion per unit length (or largest strain). For any other magnetic flux density, the strain is proportional. If the saturation magnetostriction is given as in **Table 5.5** at a saturation magnetic flux density B_m, then, assuming linear behavior, the expansion per unit length (strain) at any value of B is

$$\left(\frac{\Delta L}{L}\right)_B = \left(\frac{\Delta L}{L}\right)_{B_m} \times \left(\frac{B}{B_m}\right). \tag{5.59}$$

For a given length L_0, the value above must be multiplied by L_0 to find the absolute expansion of the device.

Some of the earliest uses of magnetostrictive materials date to the beginning of the twentieth century and include telephone receivers, hydrophones, magnetostrictive oscillators, torque meters, and scanning sonar. An early electrical sensor/actuator,

TABLE 5.5 ■ Saturation magnetostriction or magnetostrictive coefficient for some materials

Material	Saturation magnetostriction (magnetostrictive coefficient) [μm/m]	Saturation magnetic flux density [T]
Nickel	−28	0.5
49Co, 49Fe, 2V	−65	
Iron	+5	1.4–1.6
50Ni, 50Fe	+28	
87Fe, 13Al	+30	
95Ni, 5Fe	−35	
Cobalt	−50	0.6
$CoFe_2O_4$	−250	
Galfenol ($Ga_{0.19}Fe_{0.81}$)	50–320	
Terfenol-D ($Tb_{0.3}Dy_{0.7}Fe_2$)	2000	1.0
Vitreloy106A (metglas)(58.2Zr, 15.6Cu, 12.8Ni, 10.3Al, 2.8Nb)	20	1.5
Metglas-2605SC (81Fe, 3.5Si, 13.5B, 2C)	30	1.6

the first telephone receiver (the Reis telephone), tested by Johann Philipp Reis in 1861, was based on magnetostriction.

Applications for magnetostrictive devices include ultrasonic cleaners, high-force linear motors, positioners for adaptive optics, active vibration or noise control systems, medical and industrial ultrasonics, pumps, and sonar. In addition, magnetostrictive linear motors, reaction mass actuators, and tuned vibration absorbers have been designed. Less obvious applications include high cycle accelerated fatigue test stands, mine detection, hearing aids, razor blade sharpeners, and seismic detectors. Ultrasonic magnetostrictive transducers have been developed for surgical tools, underwater sonar, and chemical and material processing.

In general, the magnetostrictive effect is quite small and requires indirect methods for its measurement. There are devices, however, that use the effect directly. The operation of a basic magnetostrictive device is shown in **Figure 5.51**.

There are a number of methods by which magnetostrictive devices may be made to sense a variety of quantities. One of the simplest and most sensitive is to use magnetostrictive materials as the core of simple transformers. This is discussed below in the section on magnetometers. However, most of the applications of magnetostriction are for actuators, which will be described shortly. Nevertheless, indirect use of the magnetostrictive effect can be made to sense a variety of effects, such as position, stress, strain, and torque.

A noncontact magnetostrictive torque sensor for rotating shafts is shown in **Figure 5.52**. It consists of a sleeve of prestressed maraging steel (steel with 18% nickel, 8% cobalt, 5% molybdenum, and small amounts of titanium, copper, aluminum manganese, and silicon) tightly fitted on the shaft itself and two eddy current sensors arranged so that they are at 90° to each other and at 45° to the axis of the shaft, as shown in **Figures 5.52a** and **5.52b**. The torque sensor is based on two principles. First, the magnetostrictive steel, because it is stressed, when it is compressed its permeability decreases, since now the stress is reduced (negative Villari effect). When the steel is tensioned, its stress increases and its permeability increases (positive Villari effect). These can be seen in **Figure 5.52a**, which shows the main compression and tension lines. These are at 45° to the axis of the shaft, hence the choice for orientation of the eddy current sensors (**Figure 5.52b**). The second principle involved is the generation of eddy currents by an AC-driven coil. The eddy currents are influenced by the permeability of the material through the skin effect. A decrease in permeability will cause eddy currents to penetrate deeper into the steel sleeve, whereas an increase in permeability will cause shallower penetration (lower skin depth; see **Equation (5.35)**). Therefore, all that is now necessary is to relate these changes in eddy current induction to the torque responsible for changes in permeability.

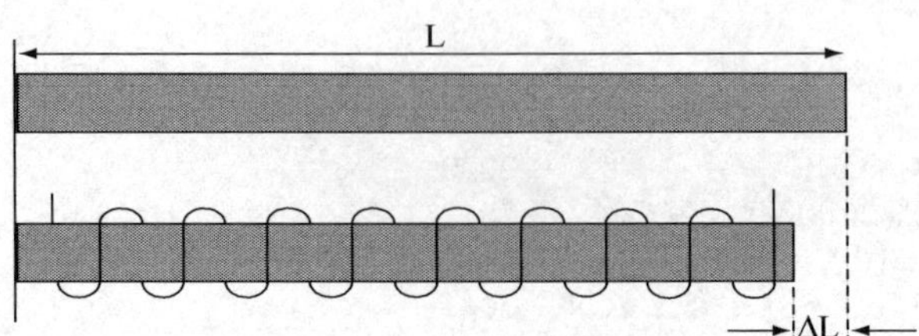

FIGURE 5.51 ■ Operation of a magnetostrictive device.

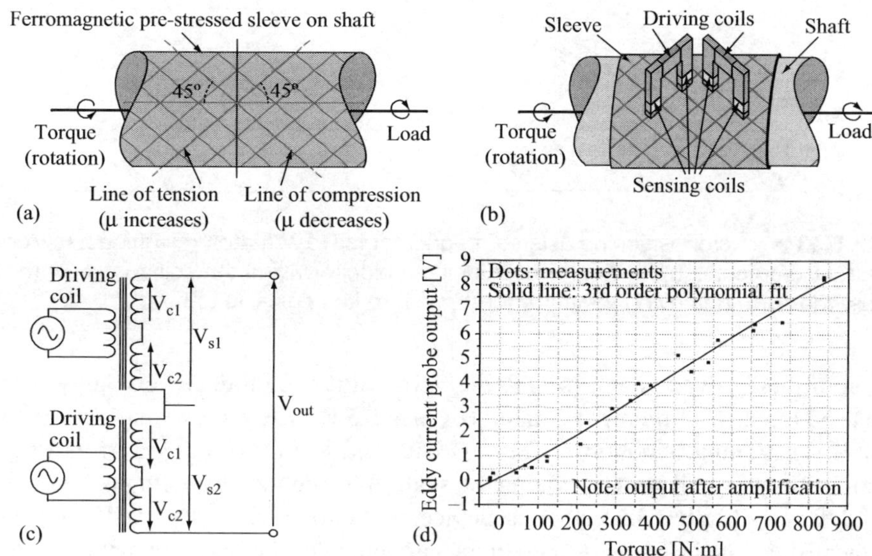

FIGURE 5.52 ■ A magnetostrictive torque sensor. (a) Prestressed ferromagnetic sleeve showing the axes of tension and compression. (b) Two c-core magnetic sensors positioned along the lines of compression and tension form a high-frequency transformer in which the flux closes through the sleeve. (c) The connection of the coils showing the differential output of the device. (d) The transfer function of the torque sensor.

The eddy current sensors consist of a driving coil at the center of the U-shaped core and two pickup coils at its tips. These act as a (high-frequency) transformer, and the voltage induced in the pickup coils depends on the stress condition in the maraging steel ring through which the flux closes (the open ends of the eddy current sensor cores are very close to the rotating sleeve to maximize output, but do not touch). The two driving coils on each sensor are connected in series so that their voltages add up. The two sensing coils are connected in a differential mode, as shown in **Figure 5.52c**.

To understand the operation, suppose first that the torque is zero. Because the two sensors are connected in differential mode, the net output is zero. As the torque increases, one sensor will experience an increase in output while the second will experience a decrease in output. Now the difference between them is the sum of the changes in voltages in the two sensors. Monitoring this voltage (which, of course, can be amplified as needed to produce a convenient span) gives a reading of torque. The experimental output of a sensor of this type is shown in **Figure 5.52d** for a shaft rotating at 200 rpm. As expected, the measurement introduces some errors, and the response of the sensor, although clearly not perfectly linear, is close to a linear response. The solid curve is shown as a polynomial fit of the experimental that can serve as the calibration curve for the sensor.

5.7.1 Magnetostrictive Actuators

Magnetostrictive actuators are quite unique. There are two distinct effects that can be used. One is the constriction (or elongation) or the torque effect produced by the Joule and Wiedemann effects discussed above. The other is due to the stress or shockwave that can be generated when a pulsed magnetic field is applied to a magnetostrictive material.

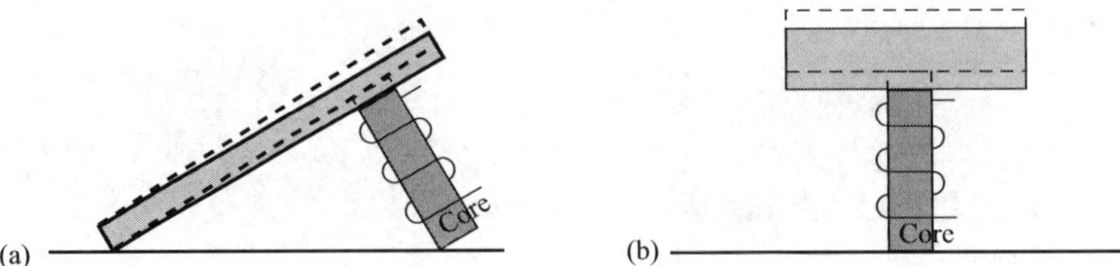

FIGURE 5.53 ■ Micropositioning using a magnetostrictive actuator. (a) Tilting a mirror for optical measurements. (b) Moving or positioning a block. Driving the coil constricts (or expands) the core. Removing the current returns it to its original length.

The first of these is very small (see **Table 5.5**), but it can produce very large forces. It can be used directly in micropositioning (**Figure 5.53**), where very small, accurate, and reversible positioning is possible. This has been used to move micromirrors to deflect light (and in other small structures), but we shall illustrate the idea with the "inchworm" motor shown in **Figure 5.54**. In this device, a bar of nickel is placed between two magnetically actuated clamps. A coil on the bar generates the requisite magnetostriction. By first clamping A, then turning on the current to the coil, end B contracts to the left. Now, clamp B is closed, clamp A is opened, and the current in the coil is turned off. End A elongates back to the original length of the bar and, in effect, the bar has now traveled to the left a distance ΔL, which depends on the magnetostrictive coefficient and the magnetic field in the bar. Whereas the motion in each step is only a few micrometers and motion is necessarily slow, this is a linear motion device that can exert relatively large forces and can be used for accurate positioning. Motion to the right is obtained by reversing the sequence of clamping and current pulses.

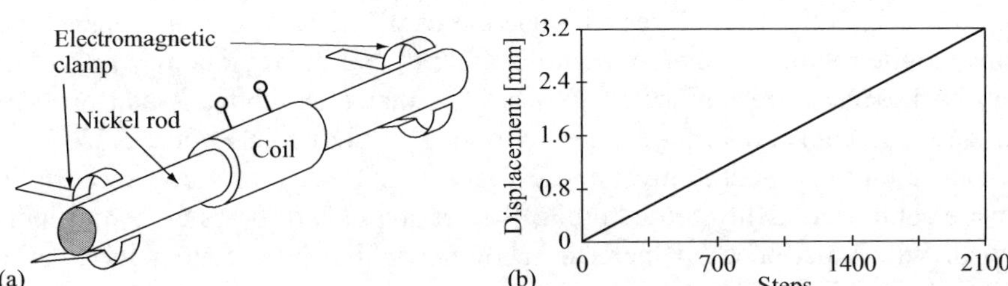

FIGURE 5.54 ■ A microstep inchworm motor. (a) Structure. (b) Response.

EXAMPLE 5.10 Linear magnetostrictive actuator

Because magnetostriction is a relatively small effect, many magnetostrictive actuators employ some means of mechanical amplification of the mechanical elongation or constriction of the magnetostrictive element. Consider the actuator in **Figure 5.55a**. The magnetostrictive bar is 30 mm long and is equipped with an elliptical ring or shell whose purpose is to amplify the horizontal magnetostrictive motion to a larger, vertical motion of the shell. The shell also closes much of the magnetic field, requiring a lower current for any given extension of the magnetostrictive bar. The magnetostrictive bar is made of Terfenol-D and the coil can generate magnetic fields varying from 0 to 0.4 T in the bar. Calculate the range of the vertical constriction of the shell.

Solution: Although the horizontal motion of the bar is immediately calculable from **Table 5.5**, we need to do some trigonometric calculations to translate this motion into the vertical motion of the shell. To do so we use **Figure 5.55b**, where we connected the vertical axis point with the horizontal axis point with a line of length l that makes an angle α with the horizontal axis. When the field is applied, the bar elongates (Terfenol-D has a positive magnetostriction coefficient) moving the horizontal axis point to the left and the vertical axis point down. The line now makes an angle β with the horizontal axis, but its length remains the same. From these two angles and the horizontal and vertical displacements we write

$$l\cos\beta = l\cos\alpha + \Delta x, \quad l\sin\beta = l\sin\alpha - \Delta y.$$

From the second relation,

$$\sin\beta = \frac{l\sin\alpha - \Delta y}{l} \rightarrow \cos\beta = \sqrt{1 - \sin^2\beta} = \frac{1}{l}\sqrt{l^2 - (l\sin\alpha - \Delta y)^2},$$

and substituting this into the first relation,

$$\sqrt{l^2 - (l\sin\alpha - \Delta y)^2} = l\cos\alpha + \Delta x.$$

Squaring both sides and arranging terms gives:

$$l^2 - (l\sin\alpha - \Delta y)^2 = (l\cos\alpha + \Delta x)^2 \rightarrow \Delta y^2 - 2l\sin\alpha\Delta y + \Delta x^2 - 2l\sin\alpha\Delta x$$
$$= l^2 - l^2\sin^2\alpha - l^2\cos^2\alpha.$$

The right-hand side of the relation is zero since $\sin^2\alpha + \cos^2\alpha = 1$ and we have a second-order equation with the unknown variable Δy in terms of known variables Δx, l, and α. Solving and taking only the positive root, we get

$$\Delta y = l\sin\alpha - \sqrt{l^2\sin^2\alpha - \Delta x(\Delta x + 2l\cos\alpha)} \quad [\text{m}].$$

In the case in **Figure 5.55a**, the angle α is

$$\alpha = \tan^{-1}\left(\frac{7.5}{15}\right) = 26.565°$$

and the length l is

$$\frac{15}{l} = \cos 26.565° \rightarrow l = \frac{15}{\cos 26.565°} = 16.77 \text{ mm}.$$

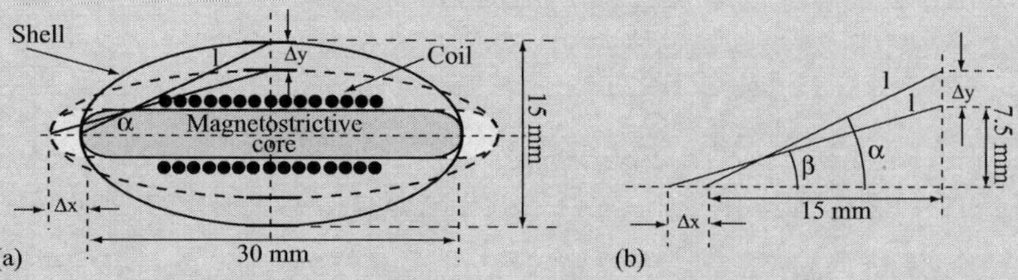

FIGURE 5.55 ■ A practical magnetostrictive actuator with mechanical amplification. (a) Structure and operation. (b) Calculation of displacement.

Now we can calculate Δx from **Table 5.5**. Since $\Delta l/l = 2000$ at a saturation flux density of 1 T, for a bar of length 30 mm and a flux density of 0.4 T we have a total expansion of

$$\Delta l = \left(\frac{\Delta l}{l}\right)_{B_m} \times \left(\frac{B}{B_m}\right) l = 2000 \times 10^{-6} \times \left(\frac{0.4}{1.0}\right) \times 30 = 2.4 \text{ mm}.$$

Now, since $\Delta l = 2\Delta x$ in **Figure 5.55b**, we can calculate Δy from the above:

$$\Delta y = l\sin\alpha - \sqrt{l^2\sin^2\alpha - \Delta x(\Delta x + 2l\cos\alpha)} = 16.77\sin 26.565^\circ$$

$$-\sqrt{(16.77)^2 \times \sin^2 26.565^\circ - 1.2 \times (1.2 + 2 \times 16.77 \times \cos 26.565^\circ)} = 3.163 \text{ mm}.$$

The shell moves twice as far since each side moves the same distance. The total motion is therefore 6.326 mm. In effect, the structure has generated an amplification of 6.326/2.4 = 2.64.

5.8 | MAGNETOMETERS

In general, magnetometers are devices that measure magnetic fields and, as such, the name can be assigned to almost any system that can measure the magnetic field. However, properly used, it refers to either very accurate sensors or low field sensing on the one hand or, on the other hand, complete systems for measuring the magnetic field, which include one or more sensors. We shall use the term as a sensor for low field sensing since it is in this capacity that magnetometers become unique devices. There are a number of methods, sometimes unrelated to each other.

5.8.1 Coil Magnetometer

To understand the fundamental methods of sensing we will start with the simplest idea of sensing a field—the small coil shown in **Figure 5.56**. In this coil the emf (voltage) measured across the coil is

$$emf = -N\frac{d\Phi}{dt} \quad [\text{V}], \text{ where } \Phi = \int_S BS\sin\theta_{BS} \quad [\text{Wb}], \tag{5.60}$$

where Φ is the flux through the coil, N is the number of turns in the coil, and θ_{BS} is the angle between the direction of the field and the normal to the area of the coil. This

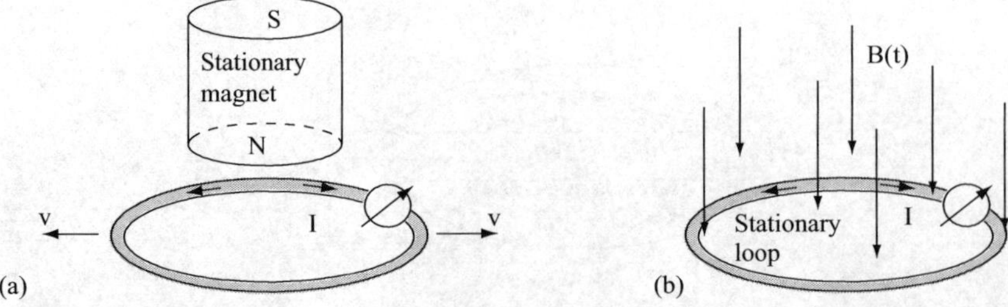

FIGURE 5.56 ■ Operation of a small search coil. (a) The coil moves in a constant magnet field. Induced emf (and hence current) depends on the motion of the coil in the DC magnetic field. (b) Induced current in a stationary coil by a time-dependent magnetic field.

relation is called Faraday's law of induction. The relations show that the output is integrating (dependent on the coil's area). This basic device indicates that to measure local fields, the area of the coil must be small, that sensitivity depends on the size and number of turns, the frequency, and, from **Figure 5.56**, that only variations in the field (due to motion in **Figure 5.56a** or due to the AC nature of the field in **Figure 5.56b**) can be detected. If the field is time dependent, it can be detected with stationary coils as well.

There are many variations on this basic device. First, differential coils may be used to detect spatial variations in the field. In other magnetometers, the coil's emf is not measured. Rather, the coil is part of an *LC* oscillator and the frequency is then inductance dependent. Any conducting and/or ferromagnetic material will alter the inductance and hence the frequency. This creates a relatively sensitive magnetometer often used in areas such as mine detection or buried object detectors (pipe detection, "treasure" hunting, etc.). While the simple coil, in all its configurations, is not normally considered a particularly sensitive magnetometer, it is often used because of its simplicity and, if properly designed and used, can be reasonably sensitive (e.g., magnetometers based on two coils have been used for airborne magnetic surveillance for mineral exploration and for detection of submarines).

EXAMPLE 5.11 The coil magnetometer

A magnetometer of surprising sensitivity can be made with a simple coil. Consider a device intended to detect and trace current-carrying wires buried in a wall or to map the magnetic field in a house or in the vicinity of power lines. The sensor itself is a simple coil, as shown in **Figure 5.57**. The emf produced by the coil based on **Equation (5.60)** is amplified and displayed or an alarm is sounded. The coil given here has 1000 turns and an average diameter of 4 cm and we wish to calculate its output due to AC magnetic fields. Assuming that a minimum output of 20 mV is needed to overcome background noise, what is the lowest magnetic flux density produced by a power line at 60 Hz that can be measured?

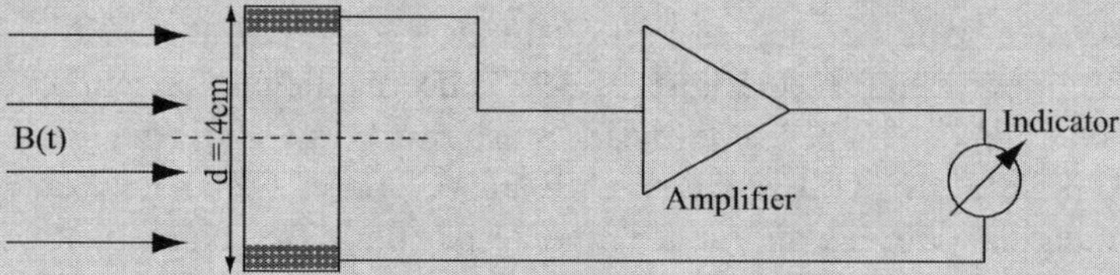

FIGURE 5.57 ■ The coil magnetometer. The amplifier is often needed unless the field is high.

Solution: We will assume that the magnetic flux density is uniform across the plane of the coil, and since the flux density is sinusoidal we write

$$emf = -N\frac{d\Phi(t)}{dt} = -NS\frac{dB(t)}{dt} = -NS\frac{d}{dt}B\sin(2\pi ft) = -2\pi fNSB\cos(2\pi ft) \quad [\text{V}].$$

For a 20 mV emf we have (neglecting the negative sign, which only indicates the phase of the emf relative to the magnetic flux density)

$$B = \frac{emf}{2\pi fNS} = \frac{20 \times 10^{-3}}{2\pi \times 60 \times 1000 \times \pi(2 \times 10^{-2})^2} = 4.22 \times 10^{-5}\ \text{T}.$$

This is on the same order of magnitude as the terrestrial magnetic flux density (approximately 60 μT).

Notes:

1. The device can only detect time dependent fields.
2. The sensitivity can be increased by increasing the number of turns, the dimensions of the coil, or if the frequency is higher. A ferromagnetic core in the coil can also help by concentrating the flux density in the coil.
3. The principle outlined here is the basis of many simple magnetometers or "field testers" ranging from "live wire detectors" in walls to "gauss meters" for general field measurement use, including electromagnetic compatibility tests and the detection of minerals 'by detecting variations in the terrestrial magnetic field using large differential magnetometers.

5.8.2 The Fluxgate Magnetometer

Much more sensitive magnetometers can be built on the basis of fluxgate sensors. The fluxgate sensor can also be used as a general purpose magnetic sensor, but it is more complex than the simple sensors described above such as the magnetoresistive sensor. It is therefore most often used where other magnetic sensors are not sensitive enough. Examples are electronic compasses, detection of fields produced by the human heart, or fields in space. These sensors have existed for many decades, but are rather large, bulky, and complex instruments specifically built for applications in scientific research. Recently they have become available as off-the-shelf sensors because of developments in new magnetostrictive materials that has allowed their miniaturization and even integration in hybrid semiconductor circuits. New fabrication techniques promise to improve these in the future, and as their size decreases, their uses will expand.

The idea of a fluxgate sensor is shown in **Figure 5.58a**. The basic principle is to compare the drive-coil current needed to saturate the core in one direction to that needed to saturate it in the opposite direction. The difference is due to the external field. In practice, it is not necessary to saturate the core, but rather to bring the core into its nonlinear range (see **Figure 5.19**). The magnetization curve for most ferromagnetic materials is highly nonlinear, so almost any ferromagnetic material is suitable. In practice, the coil is driven with an AC source (sinusoidal or square, but more often a triangular source), and under no external field, the magnetization is uniform throughout the coil and hence the sense coil will produce zero output. An external magnetic field

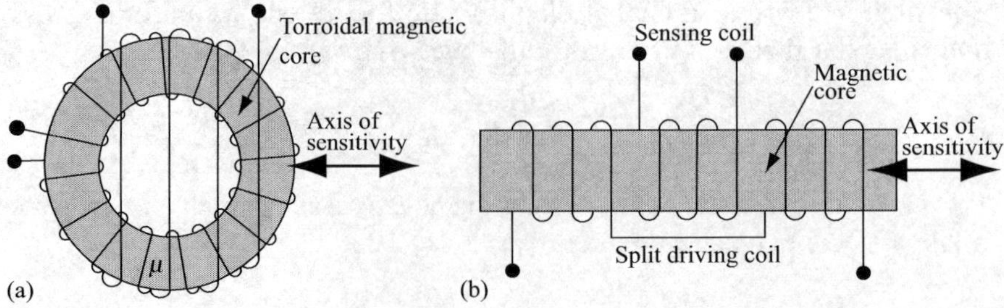

FIGURE 5.58 ■ Principle of the fluxgate magnetometer. (a) Toroidal implementation; the axis of sensitivity is shown. (b) Bar or film implementation. Sensitivity is along the long axis of the bar.

perpendicular to the sense coil changes the magnetization and, in effect, the core has now become nonuniformly magnetized, producing an emf in the sensing coil on the order of a few millivolts per microteslas [mV/μT]. The reason for the name fluxgate is this switching of the flux in the core in opposite directions. The same can be achieved by using a simple rod, as in **Figure 5.58b**. Now the two coils are wound one on top of the other and the device is sensitive to fields in the direction of the rod, but the principle is the same, the output relies on variations in permeability (nonlinearity) along the bar. A particularly useful configuration is the use of a magnetostrictive film (metglasses are a common choice), as shown in **Figure 5.58b** or a similar configuration. Because magnetostrictive materials are highly nonlinear, the sensors so produced are extremely sensitive—with sensitivities of 10^{-6} to 10^{-9} T. The sensors can be designed with two or three axes. For example, in **Figure 5.58a**, a second sensing coil can be wound perpendicular to the first. This coil will be sensitive to fields perpendicular to its area and the whole sensor now becomes a two-axis sensor. Fluxgate sensors are available in integrated circuits where permalloy is the choice material, since it can be deposited in thin films and its saturation field is low. Nevertheless, current integrated fluxgate sensors have lower sensitivities than the classical fluxgate sensors—on the order of 100 μT—but still higher than many other magnetic field sensors.

To better understand the operation of the fluxgate sensor, consider one particular and highly useful form, the so-called pulse-position fluxgate sensor. In this type of sensor, the current in the driving coil is triangular and produces a likewise triangular-shaped flux in the core. This flux may be considered a reference flux. An additional flux is produced by the external magnetic field—the field being measured. This flux either adds or subtracts from the internal, reference flux, but in any case, it produces an emf in the sensing coil. This is the measured flux. The two emfs—one from the reference flux and the other from the measured flux—are then compared. Whenever the reference field is higher than the measured field, the output is high, and whenever it is lower, the output is zero (we will see in **Chapter 11** that this can be accomplished by a simple circuit called a comparator). **Example 5.12** discusses this sensor and simulates an implementation to predict its sensitivity prior to construction.

EXAMPLE 5.12 The pulse-position fluxgate sensor: simulation results

To appreciate the value of simulation in design and to understand the operation of the fluxgate sensor, consider a fluxgate sensor as in **Figure 5.58a**, based on the pulse-position principle. The sensor and the circuits were simulated first and the results shown here are those of the simulation. This allows one to obtain results free of noise and only then build the actual circuit (shown in **Figure 5.60**). To do so, the properties of the core, the coils, and the electronic circuits are inserted into the simulation and appropriate signals are generated. In this case, a toroidal core with inner radius of 19.55 mm, outer radius of 39.25 mm, and cross-sectional area of 231 mm^2 with a saturation flux density of 0.35 T and relative permeability of 5000 is simulated. The core is wound with 12 turns for the drive coil and 50 turns for the sensing coil. The triangular reference signal oscillates at 2.5 kHz between −4.5 V and +4.5 V. The simulation produces any required number of results. We show three. **Figure 5.59a** shows the flux density in the core produced by the internal triangular generator. It shows a symmetric flux density ranging from +1 mT to −1 mT. **Figure 5.59b** shows the flux density in the core when an external flux density B_e of 500 μT

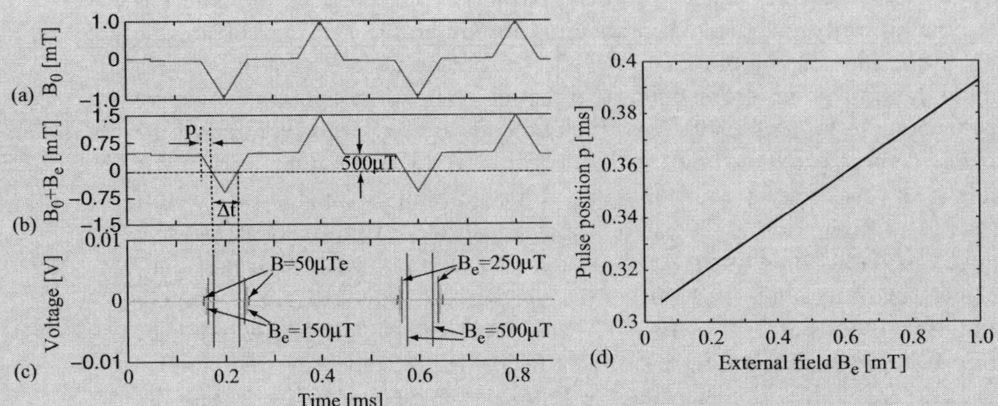

FIGURE 5.59 ■ (a) Magnetic flux density generated in the core using a triangular pulse generator. (b) The magnetic flux density in the core when an external flux density of 500 μT is measured. (c) The pulse positions obtained by detecting the zero-crossings. (d) Transfer function of the sensor given as the pulse position p versus the external magnetic flux density B_e.

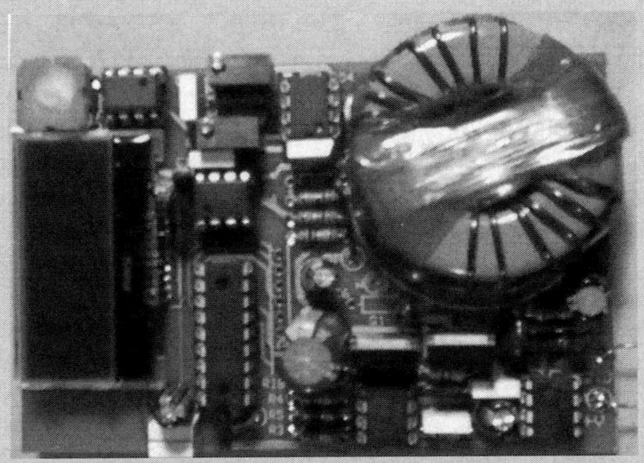

FIGURE 5.60 ■ Implementation of the fluxgate sensor. Note the toroidal coil at the upper right corner. In this implementation, based on a microprocessor, the flux density is displayed numerically on the display on the left.

is sensed. The flux density in **Figure 5.59b** is the sum of the reference (drive) flux density and the measured flux density, as can be seen by the shift upward of the zero line. The device detects the zero crossing of the flux density. The result is the output pulse in **Figure 5.59c**, showing the time width between the two comparison points. The pulse position can be taken as the distance between the pulses shown in **Figure 5.59c** and the position of the pulse (zero-crossing) when no field is applied, indicated in **Figure 5.59b** as p. Alternatively one can measure the time Δt shown in **Figure 5.59b**. **Figure 5.59d** shows the simulated transfer function indicating a linear relation and a sensitivity of 88 ms/T. This may not seem very large, but considering the fact that a time of 1 μs can be easily and accurately measured, the device can easily measure about 10 nT. The device implemented and tested based on the simulations described in this example is shown in **Figure 5.60**.

5.8.3 The SQUID

SQUID stands for superconducting quantum interference device. By far the most sensitive of all magnetometers, they can sense down to 10^{-15} T, but this kind of performance comes at a price—they operate at very low temperatures, usually at 4.2°K (liquid helium). As such, they do not seem to be the type of sensor one can simply take off the shelf and use. Surprisingly, however, higher temperature SQUIDs and integrated SQUIDs do exist. Even so, they are not as common as other types of sensors. The reason for including them here is that they represent the limits of sensing and have specific applications in the sensing of biomagnetic fields and in testing materials integrity.

The SQUID is based on the Josephson junction, formed by two superconductors separated by a small insulating gap (discovered in 1962 by Brian David Josephson). If the insulator between two superconductors is thin enough, the superconducting electrons can tunnel through it. For this purpose the most common junction is the oxide junction in a semiconductor, but there are other types. The base material is usually niobium or a lead (90%)–gold (10%) alloy with the oxide layer formed on small electrodes made of the base material that are then sandwiched to form the junction.

There are two basic types of SQUIDs. Radio frequency (RF) SQUIDs, which have only one Josephson junction and DC SQUIDs, which usually have two junctions. DC SQUIDs are more expensive to produce, but are much more sensitive.

If two Josephson junctions are connected in parallel (in a loop), electrons, which tunnel through the junctions, interfere with one another. This is caused by a phase difference between the quantum mechanical wave functions of the electrons, which is dependent upon the strength of the magnetic field through the loop. The resultant superconducting current varies with any externally applied magnetic field. The external magnetic field causes a modulation of the superconducting current through the loop, which can be measured (**Figure 5.61**). The superconducting current is set up externally by the sense loop (a single loop as in **Figure 5.61a** is used to measure fields, while two loops as in **Figure 5.61b** are used to measure the gradient in the field). The current may also be set up directly by the superconducting loop. The output is the change in voltage across the junction due to changes in the current, and since the junction is resistive, this change is measurable following amplification.

An RF SQUID operates in the same fashion except that there is only one junction and the loop is driven by an external resonant circuit that oscillates at high frequency (20–30 MHz). Any change in the internal state of the flux in the measurement loop (due to external, sensed fields) changes the resonant frequency, which is then detected and is a measure of the field.

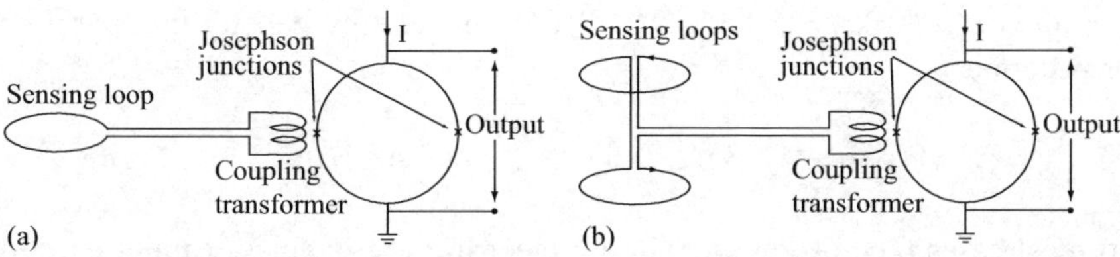

FIGURE 5.61 ■ SQUID structure and operation. (a) Measurement of magnetic fields. (b) Measurement of gradients in the magnetic field.

The main difficulty with SQUIDs is the cooling needed and the necessary bulk. Nevertheless, it is an exceedingly useful sensor where and when the cost can be justified. It is exclusively used in applications such as magnetoencephalography (measurement of magnetic fields of the brain) and in research. Many applications require that measurements of very low magnetic fields be done in shielded rooms where the terrestrial magnetic field as well as other sources, such as those from power lines, can be eliminated.

5.9 | MAGNETIC ACTUATORS

We have already discussed magnetostrictive actuators, but there are many other types of magnetic actuators and many of them are more conventional. In particular, the whole group of electric motors forms the bulk of conventional (and some less than conventional) actuation applications. The main reason that magnetic actuation is usually the best choice has to do with energy density, just as the reason for the use of the internal combustion engine has to do with energy density. An example is instructive. If we restrict ourselves to electrical actuation, there are two basic types of forces that exist. One is the Coulomb force, which acts on a charge in the electric field (see **Section 5.3.3**). The second is the magnetic force, which acts on a current in a magnetic field as defined by the Lorentz force in **Section 5.4**. These lead to energy densities that are given as follows:

Electric energy density:

$$w_e = \frac{\varepsilon E^2}{2} \quad [\mathrm{J/m^3}] \tag{5.61}$$

Magnetic energy density:

$$w_m = \frac{B^2}{2\mu} \quad [\mathrm{J/m^3}], \tag{5.62}$$

where E is the electric field intensity and B is the magnetic flux density. Now, by way of comparison, suppose we use a common value for the magnetic flux density B in an electric motor of 1 T. The permeability of iron in the motor is about $1000\mu_0 = 1000 \times 4\pi \times 10^{-7}$ H/m. This gives an energy density in the motor of

$$w_m = \frac{1^2}{2 \times 1000 \times 4\pi \times 10^{-7}} \approx 400\ \mathrm{J/m^3}. \tag{5.63}$$

On the other hand, the relative permittivity of most common dielectrics is less than 10. Taking an electric field intensity of 10^5 V/m (this is a very large electric field intensity) and a permittivity of $10\varepsilon_0 = 8.845 \times 10^{-11}$ F/m, the energy density in a purely electric actuator becomes

$$w_e = 8.854 \times 10^{-11} \frac{10^{10}}{2} \approx 0.45\ \mathrm{J/m^3}. \tag{5.64}$$

Note: In most cases the absolute maximum electric field intensity cannot exceed a few million volts per meter without breakdown, so increasing the electric field to its absolute maximum will only increase the energy density by about one order of magnitude.

Clearly then, a magnetic actuator will be capable of exerting larger forces in a smaller package than an electric actuator. In addition, it is usually easier and much safer to generate fairly large magnetic fields than to generate large electric fields. Nevertheless, it should be mentioned that electric forces and electric actuators have their own niche application in MEMS, where the required forces are small and small energy densities are acceptable (see **Chapter 10**). They are also useful in electrostatic filters and dust collectors, where the small forces are sufficient to act on charged dust particles and collect them on electrodes. Electrostatic actuation is also critical to the operation of copiers and printers, where toner particles are distributed on a printing medium using electrostatic forces.

But the inevitable consequence is that most electric actuators are based on magnetic forces, and among these, motors feature prominently. However, there are many types of motors and related devices, such as voice coil actuators and solenoids. Motors and solenoids will be discussed at length, but we start with a particular type of magnetic actuator called the voice coil actuator, because the discussion also introduces the principle upon which motors are based.

5.9.1 Voice Coil Actuators

Voice coil actuators got their name from their first and perhaps still their most widely used implementation—that of magnetically driven loudspeakers. In most applications of voice coil actuators, there is no use of voice—only the similarity in operation. The actuators are based on the interaction between the current in a coil and the magnetic field of a permanent magnet or another coil. To understand this, consider the basic structure of a loudspeaker mechanism shown in **Figure 5.62** (loudspeakers will be discussed in their own right in **Chapter 7**). The magnetic field in the gap is radial. For a current-carrying loop, the force is given by **Equation (5.26)** (Lorentz force), where L is the circumference of the loop and we assume a uniform magnetic field. With N turns, the force is $NBIL$. Of course, the field does not have to be uniform or the coil circular, but this is a simple configuration and is the one used in most speakers. The larger the current, the larger the force, and thus the larger the displacement of the speaker's cone. By reversing the current, the coil moves in the opposite direction. Before we proceed we should note the following:

1. The force is directly proportional to current for a given magnetic field. In this case (and in many voice coil actuators) it is linear with current, an important property of the device.

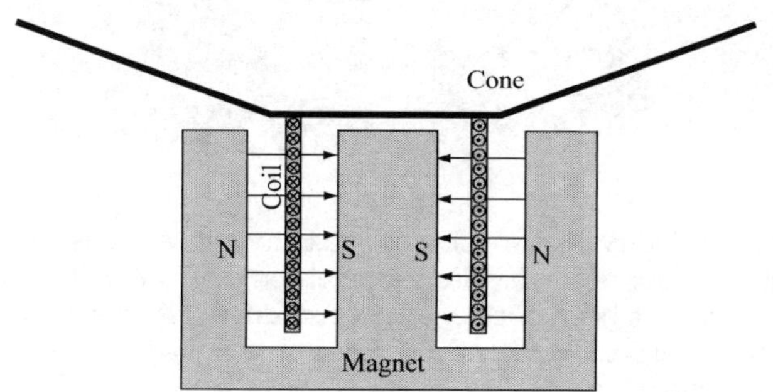

FIGURE 5.62 ■ The structure of a loudspeaker showing the interaction between the magnetic field and current in the coil.

2. The larger the coil or the magnetic field, the larger the force.
3. By allowing the coil to move, the displaced mass is small (compared with other actuators) and hence the mechanical response is rather good. For this reason, a speaker can operate at, say, 15 kHz, whereas a motor-driven actuator may take seconds to reverse.
4. It is also possible to fix the coil and allow the magnet to move.
5. The field in the actuator can be generated by an electromagnet if necessary.
6. We note here that the voice coil actuator can be turned into a sensor by simply reversing the action. If we move the coil in the magnetic field, the voltage induced in the coil will be given by Faraday's law of induction through **Equation (5.60)**. The speaker becomes a microphone and the more general voice coil actuator becomes a sensor.
7. In the absence of current, the actuator is entirely disengaged—there is no intrinsic retaining or cogging force and no friction. However, in some cases use may be made of a restoring spring to return the actuator to its rest position, as is done in speakers.
8. The motion is limited and often quite short.
9. Rotational motion can also be achieved by selecting particular coil and magnet configurations (see **Figure 5.63b**).
10. The actuator is a direct drive device.

Among these properties, the main quality that has made voice coil actuators so appealing beyond their use in speakers is that their small mass allows very high accelerations (upwards of 50 g and for very short strokes up to 300 g) at high frequencies, making them ideal candidates for fast positioning systems (e.g., in positioning read/write heads in disk drives). The forces achievable are modest in comparison to other motors, but are certainly not negligible (up to 5000 N) and the power they can handle is also significant.

Voice coil actuators are often used where very accurate positioning at high speeds is needed. Since they have no hysteresis and minimal friction, they are extremely accurate both as linear and as angular positioners. No other actuator matches their response and

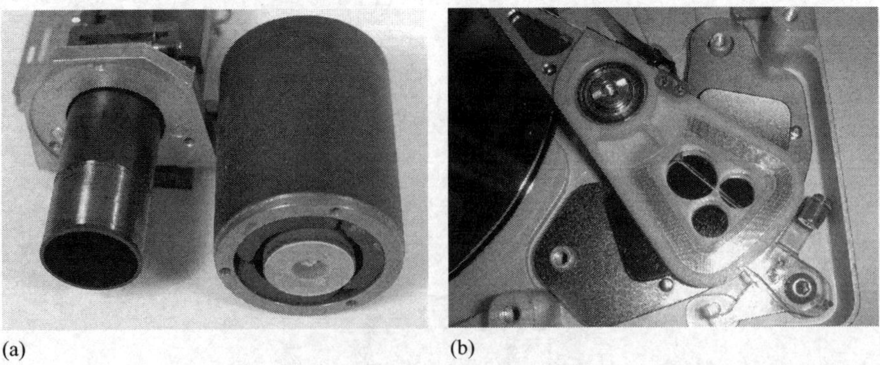

(a) (b)

FIGURE 5.63 ■ (a) A cylindrical linear voice coil actuator shown disassembled. The magnets and the gap in which the coil assembly fits are visible on the front of the assembly. (b) Rotary voice coil actuator used as a head positioner in a disk drive. Two permanent magnets can be seen under the trapezoidal coil. A steel cover closes the magnetic circuit above the coil, but it has been removed to show the coil.

acceleration. They can also be used in less critical applications, mostly in positioning and control, and also in valve actuation, pumps, and the like. Interfacing with microprocessors is usually simpler than other types of motors, and control and feedback are easily incorporated.

A large variety of voice coil actuators are available, but the cylindrical actuator in **Figure 5.63a** and the rotary actuator in **Figure 5.63b** are typical. In the cylindrical linear actuator, the magnetic field is radial, as in the loudspeaker. The coil, attached to the moving, actuating shaft, moves in and out from a center position with a maximum stroke defined by the length of the coil and the length of the cylindrical magnet. For motion to be linearly proportional to current, the coil must be within the uniform magnetic field during the entire range of motion. Ratings of these actuators are in terms of stroke, force (in newtons), acceleration, and power.

EXAMPLE 5.13 Force and acceleration a in voice coil actuator

Consider the voice coil actuator shown in **Figure 5.64a**. The coil is wound on a plastic form that rides on the inner core. Assume the coil has 400 turns, the current needed for operation is 200 mA, and that the coil never leaves the area of uniform field. The field itself is produced by a permanent magnet and equals 0.6 T anywhere within the space occupied by the moving part of the actuator. The given dimensions are $a = 2$ mm, $b = 40$ mm, and $c = 20$ mm. Calculate the force and acceleration of the actuator if the moving part weighs 80 g.

Solution: We will use **Equation (5.26)** since it gives the force on a length of wire, but we will modify it slightly to fit this configuration. Since the length of the loops vary with position in the coil (i.e., the length is $2\pi r$, where r is the radius of the loop), we first calculate a current density in the cross section of the coil by multiplying the current by the number of turns and then dividing by the cross-sectional area of the coil:

$$J = \frac{NI}{ab} = \frac{400 \times 0.2}{0.002 \times 0.04} = 1 \times 10^6 \text{ A/m}^2.$$

Now we define a ring of current of thickness dr and radius r as shown in **Figure 5.64b**. The total current in this ring is

$$dI = Jds = Jbdr \quad [\text{A}].$$

The length of the current ring is taken as its circumference since it is on this length of the current that the magnetic flux density acts. The force on this ring of current is

$$dF = BLdI = B(2\pi r)Jbdr \quad [\text{N}].$$

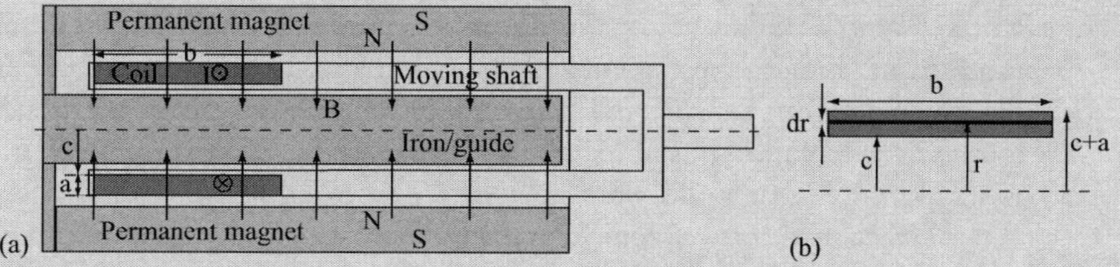

FIGURE 5.64 ■ (a) A cylindrical voice coil actuator. (b) Calculation of force on the coil.

Thus the total force on the coil is

$$F = 2\pi BJb\int_{r=c}^{c+a} rdr = 2\pi BJb\left[\frac{r^2}{2}\right]_{r=c}^{r=c+a} = \pi BJb\left[(c+a)^2 - a^2\right] \quad [\mathrm{N}].$$

Numerically this is

$$F = \pi BJb\left[(c+a)^2 - a^2\right] = \pi \times 0.6 \times 1 \times 10^6 \times 0.04\left[(0.04 + 0.002)^2 - 0.002^2\right] = 120.64\ \mathrm{N}.$$

The acceleration is calculated from Newton's law:

$$F = ma \rightarrow a = \frac{F}{m} = \frac{120.64}{0.080} = 1507.96\ \mathrm{m/s^2}.$$

With a force of 120.64 N and an acceleration of just over 150 g, this is a respectable actuator ($g = 9.81\ \mathrm{m/s^2}$).

Note: The force can be increased by increasing the radius of the moving coil, by increasing the number of turns, and/or by increasing the current in the coil. On the other hand, increasing the number of turns and physical dimensions increases the mass of the coil and hence tends to reduce acceleration, and necessarily reduces the response time of the actuator.

5.9.2 Motors as Actuators

The most common of all actuators are electrical motors in their many types and variations. It would be totally presumptuous to even try to discuss all motors, their principles, and their applications here—many volumes have been used to do so. It is important, however, to discuss some of the more salient issues associated with their use as actuators. Also, at the outset, it should be understood that motors can be used, and often are, as sensors. In fact, many motors can be used as generators, in which capacity they can sense motion, rotation, linear and angular position, and any other quantities that affect these, such as wind speed, flow velocity and rate, and much more. Some of these sensor applications will be discussed throughout this text (e.g., wind speed is commonly sensed and measured by a rotating cup arrangement that turns a small motor and the output is then proportional to wind speed), but for now we shall concentrate on their use as actuators.

Also to be recognized at this stage is that most motors are magnetic devices—they operate by attraction or repulsion between current-carrying conductors or between current-carrying conductors and permanent magnets in a manner similar to that of voice coil actuators. Unlike voice coil actuators, motors include magnetic materials (mostly iron), in addition to permanent magnets or electromagnets, to increase and concentrate the magnetic flux density and in so doing increasing power, efficiency, and available torque in the smallest possible volume. The variation in size, and power, they can deliver is staggering. Some motors are truly tiny. For example, the motors used as vibrators in cell phones are about 6–8 mm in diameter and no more than 20 mm long—some are flat, the size of a small button. On the other end, motors delivering hundreds of megawatts of power are used in the steel and mining industries. Perhaps the largest are generators in power plants—these can be 1000 MW or more. Yet there is no fundamental difference in operation between these devices.

Motors, as their name implies, deliver motion in one form or another. As such, many devices can be called motors. For example, the windup spring mechanism in a clock is a true motor.

As a general classification for actuation purposes, there are three types of motors: continuous rotational motors, stepper motors, and linear motors. Of these, the best known to the casual observer is the continuous rotational motor. However, stepper motors are much more common than one realizes, and linear motors, while not as common, are increasingly finding application in specialized systems. In a continuous motor, the shaft rotates in one direction as long as power is supplied to the motor. Some motors can be reversed (such as DC motors), while some cannot. Stepper motors provide discrete motion or steps of predetermined size. A pulse given to the motor will move it a fraction of a rotation (1°–5° is typical). To move it further, an additional pulse is necessary. This is a boon to positioning, where the accurate and repeatable steps of these motors is very useful. Linear motors are somewhere in between. First, their motion is linear rather than rotational. Because of this, they cannot be truly continuous and must be reversible. Often they are stepping motors, but they can be continuous in the range of their linear motion.

When used in actuation, some form of control is often necessary. One may need to control the speed, direction of motion, number of steps, torque applied, etc. These controls are often accomplished with microprocessors and interfacing circuits (**Chapter 12**) and become part of the actuation strategy for the system. An important class of motors is the so-called brushless DC (BLDC) motors, which is somewhere between continuous and stepping motors. Its importance comes from its control and, as a consequence, from its use in a plethora of applications, especially in data storage drives, CD drives, and in some rotary tools and toys, such as toy airplanes.

In the following sections we shall discuss, briefly, some of the important properties of motors, such as torque, power, speed, etc., and common types of motors used for actuation. Although no specific power range is implied, it should be understood that very large motors have special requirements that small motors do not (mechanical structures, power supplies, cooling, etc.). Therefore the discussion here should be viewed as relating to low-power motors.

5.9.2.1 Operation Principles

All motors operate on the principle of repulsion or attraction between magnetic poles. As an initial discussion, consider the device in **Figure 5.65a**. The two magnets are kept

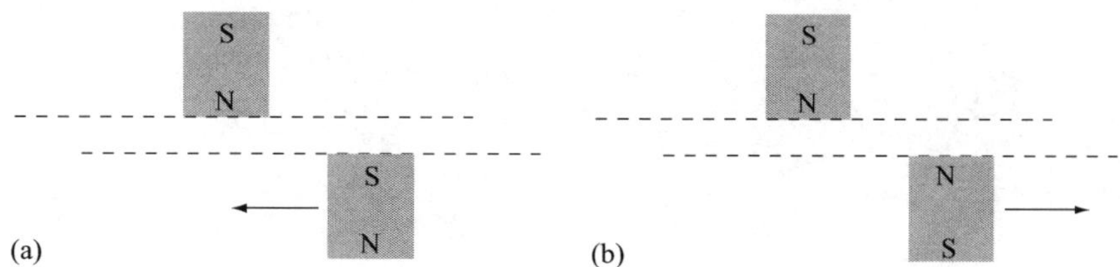

FIGURE 5.65 ■ Forces between two magnetic poles. In this schematic, the upper magnet is fixed (stator) and the lower is free to move (rotor). (a) Attraction. (b) Repulsion.

separated vertically, but the lower magnet is free to move horizontally. The two opposite poles attract and the lower magnet will move to the left until it is aligned with the upper magnet. **Figure 5.65b** is similar, but now the two magnets repel each other and the lower magnet will move to the right. In this very simple example, this is the extent of motion, but this is the first and most fundamental principle of motors. As a matter of nomenclature, the stationary pole is called a stator, whereas the moving pole is the rotor (or in the case of linear motors, a slider).

To make this into a more useful device, consider **Figure 5.66a**. Here the configuration is somewhat different, but the principle is still the same. The magnetic field (which may be produced by a permanent magnet or an electromagnet) is assumed to be constant in time and space for now. If we apply a current to the loop, and assuming the loop is initially at an angle to the field, as shown in **Figure 5.66b**, a force will exist on each of the upper and lower members of the loop equal to $F = BIh$ (Lorentz force in **Equation (5.26)**, where B is the magnetic flux density, I is the current, and h is the length of the loop). This force will rotate the loop to the right one-half turn, until the loop is perpendicular to the magnetic field. Note that the Lorentz force is always perpendicular to both the current and the magnetic field. For a motor to operate continuously, when the loop reaches this position, the current in the loop is reversed (commutated) and, assuming that the loop rotates slightly past the perpendicular position due to inertia, the force now will continue rotating it clockwise an additional half turn, and so on. Note also that the force on the loop is constant (independent of position). Obviously some additional issues have to be resolved, otherwise the loop can get stuck into a vertical position. However, without complicating the issue, it is obvious from this configuration that the motor develops a force and the force, acting on the loop, develops a torque. The latter is

$$T = 2BIhr \sin \alpha \quad [\mathrm{N \cdot m}], \tag{5.65}$$

where r is the radius of the loop. If multiple loops are used, the force, and hence the torque, is multiplied by the number of loops, N. This particular configuration requires commutation, and this can be done mechanically or electronically. **Figure 5.67a** shows the same configuration with a mechanical commutator and a permanent magnet stator producing the magnetic field. This is a simple DC motor. As the loop rotates, the commutator rotates with it. The contact with the brushes reverses the current at the right moment at the end of each turn. The number of coils can be increased, say, to two as in

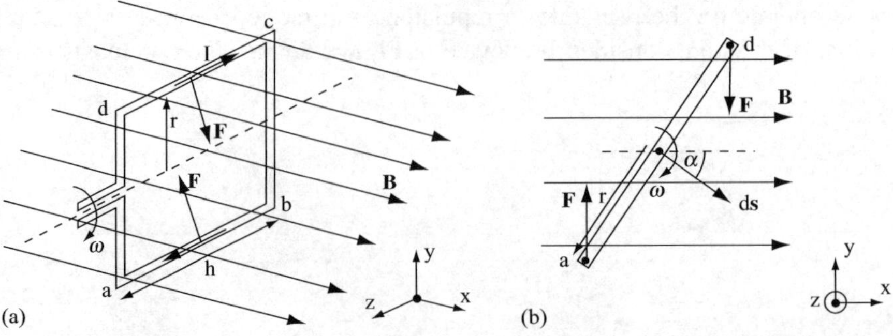

FIGURE 5.66 ■ A loop in a magnetic field showing the force on the current in the loop. (a) The force rotates the loop to the right until the loop's area is perpendicular to the field. (b) Relation between force and position of the loop.

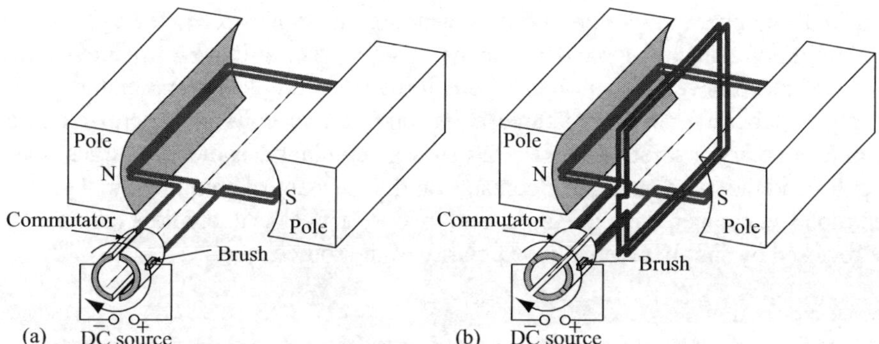

FIGURE 5.67 ■ A commutated DC motor. (a) Single coil with two commutating connections. (b) Two perpendicular coils, each with a pair of commutating connections. The brushes are stationary.

Figure 5.67b. In this case, there are four connections on the commutator so that each coil is powered in the appropriate sequence to ensure continuous rotation. In practical motors of this type, many more coils are used spaced equally around the circumference accompanied by additional commutator connections. This increases torque and makes for smoother operation due to commutation. Most small DC motors are made in this configuration or a modification of it. One particular modification is to use electromagnets for the stator and to add additional poles for the magnetic field (also spaced equally). **Figure 5.68** shows a small motor with two stator poles and eight rotating coils (note the way they are wound). The addition of iron increases force and torque. This motor is called a universal motor, and can operate on DC or AC and is perhaps most common in AC-powered hand tools. They can develop a fairly large torque but are very noisy. Typically the stator coils are connected in series with the rotor coils (series universal motors), although parallel connection is also possible. The motor in **Figure 5.68** is a high-speed universal motor as used in a hand tool. It also shows one of the problems common in these motors—damage to the commutator due to sparks developed when brushes (carbon contacts) slide over the commutator in normal operation. These brushes also wear out over time, reducing motor performance.

In many applications, and especially for low-power DC applications, a modification of this basic configuration has the magnetic field produced by a pair (or more) of permanent

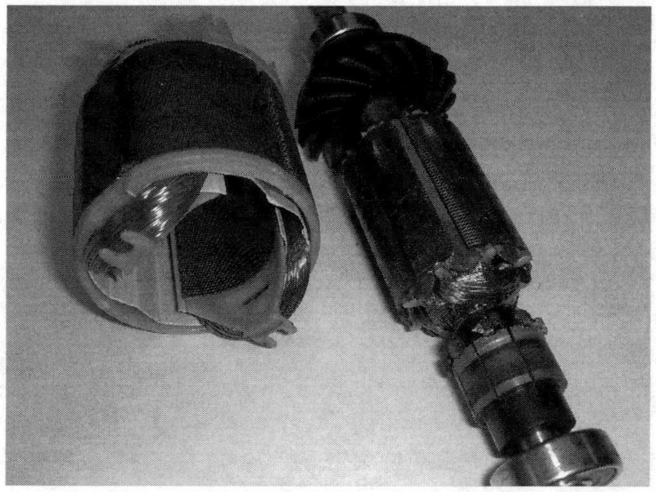

FIGURE 5.68 ■ The rotor and stator of a universal motor. Note the method of winding the coils and the damage to the commutators.

magnets and a number of poles produced by windings, as shown in **Figure 5.69**. In this case there are three poles on the rotor and two on the stator (marked as P in **Figure 5.69b**), ensuring that the motor can never get stuck in a zero force situation. The commutator operates as previously, but because there are three coils, one or two coils are energized at a time (depending on rotary position). **Figure 5.69c** shows a similar but somewhat larger motor with seven poles and the same number of contacts on the mechanical commutator. These motors are commonly encountered in tape drives and in toys, as well as in cordless hand tools. They can be reversed by simply reversing the polarity of the source.

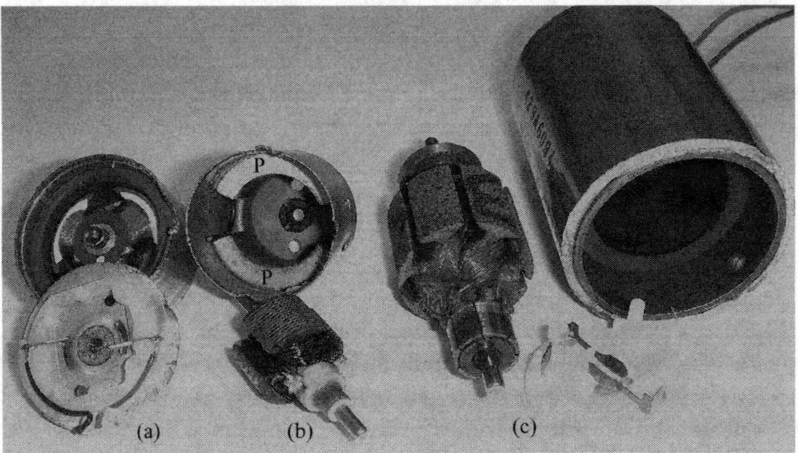

FIGURE 5.69 ■ Three small motors showing the rotors, stators, and commutators. Left and middle: three-pole rotor, two-pole stator. Right: seven-pole rotor, two-pole stator. The sliding "brushes" are seen at the bottom of the figure. In larger motors the brushes are made of carbon or graphite.

EXAMPLE 5.14 Torque in a commutated DC motor

A permanent magnet DC motor based on **Figure 5.67a** has an iron rotor on which the square coil is wound. This guarantees that the magnetic flux density in the gap between the stator and rotor where the coil resides is constant and high. Suppose the magnetic poles produce a magnetic flux density of 0.8 T. The coil is 60 mm long, has a radius of 20 mm, and contains 240 turns.

a. Calculate the maximum torque the motor can generate for a current of 0.1 A in the coils.

b. Suppose now that a second coil is added, as in **Figure 5.67b**. How does the answer in (a) change?

Solution: The fact that the flux density in the coil is constant provides a simple expression for the forces and torque.

a. The torque on the coil depends on the position of the coil, with a maximum when the flux density is parallel to the area of the coil (see **Figure 5.66b** and **Equation (5.65)**) and a minimum when perpendicular. The force on a single loop was calculated above and is available from

Equation (5.26):

$$F = BIL \quad [\text{N}].$$

In the motor discussed here, the length is $L = 0.06$ m, the radius is $r = 0.02$ m, the current is $I = 0.1$ A, and the flux density is $B = 0.8$ T. Since there are $N = 240$ turns in the coil, the total force on the coil is (the force F in **Figure 5.66b**)

$$F = NBIL = 240 \times 0.8 \times 0.1 \times 0.06 = 1.152 \text{ N}.$$

The force is constant regardless of the position of the coil. However, the torque is maximum when the coil is parallel to the field since then the force F is perpendicular to the plane of the coil. Thus the maximum torque is

$$T = 2Fr = 2 \times 1.152 \times 0.02 = 0.046 \text{ N} \cdot \text{m}.$$

b. The torque remains the same as in (a) except that whereas in (a) it varies from maximum to minimum (zero) in one-half turn, in (b) it varies from maximum to minimum in one-quarter turn, producing a smoother torque with the rotation of the motor.

5.9.2.2 Brushless, Electronically Commutated DC Motors (BLDC Motors)

DC motors may be sufficient for simple applications but their control (speed and torque control) is somewhat complex. In addition, the mechanical commutator is electrically noisy (generates sparks and therefore magnetic fields that can interfere with electronic circuits) and wears out with time. For more demanding applications, such as in disk drives, a variation of this motor is used in which the commutation is done electronically. In addition, the physical structure is often different to allow fitting in tight spaces or incorporation with integrated circuits. These motors are often flat (a popular name is flat motors) and often the rotor is a mere disk. An additional important aspect is that now, since commutation is electronic, the coils are stationary and the magnets rotate. These motors can be viewed as a type of stepper motor (we shall talk about stepper motors next), but BLDC motors are typically used for continuous rotation and hence their discussion here.

To understand their operation, consider **Figure 5.70a**. It shows a brushless DC motor with six coils forming the stator. The rotor has been taken out of its bearing and inverted to reveal the stator and the structure of the rotor. The coils are placed directly on a printed circuit board. The rotor, shown on the left, has a ring made of eight separate magnets so that the sides facing the coil (up in this figure) alternate in their magnetic field (the individual magnets can be distinguished by the brighter lines separating them (**Figure 5.70b**)). Note also the three Hall elements placed in the middle of three of the coils—these are used to sense the position of the rotating magnets for the purpose of control of speed and direction of rotation. The operation of the motor relies on two principles. First, the pitch of the stator and rotor are different (six coils but eight magnets). Second, the positions of the magnets are sensed and this information is used to drive the coils, measure the speed, and reverse the sense of rotation. By driving sequential pairs of coils, the device can be made to rotate in one direction or the other. The exact timing for switching the coils is obtained from the three Hall elements. **Figure 5.71** explains the sequence. Suppose that the initial condition is as shown in **Figure 5.71a**. The initial condition of the coils with respect to the magnets is sensed by

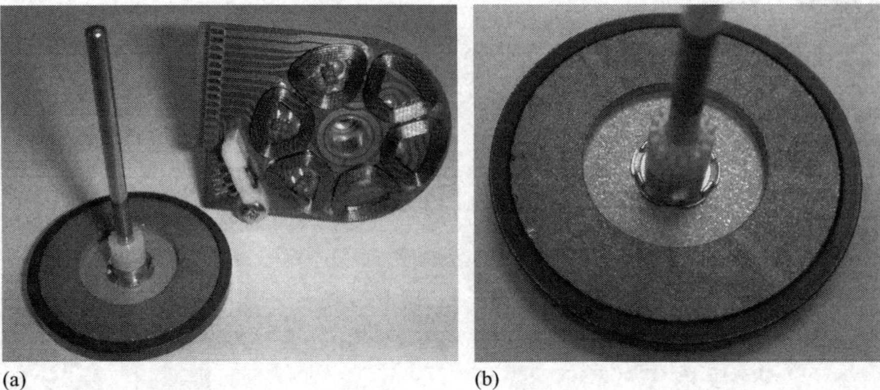

FIGURE 5.70 ■ A flat, electronically commutated DC motor (brushless DC motor or BLDC motor). (a) View of the separated rotor and stator. (b) A closer view of the rotor showing the separation between the individual magnets (lighter strips).

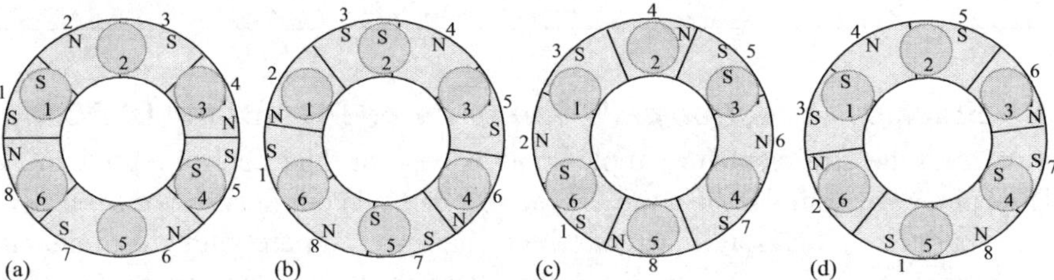

FIGURE 5.71 ■ Operation of a flat brushless DC motor. Circles represent coils, segments represent magnets with alternating polarities (marked).

the Hall elements so that a predictable direction of rotation can be defined. The magnets are shown behind the round coils and polarity is indicated on the periphery to avoid clutter. The coils are shown as grey circles and their polarity (on the side facing the magnets) is indicated in black. Now suppose coils 1 and 4 are driven so that their polarity is as shown in **Figure 5.71a**. That is, their side facing the magnets is *S*. Coil 1 will repel magnet 1 and will attract magnet 2, whereas coil 4 will repel magnet 5 and attract magnet 6. This will rotate the rotor (magnets) to the left until coil 1 is centered with magnet 2 and coil 4 is centered with magnet 6. Next, coils 2 and 5 are driven in the same way (as shown in **Figure 5.71b**). Now coil 2 will repel magnet 3 and attract magnet 4, whereas coil 5 will repel magnet 7 and attract magnet 8. Again, the magnets are forced to rotate left until coil 2 is centered with magnet 4 and coil 5 with magnet 8. This is shown in **Figure 5.71c**. Now the process repeats, but coils 3 and 6 are driven. Again rotation to the left is obtained until the coils and magnets are as in **Figure 5.71d**. This is identical to **Figure 5.71a**, and the process repeats indefinitely. This is said to be a three-phase operation and can be done digitally since all it requires is to ascertain the location of the magnets and drive the opposite coils according to the sequence above. Note that by reversing the coils currents, the north (N) poles are operating against the magnets and rotation is in the opposite direction.

This type of motor is the common choice in most digital devices such as disk drives, CD drives, video recorder heads, tape drives, and many others because it can be controlled very easily and its control is essentially digital. The speed is controlled by timing

FIGURE 5.72 ■ The stator (left) and rotor (right) of an electronically commutated DC motor (from a CD drive). The magnets are placed on the interior rim of the rotor. Note also the three Hall elements used to sense the rotor.

FIGURE 5.73 ■ A sensorless BLDC motor for electrical model aircraft applications. The speed of rotation is up to 70,000 rpm. Note the permanent magnets around the circumference of the stator.

the three phases at will. However, there are many variations in terms of the actual construction, shape, and number of magnets and coils, etc. One additional form is shown in **Figure 5.72**. In this case the magnets are placed on the inner side of the rim of the rotor and the coils are wound on an iron core to increase torque. A three-phase BLDC motor used for electrical model aircraft is shown in **Figure 5.73**. The particular motor shown can rotate at speeds up to 70,000 rpm.

Another development of BLDC motors is to dispense with the Hall elements for sensing and use the coils themselves as sensors for positioning by monitoring the induced emf in those coils that, at any given instant, are not driven. This allows for control of the motor in a sensorless mode. The control is more complicated, but there is no need for separate Hall elements to sense position. The motor in **Figure 5.73** is of this type.

5.9.2.3 AC Motors

In addition to DC motors, there is a large variety of AC motors. The most common of the conventional motors is the induction motor in its many variants. Without going into

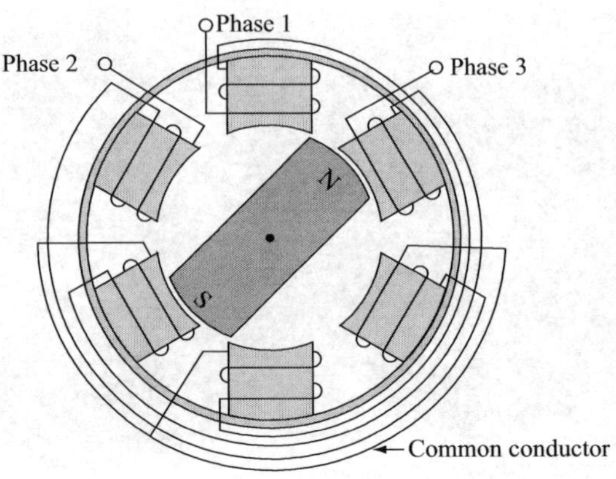

FIGURE 5.74 ■ A motor based on rotation of the magnetic field. In an induction machine, the magnet is replaced by a shorted coil.

all the details of their construction, the induction motor may be understood by first returning to **Figure 5.66**, but now the magnetic flux density is an AC field. In addition, the rotating coil is shorted (no external current connected to it). The AC field and the coil act as a transformer and an AC current is induced in the coil because it is shorted. According to Lentz's law, the current in the coil must produce an opposing field, which then forces the coil to rotate. Since now there is no commutation, continuous rotation is achieved by rotating the field. That is, suppose that an additional field is provided perpendicular to that shown in **Figure 5.66**. This can then be switched on after the loop has rotated one-half turn to keep it going. In practice this is done somewhat differently, by using the phases of the AC power supply to form a rotating field, shown schematically in **Figure 5.74** for a three-phase AC motor (a magnet is shown for the rotor, but a shorted coil acts exactly as a magnet). As the phases of the supply change with time, they generate a rotating magnetic field that drags the rotor with it, affecting the rotation.

Induction machines are very common in appliances since they are quiet, efficient, and most importantly, rotate at constant speeds that depend only on the frequency of the field and the number of poles. They are also used in control devices where constant speed is important. Control of induction motors, other than on and off, is much more involved than for DC motors, especially when variable speed is desired. A small induction motor is shown in **Figure 5.75**.

Of course, there are other types of AC motors with particular performance characteristics.

5.9.2.4 Stepper Motors

In actuation, the control of a motor for, say, exact and repeatable positioning requires, in addition to a continuous motor, some means of feedback, counting rotations, sensing position, etc. Motors that incorporate these means are called **servomotors** and they are commonly used in various control systems. In some applications servomotors have been replaced by stepper motors. Stepper motors are incremental rotation or linear motion motors. For this reason they are often viewed as "digital" motors, in the sense that each increment is fixed in size and increments are generated by a train of pulses. To understand their operation, consider first the configuration in **Figure 5.76a**. This is a two-phase stepper motor and uses a permanent magnet as the rotor, allowing a simple description of the operation. The rotor can be made to rotate in steps by properly driving the two coils, which in turn define the magnetic poles of the stator. To see the stepping

FIGURE 5.75 ■ A small induction motor.

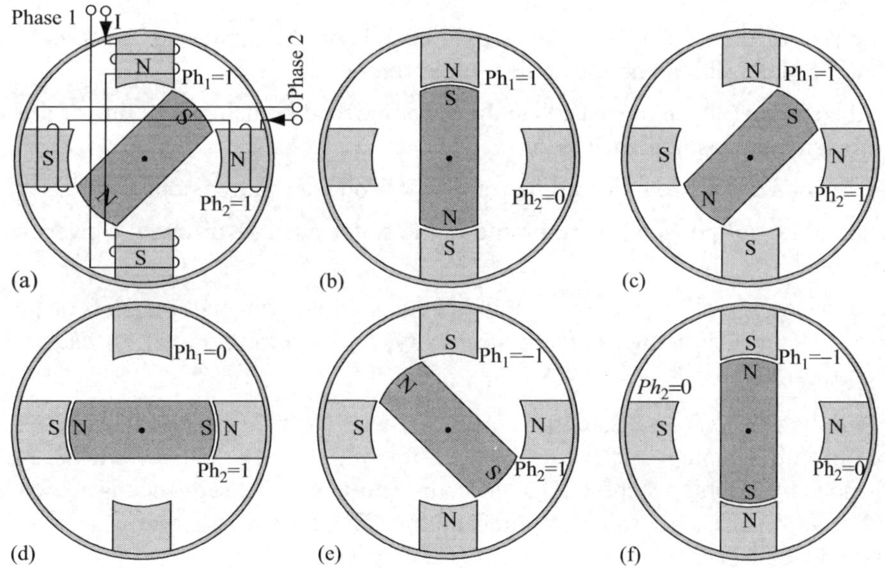

FIGURE 5.76 ■ (a) A schematic of two-phase stepping motor. The currents shown correspond to both phases being driven ($Ph_1 = 1$, $Ph_2 = 1$). (b–f) The sequence of half-stepping for the two phase motor in (a).

sequence, consider **Figure 5.76b**. By driving the two vertical coils, the magnet is held aligned vertically with coil 1. Now, if both coils are driven as in **Figure 5.76c**, the rotor will be at rest at 45°, rotating to the right. This is called a half-step and is the minimum rotation or step possible in this stepper motor. If now the vertical coil is deenergized, but the horizontal coil is kept energized, the magnet rotates an additional quarter turn to the position in **Figure 5.76d**. In the next step the current in the vertical coil is negative, in the horizontal coil it is positive, and the situation in **Figure 5.76e** is obtained. Finally, by reversing the vertical coil current and setting the horizontal coil to zero (no current), a full rotation has been completed (**Figure 5.76f**). This simple motor steps at 45° and requires eight steps to rotate one full turn. To rotate in the opposite direction the

TABLE 5.6 ■ The sequence required to turn the motor in **Figure 5.76** one full rotation (eight steps) clockwise[a]

Step	S_1	S_2
1	1	1
2	0	1
3	−1	1
4	−1	0
5	−1	−1
6	0	−1
7	1	−1
8	1	0

0 – no current, 1 current in one direction, –1 current in opposite direction.
[a]Reversing the sequence turns it counterclockwise. Shaded rows show the full step sequence.

sequence must be reversed, as shown in **Table 5.7**. The sequence above indicates the following:

1. The size of the step (number of steps) depends on the number of coils and, as we shall see later, the number of poles in the rotor.
2. Full stepping (90° in this case) can be accomplished by using only one of the stator coils (single phase) at each step.
3. More coils and more poles in the rotor will produce smaller steps.
4. The number of poles in the rotor and in the stator must be different (fewer poles in the rotor).
5. The magnetic field in the rotor can be generated by permanent magnets or by coils. We shall see that neither is really necessary. A stepper motor can be made with an iron rotor alone.

From the previous discussion it is clear that whereas the structure in **Figure 5.76** is capable of half-stepping, it can also be used for full steps (i.e., at increments of 90°). This is done by skipping steps 1, 3, 5, and 7 in **Table 5.6**. The sequence is shown in the shaded rows. That is, the same stepping motor can be used to move faster or slower through the sequence.

Consider again **Figure 5.76**, but suppose that the permanent magnet in the rotor is replaced with a piece of iron (nonmagnetized). The operation indicated above is still valid since the magnetic field produced by the stator coils will magnetize the iron (i.e., an electromagnet will attract a piece of iron). This simplifies matters considerably since now the rotor is much simpler to make. This type of stepper motor is called a **variable reluctance stepping motor** and is a common way of producing stepper motors. The principle of the motor is shown in **Figure 5.77a**. To operate it, the coils marked as 2 are first energized. This moves the rotor one step counterclockwise. Then coils marked as 3 are energized, moving one step to the left, and so on. Rotation in the opposite direction is obtained by inverting the sequence (driving coil number 3 first, then 2, and so on).

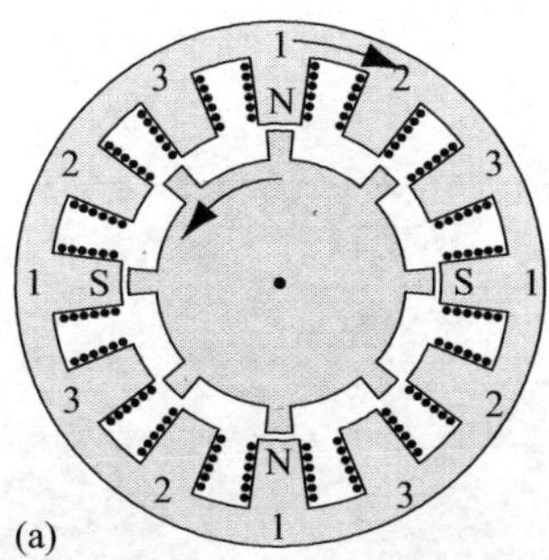

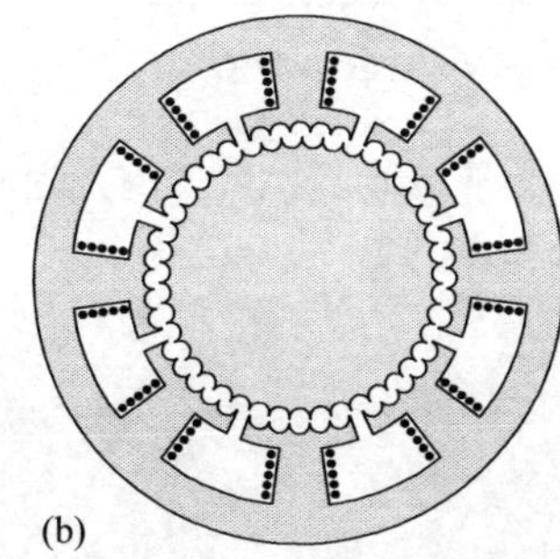

FIGURE 5.77 ■ (a) A practical stepper motor with 12 poles in the stator and 8 poles in the rotor. This motor is a three-phase motor. (b) An 8-pole (40 teeth) stator and 50-teeth rotor variable reluctance stepping motor.

Assuming there are n_s stator poles and n_r rotor poles (teeth in this case). The stator and rotor pitches are defined as

$$\theta_s = \frac{360^\circ}{n_s}, \ \theta_r = \frac{360^\circ}{n_r}. \tag{5.66}$$

The step of the stepper motor is then

$$\Delta\theta = |\theta_r - \theta_s|. \tag{5.67}$$

In the example in **Figure 5.77a** there are 12 poles in the stator and 8 in the rotor. Thus the full-stepping angle can be calculated as

$$\theta_s = \frac{360^\circ}{12} = 30^\circ, \ \theta_r = \frac{360^\circ}{8} = 45^\circ, \ \Delta\theta = |45^\circ - 30^\circ| = 15^\circ. \tag{5.68}$$

Half-stepping is possible by proper driving following the principles outlined above. For example, by driving first coil 1 we obtain the situation shown in the figure. If coils 1 and 2 are driven at the same time, the rotor will move counterclockwise one-half step (see **Example 5.16**). This stepper motor is capable of full steps of 15°. The motor is a three-phase stepper motor since the sequence to drive it in either direction repeats after three steps. Note that here the number of poles in the stator is larger than in the rotor. The opposite is just as valid and is often used in conjunction with variable reluctance stepper motors.

To simplify construction, the rotor is made of n_r teeth, as above, and the stator is made of a fixed number of poles, say eight, and each pole is toothed, as shown in **Figure 5.77b**. Note that in this case there are more teeth in the rotor (50) than in the stator (40). This produces a step of 1.8° (360/40−360/50). The motor in this figure is a four-phase motor (it requires a sequence of four pulses that repeat indefinitely).

EXAMPLE 5.15 A 200 step/revolution motor

One the most common stepper motors is the 200 steps/revolution motor. It is typically made of eight poles in the stator, each split into five teeth (a total of 40 teeth). The rotor has 50 teeth, as in **Figure 5.77b**. The full-step angle is calculated as follows:

In the stator,

$$\theta_s = \frac{360^\circ}{40} = 9^\circ.$$

In the rotor,

$$\theta_r = \frac{360^\circ}{50} = 7.2^\circ.$$

Thus the full step is

$$\Delta\theta = 9^\circ - 7.2^\circ = 1.8^\circ$$

and the number of steps per revolution is 360°/1.8° = 200.

Note: By splitting the rotor in two and shifting half of the rotor by 1/tooth, the same motor is capable of half-stepping (0.9°/step or 400 steps/revolution). It is also possible to half-step by proper driving of the coils without the need to split the rotor, as discussed above and in the following example.

EXAMPLE 5.16 Half-stepping of a variable reluctance stepper motor

Consider the variable reluctance stepper motor shown in **Figure 5.77a**. Show the driving sequence to produce half-step motion in the clockwise direction.

Solution: We start from the position shown, that is, the first step in the sequence is to drive coil 1 as shown. To move clockwise, we need to attract the tooth to the left of the aligned tooth, meaning we must drive coils 1 and 3. The condition now is shown in **Figure 5.78a**. Next we drive coil 3 to obtain the configuration in **Figure 5.78b**. Now coils 3 and 2 are driven at the same time (**Figure 5.78c**), followed by coil 2, then coils 1 and 2 together. The next step is to drive coil 1 alone to get to the starting step in **Figure 5.77a**. Thus the sequence is (1), (1 + 3), (3), (3 + 2), (2), (2 + 1).

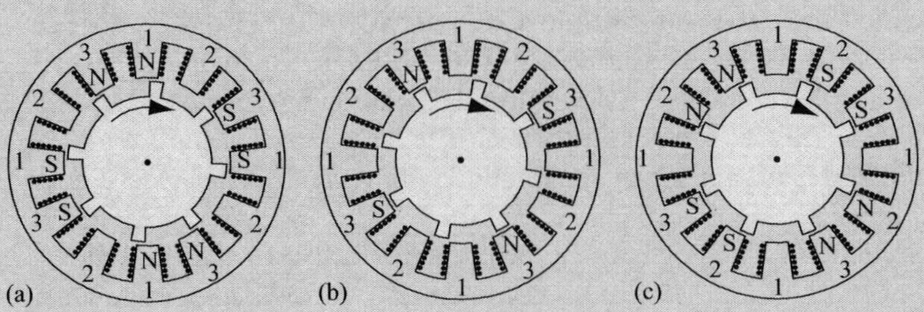

FIGURE 5.78 ■ Sequence necessary for half-stepping of the variable reluctance stepper motor in **Figure 5.77a**. Starting with the position in **Figure 5.77a,** (a–c) show the first three half-steps and the necessary coil driving sequence.

FIGURE 5.79 ■ A double-stack stepper motor with permanent magnet rotor (1.8° steps).

In general, variable reluctance stepping motors are simpler and less expensive to produce. However, when not powered, their rotor is free to move and hence they cannot hold their position. Permanent magnet rotors in stepper motors have some holding power and will maintain their position under power-off conditions.

In the description above, the stepper motor had a single rotor and a single stator. In an attempt to make motors with increasingly finer pitch (smaller steps), multiple rotors on a single shaft are used. These are called multiple stack stepper motors and they allow much finer steps by keeping one of the set of poles (on the rotor or on the stator) equal but changing the pitch on the opposite set of poles. Since now the pitch varies between stacks rather than within one stack, finer pitches are possible. The disadvantage is that the driving sequence is more complicated than in a single rotor motor. Usually the stator and each rotor have the same number of teeth but the two rotors are shifted one-half tooth apart (for a two-stack rotor). An example of an eight-pole (stator), double-stack motor is shown in **Figure 5.79**. This motor has 50 teeth on each rotor and 40 teeth on the stator. The rotors are magnetized and the motor shown has a 1.8° step but is capable of 0.9° by proper driving. We shall not pursue these motors further since they are not fundamentally different than single-stack motors.

Stepper motors come in all sizes, from tiny to very large, and are currently the motors of choice for accurate positioning and driving. They are, however, by themselves, more expensive and lower power than other motors such as DC motors. The extra cost is usually justified by their simple control and accuracy and by the fact that they can be driven from digital controllers with little more than a transistor or metal oxide semiconductor field-effect transistor (MOSFET) per phase to supply the current needed.

One can find stepper motors in industrial controls as easily as in consumer products such as printers, scanners, and cameras. In these applications, the ability of the motor to step through a predictable sequence with accurate, repeatable steps is used for fast positioning. The motors have typically low inertia, allowing them to respond quickly in both directions. In this respect they are fully capable of incorporation in fast systems while still maintaining their direct drive capabilities. Two small stepping motors are shown in **Figure 5.80**.

FIGURE 5.80 ■ Two small stepper motors. Left: A motor from a disk drive. Right: A paper advance motor from an ink jet printer.

5.9.2.5 *Linear Motors*

Conventional motors are naturally suited for rotary motion. There is, however, a need for linear actuation that cannot be met by rotary motors directly. In such cases, the rotary motion can be converted to linear motion through the use of cams, screw drives, belts, and so on. Another possibility, one that is gaining popularity, is the use of linear motors. We have in fact discussed two methods of linear motion in previous sections (voice coil actuators and the inchworm magnetostrictive motor).

A linear motor, either incorporating continuous motion or stepper motion, can be viewed as a rotary motor that has been cut and flattened so that the rotor can now slide linearly over the stator. This is shown schematically in **Figure 5.81**. Note that the slider or translator (equivalent to the rotor) may have as many poles as we wish—four are shown for clarity. Starting from the initial condition in **Figure 5.81a**, the sliding poles are driven as shown and are therefore attracted to the right. As they pass past the stator poles, they are commutated and the polarities change as in **Figure 5.81b**, again forcing motion to the right. This is merely a commutated DC machine. Motion to the left requires the opposite sequence. Based on this description, any of the motors above, including induction motors, can be built as linear motors. The stator may be very long (such as in linear drives for trains—in which the stator equals the length of the rail) or may be fairly short, depending on the application.

A variable reluctance linear stepping motor may be built as shown in **Figure 5.82**. This motor is equivalent to the rotary motor in **Figure 5.77a**. However, now the pitch is measured in units of length (so many millimeter per step). In this sequence we assume that the stator poles are driven and that the rotor is a mere toothed iron piece (variable reluctance motor). The sequence is as follows: Starting with the configuration in

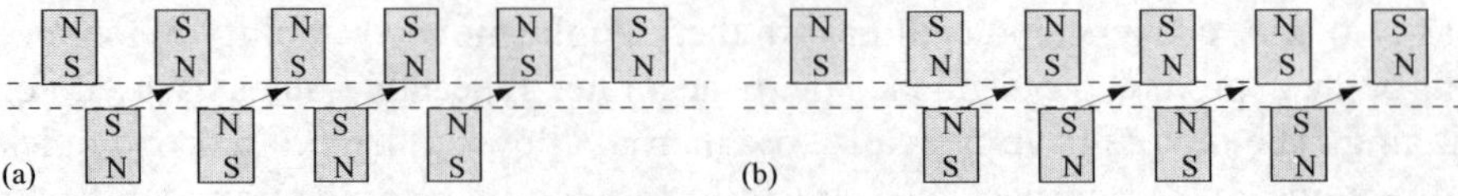

FIGURE 5.81 ■ The principle of linear motors. (a) The translator (bottom part) moves to the right until it is centered under the stator's poles. (b) The polarities of the translator are commutated and a new step can take place.

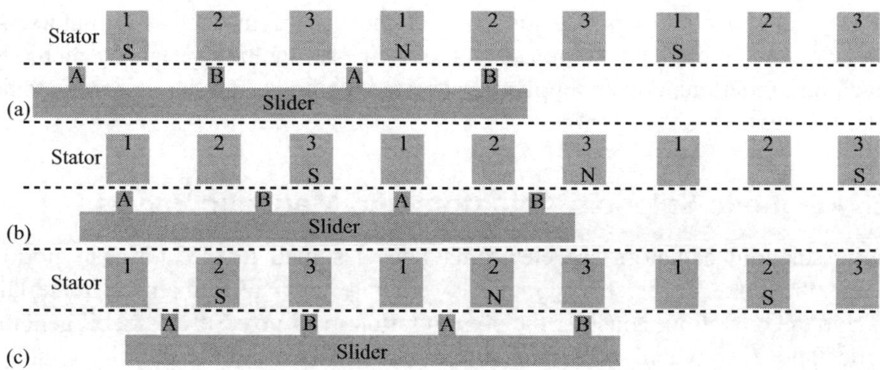

FIGURE 5.82 ■ Operation of a three-phase linear stepper motor with driven stator. Sequence is 1-3-2.

Figure 5.82a, poles marked as 1 are driven alternately as N and S, as shown. The slider moves to the right until teeth A are aligned with poles 1. This is shown in **Figure 5.82b**. Now poles 3 are driven as previously and the slider again moves to the right until teeth B are aligned with poles 3 (**Figure 5.82c**). Finally, poles 2 are driven, and the cycle completes and the relation of the slider and the stator is now as at the beginning of the sequence. The same can be accomplished with permanent magnet poles in the rotor. From **Figure 5.82** it should be noted that the pitch of the stator and slider are different. For every four poles in the stator there are three teeth in the slider. Thus each step is equal to half the pitch of the stator (i.e., in each step a tooth moves either from the middle between two poles to the center of the next pole, or vice versa). Of course, by changing the number of teeth, one can change this pitch. In the motor described here, the sequence is 1-3-2 for motion to the right. Moving in the opposite direction is accomplished by changing the sequence to (3-1-2).

In many linear stepping motors it is more practical to drive the slider rather than the stator since the stator may be very long, whereas the slider is usually short. However, the principle is the same. A variable reluctance linear stepping motor (which includes permanent magnet poles) is shown in **Figure 5.83** with the slider off the stator and inverted to reveal its poles (four poles with six teeth on each pole). These are separated 1 mm apart. The teeth on the stator are slightly smaller, allowing for the fine step motion of this device.

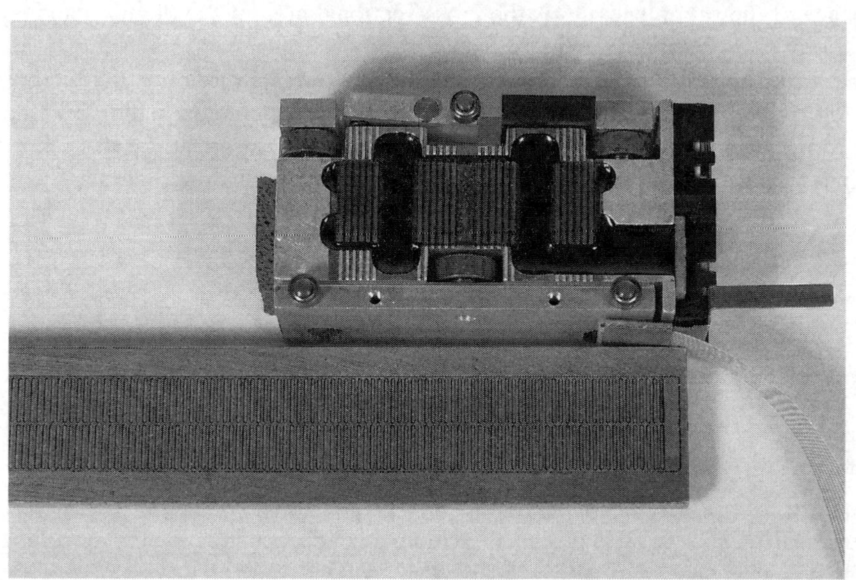

FIGURE 5.83 ■ A disassembled linear stepper motor with drive translator showing the poles of the translator and those of the stator.

There are many other issues involved in the use of motors as actuators, some mechanical, some electrical. Some aspects of their use, including starting methods for AC machines, inverters, power supplies, protection methods, and the like are outside the scope of this text.

5.9.3 Magnetic Solenoid Actuators and Magnetic Valves

Magnetic solenoid actuators are electromagnets designed to affect linear motion by exploiting the force an electromagnet can generate on a ferromagnetic material. To understand the operation, consider the configuration in **Figure 5.84a**. A coil generates a magnetic field everywhere, including in the gap between the fixed and movable iron pieces. We shall call the movable piece a plunger. In a closed magnetic path, such as the solenoid in **Figure 5.84b**, the magnetic flux density in the air gap between the plunger and the fixed iron piece is (approximately)

$$B = \frac{\mu_0 NI}{L} \quad [\mathrm{T}], \tag{5.69}$$

where N is the number of turns in the coil, I is the current in the coil, and L is the length of the gap (see **Figure 5.84a**). This is approximate because it neglects the effect of the field outside the air gap between the plunger and the fixed piece and does not apply when L approaches zero or when L is large. Nevertheless, it provides a good approximation for the magnetic flux density in many cases, especially in the case shown in **Figure 5.84b** in which the magnetic field closes through iron because of its high permeability. The method used here is known as a "virtual displacement method" and is similar to the method used to obtain the electrostatic force in **Equation (5.14)**. The force exerted on the plunger is given as

$$F = \frac{B^2 S}{\mu_0} = \frac{\mu_0 N^2 I^2 S}{L^2} \quad [\mathrm{N}], \tag{5.70}$$

where B is the magnetic flux density in the gap generated by the coil (perpendicular to the surface of the iron pieces), S is the cross-sectional area of the plunger, and μ_0 is the permeability of free space (air) in the gap.

The force on the plunger tends to close the gap and this motion is the linear motion generated by the magnetic valve actuator. As the plunger closes the gap, the force increases because L decreases. The construction shown in **Figure 5.84b** is more practical since it generates an axial field in the plunger and closes the external field so that

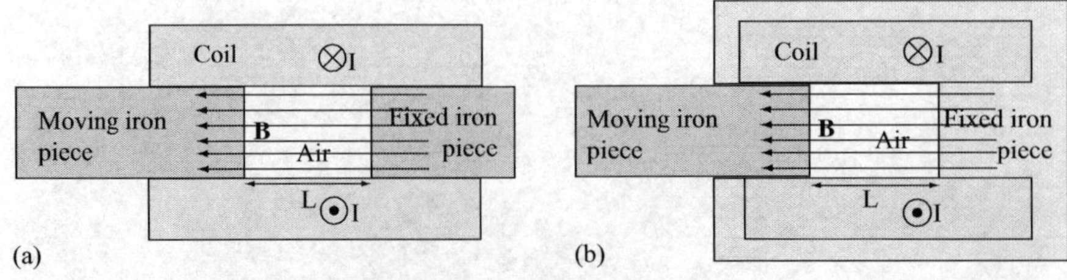

FIGURE 5.84 ■ The solenoid actuator. (a) The coil generates a magnetic field in the gap between the fixed and movable pieces. (b) A more practical construction that ensures the flux is closed and increases the magnetic flux density, and hence the force.

the total magnetic field available at the plunger is larger. In this form, the device is used as a simple go/no go actuator. That is, when energized, the gap is closed, and when deenergized it is open. This type of device is often used for electrical release of latches on doors, as a means of opening/closing fluid or gas valves, or to engage a mechanism. Examples of small linear solenoid actuators are shown in **Figure 5.85**. A modification of the linear plunger is the rotary or angular solenoid actuator, an example of which is shown in **Figure 5.86**. In this example, the rotor can move one-half turn in either direction. The rotor, which is equivalent to the plunger in the linear case, is made of a permanent magnet to increase the force.

The basic solenoid actuator is often used as the moving mechanism in valves. A basic configuration is shown in **Figure 5.87**. These valves are quite common in

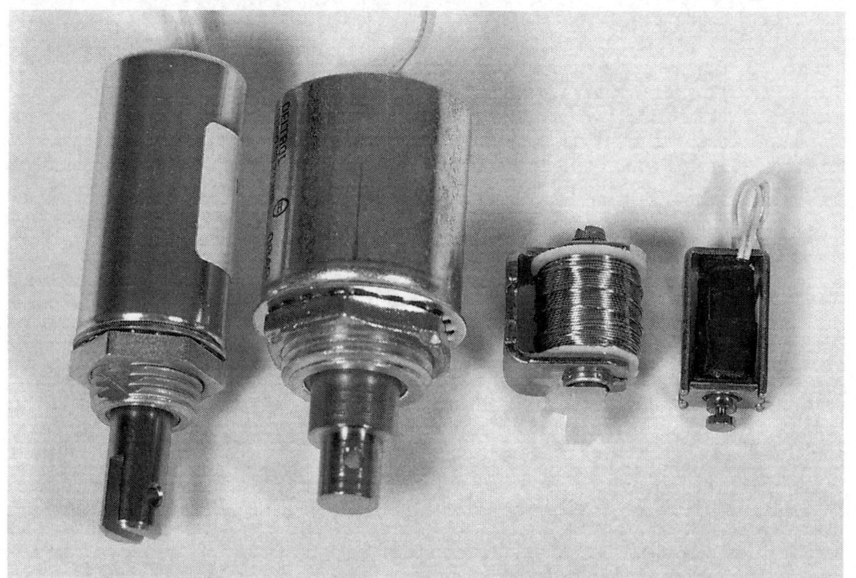

FIGURE 5.85 ■ Two linear solenoid actuators. The plungers are shown in the middle of the devices and are attached to whatever is actuated by them.

FIGURE 5.86 ■ An angular solenoid actuator. Here the rotor (equivalent to the plunger) is a permanent magnet.

control of both fluids and gases and exist in a variety of sizes, constructions, and power levels. They can be found not only in industrial processes, but also in consumer appliances such as washing machines, dishwashers, and refrigerators, as well as in cars and a variety of other products. The actuating rod (plunger) in this case acts against a spring and by properly driving the current through the solenoid its motion can be controlled as to speed and force exerted. Similar constructions can operate and control almost anything that requires linear (or rotational) motion. However, the travel of the actuating rods is relatively small, on the order of 10–20 mm, and often much less.

A magnetic valve designed for fluid flow control is shown in **Figure 5.88**. **Figure 5.89** shows a smaller valve used for control of air flow. The solenoid is about 18 mm in diameter, 25 mm long, and operates at 1.4 V and 300 mA.

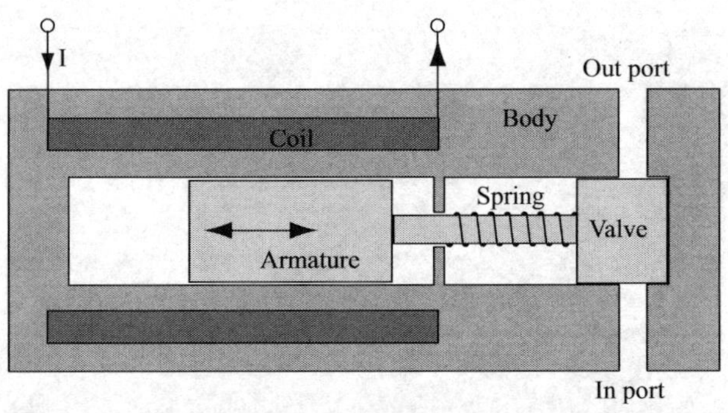

FIGURE 5.87 ■ The principle of a valve solenoid actuator, showing the coil and return spring. In this case the valve closes or opens an orifice.

FIGURE 5.88 ■ An electric valve for fluid flow control operated by a magnetic coil (28 V DC or 110 V AC).

EXAMPLE 5.17 **Force produced by a linear solenoid actuator**

A solenoid actuator has a cylindrical plunger of diameter 18 mm and a total travel (L in **Figure 5.84b**) of 10 mm. The coil contains 2000 turns and is fed with a constant current of 500 mA. Calculate the initial force the solenoid can exert (i.e., when the gap is 10 mm) and after the plunger has traveled 5 mm.

(a)

(b)

FIGURE 5.89 ■ A valve used to control air flow to a piston. The solenoid operates at 1.4 V and 300 mA. (a) Solenoid and valve. (b) Detail of the solenoid. The solenoid is 18 mm in diameter and 25 mm long.

Solution: The initial force is the most important parameter since it is this force that must act to affect action (such as opening a lock). It can be calculated directly from **Equation (5.70)** with $N = 2000$, $I = 0.5$ A, $L = 0.01$ m, and $\mu_0 = 4\pi \times 10^{-7}$ H/m. This produces a force of

$$F = \frac{\mu_0 N^2 I^2 S}{L^2} = \frac{4\pi \times 10^{-7} \times 2000^2 \times 0.5^2 \times (\pi \times 0.009^2)}{0.01^2} = 3.198 \text{ N}.$$

After a travel of 5 mm, $L = 0.005$ and the force is

$$F = \frac{\mu_0 N^2 I^2 S}{L^2} = \frac{4\pi \times 10^{-7} \times 2000^2 \times 0.5^2 \times (\pi \times 0.009^2)}{0.005^2} = 12.791 \text{ N}.$$

Not surprisingly, this force is four times larger since L is twice as small.

Notes: The forces these solenoids produce are not large, but are sufficient to open a valve, unlock a door, or pull a mechanical lever to release a device. On the other hand, they typically dissipate relatively large amounts of power in the coil and hence tend to be used intermittently. However, there are solenoids that can be turned on continuously. Note also that the calculation performed here is only valid for $L > 0$ because of the assumptions used in the development of **Equation (5.70)**. A more exact calculation can be done by taking into account the effect of the iron path and the magnetic properties of iron.

5.10 | VOLTAGE AND CURRENT SENSORS

In most cases voltage and current are measured as output of sensors or applied to actuators. But the sensing of voltage and current are important in themselves, as these are often used to affect other conditions. For example, the control and regulation of the output of a power supply requires that the output voltage and current be monitored. To maintain a constant voltage on a power line or to regulate the voltage in a car requires similar sensing of voltage and current. Fuses and circuit breakers are devices that sense the current in electric circuits and disconnect the power to the circuit when the current exceeds a preset value. Other devices protect circuits from overvoltages.

There are many mechanisms for sensing current and voltage. The most common methods are resistive and inductive, but the Hall element can be used successfully (see **Example 5.20**), as can capacitive methods. Some of the principles for DC and AC voltage and current sensing are discussed next.

5.10.1 Voltage Sensing

The potentiometer is a variable voltage divider, shown schematically in **Figure 5.90**. Although the purpose of using a potentiometer may vary, in all cases an input voltage V_{in} is divided to produce an output voltage V_{out}, which may be viewed as a "sampling" of the voltage V_{in}. That is,

$$V_{\text{out}} = \frac{V_{\text{in}}}{R} R_o \quad [\text{V}]. \tag{5.71}$$

Potentiometers come in a variety of physical implementations. The potentiometer may be a rotary or a linear device and its sampling may be linear or nonlinear, of which logarithmic potentiometers are the most common. In a rotary potentiometer, a resistance is built on a circular path and the slider rotates on a shaft to sample part of the resistance to produce the output. In a linear potentiometer, the resistance is a straight strip and the slider moves from side to side. A logarithmic-scale potentiometer's resistance is nonlinear, varying on a logarithmic scale (see **Example 2.15** for a discussion of the logarithmic potentiometer) and may be linear or rotary. Similarly, a linear-scale potentiometer may be linear or rotary. There are also potentiometers with multiturn capabilities and potentiometers without a shaft (often called trimmers) intended for one-time or occasional adjustment, usually using a screwdriver. Electronic potentiometers are devices that produce the same effect, that of sampling part of a voltage by electronic means. **Figure 5.91** shows a number of potentiometers of various types and sizes. The potentiometer allows one to sense the voltage V_{in} in **Figure 5.90**, which may be high, through use of the voltage V_{out}, which may be adjusted to a convenient level. Sometimes, when V_{in} is very high, the use of a voltage divider is a must. The potentiometer is equally suitable for DC and AC voltage sensing.

The common transformer is another means of sensing voltage, but now the voltage must be AC. The transformer was discussed in **Section 5.4.1** and is shown schematically in **Figure 5.92**. The output voltage relates to the input voltage through the turn ratio, as shown in **Equation (5.31)**. With **Figure 5.92a** as a reference, the output voltage is

$$V_{\text{out}} = \frac{V_{\text{in}}}{N_1} N_2 \quad [\text{V}]. \tag{5.72}$$

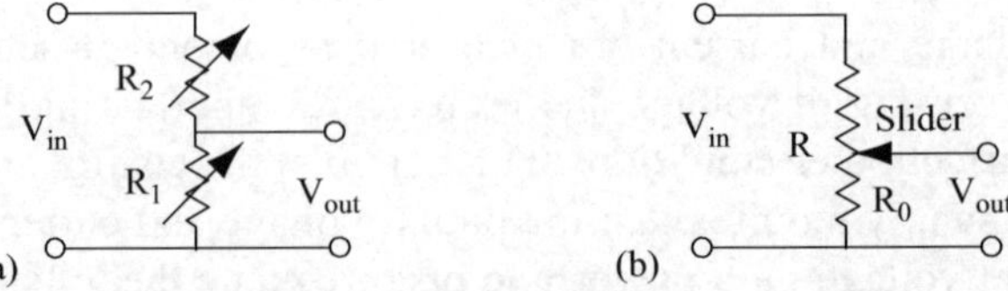

FIGURE 5.90 ■ The potentiometer as a resistive voltage divider. (a) Two variable resistors can produce any output voltage between zero and V_{in}. (b) The potentiometer does this by varying the ratio between the two resistors while their sum remains constant.

FIGURE 5.91 ■ A number of potentiometers of various types and sizes.

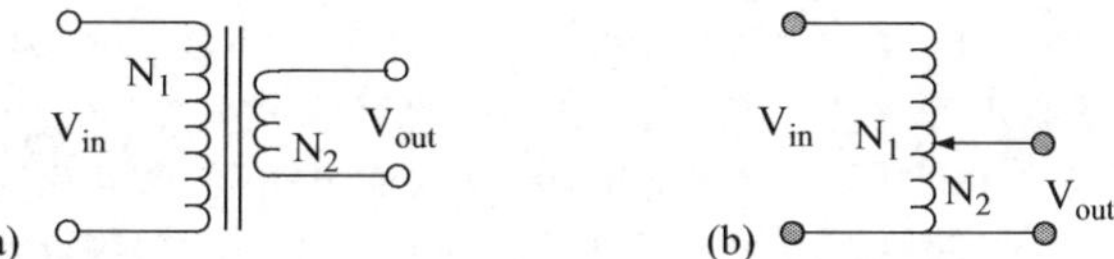

FIGURE 5.92 ■ The transformer. (a) Common isolating voltage transformer. (b) The autotransformer. Both serve as voltage sensors.

Unlike the potentiometer, the transformer also isolates the sampled voltage from the input voltage, a property that is important when mixing high and low voltages, and in particular where contact with the input voltage is to be avoided, usually for safety reasons. Although variable transformers exist (**Figure 5.92b**), they are not common, and unlike the standard transformer, which has a constant turn ratio, the variable transformer has a variable turn ratio. In fact, it is a type of potentiometer. In spite of its apparent usefulness, variable transformers are not often used because they are bulky, expensive, and most of the designs do not isolate input and output.

Voltage can also be sensed capacitively in what can be called a capacitive voltage divider, shown in **Figure 5.93a**. The output voltage is given as

$$V_{out} = \frac{C_2}{C_1 + C_2} V_{in} \quad [\mathrm{V}]. \tag{5.73}$$

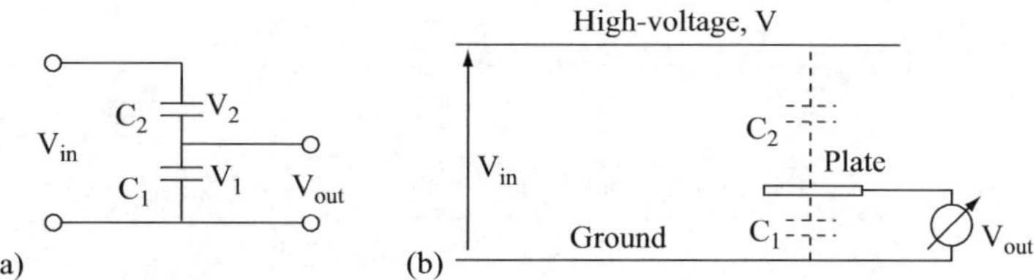

FIGURE 5.93 ■ The capacitive divider as a voltage sensor. (a) The principle. (b) An example of sensing the voltage of an overhead power line.

The capacitive method is particularly useful for measurement (sampling) of high voltages where direct measurements are not possible. It works equally well with DC or AC sources. For example, one can envision sensing the voltage of a high-voltage line in a high-voltage device as in **Figure 5.93b**. A conducting wire or a small plate at some height above the ground establishes the capacitance C_1. The capacitance between the high-voltage line and the plate establishes C_2. Once these two capacitances are measured (or are known), the voltage on the plate can be calibrated to monitor the voltage on the line. After that the line voltage variations can be sensed on the plate.

EXAMPLE 5.18 Monitoring of voltage in a high-voltage power supply

There are many installations that make use of high-voltage sources for particular applications. For example, sandpaper is often produced by attracting the abrasive particles using voltages in excess of 100 kV to the paper after applying a layer of glue. Consider a device of this type shown in **Figure 5.94**. The high-voltage supply is applied between the top and bottom surfaces, which are 2 m long, 1.2 m wide, and separated 10 cm apart, forming a capacitor. The abrasive particles are placed on the bottom surface and attracted to the top, sticking to the paper. A small plate of area S is placed at a small distance from the lower conducting surface. The potential difference between the plate and ground is connected to a microprocessor to monitor the high voltage across the plate. If the high voltage can vary between 0 and 200 kV, what must be the distance of the small plate from the bottom surface if the microprocessor operates at 5 V?

Solution: The large plates are fairly close to each other, forming a parallel plate capacitor. Therefore the electric field intensity between the plates is uniform and both the capacitor C_1 and C_2 may be considered as parallel plate capacitors in spite of the fact that the area of the small plate may not be large. Assuming the area is S, we have (as an approximation)

$$C_1 = \frac{\varepsilon_0 S}{d_1}, C_2 = \frac{\varepsilon_0 S}{d_2} \quad \text{[F]}.$$

The output must not exceed 5 V at an input voltage of 200 kV. Therefore

$$V_{\text{out}} = \frac{C_2}{C_1 + C_2} V_{\text{in}} = 5 \text{ V} = \frac{\dfrac{\varepsilon_0 S}{d_2}}{\dfrac{\varepsilon_0 S}{d_1} + \dfrac{\varepsilon_0 S}{d_2}} \times 200 \times 10^3 \rightarrow V_{\text{out}} = \frac{d_1}{d_1 + d_2} \times 200 \times 10^3 = 5 \text{ V}.$$

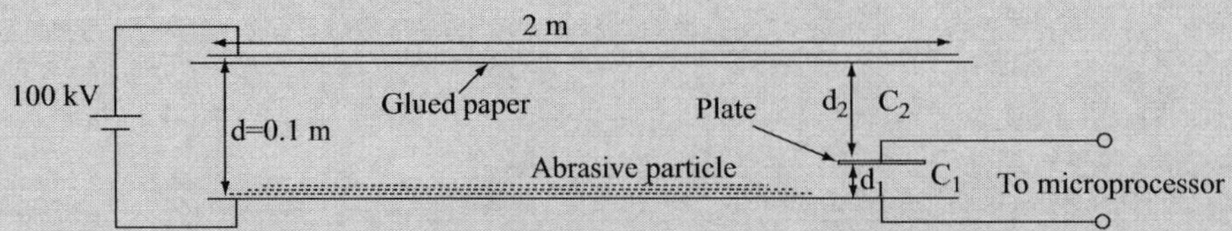

FIGURE 5.94 ■ A small capacitor used as a voltage sensor in a sandpaper production machine.

Since $d_1 + d_2 = d = 100$ mm, we can write

$$V_1 = \frac{200 \times 10^3}{100} \times d_1 = 5 \rightarrow d_1 = \frac{500}{200{,}000} = 0.0025 \text{ mm}.$$

Note that the area of the plate is immaterial. Also, the plate would have to be protected from dust. But aside from these minor difficulties, the method is viable.

5.10.2 Current Sensing

Most current sensors are in fact voltage sensors, or current to voltage converters. Here also there are a number of methods that can be used. In its simplest form, a resistor connected in series with the current to be sensed provides a voltage proportional to the current, as in **Figure 5.95**. This simple method is often used to sense current in power supplies, in electric machines, and in converters where the sensed current is used to control current or power. The sensing resistor has to be small so as not to affect the current and the voltage of the device, and is typically a fraction of an ohm, but depending on the current, it must be sufficiently large to produce a voltage drop on the order of at least 10–100 mV. This voltage can be amplified to produce the necessary control voltage.

A second method often used with AC currents is the so-called current transformer, shown in **Figure 5.96a**. In fact, it is a regular transformer with the current being sensed being the single-turn primary and with N_2 turns in the secondary. The current I produces a voltage across the primary and that voltage produces a voltage N_2 times larger in the secondary (see **Figure 5.96b**). This voltage is measured and is then an indication of the current in the conductor. There are two basic types of current transformers. One is a solid core transformer, as in **Figure 5.96a**, that requires the sensed current to pass through the

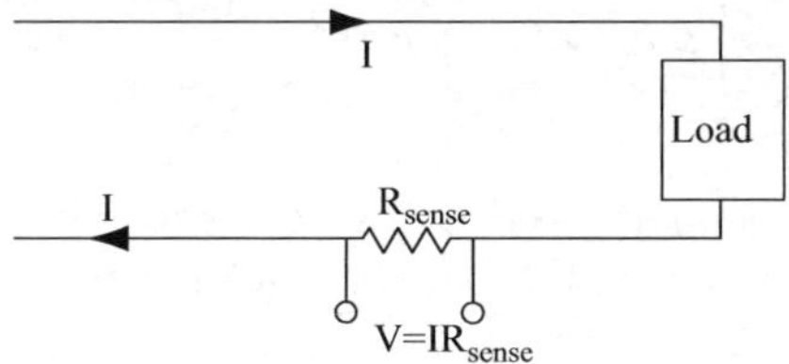

FIGURE 5.95 ■ A resistor as a current sensor. Measuring the voltage on a small, known resistor indicates the current in the load.

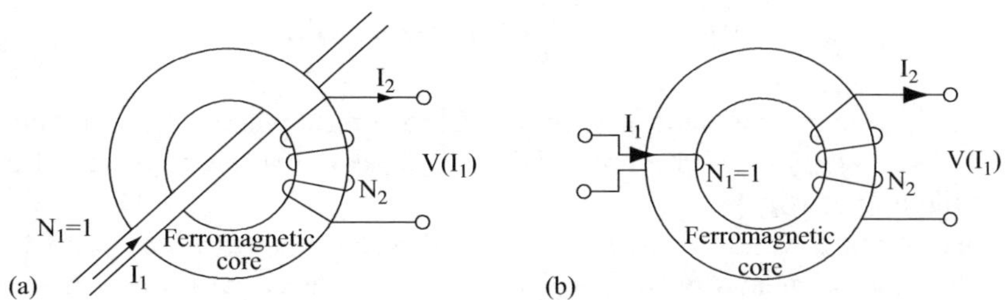

FIGURE 5.96 ■ (a) A current transformer as a current sensor. (b) The equivalent circuit with the current-carrying conductor shown as a single loop.

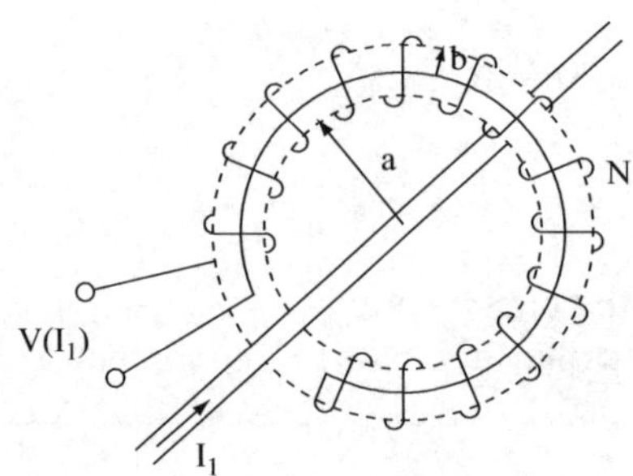

FIGURE 5.97 ■ The Rogowski coil as a current sensor. Since it is not a closed coil, it can fit around conductors more easily than the current transformer in **Figure 5.96**.

core. To facilitate use, some current transformers have hinged cores that can be opened in a manner similar to a clip and closed over the conductor through which the current is measured. The second type uses no core and is based on the Rogowski coil, shown in **Figure 5.97**. The coil is wound uniformly around a round form that is then removed and the end of the wire threaded through the coil itself so that both ends of the coil are available at one end of the coil. This means that the Rogowski coil can be placed over the conductor, facilitating measurements. The coil itself can be potted for physical protection and to maintain its form. The sensor is based on the fact that a current-carrying wire produces a magnetic flux density **B**, given in **Equation (5.17)**. The Rogowski coil has no core, therefore the permeability is μ_0 (that of air or the potting material, usually a plastic). If the coil has an average radius a, the flux density at the center of the coil is

$$B = \mu_0 \frac{I}{2\pi a} \quad [\mathrm{T}]. \tag{5.74}$$

The measured quantity is the emf in the coil. Assuming the turns of the coil are of radius b and there are N turns in the coil, the emf is calculated from **Equation (5.60)**, so we first need to calculate the flux:

$$\Phi = \int_S B ds \approx BS = B\pi b^2 = \frac{\mu_0 I b^2}{2a} \quad [\mathrm{Wb}]. \tag{5.75}$$

If the current is time dependent, and we will assume it to be of the form $I(t) = I_0 \sin(\omega t)$, where $\omega = 2\pi f$ and f is the frequency, we get from **Equation (5.60)**,

$$emf = N\frac{d\Phi}{dt} = N\frac{\mu_0 I_0 b^2}{2a}\omega \cos\omega t \quad [\mathrm{V}] \tag{5.76}$$

or

$$emf = \left(N\frac{\mu_0 b^2}{2a}\omega\right) I_0 \cos\omega t \quad [\mathrm{V}]. \tag{5.77}$$

This provides a voltage that is linear with respect to current and can be sufficiently large to be measured directly or after amplification. Note also that the higher the frequency, the higher the output voltage.

If instead of the Rogowski coil one uses the ferromagnetic core coil shown in **Figure 5.96a**, the permeability of free space, μ_0, is replaced with the permeability of the ferromagnetic core, μ, in all relations above. This produces a larger emf for the same number of turns (or fewer turns may be used to obtain the same emf) because the permeability of ferromagnetic materials is much higher. In most applications the ferromagnetic core is in the form of a toroid with an average radius a and cross-sectional radius b and the turns wound uniformly around the core. The only disadvantage of this arrangement is that the current-carrying wire whose current is measured must be threaded through the core, unless the core is hinged so it can be opened and closed around the wire as is done in hand-held clamp current probes.

EXAMPLE 5.19 Current sensor for house power monitoring

A current sensor based on the Rogowski coil is needed to sense the current entering a home. With a maximum expected current of 200 A (root mean square [RMS]), it is desired to use a sensor that produces a maximum voltage of magnitude 200 mV so that it can be connected directly to a digital voltmeter and the current read directly on the 0–200 mV scale. Design a Rogowski coil that will accomplish this. The diameter of the current-carrying conductor is 8 mm and the AC in the grid is sinusoidal at 60 Hz.

Solution: The output of the Rogowski coil in **Equation (5.77)** indicates that we can control three parameters: the average radius of the coil, a (but a must be larger than 4 mm so it can fit over the conductor), the radius of the turns b, and the number of turns. Since a is in the denominator it should be as large as practical. We will arbitrarily use a 5 cm diameter for the coil so that $a = 0.025$ m. Then we solve for b^2N and then decide on b and N so we obtain reasonable results. The peak current, by definition, is $I_0 = 200\sqrt{2}/2 = 282.84$ A. However, we will use the RMS value since the voltage is needed as RMS value.

The maximum emf is

$$emf = N\frac{\mu_0 I_0 b^2 \omega}{2a} = \frac{4\pi \times 10^{-7} \times 200 \times 2\pi \times 60}{2 \times 0.025} b^2 N = 0.2 \rightarrow b^2 N = \frac{0.02}{0.192\pi^2} = 0.1055.$$

That is, the product b^2N must be 0.1055. Since b cannot be very large we will use a 10 mm diameter for the coil ($b = 0.005$ m). This gives

$$N = \frac{0.1055}{b^2} = \frac{0.1055}{0.005^2} = 4222 \text{ turns.}$$

This is a large number of turns, but it is not impractical. Since magnet wire with diameters less than 0.05 mm is available, the coil will require about two layers of closely wound wires. A thicker wire, say 0.1 mm in diameter, would require about four layers of closely wound turns.

The parameters used here can be changed. Using a larger diameter coil would require a greater number of turns, whereas a larger turn diameter would require fewer turns. Note also that if we were to use a high-permeability core, that would reduce the number of turns by a factor equal to the relative permeability of the core, but then the coil would be closed and would not be a Rogowski coil.

Equation (5.17) (or **Equation (5.74)**) shows that the magnetic flux density B produced by a long wire carrying a current I is directly proportional to current and inversely proportional to the distance r from the wire. A current sensor can be devised by measuring the magnetic flux density produced by the current:

$$I = \left(\frac{2\pi r}{\mu}\right) B \quad [\text{A}], \tag{5.78}$$

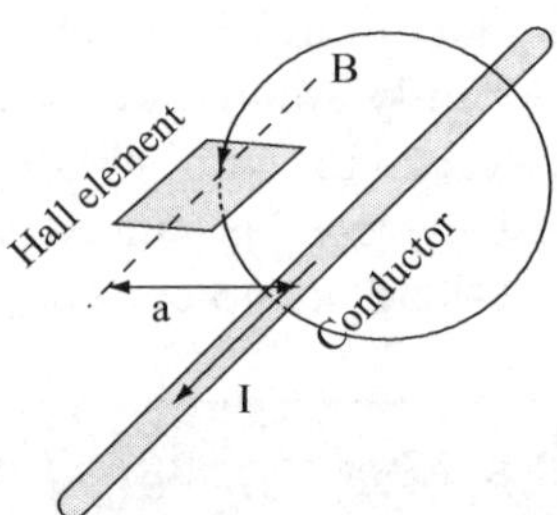

FIGURE 5.98 ■ Principle of a current sensor utilizing a Hall element. The Hall element is embedded in a plastic ring (not shown) through which the conductor passes.

where r is the distance from the conductor at which the magnetic flux density is measured and μ is the permeability at that location. The magnetic flux density may be measured using a small coil, but more often it is measured using a Hall element placed with its surface perpendicular to the magnetic flux density, as shown in **Figure 5.98**. The magnetic flux density B produced by the measured current I produces a Hall voltage according to **Equation (5.44)**:

$$V_{out} = K_H \frac{I_H B}{d} = \left(K_H \frac{I_H}{d} \frac{\mu_0}{2\pi a} \right) I \quad [\text{V}]. \tag{5.79}$$

In this relation, I_H is the bias current through the Hall element (see **Figure 5.38**), d is the thickness of the Hall element, and μ_0 is the permeability of the material in which the Hall element is embedded (assuming that it is nonmagnetic). K_H is the Hall coefficient.

EXAMPLE 5.20 A current sensor

Any current in a long conductor produces a magnetic flux density as described in **Equation (5.17)**. This magnetic flux density can be measured using a Hall element and by doing so produce a current sensor that indicates the current without the need to cut the conductor and measure the current in the conventional manner. The current sensor is shown in **Figure 5.98**. The only requirements are that the Hall element and the conductor are in a plane so that the magnetic flux density of the conductor, which is circumferential to the conductor, is perpendicular to the surface of the Hall element (or that the sensor be properly calibrated if that condition cannot be satisfied). The Hall element is embedded in a nonconducting ring that fits snuggly over the conductor and holds the Hall element in the position shown. (The ring is not shown, as it only has a mechanical function and does not affect the Hall element reading.) Using a small Hall element with a Hall coefficient of 0.01 m^3/A/s at a fixed distance of 10 mm from the center of the conductor, calculate the response of the sensor for currents between 0 and 100 A using the Hall element and biasing in **Example 5.7**.

Solution: First we calculate the range of the magnetic flux density for the given current using **Equation (5.17)**:

$$B = \frac{\mu_0 I}{2\pi r} = \frac{4\pi \times 10^{-7} I}{2\pi \times 0.01} = 2 \times 10^{-5} I \quad [\text{T}].$$

This is then introduced into **Equation (5.44)**:

$$V_{\text{out}} = K_H \frac{I_H B}{d} = K_H \frac{I_H \times 2 \times 10^{-5} I}{d} = 0.01 \times \frac{5 \times 10^{-3}}{0.1 \times 10^{-3}} 2 \times 10^{-5} I = 10^{-5} I \quad [\text{V}].$$

The same result may be obtained directly from **Equation (5.79)**. For the maximum conductor current of 100 A, this produces an output of 1 mV. Therefore the output will vary linearly from 0 to 1 mV for a current varying between 0 and 100 A. The sensitivity is 10^{-5} V/A, clearly requiring amplification for practical use.

5.11 | PROBLEMS

Capacitive sensors and actuators

5.1 **Capacitive position sensor**. A position sensor is made as follows: Two tubes with very thin walls are placed one inside the other so that they are concentric, separated by a Teflon layer, and can move in or out between two preset limits x_1 and x_2, as shown in **Figure 5.99**. The tubes are $L = 60$ mm long.

a. Calculate the minimum and maximum capacitance between the two tubes, assuming that an electric field may only exist in that area in which the tubes overlap.

b. Show how this device can be used as a sensor; that is, calculate its sensitivity for $a = 4$ mm, $b = 4.5$ mm, $x_1 = 10$ mm, $x_2 = 50$ mm, and the relative permittivity of Teflon $\varepsilon_r = 2.25$.

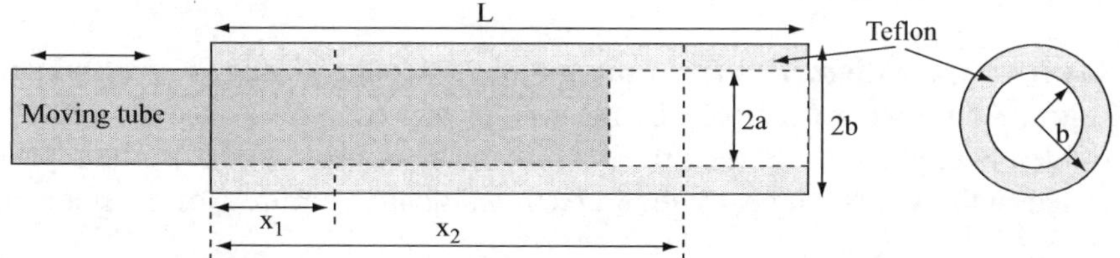

FIGURE 5.99 ■ Capacitive position sensor.

5.2 **Capacitive temperature sensor for water.** The temperature of water can be measured directly using a capacitive sensor based on the fact that the permittivity of water is highly dependent on temperature. As temperature changes from 0°C to 100°C, the relative permittivity of water, ε_w, changes from 90 to 55.

a. Given a sensor as shown in **Figure 5.100**, calculate its sensitivity assuming the permittivity change is linear with temperature.

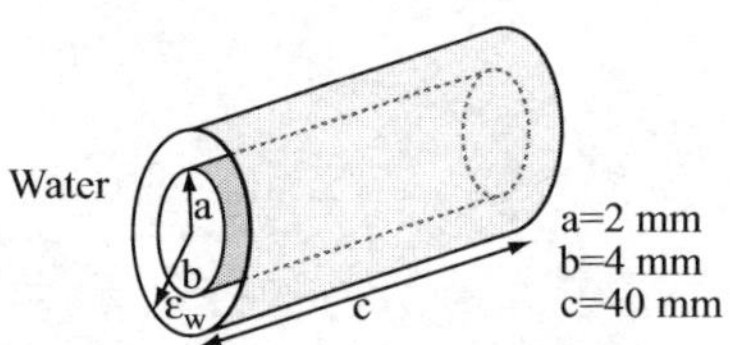

FIGURE 5.100 ■ Capacitive temperature sensor for water.

b. With a digital capacitance meter capable of a resolution of 0.2 pF, what is the resolution of the sensing system in terms of temperature?

5.3 A low-force capacitive actuator. A simple capacitive actuator can be made using the configuration in **Figure 5.101**. Two coaxial tubes form a capacitor. The space between them contains a hollow tube made of Teflon that is free to move between the two cylinders as shown. Assume the moving part is at an arbitrary location x from the edge of the capacitor and the dimensions are as shown in the figure:

a. Calculate the force the actuator can exert as a function of the applied voltage connected across the cylinders (positive to the outer cylinder, negative to the inner cylinder).

b. Show from physical considerations that the motion can only be inward, regardless of polarity of the voltage applied.

Note: The capacitance of coaxial capacitors can be calculated directly but, as a first approximation, they may be treated as parallel plate capacitors using **Equation (5.2)**, with S taken as the average between the areas of the outer and inner conductors, especially if the inner radius is not too small. The same applies to the electric field in the capacitor.

FIGURE 5.101 ■ Capacitive force actuator.

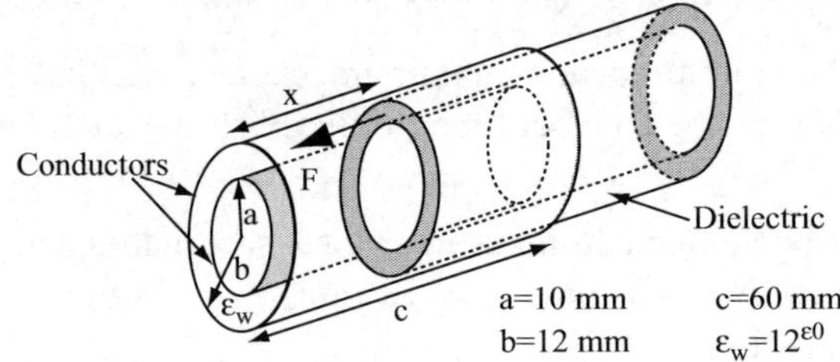

5.4 Resistive (potentiometer) fuel tank gauge. A fuel tank gauge is made as shown in **Figure 5.102**. The rotary potentiometer is linear with a resistance $R = 100\ \text{k}\Omega$ distributed over a 330° strip (i.e., the potentiometer can rotate 330°). The float is connected to the shaft of a gear with 30 teeth and that in turn rotates the potentiometer

FIGURE 5.102 ■ Resistive fuel gauge.

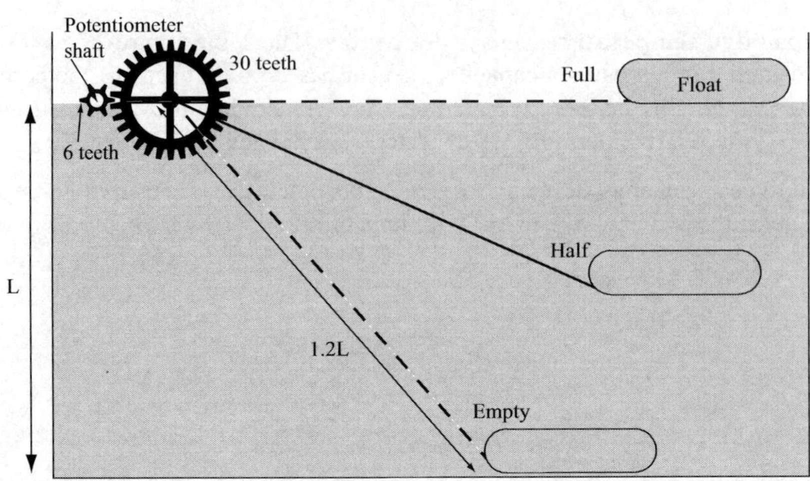

through a gear with 6 teeth. The mechanical linkage is set so that when the tank is full, the potentiometer resistance is zero. The float is shown in three positions to show how the resistance changes (increases from zero at full tank).

a. Calculate the resistance reading for empty, ¼, ½, ¾ and full tank.

b. Using the points in (a), find a linear best fit and calculate the maximum nonlinearity of the sensor. Why is the curve nonlinear?

5.5 **Corrosion rate sensor.** Structures that are subject to corrosion require a means of determining the corrosion rate so that action can be taken before the structure becomes dangerous. One way of doing so is to expose a thin wire of the same material as the structure to identical conditions by placing it in the same location as the monitored structure. The corrosion rate is often defined as corrosion depth in millimeters per year (mm/year). A corrosion rate sensor is made of a steel wire L m long and d mm in diameter. Its resistance is monitored continuously and correlated directly with the corrosion rate.

a. Find a relation between the rate of corrosion in millimeters and resistance measured. Assume corrosion is uniform around the circumference and that the corrosion products do not contribute to resistance.

b. Discuss how variations in temperature may be accounted for.

5.6 **Tide level sensor.** In an attempt to monitor maximum and minimum tide levels, a simple sensor is made as follows (**Figure 5.103**). A metal tube of inner diameter $a = 100$ mm is coated on the outside with an isolating layer of paint so that seawater cannot come in contact with the outer surface of the tube. A metal cylinder of outer diameter $b = 40$ mm is placed inside the larger tube and, using a few insulating spacers, held so that the two tubes are coaxial. Now the assembly is sunk into the sea. A series of small holes at the bottom allow seawater into the space between the tubes. A 1.5 V battery is connected through an ammeter as shown.

a. If the maximum and minimum water levels are as shown, and the conductivity of seawater is $\sigma = 4$ S/m, calculate the maximum and minimum (range) reading of the ammeter. Assume conductors are perfect conductors and both the air and sea bottom are insulators.

b. Calculate the sensitivity of the sensor.

c. Calculate the resolution for a digital ammeter that measures in steps of 1 mA.

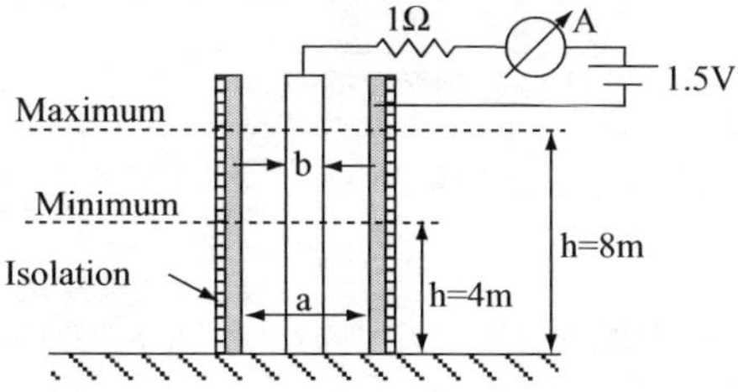

FIGURE 5.103 ■ Tide level sensor.

5.7 **Resistive position sensor.** A position sensor is made as shown in **Figure 5.104**, where the inner plate can slide back and forth over a distance of 10 cm. Both the stationary and moving sections are made of graphite and are in intimate contact. The length of the moving bar is the same as the length of the stationary section (15 cm). The conductivity of graphite is $\sigma = 10^2$ S/m. Given the dimensions in the figure,

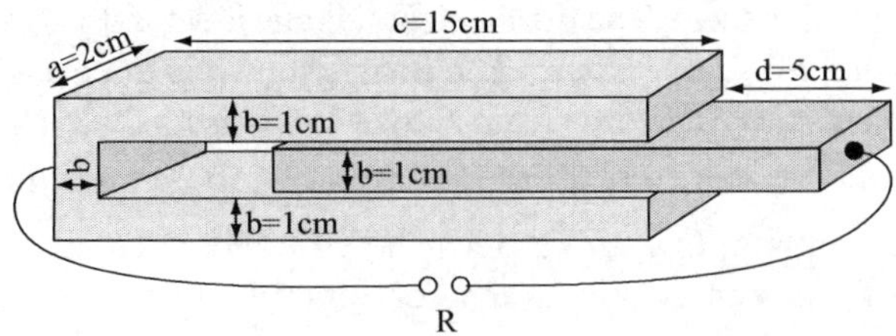

FIGURE 5.104 ■ A simple resistive position sensor.

a. Calculate the relation between position d and resistance R measured between the two.
b. Calculate the maximum and minimum resistance possible (i.e., for $d = 1$ cm and for $d = 11$ cm).
c. Calculate the sensitivity of the sensor.

Magnetic sensors and actuators

5.8 Induction proximity sensor (induced emf). In some applications it is important to sense displacement or proximity without incurring friction and losses. One possible solution is shown in **Figure 5.105**. Two solenoids of length L each are placed inside one another as shown. The radii of the solenoids are a and b and each has N turns uniformly wound over the length L. A sinusoidal current with amplitude I_0 and frequency f flows in the outer solenoid, which is stationary. $I_0 = 0.1$ A, $f = 1000$ Hz, $L = 10$ cm, $a = 10$ mm, $b = 12$ mm, and $N = 400$ turns. The coils are shown in axial cross section. The $\otimes$ in the turns indicates a current flowing in, whereas a $\odot$ indicates a current flowing out of the plane of the cut.

a. What is the emf induced in the inner, movable solenoid when the two coils overlap a distance x? Assume that $L \gg b$, which allows you to say that the field is approximately constant within the solenoid and zero outside and neglect the effects of the ends of the outer solenoid. The moving coil core is nonmagnetic (permeability equals that of space).
b. What is the sensitivity of the sensor (voltage RMS per unit length) if the core is replaced with iron with relative permeability μ_r?

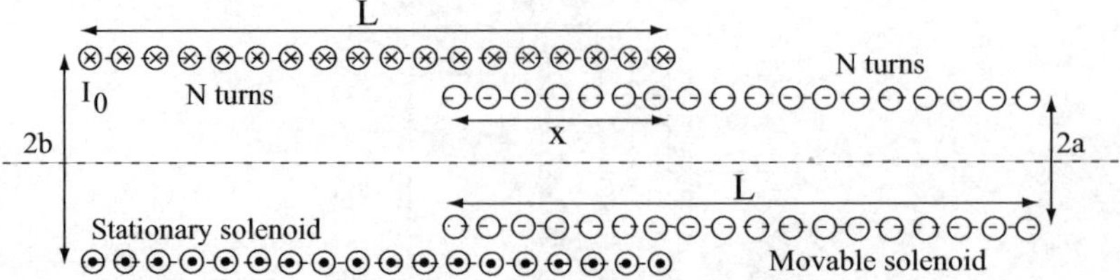

FIGURE 5.105 ■ A magnetic position sensor.

5.9 Invisible fence for dogs. In this type of system, a wire is buried under the surface and a current at a given frequency passes through the wire. The dog wears a small unit made of a pickup coil and electronics that deliver a high-voltage pulse to the dog through a couple of electrodes pressed against its skin.

The pulse is not harmful, but it is painful enough to cause the dog to keep away. In an invisible fence, the wire carries a 0.5 A (amplitude) sinusoidal current at 10 kHz. The dog carries a sensor made as a coil with 1500 turns and 30 mm in diameter.

a. If the detection level is set at 200 μV RMS (i.e., the level at which the dog will receive a correction pulse), what is the furthest distance from the wire the dog will "feel" the presence of the fence?

b. What are the conditions necessary to obtain the result in (a)?

Hall effect sensors

5.10 **Hall effect in conductors.** The Hall effect in conductors is rather small and can be calculated from **Equation (5.43)**. To see what the order of magnitudes of the Hall coefficient and Hall voltage are, consider a Hall sensor made by deposition of gold on a substrate. The sensor itself is 2 mm $\times$ 4 mm and is 0.1 mm thick. The free charge density of gold is 5.9×10^{28} electrons/m^3. Calculate the Hall coefficient and the sensitivity of the sensor in sensing magnetic flux densities. Assume the magnetic field is perpendicular to the plate and a current of 15 mA flows along the long dimension of the plate.

5.11 **Hall effect in silicon.** A Hall element is made in the form of a small silicon wafer 1 mm $\times$ 1 mm and 0.2 mm thick. An n-type silicon is used with a majority carrier density of 1.5×10^{15} carriers/cm^3, whereas the intrinsic carrier density is 1.5×10^{10} carriers/cm^3. The mobility of holes is 450 cm^2/V·s and that of electrons is 1350 cm^2/V·s. Calculate

a. The Hall coefficient of the Hall element.

b. The sensitivity of the Hall sensor in sensing magnetic fields given a fixed current of 10 mA across the sensor. Assume the magnetic flux density is applied perpendicular to the silicon plate.

c. Suppose that the intrinsic material is used to make the Hall element. Use the dimensions of the wafer to calculate the resistance of the Hall element and explain why this Hall element is not a practical device.

5.12 **Zero Hall coefficient semiconductor.** If a semiconducting device must operate in high magnetic fields, and if the purpose is not to measure the magnetic field, the Hall voltage may be detrimental to the operation of the device. Under these conditions it may be useful to dope the semiconductor so as to produce a zero Hall coefficient. Given the mobility of holes in silicon as 450 cm^2/V·s and that of electrons as 1350 cm^2/V·s, what is the ratio of n to p densities required?

5.13 **Power sensor.** A power sensor is built as shown in **Figure 5.40** as part of a DC power meter for a small electric vehicle. A small ferromagnetic core with very large permeability is used and the Hall element is placed in the small gap shown. The magnetic core saturates at a magnetic flux density of 1.4 T. The Hall element has a Hall coefficient of $K_H = 0.018$ m^3/A/s and is $d = 0.4$ mm thick. The gap is slightly larger, at $l_g = 0.5$ mm, so that the Hall element fits snuggly in the gap. Neglect the resistance of the Hall element and of the coil.

a. Assuming N turns in the coil, a resistance R, and a resistive load R_L, find the expression relating the Hall voltage and the power in the load (this is the transfer function of the sensor).

b. Find the maximum power the sensor can sense, assuming the line voltage is constant at 12 V and the coil contains $N = 100$ turns.

c. What is the sensor's reading under the conditions in (b) if the Hall element operates at a current of 5 mA?

d. What is the sensitivity of the power sensor?

5.14 **Carrier density sensor.** The Hall effect can be used in many ways, and not all of them involve sensing of magnetic fields. Consider the measurement of carrier density in a metal alloy. To perform the measurement, the metal is cut into a rectangular plate 25 mm long, 10 m wide, and 1 mm thick. The metal is placed between two neodymium-iron-boron (NeFeB) magnets, producing a constant 1 T magnetic flux density through the disk. A current of 1 A passes through the plate and the voltage across the plate is measured as 7.45 μV (**Figure 5.106**).

a. Show the polarity of the voltage given the direction of the magnetic flux density and that of the current.

b. Calculate the carrier density in the metal alloy.

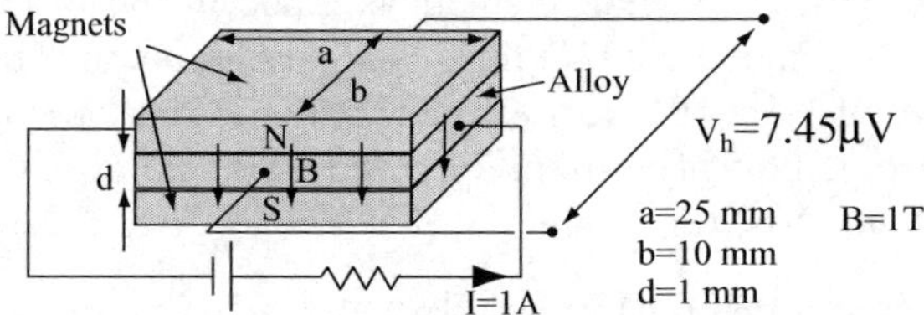

FIGURE 5.106 ■ Carrier density sensor.

5.15 **Doping concentration in *n*-type silicon**. It is required to estimate the carrier density in an *n*-type silicon sample as a means of production control. A sample 2 mm long, 1 mm wide, and 0.1 mm thick is prepared and connected as in **Figure 5.38** or **Figure 5.106**, carrying a current of 10 mA. The sample is placed between the poles of a neodymium-iron-boron (NeFeB) permanent magnet producing a magnetic flux density of 0.8 T perpendicular to the surface of the sample and a 80 μV Hall voltage is measured across the sample. Calculate the carrier concentration in the sample assuming the majority carriers dominate. Electrons in silicon have a mobility of 1350 $cm^2/V{\cdot}s$.

Magnetohydrodynamic sensors and actuators

5.16 **Magnetohydrodynamic actuator for submarine propulsion.** A submarine is designed to move using a magnetohydrodynamic pump. The pump is made in the form of a channel that is $b = 8$ m long, $d = 0.5$ m high, and $a = 1$ m wide (**Figure 5.48**). The magnetic field is constant everywhere throughout the channel in the direction shown and equal to $B = 1$ T. A current $I = 100$ kA passes through the electrodes shown and through the water between the electrodes, generating a force on the water as shown. The conductivity of seawater is 4 S/m.

a. Calculate the force produced by the pump.

b. Calculate the potential required between the two electrodes.

5.17 **Magnetohydrodynamic generator.** A magnetohydrodynamic generator is made in the form of a channel 1 m long with a cross section of 10 cm × 20 cm (**Figure 5.107**). The magnetic flux density produced between the poles on the narrow side is 0.8 T. A jet combustor is used to drive combustion gases through the channel and the gases are seeded with conducting ions to produce an effective conductivity of 100 S/m in the channel. The combustor drives the gases at 200 m/s, effectively making the gas that moves through the channel a plasma.

a. Calculate the output voltage of the generator (*emf*).

b. Calculate the maximum power it can generate if the output voltage cannot change by more than 5%.

c. Indicate qualitatively how this method might be used for sensing and what the quantities are it can sense.

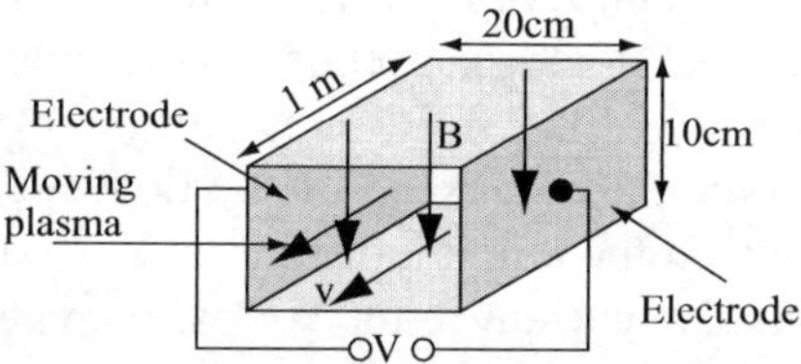

FIGURE 5.107 ■ Magneto-hydrodynamic generator.

5.18 **Magnetic flowmeter.** A magnetic flowmeter can be built as follows (**Figure 5.108**). The fluid flows in a square cross-section channel. On two opposite sides there are two coils that produce a constant magnetic flux density B_0 pointing down. The fluid contains positive and negative ions (Na^+, Cl^-, etc.). The magnetic flux density is assumed to be constant and uniform in the channel. This forces the positive and negative charges toward the two opposing electrodes.

a. Calculate the potential difference for a flowing fluid as a function of its velocity v and show its polarity. The cross-sectional area of the channel is $a \times a$ [m^2].

b. Calculate the sensitivity of the sensor to flow. Flow is measured as volume per second (i.e., m^3/s). Discuss realistic ways to improve the sensitivity of the device.

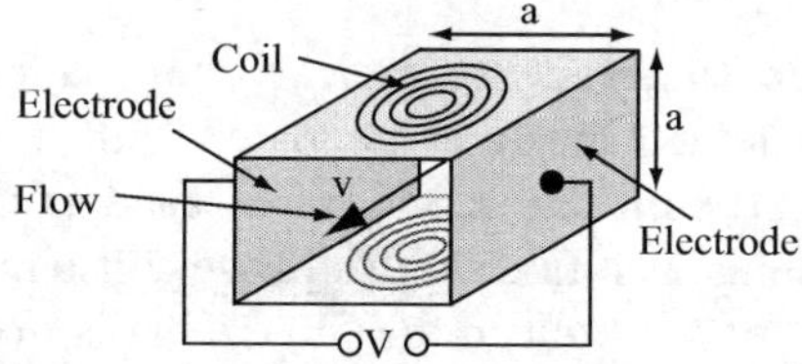

FIGURE 5.108 ■ Magnetic flowmeter.

5.19 **Magnetic gun (magnetic force).** A magnetic gun is made as shown in **Figure 5.109**. Two cylindrical conductors of diameter a (rails) and separated a distance d between centers are shorted by a conducting projectile that is free to move along the rails. When a current is applied, the magnetic flux density produced by the current in the conductors generates a force on the projectile.

a. Given $a = 10$ mm, $d = 40$ mm, and $I = 100{,}000$ A, what is the force on the projectile?
b. If the projectile has a mass of 100 g, calculate the acceleration and the exit velocity of the projectile for rails 5 m long with the projectile starting at one end and exiting the other.

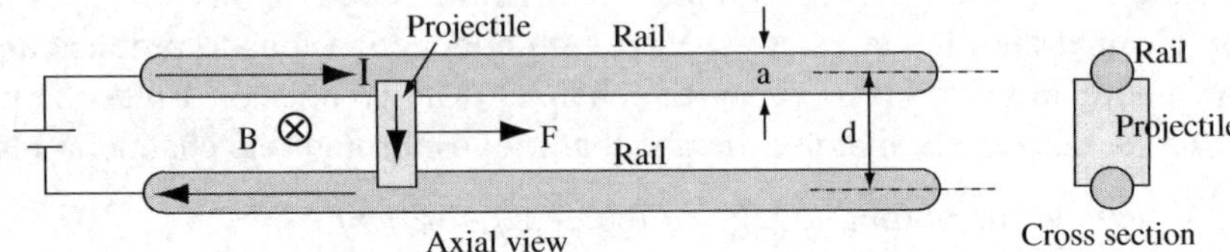

FIGURE 5.109 ■ The rail gun. (a) Axial view. (b) Cross section through the projectile.

5.20 **Magnetic density sensor.** The density of a fluid can be sensed using a magnetic sensor as follows. A sealed float (i.e., a closed container) is equipped with an amount of iron on its bottom and a coil at the bottom of the container containing the fluid is driven with a current I, as shown in **Figure 5.110**. The current in the coil is increased until the float is suspended in the fluid with its top at the surface of the fluid. The density of the float, including the iron, is known as ρ_0 and its volume is V_0. The force the coil exerts on the float equals kI^2, where k is a given constant that depends on the amount of iron in the float, the size of the coil, and distance to the float. The current in the coil is the measured quantity.

a. Find the transfer function of the sensor, that is, find the relation between the density of the fluid, ρ, and the measured current, I.
b. Calculate the sensitivity of the sensor.

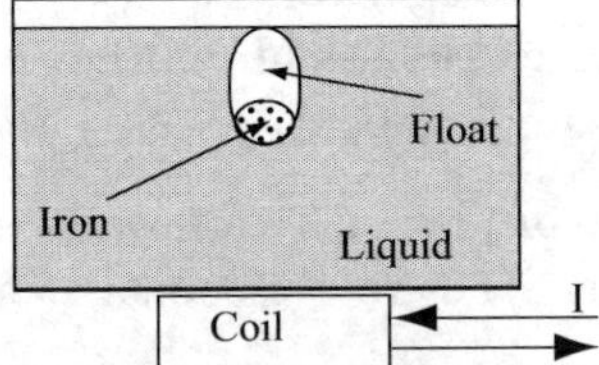

FIGURE 5.110 ■ Magnetic density sensor.

Magnetostrictive sensors and actuators

5.21 **Optical fiber magnetometer.** An optical fiber magnetometer is made of two fibers, each 100 m long. One fiber is coated with nickel and an infrared light emitting diode (LED) emitting at 850 nm is used as the source for both fibers. Light propagates along each fiber with a phase constant $\beta = 2\pi/\lambda$ [rad/m], where λ is the wavelength of the light in the fiber. That is, for every 1 m length of the fiber, the phase changes by $2\pi/\lambda$ radians. At the end of the fibers, the phases of the two signals are compared. The signal propagating through the coated fiber will have a lower phase since the sensed magnetic flux density causes the nickel coating, and hence the fiber, to contract (nickel is a magnetostrictive material; see **Table 5.5**). The device is shown in **Figure 5.111**. Assuming that the phase detector can detect a phase difference of 5°, calculate the lowest magnetic flux density detectable using this magnetometer.

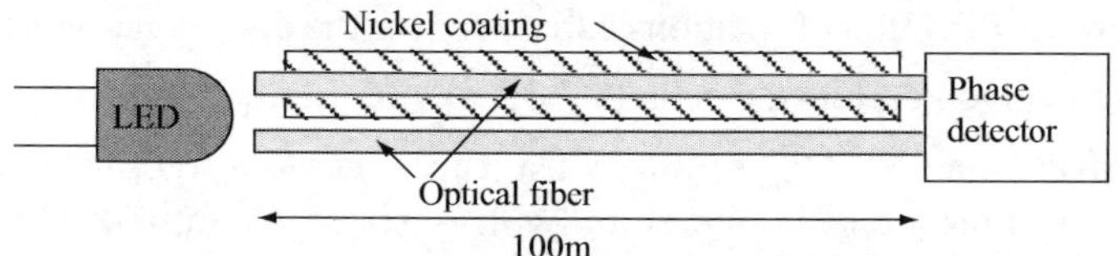

FIGURE 5.111 ■ Optical fiber magnetometer.

Voice coil actuators

5.22 **Voice coil actuator.** A cylindrical voice coil actuator is shown in **Figure 5.112**. A permanent magnet, magnetized radially, is placed on the outer surface of the gap with a free moving coil on the inner cylinder that can move back and forth. The mass of the coil is $m = 50$ g, the current in the coil has amplitude $I = 0.4$ A and frequency $f = 50$ Hz, the number of turns in the coil is $N = 240$, and the magnetic flux density due to the permanent magnet is $B = 0.8$ T. The coil has an average diameter of 22.5 mm. This device is used as a positioner or as the drive in a vibrating pump. The activating mechanism is connected directly to the coil.

a. Calculate the force on the coil.
b. If the cylinder has a restoring constant $k = 250$ N/m (provided by a spring), calculate the maximum displacement of the moving piece.
c. Calculate the maximum acceleration of the coil/cone assembly.

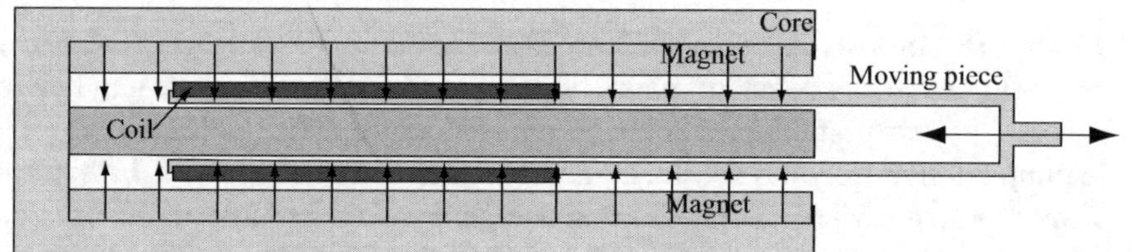

FIGURE 5.112 ■ A voice coil actuator. The magnetic field is indicated by the vertical arrows.

Motors as actuators

5.23 **A simple DC motor.** A simplified form of a motor is shown in **Figure 5.113**. The rotor and stator are separated by a gap of 0.5 mm. At a given time, the coil, which is embedded into the rotor in a groove, made of 100 turns, and carrying a current $I = 0.2$ A, is oriented as shown in the figure. The current is into the page on the top of the rotor and out of the page on the bottom of the rotor. The rotor is a cylinder of radius $a = 2$ cm and depth $b = 4$ cm. The radius of the coil in the rotor may be assumed to be the same as that of the rotor.

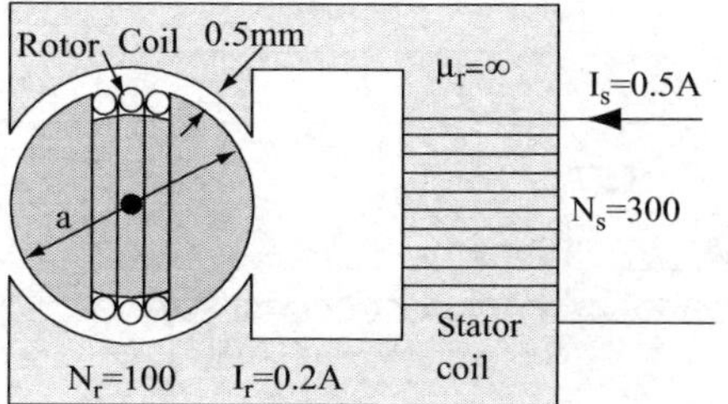

FIGURE 5.113 ■ A simplified DC motor.

a. Show the direction of rotation of the rotor for the configuration in **Figure 5.113**.

b. For the given currents, calculate the maximum torque this motor is capable of.

5.24 **Permanent magnet DC motor with three coils.** A permanent magnet DC motor has three coils in the rotor with a six-strip commutator, as shown in **Figure 5.114a**. Each of the three coils is permanently connected to a pair of strips on the commutator, as indicated. Assume the magnetic flux density in the gap is uniform and each coil is, in turn, horizontal when the brushes are at the center of the corresponding commutator strips (**Figure 5.114b** shows coil 1 in that position). Each pair of strips occupies one-third of the circle shown in **Figure 5.114b**. The radius of the rotor is 30 mm, the length 80 mm, and the permanent magnets produce a magnetic flux density of 0.75 T in the gap and throughout the rotor. The current in the coils is 0.5 A and each coil is made of 120 turns. Calculate and plot the torque as a function of time for a full cycle of the rotor.

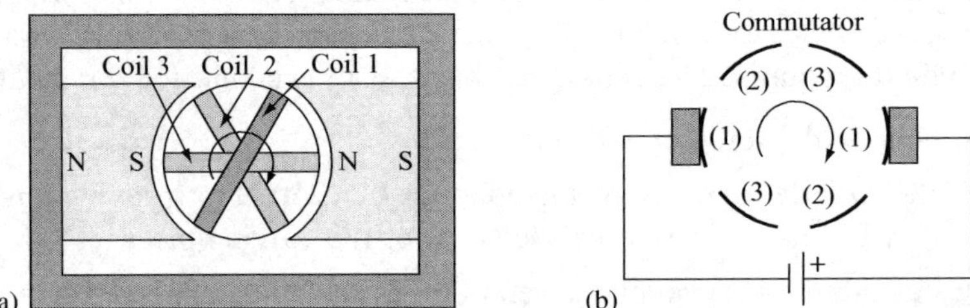

FIGURE 5.114 ■ Permanent magnet DC motor with three coils.

5.25 **Linear DC motor.** A linear DC motor is shown in **Figure 5.115**. A permanent magnet generates a constant magnetic flux density of 0.5 T in the gap between its poles. An iron plate 4 mm thick is placed between the poles of the magnets, leaving enough freedom for the magnets to move, and a coil with 1000 turns/m is wound on the iron plate. The gap on each side of the iron plate is 1 mm and the current-carrying wires shown are bonded to the plate. Wires with an ⊗ carry a current into the page, those with a ⊙ carry a current out of the page, and those without a symbol do no carry a current. As the magnet assembly moves under the influence of the magnetic forces, the currents in the conductors are shifted so that at all times the relation between currents and magnets remains the same. The plate with the conductors is stationary.

a. Calculate the force on the magnet for a current of 5 A and show the direction of motion of the magnet assembly for the current shown in the figure.

b. Show how this configuration may be changed into a stepping motor without changing the geometry and define its possible step sizes based on your configuration.

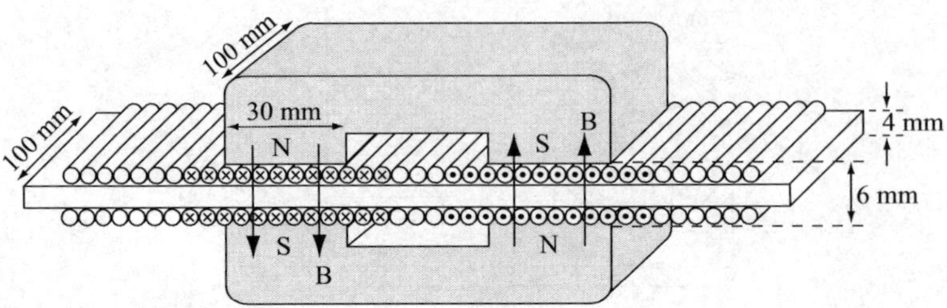

FIGURE 5.115 ■ Permanent magnet linear motor.

Stepper motors

5.26 General relations in stepper motors. The number of teeth in the rotor and stator of a stepper motor define the number of steps the motor is capable of. Show in general that if n is the number of teeth in the stator and p is the number of teeth in the rotor, then the following applies:

a. The larger n is for a fixed value p, the larger the step size.

b. The smaller the difference $n-p$, the smaller the step size.

c. The larger the numbers n and p, the smaller the step size.

d. Discuss the limitations on n and p.

5.27 High angular resolution stepper motor. A variable reluctance stepper motor is proposed with six poles in the stator and 6 teeth in each pole and 50 teeth in the rotor (usually the number of teeth in the rotor is less than in the stator because its diameter is smaller, but this is not a requirement, and in variable reluctance motors the opposite is just as practical).

a. Calculate the angle per step and number of steps per revolution.

b. Explain why a motor with the number of teeth indicated here is not likely to be produced even though it is entirely possible to do so.

5.28 Noninteger number of steps per revolution. Normally a stepper motor is designed with an integer number of steps per revolution. However, one can envision cases of noninteger number of steps, such as in cases where a specific step size is needed (say, 4.6°). Consider a motor with eight poles in the stator with 6 teeth per pole and 38 teeth in the rotor. Calculate the step size and the number of steps per revolution.

5.29 Linear stepping motor. A linear stepping motor similar to the one in **Figure 5.83** has N teeth/cm in the stator and M teeth/cm in the slider.

a. Derive a relation that allows calculation of the step size in millimeters.

b. What is the step size for a linear motor with 10 teeth/cm in the stator and 8 teeth/cm in the slider?

5.30 Force in a magnetic valve. Calculate the force on the movable part in the configuration shown in **Figure 5.116**. Edge effects in the air gaps are neglected and the flux density is assumed to be constant and perpendicular to the surfaces of the gaps. The permeability of the core may be assumed to be very large, number of turns $N = 300$, $I = 1$ A, $a = 25$ mm, $b = 10$ mm, $l = 2$ mm, and $L = 20$ mm.

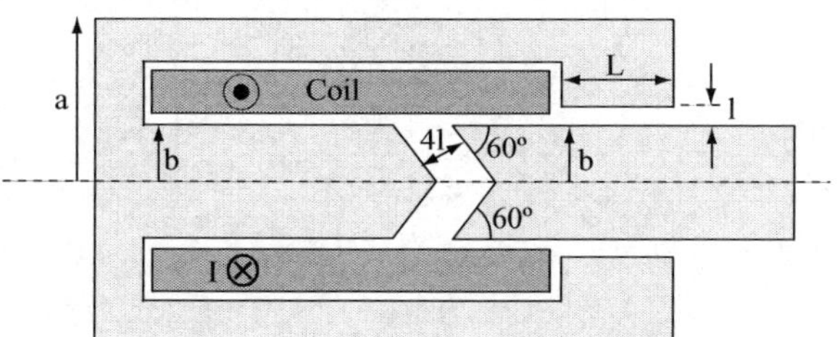

FIGURE 5.116 ■ Structure of a magnetic valve.

5.31 Solenoid actuator in a paint sprayer. A pump used to spray paint in an airless sprayer is made as shown in **Figure 5.117**. The principle is as follows: when a current is applied to the coil, the gap closes, and it opens when the current is switched off. If an alternating current is applied, the armature vibrates since the field is sinusoidal and passes through zero twice each cycle. This moves a piston back and forth a short distance, sufficient to activate a piston to pump the paint from its reservoir and spray it through an orifice. **Figure 5.117** shows a simplified form of the structure without the spray mechanism itself. In an actual device, the armature is hinged on one side and has a restoring spring. The coil contains $N = 5000$ turns and carries a sinusoidal current at 60 Hz with amplitude $I = 0.1$ A. The permeability of iron is assumed to be very large and the gaps are $d = 3$ mm. The dimensions $a = 40$ mm and $b = 20$ mm define the surface area of the poles that form the gap. Assume all flux is contained in the gaps (no leakage of flux outside the area of the poles).

a. Calculate the force exerted by the moving piece on the piston.

b. What is the force if the gap is reduced to 1 mm?

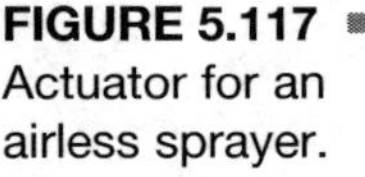
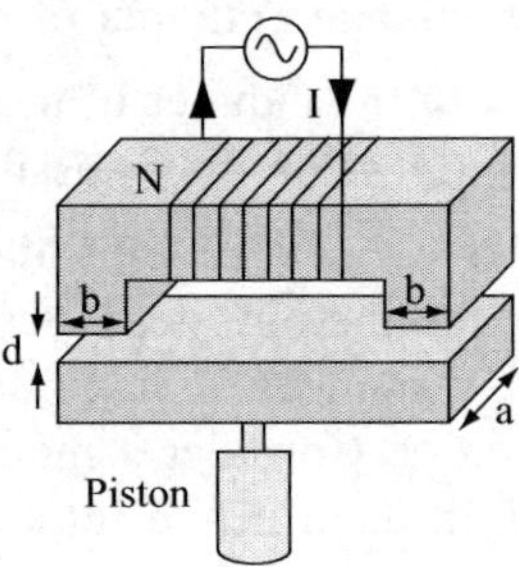

FIGURE 5.117 ▪ Actuator for an airless sprayer.

Voltage and current sensors

5.32 Ground fault circuit interrupt (GFCI). An important safety device is the GFCI (also called a residual current device (RCD)). It is intended to disconnect electrical power if current flows outside of the intended circuit, usually to ground, such as in the case when a person is electrocuted. The schematic in **Figure 5.118** shows the concept. The two conductors supplying power to an electrical socket or an appliance pass through the center of a toroidal coil or a Rogowski coil.

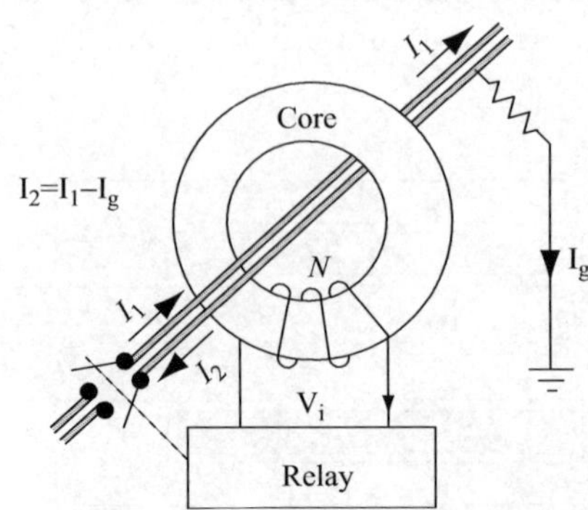

FIGURE 5.118 ▪ Principle of a GFCI sensor.

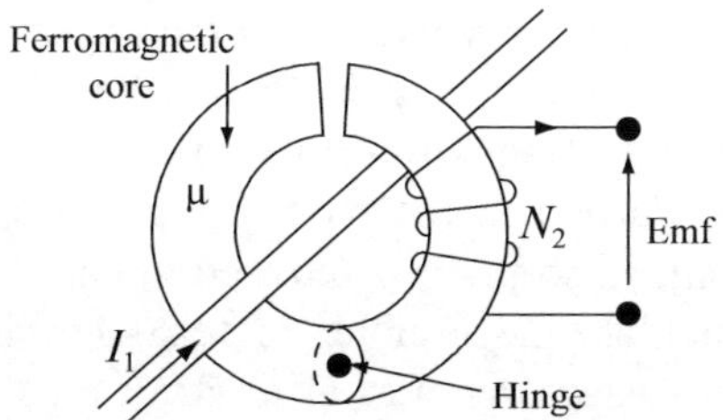

FIGURE 5.119 ■ Principle of the clamping ammeter.

Normally the currents in the two conductors are the same and the net induced voltage due to the two conductors cancel each other, producing a net zero output in the current sensor. If there is a fault and current flows to ground, say a current I_g, the return wire will carry a smaller current and the current sensor produces an output proportional to the ground current I_g. If that current exceeds a set value (typically 30 mA), the voltage induced causes the circuit to disconnect. These devices are common in many locations and are required by code in any location in close proximity to water (bathrooms, kitchens, etc.). Consider the GFCI shown schematically in **Figure 5.118**. The device is designed to operate in a 50 Hz installation and trip when the output voltage is 100 μV RMS. For a toroidal coil with average diameter $a = 30$ mm and a cross-sectional diameter of $b = 10$ mm,

a. Calculate the number of turns needed if a Rogowski coil is used and the device must trip at a ground current of 30 mA.

b. Calculate the number of turns needed to trip at a current of 30 mA if a ferromagnetic torus with relative permeability of 2000 is used as a core for the coil.

5.33 Clamping ammeter. When measuring the current in high current conductors without the need to cut the conductor and insert a regular ammeter, one uses a toroidal coil with the wire in which the current is measured passing through the toroid. To accomplish this, the torus is hinged so it can be opened and closed around the wire (see **Figure 5.119**). The *emf* in the coil on the torus is measured and related to the current in the wire.

a. Find the relation between current (RMS) and emf measured (RMS).

b. What is the maximum *emf* (peak value) for a sinusoidal current of amplitude 10 A at 60 Hz for a torus with inner radius $a = 2$ cm, outer radius $b = 4$ cm, thickness $c = 2$ cm, $N = 200$ turns, and relative permeability $\mu_r = 600$. The torus has rectangular cross section.

5.34 AC sensor. One type of commercially available current sensor is made as a simple closed magnetic core, typically rectangular, as shown in **Figure 5.120**.

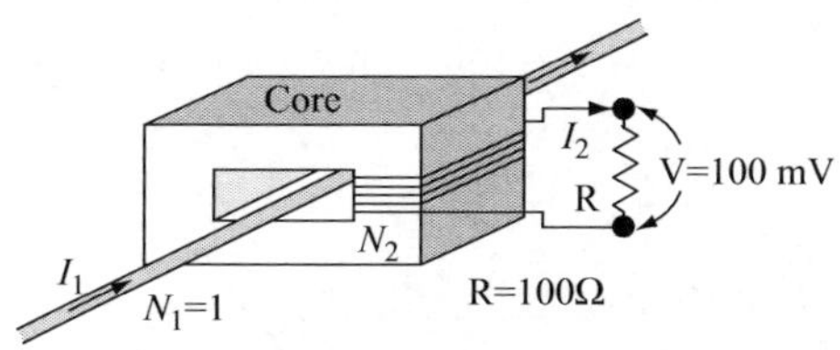

FIGURE 5.120 ■ Current sensor.

The current in a conductor is sensed by threading the wire through the central opening (this type of sensor is typically used in fixed installations) and the induced voltage in a coil wound uniformly on the core, or the current through it is used to measure the current in the wire. For a full scale of 100 A (RMS), the sensor shown is designed so that the induced voltage is 100 mV (RMS) across a load of 100 Ω. Calculate the number of turns required in the secondary coil assuming the current sensor operates as an ideal transformer.

CHAPTER 6

Mechanical Sensors and Actuators

The hand

The hand is the main body organ for interaction with the environment. An actuator as well as a sensor, it is an amazing organ when one really thinks about it. As an actuator it contains 27 bones, of which 14 make up the fingers or digital bones (3 on each finger except the thumb, which has only two), 5 are in the palm (metacarpal bones), and 8 in the wrist (carpal bones). Their structure and interconnections together with a complex series of muscles and tendons give the human hand a flexibility and dexterity not found in any other animal. Apes, monkeys, and lemurs have hands similar to humans, and other animals such as the koala have opposing thumbs, which are useful for climbing, but none are as flexible as the human hand. The hand can perform articulation of the finger bones, between the fingers and the palm, between the palm and the wrist, and between the wrist and the arm. Together with additional articulations at the elbow and shoulder, the hand is a multiaxis actuator capable of surprisingly delicate as well as gross motions. But the hand is also a tactile sensor. The fingertips in particular have the densest nerve endings in the body. They provide feedback for manipulation of objects or sense by direct touch. The hands are controlled by opposing brain hemispheres (left hand by the right hemisphere and right hand by the left hemisphere). This is true of other paired organs, including the eyes and legs.

Sensing and the skin

The skin is the largest organ in the human body, covering the whole body with a layer that averages 2–3 mm in thickness and an average area close to 2 m^2. As with other organs, it is multifunctional, serving as a protection layer for intrusion of organisms into the body, preventing loss of fluid through it, and absorbing vitamin D. It also protects the body from harmful radiation by absorbing ultraviolet radiation in melatonin as well as absorbing oxygen and excreting some chemicals. A critical function is insulation and heat regulation through sweat mechanisms and blood vessels in the dermis (the layer just below the thin, externally visible surface, called the epidermis). But of particular interest here is the function of the skin for sensing. Nerve endings on the skin sense heat, cold, pressure, vibration, and damage (injury), although the sensitivity varies from place to place. Not only can we sense with the skin, but localization of stimuli is very good and quite accurate, allowing us to detect the location of the stimulus on that large surface.

6.1 | INTRODUCTION

The class of mechanical sensors includes a fairly large number of different sensors based on many principles, but the four groups of general sensors discussed here—force sensors, accelerometers, pressure sensors, and gyroscopes—cover most of the principles involved in the sensing of mechanical quantities either directly or indirectly. Some of these sensors are used for applications that initially do not seem to relate to mechanical quantities. For example, it is possible to measure temperature through the expansion of gases in a volume (pneumatic temperature sensors are discussed in **Chapter 3**). The expansion can be sensed through the use of a strain gauge, which is a classical mechanical sensor. In this application an indirect use of a strain sensor is made to measure temperature. On the other hand, some mechanical sensors do not involve motion or force. An example of this is the optical fiber gyroscope, which will be discussed later in this chapter.

6.2 | SOME DEFINITIONS AND UNITS

Strain (dimensionless) is defined as the change in length per unit length of a sample. It is given as a fraction (i.e., 0.001) or as a percentage (i.e., 0.1%). Sometimes it is given as microstrain, meaning the strain in micrometers per meter (μm/m). The common symbol for strain is ε. Although this is the same symbol as that of electric permittivity, it should be clear from the context which quantity is implied by its use.

Stress is pressure [N/m^2] in a material. The symbol for stress is σ, and again, this should not be confused with electric conductivity, which uses the same symbol.

Modulus of elasticity is the ratio of stress to strain. This relation, often written as $\sigma = \varepsilon E$, where E is the modulus of elasticity, is called Hooke's law. The modulus of elasticity is often referred to as Young's modulus and has units of pressure [N/m^2].

The **gas constant** or **ideal gas constant** is equivalent to the Boltzmann constant. Whereas Boltzmann's constant expresses energy per temperature increment per particle, the gas constant expresses energy per temperature increment per mole. Denoted as R, its value is 8.3144621 J/mol/K.

The **specific gas constant** is the gas constant divided by the molecular mass of the gas. It is denoted as R_{specific} or R_s and equals 287.05 J/kg/K.

Pressure is force per unit area [N/m^2]. The SI derived unit of pressure is the pascal (1 pascal [Pa] = 1 newton per square meter [N/m^2]). The pascal is an exceedingly small unit and it is much more common to use the kilopascal (kPa = 10^3 Pa) and the megapascal (MPa = 10^6 Pa). Other often used units are the bar (1 bar = 0.1 MPa) and the torr (1 torr = 133 Pa). Also, for some very low pressure uses, the millibar (1 mbar = 1.333 torr = 100 Pa) and the microbar (1 μbar = 0.1 Pa) are employed. In common use one can find the atmosphere (atm). The atmosphere is defined as "the pressure exerted by a 1 m (actually the exact value is 1.032 m) column of water at 4°C on 1 cm^2 at sea level." The use of the atmosphere indicates a totally parallel system of pressure based either on a column of water or a column of mercury. In fact, the torr (named after Evangelista Torricelli) is defined as the pressure exerted by 1 mm of mercury (at 0°C and normal atmospheric pressure). Neither mmHg nor cmH_2O is an SI unit, but they still exist and in some instances are the preferred unit. For example, blood pressure is often measured in mmHg, whereas cmH_2O is used by gas utilities to measure gas pressure. One mmHg is the pressure exerted by a column of mercury 1 mm high at 0°C (the density of mercury is 13.5951 g/cm^3) and assuming the acceleration of gravity is 9.80665. Similarly 1 cmH_2O is the pressure exerted by a column of water 1 cm high at 4°C with density 1.004514556 g/cm^3 and assuming the acceleration of gravity is 9.80665 m/s^2. In the

TABLE 6.1 ■ Main units of pressure and conversion between them

	pascal	atmosphere	torr	bar	psi
pascal	1 Pa	9.869×10^{-6} atm	7.7×10^{-3} torr	10^{-5} bar	1.45×10^{-4} psi
atmosphere	101.325 kPa	1 atm	760 torr	1.01325 bar	14.7 psi
torr	133.32 Pa	1.315×10^{-3} atm	1 torr	1.33×10^{-3} bar	0.01935 psi
bar	100 kPa	0.986923 atm	750 torr	1 bar	14.51 psi
psi	6.89 kPa	0.068 atm	51.68 torr	0.0689 bar	1 psi

Note: Atmospheric (air) pressure is often given in millibars (mbar). The normal atmospheric pressure at sea level is 1013 mbar (1 atm or 101.325 kPa or 14.7 psi). However, none of these units are SI units. The proper unit to use is the pascal.

United States in particular, the common (nonmetric) unit of pressure is pounds per square inch (psi; 1 psi = 6.89 kPa = 0.068 atm).

Table 6.1 shows the main units of pressure in common use and the conversion between them. Note that only the pascal is an SI unit.

In work with pressure and pressure sensors, the concept of a vacuum is often used, sometimes as a separate quantity. Whereas vacuum means lack of pressure, it is usually understood as indicating pressure below ambient. Thus when one talks about so many pascals or psi of vacuum, this simply refers to so many pascals or psi below ambient pressure. While this may be convenient, it is, strictly speaking, not correct and the system of units does not provide for it. Therefore it should be avoided.

6.3 | FORCE SENSORS

6.3.1 Strain Gauges

The main tool in sensing force is the strain gauge. Although strain gauges, as their name implies, measure strain, the strain can be related to stress, force, torque, and a host of other stimuli, including displacement, acceleration, or position. With proper application of transduction methods, it can even be used to measure temperature, level, and many other related quantities.

At the heart of all strain gauges is the change in resistance of materials (primarily metals and semiconductors) due to a change in their length due to strain. To better understand this, consider a length of metal wire L, of conductivity σ, and cross-sectional area A. The resistance of the wire is

$$R = \frac{L}{\sigma A} \quad [\Omega]. \tag{6.1}$$

Taking the log of this expression:

$$\log R = \log\left(\frac{1}{\sigma}\right) + \log\left(\frac{L}{A}\right) = -\log\sigma + \log\left(\frac{L}{A}\right). \tag{6.2}$$

Now, taking the differential on both sides:

$$\frac{dR}{R} = -\frac{d\sigma}{\sigma} + \frac{d(L/A)}{L/A}. \tag{6.3}$$

Thus the change in resistance can be viewed as due to two terms. One is the conductivity of the material and the other (second term on the right-hand side) is due to the deformation of the conductor. For small deformations, both terms on the right-hand side are linear functions of strain, ε. Bundling both effects together (i.e., the change in conductivity and deformation) we can write

$$\frac{dR}{R} = g\varepsilon, \tag{6.4}$$

where g is the sensitivity of the strain gauge, also known as the **gauge factor**. For any given strain gauge this is a constant, ranging between 2 and 6 for most metallic strain gauges and between 40 and 200 for semiconductor strain gauges. This equation is the strain gauge relation and gives a simple linear relation between the change in resistance of the sensor and the strain applied to it.

The change in resistance due to strain increases the resistance of the strain gauge. The resistance of the strain gauge under strain is therefore

$$R(\varepsilon) = R_0(1 + g\varepsilon) \quad [\Omega], \tag{6.5}$$

where R_0 is the no-strain resistance. Before continuing, a bit more on stress, strain, and the connection between them.

Given the conductor discussed above (**Figure 6.1**), and applying a force along its axis, the stress is

$$\sigma = \frac{F}{A} = E\frac{dL}{L} = E\varepsilon \quad [\text{N/m}^2]. \tag{6.6}$$

Note: σ is used here for stress, not for conductivity. The quantities are

σ = stress [N/m^2 or Pa], which is in fact pressure.

E = modulus of elasticity or Young's modulus of the material [N/m^2 or Pa]. The modulus of elasticity is the ratio between stress and strain (Hooke's law), as can be seen from **Equation (6.6)**.

$\varepsilon = dL/L$ = strain. Strain is dimensionless and gives the change in length per unit length.

Thus strain is simply a normalized linear deformation of the material, whereas stress is a measure of the elasticity of the material.

Note: σ and ε are used here to denote stress and strain. In Chapter 5 we used the same symbols for electric conductivity and permittivity. These should not be confused.

Since strain gauges are made of metals and metal alloys (including semiconductors), they are also affected by temperature. If we assume that the resistance in **Equation (6.5)** is calculated at a reference temperature T_0, then we can write the resistance of the sensor as a function of temperature using **Equation (3.4)**:

$$R(\varepsilon, T) = R(\varepsilon)(1 + \alpha[T - T_0]) = R_0(1 + g\varepsilon)(1 + \alpha[T - T_0]) \quad [\Omega], \tag{6.7}$$

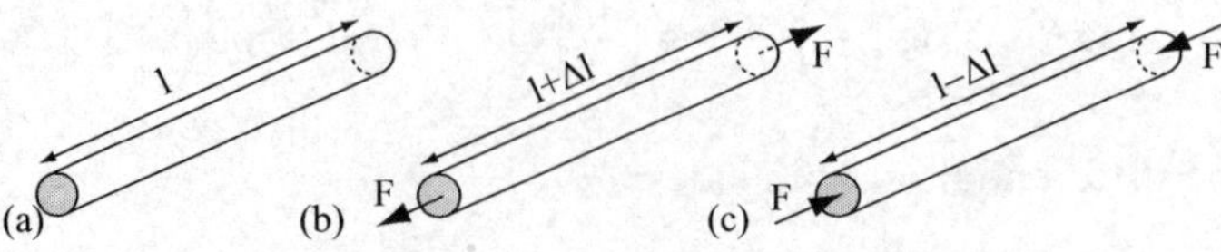

FIGURE 6.1 ■ (a) A wire of length L, cross-sectional area A, and electric conductivity σ. (b, c) Application of force to cause stress and strain in the conductor.

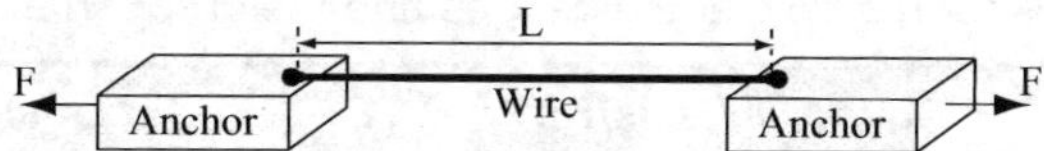

FIGURE 6.2 ▪ A rudimentary wire strain gauge (also called unbonded strain gauge).

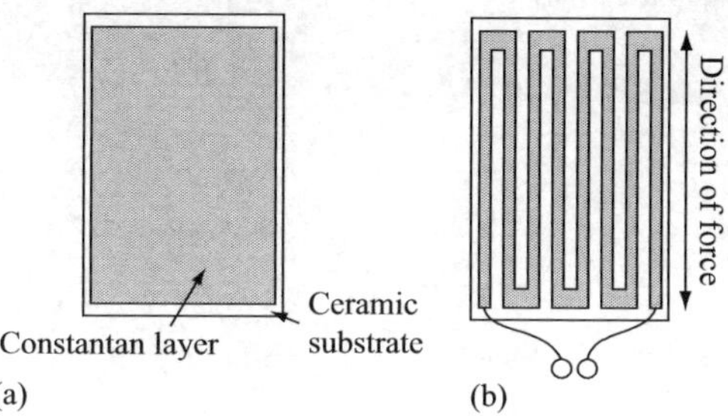

FIGURE 6.3 ▪ Common construction of a resistive strain gauge. Constantan is usually used because of its very low TCR.

where α is the TCR of the material (see **Table 3.1**). This clearly shows that the temperature and strain effects are multiplicative and indicates that strain gauges are, necessarily, sensitive to temperature variations.

Strain gauges come in many forms and types. In effect, any material, combination of materials, or physical configuration that will change its resistance (or any other property for that matter) due to strain constitutes a strain gauge. However, we will restrict our discussion here to two types that account for most of the strain gauges in use today: wire (or metal film) strain gauges and semiconductor strain gauges. In its simplest form, a metallic strain gauge can be made of a length of wire held between two fixed posts (**Figure 6.2**). When a force is applied to the posts, the wire deforms, causing a change in the wire's resistance. Although this method was used in the past and is valid, it is not very practical in terms of construction, attachment to the system whose strain needs to be measured, or in terms of the change in resistance (which is necessarily very small). Thus a more practical strain gauge is built out of a thin layer of conducting material deposited on an insulating substrate (plastic, ceramic, etc.) and etched to form a long, meandering wire, as shown in **Figure 6.3**. Constantan (an alloy made of 60% copper and 40% nickel) is the most common material because of its negligible temperature coefficient of resistance (TCR) (see **Table 3.1**). There are other materials in common use, especially at higher temperatures or when special properties are needed. **Table 6.2** shows some of the materials used for strain gauges with their properties, including the gauge factor.

Strain gauges may also be used to measure multiple axis strains by simply using more than one gauge or by producing them in standard configurations. Some of these are shown in **Figure 6.4**. **Figure 6.7** shows two commercial strain gauges.

6.3.2 Semiconductor Strain Gauges

These operate in the same way as conductor strain gauges, but their construction and properties are different. First, the gauge factor for semiconductors happens to be much higher than for metals. Second, the change in conductivity in **Equation (6.1)** due to strain is much larger than in metals. Semiconductor strain gauges are typically smaller

TABLE 6.2 ■ Materials for resistive strain gauges and their properties

Material	Gauge factor	Resistivity [Ω · mm²/m] at 20°C	TCR [10^{-6}/K]	Expansion coefficient [10^{-6}/K]	Maximum temperature [°C]
Constantan (Cu60Ni40)	2.0	0.5	10	12.5	400
Nichrome (Ni80Cr20)	2.0	1.3	100	18	1000
Manganine (Cu84Mn12Ni4)	2.2	0.43	10	17	
Nickel	−12	0.11	6000	12	
Chromel (Ni65Fe25Cr10)	2.5	0.9	300	15	800
Platinum	5.1	0.1	2450	8.9	1300
Elinvar (Fe55Ni36Cr8Mn0.5)	3.8	0.84	300	9	
Platinum-iridium (Pt80Ir20)	6.0	0.36	1700	8.9	1300
Platinum-rhodium (Pt90Rh10)	4.8	0.23	1500	8.9	
Bismuth	22	1.19	300	13.4	

Notes:

1. There are other specialized alloys often used for the production of strain gauges. These include platinum-tungsten (Pt92W08), isoelastic alloy (Fe55.5Ni36Cr08Mn05), Karma (Ni74Cr20Al03Fe03), Armour D (Fe70Cr20Al10), Manganin (Cu84Mn12Ni4), and Monel (Ni67Cu33).
2. These materials are selected for specific applications. For example, isoelastic alloy is excellent for dynamic strain/stress sensing, although it is particularly highly temperature sensitive. Platinum strain gauges are selected for high-temperature applications.
3. Many of the strain gauges must be temperature compensated.

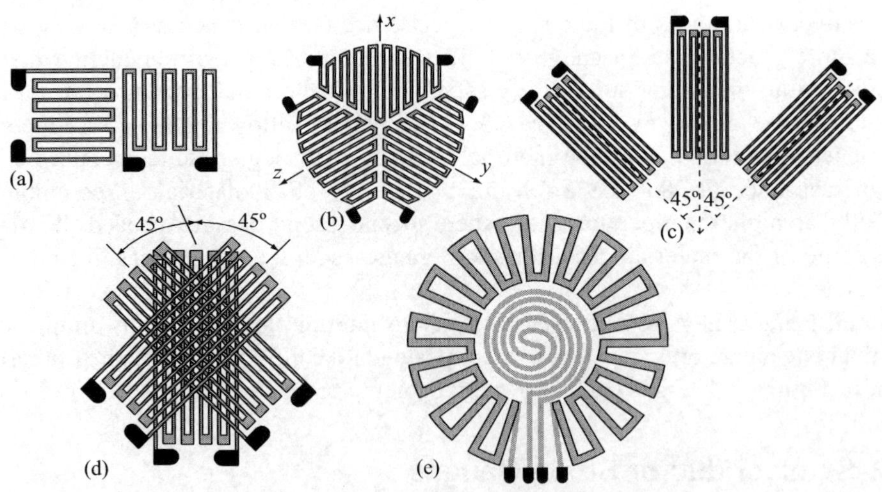

FIGURE 6.4 ■ Various configurations of strain gauges for different purposes. (a) Two-axis. (b) 120° rosette. (c) 45° rosette. (d) 45° stacked. (e) Membrane rosette.

than metal types, but are often more sensitive to temperature variations (hence temperature compensation is often incorporated within the gauge). All semiconductor materials exhibit changes in resistance due to strain, but the most common material is silicon because of its inert properties and ease of production. The base material is doped by diffusion of doping materials (usually boron for *p* type and arsenide for *n* type) to obtain a base resistance as needed. The substrate provides the means of straining the silicon chip and connections are provided by deposition of metal at the ends of the device. **Figure 6.5a** shows the construction of such a device, but a large variation in shapes and types may be found. Some of these, including multiple element gauges are shown in **Figures 6.5b–6.5f**. The range of temperatures for semiconductor strain gauges is limited to less than about 150°C.

One of the important differences between conductor and semiconductor strain gauges is that semiconductor strain gauges are nonlinear devices with typically a quadratic transfer function:

$$\frac{dR}{R} = g_1\varepsilon + g_2\varepsilon^2. \tag{6.8}$$

Although this nonlinearity is problematic in some applications, the higher sensitivity (gauge factor between 40 and 200 or more) is a boon. Also, the fact that *p* and *n* types may be used allows for PTC- and NTC-type behavior as shown in **Figure 6.6**.

The conductivity of semiconductors depends on a number of parameters, including the doping level (concentration or carrier density), type of semiconductor, temperature, radiation, pressure, and light intensity (if exposed), among others. Therefore it is essential that common effects such as temperature variations be compensated for or the errors due to these effects may be of the same order of magnitude as the effects of strain, resulting in unacceptable errors.

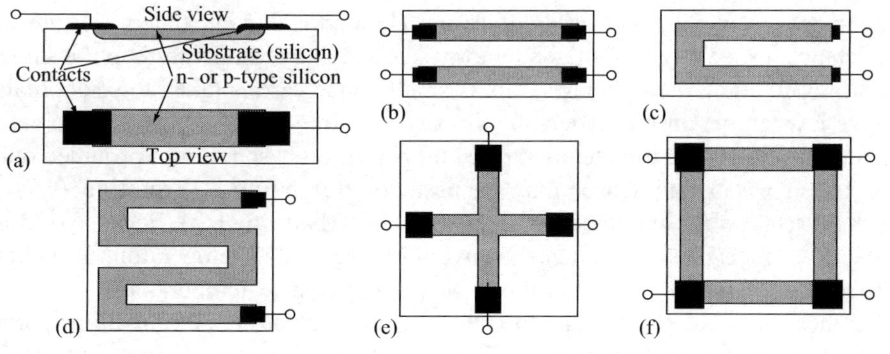

FIGURE 6.5 ■ (a) Construction of a semiconductor strain gauge. (b–f) Various configurations of semiconductor strain gauges.

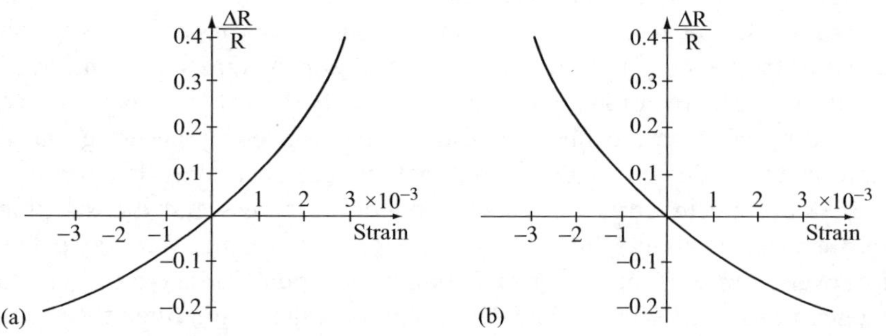

FIGURE 6.6 ■ Transfer functions for *p*- and *n*-type semiconductor strain gauges (PTC and NTC operation).

6.3.2.1 Application

To use a strain gauge as a sensor it must be made to react to force. For this to happen, the strain gauge is attached to the member in which strain is sensed, usually by bonding. Special bonding agents exist for different applications and types of materials and are usually supplied by the manufacturers of strain gauges or by specialized producers. In this mode, they are often used to sense bending strain, twisting (torsional and shear) strain, and longitudinal tensioning/deformation (axial strain) of structures such as engine shafts, bridge loading, truck weighing, and many others. Any quantity related to strain (or force), such as pressure, torque, and acceleration, can be measured directly. Other quantities may be measurable indirectly.

The properties of strain gauges vary by type and application, but most metal gauges have a nominal resistance between 100 and 1000 Ω (lower and higher resistances are available), have a gauge factor between 2 and 5, and have dimensions from less than 3 mm × 3 mm to lengths in excess of 150 mm, but almost any size may be fabricated as necessary. Rosettes (multiple-axis strain gauges) are available with 45°, 90°, and 120° axes, as well as diaphragm and other specialized configurations (see **Figure 6.4**). Typical sensitivities are 5 mΩ/Ω and deformation strain is on the order of 2–3 μm/m. Obviously much higher strains can be measured with specialized gauges.

Semiconductor strain gauges are usually smaller than most metal strain gauges and can be made with higher resistances. Because of the temperature limitations of these gauges, their use is limited to low temperatures, but they can be much less expensive than metal strain gauges and are in common use where applicable. One of the main uses of semiconductor strain gauges is as embedded devices in sensors such as accelerometers and load cells.

6.3.2.2 Errors

Strain gauges are subject to a variety of errors. The first is that due to temperatures, since the resistance, especially in semiconductors, is affected by temperature in the same way as by strain. In some metal gauges, this is small, since care is taken to select materials that have low temperature coefficients of resistance. In others, however, it can be rather large, and in semiconductors, temperature compensation is sometimes provided onboard the device or a separate sensor may be used for this purpose. **Equation (6.7)** gives the general relation for temperature effects (see also **Example 6.2**). Because of this, the nominal resistance of strain gauges is given at a reference temperature T_0 (often the reference temperature is 23°C, but it can be any convenient temperature).

Another source of error is due to lateral strains (i.e., strains in directions perpendicular to the main axis in **Figure 6.3**). These strains, and the change in resistance due to them, affect the overall reading. For this reason, strain gauges are usually built as slender devices with one dimension much larger than the other. Semiconductor strain gauges are particularly good in this respect, as their lateral sensitivity (or cross-sensitivity) is very low because of the very small dimensions of the sensor. A third source of error is due to the strain itself, which, over time, tends to permanently deform the gauge. This error can be eliminated by periodic recalibration and can be reduced by ensuring that the maximum deformation allowed is small and below that recommended for the device. Additional errors are incurred through the bonding process and through thinning of materials (or even breaking) due to cycling. Most strain gauges are rated for a given number of cycles (e.g., 10^6 or 10^7 cycles), maximum strain (3% is typical for conducting strain gauges and 1%–2% for semiconductor strain gauges), and often their temperature

FIGURE 6.7 ■ Two resistive strain gauges. Upper gauge is 25 mm × 6 mm. Lower gauge is 6 mm × 3 mm.

characteristics are specified for use with a particular material (aluminum, stainless steel, carbon steel) for optimal performance when bonded to that material. Typical accuracies when used in bridge configurations are on the order of 0.2%–0.5%.

EXAMPLE 6.1 The strain gauge

A strain gauge similar to that shown in **Figure 6.7** is made with the dimensions shown in **Figure 6.8**. The gauge is 5 μm thick. The sensor is made of constantan to reduce temperature effects.

a. Calculate the resistance of the sensor at 25°C without strain.

b. Calculate the resistance of the sensor if force is applied longitudinally causing a strain of 0.001.

c. Estimate the gauge factor from the calculations in (a) and (b).

Solution:

a. The resistance of the meander strip is calculated from **Equation (6.1)** and the data for constantan in **Table 6.2**. The conductivity at 20°C is 2×10^6 S/m (conductivity is the reciprocal of resistivity). The total length of the strip is

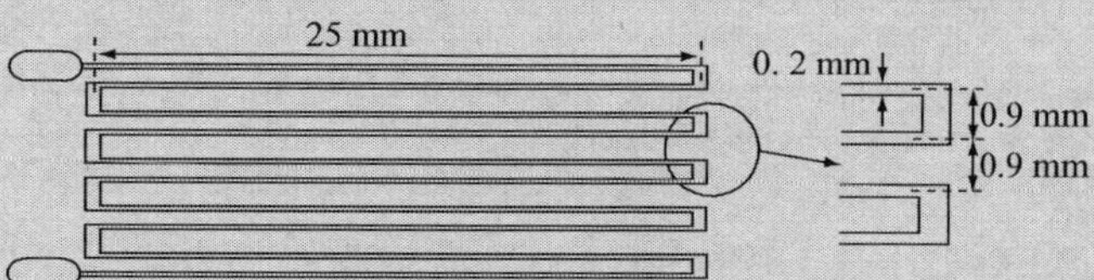

FIGURE 6.8 ■ Dimensions and structure of a strain gauge.

$$L = 10 \times 0.025 + 9 \times 0.0009 = 0.2581 \text{ m}$$

and its cross-sectional area is

$$S = 0.0002 \times 5 \times 10^{-6} = 1.0 \times 10^{-9} \text{ m}^2.$$

At 20°C,

$$R = \frac{L}{\sigma S} = \frac{0.2581}{2 \times 10^6 \times 1 \times 10^{-9}} = 129.05\ \Omega.$$

To calculate the gauge resistance at 25°C we use **Equation (3.4)** with the temperature coefficient of resistance for constantan, which equals 10^{-5} (see **Table 3.1**):

$$R(25°\text{C}) = R_0(1 + \alpha[T - T_0]),$$

where $T_0 = 20°\text{C}$ and R_0 is the resistance calculated at 20°C:

$$R(25°\text{C}) = 129.05\left(1 + 1 \times 10^{-5}[25 - 20]\right) = 129.05 \times 1.00005 = 129.05\ \Omega.$$

The resistance is virtually unchanged because of the small temperature difference and the low coefficient.

b. The strain is the change in length divided by total length:

$$\varepsilon = \frac{\Delta L}{L} = 0.001 \rightarrow \Delta L = 0.001L = 0.001 \times 0.2581 = 0.0002581\ \text{m}.$$

Thus the total length is 0.2583581 m. The cross-sectional area must also change since the volume of material must remain constant. Taking the volume v_0 as LS before deformation we get

$$S' = \frac{v_0}{L + \Delta L} = \frac{LS}{L + \Delta L} = \frac{0.2581 \times 1.0 \times 10^{-9}}{0.2583581} = 9.99 \times 10^{-10}\ \text{m}^2.$$

The resistance of the strain gauge is now

$$R = \frac{L + \Delta L}{\sigma S'} = \frac{0.2583581}{2 \times 10^6 \times 9.99 \times 10^{-10}} = 129.308\ \Omega.$$

This is a small change (0.258 Ω), but is typical of strain gauges.

c. The gauge factor is calculated from **Equation (6.4)** as an approximation:

$$g = \frac{1}{\varepsilon}\frac{dR}{R} = 1000 \times \frac{0.2583581}{129.05} = 2.0.$$

This gauge factor is as expected for conductor strain gauges.

EXAMPLE 6.2 Errors due to temperature variations

To measure strain during testing in a jet engine, a special platinum strain gauge is produced by sputtering the material into a foil and etching the gauge pattern. The sensor has a nominal resistance of 350 Ω at 20°C and a gauge factor of 8.9 (see **Table 6.2**). The platinum grade used has a temperature coefficient of resistance of 0.00385 Ω/°C. The sensor is exposed to temperature variations between −50°C and 800°C during testing.

a. Calculate the maximum resistance expected for a maximum strain of 2%.

b. Calculate the change in resistance due to temperature and maximum error due to temperature changes.

Solution:

a. From **Equation (6.4)**:

$$\frac{dR}{R} = g\varepsilon \rightarrow dR = Rg\varepsilon = 350 \times 8.9 \times 0.02 = 62.3\ \Omega.$$

The maximum resistance will be measured as $62.3 + 350 = 412.3\ \Omega$.

b. The change in resistance of the sensor due to temperature is calculated from **Equation (3.4)**:

$$R(25°\text{C}) = R_0(1 + \alpha[T - T_0]),$$

where R_0 is the resistance of the sensor at T_0. In this case, this is the resistance of the sensor at the given strain. We write

At $-50°$C, minimum strain is

$$R(-50°\text{C}) = 350(1 + 0.00385[-50 - 20]) = 255.675\ \Omega.$$

At $-50°$C, maximum strain is

$$R(-50°\text{C}) = 412.3(1 + 0.00385[-50 - 20]) = 301.185\ \Omega.$$

At 800°C, minimum strain is

$$R(800°\text{C}) = 350(1 + 0.00385[800 - 20]) = 1401.05\ \Omega.$$

At 800°C, maximum strain is

$$R(800°\text{C}) = 412.3(1 + 0.00385[800 - 20]) = 1650.44\ \Omega.$$

Clearly the change in resistance due to temperature is large. Taking the maximum resistance, the error that temperature variations will cause is

At 800°C, maximum strain is

$$error = \frac{1650.44 - 412.3}{412.3} \times 100\% = 300\%.$$

At 800°C, minimum strain is

$$error = \frac{1401.05 - 350}{350} \times 100\% = 300\%.$$

At $-50°$C, maximum strain is

$$error = \frac{301.185 - 350}{350} \times 100\% = -13.95\%.$$

At $-50°$C, minimum strain is

$$error = \frac{255.675 - 350}{350} \times 100\% = -26.95\%.$$

Therefore maximum error occurs at the higher temperature, regardless of strain. This error can be compensated for in properly designed bridge circuits (we shall see this in **Chapter 11**) and the measurement can still be done accurately. This example is extreme, but in many applications of strain gauges, temperature compensation is an essential part of sensing.

6.3.3 Other Strain Gauges

There are a number of strain gauges used for specialized applications. A very sensitive strain gauge can be made from optical fibers. In this type of gauge, the change in length of the fiber changes the phase of the light through the fiber. Measuring the light phase, either directly or by an interferrometric method, can produce readings of minute strain that cannot be obtained in other strain gauges. However, the device and the electronics necessary are far more complicated than standard gauges. There are also liquid strain gauges that rely on the resistance of an electrolytic liquid in a flexible container that can be deformed. Another type of strain gauge that is used on a limited basis is the plastic strain gauge. These are made as ribbons or threads based on graphite or carbon in a resin as a substrate and used in a way similar to other strain gauges. While they have very high gauge factors (up to about 300), they are otherwise difficult to use and inaccurate, as well as unstable mechanically, severely limiting their practical use.

6.3.4 Force and Tactile Sensors

Forces can be measured in many ways, but the simplest and most common method is to use a strain gauge and calibrate the output in units of force. Other methods include measuring acceleration of a mass ($F = ma$), measuring the displacement of a spring under action of force ($x = kF$, where k is the spring constant), measuring the pressure produced by force, and variations of these basic methods. None of these is a direct measure of force and many are more complicated than the use of a strain gauge. The transduction process may mean that the actual, measured quantity is capacitance, inductance, or, as in the case of strain gauges, resistance. The basic method of force sensing is shown in **Figure 6.9**. In this configuration one measures the tensile force by measuring the strain in the strain gauge. The sensor is usually provided with attachment holes and may also be used in compressive mode by prestressing the strain gauge. This type of sensor is often used to measure forces in machine tools, engine mounts, and the like. A common form of force sensor is the load cell. Like the force sensors in **Figure 6.9**, the load cell is instrumented with strain gauges. Usually the load cell is in the form of a cylinder (but a bewildering variety of shapes exist) placed between the two members that apply a force (e.g., the two plates of a press, between the suspension and the body of a vehicle, or between the moving and stationary parts of a truck weight scale).

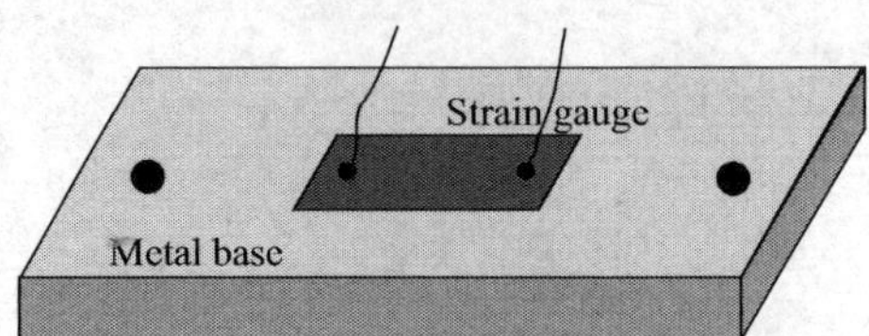

FIGURE 6.9 ■ The basic structure of a force sensor.

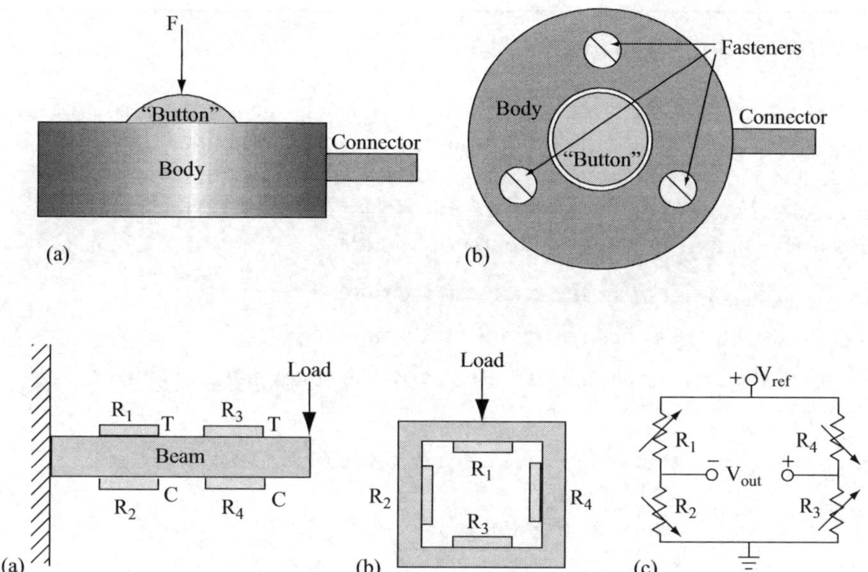

FIGURE 6.10 ■ One type of load cell—the button load cell.

FIGURE 6.11 ■ Structure of load cells. (a) Bending beam load cell. (b) "Ring" load cell. (c) The connection of the strain gauges in a bridge. Arrows pointing up indicate tension, arrows pointing down indicate compression.

Figure 6.10 shows a load cell that operates in compression mode. Note the "button," which transfers the load to the strain gauges. One or more strain gauges may be bonded to this button, which is usually a cylindrical piece, but the button may transfer the load to a beam or any other structure on which the strain gauges are bonded. The strain gauges are prestressed so that under compression the stress is lowered. However, it should be noted that in addition to the basic configuration in **Figures 6.9** and **6.10** there are literally dozens of configurations to suit any need. Load cells exist to sense forces that range from a fraction of a newton to hundreds of thousands of newtons. Although the configurations and shapes of load cells vary wildly, most use four strain gauges, two in compression mode and two in tension mode. Two common configuration are shown in **Figures 6.11a** and **6.11b**. In **Figure 6.11a**, the two gauges under the beam operate in compression mode, whereas the upper two operate in tension mode. In **Figure 6.11b**, the upper and lower members bend inward when a load is applied and hence gauges R_1 and R_3 are under tension. The side members bend outward and the gauges R_2 and R_4 are under compression. The four gauges are connected in a bridge as shown in **Figure 6.11c**. The operation of the bridge will be discussed in **Chapter 11**, including its use for load cells. At this point we simply mention that under no load gauges R_1 and R_3 have identical resistance, whereas gauges R_2 and R_4 have the same resistance (but generally different than that of R_1 and R_3). Under these conditions the bridge is balanced and the output is zero. When a load is applied, the resistance of R_1 and R_3 increases, whereas that of R_2 and R_4 decreases, taking the bridge out of balance and producing an output proportional to the load.

There are also force sensors that do not actually measure force in a quantitative way but, rather, sense the force in a qualitative way and respond to the presence of force above a threshold value. Examples are switches and keyboards, pressure-sensitive polymer mats for sensing presence, and the like.

EXAMPLE 6.3 **Force sensor for a truck scale**

A truck scale is made of a platform and four compression force sensors, one at each corner of the platform. The sensor itself is a short steel cylinder, 20 mm in diameter. A single stain gauge is prestressed to 2% strain and bonded on the outer surface of the cylinder. The strain gauges have a nominal resistance (before prestressing) of 350 Ω and a gauge factor of 6.9. The steel used for the cylinders has a modulus of elasticity (Young's modulus) of 30 GPa.

a. Calculate the maximum truck weight that the scale can measure.

b. Calculate the change in resistance of the sensors for maximum weight.

c. Calculate the sensitivity of the scale assuming the response of the strain gauges is linear.

Solution:

a. The relation between pressure and strain is given in **Equation (6.6)**:

$$\frac{F}{A} = \varepsilon E \quad [\text{Pa}],$$

where A is the cross-sectional area of the cylinder, F is the force applied, ε is the strain, and E is the modulus of elasticity. Since there are four sensors, the total force is

$$F = 4\varepsilon AE = 4 \times 0.02 \times \pi \times 0.01^2 \times 30 \times 10^9 = 753{,}982 \text{ N}.$$

This is 753,982/9.81 = 76,858 kg of force or 76.86 tons.

b. We need to relate the force to the resistance of the sensor. To do so we use **Equation (6.4)**:

$$\frac{dR}{R_0} = g\varepsilon \rightarrow dR = g\varepsilon R_0.$$

However, since the gauge is prestressed, its rest resistance is

$$R = R_0 + dR = R_0(1 + g\varepsilon) = 350(1 + 6.9 \times 0.02) = 398.3 \ \Omega,$$

where $R_0 = 350\ \Omega$ is the nominal (unstressed) resistance. As the sensor is compressed, the resistance goes down until at maximum allowable strain its resistance is R_0. Thus the change in resistance is −48.3 Ω.

c. The sensitivity is the output (resistance) divided by the input (force). For any of the sensors the force is 76.86/4 = 19.215 tons and the change in resistance is −48.3 Ω. Thus

$$S_o = -\frac{48.3}{19.215} = -2.514 \ \Omega/\text{ton}.$$

Tactile sensors are force sensors, but because the definition of "tactile" action is broader, the sensors are also more diverse. If one views a tactile action as simply sensing the presence of force, then a simple switch is a tactile sensor. This approach is commonly used in keyboards where membranes or resistive pads are used and the force is applied against the membrane or a silicon rubber layer. In applications of tactile sensing it is often important to sense a force distribution over a specified area (such as the "hand" of a robot). In such cases either an array of force sensors or a distributed sensor may be used. These are usually made from piezoelectric films that generate an electrical signal in response to deformation (passive sensors). An example is shown in **Figure 6.12**. The polyvinylidene fluoride (PVDF) film is sensitive to deformation. The lower film is driven with an AC signal and therefore it contracts and expands mechanically and periodically. This deformation is transferred to the upper film through the compression layer acting somewhat like a transformer and thus establishing a signal at the output. When the upper film is deformed by a force, its signal changes from normal and the amplitude and/or phase of the output signal is now a measure of deformation (force). Since the compression layer is thinner, when a force is applied, the output is higher and proportional (but not necessarily linearly) to the applied force. The PVDF films can be long narrow ribbons for a linear sensor or sheets of various sizes for tactile sensing over an area.

Another example is shown in **Figure 6.13**. In this case the output is normally zero. When a force is applied, the strain in the film gives rise to an output proportional to stress (force) and changing with the force. Because of this the output can be used to sense not only force, but also variations in force. This idea has been used to sense minute changes due to breathing patterns in babies, primarily in hospitals, but it can be used to sense under other conditions. In a sensor of this type, a sheet of PVDF is placed under the patient and its output is monitored for an expected pattern in the signal as the center of gravity of the body shifts with breathing. The issue of piezoelectricity and the associated piezoelectric force sensors will be revisited in **Chapter 7** in conjunction with ultrasonic sensors.

The simplest tactile sensors are made of conductive polymers or elastomers or with semiconducting polymers and are called piezoresistive sensors or force sensitive resistive (FSR) sensors. In these devices, the resistance of the material is pressure dependent and is shown schematically in **Figure 6.14**. A very simple tactile sensor can be made from conducting foam and two electrodes. The resistance of FSR sensors is a nonlinear function of force (**Figure 6.14c**), but the change in resistance is quite high (large dynamic range) and hence the sensor is rather immune to noise and easily interfaced

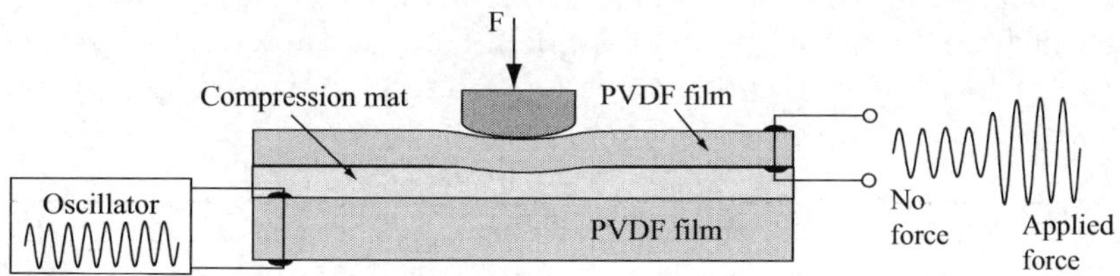

FIGURE 6.12 ▪ A piezoelectric film tactile sensor. The compression due to force changes the coupling between the lower and upper PVDF layers and hence the amplitude of the output.

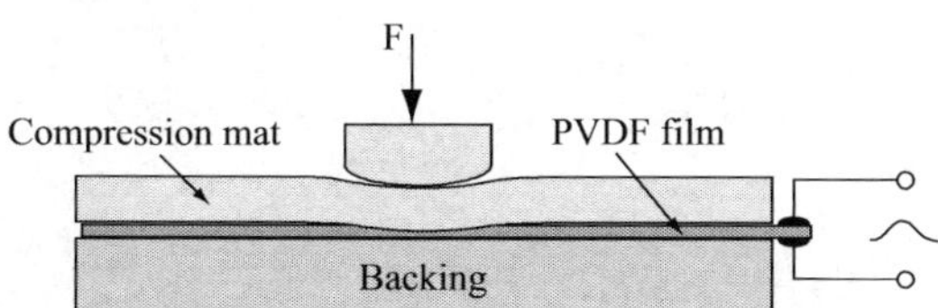

FIGURE 6.13 ▪ A piezoelectric film sensor used to detect sliding motion due to breathing. The output is monitored for a pattern consistent with the breathing pattern and the shift in the center of gravity as a consequence.

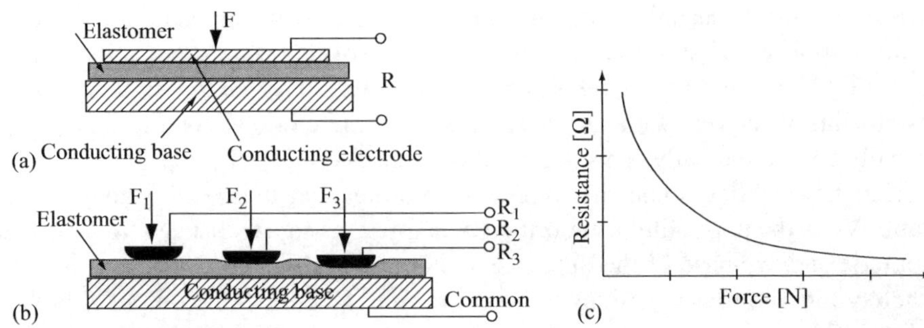

FIGURE 6.14 ■ FSR tactile sensor using conducting elastomers. (a) Principle and structure. (b) An array of tactile sensors. (c) The transfer function of FSRs.

with microprocessors. Either DC or AC sources can be used and the device can be as large or as small as needed. An array of sensors can be built by using one large electrode on one side of the film and multiple electrodes on the other side. In this configuration, sensing occurs over an area or a line (**Figure 6.14b**).

EXAMPLE 6.4 Evaluation of a force sensor

A force sensor (FSR) is evaluated experimentally. To do so, the resistance of the sensor is measured for a range of forces as follows:

F [N]	50	100	150	200	250	300	350	400	450	500	550	600	650
R [Ω]	500	256.4	169.5	144.9	125	100	95.2	78.1	71.4	65.8	59.9	60	55.9

Calculate the sensitivity of the sensor throughout its range.

Solution: Sensitivity is the slope of the resistance versus force curve and is clearly a nonlinear quantity. However, we recall from **Chapter 2** (see **Example 2.16**) that force resistive sensors have a linear relation between force (F) and conductance ($1/R$). Therefore it is simpler to first calculate the conductance.

F [N]	50	100	150	200	250	300	350	400	450	500	550	600
$1/R$ [1/Ω]	.002	.0039	.0059	.0069	.008	.01	.0105	.0128	.014	.0152	.0167	.0179

Now we have two options. We can calculate the sensitivity as a local quantity as

$$S = \frac{\Delta(1/R)}{\Delta F} \quad \left[\frac{1}{\Omega \cdot \mathrm{N}}\right]$$

or start with the resistance and calculate

$$S = \frac{\Delta R}{\Delta F} \quad [\Omega/\mathrm{N}].$$

A better approach is to find a linear fit for the conductance, then write the resistance as a function of force and take the derivative of that.

Using the linear fit in **Equation (2.19)**, we write the conductance $C = 1/R = aF + b$, where

$$a = \frac{n\sum_{i=1}^{n} x_i y_i - \left\{\sum_{i=1}^{n} x_i\right\}\left\{\sum_{i=1}^{n} y_i\right\}}{n\sum_{i=1}^{n} x_i^2 - \left\{\sum_{i=1}^{n} x_i\right\}^2}$$

and

$$b = \frac{\left\{\sum_{i=1}^{n} x_i^2\right\}\left\{\sum_{i=1}^{n} y_i\right\} - \left\{\sum_{i=1}^{n} x_i\right\}\left\{\sum_{i=1}^{n} x_i y_i\right\}}{n\sum_{i=1}^{n} x_i^2 - \left\{\sum_{i=1}^{n} x_i\right\}^2},$$

In this case, $n = 12$ is the number of points, x_i is the force at point i, and y_i is the conductance at point i. From the points in the table above, we get

$$a = 0.00014182, b = 0.0010985.$$

The conductance is

$$C = 0.00014182F + 0.0010985 \quad [1/\Omega].$$

Now we can write the resistance as

$$R = \frac{1}{0.00014182F + 0.0010985} \quad [\Omega].$$

The sensitivity is

$$\frac{dR}{dF} = \frac{dR}{dF}(0.00014182F + 0.0010985)^{-1}$$
$$= -\frac{0.00014182}{(0.00014182F + 0.0010985)^2} \quad [\Omega/\mathrm{N}].$$

Clearly sensitivity goes down with force, as can be seen from the table above as well. The negative sign simply indicates that increasing the force reduces the resistance. The method used here applies to other sensors for which a reciprocal function happens to be linear or may be approximated as such.

6.4 | ACCELEROMETERS

By virtue of Newton's second law ($F = ma$), a sensor may be made to sense acceleration by simply measuring the force on a mass. At rest, acceleration is zero and the force on the mass is zero. At any acceleration a, the force on the mass is directly proportional to mass and acceleration. This force may be sensed with any method of sensing force (see above) but, again, the strain gauge will be representative of direct force measurement.

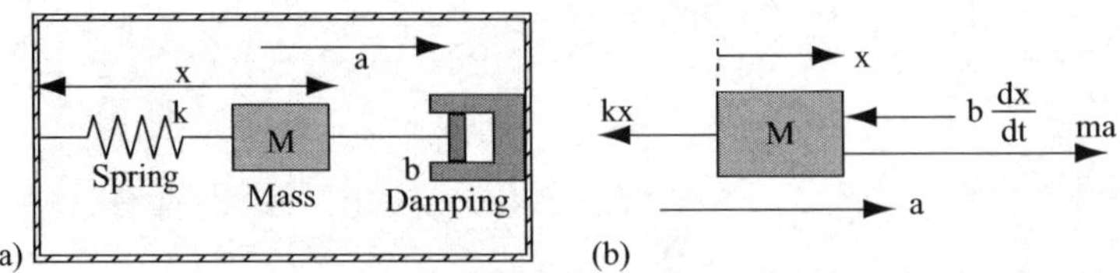

FIGURE 6.15 ■ (a) Mechanical model of an accelerometer based on sensing the force on a mass. (b) Free body diagram of the accelerometer in (a).

There are, however, other methods of sensing acceleration. Magnetic methods and electrostatic (capacitive) methods are commonly used for this purpose. In their simplest forms, the distance between the mass and a fixed surface, which depends on acceleration, can be made into a capacitor whose capacitance increases (or decreases) with acceleration. Similarly a magnetic sensor can be used by measuring the change in field due to a magnetic mass. The higher the acceleration, the closer (or farther) the magnet from a fixed surface and hence the larger or lower the magnetic field. The methods used in **Chapter 5** to sense position or proximity can now be used to sense acceleration. There are other methods of acceleration sensing including thermal methods. Velocity and vibrations may also be measured by similar methods and these will also be discussed in this section.

To understand the method of acceleration sensing it is useful to look at the mechanical model of an accelerometer based on sensing the force on a mass shown in **Figure 6.15**. The mass, which can move under the influence of forces, has a restoring force (spring) and a damping force (which prevents it from oscillating). Under these conditions, and assuming the mass can only move in one direction (along the horizontal axis), Newton's second law may be written as

$$ma = kx - b\frac{dx}{dt} \quad [\mathrm{N}]. \tag{6.9}$$

This assumes that the mass has moved a distance x under the influence of acceleration, k is the restoring (spring) constant, and b is the damping coefficient. Given the mass m and the constants k and b, a measurement of x gives an indication of the acceleration a. The mass is often called an inertial mass or proof mass.

Therefore, for a useful acceleration sensor it is sufficient to provide an element of given mass that can move relative to the sensor's housing and a means of sensing this movement. A displacement sensor (position, proximity, etc.) can be used to provide an appropriate output proportional to acceleration.

6.4.1 Capacitive Accelerometers

In this type of accelerometer, one plate of a small capacitor is fixed and connected physically to the body of the sensor. The second, which serves as the inertial mass of the sensor is free to move and connected to a restoring spring. Three basic configurations are shown in **Figure 6.16**. In these, the restoring force is provided by springs (**Figure 6.16a, c**) or by a cantilever beam (**Figure 6.16b**). In **Figures 6.16a** and **6.16b**, the distance between the plates changes with acceleration. In **Figure 6.16c**, the effective area of the capacitor plates changes while the distance between the plates stays constant. In either case, acceleration either increases the capacitance or decreases it, depending on

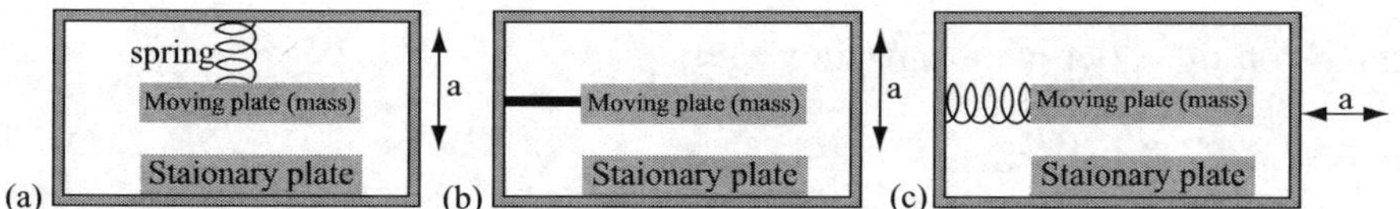

FIGURE 6.16 ■ Three basic capacitive acceleration sensors. (a) Moving plate against a spring. (b) Beam-suspended plate. (c) Sideways moving plate against a spring.

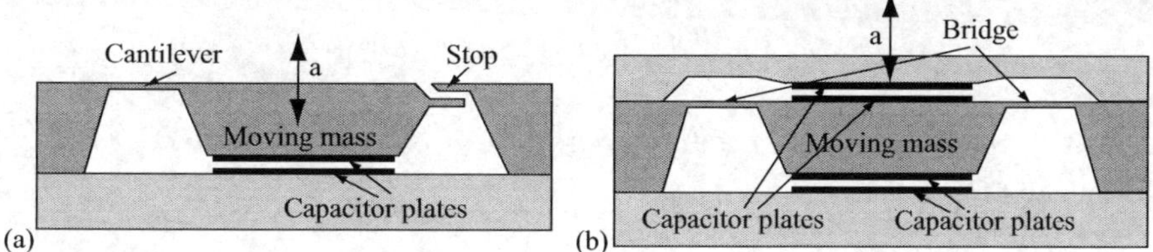

FIGURE 6.17 ■ Two basic forms of producing accelerometers. (a) Cantilever (supported on the left). (b) Bridge.

the direction of motion. Of course, for a practical accelerometer, the plates must be prevented from touching by stoppers and a damping mechanism must be added to prevent the springs or the beam from oscillating. Some of these issues are addressed in the structures in **Figure 6.17**, but regardless of the specific arrangement, the capacitance changes proportional to acceleration and the capacitance is thus a measure of acceleration. It should be noted, however, that the changes in capacitance are very small and therefore, rather than measuring these changes directly, indirect methods such as using the capacitor in an LC or RC oscillator are often used. In these configurations, the frequency of oscillation is a direct measure of acceleration. The frequency can be easily converted into a digital reading at the output. Accelerometers of this type can be produced as semiconductor devices by etching both the mass, fixed plate, and springs directly into silicon. By doing so, microaccelerometers can be produced quite easily. Two structures are shown in **Figure 6.17**. The first is a cantilever structure. The second is similar to **Figure 6.16a** and relies on etched bridges to provide the springs. In the latter structure, the mass moves between two plates and forms an upper and lower capacitor. By doing so, a differential mode may be obtained since at rest the two capacitors are the same (see **Section 5.3.1** and **Figure 5.10**). In both of these structures, limit stops are provided.

EXAMPLE 6.5 Capacitive accelerometer

Consider a simplified design of a capacitive accelerometer to be used in a car. Its ultimate function is to deploy an airbag in case of collision. Suppose the configuration in **Figure 6.16a** is used and the sensor is mounted so that in case of collision the spring elongates and the plates get closer together, increasing capacitance. Airbags are supposed to deploy when a deceleration of 60 g is detected (equivalent to crashing into a barrier at 23 km/h). The sensor has a fixed plate and a moving plate of mass 20 g. The two plates are separated a distance 0.5 mm apart, producing a capacitance of 330 pF at rest. To trigger airbag deployment, the capacitance must double. To ensure that the capacitance doubles at a deceleration of 60 g, the spring constant must be selected carefully. Find the necessary spring constant to accomplish this.

Solution: The capacitance of a parallel plate capacitor is given as

$$C = \frac{\varepsilon A}{d} \quad [\mathrm{F}].$$

This means that to double the capacitance, the distance d must be halved since everything else remains constant with acceleration. That is, the airbag will trigger when the plates are 0.25 mm apart. Under these conditions, the spring has elongated a distance $x = 0.25$ mm. The force equation now requires that the force due to deceleration equal the force on the spring:

$$ma = kx \rightarrow k = \frac{ma}{x} \quad [\mathrm{N/m}],$$

where k is the spring constant, m is the mass of the moving plate, a is deceleration, and x is the displacement of the plate. Thus we get

$$k = \frac{ma}{x} = \frac{20 \times 10^{-3} \times 60 \times 9.81}{0.25 \times 10^{-3}} = 47{,}088 \ \mathrm{N/m}.$$

Note: The calculation here is rather simple and does not address issues such as keeping the plates parallel. Nevertheless, the calculation indicates how a sensor might be designed. Some acceleration sensors used for this purpose are contact sensors, that is, when the necessary pressure has been achieved, a contact is closed (such as, e.g., the two plates touching each other). This avoids the need to actually measure capacitance and thus decreasing the response time of the sensor.

6.4.2 Strain Gauge Accelerometers

The structures in **Figures 6.16** and **6.17** can also be fitted with strain gauges to measure the strain due to acceleration. A strain gauge accelerometer is shown in **Figure 6.18**. The mass is suspended on a cantilever beam and a strain gauge senses the bending of the beam. A second strain gauge may be fitted under the beam to sense acceleration in both directions. Also, by fitting (or manufacturing) strain gauges on the bridge or cantilever beam in **Figure 6.17**, the capacitive sensor is transformed into a strain gauge sensor. In this configuration the strain gauges are usually semiconductor strain gauges, whereas in **Figure 6.18** they may be bonded metallic gauges. The operation remains the same as for capacitive accelerometers, only the means of sensing the force changes. Strain gauge sensors can be as sensitive as capacitive sensors and in some cases may be easier to work with since the measurement of resistance is typically simpler than that of capacitance. On the other hand, strain gauges are temperature sensitive and must be properly compensated.

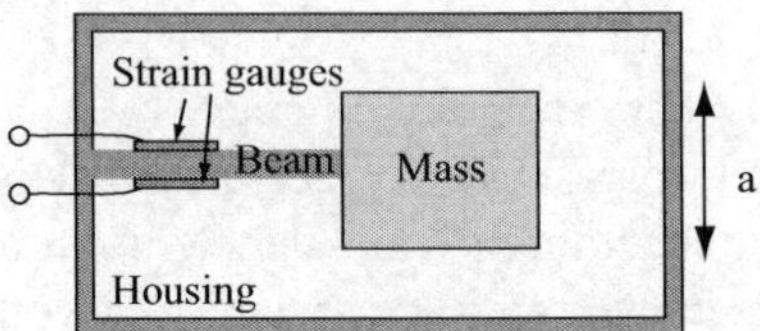

FIGURE 6.18 ■ An accelerometer in which the beam bending is sensed by two strain gauges to sense acceleration in both vertical directions.

6.4.3 Magnetic Accelerometers

A simple magnetic accelerometer can be built as a variable inductance device in which the mass, or a rod connected to and moving with the mass, links magnetically to a coil. The inductance of the coil is proportional to the position of the mass and increases the further the ferromagnetic rod penetrates into the coil (**Figure 6.19**). This configuration is a simple position sensor calibrated for acceleration. Instead of the coil, an LVDT can be used for an essentially linear indication of position. A different approach is to use a permanent magnet as a mass on a spring or cantilever beam and to sense the field of the permanent magnet using a Hall element or a magnetoresistive sensor (**Figure 6.20**). The reading of the Hall element is now proportional to the magnetic field, which is proportional to acceleration. It is also possible to bias the Hall element with a small magnet and use a ferromagnetic mass. In this configuration, the proximity of the mass changes the flux density, providing an indication of acceleration.

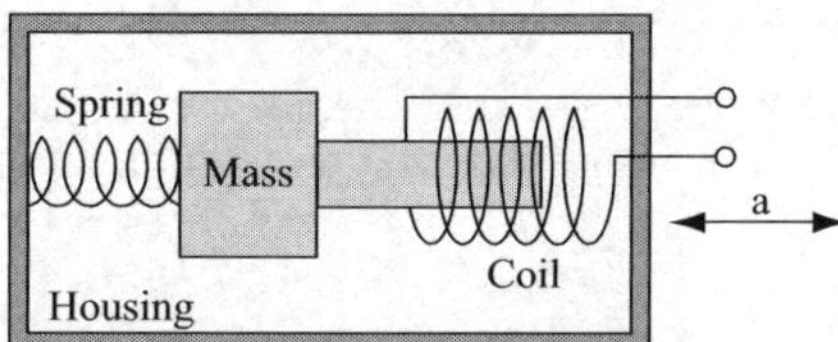

FIGURE 6.19 ■ An inductive accelerometer in which the horizontal motion of the mass is sensed by a change in the inductance of a coil.

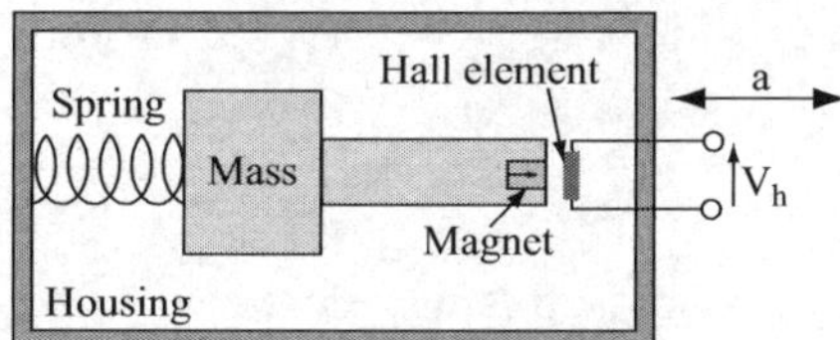

FIGURE 6.20 ■ An accelerometer in which the position of the mass is sensed by a Hall element.

EXAMPLE 6.6 **Magnetic accelerometer**

A magnetic accelerometer is built as in **Figure 6.21** with the mass being a cylinder of diameter $d = 4$ mm and some length l. The mass is 10 g and a spring with spring constant $k = 400$ N/m holds the mass in place. The mass is made of silicon steel with a relative permeability of 4000. The coil has $n = 1$ turn/mm and its inductance is sensed as a measure of the position of the mass. As the mass moves in or out of the coil, its inductance increases or decreases accordingly. In a long coil, the inductance per unit length can be approximated as (**see Equation (5.28)**)

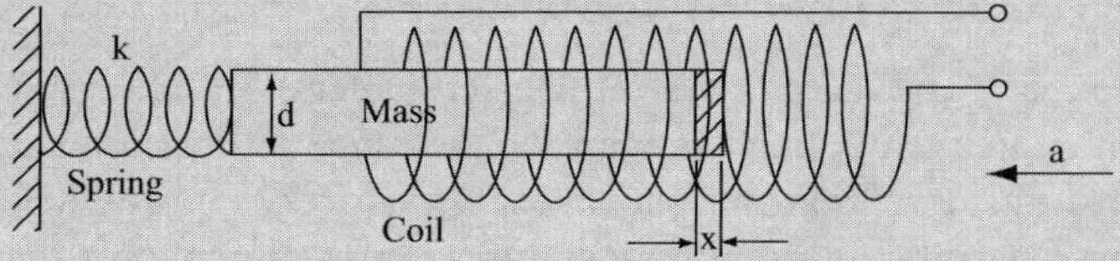

FIGURE 6.21 ■ Magnetic accelerometer.

$$L = \mu n^2 S \quad [\text{H/m}],$$

where n is the number of turns per unit length and S is the cross-sectional area of the coil. Calculate the change in voltage on the coil for an acceleration of $\pm 10\,g$ if the coil is driven with a sinusoidal current of amplitude 0.5 A and a frequency of 1 kHz.

Solution: As the mass moves a distance x into the coil, the inductance changes according to the position. As long as the distances are small, the change in inductance is linear and can be calculated as

$$\Delta L = Lx = \mu n^2 Sx \quad [\text{H}].$$

The maximum distance the mass moves in either direction is determined by the acceleration and the spring constant. That is,

$$ma = kx \to x = \frac{ma}{k} = \frac{10 \times 10^{-3} \times 10 \times 9.81}{400} = 2.4525 \text{ mm}.$$

The mass moves in or out a maximum of 2.453 mm. The change in inductance is therefore

$$\begin{aligned}\Delta L &= Lx = \mu n^2 Sx \\ &= 4000 \times 4\pi \times 10^{-7} \times 1000^2 \times \pi \times \left(2 \times 10^{-3}\right)^2 \times 2 \times 2.4525 \times 10^{-3} \\ &= 0.00031\,\text{H}.\end{aligned}$$

That is, the inductance changes by ± 310 μH.

The voltage across an inductor is related to its current as (see **Equation (5.29)**)

$$V = L\frac{dI(t)}{dt} \quad [\text{V}].$$

The change in voltage due to the change in inductance is therefore

$$\begin{aligned}\Delta V = \Delta L\frac{dI(t)}{dt} &= 310 \times 10^{-6} \times \frac{d}{dt}(0.5\sin 2\pi \times 1000t) \\ &= 310 \times 10^{-6} \times 0.5 \times 2 \times \pi \times 1000\cos 2\pi \times 1000t \\ &= 0.974\cos 2\pi \times 1000t \quad [\text{V}].\end{aligned}$$

The voltage across the coil changes by ± 0.974 V, a change sufficient for sensing.

Note that this can be increased if necessary by increasing the frequency or by increasing the current in the coil. Also, we have assumed here a linear relation based on the small travel distance and the fact that the coil is long. Friction and damping were neglected.

6.4.4 Other Accelerometers

There are many other types of accelerometers, but all employ a moving mass in one form or another. A good example of the range of principles used for this purpose is the heated gas accelerometer shown in **Figure 6.22**. In this device the gas in a cavity is heated to an equilibrium temperature and two (or more) thermocouples are provided

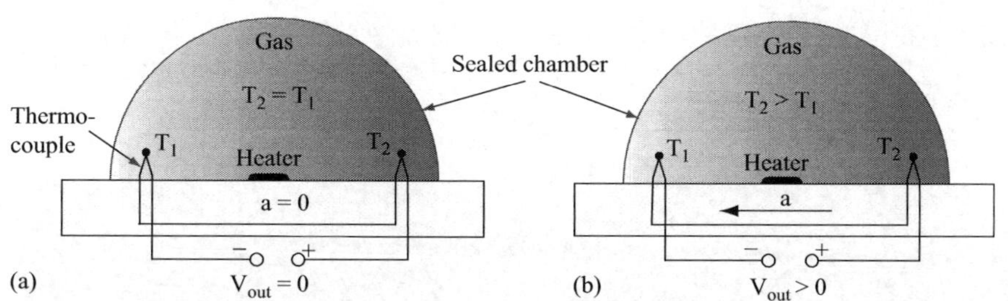

FIGURE 6.22 ■ The heated gas accelerometer.

equidistant from the heater. Under rest conditions, the two thermocouples are at the same temperature and hence their differential reading (one thermocouple is the sense thermocouple, the second is the reference thermocouple) is zero. When acceleration occurs, the gas shifts to the direction opposite of the motion (the gas is the inertial mass), causing a temperature rise that can be calibrated in terms of acceleration.

Other accelerometers use optical means (by activating a shutter by means of the moving mass), optical fiber accelerometers that use an optical fiber position sensor, vibrating reeds whose vibration rate changes with acceleration, and many more.

Finally, it should be noted that multiple-axis accelerometers can be built by essentially using single-axis accelerometers with axes perpendicular to each other. These can be fabricated as two- or three-axis accelerometers, or two or three single-axis accelerometers may be attached appropriately. Although this may seem cumbersome with regular devices, it is entirely practical to do so with microdevices. We will see that this is done routinely in MEMs (**Chapter 10**).

The uses of accelerometers are vast and include airbag deploying sensors, weapons guidance systems, vibration and shock measurement and control, and other similar applications. They can also be found in consumer devices such as telephones and computers, as well as in toys.

EXAMPLE 6.7 A seismic sensor

Detection of seismic activity, such as earthquakes, can be (and often is) undertaken through accelerometers by detecting the motion caused by earthquakes. To do so, an accelerometer is built as follows: A steel bar, 10 mm × 10 mm in cross section, is fixed in a concrete slab and extends vertically 50 cm above the slab. A 12 kg mass is welded to the top of the steel bar. To detect acceleration due to motion of the earth, a semiconductor strain gauge with a nominal resistance of 350 Ω and a gauge factor of 125 is fixed on one of the surfaces of the steel bar at the point where the bar emerges from the concrete slab. Assume the distance between the center of the mass and sensor is exactly 50 cm. Assume as well that the strain gauge is temperature compensated and the minimum change in resistance of the strain gauge that can be reliably measured is 0.01 Ω. Calculate the minimum acceleration the seismic sensor can detect. The modulus of elasticity for steel is 200 GPa.

Solution: The strain gauge relation in **Equation (6.5)** can be used to calculate the strain needed to cause a change in resistance of 0.01 Ω. Then we use the basic relations for beam bending to find the acceleration that will produce that strain. From **Equation (6.5)**:

$$R(\varepsilon) = R(1 + g\varepsilon) = 350 + 125\varepsilon \quad [\Omega].$$

Therefore

$$125\varepsilon = 0.01\ \Omega$$

or

$$\varepsilon_{\min} = \frac{0.01}{125} = 0.00008.$$

That is, an 80 microstrain will produce a change of 0.01 Ω in the resistance of the strain gauge.

As the soil moves, the acceleration a produces a force on the mass ($m = 12$ kg):

$$F = ma \quad [\text{N}].$$

The force bends the beam, causing a bending moment:

$$M = Fl = mal \quad [\text{N} \cdot \text{m}],$$

where $l = 50$ cm is the distance between the mass and the sensor.

To calculate the strain at the surface of the beam (where the strain gauge is located) we write

$$\varepsilon = \frac{M(d/2)}{EI} \quad [\text{m/m}],$$

where M is the bending moment, E is the modulus of elasticity, I is the moment of area of the beam, and d is the thickness of the beam. E is given and I is

$$I = \frac{bh^3}{12} = \frac{d^4}{12} \quad [\text{m}^4],$$

where b is the width and h is the height of the beam cross section. In this case $b = h = d = 0.01$ m and we get the strain in the bar:

$$\varepsilon = \frac{mal(d/2)}{Ed^4/12} = \frac{6mal}{Ed^3} \quad [\text{m/m}].$$

The minimum acceleration detectable is

$$a = \frac{\varepsilon Ed^3}{6ml} \quad [\text{m/s}^2].$$

For the numerical values given,

$$a = \frac{0.00008 \times 200 \times 10^9 \times (0.01)^3}{6 \times 12 \times 0.5} = 0.444\ \text{m/s}^2.$$

This is a small acceleration (about 0.045 g). Note as well that the accelerometer can be made more sensitive in a number of ways. First and foremost one can use a less "stiff" bar, that is, a bar with a lower modulus of elasticity. A bigger mass and longer bar will do the same thing. Similarly one can increase sensitivity by reducing the cross section of the bar, but of course one must come up with a reasonable compromise. For example, one cannot use a much thinner bar for the mass given, or alternatively, as one increases the mass the cross section of the bar must also be adjusted to support the mass. Finally, we note that the calculation here assumes acceleration is perpendicular to the surface on which the strain gauge is placed. Since one cannot predict the direction of acceleration in the case of earthquakes, it is necessary that at least two surfaces of the bar be equipped with strain gauges on perpendicular surfaces and the acceleration calculated from the two perpendicular components of acceleration.

6.5 | PRESSURE SENSORS

Sensing of pressure is perhaps only second in importance to sensing of strain in mechanical systems (and strain gauges are often used to sense pressure). These sensors are used either in their own right, that is, to measure pressure, or to sense secondary quantities such as force, power, temperature, or any quantity that can be related to pressure. One of the reasons for their prominence in the realm of sensors is that in sensing in gases and fluids, direct measurement of force is not an attractive option—only pressure can be measured and related to properties of these substances, including the forces they exert. Another reason for their widespread use and of exposure of most people to them is their use in cars, atmospheric weather prediction, heating and cooling, and other consumer-oriented devices. Certainly the "barometer" hanging on many a wall and the use of atmospheric pressure as an indication of weather conditions has helped popularize the concept of pressure.

The sensing of pressure, which is force per unit area, follows the same principle as the sensing of force—that of measuring the displacement of an appropriate member of the sensor in response to pressure. Therefore it is not surprising to find that the same principles used for sensing of force are used to sense pressure. Any device that will respond to pressure either by direct displacement or equivalent quantities (such as strain) is an appropriate means of sensing pressure. Thus the range of methods is quite large and includes thermal, mechanical, as well as magnetic and electrical principles.

6.5.1 Mechanical Pressure Sensors

Historically the sensing of pressure started with purely mechanical devices that did not require electrical transduction—a direct transduction from pressure to mechanical displacement was used. As such, these devices are actuators that react to pressure and, perhaps surprisingly, are as common today as ever. Some of these mechanical devices have been combined with other sensors to provide electrical output, while others are still being used in their original form. Perhaps the most common is the bourdon tube, shown in **Figure 6.23**. This sensor has been used for more than 150 years in pressure gauges, in which the dial indicator is connected directly to the tube (invented by Eugene Bourdon in 1849). This type of sensor, in different forms, is still the most common pressure gauge used today, and because it does not need additional components, it is simple and inexpensive. However, it is only really useful for relatively high pressures. It is typically used for gases, but it can also be used to sense fluid pressure.

Other methods of sensing pressure mechanically are the expansion of a diaphragm, the motion of a bellows, and the motion of a piston under the influence of pressure.

FIGURE 6.23 ■ The bourdon tube pressure sensor. The bourdon tube (C-shaped portion) expands with pressure, turning the dial (below the bezel, not seen) through a leverage arm and gear mechanism.

FIGURE 6.24 ■ The diaphragm pressure sensor.

The motion produced can be used to directly drive an indicator or can be sensed by a displacement sensor (LVDT, magnetic, capacitive, etc.) to provide a reading of pressure. A simple diaphragm pressure sensor used in wall barometers is shown in **Figure 6.24**. It is essentially a sealed metal can with relatively flexible walls. One side is held fixed (in this case by the small screw that also serves to adjust or calibrate it) while the other moves in response to pressure. This particular device is hermetically sealed at a given pressure so that any pressure below the internal pressure will force the diaphragm to expand and any higher pressure will force it to contract. While very simple and trivially inexpensive, it is easy to see its drawbacks, including the possibility of leakage and the inevitable dependence on temperature. A bellows is a similar device that can be used for direct reading or to activate another sensor. The bellows, in various forms, can also be used as an actuator. One of its common uses is in "vacuum motors," used in vehicles to activate valves and to move slats and doors, particularly in heating and air conditioning systems and in speed controls. They have survived in these roles into the modern era, mostly in vehicles, because of their simplicity, quiet operation, and the availability of a

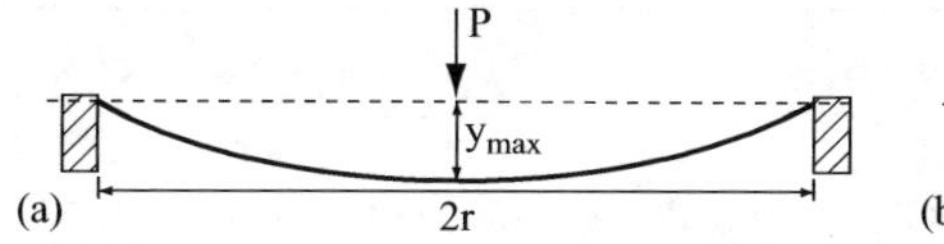

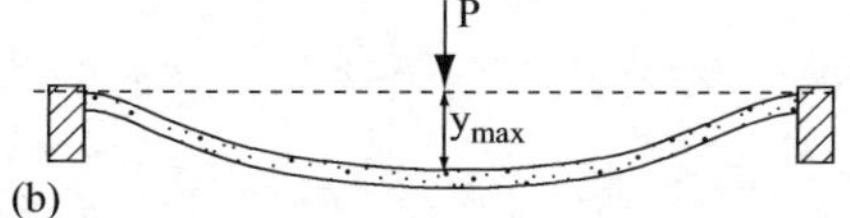

FIGURE 6.25 ■ (a) The thin plate. (b) The membrane.

source of low pressure (hence the use of the name vacuum) in internal combustion engines.

These mechanical devices indicate the need for a mechanism that can deflect under the influence of pressure. By far the most common structures used for this purpose are the thin plate and the diaphragm or membrane. In simple terms, a membrane is a thin plate with negligible thickness, whereas a plate has a finite thickness. Their behavior and response to pressure is very different. In relation to **Figure 6.25a**, the deflection of the center of a membrane (maximum deflection) that is under radial tension S and the stress in the diaphragm are given as

$$y_{max} = \frac{r^2 P}{4S} \ [\text{m}], \quad \sigma_m = \frac{S}{t} \ [\text{N/m}^2], \tag{6.10}$$

where P is the applied pressure (actually, the pressure difference between the top and bottom of the membrane), r is its radius, and t is its thickness. Strain can be calculated by dividing the strain by the modulus of elasticity (Young's modulus).

If, on the other hand, the thickness t is not negligible, the device is a thin plate (**Figure 6.25b**) and the behavior is given as

$$y_{max} = \frac{3(1 - v^2)r^4 P}{16Et^2} \ [\text{m}], \quad \sigma_m = \frac{3r^2 P}{4t^2} \ [\text{N/m}^2], \tag{6.11}$$

where E is Young's modulus and v is Poisson's ratio.

In either case, the displacement is linear with pressure, hence the widespread use of these structures for pressure sensing. The displacement y_{max} or the stress σ_m (or the equivalent strain) are measured depending on the type of sensor used. In modern sensors it is actually more common to measure strain using either a metal or, even more commonly, a semiconductor strain gauge or a piezoresistor. One advantage of using strain gauges is that the displacement needed is very small, allowing for very rugged construction and sensing of very high pressures. If displacement must be measured, this can be done capacitively, inductively, or even optically.

Pressure sensors come in four basic types, defined in terms of the pressure they sense. These are

Absolute pressure sensors (PSIA): pressure is sensed relative to absolute vacuum.

Differential pressure sensors (PSID): the difference between two pressures on two ports of the sensor is sensed.

Gauge pressure sensors (PSIG): senses the pressure relative to ambient pressure.

Sealed gauge pressure sensor (PSIS): the pressure relative to a sealed pressure chamber (usually 1 atm at sea level or 14.7 psi) is sensed.

The most common sensors are gauge sensors, but differential sensors are often used, as are sealed gauge sensors.

EXAMPLE 6.8 A piston-based mechanical pressure sensor

In a manner similar to the diaphragm, a piston acting against a spring can serve as a simple pressure sensor. This mechanical sensor/actuator is commonly used to measure pressure in tires and is shown schematically in **Figure 6.26**. A typical gauge rated at 700 kPa (100 psi) is made as a short cylinder, about 15–20 cm long and 10–15 mm in diameter. A valve at the bottom allows one-way gas entry to pressurize the cylinder. The inner stem moves against the spring and the graduations on the stem then indicate the pressure. The stem typically extends a maximum of about 5 cm so that it operates in the linear range of the spring. For a typical inner diameter of 10 mm, a pressure of 700 kPa generates a force on the piston of

$$F = PS = 700 \times 10^3 \times \pi \times \left(5 \times 10^{-3}\right)^2 = 54.978\ \text{N}.$$

The spring must compress 50 mm. This means the spring constant k, must be

$$F = kx \rightarrow k = \frac{F}{x} = \frac{54.978}{0.05} = 1100\ \text{N/m}.$$

The pressure is read on the graduation and these are at 14 kPa/mm (approximately 2 psi/mm). This, of course, is the sensitivity of the sensor.

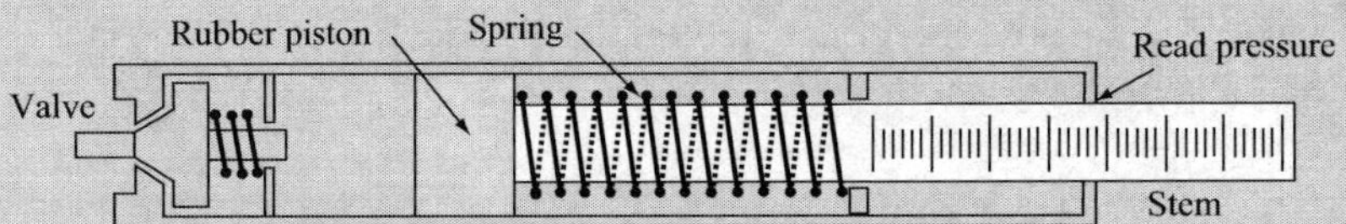

FIGURE 6.26 ■ Piston-based mechanical pressure sensor.

EXAMPLE 6.9 The sealed gauge pressure sensor as a simple barometer

The barometer shows air pressure usually by mechanical means on a simple dial. A simple sealed gauge pressure sensor similar in principle to the one shown in **Figure 6.24** can be used for this purpose. To see in more detail how this can be accomplished, consider the device in **Figure 6.27**, which consists of a chamber, sealed by a piston. The chamber is sealed at a standard atmospheric pressure, $P_0 = 101{,}325$ Pa (1.013 bar or 1013.25 mbar). The air pressure is indicated on a linear scale with reference to normal pressure. As the external pressure increases, the piston is pushed down, compressing the air in the chamber. When the external pressure decreases, the air expands, allowing the piston to move up. Air pressure in the atmosphere varies by small amounts (the highest pressure ever recorded was 1086 mbar and the lowest pressure ever recorded was 850 mbar) so a scale between 800 mbar and 1100 mbar is sufficient. Calculate the range of motion of the piston (i.e., the length of the scale).

Solution: The compression of the sealed air is governed by Boyle's law (under constant temperature conditions):

$$P_1 V_1 = P_2 V_2.$$

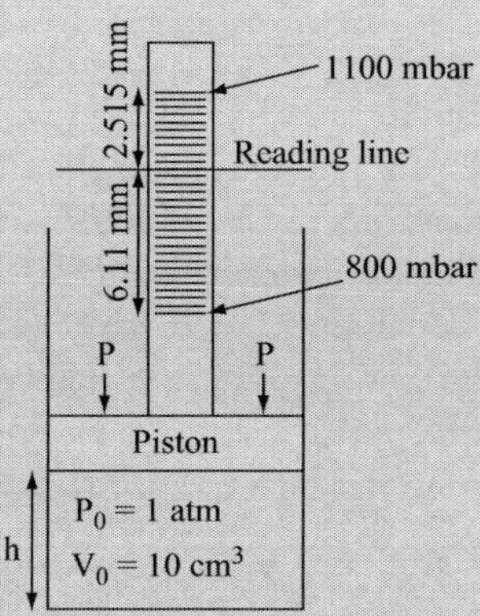

FIGURE 6.27 ■ A sealed chamber barometer.

That is, as pressure changes, the volume changes accordingly to keep the product constant. Now, taking the volume at the nominal pressure $P_0 = 1013.25$ mbar to be $V_0 = 10\ \text{cm}^3$, we write for the volumes at minimum and maximum pressure

$$P_{\min}V_{\min} = P_0V_0 \rightarrow V_{\min} = \frac{P_0V_0}{P_{\min}} = \frac{1013.25 \times 10}{850} = 11.92\ \text{cm}^3$$

and

$$P_{\max}V_{\max} = P_0V_0 \rightarrow V_{\max} = \frac{P_0V_0}{P_{\max}} = \frac{1013.25 \times 10}{1100} = 9.21\ \text{cm}^3.$$

The scale is found from the height of the displaced air column:

At low pressure,

$$V_{\min} = \pi\frac{d^2}{4}h_{\min} = 11.92 \rightarrow h_{\min} = \frac{4 \times 11.92}{\pi d^2} = \frac{4 \times 11.92}{\pi \times 2^2} = 3.7942\ \text{cm}.$$

At high pressure,

$$V_{\max} = \pi\frac{d^2}{4}h_{\max} = 9.21 \rightarrow h_{\max} = \frac{4 \times 9.21}{\pi \times 2^2} = 2.9316\ \text{cm}.$$

At nominal pressure the height is

$$h_0 = \frac{4 \times 10}{\pi \times 2^2} = 3.1831\ \text{cm}.$$

That is, given the position of the line at nominal pressure, the lowest pressure line is 6.11 mm below and the highest pressure point is 2.515 mm above the nominal pressure line. The whole range of motion is 8.625 mm.

Note that reducing the diameter by a factor of two increases the range by a factor of four (in this case, to 34.5 mm). In the fluid barometer, the piston is replaced by a fluid—usually water or oil—that serves not only as a "piston," but also as a direct indication of pressure. In other barometers the motion of the piston, or its equivalent, turns a dial.

6.5.2 Piezoresistive Pressure Sensors

Although a piezoresistor is simply a semiconductor strain gauge, and can always be replaced with a conductor strain gauge, most modern pressure sensors use it rather than the conductor-type strain gauge. Only when higher temperature operation is needed or for specialized applications are conductor strain gauges preferred. In addition, the diaphragm itself may be fabricated of silicon, a process that simplifies construction and allows for additional benefits such as onboard temperature compensating elements, amplifiers, and conditioning circuitry. The basic structure of a sensor of this type is shown in **Figure 6.28**. In this case, the two gauges are parallel to one dimension of the diaphragm. The change in resistance of the two piezoresistors is

$$\frac{\Delta R_1}{R_1} = -\frac{\Delta R_2}{R_2} = \frac{1}{2}\pi\left(\sigma_y - \sigma_x\right), \tag{6.12}$$

where σ_x and σ_y are the stresses in the transverse directions. Although other types of arrangements of the piezoresistors will result in different values for the change in resistance (e.g., R_2 in **Figure 6.29** can be placed perpendicular to R_1), this formula is representative of the expected values. In the device in **Figure 6.29**, both the piezoresistors and the diaphragm are fabricated of silicon. In this case, a vent is provided, making this a gauge sensor. If the cavity under the diaphragm is hermetically sealed and the pressure in it is P_0, then the sensor becomes a sealed gauge pressure sensor sensing the pressure $P - P_0$. A differential sensor is produced by placing the diaphragm between two chambers, each vented through a port, as shown in **Figure 6.29**.

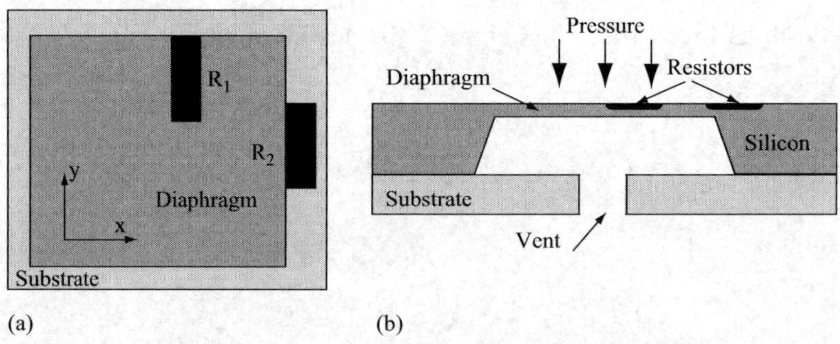

FIGURE 6.28 ■ A piezoresistive pressure sensor. (a) Placement of the piezoresistances. (b) Construction showing the diaphragm and vent hole (for gauge pressure sensors).

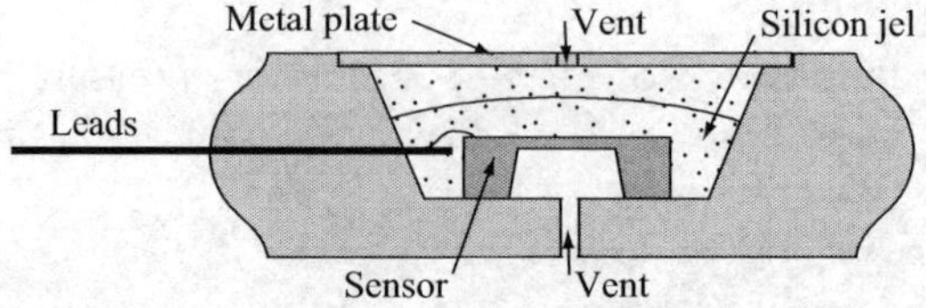

FIGURE 6.29 ■ Construction of a differential pressure sensor. The diaphragm is placed between the two ports.

A different approach is to use a single strain gauge, as in **Figure 6.30**, with a current passing through it and pressure applied perpendicular to the current. The voltage across the element is measured as an indication of the stress, and thus pressure.

There are many variations on these basic types of sensors with different materials and processes, different sensitivities, etc., but these do not constitute separate types of sensors and will not be discussed separately.

Although the most common method of sensing is through the use of semiconductor strain gauges, the construction of the body of the sensor and, in particular, that of the diaphragm varies based on applications. Stainless steel, titanium, and ceramics are used in corrosive environments and other materials, including glass, can be used for coatings.

Figures 6.31 and **6.32** show a number of pressure sensors of various constructions, sizes, and ratings.

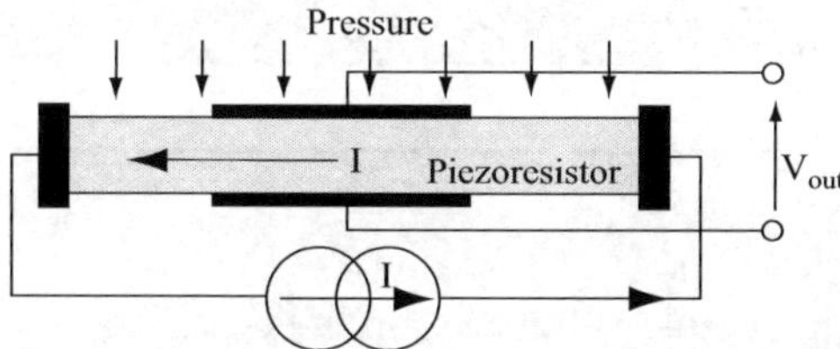

FIGURE 6.30 ■ A single piezoresistance pressure sensor. The potential across the resistor is a measure of pressure. Pressure is applied perpendicular to the current.

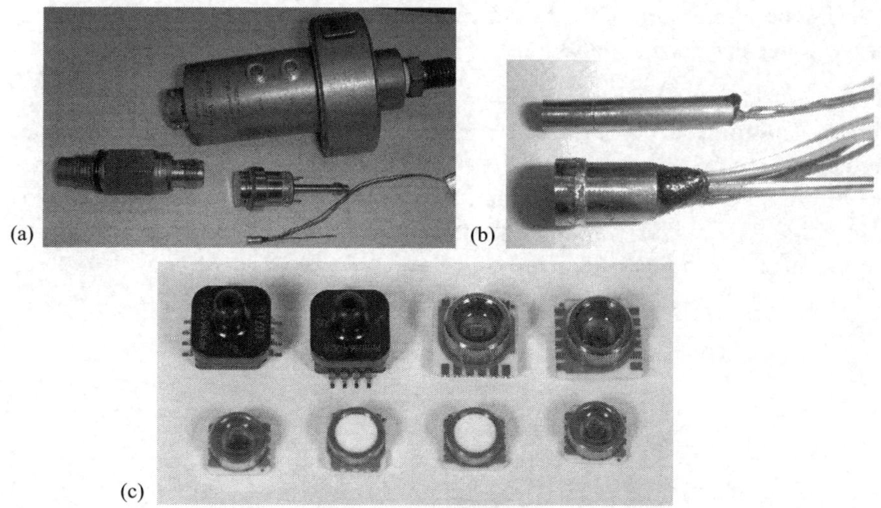

FIGURE 6.31 ■ Various pressure sensors. (a) Pressure sensors of various sizes. The smallest is 2 mm in diameter, the largest is 30 mm in diameter. Note the connectors. All are sealed gauge pressure sensors. (b) Small sensors in stainless steel housings (absolute pressure sensors). (c) Miniature surface-mount digital pressure sensors (from top left, clockwise: 14 bar, 7 bar, 1 bar, and 12 bar) sealed gauge sensors.

FIGURE 6.32 ■ Various pressure sensors. (a) A 100 psi absolute pressure sensor in a metal can. (b) A 150 psi differential pressure sensor intended for automotive applications. (c) Three 15 and 30 psi gauge pressure sensors.

EXAMPLE 6.10 Water depth sensor

A depth sensor for an autonomous diving vehicle is built as a thin disk of stainless steel of radius 6 mm, 0.5 mm thick, supported by a ring with an inner radius of 5 mm and an outer radius of 6 mm, as shown in **Figure 6.33**. The top is open to the water and the bottom is sealed at atmospheric pressure (1 atm) before lowering the vehicle into the water. A radially oriented strain gauge is attached to the lower part of the disk and senses the strain in the disk. Find a relation for the resistance of the strain gauge with depth if it has nominal resistance of 240 Ω and a gauge factor of 2.5. Assume atmospheric pressure of 1 atm, the modulus of elasticity of stainless steel is 195 GPa, and the average water density is 1025 kg/m^3.

Solution: The sensor is in effect a sealed gauge pressure sensor (PSIG) with a thin plate as a transducer. The pressure in **Equation (6.11)** is the water pressure on the top of the disk minus the sealed pressure of 1 atm. Thus we first need to calculate the pressure as a function of depth. The latter is rather simple. The pressure at the surface is 1 atm. It increases by 1 atm for every 1.032 m of depth (see section on units). Given that 1 atm equals 101.325 kPa, the pressure in water as a function of depth from the surface can be written as

$$P = \frac{d}{1.032} \times 101.325 + 101.325 \quad [\text{kPa}],$$

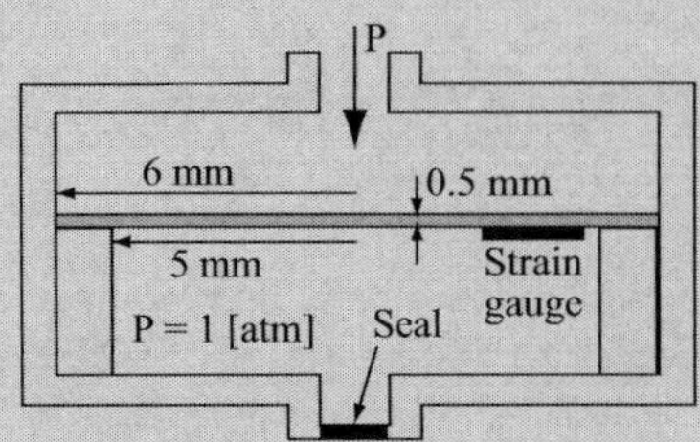

FIGURE 6.33 ■ A water depth sensor. This is in effect a sealed gauge pressure sensor measuring the difference between the pressure of the water and the pressure at the surface of the ocean (1 atm).

where d is the depth in meters. The pressure sensed by the sensor described here is

$$P = \frac{d}{1.032} \times 101{,}325 \quad [\text{Pa}]$$

because of the sealed pressure of 1 atm. This means that the sensor will measure zero pressure at the surface of the water. Therefore the depth is measured directly as

$$d = \frac{1.032P}{101325} = 10^{-5}P \quad [\text{m}].$$

To calculate the change in resistance of the sensor we use **Equation (6.5)**, which in turn requires the strain in the disk. Therefore we first calculate the stress using **Equation (6.11)** and then divide stress by the modulus of elasticity to find the strain. The stress in the disk is

$$\sigma_m = \frac{3r^2P}{4t^2} = \frac{1.032 \times 3r^2dP}{1.032 \times 4t^2} = 10^5\frac{r^2d}{t^2}\frac{3}{4} \quad [\text{N/m}^2].$$

Dividing by the modulus of elasticity we get the strain:

$$\varepsilon_m = \frac{\sigma_m}{E} = 7.5 \times 10^4\frac{r^2d}{Et^2} \quad [\text{m/m}].$$

Now we substitute this in **Equation (6.5)**:

$$R(\varepsilon) = R(1 + g\varepsilon) = R\left(1 + 7.5 \times 10^4\frac{gr^2d}{Et^2}\frac{3}{4}\right) \quad [\Omega],$$

where R is the nominal resistance of the strain gauge and g is the gauge factor. Therefore the change in resistance due to depth is

$$\Delta R = 7.5 \times 10^4\frac{gRr^2d}{Et^2}\frac{3}{4} \quad [\Omega].$$

Another way to write this is

$$d = \frac{Et^2}{7.5 \times 10^4 gRr^2}\frac{4}{3}\Delta R \quad [\text{m}].$$

In this form the depth is immediately available as a function of ΔR, the difference in resistance between the sensed resistance and the nominal resistance of the strain gauge. For the values given here,

$$d = \frac{Et^2}{7.5 \times 10^4 gRr^2}\frac{4}{3}\Delta R = \frac{195 \times 10^9 \times (0.0005)^2 \times 4}{7.5 \times 10^4 \times 2.5 \times 240 \times (0.005)^2 \times 3}\Delta R = 57.78\Delta R \quad [\text{m}].$$

That is, for every 1 m depth, the strain gauge resistance will change by $1/57.78 = 0.0173\ \Omega$. This is a simple calibration curve, and as long as the strain gauge is temperature compensated, the depth can be measured accurately.

Note: A higher gauge factor strain gauge (such as a semiconductor strain gauge) can increase the sensitivity. For example, a gauge factor of 125 (not unusual for semiconductor strain gauges) will increase the change in resistance to $0.865\ \Omega$ per meter of depth.

6.5.3 Capacitive Pressure Sensors

The deflection of the diaphragm in any of the structures described above with respect to a fixed conducting plate constitutes a capacitor in which the distance between the plates is pressure sensitive. The basic structure in **Figure 6.17** may be used or a similar configuration devised. These sensors are very simple and are particularly useful for sensing of very low pressures. At low pressure, the deflection of the diaphragm may be insufficient to cause large strain but can be relatively large in terms of capacitance. Since the capacitance may be part of an oscillator, the change in its frequency may be sufficiently large to make for a very sensitive sensor. Another advantage of capacitive pressure sensors is that they are less temperature dependent and, because stops on motion of the plate can be incorporated, they are less sensitive to overpressure. Usually overpressures two to three orders of magnitude larger than rated pressure may be easily tolerated without ill effects. The sensors are linear for small displacements, but at greater pressures the diaphragm tends to bow, causing nonlinear output.

6.5.4 Magnetic Pressure Sensors

A number of methods are used in magnetic pressure sensors. In large deflection sensors an inductive position sensor or an LVDT attached to the diaphragm can be used. However, for low pressures, the so-called variable reluctance pressure sensor is more practical. In this type of sensor the diaphragm is made of a ferromagnetic material and is part of the magnetic circuit shown in **Figure 6.34a**. The reluctance of a magnetic circuit is the magnetic equivalent of resistance in electric circuits and depends on the sizes of the magnetic paths, their permeability, and their cross-sectional areas (see **Section 5.4** for a discussion of magnetic circuits and **Equation (5.32)** for a definition of magnetic reluctance). **Figure 5.23b** shows the equivalent circuit where $\mathcal{R}_f$ indicates paths in iron, $\mathcal{R}_g$ indicates gaps and $\mathcal{R}_d$ indicates paths in the diophragm. If the magnetic core (the E-shaped path in **Figure 6.34**) and the diaphragm are made of high-permeability ferromagnetic materials, their reluctance is negligible. In this case the reluctance is directly proportional to the length of the air gap between the diaphragm and the E-core. As pressure changes, this gap changes and the inductance of the two coils changes accordingly. This inductance can be sensed directly, but more often the current in the circuit made of a fixed impedance and a variable impedance due to the motion of the diaphragm is measured, as shown in **Figure 6.34c**. The advantage of a sensor of this type is that a small deflection can cause a large change in the inductance of the circuit, making for a very sensitive device. In addition, magnetic sensors are almost devoid of temperature sensitivity, allowing operation at elevated or variable temperatures.

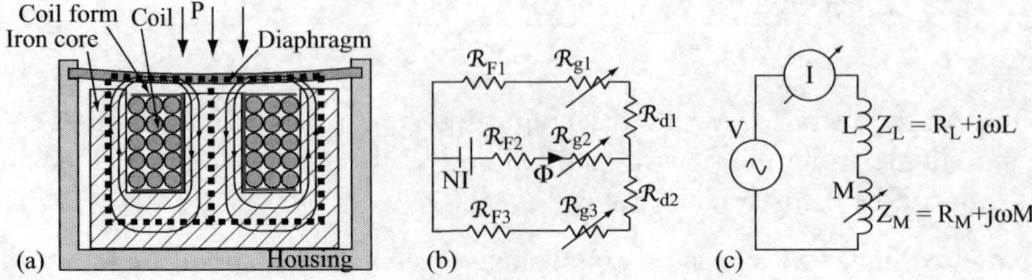

FIGURE 6.34 ■ A variable reluctance pressure sensor. (a) Structure and operation. (b) Equivalent circuit in terms of reluctances. (c) Operation with an AC source. The core and diaphragm are circular.

There are a number of other types of pressure sensors that rely on diverse principles. Optoelectronic pressure sensors use the principle of a Fabri–Perot optical resonator to measure exceedingly small displacements. In a resonator of this type, light reflected from a resonant optical cavity is measured by a photodiode to produce a measure of the sensed pressure. Another, very old method of sensing low pressures (hence they are often called vacuum sensors) is the Pirani gauge. It is based on measuring the heat loss from gases, which is dependent on pressure. The temperature of a heated element in the gas flow is sensed and correlated to pressure, usually in an absolute pressure sensor arrangement.

The properties of pressure sensors vary considerably depending on construction and on the principles used. Typically semiconductor-based sensors can only operate at low temperatures (−50°C to +150°C). Their temperature-dependent errors can be high unless properly compensated either externally or internally. The range of sensors can exceed 300 GPa (50,000 psi) and can be as small as a few pascals. Impedance is anywhere between a few hundred ohms to about 100 kΩ, again depending on the type of device. Linearity is between 0.1% and 2% and response time is typically less than 1 ms. The maximum pressure, burst pressure, and proof pressure (overpressure) are all part of the specifications of the device, as is its electrical output, which can be either direct (no internal circuitry and amplification) or after conditioning and amplification. Digital outputs are also available. As indicated above, the materials used (silicon, aluminum, titanium, stainless steel, etc.) and compatibility with gases and liquids are specified and must be followed to avoid damage and incorrect readings. Other specifications are the port sizes and shapes, connectors, venting ports, and the like. The cycling of pressure sensors is also specified, as are hysteresis (usually less than 0.1% of full scale) and repeatability (typically less than 0.1% of full scale).

6.6 | VELOCITY SENSING

Velocity sensing is actually more complicated than acceleration sensing and often requires indirect methods. This is understandable from the fact that velocity is relative and thus requires a reference. Of course, one can always measure something proportional to velocity. For example, we can infer the velocity of a car from the rotation of the wheels (or the transmission shaft—a common method of velocity measurement in cars) or count the number of rotations of a shaft per unit time in an electric motor, or, indeed, use GPS for that purpose. In other applications, such as aircraft, velocity may be inferred from pressure or from temperature sensors measuring the cooling effect of moving air. However, a free-standing sensor that measures velocity directly is much more difficult to produce. One approach that may be used is the induction of emf in a coil due to a moving magnet. However, this requires that the coil be stationary and, if the velocity is constant (no acceleration), the magnet cannot move relative to the coil since the coil must have a restoring force (spring). For changing velocity (when acceleration is not zero), the principle in **Figure 6.35** may be useful. The emf induced in the coils is governed by Faraday's law:

$$emf = -N\frac{d\Phi}{dt} \quad [\text{V}], \tag{6.13}$$

where N is the number of turns and Φ is the flux in the coil. The time derivative indicates that the magnet must be moving to produce a nonzero change in flux.

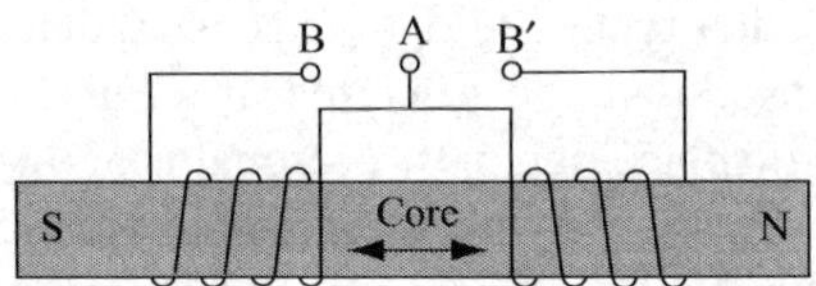

FIGURE 6.35 ■ A velocity sensor. The induced emf in the coils is proportional to the velocity of the magnet.

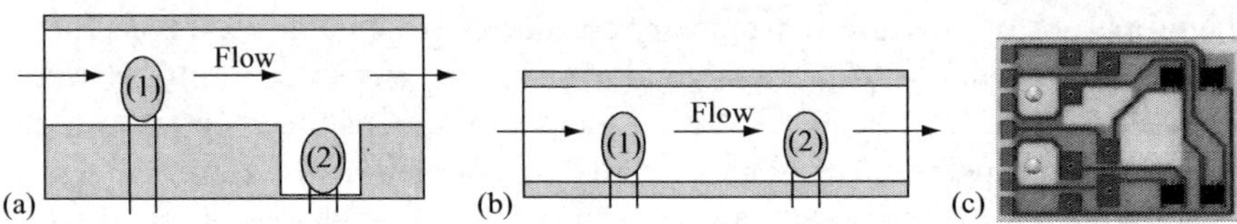

FIGURE 6.36 ■ Flow velocity sensor. (a) The downstream temperature sensor (2) is shielded from the flow but measures the air (or fluid) temperature. (b) The downstream sensor (2) is also in the flow, but is cooled less because of the heat transfer from the upstream sensor (1). (c) A fluid velocity sensor showing four deposited thermistors on a ceramic substrate (right-hand side of the picture). Flow is from top to bottom and the sensors are connected in a bridge configuration. A reference thermistor is placed on the reverse side of the substrate along with a temperature sensor.

Thus the most common approach to velocity sensing is to use an accelerometer and integrate its output using an integrating amplifier. Since velocity is the time integral of acceleration, the velocity is easily obtained, but, as before, constant velocity cannot be sensed (zero acceleration). Fortunately, in many instances, the velocity may be measured directly without the use of specific sensors. The speed of vehicles is one example, where the speed relative to the stationary ground can be measured in many ways.

The velocity of fluids and gasses can be sensed quite easily. The velocity of watercraft and aircraft can also be measured relative to stationary or moving fluids. However, the methods used to do so are all indirect. One simple method of fluid velocity sensing is to sense the cooling of a thermistor relative to a thermistor that is not exposed to the fluid flow. This is particularly useful for air flow or sensing of air velocity in an aircraft (**Figure 6.36a, b**). The downstream sensor (2) can be shielded from the flow, as in **Figure 6.37a**, or it can be in the flow, as in **Figure 6.36b**. In the first case, the downstream sensor remains at a constant temperature that only depends on the static temperature (independent of flow), whereas in the second, the upstream sensor is cooled by the flow while the downstream sensor is cooled much less because of the heated fluid downstream caused by the upstream sensor. In either case, the temperature difference can be related to fluid velocity (or, if needed, to fluid mass flow). A similar method is used to measure air mass in the intake of vehicles (will be discussed in **Chapter 10**). **Figure 6.36c** shows a fluid flow sensor that uses four thermistors, two downstream and two upstream, arranged in a bridge configuration. The sensors are deposited on a ceramic substrate about 15 mm × 20 mm in size. In essence, the temperature difference between the upstream and downstream sensors is measured and correlated with fluid velocity or fluid flow. Since the transfer function depends on the actual temperature of the fluid, an additional thermistor (on the other side of the substrate) measures the fluid temperature directly, and since this sensor is in a stagnant section of the fluid, it is not affected by the flow. In some

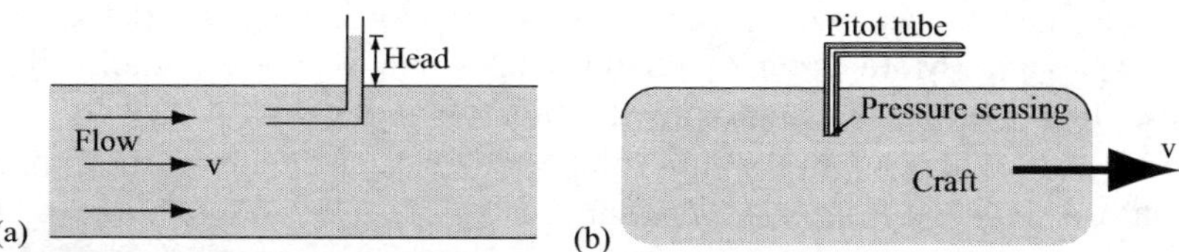

FIGURE 6.37 ■ The Pitot tube. (a) The original use was to measure water velocity and flow rates in rivers. (b) The modern adaptation to measure airspeed in an aircraft or the relative speed in a fluid. What is measured is the total (stagnant) pressure in the tube.

applications, especially at high temperatures, thermistors may be replaced with RTDs. In others, two transistors or two diodes can serve the same purpose (see **Chapter 3**).

Another common method of sensing speed is based on differential pressure: the change in pressure due to motion in the fluid gives an indication of the speed. This is a standard method of speed sensing in modern aircraft (including commercial aircraft) and is based on one of the oldest sensors in existence, the Pitot tube. The basic method, that due to Henry Pitot, dates to 1732 and was initially used to measure water speed in rivers. The principle is shown in **Figure 6.37a**. As water velocity increases, the total pressure in the tube increases and the water head rises, indicating fluid speed (or, if properly calibrated, flow rate). The modern pitot tube as used in aircraft consists of a bent tube with its opening facing forward (parallel to the body of the aircraft), shown in **Figure 6.37b**. The tube is either sealed at the interior end and the pressure at that point is measured or the pressure is allowed to act against a mechanical indicator (such as the diaphragm pressure sensor in **Figure 6.24** or even a bourdon tube) or a pressure sensor can be used. Under these conditions, the total pressure in the tube (also called the stagnant pressure since the fluid is constrained from moving) is given by Bernouli's principle:

$$P_t = P_s + P_d \quad [\text{Pa}], \tag{6.14}$$

where P_t is the total (or stagnant) pressure, P_s is the static pressure, and P_d is the dynamic pressure. In the case of the aircraft, P_s is the pressure one would measure if the aircraft were stationary (i.e., the atmospheric pressure), whereas P_d is the pressure due to the motion of the aircraft (or, in water, the motion of a boat). The latter is given as

$$P_d = \rho \frac{V^2}{2} \quad [\text{Pa}], \tag{6.15}$$

where ρ is the density of the fluid (air, water) and V is the velocity of the craft. Since the interest here is measurement of velocity:

$$V = \sqrt{\frac{2(P_t - P_s)}{\rho}} \quad [\text{m/s}]. \tag{6.16}$$

The density, ρ, can be measured separately or may be known (such as in the case of water). For flight speed purposes it is important to remember that density (and pressure) varies with altitude. The density may be deduced from pressure as (approximately)

$$\rho = \frac{P_t}{RT} \quad [\text{kg/m}^3], \tag{6.17}$$

where R is the specific gas constant (equal to 287.05 J/kg/K for dry air) and T [K] is the absolute temperature. More accurate relations take into account humidity through the use of vapor pressure, but this approximation is often sufficient in "dry air" conditions. The static pressure at height h in the atmosphere can be calculated from the altitude (or the altitude can be calculated from static pressure) using the following relation:

$$P_s = P_0\left(1 - \frac{Lh}{T_0}\right)^{\frac{gm}{RL}} \quad \text{[Pa]}, \tag{6.18}$$

where P_0 is the standard pressure at sea level (101,325 Pa), L is the temperature change with elevation, also called temperature lapse rate (0.0065 K/m), h is the altitude in meters, T_0 is the standard temperature at sea level (288.15 K), $g = 9.80665$ m/s^2 is the gravitational acceleration, m is the molar mass of dry air (0.0289644 kg/mol), and R is the gas constant (8.31447 J/mol/K). A simpler relation, often called the barometric equation, is given as follows:

$$P_s(h) = P_0 e^{-\frac{Mgh}{RT_0}} \quad \text{[Pa]}. \tag{6.19}$$

Although this formula overestimates the pressure at height h, it is commonly used, and in fact serves as the basis of many altimeters, including those in aircraft.

Because the difference between the total pressure and the static pressure is needed to measure speed, the Pitot tube has been modified to measure static pressure independently. This is known as the Prandtl tube (although it is most often called simply a Pitot tube or a pitot-static tube) and is shown in **Figure 6.38**. In this unique sensor there is an additional port open to the side of the tube to measure the static pressure. A differential pressure sensor measures the pressure difference between the total pressure (forward-facing opening) and the static pressure (side-facing opening). Now the velocity can be measured directly from **Equation (6.16)**. It is important to recognize that the sensor measures the relative fluid velocity, so in the case of aircraft, the sensor gives the aircraft velocity relative to air (the airspeed). The Pitot tube (or the Prandtl tube) has narrow openings that, particularly in aircraft, are prone to icing. This is a very dangerous situation since the engine speed is regulated using the airspeed provided by the tube. Icing has been blamed for many airplane crashes, so to minimize this possibility, the sensors are heated to prevent ice buildup. The exact same idea can be used in water for surface watercraft or underwater for submarines, again measuring the speed relative to the fluid.

Other methods of sensing speed include ultrasonic, electromagnetic, and optical methods relying on reflections from the moving object and measuring the time of flight of the wave to and from the moving object. We will encounter the ultrasonic speed

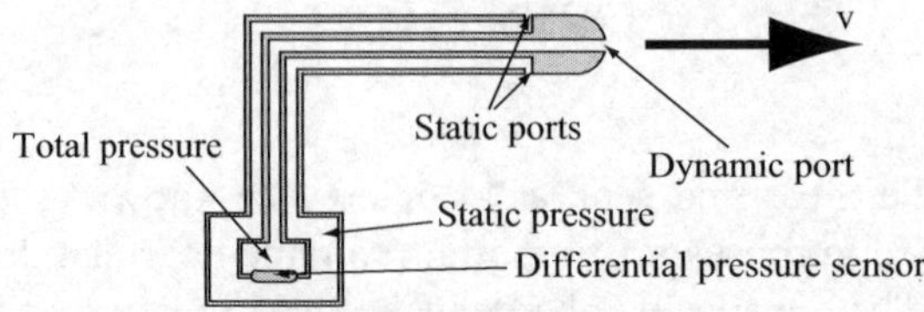

FIGURE 6.38 ■ The Prandtl tube. A differential pressure sensor measures the difference between the total pressure and the static pressure. The tube moves to the right at a velocity *v* in a fluid (air).

sensor in **Chapter 7**. Velocity can also be measured/sensed using the Doppler effect through use of ultrasonic, electromagnetic, or light waves. However, the Doppler method is not really a sensor, but rather a system that measures the change in frequency of a reflected wave due to the relative velocity between the source of the wave and the body whose velocity is being sensed. Although the system is somewhat complicated, it has important applications in weather prediction (detection and analysis of tornadoes and hurricanes), in space applications and science (measurement of moving objects, including the recession of stars), and in law enforcement (speed detection of cars, anticollision systems, and more).

EXAMPLE 6.11 Water pressure in a river

The Pitot tube can also be used to measure dynamic pressure rather than speed. Suppose a Pitot tube is immersed in the flow of a river and the dynamic pressure needs to be measured. To do so, the water head above the surface of the water is measured, as in **Figure 6.37a** (alternatively, the Prandtl tube can be used, as in **Figure 6.38**, to measure the differential pressure to which the static pressure can then be added). Assume water density is 1000 kg/m^3. Neglect the effect of temperature.

a. Given the water speed as $V_0 = 5$ m/s, and an ambient pressure equal to 101.325 kPa (1 atm), calculate the dynamic pressure due to flow just below the surface.
b. What is the dynamic pressure at a depth of 3 m?

Solution: At the surface the pressure is essentially that of the atmosphere, which we will take here as 101.325 kPa. Static water pressure under the surface increases by 101.325 kPa (1 atm) for every 1.032 m, but the dynamic pressure only depends on speed.

a. The dynamic pressure can be calculated directly from **Equation (6.15)**:

$$P_d = \rho\frac{V_0^2}{2} = 1000\frac{5^2}{2} = 12{,}500 \text{ Pa.}$$

b. The dynamic pressure remains the same as long as the speed and density remain constant. At greater depths the density changes somewhat (increases) and the dynamic pressure at a constant velocity increases as well.

6.7 | INERTIAL SENSORS: GYROSCOPES

Gyroscopes come to mind usually as stabilizing devices in aircraft and spacecraft in such applications as automatic pilots or, more recently, in stabilizing satellites so they point in the right direction. However, they are much more than that and much more common than one can imagine. Just like the magnetic compass is a navigational tool, the gyroscope is a navigational tool. Its purpose is to keep the direction of a device or vehicle or to indicate attitude. As such, they are used in all satellites, in smart weapons, and in all other applications that require attitude and position stabilization. Their accuracy has made them useful in such unlikely applications as tunnel construction and mining. As they become smaller, one can expect them to find their way into consumer products such as cars. They have already found their way into toys such as remotely controlled model aircraft.

The basic principle involved is the principle of conservation of angular momentum: "In any system of bodies or particles, the total angular momentum relative to any point in space is constant, provided no external forces act on the system."

The name gyroscope comes from concatenation of the Greek words *gyro* (rotation or circle) and *skopeein* (to see), coined by Leon Foucault, who used it to see or demonstrate the rotation of the earth around 1852. The principle was known at least since 1817, when it was first mentioned by Johann Bohnenberger, although it is not clear whether he discovered it or was the first to use it.

6.7.1 Mechanical or Rotor Gyroscopes

The mechanical gyroscope is the best known of the existing gyros and the easiest to understand, although its heyday has passed (it still exists, but in miniaturized forms). It consists of a rotating mass (heavy wheel) on an axis in a frame. The spinning mass provides an angular momentum (**Figure 6.39**). So far this is merely a rotating wheel. However, if one tries to change the direction of the axis by applying a torque to it, a torque is developed in a direction perpendicular to the axis of rotation and that of the applied torque, which forces a precession motion. This precession is the output of the gyroscope and is proportional to the torque applied to its frame and the inertia of the rotating mass. Using **Figure 6.39**, if a torque is applied to the frame of the gyroscope around the input axis, the output axis will rotate as shown. This precession now becomes a measure of the applied torque and can be used as an output to, for example, correct the direction of an airplane or the position of a satellite antenna. Application of torque in the opposite direction reverses the direction of precession. The relation between applied torque and the angular velocity of precession Ω is

$$T = I\omega\Omega \quad [\mathrm{N \cdot m}], \tag{6.20}$$

where T is the applied torque [N · m], ω is the angular velocity [rad/s], I is the inertia of the rotating mass [kg · m^2], and Ω is the **angular velocity of precession** [1/rad · s], also called the **rotational rate**. $I\omega$ is the angular momentum [kg · m^2 · rad/s]. Clearly then, Ω is a measure of the torque applied to the frame of the device:

$$\Omega = \frac{T}{I\omega} \quad [1/\mathrm{rad \cdot s}]. \tag{6.21}$$

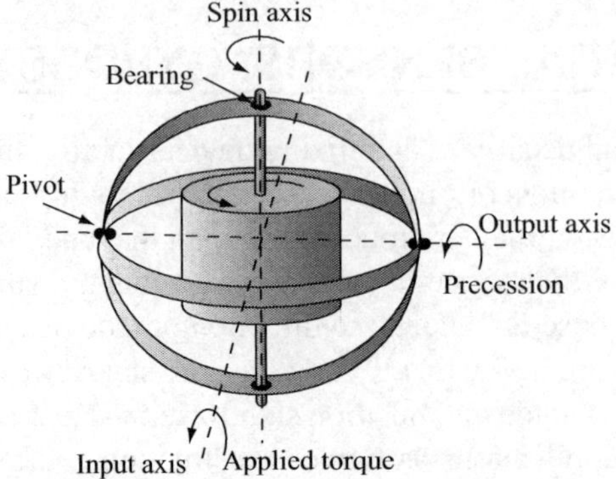

FIGURE 6.39 ■ The rotating mass gyroscope.

The device in **Figure 6.39** is a single-axis gyroscope. Two- or three-axis gyroscopes can be built by duplicating this structure with rotation axes perpendicular to each other.

This type of gyroscope has been used for many decades in aircraft, but it is a fairly large, heavy, and complex device, not easily adapted to small systems. It also has other problems associated with the spinning mass. Obviously the faster the spin and the larger the mass, the larger the angular momentum ($I\omega$) and the lower the frequency of precession for a given appiled torque. But fast rotation adds friction and requires delicate balancing of the rotating disk, as well as precision machining. This has led to many variations, including rotation in a vacuum, magnetic and electrostatic suspension, use of high-pressure gas bearings, cryogenic magnetic suspension, and more. However, none of these can ever make this device a low-cost, general-purpose sensor. Some modern gyroscopes still use the spinning mass idea, but the mass is much smaller, the motor is a small DC motor, and the whole unit is relatively small. These devices compensate for the smaller mass by using high-speed motors and sensitive sensors to sense the torque. But other types of gyroscopes designed for reliable operation at low cost have been developed. Some gyroscopes are only gyroscopes in the equivalent operation and bear no resemblance to the spinning mass gyroscope. Nevertheless, they are gyroscopes, and are highly useful at that.

Instead of the conventional gyroscope, the idea of Coriolis acceleration has been used to devise much smaller and more cost-effective gyroscopic sensors. These are built in silicon by standard etching methods and thus can be produced inexpensively. We shall discuss Coriolis acceleration-based gyroscopes in more detail in **Chapter 10**, but for completeness we outline here the basics behind Coriolis force gyroscopes. The idea is based on the fact that if a body moves linearly in a rotating frame of reference, an acceleration appears at right angles to both motions, as shown in **Figure 6.40**. The linear motion is typically supplied by the vibration of a mass, usually in a harmonic motion, and the resulting Coriolis acceleration is used for sensing. Under normal conditions the Coriolis acceleration is zero and the force associated with it is zero as well. If the sensor is rotated in the plane perpendicular to the linear vibration, an acceleration is obtained, proportional to the angular velocity, Ω.

6.7.2 Optical Gyroscopes

One of the more exciting developments in gyroscopes is the optical gyroscope, which, unlike the rotating mass gyroscope or the vibrating mass gyroscope, has no moving elements. These modern devices are used extensively for guidance and control and are based on the Sagnac effect. The Sagnac effect is based on the propagation of light in optical fibers (or in any other medium) and can be explained using **Figure 6.41**. Suppose first that the optical fiber ring is at rest and two laser beams travel the length of the ring, one in the clockwise (CW) direction, the other in the counterclockwise (CCW) direction, both produced by the same laser (so that they are at the same frequency and phase). The time it takes either beam to travel the length of the ring is $\Delta t = 2\pi Rn/c$, where n is the

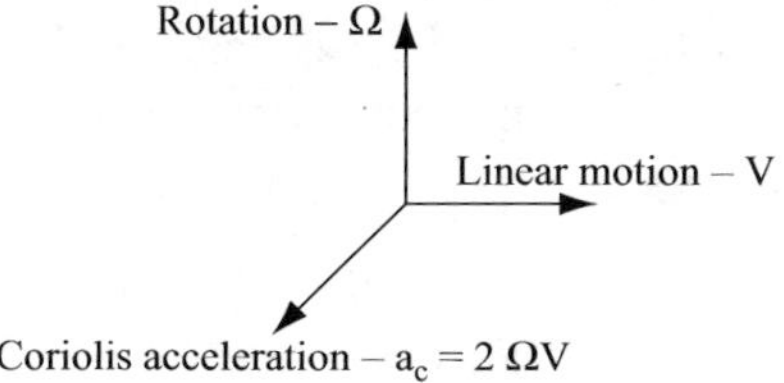

FIGURE 6.40 ■ The relation between linear velocity V, angular velocity Ω, and Coriolis acceleration a_c.

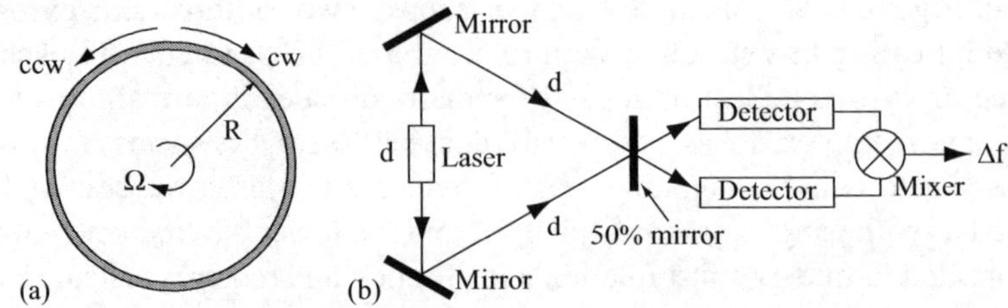

FIGURE 6.41 ■ (a) The Sagnac effect in an optical fiber ring rotating at an angular frequency Ω. (b) Implementation of the ring resonator using mirrors to "close" the ring.

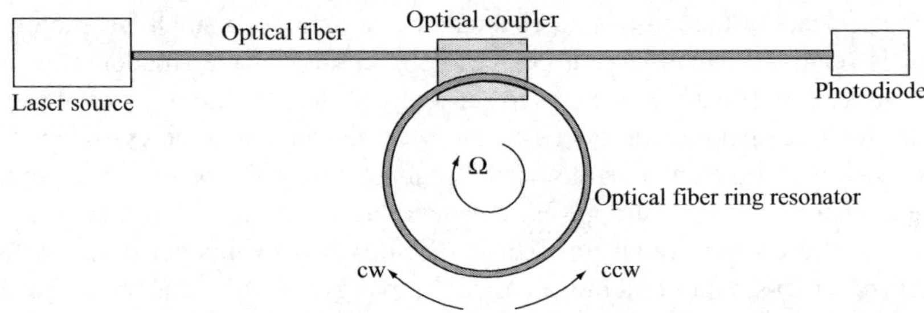

FIGURE 6.42 ■ A resonating ring optical fiber gyroscope.

index of refraction of the optical fiber and c is the speed of light in vacuum (i.e., c/n is the speed of light in the fiber).

Now suppose that the ring rotates clockwise at an angular velocity Ω [rad/s]. The beams now travel different paths in each direction. The CW beam will travel a distance $2\pi R + \Omega R\Delta t$ and the CCW beam will travel a distance $2\pi R - \Omega R\Delta t$. The difference between the two paths is

$$\Delta l = \frac{4\pi\Omega R^2 n}{c} \quad [\mathrm{m}]. \tag{6.22}$$

Note that if we divide this distance by the speed of light in the fiber, we get Δt as follows:

$$\Delta t = \frac{\Delta l n}{c} = \frac{4\pi\Omega R^2 n^2}{c^2} \quad [\mathrm{s}]. \tag{6.23}$$

Equations (6.22) and **(6.23)** provide linear relations between Ω (the stimulus in this case) and the change in the length traveled or the change in the time needed. The challenge is to measure either of these quantities. This can be done in a number of ways. One method is to build an optical resonator. A resonator is an optical path that has a dimension equal to multiple half wavelengths of the wave. In this case, a ring is built as shown in **Figure 6.42**. Light is coupled through the light coupler (beam splitter). At resonance, which occurs at a given frequency depending on the circumference of the ring, maximum power is coupled into the ring and minimum power is available at the detector. The incoming beam frequency is tuned to do just that. If the ring rotates at an angular velocity Ω, the light beams in the ring change in frequency (wavelength) to compensate for the change in the apparent length of the ring. The relation between frequency, wavelength, and length is

$$-\frac{df}{f} = \frac{d\lambda}{\lambda} = \frac{dl}{l}, \tag{6.24}$$

where the negative sign simply indicates that an increase in length decreases the resonant frequency. In effect, the wavelength of light increases in one direction and decreases in the other. The net effect is that the two beams generate a frequency difference. To show this we write

$$-\frac{\Delta f}{f} = \frac{\Delta l}{l} \rightarrow \Delta f = -f\frac{\Delta l}{l}. \tag{6.25}$$

Substituting from **Equation (6.22)**,

$$\Delta f = -f\frac{4\pi\Omega R^2 n}{lc} = -\frac{4\pi R^2}{\lambda l}\Omega \ [\text{Hz}], \tag{6.26}$$

where $\lambda = c/fn$ is the wavelength in the optical fiber. Alternatively, since $\lambda = \lambda_0 n$, where λ_0 is the wavelength in vacuum, we can write

$$\Delta f = -\frac{4\pi R^2}{\lambda_0 n l}\Omega = -\frac{4S}{\lambda_0 n l}\Omega \ [\text{Hz}], \tag{6.27}$$

where S is the area of the loop regardless of its shape. Since $l = 2\pi R$, we finally get for the circular loop,

$$\Delta f = -\frac{2R}{\lambda_0 n}\Omega \ [\text{Hz}]. \tag{6.28}$$

In all of these relations it was assumed that the detector is at the same location as the source and hence the beam travels the circumference of the ring. If the detector is, say, at the bottom of the ring and the source at the top (**Figure 6.41a**), each beam only travels half the circumference, and hence the relations must be halved. For example, the frequency shift in **Equation (6.28)** would be half of that shown.

This frequency shift is measured in the detector by mixing and filtering (we shall discuss these methods in **Chapter 11**) and is an indication of the angular velocity of precession, also called the **rotation rate**, Ω. In most cases **Equation (6.27)** is most convenient, whereas **Equation (6.28)** is only suitable for circular loops. It should be noted in particular that the larger the loop area, the larger the change in frequency, and hence the greater the sensitivity. In optical fiber gyroscopes it is a simple matter to loop the fiber N times and by doing so increase the output by a factor of N.

Figure 6.41b shows a common implementation of the Sagnac ring sensor using a set of mirrors to implement the ring. A cavity laser is used since a cavity laser generates two equal-amplitude beams traveling in opposite directions and hence the issue of splitting the beam is trivial. The two beams arriving at mirror M are directed into a detector that generates the frequency difference. This type of sensor is often called a loop or mirror gyroscope.

A different and more sensitive implementation is shown in **Figure 6.43**. Here an optical fiber is wound in a coil to increase its length and fed from a polarized light source through a beam splitter to ensure equal intensity and phase (the phase modulator adjusts for any variations in phase between the two beams). The beams propagate in opposite directions and, when returning to the detector, are at the same phase in the absence of rotation. If rotation exists, the beams will induce a phase difference at the detector that is dependent on the angular frequency of precession (rotation rate) Ω.

These devices are not cheap, but they are orders of magnitude cheaper than the spinning mass gyroscope, much smaller and lighter, and do not have the mechanical

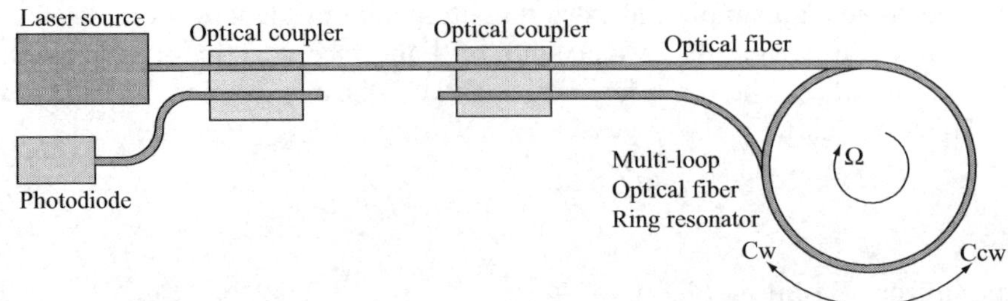

FIGURE 6.43 ■ A coil optical fiber gyroscope.

problems a rotating mass has. They have a very large dynamic range (as high as 10,000) and so they can be used for sensing rotation rates over a large span. In addition, optical fiber gyroscopes are immune to electromagnetic fields as well as to radiation and thus can be used in very hostile environments, including space. A ring gyroscope can measure the rotation of a fraction of a degree per hour. In many cases these are the devices of choice in aerospace applications, and the loop gyroscope can be produced as a microsensor.

There are other types of gyroscopes, often referred to as angular rate sensors, and some of these will be described in **Chapter 10**.

EXAMPLE 6.12 The optical gyroscope

A ring resonator is built as in **Figure 6.41a** with a radius of 10 cm. The source is a red laser operating at 850 nm in an optical fiber with an index of refraction $n = 1.516$. Calculate the output frequency for a rotation rate of 1°/h.

Solution: We first calculate the rate Ω in radians per second rather than degrees per hour:

$$\Omega = 1^\circ/\text{hr} \rightarrow \Omega = \frac{1^\circ}{180} \times \pi \times \frac{1}{3600} = 4.848 \times 10^{-6}\ \text{rad/s}.$$

The wavelength of the laser in vacuum is 850 nm, therefore we have, from **Equation (6.29)**,

$$\Delta f = -\frac{2R}{\lambda_0 n}\Omega = -\frac{2 \times 0.1}{850 \times 10^{-9} \times 1.516} \times 4.848 \times 10^{-6} = 0.752\ \text{Hz}.$$

This is not a very large shift in frequency, but it is measurable. By increasing the number of loops, say to 10, one obtains a shift of 7.52 Hz/degree/h.

6.8 | PROBLEMS

Strain gauges

6.1 Wire strain gauge. A strain gauge is made in the form of a simple round platinum-iridium wire of length 1 m and diameter 0.1 mm, used to sense strain on an antenna mast due to wind loading. Calculate the change in resistance of the sensor per pro-mil strain (1 pro-mil is 0.001 or 0.1% strain).

6.2 **NTC semiconductor strain gauge.** The following measurements are given for an NTC strain gauge:

Nominal resistance (no strain): 1 kΩ
Resistance at −3000 microstrains: 1366 Ω
Resistance at −1000 microstrains: 1100 Ω
Resistance at +3000 microstrains: 833 Ω

a. Find the transfer function of the strain gauge. Compare with **Figure 6.6b**.

b. Find the resistance of the strain gauge at 0.1% strain and at −0.1% strain.

6.3 **PTC semiconductor strain gauge:** The following measurements are given for a PTC strain gauge:

Nominal resistance (no strain): 1 kΩ
Resistance at −3000 microstrains: 833 Ω
Resistance at +1000 microstrains: 1100 Ω
Resistance at +3000 microstrains: 1366 Ω

a. Find the transfer function of the strain gauge. Compare with **Figure 6.6a**.

b. Find the resistance of the strain gauge at 0.1% strain and at −0.1% strain.

6.4 **Semiconductor strain gauges in a bridge configuration.** The strain gauge configuration in **Figure 6.5f** is used to sense the strain on a square metal sheet under tension with two opposing forces, as shown in **Figure 6.44**. Assume the material has a Young's modulus E and the strain gauges have gauge factors $g_1 = g$ and $g_2 = h$. Assume as well that the deformation of the material is elastic (i.e., the strain is not high enough to permanently deform the material) and that the strain does not exceed the maximum strain for the strain gauges used.

a. Find the resistances as a function of force of the four strain gauges of nominal resistance R_0.

b. Find the sensitivity of the four sensors to force.

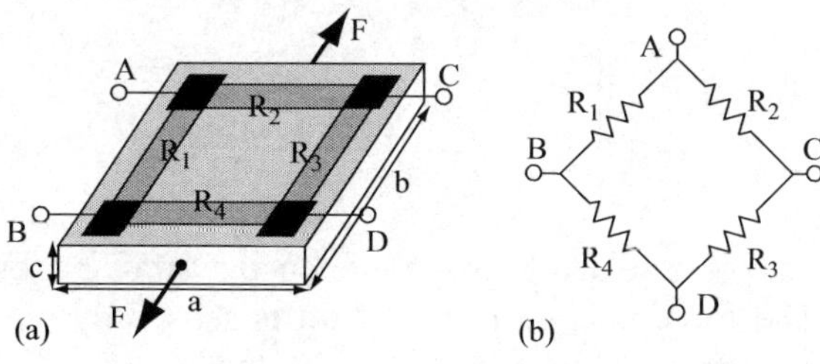

FIGURE 6.44 ■ Semiconductor strain gauge in a bridge configuration.

6.5 **Semiconductor strain gauges in a bridge configuration.** Repeat **Problem 6.4** but with the forces shown in **Figure 6.45**.

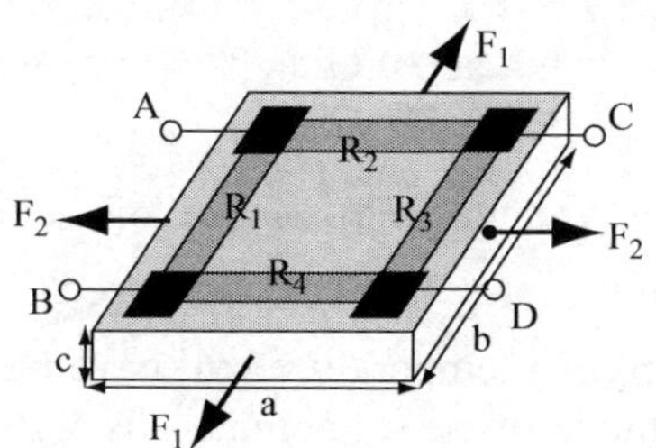

FIGURE 6.45 ■ Semiconductor strain gauges in a bridge configuration—two-axis forces.

Force and tactile sensors

6.6 Basic force sensor. The ceiling of a large building is supported by 16 vertical steel tubes, each with an inner diameter of 100 mm and an outer diameter of 140 mm. Each tube is equipped with a 240 Ω strain gauge (nominal, unstressed resistance at 20°C), prestressed to 2.5% strain. The Young's modulus for steel is 200 GPa. The strain gauge has a gauge factor of 2.2.

a. If the maximum strain measured is 2% in each tube, what is the maximum weight of the roof?

b. What is the change in the resistance of the strain gauge and what is the actual resistance reading for maximum allowable weight?

c. If the expected temperature range is 0°C–50°C, what is the error in reading of the maximum weight, assuming the sensor is not temperature compensated and it is made of constantan?

6.7 Force sensor. A force sensor is made of a strip of steel with cross-sectional area of $a = 40$ mm by $b = 10$ mm with a platinum, 350 Ω strain gauge (nominal, unstressed resistance) bonded to one surface, as shown in **Figure 6.46**. The sensor is intended for use as a compression force sensor.

a. Given the modulus of elasticity of 200 GPa, calculate the range of resistance of the sensor if it cannot exceed a strain of 3%. What is the range of forces that can be applied?

b. Calculate the sensitivity of the sensor assuming it is prestressed to 3%.

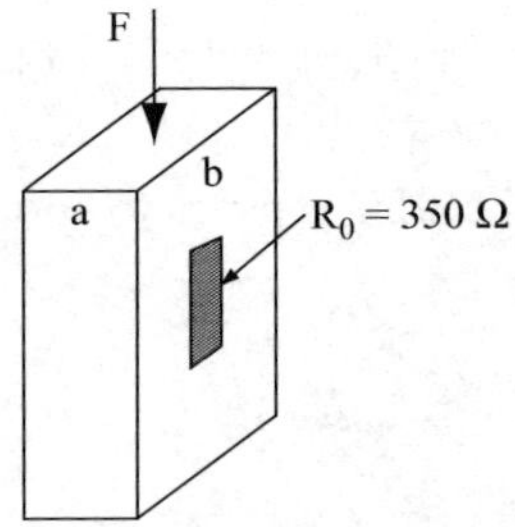

FIGURE 6.46 ■ A simple compression force sensor.

6.8 Compensated force sensor. A proposed sensing strategy for force is shown in **Figure 6.47**. The force applied to the beam is sensed by measuring the voltage across the upper sensor (marked as R_1). The lower sensor is prestressed to 3%, whereas the unstressed, nominal resistance of the strain gauges is 240 Ω at 20°C. Both strain gauges have a gauge factor of 6.4. The strain gauges are attached to a steel beam whose dimensions are given in **Figure 6.47**. The modulus of elasticity for the material used is given as 30 GPa. The two sensors are connected in series as shown.

a. Calculate the output voltage as a function of the applied force.

b. What is the maximum force that can be sensed?

c. Show that any change in temperature will have no effect on the output as long as the sensors are made of the same materials and are at the same temperature.

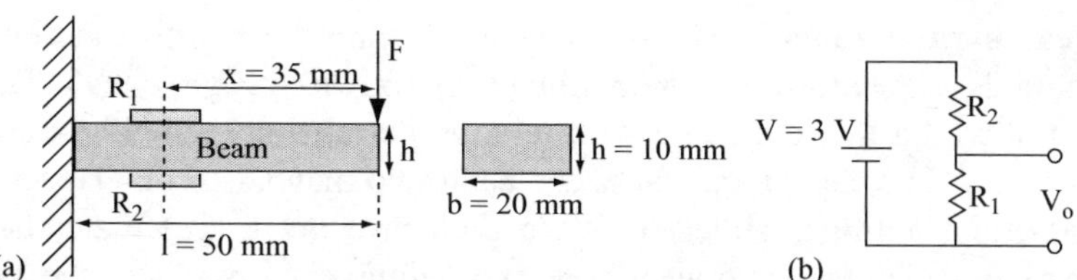

FIGURE 6.47 ■ A compensated force sensor.

6.9 **Capacitive force sensor.** A capacitive force sensor is made as shown in **Figure 6.48**. The three plates are identical, each $w = 10$ mm wide and $L = 20$ mm long. The two outer plates are fixed, whereas the center plate is suspended on a spring so that at zero force the three plates are aligned. Between the plates there is a separation sheet made of Teflon with relative permittivity of 2.25 and thickness $d = 0.1$ mm (one on each side of the center plate).

a. If the spring has a constant of 100 N/m, find the transfer function of the sensor and its sensitivity.

b. What is the maximum possible span of the sensor?

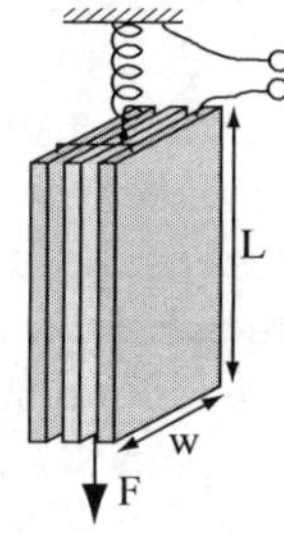

FIGURE 6.48 ■ A capacitive force sensor.

6.10 **Load sensing in a bridge.** A foot bridge is made as a simple deck, 4 m long, with a cross section 2 m (wide) and 20 cm thick, and made of wood. The deck is supported at the two ends. The maximum load allowable on the bridge is 10 tons, provided it is uniformly distributed on the deck. To sense this load, a strain gauge is placed at the center of the bridge and its resistance is monitored. If the sensor has a nominal resistance of 350 Ω and a gauge factor of 3.6, what is the reading of the strain gauge at maximum load? The modulus of elasticity for the wood used in the construction is 10 GPa.

6.11 **Overload sensing in elevators.** Most elevators are rated for a certain load, typically by specifying the number of persons allowed or by specifying maximum weight (or both). Modern elevators will not move if the maximum load is exceeded. There are many ways to sense this load, but the simplest to understand and one of the most accurate is to use a plate as the floor of the elevator and support it on load cells. Consider an elevator rated at 1500 kg (force) and its floor is supported on four load cells. Each load cell is equipped with a strain gauge with a nominal resistance of 240 Ω and a gauge factor of 5.8. Calculate the reading of each strain gauge at maximum load if the cross-sectional area of the load cell button (on which the strain gauge is mounted; see **Figure 6.10**) is 0.5 cm^2. Assume the buttons are made of steel with a modulus of elasticity of 60 GPa and the strain gauges are prestressed to 0.5%.

6.12 Linear array capacitive tactile sensor. A simple capacitive tactile sensor can be built as a linear array of simple plates as shown in **Figure 6.49**. The plates are covered with a thin dielectric with relative permittivity equal to 4 and thickness of 0.1 mm. The capacitance between each two plates is 6 pF. To sense presence or position, a finger slides over the dielectric layer. Calculate the maximum change in capacitance between any two neighboring plates as the finger passes above them, assuming the finger is a conductor and that it completely covers the two neighboring plates.

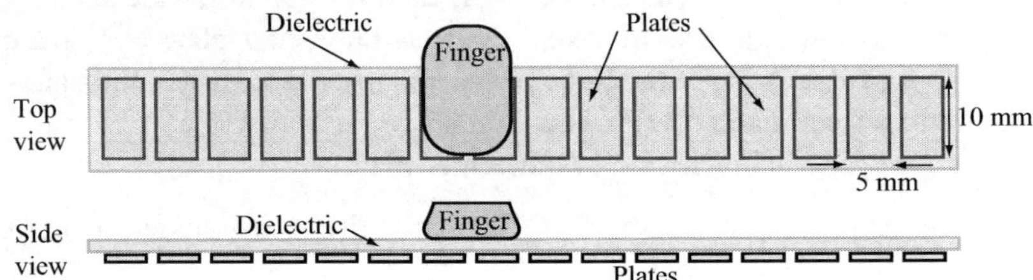

FIGURE 6.49 ■ A linear tactile sensor.

6.13 Capacitive tactile sensor: the touch pad. Touch pads can be made as two-dimensional arrays of capacitive sensors, as shown in **Figure 6.50**. A lower set of strips is covered with a dielectric and on top another set of strips is laid at 90° to the lower strips. The overlapping sections form capacitors with a capacitance that depends on the size of the strips and the material between them. The pad senses the position of a finger by the fact that as the finger slides it presses down on the top layer and pushes the strips underneath it closer together, increasing the capacitance between them. Consider a touch pad with strips 0.2 mm wide separated 0.02 mm apart with a dielectric with relative permittivity of 12. Assume that the fingers press down so that the distance between the dielectric compresses by 20%. In the process, the permittivity also increases locally by 10%. Calculate the rest capacitance at any strip intersection and the change in capacitance due to pressure from fingers.

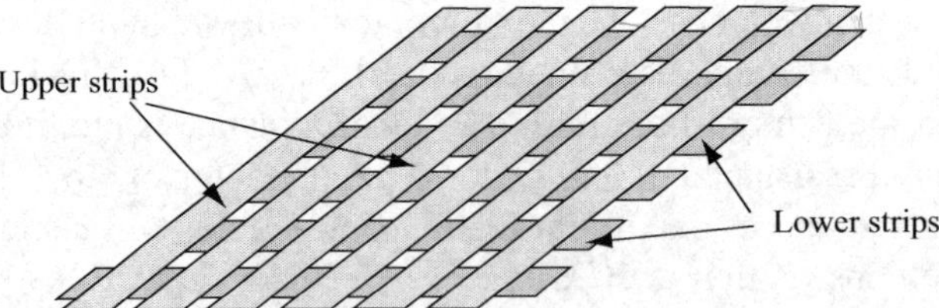

FIGURE 6.50 ■ A tactile touch pad.

Capacitive accelerometers

6.14 Force, pressure, and acceleration sensor. A force sensor is built as a capacitor with one plate fixed while the other can move against a spring with constant k [N/m]. By pressing on the movable plate, the distance the plate moves is directly proportional to the force. The plates have an area S and are separated by a distance d. The permittivity of the material between the plates is that of free space. Using **Figure 6.51**, and assuming the plate has moved a distance x from the rest position:

a. Find the relation between measured force and the capacitance of the sensor.
b. Plot a calibration curve for the following values: $k = 5$ N/m, $S = 1$ cm^2, $d = 0.02$ mm.

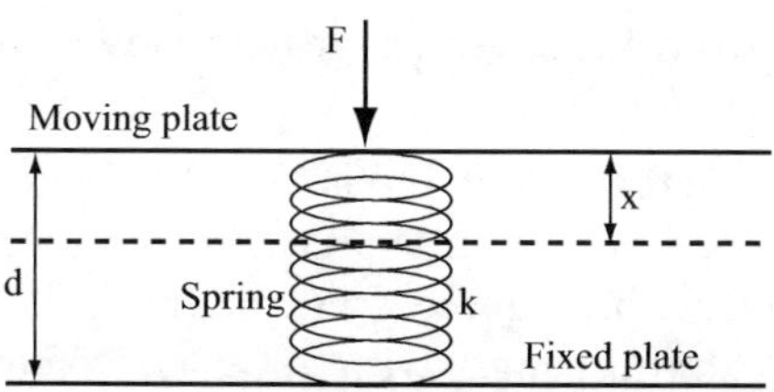

FIGURE 6.51 ■ A simple force, acceleration, or pressure sensor.

c. How can this device be used to measure pressure? Calculate the relation between pressure and capacitance.

d. How can this device be used to measure acceleration? Calculate the relation between acceleration and capacitance. Assume the mass of each plate is m [kg] and that the mass of the spring is negligible.

Note: The force necessary to compress a spring is $F = kx$. The spring in the figure represents a restoring force and is not necessarily a physical spring.

6.15 **Capacitive accelerometer.** An accelerometer is made as shown in **Figure 6.52**. The mass is small in size and together with the upper plate of the capacitor has mass of 10 g. The beam is $e = 1$ mm thick and $b = 2$ mm wide and it is made of silicon. The total length of the beam (from the center of mass to the fixed point) is $c = 20$ mm. With plates separated a distance $d = 2$ mm apart and capacitor plates 10 mm $\times$ 10 mm in area, and assuming the maximum strain in silicon cannot exceed 1%, calculate:

a. The span of acceleration the sensor is capable of. Use a modulus of elasticity of 150 GPa.

b. The range of capacitances corresponding to the span calculated in (a). Assume there is a 0.1 mm thick stop that prevents the plates from approaching each other at less than 0.1 mm.

c. The sensitivity of the sensor in pF/m/s^2.

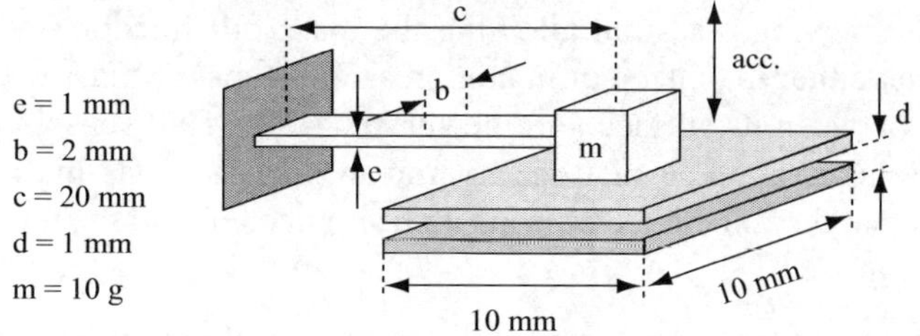

FIGURE 6.52 ■ A capacitive accelerometer.

6.16 **Strain gauge accelerometer.** The sensor in **Figure 6.52** is given again, but now the capacitor plates are removed and instead two strain gauges are placed one on the top surface of the beam and one on the bottom surface. The strain gauges are silicon gauges, very small with a nominal resistance of 1000 Ω. Both are pre-stressed to 1.5% strain and have a maximum range of 3% strain. The strain gauges are bonded in the middle of the beam (5 mm from the fixed position). For the mass, dimensions, and Young's modulus in **Problem 6.15**:

a. Calculate the span of

b. Calculate the sensitivity of the sensor if the gauge factor for the strain gauges is 50.

c. Repeat (a) and (b) if the strain gauges are moved to the fixed location of the beam.

6.17 **Two-axis accelerometer.** A two-axis accelerometer is made as shown in **Figure 6.53**. The mass, $m = 2$ g, is attached to the center of four beams, each $e = 4$ mm long and with a square cross section $b = 0.2$ mm by $c = 0.2$ mm. A semiconductor strain gauge with a gauge factor of 120 and nominal (unstressed) resistance of 1 kΩ is built onto each beam, across from the attachment point.

a. Calculate the range of the accelerometer if the strain gauges can handle ±2% strain and can work only under tension. The sensor is made of silicon with a modulus of elasticity of 150 GPa.

b. Calculate the range of resistance of the strain gauges.

c. Calculate the sensitivity of the sensor.

d. Discuss how the sensitivity of the sensor can be increased.

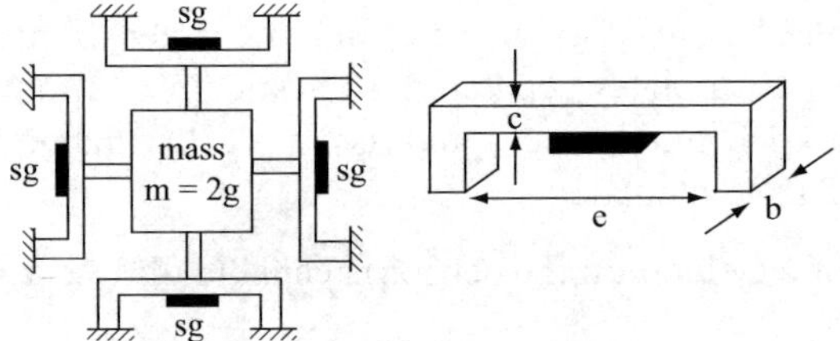

FIGURE 6.53 ■ A two-axis accelerometer.

Magnetic accelerometers

6.18 **Magnetic accelerometer.** A magnetic accelerometer is made as shown in **Figure 6.54**. A coil with $N = 500$ turns carries a current $I_c = 10$ mA, wound around the moving mass but allowing the mass full freedom to move. A spring that can act either in tension or in compression keeps the mass in place with a gap of 2 mm between its surface and the surface of the Hall sensor on the stationary pole. The latter is made of iron, 15 mm in diameter, has high relative permeability so that the reluctance of iron can be neglected, and the mass of the moving core is 10 g.

a. If the sensor must be capable of sensing accelerations up to 100 g in each direction, what is the spring constant?

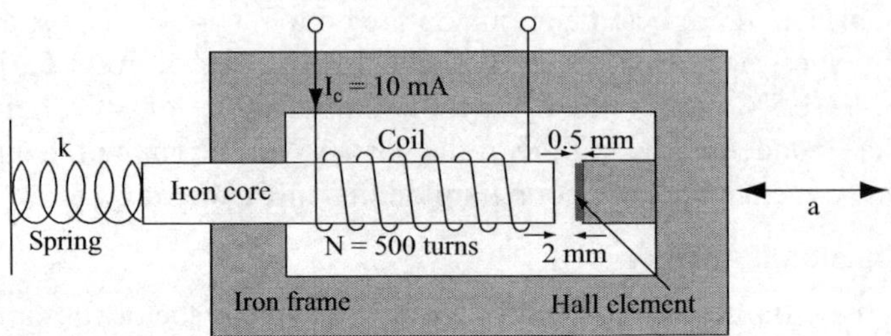

FIGURE 6.54 ■ A magnetic accelerometer.

b. If the Hall element used is connected to a current I and is d mm thick, calculate the sensitivity of the sensor as a function of acceleration and the current in the Hall element. Assume the Hall coefficient is known.

c. A silicon Hall element, 0.5 mm thick with a Hall coefficient of $-0.01\ \text{m}^3/\text{A} \cdot \text{s}$ is used, driven with a 5 mA current. Calculate the span of the output of the sensor and its sensitivity over the span.

6.19 **LVDT-based accelerometer.** LVDTs have a number of advantages over other sensors, including linearity and high output. Because of the moving core, they can be used in a number of ways, including for sensing of acceleration, although they must be modified for the purpose. Consider an LVDT designed to sense position in the range -10 mm to $+10$ mm for which it produces an output of ± 5 V (rms) (for each direction; i.e., it produces 5 V for a displacement of 10 mm). The core is free to move and a spring is attached to each end of the core to restore it to its zero position (**Figure 6.55**).

a. With a mass of 40 g, calculate the spring constant needed to sense acceleration in the range $-2\ \text{m/s}^2$ to $+2\ \text{m/s}^2$.

b. What is the sensitivity of the accelerometer?

c. The output of the LVDT is measured using a digital voltmeter with a resolution of 0.01 V. What is the resolution of the sensor?.

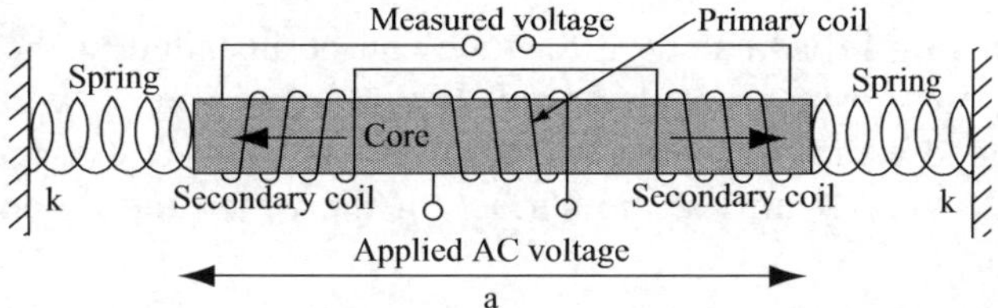

FIGURE 6.55 ■ Use of an LVDT to sense acceleration.

Pressure sensors

6.20 **The altimeter.** Most altimeters use a pressure sensor to measure height, using the barometric equation as the basis.

a. What type of pressure sensor can be used for this purpose?

b. Show how the barometric equation can be used to calibrate a pressure sensor in terms of elevation above sea level.

c. What is the required resolution of a pressure sensor to produce a 1 m resolution altimeter at sea level and at 10,000 m?

d. Calculate the span required of a pressure sensor to serve as an altimeter for mountain climbing purposes with an elevation range of 10 km (the highest mountain on earth is 8800 m high).

6.21 **The depth meter.** Depth meters are essential tools for divers and for submarines. It is proposed to use a pressure sensor and calibrate it in meters. Since pressure in water is produced by the column of water above the point at which the pressure is sensed, the relation between pressure and depth is relatively simple, and because water density can be considered constant the relation is accurate. Given the density of seawater as 1025 kg/m^3:

a. Calculate the span of a sealed pressure gauge pressure sensor to sense pressure down to 100 m. The sealed pressure is 101,325 Pa (1 atm).

b. Calculate the required resolution of the sensor for a resolution of 0.25 m of the depth meter.

c. The water density in freshwater is 1000 kg/m^3. What is the error in the reading of a depth pressure sensor if used in freshwater without recalibration?

Note: Water density does vary with temperature, but this is neglected here.

Velocity sensing

6.22 **Water speed and flow volume sensing.** To measure the speed of water in a channel and the flow volume it is suggested to use a Pitot tube and measure the head above water, as in **Figure 6.37a**.

a. Find a relation between the speed of water and the head.

b. If a difference of 0.5 cm is practical, what is the sensitivity of the device?

c. Calculate the water flow in cubic meters per second [m^3/s] as a function of velocity and as a function of the head if the cross-sectional area is S and the flow velocity is uniform throughout the channel.

6.23 **Speed sensing in a boat.** A Pitot tube can be installed in the prow or on the side of a boat to measure its speed. If a pressure sensor with a resolution of 1000 Pa and a range from 0 to 50,000 Pa is used, and assuming a pressure of 101.325 kPa (1 atm) at the surface of the water and a density of water of 1025 kg/m^3:

a. Calculate the resolution in terms of speed that can be measured, neglecting the effects of static pressure and assuming the sensor is calibrated to zero output at zero speed.

b. Calculate the range of the sensor.

6.24 **Sensing airspeed in an aircraft.** A passenger aircraft uses two Pitot tubes for speed sensing. One tube is aligned parallel to the airplane and the other is perpendicular to it, each equipped with a pressure sensor.

a. For an aircraft flying at 11,000 m, calculate the readings of the pressure sensors in each of the tubes and the differential pressure if the airplane flies at 850 km/h. The temperature at that elevation is -40°C. Neglect any effect the speed may have on air density inside the tube.

b. Suppose that the lateral tube becomes blocked with ice at 11,000 m and now the aircraft climbs to 12,000 m. What is the error in the speed reading, assuming the airplane has not changed speed and the temperature remains the same as at 11,000 m?

6.25 **Speed and depth sensing in submarines.** The Pitot and Prandtl tubes are equally effective underwater. Suppose a submarine is equipped with a forward-pointing Prandtl tube. Two independent sensors are used, one to sense static pressure and the other to sense total pressure. The density of water is assumed constant with depth, equal to 1025 kg/m^3 and independent of temperature.

a. If the submarine is expected to descend to 1000 m, what is the span required from each pressure sensor? The maximum speed of the submarine is 25 knots (1 knot = 1.854 km/h).

b. Show that these two measurements are sufficient to provide both the velocity and the depth of the submarine.

6.26 **A resistive pressure sensor.** Conducting polymers can be used to sense pressure by measuring the resistance of the polymer. A pressure sensor is proposed as follows: A small hollow spherical ball is made of a polymer with a given conductivity. The inner and outer surfaces of the ball are plated with a conducting surface to form an inner and outer electrode, as shown in **Figure 6.56**. The pressure inside the ball is sensed by measuring the resistance between the inner and outer electrodes. At a reference temperature P_0, the inner radius is $r = a$ and the thickness is $t = t_0$. Also, you may assume $a \gg t_0$ and, in general, $r \gg t$. The relation between pressure and radius is

$$r = \alpha\sqrt{P}.$$

That is, as the pressure increases, the radius increases with α a constant value. The conductivity of the material is σ. This relation holds throughout the pressure range, including the reference pressure. Find the relation between pressure and resistance between the electrodes.

Note: The fact that $r \gg t$ means that the shell is thin with respect to the radius. Use this approximation to simplify the calculations.

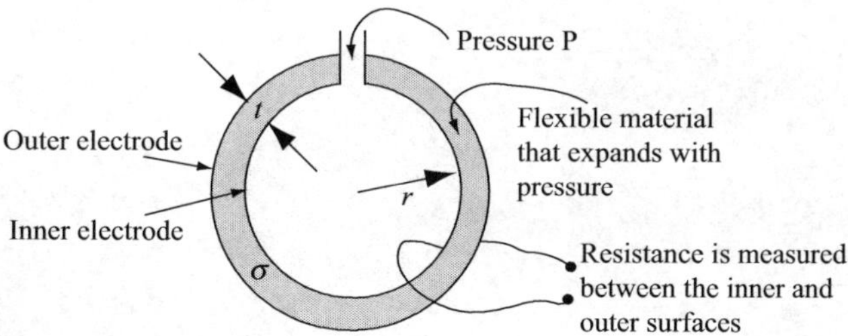

FIGURE 6.56 ■ A resistive pressure sensor.

Optical gyroscopes

6.27 **The mechanical gyroscope**. A miniature mechanical gyroscope contains a wheel of mass 50 g, radius 40 mm, and length 20 mm, rotating at 10,000 rpm.

a. Calculate the sensitivity of the gyroscope to torque perpendicular to its axis.

b. What is the lowest torque it can sense if the frequency of precession can be measured to within 0.01 rad/s?

6.28 **Ring gyroscope.** A Sagnac gyroscope is implemented as in **Figure 6.41b**.

a. With the side of the triangle $a = 5$ cm and using a green laser operating at 532 nm in a vacuum, calculate the sensitivity of the sensor (in Hz/°/s).

b. If a frequency can be reliably measured down to 0.1 Hz, what is the lowest rate that can be sensed?

6.29 Optical fiber loop gyroscope. A small optical fiber gyroscope is designed for high sensitivity. For reliable reading the output frequency resolution is set at 0.1 Hz. How many loops are required to sense a rate of 10°/h if the loop is 10 cm in diameter and an infrared LED at 850 nm is used as the source? The index of refraction of the fiber is 1.85.

6.30 Ring gyroscope. A Sagnac gyroscope is built as in **Figure 6.57**. For $a = 40$ mm and using a red laser at 680 nm, calculate the output expected for a rate of 1°/s and the sensitivity of the gyroscope.

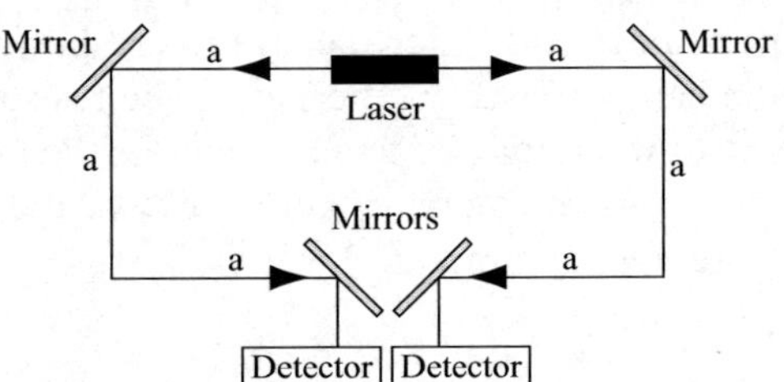

FIGURE 6.57 ■ Implementation of a ring gyroscope.

CHAPTER

Acoustic Sensors and Actuators

7

The ear

The ear is a sensor and actuator in more than one way. Essentially a mechanochemical sensor, it includes a moving mechanism on the hearing side of the structure. But the ear also features a gyroscope, the inner ear, responsible for stability and sense of position. The ear itself is made of the outer and inner ear. The external ear is no more than a means of concentrating and guiding the sound toward the tympanic membrane (eardrum). In humans, the external ear is a relatively small, static feature, but in some animals it is both large and adjustable. The fenec fox, for example, has external ears that are larger than its head. At the bottom of the ear canal, the tympanic membrane moves in response to sound and, in the process, moves an assembly of three bones, the malleus (connected to the eardrum), the incus (an intermediate flexural bone), and the stapes. The latter, the smallest bone in the body, transmits the motion to the cochlea in the inner ear. The three bones not only transmit the sound, but also amplify it through lever advantage afforded by their structure and dimensions. The cochlea is a spiral tube filled with a fluid. The stapes move like a piston, moving the fluid that in turn moves a series of hairlike structures lining the cochlea. These are the actual sensors that release a chemical onto the auditory nerve to affect hearing.

The inner ear also contains three semicircular canals arranged at 90° to each other, with two roughly vertical and one horizontal. They have a similar structure to the cochlea, including a series of hairlike structures affected by the fluid in the canals based on the position of the body. These serve to maintain balance and provide information on the position and attitude of the body. The effect of motion on these structures can be immediately seen if the body rotates, as, for example, on a merry-go-round. We temporarily loose, the ability to keep our balance.

The ear is a uniquely sensitive structure. It can sense pressures as low as 2×10^{-5} Pa (or 10^{-12} W/m^2; that is on the order of one-billionth of the atmospheric pressure) and can function at levels 10^{13} times higher. That means the dynamic range is about 130 dB. The nominal frequency response is between 20 Hz and 20,000 Hz, although most humans have a much narrower range. But the ear is also very sensitive to pitch and can distinguish very small changes in pitch and frequency. A 1 Hz difference between two sounds is easily detectable. The hearing in humans is binaural and the brain uses that to detect the direction of sources of sound. Many animals use the mechanical motion of the outer ear to accomplish the same function, but much better than we do. It should be noted as well that many animals have much more sensitive hearing than humans, with ears that respond to higher frequencies and to a wider range of frequencies.

7.1 | INTRODUCTION

The term acoustics can mean sound or the science of sound. It is in the latter sense that it is used here. Acoustics thus covers all aspects of sound waves, from low-frequency sound waves to ultrasound waves and beyond to what are simply called acoustic waves.

The distinction between acoustics and ultrasonics is based on the span of the human ear. The common, stated range of the human ear is 20 Hz–20 kHz and is based on the ability of our ears to distinguish differences in pressure (usually not only in the atmosphere, but also in water). This is called the audio or audible range or span. It should also be noted that most humans can only hear on a portion of this span (about 50 Hz–14 kHz) and that the whole span is not necessary for transmission of audio information (e.g., telephones use the range between 300 Hz and 3 kHz and an AM radio station has a frequency bandwidth of 10 kHz).

Vibrations from any source cause variations in pressure and these propagate in the substance in which they are generated at a velocity that depends on the substance. The waves are understood to be elastic waves, meaning that they can only be generated in elastic substances (gases, solids, liquids), but not in vacuum or plastic substances (plastic media absorb waves; the term plastic here indicates a material that is not rigid). Above 20 kHz, the same vibrations generate variations in pressure (in air or any other material) and these are called ultrasonic waves. Below 20 Hz, elastic waves are called infrasound. There is no specific range for ultrasound—any acoustic wave above 20 kHz qualifies, and often ultrasonic waves at frequencies well above 100 MHz can be generated and are useful in a variety of applications. Acoustic waves can be and often are generated at much higher frequencies, well above 1 GHz.

Acoustic waves, in the more general sense, cover ultrasonic and infrasonic waves and have roughly the same properties. That is, their general behavior is the same even though certain aspects of the waves change with frequency. For example, the higher the frequency of a wave, the more "direct" its propagation, that is, the less likely it is to diffract (bend) around corners and edges.

As a means of sensing and actuation, acoustic waves have developed in a number of directions. The most obvious is the use of sound waves in the audible range for the sensing of sound (microphones, hydrophones, dynamic pressure sensors) and for actuation using loudspeakers. Another direction that has contributed greatly to the development of sensors and actuators is the extensive work in sonar—the generation and detection of acoustic energy (including infrasound and ultrasound) in water, initially for military purposes and later for the study of oceans and life in the oceans, and even down to fishing aids. Out of this work has evolved the newer area of ultrasonics, which has found applications in the testing of materials, material processing, ranging, and medicine. The development of surface acoustic wave (SAW) devices has extended the range of ultrasonics well into the gigahertz region and for applications that may not seem directly connected to acoustics, such as oscillators in electronic equipment. SAW devices are important not only in sensing, especially in mass and pressure sensing, but also in a variety of chemical sensors.

Because acoustic waves are involved, it should not be surprising to find that the interest in them and their properties is not new. It is impossible to assume that ancient man did not observe that sounds propagate farther in cold dense air than in warm, thin air, or that sounds seem to be louder underwater. In fact, Leonardo da Vinci wrote in 1490 that by using a hydrophone (a tube inserted in water), he could detect noise from ships at great distances. The images from movies where someone presses his ear to the ground to detect an incoming rider are probably familiar to many. Of course, it is altogether a different issue to quantify the speed of propagation of sound and to define its relation to material properties—these came much later (starting around 1800).

7.2 | UNITS AND DEFINITIONS

Perhaps more than any other area of sensing and actuation, the issue of units, measurement, and definitions in acoustics seems confusing at times. Part of the reason is that many units have been developed from work with sound in conjunction with the span of the human ear. There are even units based on perceived quantities and, again, because of the range of human hearing, the use of logarithmic scales is very common. In the audio range, the most common way of describing the propagation of acoustic waves is through the use of sound pressure—newtons per square meter (N/m^2) or pascals (Pa)—since acoustic waves are elastic waves and their effect, especially on the human ear, manifests itself in changes in pressure. However, pressure can be directly related to power density (watts per square meter [W/m^2]), especially when the pressure acts on a diaphragm such as the eardrum, a microphone, or is generated by a loudspeaker. Thus there are two equivalent methods of describing acoustic behavior: one in terms of pressure and the other in terms of power density. Because much of the work in audio relates to hearing, the threshold of hearing holds a unique place and often pressure and power density are related to the threshold of hearing. The threshold of hearing is taken as 2×10^{-5} Pa in terms of pressure or 10^{-12} W/m^2 in terms of power density. A second point on the scale is the threshold of pain, typically taken as 100 Pa or 10 W/m^2, indicating the level beyond which damage to the ear can occur. It should be mentioned that these values are subjective and different sources will use different values for both the threshold of hearing and threshold of pain.

Because the range is so large, sound pressure level and power density are often given in decibels (dB). For the sound pressure level (SPL),

$$SPL_{dB} = 20\log_{10}\frac{P_a}{P_0} \quad [\text{dB}], \tag{7.1}$$

where P_0 is the threshold of hearing and is viewed as a reference pressure and P_a is the acoustic pressure. Thus the threshold of hearing is 0 dB and the threshold of pain is 134 dB (for the values given above). Normal speech is between approximately 45 dB and 70 dB.

For power density,

$$PD_{dB} = 10\log_{10}\frac{P_a}{P_0} \quad [\text{dB}], \tag{7.2}$$

where P_0 is 10^{-12} W/m^2 and P_a is the acoustic power density being sensed. Using the values above, the threshold of hearing is 0 dB and the threshold of pain is 130 dB. Although the numbers look similar to the SPL values, one should be very careful not to confuse the two, as they indicate different quantities.

Acoustic actuators are often specified in terms of power (e.g., the power specification of loudspeakers). The data can be given as average power, peak power (or even peak-to-peak power) based on sinusoidal excitation. It can sometimes be given as maximum power during a specific, typically short, period of time. Whereas these specifications serve mostly marketing purposes, it is important to recognize that the power specified for an actuator is almost always the electric, dissipated power, that is, the power the actuator can dissipate without being damaged. This is very different than the acoustic power the actuator can couple into the space around it. Typically the acoustic power is a very small fraction of the input electric power to the actuator. Most of that power is lost as heat in the actuator itself.

In the ultrasound range, when acoustic waves propagate in materials (other than gases) they are perceived as producing stress in materials and hence stress and strain play a significant role in analysis. Pressure and power density can still be used, but it is more common to describe behaviors in terms of displacement and strain as measures of the ultrasonic signal. When decibel scales are used, the reference pressure (or power density) is taken as 1 since now the thresholds of hearing and pain have no meaning.

EXAMPLE 7.1 Pressure and power density during normal speech

Normal speech ranges from approximately 45 dB to 70 dB, typically measured at a distance of 1 m from the speaker. What are the ranges in terms of pressure and power density at the eardrums of the listener.

Solution: Using **Equation 7.1**, we have for the lower range

$$20\log_{10}\frac{P_a}{P_0} = 45\text{ dB},$$

that is,

$$\log_{10}\frac{P_a}{P_0} = \frac{45}{20} = 2.25 \rightarrow P_a = P_0 10^{2.25} = 2\times10^{-5}\times10^{2.25} = 3.556\times10^{-3}\text{ Pa}.$$

At the higher range,

$$20\log_{10}\frac{P_a}{P_0} = 70\text{ dB},$$

or

$$\log_{10}\frac{P_a}{P_0} = \frac{70}{20} = 3.5 \rightarrow P_a = P_0 10^{3.5} = 2\times10^{-5}\times10^{3.5} = 6.325\times10^{-2}\text{ Pa}.$$

The range is between 0.0035565 Pa and 0.06325 Pa.

The power density is obtained from **Equation (7.2)**. At the lower range,

$$10\log_{10}\frac{P_a}{P_0} = 45\text{ dB},$$

or

$$\log_{10}\frac{P_a}{P_0} = \frac{45}{10} = 4.5 \rightarrow P_a = P_0 10^{4.5} = 10^{-12}\times10^{4.5} = 3.162\times10^{-8}\text{ W/m}^2.$$

At the higher range,

$$10\log_{10}\frac{P_a}{P_0} = 70\text{ dB},$$

or

$$\log_{10}\frac{P_a}{P_0} = \frac{70}{10} = 7 \rightarrow P_a = P_0 10^{7} = 10^{-12}\times10^{7} = 10^{-5}\text{ W/m}^2.$$

The range is between 31.62 nW/m^2 and 10 μW/m^2.

The properties of acoustic waves are defined by the media through which the waves propagate. Some of the properties that affect the behavior of acoustic wave include the following.

Bulk modulus (*K*) is the ratio of volume stress per unit of volume strain. It may be viewed as the ratio of the rate of increase in pressure to the resulting relative decrease in volume or the ratio of the rate of increase in pressure to the relative increase in density:

$$K = -\frac{dP}{dV/V} = \frac{dP}{d\rho/\rho} \quad [\mathrm{N/m^2}]. \tag{7.3}$$

Note that the bulk modulus has units of pressure. The bulk modulus is an indication of the resistance of the material to compression. The reciprocal, $1/K$, may be viewed as a measure of the compressibility of the material. Bulk moduli of materials are available in tables based on experimental data.

Shear modulus (*G*) is the ratio of shear stress and shear strain. It is viewed as a measure of the rigidity of the material or its resistance to shear deformation:

$$G = \frac{dP}{dx/x} \quad [\mathrm{N/m^2}], \tag{7.4}$$

where dx is the shear deformation or shear displacement. Another way to understand the definition is as the ratio of the change in pressure to the relative change in shear deformation.

The bulk and shear moduli, together with the modulus of elasticity defined in **Chapter 6**, describe the elastic properties of materials. The difference between the shear modulus and the modulus of elasticity is that the modulus of elasticity defines the linear or longitudinal deformation, whereas the shear modulus defines the transverse or shear deformation of the material.

The **ratio of specific heats** of a gas is the ratio of the specific heat capacity at constant pressure (P_c) to specific heat capacity at constant volume (V_c). It is also known as the **isentropic expansion factor** and is denoted by γ. The specific heat capacity is the amount of heat (in joules [J]) needed to raise the temperature of a unit mass (in kilograms [kg]) by 1°C (see **Section 3.1.1**).

In acoustics, some of the terms used are based on subjective measures rather than on absolute scales because of the intricate link between sound and hearing, and hence with the perception of sound by the human brain. One of these terms is **loudness**, defined as an attribute of the auditory perception that ranks sounds on a scale ranging from quiet to loud. The sensation of sound by the human brain depends on a variety of terms, including intensity (amplitude) and frequency. To measure loudness one employs two basic units: the phon and the sone.

The **phon** is a unit of loudness that measures the intensity of sound in decibels above a reference tone having a frequency of 1000 Hz and a root mean square (RMS) sound pressure of 20×10^{-6} Pa. An alternative definition is "a unit of apparent loudness, equal in number to the intensity in decibels of a 1000 Hz tone perceived to be as loud as the sound being measured."

The **sone** is a unit of perceived loudness equal to the loudness of a 1000 Hz tone at 40 dB above the threshold of hearing.

Another subjective term in common use is **tone**, which is used to describe the quality or character of a sound. Obviously it cannot be measured on any objective scale, but it is an important aspect of acoustics, especially as it relates to music.

7.3 | ELASTIC WAVES AND THEIR PROPERTIES

Sound waves are longitudinal elastic waves; that is, a pressure wave, as it propagates, changes the pressure along the direction of its propagation. Thus an acoustic wave impinging on our eardrums will push and pull on the eardrum to affect hearing. Waves, including acoustic waves, have three fundamental properties that are of special importance.

First, they have a frequency (or a range of frequencies). The frequency, f, of a wave is the number of variations of the wave per second, measured in hertz (Hz, or cycles per second). This is normally defined for harmonic waves and is understood to be the number of cycles of the harmonic wave per second. For example, if we were to count the number of crests in an ocean wave passing through a fixed point in 1 s, the result would be the frequency of that wave.

The second property is the wavelength, λ, which is related to frequency and is the distance in meters (m) a wave propagates in one cycle of the wave. In the example of the ocean waves, the wavelength is the distance between two crests (or two valleys).

The speed of propagation of a wave, c, is the speed (in meters per second [m/s]) with which the front of a wave propagates. These three quantities are related as

$$\lambda = \frac{c}{f} \quad [\mathrm{m}]. \tag{7.5}$$

The relation between frequency and wavelength can be seen in **Figure 7.1**. Although this relation may seem minor, one of the most important aspects of acoustic waves is the short wavelength they exhibit. In fact, this property is responsible for the relatively high resolution of ultrasonic tests such as tests for defects in materials or tests for medical purposes. As a rule, the resolution one can expect from any test using a wave is dependent on the wavelength. The shorter the wavelength, the higher the resolution. We will see later in this chapter that this property is taken full advantage of in SAW devices.

Waves can be transverse waves, longitudinal waves, or a combination of the two. Transverse waves are those waves that cause a change in amplitude in directions transverse to the direction of propagation of the wave. Waves produced by a tight string are of this type. When we pluck a string, it vibrates perpendicular to the length of the string while the wave itself propagates along the string. This is shown schematically in **Figure 7.2**. The figure also shows that the wave propagates away from the source, in this case in two directions. The ocean wave mentioned above and electromagnetic waves are also transverse waves.

Acoustic waves in gases and liquids are longitudinal waves. In solids they can also be transverse waves. Transverse acoustic waves are often called shear waves. To avoid confusion, and because in most cases we will encounter longitudinal waves, the discussion that follows relates to longitudinal waves. Whenever the need arises to discuss

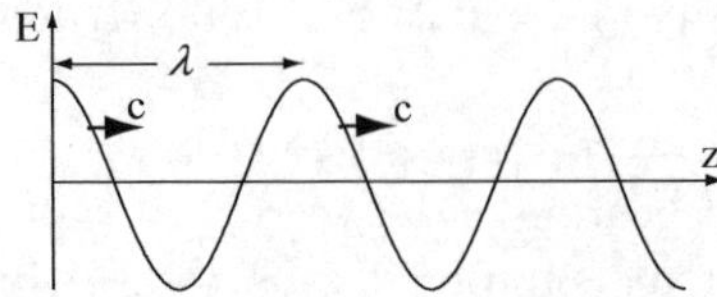

FIGURE 7.1 ■ Relation between frequency, wavelength, and speed of propagation for a generic time-harmonic wave.

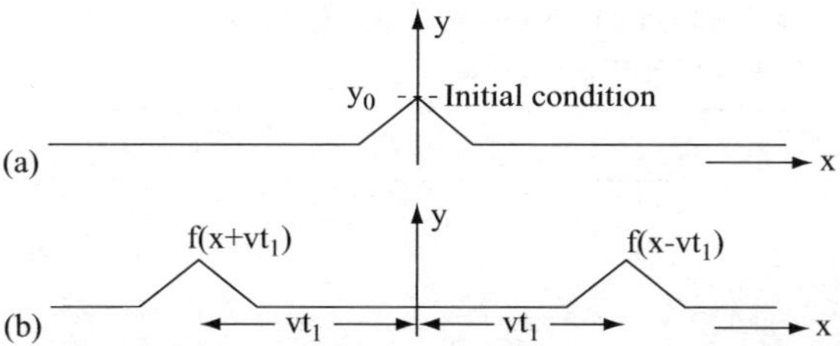

FIGURE 7.2 ■ Wave propagation along a tight string. (a) The string is plucked at $x = 0$. (b) The string and the disturbance after a time t_1.

shear waves, and later, surface waves, these will be indicated explicitly to distinguish them from longitudinal waves.

7.3.1 Longitudinal Waves

The speed of an acoustic wave is directly related to the change in volume and the resulting change in pressure (say, due to the motion of the piston in **Figure 7.3**):

$$c = \sqrt{\frac{\Delta p V}{\Delta V \rho_0}} \quad [\text{m/s}]. \tag{7.6}$$

Note that $\Delta p/(\Delta V/V)$ is in fact the bulk modulus, so one can write **Equation (7.6)** as

$$c = \sqrt{\frac{K}{\rho_0}} \quad [\text{m/s}], \tag{7.7}$$

where ρ_0 is the density of the undisturbed fluid, ΔV is the change in volume, Δp is the change in pressure, and V is the volume. In gases, this simplifies to the following:

$$c = \sqrt{\frac{\gamma p_0}{\rho_0}} \quad [\text{m/s}], \tag{7.8}$$

where p_0 is the static pressure and γ is the ratio of specific heats for the gas. Thus the speed of acoustic waves in materials is pressure and temperature dependent. In solids, the speed of sound depends on the "elasticity" of the solid—more specifically, on the shear and bulk moduli of the medium. **Table 7.1** gives the speed of sound in a number of materials for longitudinal waves. These values are experimental and will vary somewhat depending on the source. For example, the speed of sound in air at 20°C is quoted as 343 m/s to 358 m/s by various sources. The speed of sound also varies with pressure

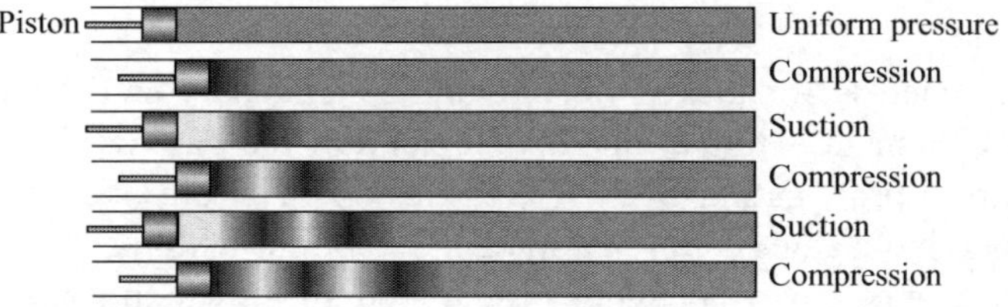

FIGURE 7.3 ■ Generation of a longitudinal wave by motion of a piston. The particles of the substance are displaced longitudinally to create local variations in pressure.

TABLE 7.1 ■ Speed of sound for longitudinal waves in some materials at given temperatures

Material	Speed [m/s]	Temperature [°C]
Air	331	0
Freshwater	1486	20
Seawater	1520	20
Muscle tissue	1580	35
Fat	1450	35
Bone	4040	35
Rubber	2300	25
Granite	6000	25
Quartz	5980	25
Glass	6800	25
Steel	5900	20
Copper	4600	20
Aluminum	6320	20
Beryllium	12,900	25
Titanium	6170	20
Brass	3800	20

and relative humidity. In solids, especially in metals, dependency on temperature is lower than in gases or liquids.

A longitudinal wave changes its amplitude along the direction of propagation. A simple example is the mechanical wave generated by a piston in a tube. As the piston moves back and forth, it compresses and decompresses the gas ahead of it. This motion then propagates along the tube, as shown in **Figure 7.3**. Acoustic waves are of this type.

Assuming for simplicity that we have a harmonic longitudinal wave of frequency f, it may be written in general terms as

$$p(x,t) = P_0 \sin(kx - \omega t) \quad [\mathrm{N/m^2}], \tag{7.9}$$

where $p(x,t)$ is the time- and position-dependent pressure in the medium, P_0 is the pressure amplitude of the wave, and k is a constant. The wave propagates in the positive x direction (in this case) and $\omega = 2\pi f$ is its angular frequency.

The amplitude of the wave is

$$P_0 = k\rho_0 c^2 y_m \quad [\mathrm{N/m^2}], \tag{7.10}$$

where y_m is the maximum displacement of a particle during compression or expansion in the wave. The constant k is called the wavenumber or the phase constant and is given as

$$k = \frac{2\pi}{\lambda} = \frac{\omega}{c} \quad [\mathrm{rad/m}]. \tag{7.11}$$

Waves carry energy. A shockwave (like that generated by an earthquake or a sonic boom) can cause damage, while a loud sound can hurt our ears or shatter a window. A wave is said to be a propagating wave if it carries energy from one point to another.

A wave can propagate in an unbounded medium with or without attenuation (losses). Attenuation of a wave depends on the medium in which it propagates and this attenuation reduces the amplitude of the wave. Attenuation of waves is exponential, with an exponent that depends on the material properties. An **attenuation constant**, α, is defined for each material and the amplitude of the wave (pressure), as it propagates, changes as follows:

$$p(x,t) = P_0 e^{-\alpha x} \sin(kx - \omega t) \quad [\mathrm{N/m^2}]. \tag{7.12}$$

This attenuation causes a loss of energy as the wave propagates and eventually dissipates all the energy in the wave. In addition, unless the wave propagates in a perfectly collimated beam, it spreads into space so its energy is spread over an increasingly larger area. Under this condition, the amplitude decreases at any point in space regardless of attenuation. The attenuation constant, α, has units of neper per meter (Np/m), where 1 Np/m = 8.686 dB/m. It should also be noted here that power, as opposed to amplitude (force, pressure, displacement), attenuates with a constant 2α.

The attenuation constant in a number of materials is given in **Table 7.2**. The attenuation constant is itself dependent on temperature, but the most striking feature of the attenuation constant in materials is its dependency on frequency. In air it also depends on relative humidity and pressure, and is roughly proportional to f^2, especially at higher frequencies. Attenuation is typically given in decibels per kilometer (dB/km), decibels per meter (dB/m), or decibels per centimeter (dB/cm). Because of the complex nature of the attenuation constant and its dependency on many parameters, its values are typically available in tables. **Table 7.3** shows some of these and their dependency on frequency for air and other materials. In some cases the properties of sound waves are available in formulas, usually based on fitting to experimental data. For example, the attenuation constant in water can be calculated from the following formula:

$$\alpha_{\mathrm{water}} = 0.00217 f^2 \quad [\mathrm{dB/cm}], \tag{7.13}$$

where f is frequency (in megahertz [MHz]). Similar formulas with differing coefficients exist for other fluids, but unfortunately the behavior in other materials is not as simple. Also, the formula for water does not apply below about 1 MHz.

TABLE 7.2 ■ Typical attenuation constants for some representative materials

Material	Attenuation constant [dB/cm]	Frequency
Steel	0.429	10 MHz
Quartz	0.02	10 MHz
Rubber	3.127	300 kHz
Glass	0.173	10 MHz
PVC	0.3	350 kHz
Water	See **Equation (7.13)**	
Aluminum	0.27	10 MHz
Copper	0.45	1 MHz

TABLE 7.3 ■ Attenuation constant (in dB/cm) and its dependency on frequency

	1 kHz	10 kHz	100 kHz	1 MHz	5 MHz	10 MHz
Air	1.4×10^{-4}	1.9×10^{-3}	0.18	1.7	40	170
Water	See **Equation (7.13)**					
Aluminum				0.008	0.078	0.27
Quartz				0.002	0.01	0.02

TABLE 7.4 ■ Acoustic impedance of some materials

Material	Acoustic impedance [kg/m²/s]
Air	415
Fresh water	1.48×10^6
Muscle tissue	1.64×10^6
Fat	1.33×10^6
Bone	7.68×10^6
Rubber	1.54×10^6
Quartz	14.5×10^6
Rubber	1.74×10^6
Steel	45.4×10^6
Aluminum	17×10^6
Copper	42.5×10^6

Another example of well-established relations is the variation of the speed of sound in pure water as a function of temperature, given as an *n*th-order polynomial and designed for specific temperature ranges. The formulas range from second- to fifth-order polynomials in which the coefficients are calculated from experimental data. An example is the following:

$$c_{\text{water}} = 1405.03 + 4.624T - 0.0383T^2 \ \ [\text{m/s}], \tag{7.14}$$

where T is the temperature (in °C). The formula is designed for the normal range of temperatures of bodies of water (10°C–40°C), but it may be used beyond this range with increased error. An approximate formula for the dependency of the speed of sound in air on temperature also exists:

$$c_{\text{air}} = 331.4 + 0.6T \ \ [\text{m/s}]. \tag{7.15}$$

The waves also posses a property called wave impedance, although in the case of acoustic waves it is often called acoustic impedance. The **wave impedance** or **acoustic impedance** is the product of density (ρ) and velocity (c):

$$Z = \rho c \ \ [\text{kg/m}^2/\text{s}]. \tag{7.16}$$

Acoustic impedance is an important parameter of the material and is useful in a number of acoustic applications, including reflection and transmission of waves, and hence in the testing of materials and the detection of objects and conditions using ultrasound. In general, elastic materials have high acoustic impedance, whereas "soft" materials tend to have low acoustic impedance. The differences can be orders of magnitude, as can be seen in **Table 7.4**. For example, the acoustic impedance of air is 415 kg/m²/s, whereas that of steel is 4.54×10^7 kg/m²/s. These large differences affect the behavior of the acoustic waves and their usefulness in various application.

EXAMPLE 7.2 Tsunami detection system

A tsunami detection and warning system consists of a number of shore stations that detect earthquakes using essentially accelerometers. The system consists of a number of basic components, including the sensors themselves and the detection stations where the accelerometers are located. A number of sensors, located at fixed positions, detect earthquakes. They determine the strength and, by triangulation, the location (epicenter) of the earthquake. This provides the distance and the likelihood that a tsunami will be generated (based on strength, location, and depth). Then the system determines how much time will elapse until the tsunami arrives at various locations. This is based on the speed of propagation of compression waves. In the earth's crust, the speed of propagation of seismic waves is approximately 4 km/s, whereas in water it is 1.52 km/s. For this reason, as well as for practical reasons of installation, the detection of seismic waves is done on land. Tsunamis travel at speeds of up to 1000 km/h.

Suppose an earthquake occurs 250 km from a city located on the seashore. The earthquake is detected at a station in a different location, 700 km from the epicenter of the earthquake. If the detection system determines that a tsunami is likely, how long do people in the city have to evacuate before the tsunami hits?

Solution: To detect the earthquake requires a time t_1:

$$t_1 = \frac{700}{4} = 175 \text{ s},$$

which is approximately 3 min.

The tsunami requires a time t_2 to travel a distance of 250 km:

$$t_2 = \frac{250}{1000} = 0.25 \text{ h},$$

which is one-quarter of an hour, or 15 min. Since one needs 3 min for detection, the city has at most 12 min to prepare. This assumes of course that the warning is issued without delay. This is one reason that tsunamis are so dangerous—the time available for preparation and evacuation is typically short except at large distances from the epicenter.

EXAMPLE 7.3 Attenuation of acoustic waves in air

Acoustic waves attenuate in air at a rate that depends on a number of factors, including temperature, pressure, relative humidity, and the frequency of the wave. All of these have a significant effect on attenuation, but to understand the propagation of ultrasound in air we will look at the effect of frequency alone. The following data are available for sound propagation in air:

Attenuation at 1 kHz, 20°C, 1 atm at sea level, 60% relative humidity, 4.8 dB/km.

Attenuation at 40 kHz, 20°C, 1 atm at sea level, 60% relative humidity, 1300 dB/km.

Attenuation at 100 kHz, 20°C, 1 atm at sea level, 60% relative humidity, 3600 dB/km.

Given a sound wave amplitude (sound pressure) of 1 Pa, calculate the sound pressure at a distance $d = 100$ m from the source at the three frequencies.

Solution: Since the attenuation is given in decibels per kilometer (dB/km), we need first to convert it into neper per meter (Np/m) so we can use **Equation (7.12)**. To do so we write the following:

At 1 kHz,

$$4.8 \text{ dB/km} = \frac{4.8}{8.686} \text{ Np/km} = \frac{4.8}{8.686 \times 1000} = 5.526 \times 10^{-4} \text{ Np/m}.$$

At 40 kHz,

$$1300 \text{ dB/km} = \frac{1300}{8.686 \times 1000} = 0.1497 \text{ Np/m}.$$

At 100 kHz,

$$3600 \text{ dB/km} = \frac{3600}{8.686 \times 1000} = 0.4145 \text{ Np/m}.$$

With these, the amplitude at a distance d, which we denote as P_d, is written in terms of the source pressure P_0 as

$$P_d = P_0 e^{-\alpha d} \quad [\text{Pa}].$$

At 1 kHz,

$$P_d = 1e^{-5.526 \times 10^{-4} \times 100} = 0.9994 \text{ Pa}.$$

At 40 kHz,

$$P_d = 1e^{-0.1497 \times 100} = 3.15 \times 10^{-7} \text{ Pa}.$$

At 100 kHz,

$$P_d = 1e^{-0.4145 \times 100} = 9.96 \times 10^{-19} \text{ Pa}.$$

These results reveal that in air, ultrasound can only be used for short-range applications. Indeed, most ultrasound applications in air use either 24 kHz or 40 kHz and are intended for ranges of less than about 20 m. The lower the frequency, the longer the range. Clearly at 1 kHz the sound has attenuated very little. It is perhaps for this reason that the human voice has evolved to use low frequencies. Lower frequencies (i.e., those sounds below our own hearing limit [infrasound]) propagate for very long distances and are used by some animals (such as elephants and whales). It should be noted that in water and in solids the attenuation is much lower and sound waves can propagate for long distances (see, e.g., **Problem 7.7**).

When a propagating wave encounters a discontinuity in the unbounded space (an object such as a wall, a change in air pressure, etc.), part of the wave is reflected and part of it is transmitted through the discontinuity. Thus we say that **reflection** and **transmission** occur, and reflected and transmitted waves can propagate in directions other than that of the original wave. The transmitted wave is understood as being the **refraction** of the wave across the discontinuity. To simplify the discussion, we define a transmission coefficient and a reflection coefficient. In the simplest case, when the propagating wave impinges on the interface perpendicularly ($\theta_i = 0$ in **Figure 7.4**), propagating from material 1 into material 2, the reflection coefficient (R) and transmission coefficient (T) are defined as

$$R = \frac{Z_2 - Z_1}{Z_2 + Z_1}, \quad T = \frac{2Z_2}{Z_2 + Z_1}, \tag{7.17}$$

where Z_1 and Z_2 are the acoustic impedances of medium 1 and medium 2, respectively.

The reflection coefficient multiplied by the amplitude of the incident wave gives the amplitude of the reflected wave. The transmission coefficient multiplied by the

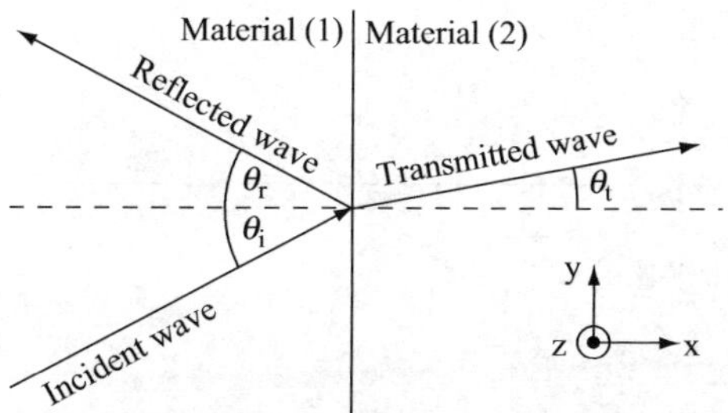

FIGURE 7.4 ▪ Reflection, transmission, and refraction of a wave.

amplitude of the incident wave gives the amplitude of the wave transmitted from medium 1 to medium 2. That is, the reflected and transmitted wave amplitudes (say, for pressure) are

$$P_r = P_i R \quad [\mathrm{N/m^2}], \quad P_t = P_i T \quad [\mathrm{N/m^2}], \tag{7.18}$$

where P_i is the incident pressure, P_r is the reflected pressure, and P_t is the transmitted pressure. Note that the reflection coefficient can be negative and varies from -1 to $+1$, whereas the transmission coefficient is always positive and varies from 0 to 2.

In acoustics, and in particular in ultrasonics, the quantities of interest are often power or energy rather than pressure. Since power and energy are related to pressure squared, the transmitted and reflected power or energy are related to the reflection and transmission coefficient squared. For example, assuming a collimated ultrasound beam with total incident power W_i, the reflected and transmitted power will be

$$W_r = W_i R^2 \quad [\mathrm{W}], \quad W_t = W_i T^2 \quad [\mathrm{W}]. \tag{7.19}$$

The refraction of the wave is defined in **Figure 7.4**. The reflected wave is reflected at an angle equal to the angle of incidence ($\theta_r = \theta_i$, defined between the direction of the propagating wave and the normal to the surface on which the wave reflects). The transmitted wave propagates in material 2 at an angle θ_t, which may be calculated from the following

$$\sin\theta_t = \frac{c_2}{c_1}\sin\theta_i, \tag{7.20}$$

where c_2 is the speed of propagation of the wave in the medium into which the wave transmits and c_1 is the speed in the medium from which the wave originates.

The reflected waves propagate in the same medium as the propagating wave and therefore can interfere with the propagating wave to the extent that their amplitude can add (constructive interference) or subtract (destructive interference). The net effect is that the total wave can have amplitudes smaller or larger than the original wave. This phenomenon is well known and leads to the idea of a standing wave. In particular, suppose that the wave is totally reflected so that the amplitudes of the reflected and incident waves are the same. This will cause some locations in space to have zero amplitudes, whereas others will have amplitudes twice as large as the incident wave. This is called a standing wave because the locations of zero amplitude (called nodes) are fixed in space, as are the locations of maxima. **Figure 7.5** shows this and also the fact that the nodes of the standing wave are at distances of $\lambda/2$, whereas maxima occur at $\lambda/4$ on either side of a node. A good example of standing waves can be seen in vibrating tight strings in which reflections occur at the locations where the strings are attached. The vibration at various

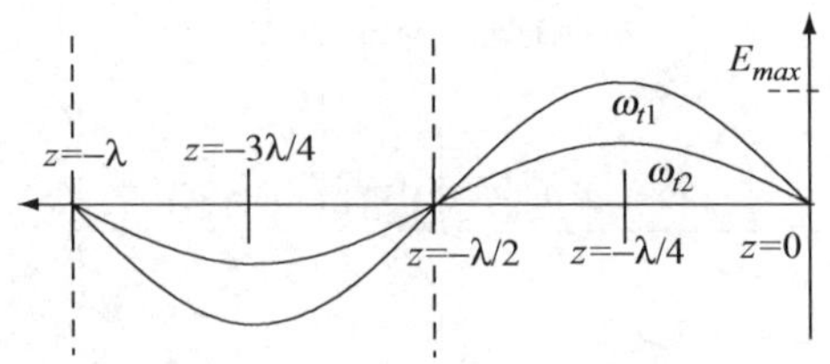

FIGURE 7.5 ■ Standing waves. The waves oscillate vertically (in time) but are stationary in space.

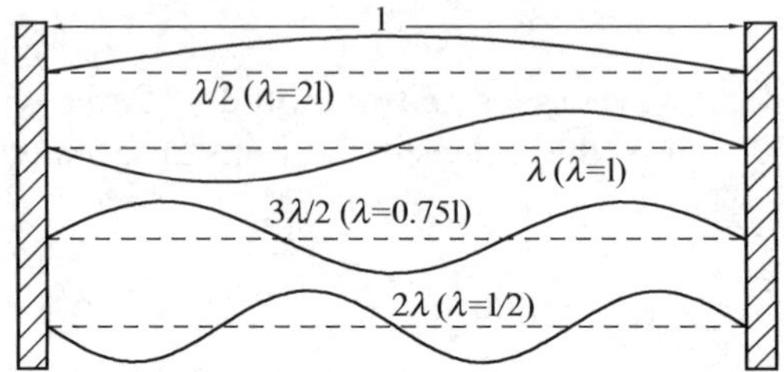

FIGURE 7.6 ■ Modes of a vibrating string. Note that the nodes of the standing wave are at equal distances from each other ($\lambda/2$) and occur at fixed locations for each mode.

wavelengths and its interaction with the air accounts for the music we perceive when a violin plays. **Figure 7.6** shows the first few modes of a vibrating string. For each mode, the nodes (zero displacement) occur at fixed physical locations.

The reflection of sound waves is also responsible for scattering. In essence, **scattering** is reflection of the waves in all directions caused by anything in the path of the waves. Dispersion of sound waves is another important property. **Dispersion** is the propagation of various frequency components at different speeds causing distortion in the received sound wave.

EXAMPLE 7.4 Properties of waves: resolution

Waves are often used for a variety of sensing and actuation functions. However, not all waves are the same and equally useful. The use of ultrasound for imaging is well established, including imaging of the body for medical purposes and use in the testing of materials. Animals as well use ultrasound in much the same way. Bats and dolphins use it for echolocation—for the identification of prey and avoidance of danger. But ultrasound is used in actuation as well. Dolphins stun fish with an intense ultrasound burst, whereas we use it to break up kidney stones, for ultrasonic cleaning, and for descaling of equipment. Electromagnetic waves, including light waves, are similarly used for imaging, as well as for echolocation, speed sensing, and a whole host of other applications. These functions are possible because of the interaction of the waves with materials, and one of the critical issues in this interaction is the wavelength. If the wavelength is long, the wave is useful in identifying large obstructions. The shorter the wavelength, the smaller the objects it can identify and hence the higher the resolution. Consider the following examples:

Ultrasound in air: A bat transmits ultrasound at 40 kHz in air. With the speed of sound in air equal to 331 m/s, the wavelength is (from **Equation (7.5)**)

$$\lambda = \frac{c}{f} = \frac{331}{40{,}000} = 8.275 \times 10^{-3}\ \text{m}.$$

This is a mere 8.275 mm, sufficiently small to hunt for insects.

Ultrasound in water: A dolphin transmits ultrasound at 24 kHz in water. With the speed of sound equal to 1500 m/s, the wavelength is

$$\lambda = \frac{c}{f} = \frac{1500}{24{,}000} = 62.5 \times 10^{-3}\ \text{m}.$$

At a wavelength of 62.5 mm, the dolphin is well equipped for fishing.

Imaging with ultrasound: An ultrasound wave at 2.75 MHz is used to monitor the condition of a human heart. Assuming the speed of sound to be the same as in water, the wavelength is

$$\lambda = \frac{c}{f} = \frac{1500}{2.75 \times 10^{6}} = 5.455 \times 10^{-4}\ \text{m}.$$

The test is capable of distinguishing features at submillimeter levels (less than 0.5 mm), sufficient for diagnostics of conditions such as deteriorating valves, blood vessel wall thickness, and more.

By way of comparison, the frequency of visible light varies between 480 THz (red) and 790 THz (violet). Its wavelength varies between 380 nm (violet) and 760 nm (red). The resolution possible with optical means is of that order of magnitude. Anything much smaller than that will not be seen using optical means (i.e., microscopes) and will require lower wavelengths (e.g., the use of electron microscopes).

7.3.2 Shear Waves

As mentioned above, solids can support shear or transverse waves in addition to longitudinal waves. In shear waves, the displacement (i.e., vibration of molecules) is perpendicular to the direction of propagation. Most of the properties defined for longitudinal waves, as well as properties such as reflection and transmission, are the same for shear waves. Other properties are different. In particular, the speed of propagation of shear waves is slower than for longitudinal waves. While the speed of propagation of longitudinal waves depends on the bulk modulus, that of shear waves depends on the shear modulus:

$$c = \sqrt{\frac{G}{\rho_0}}\quad [\text{m/s}]. \tag{7.21}$$

Since the shear modulus is lower than the bulk modulus, the speed of propagation of shear waves is lower (by about 50%).

The acoustic impedance in **Equation (7.16)** applies to shear waves as well, but since the speed is lower, so is the acoustic impedance.

7.3.3 Surface Waves

Acoustic waves can also propagate on the surface between two media and, in particular, at the interface between an elastic medium and vacuum (or air). This applies in particular to propagation on the surface of solids. Surface waves are also called Rayleigh waves and propagate on the surface of an elastic medium with little effect on the bulk of the medium and have properties that are significantly different than those of either longitudinal waves or shear waves. The most striking difference is their slower speed of propagation. The speed of propagation of a surface wave can be written as

$$c = g\sqrt{\frac{G}{\rho_0}} \quad [\text{m/s}], \tag{7.22}$$

where g is a constant that depends on the particular material but is around 0.9. This means that surface waves propagate slower than shear waves and propagate much slower than longitudinal waves.

In addition, the propagation of surface waves in ideal, elastic, flat surfaces is nondispersive, that is, their speed of propagation is independent of frequency. In reality there is some dispersion but it is lower than for other types of acoustic waves. This, combined with the slow speed of propagation, has found an important application in SAW devices. They also have uses in seismology and the study of earthquakes. The exact definition of a Rayleigh wave is "a wave that propagates at the interface between an elastic medium and vacuum or rarefied gas (air, for example) with little penetration into the bulk of the medium."

7.3.4 Lamb Waves

In addition to longitudinal, shear, and surface waves (sometimes designated as L-waves, P-waves, and S-waves, respectively), acoustic waves propagate in thin plates in a unique way dominated by modes of propagation that depend on the thickness of the plates. These are called Lamb waves (named after Horace Lamb). A plate will support an infinite number of modes that depend on the relationship between the thickness of the plate and the wavelength of the acoustic wave.

7.4 | MICROPHONES

We start the discussion on acoustic devices with the better known of these—audio sensors and actuators. Microphones and loudspeakers are familiar, at least to a certain extent. These are common devices, but like any other area in sensing and actuation, exhibit considerable variability in construction and applications. Microphones are differential pressure sensors where the output depends on the pressure difference between the front and back of a membrane. Since under normal conditions the two pressures are the same, the microphone can only sense changes in pressure and hence may be viewed as a dynamic pressure sensor. It may also be used to sense vibrations or any quantity that generates variations in pressure in air or in a fluid. Microphones designed to work in water or other fluids are called hydrophones.

7.4.1 The Carbon Microphone

The very first microphones and loudspeakers (or earphones) were devised and patented for use in telephones. In fact, the first patent of the telephone is not really a patent of a telephone, but rather, that of a microphone. Alexander Graham Bell patented the first variable resistance microphone in 1876, although in its early form it was a very inconvenient device. It was built as in **Figure 7.7** and used a liquid solution. The resistance

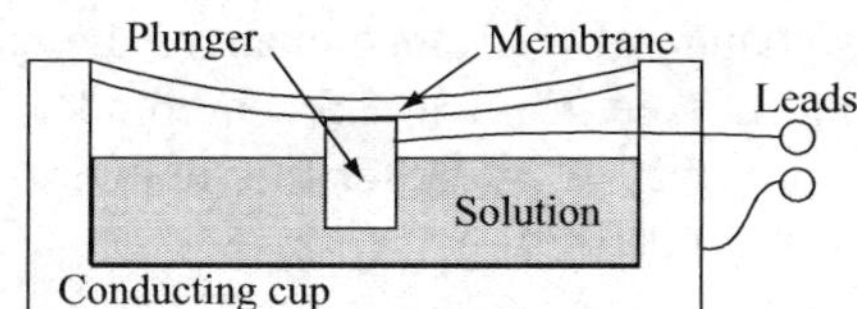

FIGURE 7.7 ■ Bell's microphone relied on changes in resistance in a solution.

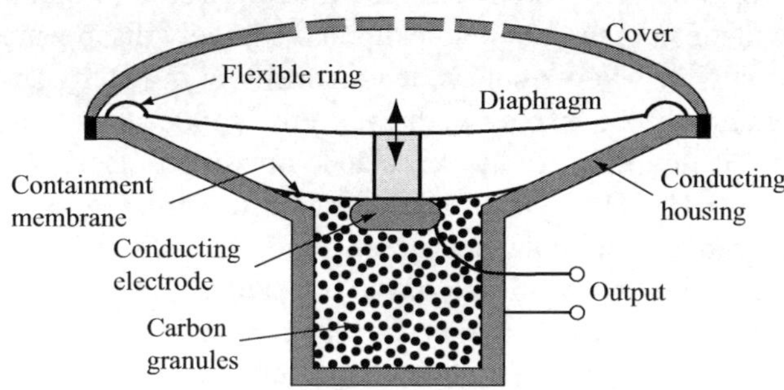

FIGURE 7.8 ■ The construction of the carbon microphone.

FIGURE 7.9 ■ The carbon microphone used in a telephone handset.

between the plunger and the body of the microphone depends on sound pressure (which pushes the plunger into the solution). This microphone worked, but was not practical and was soon replaced by others more suited for the job. The first practical microphone was invented by Thomas Edison and was essentially the same construction as Bell's microphone, but the solution was replaced with carbon or graphite particles—hence its name, the carbon microphone. Although it has many problems, it has been in continuous use in telephones ever since its invention. Because of its rather poor performance (noise, limited frequency response, dependence on position, and distortion), it has not been used since the late 1940s except in telephones. Nevertheless, it is a somewhat unique device, particularly because of the fact that it is an "amplifying" device (it can modulate large currents). In that capacity it was, and is still being used, to drive an earpiece directly without the need for an amplifier. Its structure is shown in **Figure 7.8** and a picture of a carbon microphone is shown in **Figure 7.9**. As the diaphragm moves, the resistance between the conducting electrode and the conducting housing changes, and when connected in a circuit, this change in resistance changes the current in the circuit, producing sound in the earpiece (see **Figure 1.5**). In modern telephones the carbon microphone has been largely replaced by better microphones, albeit microphones that require electronic circuits for amplification.

7.4.2 The Magnetic Microphone

The magnetic microphone, better known as the moving iron microphone, together with its cousin, the moving iron gramophone pickup, have largely disappeared and have been replaced by better devices. Nevertheless, it is worth looking at its structure since that structure is quite common in sensors (we have seen a similar device used as a pressure sensor in **Chapter 6** called the variable reluctance pressure sensor). The basic principle is shown in **Figure 7.10**. The operation is straightforward. As the armature moves (a piece of iron that moves due to the action of sound, or a needle in the case of a record pickup), it decreases the gap toward one of the poles of the iron core. This changes the reluctance in the magnetic circuit. If the coil is supplied with a constant voltage, the current in it depends on the reluctance of the circuit. Hence the current in the coil depends on the position of the armature (sound level). The moving iron microphone was a slight improvement over the carbon microphone, and perhaps the only real advantage it had was that its operation was reversible—the moving iron armature could be made to move under the influence of a current, and by doing so it could serve as an earpiece or loudspeaker. This microphone was quickly replaced by the so-called moving coil microphone, shown in **Figure 7.11**, also known as the dynamic microphone. This was the first microphone that could reproduce the whole range of the human voice, and it has survived into our time even though newer, simpler devices have been developed. The operation of the moving coil microphone is based on Faraday's law. Given a coil moving in a magnetic field, it will produce an emf as follows:

$$V = -N\frac{d\Phi}{dt} \quad [\mathrm{V}], \tag{7.23}$$

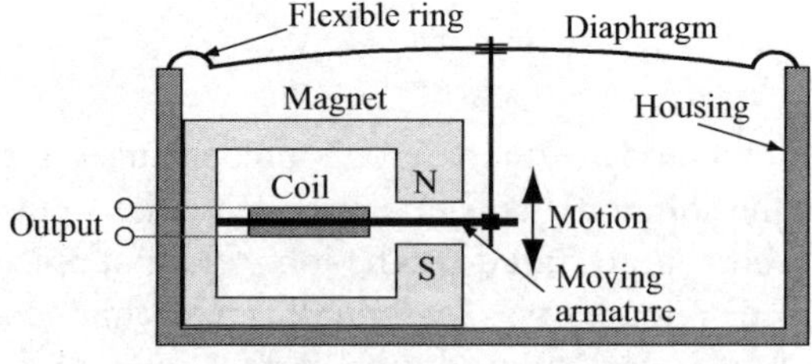

FIGURE 7.10 ■ The construction of a moving armature magnetic microphone.

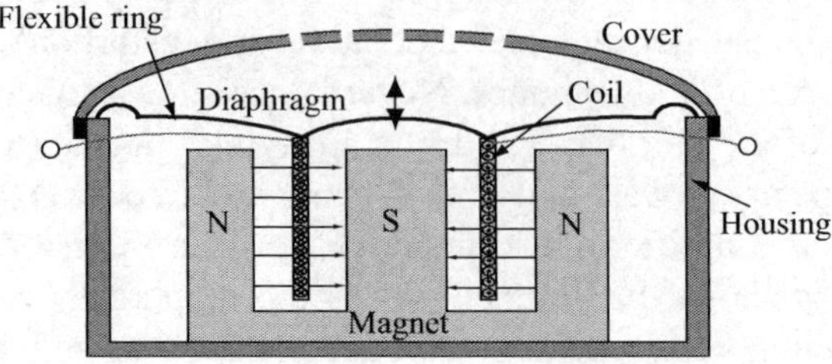

FIGURE 7.11 ■ A moving coil microphone.

where Φ is the flux in the coil and N is the number of turns. This relation also explains the term dynamic. One property that should be emphasized is that this is a passive device—it generates its own output and requires no source of power.

As the coil vibrates in the magnetic field, a voltage with appropriate polarity is generated that can then be amplified for audio reproduction. The emf, when connected in a circuit, will generate a current and both of these are proportional to the velocity of the coil. These microphones have excellent characteristics, with relatively low noise and high sensitivity. They can be connected directly to many low-input impedance amplifiers and are still in use today. Note also that the structure in **Figure 7.11** is not fundamentally different from that of a common loudspeaker or the voice coil actuator discussed in **Section 5.9.1** except that in microphones the structure is modified to increase the change in flux as the diaphragm moves and, of course, the dimensions are smaller. Therefore any small magnetic loudspeaker can serve as a dynamic microphone, and the dynamic microphone, just like the moving iron microphone, is a dual device capable of serving as a loudspeaker or earphone (with the appropriate changes in dimensions, coil size, etc.).

This also means that an alternative way of looking at the result in **Equation (7.23)** is to start with the motion of the coil in the magnetic field and the force on the charge q moving at a velocity $\mathbf{v}$ as given in **Equation (5.21)**:

$$\mathbf{F} = q\mathbf{v} \times \mathbf{B}. \tag{7.24}$$

From the fact that the force on a charge can always be written as $\mathbf{F} = q\mathbf{E}$, we conclude that $\mathbf{E} = \mathbf{v} \times \mathbf{B}$ is an electric field intensity. Integrating the field around the circumference of a loop of the coil and multiplying by the number of loops we get the emf produced in the coil:

$$emf = N\int_{loop} \mathbf{v} \times \mathbf{B} \cdot d\mathbf{l} \ [\text{V}]. \tag{7.25}$$

This emf relates to the velocity of the coil in the magnetic field but produces exactly the same result as **Equation (7.23)**.

EXAMPLE 7.5 The moving coil microphone

To understand the operation of a moving coil microphone it is sufficient to note that as the coil moves in and out of the magnetic field, the total flux through the coil changes and hence the emf induced in the coil changes. Exact calculation is not simple, as the motion of the coil under pressure depends on the mechanical properties of the diaphragm, the uniformity of the magnetic field, and the coil itself. But we can get an idea by assuming that the change in flux is proportional to the amplitude of sound and hence the position of the coil within the magnetic structure. Because of this, microphones are rated in terms of a sensitivity factor, k [mV/Pa]. For a pressure at amplitude P_0, the microphone's output emf is

$$emf = kP_0 \sin \omega t \quad [\text{mV}],$$

where $\omega = 2\pi f$ and f is the frequency of the pressure wave, indicating that only changes in signals can be detected. Sensitivities of 10–20 mV/Pa are common (much more sensitive microphones exist).

The limit of human auditory threshold is 2×10^{-5} Pa. At that pressure, a microphone with a sensitivity of 20 mV/Pa will produce an emf of 0.4 μV. This will likely be below the noise level, meaning that the signal is not usable at or near the threshold level. But at the normal speech level of about 2 Pa, the output would be 40 mV, a signal that can easily be amplified.

7.4.3 The Ribbon Microphone

Another microphone in the same class as moving armature and moving coil microphones is the ribbon microphone. This is shown in **Figure 7.12** and is a variation of the moving coil microphone. The ribbon is a thin metallic foil (aluminum, for example) between the two poles of a magnet. As the ribbon moves, an emf is induced across it based on Faraday's law in **Equation (7.23)**, except that in this case $N = 1$. The current produced by this emf is the output of the microphone. These simple microphones have wide, flat frequency responses because of the very small mass of the ribbon. However, the small mass also makes them susceptible to background noise and vibration, and often they require elaborate suspension to prevent these effects. Because of their qualities, they are often used in studio sound recordings. The impedance of these microphones is very low, typically less than 1 Ω, and they must be properly interfaced for operation with amplifiers.

7.4.4 Capacitive Microphones

Early on in the development of audio reproduction, in the early 1920s, it became apparent that the motion of a plate in a parallel plate capacitor could be used for this purpose and hence the introduction of the capacitive or "condenser" microphone (condenser is the old name for a capacitor). The basic structure in **Figure 7.13** can be

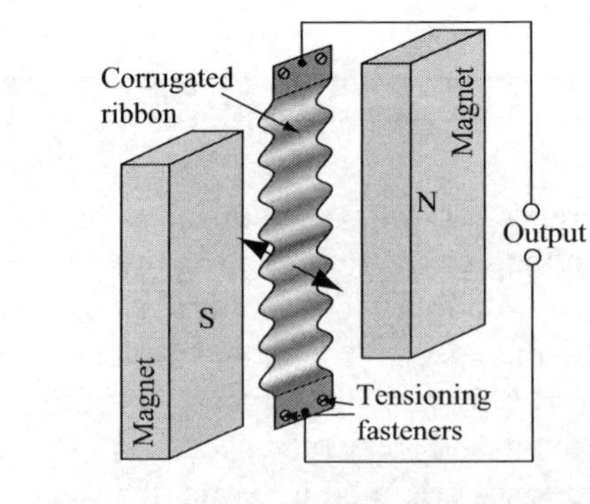

FIGURE 7.12 ■ The ribbon microphone.

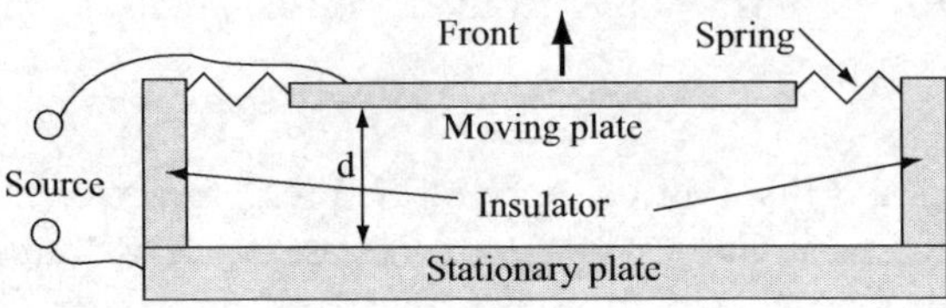

FIGURE 7.13 ■ The basic idea of a capacitive microphone.

used to understand the principle. The operation is based on the two basic equations of the parallel plate capacitor:

$$C = \frac{\varepsilon A}{d} \quad \text{and} \quad C = \frac{Q}{V} \rightarrow V = Q\frac{d}{\varepsilon A} \quad [\text{V}]. \tag{7.26}$$

This may look simple, but it also reveals a flaw in the whole idea of a simple parallel plate microphone: to produce an output voltage proportional to the distance d between the plates, a source of charge must be available. Sources of charge are not easy to come by except from external sources. Nevertheless, a solution has been found in the form of the **electret microphone**.

To understand what an electret is, it may be useful to consider first the idea of a permanent magnet. To produce a permanent magnet, a "hard" magnetic material, say, samarium-cobalt, is used and made into the shape needed. Then the material is magnetized by subjecting it to a very large external magnetic field. This moves the magnetic domains and sets up a permanent magnetization vector inside the material. When the external field is disconnected, the internal magnetization is retained by the material and that magnetization sets up the permanent magnetic field of the permanent magnet. One needs an equal or larger field to demagnetize it. An equivalent process can be done with an electric field. If a special material (it would be appropriate at this point to call it an electrically hard material) is exposed to an external magnetic field, a polarization of the atoms inside the material occurs. In these materials, when the external electric field is removed, the internal electric polarization vector is retained and this polarization vector sets up a permanent external electric field. Electrets are usually made by applying the electric field while the material is heated, to increase the atom energy and allow easier polarization. As the material cools, the polarized charges remain in this state. Materials used for this purpose are Teflon FEP (fluorinated ethylene propylene), barium titanate (BaTi), calcium titanate oxide ($CaTiO_3$), and many others. Some materials can be made into electrets by simply bombarding the material in its final shape by an electron beam.

The electret microphone is thus a capacitive microphone made of the same two conducting plates discussed above, but with a layer of an electret material under the upper plate, as shown in **Figure 7.14a**. The electret here is made of a thin film to provide the flexibility and motion necessary.

The electret generates a surface charge density σ_s on the upper metallic surface and an opposite sign surface charge density $-\sigma_s$ on the lower metal backplane. This generates an electric field intensity in the gap, s_1. The voltage across the two metallic plates, in the absence of any outside stimulus (sound), is

$$V = \frac{\sigma_s s_1}{\varepsilon_0 s + \varepsilon s_1} \quad [\text{V}]. \tag{7.27}$$

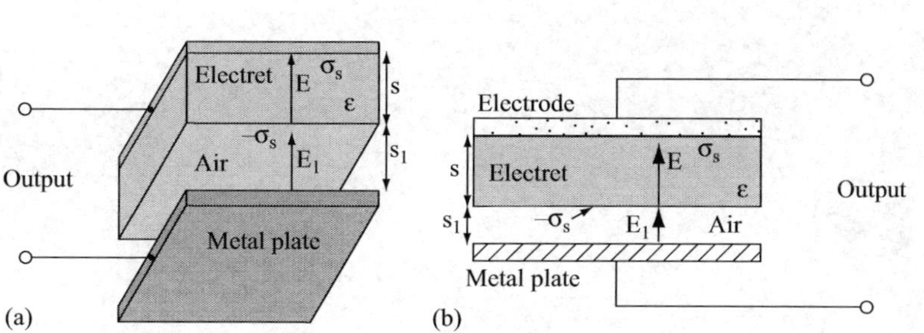

FIGURE 7.14 ■ The construction of an electret capacitive microphone.

If sound is applied to the diaphragm, the electret will move down a distance Δs and a change in voltage occurs as

$$\Delta V = \frac{\sigma_s \Delta s}{\varepsilon_0 s + \varepsilon s_1} \quad [\mathrm{V}]. \tag{7.28}$$

This voltage, which is the true output of the sensor, can be directly related to the sound pressure by first calculating the change in the gap length:

$$\Delta s = \frac{\Delta p}{\gamma p_0 s_1 + 8\pi T/A} \quad [\mathrm{m}], \tag{7.29}$$

where A is the area of the membrane, T is the tension, γ is the specific heat ratio of air, p_0 is the ambient pressure (or, in a more general sense, the pressure in the gap between the plate and the electret), and Δp is the change in pressure above ambient pressure due to sound. Thus the change in output voltage due to sound waves is obtained by substituting Δs in **Equation (7.28)**:

$$\Delta V = \frac{\sigma_s}{\varepsilon_0 s + \varepsilon s_1}\left(\frac{\Delta p}{\gamma p_0 s_1 + 8\pi T/A}\right) \quad [\mathrm{V}]. \tag{7.30}$$

This voltage can now be amplified as necessary.

Electret microphones are very popular because they are simple and do not require a source (they are passive devices). However, their impedance is very high and they require special circuits for connection to instruments. Typically a field-effect transistor (FET) preamplifier is required to match the high impedance of the microphone to the lower input impedance of the amplifier. In terms of construction, the membrane is typically made of a thin film of electret material on which a metal layer is deposited to form the movable plate.

In many ways the electret microphone is almost ideal. By proper choice of dimensions and materials, the frequency response can be totally flat from zero to a few megahertz. These microphones have very low distortions and excellent sensitivities (a few millivolts per microbar [mV/μbar]). Electret microphones are usually very small (some no more than 3 mm in diameter and about 3 mm long) and are inexpensive. Electret microphones can be found everywhere, from recording devices to cell phones. A sample of electret microphones is shown in **Figure 7.15**.

FIGURE 7.15 ■ Common electret microphones.

EXAMPLE 7.6 **The electret microphone: design considerations**

Consider the design of a small electret microphone for use in cellular phones, made in the form of a cylinder 6 mm in diameter and 3 mm long to fit in a slim telephone. Internally the designer has considerable flexibility as to the choice of materials and dimensions, as long as they fit within the external dimensions. Assuming the protective external structure requires a thickness of 0.5 mm, the diaphragm cannot be larger than 5 mm in diameter. The thickness of the diaphragm depends on the material used. Assuming a polymer, a reasonable thickness is 0.5 mm and a tension of 2 N can be easily supported by the structure. Polymers have relatively low permittivities, so we will assume a relative permittivity of 6. The gap between the electret and the lower conducting plate will be taken as 0.2 mm (the smaller the gap, the more sensitive the microphone). The ratio of specific heat in air is 1.4 (varies somewhat with temperature, but we will neglect that since the variation is rather small). The polymer can be charged at various levels, but the surface charge density cannot be very high. We will assume a charge density of 0.2 μC/m^2. With these values, and assuming an ambient pressure of 101,325 Pa (1 atm), we obtain a transfer function between the output voltage and the change in pressure using **Equation (7.30)**:

$$\Delta V = \frac{\sigma_s}{\varepsilon_0 s + \varepsilon s_1}\left(\frac{1}{\gamma p_0 s_1 + 8\pi T/A}\right)\Delta p.$$

In this relation, if p_0 is in pascals, Δp must also be in pascals. Numerically,

$$\Delta V = \frac{0.2\times 10^{-6}}{8.854\times 10^{-12}\times 0.5\times 10^{-3} + 6\times 8.854\times 10^{-12}\times 0.2\times 10^{-3}} \times\left(\frac{1}{1.4\times 0.2\times 10^{-3}\times 101{,}325 + 8\pi\times 2/(\pi\times(0.0025)^2)}\right)\Delta p$$
$$= 5.19\Delta p \quad [\mathrm{V}]$$

This is a sensitivity of 5.134 mV/Pa.

For a normal level of speech (45 dB to 70 dB) the pressure is 3.5×10^{-3} Pa to 6.3×10^{-2} Pa (see **Example 7.1**), producing an output of 18 to 327 mV, the microphone will produce a change in voltage of 10.268–20.536 mV, a level that is not only sufficient, but also rather high, for a microphone of the dimensions assumed here. Note, however, that this change is on top of the existing voltage given in **Equation (7.27)**.

The sensitivity can be improved by increasing the surface charge density or the area of the diaphragm, or decreasing the gap, the permittivity, the thickness of the electret, or the tension in the diaphragm.

7.5 | THE PIEZOELECTRIC EFFECT

The piezoelectric effect is the generation of electric charge in crystalline materials upon application of a mechanical stress. The opposite effect, often called electrostriction, is equally useful: the application of a charge across the crystal causes mechanical deformation in the material. The piezoelectric effect occurs naturally in materials such as quartz (silicon oxide) and has been used for many decades in so-called crystal oscillators. It is also a property of some ceramics and polymers, of which we have already met

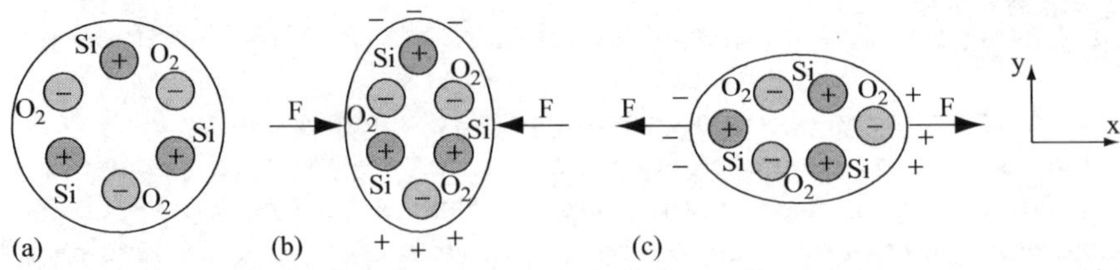

FIGURE 7.16 ■ The piezoelectric effect in a quartz crystal. (a) Undisturbed. (b) Strain applied in one direction. (c) Strain applied in the opposite direction.

the piezoresistive materials of **Chapter 5** (PZT is the best known) and the piezoresistive polymers PVF and PVDF. The piezoelectric effect has been known since 1880 and was first used in 1917 to detect and generate sound waves in water for the purpose of detecting submarines (sonar). The piezoelectric effect can be explained in a simple model by deformation of crystals. Starting with a neutral crystal (**Figure 7.16a**), a deformation in one direction (**Figure 7.16b**) displaces the molecular structure so that a net charge occurs as shown. In this case, the net charge on top is negative. Deformation on a perpendicular axis (**Figure 7.16c**) generates charges on the perpendicular axis. These charges can be collected on electrodes deposited on the crystal and measurement of the charge (or voltage) is then a measure of the displacement or deformation. This model uses the quartz crystal (SiO_2), but other piezoelectric materials behave in a similar manner. In addition, the behavior of the crystal depends on how the crystal is cut, and different cuts are used for different applications.

The polarization vector in a medium (polarization is the electric dipole moment of atoms per unit volume of the material) is related to stress through the following simple relation:

$$P = d\sigma \quad [\mathrm{C/m^2}], \tag{7.31}$$

where d is the **piezoelectric constant** and σ is the stress in the material. In reality, the polarization is direction dependent in the crystal and may be written as

$$P = P_{xx} + P_{yy} + P_{zz}, \tag{7.32}$$

where x, y, and z are the standard axes in the crystal. The relation above now becomes

$$P_{xx} = d_{11}\sigma_{xx} + d_{12}\sigma_{yy} + d_{13}\sigma_{zz}, \tag{7.33}$$

$$P_{yy} = d_{21}\sigma_{xx} + d_{22}\sigma_{yy} + d_{23}\sigma_{zz}, \tag{7.34}$$

$$P_{zz} = d_{31}\sigma_{xx} + d_{32}\sigma_{yy} + d_{33}\sigma_{zz}. \tag{7.35}$$

Now d_{ij} are the **piezoelectric coefficients** along the orthogonal axes of the crystal. Clearly then the coefficient depends on how the crystal is cut. To simplify the discussion, we will assume that d is single valued, but that depends on the type of piezoelectric material and how it is cut and excited (see the tables below for additional explanation of the indices). The inverse effect is written as

$$e = gP, \tag{7.36}$$

where e is strain (dimensionless) and g is the constant coefficient. The constant coefficient is related to the piezoelectric coefficient as

$$g = \frac{d}{e} \quad \text{or} \quad g_{ij} = \frac{d_{ij}}{e_{ij}}. \tag{7.37}$$

Normally the notation for stress is ε (see **Chapter 6**), but here the notation e is used to avoid confusion with permittivity, which is also denoted by ε. This relation also shows that the various coefficients are related to the electrical anisotropy of materials.

A third important coefficient is called the **electromechanical coupling coefficient** and is a measure of the efficiency of the electromechanical conversion:

$$k^2 = dgE \quad \text{or} \quad k_{ij}^2 = d_{ij}g_{ij}E_{ij}, \tag{7.38}$$

where E is the modulus of elasticity (Young's modulus). The electromechanical coupling coefficient is simply the ratio of the electrical and mechanical energies per unit volume. Some of these properties are listed in **Tables 7.5–7.7** for some crystals and ceramics often used in piezoelectric sensors and actuators. These tables also list some properties of polymers, materials that are becoming increasingly useful in piezoelectric (and piezoresistive) sensors.

Piezoelectric devices are often built as simple capacitors, as shown in **Figure 7.17**. Assuming that force is applied on the x-axis in this figure, the charge generated is

$$Q_x = d_{ij}F_x \quad [\text{C}]. \tag{7.39}$$

Taking the capacitance of the device to be C, the voltage developed across it is

$$V = \frac{Q_x}{C} = \frac{d_{ij}F_x}{C} = \frac{d_{ij}F_x d}{\varepsilon_{ij}A} \quad [\text{V}], \tag{7.40}$$

where d is the thickness of the piezoelectric material and A is its area. Thus the thicker the device, the larger the output voltage. A smaller area has the same effect. The output is directly proportional to force (or pressure). Note also that the pressure generates the

TABLE 7.5 ■ Piezoelectric coefficients and other properties in monocrystals

Crystal	Piezoelectric coefficient, d_{ij}, $\times 10^{-12}$ [C/N]	Relative permittivity, ε_{ij}	Coupling coefficient, k_{max}
Quartz (SiO_2)	$d_{11} = 2.31$, $d_{14} = 0.7$	$\varepsilon_{11} = 4.5$, $\varepsilon_{33} = 4.63$	0.1
ZnS	$d_{14} = 3.18$	$\varepsilon_{11} = 8.37$	0.1
CdS	$d_{15} = -14$, $d_{33} = 10.3$, $d_{31} = -5.2$	$\varepsilon_{11} = 9.35$, $\varepsilon_{33} = 10.3$	0.2
ZnO	$d_{15} = -12$, $d_{33} = 12$, $d_{31} = -4.7$	$\varepsilon_{11} = 8.2$	0.3
KDP (KH_2PO_4)	$d_{14} = 1.3$, $d_{36} = 21$	$\varepsilon_{11} = 42$, $\varepsilon_{33} = 21$	0.07
ADP ($NH_4H_2PO_4$)	$d_{14} = -1.5$, $d_{36} = 48$	$\varepsilon_{11} = 56$, $\varepsilon_{33} = 15.4$	0.1
$BaTiO_3$	$d_{15} = 400$, $d_{33} = 100$, $d_{31} = -35$	$\varepsilon_{11} = 3000$, $\varepsilon_{33} = 180$	0.6
$LiNbO_3$	$d_{31} = -1.3$, $d_{33} = 18$, $d_{22} = 20$, $d_{15} = 70$	$\varepsilon_{11} = 84$, $\varepsilon_{33} = 29$	0.68
$LiTaO_3$	$d_{31} = -3$, $d_{33} = 7$, $d_{22} = 7.5$, $d_{15} = 26$	$\varepsilon_{11} = 53$, $\varepsilon_{33} = 44$	0.47

TABLE 7.6 ■ Piezoelectric coefficients and other properties in ceramics

Ceramic	Piezoelectric coefficient, $d_{ij}, \times 10^{-12}$ [C/N]	Relative permittivity, ε	Coupling coefficient, k_{max}
$BaTiO_3$ (at 120°C)	$d_{15} = 260$, $d_{31} = -45$, $d_{33} = -100$	1400	0.2
$BaTiO_3 + 5\%CaTiO_3$ (at 105°C)	$d_{31} = 43$, $d_{33} = 77$	1200	0.25
$Pb(Zr0.53Ti0.47)O_3 + (0.5–3\%)$ La_2O_2 or Bi_2O_2 or Ta_2O_5 (at 290°C)	$d_{15} = 380$, $d_{31} = 119$, $d_{33} = 282$	1400	0.47
$(Pb0.6Ba0.4)Nb_2O_6$ (at 300°C)	$d_{31} = 67$, $d_{33} = 167$	1800	0.28
$(K0.5Na0.5)NbO_3$ (at 240°C)	$d_{31} = 49$, $d_{33} = 160$	420	0.45
PZT ($PbZr_{0.52}Ti_{0.48}O_3$)	$d_{15} = d_{24} = 584$, $d_{31} = d_{32} = -171$, $d_{33} = 374$	1730	0.46

TABLE 7.7 ■ Piezoelectric coefficients and other properties in polymers

Polymer	Piezoelectric coefficient, $d_{ij}, \times 10^{-12}$ [C/N]	Relative permittivity, ε [F/m]	Coupling coefficient, k_{max}
PVDF	$d_{31} = 23$, $d_{33} = -33$	106–113	0.14
Copolymer	$d_{31} = 11$, $d_{33} = -38$	65–75	0.28

Note: The indices *i,j* of the coefficients indicate the relation between input (force) and output (strain). Thus an index of 3,3 indicates that a force applied along the 3-axis produces a strain in that direction. An index 3,1 indicates a strain on the 1-axis when force is applied in the direction of the 3-axis of the crystal.

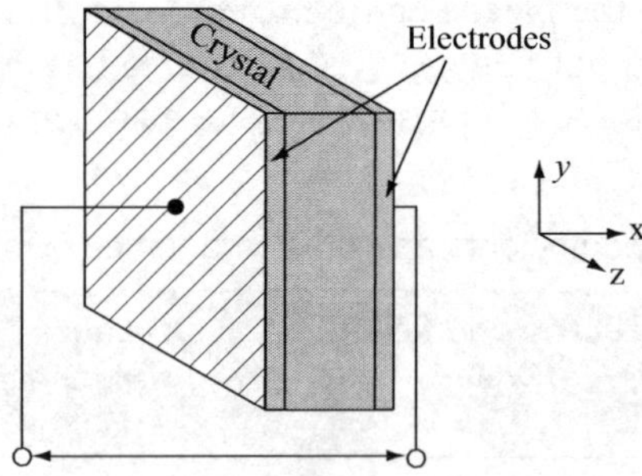

FIGURE 7.17 ■ The basic structure of a piezoelectric device.

stress in the material and hence the output may also be viewed as being proportional to stress or strain in the material. Piezoelectric sensors are often made of ceramics such as lead zirconite titanium oxide (PZT) and polymer films such as polyvinylidene fluoride (PVDF). Barium titanate ($BiTiO_3$) in crystal or ceramic form, as well as crystalline quartz, are also used for some applications.

One important development is the use of thin-film piezoelectric materials. Polymers are natural candidates for these films but they are fairly weak mechanically. Other materials such as PZT and zinc oxide (ZnO) are often used for this purpose because they have better mechanical and piezoelectric properties.

7.5.1 Electrostriction

It should be noted that the piezoelectric coefficient, with units of coulombs per newton (C/N) can also be viewed as having units of meters per volt (m/V) since 1 N/C = 1 V/m. This in turn may be viewed as follows:

$$d_{ij} \rightarrow \left[\frac{\mathrm{C}}{\mathrm{N}}\right] = \left[\frac{\mathrm{m}}{\mathrm{V}}\right] = \left[\frac{\mathrm{m/m}}{\mathrm{V/m}}\right] = \left[\frac{strain}{electric\ field\ intensity}\right].$$

Therefore the piezoelectric coefficient is strain developed per unit electric field intensity applied. This strain may be parallel or perpendicular to the applied force depending on the axes i,j involved. Note, however, that the strain produced per unit electric field intensity is small.

This gives rise to the electrostriction property. That is, when an electric field intensity is applied on a piezoelectric material, its dimensions change (strain). For example, the piezoelectric coefficient d_{33} for PZT is 374×10^{-12} C/N. That is, a sample of PZT, 1 m long will change its length by 374 pm per 1 V/m. This means that the electric field intensity must be high to produce any significant change in dimensions of the medium. Fortunately, very thin samples can be made and a large electric field intensity on the order of 1–2 million V/m can be applied to them, producing displacements on the order of hundreds of micrometers. In the example shown here, an electric field intensity of 2×10^6 V/m will produce a strain of 748 μm/m, or alternatively, 0.748 μm/mm. This is a reasonably large strain and is sufficient for many applications. Note again that this means that a 1 mm sample will increase in dimensions by 0.748 mm when an electric field intensity of 2×10^6 V/m is applied. To produce that electric field intensity across a 1 mm sample requires a voltage of 2000 V.

High voltages are typical in piezoelectric devices. For this reason, and to allow operation at more convenient, lower voltages, many applications use very thin samples. This is certainly the case for most electrostrictive actuators in which a voltage is applied to produce displacement. However, the piezoelectric effect can be used to generate high voltages by applying a strain. In this case the piezoelectric crystal should be thick to produce the required voltage (see **Problems 7.31** and **7.32**).

7.5.2 Piezoelectric Sensors

One of the most common piezoelectric sensors is the **piezoelectric microphone**, a device useful in both acoustic and ultrasound applications. The device in **Figure 7.17** can serve as a microphone by applying a force (due to sound pressure) on its surface. Given this structure, and a change in pressure Δp, the change in voltage expected (from **Equation (17.40)**) is

$$\Delta V = \frac{d_{ij}(\Delta pA)d}{\varepsilon_{ij}A} = \frac{d_{ij}d}{\varepsilon_{ij}}\Delta p \quad [\mathrm{V}]. \tag{7.41}$$

A linear relation is therefore available to sense the sound pressure. A common structure for the microphone is shown in **Figure 7.18**. The fact that capacitance is involved also indicates that piezoelectrics are high-impedance materials and thus require impedance matching networks.

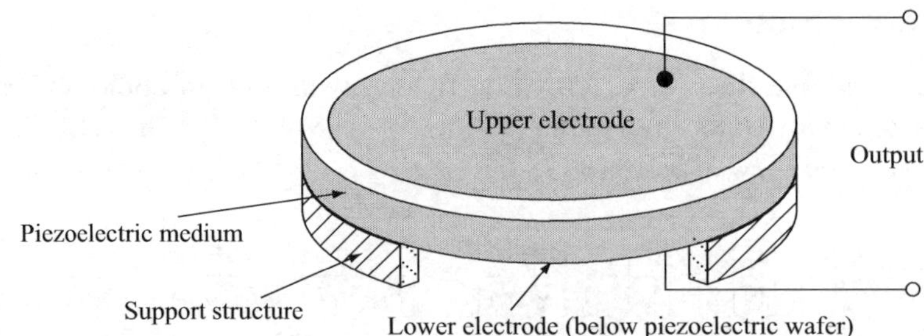

FIGURE 7.18 ■ The structure of a piezoelectric microphone.

One overriding property of these devices is that they can operate at high frequencies, hence their use as ultrasonic sensors. In addition, the piezoelectric microphone can be used as a piezoelectric actuator, and it is just as efficient. In other words, whereas there is a big difference between a magnetic (or capacitive) microphone and a loudspeaker, the piezoelectric microphone and piezoelectric actuator are essentially the same in all respects including dimensions. This complete duality is unique to piezoelectric transducers and, to a smaller extent, to magnetostrictive transducers.

Typical construction consists of films (PVDF or copolymers) with metal coatings for electrodes or disks of various piezoelectric crystals. These can be round, square, or almost any other shape. One particularly useful form is a tubelike electrode, which is usually used in hydrophones. These elements can be connected in series to cover a larger area, as is sometimes required in hydrophones.

The output of piezoelectric microphones is relatively low in the range of human speech because of the low pressures produced. Normal sensitivities are on the order of 10 μV/Pa, so in the range of normal speech one can expect voltages on the order of 100 μV, depending on the properties of the material involved and the distance of the microphone from the source of the sound.

Piezoelectric microphones have exceptional qualities and a flat frequency response. For this reason they are used in many applications, chief among them as pickups in musical instruments and for detection of low-intensity sounds such as the flow of blood in the veins. Other applications include voice-activated devices and hydrophones.

EXAMPLE 7.7 The piezoelectric microphone

A piezoelectric microphone is made of lithium titanate oxide ($LiTiO_3$) in the form of a disk that is 10 mm in diameter and 0.25 mm thick. Two electrodes, 8 mm in diameter, are coated on the opposite surfaces of the $LiTiO_3$ crystal. The crystal is cut on the 3-3 axis and is used to record speech at a distance of 1 m from a person. The sound pressure produced by normal speech at that distance is approximately 60 dB above the threshold of hearing. If the person were to shout, the sound pressure would increase to about 80 dB above the threshold of hearing. The threshold of hearing is 2×10^{-5} Pa taken at a reference of 0 dB. Calculate the range in voltages produced by the microphone under these conditions.

Solution: It is common to provide sound pressure in decibels rather than in pascals or newtons per square meter. However, we need the sound pressure in actual units so we can use **Equation (7.41)**. We therefore start by converting the given values using the relation

$$P(\text{dB}) = 20\log_{10}P \quad [\text{dB}].$$

However, because we need a zero reference at a pressure of 2×10^{-5} Pa, we must add this reference in decibels to any conversion. Thus we get

$$P_0 = 20\log_{10} 2\times10^{-5} = -94 \text{ dB}.$$

For normal level speech, $P = 60 - 94 = -34$ dB:

$$-34 \text{ dB} = 20\log_{10} P \rightarrow \log_{10} P = -1.7 \rightarrow P = 10^{-1.7} = 0.02 \text{ Pa}.$$

At an elevated level, $P = 80 - 94 = -14$ dB:

$$-14 \text{ dB} = 20\log_{10} P \rightarrow \log_{10} P = -0.7 \rightarrow P = 10^{-0.7} = 0.2 \text{ Pa}.$$

Since these pressure levels are above the ambient pressure, we can take them as changes in pressure due to speech. We use **Equation (7.41)** with the relative permittivity ε_{33}.

At normal speech,

$$\Delta V_l = \frac{d_{33}d}{\varepsilon_{33}}\Delta p = \frac{7\times10^{-12}\times0.25\times10^{-3}}{44\times8.854\times10^{-12}}0.02 = 89.84\times10^{-9} \text{ V}.$$

At an elevated level,

$$\Delta V_e = \frac{d_{33}d}{\varepsilon_{33}}\Delta p = \frac{7\times10^{-12}\times0.25\times10^{-3}}{44\times8.854\times10^{-12}}0.2 = 8.984\times10^{-7} \text{ V}.$$

The output of the microphone changes from 89.84 nV to 0.8984 μV as the voice rises from normal to shouting. This output is consistent with piezoelectric microphones that produce low outputs (because voice pressures are low).

7.6 | ACOUSTIC ACTUATORS

Among these we shall discuss two types. First is the classic loudspeaker as is used in audio reproduction. We have already discussed its basic properties in **Chapter 5** in the section on voice coil actuators. Here we will discuss other properties as these relate to the audio range. Second, we introduce the use of piezoelectric actuators for the purpose of sound generation. These devices, sometimes referred to as buzzers, are quite common in electronic equipment where audible signals (rather than voice or music) are needed. They are also much simpler, more rugged, and less expensive than classic loudspeakers. The issue of mechanical actuation using piezoelectric means will be discussed separately later in this chapter.

7.6.1 Loudspeakers

The basic structure of a loudspeaker driving mechanism is shown in **Figure 7.19**. The magnetic field in the gap is radial and acts on the coil (see **Figure 7.20**). For a

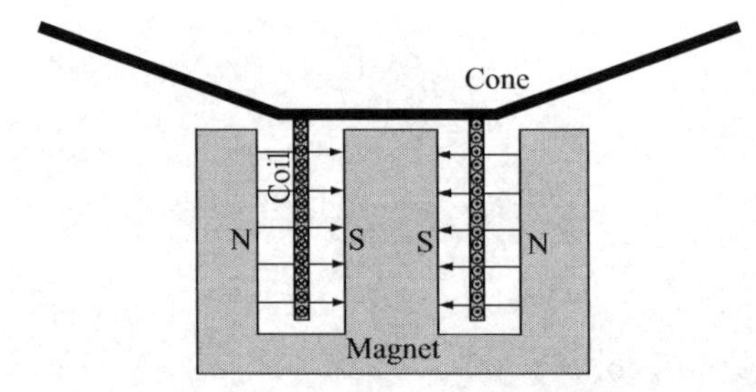

FIGURE 7.19 ■ Structure of a magnetic loudspeaker. The radial magnetic field is produced by a permanent magnet. The current in the coil is also shown.

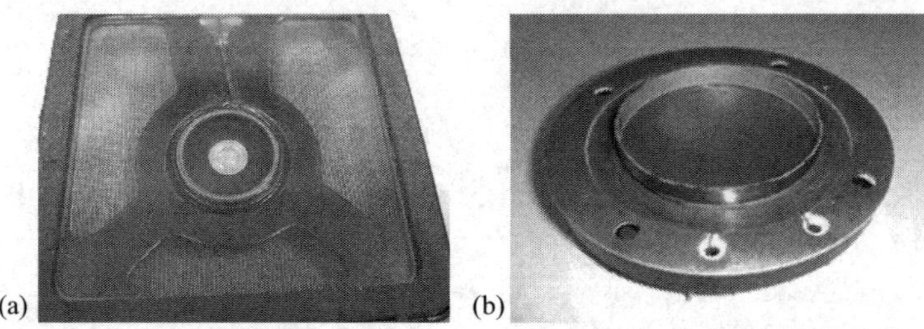

FIGURE 7.20 ■ (a) A small loudspeaker with a transparent cone (or diaphragm) showing the coil and the magnet at the center as well as the lead wires to the coil and the frame. (b) The coil of a public address loudspeaker. In this case the diaphragm is behind the coil and is made of titanium. The coil is wound on a short paper tube bonded to the diaphragm.

current-carrying loop, the force is given by the Lorentz force (see **Section 5.4**, and in particular **Equations (5.21)–(5.26)**, as well as **Section 5.9.1**). With N turns, the force is $NBIL$, where L is the circumference of the loop and we assume a uniform magnetic field. The field is only approximately uniform and the force is slightly nonlinear at the very end of travel of the coil, as was discussed in conjunction with voice coil actuators in **Section 5.9.1**. This is also the range in which most of the distortions occur.

Loudspeakers come in many varieties with various methods of construction, but as a rule the driving coil is round and the magnetic field in the gap is radial. Some old loudspeakers used electromagnets to generate the magnetic field, but all modern loudspeakers use permanent magnets for this purpose. The magnets are made as strong as possible and the gap as narrow as possible to ensure maximum force for a given current, thus reducing the power dissipated in the speaker. In most cases the coils are simple varnish-insulated copper wires wound tightly in a vertical spiral, usually in a single layer and supported by a backing of paper, Mylar, or fiberglass, depending on the speaker (see **Figure 7.20b**). The diaphragm supplies the restoring force and keeps the coil centered. The paper support for the coil and the method of leading the coils out of the device are shown as well. The cone is usually made of paper (in very small speakers they may be made of Mylar or some other reasonably stiff material; see **Figure 7.20a** and **Figure 7.22**) and is suspended on the rim of the speaker, which in turn is made as stiff as possible to avoid vibrations. The operation of a loudspeaker is essentially the motion of the coil in response to variations of current through it that in turn change the pressure in front of (and behind) the cone, thus generating a longitudinal wave in the air. The same principle can be used to generate waves in fluids or even in solids.

The power rating of a speaker is usually defined as the power in the coil, that is, the voltage across the coil multiplied by the current in the coil. This power can be specified as average or peak power, but it is not the power radiated by the cone. The radiated power is a portion of the total power supplied to the loudspeaker and is the difference between the total power and the dissipated power. The efficiency of loudspeakers is not particularly high.

The power handling capacity of the loudspeaker is the power the loudspeaker can handle without damage to its coil. The acoustic radiated power is quite different and depends on the electrical and mechanical properties of the loudspeaker. Assuming an unimpeded diaphragm connected to a coil of radius r and N turns in a magnetic field B, the radiated acoustic power is

$$P_r = \frac{2I^2B^2(2\pi rN)2Z}{R_{ml}^2 + X_{ml}^2} \quad [\mathrm{W}], \tag{7.42}$$

where Z is the acoustic impedance of air, R_{ml} is the total mechanical resistance, and X_{ml} is the total mass reactance seen by the diaphragm. However, these quantities are not easy to obtain and are often either estimated or measured for a particular loudspeaker, as they depend on the speaker and its construction. The radiated power may also be estimated from calculations of the magnetic force on the coil and the velocity of travel of the diaphragm in the magnetic field (see **Example 7.8**). However, this method is not very accurate since it does not take the mechanical properties of the loudspeaker into account.

A simplified approach to calculating the radiated power is based on the pressure generated by a piston of area A. Assuming uniform pressure across the area of the loudspeaker, the radiated acoustic power may be approximated as

$$P_{ac} = \frac{p^2{}_{ac}A}{Z} \quad [\mathrm{W}], \tag{7.43}$$

where P_{ac} is the pressure and A is the area of the loudspeaker (i.e., the circular area at the top of the cone, not the surface area of the cone). This relation may also be used to estimate the acoustic power in a buzzer, where the flat diaphragm is a better approximation of a piston.

These relations only give a rough idea of the power radiated. **Equation (7.42)** indicates that power is proportional to current, magnetic flux density, and the size (both physical and the number of turns) of the coil, whereas **Equation (7.43)** looks at power from a pressure point of view, which in turn is generated by the forces produced by the current. There are other issues that have to be taken into account, including reflections from the speaker's body, vibration of the structure, and damping due to the suspension of the cone, but the relations above are sufficient for a general understanding of radiated acoustic power.

EXAMPLE 7.8 Radiated and dissipated power in a loudspeaker

A loudspeaker is made as in **Figure 7.19**, with the following parameters: the coil is 60 mm in diameter, has 40 copper turns, and each turn is 0.5 mm in diameter, with a magnetic flux density produced by a permanent magnet equal to 0.85 T. The loudspeaker is fed with a sinusoidal current of amplitude 1 A at a frequency of 1 kHz. The coil and the diaphragm have a total mass of 25 g. Use an electric conductivity of 5.8×10^7 S/m for copper.

a. Estimate the power loss in the coil.

b. Estimate the radiated power of the loudspeaker.

c. Estimate the maximum travel of the diaphragm.

d. Discuss the approximations needed to get the results above.

Solution: The power loss can be calculated directly from the resistance of the wires (see part (d)). The power radiated by the loudspeaker is the mechanical power calculated from the product Fv, where F is the force and v is the velocity of the coil.

a. We will calculate here the DC resistance of the wires using a total length of wire:

$$L = 2\pi rn \quad [\mathrm{m}],$$

where r is the radius of the coil and n is the number of turns. The wire cross-sectional area S is

$$S = \pi \frac{d^2}{4} \quad [\mathrm{m}^2],$$

where d is the diameter of the wire. Given the conductivity of copper, the DC resistance of the coil is

$$R = \frac{L}{\sigma S} = \frac{2\pi rn}{\pi \frac{d^2}{4}} = \frac{8rn}{\sigma d^2} = \frac{8 \times 0.03 \times 40}{5.8 \times 10^7 \times (0.0005)^2} = 0.662\ \Omega.$$

To calculate the power dissipated we need the current. The current is sinusoidal at a frequency of 1 kHz:

$$I(t) = 1 \sin(2\pi \times 1000t) = 1 \sin(6283t) \quad [\mathrm{A}].$$

Power is an averaged value. Given the RMS value of the current is $I/\sqrt{2}$, where I is the amplitude (peak value) of the current, the power dissipated is

$$P = \frac{I^2 R}{2} = \frac{1 \times 0.662}{2} = 0.332\ \mathrm{W}.$$

b. To calculate power radiated we start by calculating the force the magnetic field exerts on the coil. We note that the magnet produces a uniform magnetic flux density in the gap. Therefore the loops of the coil are in a uniform radial field. Using **Equation (5.26)** for the force on a length of wire, the peak magnetic force on the coil is

$$F = B(nI)L = 2\pi rnBI = 2\pi \times 0.03 \times 40 \times 0.9 \times 1 = 6.786\ \mathrm{N},$$

where r is the radius of the coil, n is the number of turns, B is the magnetic flux density, and I is the current in the coil. This force moves the coil in or out depending on the phase of the current. We assume the speaker's diaphragm moves the coil in tandem with the current (otherwise the speaker cannot be expected to produce sound of any fidelity to the source that drives it). The time-dependent force is

$$F(t) = 6.786 \sin(6283t) \quad [\mathrm{N}]$$

From this we can calculate the acceleration of the coil:

$$F = ma \rightarrow a = \frac{F}{m} = \frac{6.786 \sin(6283t)}{30 \times 10^{-3}} = 226.2 \sin(6283t) \quad [\mathrm{m/s^2}].$$

We integrate the acceleration to obtain the velocity of the coil:

$$v(t) = \int a_0 \sin(\omega t) dt = -\frac{a_0}{\omega}\cos\omega t = -\frac{226.2}{6283}\cos(6283t) = 0.036\cos(6283t) \ \ [\text{m/s}].$$

Now we can write the instantaneous power as

$$P(t) = F(t)v(t) = 6.786\sin(6283t) \times 0.036\cos(6283t) = 0.244\sin(12566t) \ \ [\text{W}].$$

The averaged radiated power is half the amplitude of the instantaneous power:

$$P_{\text{avg}} = 0.122 \text{ W}.$$

Note: This power may not seem very high, but it is sufficient for normal listening. A higher power would necessitate a larger number of turns, a larger current, and/or a larger magnetic field. When these parameters are changed, the power dissipation changes as well. The efficiency of the loudspeaker shown here is about 73%, an excellent figure for loudspeakers.

c. The maximum displacement of the diaphragm is calculated from the equation of motion of the diaphragm. Since we know that the diaphragm is at its rest position when the current is zero (i.e., at $t = 0$), it reaches its positive maximum in ¼ cycle ($T/4$), where $T = 1$ ms. That is, the maximum force occurs after 1/4000 s. Thus we write the distance as

$$L = \int_{t=0}^{T/4} a\frac{t^2}{2} \ \ [\text{m}].$$

This would give an exact solution. A simpler way is to use the RMS value of acceleration and write

$$L = a_{\text{RMS}}\frac{t^2}{2} = \frac{226.2}{\sqrt{2}} \times \frac{(1/4000)^2}{2} = 5.7 \ \mu\text{m}.$$

The diaphragm moves only ± 5.7 μm at that frequency. The lower the frequency, the larger the displacement for the same force. The displacement obtained here may seem unreasonably small, but in fact the diaphragm only has 1 ms to move from rest to maximum displacement then to maximum displacement in the opposite direction and back to rest. For this reason smaller, lighter cones will respond to higher frequencies, whereas heavier cones will not. In the end, for any given speaker, its frequency response is to a large extent a function of this ability to follow the current.

d. We have made a number of assumptions, both explicit and implicit. The first is the use of DC resistance for the speaker. This is convenient because it is simple, but the AC resistance of conductors is frequency dependent and increases with frequency. Therefore the power loss we calculated is the minimum possible—essentially power loss at zero frequency. Second, we assumed a uniform magnetic flux density, which in actuality may not be uniform and may not be the same for loops close to the top of the magnet. More importantly, we have not taken into account mechanical issues such as forces needed to act against the restoring spring action that keeps the diaphragm at its starting position. In addition, any effects due to heat dissipation, such as a change in resistance of the coil with temperature, have been neglected.

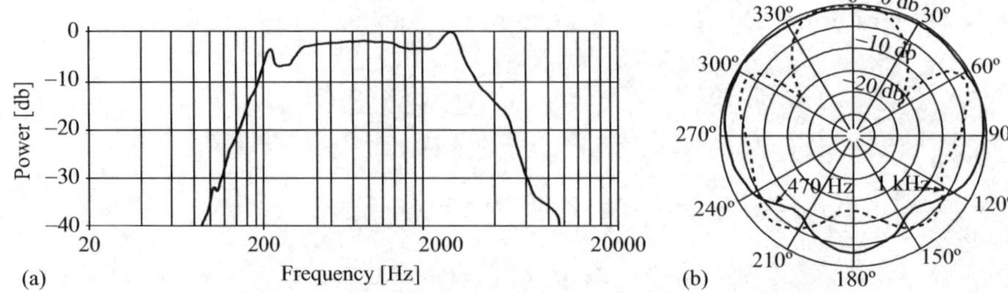

FIGURE 7.21 ■ Frequency response of a midrange loudspeaker. (a) Rectangular plot of power versus frequency. (b) Polar plot of normalized power at 470 Hz and 1 KHz.

In addition to radiated power and dissipated power, speakers are characterized by other properties such as dynamic range, maximum displacement of the diaphragm, and distortion. However, two other properties are of paramount importance. One is the frequency response of the speaker, the other its directional response (also called the radiation pattern or coverage pattern). The frequency response of a loudspeaker over its useful span is shown in **Figure 7.21a**. It shows power as a function of frequency in decibels, normalized to 1. This particular speaker shows a response between 90 Hz and 9 kHz with a bandwidth between about 200 Hz and 3.5 kHz (half power points). Also to be noted are the peaks or resonances at 220 Hz and 2.7 kHz. These are usually associated with the mechanical structure of the speaker. This speaker is obviously a general purpose speaker and others will have better responses at lower frequencies (woofers) or higher frequencies (tweeters), usually associated with the physical size of the speakers.

The directional response indicates the relative power density in different directions in space. **Figure 7.21b** shows such a plot at two frequencies indicating where in space one can expect larger or smaller power densities and the general coverage. In particular, note that the power density behind the speaker is lower than in front of it, as expected. When measuring the spatial response of loudspeakers the measured quantity may be pressure or, as in this case, power density. **Figures 7.22** and **7.23** show a number of speakers, some very small, some larger, but these only cover the "conventional" range. Many other types and shapes exist, some of them truly large.

FIGURE 7.22 ■ Some small loudspeakers. The smallest is 15 mm in diameter, the largest is 50 mm. The largest has a paper cone, whereas the others feature Mylar cones.

FIGURE 7.23 ■ A medium-size loudspeaker used for low-frequency reproduction (woofer). (a) View of the cone (front). (b) View of the cone (back) showing the magnet on top, the frame, and the connections. This loudspeaker is 16 cm in diameter.

7.6.2 Headphones and Buzzers

The loudspeakers in **Figures 7.20a**, **7.22**, and **7.23** represent the common structures of loudspeakers. Instead of moving the coil, one can conceive the opposite—moving the magnet while keeping the coil fixed. An adaptation of this idea is the moving diaphragm actuator shown in **Figure 7.24**. This is not used in loudspeakers, but has been used in the past in headphones and is in use today as earpieces in land telephones and in magnetic warning devices called buzzers. These magnetic actuators come in two basic varieties. One is simply a coil and a suspended membrane, as in **Figure 7.24**. Current in the coil attracts the membrane and variations in current move it back and forth with respect to the coil depending on the magnitude and direction of the current. A permanent magnet may also be present, as shown, to bias the device and keep the membrane in place. In this form the device acts as a small loudspeaker, but of a fairly inferior quality. It does have one advantage over conventional loudspeakers, especially in its use in telephones: because the coil is fairly large (many turns), its impedance is relatively high, so it can be connected directly in a circuit and driven by a carbon microphone without the need for an amplifier. However, for all other sound reproduction systems, it is not acceptable. Instead, modern magnetic headphones use small loudspeakers for much better sound quality.

7.6.2.1 The Magnetic Buzzer

The magnetic earpiece, mentioned above, has evolved into the modern magnetic buzzer. In this form, sound reproduction is not important, but rather the membrane is made to vibrate at a fixed frequency, say 1 kHz, to provide an audible warning for circuits, machinery, fire alarms, and the like. This can be done by driving the basic circuit in **Figure 7.24** with a square wave, usually directly from the output of a microprocessor or

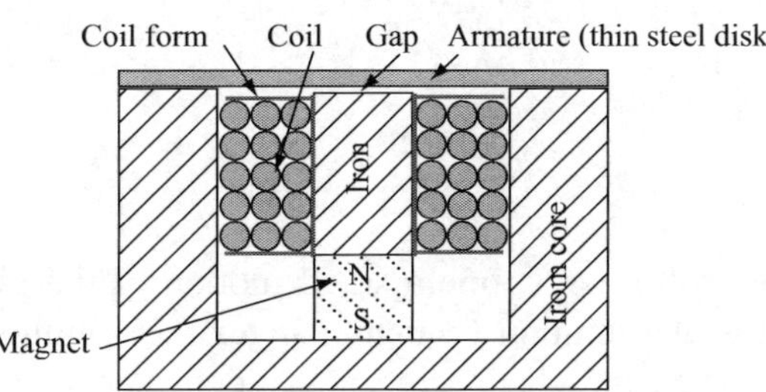

FIGURE 7.24 ■ The moving armature actuator: the buzzer.

(a) (b)

FIGURE 7.25 ■ (a) A World War II era earpiece made as a moving diaphragm element. The idea has survived in the modern magnetic buzzer. (b) Two modern magnetic buzzers based on the same principles. The one on the left is 12 mm in diameter, the one on the right is 15 mm.

through a suitable oscillator. In some devices the circuitry necessary for oscillation is internal to the device and the only external connections are to power. **Figure 7.25a** shows a World War II era earpiece showing the magnetic yoke at the center, the steel diaphragm, and the cover. **Figure 7.25b** shows two modern magnetic buzzers, 12 and 15 mm in diameter, respectively, based on the same basic structure. In the structure in **Figure 7.24**, given a coil of N turns carrying a current I, the magnetic flux density in the gap between the coil and the diaphragm can be approximated as

$$B = \frac{\mu_0 NI}{d} \quad [\mathrm{T}], \tag{7.44}$$

where d is the gap length and μ_0 is the permeability of air. For this approximation to be valid the permeability of the iron structure and the diaphragm must be large. This field generates a force on the diaphragm (see **Example 7.9**) forcing it to move. If, for example, the current were sinusoidal or a square wave, the diaphragm would move back and forth at the frequency of the signal, generating a pressure wave at that frequency. It is this property that makes the device useful as a warning device or as a means of producing simple sounds (such as the optional clicks used as feedback when typing on a keypad). The force on the diaphragm can be approximated by first calculating the energy per unit volume in the gap. This was calculated in **Equation (5.62)** as magnetic energy:

$$w_m = \frac{B^2}{2\mu_0} \quad [\mathrm{J/m^3}]. \tag{7.45}$$

Now we use the fact that a force F, moving the armature a distance dl, produces a change in volume dv and a change in work $dW_m = Fdl$:

$$dW_m = Fdl \quad [\mathrm{J}]. \tag{7.46}$$

The force becomes

$$F = \frac{dW_m}{dl} \quad [\mathrm{J}]. \tag{7.47}$$

For this relation to be useful, we define a small motion of the plate and calculate the change in energy due to that motion (which changes the volume in which the energy density exists by dv). Then the change in energy per change in distance due to that

FIGURE 7.26 ■ The piezoelectric diaphragm earpiece. The piezoelectric disk is shown in the center of the diaphragm.

motion gives the force. This method is called the virtual displacement method and is a common method of calculating forces in magnetic circuits (see **Example 7.9**).

7.6.2.2 The Piezoelectric Headphone and Piezoelectric Buzzer

Both the headphone and the buzzer also exist as piezoelectric devices in which a piezoelectric disk is physically bonded to a diaphragm. The piezoelectric element is a disk, as shown in **Figure 7.18**, and connection to a voltage source will cause a mechanical motion in the disk. When an AC source is applied, the variations in motion of the disk generate a sound at the applied source's frequency. An earphone of this type is shown in **Figure 7.26** together with its piezoelectric element, seen as the smaller disk at the center of the diaphragm.

The earpiece in **Figure 7.26** can be used as a buzzer by driving it with an AC source. However, for incorporation in an electronic circuit, these devices often come either as a device with a third connection, which when appropriately driven forces the diaphragm to oscillate at a fixed frequency, or has the necessary circuit to do so incorporated in the device. **Figure 7.27** shows a piezoelectric buzzer and, separately, its diaphragm shown from underneath. The piezoelectric element has two sections. A large circular section and a smaller finger-shaped section. The latter, when properly driven, causes local distortion in the diaphragm and the interaction of these distortions and those of the main element

FIGURE 7.27 ■ A piezoelectric buzzer showing the structure and the diaphragm with the piezoelectric disk on it.

FIGURE 7.28 ■ Piezoelectric buzzers of various sizes (13 mm–28 mm).

cause the device to oscillate at a set frequency that depends on the sizes and shapes of the two piezoelectric elements. These buzzers are very popular since they use little power, can operate down to about 1.5 V, and are rather loud, making them useful as directly driven devices in microprocessors. A device like this can be used for audible feedback or as a warning device (e.g., for a moving robot or as a backup warning in trucks and heavy equipment). **Figure 7.28** shows a number of piezoelectric buzzers of different sizes.

EXAMPLE 7.9 Pressure generated by a magnetic buzzer

A magnetic buzzer is made as shown in **Figure 7.29**. The structure is circular with an outer radius $a = 12.5$ mm and an inner radius $b = 11$ mm. The inner cylinder supporting the coil has a radius $c = 12$ mm. Assume the whole structure, including the diaphragm, is made of a high-permeability material so that any magnetic field generated by the coil is contained within the structure and the gap between the coil and the diaphragm. The gap is $d = 1$ mm. Given a coil with $N = 400$ turns and a current I at 1 kHz and an amplitude of 200 mA, calculate the maximum pressure generated by the diaphragm. Neglect mechanical losses in the system.

Solution: Since the configuration described here is essentially that of **Figure 7.24**, the magnetic flux density in the gap is (from **Equation (7.44)**):

$$B = \frac{\mu_0 NI}{d} \quad [\mathrm{T}].$$

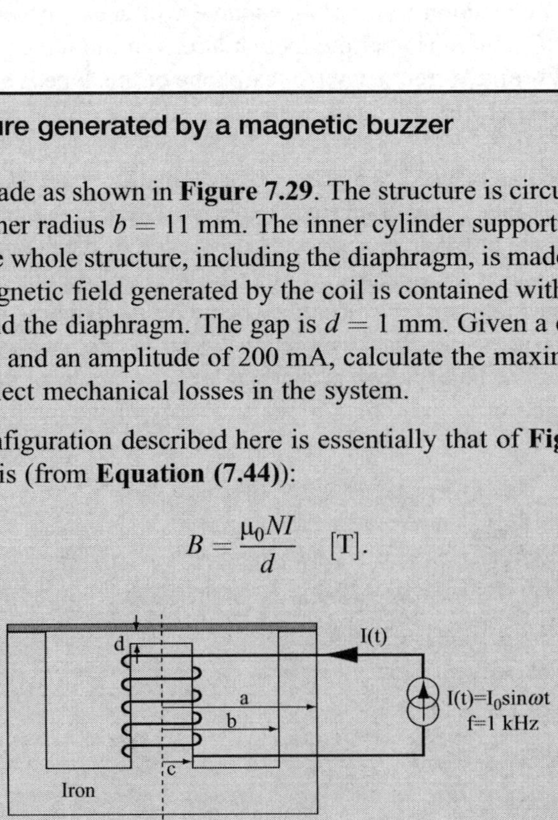

FIGURE 7.29 ■ Structure and dimensions of a magnetic buzzer.

The energy density in the gap between the core and diaphragm is

$$w_m = \frac{B^2}{2\mu_0} = \frac{\mu_0 N^2 I^2}{2d^2} \quad [\mathrm{J/m^3}].$$

Now suppose the diaphragm moves a very small distance dx so that it either reduces the gap or increases the gap. The change in energy in the gap is

$$dW = dw_m S dx = \frac{\mu_0 N^2 I^2}{2d^2} S dx \quad [\mathrm{J}],$$

where S is the cross-sectional area of the coil core. The latter is πc^2. Thus the force may be written as

$$F = \frac{dW}{dx} = \frac{\mu_0 N^2 I^2}{2d^2} \pi c^2 \quad [\mathrm{N}].$$

Since this is the force acting on the diaphragm, the pressure produced must be the force divided by the area of the diaphragm, πb^2:

$$P = \frac{F}{\pi b^2} = \frac{\mu_0 N^2 I^2 c^2}{2d^2 b^2} \quad [\mathrm{N/m^2}].$$

This is better defined as a change in pressure to indicate that it is above (or below) the ambient pressure. It should also be noted that this is dynamic pressure, that is, it only exists while the diaphragm is moving (and hence sound is produced only during that time). Once the diaphragm settles into a fixed position (such as if we apply DC instead of AC) the pressure is the ambient pressure and no sound is produced.

For the values given,

$$P = \frac{\mu_0 N^2 I^2 c^2}{2d^2 b^2} = \frac{4\pi \times 10^{-7} \times 400^2 \times 0.2^2 \times 0.006^2}{2 \times 0.001^2 \times 0.011^2} = 1196.4\ \mathrm{N/m^2}.$$

That is, 1196.4 Pa. Since buzzers are often rated in decibels, we calculate the rating of this buzzer as follows:

$$20 \log_{10} \frac{1194.6}{2 \times 10^{-5}} = 155.5\ \mathrm{dB}.$$

In other words, a very loud sound, sure to get one's attention since it is above the threshold of pain. Recall that normal speech is on the order of a few pascals or about 50 dB. Note, however, that this is the sound level at the diaphragm. At a distance from the diaphragm the sound level is reduced by attenuation and by spreading of the acoustic power.

7.7 | ULTRASONIC SENSORS AND ACTUATORS: TRANSDUCERS

Ultrasonic sensors and actuators are, in principles of operation, identical to the acoustic sensors and actuators discussed above, but they are somewhat different in construction and very different in terms of materials used and their range of frequencies. However, because the ultrasonic range starts where the audible range ends, the two, in effect, overlap. It is therefore quite reasonable to assume that an ultrasonic sensor (i.e., microphone or, as is more often the nomenclature, transducer or receiver) or actuator for the near ultrasound range should be quite similar to an acoustic sensor or actuator. In fact they are, at least at first glance. **Figure 7.30** shows an ultrasonic transmitter (left) and an ultrasonic receiver (right) designed for operation in air at 24 kHz. The first thing to note is that the two are of the same size and essentially the same construction. This is typical of piezoelectric devices, in which the same exact device can be used for both purposes, as explained above. Both use an identical piezoelectric disk similar to the one in **Figure 7.18**. The only visible difference is in the slight difference in the construction of the cone. **Figure 7.31** shows a closer view of another device, this time operating at 40 kHz, also designed to operate in air, in which the piezoelectric device is square, seen at the center below the brass supporting member (the second connecting wire is underneath). These devices operate exactly as microphones and as speakers.

These ultrasonic sensors are very common in applications in air (typical frequencies are 24 kHz and 40 kHz) for range finding and obstacle avoidance in robots. Other

FIGURE 7.30 ■ An ultrasonic transmitter (left) and an ultrasonic receiver (right) designed to operate in air at 24 kHz.

FIGURE 7.31 ■ A 40 kHz ultrasonic sensor (transducer) for operation in air.

applications are for presence detection in alarm systems and for safety in cars, where they are used for intrusion alarms and for collision avoidance when backing up. In some of these applications, higher frequencies are often used. The main difficulty with the propagation of ultrasound in air is that the attenuation of high-frequency ultrasound in air is high, so that the range of these devices is relatively short. On the other hand, the use of ultrasound is very attractive both because it is relatively simple (at these low frequencies) and because ultrasound, just like sound, tends to spread and cover a relatively large area. At higher frequencies the propagation can be much more direct and focused, but at the low frequencies used for ultrasound in air it is not. **Figure 7.32** shows a transmitter–receiver pair typically used for distance ranging in robots.

The scope of ultrasonic sensing is much wider than what is implied by the previous paragraphs. Its use for sensing and actuation is much more common and perhaps more important in materials other than air and at higher frequencies. In particular, when viewing **Table 7.1**, it is clear that ultrasound is better suited for use in solids and liquids where ultrasound propagates at higher velocities and lower attenuation. Also, solids support waves other than longitudinal, a property that allows additional flexibility in the use of ultrasonic waves: shear waves (these are transverse waves that can only exist in solids) and surface waves are two types often used in addition to longitudinal waves (see **Section 7.3**).

Ultrasonic sensors exist at almost any frequency and can certainly be made in frequencies exceeding 1 GHz. For practical applications, most sensors operate below 50 MHz, but a unique class of sensors based on SAW principles uses higher frequency to achieve a number of sensing and actuation functions. Most ultrasonic sensors and actuators are based on piezoelectric materials, but some are based on magnetostrictive materials since, in effect, what is needed is a means of converting an electrical signal into strain (for a transmitter) or strain into an electrical signal (receiver).

One particularly important property of piezoelectric materials that makes them indispensable in the design of ultrasonic sensors and actuators is their ability to oscillate at a fixed, sharply defined resonant frequency. The resonant frequency of a piezoelectric crystal (or ceramic element) depends on the material itself and its effective mass, strain, and physical dimensions, and is also influenced by temperature, pressure, and other environmental conditions such as humidity. To understand resonance, it is useful to look at the equivalent circuit of a piezoelectric device sandwiched between two electrodes, shown in **Figure 7.33a**. This circuit has two resonances: a parallel resonance and a series

FIGURE 7.32 ■ A 40 kHz transmitter–receiver pair for distance ranging in robots.

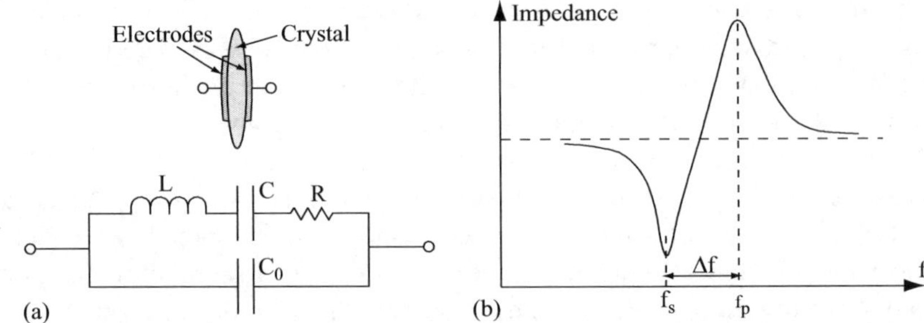

FIGURE 7.33 ■ The piezoelectric resonator. (a) The resonator and its equivalent circuit. (b) The two resonances.

resonance (also called antiresonance), shown in **Figure 7.33b**. These two resonant frequencies are given as

$$f_s = \frac{1}{2\pi\sqrt{LC}} \quad \text{[Hz] (series resonance)} \tag{7.48}$$

and

$$f_p = \frac{1}{2\pi\sqrt{LC[C_0/(C + C_0)]}} \quad \text{[Hz] (parallel resonance).} \tag{7.49}$$

In most applications a single resonance is desirable, and for these applications materials or geometries for which the two resonant frequencies are widely separated are used. To identify the frequency separation between the two resonances, a capacitance ratio is defined as

$$m = \frac{C}{C_0}. \tag{7.50}$$

With this, the relation between the two frequencies becomes

$$f_p = f_s(1 + m) \quad \text{[Hz]} \tag{7.51}$$

Thus the larger the ratio m, the larger the separation between the two resonant frequencies.

The resistance R in the equivalent circuit does not figure in the resonance, but acts as a damping (loss) factor. This is associated with the quality factor (Q-factor) of the piezoelectric device, given as

$$Q = \frac{1}{R}\sqrt{\frac{L}{C}} \quad \text{[C].} \tag{7.52}$$

The Q-factor tends to infinity for zero resistance and is, by definition, the ratio between stored and dissipated energy in the crystal.

The importance of resonance is two-fold. First, at resonance, the amplitude of mechanical distortion is highest (in transmit mode), whereas in receive mode, the signal generated is largest, meaning that the sensor is most efficient at resonance. The second reason is that the sensors operate at clear and sharp frequencies and hence the parameters of propagation, including reflections and transmissions, are clearly defined, as are other properties such as wavelength.

The construction of a piezoelectric transducer intended for operation in solids or liquids is shown in **Figure 7.34**. The piezoelectric element is rigidly attached to the front of the sensor so that vibrations can be transmitted to and from the sensor. The front of the sensor is often just a thin flat metal surface, or it may be prismatic, conical, or spherical to focus the acoustic energy. **Figure 7.34a** shows a flat, nonfocusing coupling element. **Figure 7.34b** shows a concave, focusing coupling element. The damping chamber

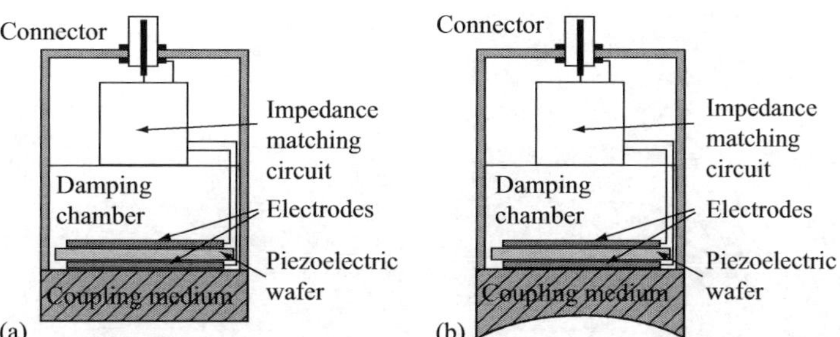

FIGURE 7.34 ■ Construction of an ultrasonic sensor. (a) Flat, nonfocusing sensor. (b) Concave, focusing sensor.

prevents ringing of the device, while the impedance matching circuit (not always present, sometimes it is part of the driving supply) matches the source with the piezoelectric element. Every sensor is specified for a resonant frequency, power, and for operational environment (solids, fluids, air, harsh environments, etc.). **Figure 7.35** shows a number of ultrasonic sensors for various applications and operating at various frequencies.

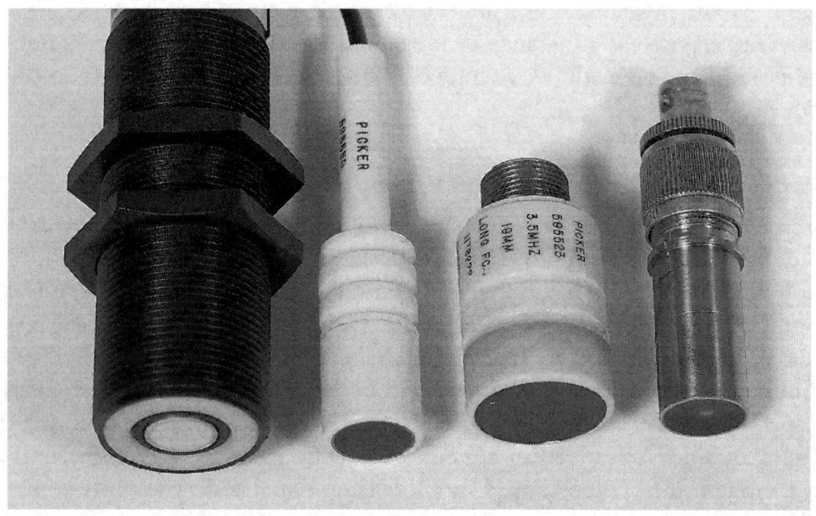

FIGURE 7.35 ■ A number of ultrasonic sensors. (Left to right) An industrial ultrasonic sensor operating at 175 kHz, a medical grade sensor operating at 2.25 MHz, an immersible sensor operating at 3.5 MHz, and a 15 MHz sensor with focusing lens used for testing of materials.

7.7.1 Pulse-Echo Operation

All ultrasonic sensors are dual—they can transmit or receive. In many applications, as, for example, in range finding, one can use two sensors (see **Figure 7.32**). In other applications the same sensor is used to transmit and to receive by switching between transmit and receive modes. That is, the sensor is driven to transmit an ultrasonic burst and then switched into receive mode to receive the echo reflected from any object the sound beam encounters. This is a common mode for operation in medical applications and in the testing of materials. The method is based on the fact that any discontinuity in the path of the acoustic wave causes reflection or scattering of the sound waves (see **Section 7.3.1**). The reflection is received and becomes an indication of the existence of the discontinuity and the amplitude of the reflection is a function of the size of the discontinuity. The exact location of the discontinuity can be found from the time it takes the waves to propagate to and from the discontinuity. This time is called time of flight. **Figure 7.36a** shows an example of finding the location/size of a defect in a piece of metal. The front and back surfaces manifest themselves as large reflections, whereas the defect usually produces a

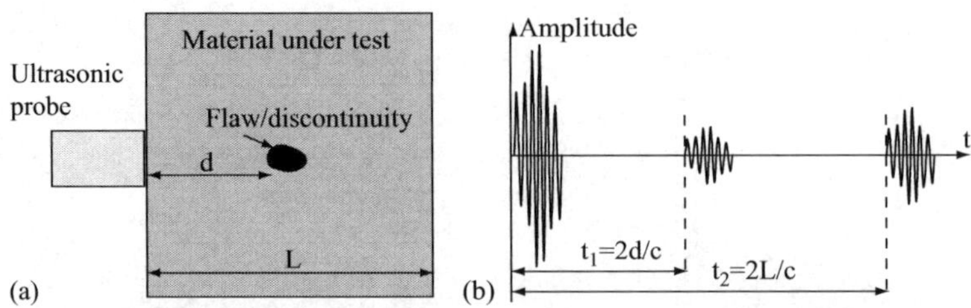

FIGURE 7.36 ■ Ultrasonic testing of materials. The echoes from various discontinuities can be detected and evaluated.

smaller signal. Its location can be easily detected. The same idea can be used to create an image of a baby in the womb, to sense a heartbeat, to measure blood vessel thickness and condition, for position sensing in industry, or in range finding. Using the configuration in **Figure 7.36**, the time it takes for the acoustic wave to reach the flaw is

$$t_1 = \frac{d}{c} \quad [\mathrm{s}]. \tag{7.53}$$

Since the ultrasonic probe receives the reflection after an additional time t_1 needed for the reflection to traverse the distance d, the signal is received back after a time $t_f = 2t_1$ and in the process the acoustic wave traverses a distance $2d$. The location of the flaw is calculated as

$$d = \frac{t_f}{2c} \quad [\mathrm{m}]. \tag{7.54}$$

Thus the location of the flaw or, for that matter, the thickness of the material may be inferred from the time of flight of the acoustic wave.

In addition to these important applications, ultrasonic sensors are useful in sensing other quantities, such as the velocity of a fluid. For this purpose there are three effects that can be used. One is the fact that sound velocity is relative to the fluid in which it travels. For example, voice carries downwind faster (by the wind velocity) than in still air. This speed difference can be measured as the time it takes the sound to get from one point to another since the speed of sound is constant and known. The second effect is based on the phase difference caused by this change in speed. The third is the Doppler effect—the frequency of the wave propagating downwind is higher than the frequency in still air or in a stagnant fluid. An example of a fluid speed sensor is shown in **Figure 7.37**. In this case, the distance and angle of the sensors are known and the transmit time downstream is

$$t = \frac{d}{c + v_f \cos\theta} \quad [\mathrm{s}], \tag{7.55}$$

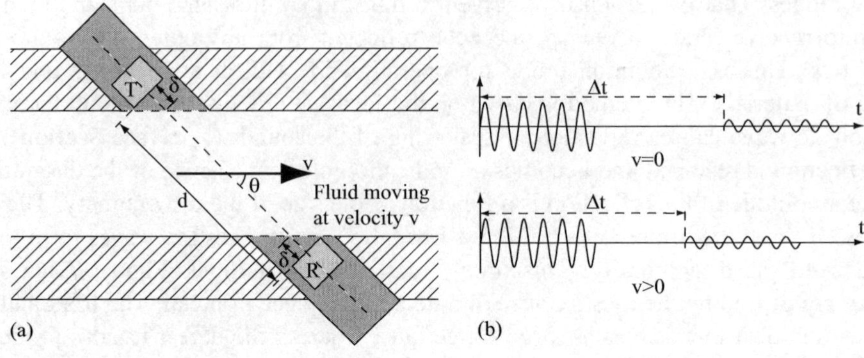

FIGURE 7.37 ■ A fluid velocity sensor. (a) The locations of the sensors. (b) The relation between transmitted and received signals.

where c is the sound velocity in the fluid and v_f is the speed of the fluid. The fluid velocity is

$$v_f = \frac{d}{t\cos\theta} - \frac{c}{\cos\theta} \quad [\text{m/s}]. \tag{7.56}$$

All terms in **Equation (7.56)** except t are known constants, so by measuring the time of flight t, the velocity is immediately available. An alternative method that is often used is based on the Doppler effect. We will discuss the Doppler effect again in **Chapter 9** in connection with radar, but the effect can be used with ultrasound as well. The basic idea is that as the wave propagates in the direction of the flow, the net speed of the wave increases by a velocity Δv. Therefore the signal arrives at the receiver sooner than it would otherwise. In effect, this means that the frequency is higher. Assuming a signal at a fixed frequency f is transmitted, the received frequency is

$$f' = \frac{f}{1 - v_f\cos\theta/c} \quad [\text{Hz}]. \tag{7.57}$$

The fluid velocity is

$$v_f = c\frac{f' - f}{f'\cos\theta} \quad [\text{m/s}]. \tag{7.58}$$

As can be seen, the change in frequency is a direct measure of fluid velocity.

Naturally, if the receiver were to be placed upstream rather than downstream, the frequency would be lower (the negative sign in **Equation (7.57)** becomes positive). The advantage of the Doppler method is in the fact that frequency is easier to measure accurately. With a constant frequency f, the method can be very accurate.

EXAMPLE 7.10 Doppler ultrasound sensing of water flow

To see the frequency levels and changes in frequency involved in a Doppler ultrasound fluid velocity sensor, consider **Figure 7.37** as a guide. The transmitter is upstream and the receiver downstream. The transmitter operates at 3.5 MHz and the sensor is at 45° to the flow. The sound velocity in water is 1500 m/s.

a. Calculate the change in frequency of the sensor for a fluid speed of 10 m/s.
b. Calculate the sensitivity of the sensor in hertz per meter per second (Hz/m/s).

Solution:

a. From **Equation (7.57)**, the change in frequency is

$$\Delta f = f' - f = \frac{f}{1 - v_f\cos\theta/c} - f = \frac{3.5\times10^6}{1 - 10\cos45^\circ/1500} - 3.5\times10^6$$

$$= 3.516577\times10^6 - 3.5\times10^6 = 16{,}577\ \text{Hz}.$$

This is a relatively large change in frequency and is easily measurable by a number of means, including a microprocessor (see **Chapter 12**).

b. The sensitivity is the change in frequency (output) over the change in fluid velocity. We write

$$\frac{df'}{dv_f} = \frac{d}{dv_f}\left(\frac{f}{1 - v_f \cos\theta/c}\right) = \frac{d}{dv_f} f\left(1 - v_f \cos\theta/c\right)^{-1}$$

$$= -f\left(1 - v_f \cos\theta/c\right)^{-2}(-\cos\theta/c) = \frac{f \cos\theta/c}{\left(1 - v_f \cos\theta/c\right)^2} \quad [\mathrm{Hz/m/s}]$$

Note that this relation looks nonlinear and seems to increase with velocity. However, the term $v_f \cos\theta/c$ is very small and hence the term in brackets in the denominator calculated for a fluid velocity of 10 m/s is

$$1 - v_f \cos\theta/c = 1 - 10\frac{\sqrt{2}}{2 \times 1500} = 0.9953.$$

This means that we can calculate a very good numerical approximation that would work for all velocities except for the unlikely case of fluid velocities that approach the sound velocity in the fluid. Taking the value above, the sensitivity is

$$\frac{df'}{dv_f} = \frac{f \cos\theta/c}{\left(1 - v_f \cos\theta/c\right)^2} = \frac{3.5 \times 10^6 \times \left(\sqrt{2}/2\right)/1500}{(0.9953)^2} = 1665.54\ \mathrm{Hz/m/s}.$$

This result is consistent with the result in (a), but it is only an approximation. By multiplying this by 10 we get 16,655 Hz instead of the 16,577 Hz we got in (a), for an error of 0.4%. Of course, the general result is more accurate than the numerical approximation.

The properties described above have also been used for other important applications. For example, the sonar used by surface ships and submarines is essentially a pulse-echo ultrasound method. The main difference is that the power involved is very large to allow long-distance sensing. It also relies on the very good propagation qualities of water. In medical applications, ultrasound is often used to sense motion, such as the motion of veins (blood pressure) or of heart valves, to detect abnormal conditions. Another useful application is to break apart kidney stones. In this case high-intensity bursts are applied to the body while it is immersed in water (the transducer is an actuator). The stones are pulverized and can then pass with the urine.

7.7.2 Magnetostrictive Transducers

For operation in air or in fluids, piezoelectric sensors seem to be the best. However, in solids there is an alternative method, based on magnetostriction, that can be much more effective. One can imagine that by applying a pulse to a magnetostrictive bar, it constricts and expands alternately to "bang" on the solid just like a hammer. These sensors are collectively called magnetostrictive ultrasonic sensors and they are used at lower frequencies (about 100 kHz) to generate higher intensity waves.

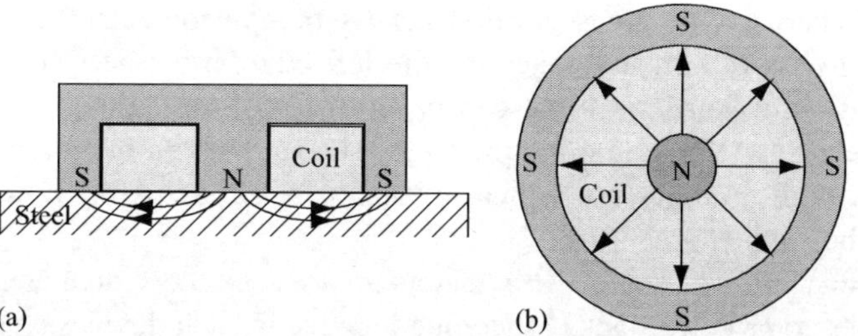

FIGURE 7.38 ■ Construction of an EMAT. (a) Side view showing the permanent magnet field. (b) Bottom view.

If, however, the ultrasound is to be coupled into a magnetostrictive material, all that is necessary is to attach a coil to the material and drive it at the required frequency. The field generated in the material itself generates stresses in the material, which in turn generates an ultrasonic wave (just like generating stress in the earth's crust generates earthquakes). The reason this type of actuator is important is that iron is magnetostrictive and hence the method can be used to generate ultrasound waves in iron and steel products for the purpose of integrity testing.

This principle is implemented as follows: A coil driven by AC (or pulses) generates induced electric currents (eddy currents) in the magnetostrictive material. A magnetic field produced by an external permanent magnet produces a force acting on these currents. The interaction between the magnetic field and the eddy currents generates stresses and an acoustic wave ensues. These devices are called electromagnetic acoustic transducers (EMATs). As with other acoustic transducers, they are dual function and can sense acoustic waves as well as generate them. **Figure 7.38** shows a schematic EMAT. EMATs are commonly used for nondestructive testing and evaluation of steel because of their simplicity, but they tend to operate at low frequencies (<100 kHz) and have relatively low efficiencies.

7.8 | PIEZOELECTRIC ACTUATORS

We have seen that piezoelectric sensors can act as actuators as used in transmitters for ultrasound. But piezoelectric devices can be used in more direct types of actuators to affect motion. One of the first types of piezoelectric actuator has been in use in analog clocks for decades. It consists of a crystal that oscillates at a fixed frequency to serve as a time reference for the clock. But other actuators have been designed that can move much larger distances and apply significant forces as well. One such device is shown in **Figure 7.39**. It is a thin, stiff plate, with the piezoelectric material bonded to it

FIGURE 7.39 ■ (a) A large-displacement, rectangular, piezoelectric actuator. (b) A cylindrical, stacked, piezoelectric actuator. It can only move about 0.05 mm but can deliver about 40 N of force. The moving shaft is seen on the left pushing against a workpiece.

(gray patch). When a voltage is applied across the piezoelectric element, one edge moves relative to the other (one edge, say the left edge, must be fixed). The motion is accompanied by force and this force can be utilized for actuation. Note, however, the high voltage needed. Although some piezoelectric sensors and actuators can operate at lower voltages, high voltages are typical of piezoelectric actuators and are one serious limitation for their widespread use.

Other approaches to piezoelectric actuators are to stack individual elements, each with its own electrodes to produce stacks of varying lengths. In such devices, the displacement is anywhere between 0.1% and 0.25% of the stack length, but this is still a small displacement. One of the advantages of these stacks is that the forces are larger than those achievable by bending plates such as the one in **Figure 7.39a**. A small actuator, capable of a displacement of about 0.05 mm and a force of about 40 N, is shown in **Figure 7.39b**.

EXAMPLE 7.11 The ultrasonic motor

The ultrasonic motor is an interesting and useful actuator, originally developed for autofocus lenses in cameras. It consists of a simple metal disk (the rotor) with a second piezoelectric disk below it. The piezoelectric disk is toothed, allowing it to flex, and as it does so it moves the rotor. To generate the wave motion of the stator disk it is necessary to generate two standing waves of equal amplitude (a standing wave is a motion, say up and down of the ring, similar to an ocean wave). A standing wave cannot generate motion just like an ocean wave, which can only move a body up and down. However, if two standing waves, 90° out of phase in space and time are generated, their sum is a travelling wave whose direction of motion depends on the frequency of the two waves and on the mode of excitation. **Figure 7.40** shows the operation in three steps, highlighting a single tooth. The wave propagates to the right (counterclockwise [CCW]) and the marked tooth first touches the rotor with the back edge since it is slightly inclined to the right (**Figure 7.40a**). As the wave propagates, the tooth straightens (**Figure 7.40b**), pushing the rotor to the left (clockwise [CW]). In **Figure 7.40c** the tooth bends to the left, pushing the disk further to the left (CW). The tooth therefore describes an elliptical path as it moves up and down, making contact with the disk during part of the cycle and causing the rotation of the disk. In this configuration, with the undulation propagating to the right, the rotor rotates in the CW direction (viewed from the top). Changing the direction of undulation reverses the direction of rotation.

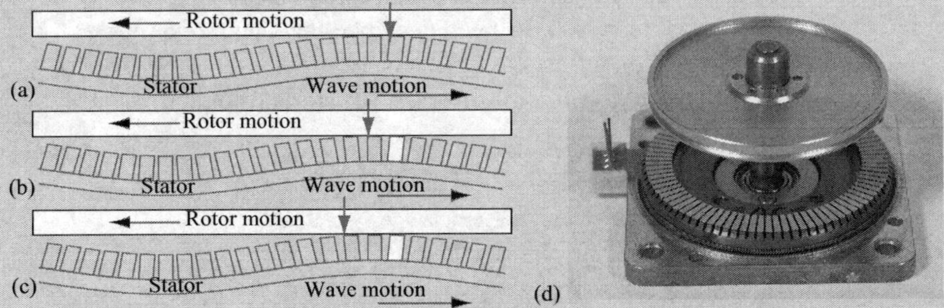

FIGURE 7.40 ■ Sequence of motion in an ultrasonic motor. (a) The leading edge of a tooth touches the rotor. (b) The motion of the rotor moves the stator to the left. (c) The trailing edge of the third tooth disengages from the rotor. The sequence has moved the rotor to the left. (d) A commercial ultrasonic motor showing the rotor (lifted above the stator) and the toothed stator. The piezoelectric segments are bonded to the bottom of the stator.

The advantages of this motor are many, including very small size, rotation speed that can be controlled directly through the propagating wave, and significant torque. It is a friction-driven motor and hence it has considerable holding torque. The ultrasonic standing waves are generated by applying high-frequency electric fields to piezoelectric strips bonded to the bottom surface of the stator. The piezoelectric strips are driven in sequence to generate the standing waves. The motor is small in size, relatively fast, does not require gearing (direct drive), and is quiet, some of the properties that made it so useful in autofocus lenses as well as other applications.

The implementation and control of motion is rather simple. The two waves are generated by applying an electric field at two or more opposite locations on the stator (i.e., the out of phase in space requirement) and, at the same time, the two locations are driven with electric fields that are 90° out of phase in time. The two waves look as follows:

$$u_1(\theta, t) = A \cos \omega t \cos n\theta$$

$$u_2(\theta, t) = A \cos(\omega t + \pi/2) \cos(n\theta + \pi/2),$$

where n is the nth mode of oscillation of the stator (n is the number of peaks in the standing wave pattern produced in the stator, can be any integer from 1 to infinity and can be controlled by the number of locations at which the stator is excited). These two waves add up to

$$\begin{aligned} u_1(\theta, t) + u_2(\theta, t) &= A \cos \omega t \cos n\theta + A \cos(\omega t + \pi/2) \cos(n\theta + \pi/2) \\ &= A \cos \omega t \cos n\theta + A[\cos(\omega t) \cos(\pi/2) - \sin(\omega t)\sin(\pi/2)] \\ &\quad \times [\cos(n\theta) \cos(\pi/2) - \sin(n\theta)\sin(\pi/2)] \\ &= A \cos \omega t \cos n\theta + A[-\sin(\omega t)][-\sin(n\theta)] \\ &= A \cos \omega t \cos n\theta + A\sin(\omega t)\sin(n\theta) = A \cos(\omega t - n\theta). \end{aligned}$$

The speed of propagation of this wave (more precisely, its phase velocity) is $v = \omega/n$ [m/s], as can be seen from **Equation (7.11)**:

$$v = \frac{\omega}{n} = \frac{2\pi f}{n} \quad [\text{m/s}].$$

The rotation of the rotor is due to the vibration of the stator teeth that make contact with the rotor and generate motion (see **Figure 7.40**). The velocity of the rotor is not the same as the phase velocity of the wave and depends on displacement of the teeth (and hence on the current in the piezoelectric elements), on the load, and on the mode of vibration. In general, the higher the mode and the larger the radius of the stator, the slower the rotational speed of the motor.

Multiplying the velocity by 1 s gives the distance travelled by the wave in 1 s. Dividing this result by the circumference of the motor gives the number of rotations per second (rps):

$$v_r = \frac{2\pi f \times (1 \text{ s})}{2\pi r n} = \frac{1}{rn} \quad [\text{rps}],$$

where r is the radius of the stator. Note that the speed of the motor is independent of the speed of the wave—it depends on the vibration speed, which tends to be constant. As an example, a motor of radius 2 cm operating in the fundamental mode ($n = 1$) at a frequency of 30 Hz will rotate at a speed of

$$v_r = \frac{1}{0.02} = 50 \text{ rps}.$$

That is, 3000 rpm. The same motor, with eight pairs of excitation locations would operate in the eighth mode and would rotate at 375 rpm.

This means that one can control the speed of rotation by changing the mode of oscillation through introduction of additional generation points on the circumference of the stator. Note also that if we change the phase from $+\pi/2$ to $-\pi/2$, the wave will propagate in the opposite direction and the motor will turn in the opposite direction.

EXAMPLE 7.12 Linear piezoelectric actuator

A simple linear actuator is made by stacking alternating conducting disks and piezoelectric disks as shown in **Figure 7.41**. There are N piezoelectric disks, each of thickness t and radius a, and $N+1$ conducting disks ($N=5$ in the figure). The conducting disk's whole purpose is to apply an external voltage to generate an electric field intensity in the piezoelectric disks. Given the properties of the piezoelectric material (relative permittivity ε_{ii} and piezoelectric constant d_{ii}),

a. Calculate the displacement of the stack for an applied voltage V.

b. Calculate the force the stack can generate for an applied voltage V.

c. Calculate the displacement and force for a 3-3 cut barium titanate oxide ($BaTiO_3$) piezoelectric with dimensions $a=10$ mm, $t=1$ mm, and $V=36$ V for a stack with $N=40$ disks.

d. What is the maximum possible displacement and force if the breakdown electric field in the crystal is 32,000 V/mm whereas the breakdown voltage in air is 3000 V/mm?

Solution: The displacement is calculated directly from the piezoelectric coefficient d_{ii}, whereas the force is calculated from **Equation (7.40)**. We start with the displacement.

a. By definition, the piezoelectric constant is the strain per unit electric field and strain is the ratio of displacement divided by length. We write for a disk of thickness t,

$$\frac{dt}{t} = d_{ii}E = d_{ii}\frac{V}{t} \quad [\mathrm{m/m}],$$

where $E = V/t$ is the electric field intensity in the disk produced by the potential difference on the disk, V. Therefore the displacement, or the change in thickness of the disk, is dt:

$$dt = d_{ii}V \quad [\mathrm{m}].$$

The total change in the length of the stack is therefore Ndt:

$$\Delta l = Ndt = Nd_{ii}V \quad [\mathrm{m}]$$

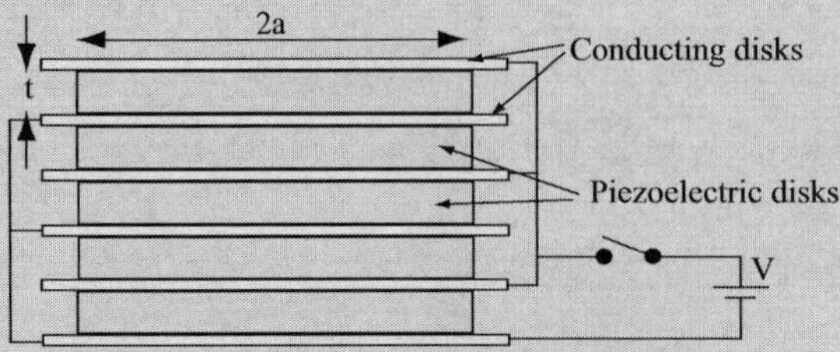

FIGURE 7.41. ■ A piezoelectric stack actuator.

b. The force is calculated from **Equation (7.40)** after rewriting the equation:

$$F = \frac{\varepsilon_{ii} A V}{t d_{ii}} \quad [\text{N}],$$

where A is the surface area of the base of the disk (πa^2) and ε_{ii} is its permittivity. This is the force produced by one disk. All other disks produce an identical force, but since the disks are in series, the total force of the N disk stack is the same as that of a single disk.

c. For the given properties and dimensions, we have

$$\Delta l = N d_{ii} V = 40 \times 100 \times 10^{-12} \times 36 = 0.144\ \mu\text{m}$$

and

$$F = \frac{\varepsilon_{ii} A V}{t d_{ii}} = \frac{180 \times 8.854 \times 10^{-12} \times \pi \times 0.01^2 \times 36}{10^{-3} \times 100 \times 10^{-12}} = 180.25\ \text{N}.$$

Note the typical characteristics of piezoelectric actuators: small displacement, but large forces.

d. The maximum electric field intensity is the breakdown electric field intensity. In this case it is 32,000 V/mm and requires a potential difference of 32,000 V. However, in air the electric field intensity is only 3000 V/mm. It is impossible to raise the voltage of the stack above 3000 V/mm because at that voltage difference there will be breakdown in air. Therefore the maximum electric field intensity is 3000 V/m or 3×10^6 V/m. We have for the maximum possible displacement,

$$\Delta l_{\max} = N d_{ii} V = 40 \times 100 \times 10^{-12} \times 3 \times 10^6 = 12{,}000\ \mu\text{m}.$$

This is 12 mm, and as a theoretical result it looks reasonable. But in fact this would require a strain of 30%, and that is not possible in a real material. Perhaps 1/10 of that or a total displacement of 1.2 mm seems reasonable (a strain of 3% is certainly possible). To obtain this will require a 300 V potential difference between the plates.

The maximum theoretical force possible is

$$F = \frac{\varepsilon_{ii} A V}{t d_{ii}} = \frac{180 \times 8.854 \times 10^{-12} \times \pi \times 0.01^2 \times 3000}{10^{-3} \times 100 \times 10^{-12}} = 15{,}020\ \text{N}.$$

Again, 1/10 of this, or 1500 N, may be more reasonable, obtainable with a 300 V potential difference.

7.9 | PIEZOELECTRIC RESONATORS AND SAW DEVICES

In **Section 7.7** we discussed ultrasonic sensors and in **Section 7.3** the theory of sound waves. Most of this was based on the idea of the generation and propagation of longitudinal waves and their interaction with materials and the environment. Sound waves in air and in fluids are essentially longitudinal waves, but under appropriate conditions other waves

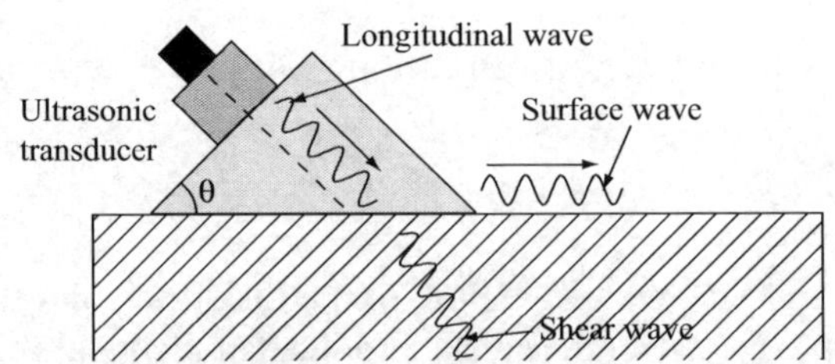

FIGURE 7.42 ■ Conversion of a longitudinal wave into a surface wave by means of a wedge.

may be generated. We saw that solids can support shear waves (**Section 7.3.2**) and the surface between solids and air can support surface waves (**Section 7.3.3**). Surface waves are of particular interest because of their slow speed of propagation and low dispersion. Under most conditions the slow propagation speed of surface waves would seem to be a disadvantage, but looking at the wavelength alone as the ratio of velocity and frequency, $\lambda = c/f$, it is clear that the lower the velocity of the wave, the shorter the wavelength in that medium. This means, for example, that if a device must be, say, one-half wavelength in size, then the same device utilizing surface waves will be physically smaller than if it were to utilize longitudinal waves. It is this property that is at the heart of SAW devices.

Surface waves can be generated in a number of ways. In a thick sample, one can set up a surface wave by a process of wave conversion. Essentially, a longitudinal wave device is used and energy coupled through a wedge at an angle to the surface. At the surface of the medium there will be both a shear wave and a surface wave (**Figure 7.42**). This is an obvious solution, but not necessarily optimal. A much more efficient method of generating surface waves, one that is almost ideally suited for fabrication is to apply metallic strips on the surface of a piezoelectric material in an interdigital fashion (a comblike structure) as shown in **Figure 7.43**. This establishes a periodic structure of metallic strips. When an oscillating source is connected across the two sets of electrodes, a periodic electric field intensity is established in the piezoelectric material, equal to the periodicity of the electrodes and parallel to the surface. (The period is equal to the distance between each two electrodes and the latter are designed so that each strip is $\lambda/4$ wide and the gap between strips is also $\lambda/4$ wide.) Because of this electric field, an equivalent, periodic stress pattern is established on the surface of the piezoelectric medium. This generates a stress wave (sound wave) that now propagates away from the electrodes in both directions. The generation is most efficient when the period of the surface wave equals the interdigital period. For example, in the structure in **Figure 7.44**, suppose the frequency of the source is 400 MHz. The speed of propagation in a piezoelectric is on the order of 3000 m/s. This gives a wavelength of 7.5 μm. Making each strip in the structure $\lambda/4$ means each strip is 1.875 μm wide and the distance between neighboring strips is 1.875 μm. This calculation shows first that the dimensions required are very small (a device at the same frequency, based on electromagnetic waves, has a

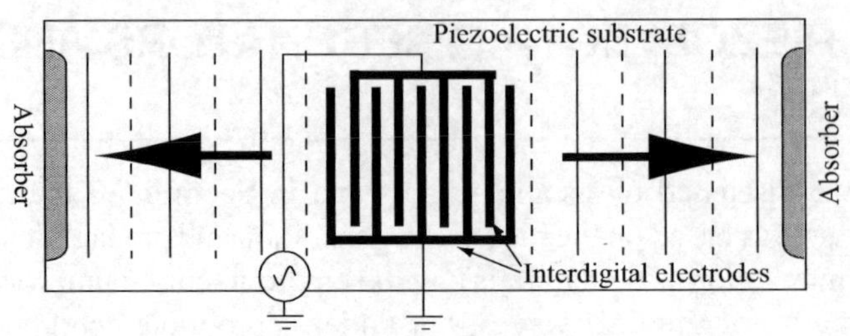

FIGURE 7.43 ■ Generation of a SAW by a series of periodic surface electrodes driven from a resonant source.

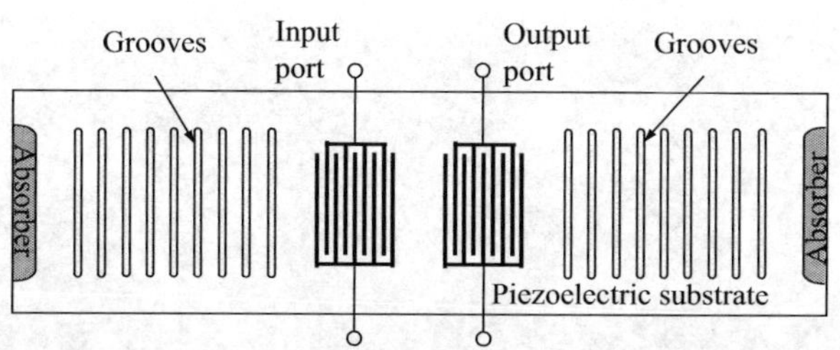

FIGURE 7.44 ■ Construction of a SAW resonator.

wavelength of 750 mm). Second, it shows that production of these devices can be accomplished using lithographic techniques that are compatible with semiconductor production methods.

Returning now to the basic structure, just as the comblike structure generates surface acoustic waves, and hence stress in the piezoelectric medium, an acoustic wave in the piezoelectric medium produces a signal in a comblike structure because of the surface stress produced by the acoustic wave. Thus the structure can be used both for generation and sensing of surface waves. That, in turn, means that the device can be used for sensing or actuation.

By far the most common use of SAW principles is in SAW resonators, filters, and delay lines. A SAW resonator is shown in **Figure 7.44**. The sections marked as *In* and *Out* are used as the input and output ports of the resonator (i.e., the external connections of the resonator). The parallel lines on each side are grooves etched in the quartz piezoelectric. The input port establishes a surface wave, which is reflected by the grooves on each side. These reflections interfere with each other, establishing a resonance that depends on the separation of the grooves in the grating. Only those signals that interfere constructively will establish a signal in the output port, the others cancel.

This device has quickly become popular as the basic element for oscillators in communication systems since a very small device can easily operate at low frequencies and at the other extreme can operate at frequencies above the limit of conventional oscillators, including crystal oscillators. **Figure 7.46** shows a numbere of SAW resonators used in low power transmitters. The device in **Figure 7.44** may also be viewed as a very narrow bandpass filter and this is in fact another of its uses.

The configuration in **Figure 7.45** is a SAW delay line. The comb on the left generates a surface wave and this is detected after a delay in the comb on the right. The delay depends on the distance between the combs and, because the wavelength is usually small, the delay can be relatively long.

In addition to these important applications, SAW devices can be used for sensing of almost any quantity, taking advantage of the properties of the piezoelectric medium. For example, the application of stress to the piezoelectric changes the speed of sound in the

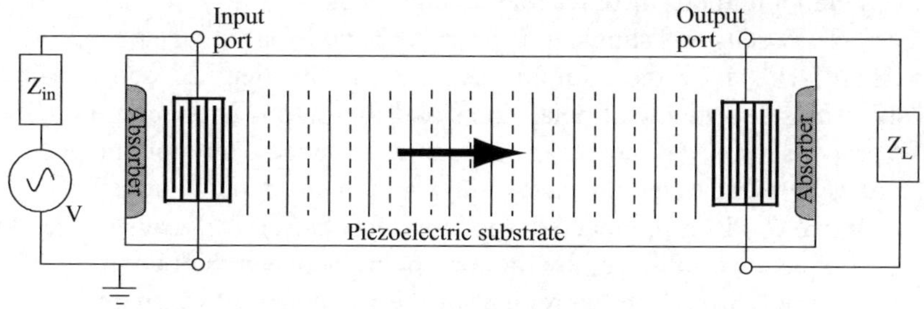

FIGURE 7.45 ■ A SAW delay line.

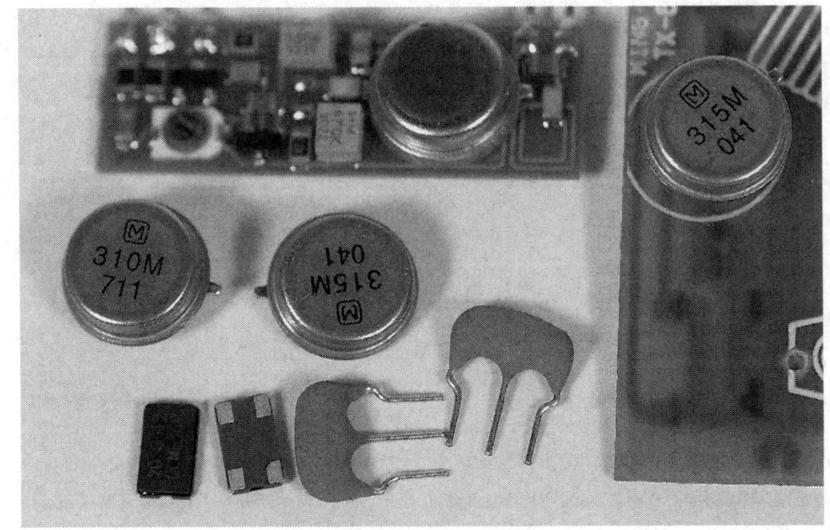

FIGURE 7.46 ■ SAW resonators used in transmitters and receivers. The device covered with the dark label resonates at 433.92 MHz (shown soldered in a transmitter). The metal can resonators are for 310 and 315 MHz applications. The surface mount devices (bottom left) and the three-pin devices (bottom center) resonate at 433.92 MHz.

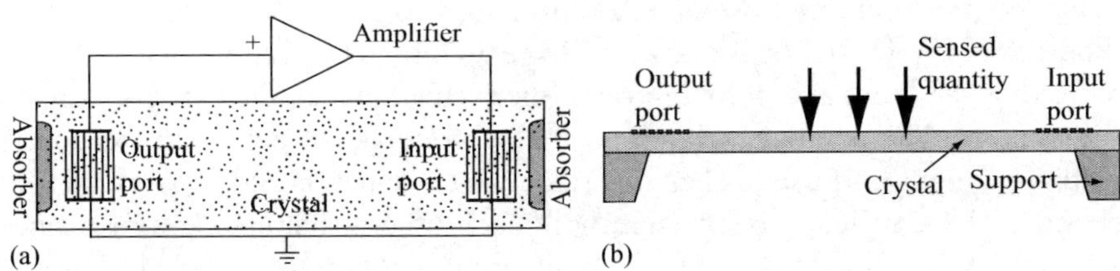

FIGURE 7.47 ■ The basic structure of a SAW sensor based on the delay line.

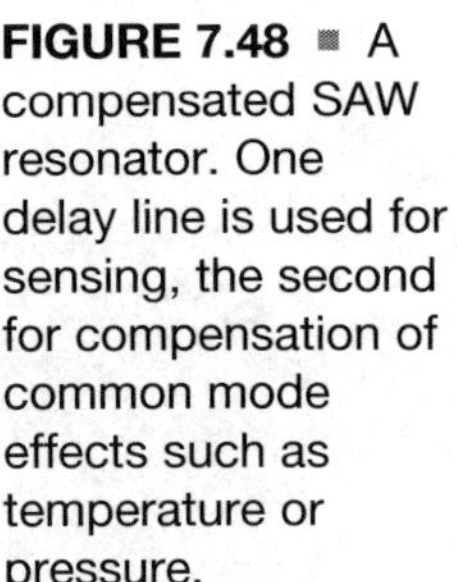

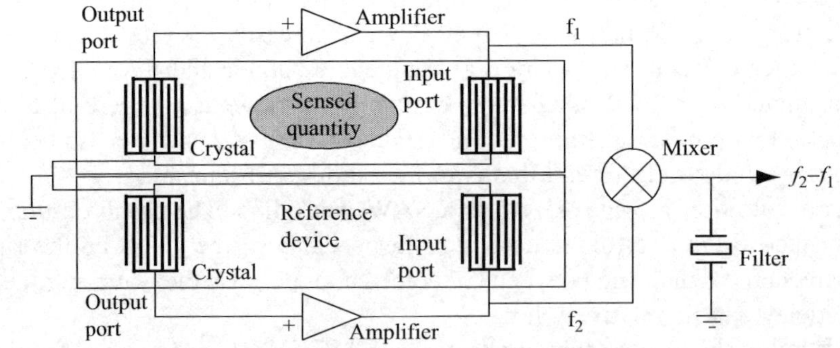

FIGURE 7.48 ■ A compensated SAW resonator. One delay line is used for sensing, the second for compensation of common mode effects such as temperature or pressure.

material. This in turn will change, say, the resonant frequency of the device in **Figure 7.44** or the delay in **Figure 7.45** and by so doing sense force, pressure, acceleration, mass, and a number of other related quantities.

The basic SAW sensor is shown in **Figure 7.47** and is based on a delay line in which the delay is influenced by the stimulus. An essentially identical sensor is shown in **Figure 7.48**. It has two identical delay lines and the output is differential. One line is used as the proper sensor, the second as a reference to cancel common mode effects such as temperature. In most cases the delay time is not measured, but rather a feedback amplifier (**Figure 7.48**) is connected (positive feedback) that causes the device to resonate at a frequency established by the time delay between the two ports. The sensed quantity is measured through measurement of the frequency of resonance.

The stimuli that can be measured are many. First, the speed of sound is temperature dependent. Temperature changes both the physical length of the delay line and the sound speed as follows:

$$L = L_0[1 + \alpha(T - T_0)] \quad [\text{m}], \quad c = c_0[1 + \delta(T - T_0)] \quad [\text{m/s}], \tag{7.59}$$

where α is the coefficient of linear expansion and δ is the temperature coefficient of sound velocity.

Both the length and the speed of sound increase with temperature, hence the delay and oscillator frequency are a function of the difference between them. In fact, the change in frequency with temperature is

$$\frac{\Delta f}{f} = (\delta - \alpha)\Delta T. \tag{7.60}$$

This is linear and a SAW sensor has a sensitivity of about 10^{-3}°C.

In sensing pressure, the delay in propagation is due to stress in the piezoelectric, as indicated above. Measurement of displacement, force, and acceleration are done by measuring the strain (pressure) produced in the sensor. Many other stimuli can be measured, including radiation (through the temperature increase), voltage (through the stress it produces through the electric field), and so on. **Equations 7.59** and **7.60** indicate a linear relationship between the change in frequency and the change in length (in this case due to temperature). That means that if the length increases, say, by 1%, the frequency must necessarily decrease by 1%. This can be used to sense any quantity that would change the length of the sensor. Given a change in length Δl, we can write

$$\frac{\Delta f}{f} = \frac{\Delta l}{l}, \tag{7.61}$$

where f is the frequency at length l. Note that the right-hand side of **Equation (7.61)** is the strain in the medium and that it can be related to pressure, force, acceleration, or mass.

We will discuss additional applications of SAW devices in chemical sensing in **Chapter 8**.

EXAMPLE 7.13 Sensitivity of a SAW pressure sensor

A pressure sensor is built with a SAW resonator serving as a beam (see **Figure 7.47b**) that resonates at 500 MHz. The maximum strain allowed in the beam is 1000 microstrain, corresponding to 10^6 Pa (9.87 atm).

a. Calculate the maximum change in frequency of the sensor.
b. Suppose that the temperature changes by 1°C. Calculate the error introduced by the change in temperature at 100 kPa if the temperature sensitivity is 10^{-4}/°C.

Solution:

a. The strain, by definition, is the change per unit length. That is,

$$\frac{\Delta l}{l} = 1000 \times 10^{-6} = 10^{-3}.$$

Therefore the change in frequency is

$$\frac{\Delta f}{f} = \frac{\Delta l}{l} = 10^{-3} \rightarrow \Delta f = 500 \times 10^6 \times 10^{-3} = 500 \times 10^3 \text{ Hz}.$$

This is a very large change in frequency with a sensitivity of 500 Hz/microstrain.

b. Since the sensor is linear and the temperature sensitivity is 10^{-3}/°C, we can write from **Equation (7.61)**:

$$\Delta f = 500 \times 10^6 \times 10^{-4} = 5 \times 10^4 \text{ Hz}.$$

At 100 kPa, the change in frequency due to pressure is only 50 kHz. That is, a change of 1°C causes a change in frequency equal to the change due to pressure. Obviously, unless the sensor is temperature compensated, it cannot be used for sensing. It is for this reason that the configuration in **Figure 7.48** is so important.

7.10 | PROBLEMS

Units

7.1 **Acoustic pressure at a distance from a source.** A jet engine on the ground generates a sound power density level of 155 dB at a distance of 10 m from the engine. Assume that the sound travels uniformly in all directions in the space above the ground. Neglect attenuation of sound waves in air.

a. What is the shortest safe distance for an operator without hearing protection?

b. What is the shortest safe distance if the operator uses hearing protection rated at 20 dB?

7.2 **Stress and strain produced by an ultrasonic actuator.** An ultrasonic actuator is used to test steel for cracks. To do so it produces a pressure of 1000 Pa on the surface of the steel. Calculate the stress and strain produced by the actuator at the contact surface if the modulus of elasticity of steel is 198 GPa. Comment on the effect this might have on the material.

Elastic waves and their properties

7.3 **Speed of propagation of sound waves.** Popular wisdom says that when you see lightning, start counting slowly (it is assumed that each count is a second). The number you reach by the time you hear the thunder is three times the distance to the lightning strike in kilometers. How accurate is this estimate?

7.4 **Ultrasonic testing of materials.** Ultrasonic testing of materials for flaws is an established method often used in industry, especially in metals, looking for flaws, cracks, thinning due to corrosion, inclusions, and other effects that can be detrimental to the functioning of a structure. It relies on reflection of ultrasonic waves from discontinuities within the structure. The resolution, that is, the smallest detail that can be "seen" by the waves, depends on frequency. For that reason, ultrasonic testing is done at relatively high frequencies. For flaws to be detectable they must be on the order of the wavelength of the acoustic wave. Consider the testing of titanium blades in jet engines for small cracks. Since these tend to initiate larger flaws followed by failure, the test must detect cracks smaller than 0.5 mm.

a. Given the speed of propagation of sound waves in titanium as 6172 m/s, what is the lowest sound frequency that will detect these flaws?

b. Discuss the consequences of using higher frequencies in ultrasonic sensors, especially for very small flaws.

7.5 Range of normal speech—the ear as a sensor. The sound intensity of human speech varies from about 10^{-12} W/m^2 (very faint whisper) to about 0.1 W/m^2 (loud scream). The human ear can detect sounds as faint as 10^{-12} W/m^2, but to understand a conversation requires at least 10^{-10} W/m^2. Normal conversation is considered to be 10^{-6} W/m^2. Suppose a person speaks normally, producing a power density of 10^{-6} W/m^2 at a distance of 1 m.

a. What is the absolute longest possible distance a person can be heard and understood assuming that the sound can be directed to the listener without spreading in space (such as speaking through a tube) and there are no losses in the path other than attenuation of sound waves in air? Use the attenuation data in **Example 7.3**.

b. What is the distance the person can be heard and understood assuming the sound waves propagate uniformly in all directions and neglecting attenuation?

7.6 Fishing sonar. The attenuation constant for sound waves in water increases with frequency. Consider the sonar used for fishing. The reflection of sound from fish schools is used for sports and commercial fishing. Fish produce relatively large reflections because of their air sack, the reflection being due to the interface between flesh and air, whereas the reflection at the interface between flesh and water is small. The signal is generated by an ultrasound actuator and detected by an ultrasound sensor. Suppose the actuator produces a power of 0.1 W at 50 kHz (a common frequency in fishing sonars) that spreads in a 30° cone as it propagates into water. Assuming that 20% of the sound reaching a fish school is reflected and that reflected sound is scattered uniformly over a half-sphere, calculate the required sensitivity of the ultrasonic sensor (in W/m^2) to detect a fish school at a depth of 20 m. Attenuation in water at 50 kHz is 15 dB/km.

7.7 Attenuation of ultrasonic waves in water. Much work with ultrasound is conducted in water or in conjunction with water (sonar, diagnostics in the body, ultrasonic cleaning, treatment for kidney stones, and others). The selection of the frequency of operation is critical for the successful implementation of ultrasonic systems and one parameter of importance is the attenuation at the selected frequency. As a rule, the higher the frequency, the higher the attenuation, and also the better the resolution. The attenuation of ultrasound in water at frequencies above 1 MHz is approximately: $\alpha = 0.217f^2$ [dB/m], where f is the frequency in megahertz. Suppose an ultrasonic test is applied for diagnostics deep into the body and requires a spatial resolution of 1 mm. The spatial resolution of ultrasound is approximately 1 wavelength. Assuming that the waves should be able to image an artifact 2 mm in diameter at 15 cm depth and the speed of sound in water is approximately 1500 m/s, calculate the lowest frequency that can be used and the minimum power the ultrasonic transmitter must transmit to receive back 10 μW. The artifact is a small bone fragment embedded in the soft tissue due to a fracture. Assume the beam produced by the transmitter is collimated in a cylinder of diameter 20 mm, equal to the diameter of the sensor, and the reflected waves scatter in all directions equally.

7.8 Water level detection. Water level can be detected and accurately measured using ultrasonic waves based on the reflection of waves at the surface. Given a fixed ultrasonic transmitter above the surface of the water, the measurement of time of flight is sufficient to sense the surface of the water (**Figure 7.49**). The properties of ultrasonic waves in air and water are given in **Table 7.1**.

a. Given the amplitude of the pulse transmitted as A, show that the amplitude of the pulse received by the receiver, B, only depends on the distance traveled in air and its properties and is not affected by the properties of water. Find the relation needed to sense water level based on the measurement of amplitude.

b. Find a relation between the water level h and the time of flight of the ultrasonic wave measuring the start of the pulse in the transmitter and that in the receiver.

c. Explain why the method in (b) is preferable.

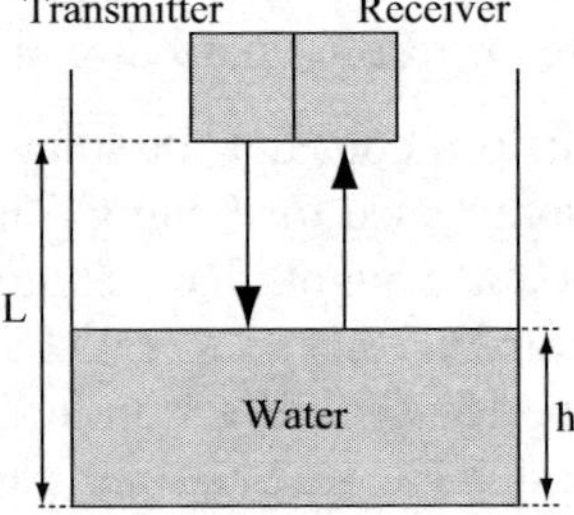

FIGURE 7.49 ■ Water level sensor.

7.9 Pulse-echo ultrasonic testing. In a pulse-echo ultrasound test, the transmitter serves as the receiver during the time the pulse is off. The transmitter sends a pulse of width Δt, at the end of which it is ready to receive. The test is applied to measure the thickness of a copper slab by detecting the reflected pulse from the far surface (**Figure 7.50**). The properties of copper are as follows: speed of propagation 4600 m/s, attenuation constant 0.45 dB/cm, acoustic impedance 42.5×10^6 kg/m^2/s. The pulse generated during transmission is 200 ns wide.

a. What is the thinnest slab that can be tested using the pulse-echo method described here and what is the maximum frequency of a train of pulses intended to repeat the process indefinitely?

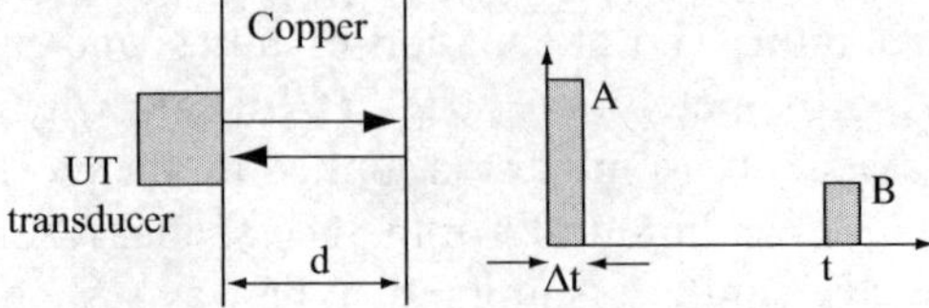

FIGURE 7.50 ■ Pulse-echo testing of materials.

b. Calculate the amplitude of the reflected pulse received by the transducer for the conditions in (a), given an amplitude of the transmitted wave, V_0.
Note: The application of a pulse to an ultrasonic transducer generates a series of sinusoidal waveforms as the transducer oscillates at its resonant frequency. However, for simplicity we will assume here that the pulse propagates and reflects as a pulse.

c. Find a relation between the thickness of the slab and the amplitude of the reflected pulse.

d. Find a relation between the thickness of slab and the time it takes to receive the pulse.

7.10 Stealth submarines. The reflection of sound waves off the hull of a submarine is one of the most important methods of their detection (another is the disturbance they create in the local terrestrial magnetic field). To avoid detection, submarines are coated with rubber (or rubberlike substances). A sonar operates at 10 kHz and produces a signal of power $P_0 = 1$ kW transmitted uniformly in a 5° circular cone, used to detect a submarine at a depth of 300 m. Assume the reflection off the submarine also propagates back in a 5° circular cone and that the power that penetrates into the rubber coating is dissipated within the coating. The top surface area of the submarine is 65 m^2.

a. Calculate the ratio of power received by the sonar from a steel hull submarine and that of a rubber-coated submarine. Is this an effective way of reducing the visibility of submarines?

b. In theory, the reflection off submarines can be reduced to zero. Explain the requirements of the coating to achieve this.

Resistive and magnetic microphones

7.11 A resistive microphone. Consider the microphone in **Figure 7.51**. It consists of conducting particles suspended in a lightweight foam. The particles are made of low-conductivity particles with a conductivity $\sigma_c = 1$ S/m, whereas the foam can be considered nonconducting. When no pressure is applied, the particles occupy 50% of the total volume. The foam has a restoring constant (spring constant) of 0.1 N/m. The conductivity of the combined foam and particles equals the conductivity of carbon multiplied by the ratio of the volume of carbon to total volume:

$$\sigma = \frac{\sigma_c V_c}{V_t} \quad [\text{S/m}].$$

To operate, a current is established through the microphone as shown and the current is then a measure of sound pressure. Normal speech produces pressures

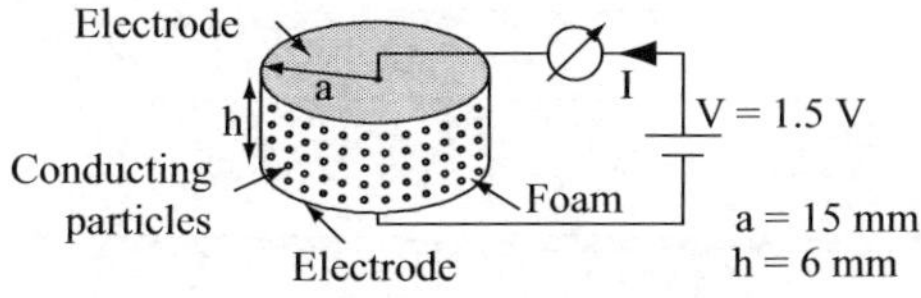

FIGURE 7.51 ■ A simple resistive microphone.

varying between 40 dB (very soft) and 70 dB (very loud). Calculate the range of current in the circuit for the pressure range in normal speech.

7.12 **Loudspeaker used as a dynamic microphone.** A small loudspeaker is used both as a microphone and as a loudspeaker in an intercom system. The loudspeaker has a cone 60 mm in diameter, an 80 turn coil of radius 15 mm, and a magnet that produces a radial magnetic flux density of 1 T (see **Figure 7.19** for the geometry and field configuration). The coil and cone have a mass of 8 g. Calculate the output of the loudspeaker as a microphone (emf) when a 60 dB sinusoidal sound pressure at 1 kHz is applied on the cone.

Capacitive microphones

7.13 **The electret microphone: sensitivity to variations in properties.** One of the parameters involved in the output of the electret microphone is the permittivity of the electret.

a. Calculate the sensitivity of the output to the relative permittivity of the electret.
b. What is the error in the expected reading if the permittivity of the electret decreases by 5% due to aging or due to variations in manufacturing? Use the data in **Example 7.6**.

7.14 **The electret pressure sensor.** The electret microphone can serve as a pressure sensor. In particular, any electret microphone can serve as a differential pressure sensor, measuring the pressure $P - P_0$, where P_0 is the air pressure, provided a way is found to apply pressure to one of the electrodes. A pressure sensor of this type is shown in **Figure 7.52**. The thickness of the electret is 1 mm and the plate is at a tension of 100 N. The distances are shown in the figure and the relative permittivity of the electret is 4.5. The ratio of specific heats in air is 1.4 and a surface charge density on the electret of 0.6 μC/m^2 may be assumed. The sensor is cylindrical, with an internal diameter of 10 mm.

a. Given the dimensions and properties, calculate and plot the output as a function of external pressure starting at 0.1 atm (assuming the metal diaphragm can withstand that pressure). The ambient pressure is 1 atm. What is the maximum pressure the sensor can respond to? What is the sensitivity of the sensor?
b. The gap between the electret and plate is now evacuated and the vent is then sealed, keeping the pressure inside the gap between the electret and the plate at 50,000 Pa (approximately 0.5 atm) when no external pressure is applied.

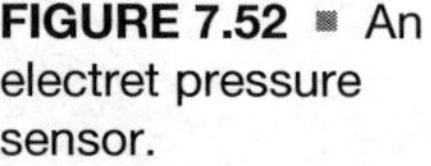

FIGURE 7.52 ■ An electret pressure sensor.

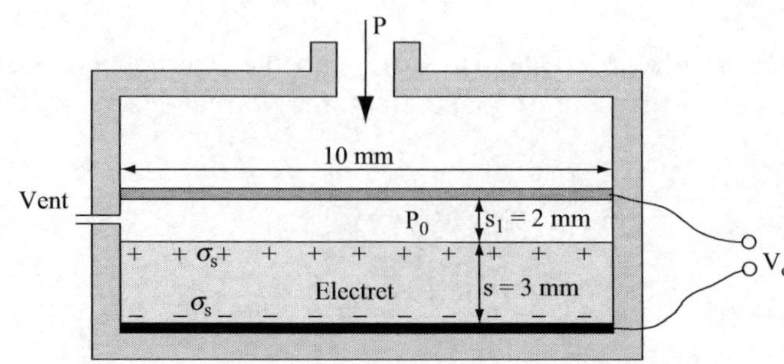

Calculate and plot the sensor output for external pressures between 10,000 Pa (0.1 atm) and the maximum pressure to which the sensor responds. Does the sensitivity change due to the change in the pressure in the gap?

The piezoelectric microphone

7.15 **Dynamic pressure sensor.** Piezoelectric devices cannot be used to measure true static pressure because once the charge has been generated on the electrodes, it will discharge through the internal impedance of the sensor and through external impedances. They are, however, well suited to measure dynamic pressure such as that due to vibrations, detonations, engine knock and ignition, and many others. In that role it is, in essence, a modified microphone. Consider a piezoelectric pressure sensor designed to sense pressure in the cylinders of a diesel engine. The normal pressure in a diesel engine at the peak of piston travel is around 4 MPa. Calculate the output expected from a lead zirconium ceramic sensor with a piezoelectric coefficient of 120×10^{-12} C/N, a capacitance of 5000 pF, and an active surface area of 1 cm^2 (i.e., the area on which the pressure acts).

7.16 **Sound intensity sensor.** Consider a flat, round piezoelectric microphone with a piezoelectric disk 25 mm in diameter and a thickness of 0.8 mm made of zinc oxide (ZnO) cut on the 3-3 axis. Two electrodes are plated on the disk, one on each side. The microphone is used to measure sound intensity to alert workers of damaging noise levels. The sound pressure the human ear responds to is in the range 2×10^{-5} Pa (0 dB) and 200 Pa (threshold of pain).

a. Calculate the output of the microphone over the entire range.
b. From a practical point of view, what is the approximate useful range of the microphone? Explain in terms of practical output voltage levels.

7.17 **The magnetic buzzer.** Small buzzers are common in portable equipment as well as fixed installations, where they provide audible feedback or warning signals. In many cases they are driven directly from microprocessors, as they require little power to be effective. A magnetic buzzer for use with a microprocessor is built as in **Figure 7.24** with an iron core of radius 6 mm containing 150 turns and with the magnet replaced by iron. The gap between the core and the diaphragm is 1 mm and the iron core as well as the diaphragm have very high permeability. The diaphragm itself has an effective radius of 12 mm. Because the coil is driven directly from a microprocessor, its maximum current is 25 mA and it cannot operate at currents below 5 mA.

a. Calculate the range of sound pressures it can generate.
b. Calculate the range of acoustic power transmitted by the buzzer and the corresponding efficiencies if the microprocessor operates at 3.3 V.

Acoustic actuators

7.18 **Force and pressure in a loudspeaker.** Consider the loudspeaker structure in **Figure 7.19** with the following specifications: $I = 2\sin 2\pi ft$, number of turns in the coil $N = 100$, radius of the coil $a = 40$ mm, and magnetic flux density $B = 0.8$ T. The radius of the cone is $b = 15$ cm. Assuming that the magnetic flux density in the coil is constant at all times,

a. Calculate the maximum force on the coil.

b. If the cone has a restoring constant $k = 750$ N/m (this restoring constant is due to the attachment of the cone to the body of the loudspeaker and acts exactly as a spring constant to return the cone to its centered position), calculate the maximum displacement of the cone.

c. Calculate the maximum pressure the cone can apply.

7.19 The electrostatic loudspeaker. Most loudspeakers are magnetic devices because the forces that can be generated are significant. However, capacitive loudspeakers, also called electrostatic loudspeakers, exist in which the diaphragm is moved by electrostatic forces. The structure of the capacitive microphone in **Figure 7.13** can be used for this purpose, with the obvious change in dimensions and the application of an AC voltage as input (see also **Section 5.3.3** on capacitive actuators). In the device in **Figure 7.13**, the moving plate is a disk 10 cm in diameter separated from the fixed, lower plate by a distance of 6 mm when the input voltage is zero. The mass of the plate is 10 g. The input is a sinusoidal voltage varying between 3000 V and 0 V at a frequency of 1 kHz (**Figure 7.53**), that is, the sinusoidal signal is centered at 1500 V.

Note: Electrostatic loudspeakers require high voltages to work and produce measurable pressures.

a. Calculate the restoring spring constant needed to ensure the moving plate does not move closer than 2 mm from the fixed plate.

b. Estimate the sound pressure generated by the loudspeaker.

c. Loudspeakers are usually characterized by their power. Estimate the average power of this loudspeaker.

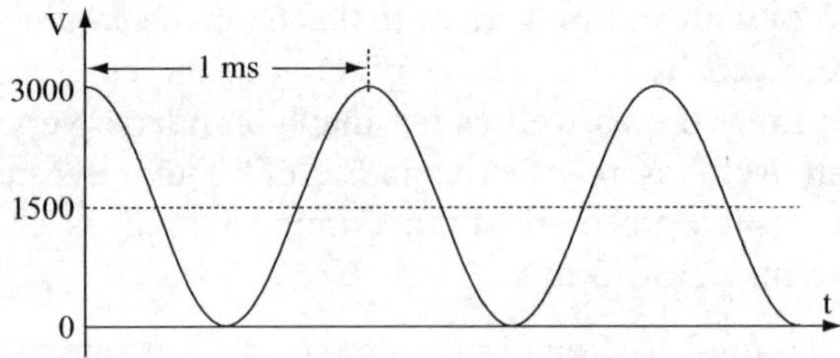

FIGURE 7.53 ■ Input voltage to the electrostatic loudspeaker.

Ultrasonic sensors

7.20 Ultrasonic evaluation of structures. In an ultrasonic test for delamination effects in steel plates, a pulse is transmitted and the signals received are detected

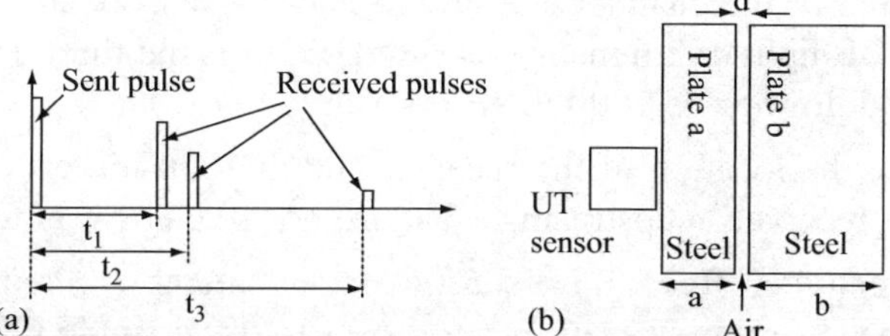

FIGURE 7.54 ■ Ultrasonic testing. (a) Signals recorded at the sensor. (b) The structure and configuration that produces the signals in (a).

on an oscilloscope. The timing of these signals is given in **Figure 7.54a**. Since delamination is suspected, it is assumed that the flaw is air-filled (a sketch of the expected configuration is shown in **Figure 7.54b**). From the signal received and the speed of sound in air, c_a, and steel, c_s, calculate the thicknesses of the two sheets and the width of the delamination.

Note: The transmitted and received signals look like those in **Figure 7.36b**, but are shown here as simple pulses for simplicity.

7.21 Doppler ultrasound sensing of fluid velocity. To measure fluid velocity in a channel it is proposed to use the configuration in **Figure 7.55**. The velocity in the channel is measured by placing both sensors at the top of the channel. The wave transmitted by the ultrasound transmitter (actuator) on the left reflects off the bottom of the channel and is received by the sensor on the right. The frequency of the ultrasound wave is 3 MHz and the speed of propagation in water is 1500 m/s.

a. Calculate the sensitivity of the system for a frequency f, fluid velocity v_f, and angle θ. Show that it is exactly the same as that for the sensor in **Figure 7.37** if the frequency, velocity, and angle are the same.

b. Calculate the frequency shift for $f = 3$ MHz and $\theta = 30°$ in water moving at a speed of 3 m/s.

c. Suppose the location of the receiver and the transmitter are interchanged so that the receiver is upstream. Now what are the answers to (a) and (b)?

7.22 Time of flight method of speed sensing. In **Problem 7.21**, instead of measuring the frequency shift, the time required for the wave to traverse the distance between the transmitter and receiver is measured.

a. If the depth of the channel is h, find a relation between the speed of flow and time. Assume a flow velocity v and a sensor angle of θ.

b. Calculate the sensitivity of the sensor.

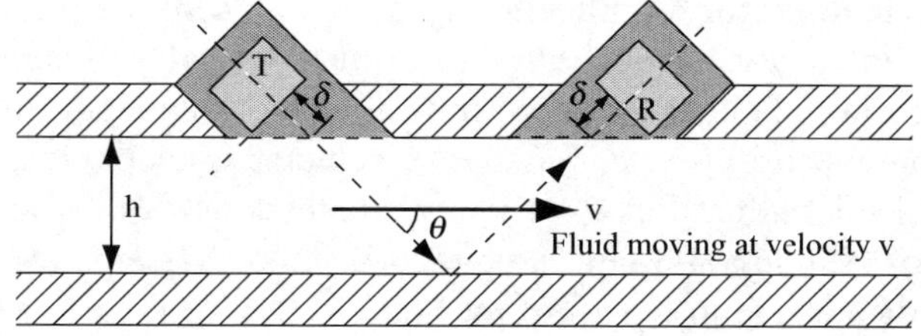

FIGURE 7.55 ■ Fluid velocity sensor adapted to sense velocity in a channel.

c. The depth of the channel is $h = 1$ m and the sensors are at an angle of 30°. A time of flight of 2.6 ms is measured. What is the fluid velocity if the speed of sound in water is 1500 m/s?

d. Suppose the location of the receiver and the transmitter are interchanged so that the receiver is upstream. What are now the answers to (a) and (b)?

7.23 **Single-transducer fluid flow sensor.** In an attempt to reduce the cost of a Doppler speed sensor an engineer proposes to use a single ultrasonic transducer and operate it in a pulsed-echo mode in which the transmitter sends a pulse then switches into receive mode to receive the reflection. That is, in **Figure 7.37a** the receiver is replaced with a reflector (a metal plate).

Note: The sensor is driven by a pulse that generates harmonics, but ultrasonic sensors are resonant, meaning they will only receive the fundamental frequency.

a. Calculate the exact shift in frequency at the receiver. Show that this is smaller than would be obtained with the configuration in **Figure 7.37a**.

b. Calculate the frequency shift for a sensor operating in water moving at a velocity $v = 5$ m/s, inclined at an angle $\theta = 60°$, and the mirror at a distance $d = 10$ cm (see **Figure 7.37a**). The sensor resonates at 3.5 MHz and the measurement is made in water, with a speed of sound of 1500 m/s. Compare this with the frequency shift that would be obtained using the configuration in **Figure 7.37a**.

c. What is the maximum pulse width the transducer can generate and still perform as intended for the sensor and fluid properties in (b)?

Piezoelectric actuators

7.24 **Sound intensity produced by a piezoelectric buzzer.** The sound intensity produced by a buzzer can be estimated from the mechanical power produced by the buzzer's piezoelectric disk. A piezoelectric buzzer is driven from a 1 kHz (square wave) source with an amplitude of 12 V, a current of amplitude 1 mA, and a 50% duty cycle. If the power efficiency of the device is 30%, and assuming the mechanical power of the buzzer is converted into sound, calculate the peak sound intensity produced by the device in power/area [W/m^2] and in decibels, considering the fact that the reference value for sound is taken as 10^{-12} W/m^2 (the threshold of hearing). The piezoelectric element is a disk 30 mm in diameter.

7.25 **Piezoelectric actuator.** Application of a voltage across a piezoelectric element causes a force to develop according to **Equation (7.40)**. This means that a strain is generated in the element, which in turn changes its length in the direction of the field produced by the potential. An actuator is made as a stack of $N = 20$ piezoelectric disks of radius $a = 10$ mm and thickness $d = 1$ mm each. The disks are made of PZT in a 3-3 cut. A voltage $V = 120$ V is applied across the stack. Calculate the change in length of the stack.

7.26 Quartz SAW resonator. A quartz SAW resonator is made as shown in **Figure 7.44**. It consists of 45 reflecting grooves on each side of the ports and the grooves are separated a distance 20 μm apart. What is the resonant frequency of the device? The speed of sound in quartz is 5900 m/s.

7.27 Strain produced by an ultrasonic actuator. An ultrasonic transducer is used to test thick aluminum billets for defects. To do so a transducer capable of generating 6 W of acoustic power operating at 10 MHz is used. The transducer is circular with a diameter of 30 mm. Assume that propagation of the ultrasonic wave is in a 15° cone and that the power density is uniform across the cross section of the cone. The acoustic properties of aluminum are given in **Tables 7.1–7.4**. The coefficient of elasticity of aluminum is 79 GPa.

a. Calculate the strain in the material at the surface at the location of the transducer.

b. Calculate the strain in the material at a depth of 60 mm.

c. Comment on the results and on the use of ultrasonic diagnostics in the body.

7.28 SAW resonator temperature sensor. The SAW resonator can be used to sense any quantity that will affect its resonant frequency, including temperature. The sound velocity of quartz is 5900 m/s at 20°C. The speed of sound is temperature dependent and increases by 0.32 mm/s/°C, and the coefficient of thermal expansion is 0.557 μm/m/°C.

a. Sketch a compensated sensor that will measure temperature but not be affected by other quantities such as ambient pressure. Can this be done in practice?

b. Calculate the sensitivity to temperature for a SAW sensor operating at 400 MHz. Comment on the practicality of this sensor.

7.29 SAW resonator as a pressure sensor. The structure in **Figure 7.56** is used as a pressure sensor. The sensor is made of quartz and consists of a number of grooves separated 10 μm apart. The area on which the pressure operates is $w = 2$ mm wide, $L = 10$ mm long, and $d = 0.8$ mm thick. Assume the quartz chip bends under pressure as a simply supported beam of thickness 0.8 mm with its supports at the edges of the device. Quartz has a modulus of elasticity of 71.7 GPa.

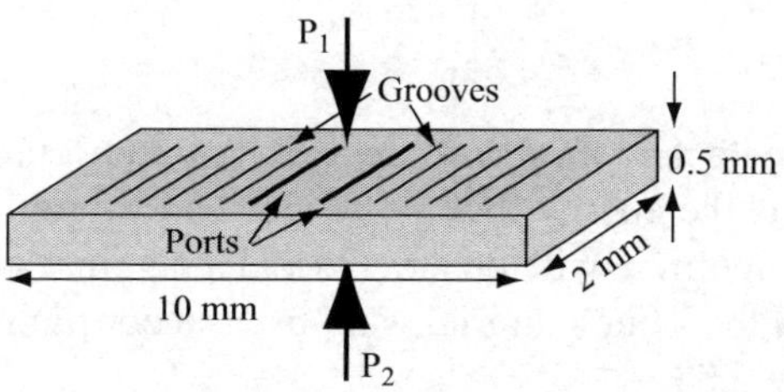

FIGURE 7.56 ■ The SAW resonator as a pressure sensor.

a. Calculate the sensitivity to pressure and the shift in resonant frequency for a pressure of 1 atm above ambient if the pressure P_1 is applied on the upper surface ($P_2 = 1$ atm).

b. Calculate the resonant frequency of the sensor for $P_1 = 120{,}000$ Pa and $P_2 = 150{,}000$ Pa.

7.30 **SAW mass sensor.** A SAW resonator made as in **Figure 7.48** has a length $a = 4$ mm, width $w = 2$ mm, and thickness $t = 0.2$ mm. The sensor is made of fused quartz and oscillates at 120 MHz without a stimulus. The modulus of elasticity for quartz is 71.7 GPa.

a. If the sensor is used to sense mass, what is the sensitivity of the sensor (in g/Hz)? Assume the mass is uniformly distributed on the sensor.

b. The maximum strain allowable in quartz is 1.2%. Calculate the range and span of the sensor.

c. If the frequency is measured by a frequency counter and the lowest frequency change that can be distinguished is 10 Hz, what is the resolution of the instrument?

7.31 **Piezoelectric ignition device.** A unique and common actuator makes use of a piezoelectric device to generate sufficiently high voltages that can generate sparks to ignite gas. This device can be found in all cigarette lighters and in ignition switches for gas kitchen stoves, furnaces, and other applications. The device uses a small, typically cylindrical crystal and a spring-loaded hammer that delivers a fixed, known force when hitting the crystal. Consider the device used in cigarette lighters. The crystal is 2 mm in diameter and 10 mm long (**Figure 7.57**).

a. Assuming the crystal is $BaTiO_3$ with a 3-3 cut and the required voltage to produce a spark of the appropriate size is 3200 V, calculate the impact force required. Describe the approximations needed and their validity.

b. Describe how that force can be generated using a spring with a spring constant $k = 2000$ N/m.

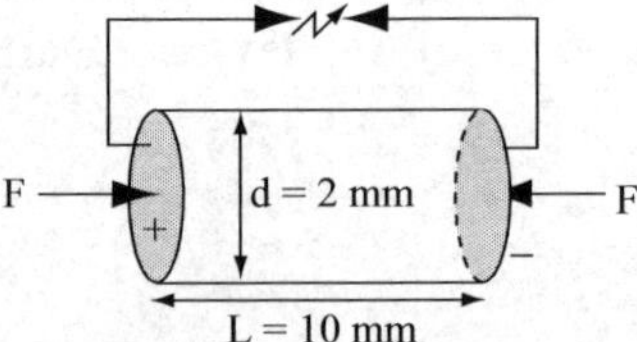

FIGURE 7.57 ■ Piezoelectric gas ignition device.

7.32 **Improved piezoelectric ignition device.** An important parameter in piezoelectric igniters is the energy that the device can supply. The higher the energy, the more likely ignition will occur. Consider again **Problem 7.31**. Using the same crystal and the same material, suppose the configuration is changed to that shown in **Figure 7.58**.

a. Using the dimensions shown and assuming a $BaTiO_3$ crystal with a 3-3 cut, calculate the force required to produce a voltage of 3200 V across the spark gap.

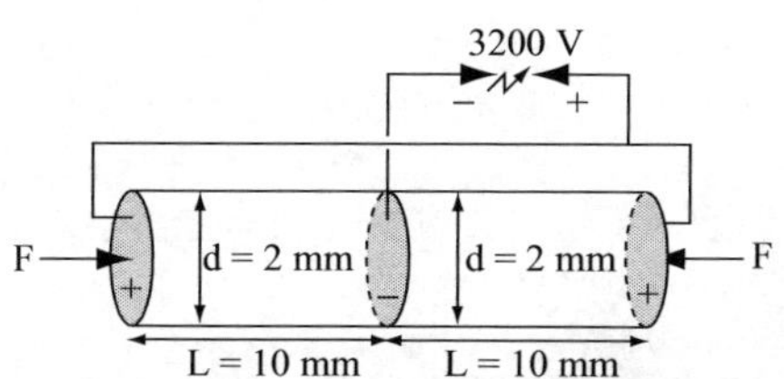

FIGURE 7.58 ■ Improved piezoelectric gas ignition device.

b. Show in general terms that the energy supplied by this device is twice the energy supplied by the device in **Figure 7.57**.

c. Where does the extra energy come from?

d. Show how the energy available from the ignition device can be further increased without increasing the force and discuss the practicality of your approach.

a.) $V = \frac{Q}{C} = \frac{d_{ij} F}{C} = \frac{d_{ij} F d}{\varepsilon_{ij} A}$

$F = V \frac{\varepsilon_{ij} A}{d_{ij} d} = \frac{V \varepsilon_{ij} A}{d_{33} d}$

$F = 16N$ ← a lot of force for small area

$d = 10 \times 10^{-3} m$

$d_{33} = 100 \times 10^{-12}$

$V = 3200 V$

$\varepsilon = 180$

$\varepsilon_0 = 8.85 \times 10^{-12}$

$A = \pi r^2$

$A = \pi (1 \times 10^{-3})^2$

CHAPTER 8

Chemical Sensor and Actuators

The tongue and the nose

Two of our most important chemical sensors, the tongue and the nose not only share a close and connected space but also cooperate in determining taste. The tongue is a multifunction muscle, perhaps the most flexible in the body. Taste, the chemical analysis of substances that come in contact with the tongue, is processed by taste buds or sensors and can detect five distinct flavors: salty, sour, bitter, sweet, and savory. Although taste buds are found mostly on the tongue, some can be found as well on the soft palate, upper esophagus, and epiglottis (the area in the back of the mouth between the tongue and the larynx). Most taste buds reside in protrusions on the surface of the tongue and open toward its upper surface, an opening through which food elements come in contact with it (gustatory pore). The human tongue may contain upwards of 8000 taste buds or as few as 2000, depending on individual variations and on age. Taste is transmitted through nerves to the gustatory section of the brain.

The tongue has other functions as well. In humans it is an integral part of processing food and cleaning the mouth and, significantly, of speech. As such it serves as a mechanical organ. In some animals it is part of the heat regulation mechanism (as, e.g., in dogs). In many animals it serves as an indispensable hygienic function in cleaning fur or drinking (e.g., in cats) and the cleaning of soft organs (such as cleaning the eyes in some reptiles or the muzzle in bovines). Specialized functions of the tongue can be found, an example being the prehensile tongue of the chameleon or the elongated tongue of the giraffe serving as a hook for feeding purposes.

The second chemical organ is the nose. It consists of a relatively simple structure with its external, visible protrusion and its two nostrils. Internally it has a number of functions. Immediately behind the nostrils are three bony surfaces called conchae that force and regulate the air flow downward toward the lungs. These also warm the air and, together with a mucous surface and hairs, filter the air of debris and dust. Soft tissue on their sides also controls the amount of air and its speed by constricting or enlarging the opening. Above, in the upper part of the nose cavity and out of the main airstream, a separate cavity contains the olfactory organ, the cells that are responsible for smell. This cavity is open toward the airstream, sampling the air, but because air does not flow through it, the molecules linger in it long enough to accomplish the smelling function. It is for this reason that smells sometimes seem to linger long after their causes have disappeared. The olfactory cells are connected to the olfactory section of the brain. The sense of smell is usually not considered as critical as that of sight or hearing, but it is somehow connected with long-term memory. Long after the sights or sounds of an event have faded, the odors of a place or a situation linger in the brain, still vivid and evoking. The nose also has certain adaptations. In most mammals the nose has a secondary olfactory bulb called the vomeronasal organs that sense certain chemical messages associated with social and sexual conditions. These organs bypass the cerebral cortex and link to sections in the brain responsible for reproduction and maternity and also affect aggressiveness in males. Another adaptation in some reptiles (snakes, lizards) is the combination of a forked tongue that samples the air and deposits molecules into an organ (called the Jacobson organ) on the roof of the mouth to chemically sense the environment.

8.1 | INTRODUCTION

To most, chemical sensors are likely to be the least understood of sensors and least known, even though they are common in the home, transportation, and places of work. The principles involved are often very different than those applied to other sensors and the method of sensing may be different as well. In many chemical sensors, sampling of a substance occurs. The sample is then allowed to interact in some fashion with elements of the sensor and usually an electric output is obtained from this reaction. Some sensors perform a complete analysis of the substance, while in others a direct output is obtained simply from the presence of the substance. Even the units involved are sometimes obscure to all but those versed in chemistry or chemical engineering.

However, chemical sensing is no different than other areas of sensing in that stimuli are involved, a physical sensor is employed, and the output is used to affect an appropriate action. It is worth reiterating that chemical sensors are very common and the role of chemical sensing is only likely to increase as the need for more stringent environmental monitoring and protection increases. An important role of sensors is in environmental monitoring, protection and tracking of hazardous materials, as well as the use of chemical sensors to track natural and man-made occurrences, including pollution, waterways infestation, migration of species, and, of course, weather prediction and tracking. In the sciences and in medicine, the sampling of substances such as oxygen, blood, and alcohol is well known. The food industry relies on them heavily in monitoring food processing and food safety and the military has been using chemical sensors at least since World War I to track chemical agents used in chemical warfare. Pollution control in vehicles is done on a vast scale with literally billions of chemical sensors in use. And just as important are the uses around the home: carbon monoxide (CO) detectors, smoke alarms, pH meters, and many more.

And chemical actuators also exist. We tend to think of actuators in terms of mechanical actuation, although by now it should be clear that any action taken by a system may be viewed as an output of that system and hence qualifies as actuation. In that sense, chemical actuators are those devices and processes that perform a chemical reaction or process to affect a specific outcome. For example, chemical scrubbers, whose role is to remove a substance or substances (usually for purposes of pollution control), are an important class of chemical actuators. Also used for pollution control is the catalytic converter, whose best-known use is in vehicles. And if mechanical actuation is easier to conceptualize, then the internal combustion engine or the deployment of an airbag during an accident may be good examples of actuators that may be termed chemical (although they may equally well be called mechanical actuators).

The plethora of applications and devices presents another problem—how to classify chemical sensors and actuators and the proper approach for their presentation. It seems that the first level of distinction between chemical stimuli is between a direct and indirect output. In a direct sensor, the chemical reaction or the presence of a chemical produces a measured electrical output. A simple example is the capacitive moisture sensor—the capacitance is directly proportional to the amount of water (or other fluid) present between its plates. An indirect (also called complex) sensor relies on a secondary, indirect reading of the sensed stimulus. For example, in an optical smoke detector, an optical sensor such as a photoresistor is illuminated by a source and establishes a background reading. Smoke is "sampled" by allowing it to flow between the source and sensor and alter the light intensity, its velocity, its phase, or some other measurable property. Some chemical sensors are much more complex than that and may involve more transduction steps. In fact, some may be viewed as complete instruments or processes.

Another distinction that can be made is on the basis of the stimulus itself. For example, the sensing of stimuli such as acidity, conductivity, and oxidation–reduction potential may form the basis of a classification.

We will avoid a rigid classification and will concentrate on those chemical sensors that are most important from a practical point of view while trying to cover the principles involved in chemical sensing. In doing so, we will try to steer clear of most chemical reactions and the formulas associated with them, replacing these with physical explanations that convey the process and explain the results without the baggage of analytic chemistry. We will start with the class of electrochemical sensors. This class includes those sensors that convert a chemical quantity directly into an electrical reading and follows the definition above for direct sensors. The second group is those sensors that generate heat and where heat is the sensed quantity. These sensors, just like the thermo-optical sensors in **Chapter 4**, are indirect sensors, as are the optical chemical sensors. Following these are some of the most common sensors, such as pH and gas sensors. Humidity and moisture sensors are included here even though their sensing is not truly chemical, but because the sensing methods and materials relate to chemical sensors.

8.2 | CHEMICAL UNITS

Most units used in conjunction with chemical sensors and actuators are the same as in other disciplines, but there are a few that are unique. These are defined here before we use them.

Mole (mol): the only chemical base SI unit, defined as an amount of substance equal to approximately 6.02214×10^{23} (Avogadro's number) molecules of that substance. Units of millimole (mmol), kilomole (kmol), etc., are sometimes employed.

Molar mass (g/mol): the mass in grams of a mole of a substance.

Gram-equivalent (g-eq): the mass of 1 equivalent, that is, the mass of a given substance that will either (1) supply or react with 1 mole of hydrogen cations (H^+) in an acid/base solution or (2) supply or react with 1 mole of electrons in an oxidation–reduction (redox) reaction. The gram-equivalent is more general than that and for general purpose use is equal to the molar mass (mass/mole) divided by the valence of the atom or molecule being considered.

Parts per million (ppm) and **parts per billion (ppb):** dimensionless quantities that in their most common usage represent a fraction of a quantity, such as a mass fraction (1 mg/kg = 1 ppm or 10 μg/kg = 10 ppb). However, it can represent any other fraction. These notations are used in the same fashion as the percent (%) to indicate the fraction of one species into the whole. Although not strictly correct, it is sometimes used to represent a change with respect to a variable.

For example, we may say that the change is 1 ppm/°C. Or we may say that the change in volume of a material is 100 ppm/°C, meaning that the change in volume is 100 $\mu m^3/m^3/°C$. The units ppm and ppb are not part of the SI system, but are universally accepted and commonly used in chemistry and medicine. Any use of ppb should be undertaken carefully, as the billion has two different meanings. In the United States the billion is used as the so-called short scale: 1 billion = 10^9. The traditional value for billion is the long scale: 1 billion = 10^{12}. The unit ppm refers to the short scale (1 ppm = one part in 10^9).

EXAMPLE 8.1 Conversion between mole and mass

The mole is not a fixed quantity, that is, a mole of one substance represents a different mass than a mole of another substance. Consider oxygen, hydrogen, and water. A mole of each has the same number of molecules or atoms, but the masses are different. To convert from moles to mass (or vice versa) we use the atomic units of the substance.

Oxygen has an atomic mass equal to 16 amu (atomic mass units). Therefore 1 mole represents a mass of 16 g. Hydrogen has an atomic mass equal to 1.008 amu. Thus 1 mole of hydrogen has a mass of 1.008 g. Water (H_2O) has atomic mass of $2 \times 1.008 + 16 = 18.016$ amu. The mass of 1 mole of water is 18.016 g. The molar mass is calculated from the atomic mass of the constituents of the substance by adding the masses of all constituents in 1 mole.

8.3 | ELECTROCHEMICAL SENSORS

An electrochemical sensor is expected to exhibit changes in resistance (conductivity) or changes in capacitance (permittivity) due to substances or reactions. These may carry different names. For example, potentiometric sensors are those that do not involve current, only measurement of capacitance and voltage. Amperimetric sensors rely on measuring current, whereas conductimetric sensors rely on measurement of conductivity (resistance). These are different names for the same properties since voltage, current, and resistance are related by Ohm's law.

Electrochemical sensors include a large number of sensing methods, all based on the broad area of electrochemistry. Many common devices, including fuel cells (an actuator), surface conductivity sensors, enzyme electrodes, oxidation sensors, and humidity sensors, belong to this category. We shall start with some of the simplest and most useful sensors available, the metal oxide sensors.

8.3.1 Metal Oxide Sensors

Metal oxide sensors rely on a very well-known property of metal oxides at elevated temperatures to change their surface potential, and therefore their conductivity, in the presence of various reducible gases such as ethyl alcohol, methane, and many others, sometimes selectively, sometimes not. Metal oxides that can be used are oxides of tin (SnO_2), zinc (ZnO), iron (Fe_2O_3), zirconium (ZrO_2), titanium (TiO_2), and wolfram (WO_3). These are semiconductor materials and may be either *p*- or *n*-type, with preference toward *n*-type materials. The fabrication is relatively simple and may be based on silicon processes or other thin or thick film technologies. The basic principle is that when an oxide is held at elevated temperatures, the surrounding gases react with the oxygen in the oxide, causing changes in the resistivity of the material. The essential components are the high temperature, the oxide, and the reaction in the oxide.

As a representative sensor, consider the carbon monoxide (CO) sensor shown in **Figure 8.1a**. It consists of a heater and a thin layer of tin dioxide (SnO_2) above it. In terms of construction, a silicon layer is first created to serve as a temporary support for the structure. Above it, a silicon dioxide (SiO_2) layer is thermally grown. This layer must be capable of withstanding high temperatures. A layer of gold is sputtered on top of the SiO_2 layer and etched to form a long meandering wire that serves as the heating element by driving it with a sufficiently high current. A second layer of SiO_2 is deposited on top, sandwiching the gold heating element. Then the SnO_2 layer is sputtered on top and patterned with grooves to increase its active surface. The original silicon material is finally etched away to decrease the heat capacity of the sensor. The sensing area can be quite small: 1–1.5 mm^2. The device is heated to 300°C to operate, but because the size is very small and the heat capacity is small as well, the power needed is typically small—on the order of 100 mW. The conductivity of the oxide can be written as

$$\sigma = \sigma_0 + kP^m \quad [\mathrm{S/m}], \tag{8.1}$$

where σ_0 is the conductivity of the SnO_2 at 300°C but without CO present, P is the concentration of the CO gas [ppm], k is a sensitivity coefficient (determined experimentally for various oxides), and the exponent m is again an experimental value, which for SnO_2 is about 0.5. Thus the conductivity increases with an increase in concentration, as shown in **Figure 8.1b**. The resistance is proportional to the inverse of conductivity, so it may be written as

$$R = aP^{-\alpha} \quad [\Omega], \tag{8.2}$$

where a is a constant defined by the material and construction, α is an experimental quantity for the gas, and P is its concentration [ppm]. This simple relation defines the response of the sensor to various gases, but only on a range of concentrations since it cannot define the resistance at zero concentration. The response is exponential (linear on a log scale) and a transfer function of the type shown in **Figure 8.1b** must be defined for each gas and each type of oxide. SiO_2-based sensors as well as ZnO sensors can also be used to sense carbon dioxide (CO_2), toluene, benzene, ether, ethyl alcohol, and propane with excellent sensitivity (1–50 ppm).

A variation of the structure above is shown in **Figure 8.2**. It consists of an SnO_2 layer on a ferrite substrate. The heater here is provided by a thick layer of rubidium dioxide (RuO_2) fed through two gold contacts (C and D). The resistance of the very thin SnO_2 layer (less than about 0.5 μm) is measured between the two gold contacts on top (A and B). This sensor, which operates as above, is sensitive primarily to ethanol and CO.

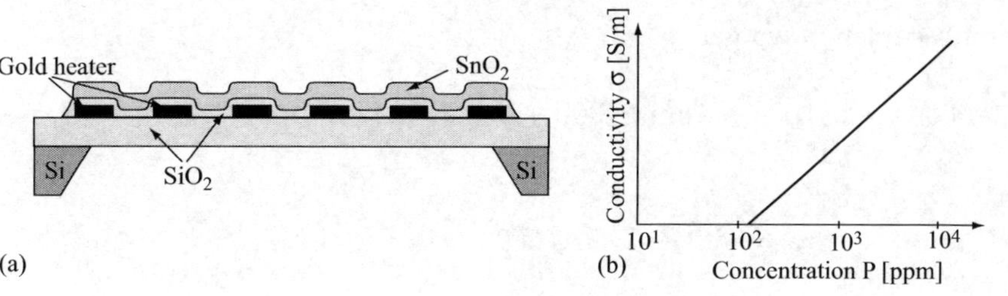

FIGURE 8.1 ■ A metal oxide CO sensor. (a) Construction. (b) Transfer function.

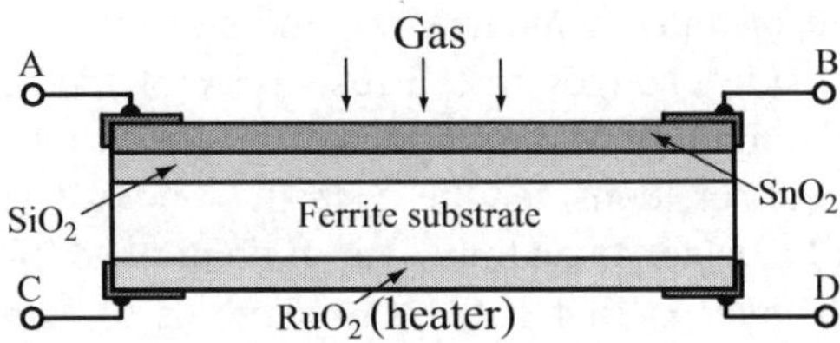

FIGURE 8.2 ■ An ethanol and CO sensor.

EXAMPLE 8.2 Alcohol sensor

An alcohol sensor is made by deposition of a thin layer of tungsten trioxide (WO_3) nanoparticles on a substrate. To evaluate its performance, the resistance of the sensor is measured at two concentrations. At 100 ppm the resistance is 161 kΩ, while at 1000 ppm its resistance is 112 kΩ. The resistance of the sensor in air in the absence of alcohol is 320 kΩ. Calculate the sensitivity of the sensor in ohms/ppm.

Solution: We can use **Equation (8.2)** to evaluate the constants a and α. Then we can calculate the sensitivity based on the definition.

At 100 ppm,

$$R_1 = 161{,}000 = a100^{-\alpha} \quad [\Omega].$$

At 1000 ppm,

$$R_2 = 112{,}000 = a1000^{-\alpha} \quad [\Omega].$$

To evaluate the constants we take the natural logarithm on both sides of both relations:

$$\ln 161{,}000 = \ln a - \alpha \ln(100)$$

$$\ln 112{,}000 = \ln a - \alpha \ln(1000).$$

Subtracting the second relation from the first we get

$$\ln 161{,}000 - \ln 112{,}00 = \alpha \ln(1000) - \alpha \ln(100),$$

or

$$\alpha = \frac{\ln 161{,}000 - \ln 112{,}00}{\ln 1000 - \ln 100} = \frac{\ln\left(\frac{161}{112}\right)}{\ln 10} = 0.1576.$$

Substituting back into either relation we get

$$\ln a = \ln 112{,}000 + \alpha \ln(1000) = \ln 112{,}000 + 0.1576 \ln(1000) = 12.7149.$$

Therefore

$$a = e^{12.7149} = 332{,}667.$$

The relation for resistance now becomes

$$R = 332{,}667P^{-0.1576} \quad [\Omega]$$

The sensitivity may be written as

$$S = \frac{dR}{dP} = -0.1576 \times 332{,}667P^{-1.1576} = -52{,}428P^{-1.1576} \quad [\Omega/\text{ppm}].$$

The sensitivity varies along the curve as expected. For example, at 500 ppm the sensitivity is 39.37 Ω/ppm, whereas at 200 ppm it is 113.73 Ω/ppm.

As mentioned above, the reaction is with oxygen, and hence any reducible gas (a gas that reacts with oxygen) will be detected. This lack of selectivity is a common problem in metal oxide sensors. To overcome this problem, one can select temperatures at which the required gas reacts but not others or the particular gas may be filtered. These sensors are used in many applications, from CO and CO_2 detectors to oxygen sensors in automobiles. The latter, for example, uses a TiO_2 sensor built as above in which resistance increases in proportion to the concentration of oxygen. This is commonly used in other applications, such as the sensing of oxygen in water (for pollution control purposes). The process can also be used to determine the amount of available organic material in water by first evaporating the water and then oxygenating the residue to determine how much oxygen is consumed. The amount of oxygen consumed in the reaction is then an indication of the amount of organic material in the sample.

8.3.2 Solid Electrolyte Sensors

Another important type of sensor that has found significant commercial applications is the solid electrolyte sensor, most often used in oxygen sensors, including those in automobiles. In these sensors a solid galvanic cell (battery cell) is built that produces an emf across two electrodes based on the oxygen concentrations at the two electrodes under constant temperature and pressure. A solid electrolyte, usually made of zirconium dioxide (ZrO_2) and calcium oxide (CaO) in a roughly 90%:10% ratio, is often used because it has high oxygen ion conductivity at elevated temperatures (above 500°C). The electrode is made of sintered ZrO_2 powder (sintering makes the powder into a ceramic). The inner and outer electrodes are made of platinum that act as catalysts and absorb oxygen. The structure is shown in **Figure 8.3** for an exhaust oxygen sensor in an automobile. The potential across the electrodes is

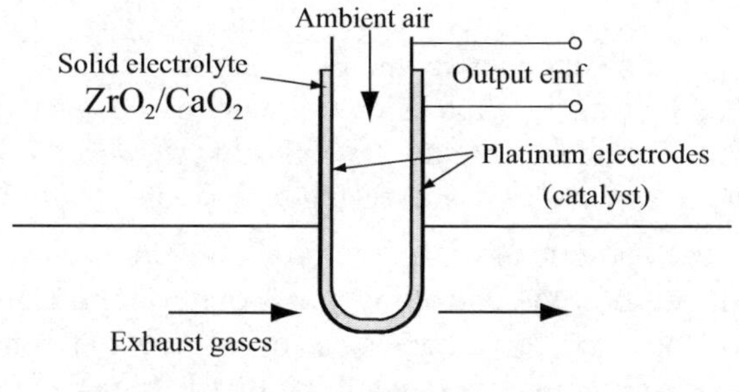

FIGURE 8.3 ■ A solid electrolyte oxygen sensor for car engines used as an active sensor.

$$emf = \frac{RT}{4F}\ln\left(\frac{P^1_{O2}}{P^2_{O2}}\right) \quad [V], \tag{8.3}$$

where R is the universal gas constant (8.314472 J/K/mol), T is the temperature (K), and F is the Faraday constant (96,487 C/mol). P^1_{O2} is the concentration of oxygen in the atmosphere, P^2_{O2} is the concentration of oxygen in the exhaust, both heated to the same temperature. There is also a small constant added to the relation in **Equation (8.3)** that represents the emf when both concentrations are the same. Ideally this constant should be zero, but in practice it is not. We will leave it out, however, because it is small and because it depends on the sensor (construction, materials, etc.) and hence is taken care of in the calibration of the sensor. The oxygen sensor is used to adjust the fuel ratio at the most efficient rate at which pollutants NO and NO_2, known together as NO_x and CO are converted into N_2, CO_2, and H_2O, all of which are natural constituents in the atmosphere and hence considered nonpollutants. In a heated oxygen sensor, the oxygen concentrations in the exhaust stream produce emfs between about 2 mV for atmospheric concentration (20.6%) and about 60 mV for oxygen concentrations around 1%.

EXAMPLE 8.3 Strategies for emission control in internal combustion engines

There are a number of strategies that can be followed to reduce emissions of noxious gases from vehicles equipped with internal combustion engines. All involve a catalytic converter and oxygen sensors (see **Section 8.9.1** for the operation of the catalytic converter) and convert CO as well as NO to benign forms such as CO_2, N_2, and H_2O_2. These are as follows:

1. The oxygen level is sensed before gasses get into the catalytic converter. This strategy is based on the premise that given enough oxygen in the exhaust stream most (if not all) gases will be converted to benign forms. This method has the advantage of simplicity—it senses relatively large concentrations of oxygen. Based on known fuel needs, the oxygen level can be increased or decreased to attain minimum emissions. However, one never knows if the oxygen is actually being used to oxidize pollutants or if it is simply released with the exhaust. A perfectly functioning or totally nonfunctioning catalytic converter would be treated the same by the control system.
2. It is possible to measure only the oxygen in the exhaust after the catalytic converter, assuming that if there is some oxygen in the exhaust stream, the catalytic converter converted all the noxious gases that could have been converted. The advantage of this method is that it indicates how much oxygen has not been consumed and can increase the oxygen levels in the intake to keep a minimum level of oxygen in the exhaust. Similarly, if too much oxygen is present it can reduce the oxygen levels, since an engine that runs too lean also runs too hot. The problem with this method is that an inefficient or nonfunctioning catalytic converter would leave too much oxygen in the exhaust, forcing the control system to reduce oxygen at the intake and forcing the engine to run too rich, increasing emissions.
3. By placing an oxygen sensor before and one after the catalytic converter, a true closed-loop control may be accomplished. In addition to controlling emissions, comparison of the readings of the two sensors can also be used for diagnostics to indicate the condition of the catalytic converter (or the absence of a converter, something that is illegal) and the sensors

themselves. Cars equipped with an onboard diagnostics (OBDII) system have two sensors and the system issues diagnostics codes covering the status of the sensors and of the catalytic converter (see **Example 1.1** for some of the sensors covered by the OBDII system). This method is optimal in the sense that it avoids the main problems associated with single-sensor systems.

A common strategy for pollution control in an internal combustion engine is to sense the oxygen concentration before and after the catalytic converter and use the readings of the oxygen sensors to adjust the fuel:air ratio in the engine to ensure full burning of noxious gases such as CO. By measuring both the input and output oxygen concentration one gets an indication of the effectiveness of the catalytic converter itself (see **Example 8.4**).

In engines that operate in a much leaner mode, the solid electrolyte sensor is not sufficiently sensitive, since in these high-efficiency engines the amount of oxygen in the exhaust is high and the reading of the electrolytic cell is insufficient. In such engines, the same basic sensor is used, that is, a solid electrolyte between two platinum electrodes, as shown in **Figure 8.4**, but a potential is applied to the cell. This arrangement forces (pumps) oxygen across the electrolyte and a current is produced proportional to the oxygen concentration in the exhaust. This sensor is called a diffusion oxygen sensor or a diffusion-controlled limiting current oxygen sensor.

Another important application of solid electrolyte sensors is in oxygen sensing in the production of steel and other molten materials, since the quality of the final product is a direct result of the amount of oxygen in the process. The sensor is shown in **Figure 8.5**. The molybdenum needle is used to keep the device from melting when it is inserted in the molten steel. A potential difference is developed across the cell (between the molybdenum and the outer layer). The voltage is measured between the inner electrode and the outer layer through an iron electrode dipped into the molten steel. The voltage developed is directly proportional to the oxygen concentration in the molten steel.

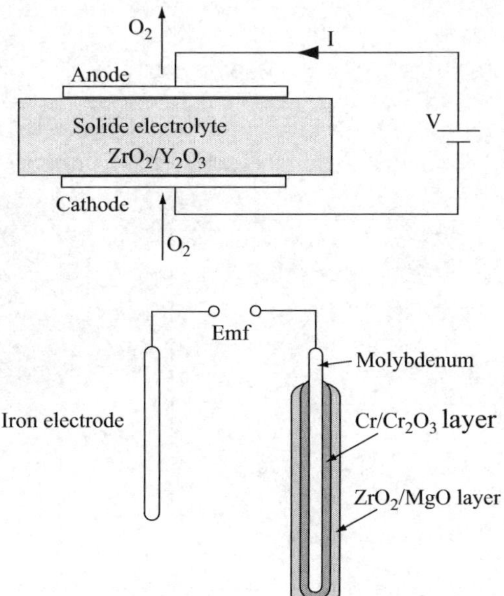

FIGURE 8.4 ■ A diffusion-controlled limiting current oxygen sensor used as a passive sensor.

FIGURE 8.5 ■ Oxygen sensor for molten metals.

EXAMPLE 8.4 Oxygen sensors in cars: efficiency of catalytic converters

To monitor the efficiency of a catalytic converter, one can use an oxygen sensor before and another after the catalytic converter. Taking the difference between the potentials of the two sensors makes a differential sensor and gives an indication of the conversion efficiency of the converter. The larger the difference, the more oxygen is used in the converter, and hence the higher its conversion efficiency. Suppose the oxygen content entering the converter is 10% and the minimum oxygen concentration required in the exhaust is 1%. Calculate the transfer function of the differential sensor assuming both sensors are at a temperature of 750°C.

Solution: The output of each sensor is calculated using **Equation (8.3)**. We denote P_0, the oxygen concentration in air; P_{in}, the concentration in the exhaust before entering the converter; and P_{out}, the concentration after the converter. The potentials of the two sensors are

For the sensor in front of the converter:

$$emf_1 = \frac{RT}{4F}\ln\left(\frac{P_0}{P_{in}}\right) \quad [V].$$

The emf in the exit sensor:

$$emf_2 = \frac{RT}{4F}\ln\left(\frac{P_0}{P_{out}}\right) \quad [V].$$

Since emf_2 is necessarily larger than emf_1, we calculate the difference as

$$emf_2 - emf = \frac{RT}{4F}\ln\left(\frac{P_0}{P_{out}}\right) - \frac{RT}{4F}\ln\left(\frac{P_0}{P_{in}}\right) = \frac{RT}{4F}\ln\left(\frac{P_{in}}{P_{out}}\right) \quad [V].$$

For the given input and exhaust concentrations we get

$$\Delta emf = emf_2 - emf = \frac{RT}{4F}\ln\left(\frac{P_{in}}{P_{out}}\right) = \frac{8.314472 \times 1023.15}{4 \times 96487}(\ln P_{in} - \ln P_{out})$$
$$= 0.02204(\ln 0.1 - \ln P_{out}) = -0.02204 \ln P_{out} - 0.05075 \quad [V].$$

That is,

$$\Delta emf = -0.02204 \ln P_{out} - 0.05075 \quad [V].$$

This transfer function is shown in **Figure 8.6** for concentrations between 0.1 (10%) and 0.01 (1%). The potential at 1% oxygen in the exhaust (after the converter) is approximately 50 mV.

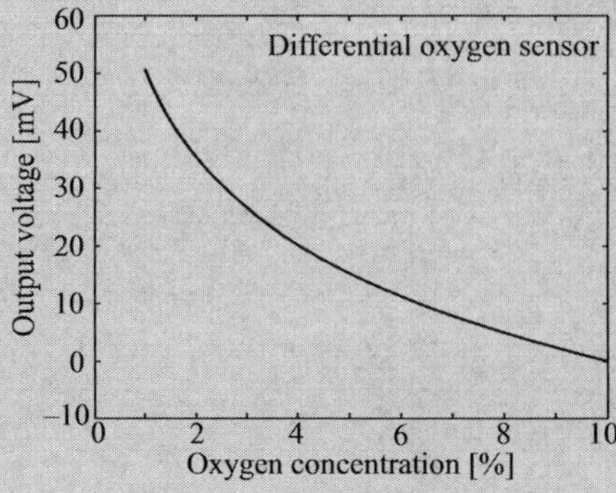

FIGURE 8.6 ■ Transfer function of a differential oxygen sensor.

This decreases with an increase in concentration until at 10% oxygen it is zero (both sensors are at the same potential since the oxygen concentrations are the same). The higher the output, the higher the efficiency of the converter (i.e., the lower the oxygen concentration in the exhaust).

8.3.3 The Metal Oxide Semiconductor (MOS) Chemical Sensor

A unique development in sensors is the use of the basic metal oxide semiconductor field-effect transistor (MOSFET) structure, commonly used in electronics, as a chemical sensor. The basic idea is that of the classic MOSFET transistor, in that the gate serves as the sensing surface. The advantage of this is that a very simple and sensitive device is obtained that controls the current through the MOSFET. The interfacing of such a device is simple and there are fewer problems (such as heating, temperature sensing, and compensation, etc.) to overcome. It is therefore not surprising that the basic MOS structure has been developed into an array of sensors for diverse applications.

For example, by simply replacing the metal gate in **Figure 8.7** with palladium, the MOSFET becomes a hydrogen sensor, since the palladium gate absorbs hydrogen and its potential changes accordingly. Sensitivity is down to about 1 ppm. Other similar structures can sense gases such as H_2S and NH_3. Palladium MOSFETs (Pd-gate MOSFETs) can also be used to sense oxygen in water, relying on the fact that the absorption efficiency of oxygen decreases in proportion to the amount of oxygen present.

We shall say more about MOSFET sensors in the subsequent section on pH sensing, since MOSFETS have been very successful in sensing pH.

8.4 | POTENTIOMETRIC SENSORS

A large subset of electrochemical sensors is the so-called potentiometric sensors. These are based on an electric potential that develops at the surface of a solid material when immersed in a solution containing ions that can be exchanged at the surface. The potential is proportional to the number or density of ions in the solution. A potential difference between the surface of the solid and the solution occurs because of charge separation at the surface. This contact potential, analogous to that used to set up a voltaic cell, cannot be measured directly. However, if a second, reference electrode is provided, an electrochemical cell is set up and the potential across the two electrodes is directly measurable. To ensure that the potential is measured accurately, and therefore that the ion concentration is properly represented by the potential, it is critical that the current

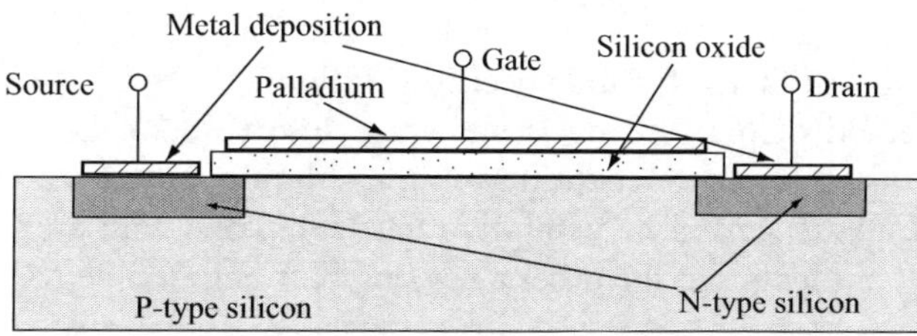

FIGURE 8.7 ■ The MOSFET structure used as a chemical sensor by replacing the gate with a species-sensitive material.

drawn by the measuring instrument is as small as possible (any current is a load on the cell and therefore reduces the measured potential).

For a sensor of this type to be useful, the potential generated must be ion specific—that is, the electrodes must be able to distinguish between solutions. These are called ion-specific electrodes or membranes. There are four types of membranes:

Glass membranes: selective for H^+, Na^+, $NH_4{}^+$ and similar ions.

Polymer-immobilized membranes: in this type of membrane, an ion-selective agent is immobilized (trapped) in a polymer matrix. A typical polymer is polyvinyl chloride (PVC).

Gel-immobilized enzyme membranes: the surface reaction is between an ion-specific enzyme, which in turn is either bonded onto a solid surface or immobilized into a matrix, and the solution.

Soluble inorganic salt membranes: either crystalline or powdered salts pressed into a solid are used. Typical salts are LaF_3 or mixtures of salts such as silver sulfide (Ag_2S) and silver chloride (AgCl). These electrodes are selective to fluoride (F^-), sulfur (S^{2-}), chlorine (Cl^-), and similar ions.

8.4.1 Glass Membrane Sensors

By far the oldest of the ion-selective electrodes, the glass membrane has been in use for pH sensing since the mid-1930s and is as common as ever. The electrode is a glass made with the addition of Na_2O and aluminum oxide (Al_2O_3), made into a very thin tubelike membrane. This results in a high-resistance membrane, which nevertheless allows transfer of ions across it. The pH sensor measures the concentration of H^+ ions in a solution as follows:

$$pH = -\log|H^+|, \tag{8.4}$$

where the concentration of hydrogen atoms is in terms of gram-equivalent per liter (g-eq/L). A concentration of 1 g-eq/L means a pH of 0, a concentration of 10^{-1} means a pH of 1, and so on. The normal pH scale is between 0 and 14, which corresponds to concentrations of 10^0 to 10^{-14} g-eq/L. However, pH can and is defined beyond that scale. A concentration of 10 g-eq/L will produce a pH of -1, whereas a concentration of 10^{-18} will produce a pH of 18. The higher the concentration, the more acidic the solution, and vice versa. A pH of 7 is considered neutral simply because that is the normal pH of water. In many instances, when the solutions are weak (weak acids or bases), **Equation (8.4)** is accurate and pH measures the actual ion concentration. In strong solutions the concentration of H^+ must be multiplied by an activity factor to take into account the contribution of other ions to the pH reading. The activity factor takes into account the interaction between all ions in the solution. In this case the pH is given as

$$pH = -\log|\gamma_H H^+|, \tag{8.5}$$

where γ_H is the activity factor for the specific solution.

The basic method of pH sensing is shown in **Figure 8.8a**. In principle all that is needed is to measure the ion concentration in the solution. However, that is difficult to do directly, hence pH is sensed by using two half-cells, one with a known pH, called a reference half-cell or electrode, the other a sensing half-cell or electrode. In **Figure 8.8a**

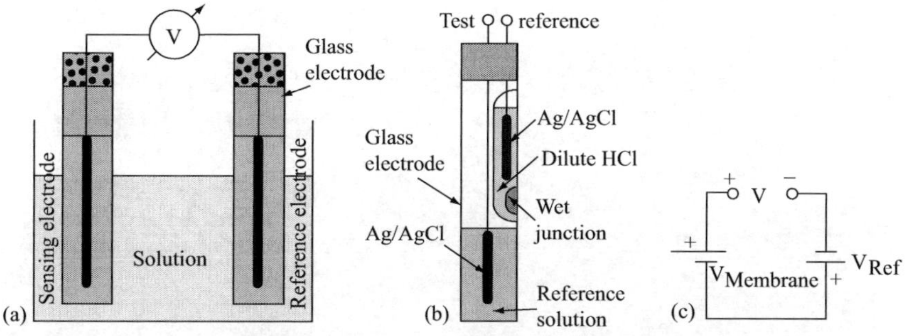

FIGURE 8.8 ■ (a) Basic method of pH sensing using a glass membrane. (b) A glass membrane pH probe incorporating the reference electrode in a single unit. (c) Equivalent circuit.

the sensing glass membrane electrode is shown on the left and a reference electrode on the right. The reference electrode is typically a silver/silver chloride (Ag/AgCl) electrode in a potassium chloride (KCl) aqueous solution or a saturated calomel electrode (Hg/Hg_2Cl_2 in a KCl solution). The reference electrode is normally incorporated into the test electrode so that the user only has to deal with a single probe, as shown in **Figure 8.8b**. Since what is actually measured is the difference between the electrode potential and the reference potential, it is easier to understand the measurement in the equivalent circuit in **Figure 8.8c**. The potential measured by the instrument is

$$V = V_{ref} + V_{membrane} \quad [V], \tag{8.6}$$

where V_{ref} is a constant value and $V_{membrane}$ depends on the ion concentration in the solution. The latter is given by the Nernst equation (which gives the potential of any half-cell) as

$$V_{membrane} = \frac{RT}{nF}\ln(a) = \frac{2.303RT}{nF}\log_{10}(a) = \frac{2.303RT}{nF}\text{pH} \quad [V], \tag{8.7}$$

where R is the universal gas constant (equal to 8.314462 J/mol/K), F is Faraday's constant (equal to 96,485.309 C/mol), n is the net number of negative charges transferred in the reaction, a is the activity of the ions involved in the reaction, and T is the temperature of the solution [K]. The term 2.303 comes from the fact that $\ln(a) = \log_{10}(a)/\log_{10}e$, that is, $2.303 = 1/\log_{10}e$. For H^+ ions, $n = 1$ (one electron transferred) and $\log_{10}(a)$ is the pH, leading to the following relation for pH by substituting **Equation (8.7)** into **Equation (8.6)**:

$$\text{pH} = \frac{(V - V_{ref})F}{2.303RT}. \tag{8.8}$$

Note that the activity a is an effective concentration, that is, the equivalent concentration that takes into account all interactions between ions. As indicated above, in weak acids or bases a represents the actual concentration. When the activity is less than 1, it is given as a fraction (e.g., 0.9).

In a measurement, V is the actual quantity measured—the rest of the quantities are taken into account internally. For this reason it is important for the voltage to be constant and stable and that the temperature is either taken into account or is compensated in the circuit itself. The voltage of the reference electrode is typically known or can be calculated from **Equation (8.7)**. For example, the saturated calomel (Hg/Hg_2Cl_2) (see **Figures 8.8** and **8.9**) electrode mentioned above has a potential of +0.244 V.

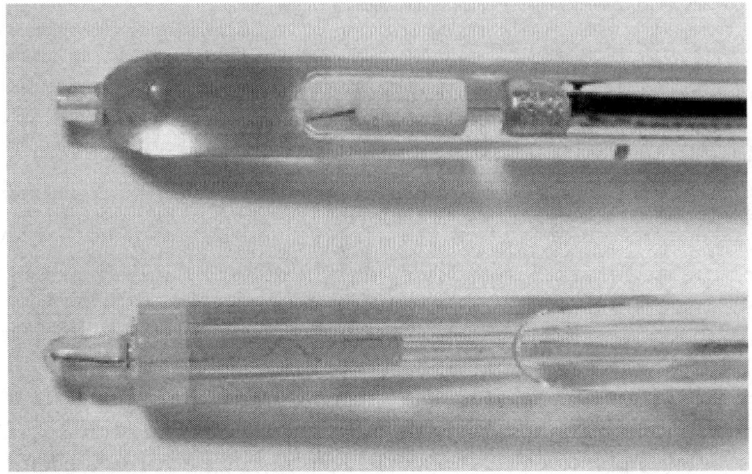

FIGURE 8.9 ■ A saturated calomel reference electrode (top) and a pH electrode (bottom).

The Ag/AgCl electrode has a potential of +0.197 V. The copper/copper sulfate ($Cu/CuSO_4$) electrode has a potential of +0.314 V. There are of course other reference electrodes that can be used. **Figure 8.9** shows a saturated calomel and a pH electrode.

The pH sensor is used by first immersing the electrode into a conditioning solution of hydrochloric acid (HCl) (0.1 mol/L) and then immersing it into the solution to be tested. The electric (voltage) output is calibrated directly in pH.

Modifications of the basic configuration, both in terms of the reference electrode (filling) and the constituents of the glass membrane, lead to sensitivity to other types of ions as well as to sensors capable of sensing concentrations of dissolved gas in solutions, particularly ammonia, but also CO_2, sulfur dioxide (SO_2), hydrogen fluoride (HF), hydrogen sulfide (H_2S), and hydrogen cyanide (HCN). In essence a pH-like electrode sensitive to the desired ions is used to sense the concentration of the respective ions in solutions. These sensors are important devices in industrial processes, in pollution control, and in environmental sensing.

EXAMPLE 8.5 Basic pH measurement

An uncompensated pH sensor is used to sense the pH in a fish tank.

a. Calculate the voltage measured using a saturated calomel reference electrode if the water is neutral (pH = 7) and the sensor is calibrated at 20°C.

b. Calculate the error in the voltage reading of the sensor if the temperature increases by 15°C. What is the expected pH reading at 35°C?

Solution:

a. We can use **Equation (8.8)** directly to calculate V since the reference voltage of the saturated calomel electrode is 0.244 V and the calibrated pH is 7. Thus we have

$$7 = \frac{(V - 0.244)F}{2.303RT} \to V = \frac{7 \times 2.303 \times 8.314462 \times 293.15}{9.64 \times 10^4} + 0.244 = 0.6516 \text{ V}.$$

b. To calculate error we write

$$\text{pH} = \frac{(V - 0.244) \times 9.64 \times 10^4}{2.303 \times 8.314462} T^{-1}.$$

The voltage read by the instrument may be written as

$$V = \frac{7 \times 2.303 \times 8.314462 \times T}{9.64 \times 10^4} + 0.244 = 13.9043 \times 10^{-4} T + 0.244$$
$$= 13.9043 \times 10^{-4} \times 308.15 + 0.244 = 0.6725 \text{ V}.$$

This is an error of $0.6725 - 0.6516 = 0.0209$ V or

$$e = \frac{0.6725 - 0.6516}{0.6516} \times 100\% = 3.2\%.$$

The expected pH can again be calculated from **Equation (8.8)** since we have the potentials at the given temperature:

$$\text{pH} = \frac{(V - V_{\text{ref}})F}{2.303RT} = \frac{(0.6725 - 0.244) \times 9.64 \times 10^4}{2.303 \times 8.314462 \times 308.15} = 7.00064.$$

The error is rather small because of the logarithmic nature of the pH reading.

8.4.2 Soluble Inorganic Salt Membrane Sensors

These membranes are based on soluble inorganic salts, which undergo ion exchange interaction in water and generate the required potential at the interface. Typical salts are lanthanum fluoride (LaF_3) and silver sulfide (Ag_2S). A membrane made of these materials may be either a single crystal membrane, a disk made of sintered powdered salt, or the powdered salt may be embedded into a polymer matrix, each of these leading to sensors of similar operation but different properties and sensitivities.

The structure of a sensor used to sense the fluoride concentration in water is shown in **Figure 8.10**. The sensing membrane is made in the form of a thin disk of LaF_3 grown as a single crystal. The reference electrode is created in the internal solution (in this case a sodium fluoride/sodium chloride [NaF/NaCl] solution in 0.1 mol/L solution). The sensor shown can detect concentrations of fluoride in water between 0.1 and 2000 mg/L. This sensor is commonly used to monitor fluoride in drinking water (normal concentration of about 1 mg/L).

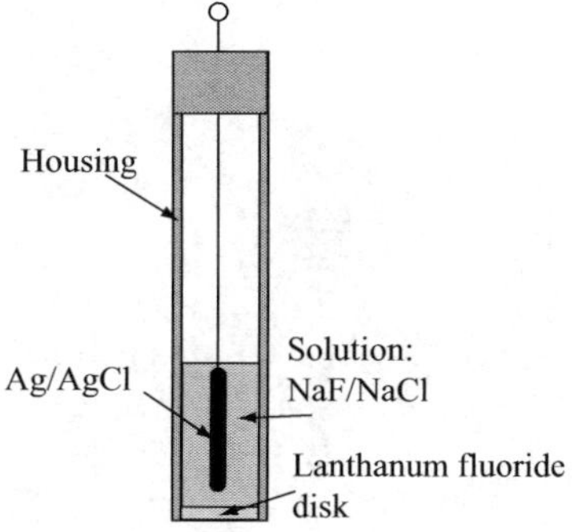

FIGURE 8.10 ■ Soluble inorganic salt membrane sensor for fluoride.

Membranes may be made of other materials, such as Ag_2S, which is easily made into thin sintered disks from powdered material and may be used in lieu of the single crystal. Also, in this form, other compounds may be added to affect the properties of the membrane, and hence its sensitivity to other ions. This leads to selective sensors sensitive to ions of chlorine, cadmium, lead, and copper, which are often used to sense for dissolved heavy metals in water.

Polymeric membranes are made by using a polymeric binder for the powdered salt in a ratio of about 50% salt and 50% binding material. The common binding materials are PVC, polyethylene, and silicon rubber. In terms of performance, these membranes are similar to sintered disks.

8.4.3 Polymer-Immobilized Ionophore Membranes

A development in the inorganic salt membrane is the use of polymer-immobilized membranes. In these, ion-selective organic reagents are used in the production of the polymer by including them in the plasticizers, particularly for PVC. A reagent, called an ionophore (or ion exchanger), is dissolved in the plasticizer at a concentration of about 1%. This produces a polymer film, which can then be used as the membrane, replacing the crystal or disk as the membrane in sensors. The construction of the sensor is simple and is shown in **Figure 8.11**. The sensor shown includes an Ag/AgCl reference electrode. The resulting sensor is a relatively high-resistance sensor. A different approach to building polymer-immobilized ionophore membranes is shown in **Figure 8.12**. It is made of an inner platinum wire on which the polymer membrane is coated and the wire is protected with a coating of paraffin. This is called a coated wire electrode. To be useful a reference membrane must be added.

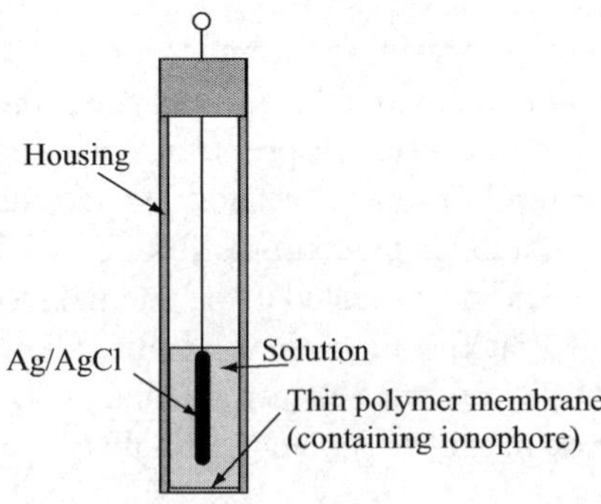

FIGURE 8.11 ■ Polymer-immobilized ionophore membrane sensor.

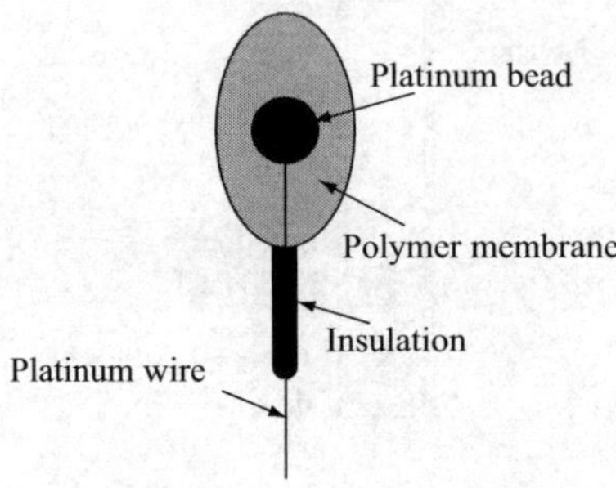

FIGURE 8.12 ■ Coated wire electrode.

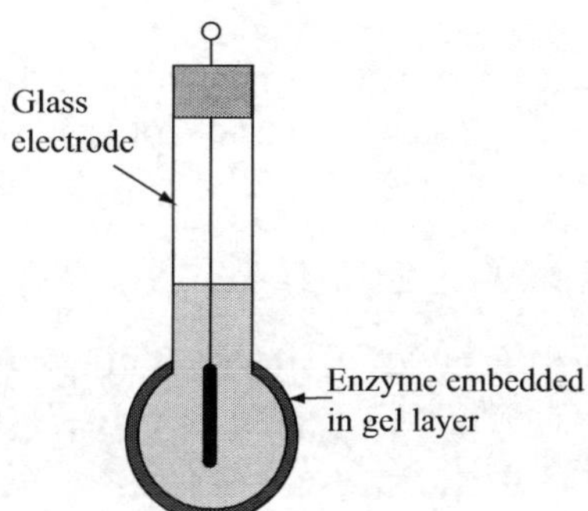

FIGURE 8.13 ■ Gel-immobilized enzyme membrane sensor.

Polymer membranes of this type can be made selective to many ions by the use of different ionophores. Sensitivity to calcium and potassium can be designed and these two types of sensors are routinely used to sense calcium in the blood and potassium in seawater. Nitrate-selective membranes are also available for sensing of nitrates in soil (fertilizers) and in runoff from fertilized fields.

8.4.4 Gel-Immobilized Enzyme Membranes

These sensors are similar in principle to polymer-immobilized ionophore membranes, but instead a gel is used and the ionophore is replaced by an enzyme designed to be selective to a particular ion. The enzyme, a biomaterial, is immobilized in a gel (polyacrylamide) and held in place on a glass membrane electrode, as shown in **Figure 8.13**. The choice of the enzyme and the choice of the glass electrode define the selectivity of the sensor. These sensors exist for the sensing of a variety of important analytes, including urea and glucose, L-amino acids, penicillin, and others. The operation is simple: The sensor is placed in the solution to be sensed, which diffuses into the gel and reacts with the enzyme. The ions released are then sensed by the glass electrode. Although these sensors are slow in response because of the need for diffusion, they are very useful in analysis in medicine, including the analysis of blood and urine.

EXAMPLE 8.6 **Sensing of fluoride in water**

Fluoride is an important additive to water, particularly useful for dental health in children. In addition to its use in drinking water, it is often added to toothpaste to strengthen the enamel on teeth. Sensing of the concentration of fluoride in solutions is typically done with the configuration in **Figure 8.10**. The LaF_3 disk serves as the sensing membrane sensitive to F^- ions, whereas the reference is the Ag/AgCl electrode with a potential of 0.199 V. To test for concentrations of fluoride one assumes the concentration is very low, otherwise the test shows activity rather than actual concentration. In many cases the concentration is given in ppm or ppb. The potential of the electrode can be calculated using **Equation (8.7)**, but since fluoride is a negative ion, $n = -1$. We will calculate it at 25°C:

$$V_{\text{membrane}} = \frac{2.303RT}{nF}\log_{10}(a) = -\frac{2.303 \times 8.314462 \times 298.15}{1 \times 9.64 \times 10^4}\log_{10}(a)$$

$$= 0.05922\log_{10}(a) \quad [\text{V}].$$

As mentioned above, a represents the concentration when it is small and the membrane voltage represents the half-cell voltage (across the LaF_3 crystal). The voltage measured is given in **Equation (8.6)**:

$$V = V_{ref} + V_{membrane} = 0.199 - 0.05922 \log_{10}(a) \quad [V].$$

This relation allows immediate calculation of the concentration from the measured voltage, V:

$$\log_{10}(a) = -\frac{V - 0.199}{0.05922}.$$

If this voltage were to be measured with a pH meter, the pH reading would then represent the concentration of a.

For example, a reading of 0.35 V represents a concentration of

$$\log_{10}(a) = -\frac{0.35 - 0.199}{0.05922} = -\frac{0.151}{0.05922} = -2.5498 \rightarrow a = 0.002819.$$

Note that the higher the concentration, the lower the measured voltage.

Of course, an instrument used to sense fluoride would typically be calibrated in ppm or in any other convenient representation such as percentage.

8.4.5 The Ion-Sensitive Field-Effect Transistor (ISFET)

Also called the chemFET, this is essentially a MOSFET in which the gate has been replaced by an ion-selective membrane. Any of the membranes discussed above may be used, but the glass and polymeric membranes are most common. In its simplest form, a separate reference electrode is used, but a miniaturized reference electrode may be easily incorporated within the gate structure as shown in **Figure 8.14a**. The gate is then allowed to come in contact with the sample to be tested and the drain current is measured to indicate the ion concentration. The most important use of this device is for measurements of pH, in which capacity the chemFET replaces the glass membrane. Other applications are in the sensing of ions such as calcium (Ca^{++}), manganese (Mn^{++}), and potassium (K^+) through the use of immobilized ionophore membranes. ChemFET pH sensors are sold commercially and in many applications are considered more appropriate than glass pH sensors, if for no other reason than they are sturdier. However, they are relatively expensive.

Figure 8.14a shows the basic structure of the ISFET. The electronic circuit in **Figure 8.14b** and the equivalent circuit in **Figure 8.14c** explain the operation. The

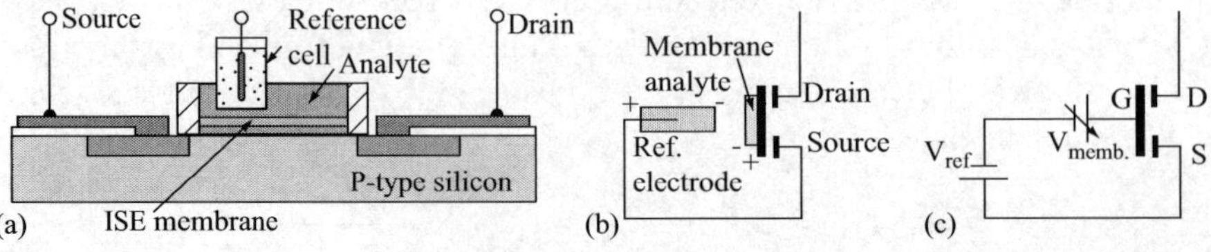

FIGURE 8.14 ■ (a) Structure of the ISFET. Note the reference electrode built into the sensor. (b) The reference and sensing membranes. (c) Equivalent circuit.

reference electrode produces a potential based on **Equation (8.7)**. This is a fixed potential that depends on the reference electrode, which is typically a saturated calomel electrode. The membrane also develops a potential based on **Equation (8.7)**, but that potential varies based on either the pH of the solution or the ion concentration in the solution. The variable voltage provided by the membrane defines the current in the MOSFET, and hence the output of the sensor. The typical sensitivity at the gate is 30–60 mV per pH unit or per concentration unit of ions (i.e., $\log_{10}a$).

8.5 | THERMOCHEMICAL SENSORS

Thermochemical or calorimetric sensors form a class of sensors that rely on the heat generated in chemical reactions to sense the amount of particular substances (reactants) involved. There are three basic sensing strategies, each leading to sensors for different applications. The most obvious is to sense the temperature increase due to a reaction using a temperature sensor such as a thermistor or thermocouple. The second type is a catalytic sensor used for sensing of flammable gases. The third measures the thermal conductivity in air due to the presence of the sensed gas.

8.5.1 Thermistor-Based Chemical Sensors

The basic principle is to sense the small change in temperature due to the chemical reaction. Since the increase in temperature is due to the reaction, a reference temperature sensor is usually employed to sense the temperature of the solution, then the difference in temperature is related to the concentration of the sensed substance. The most common approach is to use an enzyme-based reaction because enzymes are highly selective (so that the reaction can be ascertained) and because they generate significant amounts of heat. A typical sensor is made by coating the enzyme directly on the thermistor. The thermistor itself is a small bead thermistor, making for a very compact, highly sensitive sensor. The construction is shown in **Figure 8.15**. This sensor has been used to sense the concentration of urea and glucose, each with its own enzyme (urease or glucose enzymes, respectively). The amount of heat generated is proportional to the amount of the substance sensed in the solution. The temperature difference between the treated thermistor and the reference thermistor is then related to the concentration of the substance. In general, the heat lost or gained depends on the change in enthalpy in the reaction. Only part of that energy contributes to the change in temperature of the

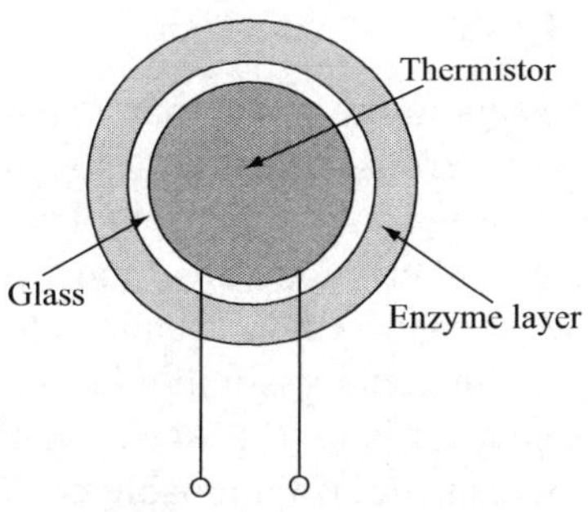

FIGURE 8.15 ■ A thermochemical sensor as used to sense urea or glucose (depending on the enzyme used), making use of a small bead thermistor.

sensor, depending on the environment and the heat capacity of that environment. In air and other gases the specific heat capacity is rather high and most if not all of the heat generated contributes to the temperature change in the sensors. In solutions, particularly aqueous solutions, some of the heat is absorbed by the solution and does not contribute to the change in temperature. This also depends on the speed of the reaction. Fast reactions tend to be more accurate, as less heat is conducted away from the sensors. The heat produced by the reaction increases the temperature of the sensor through the self-heat relation of the thermistor. Given a change in enthalpy ΔH [J], the change in temperature of the sensor (the thermistor, enzyme layer, and the immediate vicinity) with specific heat capacity C_p is

$$\Delta T = \frac{\Delta H}{C_p} n \quad [^{\circ}\mathrm{C}], \tag{8.9}$$

where ΔH is typically given in J/mol, C_p in J/mol/K, and n is the number of moles of analyte taking part in the reaction (dimensionless). In some cases ΔH is available in J/g or kJ/kg. In these cases C_p must also be in J/g/K or in kJ/kg/K. It should be remembered also that one can convert from mass to moles and vice versa through use of the atomic masses of the substances involved. Clearly each substance involved, including the sensor, the enzyme, and the solution (or air), will have its own heat capacity, but it is likely that the heat capacity of the sensor will dominate (i.e., it is typically lower than the surroundings), or some average heat capacity can be used, perhaps from a calibrated test. Given the change in temperature, $\Delta T = T - T_0$, the change in resistance of a thermistor is given by **Equation (3.12)**:

$$R(T) = R(T_0)e^{\beta(1/T - 1/T_0)} \quad [\Omega], \tag{8.10}$$

where the resistance now changes from $R(T_0)$ to $R(T) = R(T_0 + \Delta T)$. These relations are useful provided the change in enthalpy can be captured and the heat capacity is known. Otherwise the response of the temperature sensor (in this case a thermistor) must be established experimentally for every reaction and environment (solution). The temperature T can be measured with a sensor identical to that in **Figure 8.15**, but without the enzyme layer, and hence not taking part in the reaction.

Although some thermistors can measure temperature differences as low as 0.001°C, most are less sensitive than that, and the overall sensitivity of the sensor depends on the amount of heat generated. In the examples above, the glucose reaction is much more sensitive than the urea reaction because glucose has a much higher enthalpy.

8.5.2 Catalytic Sensors

These are true calorimetric sensors in the sense that a sample of the analyte is burned and the heat generated in the process is measured through a temperature sensor. This type of sensor is common and serves as the main tool in the detection of flammable gases such as methane, butane, carbon monoxide, and hydrogen, fuel vapors such as gasoline, as well as flammable solvents (ether, acetone, etc.). The basic principle is the sampling of air containing the flammable gas into a heated environment that combusts the gas and generates heat. A catalyst is used to speed up the process. The temperature sensed is then indicated as a percentage of flammable gas in air.

The simplest form of the sensor is to use a current-carrying platinum coil. The platinum coil has two purposes. It heats up due to its own resistance and it serves as a catalyst for hydrocarbons (this is the reason why it is the active material in catalytic converters in automobiles). Other, better catalysts are palladium and rhodium, but the principle is the same. The gas combusts and releases heat, raising the temperature, and hence the resistance, of the platinum coil. This change in resistance is then a direct indication of the amount of flammable gas in the sampled air. One such sensor, called a "pellistor" (the name comes from a combination of pellet and resistor), is shown in **Figure 8.16**. It uses the same heater and temperature-sensing mechanism (platinum coil) but uses a palladium catalyst either external to the ceramic bead or embedded in it. The second is better because there is less of a chance of contamination by noncombustible gases (an effect called poisoning, which reduces sensitivity). The advantage of these devices is that they operate at lower temperatures (about 500°C as opposed to about 1000°C for the platinum coil sensor). A sensor of this type will contain two beads, one inert and one sensing bead, the former serving as a reference in the sensing head shown in **Figure 8.17**. The sample of air being tested diffuses through a metal membrane (slowly) and activates the sensor. This generates a reaction in a few seconds. The operation is governed by **Equation (8.9)**, that is, the temperature of the pellistor changes by an amount ΔT depending on the change in enthalpy, but now, because the reaction occurs in air, the specific heat capacity is that of the sensor itself, which tends to be much smaller than that of the gas, that is, the change in temperature is dominated by the reaction and little of the heat is lost into the air during the reaction. This change in temperature changes the resistance of the platinum coil as described in **Equation (3.4)**:

$$R(T) = R(T_0)(1 + \alpha[T - T_0]) \quad [\Omega]. \tag{8.11}$$

The change in resistance is small since the temperature coefficient of resistance (TCR) for platinum is relatively small (see **Table 3.1**), as is the amount of gas sampled.

In common applications, these sensors are used in mines to detect methane and in industry to sense solvents in air. For practical purposes, the most important issue is the concentration at which a flammable gas explodes. This is called the **lower explosive**

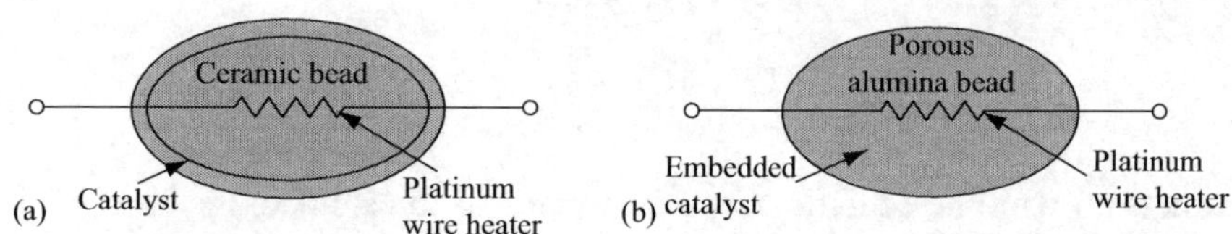

FIGURE 8.16 ■ Catalytic sensors (pellistors). (a) The catalyst is coated on the ceramic bead. (b) The catalyst is embedded in alumina.

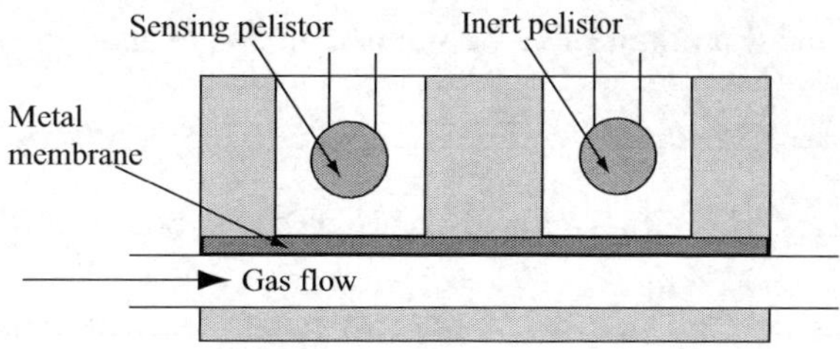

FIGURE 8.17 ■ Construction of a catalytic sensor with a reference sensor using pellistors.

limit (LEL), below which a gas will not ignite. For methane, for example, the LEL limit is 5% (by volume, in air). Thus a methane sensor will be calibrated in percentage of LEL, where 100% LEL corresponds to 5% methane in air.

EXAMPLE 8.7 Detection of carbon monoxide

A pellistor is used to sense CO (assuming a CO-specific ceramic bead is used). The pellistor samples 1 mg of air that contains 1% (by mass) CO. The pellistor operates at a steady-state temperature of 700°C. Its heater is made of platinum and at that temperature has a resistance of 1200 Ω. The TCR of the platinum alloy used is 0.00362/°C at the temperature employed by the sensor. CO has a specific heat capacity of 29 J/mol/K and an enthalpy of combustion in oxygen of 283 kJ/mol. The specific heat capacity of the pellistor itself is 0.750 J/g/K (similar to glass). What is the expected change in resistance of the pellistor due to the combustion of the CO?

Solution: This example shows some of the difficulties involved in this type of calculation. We have the specific heats for both CO and the sensor, but which of them should we use? On top of that, perhaps we should use the specific heat of air. Also, we need to deal with mixed units since the specific heat capacity of the pellistor is, necessarily, given per gram per kelvin. Nevertheless, with a little care and some approximations, meaningful results can be obtained.

First, we note that the fraction of mass sampled is 0.01 mg or 10^{-5} g. We can use $n = 10^{-5}$ in **Equation (8.9)** provided that the enthalpy is written in units of J/g and heat capacity is written in units of J/g/K. In general, gases have a higher heat capacity than most other materials so that the proper quantity to use for the specific heat capacity is that of the pellistor, since it is its temperature that is being raised, whereas the temperature of air changes very little. In any case, CO is a very small fraction of the volume involved. But we need to convert the specific heat capacity to J/g/K. To do so we note that the CO atom is made of one carbon atom and one oxygen atom. The molar mass (MM) of CO is

$$MM(CO) = 1 \times 12.01 + 1 \times 16 = 28.01 \text{ g}.$$

That is, 1 mole of CO has a mass of 28.01 g. Therefore the enthalpy is

$$\Delta H = 283\left(\frac{\text{kJ}}{\text{mol}}\right) = \frac{283}{28.01}\left(\frac{\text{kJ}}{\text{g}}\right)$$

Now we can use **Equation (8.9)** with the specific heat of the pellistor to calculate the change in temperature due to combustion of the CO:

$$\Delta T = \frac{\Delta H}{C_p} n = \frac{283 \times 10^3/28.01}{0.750} \times 10^{-5} = 0.1347 \text{ K}.$$

The change is small because of the small amount of CO combusted. The change in resistance is

$$R(T) = R(T_0)(1 + \alpha[T - T_0]) = 1200 \times (1 + 0.00362 \times 0.1347) = 1200.585\ \Omega.$$

The resistance changes by 0.585 Ω. This is a small change (0.05%), but nevertheless it is measurable.

8.5.3 Thermal Conductivity Sensor

Thermal conductivity sensors do not involve any chemical reaction, but rather use the thermal properties of gases for sensing. A sensor of this type is shown in **Figure 8.18**. It consists of a heater set at a given temperature (around 250°C) in the path of a gas. The heater looses heat to the surrounding area, depending on the gas with which it comes in contact. As the gas concentration increases a larger amount of heat is lost compared to the loss in air, and the temperature of the heater as well as its resistance decreases. This is particularly so with gases that have high thermal conductivities. This change in resistance is sensed and calibrated in terms of gas concentration. Unlike the previous two types of sensors, this sensor is useful for high concentrations of gas. It can be used for inert gases such as nitrogen, argon, and CO_2, as well as for volatile gases, provided of course that the concentration is below the LEL. The sensor is in common use in industry and is a useful tool in gas chromatography in the lab. Based on Fourier's law, it combines the law of heat transfer and Ohm's law in that the heat loss (or gain) of the sensor is sensed through the change in a heated resistor. As with many other sensors, a comparative measurement against a reference sensor, one that is not exposed to the gas being sensed, provides better resolution.

However, the relation between the temperature change and heat loss is relatively complex. Because of that, thermal conductivity sensors must be calibrated, but accurate computation of a sensor's response is difficult and requires specific information for the sensor (dimensions, thermal properties, etc.). Nevertheless, in spite of these difficulties, thermal conductivity sensors exist commercially and are an important tool in the evaluation of gases.

8.6 | OPTICAL CHEMICAL SENSORS

The propagation of light, and in a broader sense that of any electromagnetic radiation in any medium, is governed by the properties of the medium. The transmission, reflection, and absorption (attenuation) of light in a medium, its velocity, and hence its wavelength are dependent on these properties. The optical properties of the medium can all serve as the basis for sensing either by themselves or in conjunction with other transduction mechanisms and sensors. For example, optical smoke detectors use the transmission of light through smoke to detect its presence. Other substances may be sensed in this way, sometimes by adding agents to, for example, color the substance being tested. However, much more complex mechanisms can be used to obtain highly sensitive sensors for a variety of chemical substances and reactions.

In many optical sensors, use is being made of an electrode whose properties change according to the substance being tested. An electrode of this type is called an "optode," a name that parallels the name "electrode" for electrical properties. The optode has an

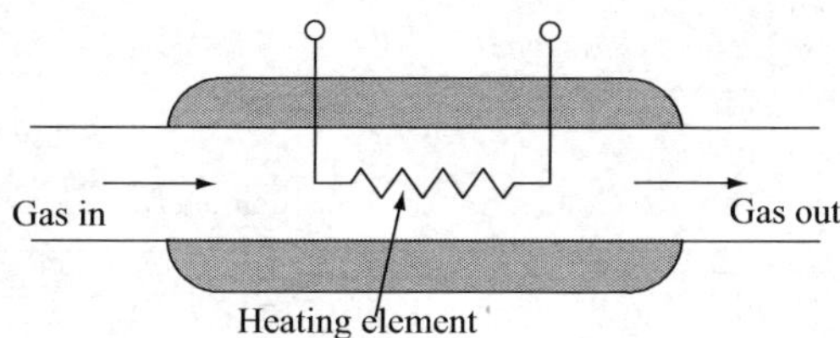

FIGURE 8.18 ■ The construction of a thermal conductivity sensor.

important advantage in that no reference is needed and it is well suited for use with optical guiding systems such as optical fibers.

Other options for optochemical sensing are the properties of some substances to fluoresce or phosphoresce under optical radiation. These chemiluminescence properties can be sensed and used for indications of specific materials or properties. Luminescence can be a highly sensitive method because the luminescence is typically at a different frequency (wavelength) than the frequency (wavelength) of the exiting radiation. Luminescence occurs more often with ultraviolet (UV) radiation, but it can occur in the infrared (IR) or visible range as well, and is often used for detection.

Most optical sensing mechanisms, including luminescence, rely at least in part on the absorption of light by the substance through which it propagates or on which it impinges. This absorption is important in sensors based on the transmission of light and is governed by the Beer–Lambert law, stated as follows:

$$A = \varepsilon b M, \tag{8.12}$$

where ε is the absorption coefficient characteristic of the medium [10^3 cm^2/mol], b is the path length [cm] traveled, and M is the concentration [mol/L]. A is the absorbance $A = \log(P_0/P)$, where P_0 is the incident and P is the transmitted light intensity. This linear relationship only applies to monochromatic radiation.

Perhaps the simplest optochemical sensors are the so-called reflectance sensors, which rely on the reflective properties of a membrane or substance to infer a property of the substance with which it is in contact. In many instances an optical fiber is used as an optical waveguide. The basic structure is shown in **Figure 8.19**. A source of light (LED, white light, laser) generates a beam that is conducted through the optical fiber to the optode. The optical properties of the optode are altered by the substance with which it interacts and the reflected beam is then a function of the concentration of the analyte or its reaction products in the optode. It is also possible to separate the incident and reflected beams by separate optical guides, but usually this is not necessary.

An alternative way of sensing is to use an uncladded optical fiber so some light is lost through the walls of the fiber. This is called an **evanescent loss** and depends on what is in contact with the walls of the fiber. The principle is shown in **Figure 8.20**. In this

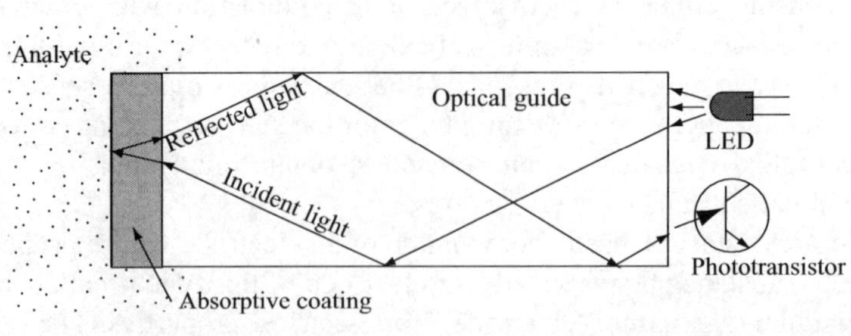

FIGURE 8.19 ■ Principle and structure of a reflectance sensor.

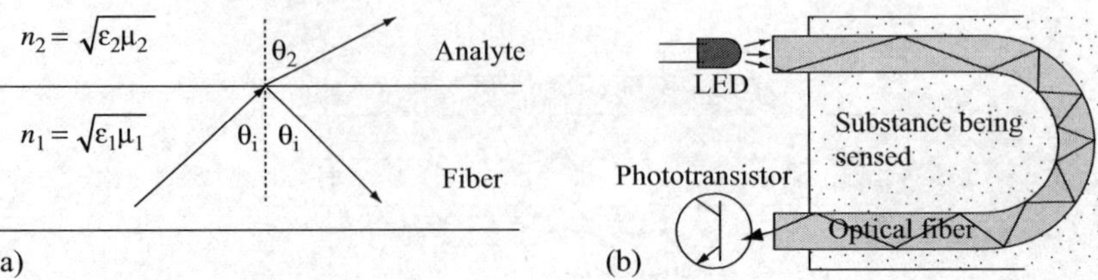

FIGURE 8.20 ■ Evanescent loss sensor. (a) Principle. (b) Structure.

type of sensor the coupling to the analyte is through the walls of the fiber rather than its end. This also means that rather than reflection, the power transmission through the fiber is sensed—a quantity influenced by the power loss through the walls of the fiber. The transmitted wave is then dependent on the amount of light absorbed in the analyte and is therefore a function of the analyte optical properties (primarily the index of refraction) on the surface of the optical fiber. Sensing using this method is based on the transmission properties between a high-permittivity dielectric and a lower-permittivity dielectric such as between glass (or plastic) and air. Given a light wave propagating inside a fiber and impinging from within on the interface between the fiber and analyte, the wave is reflected back into the fiber but part of the power is transmitted across the interface into the analyte. The principle is shown in **Figure 8.20a**. A light beam (a laser beam, for example) is incident on the interface between medium 1 and medium 2 at an angle θ_ι. The beam is reflected at the same angle ($\theta_r = \theta_\iota$). The transmitted (refracted) wave that will penetrate into medium 2 follows from the Snell law of refraction:

$$\frac{\sin\theta_i}{\sin\theta_t} = \frac{n_2}{n_1}, \tag{8.13}$$

where n_1 is the optical index of refraction in medium 1 and n_2 is the optical index of refraction in medium 2. The index of refraction in a dielectric (such as glass or air) is given as

$$n = \sqrt{\varepsilon_r \mu_r}, \tag{8.14}$$

where ε_r and μ_r are the relative permittivity and relative permeability of the medium, respectively. If the angle of refraction θ_t is equal to or greater than 90°, the wave does not penetrate into medium 2 and is totally contained within medium 1. The critical angle is the incident angle θ_i, for which the refraction angle is 90°, and is given as

$$\theta_i = \sin^{-1}\left(\frac{n_2}{n_1}\right) = \sin^{-1}\sqrt{\frac{\varepsilon_{r2}\mu_{r2}}{\varepsilon_{r1}\mu_{r1}}}, \text{provided } \varepsilon_{r2}\mu_{r2} < \varepsilon_{r1}\mu_{r1}. \tag{8.15}$$

As long as medium 2 has a lower index of refraction, a critical angle exists. The importance of this is that for incident angles larger or equal to the critical angle, all power in the incident wave is reflected back into the volume of the fiber and none is "lost" by propagating into medium 2. Therefore all power is available at the detector (neglecting any possible losses in the fiber itself). At angles of incidence less than the critical angle, the power available at the detector is reduced because of the power transmitted across the interface between the fiber and analyte.

There are two methods that can be used to exploit the critical angle for sensing purposes. The first relies on an increase in the reflected wave power because less power propagates into medium 2, whereas the second does exactly the opposite. To understand this, suppose medium 2 is air and medium 1 is glass. The incident angle is small enough so that the wave penetrates into medium 2, that is, the wave impinges on the surface of the fiber below the critical angle. A portion of the wave is reflected and a portion propagates into air. Now suppose air is replaced with a dielectric such as water vapor or any other substance with a higher permittivity but still satisfying the condition $\varepsilon_{r2}\mu_{r2} < \varepsilon_{r1}\mu_{r1}$. The index of refraction n_2 increases and less power (or none) propagates into medium 2. Necessarily the proportion of the reflected power increases with the permittivity of medium 2. Therefore, in this case, the higher the index of refraction of medium 2, the larger the power measured in the reflected wave and a sensor can then be calibrated to sense, for example, relative humidity or the concentration of a substance.

The second method assumes the permittivity of medium 2 is larger than that of the fiber and hence **Equation (8.15)** cannot be satisfied. It that case, the higher the index of refraction of the medium, the lower the reflected power. This method is useful when using the sensor in solutions such as water or when detecting fluids. In practice, the angle of incidence is not a fixed value, but rather a range of angles. Some rays may refract, some not, so that the relation between detected power and the index of refraction is not as simple as the description above. Nevertheless, the output can be calibrated to detect and, to a certain extent, to measure variations in the index of refraction of media on the surface of the sensor.

An evanescent loss sensor capable of detecting and differentiating between various types of fuels (from heavy oil to gasoline) is shown in **Figure 8.21**. It can be used to detect fuel leaks from pumping systems.

pH sensing can be done optically by using special optodes that change color with changes in pH. However, in these systems only about 1 pH unit on either side of the pH of the optode (before the analyte interaction) can be sensed. Although this is a narrow range, it is sufficient for some applications. A sensor of this type is shown in **Figure 8.22**. In this sensor a hydrogen-permeable membrane is used in which phenol red is immobilized on polyacrylamide microspheres. The membrane is a dialysis tube (cellulose acetate) and the optode thus created is attached to the end of an optical fiber. When immersed in the analyte, it diffuses into the optode. Phenol red is known to absorb light at a wavelength of 560 nm (yellow-green light). The amount of light absorbed

FIGURE 8.21 ■ An evanescent loss sensor for detection of fluids. The optical waveguide can be seen at the center of the picture.

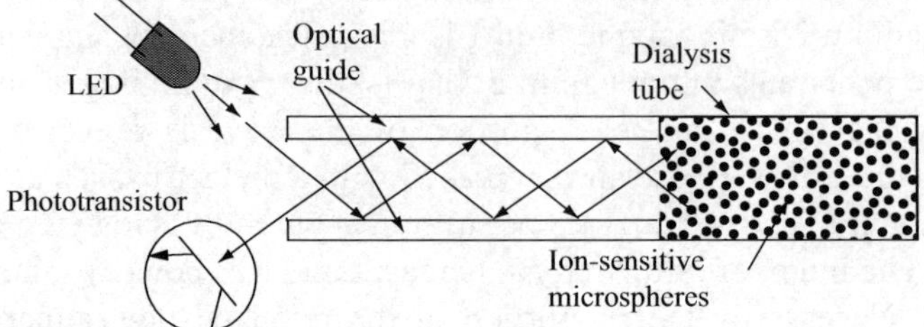

FIGURE 8.22 ■ Structure and principle of an optical pH meter.

depends on the pH and hence the reflected light will change with the pH. The difference between the incident and reflected intensities is then related to pH.

Either the arrangement in **Figure 8.19** or **Figure 8.22** can be used to sense fluorescence. Since fluorescence occurs at different wavelengths than the exciting radiation, the separation between the incident and reflected waves is based on filtering.

A sensor similar to the pH sensor in **Figure 8.22** uses the fluorescent properties of 8-hydroxypyrene-1,3,6-trisulphonic (HPTS), a weak acid. This substance fluoresces when excited by UV light at 405 nm. The intensity of fluorescence is then related to the pH. This material is particularly useful since its normal pH is 7.3, so measurements around the neutral point can be made and, in particular, physiological measurements.

Optodes can also be used to sense ions. Metal ions are particularly easy to sense because they can form highly colored complexes with a variety of reagents. These reagents are embedded in the optode and the reflective properties of the optode–analyte interface are then related to the concentration of the metal ions. Fluorescence is also common in metal ions, a method that is used extensively in analytical chemistry, primarily by use of UV light with fluorescence in the visible range. These methods have been used to sense a variety of ions, including oxygen in water, penicillin, and glucose in blood.

8.7 | MASS SENSORS

Another method of chemical sensing is to detect changes in the mass of a sensing element due to absorption of an analyte. The idea is obvious, but it should be immediately realized that the masses involved in absorption, say, of a gas or water vapor, are minute and a method must be found that is sufficiently sensitive to these minute mass changes. For this reason, mass sensors are also called **microgravimetric** or **microbalance sensors**. In a practical sensor it is not possible to sense this change in mass directly and therefore indirect methods must be used. This is done by using piezoelectric crystals such as quartz and setting them into oscillation at their resonant frequency (see **Section 7.7**). This resonant frequency is dependent on the way the crystal is cut and on dimensions, but once these have been fixed, any change in mass of the crystal will change its resonant frequency. The sensitivity is generally very high, on the order of 10^{-9} g/Hz and the limit sensitivity is about 10^{-12} g. Since the resonant frequency of crystals can be very high, the change in frequency due to a change in mass is significant and can be accurately measured digitally. As a result, these sensors are hihgly sensitive and relatively simple. An equivalent approach can be taken with SAW resonators, which are piezoelectric resonators, but because they operate on the propagation of acoustic waves in the piezoelectric material, the wavelengths are shorter and they can resonate at even higher frequencies than crystals resonators (see **Section 7.9**), hence they offer even higher sensitivities.

The shift in resonant frequency of a crystal due to a change in mass Δm can be written as

$$\Delta f = -S_m \Delta_m \quad [\text{Hz}], \tag{8.16}$$

where Δf is the change in the resonant frequency, S_m is a sensitivity factor that depends of the crystal (cut, shape, mounting, etc.), and Δm is the change in mass per unit area (typically given in g/cm^2). The sensitivity factor is given in $Hz{\cdot}cm^2/\mu g$. Since the sensitivity is more or less constant with the change in mass (for small mass changes), the shift in resonant frequency is linear. On the other hand, the sensitivity factor depends on frequency, so the factors are given for given crystals resonating at given frequencies. Note also that the change in frequency is negative, that is, the frequency decreases as the mass increases. Typical values for the sensitivity factor are 40–60 $Hz{\cdot}cm^2/\mu g$.

Equation (8.16) may be written in an inverse form:

$$\Delta m = C_m \Delta f \quad [g/cm^2]. \tag{8.17}$$

In this form C_m is a mass factor or mass sensitivity factor and is given in $ng/Hz/cm^2$ and Δm is given in g/cm^2. Typical values for C_m are 4–6 $ng/Hz/cm^2$. This form gives the added mass for the measured change in frequency, Δf.

The mass due to the analyte may be absorbed directly into the crystal (or any piezoelectric material) or it may be absorbed in a coating on the crystal. All in all, these are simple and efficient sensors. The main problem is that selectivity is poor since crystals and coatings tend to absorb more than one species, confounding discrimination between species. Also, a basic requirement is that the process be reversible, that is, the absorbed species must be removable (e.g., by heating) without any hysteresis. Although sensors more or less specific to various gases have been developed, the most common is the water vapor sensor.

EXAMPLE 8.8 Sensitivity of a mass sensor

Consider a quartz crystal oscillating at 10 MHz. The crystal is a disk with a diameter of 20 mm. The sensor is used to sense pollen concentrations in air for purposes of warning the public when the pollen count is high. For this purpose the surface of the crystal is coated with a sticky substance to trap pollen particles. Assume that a frequency shift of 100 Hz can be reliably and accurately measured and that the crystal has a mass sensitivity of 4.5 $ng/Hz/cm^2$. On average, a grain of pollen weighs 200 ng. What is the minimum number of pollen grains that can be reliably detected?

Solution: The mass per unit area that can be detected is

$$\Delta m = C_m \Delta f = 4.5 \times 10^{-9} \times 100 = 450 \times 10^{-9}\ g/cm^2.$$

The area of the sensor is $\pi \times 1^2 = \pi\ cm^2$, so the total mass detectable is

$$\Delta M = \Delta m S = 450 \times 10^{-9} \times \pi = 1413.7 \times 10^{-9}\ g.$$

This is 7 grains of pollen (1413.7/200 = 7.07).

Note: Although the values used here are realistic, there are other issues to consider in this type of sensor, including frequency drift with temperature and the frequency stability of the crystal oscillator. These can change the measurement by a few hertz. Typical values are about 5–10 Hz/°C, meaning that due to temperature variations alone, the measurement may be off by as much as 10%. At lower sensitivities (larger frequency variations) this may be negligible. If, for example, we were to assume a minimum measurable frequency of 1 kHz, the error would only be 1%, but sensitivity would decrease to 70 grains of pollen. This reduction in sensitivity can be compensated for by allowing a longer collection time. Temperature itself may be controlled quite accurately, reducing the errors in measurement.

8.7.1 Mass Humidity and Gas Sensors

A mass humidity sensor is made by simply coating the resonating crystal with a thin layer of hygroscopic material, which can then absorb water vapor. With hygroscopic coatings and resonators, including SAW resonators, an appropriate medium can be employed since it is not necessary that the crystal itself absorb the vapor. There are many hygroscopic materials that can be used, including polymers, gelatins, silica, and fluorides. The moisture is removed after sensing by heating the absorbing layer. Although a sensor of this type can be quite sensitive, its response time is slow. It may take many seconds (20–30 s) for sensing and many more for regeneration of the sensor (30–50 s).

Nevertheless, the method is useful and has been applied to sensing of a large variety of gases and vapors, some being sensed at room temperatures, some at elevated temperatures. The main difference between sensing one gas and another is in the coating, in an attempt to make the sensor selective. The applications are mostly in sensing of noxious gases and dangerous substances such as mercury. The sensing of SO_2 (mostly due to the burning of coal and fuels) has been accomplished by amine coatings, which react with SO_2. Concentrations as low as 10 ppb are detectable.

When detecting ammonia (in testing of the environmental effects of wastewater and sewage), the coating is ascorbic acid or pyridoxine hydrochloride (and other similar compounds) with sensitivities down to microgram per kilogram.

Hydrocarbon sulfide is similarly detected by using acetate coatings (silver, copper, and lead acetates are used, as well as others). Mercury vapor is sensed by using gold as a coating, since the two elements form an amalgam that increases the mass of the gold coating. Other applications are in sensing hydrocarbons, nitrotoluenes (emitted by explosives), and gases emitted by pesticides, insecticides, and other sources.

8.7.2 SAW Mass Sensors

The use of crystal resonators has been shown to be a useful and sensitive method of sensing, partly owing to the high frequency of resonance. SAW resonators, based on the use of delay lines (**Section 7.9**), resonate at much higher frequencies and the resonant frequency is highly dependent on the speed of sound in the piezoelectric material. A SAW mass sensor is therefore made as a delay line resonator. The delay line itself is coated with the specific reactive coating for the gas to be sensed. This is shown in **Figure 8.23**. Air containing the gas is sampled (drawn above the membrane) and the resonant frequency measured. The same method can be used to sense solid particles such as pollen or pollutants by replacing the membrane with a sticky substance. Of course, the problem then is regeneration—cleaning the surface for the next sampling.

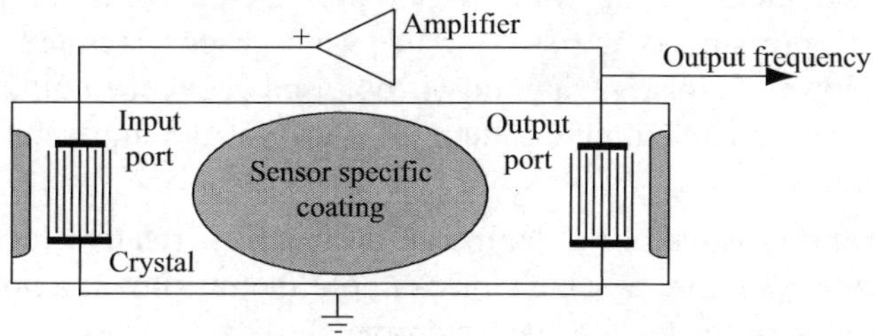

FIGURE 8.23 ■ A SAW mass sensor based on a delay line oscillator. Output frequency is related to the mass of the coating.

TABLE 8.1 ■ Some sensed substances and the coatings used for that purpose

Compound	Chemical coating	SAW material
SO_2	TEA (triethanolamine)	Lithium niobate
H_2	Pd (palladium)	Lithium niobate, silicon
NH_3	Pt (plutonium)	Quartz
H_2S	WO (Wolfram oxide)	Lithium niobate
Water vapor	Hygroscopic material	Lithium niobate
NO_2	PC (phthalocyanine)	Lithium niobate, quartz
NO_2, NH_3, SO_2, CH_4	PC (phthalocyanine)	Lithium niobate
Explosives vapor, drugs	Polymer	Quartz
SO_2, methane	None	Lithium niobate

Some sensors of this type are thus single-use sensors. As was the case with crystal microbalance sensors, the choice of coating determines the selectivity of the sensor. **Table 8.1** shows some sensed substances and the appropriate coatings.

The sensitivities of SAW resonators can be much higher than those of crystal resonators, with limit sensitivities of approximately 10^{-15} g. Sensitivities expected are on the order of 50 μHz/Hz. This means that for a 500 MHz resonator, the frequency shift sensitivity is 25 kHz. This is more than sufficient for accurate sensing.

8.8 | HUMIDITY AND MOISTURE SENSORS

The use of a mass sensor to sense humidity has been described previously and the extension of this method to SAW sensors is indicated in **Table 8.1**. There are other methods of sensing humidity, but all involve some type of hygroscopic medium to absorb water vapor. These can take many forms, of which capacitive, conductive, and optical are the most common.

The terms humidity and moisture are not interchangeable. Humidity refers to the water content in gases, such as in the atmosphere. Moisture is the water content in any solid or liquid. Other important, related quantities are the dew point temperature, absolute humidity, and relative humidity. These are defined as follows:

Absolute humidity is the mass of water vapor per unit volume of wet gas, expressed in g/m^3.

Relative humidity is the ratio of the water vapor pressure of the gas (usually air) to the maximum saturation water vapor pressure in the same gas at the same temperature. Saturation is the water vapor pressure at which droplets form. The atmospheric pressure is the sum of the water vapor pressure and the dry air pressure. However, relative humidity is not used above the boiling point of water (100°C) since the maximum saturation above that temperature changes with temperature.

Dew point temperature is the temperature at which relative humidity is 100%. This is the temperature at which air can hold the maximum amount of moisture. Cooling below it creates fog (water droplets), dew, or frost.

8.8.1 Capacitive Moisture Sensors

The simplest moisture sensor is the capacitive sensor since it simply relies on the change in permittivity due to moisture. The permittivity of water is rather high ($80\varepsilon_0$ at low frequencies). Humidity is different than liquid water, and hence the permittivity of humid air is either given in tables as a function of relative humidity or can be calculated from the following empirical relation:

$$\varepsilon = \left(1 + \frac{1.5826}{T}\left(P_{ma} + \frac{0.36P_{ws}}{T}RH\right)10^{-6}\right)\varepsilon_0 \quad [\mathrm{F/m}], \tag{8.18}$$

where ε_0 is the permittivity of vacuum, T is the absolute temperature [K], P_{ma} is the pressure of moist air [Pa], RH is the relative humidity [%], and P_{ws} is the pressure of saturated water vapor pressure [Pa] at the temperature T. These quantities may seem a bit confusing. The pressure of moist air is the partial pressure exerted by water vapor in the atmosphere. It is temperature dependent and reaches the ambient pressure at 100°C (that's the reason water boils at 100°C). Saturated water vapor pressure is the water vapor pressure at 100% humidity, and is also temperature dependent. Both of these quantities can be calculated and are available in tables. The pressure of saturated water vapor and the pressure of moist air can be calculated from the following experimental formulas:

$$P_{ws} = 133.322 \times 10^{0.66077+7.5t/(237.3+t)} \quad [\mathrm{Pa}] \tag{8.19}$$

$$P_{ma} = 133.322e^{20.386-5132/(t+273.15)} \quad [\mathrm{Pa}], \tag{8.20}$$

where t is given in °C and P_{ws} in Pa.

The capacitance of a parallel plate capacitor is $C = \varepsilon A/d$, and this then establishes a relation between capacitance and relative humidity (A is the area of the capacitor, d is the distance between the plates, and ε is the permittivity of the substance between the plates):

$$C = C_0 + C_0\frac{1.5826P_{ma}}{T}10^{-6} + C_0\frac{75.966P_{ws}}{T}10^{-6}RH \quad [\mathrm{F}], \tag{8.21}$$

where C_0 is the capacitance of the capacitor in vacuum ($C_0 = \varepsilon_0 A/d$). This relation is linear at any given temperature. The pressure of moist air adds a fixed component, whereas the variable component is entirely due to humidity. However, for a practical capacitor, the capacitance is rather small (the capacitor plates cannot be too large for practical reasons and the distance between the plates must be reasonable—at least a few micrometers—to allow air motion). In practical designs, means of increasing this capacitance are used. One approach is to use a hygroscopic material between the plates, both to increase the capacitance at no humidity and to absorb the water vapor. These materials can be hygroscopic polymer films. The metal plates can be made of gold. In a device of this type the capacitance can be approximated as

$$C = C_0 + C_0\alpha_h H \quad [\mathrm{F}], \tag{8.22}$$

where α_h is a moisture coefficient. α_h is not necessarily constant, and in general may depend on temperature and on the relative humidity itself. This method assumes that the moisture content in the hygroscopic polymer is directly proportional to relative humidity and that as the humidity changes, the moisture content changes (i.e., the film does not retain water). Under these conditions the sensing is continuous, but, as expected, changes are slow and the sensor's output necessarily lags behind, especially if changes in moisture are quick. A sensor of this type can sense relative humidity from about 5% to 90% at an accuracy of 2%–3%.

EXAMPLE 8.9 Capacitive relative humidity sensor

A capacitive relative humidity sensor using a hygroscopic polymer is given. To evaluate its properties, the capacitance is measured at 20% RH and at 80% RH. The results are $C = 448.4$ pF at 20% RH and $C = 491.6$ pF at 80% RH. The capacitor is a simple parallel plate capacitor.

a. Calculate the moisture coefficient of the sensor, its output full scale (OFS), and its sensitivity.
b. Calculate the range of relative permittivities of the sensor for the OFS in (a).

Solution: Using **Equation (8.22)** we can determine the dry capacitance C_0 (i.e., the capacitance at zero relative humidity) and the moisture coefficient. Then using the formula for the parallel plate capacitor, the permittivity is calculated directly.

a. At 20% RH

$$448.4 = C_0 + \alpha C_0 \times 20 \text{ pF}$$

and at 80% RH

$$491.6 = C_0 + \alpha C_0 \times 80 \text{ pF}.$$

Subtracting the first relation from the second:

$$491.6 - 448.4 = \alpha C_0 \times (80 - 20) \rightarrow 43.2 = \alpha C_0 \times 60 \rightarrow \alpha C_0 = \frac{60}{43.2} = 1.3889.$$

Substituting this back into the first relation:

$$448.4 = C_0 + 1.3889 \times 20 \rightarrow C_0 = 448.4 - 1.3889 \times 20 = 420.62 \text{ pF}.$$

The moisture coefficient is

$$\alpha C_0 = 1.3889 \rightarrow \alpha = \frac{1.3889}{C_0} = \frac{1.3889}{420.62} = 0.003302.$$

The OFS is the capacitance of the sensor for the input full scale (IFS). Necessarily the IFS is 0%–100%. The relation for capacitance is

$$C = 420.62 + 1.3889H \quad [\text{pF}].$$

At 0% RH we already obtained $C = C_0 = 420.62$ pF. At 100% RH we have

$$C = 420.62 + 1.3889 \times 100 = 559.51 \text{ pF}.$$

That is, the OFS is 420.62 pF to 559.51 pF, or 138.89 pF.

Since the output is linear, it suffices to divide the OFS by the IFS to obtain the sensitivity. In this case this is 1.3889 pF/% RH.

b. The permittivity is calculated from the capacitance of a parallel plate capacitor. We write

$$C = \varepsilon \frac{A}{d} \quad [\mathrm{F}].$$

Because we do not know the plate area or the distance between them, we calculate the ratio A/d from the empty capacitor:

$$\frac{A}{d} = \frac{C_0}{\varepsilon_0} = \frac{429.88 \times 10^{-12}}{8.854 \times 10^{-12}} = 45.552 \text{ m}.$$

This ratio remains the same no matter what the relative humidity is. At 100% humidity, the capacitance is 522.47 pF and the permittivity is

$$\varepsilon = \frac{C}{A/d} = \frac{522.47 \times 10^{-12}}{45.552} = 11.4697 \times 10^{-12} \text{ F/m}.$$

The relative permittivity is

$$\varepsilon_r = \frac{\varepsilon}{\varepsilon_0} = \frac{11.4697 \times 10^{-12}}{8.854 \times 10^{-12}} = 1.2954.$$

The relative permittivity varies from 1 to 1.2954. This is a reasonably large change (almost 30%) in relative permittivity as humidity varies from 0% to 100%.

One of the difficulties with parallel plate capacitor humidity sensors is that the hygroscopic film must be thin and moisture can only penetrate from the sides. They are therefore slow to respond to changes in moisture because of the time it takes for moisture to penetrate throughout the film. A different approach is shown in **Figure 8.24**. Here the capacitor is flat and built from a series of interdigitated electrodes to increase capacitance. The hygroscopic dielectric may be made of SiO_2 or phosphorosilicate glass. The layer is very thin to improve response. Because the sensor is based on silicon, temperature sensors are easily incorporated to allow for compensation, as are other components such as oscillators. The capacitance of the device is low and therefore it will normally be used as part of an oscillator and the frequency sensed. However, the permittivity of the dielectric is frequency dependent (goes down with frequency). This means that the frequency cannot be too high, especially if low humidity levels are sensed.

8.8.2 Resistive Humidity Sensor

Humidity is known to change the resistivity (conductivity) of some materials. This can be used to build resistive relative humidity sensors. To do so, a hygroscopic conducting layer and two electrodes are provided. Often the electrodes will be interdigitated to

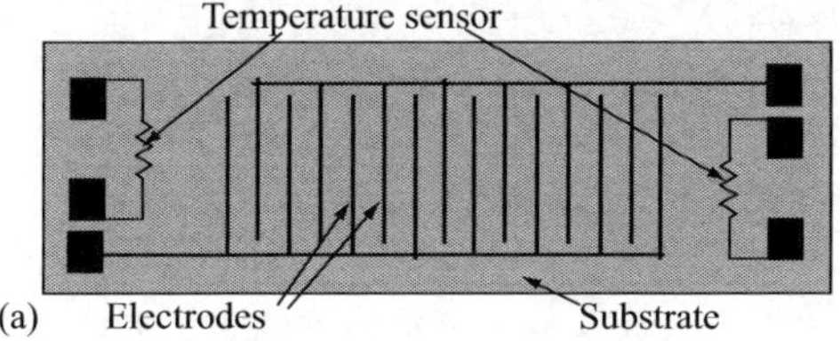

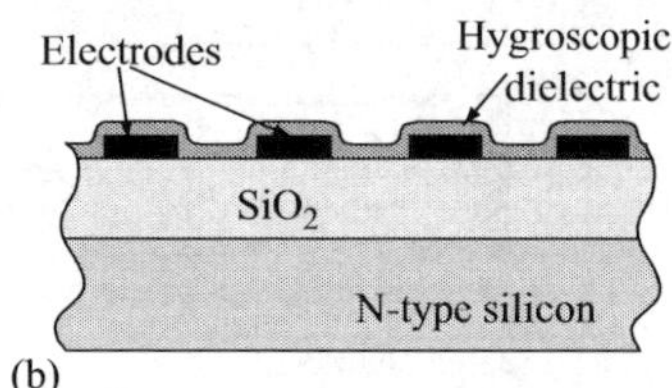

FIGURE 8.24 ■ A capacitive sensor with interdigitated electrodes.

increase the contact area, as shown in **Figure 8.25**. The hygroscopic conductive layer must have a relatively high resistance, which decreases with humidity (actually absorbed moisture). There are some materials that can be used for this purpose, including polystyrene treated with sulfuric acid and solid polyelectrolytes, but a better structure is shown in **Figure 8.26**. It operates as above but the base material is silicon. An aluminum layer is formed on the silicon (highly doped so its resistivity is low). The aluminum layer is oxidized to form a layer of Al_2O_3, which is porous and hygroscopic and has a low conductivity that increases with relative humidity. An electrode of porous gold is deposited on top to create the second contact and to allow moisture absorption in the Al_2O_3 layer. The resistance between the upper gold electrode and the substrate electrode is then a measure of relative humidity.

8.8.3 Thermal Conduction Moisture Sensors

Humidity may also be measured through thermal conduction, as higher humidity will increase thermal conduction. This sensor senses absolute humidity rather than relative humidity. The sensor makes use of two thermistors connected in a differential or bridge connection (bridge connection is shown in **Figure 8.27a**). The thermistors are heated to an identical temperature by the current through them so that the differential output is zero in dry air. One thermistor is kept in an enclosed chamber as a reference and its resistance is constant. The other is exposed to air and its temperature changes with humidity. As humidity increases, the thermistor temperature decreases and hence its resistance

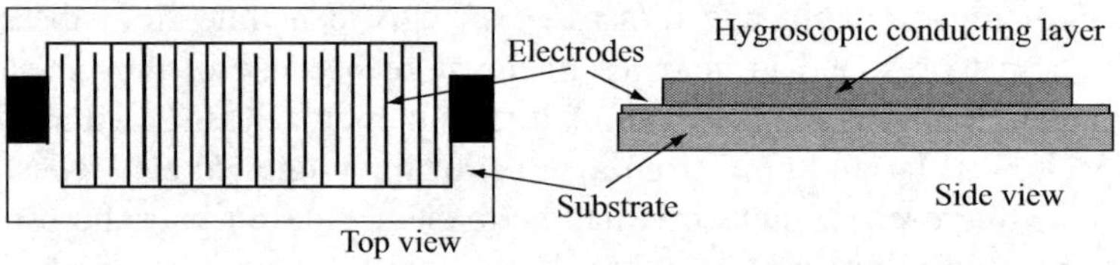

FIGURE 8.25 ■ A humidity sensor based on the interdigitated electrode capacitor.

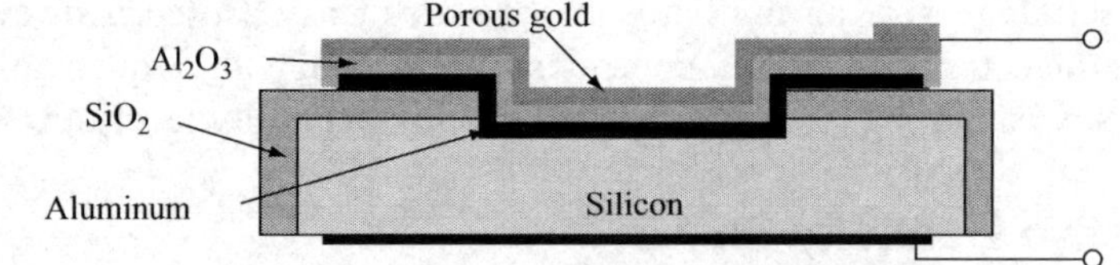

FIGURE 8.26 ■ A porous (hygroscopic) layer humidity sensor.

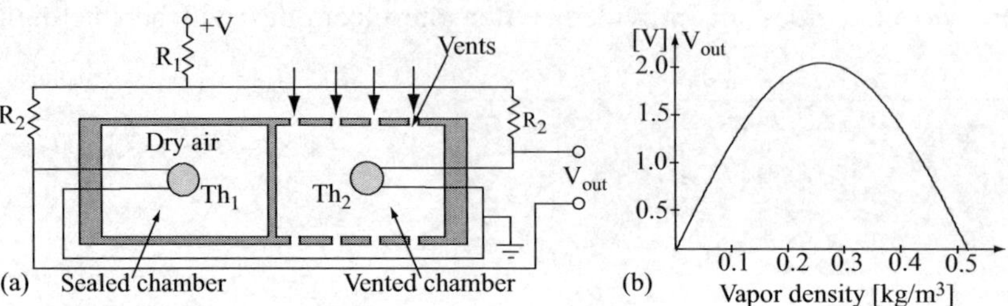

FIGURE 8.27 ■ A thermal conductivity moisture sensor. (a) Structure. (b) Response.

increases (for NTC thermistors). At saturation the peak resistance is reached, beyond which the output drops again as the thermal conductivity decreases (**Figure 8.27b**).

8.8.4 Optical Humidity Sensor

By far the most accurate humidity sensing method is optical and is based on duplicating the dew point by regulating the temperature of a mirror. When the dew point is reached, the relative humidity is 100%. The relative humidity (*RH*) is obtainable from the dew point temperature and the saturation water vapor pressure relation:

$$DPT = \frac{237.3\left(0.66077 - \log_{10}\left(\frac{P_{ws} \cdot RH/100}{133.322}\right)\right)}{\log_{10}\left(\frac{P_{ws} \cdot RH/100}{133.322}\right) - 8.16077}, \quad (8.23)$$

where P_{ws} [Pa] is the saturation water vapor pressure given in **Equation (8.19)**.

It should be noted that the higher the relative humidity at any temperature, the higher the dew point temperature until, at 100% relative humidity, the dew point temperature equals the temperature of the air (see **Example 8.10**). By measuring the ambient temperature t and then evaluating the dew point temperature (*DPT*), the relative humidity (*RH*) can be calculated from **Equation (8.23)**. Thus the basic idea is to use a dew point sensor built as shown in **Figure 8.28**. The sensor is based on detecting the dew point on the surface of a mirror. To do so, light is reflected off the mirror and the light intensity monitored. A Peltier cell is used to cool the mirror to its dew point. When the dew point temperature is reached, the controller keeps the mirror at the dew point temperature by regulating the current in the Peltier cell. The reflectivity of the mirror decreases since water droplets form on the mirror (the mirror fogs up). This temperature is measured and is the dew point temperature in **Equation (8.23)**. Although this is a rather complex sensor and includes a reference cell (which is kept at the same temperature) for balancing, it is very accurate, capable of sensing the dew point temperature at accuracies of less than 0.05°C.

The same measurement can be done with the crystal microbalance sensor described in the previous section. In that case, the resonant frequency of a crystal covered with a water-selective coating is used and its resonant frequency sensed while the sensor is cooled. At the dew point, the sensor's coating is saturated and the frequency drops to its lowest value. Equally well, a SAW mass sensor can be used with even greater accuracy. The heating/cooling is achieved as in **Figure 8.28** by use of a Peltier cell.

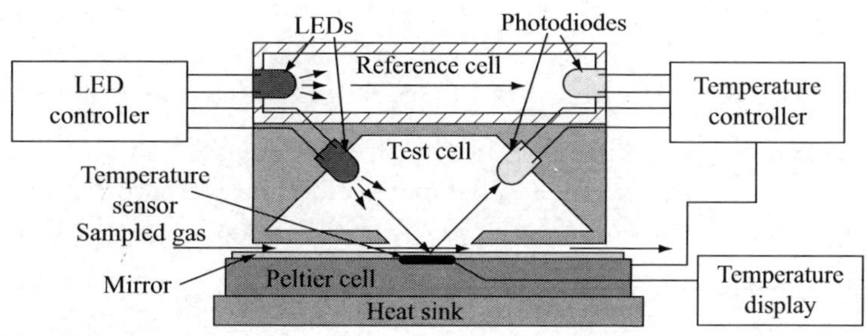

FIGURE 8.28 ■ An optical dew point sensor based on measurement of the dew point.

EXAMPLE 8.10 Calculation of dew point temperature

Calculate the dew point temperature in air at 60% relative humidity at 25°C. Show that at 100% humidity, the dew point temperature must equal the ambient temperature of 25°C.

Solution: The saturation water vapor is calculated first from **Equation (8.19)** as follows:

$$P_{ws} = 133.322 \times 10^{0.66077+7.5t/(237+25)} = 3165.94 \text{ Pa}.$$

The dew point temperature is calculated from **Equation (8.23)**:

$$DPT = \frac{237.3\left(0.66077 - \log_{10}\left(\dfrac{P_{ws} \cdot RH/100}{133.322}\right)\right)}{\log_{10}\left(\dfrac{P_{ws} \cdot RH/100}{133.322}\right) - 8.16077}$$

$$= \frac{237.3 \times \left(0.66077 - \log_{10}\left(\dfrac{3165.94 \times 60/100}{133.322}\right)\right)}{\log_{10}\left(\dfrac{3165.94 \times 60/100}{133.322}\right) - 8.16077}$$

$$= 16.69°\text{C}.$$

That is, water droplets (condensation) will form at any temperature below 16.69°C.

At 100% humidity we get

$$DPT = \frac{237.3\left(0.66077 - \log_{10}\left(\dfrac{3165.94 \times 100/100}{133.322}\right)\right)}{\log_{10}\left(\dfrac{3165.94 \times 100/100}{133.322}\right) - 8.16077}$$

$$= 25.0°\text{C}$$

as expected.

EXAMPLE 8.11 Absolute humidity sensor

Humidity is intricately linked with temperature and pressure and any attempt at sensing humidity must, at the very least, take these into account. However, given the pressure and temperature, measurement of humidity can be relatively simple. Consider the measurement of absolute humidity (amount of water) in air using a capacitive sensor. At 30°C and 1 atm of pressure, the amount of water in air varies between 0 and 30 g/m^3. Given a parallel plate capacitor with a plate area of 10 cm^2 and separation between the plates of 0.01 mm, estimate the range of capacitance of the sensor.

Solution: As the humidity increases, the relative permittivity of air increases. The relative permittivity of water at that temperature is approximately 80, whereas the relative permittivity of air is 1. One way to estimate the permittivity of the mixture is to use a volume average, as follows:

$$\varepsilon_r = \frac{\varepsilon_{rw} \times v_w + \varepsilon_{ra} \times v_a}{v_w + v_a} = \frac{80 \times 30 + 1 \times 10^6}{10^6} = 1.0024,$$

where 30 g of water was taken as 30 cL in volume and 1 m^3 of air is 10^6 cL. ε_{rw} is the relative permittivity of water, ε_{ra} is the relative permittivity of air, v_w is the volume of water, and v_a is the volume of air.

The relative permittivity varies between 1 (no moisture) and 1.0024 when the air is saturated. In terms of capacitance (see **Equation (5.2)**),

$$C_{max} = \frac{\varepsilon_0 \varepsilon_r S}{d} = \frac{8.854 \times 10^{-12} \times 1.0024 \times 10 \times 10^{-4}}{0.01 \times 10^{-3}} = 8.875 \times 10^{-10}\ \text{F}$$

and

$$C_{min} = \frac{\varepsilon_0 \varepsilon_r S}{d} = \frac{8.854 \times 10^{-12} \times 1 \times 10 \times 10^{-4}}{0.01 \times 10^{-3}} = 8.854 \times 10^{-10}\ \text{F},$$

where S is the area of each plate and d is the separation between them. ε_0 is the permittivity of vacuum. The capacitance changes between 885.4 pF and 887.5 pF. This is a small change (about 0.34%), but nevertheless it is measurable, especially if the capacitor is part of an oscillator and the frequency is measured (see **Chapter 11**). The change in capacitance can be increased by adding a hygroscopic material between the plates, but the advantage of the simple capacitor is that its response is faster and one does not have to worry about "drying" the hygroscopic material before measurements. It is also obvious from this example why this type of sensor is not the best—the change in capacitance is small and, since that depends on pressure and temperature, sensing is likely to be inaccurate.

8.9 | CHEMICAL ACTUATION

Chemical actuation can take many forms. The most obvious is a chemical reaction whose purpose is to affect an outcome. But even here there are many forms of reactions that are being used. One type of reaction is the conversion or oxidation processes that take place in the catalytic converter of a vehicle. The purpose, of course, is to reduce polluting constituents in the exhaust stream. Another type is the explosive inflation of an airbag. Although one may argue that this is a purely mechanical action, it is in fact the explosion of a charge that generates sufficient gas, sufficiently fast for the airbag system to be effective. And the whole idea of the internal combustion engine is based on what can be properly called chemical actuation, where the combustion converts hydrocarbons into gases (mostly CO_2, but also CO, NO_x, and SO_4, of which only CO_2 is considered a nonpollutant). A third example of chemical actuation is the electroplating process and cell.

There are many more chemical actuators, including chemical scrubbers, galvanic cells (batteries), and electrolytic cells, but we will concentrate here on only four: the catalytic converter, the airbag or explosive actuator, electroplating, and cathodic protection against corrosion.

8.9.1 The Catalytic Converter

The catalytic converter as used in vehicles has become one of the primary tools in pollution control and is universally used in gasoline fueled cars. With some modifications it is also applicable to diesel vehicles. The catalytic converter is used for three purposes:

1. Oxidation of CO to CO_2 to reduce the presence of this pollutant in air:

$$2CO + O_2 \rightarrow 2CO_2. \tag{8.24}$$

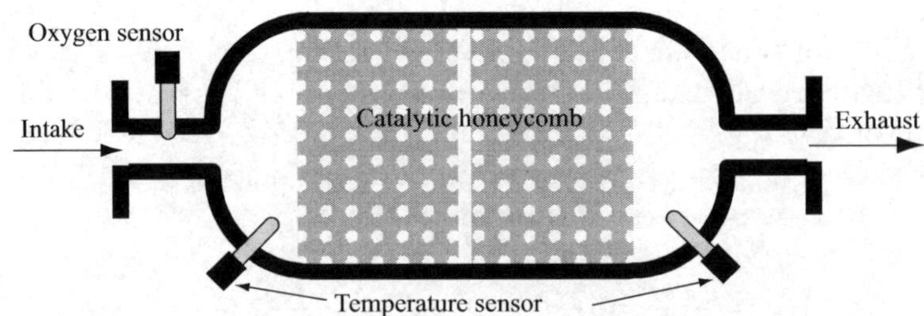

FIGURE 8.29 ■ The catalytic converter.

2. Oxidation of unburned hydrocarbons (HCs) to (CO_2) and water (H_2O):

$$C_xH_{2x+2} + [(3x+1)/2]O_2 \rightarrow xCO_2 + (x+1)H_2O. \tag{8.25}$$

3. Reduction of nitrogen oxides (NO and NO_2, together referred to as NO_x) to free nitrogen and oxygen:

$$2NO_x \rightarrow xO_2 + N_2. \tag{8.26}$$

These pollutants are produced either through incomplete combustion (CO) or as a high-temperature reaction (NO_x). The catalytic converter also produces by-products, including hydrogen sulfide (H_2S) and ammonia (NH_3). H_2S is controlled through the reduction of sulfur in gasoline and by the catalytic converter to eliminate by-products.

The converter is actually quite simple. It is a chamber enclosing a honeycomb structure made of Al_2O_3 or a mesh structure to increase surface contact with the exhaust gases. The structure is coated with a catalyst, typically platinum (but other catalysts such as palladium, rhodium, cerium, manganese, and nickel may be used for specific purposes). The whole structure is heated to a temperature between 600°C and 800°C by the exhaust gases. The catalyst only stimulates the chemical reaction, it does not take part in it. For this to happen the converter must first reach a minimum temperature (400°C–600°C). Beyond that the efficiency increases until at normal operating temperatures it reaches 90% or higher. The basic structure is shown in **Figure 8.29**, which also shows temperature sensors and at least one oxygen sensor. The oxygen sensor is necessary to control the reaction since oxygen in sufficient quantities must be present for the conversion to occur. This is supplied through the combustion process by increasing or decreasing the amount of oxygen in the mixture. The temperature sensors monitor the operation of the converter. For example, as CO oxidizes, the temperature increases, showing a higher temperature at the exhaust of the converter.

EXAMPLE 8.12 Overheating of catalytic converters

The catalytic converter in a car can overheat or even melt due to excess heat produced by the reactions occurring in it. Specifically, the conversion of CO produces additional heat that elevates the temperature of the converter. If the engine produces large quantities of CO (due to incomplete burning of fuel), the catalytic converter may suffer permanent damage. To understand this better consider a four-stroke, six-cylinder, internal combustion engine with a total displacement of 2400 cc running at 2000 rpm. Typical concentrations of CO in the exhaust of an engine before the catalytic converter are about 5000 ppm, whereas after the catalytic converter they drop to less than 100 ppm. Estimate the heat generated from the conversion of CO to CO_2 in the catalytic converter of the engine per minute. The density of air is 1.2 kg/m^3 (at 20°C) and the properties of CO are a specific heat capacity of 29 J/mol/K and an enthalpy of combustion in oxygen of 283 kJ/mol.

Solution: The total mass of gasses in the exhaust equals the air mass plus the mass of the fuel. Since the mass of fuel is rather small compared with the air mass, we shall neglect it in this calculation. We first calculate the mass of the air and then estimate the molar mass of the CO being converted.

The displacement of an engine is the volume of all of its cylinders, and each cylinder has a displacement volume of 400 cc. In a four-stroke engine, one cylinder is filled each half-rotation, so that all four cylinders are filled in two rotations. That is, the engine takes in 1.2 L of air per rotation or $1.2 \times 2000 = 2400$ L/min. This translates into $2400/1000 = 2.4$ m^3, or a mass of $2.4 \times 1.2 = 2.88$ kg of air per minute. Note that since the mass of fuel consumed per minute is only about 10 g, its contribution to the total exhaust mass is small. The concentration of CO is 5000 ppm, meaning that the mass of CO in the exhaust is

$$Mass_{CO} = 2.88 \times 5{,}000 \times 10^{-6} = 0.01414 \text{ kg/min},$$

or 14.14 g/min. The molar mass of CO is (see **Example 8.7**)

$$MM(CO) = 1 \times 12.01 + 1 \times 16 = 28.01 \text{ g}.$$

That is, 1 mole of CO has a mass of 28.01 g. Therefore the heat generated per minute is

$$H = 283 \times \frac{14.14}{28.01} = 142.864 \text{ kJ}.$$

The conversion generates about 143 kJ of heat.

This heat raises the temperature of the catalytic converter. The change in temperature can be calculated given the thermal properties of the converter and ambient temperature. In a car that is a difficult task, since air motion as the car moves changes the conditions dynamically. Nevertheless, this is a concern, and high levels of CO can lead to overheating of the converter and its possible failure.

8.9.2 The Airbag

The airbag system in vehicles is used as a safety device to protect occupants in case of collision. A number of sensors (accelerometers, wheel speed sensors, impact sensors, and others) are monitored to determine if a collision occurred and the airbag needs to be deployed. A small explosive charge is set electrically that then initiates the reaction and starts the gas-generation process. A variety of materials have been and are being used, but most of them release nitrogen as the primary gas. For example, in early systems, sodium azide (NaH_3) was used as a propellant. When ignited it produces sodium and nitrogen:

$$2NaH_3 \rightarrow 2Na + 3N_2. \quad (8.27)$$

There are other propellants that are less toxic, some are organic, some inorganic, and some airbag systems use compressed nitrogen or argon for the same purpose. Some of the alternatives are triazole ($C_2H_3N_3$), tetrazole (CH_2N_2), nitroguanidine ($CH_4N_4O_2$), nitrocellulose ($C_6N_7(NO_2)_3O_5$), and others, usually with stabilizers and reaction modifiers added to increase stability over time and to control the rate of the reaction (many of these materials are explosives and unstable, hence the need for additives). A typical airbag will contain between 50 and 150 g of propellant depending on the volume of the airbag. This generates a high volume of nitrogen that allows fast inflation of the bag and sufficient pressure to absorb the impact of the body.

EXAMPLE 8.13 Inflation of an airbag

Estimate the pressure in an airbag that uses 100 g of NaH_3, assuming the airbag has a volume of 50 L. Assume as well that the temperature of the gas increases to 50°C and that no gas escapes from the airbag. Naturally not all these conditions are satisfied in reality, but they allow an estimate of the process. For example, the airbag has vents to deflate the gas, but the inflation is so fast that initially the approximation here is valid.

Solution: From the reaction in **Equations (8.27)**, 2 moles of NaH_3 generates 3 moles of nitrogen. At standard temperature and pressure (STP) a mole of gas (any gas) is 22.4 L. Thus we need to calculate the number of moles generated and for that we must evaluate the molar mass of NaH_3. Using the periodic table we write

$$MM = 22.9897 + 3 \times 14.0067 = 65.0099\ \text{g/mol}.$$

Thus 100 g of NaH_3 will produce n moles:

$$n = \frac{100}{65.0099} = 1.5382\ \text{mol}.$$

However, since 2 moles of NaH_3 produces 3 moles of nitrogen gas, n must be multiplied by 3/2. Now we can use the ideal gas relation as follows:

$$PV = nRT,$$

where P is pressure [N/m], V is volume [m^3], n is the number of moles, R is the gas constant, equal to 8.3144621 J/mol/K, and T is the temperature [K]. The pressure inside the bag is therefore

$$P = \frac{nRT}{V} = \frac{1.5382 \times 1.5 \times 8.3144621 \times 323.15}{0.050} = 123{,}988\ \text{N/m}^2.$$

This pressure is somewhat low. Airbags for adults require a pressure of about 150–200 kPa.

Notes: The temperature used here is an estimated value and, because the pressure builds up quickly, it may not be uniform within the bag and may be higher. The bag deflates through vents in about 2 s. The inflation time is typically 40–50 ms.

8.9.3 Electroplating

Electroplating is an electrodeposition process through which a metal is coated with a thin layer of another metal to affect a desired property. In many cases this deposition is decorative, in others it is protective, and in still others it may be structural. In effect, metal ions in a solution are moved by means of an electric field from the solution to the

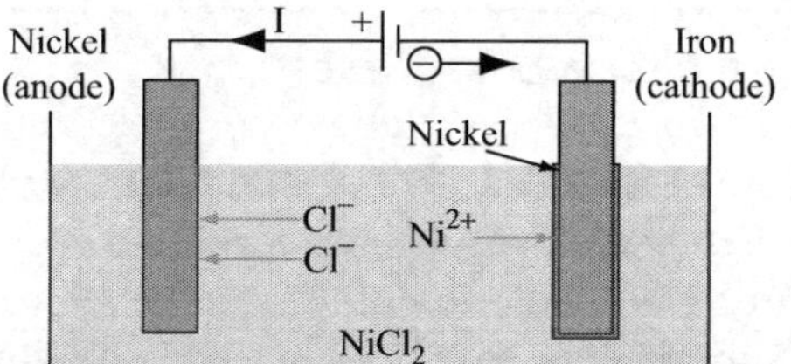

FIGURE 8.30 ■ Electroplating of nickel on iron.

medium being coated. To maintain the process, a sacrificial electrode of the metal used for coating is usually (but not always) used to supply the ions. The process, in its simplest form, is shown in **Figure 8.30**. In this case an iron piece connected as the cathode is coated with nickel. The electrolyte is usually an aqueous solution of a salt of the metal used for coating. In the case shown here the solution is nickel chloride ($NiCl_2$) and the anode is nickel to supply the ions. The $NiCl_2$ ionizes in water to nickel cations (ions with excess protons or positive ions; Ni^{2+}) and chlorine anions (ions with excess electrons or negative ions; Cl^{1-}). When the cations reach the cathode they reduce to metallic nickel by gaining two electrons. At the same time, the chlorine anions give up their electron and reduce to chlorine. This is released as gas at the anode. The role of DC is particularly important in the process. First, the metal mass being electroplated is directly proportional to the current since the extra electrons needed for reduction are supplied by the current. This is usually stated through Faraday's law as follows:

1. The mass of the metal coating is proportional to the quantity of electricity passed through the cell, and
2. The mass of the material liberated is proportional to its electrochemical equivalent. This can be stated as follows:

$$W = \frac{Ita}{nF} \quad [\text{g}], \tag{8.28}$$

where W is the mass [g], I is the current [A], t is time [s], a is the atomic weight of the metal, n is the valence of the dissolved metal [g-eq/mole], and $F = 96{,}485.309$ C/g-eq is Faraday's constant. In this particular case $n = 2$ (the reduction requires two electrons). Faraday's constant means that to deposit 1 g of metal one needs a charge equal to nF coulombs ($\text{A} \cdot \text{s}$). This charge is supplied by the current over time. Faraday's constant also indicates one specific problem with electroplating: it requires very large currents or, alternatively, it can be a very slow process. The voltages used are typically low, on the order of only a few volts, but nevertheless the energy needed is significant.

There are many variations of the basic process and many different solutions are used, each with its own properties and each adapted for a particular use, but these issues are more technological than fundamental. For example, in gold plating one does not use a sacrificial gold anode, but rather a carbon or lead anode. All ions are supplied by the solution (usually a gold-cyanide solution) and this solution must be replenished to sustain the plating process. Depending on the materials used, the process may release gases, some of which need to be processed and may produce dangerous substances that again need to be treated properly.

Electroplating is a common process in use since the very discovery of the electric cell by Alessandro Volta in 1800. It was first reported in 1805 following Volta's invention, but there is some intriguing speculation that the process was known in ancient times.

EXAMPLE 8.14 **Gold plating of printed circuit board traces**

Printed circuit boards are made of copper over fiberglass, but certain parts are often gold plated to improve contact and prevent corrosion. These include connector traces and pads. To get an idea of the issues involved, consider the plating of a printed circuit board on which a total of 8 cm^2 are gold plated to a thickness of 25 μm. Plating is done in a gold-cyanide solution, $Au(Cn)_2^{-1}$, at a relatively low current density of 10^4 A/m^2 to ensure a smooth coating. The gold-cyanide ion dissociates into a gold cation (Au^+) and two anions ($2Cn^-$). Calculate the time needed for plating.

Solution: In this case the reduction requires a single electron, $n = 1$. The atomic weight of gold is $a = 196.966543$. The total mass that needs to be plated is calculated from the volume needed and the atomic mass. The volume is

$$vol = area \times thickness = 8 \times 10^{-4} \times 25 \times 10^{-6} = 2 \times 10^{-8}\ \text{m}^3.$$

The density of gold is 19,320 kg/m^3. Therefore the mass is

$$M = vol \times density = 2 \times 10^{-8} \times 19320 = 3.864 \times 10^{-4}\ \text{kg}.$$

Equation (8.28) requires the mass in grams. The total mass needed is 0.3864 g. The current needed is the given current density multiplied by the area being plated:

$$I = area \times current\ density = 8 \times 10^{-4} \times 10^4 = 8\ \text{A}.$$

The time needed is found from **Equation (8.28)**:

$$t = \frac{nFW}{Ia} = \frac{1 \times 96,485.309 \times 0.3864}{8 \times 197} = 23.66\ \text{s}.$$

It takes just under 24 s to affect the plating.

8.9.4 Cathodic Protection

Corrosion of metals occurs when the metal transfers electrons in the presence of oxygen, starting a reaction that ends in any of a number of corrosion products. The best known of these products is iron oxide (Fe_2O_3), but other products exist and are quite common. The reactions are facilitated by the presence of water and oxygen and accelerated by acids. Therefore one can say that corrosion takes place in an electrolytic cell. The reactions that lead to the formation of Fe_2O_3 (rust) are as follows: iron oxidizes in the presence of oxygen by transferring electrons to oxygen,

$$Fe \rightarrow Fe^{2+} + 2\ e^-. \tag{8.29}$$

If water is present, the excess electrons, oxygen, and water form hydroxide ions:

$$O_2 + 4\ e^- + 2\ H_2O \rightarrow 4\ OH^-. \tag{8.30}$$

The iron ions react with oxygen:

$$4\ Fe^{2+} + O_2 \rightarrow 4\ Fe^{3+} + 2\ O^{2-}. \tag{8.31}$$

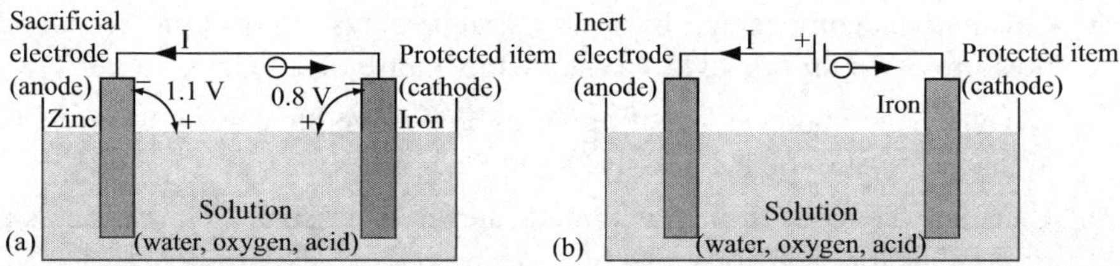

FIGURE 8.31 ▪ Cathodic protection. (a) Passive or sacrificial cathodic protection. (b) Active or impressed current cathodic protection.

The reaction that leads to the formation of Fe_2O_3 is

$$Fe^{3+} + 3\,H_2O \rightleftharpoons Fe(OH)_3 + 3\,H^+. \tag{8.32}$$

The $Fe(OH)_3$ product dehydrates and leads to the formation of Fe_2O_3 as follows:

$$Fe(OH)_3 \rightleftharpoons FeO(OH) + H_2O. \tag{8.33}$$

$$2\,FeO(OH) \rightleftharpoons Fe_2O_3 + H_2O. \tag{8.34}$$

As mentioned, there are many iron corrosion products whose formation depends on the availability of oxygen and water as well as other salts and acids.

Protection against corrosion, short of preventing contact with water and oxygen (through paints, coatings, or plating), must eliminate the oxidation of iron. That is, if one can prevent the transfer of electrons from iron to oxygen, the process stops and iron is protected against corrosion. This is the role of cathodic protection. The method is shown schematically in **Figure 8.31**. There are two methods that are commonly used. One, shown in **Figure 8.31a**, consists of setting up a galvanic cell where the anode is any metal with a contact potential that is more negative than the contact potential of the protected metal. This forces electrons to flow from the anode to the cathode (iron) in opposition to the oxidation process above. In the process, the anode is consumed (it is sacrificial) and eventually must be replaced. In the cathodic protection of iron, the most common anode is zinc. It has a contact potential of -1.1 V, whereas the contact potential of iron can vary between -0.2 and -0.8 V, depending on the composition and treatment (e.g., steel is less active and hence has a less negative contact potential than, say, cast iron). There are other materials that can be used, most notably magnesium alloys (contact potential of -1.5 to -1.7 V) and aluminum (contact potential of -0.8 V). The second method is an active method shown in **Figure 8.31b**. It consists of a nonsacrificial anode and a power supply that produces the countercurrent. The anode may be an iron alloy, but can be graphite or, in some cases, a platinum-coated wire. The current is adjusted to counter the oxidation electron flow, usually by measuring the contact potential and ensuring it is below -1.0 to -1.1 V.

8.10 | PROBLEMS

Units

8.1 **Composition of air.** The approximate composition of dry air by volume at 20°C is as follows: nitrogen (N_2) 78.09%, oxygen (O_2) 20.95%, argon (Ar) 0.93%, and CO_2 0.03%.

a. Calculate its composition by mass. Assume air is an ideal gas at an ambient pressure of 1 atm (101,325 Pa) and temperature of 20°C.

b. Calculate the composition of air in mol/m^3. The air density at 20°C and an ambient pressure of 1 atm is 1.2 kg/m^3.

c. Calculate the number of atoms (molecules in the case of CO_2) per cubic meter (m^3) of air for each constituent.

Note: There are many other constituents in air, but the four included here are the most important in terms of quantities.

8.2 **Burning of natural gas.** The reaction that takes place during the burning of natural gas (methane [CH_4]) is as follows:

$$CH_4 + 2(O_2) \rightarrow CO_2 + 2(H_2O).$$

In the process, the reaction produces heat at a rate of 890 kJ/mol. Air is at an ambient pressure of 1 atm and a temperature of 20°C. Under these conditions, air contains 21% oxygen by volume or 23% oxygen by mass.

a. Calculate the necessary proportions of gas and air by volume to produce complete burning of the gas.

b. Calculate the necessary proportions of gas and air by mass to produce complete burning of the gas.

c. Natural gas is delivered to a furnace at a pressure of 2 atm at 20°C. Calculate the heat generated per cubic meter of natural gas.

8.3 **Molar mass and gram-equivalent.**

a. Calculate the molar mass of CO_2.

b. Calculate the molar mass of magnesium.

c. Calculate the gram-equivalent of CO_2 dissolved in water. The solution of CO_2 in water is as follows:

$$CO_2 + H_2O \rightarrow H^+ + HCO_3{}^-.$$

d. Calculate the gram-equivalent of magnesium ions in water (Mg^{++}).

8.4 **Conversion between units.** A total of 0.01 mol of sulfuric acid (H_2SO_4) is mixed into 1 L of distilled water (H_2O). The density of water is 1 g/cm^3 and that of sulfuric acid is 1.84 g/cm^3. Calculate the concentration of sulfuric acid in ppm as

a. A mass fraction.

b. A volume fraction.

Metal oxide sensors

8.5 **Oxygen sensor in internal combustion engines.** The use of oxygen sensors to reduce emission of noxious gases in internal combustion engines is mandated by the need to comply with pollution regulations. The sensor is used to sense the ratio between the oxygen concentration in air and in the combustion stream and control the intake of oxygen to reduce emissions. The concentration of oxygen in air is approximately 20.9% (by volume). Calculate the range of readings of the oxygen sensor in the exhaust stream from conditions of no combustion (20.9% oxygen in the exhaust stream) to 4% oxygen in the stream. The exhaust is at a temperature of 600°C.

Note: It is desirable that a certain percentage of oxygen remains in the stream to allow the catalytic converter to operate and remove some of the combustion by-products such as CO. However, too much oxygen leads to lean combustion and can cause overheating of the engine.

8.6 Carbon monoxide sensor. The CO sensor in **Figure 8.1** is used to detect CO in a home and provide an alarm for concentrations greater than 50 ppm (the maximum allowable long-time workplace exposure level in the United States). To calibrate the sensor, its resistance is measured at 10 ppm and 100 ppm CO. The measured values are 22 kΩ and 17 kΩ, respectively. Calculate the sensor reading at which the alarm is triggered and the sensitivity of the sensor.

8.7 Metal oxide sensors and temperature variations. The resistance of metal oxide sensors depends on, among other things, temperature since the conductivity of the metal oxide is temperature dependent. The change in resistance as a measure of the concentration of the analyte relies on the fact that the temperature of the sensor is constant. Consider a thin film tin oxide sensor operating at 300°C used to sense CO. The calibration values are 16.5 Ω at a concentration of 75 ppm and 492 Ω at a concentration of 15 ppm. The conductivity of tin oxide is 6.4 S/m at 20°C and its temperature coefficient of resistance is −0.002055/°C.

Note: Tin oxide, unlike many other metal oxide materials, has relatively high electric conductivity.

a. Calculate the sensitivity of the sensor throughout its range and at the two given calibration points.

b. Calculate the relative error in the base conductivity of the material due to variations of temperature around the base temperature of 300°C.

c. Discuss the implications of the result in (b).

8.8 Sensing of oxygen in molten steel. An oxygen sensor similar to that shown in **Figure 8.5** is used to sense the oxygen concentration during the production of steel. The temperature of molten steel is kept a little above the melting point at 1550°C to ensure its flowing properties. The concentration of oxygen in air at that temperature is 18.5%. Oxygen is required in steel processing to react with carbon and hence produce low carbon steel. At the end of the process, excess oxygen must be removed, with residual oxygen in steel typically at about 100 ppm (by volume).

a. Calculate the emf expected from the sensor at a concentration of 100 ppm oxygen assuming the output is zeroed for equal concentrations for oxygen in air and steel.

b. Calculate the sensitivity of the sensor for oxygen concentrations in steel.

Solid electrolyte sensor

8.9 Pollution control in a wood-burning stove. Wood-burning stoves and hearths are a pleasant source of heat on a cold winter day, but they are highly polluting, and unless properly ventilated they can be dangerous. In an attempt to control pollution an oxygen sensor is placed in the flue and used to control a fan that supplies the necessary additional air to properly burn the wood and reduce pollution. The temperature of the flue is 470°C and the oxygen level in the flue should not go below 8%. The normal oxygen level in the interior of a house is 20%. The system is set to keep the oxygen level between 8% and 12%. To ensure that the temperature of the flue does not rise too much, the fan is turned on when the oxygen level decreases to 8% and is turned off when it reaches 12%. Calculate the sensor output voltages at which the fan turns on and off.

Glass membrane sensors

8.10 pH measurements. In a pH meter the reading of the device is calibrated in pH values from 1 to 14. The actual meter is a high-impedance voltmeter and measurements are done at an ambient temperature of 24°C.

a. Calculate the range of voltages the voltmeter must be capable of displaying for the range of pH between 1 and 14 given a Ag/AgCl reference electrode.

b. Calculate the range of errors per °C for the range of pH between 1 and 14.

8.11 Effect of CO_2 absorption on the pH of water. Water absorbs CO_2 with a maximum concentration of 1.45 g/L. If neutral water (pH = 7) is left for long periods of time exposed to air it will absorb CO_2 from the air and become increasingly acidic, although the absorption rate is slow. The reaction responsible for the increase in acidity of water is the following:

$$CO_2 + H_2O \rightarrow H^+ + HCO_3^-.$$

Calculate the pH of initially neutral water after it has absorbed 1.45 g of CO_2 per liter.

8.12 pH and acid rain. Although rainwater is in itself slightly acidic (with a pH between 5 and 6), whenever the pH is below 5 the rain is considered to be acidic, and hence detrimental to the environment. The causes of acid rain are mostly emissions from power plants, vehicles, and other chemical pollutants. CO_2 is one source (see **Problem 8.11**), but the source of most concern is sulfur dioxide (SO_2), whose main source is the burning of coal, primarily in coal-fired power plants, but also from vehicles and from volcanic eruptions. The reaction that occurs in the atmosphere is the following:

$$2(SO_2) + O_2 \rightarrow 2SO_3$$

followed by

$$SO_3 + H_2O \rightarrow H_2SO_4.$$

Sulfuric acid in water produces hydrogen cations and SO_4 anions:

$$H_2SO_4 \rightarrow 2H^+ + SO_4^{2-}.$$

To understand the problem of acid rain, consider an atmospheric concentration of 2 ppm SO_2. Assuming that 0.5 ppm SO_2 is absorbed by falling rain that in the absence of SO_2 in the atmosphere would have a pH of 5.3, what is the pH of the rainwater after absorption of SO_2?

Soluble inorganic salt membrane sensors

8.13 **Chloride ion sensor.** A silver chloride mixed with silver sulfide ($Ag_2S/AgCl$) membrane is used to sense low concentrations of chloride in water by detecting the Cl^- ion. With a Ag/AgCl reference electrode, the instrument measures a potential of 0.275 V at 32°C. Calculate the concentration of chloride in the water.

8.14 **Lead sensor and errors.** To sense lead in water, one can use a Ag_2S membrane mixed with lead sulfide (PbS). The membrane senses the Pb^{2+} ion. Suppose a concentration of 100 ppm is measured, calibrated at 25°C using a saturated calomel reference electrode in a normal pH meter (intended to sense hydrogen in water).

a. Calculate the potential expected across the electrodes.

b. What is the error in reading of the concentration if the temperature rises to 30°C and if no compensation for temperature is provided?

Thermistor-based chemical sensors

8.15 **Blood glucose sensor.** To sense the concentration of glucose in blood for the purpose of monitoring diabetes, one can use a thermistor coated with the enzyme glucose oxidase and sense the temperature of the thermistor. Normal glucose levels in blood are between 3.6 and 5.8 mmol/L. Glucose has the formula $C_6H_{12}O_6$ and has an enthalpy of 1270 kJ/mol. The thermistor has a heat capacity of 24 mJ/K (the heat capacity for thermistors is typically given in mJ/K, a quantity that takes into account its mass) and a nominal resistance of 24 kΩ at 20°C. Assume the enzyme samples 0.1 mg of blood. Assume as well that blood is mostly water. The sensing is done at the normal blood temperature of 36.8°C.

a. Calculate the increase in temperature of the thermistor over the range of normal glucose levels in blood.

b. Calculate the sensitivity of the sensor over the span given.

c. If, in addition, one measures the resistance of the thermistor as 18.68 kΩ at 30°C, calculate the range and span of the resistance of the thermistor and its sensitivity in terms of the resistance measured.

8.16 **Sugar (sucrose) sensor for sugar production.** In the production of sugar from sugar cane, the stalks are first chopped and pressed to yield a juice from which sugar is refined. Sugar cane produces a juice with a typical sugar concentration in water of 13% by mass. To sense the concentration the sucrose phosphate synthase enzyme can be used in a thermistor-based sensor to catalyze the sugar. Sucrose has the formula $C_{12}H_{22}O_{11}$ and an enthalpy of 5644 kJ/mol. A 12 kΩ thermistor (at the solution temperature) with a heat capacity of 240 mJ/K and a self-heat of 0.05 °C/mW is employed for this purpose.

a. If the sensor samples 0.2 mg of the solution, calculate the sensitivity of the sensor in °C per percent sugar.

b. If a current of 1.8 mA through the thermistor is required to operate the sensing circuitry, what is the error expected due to self-heat, neglecting the change in resistance of the sensor due to the change in temperature of the sensor when sugar is sampled?

Catalytic sensors

8.17 **Methane detector in mines.** A catalytic sensor based on a pellistor can be used to detect the methane concentration in mines to alert miners when the concentration is too high. The sensor is calibrated in LEL showing the percentage of LEL. Methane has the formula CH_4 and an enthalpy of 882 kJ/mol. The pellistor is based on alumina with a heat capacity of 775 J/kg/K, a mass of 0.8 g, and the platinum heater has a resistance of 1250 Ω at the operating temperature of 540°C. If the sensor samples a volume of 0.75 cm^3 of air, calculate the sensitivity of the sensor in ohms per percent LEL. The air and methane before sampling are at a temperature of 30°C and the ambient pressure is 101,325 Pa (1 atm).

Optical chemical sensors

8.18 **Water leakage sensor.** An evanescent loss sensor uses an optical fiber to sense water leakage into the bottom of a boat. The fiber is strung close to the interior surface of the boat, but not touching it, so it does not detect condensation of water at the bottom. To ensure that the sensor will only detect water, the angle of incidence of the laser beam (see **Figure 8.32**) is adjusted so that the total reflection occurs at all permittivities lower than that of water. The relative permittivity of glass at optical frequencies is 1.65 and that of water is 1.34. Detection occurs when the power transmitted along the optical fiber decreases because of transmission into the water through the interface.

a. Calculate the angle of incidence, θ_i, of the light beam to ensure that water will be detected.

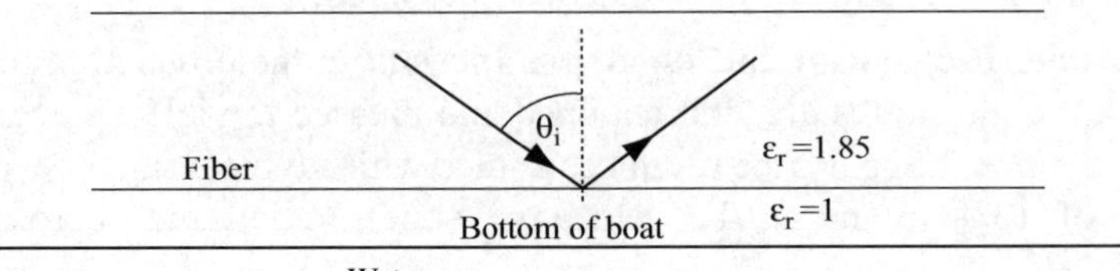

FIGURE 8.32 Water leakage sensor.

b. If one wishes to detect any medium with relative permittivity $\varepsilon_r \leq 1.65$ in contact with the fiber rather than limiting this to water, what is the incidence angle θ_i?

8.19 **Petroleum leakage sensor.** The sensor in **Figure 8.20** is used to detect petroleum leakage or water leakage from petroleum-transport hoses. These hoses are made with a double wall, as shown in **Figure 8.33**, with the sensor between the two walls. The purpose of the sensor is to detect oil leaking out through the inner hose or water leaking in through the outer hose. One simple method of detecting both oil and water is to use two sensors, one set to detect oil, the other set to detect water (a single sensor can be used as well with appropriate detection electronics). The relative permittivity of seawater at optical frequencies is 1.333, the relative permittivity of oil is 1.458, and the relative permittivity of the polycarbonate fiber used for the sensor is 1.585 (at the infrared frequency used for detection).

a. Calculate the range of angles of incidence required in sensor 1 so it will detect water.

b. Calculate the angle of incidence required in sensor 2 so it will detect oil.

c. Will sensor 1 detect oil as well?

d. Will sensor 2 detect water as well?

e. If the answer to (c) or (d) is yes, show what the two sensors will show under various conditions (no leakage, oil leakage, or water leakage) and how detection of water and oil can be guaranteed to be positive.

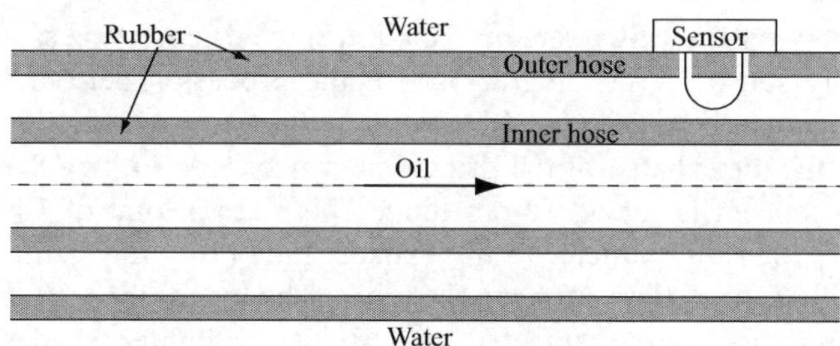

FIGURE 8.33 ■ A leakage sensor in the space between two rubber walls.

Mass sensors

8.20 **Measurement of corrosion rate using a crystal microbalance.** The crystal microbalance is an important analytical tool capable of high sensitivities and is commonly used in the laboratory. In one of its common forms a crystal disk is coated with opposite electrodes, typically of gold, and designed to resonate at a given frequency, typically between 6 and 18 MHz. The disk is connected to an oscillator and oscillates at its fundamental frequency. Any change in the mass of the disk changes the frequency. To measure the corrosion rate of iron in moist air, a crystal microbalance is used after one or both electrodes is coated with iron. In **Figure 8.34** the gold electrodes are 8 mm in diameter and the crystal disk is designed to oscillate at 10 MHz. A coating of 0.5 mg of iron is deposited on each of the gold disks, covering them entirely. The crystal used has a sensitivity factor of 54 Hz·cm^2/μg.

a. Calculate the resonant frequency before corrosion occurs.

b. The rate of corrosion is measured in mm/year, that is, the thickness of material in millimeters corroded in 1 year. Assuming the instrument can reliably detect a change in frequency of 10 kHz, calculate the sensitivity of the microbalance in this application. The density of iron is 7.87 g/cm^3.

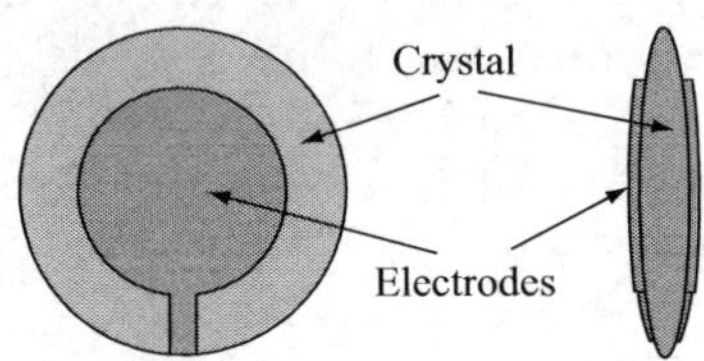

FIGURE 8.34 ■ A gold-plated crystal used for mass sensing.

Humidity and moisture sensors

8.21 **Capacitive humidity sensor.** A simple humidity sensor, although not the most sensitive, may be obtained by measuring the capacitance of an air-filled parallel plate capacitor. Consider a capacitor with two plates, each 4 cm^2 in area with a 0.2 mm separation between plates.

a. Calculate and plot the capacitance expected from the sensor for the range between 10% RH and 90% RH at an ambient temperature of 25°C.

b. Calculate the sensitivity of the sensor.

8.22 **Clothes drying humidity sensor.** To control the drying process in a clothes drier a humidity sensor is an integral part of the process. There are many types of sensors that can be used. Consider the possibility of a capacitive sensor built in line with the air exhaust of the drier. The sensor (see **Figure 8.35**) is made of a series of concentric tubes, 12 cm long. There are a total of 14 tubes, with the outer tube equal in diameter to the exhaust tube (100 mm) and separated 2 mm apart to allow air to flow between the tubes. The inner tube is 76 mm in diameter. Alternating tubes are connected together (i.e., lighter shaded tubes are all

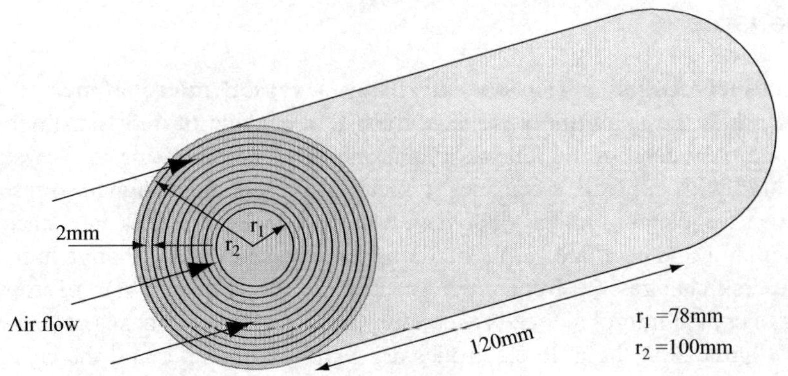

FIGURE 8.35 ■ A capacitive humidity sensor.

connected together and darker shaded tubes are connected together to form a multiconductor coaxial capacitor). Approximate the capacitance of the cylindrical capacitor by the parallel plate equivalent, where the area of the plates is the average between the area of the outer and inner plates.

a. Calculate the sensitivity of the sensor in pF/%RH.

b. If the temperature in the exhaust varies during the drying process from 50°C to 75°C, what is the variation in sensitivity expected?

8.23 **Relative humidity sensing.** The dew point temperature is measured using the device in **Figure 8.28** at an ambient temperature of 32°C and was found to be 22.6°C. Calculate the relative humidity in air.

8.24 **Dew point versus relative humidity.** Calculate and plot the dew point temperature at 27°C as the relative humidity varies from 0% to 100%. Use increments of 10°C.

8.25 **Capacitive humidity sensor.** The following data were collected for a capacitive humidity sensor based on an alumina hygroscopic layer at 20°C and at 60°C. The capacitance of the sensor at 0% humidity is 303 pF and the relative permittivity of (dry) alumina is 9.8. Assume a parallel plate capacitor structure.

Relative humidity [%]	0	10	20	40	60	80	90
Capacitance at 20°C [pF]	303	352	432	608	858	1216	1617
Capacitance at 60°C [pF]	303	345	394	508	655	845	963

a. Calculate the amount of water absorbed by the alumina layer at 20°C if its volume is 0.8 mm^3. Plot the amount of water (mass) absorbed as a function of relative humidity. The density of water is 1 g/cm^3 and the relative permittivity is 80.

b. At 60°C the relative permittivity of water reduces to 72. Calculate the amount of water (mass) absorbed and compare it with that at 20°C. Plot both.

c. Discuss the consequences of the results in (a) and (b) on the performance of the sensor. In particular, address the issues of response time (including the time needed to remove the moisture) and sensitivity to temperature variations.

8.26 **Dew point humidity sensor.** The dew point humidity sensor is one of the most accurate methods of sensing relative humidity, even though it is not the most convenient. In applications where accuracy is important this inconvenience is a minor issue. In a measurement the dew point sensor temperature is found to be 37°C at an ambient temperature of 90°C. Calculate the relative humidity of air.

8.27 **Dew point humidity sensor.** Show that the dew point temperature cannot be higher than the ambient temperature in two ways:

a. Theoretically.

b. By using a dew point temperature of 30°C at an ambient temperature of 25°C.

8.28 Sensitivity and resolution of the dew point sensor.

a. Calculate the sensitivity of the dew point sensor as a relative humidity sensor.

b. If the temperature sensor in **Figure 8.28** is capable of a resolution of ΔT_d [°C] at an ambient temperature T_a for which the dew point temperature is T_d, what is the resolution of a humidity sensor based on the dew point sensor?

8.29 Relative humidity as a function of dew point temperature. Calculate and plot the relative humidity at ambient temperatures $T = 20°C$, 25°C, and 30°C as the dew point temperature varies from –20°C to T. Use increments of 1°C.

Chemical actuation

8.30 Pollution and power loss in diesel engines. A small diesel engine is used to generate power. The generator output is 10 kW. The engine has a rated efficiency of 50% and consumes a common diesel fuel with rated energy of 32 MJ/L. The four-cylinder, four-stroke engine runs at a constant 1200 rpm and has a displacement of 850 cc. A catalytic converter is used to reduce CO emissions from 7000 ppm to 25 ppm as required by law. The density of air is 1.2 kg/m^3 (at 20°C) and the properties of CO are a specific heat capacity of 29 J/mol/K and an enthalpy of combustion in oxygen of 283 kJ/mol.

a. Calculate the power generated by the conversion of CO in the catalytic converter (power is energy per unit time [J/s]). Assume air intake is at the ambient temperature (20°C).

b. Calculate the efficiency of power generation and estimate the reduction in fuel consumption (in percent) if the engine emitted no CO at all, assuming the energy that otherwise would be generated in the converter is recovered in the engine itself. In estimating the reduction in fuel consumption, assume that fuel consumption is linearly related to power output, that is, if the output power is reduced by x%, fuel consumption is reduced by the same percentage.

Note: We usually think of pollution in terms of its negative effects on us and the environment, and of pollution control as a necessary, sometimes costly process. However, pollution has other costs, as this example shows, and elimination of pollution through the use of clean fuels and complete combustion has significant benefits.

8.31 Design of an airbag. A 75 L airbag must inflate to a peak pressure of 180 kPa. The nominal temperature for the design is 20°C.

a. Calculate the amount (mass) of NaH_3 propellant needed to achieve that. Neglect the increase in temperature of the gas during the reaction and assume inflation is entirely due to nitrogen gas generated during the reaction.

b. Given the amount (mass) of propellant calculated in (a) with the bag deploying at 0°C and no increase in the temperature of the gas during the reaction, what is the pressure expected in the bag?

c. The reaction that generates nitrogen gas does increase the temperature of the gas. Suppose the temperature rises by 50°C above ambient. What are the answers to (a) and (b) now?

Note: Pressure regulation in airbags is rudimentary but is an essential part of safety. Too high a pressure and injury might occur due to impact with the bag, whereas too low a pressure defeats the purpose of the airbag and can result in injury. For this reason, most airbags have some means of controlling pressure.

8.32 **Compressed nitrogen airbag system.** In principle, one can use compressed nitrogen to inflate airbags, avoiding the need for explosives and unstable materials. However, this is not as simple as it sounds, primarily because of the volume and pressure needed. Suppose a container capable of withstanding a pressure of 2.5 MPa is used to inflate a 105 L airbag (a typical airbag volume) to a pressure of 175,000 Pa.

a. What must be the volume of the container, assuming an ambient temperature of 30°C and that there is no increase in temperature during inflation and the gas container is also at 30°C.

b. Expanding gas cools during expansion. If one can assume a decrease in gas temperature of 30°C during inflation, what is the necessary volume of the container, assuming the same pressure? The ambient temperature is 30°C.

c. Because vehicles are designed to operate and be stored at elevated temperatures, the container designed in (a) or (b) must withstand the additional pressure due to changes in temperature. Assuming the container must be able to withstand temperatures between −60°C and +75°C (to allow for a reasonable margin of safety), what is the minimum and maximum pressure expected in the container?

d. What are your conclusions from the results above?

8.33 **Resistive electroplating coupon.** In an attempt to control plating thickness one can use a resistive coupon—a simple wire or strip made of the base material on which electroplating is performed. The resistance of the coupon changes with the coating thickness, and by measuring this resistance one can stop the coating process at the right time. Suppose electroplating of nickel on iron is performed and the nickel thickness required is 10 μm. A coupon is made as a very thin strip 4 cm long, 1 cm wide, and 0.5 mm thick of the same iron grade.

a. Calculate the change in resistance of the coupon from no plating to 10 μm nickel plating. The conductivities of iron and nickel are 1.12×10^7 S/m and 1.46×10^7 S/m, respectively.

b. The density of nickel is 8900 kg/m^3. If the coating thickness is achieved in 8 min, 35 s, what is the current density used in the plating process?

Note: The measurement of resistance must be made out of the solution, otherwise the resistance is affected by the solution itself, which is conductive. The coupon can be reused, but the calibration (zero coating resistance) changes and must be measured before electroplating begins.

8.34 **Aluminum production.** Aluminum is produced in an electrolytic process essentially identical to the electroplating process except that the electrodes used are carbon (graphite) and the process is done at high temperatures so that aluminum is in its liquid state. The process starts with alumina (Al_2O_3) in molten cryolite (Na_3AlF_6). The latter serves to conduct electricity. The process is called the Hall process and is as follows:

$$2Al_2O_3 + 3C \rightarrow 4Al + 3CO_2,$$

with carbon coming from the graphite electrodes and CO_2 being emitted as gas. To operate, a typical voltage of 4.5 V is used to generate a current of 100 kA.

a. Calculate the time it takes to produce one ton of aluminum.

b. Calculate the energy needed per ton of aluminum.

c. Calculate the mass of CO_2 released per ton of aluminum.

d. Calculate the mass of carbon needed for the process per ton of aluminum.

CHAPTER 9

Radiation Sensors and Actuators

Background radiation

The modern world has an almost innate fear of nuclear radiation. It may be the heritage of Hiroshima and Nagasaki or it may be that we just fear the unknown, the invisible, and of course there are some very good reasons to be careful. Nuclear radiation can cause damage to cells and in high doses is known to cause cancer or even death. However, radiation comes in many shades and forms. All electromagnetic waves fall in the same general category of radiation, the difference being only in frequency (and with it in energy). If one were to imagine an instrument with a dial that can change the frequency from zero to infinity, then as the frequency would rise, it would first generate low-frequency fields, first in the audio range, then into ultrasonics, then above about 200 kHz, what colloquially is called radio waves would appear. Further up, the instrument will pass through VHF (very high frequency), UHF (ultra high frequency), and then into the microwave region. Beyond that lies millimeter waves and then infrared (IR) radiation, followed by visible light and ultraviolet (UV), then into X-rays, α, β, and γ rays, and further up into cosmic rays. As the frequency increases, the energy associated with the waves increases and the radiation effects become more pronounced. As is generally known, UV and X-rays are harmful radiation and are part of the cumulative effect of radiation on our lives and health. It is expected that people working with X-rays will naturally be exposed to more radiation than those that may only have a scan in a lifetime. Pilots and frequent fliers will necessarily be affected by cosmic rays, as are astronauts in space. But beyond these, there is a background radiation level more or less constant over the globe. It is a low-level radiation caused by radioactive isotopes in rocks and soils on the order of 20–50 becquerel/minute (Bq/min) that can be detected with Geiger counters. This radiation is of no consequence to health, as it is too low to do any damage. The exposure level is, on an average, about 2.4 millisievert/year (mSv/yr). But there are locations and conditions in which the background radiation can be higher and of more concern. Granite rocks and hot springs tend to have higher radiation levels and certain areas around the globe have naturally occurring high radiation levels, as high as 250 mSv/yr or higher. On the other hand, sedimentary rocks and limestone have lower levels. Underground locations, including quarries, mines, or even basements, can have higher levels primarily from radon (a decomposition by-product of naturally occurring uranium and its isotopes), and radon can be found in the atmosphere as well as in water. However, beyond reasonable caution it should be remembered that these are natural sources that have been there from time immemorial and will be with us for any imaginable future.

9.1 | INTRODUCTION

We discussed radiation in **Chapter 4** when talking about light sensors. The particular emphasis there was on the general range occupied by infrared (IR), visible, and ultraviolet (UV) radiation. Here we will concern ourselves with the ranges below and above these. Specifically, the range above UV is characterized by ionization—that is, the frequency is sufficiently high to ionize molecules based on Planck's equation (**Equation (9.1)**). The frequencies are so high (above 750 THz) that many forms of radiation can penetrate through materials, and therefore the methods of sensing must rely on different principles than at lower frequencies. On the other hand, below the IR region the electromagnetic radiation can be generated and detected by simple antennas. Thus we will also discuss the idea of an antenna and its use as a sensor and an actuator.

All radiation can be viewed as electromagnetic radiation. Hence many of the sensing strategies, including those discussed in **Chapter 4**, may be viewed as radiation sensing. We will, however, follow the conventional nomenclature and call low-frequency radiation "electromagnetic" (electromagnetic waves, electromagnetic energy, etc.) and high-frequency radiation simply "radiation" (as in X-rays, α, β, γ, or cosmic radiation).

An important distinction in radiation is based on the Planck equation and uses photon energy to distinguish between types of radiation based on their respective energies:

$$e = hf \quad [\mathrm{J}], \tag{9.1}$$

where $h = 6.6262 \times 10^{-34}$ J/s or $h = 4.135667 \times 10^{-15}$ eV/s is Planck's constant, f is the frequency [Hz], and e is the photon energy, measured in joules [J] or electron-volts [eV]. At high frequencies, one can view radiation either as particles or as waves. The energy in these waves is also given by Planck's equation. Their wavelength is given by the de Broglie equation:

$$\lambda = \frac{h}{p} \quad [\mathrm{m}], \tag{9.2}$$

where h is Plank's constant and p is the momentum of the particle(s) given as $p = mv$ (m is mass [kg] and v is velocity [m/s]).

The higher the frequency, the higher the photon energy. At high frequencies the photon energy is sufficient to strip electrons from atoms—this is called **ionization**, and the radiation is said to be **ionizing radiation**. At low frequencies ionization does not happen, and hence these waves are said to be nonionizing and the radiation **nonionizing radiation**. The highest frequency in the microwave region is 300 GHz, resulting in a photon energy of 0.02 eV. This is considered to be nonionizing. The lowest frequency in the X-ray region is approximately 3×10^{16} Hz and the photon energy is 2000 eV, clearly an ionizing radiation. From a safety point of view, ionizing radiation is much more dangerous, but from a sensing point of view, this property opens new ways of sensing based on the ionization properties of radiation.

One thing must be made clear: some view radioactive radiation as something different than, say, X-ray radiation or microwaves—it is often viewed as particle radiation. Indeed, one can take this approach based on the duality of electromagnetic radiation, just as we can view light as an electromagnetic wave or as particles—photons. For consistency, we will base most of the discussion on the photon energy of radiation and will not emphasize the particle aspects. Nevertheless, in some cases it will be convenient to talk about particles. For example, in ionization sensors such as the Geiger–Muller counter, it is customary to talk about "counting" particles. In such cases it will be more convenient to discuss particles, even though the same can be accomplished from a wave propagation point of view. Thus the term "radiation" can mean either waves or particles.

EXAMPLE 9.1 Radiation and radiation safety

To see what is meant by ionizing and nonionizing radiation and the relation of ionization to radiation safety, consider two radiating sources: one is visible blue light, the other a source of X-rays. Blue light has a frequency of 714 THz (714×10^{12} Hz). Its photon energy is

$$e = hf = 6.6262 \times 10^{-34} \times 714 \times 10^{12} = 4.731 \times 10^{-19}\ \text{J}.$$

It is customary to give photon energy in units of electron-volts (eV): 1 eV $= 1.602 \times 10^{-19}$ J. Thus we can write

$$e = \frac{4.731 \times 10^{-19}}{1.602 \times 10^{-19}} = 2.953\ \text{eV}.$$

X-rays range from 30 pHz (30×10^{15} Hz) to 30 eHz (30×10^{18} Hz). Taking the lower limit,

$$e = hf = 6.6262 \times 10^{-34} \times 30 \times 10^{15} = 1.988 \times 10^{-17}\ \text{J}$$

or

$$e = \frac{1.988 \times 10^{-17}}{1.602 \times 10^{-19}} = 124.1\ \text{eV}.$$

Clearly, visible light cannot be considered "dangerous," and we know that it is not ionizing. X-ray radiation, especially at higher frequencies, is orders of magnitude more energetic and is ionizing. Hence we consider X-ray radiation in the same category as radioactive radiation and must be protected from it as much as possible.

Many of the radiation sensors based on ionization are used to sense the radiation itself, that is, to detect and quantify radiation from sources such as X-rays and from nuclear sources (α, β, γ, and neutron radiation). There are exceptions, however, such as smoke detection and the measurement of material thickness through α, β, or γ radiation. On the other hand, in the lower range the sensing of a variety of parameters through microwaves is the most practical method, while sensing of the microwaves themselves is not (however, we discussed in **Chapter 4** the use of bolometers to sense microwave power).

9.2 | UNITS OF RADIATION

The units for radiation, except for low-frequency electromagnetic radiation, are divided into three types and relate to radioactivity as well as to X-rays. The three sets of units are the units of activity, exposure, and absorbed dose. In addition there is a set of units for dose equivalent.

The basic unit of activity is the becquerel (Bq), which is defined as one transition (disintegration) per second. It indicates the rate of decay of a radionuclide. An older unit of activity was the curie (1 curie $= 3.7 \times 10^{10}$ Bq). The becquerel is a small unit, so mega-, giga-, and terabecquerels (MBq, GBq, and Tbq) are often used.

The basic unit of exposure is the coulomb per kilogram (C/kg), which is equivalent to the ampere second per kilogram (A·s/kg). The older unit was the roentgen

TABLE 9.1 ■ Summary of units of radiation

	Current unit	Old unit
Activity	becquerel (Bq)	curie (Ci), 1 Ci = 3.7×10^{10} Bq
Exposure	coulomb/kilogram (C/kg)	roentgen, 1 roentgen = 2.58×10^{-4} C/kg
Absorbed dose	gray (Gy)	rad, 1 rad = 100 Gy
Dose equivalent	sievert (Sv)	rem, 1 rem = 100 Sv

(1 roentgen = 2.58×10^{-4} C/kg). The coulomb per kilogram is a very large unit and units of milli-, micro-, and picocoulombs per kilogram (mC/kg, μC/kg, and pC/kg) are often used.

Absorbed dose is measured in grays (Gy). The gray is energy per kilogram, that is, 1 Gy = 1 J/kg. The old unit of absorbed dose was the rad (1 rad = 100 Gy).

The unit for dose equivalence is the sievert (Sv), which is measured in joules per kilogram (J/kg) as well. The old unit is the rem (1 rem = 100 Sv). Note that the sievert and the gray are the same. This is because they measure identical quantities in air. However, the dose equivalent for a body (like the human body) is obtained by multiplying the absorbed dose by a quality factor to obtain the dose equivalent. When people are exposed to radioactive radiation, their exposure is measured in sieverts. For example, the allowed exposure for workers in nuclear power plants in the United States is 50 mSv/yr. The units of radiation are summarized in **Table 9.1**.

Although the SI units for radiation and radiation exposure are clearly defined, older units like the curie (Ci), the rad, and even the roentgen persist in some industries and some devices. It is common, for example, to find the radioactive isotope in smoke detectors to be rated in microcuries (μCi) rather than becquerels. Similarly, radiation badges in the United States are typically rated in millirem (mrem) rather than sieverts. It should be noted as well that commonly used units for energy, like the calorie (cal) and the electron-volt (eV) are not SI units, but they are commonly found in practical use.

There are other derived and customary units used in conjunction with radiation. One example is the mass attenuation coefficient, measured in square centimeters per gram (cm^2/g). By multiplying the mass attenuation coefficient by the density of the medium one obtains a linear attenuation coefficient [1/m]. Therefore the mass attenuation coefficient is a normalized value convenient in comparing various media. Another derived unit often used is the stopping power of a medium. When high-energy radiation propagates through a medium the energy loss in the medium is defined in terms of its "linear stopping power." It is, in effect, the energy loss per unit length, typically in MeV/m or MeV/cm, normalized with respect to the density of the medium. The units are $(\text{MeV/cm})/(\text{g/cm}^3) = \text{MeV}\cdot\text{cm}^2/\text{g}$ or $(\text{MeV/m})/(\text{kg/m}^3) = \text{MeV}\cdot\text{m}^2/\text{kg}$. This designation allows comparisons of energy loss without considering the density of the medium. To obtain the energy loss per unit length of the medium, one must multiply the stopping power of the medium by its density.

9.3 | RADIATION SENSORS

We will start the discussion with ionization sensors (more often called detectors), and only then will we discuss the much lower frequency methods based on electromagnetic radiation—antennas. There are three basic types of radiation sensors: ionization sensors, scintillation sensors, and semiconductor radiation sensors. Some of these sensors may be simple detectors—that is, they simply detect the presence of radiation with no quantification, whereas others quantify the radiation in some way.

9.3.1 Ionization Sensors (Detectors)

In an ionization sensor, the radiation passing through a medium (gas or solid) generates electron–proton pairs whose density and energy depend on the energy of the ionizing radiation. These charges can then be attracted to electrodes and measured or they can be accelerated through the use of electric or magnetic fields for further use.

9.3.1.1 Ionization Chambers

The simplest and oldest type of radiation sensor is the ionization chamber. The chamber is a gas-filled chamber, usually at low pressure, with predictable response to radiation. In most gases the ionization energy for the outer electrons is rather small, 10–20 eV. Nevertheless, a somewhat higher energy is required since some energy may be absorbed without releasing charged pairs (by moving electrons into higher energy bands within the atom). For the purpose of sensing, the important quantity is the ***W* value**. It is an average energy transferred per ion pair generated. **Table 9.2** gives the W values for a few gases used in ion chambers. Clearly ion pairs can also recombine. Therefore in an ionization chamber, the current generated is due to an average rate of ion generation. The principle is shown in **Figure 9.1**. When no ionization occurs, there is no current, as the gas has negligible resistance. The voltage across the cell is relatively high and attracts the charges, reducing recombination. Under these conditions the steady-state current is a good measure of the ionization rate. The chamber operates in the saturation region of the I–V curve. The saturation current in an ion chamber can be calculated from the ionizing energy of the radiation that produces the ionization and their activity. Given a source of particles with energy E_s and activity A, the current in a chamber is

$$I_s = q\left(\frac{E_s}{E_i}\right)A\eta \quad [\mathrm{A}], \tag{9.3}$$

TABLE 9.2 ■ W values for various gases used in ionization chambers [eV/ion pair]

Gas	Electrons (fast)	Alpha particles
Argon (A)	27.0	25.9
Helium (He)	32.5	31.7
Nitrogen (N_2)	35.8	36.0
Air	35.0	35.2
Methane (CH_4)	30.2	29.0
Xenon (Xe)		23.0

Note: Fast electrons typically mean β radiation.

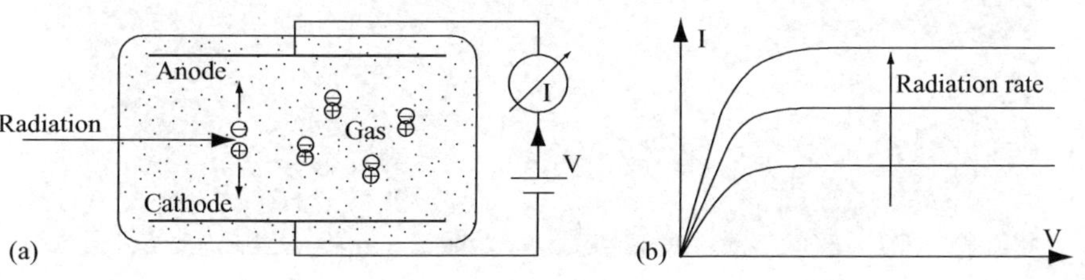

FIGURE 9.1 ■ (a) Ionization chamber. (b) I–V curve for an ionization chamber.

where E_i is the electron–proton pair energy and depends on the gas in the chamber, A is the activity of the source (the number of disintegrations per second), and η is an efficiency term that takes into account any masking or recombination within the chamber. Clearly, the higher the particle energy, the higher the current across the chamber. Alternatively, looking at it as an electromagnetic radiation, the higher the radiation frequency and the higher the voltage across the electrodes, the higher the current across the chamber.

The most common practical use for ionization chambers is in smoke detectors. In these, the chamber is open to the air and ionization occurs in air. A small radioactive source (usually americum-241) ionizes the air at a constant rate causing a small, constant ionization current between the anode and cathode of the chamber. The source emits mostly α particles. These are heavy particles and are easily blocked. In air they only propagate a few centimeters, but that is sufficient to establish an ionization current (called the saturation current) in the chamber. Combustion products such as smoke that enter the chamber are much larger and heavier than air molecules and form centers around which positive and negative charges recombine (some particles become positive, some negative through collisions with charged air molecules). This reduces the ionization current and triggers an alarm. In most smoke detectors there are actually two chambers. One is as described above, but because it can be triggered by humidity, dust, and even by pressure differences or small insects, a second, reference chamber is provided in which the openings to air are too small to allow large smoke particles, but they will allow water vapor. The trigger is now based on the difference between these two currents. **Figure 9.2** shows the ionization chambers of a residential smoke detector. The black chamber is the reference chamber and the white chamber is the sensing chamber.

Another example is shown in **Figure 9.3**. This is a fabric density sensor. It is made of two sections with the fabric between them. One part contains a low-energy radioactive isotope (typically krypton-85) while the second part is an ionization chamber. The ionization current established in the chamber is reduced by the fabric. The denser the fabric, the lower the ionization current. The ionization current is calibrated in terms of density (i.e., weight per unit area). Similar devices are calibrated in terms of thickness (rubber, for example) or other quantities that affect the amount of radiation that passes

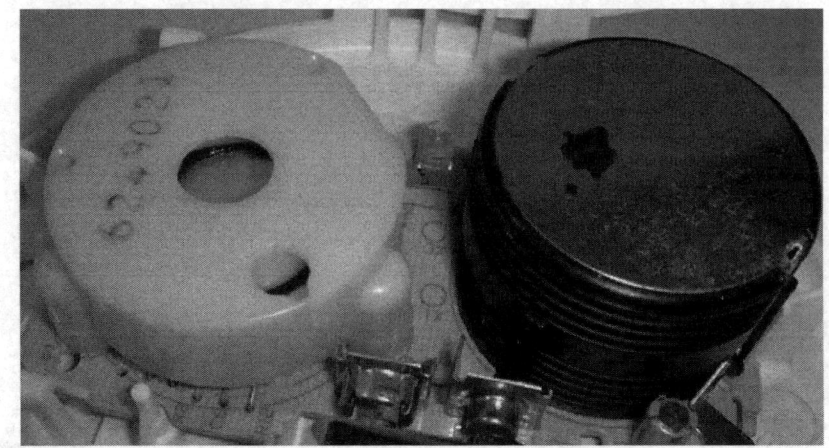

FIGURE 9.2 ■ Ionization chambers in a residential smoke detector. On the left is the sensing chamber (see opening). On the right is the reference chamber.

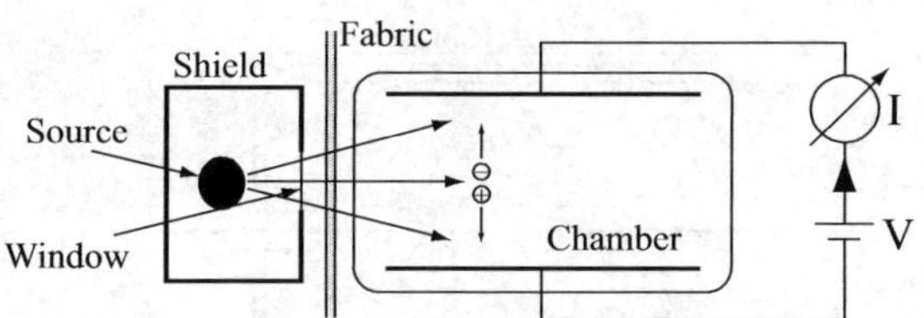

FIGURE 9.3 ■ Nuclear fabric density sensor.

through, such as moisture. Because radiation passes across the fabric, lighter particles are used (β particles). In some isotopes, like krypton-85, the radiation is mostly β particles.

Similar devices are used for radiology and for nondestructive testing of materials in a fashion similar to X-rays. However, in such applications high-energy γ radiation is typically used, generated from isotopes such as iridium-192 or cobalt-60 because these need to penetrate either through thicker sections or through more absorbing materials such as metals.

Whereas the chamber in **Figure 9.1** is sufficient for high-energy radiation, for low-energy X-rays, or for lower activity sources, a better approach is needed. The answer is the proportional chamber.

9.3.1.2 Proportional Chamber

A **proportional chamber** is essentially a gas ionization chamber, but the potential across the electrodes is high enough to produce an electric field in excess of 10^6 V/m. Under these conditions, the electrons are accelerated and in the process they collide with atoms, releasing additional electrons (and protons) in a process called the Townsend avalanche. These charges are collected at the anode, and because of this multiplication effect they can be used to detect lower intensity radiation. The device is also called a proportional counter or multiplier. If the electric field is increased further, the output becomes nonlinear due to the increased number of protons, which, being heavier than electrons, cannot move as fast, causing accumulation of a space charge. The operation now is in what is called the limited proportional zone or region. **Figure 9.4** shows the regions of operation of the various types of gas chambers.

9.3.1.3 Geiger–Muller Counters

When the voltage across an ionization chamber is sufficiently high, the output is not dependent on the ionization energy, but rather is a function of the electric field in the chamber. Because of this, the chamber can "count" single particles, whereas this would be insufficient to trigger a proportional chamber. This device is called a Geiger–Muller (G–M) counter. The very high voltage can also trigger a false reading immediately after a valid reading because of the ionized atoms in the chamber. To prevent this, a

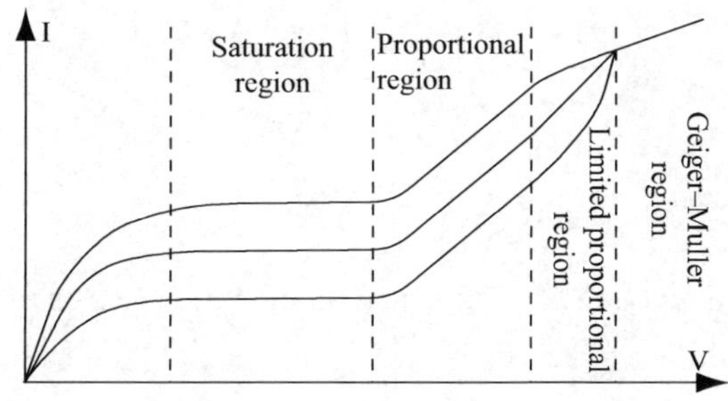

FIGURE 9.4 ■ Regions of operation of the various types of ionization chamber sensors.

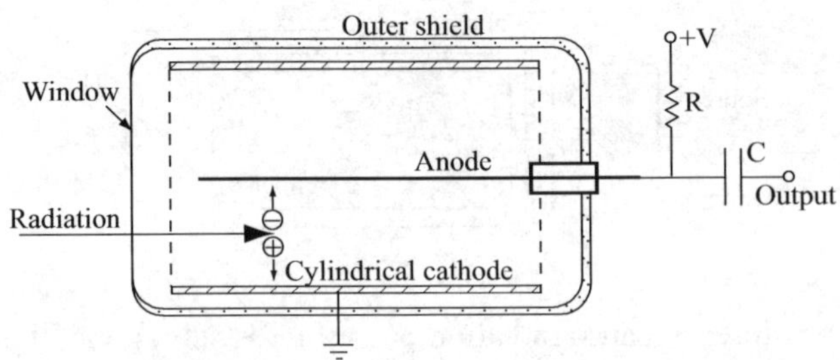

FIGURE 9.5 ■ The Geiger–Muller sensor.

quenching gas is added to the noble gas that fills the counter chamber. The G–M counter is made as a tube, about 10–15 cm long and about 3 cm in diameter. A window (transparent to radiation) is provided to allow penetration of radiation. The tube is filled with argon or helium with about 5%–10% ethyl alcohol to quench triggering. The operation relies heavily on the avalanche effect and in the process UV radiation is released, which adds to the avalanche process. These processes result in an output that is about the same no matter what the ionization energy of the input radiation is (as long as it is sufficient to produce ionization). Because of the very high voltage, a single particle can release 10^9–10^{10} ion pairs. This means that a G–M counter is essentially guaranteed to detect any ionizing radiation through it.

The efficiency of all ionization chambers, including G–M counters, depends on the type of radiation. The cathodes also greatly influence this efficiency. In general, high atomic number cathodes are used for higher energy radiation (γ rays) and lower atomic number cathodes for lower energy radiation.

The structure of a G–M counter is shown in **Figure 9.5**.

EXAMPLE 9.2 Geiger–Muller and background radiation

To evaluate the performance of a G–M tube, it was installed on a fixed stand in a stone quarry to measure the background radiation in counts per minute. The granite stone quarry was selected because it was expected to have radiation levels higher than normal. The voltage across the tube was varied from 100 V and 1000 V in increments of 50 V. Because background radiation is not constant in time, 12 readings at each voltage value were averaged to obtain the result shown in **Figure 9.6**. Each reading is the counts (clicks) in 1 min.

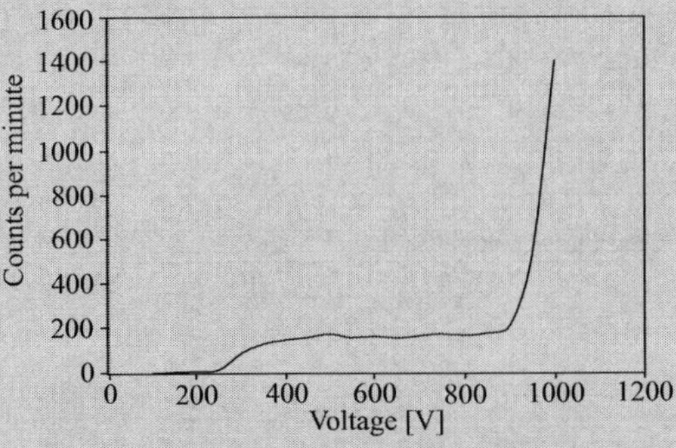

FIGURE 9.6 ■ Geiger–Muller count as a function of voltage across the tube. Source of radiation: background radiation in a quarry.

This test shows the typical voltage characteristics of a G–M tube. Initially the voltage is not sufficiently high to ionize the gas, therefore the count is very low. Above about 300 V the number of counts increases. Between about 400 V and 850 V the count is fairly constant—in the case shown here it averages 153 counts/min. This is the range of operation of the tube, that is, it must operate at a voltage between 400 V and 850 V if the readings are to represent the radiation. Above 850 V the count increases very fast and the tube enters a state of continuous discharge, that is, the avalanche effect takes over and the count does not represent the incoming radiation. The results shown here are specific for a particular tube and will vary for other tubes. Even two specimens of the same type will have slightly different curves.

Notes:

1. The background radiation in most locations on the planet is less than 20 counts/min. In mines and quarries the counts rise to about 150, depending of course on the type of stones in the mine or quarry. Clays and granite tend to have higher radiation levels than sandstone.
2. The exposure level is measured in microsieverts per hour [μSv/h] or millisieverts per hour [mSv/h]. There is no direct relation between counts and the sievert, although one can find approximate values, typically equating 100 counts/min to 1 μSv/h. With the result above, a person that spent 8 hours a day working in a quarry would absorb about 4400 μSv/yr. By comparison, nuclear power plant workers are allowed a maximum of 50 mSv/yr. Any dose larger than 100 mSv is considered carcinogenic. The normal dose from background radiation is less than about 2 mSv/yr.

9.3.2 Scintillation Sensors

A relatively simple method of sensing radiation is to take advantage of the radiation to light conversion (scintillation) that occurs in certain materials. The light intensity generated is then a measure of the radiation's kinetic energy. Some scintillation sensors are used as detectors in which the exact relationship to radiation is not critical. In others it is important that a linear relationship exists and that the light conversion be efficient. Also, the materials used should exhibit fast light decay following irradiation (photoluminescence) to allow fast response by the detector. The most common material used for this purpose is sodium iodine (other alkali halide crystals can be used and activation materials such as thallium are added), but there are also organic materials and plastics that can be used for this purpose. Many of these have faster responses than the inorganic crystals.

Light conversion is fairly weak because it involves inefficient processes. Therefore the light obtained in these scintillating materials is of low intensity and requires "amplification" to be detectable. To increase the sensitivity, a photomultiplier or a charge-coupled device (CCD; see **Sections 4.6.2** and **4.7**) can be used as the detector mechanism, as shown in **Figure 9.7**. The large gain of photomultipliers is critical in the success of these devices. The reading is a function of many parameters. First, the energy of the particles and the efficiency of conversion (about 10%) define how many photons are generated. Part of this number, say k, reaches the cathode of the photomultiplier. Then the cathode of the photomultiplier has a quantum efficiency (about 20%–25%).

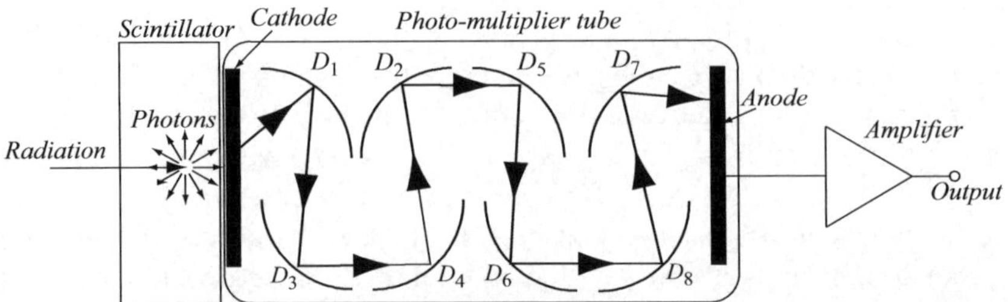

FIGURE 9.7 ■ A scintillation sensor making use of a photomultiplier to detect the feeble light emitted by the scintillator.

This number, say k_1, is now multiplied by the gain of the photomultiplier G, which can be on the order of 10^6–10^8.

EXAMPLE 9.3 Detection of cosmic radiation

One of the simplest methods of detecting cosmic radiation is through the use of two scintillator layers and two photomultiplier tubes, as shown in **Figure 9.8**. The detector is placed parallel to the surface of the earth because muons produced by cosmic radiation move more or less perpendicular to the surface of the earth at relativistic speeds (about $0.95c$). The scintillators and the

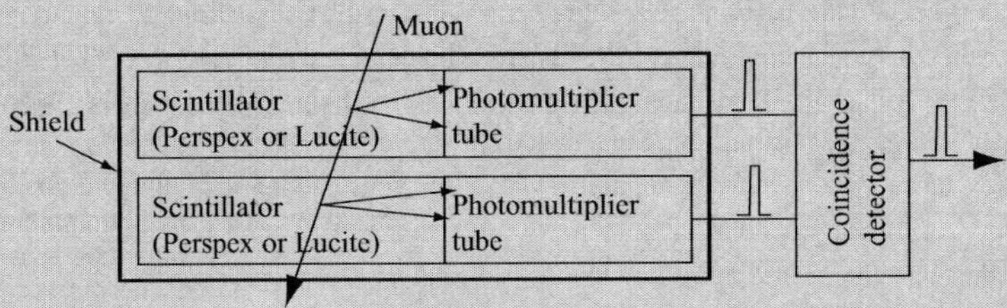

FIGURE 9.8 ■ Coincidence detection of muons produced by cosmic radiation using two scintillators and photomultiplier tubes.

photomultiplier tubes are shielded. The scintillators are simple sheets of Perspex or Lucite. The reason for this arrangement is that almost any radiation source will cause scintillation. With two scintillators, only if both tubes detect scintillation at the same time can one be sure that the source is a muon, since lower energy radiation will be blocked by the shield and if it arrives at one scintillator it will not arrive at the other. Therefore one must detect both signals and correlate them. This is relatively easy with two tubes, but it can be done with one tube as well with appropriate electronics. The method is often called coincidence detection because one tries to ascertain whether both detectors respond to the same source.

9.3.3 Semiconductor Radiation Detectors

Just as light radiation can be detected in semiconductors through the release of charges across the bandgap of the semiconductor, higher energy radiation can be expected do so as well. In principle, any semiconductor light sensor will also be sensitive to higher energy radiation, but in practice there are a few issues that have to be resolved. First, because the energy is high, the lower bandgaps are not useful since they would produce currents that are too high. Second, high-energy radiation can easily penetrate through the thin semiconductor layers without releasing charges. Thus thicker devices and heavier materials are needed. Also, in the detection of low radiation levels, the background noise, due to the "dark" current (current from thermal sources), can seriously interfere with the detector. Because of this, some semiconducting radiation sensors can only be used at cryogenic temperatures, and those that are used at room temperature must be made of high-purity materials.

When an energetic particle penetrates into a semiconductor, it initiates a process that releases electrons (and holes) through direct interaction with the crystal and through secondary emissions by the primary electrons, which are usually of much higher energy. The net effect is that to produce a hole–electron pair, a specific amount of ionization energy, on the order of 3–5 eV, is required. Because this is only about 1/10 of the energy required to release an ion pair in gases, the basic sensitivity of semiconductor sensors is an order of magnitude higher than in gases. In addition, the efficiency is typically higher because of the higher density of semiconductors. The relevant properties of some common semiconductors are listed in **Table 9.3**.

Semiconductor radiation sensors may be divided into two types. The first type is a simple intrinsic material with two electrodes. The second type is based on the sensitivity of a normal diode to detect radiation of any kind, from IR to γ radiation.

9.3.3.1 Bulk Semiconductor Radiation Sensor

The sensor, made of a volume of intrinsic semiconductor and two electrodes across which a voltage is applied (see **Figure 9.9a**), parallels the idea of photoresistors discussed in **Section 4.5.1**. Another way to look at it is as an "ionization chamber" in which the gas is replaced with a solid semiconductor material, but necessarily of much smaller dimensions. It relies on the change in resistance of the semiconductor due to the generation of charges by the incoming radiation. For this reason it is sometimes called a **bulk resistivity radiation sensor.**

TABLE 9.3 ■ Properties of some common semiconductors

Material	Operating temp [°K]	Atomic number	Bandgap [eV]	Energy per electron–hole pair [eV]
Silicon (Si)	300	14	1.12	3.61
Germanium (Ge)	77	32	0.74	2.98
Cadmium telluride (CdTe)	300	48, 52	1.47	4.43
Mercury iodine (HgI_2)	300	80, 53	2.13	6.5
Gallium arsenide (GaAs)	300	31, 33	1.43	4.2

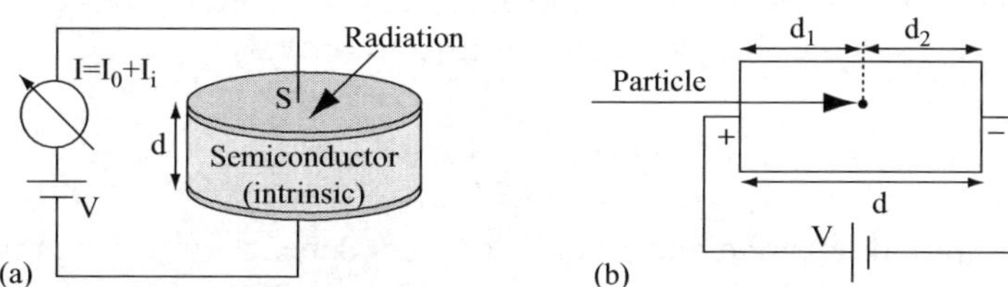

FIGURE 9.9 ■ (a) Bulk semiconductor sensor. (b) The process of generating the ionization current.

Depending on the type of base material, additional restrictions must be imposed. Unlike silicon, germanium can only be used at cryogenic temperatures. On the other hand, silicon is a light material (atomic number 14) and is therefore very inefficient for energetic radiation such as γ rays. For this purpose, cadmium telluride (CdTe) is most often used because it combines heavy materials (atomic numbers 48 and 52) with relatively high bandgap energies. Other materials that can be used are mercuric iodine (HgI_2) and gallium arsenide (GaAs). The surface area of these devices can be quite large (some as large as 50 mm in diameter) or very small (1 mm in diameter) depending on applications. Conductivity under dark conditions is on the order of 10^{-8} to 10^{-10} S/cm depending on the construction and on doping, if any (intrinsic materials have higher resistivity).

The simplest way to view the behavior of these devices is in terms of their conductivity, in the same way we looked at photoresistors. A semiconductor has a conductivity that depends on doping and temperature (see **Section 4.5.1**). Radiation increases the conductivity of the medium (decreases resistivity) by releasing additional carriers. This increases the current, and this change in current is then a measure of radiation.

A bulk resistivity sensor is shown in **Figure 9.9**. The current through the device is composed of two terms. One is the current in the absence of radiation based on the conductivity of the material. The other is the intrinsic conductivity of the material (see **Equation (4.4)**):

$$\sigma = e(\mu_e n + \mu_p p) \quad [\text{S/m}], \tag{9.4}$$

where μ_e and μ_p are the mobilities of electrons and holes, respectively; n and p are the concentrations of electrons and holes, respectively; and e is the charge of the electron. This produces a resistance R_0:

$$R_0 = \frac{d}{\sigma S} = \frac{d}{e(\mu_e n + \mu_p p)S} \quad [\Omega]. \tag{9.5}$$

The current in the absence of radiation is entirely due to the resistance of the device:

$$I_0 = \frac{V}{R_0} = \frac{V}{d} e(\mu_e n + \mu_p p)S \quad [\text{A}], \tag{9.6}$$

where the term $E = V/d$ is the electric field intensity produced by the source across the semiconductor. The current I_0 is present as long as the device is connected to a voltage source. When radiation is detected, the additional charges produced in the bulk of the material increase the current by generating an ionization current that in effect decreases the resistance of the device.

The ionization current through the semiconductor is calculated as the ratio of the charge generated and the time it takes the charge to reach the electrodes (transition time). Referring to **Figure 9.9b**, the charge generated per interaction (i.e., per particle or per photon) is given in the first two terms in **Equation (9.3)**:

$$Q = e\left(\frac{E_s}{E_i}\right) \quad [\text{C}]. \tag{9.7}$$

The transition time of negative and positive carriers depends on the mobilities of these carriers and hence their velocities (often called drift velocities). The latter are

$$v_e = \mu_e E, \quad v_p = \mu_p E \quad [\text{m/s}]. \tag{9.8}$$

As the positive and negative carriers move toward the opposite plates, they generate a current equal to charge divided by transition time. Depending on where the charges are generated and on their velocities (the negative carriers, electrons, move much faster than the positive carriers), the contribution of the negative carriers to current is higher. Suppose the carriers are generated at a distance d_1 from the positive plate and d_2 from the negative plate. The transition times through the device for electrons and protons are

$$t_e = \frac{d_1}{v_e} = \frac{d_1}{\mu_e E} = \frac{d_1 d}{\mu_e V}, \quad t_p = \frac{d_2 d}{\mu_p V} \quad [\text{s}]. \tag{9.9}$$

The total transition time, that is, the total time needed for the charges to collect on the opposite plates, is

$$t = t_e + t_p = \frac{d_1 d}{\mu_e V} + \frac{d_2 d}{\mu_p V} = \frac{d}{V}\left(\frac{d_1\mu_p + d_2\mu_e}{\mu_e\mu_p}\right) \quad [\text{s}]. \tag{9.10}$$

This is an approximation since the charges move at the same time and therefore times are not strictly additive, but since electrons move faster than holes, the error in this approximation is small. The ionization current through the semiconductor is the charge collected divided by time:

$$I_i = \frac{Q}{t} = \frac{e}{t}\left(\frac{E_s}{E_i}\right) = e\left(\frac{E_s}{E_i}\right)\frac{V}{d}\left(\frac{\mu_e\mu_p}{d_1\mu_p + d_2\mu_e}\right) \quad [\text{A}]. \tag{9.11}$$

The total current through the device is $I_0 + I_i$:

$$I = I_0 + I_i = \frac{V}{d}e(\mu_e n + \mu_p p)S + e\left(\frac{E_s}{E_i}\right)\frac{V}{d}\left(\frac{\mu_e\mu_p}{d_1\mu_p + d_2\mu_e}\right) \quad [\text{A}]. \tag{9.12}$$

The materials used are typically intrinsic semiconductors with $n = p = n_i$, where n_i is the intrinsic carrier concentration. Therefore

$$I = I_0 + I_i = e\frac{V}{d}\left[n_i(\mu_e + \mu_p)S + \left(\frac{E_s}{E_i}\right)\left(\frac{\mu_e\mu_p}{d_1\mu_p + d_2\mu_e}\right)\right] \quad [\text{A}]. \tag{9.13}$$

One can assume further that, on average, the charges are generated at the center of the device with $d_1 = d/2$, $d_2 = d/2$ and get

$$I = e\frac{V}{d}\left[n_i(\mu_e + \mu_p)S + \frac{2}{d}\left(\frac{E_s}{E_i}\right)\left(\frac{\mu_e\mu_p}{\mu_p + \mu_e}\right)\right] \quad [\text{A}]. \tag{9.14}$$

The resistance can now be calculated if necessary. Note that even though it does not appear explicitly, the current and resistance are temperature dependent through the dependence of the mobilities as well as charge densities on temperature (see **Chapter 4**). The model used here is rather simple. It does not take into account the fact that as radiation penetrates deeper into the material, its energy reduces exponentially. Alpha particles create virtually surface interactions (very little penetration in the medium) and hence it is entirely absorbed in the material. A simple model of the absorption of β, γ, and X-ray radiation can be written as follows:

$$E_s(x) = E_s(0)e^{-kx} \quad [\text{eV}], \tag{9.15}$$

where x is the distance the radiation penetrates into the material, $E_s(0)$ is its energy at the surface of the material, and k [1/m] is a linear attenuation coefficient based on the probability that an interaction takes place and releases carrier pairs. This coefficient depends on the radiation energy and on the density of the material used in the device. Gamma and X-ray radiation are not charged and hence they are viewed as photons (or the equivalent wave form), whereas β radiation is viewed as radiation of charged particles. The attenuation coefficients for the various types of radiation are available in tables for most materials and energy levels.

The calculations shown here also assume the electric field intensity to be uniform and that all energy is absorbed in the detector. In a way, these are contradictory. To neglect the energy absorption with depth means to assume thin materials (d is small), but that means that not all energy can be absorbed, requiring the use of an absorption coefficient that may be small (and hence the number of carriers generated will be small), leading to low sensitivity. In addition, because the mobilities of electrons and holes are different, the current will depend on where the charges are generated. The current due to each interaction will look like a short pulse that evolves over a short period of time as the electrons and holes propagate toward the electrodes. This results in an output that looks like a series of pulses, and the number of counts is then a measure of the radiation level. Nevertheless, the simple model used here is useful in understanding the phenomena and in approximating the currents in the detector.

The relations above are useful when particles are involved since the current can then be related to carriers generated and their transition times through the semiconductor. However, when a radiation flux is involved, the current in the semiconductor cannot be related to single particles and one has to start with incident power (also called energy rate [J/s]). If the energy per unit time (power) P_s absorbed in the sensor is given, the ionization current can be calculated directly as

$$I_i = \frac{Q}{t} = e\left(\frac{P_s}{E_i}\right) \quad \text{[A]}, \tag{9.16}$$

where E_i is the ion-pair energy. The power absorbed may not necessarily be continuous, but rather may be a radiation event that may last a time Δt. The current during that time is given by **Equation (9.16)**. However, it is implicit that the event is longer than the transition time of electrons in the sensor. It should be noted that **Equation (9.16)** is independent of the voltage across the sensor, but the sensor must be biased for the carriers to migrate to the surface. An insufficient bias may reduce the current through the device and consequently the charge on the electrodes. The relation in **Equation (9.16)** assumes that all charges are collected at the electrodes at a constant rate, resulting in a constant current. The current I_0 remains unchanged as it has nothing to do with the radiation—it is only dependent on the voltage applied across the sensor.

In many cases the power density rather than power may be given. To obtain the power absorbed, the power density is multiplied by the area of the sensor.

9.3.3.2 Semiconducting Junction Radiation Sensors

The second type of semiconductor radiation sensor consists essentially of diodes in reverse bias. This ensures a small (ideally negligible) background (dark) current as for any diode. The reverse current produced by radiation is then a measure of the kinetic energy of the radiation. In practical devices, the diode must be thick to ensure absorption of the energy due to fast particles. The most common construction is similar to the PIN diode and is shown in **Figure 9.10**. In this construction, a normal diode is built, but with a much thicker intrinsic region producing lower reverse currents than normal diodes. This region is doped with balanced impurities so that it resembles an intrinsic material. To accomplish that and to avoid the tendency of drift toward either an n or p behavior, an ion-drifting process is employed in which a compensating material is diffused throughout the layer. The material of choice for this purpose is lithium.

In a reverse-biased diode, the only current in the absence of radiation is the dark current produced by thermal effects, and that is typically very small. Therefore the current in the diode can be viewed as being entirely due to radiation. For this reason also, when materials with low bandgaps such as germanium are used, they must be cooled to cryogenic temperatures (typically using liquid nitrogen to 77°K).

To better understand the detection process, consider **Figure 9.10**. The intrinsic region of the PIN structure is wide, as indicated qualitatively in the figure and sandwiched between n and p materials. An ionization source is assumed to generate charges Q and $-Q$ (holes and electrons) at a given point in the medium according to **Equation (9.7)**. The carriers now move to the opposite polarity surfaces under the effect of the electric field intensity. The latter will be assumed here to be constant and equal to V/d. The current is as indicated in **Equation (9.11)**. If one chooses to use the average value in **Equation (9.14)**, the first term should be set to zero since in the absence of radiation the diode current is assumed to be zero. It should again be noted that the charge generation depends on the depth within the diode where the charge is generated, but otherwise the only advantage of the diode over the bulk intrinsic sensor is in the fact that the diode can be reverse biased, eliminating the ohmic current I_0 in **Equation (9.6)** and in **Equations (9.12)–(9.14)**. If the diode absorbs a certain power (energy per unit time) rather than single particles or photons, **Equation (9.16)** is appropriate for use. However, based on the considerations above, the results given here should be used only as representative estimates as exact calculations require much more complex models that take into account recombination and secondary generation of carriers, nonuniform electric fields within the device, attenuation of energy with depth, and the absorption efficiency of the material. The latter, in particular, is fairly low in low atomic number materials like silicon and higher in materials like germanium or GaAs.

Just as with diode light sensors, the idea of avalanching can be used to increase the sensitivity of semiconductor radiation detectors, especially at lower energy levels.

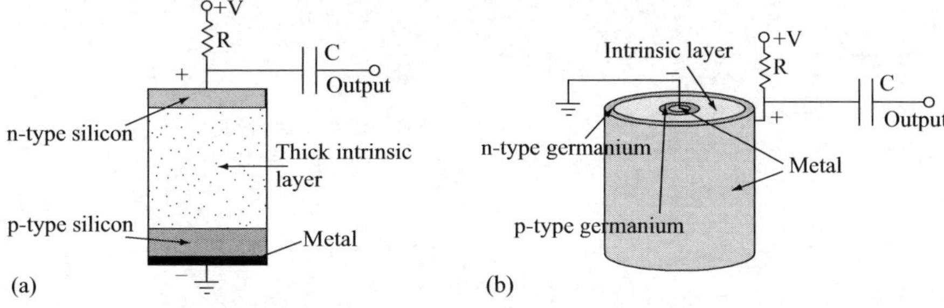

FIGURE 9.10 ■ Semiconductor radiation sensor. (a) A typical silicon sensor built as a regular, planar diode with a thick intrinsic layer. (b) A coaxial, germanium diode intended for higher energy radiation levels.

These are called avalanche detectors and operate similarly to the proportional chamber detectors discussed above. While this can increase the sensitivity by about two orders of magnitude, it is important to use these only for low energies or the barrier can be easily breached and the sensor destroyed.

The diode may be forward or reverse biased, but the preferred mode is reverse biasing (see **Figure 9.10**), where the change in current is high against a very low "dark" current. The relations in **Sections 4.4** and **4.5** defining dark current and the change in conductivity due to radiation apply here as well. The main difference is in the much higher energy of the radiation and the lower efficiency of the device.

Semiconducting radiation sensors are sensitive and versatile radiation sensors, but they suffer from a number of limitations. Chief among them is damage that can occur when exposed to radiation over time. The damage can occur in the semiconductor lattice, in the packaging, or in the metal layers and connectors. Prolonged radiation may also increase the leakage (dark) current, resulting in a loss of energy resolution of the sensor. In addition, the temperature limits of the sensor must be taken into account (unless a cooled sensor is used).

EXAMPLE 9.4 Germanium semiconductor sensor and its sensitivity to radiation

A germanium diode is used to detect radiation with an energy of 1.5 MeV. To do so one has the option of exposing either the anode or the cathode to the incoming radiation (see **Figure 9.11**). Assume that the energy is absorbed entirely at the point of entry (i.e., at the cathode or at the anode). With mobilities of 1200 cm^2/V·s for holes and 3800 cm^2/V·s for electrons, calculate the current through the diode for the two configurations with reverse bias: $V = 24$ V and $d = 20$ mm. Explain the difference and draw conclusions regarding sensitivity and the shape of the pulse expected if the radiation is absorbed uniformly across the intrinsic layer. Neglect the effects of electrodes and of the n and p layers and assume a single radiation event, that is, a single particle or a short burst of radiation.

Solution: The current through the diode is the charge collected divided by the transit time between the electrodes. The general relation for transit time is given by **Equation (9.10)**. Using the general relation and **Figure 9.9b**, we write the transition time for **Figure 9.11a**:

$$t = t_e + t_p = \frac{d_1 d}{\mu_e V} + \frac{d_2 d}{\mu_p V} = \frac{d}{V}\left(\frac{d_1\mu_p + d_2\mu_e}{\mu_e\mu_p}\right) \quad [\text{s}].$$

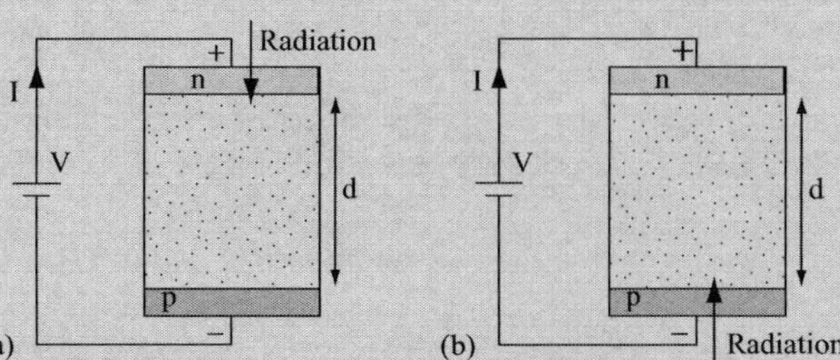

FIGURE 9.11 ■ Radiation sensor. (a) Radiation is absorbed next to the cathode. (b) Radiation is absorbed next to the anode.

However, in this case the transition time for electrons is zero since they are generated at the anode. The only current is due to holes propagating toward the cathode. Hence $d_1 = 0$, $d_2 = d$ and we have

$$t = t_p = \frac{d^2}{\mu_p V} = \frac{0.02^2}{1200 \times 10^{-4} \times 24} = 0.139 \times 10^{-3} \text{ s}.$$

The charge collected is found from the energy absorbed and the energy per electron–hole pair for germanium. The latter is available in **Table 9.2** as 2.98 eV. The charge generated is therefore

$$Q = e\frac{E_s}{E_i} = 1.61 \times 10^{-19} \times \frac{1.5 \times 10^6}{2.98} = 8.104 \times 10^{-14} \text{ C},$$

where e is the charge of the electron, E_s is the absorbed energy, and E_i is the energy needed to generate an electron–hole pair. The current is given (see **Equation (9.11)**) as

$$I_p = \frac{Q}{t_p} = \frac{8.104 \times 10^{-14}}{0.139 \times 10^{-3}} = 5.83 \times 10^{-10} \text{ A}.$$

The current is 0.583 nA. This is a small current and its measurement requires a very low leakage current (dark current) in the diode. The index p indicates this is a current due to holes.

In the case shown in **Figure 9.11b**, the current is entirely due to electrons since the holes are captured immediately on the cathode. The current can be calculated directly from **Equation (9.11)**, but now, referring again to **Figure 9.9b**, $d_2 = 0$, $d_1 = d$ and we get

$$I_e = e\left(\frac{E_s}{E_i}\right)\frac{V}{d}\left(\frac{\mu_e\mu_p}{d_1\mu_p + d_2\mu_e}\right) = e\left(\frac{E_s}{E_i}\right)\frac{V}{d}\left(\frac{\mu_e}{d}\right)$$

$$= 1.61 \times 10^{-19} \times \frac{1.5 \times 10^6 \times 24 \times 3800 \times 10^{-4}}{2.98 \times 0.02^2} = 1.848 \times 10^{-9} \text{ A}.$$

The current is now approximately three times larger because the transition time is about three times shorter.

Clearly the device is more sensitive to radiation close to its cathode because the electrons have higher mobility and hence generate a larger current. The transition time may be viewed as the delay in the onset of the pulse, that is, one only gets an indication of the radiation after the electrons or holes (or both) have arrived at the appropriate electrodes. In reality, pairs will be generated throughout the volume and the current will vary with time depending on where the charges are generated. This produces a varying width pulse because of charges arriving at different times. The pulse will start rising at the moment radiation reaches the device, and will increase to a peak when the charge arriving at the electrodes peaks, and then diminish until all charges generated have been captured. This applies to single events. If radiation is constant over time, the current will also be constant as the charge arriving at the electrodes will be in a steady state.

9.4 | MICROWAVE RADIATION

Microwaves are often employed in the sensing of a variety of stimuli because of the relative ease of generating, manipulating, and detecting microwave radiation. Certainly their use in speed measurements and in sensing of the environment (radar, Doppler radar, weather radar, mapping of the earth and planets, etc.) should be well known. All of these applications and sensors are based on the properties, especially the propagation properties, of electromagnetic waves at any frequency, including at optical frequencies.

We have discussed most of the properties of waves in conjunction with the propagation of sound waves in **Section 7.3** and the frequency ranges of various types of radiation, including microwaves, in **Section 4.1**. Although all properties of waves discussed in **Chapter 7** apply here as well, electromagnetic waves are different from acoustic waves in three fundamental ways:

- The electromagnetic wave is a transverse wave.
- The electromagnetic wave is the variation in space and time of the electric and magnetic field intensities.
- The electric field intensity E and the magnetic field intensity H are transverse to the direction of propagation of the wave (in most cases of practical interest) and perpendicular to each other. The wave is then called a transverse electromagnetic (TEM) wave. The electric and magnetic field intensities can exist in matter as well as in vacuum. Therefore electromagnetic waves propagate in vacuum, whereas sound waves do not. In fact, vacuum is ideal for wave propagation because there are no losses and hence no attenuation of the wave. Although there are other types of electromagnetic waves, we will restrict our discussion here to TEM waves, with very little loss of generality.

A visual interpretation of how a TEM wave propagates is shown in **Figure 9.12**.

The properties of the electromagnetic wave are significantly different from those of the acoustic wave numerically as well. Most important is the speed of propagation of the wave (also called the phase velocity). This is given as

$$v_p = \frac{1}{\sqrt{\varepsilon\mu}} \quad [\mathrm{m/s}], \tag{9.17}$$

where ε is the permittivity and μ is the permeability of the medium in which the wave propagates. Naturally, any relation, such as the wavelength λ and wave number k, which depends on phase velocity, also changes appropriately. The phase velocity of electromagnetic

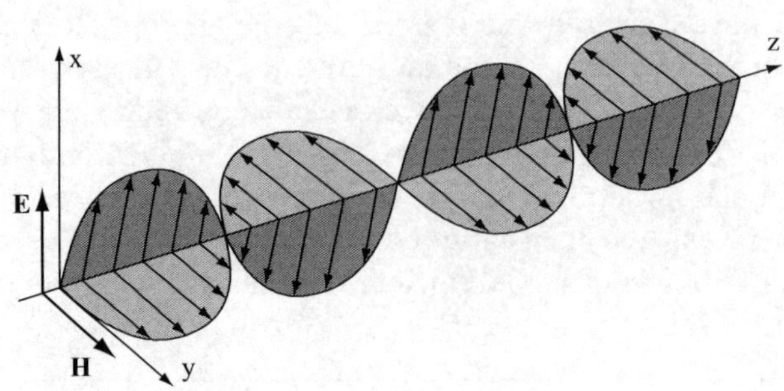

FIGURE 9.12 ■ Propagation of a TEM wave. The electric and magnetic fields are perpendicular to each other and to the direction of propagation.

waves in vacuum is 3×10^8 m/s, but it is lower in all other media. Losses also change the phase velocity, but we will neglect this effect as it is not fundamental to the discussion. As the wave propagates its phase changes. Given a source that radiates electromagnetic waves (e.g., an antenna) with an electric field intensity amplitude E_0, the electric field intensity at a position in space a distance R from the source will be

$$E = E_0 e^{-j\beta R} \quad [\mathrm{V/m}]. \tag{9.18}$$

Note that E_0 is written here as a phasor (i.e., a term $e^{j\omega t}$, where $\omega = 2\pi f$ and f is the frequency of the wave, is implied in the notation) and it may be a function of position (or coordinates). This simple model is valid for TEM waves and indicates that as the wave propagates its phase changes. This is easier to see by writing the equation in the time domain:

$$E = E_0 \cos(\omega t - \beta R) \quad [\mathrm{V/m}], \tag{9.19}$$

where β is the phase constant of the materials. In lossless materials, the phase constant is equal to the wave number and is defined as

$$\beta = \frac{\omega}{v_p} = \omega\sqrt{\mu\varepsilon} \quad [\mathrm{rad/m}]. \tag{9.20}$$

In addition to the change in phase, the wave's amplitude may be attenuated because of losses in the medium through which the wave propagates. The attenuation of electromagnetic waves is exponential and material dependent. Attenuation is low in low-conductivity materials such as dielectrics, but is high in conducting materials. Attenuation is zero in vacuum and in perfect dielectrics (lossless materials in which the conductivity is zero or is so low as to be negligible). Each medium is characterized by an attenuation constant that, in general, depends on frequency. Incorporating the attenuation constant α [Np/m] as indicated in **Section 7.3.1**, the more general form of a propagating wave is

$$E = E_0 e^{-\alpha R} e^{-j\beta R} \quad [\mathrm{V/m}], \tag{9.21}$$

or in the time domain,

$$E = E_0 e^{-\alpha R} \cos(\omega t - \beta R) \quad [\mathrm{V/m}]. \tag{9.22}$$

In TEM waves, the magnetic field intensity is perpendicular to the electric field intensity, and its magnitude is related through the wave impedance of the medium:

$$\frac{|E|}{|H|} = \eta \quad [\Omega]. \tag{9.23}$$

The wave impedance is characteristic of the medium. In free space (and, as a very good approximation in air) the wave impedance equals 377 Ω.

The magnetic field intensity can now be written as

$$H = \frac{E_0}{\eta} e^{-\alpha R} e^{-j\beta R} \quad [\mathrm{A/m}] \quad \text{or} \quad H = \frac{E_0}{\eta} e^{-\alpha R} \cos(\omega t - \beta R) \quad [\mathrm{A/m}]. \tag{9.24}$$

That means that the electromagnetic wave propagates power with a power density,

$$P_{av} = \frac{E_0^2}{2\eta} e^{-2\alpha R} \quad \left[\mathrm{W/m^2}\right]. \tag{9.25}$$

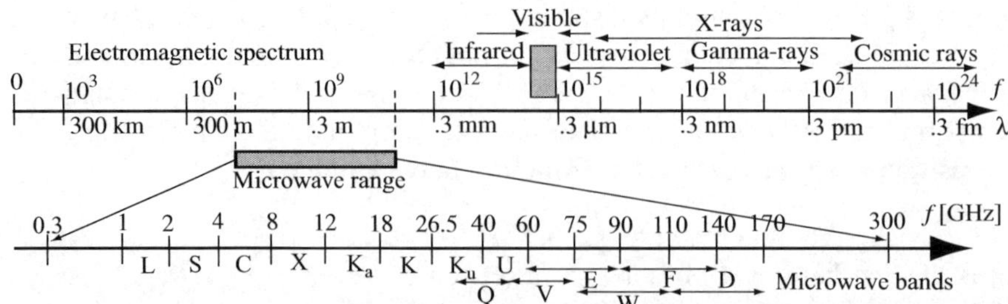

FIGURE 9.13 ■ The electromagnetic spectrum with details of the microwave portion, including microwave band designations.

The whole spectrum of electromagnetic waves, from very low to very high frequencies, can be used for sensing, but we will restrict the discussion here to microwaves. The microwave spectrum is defined broadly from about 300 MHz to 300 GHz (wavelengths from 1 m to 1 mm). The band above this is sometimes called millimeter waves and overlaps with the low IR band. This spectrum is divided into bands for purposes of identification, as shown in **Figure 9.13**. Although microwaves extend to 300 GHz, most applications are below 50 GHz. Part of the reason is regulatory, based on frequency allocations, and part is the fact that electronic circuits at higher frequencies are harder to come by, more difficult to design with, and their performance is reduced.

9.4.1 Microwave Sensors

Sensing with microwaves is based on four distinct methods, each with its advantages, disadvantages, and areas of application:

- Propagation of waves.
- Reflection and scattering of waves.
- Transmission of waves.
- Resonance.

These can be combined in a sensor to affect a particular function.

9.4.1.1 Radar

The best known method of sensing with microwaves is radar (**ra**dio **d**etection **a**nd **r**anging). In its simplest form it is not much different than a simple flashlight (source) and our eyes (detector), and is shown schematically in **Figure 9.14**. Clearly, the larger the target and the more intense the source of waves, the larger the signal received back from the target. Reception may be by the same antenna that serves as the source (pulsed-echo radar or a-static radar) or it may be by a second antenna (continuous or bistatic radar). Both are shown in **Figure 9.15**. The operation of radar is based on the scattering of waves by any target the incident waves encounter. For any object in the path of

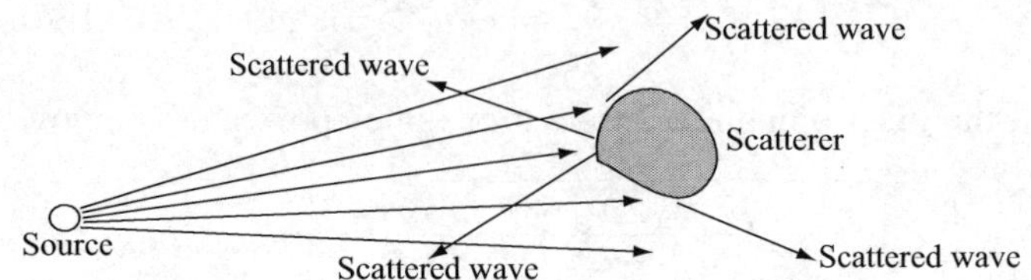

FIGURE 9.14 ■ Scattering of electromagnetic waves by an object.

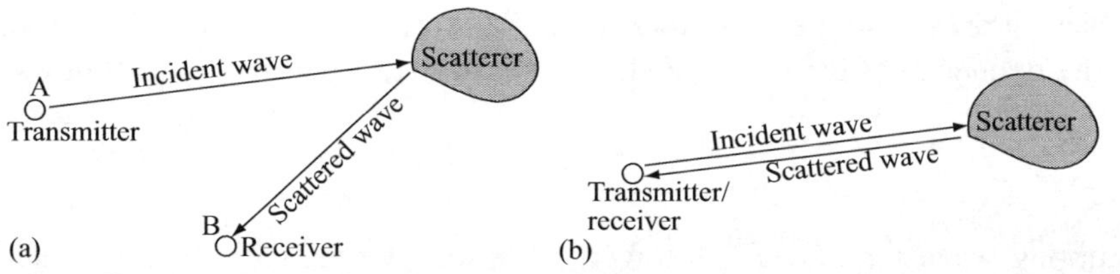

FIGURE 9.15 ■ (a) Bistatic and (b) a-static radar principles.

electromagnetic waves, the scattering coefficient σ, called the scattering cross section or radar cross section, is

$$\sigma = 4\pi R^2 \frac{P_s}{P_i} \quad [\mathrm{m}^2], \tag{9.26}$$

where P_s is the scattered power density [W/m^2], P_i is the incident power density [W/m^2], and R is the distance from the source to the target [m]. With this, the power received is calculated from the radar equation:

$$P_r = P_{rad}\sigma \frac{\lambda^2 D^2}{(4\pi)^3 R^4} \quad [\mathrm{W}], \tag{9.27}$$

where λ is the wavelength, σ is the radar cross section, P_r the total received power [W], P_{rad} is the total radiated power [W], and D is the directivity of the antenna. The latter is a property of the antenna and is an indication of how directive the radiation is. It depends on the type and construction of the antenna.

Although numerical values of power received by a radar antenna will vary, it is clear that this is a short-range device because of dependency on $1/R^4$. Nevertheless, radar is one of the most useful sensing systems, capable of sensing the distance as well as the size (radar cross section) of objects. In more sophisticated systems the position (distance and attitude) may be sensed as well as the speed of the target, but this is obviously as much a function of the signal processing involved as of the radar itself. Radar can also sense material properties. In that capacity it can detect precipitation, the composition of structures, the depth of ice and snow, and a myriad of other properties from insect swarms to the existence of water on distant planets.

A different approach to radar sensing is based on the Doppler effect. In this type of radar the amplitude and power involved are not important (as long as a reflection is received). Rather, the Doppler effect is used. This effect is simply a change in the frequency of the reflected waves due to the speed of a target (see also **Section 7.7.1** for the use of Doppler in ultrasound sensing). Consider a vehicle moving away from a source at a velocity v, as shown in **Figure 9.16**. The source transmits a signal at frequency f. In a time Δt, the vehicle has moved a distance ΔS. Because of the motion, the reflected signal

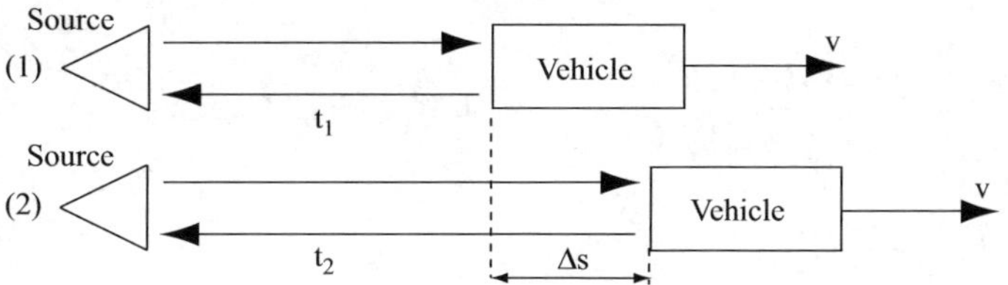

FIGURE 9.16 ■ Speed sensing with radar: either time of flight or Doppler shift may be used.

arrives back at the transmitter after a delay of $2\Delta t$, where $\Delta t = \Delta S/v$. This delay causes a shift in the frequency of the received signal. The shifted frequency is as follows:

$$f' = \frac{f}{1 + 2v/c} \quad [\text{Hz}]. \tag{9.28}$$

The returning wave's frequency is lower the higher the velocity of the vehicle. If the motion is toward the radar source, the frequency increases (velocity is negative). Measuring this frequency gives an accurate indication of the speed of the vehicle. This is how a police speed detector operates, but the same can be used to detect aircraft or tornadoes. On the other hand, Doppler radar is totally blind to stationary objects. Doppler radar is actively developed for anticollision systems in vehicles and for active cruise control.

Radar relies heavily on good antennas and, in particular, on the high directivity of these antennas. Therefore practical radar sensors operate at relatively high frequencies—from about 2 GHz to 30 GHz with anticollision systems intended to operate in excess of 80 GHz.

There are many other types of radar. One is the into-the-ground radar (also called ground penetrating radar). This system operates at lower frequencies for the purpose of penetrating and mapping underground objects. For space exploration and for mapping of planets, synthetic aperture radar (SAR) has been developed. This method makes use of moving antennas and signal processing to increase the effective range, sensitivity, and apparent power of the radar.

EXAMPLE 9.5 Speed detection by radar

The use of Doppler radar for speed detection and enforcement on the roads is very common and has been in use for a long time. Most speed radars operate in the X-band (8–12 GHz), Ka-band (27–40 GHz), and K-band (18–26 GHz). Suppose a 10 GHz radar (also called a radar gun or speed gun) measures the speed of a car at 100 km/h moving toward the radar gun.

a. What is the change in the frequency of the reflected wave due to the speed of the car?

b. Calculate the sensitivity of the device in Hz/km.

Solution: The radar transmits a signal at a frequency f and receives a reflection at a frequency f'. The internal circuitry subtracts the two frequencies to obtain the frequency difference. This is calibrated in speed, giving the operator a direct indication of the speed of the vehicle.

a. The reflected signal frequency is calculated from **Equation (9.28)**, but first we must calculate the speed of the vehicle in meter per second:

$$v = \frac{100000}{3600} = 27.78 \text{ m/s}.$$

The reflected signal frequency is

$$f' = \frac{f}{1 - 2v/c} = \frac{10 \times 10^9}{1 - 2 \times 27.78/3 \times 10^8} = 10{,}000{,}001{,}852 \text{ Hz}.$$

Thus the change in frequency is 1852 Hz.

Note that the speed was taken as negative because the car moves toward the observer.

b. To calculate the sensitivity we simply substitute $v = 1$ km/h $= 1000/3600$ m/s:

$$f' = \frac{f}{1 - 2v/c} = \frac{10 \times 10^9}{1 - 2 \times 0.2778/3 \times 10^8} = 10{,}000{,}000{,}018.5 \text{ Hz}.$$

Thus the change in frequency is 18.5 Hz and sensitivity is 18.5 Hz/km.

Note: These frequencies seem small, but in fact the speed Doppler radar is very accurate even if the oscillator that generates the frequency f is not perfect, since the subtraction (done in a circuit called mixer) uses the frequency f and does not assume a fixed value. As long as the fundamental frequency does not vary significantly during the measurement, the speed can be inferred accurately.

9.4.1.2 Reflection and Transmission Sensors

A somewhat different approach, applicable at very short ranges, is to send an electromagnetic wave and sense the reflected waves, but unlike in radar, the propagation effect is negligible since the distance is very short. This is shown schematically in **Figure 9.17**. The reflection coefficient of an electromagnetic wave depends on the wave impedance of the materials involved. Assuming that the source is in air, and it propagates into a material, which we will denote as 1, the wave impedances of the materials are

$$\eta_0 = \sqrt{\frac{\mu_0}{\varepsilon_0}} \quad [\Omega], \quad \eta_1 = \sqrt{\frac{j\omega\mu_1}{\sigma_1 + j\omega\varepsilon_1}} < \eta_0 \quad [\Omega], \tag{9.29}$$

where σ_1 is the conductivity of medium 1. If the medium is nonconducting (a perfect or lossless dielectric), the latter reduces to $\eta_1 = \sqrt{\mu_1/\varepsilon_1}$ and becomes a real number. Air will be considered a lossless dielectric for the purpose of reflection sensors because its losses are small and the distances involved are short.

To define reflection and transmission of electromagnetic waves, we define first a reflection coefficient (Γ) and a transmission coefficient (T) as follows:

$$\Gamma = \frac{E_r}{E_i}, \quad T = \frac{E_t}{E_i}, \tag{9.30}$$

where E_i is the amplitude of the incident electric field intensity, E_r is the amplitude of the reflected electric field intensity, and E_t is the amplitude of the transmitted electric field intensity (see **Figure 9.17**). The reflection and transmission coefficients depend on the angle of incidence as well. Further, these depend on the direction of the electric field

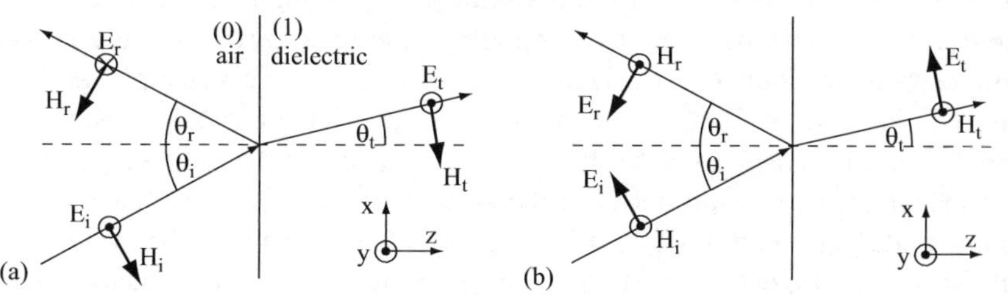

FIGURE 9.17 ■ Reflection of electromagnetic waves from a dielectric. (a) Perpendicular polarization. (b) Parallel polarization.

intensity—this is designated as the polarization of the electric field intensity. One distinguishes two polarizations: parallel polarization refers to the electric field intensity being in the plane of the incidence (the plane defined by the direction of propagation of the wave and the normal to the interface at the location of incidence). The second is perpendicular polarization, in which the electric field intensity is perpendicular to the plane of incidence. The wave shown in **Figure 9.17a** is polarized perpendicular to the plane of incidence, whereas the wave in **Figure 9.17b** is polarized parallel to the plane of incidence. The coefficients are given as follows:

For parallel polarization (denoted with $\|$):

$$\Gamma_{\|} = \frac{E_r}{E_i} = \frac{\eta_1 \cos\theta_t - \eta_0 \cos\theta_i}{\eta_1 \cos\theta_t + \eta_0 \cos\theta_i}, \quad T_{\|} = \frac{E_t}{E_i} = \frac{2\eta_1 \cos\theta_i}{\eta_1 \cos\theta_t + \eta_0 \cos\theta_i}. \tag{9.31}$$

For perpendicular polarization (denoted with $\perp$):

$$\Gamma_{\perp} = \frac{E_r}{E_i} = \frac{\eta_1 \cos\theta_i - \eta_0 \cos\theta_t}{\eta_1 \cos\theta_i + \eta_0 \cos\theta_t}, \quad T_{\perp} = \frac{E_t}{E_i} = \frac{2\eta_1 \cos\theta_i}{\eta_1 \cos\theta_i + \eta_0 \cos\theta_t}. \tag{9.32}$$

In addition, the incidence and transmission angles are related through Snell's law of refraction:

$$\frac{\sin\theta_t}{\sin\theta_i} = \frac{n_0}{n_1}, \tag{9.33}$$

where n_0 and n_1 are the indices of refraction of air and medium 1, respectively. The latter are given as

$$n_0 = \sqrt{\varepsilon_{r0}\mu_{r0}} = 1, \quad n_1 = \sqrt{\varepsilon_{r1}\mu_{r1}} > 1, \tag{9.34}$$

where ε_r and μ_r are the relative permittivity and relative permeability of the respective medium. The use of Snell's law allows one to calculate the reflection and transmission coefficients in terms of the angle of incidence and material properties alone.

In the particular case of perpendicular incidence ($\theta_i = 0$), the reflection and transmission coefficients become

$$\Gamma = \frac{\eta_1 - \eta_0}{\eta_1 + \eta_0}, \quad T = \frac{2\eta_1}{\eta_1 + \eta_0}. \tag{9.35}$$

The reflection coefficient varies between -1 and $+1$ depending on the properties of the materials (can be complex as well, since permittivities of lossy materials are complex). The transmission coefficient varies between 0 and 2. Thus for an incident amplitude E_0, the reflected amplitude is ΓE_0 and the transmitted amplitude is TE_0.

In a reflection sensor, the amplitude of the reflected wave, ΓE_0, is measured and can be directly linked to the permittivity in material 1. The reflection coefficient depends on permittivity, which depends on many parameters, the most obvious being moisture, but also composition and density. Reflection sensors can be very simple and effective. **Example 9.7** discusses such a sensor intended for the detection of mines.

A transmission sensor may be built equally easily and is shown in principle in **Figure 9.18**. The transmission between the source and the detector is a function of the intervening material. This can be calibrated in terms of any of the properties of the material. Moisture content is most often the stimulus, since water has a high permittivity, can be sensed easily, and is important to a wide range of industries (paper, textile,

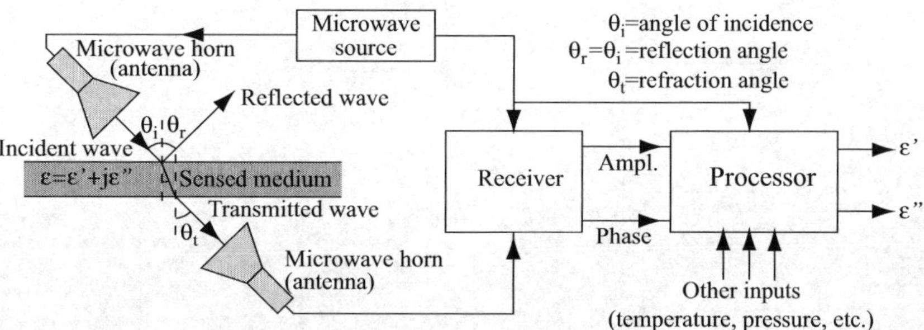

FIGURE 9.18 ■ A transmission sensor. The output is a function of the transmission through the material tested and is affected by a number of parameters, most notably moisture.

food, etc.). A sensor of this type might be used to monitor the drying of grain before storage, the production of dough for baking, or the thickness of paper on the production line. The sensor in **Figure 9.18** indicates the measurement of real (ε') and imaginary (ε'') parts of the permittivity (both may be affected by, for example, water content), but it can be calibrated to indicate mass, moisture content, density, or any other quantity that affects permittivity. It should also be noted that it is often difficult to distinguish between different quantities that may affect permittivity in similar ways.

EXAMPLE 9.6 Transmission of microwaves and detection of wildlife

A common hazard along highways is collisions with animals that wander into the roadway. Of particular concern are larger animals such as deer. A transmission-type microwave system is in use in the United States in areas of particular concern, such as highways along forests and marshes where it is common for animals to cross roads as part of their natural pattern of movement. The system transmits a microwave beam around 35 GHz to a receiver at some distance (typically 200–400 m) away. This establishes the normal signal received in a system similar to that shown in **Figure 9.18**. When a larger animal crosses the beam, the signal received decreases in amplitude. The system determines the trigger level based on a number of parameters and warns drivers about the presence of animals. The warning is typically in the form of a flashing yellow light. The driver is supposed to exercise caution by slowing down to avoid collisions, or if a collision does happen, to mitigate its consequences.

EXAMPLE 9.7 Microwave detection of buried dielectric objects

Metal detectors are very useful in detecting metals buried underground, at least to a certain depth. However, dielectrics such as pipes or even nonmetallic mines buried just underneath the surface are difficult to detect. A microwave transmission will penetrate to some depth under the soil and will reflect from any discontinuity in the soil properties, be it a rock or a piece of plastic. To increase sensitivity and resolution, a differential sensor may be employed as shown in **Figure 9.19a**. A single transmitter transmits a beam (at 10 GHz in this case). Two receiving antennas, symmetrically spaced about the transmitter and above it, receive any reflections from the target. As long as the soil is uniform, the reflections into each receiving antenna are identical (or nearly so). The two signals are amplified after down-converting to a convenient frequency for

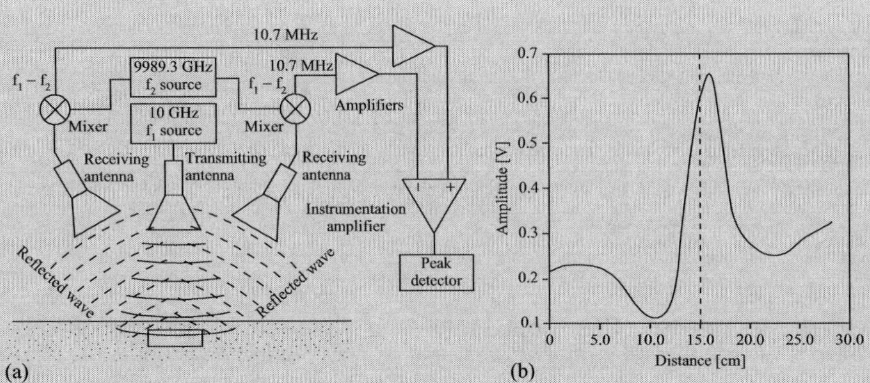

FIGURE 9.19 ■ Differential reflection sensor used to detect buried objects such as mines.

amplification (down-converting is a process of reducing the frequency by mixing two frequencies to obtain the difference between them). The amplified signals are then fed to the differential inputs of an instrumentation amplifier (more on this in **Chapter 11**). Under normal conditions the output is zero. If one receiver receives a larger signal, the output will shift from zero, indicating a discontinuity in the medium. **Figure 9.19b** shows the output detected as the three antennas move in tandem across a buried Perspex box to simulate a buried mine.

9.4.1.3 Resonant Microwave Sensors

A third important method of sensing with microwaves is based on microwave resonators. A microwave resonator may be thought of as a box or cavity with conducting walls that confines the waves. In essence, standing waves are generated (provided that energy is coupled into the structure) in each dimension of the cavity. The standing waves the cavity can support must be a multiple integer of half-wavelengths in any dimension or a combination of these. These are the resonant frequencies. For a rectangular cavity of dimensions a, b, and c, the resonant frequencies are

$$f_{mnp} = \frac{1}{2\pi\sqrt{\mu\varepsilon}}\sqrt{\left(\frac{m}{a}\right)^2 + \left(\frac{n}{b}\right)^2 + \left(\frac{p}{c}\right)^2} \quad \text{[Hz]}, \tag{9.36}$$

where m, n, and p are integers (0, 1, 2,...) and can take different values. These define the modes of resonance of the cavity. For example, in an air-filled cavity, for $m = 1$, $n = 0$, and $p = 0$, the 100 mode is excited and its frequency in a cavity of dimensions $a = b = c = 0.1$ m is 477.46 MHz. Not all values of m, n, and p result in valid modes, but for simplicity's sake the discussion here should be sufficient. Also, cavities do not need to be rectangular, they can be cylindrical or of any complex shape, in which case the analysis is much more complicated.

The importance of this result is that at resonance the fields in the cavity are very high while off resonance they are very low. The cavity acts as a sharp bandpass filter at resonance. From the sensing point of view it is important to note that the resonant frequency depends on the electrical properties of the material in the cavity—its permittivity and its permeability, in addition to physical dimensions. Thus any material

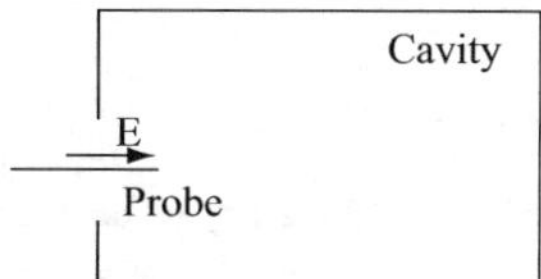

FIGURE 9.20 ■ Coupling energy into a cavity resonator.

inserted in the cavity will reduce its resonant frequency since air (actually vacuum) has the lowest permittivity. Because resonance is sharp, the change in resonant frequency is easily measured and can be correlated to the sensed quantity. Sensors built based on cavity resonance are simple and very sensitive.

The resonant frequency in **Equation (9.36)** depends on the permeability μ and permittivity ε of the medium in the cavity. In virtually all practical applications the permeability is that of free space. The permittivity is typically due to a mixture of materials. For example, if the cavity is used to sense humidity, it will contain air and water vapor. In other cases it may contain a mixture of materials of very different permittivities. In such cases ε is replaced with an effective permittivity based on the volume of each material and its permittivity. There are many mixing formulas for various conditions, but the simplest is the following:

$$\varepsilon_{eff} = \frac{\sum_{i=1}^{N} \varepsilon_i v_i}{\sum_{i=1}^{N} v_i}, \tag{9.37}$$

where we assume N media, each with its own permittivity and volume. The sum of volumes is the total volume of the cavity. The relation in **Equation (9.37)** is useful in many instances, particularly when the substances are uniformly mixed (we used this relation in **Example 8.11** to calculate an effective permittivity of moist air). But, as an approximation, it can also be used when the substances are separated, such as, for example, when the cavity contains separate distinct objects.

There are two necessary conditions to produce a cavity resonator sensor. First, the property sensed must somehow alter the material permittivity in the cavity or its dimensions. Second, a means of coupling energy into the cavity must be found. The resonant frequency is then measured and, provided a transfer function can be established, the stimulus is sensed directly. Energy can be supplied to a cavity in many ways, the simplest way being to insert a probe (a small antenna) that radiates fields into the cavity. This is shown in **Figure 9.20**. Those fields at the right frequency are amplified as standing waves, the others are negligible. To sense a quantity, the permittivity must change with this quantity. This can be accomplished in a number of ways. For gases, it is sufficient to provide holes in the walls of the cavity to allow them to penetrate as shown in **Figure 9.21**. In this form, the cavity can sense gases emitted by explosives, fumes from chemical processes, smoke, moisture, and almost anything else that has a permittivity greater than air. These "sniffers" can be extremely sensitive, but it is difficult to separate the effects of, say, smoke and moisture, and the measurement of resonant frequency at the frequencies involved is not a trivial issue. Nevertheless, resonant methods are some of the most useful in the evaluation of gases. Solids may be equally sensed for variations in permittivity, provided they can be inserted into the cavity. The change in resonant frequency is usually quite small, on the order of a fraction of a percent, but since the frequencies are high, it is sufficient for detection.

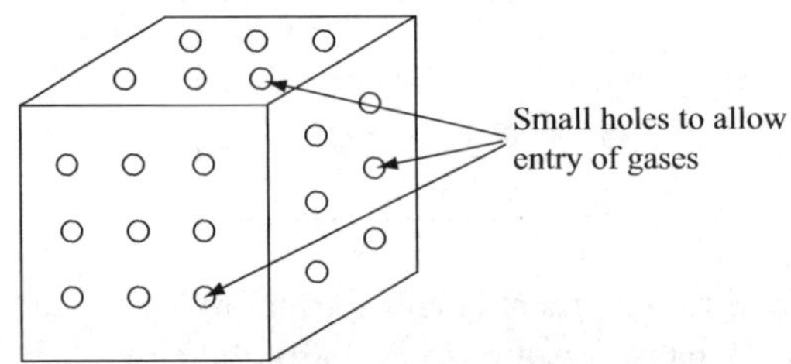

FIGURE 9.21 ■ A cavity resonator with openings for gas sampling. The diameter of the holes must be much smaller than the wavelength at the resonant frequency.

EXAMPLE 9.8 Microwave sensing of moisture content

Sensing in a microwave cavity relies on the change in relative permittivity or the change in volume of the cavity. Suppose a rectangular cavity of dimensions $a = 20$ mm, $b = 20$ mm, and $c = 40$ mm is used to measure the moisture content in the air of a large industrial drier by measuring the relative humidity. The drier operates by supplying an airstream at 70°C and removing moisture by drawing the air out. The relative permittivity of air at saturation humidity (100% humidity) at 70°C is 1.00213, whereas at zero humidity it is 1.0. The product in the drier is considered dry when the relative humidity in air is 20% or less.

a. Calculate the resonant frequency of the cavity in the extremes of humidity and at 20% humidity, assuming permittivity varies linearly with relative humidity. Use the 100 ($m = 1$, $n = 0, p = 0$) mode.
b. If frequency increments of 1 kHz can be accurately measured, what is the resolution of the sensor in terms of relative humidity, assuming permittivity varies linearly with humidity?

Solution:

a. The mode selected defines the resonant frequency for any given permittivity from **Equation (9.36)**:

$$f_{mnp} = \frac{1}{2\pi\sqrt{\mu\varepsilon}}\sqrt{\left(\frac{1}{a}\right)^2 + \left(\frac{0}{b}\right)^2 + \left(\frac{0}{c}\right)^2} = \frac{1}{2\pi a\sqrt{\mu\varepsilon}} \quad \text{[Hz]}.$$

Air has the permeability ($\mu = \mu_0$) of free space.

At 0% humidity, the resonant frequency is

$$f_{100} = \frac{1}{2\pi a\sqrt{\mu_0\varepsilon_0}} = \frac{1}{2\pi \times 0.02\sqrt{4\pi \times 10^{-7} \times 8.854 \times 10^{-12}}} = 2{,}385{,}697{,}883 \text{ Hz}.$$

At 100% humidity, the resonant frequency is

$$f_{100} = \frac{1}{2\pi a\sqrt{\mu_0 \times 1.00213\varepsilon_0}} = \frac{1}{2\pi \times 0.02\sqrt{4\pi \times 10^{-7} \times 1.00213 \times 8.854 \times 10^{-12}}}$$

$$= 2{,}383{,}161{,}166 \text{ Hz}.$$

At 20% relative humidity, the relative permittivity is

$$\varepsilon_r = 1 + \frac{0.00213}{100} \times 20 = 1.000426$$

and therefore the resonant frequency is

$$f_{100} = \frac{1}{2\pi a\sqrt{\mu_0 \times 1{\cdot}00213\varepsilon_0}} = \frac{1}{2\pi \times 0.02\sqrt{4\pi \times 10^{-7}} \times 1.000426 \times 8.854 \times 10^{-12}}$$

$$= 2{,}385{,}189{,}892 \text{ Hz}.$$

Note that the frequency has increased from 100% humidity to 20% humidity by 2.028 MHz. This is an easily measurable quantity.

b. An increment in frequency of 1 kHz leads to the following:

$$\frac{1}{2\pi a\sqrt{\mu_0\varepsilon}} = \frac{1}{2\pi a\sqrt{\mu_0(\varepsilon + \Delta\varepsilon)}} = 1000 \text{ Hz},$$

from which we can calculate $\Delta\varepsilon$. The relation between frequency and permittivity is nonlinear, but only very slightly, so the change in frequency is relatively small. Therefore we may safely assume that the change in frequency with the change in relative humidity is more or less linear and we argue as follows: The change in frequency from zero humidity to 100% humidity is

$$\Delta f = f(0\%) - f(100\%) = 2{,}385{,}697{,}883 - 2{,}383{,}161{,}166 - 2{,}536{,}717 \text{ Hz}.$$

Since we can measure accurately 1 kHz, the measurement range can distinguish 2536.7 levels of humidity. Therefore the resolution is 100%/2537 = 0.039% humidity. That is, the resolution is 0.039%.

Note: The system described here is very simple and does not take into account errors such as drift in resonant frequency and the effects of changes in temperature on the permittivity. However, the method is sound and may be easily employed in applications where the cost is justified. In the end the resolution would be much lower, but even a 1% relative humidity resolution may be more than sufficient in practice.

To allow measurements on solids, the idea of the cavity sensor can be extended by partially opening the cavity and allowing the solids to move through it. An example of this type of sensing is shown in **Figure 9.22**. Here the resonance is established by the two strips acting as a transmission line between the two plates. Resonance depends on the lengths of the strips as well as the location and size of the outer plates. The material to be sensed for variations in permittivity passes between the strips. This method has been successfully used to sense moisture content in paper, wood veneers, and plywood and to monitor the curing process in rubber and polymers. To improve performance, the plates are bent down to partially enclose the cavity. This improves sensitivity and reduces influences from outside. **Figure 9.23** shows an open-cavity resonator operating at 370 MHz in air and designed to monitor the water content in drying latex in a continuous industrial coating process. The change in resonant frequency is only about 2 MHz (from wet to dry),

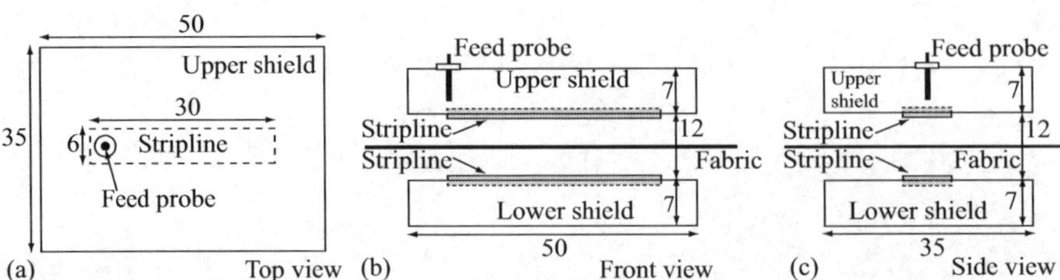

FIGURE 9.22 ■ A stripline cavity resonator. The open cavity allows testing of continuous production materials such as paper. Dimensions are in centimeters.

FIGURE 9.23 ■ An open stripline resonator operating at 370 MHz used for moisture sensing in latex-coated fabric. One stripline and antenna (brass rod above the left edge of the stripline) are seen in the top half of the resonator.

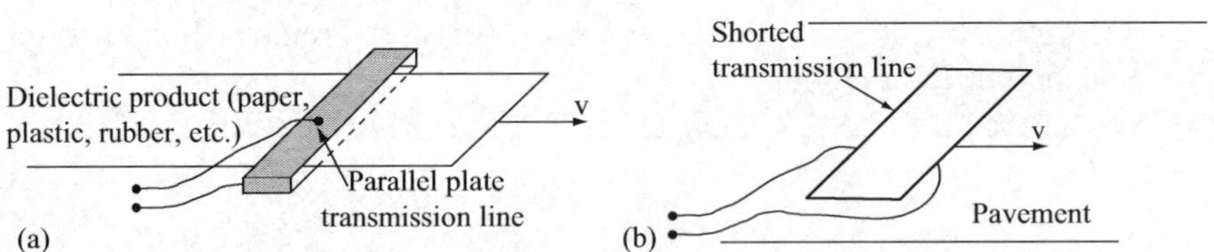

FIGURE 9.24 ■ (a) An open transmission line resonator for thickness, density, or moisture content sensing of flat products. (b) A transmission line sensor for road pavement thickness or density sensing. The sensor moves (often attached to a vehicle) on the pavement to monitor its condition.

which represents about 0.5% change in frequency. However, with the use of a commercial network analyzer, changes on the order of less than 1 kHz are easily measured, making this a very sensitive device.

A variation of the open resonator is the transmission line resonator shown in **Figure 9.24**. This is made of two strips at a fixed distance from each other and shorted at both ends. Connections are made to each strip and fed by a source. The resonant frequency depends on the dimensions and locations of the feed wires and, of course, on the permittivity of the medium between them. A similar device is commonly used to sense the thickness or density of asphalt on roads. The resonator is driven above and close to the road and the resonant frequency is monitored. Any increase indicates a thinning of the asphalt layer, while any decrease indicates a thicker layer.

9.4.1.4 Propagation Effects and Sensing

Perhaps the simplest method that can be used for sensing with microwaves, and in a more general sense with any electromagnetic wave, takes advantage of the fact that electromagnetic waves, as they propagate in space, are attenuated and the field spreads in space depending on the properties of the source. That allows one to sense distance by simply measuring the electric (or magnetic) field amplitude, given knowledge of the amplitude at the source or at some other location in space between the sensing location and the source, based on **Equations (9.21)** and **(9.22)**. Similarly, if the distance is known, one can sense the material properties (primarily permittivity) in that space since the amplitude depends on the attenuation constant. These methods are very simple, but not very accurate, since there are many effects that can change the amplitude. These include moisture, air density, the presence or proximity of conducting bodies such as earth, and many others. Nevertheless, position sensors based on the measurement of amplitude in receivers given a known transmitter do exist, and for some applications these are sufficiently accurate. One can also imagine time-of-flight measurements, as were discussed in **Chapter 7** with acoustic waves. For example, electromagnetic waves require 3 ns to travel a distance of 1 m. It is therefore quite possible to measure, say, a distance of 100 m (time of flight of 300 ns), but given current electronic components, it would be difficult to do so economically. However, longer distances, on the order of kilometers, can be measured accurately and economically.

9.5 | ANTENNAS AS SENSORS AND ACTUATORS

Antennas are unique devices and are not normally thought of as sensors since they are usually associated with transmitters and receivers. But they are true sensors—sensing the electric field or the magnetic field in the electromagnetic wave. Thus one can say that the receiver and transmitter are in fact transducers and the antenna is the sensor (in a receiver) or the actuator (in a transmitter). In microwave work, antennas are often referred to as "probes" because of their use as sensors and actuators (receiving and transmitting antennas).

All antennas are based on or can be related to one of two related fundamental or elementary antennas. These are called the electric and magnetic dipoles, or sometimes the elementary electric and magnetic dipoles.

9.5.1 General Relations

The electric dipole is simply a very short antenna, made as shown in **Figure 9.25a**. It consists of two short conducting segments carrying a current I_0 fed by a transmission line. The magnetic dipole, shown in **Figure 9.25b**, is a loop of small diameter that is fed by a transmission line. Their names are related to the fields they produce, which look like the fields of an electric dipole and a magnetic dipole, respectively. In all other

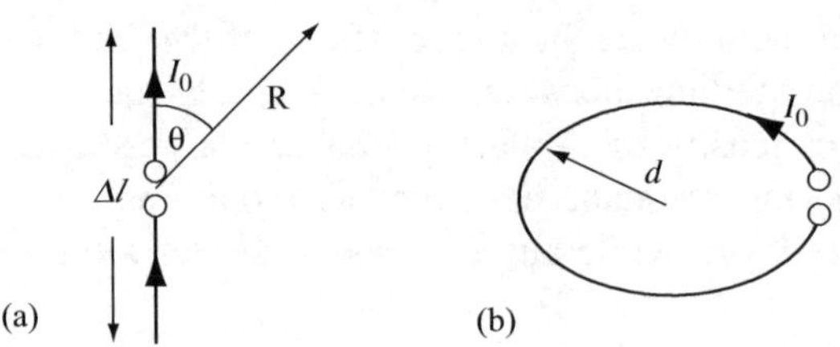

FIGURE 9.25 ■ (a) An elementary electric dipole antenna. (b) An elementary magnetic dipole antenna.

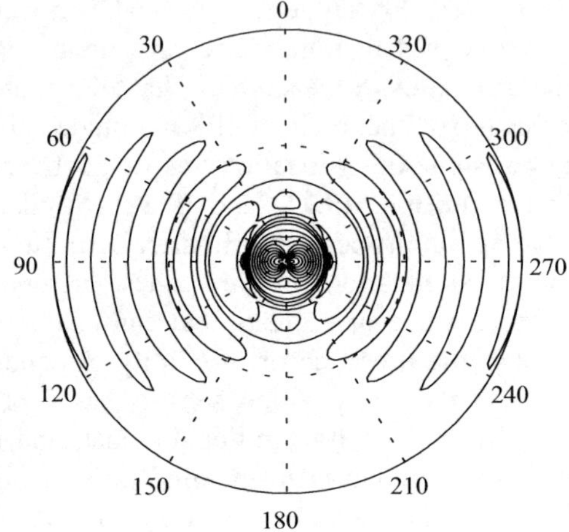

FIGURE 9.26 ■ Radiation from an electrical dipole. Note the pattern: maximum radiation in the horizontal direction ($\theta = 90°$) and zero radiation in the perpendicular direction ($\theta = 0°$ or $\theta = 180°$).

respects, the two antennas are very similar, and in fact the two produce identical field distribution in space except that the magnetic field of the electric dipole is identical (in shape) to the electric field of the magnetic dipole and vice versa. The field radiated from a small dipole (electric or magnetic) is shown in **Figure 9.26**. It shows that near the antenna the field is essentially the same as for an electrostatic dipole (two point charges of opposite polarity at a very short distance from each other). For this reason, it is called the electrostatic field or the near field. When electric dipole antennas are very close to a source (less than about one wavelength), they behave more or less like capacitors. At larger distances, the antennas radiate (or receive radiation) in what is called the far field. Usually antennas are used in the far field. The electric and magnetic field intensities of a dipole in the far field are

$$H = \frac{I\Delta l}{2\lambda R} e^{-j\beta/R} \sin\theta_{IR} \quad [\text{A/m}], \quad E = \eta H \quad [\text{V/m}], \tag{9.38}$$

where Δl is the length of the dipole, λ is the wavelength, R is the distance from the antenna to the location where the field is measured, and θ_{IP} is the angle between the antenna and the direction to the point at which the field is needed (a spherical system of coordinates is usually assumed in orienting antennas). η is the wave impedance and β is the phase constant (see **Equation (9.20)**).

It should be noted from **Equation (9.38)** that the ratio between the electric field and magnetic field is constant and equals the wave impedance (see **Equation (9.23)**). This impedance is only dependent on material properties, as follows:

$$\eta = \frac{|E|}{|H|} = \sqrt{\frac{\mu}{\varepsilon}} \quad [\Omega]. \tag{9.39}$$

Since permittivity is in general a complex number, the wave impedance is also a complex number. The electric and magnetic fields are perpendicular to each other and both are perpendicular to the direction of propagation of the wave (radial direction). The relation above also shows that maximum fields are obtained when $\theta = 90°$, that is, perpendicular to the current. A plot of the relation will reveal that the fields diminish as the angle becomes smaller or larger, and at $\theta = 0°$ the field is zero. This plot is called the radiation pattern of the antenna and gives the distribution of the field over a plane that contains the dipole (other planes may also be selected and similarly described). The radiation pattern changes with the length and type of the antenna. Another important quantity is the directivity of the antenna, which can also be obtained from the radiation pattern or the formulas for the electric field or magnetic field. It simply indicates the relative power density in all directions in space. If the antenna is not short, it can be viewed as an assembly of elemental antennas. The fields of long wire antennas are obtained by integrating the fields in **Equation (9.38)** over the length of the wires after replacing Δl by dl (i.e., a differential length dipole). As mentioned above, the exact same considerations apply to magnetic dipole antennas (loop antennas) except that the behavior of the magnetic field intensity of the magnetic dipole is the same as the magnetic field intensity of the electric dipole and vice versa. Antennas are dual elements—they are equally suitable for transmission and for reception.

9.5.2 Antennas as Sensing Elements

From a sensing point of view, the electric dipole may be viewed as an electric field sensor. Of course, the magnetic dipole senses a magnetic field, but since the relation between the electric field and magnetic field is known everywhere, sensing one field or the other amounts to the same thing. To see how sensing occurs, consider **Figure 9.27**, which shows a propagating wave at the location of the antenna, making an angle θ with it. The electric field intensity of the wave is E and it is perpendicular to the direction of propagation of the wave. The voltage of the antenna due to this field (assuming l is small) is

$$V_d = El \sin\theta \quad [\mathrm{V}]. \tag{9.40}$$

Thus a linear relation between the electric field intensity in the wave and the voltage on the antenna is obtained. Thus **Equation (9.38)** establishes the relation for actuation, whereas **Equation (9.40)** establishes the relation for sensing.

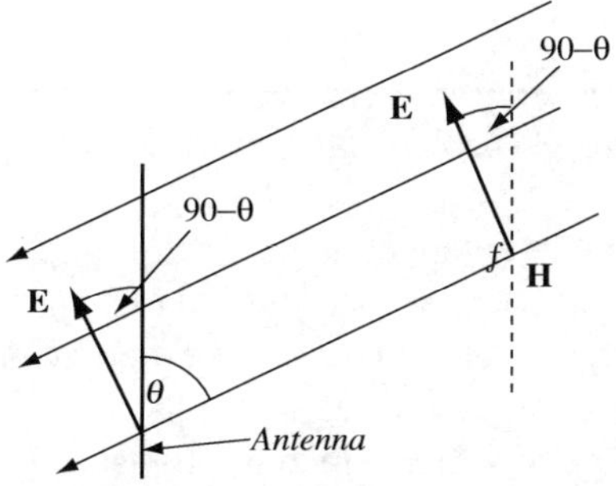

FIGURE 9.27 ■ A small dipole as a sensing element.

More practical antennas are made of various lengths (or, in the case of loops, various diameters), may have different shapes, and may in fact be an array of antennas, but these changes are not fundamental. In general, the "larger" the antenna, the higher the power it can transmit or receive (not always and not linearly proportional to antenna size). Also, the size of the antenna changes the radiation pattern of the antenna, but again, the change is not a fundamental change. Antennas are very efficient sensors/actuators with conversion efficiencies that can easily exceed 95%.

In practical applications, certain antennas have been shown to be better than others in some respects. Most applications try to use a $\lambda/2$ antenna if possible, primarily because its input impedance can be shown to be 73 Ω and the antenna has a good omnidirectional radiation pattern, whereas other antenna lengths have higher or lower impedances and different radiation patterns. Dipole antennas may sometimes be replaced by monopoles (half a dipole—similar to the car antenna or the telescoping antenna in some radios), with appropriate changes in properties (half the impedance, half the total radiated power, etc.).

Some antennas are more directive than others, meaning that they will radiate (or receive) with preferential directions in space. The dipole in **Equation (9.38)** has its highest amplitude at 90° with respect to the axis of the antenna, but others, such as reflector antennas (dish antennas), are much more directive, often with a narrow radiating/receiving beam. But a highly directive antenna does not have to be a reflector antenna. One of the more common, highly directive antenna is the Yagi antenna. It may be recognized as the common rooftop TV antenna, but it can serve in other applications, such as point-to-point communication or data transmission, in Wi-Fi repeaters, and in remotely controlled installations. In some applications, such as broadcasting, remote control of devices, and data transmission, a nondirective antenna allows coverage of large spaces, but in others such as radar, directivity is essential not only to "concentrate" the power, but also to allow direction identification.

Antennas come in an array of sizes, from the tiny to the truly gigantic. Some integrated antennas are only millimeters in length while others can be a few kilometers long. Some are massive structures, such as antennas for radio telescopes or deep space communication (these are typically reflector antennas—the antenna itself is much smaller and located at the focal length of the parabolic dish). In other applications, antennas are built as part of the structure, either on a printed circuit board, such as in mobile phones and remote door openers, or on the surface of structures such as aircraft or missiles.

Antennas can be used to detect and quantify electromagnetic sources ranging from lightning strikes to radiation from distant galaxies, and anything in between. As actuators they may be used to open a door remotely, identify a key in a car, treat tumors, or warm food, and a myriad of other applications.

EXAMPLE 9.9 Position location and the Global Positioning System (GPS)

An important use of high-frequency waves and antennas is in positioning, that is, the identification of one's position on the planet or in space. There are a number of variations of the method, but they all involve measuring distance, directly or indirectly. We start with the oldest and least accurate method, called triangulation. It involves two fixed-position transmitters, a receiver with adjustable antenna, and a map. Consider **Figure 9.28a**. Since the transmitter transmits equally in

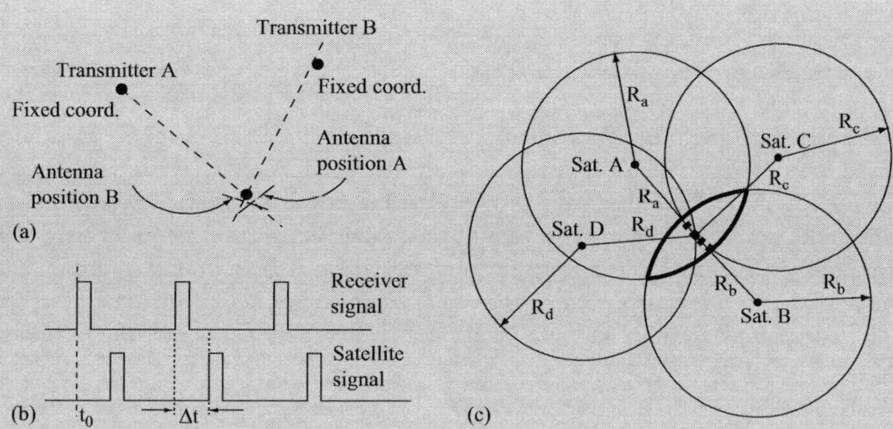

FIGURE 9.28 ▪ (a) The concept of triangulation on a surface (i.e., on a map). (b) Timing pulses in the GPS. The time difference between the synchronized transmitter and receiver is due to the time needed for the waves to travel from the transmitter to the receiver. (c) In space, a minimum of four satellites are needed to pinpoint the receiver.

all radial directions, by adjusting the antenna angle θ so that reception is maximum, the direction relative to the transmitting antenna can be identified. **Figure 9.28a** shows the classic concept of triangulation. The receiving antenna is adjusted for maximum reception and the direction is marked. The line on the map shows that direction, but not the distance, so the receiver could be very close or very far from transmitter A. Now the receiving antenna is adjusted again until maximum reception is received from transmitter B. The two directions intersect at one and only one point on the map. Since the position of the two transmitters is known, the intersection point on the map shows the exact position of the receiver. Triangulation was used extensively for navigation of ships, with shore stations providing the fixed positions. The system was called Loran (**lo**ng **ra**nge **n**avigation), but its use declined after the introduction of the Global Positioning System (GPS) and it was discontinued in late 2010. It worked a bit differently than described above, and rather than measuring signal strength, it measured the time needed for the signal to reach the receiver (since that does not require moving antennas). From time and speed of propagation in air, one could calculate the distance from two (or more) fixed positions intersecting at one point—the location of the receiver.

The same idea is used in the GPS. The GPS consists of 24 fixed satellites transmitting timing and identification pulses based on an atomic clock in each satellite. The information consists of the satellite position and the clock time (i.e., the exact time or "wall time"). The GPS antenna receives these pulses based on which the GPS receiver performs two fundamental tasks. First, it synchronizes its clock with that of the GPS satellites it receives. To do so it checks a number of satellites (minimum of four). It now resets its own clock to be identical with the clocks in the satellites (satellite clocks are synchronized among themselves using signals received from earth). Now the receiver generates a series of pulses identical to those the satellite sends, separated by a known time t_0, both starting at the same "wall time." The pulse received from each satellite is delayed a time Δt_n because of the distance to the satellites (n indicates the satellite ID). This is shown schematically in **Figure 9.28b**. Since the speed of electromagnetic waves is known exactly, the distance to each satellite is calculated as $R_n = v\Delta t_n$. Now the receiver "intersects" these distances at one point in space—the position of the receiver. The best way to understand how this is possible is to imagine one satellite, marked as A in **Figure 9.28c**. The receiver knows the distance, say, R_a (from the time delay), so it can be at any point on a sphere of radius R_a (the center is the location of the satellite, information that is known because it is part of the transmission from that satellite). Now, using a second satellite, B, the intersection between the two spheres is a circle on which the receiver is located (indicated as thick solid arcs). A third satellite's sphere (C) intersects the circle at two possible points (the thick dashed arc's intersection with the two thick arcs). The fourth satellite, D, determines a single point between the two points on the dashed arc. Hence the need for four satellites as a minimum. Additional satellites decrease errors, since with four satellites the receiver can only determine an intersection point between the two points determined by the third satellite and not necessarily the exact location of the GPS receiver.

Given the coordinates of the four satellites as (x_i, y_i, z_i), $i = 1,2,3,4$, the coordinates of the receiver as (x,y,z), and the time differences in the signals due to the four satellites as Δt_i, $i = 1,2,3,4$, the coordinates of the receiver (x,y,z) can be calculated as follows: Suppose first that a reference satellite exists at the origin of the system of coordinates and its signal is received with a zero time delay. Taking one of the satellites, say satellite m, we can write for the distance of the receiver to that satellite

$$R_m^2 = (x - x_m)^2 + (y - y_m)^2 + (z - z_m)^2.$$

The distance to the reference satellite is

$$R_r^2 = x^2 + y^2 + z^2.$$

We can write the distance R_m in terms of R_r as follows:

$$R_m^2 = (R_r + v\Delta t_m)^2 = R_r^2 + 2R_r v\Delta t_m + (v\Delta t_m)^2,$$

where Δt_m is the time delay of the signal received from satellite m and v is the speed of propagation of the signal (the speed of light in this case). Rewriting the relation we get

$$R_r^2 - R_m^2 + 2R_r v\Delta t_m + (v\Delta t_m)^2 = 0.$$

Dividing by $v\Delta t_m$ we get

$$\frac{R_r^2 - R_m^2}{v\Delta t_m} + 2R_r + v\Delta t_m = 0.$$

Now, writing this equation for satellite 1 we have ($m = 1$)

$$\frac{R_r^2 - R_1^2}{v\Delta t_1} + 2R_r + v\Delta t_1 = 0.$$

Subtracting this from the previous equation we get

$$\frac{R_r^2 - R_m^2}{v\Delta t_m} - \frac{R_r^2 - R_1^2}{v\Delta t_1} + v(\Delta t_m - \Delta t_1) = 0.$$

Now we write

$$R_r^2 - R_m^2 = x^2 + y^2 + z^2 - (x - x_m)^2 - (y - y_m)^2 - (z - z_m)^2$$
$$= -x_m^2 - y_m^2 - z_m^2 + (2x_m)x + (2y_m)y + (2z_m)z$$

and

$$R_r^2 - R_1^2 = -x_1^2 - y_1^2 - z_1^2 + (2x_m)x + (2y_1)y + (2z_1)z.$$

Therefore we have

$$\frac{-x_m^2 - y_m^2 - z_m^2 + (2x_m)x + (2y_m)y + (2z_m)z}{v\Delta t_m} - \frac{-x_1^2 - y_1^2 - z_1^2 + (2x_1)x + (2y_1)y + (2z_1)z}{v\Delta t_1}$$
$$+ v(\Delta t_m - \Delta t_1) = 0.$$

Collecting terms we get

$$-\frac{x_m^2 + y_m^2 + z_m^2}{v\Delta t_m} + \frac{x_1^2 + y_1^2 + z_1^2}{v\Delta t_1} + v(\Delta t_m - \Delta t_1)$$
$$+ \frac{2}{v}\left(\frac{x_m}{\Delta t_m} - \frac{x_1}{\Delta t_1}\right)x + \frac{2}{v}\left(\frac{y_m}{\Delta t_m} - \frac{y_1}{\Delta t_1}\right)y + \frac{2}{v}\left(\frac{z_m}{\Delta t_m} - \frac{z_1}{\Delta t_1}\right)z = 0.$$

Now, writing this equation for $m = 2$, $m = 3$, and $m = 4$ provides three equations in the unknowns x, y, and z:

$$-\frac{x_2^2 - y_2^2 - z_2^2}{v\Delta t_2} + \frac{x_1^2 + y_1^2 + z_1^2}{v\Delta t_1} + v(\Delta t_2 - \Delta t_1) + \frac{2}{v}\left(\frac{x_2}{\Delta t_2} - \frac{x_1}{\Delta t_1}\right)x + \frac{2}{v}\left(\frac{y_2}{\Delta t_2} - \frac{y_1}{\Delta t_1}\right)y$$
$$+ \frac{2}{v}\left(\frac{z_2}{\Delta t_2} - \frac{z_1}{\Delta t_1}\right)z = 0,$$

$$-\frac{x_3^2 + y_3^2 + z_3^2}{v\Delta t_3} + \frac{x_1^2 + y_1^2 + z_1^2}{v\Delta t_1} + v(\Delta t_3 - \Delta t_1) + \frac{2}{v}\left(\frac{x_3}{\Delta t_3} - \frac{x_1}{\Delta t_1}\right)x + \frac{2}{v}\left(\frac{y_3}{\Delta t_3} - \frac{y_1}{\Delta t_1}\right)y$$
$$+ \frac{2}{v}\left(\frac{z_3}{\Delta t_3} - \frac{z_1}{\Delta t_1}\right)z = 0,$$

$$-\frac{x_4^2 + y_4^2 + z_4^2}{v\Delta t_4} + \frac{x_1^2 + y_1^2 + z_1^2}{v\Delta t_1} + v(\Delta t_4 - \Delta t_1) + \frac{2}{v}\left(\frac{x_4}{\Delta t_4} - \frac{x_1}{\Delta t_1}\right)x + \frac{2}{v}\left(\frac{y_4}{\Delta t_4} - \frac{y_1}{\Delta t_1}\right)y$$
$$+ \frac{2}{v}\left(\frac{z_4}{\Delta t_4} - \frac{z_1}{\Delta t_1}\right)z = 0.$$

Solving these three equations provides the coordinates (x,y,z) of the receiver. Note, in particular, that the "reference" satellite is not part of these relations and was only needed to derive the relations. The reason why it is not needed is that the reference timing is provided by the synchronized clock. Note, however, that the positions of the four satellites and the delays from the four satellites are needed to solve for x, y, and z.

Although we use the term triangulation, the method is much more general and falls under the term of multilateration (or trilateration in the case of three sources). It can also be used to detect the location of a source using N receivers, with a minimum $N = 3$ for a location in a plane or $N = 4$ for a location in space.

9.5.3 Antennas as Actuators

So far we have discussed antennas and their role in sensing, that is, the antenna as a means of detecting or sensing an electric field (and therefore the magnetic field) at its location. This was, in essence, the role of the receiving antenna. However, the antenna is equally important as a transmitting element, an element that couples power from a source. Based on our definition of sensors and actuator, the antenna in this mode is an actuator. Antennas are very efficient as transmitters and the power loss in antennas is minimal (radiation efficiency is high). What is perhaps unique is that any antenna can both receive (sensor) and transmit (actuator). Not only can the antenna do both, but there is no distinction between transmitting and receiving antennas. The properties of antennas as receiving and transmitting are summarized in the reciprocity theorem, which says that "if antenna A transmits a signal and antenna B receives the signal, then by interchanging between the two antennas the signal received is identical." This property is fundamental in communication and is at the heart of systems that transmit and receive. Transmitting antennas have many specific actuation tasks. For example, in the discussion on cavity resonators above, the cavity was driven into resonance by an antenna of one type or another. It should also be clear that the two basic antennas discussed above are the simplest and most fundamental. A large variety of other antennas exist, as can be verified by any cursory inspection of rooftops or a communication mast. But all antennas can be analyzed based on the line (electric) dipole or the loop (magnetic) dipole.

The actuation function of a transmitting antenna is to generate an electric (and a magnetic) field; in other words, to couple power into the space around the antenna. This power can then perform actions of considerable importance. A common use is in heating food using microwaves. In that function, microwave energy is coupled from a microwave generator (magnetron) into the space of the oven and into the food in that space. The energy excites water molecules into motion, a process that generates heat. Heating in a microwave oven is the heating of water molecules, and in fact the frequencies used for microwave heating (e.g., 2.45 GHz for food heating and cooking in the home, 13.52 MHz for industrial heating and processing) were selected so that water absorbs maximum energy. Although microwave heating is important for consumers, it is also an important industrial process. Because microwave heating can be fast, it has found considerable applications in the freeze-drying of foods. To accomplish this, the space of the oven is held under vacuum while the food is processed. Under these conditions, water sublimates and is removed from the food with very little damage to the remaining tissue.

The heating effects of microwaves are also used with considerable success in medicine. One of these is in hyperthermia, especially for the treatment of tumors, but also in surgery. Tumors can be treated based on two fundamental properties. One is that when local heating occurs in the body, blood flow to the area increases to cool it to avoid damage. However, tumors are relatively poor in blood vessels, meaning that the body can cool down healthy tissue but cannot cool down tumors. Thus microwave heating will affect the tumor while leaving the healthy tissue unaffected or affected much less. Microwave application can be local by the insertion of antennas in the immediate vicinity of the tumor or can be a more general volume application.

9.6 | PROBLEMS

Radiation safety

9.1 Radiation safety and microwaves. Many people are concerned with the radiation effects of microwaves. Although there are effects other than radiation associated with microwaves, one way to quantize the effect is through the photon energy. Microwaves are considered to extend from 300 MHz to 300 GHz. Calculate the photon energy and indicate if the radiation in that frequency range is ionizing or nonionizing.

9.2 Flying and radiation exposure. Pilots, astronauts, and frequent fliers are exposed to hazardous radiation in the form of cosmic rays. These high-energy particles can be characterized by frequencies from about 30×10^{18} to 30×10^{34} Hz. X-rays range between 30×10^{15} and 30×10^{18} Hz. Write the photon energy associated with cosmic rays and compare them with that of X-rays.

9.3 UV radiation and cancer. It is universally accepted that prolonged exposure to UV light can contribute to the development of skin cancer. Nevertheless, UV radiation is a normal part of living on earth and is important in maintaining good health. Only excessive exposure is considered harmful. The sun's radiation reaching the surface of the earth is about 1200 W/m^2. Of that, roughly 0.5% is UV radiation (most of the UV radiation, about 98%, is absorbed in the ozone layer). An average person tanning on the beach weighs 60 kg and exposes a skin surface area of about 1.5 m^2. If about 50% of the UV radiation is absorbed in the skin, what is the dose of UV energy absorbed per hour of tanning?

Ionization sensors (detectors)

9.4 The residential smoke detector. Residential smoke detectors use a simple ionization chamber, open to the air, and a small radioactive pellet that ionizes the air inside the chamber at a constant rate. The source is americium-241 (Am-241), which produces mostly heavy α particles (these are easily absorbed in air and can only propagate about 3 cm). Smoke detectors contain approximately 0.3 μg of Am-241. The activity of Am-241 is 3.7×10^4 Bq and the ionization energy of the α particles it emits is 5.486×10^6 eV.

a. Assuming the efficiency is 100%, calculate the ionization current that will flow in the chamber if the potential across the chamber is high enough to attract all charges without recombination.

b. If the smoke detector circuit is fed by a 9 V battery with a capacity of 950 mAh and the electronic circuits consume an average of 50 μA in addition to the current through the chamber, what is a reasonable recommendation as to the frequency of battery change?

9.5 Industrial smoke detectors. Industrial smoke detectors are similar to residential smoke detectors but usually contain a larger amount of radioactive substance to generate a higher saturation current. The amount of radioactive substance is usually written on the smoke detector and given in microcurie (μCi). An industrial smoke detector is rated as having 45 μCi of Am-241. The α particles emitted by Am-241 have an energy of 5.486×10^6 eV. Calculate the saturation current in the chamber assuming 100% efficiency.

9.6 Fabric density sensor. Fabric density sensors come in many forms and one of these is based on measurement of the current in an ionization chamber. The saturation current is established between two plates using a krypton-85 isotope in a separate canister. The chamber is placed on one side of the fabric and the β-source on the other, as shown in **Figure 9.3**. The ionization current is measured as an indication of fabric density, the latter being in gram per square meter (g/m^2). The quantity of isotope used generates 3.7 GBq and the particles have an energy of 687 keV. The chamber is sealed, evacuated, and filled with xenon. This establishes an energy of 23 eV per ion-pair generated. During calibration of the device it is found that the highest fabric density measurable is 800 g/m^2 and results in zero ionization current. The relation between density and current may be assumed to be linear (approximately). Find the sensitivity of the sensor and its theoretical resolution. Assume 100% efficiency of the source and, in the absence of the fabric, all radiation emitted by the source passes through the chamber.

9.7 Geiger–Muller counters and interpretation of counts. Geiger–Muller counters are popular because they provide immediate feedback, often in the form of audible clicks. Counters also come with adjustable sensitivity. Sensitivity is adjusted by increasing or decreasing the voltage on the anode. However, interpretation of the results is poor and subjective.

a. Suppose the output is a continuous sound, that is, one cannot distinguish individual clicks. What are the possible interpretations of this output?

b. Suppose the sensor emits two clicks per second. What are the possible interpretations of this result?

c. Discuss how the counter may be used successfully without confusion in interpretation.

Semiconductor radiation sensors

9.8 Absorption of energy in radiation sensors. One of the main problems associated with radiation sensors is the amount of energy absorbed by the sensor itself. Consider four hypothetical radiation sensors for X-ray radiation as follows:

1. Bulk resistance silicon sensor, 1 mm thick with negligible electrodes thickness.
2. Same as in (a) with gold electrodes 10 μm thick.

3. Cadmium telluride sensor, 1 mm thick with negligible electrodes thickness.
4. Same as in (c) but with 10 mm thick gold electrodes.

The X-ray radiation is at 100 keV perpendicular to one of the electrodes. The linear attenuation coefficients for the materials involved, at 100 keV, are as follows:

Material	Silicon	Cadmium telluride	Gold
Linear attenuation coefficient	0.4275/cm	10.36/cm	99.55/cm

a. Calculate the fraction of the incoming X-ray energy absorbed by the four sensors, that is, the energy that will generate carriers.

b. Calculate the number of charges generated per photon for the four sensors.

9.9 Alpha particle radiation sensor. A silicon diode radiation sensor is used to check the radioactive source in a smoke detector. According to the label, the source is given as 1 μCi. The source is Am-241 with an energy of 5.486×10^6 eV and an activity of 12.95×10^{10} Bq. The energy needed to generate an electron–hole pair is 3.61 eV. The diode is connected to an inverse potential of 12 V, has a thickness of 0.5 mm, and the mobilities are 450 $cm^2/V{\cdot}s$ for holes and 1350 $cm^2/V{\cdot}s$ for electrons. The radiation from Am-241 is mostly α particles that penetrate very little into the silicon and hence may be assumed to be absorbed at the surface of the diode. For this reason the junction is exposed directly so radiation does not pass through the electrodes. What is the current through the diode if all particles emitted are trapped by the diode?

9.10 Detection of γ radiation. A bulk resistance germanium sensor is used to detect γ radiation at a frequency of 10^{20} Hz. The sensor is 6 mm thick, has a diameter of 12 mm, and the electrodes are 50 μm thick. The sensor is connected as in **Figure 9.29**. Calculate the output pulse for single interaction (photon). The linear attenuation coefficients for germanium and gold at the energy levels of the γ rays given here are 8.873/cm and 42.1/cm, respectively. The mobilities of protons and electrons in germanium are 1200 $cm^2/V{\cdot}s$ for holes and 3800 $cm^2/V{\cdot}s$ for electrons. The intrinsic carrier concentration is $2.4 \times 10^{13}/cm^3$.

a. Calculate the pulse expected from the circuit output (amplitude and sign) as measured. Assume radiation enters through the anode.

b. Calculate the current through the diode in the absence of radiation.

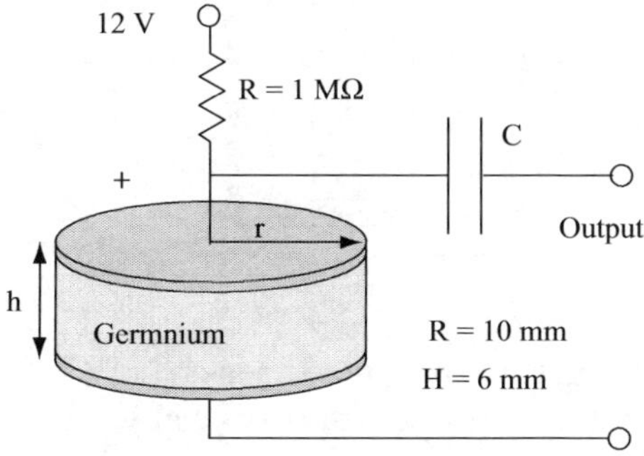

FIGURE 9.29 ■ Bulk resistivity sensor for γ radiation.

9.11 Detection of high-energy cosmic rays. Cosmic rays are rather difficult to detect with semiconductor sensors for two main reasons. First, cosmic rays can have energies on the order of 10^{18} eV or higher. Second, they are not a continuous flux, but rather individual events. But even at the lower end of their energy spectrum, cosmic rays are likely to penetrate through a sensor without generating carriers or generating too few for detection. In fact, detection is almost always indirect by detecting muons generated by cosmic rays as they collide with air or other matter. (Muons are high-energy charged particles with the charge of the electron but about 200 times heavier, and only exist for a period of a few microseconds.) Most muons have energies of about 4 GeV at the surface of the earth and about 6 GeV in the upper atmosphere, but some can have energies in excess of 100 GeV. Suppose one were to attempt to detect high-energy muons in the 100 GeV range using semiconductor sensors. Since semiconductor sensors would absorb little or no energy, it is essential to absorb the excess energy before the muons reach the sensor. This is done by placing the sensors deep underground, underwater, or by using thick layers of high-density materials in front of the sensor. Assuming that a germanium semiconducting sensor operates best below 10 MeV and the incoming muons have energy of 4 GeV:

a. Calculate the depth at which the sensor should be placed underwater if the stopping power of water is 7.3 MeV·cm^2/g. Water has density of 1 g/cm^3.

b. Calculate the thickness of a lead shield with a stopping power of 3.55 MeV·cm^2/g and density of 11.34 g/cm^3.

9.12 Silicon radiation detector. A silicon diode is used to detect radiation with an energy of 2.8 MeV. To do so the diode is reverse biased with an 18 V source and a window is provided so that radiation can only enter the device at its center (see **Figure 9.30**). Assume that the energy is absorbed entirely at the point of entry. The diode is 2 mm thick.

a. With mobilities of 350 cm^2/V·s for holes and 1350 cm^2/V·s for electrons, calculate the current through the diode. Neglect the effects of electrodes and of the n and p layers.

b. What is the sensitivity of the diode?

c. Sketch the current pulse expected if the radiation can be considered as a single event occurring at a time t_0.

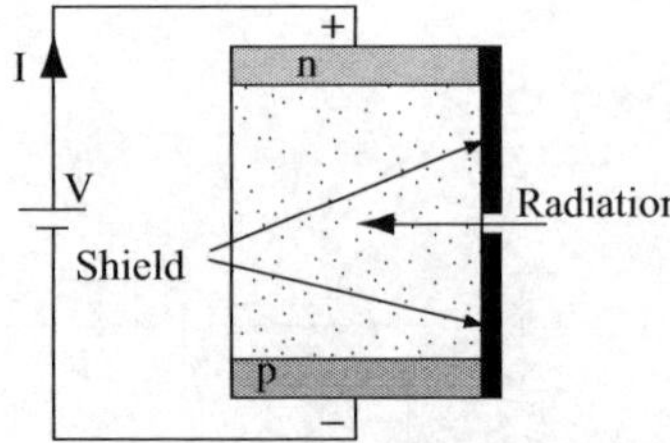

FIGURE 9.30 ■ Silicon radiation sensor.

9.13 UV sensing with a germanium diode sensor. UV is low-energy radiation that occurs naturally but is also generated artificially for a variety of industrial and medical purposes. In an industrial curing process, a source produces UV at a power density of 250 mW/cm^2. Curing is continuous and lasts a few seconds. A small silicon diode in reverse mode with an exposure area of 10 mm^2 is used to sense the radiation level for purposes of controlling its intensity. Because the diode is small, neglect the transition time.

a. Calculate the current in the silicon diode.

b. Calculate the sensitivity of the sensor.

c. Is this a practical way of sensing UV radiation? If not, how can this be done better?

Microwave radiation sensors

9.14 The red shift and speed of expansion of the universe. One of the more interesting applications of indirect sensing is that of the speed of expansion of the universe, often called the "red shift." The idea is based on the measurement of spectral emissions of elements (typically hydrogen, which emits electromagnetic radiation at regular, well-defined wavelengths). When light from distant sources passes through hydrogen, the spectral lines shift either to longer wavelengths (hydrogen moving away from the observer) or shorter wavelengths (hydrogen moving toward the observer). When the wavelengths increase, the phenomenon is called "red shift," since the shift is toward longer wavelengths (red). A contracting universe would cause a "blue shift," or a shift toward shorter wavelengths. One of the more common measures for red shift is the shift of the spectral wavelength for hydrogen, specifically the spectral line at 486.1 nm (blue spectral line). An observation on radiation from a galaxy shows that the spectral line has shifted to 537.5 nm. Assuming the shift is due to the classic Doppler effect, calculate the recession speed of the galaxy.

Notes:

1. At speeds close to the speed of light the classic Doppler effect is not a good measure of the shift of frequency due to speed. The relativistic Doppler effect is then used.
2. Galaxies further away from us move at increasing speeds, indicating an expanding universe.
3. Hydrogen is often used for red shift measurements because of its abundance in the universe. The most common spectral lines used are the violet (436 nm), the blue (486.1 nm), and the red (656.3 nm) lines.

9.15 Doppler radar. The ability to sense and measure speed remotely has found applications in diverse areas from speed control to flight control to weather prediction. The use of frequency shift information due to the motion of clouds allows the prediction of storm evolution in real time and provides a warning of severe weather phenomena such as tornadoes, thunderstorms, and others. Consider a storm front approaching a Doppler radar at a velocity v the radar, operating at 10 GHz, scans 360° and has a range of 100 km. Beyond this range the radar cannot detect. At the time of the scan, the nearest point of the storm front is 40 km from the radar antenna. Calculate the frequency output of the radar and the frequency shift as a

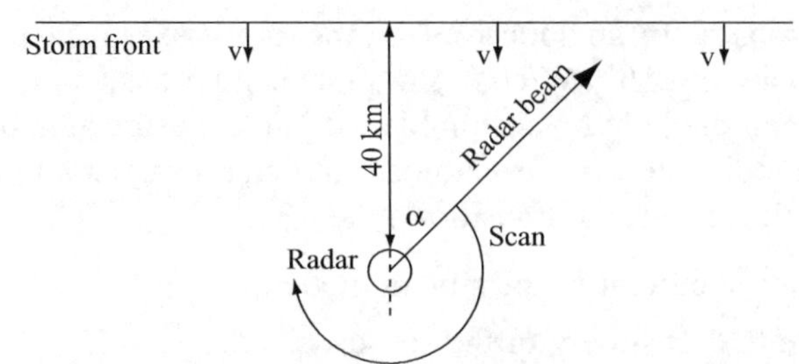

FIGURE 9.31 ■ A radar sensing an approaching storm front.

function of the scan angle, taking the scan angle with respect to the direction perpendicular to the storm front. Plot the output. Assume the storm front is very wide. **Figure 9.31** shows the situation schematically.

9.16 **Sensing of moisture during the production of plywood.** One of the parameters that must be controlled during the production of plywood is the moisture content of the product. To do so, the reflection and transmission of an electromagnetic wave is monitored during the drying process, and drying is stopped when moisture falls below 12%. Moisture content is defined as the ratio of the mass of water to the mass of dry product. The relative permittivity of the dry plywood (0% moisture content) at the frequency used for testing is 2.8, whereas that of water is 56. The density of plywood is 600 kg/m^3 and that of water is 1000 kg/m^3. Assume the total volume of the plywood increases by the volume of water absorbed. Given an incident electromagnetic wave with an electric field intensity of amplitude E_0 at the surface of the plywood, propagating perpendicular to the surface, assuming uniform moisture distribution throughout the fibers of wood and neglecting attenuation:

a. Calculate the amplitudes of the reflected and transmitted electric field intensity signals as a function of moisture content.

b. Calculate the amplitudes of the reflected and transmitted electric field intensity signals at 12% moisture content.

c. Explain which of the signals is a better representation of the moisture content.

Note: Attenuation within the plywood will affect the field, as will the existence of multiple reflections, but these are neglected here to simplify calculations and get some idea of how reflection and transmission may be used for sensing.

9.17 **Laser speed detection.** A laser beam can be used in a manner similar to radar to sense distance and speed. The optical equivalent to radar is called LIDAR (**li**ght **d**etection **a**nd **r**anging). The system works by calculating the time of flight of the laser beam to the object and back. This gives a direct indication of the distance of the object. By taking two samples, the time of flight difference between the two samples is an indication of the distance the target has moved between the two samples. An IR laser speed detector sends a series of pulses at intervals of 7.5 ms. Taking one particular pulse, it is received with a delay of 2.1 μs (i.e., the pulse is received 2.1 μs after the pulse was sent). The next transmitted pulse is received with a delay of 2.101 μs.

a. Calculate the distance from the laser source and target.

b. Calculate the speed of the target.

Note: A typical measurement uses 40–80 pulses, averaging the speeds calculated to ensure accuracy. The whole sequence takes about 300 ms.

9.18 **Radar speed detection.** The common radar speed detector used for traffic control is a Doppler radar measuring the shift in frequency due to an approaching vehicle. The typical frequencies used are around 10 GHz (X-band), 20 GHz (K-band), and 30 GHz (Ka-band).

a. Suppose a Ka-band radar speed detector is used to measure the speed of a vehicle by placing the radar directly in the path of the vehicle. Calculate the shift in frequency measured by the radar gun at a vehicle approaching at 130 km/h.

b. In practice, the radar gun cannot be placed in the path of the vehicle. Calculate the error in the speed reading if the radar gun is placed at an angle of 15° with respect to the direction of the vehicle.

Reflection and transmission sensors

9.19 **Microwave moisture sensor.** In the production of dough for baking goods, such as cookies, it is very important that the moisture in the product be tightly controlled to ensure consistent quality. Consider the sensor in **Figure 9.32** intended to measure the water content in a continuous sheet of dough before it enters the baking process. The permittivity of the dry products in dough is low and we will assume here it has a relative permittivity of 2.2. The relative permittivity of water is 24 at the frequency at which it is measured. Both have the permeability of free space. The ideal moisture content is 28% by volume. To measure the moisture content, an electromagnetic wave of amplitude E_0 is emitted by antenna A and received by antenna B. Assume air has a relative permittivity of 1 and no attenuation. Neglect any attenuation in the dough.

a. Calculate the amplitude of the electric field intensity received by antenna B, neglecting any internal reflections in the dough itself (**Figure 9.32a**).

b. Calculate the amplitude received if one takes into account one internal reflection in the dough (**Figure 9.32b**).

c. Calculate the sensitivity of the sensor (i.e., the change in amplitude for a given [small] change in moisture content, say, 1%) using the basic result in (a). Is the sensitivity constant, that is, do you expect the transfer function to be linear?

d. Under the conditions in this problem, does the dough thickness have any influence on the amplitude or the sensitivity?

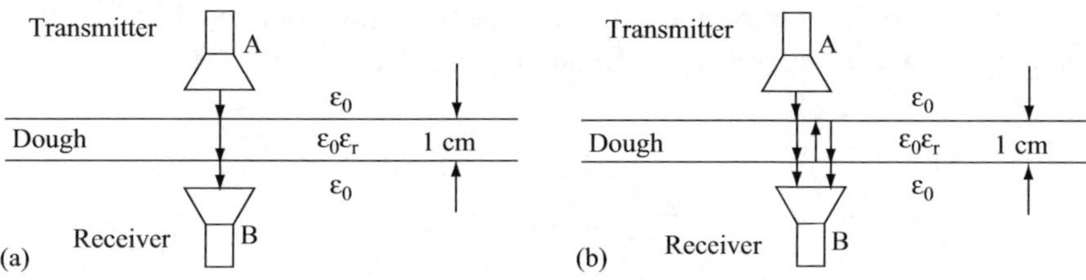

FIGURE 9.32 ■ Sensing of the water content in dough. (a) Taking into account single transmission through the dough. (b) Taking into account internal reflections in the dough.

9.20 **Reflection sensing of grain permittivity.** Bulk grain has a permittivity that depends on the type of grain, its moisture, and its density. A microwave reflection sensor can be used to sense the type of grain and its moisture content. Consider the sensor in **Figure 9.33**. The transmitting antenna transmits a perpendicularly polarized wave toward a layer of wheat grain moving on a conveyor belt during the drying process. The belt is a tight, thin mesh to allow hot air to move between the grains. The amplitude of the electric field intensity at the transmitter is given as E_0 and the amplitude of the received electric field intensity is measured as a means of determining the moisture content of the grain. Wheat in bulk has a relative permittivity of 3.1 at 5% moisture content and 6.4 at 25% moisture content, changing linearly between these two values. For storage purposes the grain should have a moisture content of about 8%.

a. Calculate the measured electric field intensity at the receiving antenna for the range of moisture contents given. Neglect any effect due to the belt and neglect losses in the grain or air.

b. Plot the output as a function of moisture content and plot the sensitivity of the sensor. Normalize the amplitude to 1 (i.e., plot the received electric field intensity divided by E_0).

c. Calculate the output at 8% moisture content.

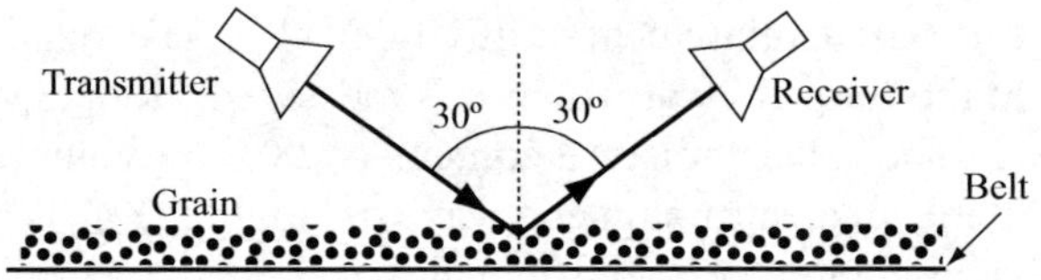

FIGURE 9.33 Reflection sensing of the moisture content in grain.

Resonant microwave sensors

9.21 **Sensitivity of a resonant humidity sensor.** Calculate the sensitivity of the sensor in **Example 9.8**.

9.22 **Thickness gauging using microwave resonant sensors.** The resonant frequency of microwave cavities is directly proportional to the amount of dielectric material present in the cavity. This allows for simple gauging of sheet products such as rubber, paper, plastics, and fabrics. Consider a cavity that has been "opened" to allow a sheet of rubber to pass between the two halves of the metal cavity, as shown in **Figure 9.34** (the height of the cavity is $2a + t_{max}$, where t_{max} is the opening size through which the rubber sheet passes and is the maximum thickness of rubber expected). The connections to couple energy into the cavity and to measure the resonant frequency are not shown. The lowest resonant

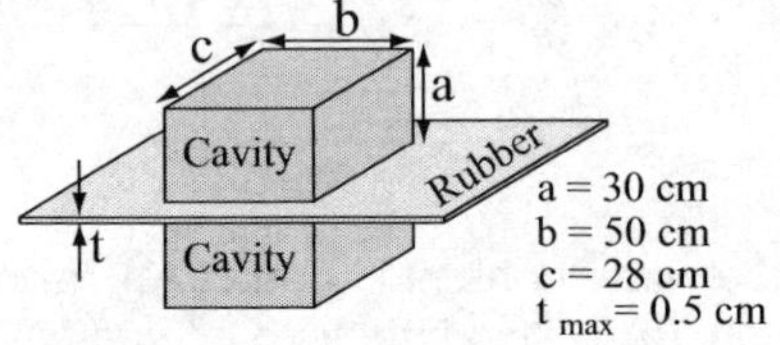

FIGURE 9.34 Rubber thickness gauging in an open cavity resonator.

frequency is used for measurements. Assume that the resonant frequency acts as if the interior of the cavity is a mixture of rubber and air (by volume):

a. Calculate the resonant frequency as a function of rubber thickness. The relative permittivity of rubber is 4.5 and that of air is 1 (dry air).

b. Calculate the sensitivity of the sensor to thickness variations.

c. If the frequency of the sensor can be reliably measured down to 250 Hz, what is the resolution of the sensor for thickness measurements at a given rubber thickness of 2.5 mm?

d. What is the error in the resonant frequency when measuring a 2.5 mm thick rubber sheet if the air inside the cavity is at 100% relative humidity (the relative permittivity of air at 100% humidity is 1.00213)?

Note: Although the cavity is open to allow passage of the rubber sheet, as long as the opening is small with respect to the wavelength, the cavity will resonate as if it were a closed cavity.

9.23 **Resonant fluid level sensor.** A high-resolution microwave cavity resonator level sensor is shown in **Figure 9.35**. The cavity resonates in the lowest mode. The dimensions of the cavity are $a = c = 25$ mm, $b = 60$ mm. The connections to couple energy into the cavity and to measure the resonant frequency are not shown. Assume the fluid is conducting (i.e., seawater).

a. Calculate the range of frequencies of the resonator as water level increases from $h = 0$ to $h = b/2$.

b. Calculate the sensitivity of the sensor.

c. Assume the fluid is oil with a relative permittivity of 3.5. What is the range of frequencies for the same level and the sensitivity of the sensor?

Antennas as sensors

9.24 **Lightning detection and location.** Detection of lightning and sensing its intensity is an important activity in weather prediction. National and global networks of lightning sensors fulfill this need. The electromagnetic radiation emitted during the lightning strike propagates at the speed of light and is detected by relatively simple antennas. In ground-based systems, the coordinates of a lightning strike are found by trilateration. The simplest algorithms rely on the coordinates of fixed antennas and record of the time a lightning signal is

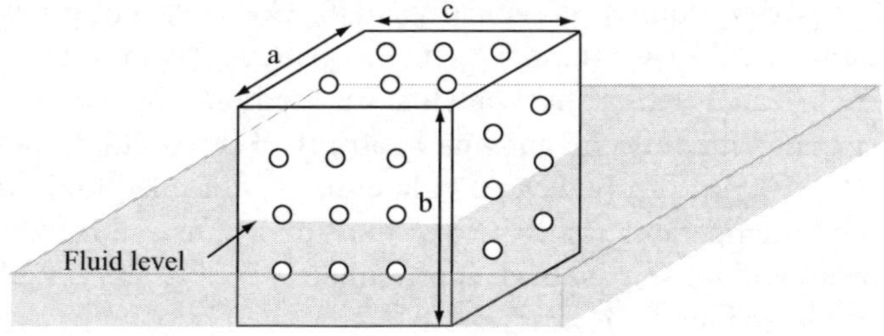

FIGURE 9.35 ■ Microwave water level sensor.

detected. Consider a system that uses the signals from three receiving antennas as follows: antenna 1 at coordinates (x_1,y_1), antenna 2 at coordinates (x_2,y_2), and antenna 3 at coordinates (x_3,y_3). The electromagnetic signal from a lightning strike is received by antenna 1 at time t_1, by antenna 2 at time t_2, and by antenna 3 at time t_3. Assume that another receiving antenna is located at a reference point (0,0,0) and the signal arrives at the reference antenna at time $t_0 = 0$.

a. Calculate the coordinates of the lightning strike from these data.

b. Show that a fourth antenna can provide the location in three dimensions (i.e., can provide the elevation of the lightning strike).

9.25 Detection of electromagnetic radiation sources using directional antennas. The location of source of electromagnetic radiation, such as an emergency transmitter of a downed airplane, a lost hiker equipped with a transmitter, a ship at sea, or an animal equipped with a transmitting collar, can be done relatively easily using two fixed receivers, provided that the antennas are directive (i.e., the maximum reception is in a distinct direction in space). Typical antennas of this type are the multielement Yagi antenna and the common dish antenna (among others). In addition, the location of the two antennas must be known precisely.

a. Show how the coordinates of the radiating source can be obtained assuming the two receiving antennas and the source are on a plane, and the directions can be determined with respect to the line connecting the two receivers. The locations of the receiving antennas are known from GPS readings.

b. Calculate the GPS coordinates of a source if receiver A, located at GPS position (N22°38′25″, E54°22′0″), determines the reception to be at $\alpha = 68°$ to the line connecting it to receiver B. Receiver B is at GPS position (N21°32′20″, E55°44′12″) and determines the angle with respect to the line connecting A and B as $\beta = 75°$. One minute of latitude or longitude is 1.8 km in the region where these measurements are made. Is the solution unique? Explain.

Antennas as actuators

9.26 Microwave hyperthermia cancer treatment. A useful application of small antennas is in the local heating of tumors to shrink or destroy them. An antenna is inserted into or very near the tumor and the area is heated to about 42°C (sometimes higher). Since tumors have fewer blood vessels they cannot dissipate heat and hence they tend to heat to temperatures above that of the surrounding tissue. Typical malignant tissue will absorb up to four times more heat than healthy tissue. This damages the tumor cells with very little damage to healthy tissue. The power required is typically small. Consider the following (hypothetical) example. A breast tumor 1 cm in diameter (it may be assumed to be spherical) is heated with a short antenna inserted into the tumor. Assuming that all power transmitted by the antenna is absorbed by the tumor and the antenna radiates $P = 100$ mW of power, calculate the time needed to destroy the tumor. Normal body temperature is 38°C and the tumor must be heated to 43°C. The heat capacity of body tissue is approximately $C = 3500$ J/kg/°K and tissue density is approximately $\rho = 1.1$ g/cm^3.

9.27 **Microwave cooking.** One of the most ubiquitous examples of a radiation actuator is the microwave oven. Although the physical process of microwave heating is not very complex, it often suffices to discuss its heating effects in terms of the absorption of energy by the substance being heated. An average size microwave oven in the home is rated at 800 W. It actually consumes almost twice that amount of power, but that is the amount of power that it can couple into food. Its absorption efficiency is almost 100% (whereas the overall efficiency, that is, the ratio of power used to cook to the input power, is between 50% and 60%). Consider the heating of a 200 g cup of water at 20°C to boiling (100°C). Water has a specific heat of $C = 4185$ J/kg/°K and a heat of vaporization of $C_v = 2.27$ MJ/kg. The microwave oven has a conversion efficiency of 97%.

a. Calculate the time needed to boil the water.
b. Calculate the time from turning on the oven until all the water has been vaporized.

CHAPTER 10

MEMS and Smart Sensors

Robotics and Mechatronics

Robotics draws on electrical, mechanical, and computer engineering to design, construct, and operate robots for a variety of purposes. Although the term is somewhat vague in that it can include trivially simple or devilishly complicated autonomous systems, the term is generally understood to mean automation and control in response to inputs. In the popular domain, robots are often thought of as systems that embody artificial intelligence, but in practice these are devices and systems that accomplish a task or a series of tasks based on a preset program and, most importantly, in response to inputs. Some robots, of course, are just automata—they perform a task or a series of tasks indefinitely. To this class belongs any number of industrial robots. Others are more sophisticated in that they include a variety of sensors and actuators to interact with their environment in a more "intelligent" way. The sensors in a robot allow it to "understand" its environment, and the processors allow it to process that data and act on it as necessary. Thus a robotic arm may be able to pick up an egg without crushing it or to lift a car engine and place it in a car during production, provided that it has the proper sensors and actuators and these have been programmed to handle the task. Surprisingly perhaps, robotics and the idea of robots is not very new. Automata have been designed and built at least as far back as the first century CE (Heron of Alexandria lived between 10 and 70 CE, and in his *Pneumatica and Automata* described dozens of automatons with self-regulating mechanisms ranging from water clocks to special effects for the theater). Even the concept of humanoid automata goes back to at least the thirteenth century. Leonardo de Vinci's mechanical knight is well known among these. In art, literature, and legend, robots are even more common. The Golem of Prague was built out of clay and breathed with life through prayer by the Rabi of Prague to protect his people only to find out that it had a will of its own. In *Coppelia*, a classic ballet, Dr. Coppelius, a kind of "mad inventor," builds a beautiful windup life-size doll that can dance and charm everybody until, of course, she comes to a screeching halt when the spring unwinds. Even Pinocchio, who started as a wooden doll, progresses to a "robotic" stage before he gets a soul. The term robotics itself originated in a theater play (Rossum's *Universal Robots*, 1920) before it appears in science fiction in the 1940s and before its use in the context we know. Mechatronics is a concatenation of mechanical and electronics and is an approach aimed at the integration of mechanics, electronics, control, and computer science/engineering in the design of products to improve and optimize their functionality. As such, it is not limited to robotics, although, mostly through fiction and movies, it sometimes takes the connotation of science fiction or, more often, that of the integration of mechanical systems with living things.

10.1 | INTRODUCTION

In the previous chapters we talked about many sensors and actuators of various types. The discussion focused on the principles of operation and on some of the applications of these sensors and actuators.

In this chapter we look at some additional aspects of sensors and actuators, aspects that could not have been discussed in conjunction with the principles of conventional devices. First, we discuss a class of devices called microelectromechanical systems (MEMS). The term MEMS relates more to the method of production of sensors and actuators, whereas the sensors and actuators themselves are some of the devices discussed previously as well as others. We discuss them here because they are unique not only in the methods used to produce them, but at least some of them can only be produced as MEMS. One can imagine an electrostatic actuator, at least in principle. But only as an MEMS device does it become a useful, practical device. Then there is the issue of scale of fabrication. Using techniques borrowed from semiconductor production, enhanced by micromachining techniques, it became possible to mass produce sensors such as accelerometers and pressure sensors, and actuators such as microvalves and pumps. Many of these devices have been developed for the automotive industry, but they have found their way into others areas, including medicine.

Following the discussion on MEMS we tackle the issue of smart sensors, which again includes many methods and many types of sensors, but in general, smart sensors imply that additional electronics have been incorporated with the sensors. This may mean, for example, that a processor, an amplifier, or some other type of circuitry has been incorporated with the sensor. The link with MEMS is in the production methods—the electronics needed to make the sensors smart are produced in silicon using some of the same semiconductor fabrication techniques. Smart sensors (or actuators) are not always very "smart" in the sense of what they can do, but they are a step above regular active or passive sensors. Sometimes a smart sensor is a true necessity and evolved from the need to solve a problem. For example, a sensor may need to be physically close to the processing circuit. It is then not a far stretch to bring the processor to the sensor and package them together. In other cases, the production method, especially with silicon-based sensors, happens to be the same and therefore it is only natural that the two should be combined to produce a device with enhanced properties and overall lower cost. The levels of "smartness" vary and may be trivial or very sophisticated. In the limit, the sensor or actuator can become a whole system requiring little else to operate and may include wireless transmitters, receivers or transceivers, microprocessors, power supplies and power management circuitry, etc.

A third topic to be discussed is the issue of wireless sensing. Although a sensor or actuator cannot be "wireless" by itself, the term wireless sensors and actuators is commonly used for devices that communicate with the outside world through a wireless link. This approach is becoming more common and hence we will discuss not only issues associated with sensors, but also wireless issues including frequencies, methods of modulation, and issues of antennas and coverage, including range. The final topic in this chapter is sensor arrays. We have already discussed sensor arrays, especially optical arrays, but also arrays of thermocouples and magnetoresistors. The meaning here is different. Whereas an optical sensor array is a sensor made of a number of individual sensors put together to perform a function (usually imaging), sensor arrays here will mean an array of sensors, spatially separated, where each sensor performs its function and sends the data to a central location for processing. Thus the main distinction is in the spatial distribution of the sensors or actuators.

10.2 | PRODUCTION OF MEMS

Microelectromechanical sensors and actuators form a class of devices that use two distinct properties. First, the devices are produced using micromachining methods borrowed from and expanding on semiconductor production methods. Second, they contain mechanical members such as flexural beams, diaphragms, machined channels, or chambers, or, indeed, truly moving parts such as wheels and cogs. In a wider view, any sensor/actuator that is micromachined may be termed a MEMS device. This means that it is structured or sculpted out of a base material such as silicon by various means but does not necessarily include moving elements.

Here we will take a much narrower view, primarily for the purpose of narrowing down the subject and to avoid overlap with previously discussed material such as that in **Chapters 6** and **7**, and to a lesser extent in **Chapter 5**, where various sensors such as semiconductor-based pressure sensors as well as accelerometers and others that are commonly micromachined were discussed. To do so we will concentrate here strictly on those sensors and actuators that employ moving members in the true sense, including micromotors, micropositioners, valves, and the like (all are actuators), and devices that employ beams, diaphragms, vibrating elements, and others for sensing purposes. However, before we do so it is well worth discussing some topics in micromachining and semiconductor processing since these methods are at the heart of all MEMS.

The production of MEMS is based on a series of techniques, including deposition, patterning, oxidation, etching, doping, and others, all familiar from the production of integrated circuits in silicon and other semiconducting materials. A few of the methods used for the construction of MEMS are defined next, together with their functions:

Oxidation. A layer that can vary in thickness up to a few micrometers is created on the surface of semiconductors at high temperatures in preparation for processing and to create insulating layers as necessary. The process may be applied many times during a process.

Patterning. During various stages of production, various patterns need to be defined (conducting pads, areas of doping, shapes of substructures such as cantilever beams, rotors, strain gauges, temperature sensors, diodes, transistors, and others). These are made using lithographic techniques in which a photoresist is placed on the silicon and exposed through a proper mask using ultraviolet (UV) sources. Following development, the pattern is created and then hardened by baking the remaining photoresist. This is followed by etching to create a pattern of (in this case) the oxide layer. **Figure 10.1** shows the basic steps involved in patterning.

Etching. Following patterning, sections may be removed using various types of etchants. For example, the pressure chambers in pressure sensors or the beam and mass in accelerometers are produced using this process. Various methods of etching and various etchants are used for specific purposes. **Figure 10.2** shows some of the "wet" etching methods often employed. The etching can be uniform, as in **Figure 10.2a**, or may be done along preferential directions (usually crystal boundaries that exhibit various levels of

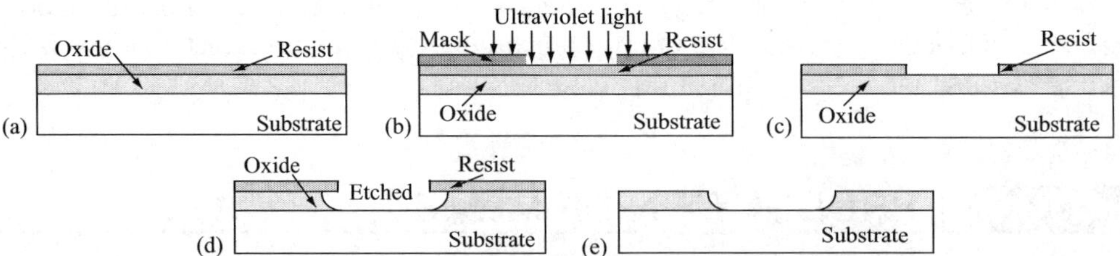

FIGURE 10.1 ▪ Basic steps in patterning. (a) Growing an oxide layer and coating it with a photoresist. (b) A mask is used to create the pattern for etching. (c) The resist is "fixed." (d) Etchant is used to etch away the oxide layer in selected areas. (e) The final result after cleaning away the resist.

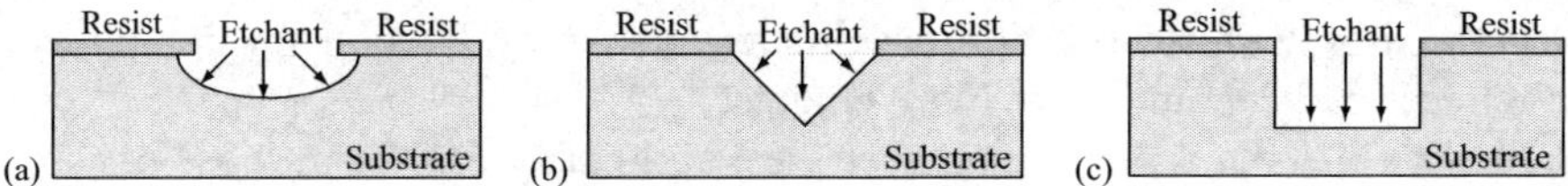

FIGURE 10.2 ▪ Methods of wet etching. (a) Uniform or isotropic etching. (b) Etching along crystal boundaries taking advantage of the varying rates of etching along these boundaries. (c) Anisotropic etching, stopped at vertical walls by etch-stop layers.

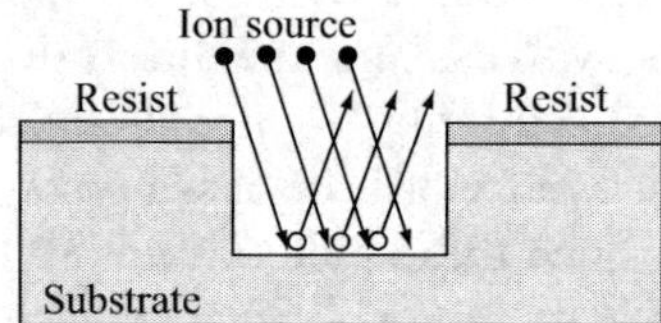

FIGURE 10.3 ▪ Methods of plasma (dry) etching.

resistance to etching), as in **Figure 10.2b**. It can also use various means of stopping the etching to achieve specific results, such as that shown in **Figure 10.2c**. Other methods, including plasma etching (dry etching), shown in **Figure 10.3**, are used for specific applications. In this method, rather than just a chemical process, ions are used to bombard the exposed areas and remove the required material. Here again there are many variations in use that vary from purely physical to entirely chemical or any combination of the two.

Doping. The production of various types of silicon (*n*, *p*, and intrinsic types at various levels of doping and hence conductivities) are also generated through the patterning masks as necessary. Again, there are various methods, such as diffusion of dopants in the atmosphere around the wafer or ion implantation. Common materials are gases of phosphorus (*n*-type) or boron (*p*-type). Another method is ion implantation using *p*-type elements such as boron and *n*-type elements such as arsenic. Following ion implantation, the material must be annealed to allow its atomic structure to relax into its final position. Doping is also used as a mechanism for controlling the etching process, since heavily doped silicon etches much slower (or not at all, depending on the etchant and level of doping) and hence layers of doped silicon are often used as etch stops.

Deposition. In the production process it is often necessary to deposit various layers of materials. Thin films of polycrystalline silicon and other materials, including metals, are deposited in various thicknesses and in various layers as necessary. There are many methods of deposition, but the most common methods are based on chemical vapor deposition whereby the wafer is placed in an atmosphere containing the vapor of the material to be deposited (usually at elevated temperatures). Metal deposition includes aluminum, gold, nickel, and others.

Bonding. A range of bonding techniques may be employed at various stages of the production process. In some cases bonding simply means bonding the silicon wafer to a substrate or package, but in others bonding is used to seal chambers (such as in gauge pressure sensors or absolute pressure sensors). Bonding can be done by using bonding agents, fusing silicon to silicon, bonding to glass, and many others, at both low and high temperatures.

Testing and packaging. Once the device is complete, usually made on a wafer that may contain hundreds or thousands of individual devices, it is tested, cut, bonded to a substrate in the package; electrical connections are made, and the device is incorporated in its package, which may be a sealed package or may have openings (such as in pressure sensors or microchannel fluid flow sensors).

The processes described above are common integrated circuit techniques and form the basis of semiconductor production. However, for MEMS production, additional techniques are needed. These are termed micromachining techniques. Many of these are methods that allow construction or sculpting of structural and/or moving members of the MEMS device. Some of the most common micromachining methods are as follows:

Bulk micromachining. The wafer is deep etched, taking advantage of special etchants and etching techniques for different sections and layers as necessary. One approach is to use etchants whose rate of etching depends on the orientation of the silicon crystal. Also, various methods of stopping the etching at the desired depth are used to control the structure. These methods can vary from simply timing the etching process to inserting specific materials, such as doped layers, to stop the process at the required depth. This method allows the creation of deep chambers as well as structural members such as diaphragms, beams, etc. Deep etching of a membrane (for use in a pressure sensor) is shown in **Figure 10.4** (see also **Figures 6.28** and **6.29**). The figure also indicates the different rates of etching on different crystal cuts. The slanted etch boundary is a result of the slower rate of etching on these crystal boundaries. A more complex example of this process is shown in **Section 6.4.1** (see **Figure 6.17**), where we discussed capacitive accelerometers.

Surface micromachining. Here the process takes place on the surface of the wafer. The process can also be called a layering or sculpting process. Often the structure of interest is built up layer by layer and in each step one or more of the techniques described above may be used. An important part of this technique is the use of sacrificial layers—layers of silicon dioxide (SiO_2) and the use of polycrystalline silicon as the structural material. Of course, other materials may be used and, in addition, metals such as aluminum and gold are used for conduction and contacts, and ferromagnetic materials including iron alloys and nickel may be used when necessary (such as in the production of microcoils). This method of production can produce free-standing elements such as

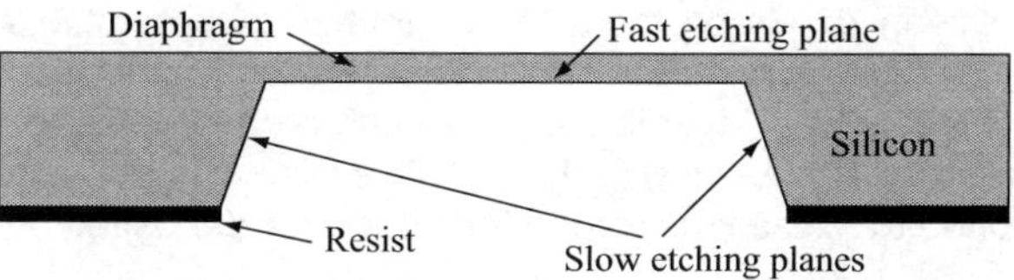

FIGURE 10.4 ■ Deep etching of a membrane from a silicon wafer.

cantilever beams in accelerometers or even moving parts of considerable complexity. **Figure 10.5** shows a micromotor. It consists of a free rotor held in place by a bearing and stator through which the motor is driven. In this case the rotor has three poles and the stator has nine. The rotation of the motor is produced by applying potentials to the stator poles to the left of the rotor pole to turn it counterclockwise or to the right for clockwise rotation. The stator voltage induces opposite charges on the nearby rotor pole and the attraction between the stator and rotor moves the rotor one position (one-ninth of a turn). A motor of this sort may have any number of poles and may be 100–500 μm in diameter. Obviously it cannot produce high power or torque, but on the other hand, it can rotate extremely fast. **Figure 10.6** shows the basic steps involved in producing an electrostatic micromotor as an example. First, a thin SiO_2 sacrificial layer is built on a silicon substrate (**Figure 10.6a**). The layer is patterned and etched to accept the rotor and stator components (**Figure 10.6b**). The space that will form the rotor bushing is etched first. A polycrystalline layer is deposited on top, patterned, and etched to create the rotor and stator (**Figure 10.6c**; the top view is that in **Figure 10.5**). Note the small, triangular extension of the rotor into the sacrificial layer. It forms a bushing that allows it to ride on the substrate with minimum friction and keeps it aligned with the stator once the sacrificial layer has been removed (**Figure 10.6b**). A second, thin SiO_2 layer is deposited on top, patterned, and etched (**Figure 10.6d**) to accept the bearing. On top of

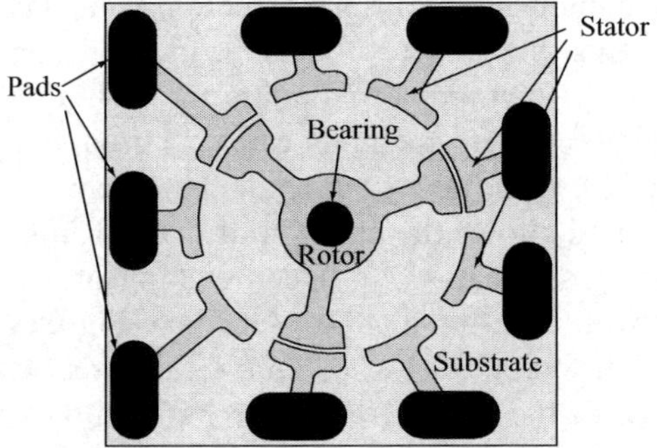

FIGURE 10.5 ■ An electrostatic micromotor.

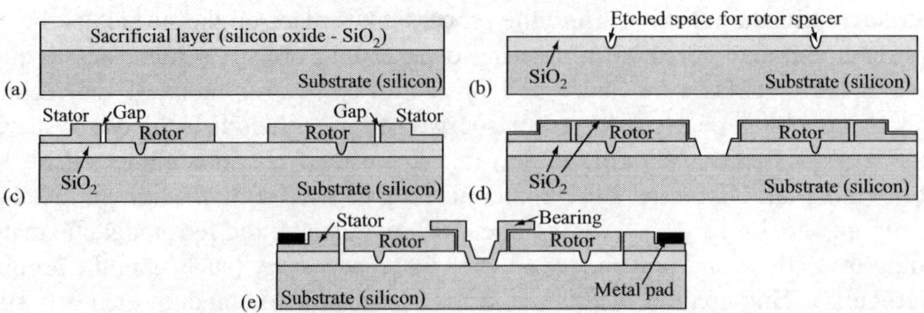

FIGURE 10.6 ■ Surface micromachining of an electrostatic micromotor. (a) A sacrificial layer of SiO_2 is grown on silicon. (b) Grooves are etched that will become the support and spacers for the rotor. (c) A layer of polycrystalline silicon is deposited, patterned, and etched to form the rotor. (d) A second sacrificial layer is deposited and etched down to the substrate at the location of the bearing. (e) The bearing (polycrystalline silicon) is deposited and most of the sacrificial layer is etched to release the rotor. Metal pads are deposited on the stator elements (see **Figure 10.5**).

that a layer of polycrystalline silicon is again deposited, patterned, and etched to form the bearing (**Figure 10.6e**). Now the sacrificial layers are all etched away to free the rotor. Aluminum is deposited on the stator connections (**Figure 10.6e**), and these are connected to pins on a package so that the motor can be driven externally.

The method can be used to produce a variety of sensors and actuators and is particularly well suited for the production of free-standing components such as cantilevered beams in accelerometers, bending mirrors in optical actuators, diaphragms in pressure sensors, and, of course, free-standing components in motors and other moving actuators. As can be imagined, there are many variations and modifications of the basic method described here, as well as other materials and material combinations.

In contrast with surface micromachining, there are also volume methods of sculpting that can produce rather intricate structures. In various forms these methods, called microstereolithographic methods, use a thick UV-curable photoresist. The structure to be built is defined by a focused beam that creates multiple layers of solidified resist to build the structure. The result is typically a three-dimensional (3D) structure with resolutions down to a few micrometers and dimensions that can vary from a few micrometers to hundreds of micrometers.

In addition to fabrication methods borrowed from silicon processing, there are so-called nonsilicon technologies. Some of these are capable of producing very slender, high aspect ratio structures, such as cogs and wheels, at a scale that is typically larger than common semiconductor scales (a few millimeters).

The devices produced using any or all of these methods can be, and often are, integrated with semiconductor circuits to produce smart sensors/actuators. An example may be a pressure sensor produced using bulk micromachining followed (or preceded) by production of semiconductor strain gauges and/or amplifiers or other processing elements.

EXAMPLE 10.1 MEMS production of a magnetic sensor

In general, a magnetic sensor will require at the very least a coil and a magnetic core. Consider the fabrication of a MEMS sensor that has the following characteristics: a core made of a magnetic material (such as permalloy) and a 5 turn coil symmetrically wound around the core, connected to pads for external current supply. The coil is made of aluminum. Other materials must be chosen to accomplish the required functionality of the device. The dimensions are not important, but they must be compatible with MEMS fabrication methods. Show the steps necessary to produce the device.

Solution: The process starts by selecting the materials, from the substrate through the sacrificial layers, and elements of the geometry. The coil is conducting, as is the core. That means that they cannot touch—there must be an insulating layer such as SiO_2 or air between them. The coils as well as the core must be supported, meaning that the support must be nonconductive. Finally, it should be clear that the process can be accomplished in various ways and the final device will depend on how it was made. The following is a possible process:

1. The substrate, on which the whole device is built, may be chosen as silicon for ease of production.
2. Grow a thick layer of SiO_2 on top of the substrate. This will serve as an insulating layer on which the coil will be built (silicon itself is conducting, hence the need for the insulating layer).

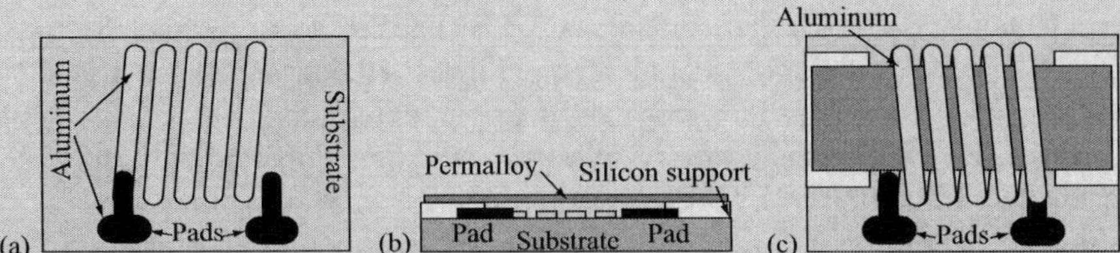

FIGURE 10.7 ■ Steps in the construction of a MEMS coil. (a) Deposition of the lower part of the coil. (b) Deposition of the permalloy core above a sacrificial SiO_2 layer (shown in profile). (c) Deposition of the upper part of the coil above a new SiO_2 sacrificial layer.

3. An aluminum layer is deposited on the SiO_2 layer and etched, as shown in **Figure 10.7a**, to create the lower layer of the coil as well as the connection pads. Note that the pad on the right will connect to the end of the upper layer of the coil.
4. A second layer of SiO_2 is grown on top and etched to create a support for the core and to insulate it from the coil.
5. A layer of magnetic material (permalloy is often used for this purpose) of the desired thickness is deposited on top and etched to the required length and width. The sacrificial SiO_2 underneath is etched away to leave the permalloy suspended above the lower part of the coil. Stop layers may be used to keep the SiO_2 under the ends of the permalloy strip from etching away so that they support it. **Figure 10.7b** shows a side view of the structure (dimensions are exaggerated).
6. A new layer of SiO_2 is deposited to support the upper part of the coil. Holes are etched in this to expose the ends of the bottom layer coil strips as well as the right-hand pad. This will allow the upper strips of the coil to connect to the lower strips.
7. An aluminum layer is deposited on top and through the holes. This is etched as shown in **Figure 10.7c** to form the upper part of the coil. The coil is now complete, surrounds the core, and the intervening material is air.

Notes: There are many issues that have been neglected here. First, for a coil to be rigid enough to stay suspended, the aluminum deposit must be sufficiently thick. An alternative is a coil that can be made of polycrystalline silicon, which is naturally more rigid than aluminum. The insulating layers between the coil layers and the core may be left in place to support the coil, but that may mean the coil can overheat because of the thermal insulation properties of the SiO_2. For these reasons, and others that are less obvious, construction of a device requires considerable expertise and knowledge of the processes involved. Often, especially in the prototyping process, it may be viewed as an art rather than an exact science. Once the process is successful, however, its replication is relatively easy and is based on the experience gained in prototyping. Software tools of considerable sophistication allow design and simulation before prototyping and testing after prototyping, and all that prior to any production run.

10.3 | MEMS SENSORS AND ACTUATORS

As discussed previously, sensors and actuators, in themselves, cannot be said to be MEMS devices. Rather, they are devices built, or indeed enabled, by the use of MEMS techniques to produce them. In that sense one cannot identify a new class of devices. However, there are sensors, and of course actuators, that implement existing sensing technologies in ways that can only be done in the context of microfabrication. For example, a flow sensor can be built by measuring the temperature difference between two temperature sensors, one upstream and the other downstream (see **Chapter 3**). However, if one needs to do so in a microchannel, the only way to produce this is through MEMS methods. Similarly, an inkjet in an inkjet printer is merely a nozzle that sprays a droplet of ink onto a page. Early nozzles were essentially miniaturized nozzles, activated by heat. Others used nozzles drilled by lasers. The extension to MEMS-based inkjet nozzles is a natural way to achieve proper control over the amount of ink projected and at reasonable speeds. In essence, the dimensions and the fabrication techniques give these devices their unique properties and advantages. Then, of course, because MEMS are produced in semiconducting materials and with techniques compatible with semiconductors, the devices so produced can be integrated with electronic devices to produce smart sensors and actuators in ways that are not possible otherwise. It is therefore useful to discuss here some sensors and actuators that have benefited immensely from MEMS techniques even though they are based on one or more of the principles discussed in the previous chapters.

In fact, many of the sensors and actuators we discussed in previous chapters can be, and some are, made using MEMS techniques. Some of these attempts have resulted in improved performance and some have seen commercial success. It should be remembered, however, that while attempts to construct MEMS sensors and actuators of a bewildering variety have been undertaken, not all of these attempts are practical and not all of them will see the light of day as production level devices. Nevertheless, MEMS offers exciting paths to future sensors and actuators and, by their very nature, to the emergence of advanced smart sensors.

10.3.1 MEMS Sensors

10.3.1.1 Pressure Sensors

One of the first sensors to be produced and marketed as a MEMS device was the pressure sensor, fueled mostly by needs in the automotive industry. The methods of producing freestanding beams and mass as well as diaphragms and the inclusion of piezoresistive strain gauges have facilitated the production of pressure sensors, and the miniaturization afforded by the process and the addition of electronics either on the dye or on separate dyes, but still within the package, led to smart sensors of considerable utility and flexibility. The capacitive pressure sensors and the piezoresistive pressure sensors in **Chapter 6** (see, e.g., **Figures 6.18**, **6.28**, and **6.29**) are of this type. Bridge sensors, compensation mechanisms, as well as high-temperature sensors have been implemented as well. One of the main advantages of implementation of these sensors in MEMS is the lower production cost and repeatability of the process.

10.3.1.2 Mass Air Flow Sensors

Mass air flow (MAF) sensors are common devices in the automotive industry, used primarily to sense the air intake of engines for the purpose of controlling combustion. The sensors are based on the fact that heat loss from a heated element is proportional to air mass flow over the element. In its simplest form, a sensor consists of a heated wire in the path of the flow (**Figure 10.8a**). The device in **Figure 10.8a** is known as a hot-wire anemometer and

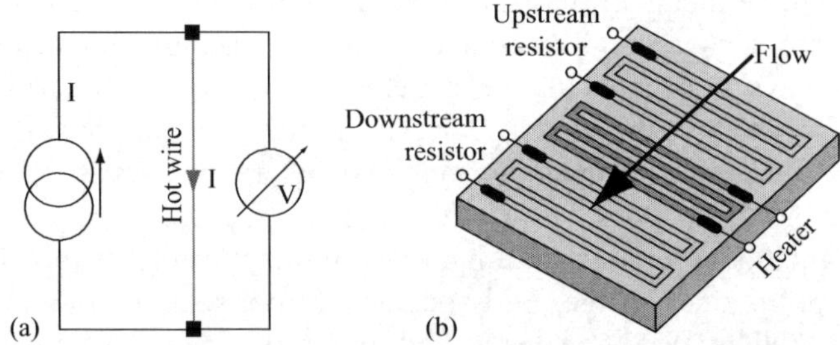

FIGURE 10.8 ■ Mass flow sensor. (a) Basic principle of a heated wire mass flow sensor. (b) A MEMS implementation using the temperature difference between an upstream (cold) and a downstream resistor (hot). Under no flow, both the downstream and upstream sensors will be at the same temperature.

measures airspeed. However, the air mass flow can be equally well measured by proper calibration. By keeping the current (or voltage) constant, the temperature of the wire, and hence its resistance, is fixed. Air flowing over the element reduces its temperature, decreasing its resistance (for most metals). Direct measurement of voltage (for constant current) or, in general, measurement of the resistance of, or power dissipated in the wire gives an output indicating the mass flow. However, direct measurement is not very practical because the resistance is very small (the wire is short and made of platinum or tungsten, both of which have high conductivity). Instead, many hot-wire sensors rely on direct or indirect temperature sensing. One approach is to change the current in the wire so as to keep the temperature constant. The measurement is typically that of the current needed to restore the temperature steady state (for constant voltage), since air mass flow is directly proportional to power. The air mass flow sensor based on the wire anemometer is particularly useful because it is independent of air pressure and air density and has a particularly fast response.

MEMS air mass flow sensors operate on a variation of this principle: A heater element raises the temperature of two resistor elements, one upstream and one downstream of the heater (**Figure 10.8b**). Mass flow over the sensor cools the upstream resistor and heats the downstream resistor. The temperatures of the two resistors can be measured directly. The output is the difference between the two temperatures and is proportional to the mass flow. The advantage of this configuration is that the output is zero at zero flow and common effects, such as the temperature of the housing, have no effect on the output because of the differential nature of the sensor. This is particularly important in motor vehicles, which may operate over a wide temperature range. The resistors must be well insulated so that there is little or no heat loss to the body of the sensor. This is done by constructing the device over a pit, using the gas in the pit as an insulator.

EXAMPLE 10.2 Hot-wire mass flow sensor

A mass flow sensor used in a vehicle needs to be calibrated for voltage outputs between 0 and 5 V. To produce a zero output voltage for zero mass flow, the sensor shown in **Figure 10.8b** is connected in a bridge configuration (discussed in **Chapter 11**). The flow is adjusted using a variable speed fan and measured for specific voltages as shown in the table below using a separate, calibrated flow sensor. Nominally the sensor should produce 5 V at a mass flow of 80 kg/min. The calibration curve is the plot of these values and is shown in **Figure 10.9**.

Mass flow [kg/min]	0	0.4	0.63	1.66	3.31	6.64	9.97	12.4	14.69	17.28
Voltage [V]	0	0.1	0.3	0.8	1.3	1.9	2.3	2.6	2.8	3.0

20.22	21.85	25.46	29.63	34.48	40.15	43.45	50.59	59.11	69.16	81.01
3.2	3.3	3.5	3.7	3.9	4.1	4.2	4.4	4.6	4.8	5.0

Note the nonlinear curve with much higher sensitivity at lower flows. Note also from the table that the mass measured at 5 V is slightly larger than the nominal value (by 1.01 kg/min, or 1.26% of full scale). The calibration curve can now be used to derive the mass flow for the engine. It is perhaps surprising at first to see an air mass of 80 kg/min, since we don't usually think of air as a mass, but it should be remembered that air has a density of 1.294 kg/m^3 (at 0°C and 101.325 kPa). As an example, a 2000 cc engine running at 4000 rpm needs approximately 10 m^3 of air/min or about 13 kg/min. Obviously, larger engines in trucks will require much more than that.

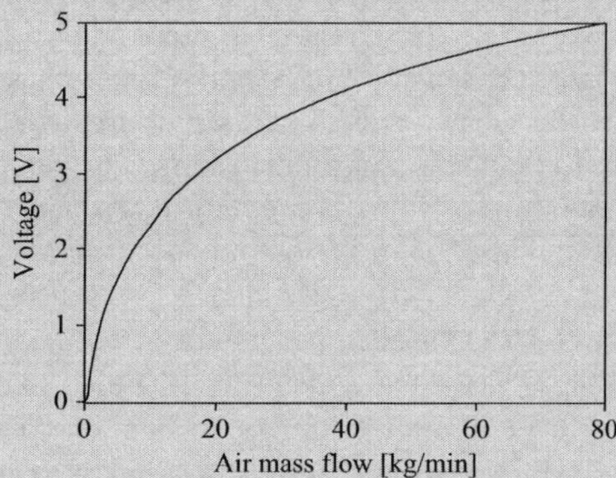

FIGURE 10.9 ■ Transfer function of an air mass flow sensor.

10.3.1.3 Inertial Sensors

Another area where MEMS have been successful is in the development and production of inertial sensors, specifically accelerometers and gyroscopes. These developments were driven by the increased use of handheld and battery-operated devices such as phones and GPS and the needs of autonomous or assisted navigation systems. The automotive industry was the initial driving force and beneficiary of some of these devices, particularly acceleration sensors for use with airbags, antilock braking, and in active suspension systems. Some of the sensors we saw in **Chapter 6** (see, e.g., **Figure 6.29**) were in fact MEMS. The fabrication of cantilever beams, bridges, and the moving mass needed for accelerometers (also called proof mass, inertial mass, or seismic mass) are natural uses for MEMS techniques, as are integrated sensors, be they capacitive, semiconductor, or piezoresistive strain gauges needed to translate the motion (strain) of the beam or mass into a useful reading. Early in the development of MEMS accelerometers, the accelerometer itself (i.e., the mass), diaphragm or beams, and strain gauges were produced separately, whereas the electronics needed were added externally as independent circuits. Later sensors incorporated the electronics either on a separate dye in the same package or integrated on the same substrate as the sensor. MEMS production allows relatively easy production of one-axis or two-axis accelerometers. Three-axis accelerometers can also be made and can be obtained commercially. In the simplest configuration they can be made of three one-axis accelerometers with their masses moving at right angles to each other. Alternatively, a two-axis accelerometer and a separate one-axis accelerometer mounted (possibly in the same package) with its response axis perpendicular to the two-axis accelerometer can be built. However, these are difficult to produce in a single device. The more common method of producing three-axis accelerometers is to use a two-axis accelerometer and extract the third-axis signal from the two axes. A simple configuration that allows this is shown in **Figure 10.10**. The mass is suspended on four beams and each of the beams is attached to a flexural member that can be flexed in the plane as well as perpendicular to the plane. Now the mass can move up–down, left–right, and in–out. A number of piezoresistive sensors are produced on each flexural member, typically two on each, but more can be added as needed. The components of acceleration are extracted as follows: If the motion is up–down or left–right, the signals on the horizontal or vertical flexural members are used, respectively. The third axis is extracted from these two. When, for example, the mass moves out, all sensors are equally stressed, something that cannot occur with acceleration in the plane. The signal is then extracted from all sensors. Of course, the mass can flex the members in a much more complicated fashion, necessitating a more complex extraction algorithm, something that can be done on-chip (a smart sensor) or can be done externally given the outputs of all piezoresistors.

Multiple-axis acceleration can also be sensed based on temperature variations, as shown in **Figure 10.11**. A chamber is fabricated in silicon and a heating resistor is built at the center. This heats the gas inside the chamber to a temperature above ambient. Four temperature sensors placed at the corners of the chamber measure the temperature at their locations. If the sensor is stationary, all four temperature sensors (semiconducting thermocouples or *pn* diodes) are at the same temperature. As an accelerator, it operates like any gas accelerometer in which the gas serves as the moving mass (see **Figure 6.22** for a one-axis conventional gas accelerometer). The four temperature sensors make this device a two-axis accelerometer. This type of sensor can also detect static tilting, since the tilting on a particular axis will allow the hot gas to move up the chamber, creating a

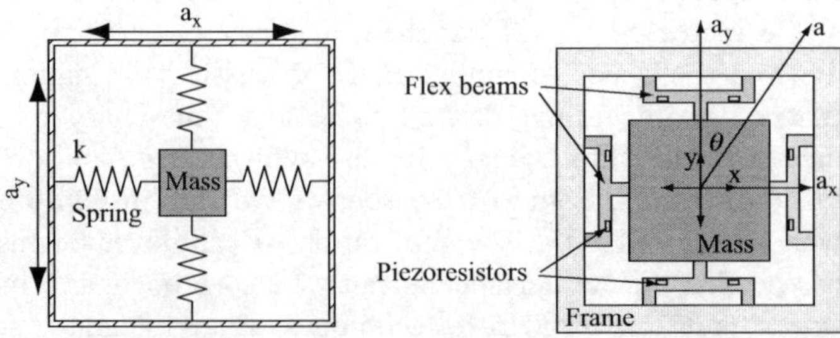

FIGURE 10.10 ■ (a) A two-axis accelerometer. (b) A three-axis accelerometer. The mass is free to move in the plane as shown, but also vertically. The signal for the third axis is extracted from the piezoresistors of the two axes.

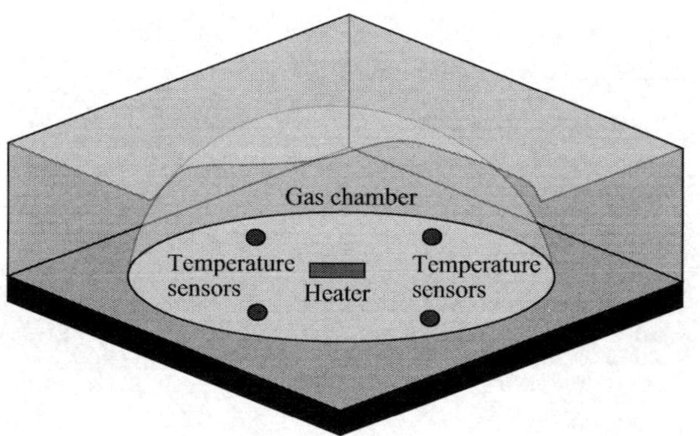

FIGURE 10.11 ■ Principle of operation of a dual-axis heated gas accelerometer.

temperature difference between the sensors. Additional sensors can be used to make it a true three-axis accelerometer. Although this type of sensor can be implemented in a conventional fashion, as in **Figure 6.22**, the MEMs implementation has distinct advantages. The small size of the sensors makes it much more responsive and integration with additional onboard electronics can make it a simple device to use. The small size of the chamber means that the power needed to heat the gas is low and additional temperature sensors can sense the ambient temperature and regulate the chamber temperature to optimize the performance. Sensors of this type are available commercially with varying degrees of integration.

10.3.1.4 Angular Rate Sensors

If there is one sensor that can exemplify the dramatic difference between conventional and MEMS sensors, it must be the gyroscope. Not only is the conventional gyroscope a relatively large device, it is also one of the more expensive devices available. Even if we leave alone the classic mechanical gyroscope, originally developed for use in aircraft and ships (see **Figure 6.39**), and concentrate on the Coriolis force gyroscopes and, in particular, on the optical fiber gyroscopes, the cost of these, as well as their reliance on exact and expensive optical components, make them out of reach for most applications. The need for miniature inertial sensors for navigation and for consumer products has led to the development of a number of MEMS configurations, all based on the Coriolis force. In addition to the optical methods discussed in **Chapter 6**, there are other basic methods that have been tested for this purpose, and some have had commercial success and have been incorporated in a number of products, including motor vehicles and handheld consumer devices, not to mention military applications. It should be noted here that the term gyroscope, which literally means "measurement of rotation," is not the proper name for vibrating devices. These should be called angular rate sensors, since nothing is rotating. There are a number of configurations that are possible, including the tuning fork and the ring angular rate sensor, and they come in many variations.

EXAMPLE 10.3 Strain in an accelerometer

Consider the accelerometer in **Figure 10.10**. Assume that the mass is 1 g and all parts of the sensor are made of silicon. The four flexural members are 100 μm long and their cross section is a square, 5 μm × 5 μm. Silicon has a modulus of elasticity of 150 GPa. Find the relation between the acceleration and the strain measured by any of the piezoresistive sensors on the vertical flexural

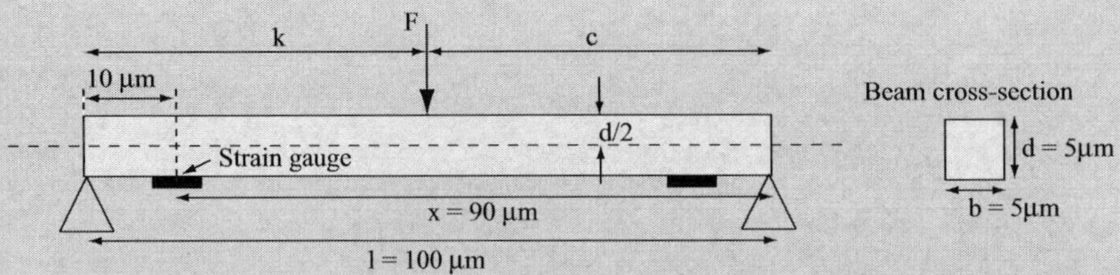

FIGURE 10.12 ■ Beam dimensions and cross section used to calculate strain.

members and calculate the strain at an acceleration of 2 g if the strain gauges are small and placed 40 μm from the center of the flexural member.

Solution: The strain in the flexural member is found from the force produced by the acceleration, which acts on the flexural beam shown in **Figure 10.12** as a simple supported beam. We start with the force due to acceleration:

$$F = ma \quad [\mathrm{N}].$$

This force acts on the center of the flexural beam. To calculate the strain at the surface of the beam (where the piezoresistive sensors are located) we write

$$\varepsilon = \frac{M(x)}{EI}\frac{d}{2},$$

where $M(x)$ is the bending moment at the location of the strain gauge, E is the modulus of elasticity, I is the moment of area of the beam, and d is the thickness of the beam. E is given and M and I are as follows:

$$M(x) = \frac{Fk}{l}(l - x) \quad [\mathrm{N \cdot m}]$$

$$I = \frac{bh^3}{12} \quad [\mathrm{m}^4],$$

where l is the length of the beam, b is the width, and h is the height of the beam cross section (see **Figure 10.12**). In the relation for the bending moment, $a < x < l$, that is, the distance x must be taken from the "far" support as shown in the figure. Since $k = c = l/2$, we have

$$M(x) = \frac{F}{2}(l - x) \quad [\mathrm{N \cdot m}].$$

With these, and noting that $b = h = d$, we get

$$\varepsilon = \frac{ma/2}{Ebh^3/12}(d/2)(l - x) = 6\frac{ma}{Ed^3}(l - x).$$

A quick check shows that the result is dimensionless, as required. Therefore the relation is

$$\varepsilon = \left[6\frac{m}{Ed^3}(l - x)\right]a.$$

This is a linear relationship, with a sensitivity (slope) of

$$\frac{\varepsilon}{a} = 6\frac{m}{Ed^3}(l - x) \quad \left[\frac{\mathrm{m/m}}{\mathrm{m/s^2}}\right].$$

At an acceleration of 2 g ($2 \times 9.81 = 19.62$ m/s^2) and with the given values, the strain is

$$\varepsilon = 6 \times \frac{1 \times 10^{-3} \times 2 \times 9.81}{150 \times 10^9 \times \left(5 \times 10^{-6}\right)^3}\left(100 \times 10^{-6} - 90 \times 10^{-6}\right) = 0.010464 \text{ m/m}.$$

This is a reasonable strain (1%) for silicon. In fact, the strain is half that since the force operates simultaneously on two beams (lower and upper), an effect that reduces the strain in each beam (as if the beam were twice as thick). Depending on the gauge factor of the sensors, the acceleration will produce a significant change in their resistance. Note that the strain can be increased by moving the strain gauges closer to the center. However, the purpose of the two strain gauges per beam is to resolve arbitrarily directed accelerations, and functionality would be diminished with strain gauges close to the center of the beam.

a. Tuning fork angular rate sensor. The principle of the tuning fork sensor is shown in **Figure 10.13**. It consists of a structure similar to the tuning fork used, for example, to tune pianos. The tines of the fork are set into mechanical oscillations using piezoelectric devices built onto the fork. In MEMS, the piezoelectric plate is made during the production process and integrated with the tines. It should also be noted that the tuning fork produced in MEMS has tines that are rather short and, in comparison, wide, whereas the common tuning fork is long and slender. The tines oscillate in the basic mode, meaning that at any given moment the tines move in opposite directions, as shown. If the fork is rotated, the tines experience a Coriolis force (or acceleration) perpendicular to the oscillatory motion. This opposite motion of the tines torques the fork and a piezoresistive sensor on the stem of the fork measures the strain associated with this torsional motion to determine the angular rate. The fork can also be set into oscillation by electrostatic forces, but piezoelectric actuators built onto the tines of the fork provide larger forces. A possible implementation of the tuning fork angular rate sensor is shown in **Figure 10.14**. Note the aspect ratio of the tines of the fork and the

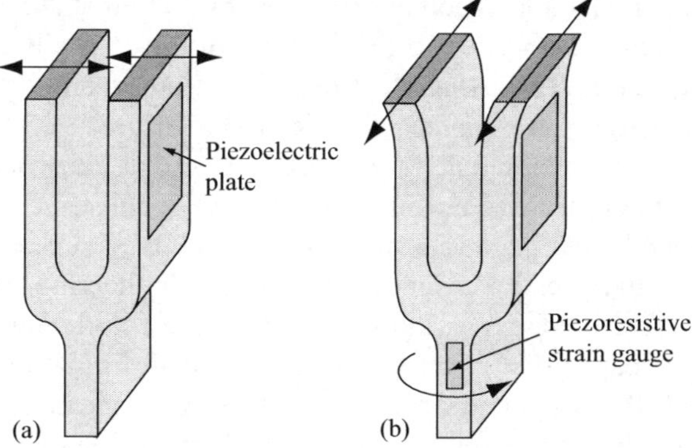

FIGURE 10.13 ■ The principle of the tuning fork angular rate sensor. (a) The oscillation of a tuning fork. (b) If the fork is torqued, the Coriolis force produces displacement of the tines perpendicular to the axis of vibration, straining the stem of the fork. Measurement of the strain in the stem is a measure of the angular rate.

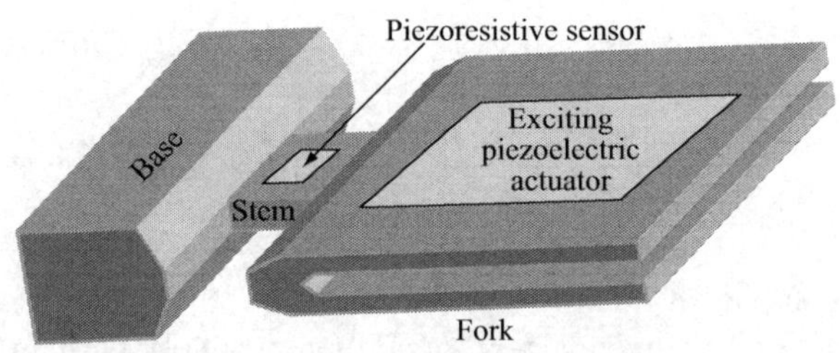

FIGURE 10.14 ■ MEMS tuning fork angular rate sensor. Piezoelectric excitation is shown, but excitation can also be capacitive.

excitation method. There are a number of variations on the basic sensor, including in the methods of excitation. Angular rate is measured in degrees/second or degrees/hour and a good sensor might measure a rate smaller than 0.001 degrees/h.

b. Vibrating ring angular rate sensor. The ring sensor is based on the known fact that a ring, when set into resonance, will distort its shape into elliptical shapes around stationary points or nodes. This phenomenon is called the wine glass oscillation after its discovery in glasses by G. H. Brian in 1890. He noticed that a ringing wine glass (e.g., by lightly tapping a wine glass with a knife) changed tone if the wine glass was spinning. The explanation is very simple: A wine glass, or more specifically its lip (a ring) will oscillate by deforming its shape into elliptical forms in a very specific way. The ring distorts from its original circular shape into an ellipse and then back to a circle. The oscillation continues, but now the ellipse that forms is rotated at 90°. Thus the nodes of the resonant mode are at 90° from each other around the ring. Because of the circular symmetry of the ring, the second mode is rotated by 45° to the first mode, meaning that the maximum in the first mode (antinode) corresponds to the minimum in the second mode (node). These modes are shown in **Figure 10.15a**. The dashed circle represents the undisturbed ring. The ellipses represent the oscillation modes, with the arrows showing the antinodes of the mode (maximum distortion). If the ring is stationary (i.e., it does not rotate), only the first mode is excited. If the ring is now rotated, the second mode is excited as well, and the overall oscillation is a linear combination of the first and second modes, and now the nodes and antinodes shift from their original location (which is determined by the location of the impulse). The shift in tone due to rotation can be heard and is proportional to the rate of rotation. The ring sensor is an implementation of this idea whereby a ring is set into oscillation either by electrostatic or electromagnetic means and the nodes are established in the primary mode of oscillation. This is shown schematically in **Figure 10.15b** using electrostatic excitation and capacitive sensing of the vibration of the ring. Some of the electrodes shown are used to set the ring into oscillation by applying a voltage that attracts the ring to that electrode. Since the ring is anchored at its axis and otherwise not in contact with the substrate, it will oscillate at a frequency that depends on its dimension, mass, and mechanical properties of the material of which it and the flex members are made. If the ring experiences angular motion, the Coriolis force will excite the secondary mode and the nodes (and antinodes) will shift. A series of capacitive electrodes measure the deformation of the ring (capacitance changes with the distance of the ring surface from the sensing electrodes) and that shift is used to determine the angular rate. The torsional members supply the restoring force for oscillation.

c. MEMS fluxgate magnetic sensor. It was mentioned before that MEMS are not limited to semiconductors and other materials may be used, including ferromagnetic materials.

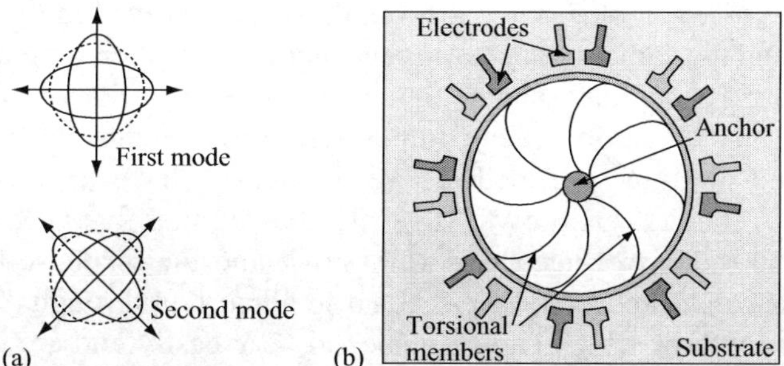

FIGURE 10.15 ■ The ring angular rate sensor. (a) Modes of oscillation of the ring indicating the antinodes (maximum distortion; arrows). (b) The ring is set into oscillation by some of the electrodes through capacitive coupling and the position of the nodes and antinodes of oscillation sensed by the other electrodes. The shift in location of the nodes is a measure of the angular rate.

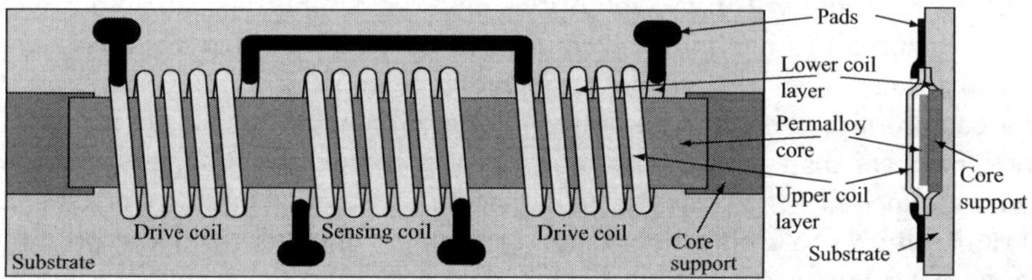

FIGURE 10.16 ■ Schematic of a MEMS fluxgate sensor demonstrating the use of ferromagnetic materials and magnetic coils.

An example of a complex microstructure for use in magnetic sensing is the microfluxgate sensor, shown schematically in **Figure 10.16**. It consists of a substrate that may be a ceramic or silicon with a SiO_2 layer on top. The sensor itself consists of a strip of permalloy (see **Section 5.8.2** and **Figure 5.58b** for an explanation) and two coils wound on it. As in **Figure 5.58b**, the device is sensitive to magnetic fields along the strip. In MEMS, this simple structure requires production of the coils in two layers, one below and one above the permalloy core, and connection between them during the deposition of the upper coil layer (see **Example 10.1**). The core must be ferromagnetic, hence the choice of permalloy, an anisotropic material with high permeability and one that can be deposited by means used in semiconductors production. The coils can be made of aluminum, but more often they are made of doped silicon or polysilicon to ensure high conductivity and sufficient stiffness. The use of MEMS offers options that are difficult to obtain with conventional sensors. For example, two or more sensors can be produced on the same substrate either to produce differential readings, to sense spatial variations in fields, or, by producing two sensors with the axes perpendicular to each other, to sense two (or, indeed, three) components of the field, with little additional effort or cost.

10.3.2 MEMS Actuators

Whereas many MEMS sensors are essentially adaptations or miniaturizations of conventional sensors to the scale afforded by MEMS, actuators are not. The methods of actuation are the same as in conventional actuators, relying on electrostatic, magnetic, thermal, piezoelectric, and other methods, but the scale involved in these devices has a profound effect and the choices of actuation mechanism are often dramatically different than in conventional actuators, sometimes even surprising. For example, the use of heat in conventional actuation is very limited, primarily because of the response time of the device and the power needed. These restrictions have to do mostly with the physical size of the device. On the MEMS scale, the volumes that need to be heated are minute and hence the power needed is very small and the response time is small as well. The heating can be achieved by passing a small current through the device or by a separate resistor, both of which can be achieved very easily and are compatible with MEMS methods. Similarly, electrostatic actuation is extremely important, both because the forces needed are minute and because capacitive actuation is easy to produce and control. On the other hand, magnetic actuation, which forms the bulk of conventional actuators, is muted in MEMS precisely because it is well suited to produce large forces and because it relies on coils and permanent magnets to do so. Nevertheless, coils can be produced in MEMS and permanent magnets as well as ferromagnetic materials can be deposited and integrated in a device to produce magnetic actuators. These are in the minority, however, simply because other methods, including piezoelectric methods, are more efficient and simpler on this scale. The forces and torques one can obtain, or in more general terms, the power a MEMS actuator can produce, are commensurate with the dimensions of the device. Hence the actuation mechanisms adopted are those that are compatible with these restrictions. One cannot expect a MEMS device to replace a DC motor, but it can, for example, move a micromirror to shift a light beam or produce droplets for inkjet printing, and can do so faster than a macrodevice. Thus it is important to match expectations and devices and apply MEMS to those areas in which they can be used effectively. As with sensors, many configurations have been proven to work, but only a few have been developed into successful products, a situation likely to change in the future. The examples of actuators that follow are representative of both commercial devices and ideas that may or may not become commercially viable. In all cases, however, they are interesting, and in many cases, strangely beautiful.

10.3.2.1 Thermal and Piezoelectric Actuation

Inkjet nozzles can be micromachined onto a substrate, as shown in **Figure 10.17**. The operation is rather simple. A small ink reservoir, connected to the main ink reservoir, is heated using a thin film resistor. The ink is heated rapidly (within a few microseconds) to temperatures as high as 200°C–300°C, increasing the pressure in the reservoir to 1–1.5 MPa. A jet of ink is expelled which, when the heat is removed, collapses back into the reservoir, leaving behind a droplet that continues toward the printed page. One of the advantages of a device like this is that multiple nozzles can be built on the same substrate, printing a whole line of dots as the device advances and thus forming the required images. Typically a linear array is built, with the distance between nozzles defining the

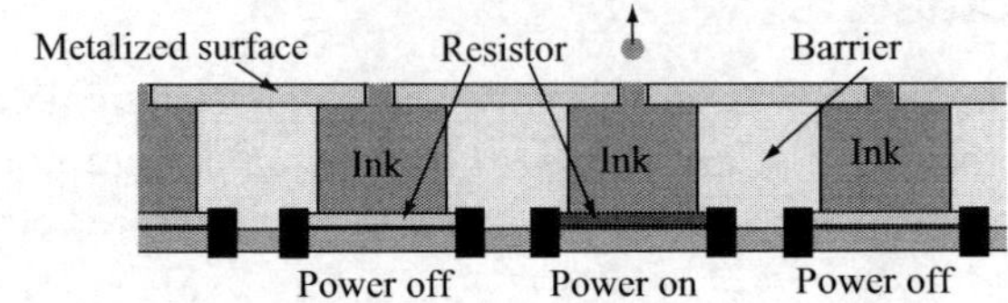

FIGURE 10.17 ■ The principle of a MEMS thermal inkjet element. A section of a linear array is shown.

resolution. For example, at a resolution of 2400 dots/in. (95 dots/mm), the distance between the centers of two nozzles in the array is approximately 10 μm. Although thermal devices tend to be slow, the small sizes involved means that little energy is needed to raise the temperature, and in fact, a droplet in a MEMS inkjet can be generated in less than 50 μs. The use of MEMS can extend the idea above to print a whole line at one time or even a whole page without the need to move the nozzles, hence achieving much faster speeds than are possible with moving print heads. It should be noted that the ink droplets can be generated in other ways, most notably by ultrasound. The idea is similar—the generation of high pressure in a short time to expel the ink. In a device of this type, the resistor is replaced by a piezoelectric element.

EXAMPLE 10.4 The piezoelectric inkjet

To produce an inkjet as part of an inkjet printing cartridge, a small cylindrical chamber is made and the ink is expelled through the action of a piezoelectric device. The arrangement is shown in **Figure 10.18**, where the bottom of the chamber is sealed with a diaphragm and a piezoelectric disk is placed between the substrate and diaphragm. When a voltage V is applied, the piezoelectric disk expands (positive piezoelectric coefficient), decreasing the volume of the chamber and expelling the ink. The piezoelectric disk is made of PZT with a piezoelectric coefficient of 374×10^{-12} C/N, a relative permittivity of 1700, and a thickness of 25 μm. A voltage of 3.6 V is applied across the disk when operated. The upper and lower surfaces are coated with aluminum to allow connection of the voltage and to produce a uniform electric field intensity in the material. Assume the ink is water based and hence has the density of water (1 g/cm^3).

a. Calculate the amount (mass) of ink expelled.
b. Calculate the maximum force produced by the piezoelectric device and the peak pressure in the ink chamber.
c. How many droplets will an ink cartridge that contains 10 g of ink produce?

Solution:

a. The ink is expelled by the expansion of the piezoelectric element in **Figure 10.18**. This is calculated as follows: the piezoelectric coefficient is the strain per unit electric field intensity, that is,

$$d = 374 \times 10^{-12} \ \frac{\mathrm{m/m}}{\mathrm{V/m}}.$$

The electric field intensity is that of a parallel plate capacitor:

$$E = \frac{V}{t} = \frac{3.6}{10 \times 10^{-6}} = 3.6 \times 10^{5} \ \mathrm{V/m}.$$

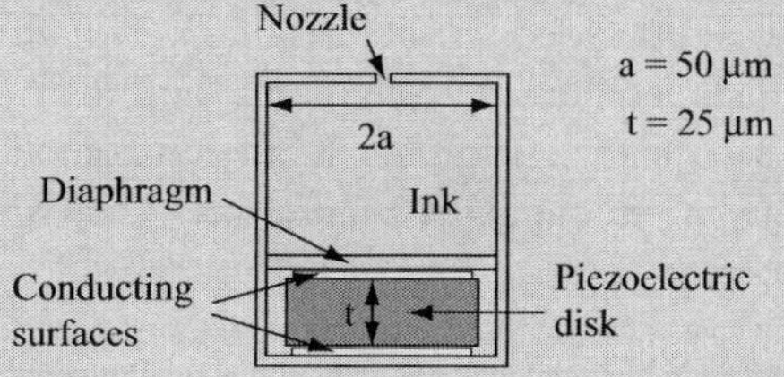

FIGURE 10.18 ■ A piezoelectric-actuated inkjet.

Thus the strain is

$$e = Ed = 3.6 \times 10^5 \times 374 \times 10^{-12} = 1346.4 \times 10^{-5} \text{ m/m}.$$

However, since the disk is only 25 μm thick, the total vertical displacement of the disk is

$$dl_{\text{disk}} = Edt = 1346.4 \times 10^{-5} \times 25 \times 10^{-6} = 0.003366 \times 10^{-6} \text{ m}.$$

The volume of ink displaced is

$$dv = \pi a^2 dl_{\text{disk}} = \pi \times \left(50 \times 10^{-6}\right)^2 0.003366 \times 10^{-6} = 26.44 \times 10^{-18} \text{ m}^3.$$

This is 26.44×10^{-18} m^3 or 26.44 μm^3.

b. To calculate the force produced we use **Equation (7.40)**. Rewriting it in terms of force, we have

$$F = \frac{\varepsilon A V}{td} \quad [\text{N}],$$

where A is the surface area of the disk, ε is its permittivity, t is the thickness, d is the piezoelectric coefficient, and V is the applied voltage. With the given values,

$$F = \frac{\varepsilon A V}{td} = \frac{1700 \times 8.854 \times 10^{-12} \times \pi \times \left(50 \times 10^{-6}\right)^2 \times 3.6}{25 \times 10^{-6} \times 374 \times 10^{-12}} = 0.0455 \text{ N}.$$

The pressure is found by dividing this by the area of the disk:

$$P = \frac{F}{\pi a^2} = \frac{0.0455}{\pi \times \left(50 \times 10^{-6}\right)^2} = 5.8 \times 10^6 \text{ Pa}.$$

Note that even though the force seems minute, the pressure is very high (over 60 atm).

c. With the volume calculated in (a), the mass of ink in a droplet is

$$w = dv\rho = 26.44 \times 10^{-18} \times 1000 = 26.44 \times 10^{-15} \text{ kg},$$

where ρ is the density of water. Thus the number of droplets the inkjet can produce from a single cartridge is

$$N = \frac{W}{w} = \frac{10 \times 10^{-3}}{26.44 \times 10^{-15}} = 3.7810^{11} \text{ droplets}.$$

10.3.2.2 Electrostatic Actuation

Electrostatic actuators are based on the attraction force between the plates of a capacitor, a subject that was discussed in **Section 5.3.3**. This force is proportional to the area of the capacitor, the distance between the plates, and the potential applied across the plates. In fact, there are two basic configurations to consider. The first, shown in **Figure 10.19a**, is a classic parallel plate capacitive actuator. As a first approximation, the electric field intensity between the plates is

$$E = \frac{V}{d} \quad [\text{V/m}]. \tag{10.1}$$

The energy per unit volume is

$$w = \frac{\varepsilon E^2}{2} = \frac{\varepsilon V^2}{2d^2} \quad [\mathrm{J/m^2}]. \tag{10.2}$$

Now, assuming that the plates move closer together a distance *dl*, the change in energy is the energy density multiplied by the change in volume:

$$dW = wdv = \frac{\varepsilon V^2}{2d^2} abdl \quad [\mathrm{J}]. \tag{10.3}$$

The force is defined as the change in energy per unit length:

$$F = \frac{dW}{dl} = \frac{\varepsilon V^2}{2d^2} ab \quad [\mathrm{N}]. \tag{10.4}$$

This relation shows that the force is proportional to voltage squared and the area of the plates (*ab*) and inversely proportional to the distance between the plates, *d*.

The second useful configuration is shown in **Figure 10.19b**. In this case the distance between the two plates remains the same, but the plates can slide relative to each other (see also **Section 5.3.3**). The lateral force, that is, the force that tends to move the upper plate to the left, is the rate of change in energy with respect to distance. Assuming the upper plate moves a differential length to the left, the change in volume between the plates is *bddl* and we have

$$dW = wdv = \frac{\varepsilon V^2}{2d^2} bddl \quad [\mathrm{J}]. \tag{10.5}$$

The force is now

$$F = \frac{dW}{dl} = \frac{\varepsilon V^2}{2d} b \quad [\mathrm{N}]. \tag{10.6}$$

There is a third possibility, that of both plates remaining fixed and the material between the plates being free to move. The result, however, is exactly as in the case of the lateral moving plate (**Equation 10.6**) and hence is not shown here (but see **Section 5.3.3**).

In MEMS the dimensions are very small. To increase this force to a useful value, the structure in **Figure 10.19b** is often modified into the comb structure shown in

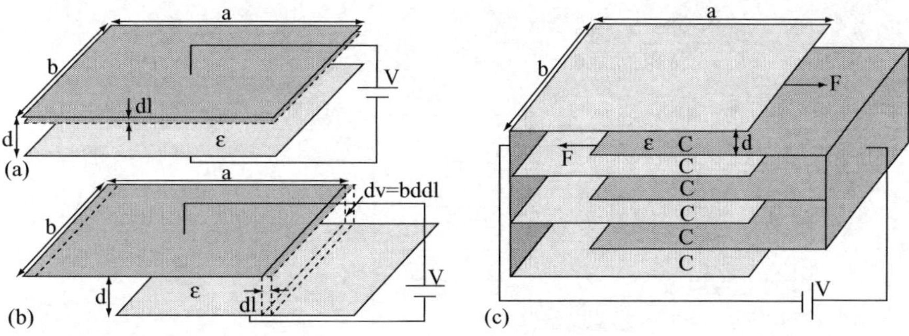

FIGURE 10.19 ■ Forces in capacitive actuators. (a) Attraction between two plates connected to a voltage source. (b) Force due to a change in energy as the upper plate slides above the lower plate. (c) Comb structure that multiplies the force of the structure in (b).

Figure 10.19c. A capacitor is thus created between each pair of plates. In this case there are a total of six capacitors, with two plates on one side and three plates on the other. Assuming N plates to be on one side and $N+1$ plates on the other, the number of capacitors is $2N$, and the force produced by a comb drive with air between the plates and with an applied voltage V is

$$F = 2N\frac{\varepsilon_0 V^2}{2d}b \quad [\text{N}]. \tag{10.7}$$

It should be noted, however, that the forces are still very small. With b and d only a few micrometers, V of only a few volts, and ε_0 on the order of 10^{-12} F/m, one cannot expect very large forces. But at the same time, large forces are not needed for the type of actuators encountered on this level. Note also that the dimension a defines the stroke of the drive, but has no effect on the force. Neglecting edge effects, the force is constant along the stroke.

Figure 10.20 shows a comb drive that operates in a pull-pull mode. One can understand the operation of the comb drive as the attraction between the plates of a parallel plate capacitor. In practice, the drive signal is applied to one side of the comb (usually the stationary or fixed plates), whereas the charge on the opposite side of the comb is generated by induction. In the configuration shown in **Figure 10.20**, the left stationary comb and the right stationary comb are driven alternately to create a back-and-forth motion that can be used for various applications, such as ratchet mechanisms or to set a structure in resonance. The thin vertical beams serve as springs to restore the device to its idle position at the center and keep the moving plates centered between the stationary plates. In practical applications, these structures can be much more complex than what is shown here. In some cases they are long and slender, in others they are folded, but their purpose is the same.

EXAMPLE 10.5 Forces in a comb drive

A comb drive is made of 40 plates on each side, and each plate is 30 μm long and 10 μm deep, separated 2 μm apart. Calculate the force for a 5 V source applied across the comb.

Solution: Since each side contains $N = 40$ plates, the total number of capacitors is $2N-1 = 79$. The force is

$$F = (2N-1)\frac{\varepsilon_0 V^2}{2d}b = 79\frac{8.854 \times 10^{-12} \times 5^2}{2 \times 2 \times 10^{-6}} 10 \times 10^{-6} = 4.372 \times 10^{-8}\ \text{N}.$$

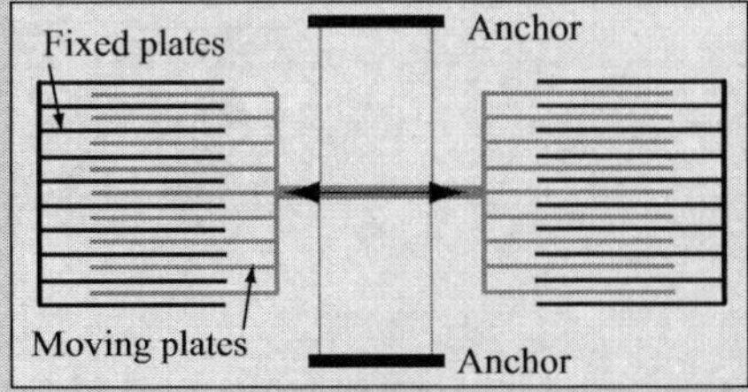

FIGURE 10.20 ■ A typical comb actuator complete with restoring springs and centering control. The comb moves from side to side by alternately powering the left and right sections.

This is a mere 43.7 nN, but is sufficient for many applications. The stroke of this device is close to the length of the plates, or almost 30 μm. Typically only about half of that or less is used. Note that the force is independent of the stroke.

EXAMPLE 10.6 Torque in a micromotor

To get an idea of the order of magnitude of the torque a micromotor can produce, consider the motor shown in **Figure 10.5**. Assume the radius of the rotor is 50 μm and both the rotor and stator are 6 μm high. The gap between the rotor and stator is 2 μm and each of the stator elements forms a 30° arc. A 5 V voltage is applied to the motor.

Solution: We can use the force calculated in **Equation (10.6)** directly by imagining the rotor in **Figure 10.5** displaces slightly. Each pair of rotor–stator plates will then look as in **Figure 10.19b**, with $d = 2$ μm being the distance between the plates and $b = 6$ μm being the height of the rotor. Note again that the width of the plates plays no role in the calculation of force. The force on each of the rotor plates is circumferential and equal to

$$F = \frac{\varepsilon V^2}{2d} b = \frac{8.854 \times 10^{-12} \times 5^2}{2 \times 2 \times 10^{-6}} \times 6 \times 10^{-6} = 3.32 \times 10^{-10}\ \text{N}.$$

The torque is the force multiplied by the radius and multiplied by 3 since there are three rotor elements activated at the same time:

$$T = 3Fr = 3 \times 3.32 \times 10^{-10} \times 50 \times 10^{-6} = 4.98 \times 10^{-14}\quad \text{N} \cdot \text{m}.$$

Clearly the torque is low, as expected from an actuator of this scale, as one would not expect micromotors to excel in torque production.

10.3.3 Some Applications

10.3.3.1 Optical Switches

An example of an application in which MEMS excel is the optical switch, an important component in optical fiber communication. In electronics, switching is done using transistors, but in optics it is usually done by diverting mirrors. Since optical fibers are very thin, the beamwidth is small and a micromirror is sufficiently large to deflect the beam. **Figure 10.21** shows a simple device that can switch the beam between two input optical fibers to two output optical fibers. In the configuration shown, the comb driver retracts the mirror when the driver is activated electrostatically. When retracted, the beam from fiber 1 (input) is coupled with fiber 4 (output) and fiber 2 (input) couples with fiber 3 (output). When not activated, fiber 1 couples with fiber 3 and fiber 2 with fiber 4. Obviously a very simple device, it is also very effective and responsive. Of course, much more complex configurations can be devised, including arrays of switches and two-way switches.

10.3.3.2 Mirrors and Mirror Arrays

Mirrors for various purposes, including the projection of images and for displays, have been an early target of MEMS development because of their simplicity and the fact that the manipulation of light requires little energy. Some, like the optical switches discussed above as well as projection systems, have succeeded in the marketplace. The object of micromirrors can be broadly divided into two areas: shifting the direction of light, such as in the projection of images or in light switches, and the modification or modulation of surface reflectivity, something that is useful in displays. In the first case, single mirrors may be used to deflect the direction of a laser beam or an array may be used to create a larger reflective surface whose direction of reflection is controllable. In surface-altering mirrors, an array is used to change the reflectivity of the surface. **Figure 10.22a** shows an example of a reflective system. The flat mirror, produced by deposition of aluminum on polysilicon, is electrostatically actuated by applying a potential between the mirror itself and a stationary electrode on the substrate. A hinge creates a restoring force that keeps the mirror tilted at a fixed, maximum angle. When the voltage is applied, the mirror is attracted to the fixed electrode and its angle determined by the magnitude of the voltage and the restoring force of the hinge. The mirror may be operated as a digital device, switching between two angles, or, in principle, it may be actuated as an analog device moving between a maximum and a minimum angle. The mirrors are typically produced by surface micromachining techniques. An example of surface-modifying mirrors is shown in **Figure 10.22b**. Here the mirror is flat under no power condition, but the individual mirrors, perhaps 50 μm on the side, will become convex when the potential is applied between the mirror and the fixed electrode below it. The surface reflectivity is thus altered, modifying the optical properties of the surface as a whole.

10.3.3.3 Pumps

The electrostatic actuation of micropumps is a simple example of actuation by attraction of two parallel plates of a capacitor. The idea is shown in **Figure 10.23a** as a miniaturized implementation of the flap pump (conventional flap pumps are often used to pump air in small aquariums). The diaphragm serves as a movable plate that is attracted to the fixed plate when a potential is applied across the two plates. The plate distorts

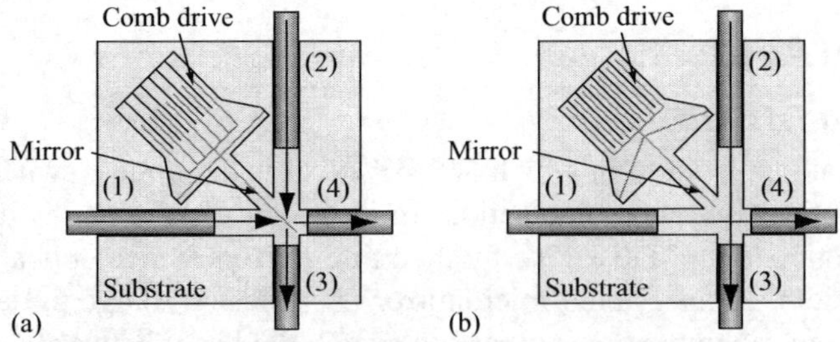

FIGURE 10.21 ■ A 2 × 2 (two inputs, two outputs) optical switch. (a) The mirror in its resting, nonactivated position. (b) The comb drive is activated, retracting the mirror and switching the beams.

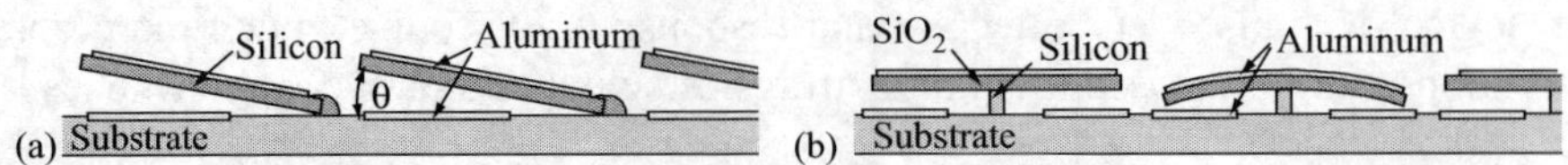

FIGURE 10.22 ■ (a) Tilting mirror activated by electrostatic forces and used in a projection system as an array of mirrors. (b) Another method of deforming a mirror to affect the surface reflectivity.

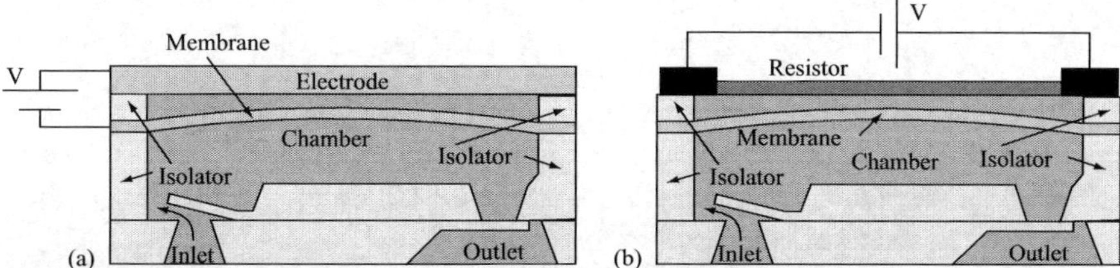

FIGURE 10.23 ▪ Two pumping configurations. (a) A flap pump driven by electrostatic forces. (b) A heat-activated flap pump. Both pumps are shown in the suction mode. When the source is removed the diaphragm relaxes and expels the fluid through the outlet flap.

upward, creating a suction action that opens the intake flap or poppet, filling the chamber. When the potential is removed, the diaphragm moves downward and forces the input flap to close and the outlet flap to open, allowing the gas or fluid to exit. Although the amount of fluid is necessarily small, this type of device can be used exactly because of that—to meter small doses of fluid or dispense drugs in precise quantities. A similar action can be obtained by heat expansion, as shown in **Figure 10.23b**. A heating element on top of the diaphragm heats it, forcing it to expand. The expansion moves it upwards, pulling in fluid past the inlet check valve. When the heat is removed the diaphragm moves downward, forcing the fluid past the outlet valve. The diaphragm can be made as a bimetal structure (aluminum–silicon or nickel–silicon can be used) and can be insulated from the fluid with a thin layer of SiO_2.

10.3.3.4 Valves

Valves of various shapes and for various purposes can be made in MEMS and actuated either electrostatically, by heat, or magnetically. An example of the use of bimetallic actuation is shown in **Figure 10.24**. When heat is applied, the poppet moves upward, opening the valve. In its normal state the valve is closed. Clearly the valve can be activated electrostatically by removing the heater and adding a fixed conducting plate above the poppet, similar to **Figure 10.23a**. A normally open or normally closed valve can also be devised using a comb actuator or using a magnetic actuator. As an example of magnetic actuation in MEMS, consider the structure in **Figure 10.25**. This is

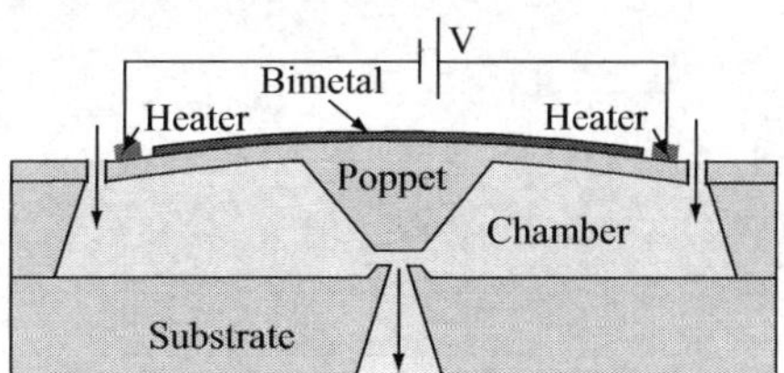

FIGURE 10.24 ▪ A normally closed valve activated by heat using a bimetal driver.

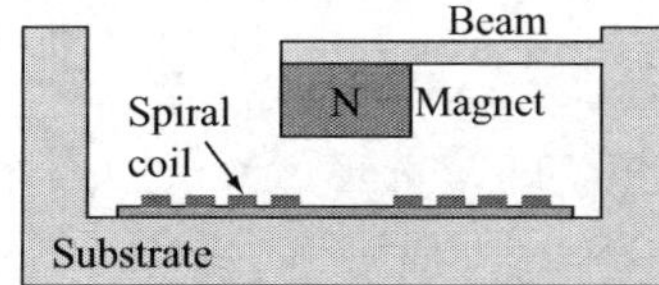

FIGURE 10.25 ▪ A direct magnetic drive akin to the voice coil actuator.

EXAMPLE 10.7 Heat-activated microvalve

Consider the heat-activated valve shown in **Figure 10.24**. To get an idea of the principles involved, assume that the bimetal is a strip 2 mm long and 50 μm thick, made of copper–nickel, and perfectly flat at 20°C. The bimetal strip is heated to 150°C. Calculate the distance the poppet lifts above the substrate. Is that sufficient to open the valve?

Solution: To calculate the lift of the poppet, recall that a bimetal strip fixed at one end bends into a circular strip of radius r as the free end bends, say, downward. This was discussed in **Section 3.5.2**, and more specifically in **Equation (3.38)**:

$$r = \frac{2t}{3(\alpha_u - \alpha_l)(T_2 - T_1)} \quad [\mathrm{m}],$$

where t is the thickness of the bimetal strip, T_1 is the reference temperature at which the strip is flat, T_2 is the actual temperature of the strip, and α_u and α_l are the coefficients of expansion of the upper and lower conductors.

In the case discussed here, both ends are constrained, forcing the center of the strip to rise, but the radius of the strip at the temperature $T_2 = 150°\mathrm{C}$ is the same as if one end were free (approximately). From **Table 3.10** we have the coefficients of expansion for copper (upper conductor) and nickel (lower conductor) as $\alpha_u = 16.6 \times 10^{-6}/°\mathrm{C}$ and $\alpha_l = 11.8 \times 10^{-6}/°\mathrm{C}$. This gives a radius of

$$r = \frac{2 \times 50 \times 10^{-6}}{3 \times (16.6 - 11.8) \times 10^{-6} \times (150 - 20)} = 5.342 \times 10^{-2} \text{ m},$$

or about 53.42 mm. To calculate the lift of the center of the strip we use the sketch in **Figure 10.26**. The angle α is

$$\alpha = \sin^{-1}\left(\frac{c}{r}\right) = \sin^{-1}\left(\frac{1}{53.42}\right).$$

The distance x then becomes

$$x = r\cos\alpha = 53.42\cos\left[\sin^{-1}\left(\frac{1}{53.42}\right)\right]$$

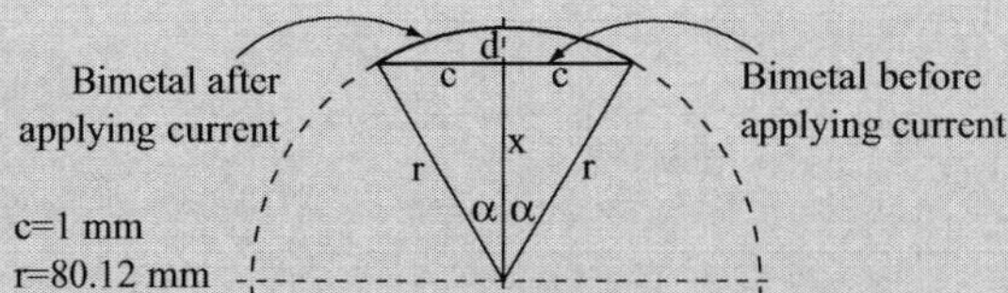

FIGURE 10.26 ■ Calculation of the lift of the poppet in **Figure 10.24**.

and the lift, d, is

$$d = r - x = 53.42\left(1 - \cos\left[\sin^{-1}\left(\frac{1}{53.42}\right)\right]\right) = 0.00936 \text{ mm}.$$

The poppet lifts 9.36 μm. This may not seem like a large distance, but this is a microvalve and, for the dimensions involved, it is sufficient to open the valve.

a direct, simple drive whereby a fixed spiral coil, when a current flows through it, attracts a movable permanent magnet. This motion can be used to close or open a valve, move a mirror, or move a diaphragm exactly as in a voice coil actuator. For example, this structure can be used as a micromicrophone or a microspeaker (or as a dynamic pressure sensor). There are many variations on this simple structure and a variety of magnetic materials that allow production of closed and open magnetic circuits (permalloy, nickel, nickel–iron, and others). Micromagnets can be produced from a composition of cobalt–platinum and other magnetic materials.

10.3.3.5 Others

The MEMS sensors and actuators mentioned above are but a sampling of the variety of devices that have been or can be implemented. In the early days of development of MEMS, a variety of very interesting devices were built, including rotating motors, grippers, latches, ratcheting mechanisms, initially intended to demonstrate the technology. With the success of some of the inertial sensors came the push toward more complex and more varied sensors and actuators. Out of these early steps came many more applications, including microchannels, micromotors, and, indeed, devices that may not seem like MEMS devices, such as the common SAW resonator, delay lines, and filters discussed in **Chapters 7** and **8**. Other areas where MEMS have made progress are in the production of low- and high-pressure sensors, projection and display devices, and biomedical sensors.

10.4 | SMART SENSORS AND ACTUATORS

A smart sensor or actuator is any sensor or actuator in which some level of "intelligence" has been introduced. This simply means that in addition to the normal function of the device, circuitry has been added to take on additional functions, such as communication, power management, local signal and data processing, and sometimes even decision making. That also means that additional power for the electronic circuits must be available for the purpose. For example, a microprocessor can be added to the sensor/actuator to analyze data, perhaps to digitize the output from the sensor, to compensate for unwanted stimuli, to "linearize" the response, to remove offset, and, of course, to communicate with the main processor to which the sensor is ultimately connected. The level of intelligence, or how "smart" the sensor is, may vary from the trivial to the truly complex. At the lower end, this may include such simple circuits as current and voltage limiters, active filters, and protection and compensation circuits, or perhaps a temperature-sensing circuit. On the high end of the spectrum, all functions of processing, including digitization, data transmission (wire or wireless), data logging, and any other conceivable function, can be included onboard, making the sensor a free-standing sensing system. In actuators, protection circuits such as thermal protection, overvoltage, and overcurrent protection, as well as motion and limit functions, can be added. Other options are counters, alarms, data loggers, and many others. The integration of electronics with sensors, particularly silicon-based sensors, is always possible, but it is based on the commercial viability of the sensor. Sometimes this integration makes perfect sense, especially when they are used for mass market applications (the automobile or toy industries, for example). In other cases it is better to leave the sensor as a general purpose sensor and allow the designer to integrate it in the design as necessary.

It should be noted here that there are two types of devices that can be called smart. One is a truly integrated device in which the sensor or actuator and the electronics are packaged into a single die or an integrated circuit package. This is usually the case with silicon-based devices. When this is done, the resulting device is mass produced and available to designers as a component in a standard footprint. Another type is a packaged system that can include multiple components in a single unit. This can take the form of a small board, a plug-in package, or a "box." Usually the package is larger and the device is produced in smaller quantities, sometimes even made as a "custom" device and often intended for specific applications or industries.

Figure 10.27 shows a general schematic of a smart sensor. The various blocks shown are representative of what a smart sensor may consist of. The sensor may be almost any sensor, although the circuitry will vary for different types of sensors, as we shall see in the following two chapters. The block marked as interface circuitry may consist of a variety of functions. This may simply mean a filter or a voltage matching circuitry. It may match impedances or isolate the sensor from the microprocessor. Then again it may consist of a signal processing unit or a digital-to-analog (D/A) converter or any combination of these and other functions. The microprocessor, by its nature, can record data; measure parameters such as time, frequency, voltage, and current; store data, parameters, and commands; and so on. The power supply may be fed by the power grid, may be battery based, or may be scavenged from various sources, and the circuit, together with the microprocessor, can also control power usage, for example, by scheduling power, switching into low power modes, providing warnings such as low battery, and more. The "communication end" module may again include many functions. It may be a modulator, encoder, and transmitter to send data over a wireless or a wired link. It may include drivers and communications protocols, and it may even be used to supply the sensor with power over the communication link.

The communication link may be wired or wireless, or may be dual purpose. It may be a two-way communication link to allow access, remote control, and programming of the microprocessor or it may be simpler, allowing the transfer of data only from the sensor to the base unit.

On the other end of the link, in the base unit, the roles are somewhat reversed. The communication end will now receive the signals and again fit them to the microprocessor for any further processing, storage, and data logging, as necessary. The data from the sensor must be acted upon. This may mean displaying the data or initiating other actions, such as switching on various actuators, all depending on what the system is supposed to do.

Although many of these functions and circuits will be discussed **Chapter 11**, it should be clear at this point that almost any conceivable function of almost any complexity can be implemented as part of the sensor unit.

A very similar diagram can be used to operate a smart actuator. In this configuration, it is usually the base station that initiates processes by sending commands and data to the actuator unit and the local microprocessor has to be interfaced with the actuator.

We have alluded repeatedly to circuit integration with sensors and actuators, especially in conjunction with the MEMS devices described above. In the following sections, and in the next two chapters, the issues of smart sensors and actuators will take center stage. We start with wireless sensors in the following section, followed by sensor

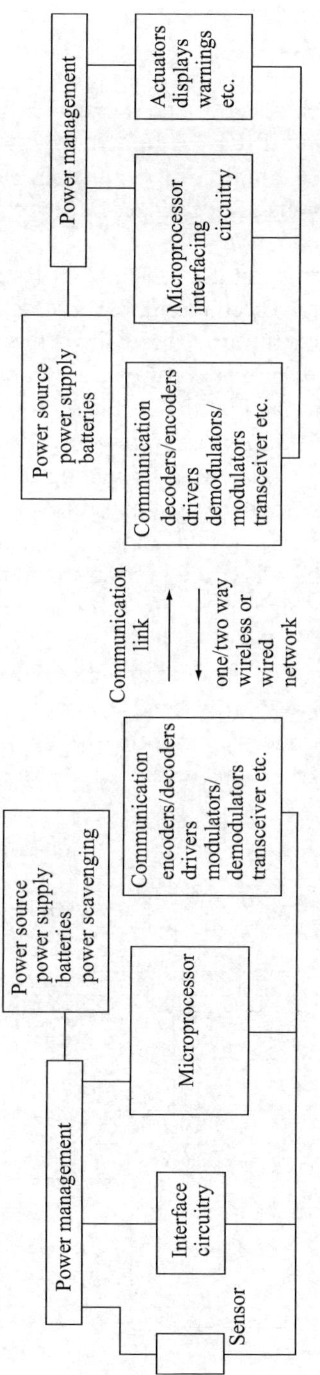

FIGURE 10.27 ■ A general smart sensor.

networks. In the next two chapters we will discuss the circuits and systems needed to interface sensors and actuators to microprocessors.

EXAMPLE 10.8 A smart remote pressure and temperature sensor in vehicle tires

A sensor designed to reside inside the tire of a vehicle and send the driver information on pressure and temperature is shown in **Figure 10.28**. Many of the components shown will be described in **Chapter 11**, but for now this should serve as an example of what a smart sensor might be. In this particular case the sensors and various components are separate entities, but all of these can be integrated together in a single chip for mass production.

The data from the analog pressure and temperature sensors is digitized internally in the microprocessor using internal analog-to-digital (A/D) converters. These data are stored internally as well as encoded and sent over a wireless link through the wall of the tire to the dashboard, where it is received and displayed. A very low power duty cycle switching circuit turns the whole system on and off on a scheduled sequence so that the unit is on for about 2 s every 2 min, sending the information and then switching off. This arrangement allows the whole unit to operate for about 5 years on an internal battery, consuming an average of about 25 μA. This is critical for a system that cannot be easily accessed for battery replacement. The unit in the vehicle reverses the process. After receiving the data, it decodes the information and displays pressure and temperature and issues alarms (i.e., when the pressure is dangerously low or high or when the temperature is too high, indicating imminent tire failure). The system can be programmed to send information and warnings over a wireless link, such as a cell phone or satellite to service centers, something that may be important for fleet vehicles for repair and tracking. A system like this can also be linked with a GPS system to send location information.

In this arrangement, the communication is one way and the driver has no control over the sensor since the purpose is simply to send data to the driver. If communication in the other

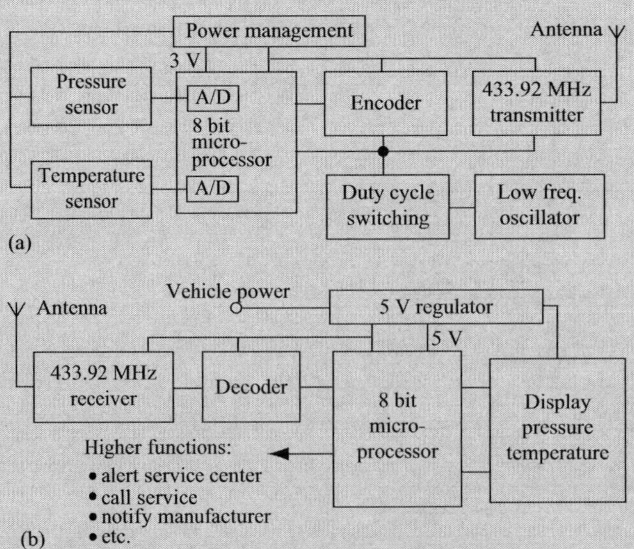

FIGURE 10.28 ■ Remote pressure and temperature sensing for tires in vehicles. (a) In-tire smart sensor. (b) In-vehicle monitor.

direction is deemed necessary, the receiver and transmitter must be replaced with transceivers for two-way communication. Of course, other modifications must be made, but since we limit ourselves to block diagrams, there is no need to go into further details. Note also that power management is important. The sensor uses a 3.3 or 3.6 V battery (i.e., a lithium-ion battery), whereas the unit in the vehicle uses the 12 V source from which it derives the 5 V necessary to drive the microprocessor and other components. Tire pressure monitoring systems (TPMS) are standard in passanger vehicles in the US.

EXAMPLE 10.9 A smart remote tire pressure controller

In most vehicles the pressure in tires is maintained by the driver by occasionally checking the pressure and adjusting it. However, proper tire pressure is critical to safety as well as to efficiency (fuel consumption). It also has an impact on traction and may need to be adjusted under various driving conditions. Hard surfaces usually favor higher pressures, whereas soft surfaces (snow, sand) require lower pressures. Similarly, high temperatures increase tire pressure, increasing the possibility of tire failure, whereas low temperature decreases pressure, causing excessive wear of the tire and poor fuel efficiency. A smart actuator system that allows automatic adjustment of pressure to set conditions as well as intervention by the operator is shown in **Figure 10.29**. It consists of an in-tire unit with a three-way valve operated by a small motor. The valve has three ports. Port 1 is connected to a source of air pressure. This may be as simple as a carbon dioxide (CO_2) canister (built into the wheel and replaceable from the outside) or a chemical reactor that generates gas on demand. Port 2 opens inside the tire and allows inflation by connecting ports 1 and 2 through the valve. Port 3 vents to the outside of the tire and allows deflation by connecting ports 2 and 3 through the valve. The valve is controlled by the microprocessor and a feedback link is provided so that the microprocessor can ascertain the position of the valve. Two sensors,

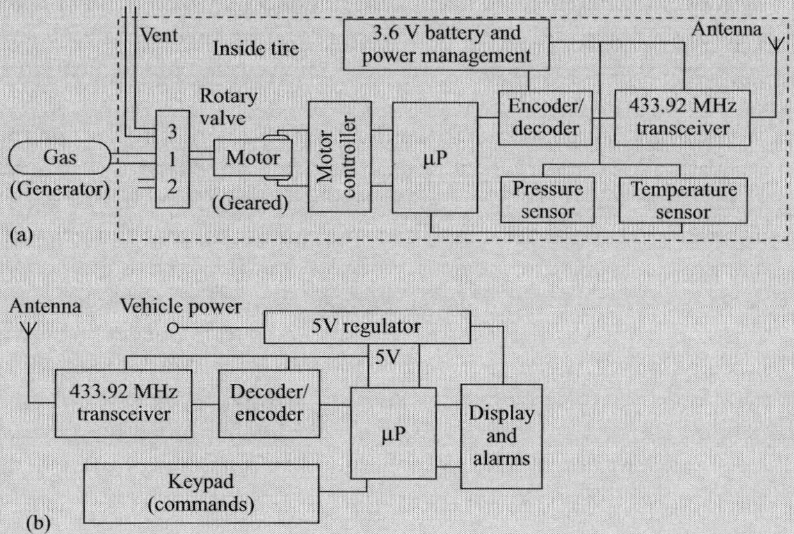

FIGURE 10.29 ■ A remote tire pressure controller. (a) In-tire smart actuator. (b) In-vehicle monitor and controller.

a pressure and a temperature sensor, provide feedback used to affect inflation or deflation. The pressure and temperature are sent to the operator over the two-way wireless link. The smart actuator also includes a power management system to ensure minimum power consumption and battery operation over many years.

In the vehicle, the operator can monitor the tires and can override the conditions in the tire by resetting the optimal conditions to match the terrain. The display may also include alarms such as low pressure, high temperature, failure of the valves, low battery, low pressure in the gas generator, and others. It should be noted as well that part of the "intelligence" of this actuator is the use of pressure and temperature sensors. To this may be added other sensors. For example, a vibration sensor may give the system information on surface conditions.

10.4.1 Wireless Sensors and Actuators and Issues Associated with Their Use

Sensors or actuators are not, by themselves, wireless. The term wireless refers more to the link with the device than to the device itself. In most cases a communication link is made available to send data and control signals out, and sometimes to receive as well. This means that the sensor must, by necessity, be a smart sensor, since most sensors do not produce data that can be sent directly over a wireless link. For example, a thermocouple produces a DC signal (or a slowly varying signal). This signal must first be digitized and then used to modulate a carrier before it can be transmitted. On the other end, at the processor, the opposite process must take place. Some sensors are truly digital and produce, for example, a signal whose frequency is proportional to the stimulus. These are usually easier to interface with wireless systems. All this is fairly obvious and represents essentially replacement of the physical link with a wireless link. However, it also means that remote sensing in the true sense can now take place. The sensor (or for that matter an actuator, or both) may be far from the processor, perhaps on a different continent, in space, or on a different planet. The communication link itself may be a short-range link or may use microwave communication links, wireless communication systems, or satellites, depending on the need.

Many sensing systems make use of short-range wireless communication, often in a dedicated link. These typically use one of the ISM (Industrial, Scientific, and Medical) or the short range devices (SRD) bands allocated for short-range unlicensed communication. The frequencies are used in a variety of applications, such as remote control (garage door openers, keyless entry in vehicles and buildings, driverless vehicles), hobbies, entertainment (model aircraft, vehicles, microphones), and data transfer. Operation in ISM bands is strictly enforced as to frequency, bandwidth, and transmitted power allowed. This also means that the range is short—typically less than 100 m, and often much less. Nevertheless, the range is often sufficient for remote sensing within a building, around the factory floor in a vehicle, or around the home. In many cases the necessary range is so small that even an induction link is sufficient, but nevertheless, this is a wireless link.

10.4.1.1 The ISM and SRD Bands

In the United States, the Federal Communications Commission (FCC) is responsible for the allocation of frequencies for general use by industry and the public. In Europe and many other countries, the European Telecommunications Standards Institute (ETSI), the European Radio Communication Office (ERO), and the Comité International Spécial des Perturbations Radioélectriques (CISPR) of the International Telecommunications Organization (ITO) regulate the use of frequencies. The allocations in the United States (and Canada) and those in Europe and other countries are not the same but somewhat overlap.

The ISM band was originally allocated for use by industry for such applications as microwave ovens, dielectric welders, and similar uses, as well as in medical applications, including microwave treatment of tumors. These frequencies are shown in **Table 10.1**. The low frequencies are commonly used in industrial microwave heating, welding, and cooking, but also in radio frequency identification (RFID) tagging and short-range communication. Other frequencies, such as the 2.45 GHz band, are used in consumer products such as microwave ovens, as well as in cellular phones, Wi-Fi systems, and many others. The 915 MHz band is universally used for communication and control, as well as for RFID applications. Some of the allocated frequencies have no current use, but have been incorporated for future use.

The short-range devices (SRD) frequencies have been allocated for what is often called "unregulated" use. Unregulated is a misnomer because they are actually very tightly regulated, but they are available for use without special licenses by any product as long as they conform to the provisions of the regulation in terms of frequency, bandwidth, power, and often, duty cycle. The SRD frequencies are shown in **Table 10.2**. Of these, the only truly internationally accepted frequency is 433 MHz. This is almost universally used for short-range control (keyless entry systems, garage door openers). In the United States, other bands have been used in the past for this purpose (290 MHz,

TABLE 10.1 ▪ ISM allocations, uses, and allowable power

Frequency	Some typical applications	Power/field strength
124–135 kHz	Low frequency, inductive coupling, RFIDs, tire pressure sensing	72 dBμA/m
6.765–6.795 MHz	Inductive coupling, RFIDs	42 dBμA/m
7.400–8.800 MHz	Article surveillance	9 dBμA/m
13.553–13.567 MHz	Inductive coupling, contactless smartcards, smartlabels, item management, dielectric welding, short-range communication	42 dBμA/m
26.957–27.283 MHz	Industrial microwave ovens, dielectric welding	42 dBμA/m
40.660–40.700 MHz	Industrial microwave ovens, dielectric welding	42 dBμA/m
433.050–434.79 MHz	Remote entry, wireless control	10–100 mW
2.400–2.483 GHz	Remote control, vehicle identification, microwave ovens, LANs, Bluetooth, WLAN, ZigBee, cordless telephones	4 W spread spectrum, United States/Canada only; 500 mW, Europe
5.725–5.875 GHz	Wireless video cameras for security, wireless communication, control, WiMAX, future use	4 W United States/Canada, 500 mW Europe
24.000–24.25 GHz	Future use	4 W United States/Canada, 500 mW Europe

Note: 1. The last three bands are divided into channels, each 0.5 MHz in bandwidth.
2. The power/field values are for communication applications. In microwave ovens the power is much higher since the system is enclosed.

TABLE 10.2 ■ SRD allocations, uses, and allowable power

Frequency [MHz]	Some applications	Power
433.050–434.79	See **Table 10.1**	10 mW
863.0–870.0	Various, including wireless audio, alarms, RFIDs, cellular communication (telephones)	5–500 mW depending on band
902.5–928	Same as above	4 W spread spectrum, United States/Canada only

310 MHz, 315MHz, and 418 MHz), but the tendency is to conform with international allocations. The higher bands (860–928 MHz) are still separate and there does not seem to be any convergence to common bands.

It should be noted that each frequency band comes with its own constraints. For example, the higher frequencies (860 MHz and higher) usually incorporate channels. One can operate within one channel or hop between them, but one cannot use a bandwidth that spans two channels. In the 433 MHz range, the bandwidth is fixed, there are no channels (i.e., a single band), power is constrained to 10 mW (or up to 100 mW under special license), and the duty cycle is no larger than 10%. That is, one is allowed to transmit for up to 1 s at a time with a minimum of 10 s between transmissions.

Sensors and actuators often operate in the ranges mentioned above either separately or in conjunction with other systems. For example, RFIDs are being used extensively for identification and tagging of items ranging from consumer products to pets. In that capacity they identify the item and contribute to efficient distribution and tracking of products. However, they are also very important in sensing. For example, many cars currently employ a key identification system whereby only the key that has been properly programmed can be used. The system employs an RFID in the key and a transceiver in the car dashboard or steering column that senses the presence and identifies the key. What is unique here is the very short range (usually below 1 m) and the wireless communication at low frequencies (typically 13 MHz). Other RFIDs employ true sensors to monitor a function, such as temperature or the health of a farm animal. As was stated above, there is nothing particularly ingenious in incorporating sensors in wireless systems once the special needs of signal processing imposed by the wireless system have been taken care of.

10.4.1.2 The Wireless Link and Data Handling

The wireless link in a sensor or actuator is similar to any conventional wireless link. The only unique aspect is the information being transferred over that link. It is exactly for this reason that the use of wireless links in conjunction with sensors is common and started early. But whereas the wireless link itself is conventional (for the most part), the handling of data is not. In particular, because of the diversity of sensors and actuators, and the data they produce or require, this is likely to be a more complex issue in the development of wireless sensors and actuators and in their successful use. Some sensors produce outputs that can be handled rather easily. Examples are SAW devices that produce a frequency proportional to the stimulus. Others, such as thermocouples, produce a DC output at very low amplitudes. This usually means that additional circuitry must be added before the signal can be transmitted over the wireless link. Of course, at the receiving end the signal must be restored to produce the desired output, again necessitating additional circuitry. In actuators the issues are similar, and in addition to wireless devices there will be a need for signal conditioning, amplifiers, and the like. Since the signals are processed for the purpose of matching the wireless requirements, we also need to worry about signal integrity, including noise and interference, and must take steps to overcome any deficiency introduced by the wireless link. The wireless link is but a tool and it cannot be allowed to alter the signal. For this reason, the information from most sensors and actuators is transferred using digital signals.

To accomplish all that, there are a number of components and methods that need to be taken into account. First and foremost are the transmitters, receivers, and transceivers that perform the actual wireless transfer of data. The sensors and actuators, together with any necessary interfacing circuitry, must match these in terms of needs—signal levels, frequencies (of data), bandwidth, data rate, etc. The signals cannot be transmitted directly. They must be modulated over the wireless carrier at the transmitter and demodulated at the receiver end to recover the data. Often, too, the signals must be encoded either for security in transmission (against eavesdropping), signal integrity (interference and corruption by signals from sources outside the link), or for safety (to ensure that the signal received in fact comes from the intended sensor or reaches the intended actuator). This is particularly important since many devices must share a single frequency or a small number of channels in a narrow frequency range. Additional considerations must be given to antennas, environmental conditions affecting the link, power requirements and power sources, and a host of other issues affecting the performance of the entire system.

A wireless link consists of at least a transmitter and a receiver, and each has an antenna. Many of the antennas typically used with sensors and actuators are simple quarter-wavelength monopole antenna. Others are loop antennas on printed circuit boards or miniature integrated antennas. Most antennas are omnidirectional, with maximum gain on the plane perpendicular to the antenna. Each antenna has its own properties, including gain, efficiency, and critically important, impedance. One usually assumes that the impedance of the antenna is matched to the transmitter or receiver, but if it is not, the transmitter output power is reduced, sometimes significantly, and so is the resulting range of transmission. Mismatched antennas also contribute to problems of data integrity.

Finally, and equally important, is the communication path. In most wireless systems this assumes communication in straight lines of sight (i.e., in lines on which the receiver and transmitter can "see" each other). Under these conditions the signal experiences two basic effects. One is the spread of the transmitted power onto an increasingly larger area, and hence diminishing power densities at larger distances. One can imagine this process in a simplistic manner assuming an isotropic antenna (an antenna that transmits uniformly in all directions in space) transmitting at power P. At a distance R from the antenna, the power density is $P/(4\pi R^2)$. Since the power received by a receiving antenna depends on the power density at its location, the power received depends on distance. The second effect is attenuation in air. Air is a lossy dielectric, that is, in addition to its permittivity and permeability it also has conductivity. The latter causes losses and the signal available at the receiver diminishes because of these losses. The signal diminishes exponentially, the exponent being dependent on the conductivity in the path of the transmitted signal through the attenuation constant (see **Section 9.4**).

But the path is often more complicated than this. For one, the signal cannot spread uniformly in all directions in space because of the presence of the earth (unless the link is in outer space). The presence of the earth also introduces additional losses (path losses). Any obstruction in the path of communication scatters part of the power and almost always introduces losses. Some of these issues were discussed in **Chapter 9** in conjunction with electromagnetic waves, and these apply here as well.

10.4.1.3 Transmitters, Receivers, and Transceivers

Figure 10.30 shows possible configurations for a wireless link for sensors and actuators. The transmitter and receiver are electronic circuits that can include oscillators, amplifiers, modulators, and demodulators (**Figure 10.30** shows these components separately). They can be rather simple or can be quite complex depending on what needs to be accomplished. We will view these as components rather than looking at their internal operation. Transmitters and receivers at specific frequencies and with specific properties (type of modulation, data rate, power, sensitivity, etc.) can be obtained as off-the-shelf devices or can be integrated within the sensor's package. Transceivers are devices that include a transmitter and a receiver within the same package to allow either two-way communication or data relaying. Although the principles involved in building receivers, transmitters, and transceivers are not very complicated, the devices themselves are highly specialized, requiring special high-frequency design methods to ensure proper and efficient operation, stable frequencies, and impedance matching. For this reason, with very few exceptions, these devices are individual components in the overall design and are rarely, if ever, integrated with sensors and actuators. In addition, all of them need an antenna, either internal to the device or external. The antenna can be printed on the printed circuit board, a dangling wire, an antenna screwed onto a special connector, or an integrated antenna. In most practical cases involving sensors the antenna is a monopole antenna one-quarter wavelength long or a loop antenna on a printed circuit board, but integrated antennas are also commonly used because of their small size.

10.4.2 Modulation and Demodulation

The carrier signal, that is, the signal transmitted by the transmitter, must be at the frequency allocated for the particular service. For example, if one were to use a link at 915 MHz, the frequency of the transmitted signal is 915 MHz with a bandwidth that is defined by the allocation. The information, that is, the signal from a sensor, must be carried by the carrier and fit within the bandwidth of the channel. To do so the carrier is modulated by the information (i.e., changed, modified) in some distinct way before transmission and then demodulated at the receiver to restore the information. There are a

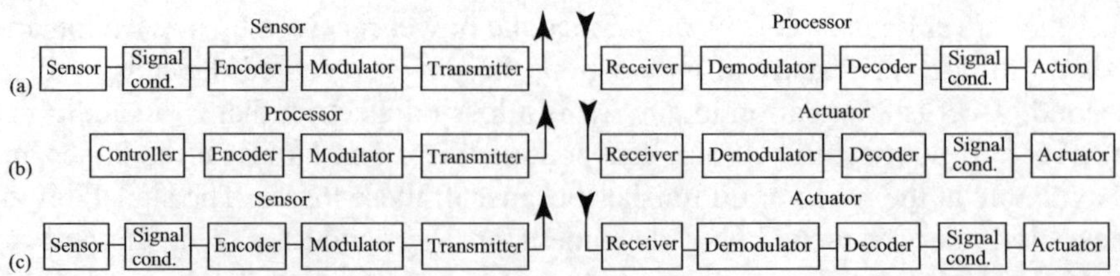

FIGURE 10.30 ■ Communication links. (a) Communication link between a sensor and a base station. (b) Communication link between a base station and an actuator. (c) Communication link between a sensor and an actuator.

number of methods of modulation and demodulation that are important in the context of sensors and actuators. If the sensor's output or if the signal to be transmitted to an actuator is analog, then the modulation can be analog modulation. If the signal is digital or if it is digitized before transmission, the modulation will be digital. In all cases, the carrier is an analog signal modulated by an analog or digital signal, that is, the transmitter transmits an analog signal.

10.4.2.1 Amplitude Modulation

The three most common analog modulation methods are amplitude, frequency, and phase modulation. Amplitude modulation (AM) is shown in **Figure 10.31** in a block diagram and in **Figure 10.32** as signals. It consists of modifying the amplitude of the carrier signal by the information signal. The depth of modulation can be controlled. A carrier signal of amplitude A_c and frequency f_c is generated and an information signal of amplitude A_m and frequency f_m is available (from a sensor or from any other source). The latter is also called the modulating signal. The carrier signal can be written as

$$A(t) = A_c \sin(2\pi f_c t). \tag{10.8}$$

The modulating signal will depend on its source, but for simplicity we will assume here it is also sinusoidal,

$$M(t) = A_m \cos(2\pi f_m t + \phi), \tag{10.9}$$

where a phase angle ϕ is added for generality. The frequency of the information signal must be much lower than the carrier frequency ($f_m \ll f_c$) and the amplitude of the carrier signal larger or equal to the amplitude of the information signal ($A_c \geq A_m$). In AM, the modulated signal transmitted has the following form:

$$S(t) = [A_c + A_m \cos(2\pi f_m t + \phi)]\sin(2\pi f_c t). \tag{10.10}$$

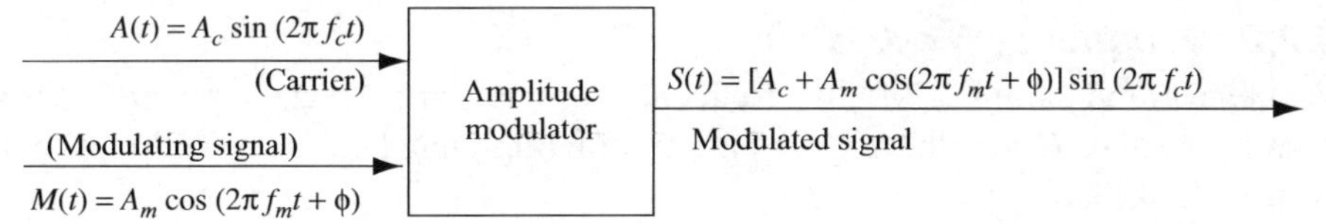

FIGURE 10.31 ▪ Amplitude modulation.

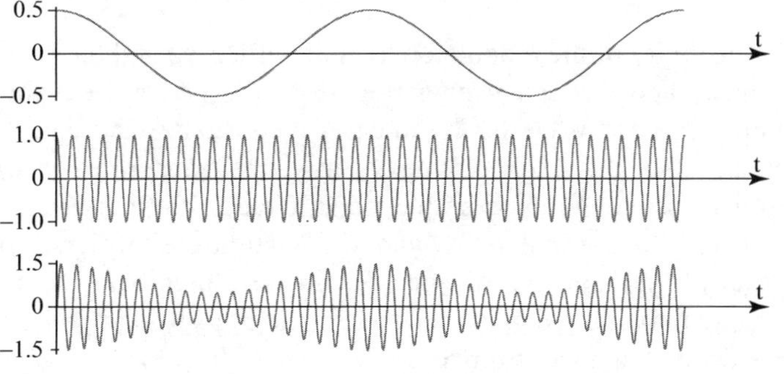

FIGURE 10.32 ▪ Amplitude modulation of an analog signal.

Clearly the amplitude of the modulated signal contains the information necessary: the amplitude and frequency of the information. This is shown in **Figure 10.32**. The ratio $m = A_m/A_c$ is called the modulation index or modulation depth, and assuming $A_m \leq A_c$, it varies between zero (no modulation) and 1 (100% modulation). In **Figure 10.32** the modulation index is 0.5, or 50% modulation. Note that the product of the two signals produces two terms: one with frequency $f_c - f_m$ and one with frequency $f_c + f_m$, and a third signal with frequency f_c:

$$S(t) = A_c\sin(2\pi f_c t) + A_m\cos(2\pi f_m t + \phi)\sin(2\pi f_c)$$
$$= A_c\sin(2\pi f_c) + \frac{A_m}{2}\sin[2\pi(f_c + f_m)t + \phi] + \frac{A_m}{2}\sin[2\pi(f_c - f_m)t + \phi]. \quad (10.11)$$

This means that the bandwidth necessary to transmit the information is $2f_m$ (i.e., from $f_c - f_m$ to $f_c + f_m$). This must not exceed the width of the channel allocated for the transmission.

There are many variations on basic AM. For example, if we set $A_c = 0$ in **Equation 10.10** or **10.11**, that is, if we suppress the carrier (by filtering it out prior to transmission), the information is still available in the two terms called sidebands. In fact, the same information—amplitude, frequency, and phase—is available twice, in both the upper and lower sidebands. One then has the option of transmitting all three signals (conventional AM signal), transmitting both sidebands but no carrier (double sideband [DSB] modulation), or transmitting one sideband (single sideband [SSB] modulation). Although in most analog applications in sensors and actuators conventional AM is used, it is well worth noting that the other options exist and that these have some value, as well as consequences. For example, SSB transmission reduces the required bandwidth to half that of the AM transmission and the amount of power needed for the same signal at the demodulator is lower, as can be seen from **Equation 10.11**. On the other hand, the circuitry on both the transmitter and receiver side is more complex. Because of that, these methods are not used with sensors, but one can envision special cases in which these may be advantageous.

10.4.2.2 Frequency Modulation

In frequency modulation (FM) the frequency of the carrier varies linearly with the information signal. Given the carrier and modulating signals as above, the modulated signal now looks as follows:

$$S(t) = A_c\cos(2\pi f_c t + 2\pi k_f \int_0^t A_m\cos(2\pi f_m t + \phi)dt), \quad (10.12)$$

where k_f is the sensitivity of the modulator (in units of Hz/V) and the product $\Delta f = k_f A_m$ represents the maximum frequency deviation from the center frequency (carrier frequency), assuming that the value of the integral is normalized to ± 1. The value of k_f must be selected so that the frequency deviation to be transmitted is within the band available for transmission. As an example, in FM radio, the bandwidth available per channel is 200 kHz. The maximum frequency deviation therefore is a measure of the amplitude of the information, A_m. It should also be remembered that for sinusoidal signals, $2A_m k_f$ is the maximum bandwidth used in the transmission, but for other signals it may be considerably higher. **Figure 10.33** shows the block diagram for FM and

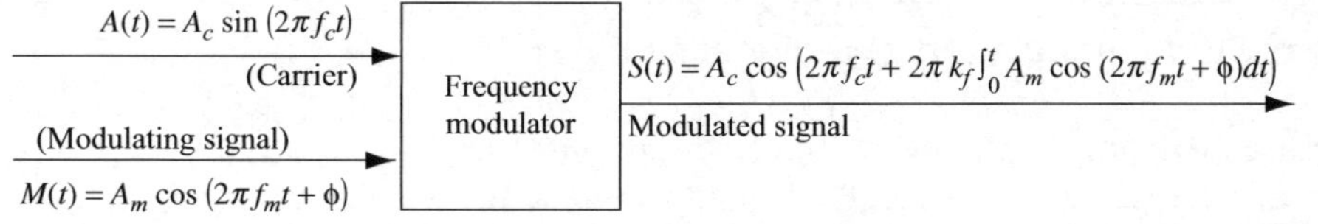

FIGURE 10.33 ■ Block diagram of a frequency modulator.

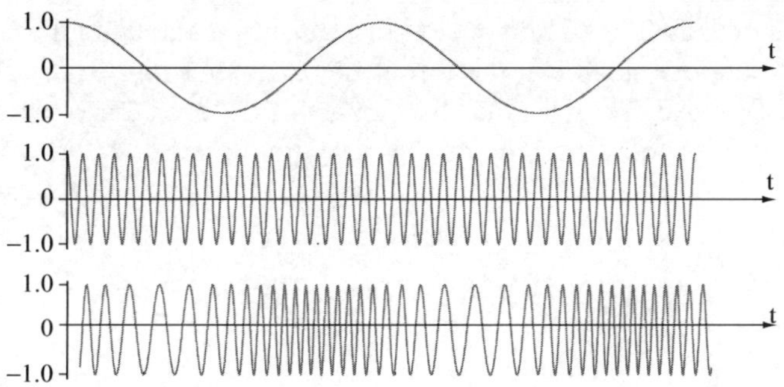

FIGURE 10.34 ■ Frequency modulation of an analog signal.

Figure 10.34 shows the expected signal for a sinusoidal modulating signal. As with AM, a modulation index $m = \Delta f / f_m$ may be defined, but unlike AM, m can be larger than 1. If $m \ll 1$, the modulation is called narrowband modulation, while $m \gg 1$ results in wideband modulation.

For sinusoidal signals the integral in **Equation (10.12)** can be evaluated analytically and the modulated signal becomes

$$S(t) = A_c \cos\left(2\pi f_c t + k_f A_m \frac{\cos(2\pi f_m t + \phi)}{f_m}\right). \tag{10.13}$$

10.4.2.3 Phase Modulation

In phase modulation (PM), the phase of the carrier varies linearly with the information signal. The output of the modulator now looks as follows:

$$S(t) = A_c \cos\left(2\pi f_c t + k_p A_m \cos(2\pi f_m t + \phi)\right), \tag{10.14}$$

where k_p is the phase sensitivity of the modulator (in units of rad/V) and $\Delta_p = k_p A_m$ is the maximum phase deviation due to the signal. This deviation represents the amplitude of the information signal, A_m. Since a change in frequency or in phase will have the same effect on the signal, FM may be viewed as a particular case of PM.

There are other methods of modulation and many variants on the methods described here, but in all cases one has to find a way to modulate a fixed-frequency carrier signal with an information signal.

When the signals are digital the modulation is somewhat different. The carrier is still the same, a sinusoidal signal of constant frequency and constant amplitude. But since the digital signal has only two states, the modulation and the representation of the modulated signal are simplified. The digital equivalents for amplitude, frequency, and phase modulation are amplitude shift keying (ASK), frequency shift keying (FSK), and phase shift keying (PSK).

EXAMPLE 10.10 Amplitude modulation of a digital signal

A train of pulses with a 50% duty cycle and a frequency of 1000 Hz is to be transmitted over an AM radio at a carrier frequency of 1.2 MHz. Each channel on the AM band has a bandwidth of 10 kHz. Calculate and plot the shape of the signal received assuming demodulation does not introduce any errors in the signal but limits it to ±5 kHz in bandwidth.

Solution: Because the transmission is that of square pulses, one must first calculate the frequency content of the pulses using the Fourier transform, then limit the transmission of the carrier frequency to 5 kHz on each side of the carrier. All other harmonics of the signal are removed and reconstruction in the receiver is due to those harmonics that lie within the 5 kHz bandwidth.

We start with the general Fourier series representation of the function $f(t)$ shown in **Figure 10.35a**,

$$f(t) = \frac{1}{2}a_0 + \sum_{n=1}^{\infty} a_n \cos\left(\frac{n\pi t}{T}\right) + \sum_{n=1}^{\infty} b_n \sin\left(\frac{n\pi t}{T}\right),$$

where

$$a_0 = \frac{1}{T}\int_0^{2T} f(t)dt,$$

$$a_n = \frac{1}{T}\int_0^{2T} f(t)\cos\left(\frac{n\pi x}{T}\right)dt,$$

and

$$b_n = \frac{1}{T}\int_0^{2T} f(t)\sin\left(\frac{n\pi x}{T}\right)dt.$$

Because the function $f(t)$ is odd, $a_0 = 1$, $a_n = 0$ and the pulse can be represented as

$$F(t) = \frac{1}{2} + \sum_{n=1}^{\infty} b_n \sin\left(\frac{n\pi t}{T}\right).$$

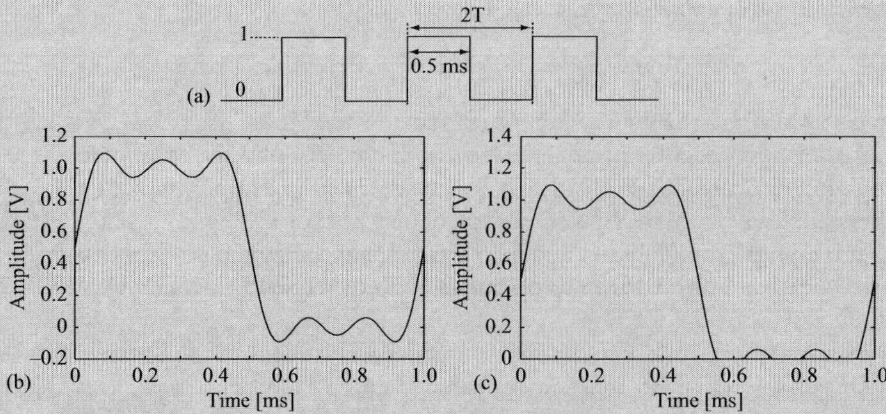

FIGURE 10.35 ■ (a) The digital signal before modulation and transmission. (b) The reconstituted signal as the Fourier transform generates it. (c) The signal after demodulation.

$a_0/2$ is clearly the DC level of the signal, and in this case it is equal to ½. To calculate the coefficient b_n we write

$$b_n = \frac{1}{T}\int_{t=0}^{T} f(t)\sin\left(\frac{n\pi t}{T}\right)dt = \frac{1}{T}\int_{t=0}^{T} 1\sin\left(\frac{n\pi t}{T}\right)dt = \begin{cases} 0 & \text{for } n \text{ even} \\ \dfrac{2}{n\pi} & \text{for } n \text{ odd} \end{cases}.$$

Thus the representation is

$$F(t) = \frac{1}{2} + \frac{2}{\pi}\sum_{n=1,3,5...}^{\infty} \frac{1}{n}\sin\left(\frac{n\pi t}{T}\right).$$

Now, since the frequency of the signal is $f = 1/2T$, we have $1/T = 2f$ and we can write

$$F(t) = \frac{1}{2} + \frac{2}{\pi}\sum_{n=1,3,5...}^{\infty} \frac{1}{n}\sin(2\pi nft) = \frac{1}{2} + \frac{2}{\pi}\left(\sin 2\pi ft + \frac{1}{3}\sin 6\pi ft + \frac{1}{5}\sin 10\pi ft + ...\right).$$

In our case, $f = 1$ kHz and the bandwidth is 5 kHz. Therefore the only terms that can exist are the first three terms in the expansion (the fundamental frequency is 1 kHz, the third harmonic is at 3 kHz, and the fifth harmonic is at 5 kHz. Thus the signal transmitted (and hence the signal at the receiver) is

$$F(t) = \frac{1}{2} + \frac{2}{\pi}\left(\sin 2\pi \times 10^3 t + \frac{1}{3}\sin 6\pi \times 10^3 t + \frac{1}{5}\sin 10\pi \times 10^3 t\right), \quad 0 \le t \le 10^{-3} s,$$

where the amplitude was assumed to be unity since the signal can be amplified as needed.

The plot of $F(t)$ is shown in **Figure 10.35b**. After demodulation only the positive part of the pulse will be available, as shown in **Figure 10.35c**. Obviously this figure represents only one cycle of the signal.

Note that the reason for the rounded-off signal is the lack of the higher harmonics because of the narrow bandwidth of the transmitter. Nevertheless, the signal is fully recoverable after some signal conditioning.

10.4.2.4 Amplitude Shift Keying

Amplitude shift keying (ASK) is a common method of modulation for digital signals and may be viewed as equivalent to AM. In this method the carrier is modulated by introducing a shift in the amplitude of the carrier to correspond to the digital signal. The amplitude of the carrier shifts with the bitstream so that a "1" corresponds to one amplitude level whereas a "0" corresponds to a second amplitude level:

$$A(t) = mA_c\sin(2\pi f_c t), m = [a, b], \tag{10.15}$$

where m takes two different values a and b (e.g., $m = 0.2$ for logical "0" and $m = 0.8$ for logical "1"). The modulation of the [1,0,0,1,0,1,1,0,0,1] stream using these two levels is shown in **Figure 10.36a**.

A particularly common ASK modulation method uses $m = [0,1]$ and is called on/off keying (OOK) and is commonly used. In this method the carrier is switched on for

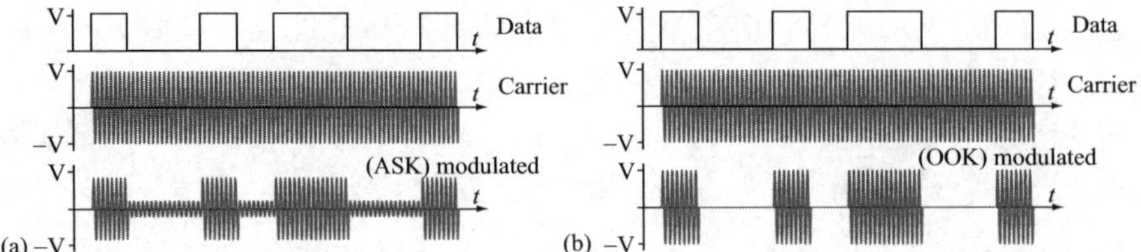

FIGURE 10.36 ■ (a) Amplitude shift keying. (b) On/off keying.

the "1" level and off for the "0" level, as shown in **Figure 10.36b**. One of the advantages of this method is that power is not transmitted during zero periods and hence the power requirements of the transmitter are minimized, an important issue in low-power systems and, in particular, in battery-operated devices. The OOK method may be described as

$$A(t) = mA_c\sin(2\pi f_c t), m = [0, 1]. \tag{10.16}$$

10.4.2.5 Frequency Shift Keying

As the name implies, in frequency shift keying (FSK) the frequency of the carrier is switched between two frequencies, one representing the "0" level, the other the "1" level. This is shown in **Figure 10.37a**. The representation is

$$S(t) = A_c\sin(2\pi f_i t), f_i = [f_1, f_2]. \tag{10.17}$$

10.4.2.6 Phase Shift Keying

In phase shift keying (PSK) the phase of the carrier is shifted to represent the digital signal. There are many variations, but possibly the simplest (also called binary phase shift keying [BPSK]) uses a zero phase for, say, "1", and shifts it by π for "0" or vice versa. The representation is

$$S(t) = A_c\sin(2\pi ft) \text{ for } 1,\ A_c\sin(2\pi ft + \pi) \text{ for } 0. \tag{10.18}$$

The choice above is arbitrary and other phase values may be used. This form is shown in **Figure 10.37b**.

There are many more methods of modulation, both for analog (quadrature amplitude modulation [QAM], space modulation [SM], etc.) and for digital modulation (minimum shift keying [MSK], pulse position modulation [PPM], continuous phase modulation [CPM], etc.). There are also methods for modulation for spread spectrum applications and some special, limited use methods. Again, as indicated above, the first important issue is to be able to represent the data so that it can be detected in the receiver. Of course, each method has its own advantages and disadvantages. Some, such as AM and OOK, are very simple but may not be the most efficient. Others require a wider bandwidth (FM, PSK) or are less susceptible to noise and interference (FM, FSK).

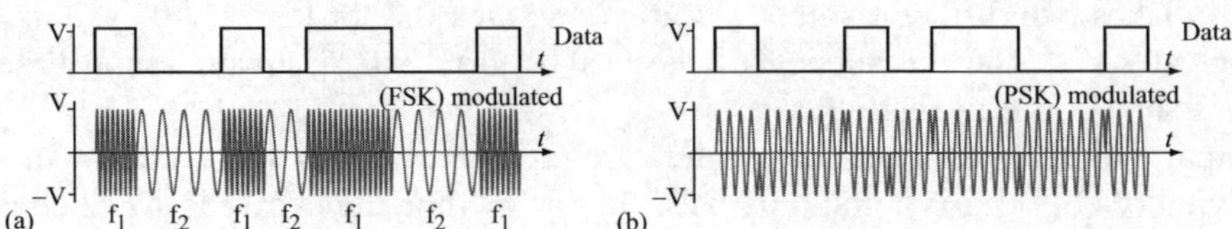

FIGURE 10.37 ■ (a) Frequency shift keying. (b) Phase shift keying.

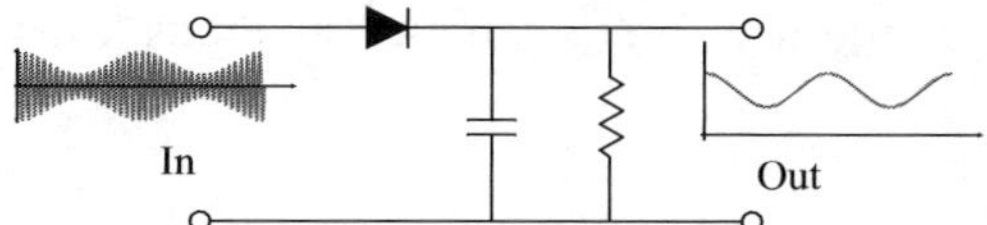

FIGURE 10.38 ▪ The principle of amplitude demodulation.

10.4.3 Demodulation

Any modulated information must be demodulated after reception to restore it to usable form. The principles are rather simple even though the circuits may not be. A demodulator must accept the modulated signal and produce a signal identical to the original. Since a signal, such as a digital signal, is represented by an amplitude and a frequency, the demodulator must reproduce the amplitude without modifying the frequency of the original signal. In amplitude methods of modulation, the signal maintains its amplitude but rides on the carrier. Therefore demodulation simply means filtering out the high-frequency carrier while retaining the low-frequency amplitude or envelope. In frequency and phase demodulation, as in amplitude demodulation, the modulation process is reversed to retrieve the original information.

10.4.3.1 Amplitude Demodulation

An AM demodulator is simply a rectifier that removes the negative part of the modulated signal (see **Figure 10.32**), followed by a capacitor of sufficient size to have low impedance at the frequency of the carrier but sufficiently high impedance at the frequency of the signal. **Figure 10.38** shows the fundamental AM demodulator with the signals before and after rectification and filtering. The circuit is an envelope detector. At the end of the process the signal in **Equation (10.10)** results in

$$M(t) = A_m \cos(2\pi f_m t + \phi). \tag{10.19}$$

This is the original modulating signal in **Equation (10.9)**.

10.4.3.2 Frequency and Phase Demodulation

The two methods are rather similar, as can be seen from **Equations (10.12)** and **(10.14)**. Conceptually, a frequency demodulator requires a differentiator to remove the integration in **Equation (10.12)** followed by an AM demodulator to remove the high-frequency components. A phase demodulator is essentially the same, but since **Equation (10.14)** does not contain the integral term, that must be added. Hence a phase demodulator includes an integrator. These are shown in **Figure 10.39**. It should be noted that the methods here are schematic only. In practice, demodulation can be done in a number of ways. One of the simplest will be discussed in **Chapter 11** and is called a frequency-to-

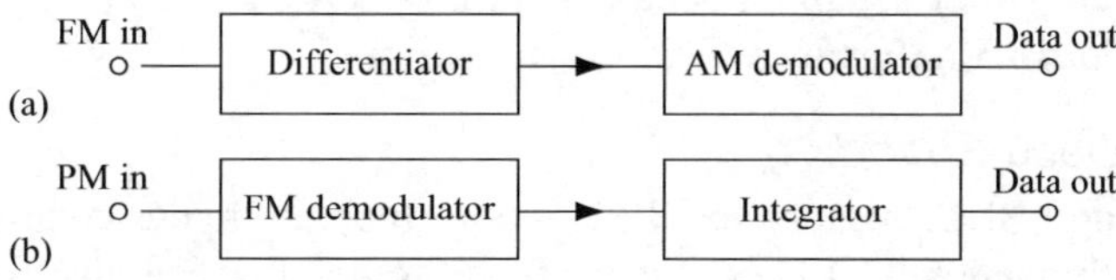

FIGURE 10.39 ▪ (a) The principle of frequency demodulation. (b) The principle of phase demodulation.

voltage converter. There are other, more complex methods in use, but at this point the principles will suffice.

10.4.4 Encoding and Decoding

Encoding of digital signals is important for a number of reasons. First, it helps in preventing data loss and data corruption by creating a "common language" between the transmitter and receiver, that is, the receiver "knows" what to look for when decoding the message. For example, the encoded signal may contain clock synchronization information that can be used in the receiver to restore the clock and to detect pulse start and end, pulse width, etc. Encoding can also separate between uplink and downlink data streams and facilitate higher data rates. Combined with additional information such as identification numbers, encoding allows secure communication and allows multiple links to share a single channel without interfering with each other. Error correcting codes and secure communication codes are also of significant concern in sensors and actuators, but these are out of scope of this discussion.

There are literally dozens of codes that are being used for encoding, ranging from the very simple to very complex, all with specific properties and applications. We will look at a few simple codes and their characteristics as examples of what can be done. Once a code is defined by its properties, encoding and decoding are implemented in software, although hardware modules can also be built for specific encoding and decoding applications. In a practical application the need for, properties, and cost of encoding must be evaluated and the selection must be made based on these and other requirements of the system. Also note that the encoding methods used in practice are described in standards.

10.4.4.1 Unipolar and Bipolar Encoding

The simplest and most obvious encoding is unipolar code. It represents a logical "1" with a positive voltage and a logical "0" with a zero voltage, thereby creating an obvious and direct representation of a data stream. The coded signal is clocked, but because no information about the clock is encoded, the decoder cannot reconstruct the clock or synchronize with it. An additional disadvantage of the method is that the average of the signal (DC level) is approximately half the maximum value of the voltage (logic "1"). This can be reduced to approximately half of its value by forcing the signal to go to zero in the middle of the bit. However, whereas the output returns to zero for the logical "1", the zero remains at its normal position (which happens to be zero). In a bipolar code, that is, one that uses a positive voltage to represent one state and a negative voltage to represent a zero, the return to zero in the middle of either pulse allows for clock synchronization, and we say that the code is self-clocking. Also bipolar codes have zero DC levels, helping to reduce the power needed for transmission. However, in bipolar codes there is not usually a need to return to zero. These are non-return-to-zero (NRZ) codes. **Figure 10.40** shows these considerations.

10.4.4.2 Biphase Encoding

Biphase encoding (BPC, also called FM1 code, biphase mark code [BMC] or frequency–double frequency [F2F] code) starts with a clock that is twice the data rate (see **Figure 10.41**). Each logical state of the digital signal is represented by two bits, delimited by the clock. The logical level in the stream produced by the code changes

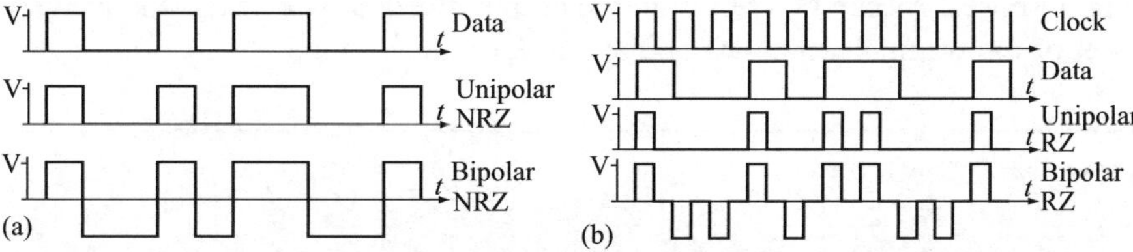

FIGURE 10.40 ■ Unipolar and bipolar codes. (a) Unipolar and bipolar NRZ codes. (b) Unipolar and bipolar return-to-zero (RZ) codes. Note that these require a clock.

state at the end of a clock cycle and also changes in the middle of any logical "1" (but not in the middle of logical "0"). As a result, logical "1" in the data stream is represented by two different bits (10 or 01), whereas logical "0" is represented by two identical consecutive bits (00 or 11). This is shown in **Figure 10.41**. To be able to detect these combinations it is absolutely necessary to reconstruct the clock in the receiver. This is done simply by noticing that the signal output stream changes (from 0 to 1 or from 1 to 0) every one or every two clock cycles. This establishes synchronization with the local clock in the receiver. Now it is only a question of comparing the first and second consecutive bits in every cell composed of two clock cycles. Biphase coded signals have zero average DC voltage, a feature that helps reduce power in the transmitter as well as noise. Of course, the price to pay is a clock rate twice as high and hence a data rate that is twice as high as the original data.

10.4.4.3 Manchester Code

Manchester code (also known as phase encoding [PE]) is one of the most common methods of encoding for digital signals. Each data bit occupies the same time slot—one clock cycle (see **Figure 10.42**). The output changes state on the negative edge of the clock, meaning that the transition occurs in the middle of the data bit. The direction of this midbit transition indicates the bit, whereas any other transitions do not carry information. A transition from logical "0" to logical "1" represents a 0, whereas a transition from logical "1" to logical "0" represents a 1 (a variation of this, called the inverse Manchester code inverses these transitions). The transitions in the Manchester code allow the receiver to recover the clock and align the bits with the clock for

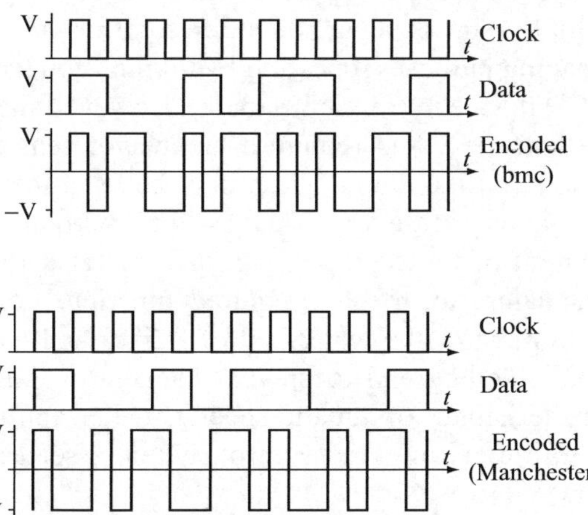

FIGURE 10.41 ■ Biphase mark code or biphase encoding.

FIGURE 10.42 ■ Manchester encoding.

decoding purposes. As can be seen from the output, the data rate of the encoded signal is twice that of the original data. The signal has zero DC voltage.

EXAMPLE 10.11 DTMF encoding

Dual tone multiple frequency (DTMF) encoding was originally developed for tone dialing on land telephones using numeric keypads, but it is also used for control in communication networks and other applications. The system transmits two tones (sinusoidal signals at defined frequencies) for each data bit being sent. These tones are detected and decoded in the decoder to identify the information.

Figure 10.43 shows the standard used for telephone dialing as well as the encoded and decoded information for the number 6027 (for dialing interior extension numbers within an organization). The encoder transmits the two frequencies associated with each number mixed together. Pressing the number 6 transmits a combined signal at 770 and 1447 Hz for 70 ms. The encoder then introduces a break of 40 ms before sending the second number by mixing 941 and 1336 Hz signals for 70 ms, and so on until the required number has been dialed. In the decoder, the frequencies are separated into two streams to identify the number dialed. The frequencies used were selected so that none is a harmonic of any other and their sum or difference do not result in any of the eight frequencies used.

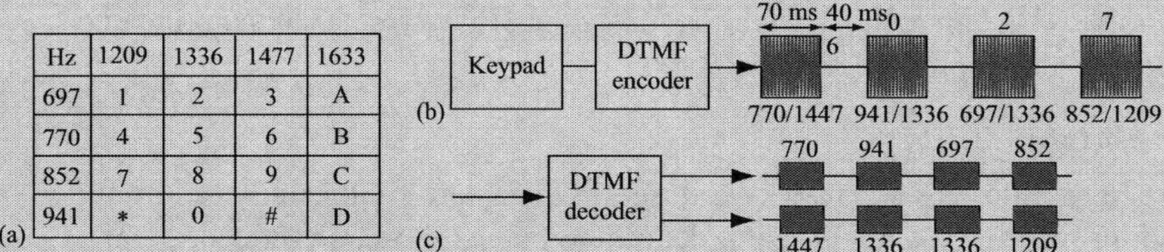

Hz	1209	1336	1477	1633
697	1	2	3	A
770	4	5	6	B
852	7	8	9	C
941	*	0	#	D

FIGURE 10.43 ■ (a) DTMF frequencies as used in the United States. (b) Encoded signal. (c) Decoded signals for dialed number 6027.

10.5 | SENSOR NETWORKS

Our discussion so far has concentrated on sensors and actuators operating independently, or in some cases, multiple sensors being connected to affect an output (thermocouples, for example) or for imaging purposes (one- and two-dimensional optical arrays such as a scanner sensor or a CCD device in a digital camera). However, there are systems that are much more complex than these and require a number of sensors (sometimes a large number) in a distributed configuration. These may be used to sense a distributed stimulus, sometimes over a very large area, such as, for environmental monitoring or for traffic control. The output of these sensors can then be used to make decisions or to operate appropriate actuators to affect a required function. For instance, a series of sensors may be used to monitor the watershed of a river and automatically open spill gates or dams to avoid flooding and to prevent damage or, alternatively, to regulate water flow to protect the habitat of aquatic species. Other applications are in sensing hazardous materials, traffic sensing and control, wireless security networks in public areas, and many others.

In a system of this type, the network is relatively simple. Each sensor performs its sensing function and transmits the data to a processor or a central node by an appropriate communication link, including wireless. Whereas the transmission may go through intermediate stations, the final goal is to get the data from all sensors to a processing center where it can be analyzed, verified, and then used to make appropriate decisions.

The discussion of sensor networks does not preclude the inclusion of actuators in the network or, indeed, networks of actuators. In fact, based on the principle that networks are normally used as inputs to a system, actuators must, in some way, be present. In some cases they will be integral with the network, whereby the data collected within the network may be used to act locally. For example, a traffic control system may be using wire loops to detect the presence of vehicles and activate traffic lights at an intersection, or the sensors may follow an emergency vehicle and activate appropriate traffic lights along its path. In other networks the actuators may be at a central location, such as in the spillway of a dam or in an air conditioning system in a building. There are differences, of course, especially in power requirements and interfacing of actuators, but in terms of the communication requirements over the network, sensors and actuators are similar.

A network of sensors is a distributed system in which a sensor (or actuator) is physically placed at a node of the network to perform a stated function. Data must be transmitted from the node to a processor (one-directional communication), from the processor to the node, or both (two-way communication). The nodes may be distributed uniformly or nonuniformly and each node may be of a different nature (i.e., a different type of sensor or actuator). In many cases nodes are in fixed locations, but not always. A sensors network may be used to monitor vehicles or the migration of birds. Communication may be by wire or wireless, or a combination of the two, depending on the needs and cost of the system. For example, temperature control sensors and actuators in a building are often wired since they are in fixed locations and wiring is usually built-in during construction. On the other hand, an ad hoc network may be added to monitor, say, traffic in a library, for which a simple wireless network may be more appropriate. Some applications require more complex communication paths. For example, monitoring the migration of marine species may require wireless links, fixed stations on buoys and ashore, moving nodes on ships, satellite links, and wired links on land, with all the difficulties these entail. Networks can also take advantage of land and cell communication systems and, of course, the Internet. For example, home automation and monitoring systems can use all of these means to monitor and control appliances in a home remotely or locally. Local control may be by wire or wireless links, whereas remote control may use telephone links or Internet-controlled appliances. Indeed, the variety in a sensor network is as wide as the imagination allows.

The simplest structure of a sensor network is afforded by direct link between the nodes and the center node, either by wire or through wireless links. This is shown in **Figure 10.44a** and is sometimes called a star network. This structure is appropriate when the number of nodes is small, the distances short, or when some feature in the network justifies this approach. This may be the case in a building where, for example, each temperature sensor is wired independently and all sensors connected to a processor. Similarly, if the nodes cover a small area, each node may be equipped with a wireless unit, and since all units are within range of the hub, communication can take place directly. The communication may be one-directional or bidirectional, as needed. The network can be extended by connecting the hub to other hubs or to points of service such

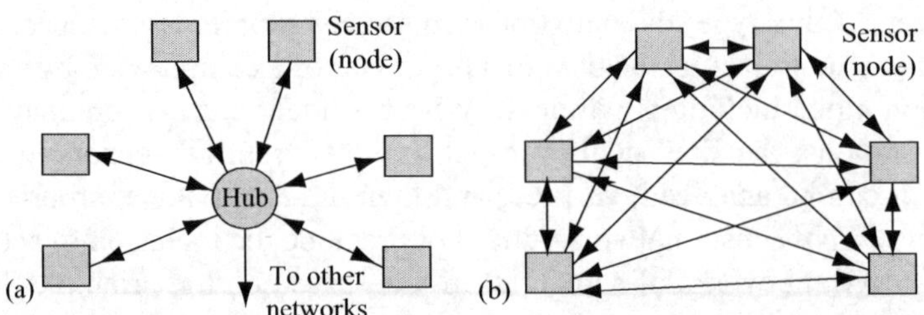

FIGURE 10.44 ■ (a) A star network. (b) A fully connected network. A hub or a central node may also be included.

as Ethernet, a bus, satellites, and so on. The nodes in the network, especially in a wireless network, do not have to be fixed and may represent moving vehicles, for example, or nodes that only turn on following a schedule or turn on randomly. Another possible network is the fully connected network shown in **Figure 10.44b**. Here each sensor can communicate with every other sensor in the network. The system is complex and protocols are needed to arbitrate the various communication paths. Also, as the number of nodes increases, the complexity increases rapidly. Unless there is a good reason for a network of this type, it should be avoided.

Another type of network is shown in **Figure 10.45**. Here the sensors connect to a bus through an appropriate protocol. This system is employed in vehicles, for example, using a special bus (the Controller Area Network [CAN] bus is most often used in vehicles) or in computer systems (using, e.g., the Universal Serial Bus [USB] protocol). Each sensor/actuator connects to the bus through a bus interface and communication on the bus is controlled through a protocol that includes node identification. The communication, as in any network, may be one or two directional, and different nodes may have different priorities on the network.

In many cases a direct link is not practical, either because of distances (range), cost, or other constraints. In these cases communication takes place in a leapfrog fashion—each node communicates with its nearest neighbor or with a local center by wire or, through a wireless link. Two possibilities are shown in **Figure 10.46**. This type of communication is more complex and requires node identification, acknowledgement of data transfer, and

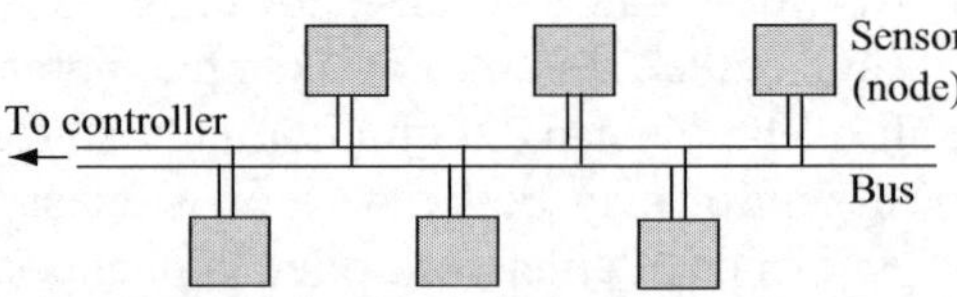

FIGURE 10.45 ■ A bus network. The communication is controlled by the bus—nodes do not communicate directly between each other.

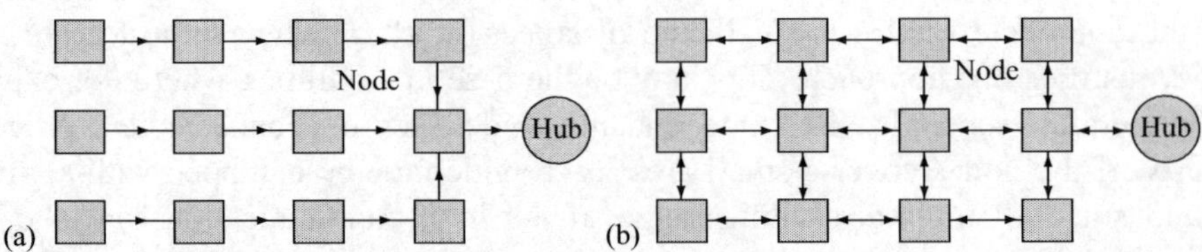

FIGURE 10.46 ■ Distributed sensor network showing neighbor-to-neighbor communication. (a) One-way communication from sensors to the center. (b) Two-way communication.

some means to control the reliability of data transfer. Software protocols need to be established to take care of all this. It is also more complex in terms of the hardware involved. In addition to the need for power, each node must contain at the very minimum a transceiver and some means for local processing of data. Likely the nodes of a network will contain a microprocessor and be controlled by an appropriate operating system. In any of these networks there may be one or more hubs or central nodes, and the network may communicate remotely through a wired link, a satellite, or the cell communication system.

The cell telephone system is the largest wireless network in existence and it is only natural that sensors and actuators take advantage of it. Cellular communication nodes similar to those in cell phones are used to connect and transfer data. The network already has in place all the means of connecting between any two nodes in the system and bidirectional data transfer. It can also take care of moving nodes through roaming protocols and can be used for both sensors and actuators. Although the system may be costly, it is entirely flexible and spans the globe. It allows for a distributed and reconfigurable structure. **Figure 10.47** shows a depiction of a network of this kind. Needless to say, a cell phone-based network can be combined with wired and wireless nodes or with satellite-linked nodes.

Finally, one should remember that the Internet offers some unique possibilities for networks. There are existing protocols and hardware that allow one to use the Internet as the basis of sensor and actuator networks as well as control and automation. In industry or in the home, the Web offers simple, reconfigurable ways of networking and remote data collection and activation. Even when the network itself is not Web based, connection to the Web at one or multiple points in the Web is common.

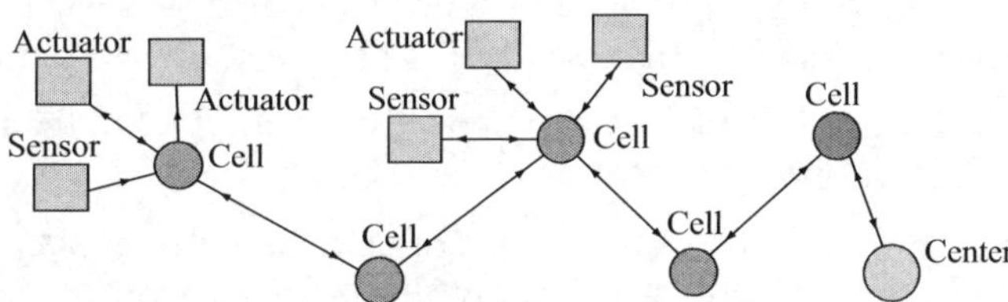

FIGURE 10.47 ■ A cell telephone-based network of sensors and actuators. Some nodes may only support or require one-way communication, others may require two-way communication.

EXAMPLE 10.12 A simple linear wireless network

When oil needs to be pumped to and from large oil supertankers, it is commonly done through floating double-walled hoses. The double-walled rubber hose is designed so that leakage of one wall (either leakage of oil through the inner hose or water through the outer hose) causes accumulation of the fluid between the two hoses. A sensor between the two layers senses the presence of the fluid and alerts the operator to the need to replace the section before both walls leak and contamination occurs. The hose is made of sections about 10–12 m long with an inner diameter of 15–60 cm. The hoses can be quite long, sometimes 1 km or longer. The detection of leaking hoses can be done by inspection by divers or the information can be transferred through wireless links. A linear network suitable for this purpose is shown below. Each node is made of a sensor that

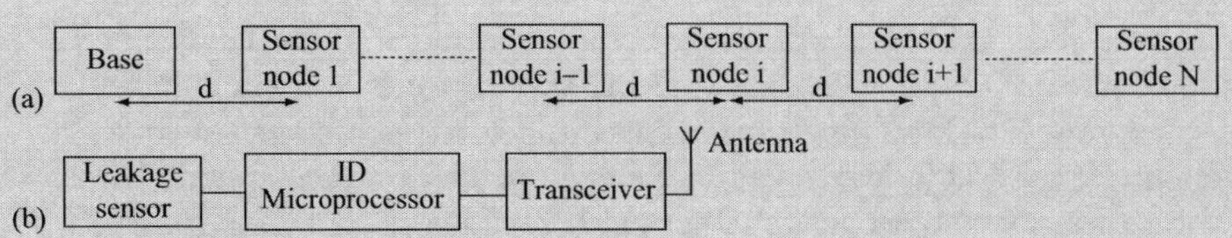

FIGURE 10.48 ■ (a) The structure of the linear sensor network for leakage detection. (b) The main components of a node in the network.

detects the presence of water or oil and a transceiver (see **Figure 10.48a**). A possible algorithm for a system of this type is as follows:

1. All sensor nodes are in sleep mode and never transmit (to conserve battery power):
   a. Unless a sensor detects a leak.
   b. In response to interrogation by the base to check the functioning of the network.
   c. If the battery in a sensor is below a certain voltage, indicating the node must be replaced.
2. Detection of a leak is transmitted by the sensor. This transmission is received by at least the two closest (neighboring) nodes. The receiving node retransmits the information and identification information.
3. Each sensor node that receives a transmission retransmits it to the nearest two sensors with additional information on itself (such as an identification number).
4. Any sensor that has already transmitted that information will not retransmit it within a given time to allow the network to transmit the data to the base and revert to sleep mode.
5. Once the information is received by the base, an acknowledgement is sent back to the network to acknowledge receipt of the information.
6. If no acknowledgement is received by the initiating node within a set time, the information is retransmitted, repeating the process.
7. The information path can be interrupted if a node is not operational (e.g., due to a damaged sensor or a low battery). To mitigate this possibility, the base interrogates the network at given intervals to ensure the viability of the network and to identify inoperative nodes that can then be replaced.

Note: Wireless transmission in water, especially seawater, is very limited in range and even 10 m may be too long. If that happens to be the case, transmission can take place internal to the hose, through oil (which is a dielectric). Antennas may also be placed between the two walls of the hose since at least part of the hose is above water. Other possibilities are to use small floats for antennas. A system like this requires the use of extremely low power devices to ensure a battery life measured in years.

10.6 | PROBLEMS

Microelectromechanical system (MEMS) sensors and actuators

10.1 The MEMS transformer. In some MEMS devices, such as MEMS fluxgate magnetometers or magnetic actuators, it becomes necessary to build inductors, and in transformers the magnetic core must be closed.

a. Show in a few steps how a magnetic transformer can be built as a MEMS device.

b. Describe the steps, methods, and materials needed.

c. Comment on the challenges of magnetic structures within MEMS.

10.2 Capacitive accelerometer. Capacitive devices, including accelerometers, are relatively easy to build with MEMS techniques.

a. Show the steps needed to build a capacitive accelerometer such as the one shown in **Figure 6.17a**.

b. Show the steps needed to build a capacitive accelerometer such as the one shown in **Figure 6.17b**.

10.3 Piezoresistive accelerometer. The basic structure of a piezoresistive accelerometer is shown in **Figure 10.10**.

a. Describe the steps needed to build such a device.

b. Describe the materials needed and the challenges of a device of this complexity.

10.4 Activation of a tuning fork sensor. In **Figures 10.13** and **10.14**, piezoelectric plates are used to drive the tines of a tuning fork into resonance. This is done by applying voltage pulses to the piezoelectric plate, which in turn expands and contracts rapidly, applying a pulse of force to the tines, driving them into oscillations. Consider one of these plates, 100 μm × 100 μm in area and 8 μm thick, made of SiO_2 with the following properties: piezoelectric coefficient 2.31 C/N, relative permittivity 4.63 (see **Figure 10.49**), and coefficient of elasticity 110 MPa. To apply the voltage the piezoelectric plate is plated on both sides by deposition of aluminum. A voltage pulse of 12 V is applied to the piezoelectric plate:

a. Calculate the maximum displacement the piezoelectric plate can provide.

b. Calculate the maximum force it can apply.

c. Discuss the force and displacement and their adequacy for the actuation of the tines.

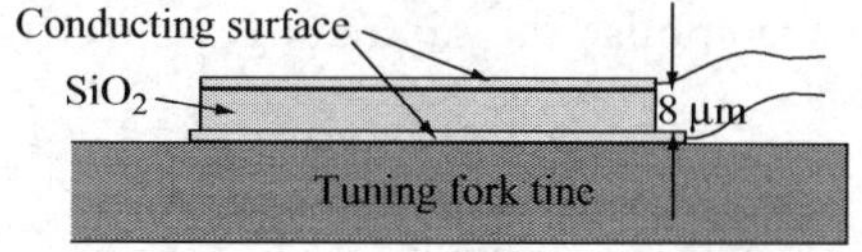

FIGURE 10.49 ■ Piezoelectric plate on the tine of a tuning fork sensor.

10.5 Two-axis accelerometer. Consider again the accelerometer in **Example 10.3**. Assume that the mass is 1 g and all parts of the sensor are made of silicon. The four flexural members are 100 μm long and their cross section is a square, 5 μm × 5 μm. Silicon has a modulus of elasticity of 150 GPa. An acceleration of 10 g is applied at an angle θ to the vertical axis (y-axis) in the x–y plane in **Figure 10.10**.

a. Calculate the strain measured by each of the eight strain gauges if they are placed 40 μm from the centers of the flexural members.

b. Suppose the acceleration is now at the same angle to the y-axis but also makes an angle ϕ with the x–y plane. If one assumes that the strain gauges can only measure bending strain, what is the error as a function of ϕ in the reading of the strain?

10.6 MEMS accelerometer. The accelerometer in **Figure 10.50** is designed to have unequal response in the two axes. The flexural members have a cross section of 5 μm × 5 μm and the mass is 2 g. The eight strain gauges are placed 10 μm from the ends of the flexural beams. Given an acceleration of 4 g at 45° to the x-axis in the x–y plane, calculate the strains measured by each of the eight strain gauges. The flexural members are silicon with a modulus of elasticity of 150 GPa.

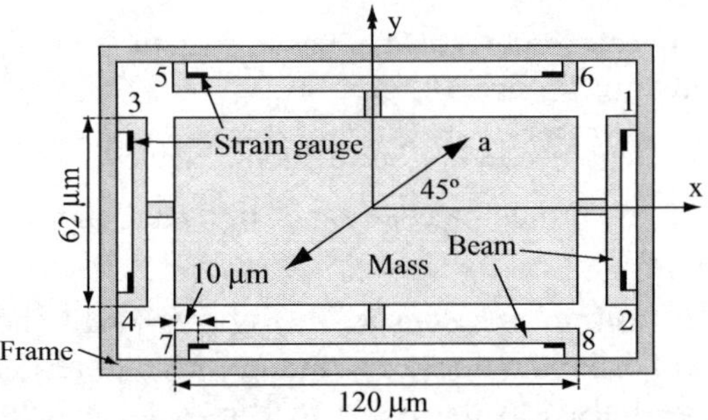

FIGURE 10.50 ■ Two-axis accelerometer with a different response on the two axes.

10.7 One-axis MEMS accelerometer. A one axis accelerometer is built as shown in **Figure 10.51** in silicon carbide (modulus of elasticity = 600 GPa). The dimensions are shown in the figure. The mass is 1.8 g. Assume that under acceleration the mass remains rigid and only the beam strains. The sensors are very small and for practical purposes they are placed at the location where the beam is fixed to the frame.

a. If silicon carbide can withstand a maximum strain of 2.2%, what is the range of the accelerometer and what is its span?

b. The accelerometer needs to be redesigned to have a range of ±100 m/s^2. Calculate the required mass to accomplish this range.

c. Suppose the mass in (a) must be reduced to 500 mg. What must be the length of the beam to accomplish the same range as in (b)?

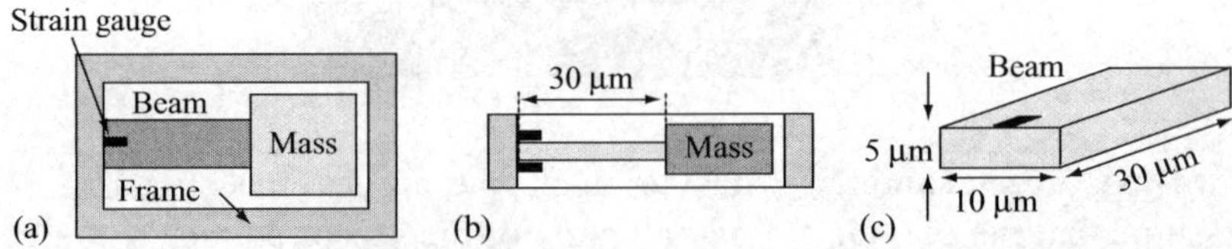

FIGURE 10.51 ■ MEMS accelerometer. (a) Top view. (b) Side view. (c) Details of the beam and location of the strain gauge.

10.8 Increased forces and torque in a MEMS motor. Consider the basic micromotor in **Figure 10.5**. In an attempt to increase force and torque, each stator and rotor section is built as a comb, as shown in **Figure 10.52** (shown in cross section). The stator combs are made of seven teeth and the rotor of six teeth. The radius of the rotor is $r = 120$ μm and the overlap between the stator and rotor fins is $e = 6$ μm. The gap between stator and rotor fins is $d = 2$ μm. If three stator sections are driven at any given time by applying a 5 V source,

a. Calculate the force produced between stator and rotor.

b. Calculate the torque of the motor.

c. What are the difficulties in producing and operating a motor of this type, especially if the number of fins is increased?

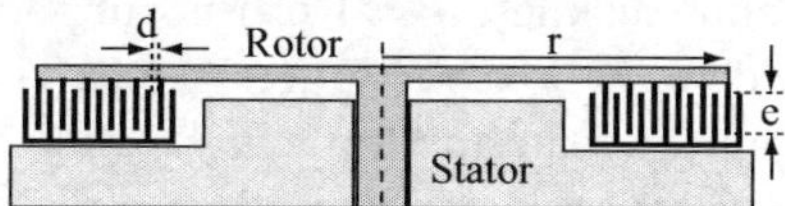

FIGURE 10.52 ■ Increased force and torque micromotor (cross-section view).

MEMS actuators

10.9 **The inkjet printer.** To appreciate the operation of the inkjet printer, consider the following example. In an inkjet, the ink is stored in a container with a nozzle as shown in **Figure 10.17**. Suppose the container is 100 μm × 100 μm × 100 μm in dimensions and is full of ink. To expel a droplet, a resistor is located at the bottom of the container. When a voltage is applied to the resistor, it heats up the ink and the volume expansion due to heat expels a droplet. Since ink is water based, we will assume the properties of water: density 1 g/cm^3, specific heat $C = 4.185$ kJ/kg/K, and coefficient of volume expansion 214×10^{-6}/°C (see **Table 3.10**).

a. If the inkjet is operated from a 5 V source, calculate the resistance needed to raise the temperature of the inkwell from ambient temperature (20°C) to 200°C in 40 μs (this is the time needed to generate a droplet).

b. Calculate the volume and mass of the droplet expelled through the nozzle.

10.10 **Electrostatically driven inkjet.** An electrostatically driven inkjet is shown in **Figure 10.53**. It consists of an ink chamber in the form of a cylinder 40 μm in diameter and 40 μm high with a nozzle to allow the ink out. The bottom of the ink chamber is a thin disk coated with a conducting layer on the underside. A second conducting layer on the substrate together with the conducting layer on the upper disk form a capacitor. To operate the inkjet, a voltage $V = 12$ V is applied to the capacitor as shown. This pulls the disk into a concave form, drawing an additional amount of ink as the volume of the chamber increases.

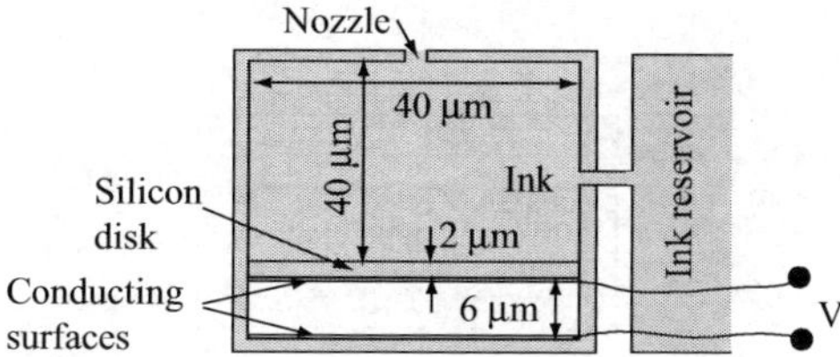

FIGURE 10.53 ■ Electrostatically actuated ink jet.

When the voltage is disconnected the disk returns to its original flat form, ejecting the extra ink. The disk is made of silicon with a coefficient of elasticity (Young's modulus) of 150 GPa and a Poisson ratio of 0.17.

a. Calculate the volume of the droplet being ejected by the inkjet.

b. Calculate the maximum pressure in the ink chamber.

Optical actuators

10.11 **A micromirror optical actuator.** A micromirror in the shape of a disk of radius $r = 50$ μm and thickness $t = 1$ μm is supported at its center by a non-conducting post made of SiO_2 (see **Figure 10.54**). The disk is coated with a thin aluminum deposition to create the reflecting surface and an aluminum layer is deposited on the substrate. The bottom of the disk is coated as well to create a continuous conducting surface. The actuation is done electrostatically by applying a voltage V between the two aluminum surfaces. The distance between the two conducting layers is $d = 3$ μm.

Viewing the mirror as a thin disk made of silicon with a modulus of elasticity of 150 GPa and a Poisson ratio of 0.17, calculate the deflection of the edges of the mirror as a function of applied voltage. Neglect the thickness of the aluminum deposition.

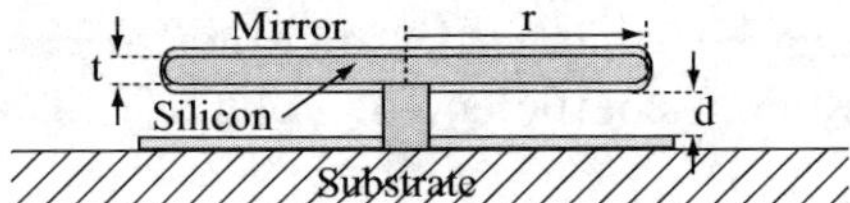

FIGURE 10.54 ■ Structure of an electrostatically actuated micromirror.

Pumps and valves

10.12 **Normally closed magnetic valve**. A magnetic valve is built as shown in **Figure 10.55**. The poppet is suspended on a series of thin arms that provide it with a spring constant k that keeps the valve closed. The spiral coil is fed with a current I that generates a magnetic flux density that is approximately linear, given as

$$B = (a - r)CI \quad [\mathrm{T}],$$

where r is the radial distance from the axis, a is the outer radius of the coil, and C is a measured constant (given) that depends on the number of turns, their

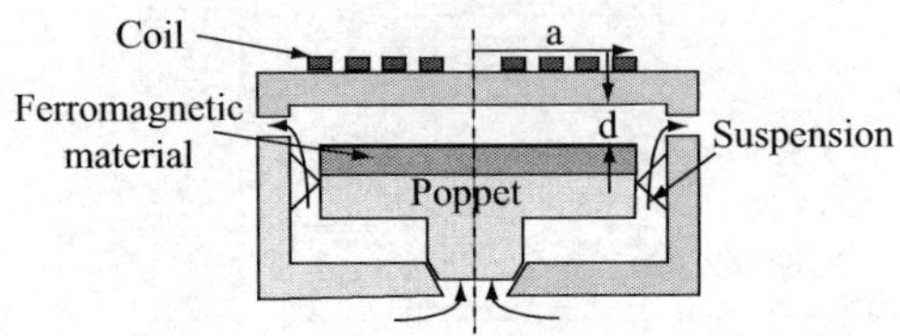

FIGURE 10.55 ■ Magnetically actuated microvalve.

density, and the ferromagnetic material's permeability. Estimate the current needed to completely open the valve, that is, to move the poppet up a distance d. Neglect gravitational forces and assume that the magnetic flux density in the gap remains unchanged as the poppet moves up because the motion is so small.

10.13 **A nonlinear force microactuator.** Consider the comb drive in **Figure 10.56**. Rather than moving sideways (i.e., the fins moving in and out), the fins move up and down. The figure shows the actuator position before the application of power. The separation between the plates is $d = 4$ μm and the plates are 60 μm long and 20 μm deep. If the actuator is required to have an upward travel of 3.5 μm (a stopping block, not shown, prevents the plates from getting closer than 0.5 μm from each other),

a. Calculate the maximum voltage that can be applied to the actuator given that the breakdown voltage in air is 3000 V/mm.

b. Calculate the range of forces corresponding to the position of the moving piece.

c. What would happen if the dimension marked as 4d were made equal to d (i.e., the moving plate is 4 μm from each of the stationary plates?

d. Suppose a controller is used so that the voltage can be adjusted with the position of the moving part to obtain larger forces at the initial position of the moving part. What is the range of voltages and the forces expected? Do you expect the force to be linear with position?

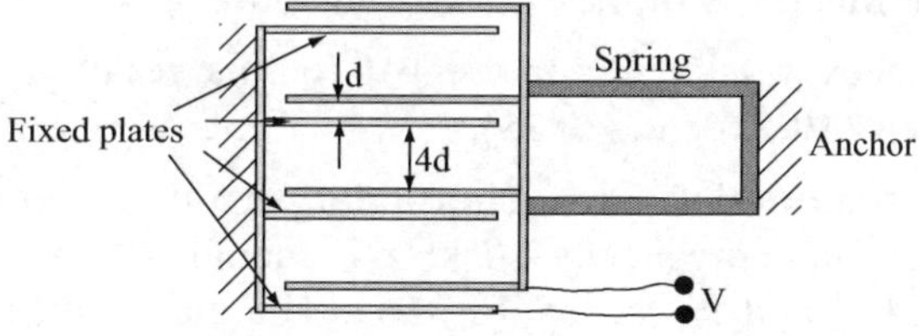

FIGURE 10.56 ■ Comb microactuator.

10.14 **Continuous optical actuator.** Consider a mirror actuated by an electrostatic comb as shown in **Figure 10.57**. The laser beam must be scanned over a 24° range from the position shown. The length of the mirror is 40 μm and it is at a 12° angle to the surface of the device. The plates of the comb drive are separated 2 μm from each other (there are 11 plates in the moving part and 10 in the stationary part of the drive) and the depth of the structure (perpendicular to the page) is 35 μm. Assume the upper tines of the comb are connected to the positive pole of the source and the lower tines to the negative pole with air

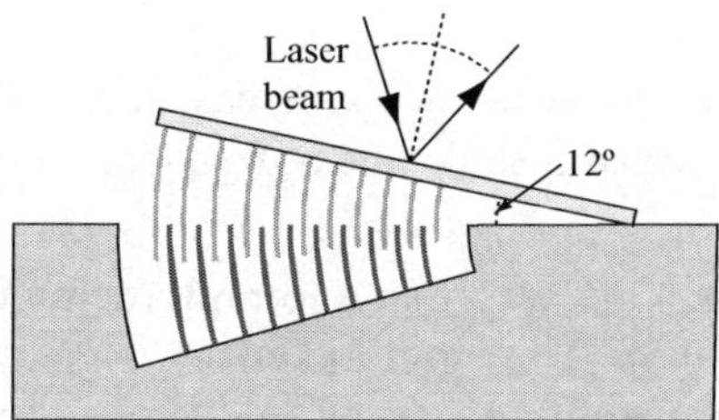

FIGURE 10.57 ■ A comb-actuated mirror.

between the tines. Neglect the thickness of the tines. A torsional spring restores the comb to its initial condition when voltage is removed. The spring is designed so that at an applied voltage of 5 V the comb is closed (maximum displacement).

a. Calculate the maximum force on the comb.

b. What is the resolution of the sensor (the smallest angle the beam can be moved or incremented) if the voltage can be accurately controlled in increments of 0.5 mV?

Modulation and demodulation

10.15 Amplitude modulation of a digital signal. A train of pulses of amplitude 5 V, frequency 10 kHz, and a 50% duty cycle is modulated using an amplitude modulator and a carrier frequency of 1.6 MHz:

a. Calculate and plot the signal with a modulation depth of 30%.

b. What is the necessary bandwidth to transmit the signal so that the signal can be fully recovered in the receiver? Discuss.

10.16 Frequency modulation of a digital signal. The signal in **Problem 10.15** is frequency modulated on a 100 MHz carrier whose amplitude is 12 V:

a. What is the necessary bandwidth to fully recover the signal in the receiver?

b. In FM radios, the bandwidth per channel is 200 kHz. Assuming that the bandwidth allowed for the signal is 100 kHz, compute the signal that will be received. Make use of the Fourier transform.

c. Write the expression for the signal for the result in (b) making use of **Equations (10.12)** and **(10.13).**

10.17 Frequency modulation of a sinusoidal signal. A cosinusoidal signal of amplitude 5 V and frequency 10 kHz is modulated over a carrier signal of amplitude 12 V and frequency 10 MHz. The maximum allowable frequency deviation is 500 kHz (wideband modulation). Assume zero phase in the signals.

a. Calculate and plot the modulated signal.

b. What is the modulation index of the modulated signal and what is the sensitivity of the modulator?

10.18 Phase modulation and distortions. Music on a broadcast transmission is modulated using phase modulation with a carrier frequency of 98 MHz:

a. If the bandwidth available for the transmission channel is 40 kHz and the sound transmitted varies in frequency between 20 Hz and 16 kHz, with an amplitude of 4 V, what is the maximum phase sensitivity the modulator can have?

b. Suppose it becomes necessary to increase the phase sensitivity by 20% over the value in (a). What is the range of frequencies that will be distorted due to this change?

10.19 FSK demodulator. One way of performing demodulation of FSK signals is shown in **Figure 10.58**. The input modulated signal is first delayed by a fixed time Δt and then the original and delayed signals are multiplied in the mixer.

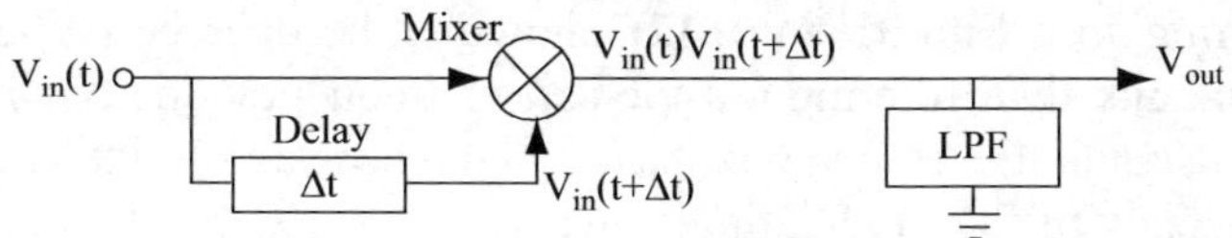

FIGURE 10.58 ■ Implementation of an FSK modulator.

Following the mixer is a low-pass filter that eliminates high-frequency signals. Suppose an FSK signal in which a zero is represented by $f_1 = 10$ kHz and a one is represented by $f_2 = 20$ kHz is received. That is, the signals is $A_c\cos(2\pi f_1 t)$ for a "0" and $A_c\cos(2\pi f_2 t)$ for a "1". These are used to encode a pulse train of amplitude 1 V and frequency of 1 kHz.

a. Show that for a fixed frequency f, the output of the demodulator is a DC level signal that only depends on the amplitude A, the frequency f, and the time delay Δt.

b. For the values of f_1 and f_2 above, calculate the output of the demodulator for a unit amplitude ($A_c = 1$). Use a fixed delay equal to half the cycle of f_1.

c. Sketch the modulated and demodulated signal.

10.20 **Digital modulators.** Digital modulators are relatively simple. They require only oscillators and electronic switches. **Figure 10.59** shows the principle of an FSK modulator (the implementation of electronic circuits and electronic switches will be discussed in **Chapter 11**). Consider the switches to be ideal and the controlling circuit to allow the switches to close and open according to the digital input. The switch position shown is for data "1".

a. Explain the circuit and sketch the output for a digital sequence 10011011.

b. Following the idea of switched oscillators, "design" an OOK modulator. Sketch the output for the digital sequence 11001011.

c. How would you implement a PSK modulator based on oscillators and switches? Draw a circuit to demonstrate the principle. Sketch the output for the digital sequence 01001110.

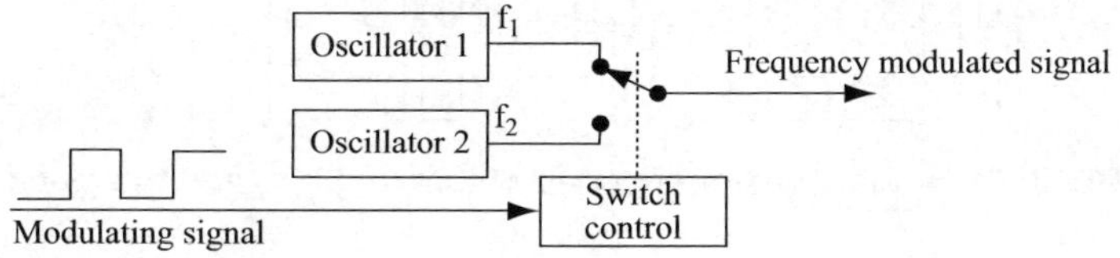

FIGURE 10.59 ■ Implementation of a digital FSK modulator.

10.21 **Amplitude demodulation**. Starting with the AM signal in **Equation (10.10)**:

a. Show that filtering out the carrier using a low-pass filter is sufficient to obtain a signal whose amplitude is proportional to the amplitude of the modulating signal.

b. What is the purpose of the rectification of the signal as part of the demodulation? Can demodulation be done without rectification?

10.22 **Frequency demodulation. Figure 10.39a** shows the block diagram for FM demodulation.

a. Starting with **Equation (10.12)**, show that by differentiating the modulated signal and then filtering out the carrier frequency one obtains the demodulated signal, that is, one obtains a signal whose amplitude is a function of the frequency of the modulating signal.

b. What is the frequency of the demodulated signal?

10.23 **Phase demodulation. Figure 10.39b** shows the block diagram for PM demodulation. Starting with **Equation (10.14)**, show that by first performing frequency demodulation as indicated in **Figure 10.39a** and then integrating the signal one obtains an output whose amplitude is proportional to the phase modulating the signal.

Encoding and decoding

10.24 **Unipolar and bipolar encoding.** A 16-bit sequence of digital information is given as D8FF in hexadecimal notation. The bits of the information are as follows: each bit consists of three clock pulses; a "0" is defined as a high for one clock cycle, followed by two low clock cycles; a "1" is defined by two high clock cycles, followed by a low clock cycle (see **Figure 10.60**).

a. Encode and sketch the information after encoding using NRZ unipolar code.

b. Encode and sketch the information after encoding using NRZ bipolar code.

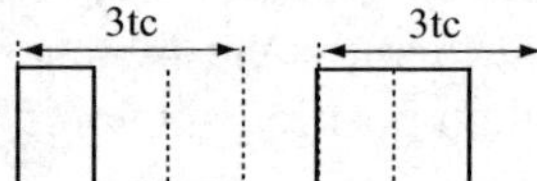

FIGURE 10.60 ■ A representation of "0" and "1" in a digital signal.

10.25 **Manchester decoding.** A 24-bit sequence of digital Manchester encoded datum is given together with the clock in **Figure 10.61**. Decode the information, sketch it, and give the datum in digital and hexadecimal formats.

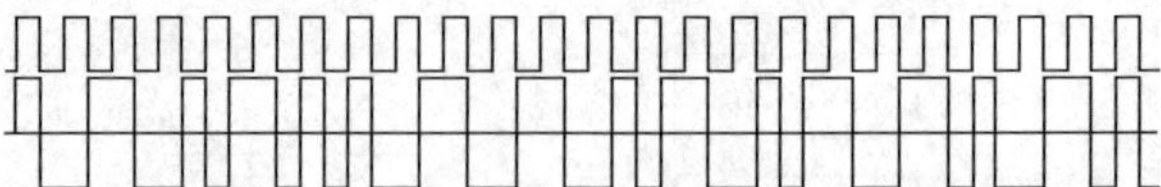

FIGURE 10.61 ■ Manchester encoded datum (bottom) and the clock (top).

10.26 **Pulse width modulation encoding.** In digital systems where noise may be a problem, and the distinction between logical "0" and logical "1" may be compromised, one may opt for a PWM method of encoding in which both the 0 and the 1 are encoded as pulses, the width of which indicates the state. An example is shown in **Figure 10.62** in which a "0" is indicated by a pulse two clock cycles wide and a "1" as a pulse one clock cycle wide, each digit being formed by three clock cycles.

a. Using this method, encode the decimal number 39,572. Sketch the output.

b. With the clock available at the receiver, describe an algorithm that can decode the PWM encoded data.

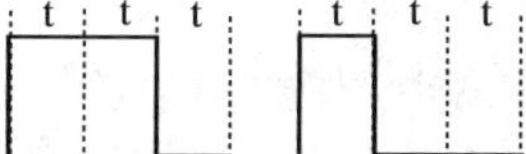

FIGURE 10.62 ■ A possible pulsewidth representation of logical "0" and logical "1".

10.27 **CAN bus NRZ encoding and decoding.** The CAN bus is a common bus used in vehicles to connect sensors and actuators and affect data communication with them. It uses an NRZ encoding scheme, meaning that at the end of a clock pulse the signal remains in its previous state unless the signal itself changes state. For the purpose of this problem, assume a "1" is +5 V and a "0" is −1.5 V and transitions occur at the trailing edge of the clock (the CAN bus specification is based on differential voltages and is much more complex than what this problem assumes). Consider a clock with pulses $t/2$ μs high and $t/2$ μs low (i.e., the clock period is t). A sensor sends the 16-bit (2 byte) digital data 1011 0011 1001 1011, where each bit is $5t$ long. The initial state (before data arrives) is "0".

a. Draw the signal produced and transmitted on the bus.

b. Suppose the length of each data bit is $4.6t$. How does the signal look now? Draw the signal and compare to (a), and in particular the locations at which the pulses change state.

Sensor networks

10.28 **Water quality sensor network.** Design a network to monitor a national park for water quality, defining the types of sensors and communication protocols. Discuss the options for network connectivity that are practical for the design. Estimate the number and types of nodes needed, the power supply, etc. Sketch a diagram of the components of the network.

10.29 **National lightning strike sensor network.** Design a sensor network to monitor lightning strikes over the country as part of the weather prediction system. Define the sensors, type of network, and communication needed. Estimate the sensor density given that lightning strikes can reliably be detected over distances of up to 150 km and the location of a particular lightning strike requires a minimum of three sensors within that range. Estimate the number of sensors needed to detect every strike within the United States (approximate the area as a rectangle about 4000 km × 3000 km). Define the communication methods necessary to accomplish detection, location, and reporting to a central location.

10.30 **Fire detection/suppression network: a network of sensors and actuators.** Address and define the following components of the network within a large complex that includes housing, offices, retail, and public spaces.

a. Detection of fire/smoke: type of sensors, sensor density, and errors in sensing.

b. Local suppression: methods of suppression and safety.

c. Alarms: type of alarms and actuators.

d. Communication to emergency services.

e. Communication to centers.

f. Communication to neighboring nodes.

10.31 **The CAN bus.** The CAN bus, designed specifically for vehicles, is a linear, two-wire network designed to connect sensors and actuators to a central processor unit. Discuss the following issues associated with such a bus:

a. The connection of a sensor or actuator on the single two-wire bus. How can one connect multiple devices and communicate to/from them?

b. Discuss possible protocols to avoid collisions between messages.

c. What are the specific properties necessary for a network of this type to operate in a vehicle environment?

CHAPTER 11

Interfacing Methods and Circuits

The nervous system

The nervous system in the body is a network of neurons and is responsible for the transmission of signals between the various sensors and the brain and from the brain to the various parts of the body to affect actuation through muscles. The system is made of a number of components and divided into two main parts. One is the central nervous system, consisting of the brain and the spinal cord (the retina is also considered part of the central nervous system). In general, the central nervous system is associated with processing sensory as well as other signals. The second is the peripheral nervous system, made of sensory neurons and glial cells that interconnect the neurons in rather complex patterns. Some neurons are clustered in ganglia (the largest ganglion in the body is located in the spinal cord and is responsible for many motor functions of the body). The neurons are specialized cells than can take specific functions such as sensing, neurotransmission, or connectivity and communicate with other neurons or with specific parts in the body. Most signals are communicated between neurons as electrical impulses over axons, elongated structures that connect between neurons. Neurons connect to cells through membrane junctions called synapses that allow transmission of electrical or chemical signals. Some signals are broadcast through release of hormones. Bundles of axons are called nerves.

Within the peripheral nervous system the nerves have three major functions. The first is sensory, conducting sensory signals from various receptors in the body and on the skin to the central nervous system (most receptors are on the skin, but also in sensing organs including the ears, nose, tongue, and eyes). The second function is the motor function, transmitting signals to muscles and organs to affect the motor functions of the body. A third function is to control autonomous actions of the body such as breathing and the heart, as well as involuntary reactions such as closing of the eyelids, flight from danger, and the conservation of body energy when needed.

From a sensing and actuation point of view, the nervous system provides a means of connecting the sensory and actuation functions in the body with the central nervous system (the brain) and transfers feedback to control body action and perception of its environment.

11.1 | INTRODUCTION

A sensor or an actuator can rarely operate on its own. Exceptions exist, such as bimetal sensors, which can both sense and directly actuate a switch or a dial. However, in the majority of cases an electric circuit of some sort is involved. The circuit can be as simple as adding a power source or a transformer, but more often it involves amplification, impedance matching, signal conditioning, and other such functions. In other cases, a digital output is required or desirable so that an analog-to-digital (A/D) conversion may be needed, or in some cases simpler A/D methods of conversion may be used. Often, too, the circuit is a microprocessor or programmer of some sort. The same general considerations apply to actuators, except that now amplifiers are likely to involve larger powers and conversion might be from digital to analog (D/A) since many actuators are analog in nature. Thus there are considerations that seem to be secondary to sensing and actuation but which are necessary for the success of the sensing/actuation strategy in which the sensor or actuator is only one part, and not always the critical part. These have to do with the circuits necessary to interface the device and hence make the sensor or actuator useful. The considerations of interfacing should be integral to the design process since selection of a device for a particular application can thereby be simplified considerably. If a digital device exists, it would be wasteful to select an equivalent analog device and add the required circuitry to convert its output to a digital format. The likely outcome is a more cumbersome and expensive system that may take more time to produce. Alternative sensing strategies and alternative sensors should always be considered before settling on a particular solution, since many different sensor types and many different sensors can accomplish the same task and it is not always obvious a priori which strategy is best overall. In these considerations, the issue of interfacing must be addressed, regardless of the constraints on the design. If, for example, cost is the overriding consideration, the choice of the simplest, least expensive sensor is not always the one that will produce the overall least expensive design. Similarly, it makes no sense to get an expensive sensor that, say, can sense down to 0.001°C only to find out that the A/D conversion will limit the accuracy to, say, 0.01°C or that the display envisioned as the output to the system can only display increments of 0.1°C.

Although there are many types of sensors and actuators based on very different principles, from an interfacing point of view there are commonalities between them that must be considered. First, most sensors' outputs are electric; their output might be resistance, voltage, or current. These can be measured directly after proper signal conditioning and, perhaps, amplification. In other cases, the output is a capacitance or an inductance. These usually require additional circuitry, such as the construction of an oscillator and then measurement of the frequency of the oscillator. In some cases the output of a sensor is frequency, whereas the input to an actuator may be pulses of varying widths.

Another important consideration is the signal level involved, as there is a large range of signal levels in sensors and actuators. A thermocouple's output may be on the order of a few microvolt DC, whereas that of an LVDT may easily produce 5 V AC. In actuators, voltages and currents may be quite high. A piezoelectric actuator may require a few hundred volts to operate (very little current), whereas a solenoid valve usually operates at perhaps 12–24 V with currents that may exceed a few amperes. The circuitry required to drive these and to interface them to, say, a microprocessor are vastly different and require special attention on the part of the engineer. In addition, one has to consider such issues as response (both electrical and mechanical), spans, power dissipation, as well as considerations of power quality and availability. Systems connected to the grid and cordless systems will have different requirements and considerations both in terms of operation and in terms of safety.

The purpose of this chapter is to discuss the general issues associated with interfacing and to outline the more general interfacing circuits the engineer is likely to be exposed to. However, no general discussion can prepare one for all eventualities and it should be recognized that there are both exceptions and extensions to the methods discussed here. For example, an A/D is a simple—if not inexpensive—method of digitizing a signal for the purpose of interfacing with a microprocessor. However, this approach may not be necessary, or may be too expensive, in some cases. A case in point: Suppose that a Hall element is used to sense the teeth on a rotating gear. The signal from the Hall element is an AC voltage (more or less sinusoidal) and only the peaks are necessary to sense the gears. In this case a simple peak detector, followed perhaps by simple signal conditioning may be adequate. An A/D converter will not provide any additional benefit and is a much more complex and expensive solution. On the other hand, if a microprocessor is used and an A/D converter is available onboard, it may be acceptable to use it for this purpose in lieu of adding circuitry.

We will start with a discussion of amplifiers, particularly operational amplifiers (or op-amps), since these offer excellent, simple solutions to amplification in virtually the whole range of signals and frequencies encountered in sensing and actuation. They offer equally useful possibilities of signal conditioning and filtering, as well as impedance matching. Power amplifiers are more important in actuation, but they involve some common principles with operational amplifiers and thus are discussed together. A/D and D/A conversion circuits are discussed next, as these are essential in interfacing with digital devices such as microprocessors. We start with simple threshold methods, followed by more sophisticated voltage-to-frequency conversion circuits and true A/D converters. The subject of bridge circuits is introduced next with a discussion of sensitivity and concluding with amplified bridges, combining them with some of the operational amplifier circuits discussed earlier. Following that, we deal with data transmission, emphasizing the need for accuracy with low-level signals and the methods appropriate for their transmission. As a necessity of any circuit, but more so in sensors and actuators, it is important to understand the need for excitation of the circuits, including power supplies and their effect on the circuit. These include DC power supplies as well as AC sources of excitation. Sinusoidal and square wave oscillators are discussed in this context as well. The final section of the chapter discusses noise and interference in sensors and general methods of addressing them.

A word on circuit and circuit choices. Modern electronics offers a bewildering variety of components and circuits, some general purpose, some specialized. The choice of a particular component for an application is both easy and extremely difficult. It is easy because of the variety of components available and the fact that one is likely to find a circuit that will do the job. It is difficult because each choice offers particular advantages and constraints. Sometimes it is frustrating because what seems to be the natural choice cannot or should not be used for a variety of reasons. For example, suppose that one needs to amplify a signal. A transistor, with proper biasing will do just fine. However, an operational amplifier, which is a much more complex circuit may offer a better solution overall (in performance, design time, and often even in cost) even though the operational amplifier is really much more than is needed for the application. Similarly, one may only need to read a sensor's signal and turn on a light, say, when the temperature has exceeded 75°C (as a warning). Any electrical engineer can design a simple circuit to do that. However, it may be cheaper to use a microprocessor for this purpose even though the use of a microprocessor is clearly "overkill." Simply put, from the point of view of cost, the number of components (and hence space), and possible future changes, the microprocessor is a much more attractive solution in spite of the fact that programming is needed.

11.2 | AMPLIFIERS

An amplifier is a device that amplifies a signal—almost always a voltage—from a low level to the required level (current and power amplifiers also exist). In the case here, the low voltage output of a sensor, say, a thermocouple, may be amplified to a level required by a controller or a display. The amplification may be quite large—sometimes on the order of 10^6—or it may be quite small, depending on the need of the interfacing circuitry. Amplifiers can also be used for impedance matching purposes even when no amplification is needed or they can be used for the sole purpose of signal conditioning, signal translation, or for isolation between the sensor and the controller to which it connects. Equally well, power amplifiers, which usually connect to actuators, serve similar purposes beyond providing the power necessary to drive the actuator.

Amplifiers can be very simple—a transistor with its associated biasing network—or they can be more complex circuits that involve many amplification stages of varying complexity. However, since our interest here is the function rather than the design details of circuits, we will use the operational amplifier as the basic building block for amplification. This is not merely a convenience—operational amplifiers are basic devices and may be viewed as components. An engineer, especially when interfacing sensors, is not likely to delve into the design of electronic circuits below the level of operational amplifiers. Although there are instances where this may be done to great advantage, operational amplifiers are almost always a better, less expensive, and higher performance choice.

11.2.1 The Operational Amplifier

An operational amplifier is a fairly complex electronic circuit that it is based on the idea of the differential voltage amplifier shown in **Figure 11.1**. In this circuit, which is based on simple transistors, the output is a function of the difference between the two inputs. Assuming the output to be zero when both inputs are at zero potential, the operation is as follows: When the voltage on the base of Q_1 increases, its bias increases while that on Q_2 decreases because of the common emitter resistance. Q_1 conducts more than Q_2 and the output is positive with respect to ground. If the sequence is inverted, the opposite occurs. If, however, both inputs increase or decrease equally, there will be no change in output (zero difference between inputs).

The difference amplifier serves as the front end or the input of the operational amplifier and is followed by additional circuitry (additional amplification stages,

FIGURE 11.1 ■ The differential amplifier forms the basis for all operational amplifiers.

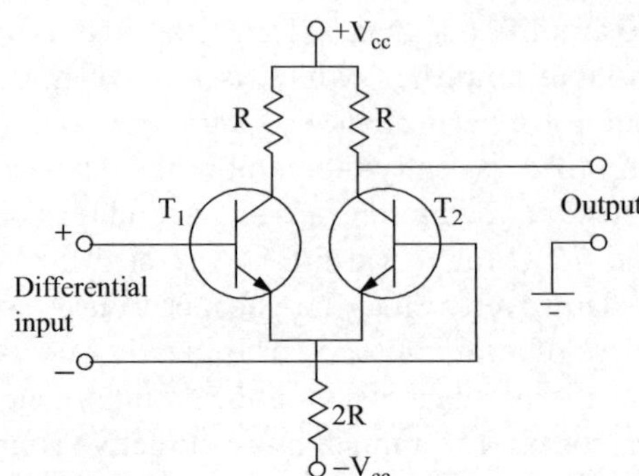

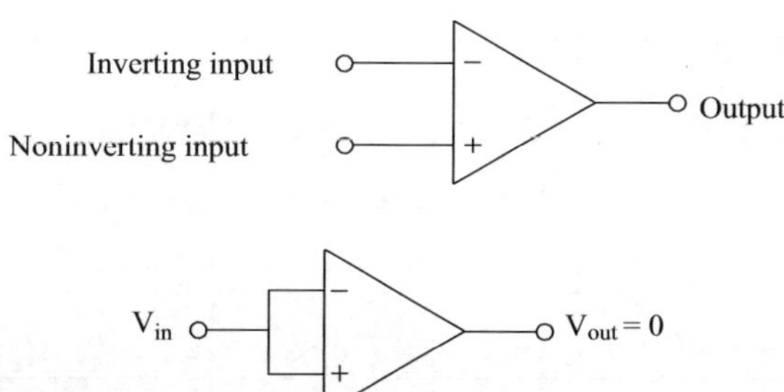

FIGURE 11.2 ■ The symbol for an operational amplifier.

FIGURE 11.3 ■ Common-mode signal and output.

temperature and drift compensation, output amplifiers, etc.), but these are of no interest to us other than the fact that they affect the specifications of the operational amplifier. There are also various modifications that allow operational amplifiers to operate under certain conditions or to perform specific functions. Some are "low noise" devices, others can operate from a single polarity source, still others can operate at higher frequencies or are particularly suited to amplify low signals. If the input transistors are replaced with field-effect transistors (FETs), the input impedance increases considerably, requiring even lower input currents from the sensors connected to it. All these are important but are mere variations of the basic circuit. To understand the properties of amplifiers, we will consider it as a simple block diagram, shown in **Figure 11.2**, and discuss its general properties based on this diagram. The more salient properties of operational amplifiers are discussed below.

11.2.1.1 Differential Voltage Gain

This is the amplification of the difference between the two inputs:

$$V_o = V_i A_d, \tag{11.1}$$

where V_i is the differential input voltage and A_d is the **open loop gain**, and in a good amplifier it should be as high as possible. Open loop gains of 10^6 or higher are common. An ideal amplifier is said to have infinite open loop gain.

11.2.1.2 Common-Mode Voltage Gain

By virtue of the differential nature of the amplifier, the common-mode gain should be zero. Practical amplifiers may have some common-mode gain because of the mismatch between the two inputs, but this should be small. The common-mode voltage gain is indicated as A_{cm}. The concept is shown in **Figure 11.3**. In the specification of operational amplifiers, it is more common to find the term common-mode rejection ratio (CMRR), defined as the ratio between A_d and A_{cm}:

$$CMRR = \frac{A_d}{A_{cm}}. \tag{11.2}$$

In an ideal amplifier this is infinite. A good amplifier will have a CMRR that is very high.

11.2.1.3 Bandwidth

Bandwidth is the range of frequencies that can be amplified. Usually the amplifier operates down to DC and has a flat response up to a maximum frequency (this is device specific) at which the output power decreases by 3 dB. An ideal amplifier will have an infinite bandwidth. The open gain bandwidth of a practical amplifier is fairly low, so a

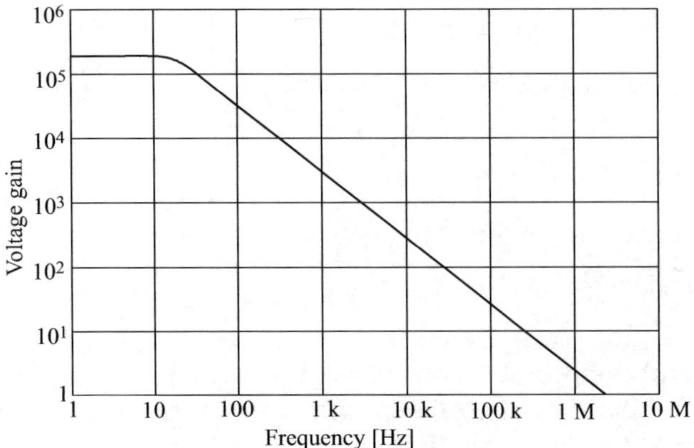

FIGURE 11.4 ■ Bandwidth of an operational amplifier.

more important quantity is the bandwidth at the actual gain at which the operational amplifier operates. This may be seen in **Figure 11.4**, which shows that the lower the gain, the higher the bandwidth. Data sheets therefore cite what is called the gain–bandwidth product. This indicates the frequency at which the gain drops to 1, and is also called the **unity gain frequency** (or **0 dB gain frequency**). For example, in **Figure 11.4**, the open loop bandwidth at a gain of 1000 is approximately 2.5 kHz and the unity gain frequency is approximately 5 MHz.

11.2.1.4 Slew Rate

Slew rate is the rate of change of the output in response to a step change in input, typically given in volts per microsecond (V/μs). The practicality of this is that if a signal at the input changes faster than the slew rate, the output will lag behind it and a distorted signal will be obtained. This limits the usable frequency range of the amplifier. For example, an ideal square wave will have an increasing and decreasing slope at the output defined by the slew rate. Consider the schematic in **Figure 11.5**. It shows an amplifier with a slew rate of 2 V/μs (a very low slew rate). It takes the amplifier's output 7.5 ms to increase from 0 to 15 V (**Figure 11.5a**). The pulse is still recognizable as a square wave, although its width and shape have changed. Above some signal frequency, the square pulse is not recognizable simply because of the slew rate. In **Figure 11.5b** the pulse width is only 10 ms. The pulse is high for 5 ms and the output increases linearly to 10 V. During the period the input pulse is low, the output falls to zero. Clearly this is not a square pulse and its amplitude has decreases. At 1 MHz the amplitude of the pulse will only be 1 V, and that is assuming the amplifier can operate at 1 MHz. Therefore the slew rate limits the usable frequency at which the amplifier can operate, typically separate from the bandwidth, that is, even if the bandwidth is sufficiently high, if the slew rate is low, distortion of signals will occur.

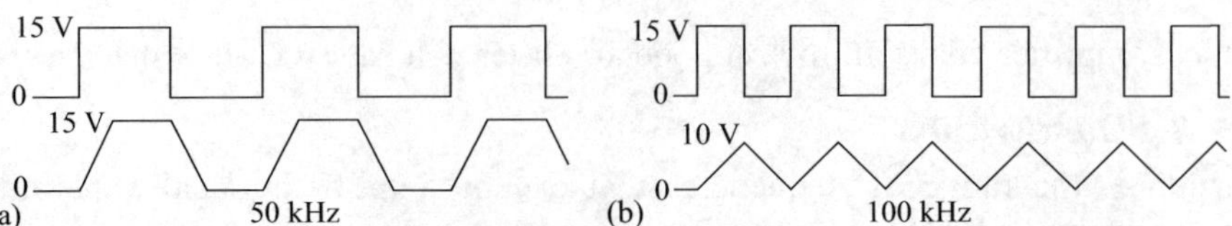

FIGURE 11.5 ■ Effect of slew rate. (a) The slew rate limits the increase and decrease rate of the signal. (b) As the frequency increases, the pulse shape and amplitude are modified.

11.2.1.5 Input Impedance

Input impedance is the impedance seen by the sensor when connected to the operational amplifier in open loop mode. Typically this impedance is high (ideally infinite), but it varies with frequency and the way the amplifier is configured. Typical input impedance for conventional amplifiers is on the order of 1 MΩ, but it can be on the order of hundreds of megaohms for FET input amplifiers. The closed loop impedance can be much lower, as will be seen shortly. This impedance defines the current needed to drive the amplifier, and hence the load it represents to the sensor.

11.2.1.6 Output Impedance

This is the impedance seen by the load. Ideally this should be zero since then the output voltage of the amplifier does not vary with the load, but in practice it is finite and depends on gain. Usually the output impedance is given for open loop, whereas at lower gains the impedance is lower. A good operational amplifier will have an open loop output resistance of only a few ohms.

11.2.1.7 Temperature Drift and Noise

These refer to variations of output with the temperature and noise characteristics of the device. These are provided by the data sheet and are usually very small. Nevertheless, for low signals, noise can be important, whereas temperature drift, if unacceptable, must be compensated through external circuits.

11.2.1.8 Power Requirements

The classic operational amplifier is designed so that its output is $\pm V_{cc}$, or rail to rail. This is in accordance with its differential input. This dual supply operation is common to many operational amplifiers, although the limits can be as low as ±5 V (or lower) and as high as ±35 V (sometimes higher). Many operational amplifiers are designed for single supply operation, varying from less than 3 V to more than 30 V, and some can be used in either single supply or dual supply modes. The current through the amplifier is an important consideration in the use of an operational amplifier, especially the quiescent current (no load), since it gives a good indication of the power needed to operate the device. This is particularly important in battery-operated devices. The current under load will depend on the application, but it is usually fairly small—a few milliamps (in some cases much less). In selecting a power supply for operational amplifiers, care should be taken with the noise that the power supply can inject into the amplifier. The effect of the power supply on the amplifier is specified through the **power supply rejection ratio (PSRR)** of the specific amplifier.

11.2.2 Inverting and Noninverting Amplifiers

From the specifications above, it is clear that the performance of the amplifier depends on how it is used and, in particular, on the gain of the amplifier. In most practical circuits, the open loop gain is not useful and a specific, lower gain must be established. For example, we might have a 50 mV output (maximum) from a sensor and require this output to be amplified, say, by 100, to obtain 5 V (maximum) for connection to an A/D converter. This can be done with one of the two basic circuits shown in **Figure 11.6**, both of which establish a means of negative feedback to reduce the gain from the open loop gain to the required level.

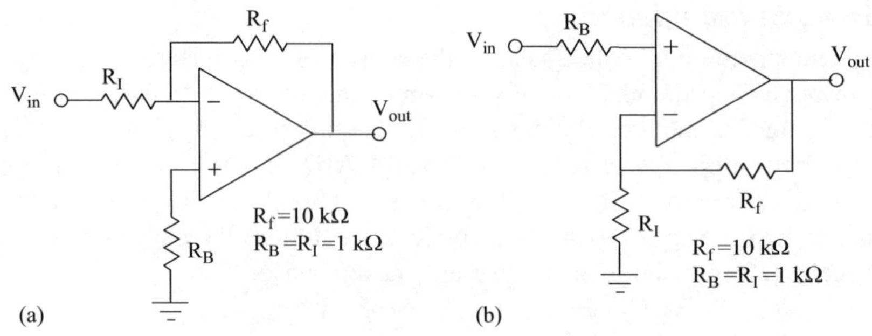

FIGURE 11.6 (a) Inverting operational amplifier. (b) Noninverting operational amplifier.

11.2.2.1 The Inverting Amplifier

In the amplifier in **Figure 11.6a**, the output is inverted with respect to the input (180° out of phase) because the input is connected to the inverting input. The feedback resistor, R_f, feeds back some of this output to the negative input, effectively reducing the gain. The gain of the amplifier is now given as

$$A_v = -\frac{R_f}{R_I}. \tag{11.3}$$

In the case shown here this is exactly −10.

The input impedance of the inverting amplifier is given as

$$R_i = R_1 \tag{11.4}$$

and is equal to 1 kΩ. Clearly, if a higher input resistance is needed, larger resistances might need to be used, or perhaps a different amplifier will be needed as well, or more likely, a noninverting amplifier will have to be used (see below).

The output impedance of the inverting amplifier is somewhat more complex:

$$R_o = \frac{(R_I + R_f)\,(open\ loop\ output\ impedance)}{R_I A_{OL}} \quad [\Omega], \tag{11.5}$$

where the open loop output impedance is as listed on the data sheet and the open loop gain is the open loop gain at the frequency at which the device is operated (see **Figure 11.4**) and may not be the maximum open loop gain. For example, a general purpose operational amplifier has an open loop output impedance of 75 Ω and the open loop gain at 1 kHz is about 1000. This gives an output impedance of

$$R_o = \frac{(1{,}000 + 10{,}000)(75)}{1{,}000 \times 1{,}000} = 0.825\ \Omega.$$

The bandwidth of the amplifier is also influenced by the feedback and is

$$BW = \frac{(unity\ gain\ frequency)\ R_I}{R_I + R_f} \quad [\text{Hz}]. \tag{11.6}$$

These values show how the amplifier can be used to amplify and to change the input and output impedances to match a sensor to a controller. Note, however, that both the input and output impedance in this case have gone decreased, something that may or may not be acceptable. In any case, this must be considered when interfacing.

11.2.2.2 The Noninverting Amplifier

If the noninverting amplifier in **Figure 11.6b** is used, the relations above change as follows:

Gain:

$$A_v = 1 + \frac{R_f}{R_I}. \tag{11.7}$$

For the circuit shown, this is 11.

Input impedance:

$$R_i = R_{op}A_{ol}\frac{R_I}{R_I + R_f} \quad [\Omega], \tag{11.8}$$

where R_{op} is the open loop input impedance of the operational amplifier as given in the spec sheet and A_{ol} is the open loop gain of the amplifier. Assuming an open loop impedance of 1 MΩ (modest value) and an open loop gain of 10^6, we get an input impedance of 10^{11} Ω. This is close to the ideal impedance an amplifier should have. The output impedance and the bandwidth are the same as for the inverting amplifier given in **Equations (11.5)** and **(11.6)**, respectively. It should be noted that the main reason for using a noninverting amplifier is that its input impedance is very high, making it almost ideal for many sensors.

There are other properties that need to be considered, such as output current and load resistance. A proper design will take these properties as well as slew rate, noise, temperature variations, etc., into consideration.

EXAMPLE 11.1 Design of an amplifier

The output of a piezoelectric microphone varies from −10 μV to 10 μV (see **Equation (7.41)** for the relation between the change in voltage and the change in pressure in a piezoelectric device) for the normal range of human speech. It is generally assumed that humans hear in the range between 20 Hz and 20 kHz. The output from the microphone must be amplified to be between +2 V and −2 V when the input varies between −10 μV and 10 μV, flat over the frequency span. To do so it is proposed to use an operational amplifier whose frequency response is given in **Figure 11.4**. The following data are available for the operational amplifier: unity gain bandwidth = 5 MHz, open loop gain = 200,000, open loop input impedance = 500 kΩ, and open loop output impedance = 75 Ω.

a. Design an amplifier circuit that will interface the microphone and provide the required output.

b. Calculate the input impedance and the output impedance of the circuit.

Solution: Before proceeding it should be noted that the required amplification is 200,000 (2 V/10 μV). Although the open loop gain of the operational amplifier is 200,000, the bandwidth at that gain is only about 20 Hz. Clearly then, more than one amplifier will be needed, the number being dictated by the required bandwidth. In addition, the output is inverted—the microphone must be connected to a noninverting amplifier since piezoelectric microphones have high impedance.

a. The required bandwidth is 20 kHz. From **Figure 11.4**, at that bandwidth, the gain is about 110. To ensure proper frequency response we will assume a maximum gain of 100. Therefore we need

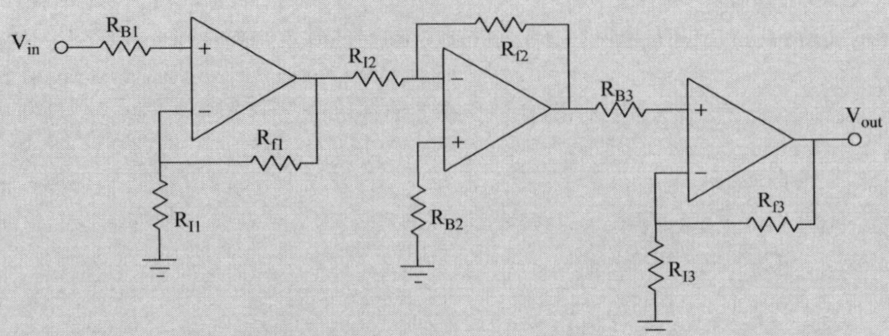

FIGURE 11.7 ■ A three-stage inverting amplifier.

three amplifiers. We have many options here: we could select a gain of 100 for the first stage, 100 for the second, and 20 for the third, or perhaps 50, 50, and 80, or any other combination within the requirements of maximum gain for the bandwidth. We will select here the first option (arbitrarily) with the first stage being noninverting, the second inverting, and the third noninverting to ensure the output is inverted and the input is high impedance. The circuit is shown in **Figure 11.7**. For the first amplifier we have, from **Equation (11.7)**:

$$A_1 = 1 + \frac{R_{f1}}{R_{I1}} = 100 \rightarrow \frac{R_{f1}}{R_{I1}} = 99.$$

Selecting $R_{I1} = 1$ kΩ, $R_{f1} = 99$ kΩ is an obvious choice (but see the note below). R_{B1} is selected equal to R_{I1}.

For the second stage we have, from **Equation (11.3)**:

$$A_2 = -\frac{R_{f2}}{R_{I2}} = 100 \rightarrow \frac{R_{f2}}{R_{I1}} = 100.$$

Again we select convenient values: $R_{I2} = 1$ kΩ, $R_{f2} = 100$ kΩ, and $R_{B2} = 1$ kΩ.

The third stage is again noninverting:

$$A_3 = 1 + \frac{R_{f3}}{R_{I3}} = 20 \rightarrow \frac{R_{f3}}{R_{I3}} = 19.$$

The resistors are $R_{I3} = 1$ kΩ, $R_{f3} = 19$ kΩ, and $R_{B3} = 1$ kΩ.

Note: The 19 kΩ and 99 kΩ resistors are not standard values and may be difficult to obtain. One may select 20 kΩ and 100 kΩ, but now the total amplification will be 101 × 100 × 21 = 212,100.

b. The input impedance is that of a noninverting amplifier given in **Equation (11.8)**:

$$R_I = R_{op}A_{ol}\frac{R_{I1}}{R_{I1} + R_{f1}} = 500{,}000 \times 200{,}000\frac{1{,}000}{1{,}000 + 99{,}000} = 10^9\ \Omega.$$

This is 1000 MΩ and should be sufficiently high for a piezoelectric microphone.

The output impedance is calculated from **Equation (11.5)**:

$$R_o = \frac{(R_{I3} + R_{f3})(open\ loop\ output\ impedance)}{R_{I3}A_{OL}} = \frac{(1{,}000 + 19{,}000) \times 75}{1{,}000 \times 110} = 13.6\ \Omega.$$

Note: Strictly speaking, the output impedance varies with frequency since A_{ol} in this relation is the open loop gain at the frequency at which the amplifier operates. We have used the open loop gain at the highest frequency (20 kHz). At the lowest frequency (20 Hz), the open loop gain is 200,000 and the output impedance is only 0.0075 Ω.

11.2.3 The Voltage Follower

If the feedback resistor in the noninverting amplifier is set to zero, the circuit in **Figure 11.8** is obtained, called a voltage follower. The first thing to notice is that the gain is 1 because of the 100% negative feedback. This circuit does not amplify, but the input impedance now is very large and equal to

$$R_i = R_{op}A_{ol} \quad [\Omega], \tag{11.9}$$

whereas the output impedance is very low and equal to

$$R_o = \frac{(open\ loop\ output\ impedance)}{A_{ol}} \quad [\Omega]. \tag{11.10}$$

Therefore the value of the voltage follower is to serve in impedance matching. One can use this circuit to connect, say, a capacitive sensor or perhaps an electret microphone. If amplification is necessary, the voltage follower can be followed by an inverting or noninverting amplifier.

11.2.4 The Instrumentation Amplifier

The instrumentation amplifier is different than a regular operational amplifier in that its gain is finite and both inputs are available to signals. These amplifiers are available as single devices, but to understand how they operate, one should view them as being made of three operational amplifiers (it is possible to make them with two operational amplifiers or even with a single amplifier, but it is best understood as a three operational amplifier device), as shown in **Figure 11.9**. The gain of an amplifier of this type is

$$A_v = \left(1 + \frac{2R_1}{R_G}\right)\left(\frac{R_3}{R_2}\right). \tag{11.11}$$

In a commercial instrumentation amplifier, all resistances except R_G are internal and produce a gain usually around 100 for the three amplifiers. R_G is external and can be set by the user to obtain the gain required for the instrumentation amplifier within certain limits. In most cases $R_3 = R_2$, and in many instrumentation amplifiers all internal resistors are the same: $R_2 = R_3 = R_1 = R_0$. This allows better control of the accuracy of the resistors and better overall performance. In that case, **Equation (11.11)** becomes

$$A_v = \left(1 + \frac{2R_0}{R_G}\right). \tag{11.12}$$

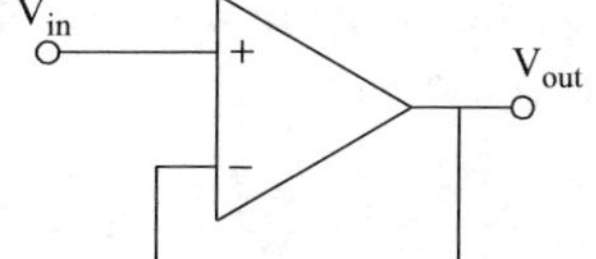

FIGURE 11.8 ■ The voltage follower.

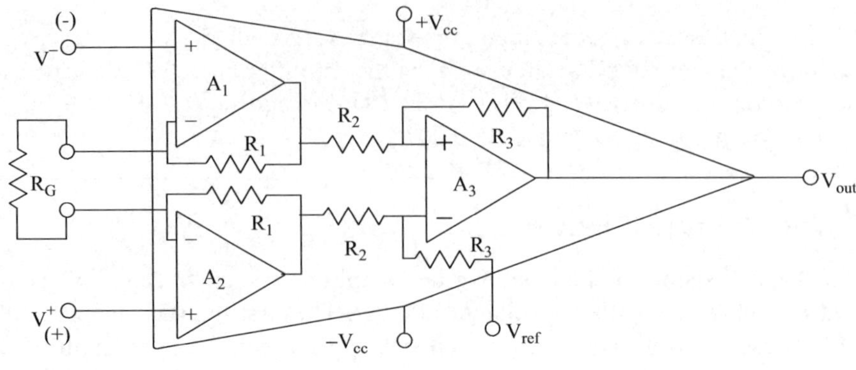

FIGURE 11.9 ■ The instrumentation amplifier. Note that the upper input is inverting.

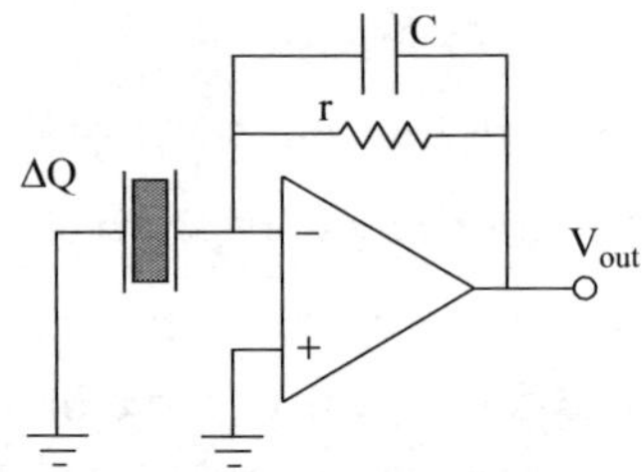

FIGURE 11.10 ■ Charge amplifier.

As a result, the gain is defined entirely by the external resistor R_G. The output of the instrumentation amplifier is

$$V_o = A_v(V^+ - V^-) \quad [\text{V}]. \tag{11.13}$$

Thus the main use of this type of amplifier is to obtain an output that is proportional to the difference between the inputs. This is important in differential sensors, especially when one sensor is used to sense the stimulus and an identical sensor is used for reference (such as when temperature compensation is needed).

Each of the inputs has the high impedance of the amplifier to which it is connected and the output impedance is low, as we have seen above. The main problem in a circuit of this type is that the CMRR depends on matching the resistances (R_1, R_2, and R_3) in each section of the circuit, that is, the resistances in the upper amplifier must match the corresponding resistances in the lower amplifier. Since these are internal, they are adjusted during production to obtain the required CMRR, and in most cases they are identical resistors, as mentioned above.

11.2.5 The Charge Amplifier

Whereas the basic circuits are the inverting and noninverting amplifiers shown above, the operational amplifier can be made to produce other desirable functions, depending on the feedback circuit. A useful example is the so-called charge amplifier shown in **Figure 11.10**. Of course, charge cannot be amplified, but the output voltage can be made proportional to charge. Since this is an inverting amplifier, the gain is given in **Equation (11.3)**, except that the feedback resistor is replaced with the impedance of a capacitor. The latter is $1/j\omega C$ and the output of the inverting amplifier is

$$A_v = -\frac{R_f}{R_I} = -\frac{1/j\omega C}{1/j\omega C_0} = -\frac{C_0}{C}, \tag{11.14}$$

where C_0 is the capacitance connected across the inverting input. Now assuming that a change in charge occurs on the capacitor, equal to $\Delta Q = C_0\Delta V$, the output voltage can be written as

$$V_o = -\Delta V\frac{C_0}{C} = -\frac{\Delta Q}{C} \quad [\text{V}]. \tag{11.15}$$

In effect, the charge generated at the input is amplified. If C is small, a small change in charge at the input can generate a large voltage swing in the output. This is a useful method of connecting capacitive sensors such as pyroelectric sensors and other capacitive sensors whose output is low. To do so, it is necessary for the input impedance to be very high and care must be taken in connections (such as the use of very good capacitors). Commercial charge amplifiers use FETs for the difference amplifier to ensure the necessary high-input impedance. The resistance r in **Figure 11.10** is added to ensure the capacitor C is discharged at a low rate as the input charge decreases (otherwise the higher reading may persist).

EXAMPLE 11.2 Interfacing of a pyroelectric sensor

In **Example 4.11** we discussed a pyroelectric sensor used for motion detection. The sensor produced a change in charge of $\Delta Q = 3.355 \times 10^{-10}$ C for a change in temperature of the pyroelectric chip of 0.01°C due to motion of a person. The change in voltage across the sensor was calculated as 0.0296 V. Calculate the capacitor needed to produce a rail voltage output (output equal to the power supply level) in the charge amplifier shown in **Figure 11.10**. Calculate the gain of the amplifier as well. The operational amplifier operates at ±15 V. Discuss.

Solution: Equation (11.15) provides the necessary relation to calculate the capacitor. The rail voltage of the amplifier is +15 V or −15 V:

$$C = -\frac{\Delta Q}{V_o} = -\frac{3.355 \times 10^{-10}}{-15} = 22.37 \times 10^{-12}\ \text{F}.$$

This is a 22.37 pF capacitor. Note that for a positive change in charge across the sensor (i.e., for an increase in temperature), the output is negative, and vice versa. For that reason the denominator was entered as negative since capacitance must be positive.

The gain of the amplifier is given in **Equation (11.14)**, but we first need to calculate C_0, the capacitance of the sensor. Typically this will be given in its specifications, but since we know the voltage produced due to the change in charge we can write

$$C_0 = \frac{\Delta Q}{\Delta V} = \frac{3.355 \times 10^{-10}}{0.0296} = 1.1334 \times 10^{-8}\ \text{F}.$$

The gain is

$$A_v = -\frac{C_0}{C} = -\frac{1.1334 \times 10^{-8}}{22.37 \times 10^{-12}} = 506.68.$$

This is a fairly high gain and may require a two-stage amplifier. The first of the two stages must be a charge amplifier, the second can be a noninverting voltage amplifier. In principle, any amplifier can be configured as a charge amplifier, but the requirement of very high input impedance (to prevent "discharging" the sensor) limits these to FET input amplifiers. These are special operational amplifiers, but are not uncommon. There are also operational amplifiers that have been specifically built as charge amplifiers. When it becomes necessary to use a charge amplifier, care should be taken not only in selection of the amplifier and the feedback capacitor but also connections and, in particular, the choice of the printed circuit board. Parasitic capacitances can change the gain, and losses in the printed circuit board can change the effective input impedance of the amplifier.

11.2.6 The Integrator and the Differentiator

The operational amplifier integrator circuit is another fundamental circuit often used in the interfacing of sensors. As the names imply, the output of the circuit is the integral of the input voltage for an integrator and the derivative of the input for a differentiator. The basic circuits are shown in **Figure 11.11**. The integrator consists of an inverting amplifier (see **Figure 11.6a**) with a capacitor across the feedback resistor. The operation of the integrator can be understood by recalling that the open loop amplification of the amplifier is very high, as is its input impedance. That means that the potential difference between the negative and positive inputs and the current into the negative input are negligible and the negative input of the amplifier is essentially at ground potential.

Consider now **Figure 11.11a**. Given an input voltage V_i, the current through the resistance R_I is $I = V_i/R_I$. This current cannot flow into the input of the amplifier and therefore must flow into the capacitor. By definition, the current in the capacitor is

$$I_C = C\frac{dV_C}{dt} \quad [\text{A}]. \tag{11.16}$$

Since this must be equal to the current through R_I we have

$$C\frac{dV_C}{dt} = \frac{V_i}{R_I} \rightarrow dV_C = \frac{V_i}{CR_I}dt \quad [\text{V}]. \tag{11.17}$$

The voltage across the capacitor is

$$V_C = \int_0^t \frac{V_i}{CR_I}dt \quad [\text{V}]. \tag{11.18}$$

The output voltage is the negative of the capacitor voltage:

$$V_o = -V_C = -\int_0^t \frac{V_i}{CR_I}dt \quad [\text{V}]. \tag{11.19}$$

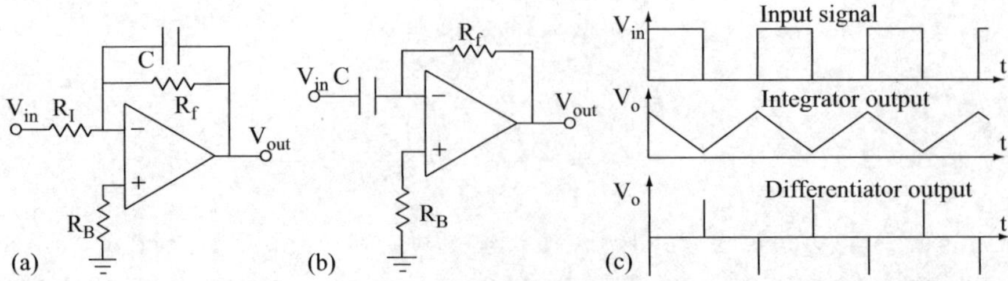

FIGURE 11.11 ■ (a) The operational amplifier integrator. (b) The operational amplifier differentiator. (c) The output of the integrator and of the differentiator for a train of pulses.

If the output is not zero when the input voltage is applied, that is, if it is at an initial value V_{initial}, this must be added to the output produced by the capacitor:

$$V_o = -\int_0^t \frac{V_i}{R_I C} dt + V_{\text{initial}} \quad [\text{V}]. \tag{11.20}$$

The output of the amplifier is the integral of the input. With constant values of V_i, R_I, and C, V_o is a linear function with a negative slope for V_i positive and positive slope for V_i negative. **Figure 11.11c** shows the output of the integrator for a square wave at the input. The input is shown as a positive signal, but it may also be a bipolar signal.

By interchanging between capacitor C and the input resistance R_I, we obtain the opposite function, that of differentiation. The circuit is shown in **Figure 11.11b**. Following the same arguments as above, we note that the current through the feedback resistor is $I_f = V_o/R_f$. This must also be the current through the capacitor since there can be no current into the negative input of the amplifier. The current through the capacitor is

$$I_C = -C\frac{dV_i}{dt} = \frac{V_o}{R_f} \quad [\text{A}]. \tag{11.21}$$

The negative sign indicates that the current is from output to input. The output voltage is

$$V_o = -R_f C\frac{dV_i}{dt} \quad [\text{V}]. \tag{11.22}$$

Clearly the input must be a function of time. The output for a square wave is shown in **Figure 11.11c**. Note that an ideal differentiator in this case should produce a very narrow positively or negatively going pulse. In reality it will be more of a narrow triangular pulse since the square wave is never ideal and the differentiator will reflect the finite increase/decrease time of the square wave.

11.2.7 The Current Amplifier

Another example of the use of an operational amplifier to a specific end is the current amplifier shown in **Figure 11.12**. The input voltage at the inverting input is $V_i = ir$. As in any inverting amplifier, the output is

$$V_o = -V_i\frac{R}{r} = -iR \quad [\text{V}]. \tag{11.23}$$

This is a very useful device when sensing with very low impedance sensors. For example, this circuit may be used with thermocouples whose impedance can be trivially low. They may be connected directly (r then represents the resistance of the

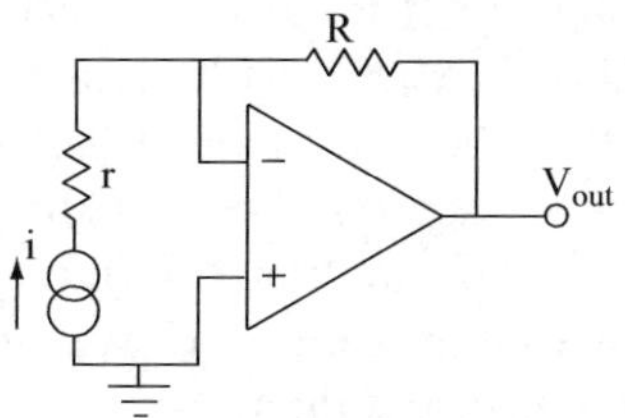

FIGURE 11.12 ■ Current amplifier.

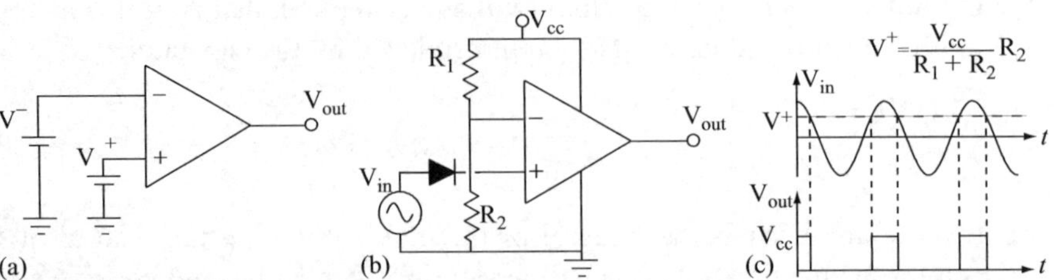

FIGURE 11.13 ■ The comparator. (a) Principle. (b) A practical, single-polarity circuit. The diode in the input prevents negative voltages from damaging the amplifier. (c) Output as a function of input voltage.

thermocouple). The output is a direct function of the current the thermocouple produces, which can be fairly large as opposed to its low voltage.

11.2.8 The Comparator

The operational amplifier can be used in open loop mode but, because its gain is so high, a very small signal at the input will saturate the output. That is, for practically any given input, the output will be either $+V_{cc}$ or $-V_{cc}$ depending on the polarity of the input signal. This property has a useful application—that of a comparator. Consider the circuit in **Figure 11.13a**. The negative input is set at a voltage V^- and the positive input at $V^+ = 0$. Therefore the output is $-A_{ol}V^- = -V_{cc}$. Now suppose we increase V^+. The output is $(V^+ - V^-)A_{ol}$. As long as $V^+ < V^-$, the output remains $-V_{cc}$. If $V^+ > V^-$, the output changes to $+V_{cc}$. Thus the function of this device is to compare the two inputs and to indicate which one is higher.

The comparator is useful beyond simple comparison. It is used extensively in A/D and D/A conversion of signals and in many other aspects of sensing and actuation. **Figure 11.13b** shows a practical connection of the comparator designed to operate from a single-polarity voltage. Resistors R_1 and R_2 form a voltage divider setting the negative input to a fixed voltage:

$$V^- = V_{cc}\frac{R_2}{R_1 + R_2} \quad [\mathrm{V}]. \tag{11.24}$$

This voltage now serves as the comparing or reference voltage. The output will flip to $+V_{cc}$ when $V^+ > V^-$ and to zero when $V^+ < V^-$. Of course, the reference voltage can be set on the positive input and the signal to be compared entered on the negative input. Also, a practical amplifier with a single-polarity power supply cannot tolerate a negative input, hence the diode in the input circuit.

One particular problem with comparators is chatter, which can occur if the two inputs are very close to each other or if $V^+ - V^-$ is very close to zero. The output will then flip between negative and positive V_{cc} as long as the condition persists. To avoid this, a small hysteresis is added to the input so that the change to $+V_{cc}$ occurs slightly above zero and the change to $-V_{cc}$ (or zero) occurs slightly below zero (see **Example 11.3**). Hysteresis is accomplished by adding a relatively large resistor between the output and the positive input of the comparator. However, in doing so it is necessary to shift the reference voltage to the positive input.

EXAMPLE 11.3 Derivation of a clock signal from the mains

In an electric clock the reference frequency is derived from the mains voltage since the frequency of the grid is tightly controlled and sufficiently accurate for most purposes. A simple means of doing so is to feed a sinusoidal voltage derived from the grid to the negative input of a comparator and set the reference voltage at some appropriate level on the positive input. **Figure 11.14a** shows a possible design together with the signals involved. The 240 V, 50 Hz (European grid) signal is first stepped down through a transformer to a ±6 V RMS (or about ±8.4 V peak). The input is $V(t) = 8.4\sin(2\pi \times 50t) = 8.4\sin(314t)$. The diode is added so that the positive input does not go below zero since the comparator shown operates between 0 and +12 V and will not tolerate negative voltages. The reference voltage is set at 6 V by selecting R_1 and R_2. When the input rises above 6 V, the output decreases to zero, and when it goes below 6 V, the output increases to +12 V. The output is shown in the figure.

When the input is exactly 6 V, the output is indeterminate and likely it will very rapidly oscillate between 0 and +12 V. To avoid this, a resistor is added between the output and the positive input, as shown in **Figure 11.14b**. This resistor changes the reference voltage as follows:

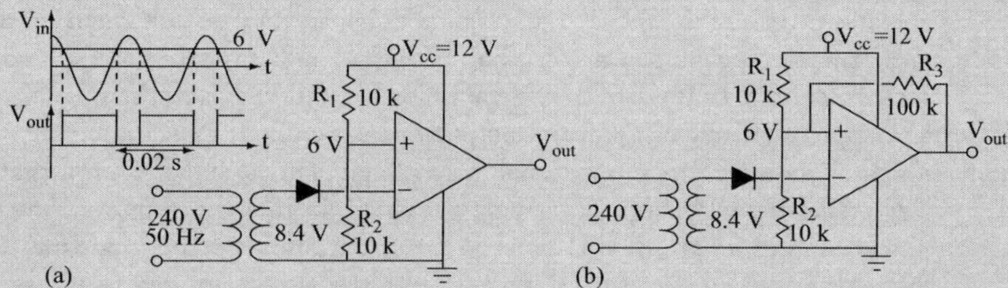

FIGURE 11.14 ■ (a) A comparator used to generate a 20 ms clock signal. (b) The addition of hysteresis by adding R_3 to eliminate output chatter.

When the output is high ($V_o = V_{cc}$), R_3 is in parallel with R_1. The voltage at the positive input is

$$V^+ = V_{cc}\frac{R_2}{R_2 + R_1 \,||\, R_3} = 12\frac{10^4}{10^4 + \left(\dfrac{10^4 \times 10^5}{10^4 + 10^5}\right)} = 12\frac{10^4}{10^4 + 9.09 \times 10^3} = 6.28 \text{ V}.$$

When the output is zero, the output is at ground level, R_3 is in parallel with R_2 and the reference voltage is

$$V^+ = V_{cc}\frac{R_2 \,||\, R_3}{R_1 + R_2 \,||\, R_3} = 12\frac{\left(\dfrac{10^4 \times 10^5}{10^4 + 10^5}\right)}{10^4 + \left(\dfrac{10^4 \times 10^5}{10^4 + 10^5}\right)} = 12\frac{9.09 \times 10^3}{10^4 + 9.09 \times 10^3} = 5.71 \text{ V}.$$

Now the shift to output zero occurs when the input voltage is 6.28 V and to output high when the input signal goes below 5.71 V. This eliminates the chatter in the output entirely.

Notes:

1. The smaller the hysteresis resistor (R_3), the larger the difference between the two voltages and hence the larger the hysteresis, and vice versa.

2. Using the method shown here, hysteresis can only be introduced on the positive input, but there are other methods of adding hysteresis.
3. The pulse width can be calculated from the reference voltages since the input signal is sinusoidal (see **Problem 11.10**).
4. The actual values of R_1 and R_2 are not important. However, they should not be too small so as not to dissipate too much power.

11.3 | POWER AMPLIFIERS

A power amplifier is an integrated or discrete components circuit whose power output is the input power multiplied by a power gain:

$$P_o = P_i A_p \quad [\mathrm{W}]. \tag{11.25}$$

That is, the amplifier is capable of boosting the power level of a signal to match the needs of, for example, an actuator. The obvious use of power amplifiers is in driving actuators, especially those that require considerable power, such as audio amplifiers used to drive speakers and voice coil actuators and amplifiers for solenoid actuators and motors. In spite of the fact that the amplifier is called a power amplifier, it is really either a voltage amplifier or a current amplifier (also called transconductance amplifier). In a voltage amplifier, the input signal is a voltage. This voltage is amplified and in the final stage a sufficiently high current is provided so that the required power is met. Most power amplifiers are of this type. In a current amplifier the opposite occurs. In effect, this can be viewed as essentially boosting the signal voltage to the required level and allowing the load to draw the necessary current.

Power amplifiers are divided into linear and pulse width modulated (PWM) amplifiers. In a linear amplifier, the output (voltage) is a linear function of the input and can be anything between $-V_{cc}$ and $+V_{cc}$. In a PWM amplifier the output is either V_{cc} or zero and the power delivered is set by the time the output is on.

11.3.1 Linear Power Amplifiers

As mentioned, the first step is to amplify the signal to the required output level. This can be done using any amplifier, but we shall assume an operational amplifier is used for this purpose. Then this voltage is applied to an "output stage" that does not need to amplify, but, rather, supplies the necessary current. A simple example is shown in **Figure 11.15**.

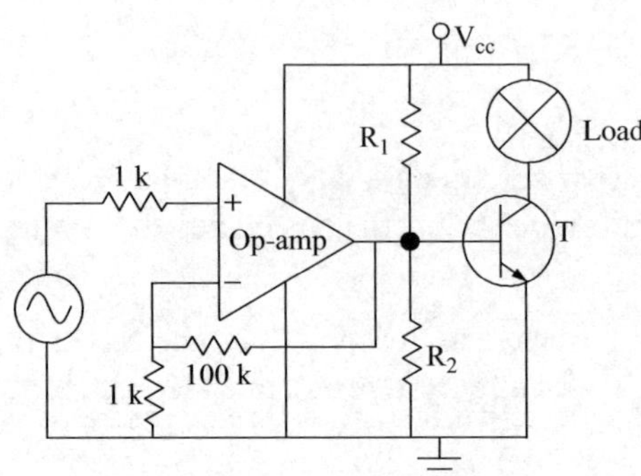

FIGURE 11.15 ■ A linear power amplifier.

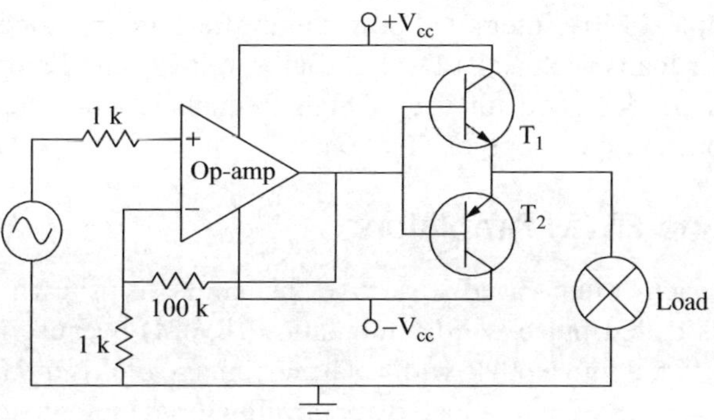

FIGURE 11.16 ■ The class B push-pull amplifier.

It shows the so-called class A power amplifier using an *npn* bipolar junction transistor (BJT). The amplifier is set for a gain of 101 (noninverting amplifier). The output then drives the transistor, whose output will swing, at most between 0 and V_{CC}, and supplies a current that is V/R_L, where R_L is the resistance of the load. The transistor may be viewed as a current amplifier since its collector current is the base current multiplied by the amplification of the transistor (see **Equation (4.18)**). The class A designation indicates amplifiers for which the output stage is always conducting, as in the case above. The transistor in **Figure 11.15** can be replaced with a MOSFET for higher currents. It is also assumed that the output does not saturate—in saturation, the output voltage is constant and not a function of the input and hence the amplifier does not operate in its linear range. This type of amplifier is sometimes used to drive relatively small loads such as light indicators, small DC motors, relays, and some solenoid valves. In some cases the amplification is set high enough to saturate the amplifier, in which case the amplifier operates as an on–off circuit, which again is useful in driving indicators, opening or closing solenoids, etc.

A better approach, and one that is often used, especially in audio amplifiers, is the class B or push-pull amplifier shown in **Figure 11.16**. This operates exactly as in the previous case except that under no input, the output is zero and there is no conduction in the transistors (or MOSFETs). When the input is positive, the upper transistor conducts, supplying the load, and when the input is negative, the lower transistor supplies the load. The voltage in the load can swing between $+V_{CC}$ and $-V_{CC}$ and the current is again defined by the load. The output stage is made of a pair of power transistors, one *pnp* and one *npn* (or of *p*- and *n*-type MOSFETs).

There are many variations of these basic amplifiers. For example, feedback may be added from output to input or to an intermediate stage. Similarly, it is common to protect the output stage from short circuits as well as from spikes due to inductive and capacitive loads. However, these are details beyond the scope of this chapter.

In terms of performance characteristics of linear amplifiers, the obvious are the power output and the type and level of input. For example, an amplifier may be specified as supplying 100 W for a 1 V input. Next is the distortion level. Usually the distortions are specified as a percentage of output. The most common specification is the total harmonic distortion (THD) as a percentage of output. A good audio amplifier will have less than, say, 0.1% THD. Although THD is more important in audio reproduction, it can also affect the performance of actuators. Other specifications are the temperature increase due to power dissipation in the transistors and the output impedance of the amplifier. The latter must match the impedance of the load for maximum power transfer and hence maximum efficiency.

Power amplifiers of various power levels exist either as integrated circuits or as discrete component circuits. Usually the discrete circuits can supply higher powers, but they are more complex. Most integrated power amplifiers are intended for audio use, but they can drive other loads such as LEDs, lightbulbs, relays, small motors, etc. There are also power amplifiers designed for use at high frequencies, but these are highly specialized circuits, outside the scope of this text.

11.3.2 PWM and PWM Amplifiers

A different approach often used to drive actuators is shown schematically in **Figure 11.17**. It is called pulse width modulation (PWM) because the amplitude of a given signal is translated into a pulse width. The advantage of this method is that now one can control the power supplied to a load by controlling how long the load is connected to the power (pulse width) rather than by controlling the amplitude. Therefore the amplifier in the previous circuits is replaced with a simple (electronic) switch and the voltage across the load is either zero or V_{cc}. To understand how this works, consider first **Figure 11.18**. The oscillator in **Figure 10.17** generates a triangular wave of constant amplitude and frequency. This is fed to the negative input of the comparator. The signal to be represented as a PWM signal is fed directly to the positive input. In this example we assume the comparator swings between V^+ and zero (many comparators swing between V^+ and V^-). The output of the comparator is positive when the signal is higher than the triangular wave and zero when it is lower. The result is a pulse with a width proportional to the amplitude of the signal. Note that if the sinusoidal signal swings negative (i.e., if its average is zero), only the positive part of the signal will be represented. In that case one can choose comparators that will swing negative and obtain a PWM signal with positive pulses for the positive part of the signal and negative pulses for the negative part of the signal. It should be noted that for proper representation the triangular wave frequency must be much higher than the signal being represented, but this depends as well on the application. For example, if this is used to dim a lightbulb using a 60 Hz source, the PWM must be on the order of 10–20 times that (i.e., 600–1200 Hz) to properly represent the signal.

PWM can be used to control the power into a load, as shown in **Figure 11.19**. Here the power transistor is driven on and off so that the voltage on the load can only be zero

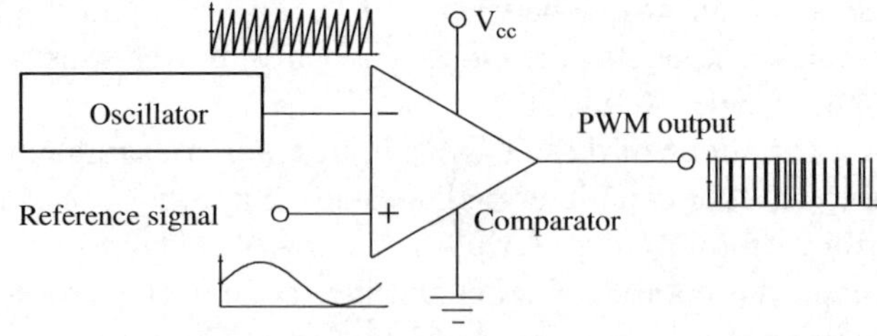

FIGURE 11.17 ■ Schematic of the PWM principle.

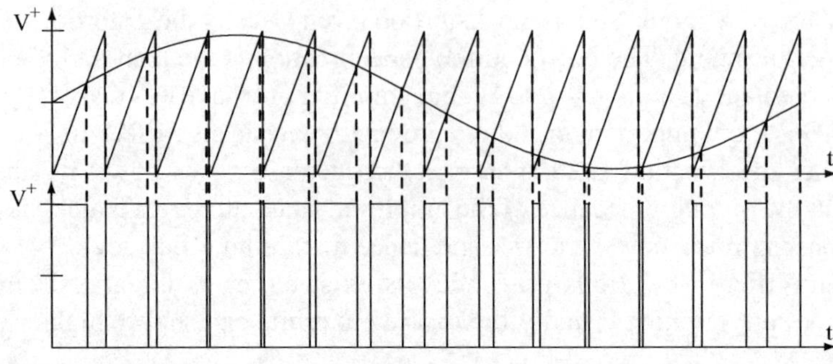

FIGURE 11.18 ■ Generation of a PWM signal. The sinusoidal signal's amplitude is represented as pulse widths.

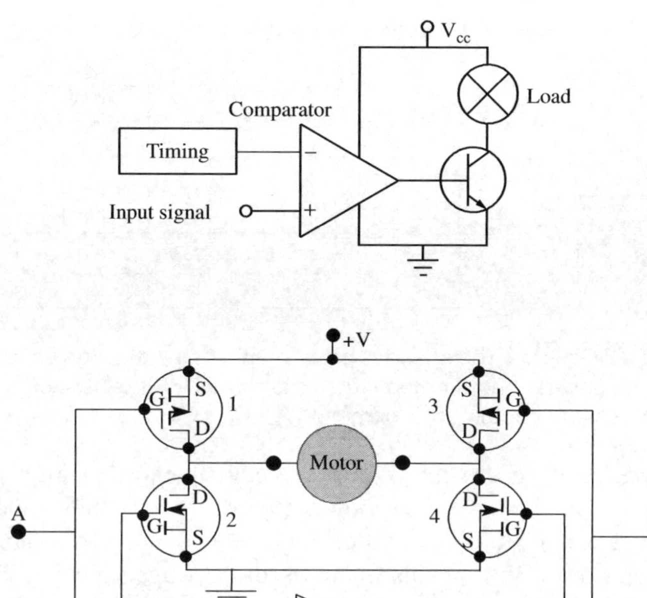

FIGURE 11.19 ■ A load driven by a PWM circuit. The transistor serves as a switch and power is controlled by the average of the PWM signal.

FIGURE 11.20 ■ Driving an H-bridge from a PWM source. Both inputs (A and B) are PWM signals to control the speed of the motor in both directions.

or V_{cc}. The width of the pulse defines the average power in the load since the total pulse length is defined by the timing signal (clock) and is fixed. This circuit can drive a small DC motor or change the intensity of a small lightbulb, but it is not an efficient method because of losses in the transistor.

Figure 11.20 shows an example often used to control both the speed and direction of DC motors. It is called an H-bridge, for obvious reasons. A pulse of constant amplitude but varying duty cycle connected to point A will drive MOSFETs 1 and 4, turning the motor in one direction. The duty cycle defines the average current in the motor and hence its speed. Connecting to point B turns on MOSFETs 2 and 3, reversing the direction of rotation. The bridge can also brake the motor by shorting both terminals to ground or to V_{cc}. To do so, each input must be available independently or additional circuitry must be employed. The two inverters used in this circuit ensure that only two diagonally opposite MOSFETs conduct at any given time. The inverters are digital circuits whose output is zero when the input is high, and vice versa. Inverters will be explained in the following section as part of the discussion on digital circuits.

Although some precautions must be taken to ensure that only opposite MOSFETs conduct (e.g., if MOSFETs 1 and 2 conduct at the same time, the power supply is shorted and the current in MOSFETs 1 and 2 is only limited by their internal resistance, resulting in instant damage), this is one of the most common circuits used for bidirectional control of motors and other actuators. The controller for H-bridges can be a small microprocessor or a dedicated logic circuit. Integrated PWM circuits and controllers are available commercially and are integrated within some microprocessors.

EXAMPLE 11.4 Speed control of a DC motor

PWM can be used with any signal, DC or AC. As an example of conversion of a DC signal and its use to control the speed of a DC motor, consider the circuit in **Figure 11.21a**. The motor's speed

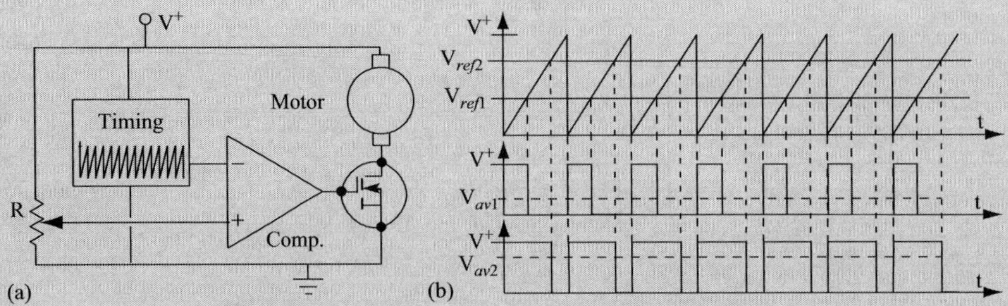

FIGURE 11.21 ■ Speed control of a DC motor. The speed is proportional to the average voltage across the motor. (a) The circuit showing the speed control potentiometer and the PWM generator. (b) The generation of PWM signals for two positions of the potentiometer.

is controlled by controlling its average power. As the voltage at the comparator's input increases, so does the pulse width and hence the power in the motor. **Figure 11.21b** shows the representation of two values of the input voltage to the transistor (and hence the power in the motor). The dashed horizontal line through the PWM signals indicates the average voltage across the motor (and hence the relative speed of the motor).

11.4 | DIGITAL CIRCUITS

Although the focus of this chapter is on interfacing and most digital functions can be accomplished through the microprocessor, it is useful to discuss here a few fundamental digital circuits. Not only are these useful in interfacing, but they can also accomplish important functions on their own, as we shall see later in this chapter. In fact, one digital circuit has already been mentioned in the previous section. The inverter is one of the simplest digital circuits and one of the most useful.

The first class of digital circuits we look at is the class of logic gates. These are simple circuits that accomplish logical functions such as OR, AND, NOR (not OR), NAND (not AND), XOR (exclusive OR), and others. These are available as integrated circuits in various logic families.

Figure 11.22 shows the symbols of the most common logic gates and **Table 11.1** shows the "truth table" for these gates. The truth table is simply the output of the gate for all possible combinations of inputs. A "0" indicates ground level voltage (0 V) and a "1" indicates a high output (V_{cc}). The high voltage depends on the type of circuit (often called a logic family) and can be as low as 1 V (or even lower) and as high as 15 V (or higher). Logic circuits can be designed for any voltage level, but the standard logic families use voltages between 1 and 15 V, with most operating at 3.3 V or 5 V.

Internally all gates only use NOR or NAND gates. There are two reasons for this. The most important reason is that any logical function can be accomplished with NOR

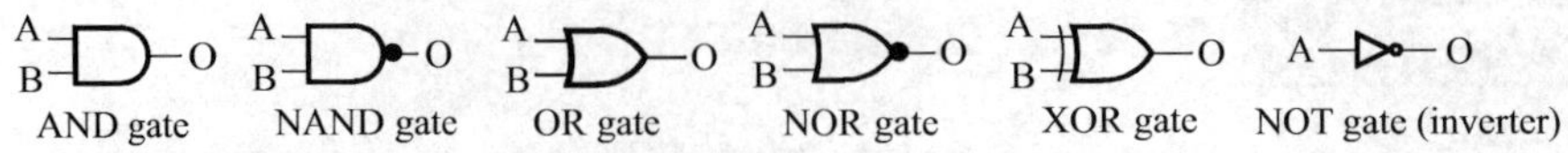

A B O AND gate · A B O NAND gate · A B O OR gate · A B O NOR gate · A B O XOR gate · A O NOT gate (inverter)

FIGURE 11.22 ■ Various logic gates. Their truth tables are shown in **Table 11.1**.

TABLE 11.1 ■ Truth tables for common logic circuits

A	B	OR	AND	NOR	NAND	XOR	Inverter	
0	0	0	0	1	1	1	**I/P**	**O/P**
0	1	1	0	0	1	0	0	1
1	0	1	0	0	1	0	1	0
1	1	1	1	0	1	1		

gates alone or with NAND gates alone. Therefore, although all of the above gates exist as components, internally they are implemented with NOR or NAND gates. The second reason this is done is that the integration of components is more efficient when it uses a few repetitive structures rather than many different structures. Thus, even though implementation with, say, NAND gates may actually increase the number of components in a particular integrated circuit, its implementation is more effective and less expensive. The use of NAND and NOR gates as "universal gates" to build other circuits is based on the DeMorgan theorems:

$$\overline{AB} = \overline{A} + \overline{B} \tag{11.26}$$

$$\overline{A+B} = \overline{AB}. \tag{11.27}$$

The various logic operations are denoted in simple mathematical notation. The AND operation of inputs A and B is denoted as AB, whereas the OR operation is denoted as $A+B$. NAND and NOR are negated AND and negated OR operations and are indicated by a bar above: $\overline{AB}$ for NAND and $\overline{A+B}$ for NOR. Similarly, if the input of an inverter is A, its output is $\overline{A}$. Thus the DeMorgan theorems say that A NAND B equals NOT A plus NOT B and that A NOR B equals NOT A AND NOT B. These two rules can be used to build other circuits, as is shown in **Example 11.5**.

The gates discussed so far were two-input gates, but multiple-input gates can be built and exist in integrated devices. Much more complex logic operations can be defined by combining various gates. It should also be noted here that the transition of signals through gates takes time. Some logic families are faster than others, but these delays must be taken into account when using gates as part of interfacing.

The various gates serve many purposes. They can, of course, perform the logic operations implied by their structure, but they can also serve circuit functions, and they are the basis of many more complex digital circuits, including microprocessors and computers. For example, an AND gate can be used as a simple switch or gate for a digital signal. Suppose the signal, consisting of a sequence of pulses, is connected to input A of the gate. If input B is "1", the signal will appear at the output. But if B is "0", the signal is blocked. Another simple example is the use of the XOR gate to compare the signals at inputs A and B. The output of the XOR gate is "1" when the two inputs are the same but "0" when they differ. One can easily imagine the usefulness of such a function in comparing signals or logic states, especially with multiple-input gates. **Figure 11.23** shows a simple implementation of the NOT, NOR, and NAND gates using transistors. Many logic families use MOSFETs to accomplish the same functions, but with very different properties (MOSFETs can operate at lower currents, leading to lower power devices).

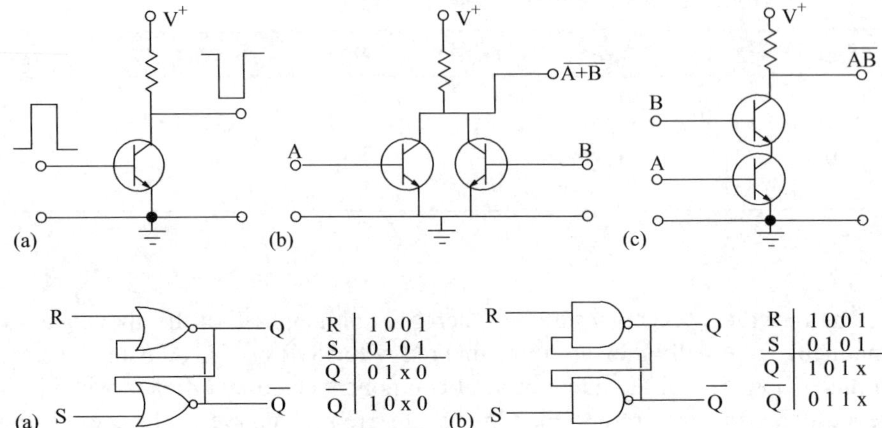

FIGURE 11.23 ■ Implementation of gates using transistors. (a) The NOT gate. (b) The two-input NOR gate. (c) The two-input NAND gate.

FIGURE 11.24 ■ The SR latch or flip-flop. (a) Implementation with NOR gates. (b) Implementation with NAND gates. Note that the truth tables are different. x indicates no change.

Logical gates are the building blocks of much more complex circuits, some of which we will see later. But it is instructive to discuss here one particular digital circuit built with NAND or NOR gates called the flip-flop. The most fundamental implementation of the flip-flop is shown in **Figure 11.24**. **Figure 11.24a** shows an implementation using NOR gates, together with its truth table. This device is called a set-reset (or SR) flip-flop or latch since a "1" on the S input forces an output $Q = 1$ and a "1" on the R input forces an output $Q = 0$. If both inputs are "0", the output stays at its previous state (hold), whereas if both inputs are "1", the output is indeterminate (this is a restricted state and should not be allowed to occur). **Figure 11.24b** shows an implementation with NAND gates in which the restricted state is $R = 0$ and $S = 0$, whereas no change occurs if both inputs are "1". The truth tables show one of the basic functions of a flip-flop—that of a storage device. The data on the output can be changed and then stored as necessary by proper control of the inputs. There are other types of flip-flops, making use of additional gates to eliminate restricted states and to accomplish specific functions. Some are designed to be "clocked," that is, to change state at clocked times, such as at the rising or the trailing edge of the clock. Others include set (sometimes called preset) and/or reset (often called clear) functions that force the output to a predetermined state.

An example is shown in **Figure 11.25** showing a D-latch. Its main function is to move the data on the input to the output at the rising edge of the clock. The D-latch is in fact a development of the basic SR latch. First, the SR latch is clocked as shown in **Figure 11.25a**. That means that the inputs S and R are transferred to the fundamental SR latch (indicated here with inputs s and r) only when the clock is high. When the clock is low, the s and r inputs are zero and the output remains unchanged. The main effect of the clock is to allow the latch to settle into its prescribed output. Inputs $S = 1$ and $R = 1$ are still not allowed since the output is indeterminate. To avoid this situation, the circuit is modified as in **Figure 11.25b**. Now the input, indicated as D, sets the output to "1" if $D = 1$, and the clock changes from 0 to 1. If $D = 0$ and the clock goes from high to low, the output goes low. That is, the input is stored on the output only during the clock cycle. To make this into a useful device, two D-latches are used as shown in **Figure 11.25c**. In this circuit, the D flip-flop toggles the output as follows: if the input is high, the output

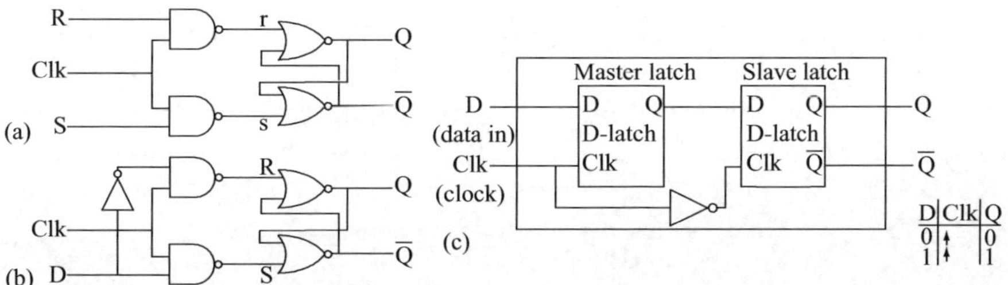

FIGURE 11.25 ■ The evolution of the D flip-flop from the SR latch. (a) The clocked SR latch. (b) The basic D-latch. (c) The D flip-flop is made of two D-latches and an inverter together with its truth table.

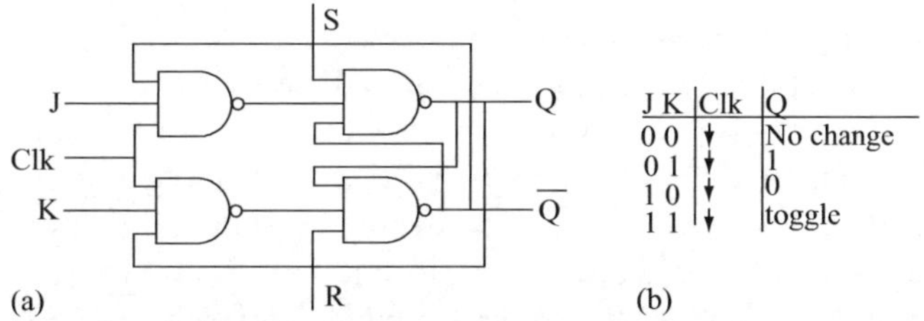

FIGURE 11.26 ■ The J-K flip-flop including preset (S) and clear (R) functions. (a) Implementation with NAND gates. (b) Truth table.

goes high if it was previously low, when the clock goes low but does not change state if the output was previously high. When D goes low, the output goes low when the clock goes low. In effect, the D flip-flop toggles the output between $Q = 0$ and $Q = 1$ by changing the input. D-type flip-flops are often used in shift registers to store and retrieve data (see **Example 11.6**).

Another useful device is the J–K flip-flop shown in **Figure 11.26**. Its truth table shows that when $J = 1$ and $K = 1$ the output toggles (changes state), whereas when $J = 1$ and $K = 0$ the output is 1 and when $J = 0$ and $K = 1$ the output is zero. The output changes on the falling edge of the clock. If both inputs are 0, there is no change in the output. If $S = 0$ (set or preset), the output is set to 1 and if $R = 0$ (clear), the output is set to 0. The toggle function of the flip-flop is often used to divide the frequency of an incoming signal by two and hence to build dividers and counters (see **Example 11.7**).

EXAMPLE 11.5 Three-input AND gates

Show how one can build a three-input AND gate using only NAND gates and only NOR gates, each with only two inputs.

Solution: The simplest way of building the required gates is to start with the DeMorgan theorems. We require an output ABC for inputs A, B, and C.

For NAND gates we only have two inputs, so we first write the output as (AB)C. Since we use NAND gates, to obtain this output the input must be AB and C. The output will be $\overline{ABC}$ and by inverting that we get the required output as ABC. This partial step is shown in **Figure 11.27a**. To obtain AB we use another NAND gate with inputs A and B that will produce $\overline{AB}$. This is inverted to obtain AB. The three-input AND gate requires four NAND gates and is shown in **Figure 11.27b** together with its truth table.

For NOR gates we can use the second DeMorgan theorem (**Equation (11.27)**), but again we recall that since we have only two inputs for each NOR gate we write the required output as (AB)C. Since $\overline{A+B} = \overline{AB}$, we first produce the following product:

$$\overline{\overline{A}+\overline{B}} = \overline{\overline{A}\,\overline{B}} = AB$$

That is, we first negate A and B to produce AB. Now again applying the theorem we write $\overline{AB+C} = \overline{ABC}$. Negating AB and C before the NOR gate produces the correct output:

$$\overline{\overline{AB}+\overline{C}} = \overline{\overline{AB}\,\overline{C}} = ABC$$

The implementation together with its truth table is shown in **Figure 11.27c**.

The two implementations are functionally identical and similar in implementation but there are differences as well. In particular, the NAND implementation requires fewer gates than the NOR implementation. In both implementations, the passage of signals is not symmetric. Signals AB pass through four gates, whereas C passes through only two gates. The total delay is twice as long for signals A and B as for signal C. This asymmetry means that the various signals may reach the output at different times, possibly modifying the expected signal. To fix this one can add two NOT gates (inverters) in the path of C, as shown in **Figure 11.28**. The delay is now four gate delays for each of the signals. Note also that the NAND implementation required four gates (or six gates for the symmetric form), whereas the NOR implementation required six (or eight) gates. This, of course, depends on the function being implemented and does not mean that NAND implementation is always more economical.

Note also that the implementation is not necessarily unique or optimal in terms of the number of gates. Often considerations such as delays or symmetry are more important than the number of gates.

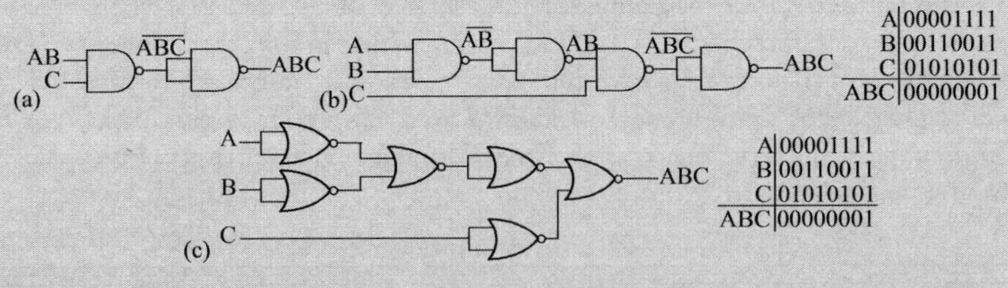

FIGURE 11.27 ■ Three-input AND gate. (a) Starting point with NAND gates. (b) The three-input AND gate implemented with NAND gates. (c) The three-input AND gate implemented with NOR gates.

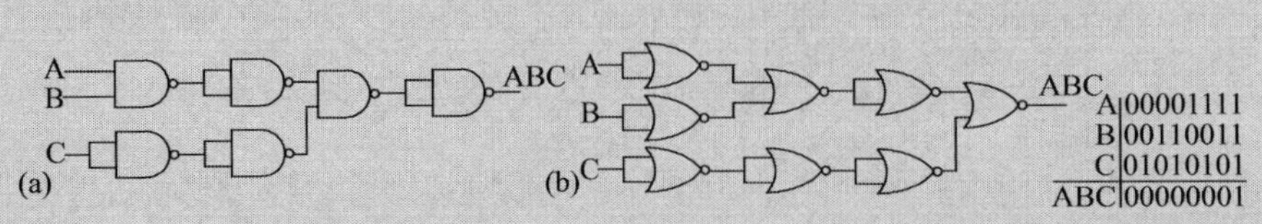

FIGURE 11.28 ■ Symmetric implementations of a three-input AND gate that produces equal delays for all three inputs. (a) Using NAND gates. (b) Using NOR gates.

EXAMPLE 11.6 The shift register

The shift register is a device made of a number of flip-flops that allows shifting of data into the device for storage purposes and the shifting of data out to retrieve it. The data can be shifted in or out serially or in parallel depending on how the shift register is designed. Consider the 4-bit shift register in **Figure 11.29**. It consists of four D flip-flops connected as shown. The clock inputs of all four flip-flops are connected in parallel and serve as a "shift" command. Given input data 1101 (as a string of pulses):

a. Show what the sequence is to enter the data into the shift register serially.

b. Show how the data stored in (a) can be shifted out serially.

c. Show how the data can be taken out as parallel data.

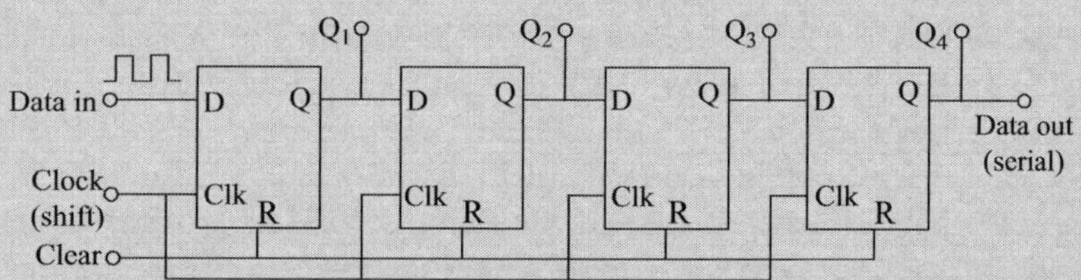

FIGURE 11.29 ■ Four-bit serial-in shift register. Output can be serial or parallel.

Solution: The input data is streamed in as it arrives using the clock input as a shift command.

a. First, the shift register can be cleared, although this is not necessary, since as the data are shifted in, the existing data are shifted out of the register and replaced with the input data. Assuming (for simplicity) that the register has been cleared, the register will contain 0000. In the first step, the first bit is shifted in and the register will show 1000. In the second step, the data in the shift register are shifted one step to the right and the second bit enters the first flip-flop, which now shows 0100. In the next two steps, the register will show 1010 and 1101, respectively, completing the data shift in four serial steps. Note that the output Q_4 is "lost" as the next bit is shifted in unless it is saved elsewhere.

b. The shift out occurs exactly as the shift in. Initially the shift register contains data 1101. The first shift moves all bits to the right. The content of Q_4 is 1 and this bit is shifted out. In the next step, $Q_4 = 0$ is shifted out followed by $Q_4 = 1$ and again $Q_4 = 1$. The data shifted out is 1101 as required. These data are "lost" unless stored elsewhere. This then constitutes a serial-in, serial-out (SISO) shift register.

c. After the shift in of the data, the four bits—$Q_1 = 1, Q_2 = 1, Q_3 = 0$, and $Q_4 = 1$—can be "taken" out by connecting to the four outputs. This constitutes a serial-in, parallel-out (SIPO) shift register.

Notes:

1. There are also parallel-in, serial-out (PISO) and parallel-in, parallel-out (PIPO) shift registers.
2. The size of the shift register can be arbitrarily long.
3. The time needed to shift in data depends on the data itself, the clock, and the size of the register, and for serial-in registers it can be long.
4. The type of shift register is often dictated by the type of data to be entered. If the data are inherently serial, an SISO or SIPO register must be used.

EXAMPLE 11.7 The digital counter or divider

Counters are just that, they count pulses. Their importance comes from the fact that they can be used to divide data by a convenient number, such as 10 or 16, for timing purposes. For example, one may use the frequency (50 or 60 Hz) to produce a clock. This is done by first digitizing the sinusoidal input to produce a square wave. Then, dividing the input by 50 or 60, one obtains a pulse per second. Minutes are obtained by dividing this by 60 again and hours by again dividing by 60. Days are marked by dividing again by 24 and so on.

As part of an electronic counter it is necessary to divide a clock signal by 10. The signal is a sequence of pulses. Design a counter/divider to do so and show the output signal obtained from the counter.

Solution: Dividing and counting is best obtained using J–K flip-flops because of the toggling obtained when $J = 1$ and $K = 1$. The signal is fed to the clock input. A flip-flop divides by two. That means that two flip-flops can divide a signal by 4, three flip-flops can divide by 8, and four flip-flops can divide by 16. To obtain a divide by 10 we must use four flip-flops and force them to reset when the count of 10 is reached. This is done by using additional gates to "intercept" the count "10" and reset the flip-flops. By doing so, the counter will count from 0000 to 1001 (zero to nine) and then reset back to zero as soon as the outputs become 1010. Resetting is done by setting the clear (R) to zero. The counter with its reset function is shown in **Figure 11.30a**.

Figure 11.30b shows the input and the outputs Q_1 through Q_4. Note that the reset occurs at the very beginning of the eleventh count and the sequence repeats. The frequency of the output at Q_4 is the frequency of the input divided by 10, that is, for every 10 input pulses, Q_4 produces one output pulse.

There are, of course, many more digital circuits and many of them are very complex, but they are all based on the fundamentals described here. In the limit, a digital computer is in fact an assembly of gates, gatelike circuits, and circuits built out of gates to perform all digital functions. Needless to say, modern digital devices take advantage of various types of circuits in addition to

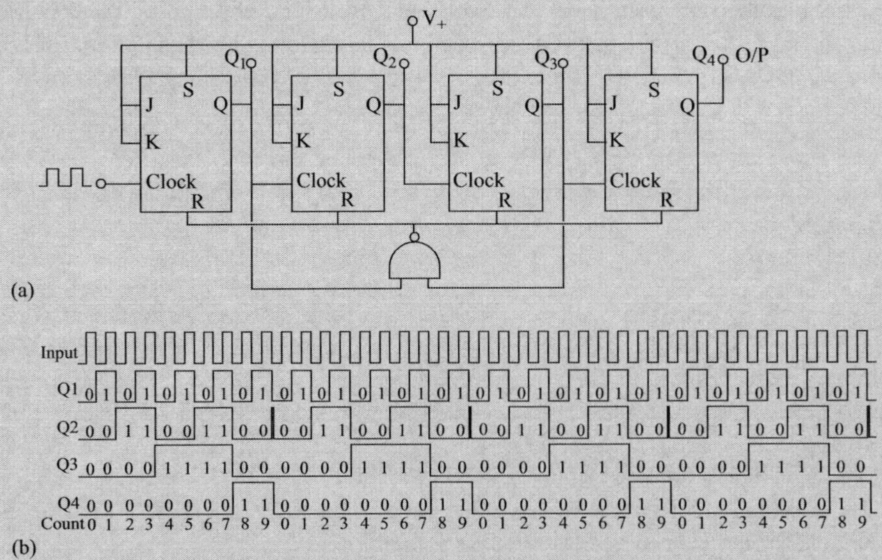

FIGURE 11.30 ■ (a) A divide-by-ten counter. (b) The signals obtained at the Q outputs of the flip-flops. The output on Q_4 is the input signal divide by 10.

the basic circuits shown here. For example, although the flip-flop is the basic "memory" cell, modern memories in digital devices have evolved beyond this basic form to metal oxide semiconductor (MOS) memory (see **Chapter 4**). Similarly, the level of integration tends to be much higher than is shown in the discussion above. Whereas one can get individual gates or a number of gates in a package as standard components, it would be impractical (and expensive), to build, for example, a clock using individual gates and flip-flops. It would be more practical to use a "clock" component or even a microprocessor and program it to operate as a clock (see **Chapter 12**). The purpose of the discussion here was to introduce the principles so that one can get a better feel for what is involved in the design and application of digital circuits in conjunction with interfacing.

11.5 | A/D AND D/A CONVERTERS

Analog-to-digital (A/D) converters and digital-to-analog (D/A) converters are the means by which a signal can be converted from analog to digital or from digital to analog as necessary. The idea is obvious, but the devices can be rather complex. However, there are certain types of D/A and A/D converters that are trivially simple. We will start with these and only then discuss some of the more complex schemes. Of course, in certain applications one of these simple methods is sufficient.

A/D and, to a lesser extent, D/A conversion are common in sensing systems since most sensors and actuators are analog devices. However, A/D converters usually require a high level voltage, much above the output of some sensors. Often the output from the sensor must be amplified first and only then converted. This leads to errors and noise and has resulted in the development of direct digitization methods based on oscillators (to be discussed below). A/D and D/A converters are available as components and are often integrated in microprocessors.

11.5.1 A/D Conversion

11.5.1.1 Threshold Digitization

In some cases an analog signal represents simple data, such as the position of items, counting of items on a production line, or monitoring passing vehicles. For example, in most car ignition systems, the ignition signal is obtained from a Hall element. The signal obtained is quite small and looks more or less sinusoidal, with the peaks indicating firing timing. In such a case it is sufficient to use a threshold detector to produce a digital output. An example is shown in **Figure 11.31a**. The output from the Hall element varies from 100 to 150 mV. This signal can be fed into a comparator, as shown in **Figure 11.32**, and the negative input set through the resistors to 130 mV. Normally the

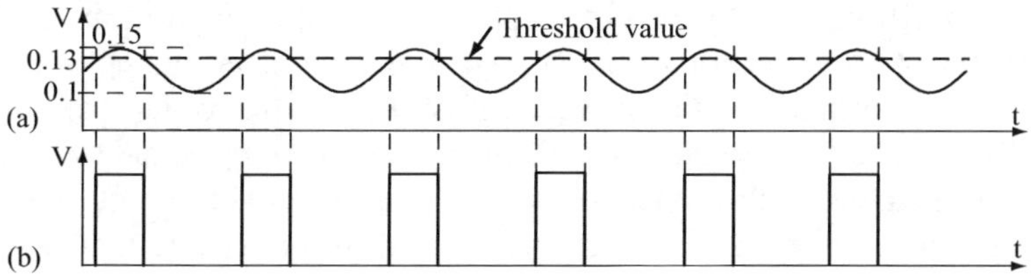

FIGURE 11.31 ■ Threshold digitization. (a) Original signal. (b) Digitized signal.

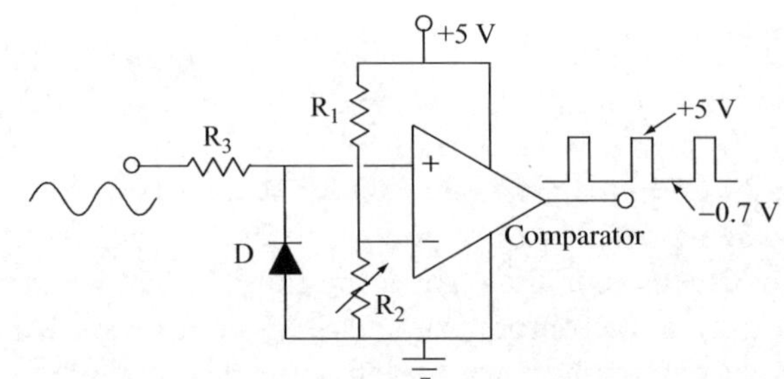

FIGURE 11.32 ■ Use of a comparator for threshold digitization. The diode eliminates negative voltages.

output is zero until the voltage on the positive input rises above the threshold of 130 mV. When the input dips below 130 mV, the output goes back to zero. The output in **Figure 11.31b** is obtained and now each pulse represents a tooth on the gear. Counting the teeth in a given time can give the speed of rotation of the gear or other data (such as when to fire a cylinder in an engine). Note also that if a tooth is missing, the corresponding pulse will be absent, whereas if the distance between teeth is not fixed, the distance between pulses will be variable. This method is effective when voltages at the input change across the comparison point but, at the comparison point itself, the output of the comparator is not properly defined and the output can change states back and forth, creating spurious pulses. To avoid this, hysteresis is added to the comparator so that the transition from low to high occurs at, say, $V_0 - \Delta V$ and the transition from high to low occurs at $V_0 + \Delta V$ (see **Example 11.3**).

Another approach to digitization of signals that can be used is the direct use of a Schmitt trigger. The Schmitt trigger is essentially a digital comparator with a built-in hysteresis, as described above, whose transition is around $V_{cc}/2$. This is a simple method of digitization and is sufficient for many applications. It is commonly used for the purpose above, but also in flow meters in which a rotating paddle operates a Hall element or another magnetic sensor, and it is also useful for optical sensors that use the idea of interruption of a beam (often used to count people passing a location or items on a production line). It is not suitable, however, for measuring the level of a signal, such as the voltage from a thermocouple.

11.5.1.2 Threshold Voltage-to-Frequency Conversion

In many sensors the output is too small to use the method above, or for that matter, to be sent over normal lines over any distance. In such cases a voltage-to-frequency conversion can be performed at the location of the sensor and the digital signal is then transferred over the line to the controller. The output now is not voltage, but rather a frequency, that is directly proportional to voltage (or current). These voltage-to-frequency (V/F) converters or voltage-controlled oscillators are relatively simple and accurate circuits and have been used for other purposes. Their main advantage over the threshold method above is that lower level signals may be involved and the problems with noisy transitions around the comparison voltage are eliminated. A circuit of this type is shown in **Figure 11.33**, as used with a thermistor. The circuit is an operational amplifier integrator. The voltage across the capacitor is the integral of the voltage at the noninverting leg of the amplifier. This voltage is proportional to the voltage across R_2. As the voltage on the capacitor increases, a threshold circuit checks this voltage and when the threshold has been reached, an electronic switch shorts the capacitor and discharges it. The switch then opens and allows the capacitor to recharge. The voltage on the capacitor is a triangular shape whose

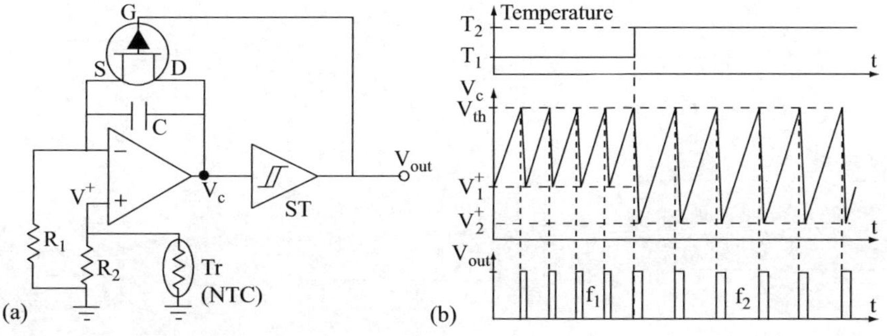

FIGURE 11.33 ■ Direct V/F conversion. (a) The circuit using a thermistor. (b) Output as a function of temperature.

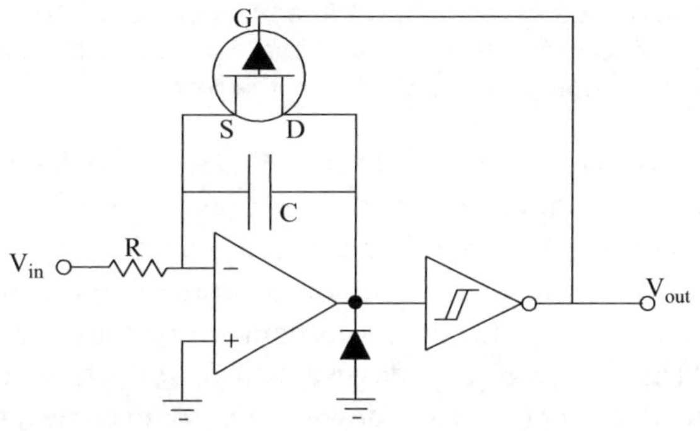

FIGURE 11.34 ■ A simple V/F converter based on an integrator and a Schmitt trigger.

width (i.e., the integration time) depends on the voltage at the noninverting input. At low temperatures the resistance is high and the voltage at the noninverting input will have a certain value. The output of the amplifier changes at a frequency f_1. If the temperature increases (to T_2), its resistance diminishes and the total resistance at the noninverting input decreases. This reduces the input voltage and hence the integration time until the capacitor reaches the threshold level (i.e., it reduces the time constant *RC*). The result is that the amplifier changes state slower and the output is a lower frequency, f_2. Since small changes in frequency can be easily detected, this can be a very sensitive method of digitization for small signal sensors. The method relies on the hysteresis of the Schmitt trigger and on the discharge time of the capacitor through the MOSFET, since that controls the width of the output pulses and the charge/discharge times, as can be seen from **Figure 11.33b**. The method can be used with optical sensors as well (such as photoresistors). Similar methods exist for capacitive sensors and the method can be adapted for other applications.

In the method shown in **Figure 11.33**, the sensor is an integral part of the V/F converter. However, the same basic circuit can be used as a V/F converter for any DC or slowly varying signal from any source, including sensors. The circuit is shown in **Figure 11.34**. The operation is as above except that the signal is fed at the inverting input. The circuit is particularly effective if the input signal level is relatively high.

Another simple and effective V/F method is shown in **Figure 11.35**. It consists of a square wave oscillator (called a multivibrator) and a control circuit. The on and off times of the waveform (hence frequency) are controlled by the charging and discharging times of the capacitors (**Figure 11.35a**). To control its frequency, the voltage to be converted is amplified and fed as currents to the bases of the two transistors. The larger the base current, the larger the collector current and the faster the charge/discharge and hence the

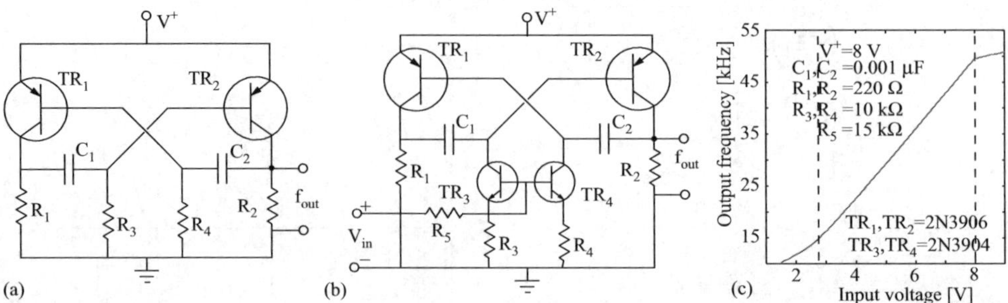

FIGURE 11.35 ■ A simple V/F converter based on a square wave multivibrator. (a) A simple multivibrator generating a square wave. (b) A V/F converter based on the multivibrator. The charge/discharge times are controlled by the input voltage V_{in}. (c) The transfer characteristics of the V/F converter for the components shown on the plot.

higher the frequency of the multivibrator. **Figure 11.35b** shows the V/F arrangement and **Figure 11.35c** shows the transfer curve for a particular set of components in **Figure 11.35b**. In spite of the fact that this is a trivially simple circuit, the relation between input voltage and output frequency is quite linear in the range between 2.75 and 8 V. Below 2.5 V and above 8 V the output becomes nonlinear. With a resolution of approximately 6600 Hz/V it can resolve down to below 1 mV. More complex circuits can both widen the range of the input voltage down to zero or even negative values, increase resolution, and stabilize the circuit against variations in temperature, power supply, etc.

11.5.1.3 True A/D Converters

The threshold methods of V/F conversion discussed so far are effective and useful and have the distinct advantage of simplicity. But these are not universally applicable and their performance may be limited. For example, the circuit in **Figure 11.35** has limitations at low input voltages and it is not perfectly linear.

There are, however, A/D converters that eliminate these issues. These have evolved into off-the-shelf components or have been integrated within microprocessors, providing the designer freedom from the details of the circuitry through well-defined transfer functions, linear conversion, and clearly defined conversion limits. We will look here at two of the more common A/D converters as representative devices. These are the dual-slope A/D converter based on methods similar to those of threshold digitization and the successive approximation A/D converter based on comparison of voltages.

11.5.1.4 Dual-Slope A/D Converter

The dual-slope A/D converter is perhaps the simpler (and slower) of the true A/D converters. The circuit is shown in **Figure 11.36a**. Its operation is based on the following

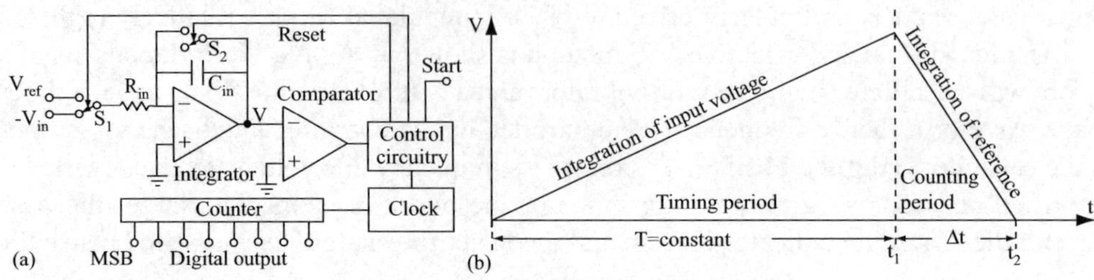

FIGURE 11.36 ■ Dual-slope A/D conversion. (a) The circuit. (b) Charging and discharging times.

principle: Suppose a capacitor is charged from the voltage to be converted through a resistor for a fixed, predetermined time T. The capacitor reaches a voltage V_T that is

$$V_T = V_{in}\frac{T}{RC} \quad [\mathrm{V}]. \tag{11.28}$$

At time T, V_{in} is disconnected and a negative reference voltage of known magnitude is connected to the capacitor through the same resistor. This discharges the capacitor down to zero in a time ΔT:

$$-V_T = -V_{ref}\frac{\Delta T}{RC} \quad [\mathrm{V}]. \tag{11.29}$$

Since these two voltages are equal in magnitude we have

$$V_{in}\frac{T}{RC} = V_{ref}\frac{\Delta T}{RC} \rightarrow \frac{V_{in}}{V_{ref}} = \frac{\Delta T}{T}. \tag{11.30}$$

In addition to this, a fixed-frequency clock is turned on at the beginning of the discharge cycle and off at the end of the discharge cycle and the number of pulses are counted by a pulse counter. Since ΔT and T are known and the counter knows exactly how many pulses have been counted, this count becomes the digital representation of the input voltage. The schematic diagram of a dual-slope converter based on these principles is shown in **Figure 11.36a**. The method is rather slow, with approximately $1/2T$ conversions per second. It is also limited in accuracy by the timing measurements, accuracy of the analog devices, and, of course, by noise. High-frequency noise is reduced by the integration process and low-frequency noise is proportional to T (the smaller T is, the less low-frequency noise). The dual-slope A/D converter is the method of choice for many sensing applications, in spite of its rather slow response, because it is simple and readily built from standard components. For most sensors its performance and noise characteristics are quite sufficient, and because of the integration involved, it tends to smooth variations in the signal during the integration. The method is also used in digital voltmeters and other digital instruments.

EXAMPLE 11.8 A 200 mV, 3½-digit voltmeter

A 3½-digit voltmeter can display full three digits whereas the fourth is either 0 or 1 (hence the term 3½ digit). Thus the display range is 0–1999. A voltmeter capable of measuring up to 200 mV (actually 199.9 mV) is required and an A/D converter based on the dual-slope method needs to be designed. Assume a 1.2 V reference source and a 32 kHz oscillator are available for this purpose.

a. Design a reasonable A/D converter including values for the integrator capacitor C_{in} and resistor R_{in}.

b. What are the internal resolution of the A/D converter and the overall resolution of the voltmeter?

Solution:

a. The design starts with the discharge time Δt, since that defines the number of pulses counted by the counter and displayed. In this case it is convenient to count 2000 pulses corresponding to the display of 199.9. For the 32 kHz oscillator we have

$$\frac{1}{\Delta t} = \frac{32{,}000}{2{,}000} = 16 \rightarrow \Delta t = \frac{1}{16} = 0.0625\ \mathrm{s}.$$

That is, by ensuring a discharge time of 62.5 ms, the output will show 200.0 for a 200 mV input.

Now we have, from **Equation (11.30)**:

$$\frac{V_{\text{in}}}{V_{\text{ref}}} = \frac{\Delta t}{T} \rightarrow \frac{\Delta t}{T} = \frac{0.2}{1.2} \rightarrow T = 6\Delta t = 6 \times 0.0625 = 0.375 \text{ s.}$$

This will ensure about 2 measurements/s, a figure typical for digital voltmeters.

To select the integrating capacitor C_{in} and the charging resistor R_{in}, we must define the maximum voltage V_t to which the capacitor will charge. This clearly cannot be larger than the measured voltage or the reference voltage, but it should be as large as possible. We will select it arbitrarily to be 150 mV (but it must be lower than the full scale of 200 mV). Thus from **Equation (11.28)** we have

$$V_t = V_{\text{in}} \frac{T}{R_{\text{in}} C_{\text{in}}} \rightarrow R_{\text{in}} C_{\text{in}} = \frac{V_{\text{in}}}{V_t} T = \frac{0.2}{0.15} \times 0.375 = 0.5 \text{ s.}$$

Now, selecting a standard capacitance of 1 μF, the required resistor is 500 kΩ.

b. The internal resolution of the counter is 1 bit or, in terms of the input, 0.2/2000 = 0.1 mV. Assuming proper design of the integrator, the internal resolution should be 0.1 mV. This is also the display resolution. One can therefore say that the overall resolution is 0.1 mV. This can then be taken as the resolution of the system.

Note, however, that if we used, say, a 64 kHz oscillator, the internal resolution would have been 0.05 mV since the counter would have counted 4000 counts in the time Δt (assuming the timing stays the same). However, the display still cannot increment by more than 0.1 mV, so the overall resolution stays 0.1 V.

11.5.1.5 Successive Approximation A/D

This is often the method of choice in A/D converter components and in many microprocessors. It is available in many off-the-shelf components with varying degrees of accuracy, and depending on the number of bits of resolution, it may resolve down to a few microvolts. The basic structure is shown in **Figure 11.37** for an 8-bit A/D converter. It consists of a precision comparator, a shift register, a D/A converter (D/A conversion will be discussed in the following section), and a precision reference voltage V_{ref}. The operation is as follows: First, all registers are cleared, which forces the comparator to HIGH since the output of the D/A converter is zero. This forces a 1 into the most significant bit (MSB) of the register. The D/A converter generates an analog voltage V_a, which for MSB = 1 is half

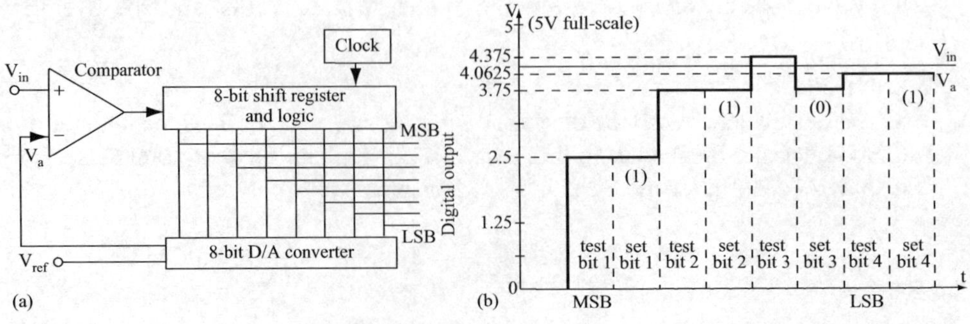

FIGURE 11.37 ■ An 8-bit successive approximation A/D converter. (a) The circuit. (b) The output sequence for input voltage V_{in}.

the full-scale input. This is compared to V_{in}. If V_{in} is larger than V_a, the output stays high and the clock shifts this into the next bit in the register. The register now shows 11000000. If this is smaller than V_{in}, the output goes low and the register shows 010000000. Assuming that the input is still higher, the D/A converter generates a voltage $V_a = (1/2 + 1/4)V_{ref}$. If this is higher than the input, the register will show 011000000, but if it is lower, it will show 11100000, and so on, until, after *n* steps, the final result is obtained. The data are read from the shift register and represent the voltage digitally. **Figure 11.37a** shows a parallel output. In some cases the register data are shifted out serially, requiring *n* clock steps to do so, where *n* is the number of digits ($n = 8$ in this figure).

A/D converters of this type exist with much higher resolutions. Resolutions of up to 14 bits (see **Example 11.9**) are common, with 8 bits being the lower limit. An 8-bit A/D converter has a resolution of $V_{in}/2^8 = 0.004V_{in}$. For a 5 V full scale, the resolution is 20 mV. This may not be sufficient for low-level signals, in which case a 10-, 12-, or 14-bit A/D converter may be used (a 14-bit A/D converter has a resolution of 0.3 mV for a 5 V full scale). Higher resolution A/D converters also exist and are sometimes necessary for high-accuracy digitization of signals (such as the digitization of audio signals for music reproduction). A critical component of the method is the reference voltage, as it represents the full-scale value and must be constant. The fact that the full scale of A/D converters is typically the power supply voltage means it is almost always necessary to amplify signals from devices such as thermocouples if they must be digitized.

The advantage of the successive approximation A/D converter is that the conversion is done in *n* steps (fixed) and is much faster than other methods. On the other hand, the accuracy of the device depends heavily on the comparator, the D/A converter, and the reference voltage. Commercial devices are relatively expensive, especially if more than 12 or 14 bits are needed. On the other hand, this type of A/D converter has been incorporated directly into microprocessors and can sometimes be used for sensing as part of the overall circuitry. Some microprocessors have multiple A/D channels and some use a different type of A/D converter.

EXAMPLE 11.9 A 14-bit successive approximation A/D converter

A 14-bit successive approximation A/D converter in a microprocessor operates at 5 V and is required to measure a 4.21 V input. The clock is set to 1 MHz.

a. Following **Figure 11.37**, define the basic parameters and the hardware needed.

b. What is the digital output of the converter?

c. How accurate is the representation?

Solution:

a. The converter requires a 14-bit D/A converter and a 14-bit shift register. The clock imposes a minimum cycle of 1 μs, meaning that each test and set step takes at least one cycle. Thus the conversion requires at least 28 cycles, or 28 μs. In fact, it takes more, since, for example, the output may need to be clocked at the end of the conversion and each step may take more than one cycle, depending on how the algorithm is implemented. The resolution is $5\text{ V}/2^{14} = 5\text{ V}/16{,}384 = 0.305176$ mV.

b. Following the steps in **Figure 11.37**, the first comparison is with $5\text{ V}/2^1 = 2.5$ V. Since this is smaller than 4.21 V, the most significant bit is 1. Next we compare with $5\text{ V}/2^2 = 1.5$ V.

Thus 2.5 V + 1.25 V = 3.75 V < 4.21 V, hence the second bit is 1. Continuing this, we can write for the output:

$$4.21V \approx 5V\left(\frac{[1]}{2^1}+\frac{[1]}{2^2}+\frac{[0]}{2^3}+\frac{[1]}{2^4}+\frac{[0]}{2^5}+\frac{[1]}{2^6}+\frac{[1]}{2^7}+\frac{[1]}{2^8}+\frac{[1]}{2^9}+\frac{[0]}{2^{10}}+\frac{[0]}{2^{11}}+\frac{[0]}{2^{12}}+\frac{[1]}{2^{13}}+\frac{[1]}{2^{14}}\right)$$

$$= 4.2098999\,\mathrm{V}.$$

Where any contribution that takes the output above 4.21 V is a "0" digit and any contribution that keeps the output below 4.21 V is a "1" digit. Taking the values in square brackets, the digital output is 11010111100011.

c. The resolution of the A/D converter is 1 bit, or 0.305176 mV, as indicated in (a). However, the difference between the analog input (4.21 V) and the represented output (4.2098999) is only 0.1 mV. This gives an error of

$$error = \frac{4.21 - 4.2098999}{4.21} \times 100 = 0.00238\%.$$

A/D converters with up to 24 bits are readily available. In principle, higher resolutions are possible but are difficult to realize. There are, of course, other methods of A/D conversion not discussed here. Of these, direct conversion or flash A/D conversion is particularly useful, as it uses a bank of comparators and hence achieves parallel conversion in a shorter time. The device is fast, but expensive and difficult to make in resolutions above 8 or 10 bits. Another type, called the sigma-delta A/D converter, relies heavily on sampling and filtering of the signal to be digitized. There are other methods—some specialized, some general purpose. Nevertheless, the methods shown here are representative and most often used within microprocessors and hence are relevant to the discussion in the context of sensors and actuators.

11.5.2 D/A Conversion

D/A conversion is not often used with sensors but is sometimes used with actuators. This occurs when a digital device, such as a microprocessor, must provide an analog output. A good example of this is in audio reproduction, where the audio signal may well be handled digitally but our ears are analog and the audio signal needs to be converted back to analog form. Another example is in the detection of a train of pulses sent from a sensor where the analog value may be used to, for example, turn on a device. D/A converters are important components in A/D conversion (see the successive approximation A/D converter in the previous section). In general, D/A conversion should be avoided if possible by the use of digital actuators (such as brushless DC motors and stepper motors), but there will be cases in which D/A conversion becomes necessary. As with A/D converters, there are different ways of accomplishing A/D conversion.

11.5.2.1 Resistive Ladder Network D/A Conversion

The most common method used in simple converters is based on the ladder network shown in **Figure 11.38**. It consists of a voltage follower so that the input impedance is very high

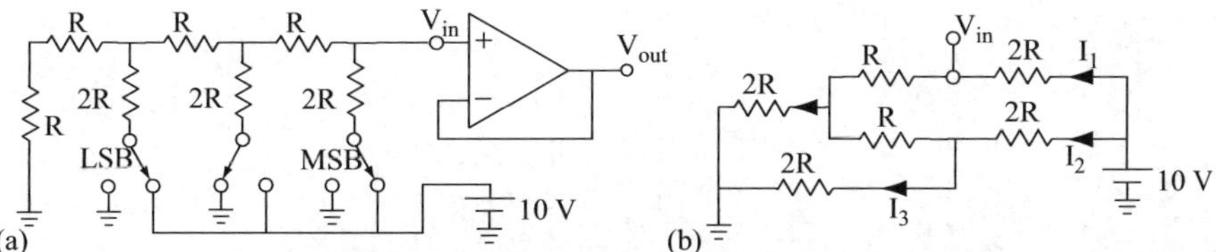

FIGURE 11.38 ■ D/A conversion based on the resistive ladder network. (a) The circuit for a 3-bit D/A converter. (b) Equivalent circuit for the ladder network for digital input 101.

and the output of the follower equals the voltage at its noninverting input. This voltage is generated by a resistance network. The latter is chosen so that the combination of series and parallel resistances represents the digital input as a unique voltage. The switches are digitally controlled analog switches (MOSFETs). Depending on the digital input, various switches connect resistances in series or in parallel. For example, suppose that the digital value 101 is to be converted. The switches will be as in **Figure 11.38**. The MSB is "1" and hence the switch for this bit is connected to the reference voltage (10 V in this case). The next bit is "0", hence its switch is connected to ground (zero voltage). The LSB switch is connected to 10 V. The switches reconfigure the resistive network as shown in **Figure 11.38b**, producing exactly 6.25 V at the amplifier's input as can be verified from **Figure 11.38b**. The ladder can be extended as necessary for any number of bits.

The accuracy and usefulness of a D/A converter depends on the quality and accuracy of the ladder network and the reference voltage used, as well as the quality and resistance of the switches. The resolution of the circuit is 1 bit, that is, the analog output can only change in the equivalent of 1 bit. In this case, 1 bit represents 10 V/2^3 = 10 V/8 = 1.25 V. Thus the analog voltage can only increment in steps of 1.25 V.

EXAMPLE 11.10 An 8-bit D/A converter based on the ladder network

Build an 8-bit D/A converter based on the resistive ladder network. Assume the available reference voltage is 10 V.

a. Calculate the analog output for digital input 11010010.

b. Does it matter what the resistance R is?

c. What is the resolution of the D/A?

Solution: The ladder network is shown in **Figure 11.39a** (representing digital 11010010). The resistance $R = 10$ kΩ.

a. To calculate the output we use the equivalent circuit in **Figure 11.39b** and compute the loop currents shown (resistances are in kΩ):

$$30I_1 + 10(I_1 + I_2) + 20(I_1 + I_2 + I_3) = 10$$
$$20I_2 + 10(I_1 + I_2) + 20(I_1 + I_2 + I_3) = 10$$
$$20(I_3 + I_4) + 10I_3 + 20(I_1 + I_2 + I_3) = 10$$
$$20(I_4 + I_5) + 10I_4 + 20(I_3 + I_4) = 10$$
$$20(I_5 + I_6 + I_7) + 10(I_5 + I_6) + 10I_5 + 20(I_4 + I_5) = 10$$
$$20(I_5 + I_6 + I_7) + 10(I_5 + I_6) + 20I_6 = 10$$
$$20(I_5 + I_6 + I_7) + 20I_7 = 10,$$

or as a system of equations:

$$\begin{bmatrix} 60 & 30 & 20 & 0 & 0 & 0 & 0 \\ 30 & 50 & 20 & 0 & 0 & 0 & 0 \\ 20 & 20 & 50 & 20 & 0 & 0 & 0 \\ 0 & 0 & 20 & 50 & 20 & 0 & 0 \\ 0 & 0 & 0 & 20 & 60 & 30 & 20 \\ 0 & 0 & 0 & 0 & 30 & 50 & 20 \\ 0 & 0 & 0 & 0 & 20 & 20 & 40 \end{bmatrix} \begin{Bmatrix} I_1 \\ I_2 \\ I_3 \\ I_4 \\ I_5 \\ I_6 \\ I_7 \end{Bmatrix} = \begin{Bmatrix} 10 \\ 10 \\ 10 \\ 10 \\ 10 \\ 10 \\ 10 \end{Bmatrix}.$$

Solving this system gives

$$I_1 = 0.08984375 \text{ mA}$$
$$I_2 = 0.135765625 \text{ mA}$$
$$I_3 = 0.0283203125 \text{ mA}$$
$$I_4 = 0.20458984375 \text{ mA}$$
$$I_5 = -0.039794921875 \text{ mA}$$
$$I_6 = 0.1448974609375 \text{ mA}$$
$$I_7 = 0.19744873046875 \text{ mA}.$$

Now the voltage at the input to the voltage follower (and hence the output of the converter) is

$$V_{out} = V_{in} = 10 - 20I_1 = 10 - 20 \times 0.08984375 = 8.203125 \text{ V}.$$

b. The value of the resistor is important only in terms of the current consumption of the converter. It should not be too small so the current consumption is reasonable, but it should not be too high either, since then noise may be significant compared to the actual current. The resistor is usually on the order of 1–10 kΩ.

c. The digital resolution is 1 bit. Since 8 bits represent $2^8 = 256$ states, the analog resolution is $10/256 = 39.06 \times 10^{-3}$ V or 39.06 mV.

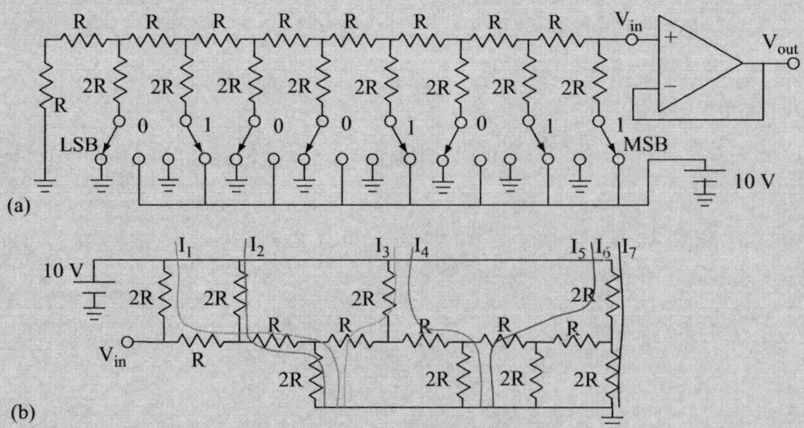

FIGURE 11.39 ■ (a) An 8-bit A/D converter with digital input 11010010. (b) The equivalent ladder network showing loop currents used to calculate the analog output voltage.

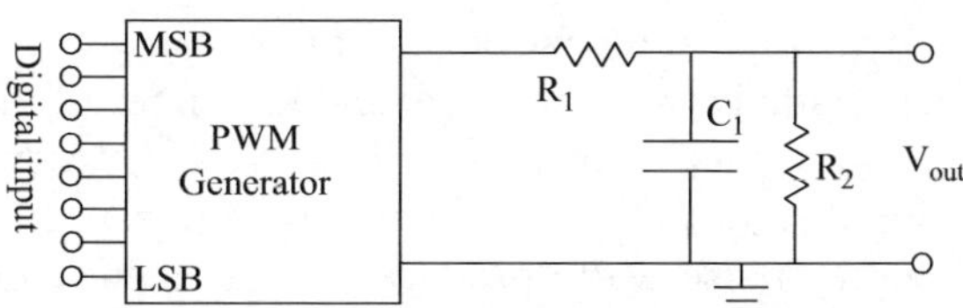

FIGURE 11.40 ■ The PWM D/A converter.

11.5.2.2 PWM D/A Conversion

D/A conversion can also be implemented using a PWM technique in conjunction with a low-pass filter. This is perhaps the simplest method of D/A conversion. It can be used in noncritical applications and is commonly used in the conversion of digital sound. The method is shown schematically in **Figure 11.40**. First, the digital data are converted into a PWM format. In PWM, the higher the digital value, the wider the pulse of a train of pulses generated by a PWM generator. Although this may seem a complicated task, PWM generators are common and available either as components or as peripherals in microprocessors. The PWM train is applied to the filter that in **Figure 11.40** is represented as the simple R_1–C_1 low-pass filter. The capacitor is charged during the on time of the pulse and discharged during the off time. The wider the pulse (i.e., the lower the duty cycle), the higher the voltage across the capacitor. This voltage is the analog output of the D/A converter. This method of conversion is ideally suited for audio reproduction, but it is also suited for motor control in which the motor itself is part of the filter, as well as for other applications. R_2 is added to ensure that the capacitor does not retain its state as the digital input value changes. The choice of components is important in that a large capacitor, that will smooth the output, will also reduce the response time of the circuit.

11.5.2.3 Frequency to Voltage (F/V) D/A Conversion

A form of D/A conversion is frequency to voltage (F/V) conversion. Unlike the PWM D/A converter, the output of an F/V converter is proportional to frequency. In that sense it is not, strictly speaking, a D/A converter, but rather a method of frequency detection or frequency demodulation (see **Chapter 10**). It is, however, closely related to the PWM D/A method and shares with it the use of the low-pass filter. If the frequency of the digital signal represents the digital output of a sensor, then the output of the F/V converter represents that data in analog form. For example, many capacitive and inductive sensors are connected as part of oscillators and the frequency is used to indicate the stimulus. A schematic F/V converter is shown in **Figure 11.41a**. The incoming square wave passes through a monostable multivibrator that at every pulse rise produces a fixed-width output pulse (a monostable multivibrator generates a single pulse each time its input changes and is often called a one-shot multivibrator). This modified signal has

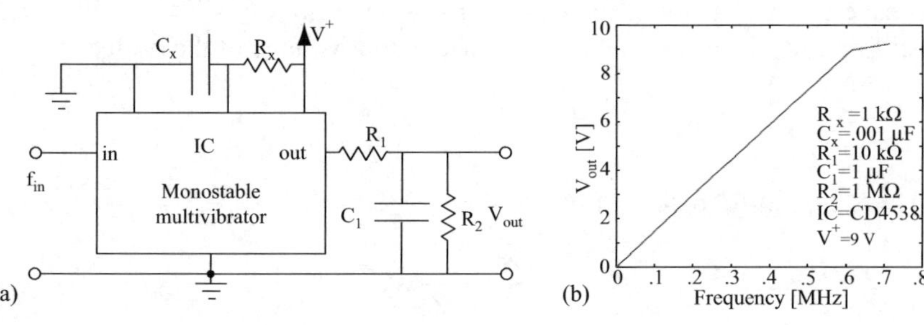

FIGURE 11.41 ■ (a) Schematic of an F/V converter. (b) The output for specific values showing a linear analog output up to a maximum of 613 kHz.

the same frequency as the original signal but a different duty cycle. As the frequency increases, the duty cycle (ratio between the high and low times of the pulses) decreases. Hence the output on the capacitor increases linearly with frequency, as shown in **Figure 11.41b**. The circuit can handle frequencies down to zero (theoretically), but the upper range depends on the pulse width produced by the monostable multivibrator. In effect, when the pulse width (Δt) equals the cycle time of the incoming signal (i.e., $\Delta t = 1/f$), the output remains constant at its maximum value, as can be seen in **Figure 11.41b** (see also **Example 11.11**). Nevertheless, the dynamic range of the circuit is typically large. The range can be increased by reducing the pulse width (by reducing the value of R_x, C_x, or both) or reduced by increasing the pulse width (by increasing the value of R_x, C_x, or both). Reducing the range increases sensitivity, that is, the change in output for a given change in frequency is larger, whereas increasing the range allows a larger span. These properties of the circuit are dictated by the expected frequency range in the application. The low-pass filter is made of R_1 and C_1 and its function and operation are identical to that in **Figure 11.40**.

EXAMPLE 11.11 **F/V converter**

The circuit in **Figure 11.41** is used with the following values: $R_1 = 10\ \text{k}\Omega$, $R_2 = 1\ \text{M}\Omega$, $R_x = 1\ \text{k}\Omega$, $C_1 = 1\ \mu\text{F}$, and $C_x = 0.001\ \mu\text{F}$. The monostable multivibrator is a CMOS device. For the values given for R_x and C_x, the monostable multivibrator produces a fixed pulse width of approximately 0.8 μs. As the frequency of the input signal varies from 1 to 613 kHz, the output across the capacitor varies from 0.01 to 8.92 V, as shown **Figure 11.41b**. Above 613 kHz the output remains essentially flat, indicating that the pulse width in the incoming signal equals that of the monostable multivibrator (i.e., $1/(2 \times 613{,}000) = 0.815\ \mu\text{s}$). The sensitivity of this circuit is

$$s_o = \frac{V_{\text{out}}}{F_{\text{in}}} = \frac{8.92 - 0.01}{(613 - 1) \times 10^3} = 1.456 \times 10^{-5}\ \text{V/Hz},$$

or 14.56 mV/kHz.

The curve in **Figure 11.41b** was obtained experimentally with the circuit and components shown.

11.6 | BRIDGE CIRCUITS

Bridge circuits are some of the oldest circuits used in conjunction with sensors as well as other applications. The bridge is known as the Wheatstone bridge, but variations of the bridge exist and have different names. The basic Wheatstone bridge is shown in **Figure 11.42**. It consists of four impedances $Z_i = R_i + jX_i$. The output voltage of the bridge is

$$V_o = V_{\text{ref}}\left(\frac{Z_1}{Z_1 + Z_2} - \frac{Z_3}{Z_3 + Z_4}\right) \quad [\text{V}]. \tag{11.31}$$

The bridge is said to be balanced if

$$\frac{Z_1}{Z_2} = \frac{Z_3}{Z_4}. \tag{11.32}$$

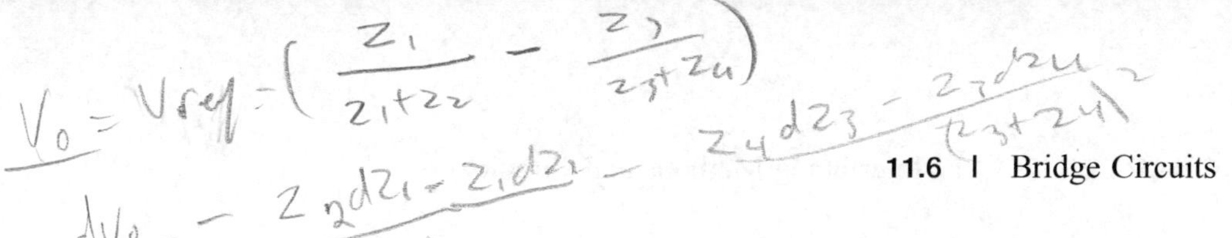

FIGURE 11.42 ■ An impedance bridge. Any of the impedances may be a sensor or a fixed impedance.

Under this condition, the output voltage is zero. This, in fact, is one of the main attractions for its use in sensors. If, for example, Z_1 represents the impedance of a sensor, by proper choice of the other impedances the output can be set to zero at any given value of Z_1 (and hence of the measurand). Any change in Z_1 will change the value of V_{out}, indicating the change in stimulus. Of course, one can do much more than that, and bridges can be used for signal translation and for temperature compensation, among other things.

11.6.1 Sensitivity

The sensitivity of the output voltage to change in any of the impedances can be calculated by first taking the derivatives:

$$\frac{dV_o}{dZ_1} = V_{\text{ref}}\frac{Z_2}{(Z_1+Z_2)^2}, \quad \frac{dV_o}{dZ_2} = -V_{\text{ref}}\frac{Z_1}{(Z_1+Z_2)^2} \quad [\text{V}/\Omega] \tag{11.33}$$

and

$$\frac{dV_o}{dZ_3} = -V_{\text{ref}}\frac{Z_4}{(Z_3+Z_4)^2}, \quad \frac{dV_o}{dZ_4} = V_{\text{ref}}\frac{Z_3}{(Z_3+Z_4)^2} \quad [\text{V}/\Omega]. \tag{11.34}$$

Summing and cross-multiplying gives the bridge sensitivity:

$$\frac{dV_o}{V_{ref}} = \frac{Z_2dZ_1 - Z_1dZ_2}{(Z_1+Z_2)^2} - \frac{Z_4dZ_3 - Z_3dZ_4}{(Z_3+Z_4)^2} \quad [\text{V/V}]. \tag{11.35}$$

This relation reveals that if $Z_1 = Z_2$ and $Z_3 = Z_4$, the bridge is balanced, and if the change is, say, $dZ_1 = dZ_2$ and $dZ_3 = dZ_4$, the change in output is zero. This is the basic idea used in compensating a sensor for temperature variation and any other common-mode effect. For example, suppose that a pressure sensor has an impedance $Z_1 = 100\ \Omega$ and a sensitivity to temperature $dZ_1 = 0.5\ \Omega/°\text{C}$. Two identical sensors may be used as Z_1 and Z_2, but sensor Z_2 is not exposed to pressure (it is only exposed to the same temperature as Z_1). Z_3 and Z_4 are equal and are made of the same material—these are typically simple resistors. Under these conditions, there will be no output due to temperature changes and the sensor is properly compensated for temperature variations. If, however, pressure changes, the output changes based on **Equation (11.31)** since the impedance Z_1 changes with pressure.

If all impedances in the bridge are fixed and only Z_1 varies (this is the sensor), then $dZ_2 = 0$, $dZ_3 = 0$, and $dZ_4 = 0$ and the bridge sensitivity becomes

$$\frac{dV_o}{V_{ref}} = \frac{Z_2 dZ_1}{(Z_1 + Z_1)^2} \text{ or } \frac{dV_o}{V_{ref}} = \frac{dZ_1}{4Z_1} \quad [\text{V/V}], \text{ if } Z_2 = Z_1. \tag{11.36}$$

This type of bridge, especially with resistive branches, is the common method of sensing with strain gauges, piezoresistive sensors, Hall elements, thermistors, force and pressure sensors, and others. Use of the bridge allows a convenient reference voltage (nulling), temperature compensation, and compensation of other sources of common-mode noise. It is also very simple and can be easily connected to amplifiers for further processing.

The sensitivity of the bridge can be increased by using two diagonal elements as sensors. For example, suppose we wish to measure strain using strain gauges. Instead of using a single strain gauge (say, Z_1 in **Figure 11.42**), we may use two strain gauges and place them on diagonal arms of the bridge, in this case Z_1 and Z_4. The two strain gauges are exposed to the same measurand—that is, they sense exactly the same strain. If the measurement starts with a balanced bridge, at a given strain, Z_1 and Z_4 will increase to $Z_1 + dZ$ and $Z_4 + dZ$ (since the strain gauges are identical, the change will be identical as well). Z_2 and Z_3 are resistors and do not change with strain. Now, with $Z_1 = Z_2 = Z_3 = Z_4 = Z_0$, we have from **Equation (11.35)**:

$$\frac{dV_o}{V_{ref}} = \frac{dZ}{2Z_0} \quad [\text{V/V}]. \tag{11.37}$$

Note that this approach produces a sensitivity twice as high as that of the single sensor given in **Equation (11.36)**. This method of sensing is shown in **Figure 11.43**. The two

FIGURE 11.43 ■ Increased sensitivity bridge. Z_1 and Z_4 sense the strain in a beam, Z_2 and Z_3 are resistors selected to balance the bridge. Note the bonding substance.

strain gauges on the left are both bonded to the steel beam and measure the same strain. The two resistors on the right are also bonded to the beam, but, of course, they do not measure anything. They are, however, at the same temperature as the sensors.

Note also that the increase in sensitivity afforded by this configuration introduces a new problem: the sensitivity to temperature variations has also doubled. But what is more critical is that the common method of temperature compensation, that is, using two identical sensors, one of which is not sensing, cannot be used. In fact, temperature compensation using that principle requires that all four elements in the bridge be sensors (in this case, four identical strain gauges), with Z_1 and Z_4 sensing strain and Z_2 and Z_3 only sensing the temperature common to all four sensors.

An even more sensitive method, often used in load cells, is to use four identical sensors and use the change in resistance of all four sensors in the bridge. This produces a sensitivity four times as high as that of a single sensor (see **Problem 11.34** for an example) as well as producing temperature compensation, since all sensors vary identically with temperature, so that temperature does not change the bridge output.

EXAMPLE 11.12 Compensation of temperature variations in sensors

In **Example 6.2** we calculated the temperature effects on a platinum strain gauge (350 Ω at 20°C) used in a variable temperature environment. The results obtained there were as follows: At the reference temperature (20°C) the resistance of the strain gauge varied from 350 Ω for zero strain to 412.3 Ω for 2% strain. As the temperature varied from −50°C to 800°C, the resistance without strain applied varied from 255.675 to 1401.05 Ω, whereas when a 2% strain was present, the resistance varied from 301.185 to 1650.44 Ω.

a. Show how the temperature effects can be compensated using a bridge.

b. Show that if the bridge is correctly set up, the output is not affected by temperature.

c. Find the output of the bridge with a reference voltage of 10 V for the range (0%–2% strain).

Solution:

a. To compensate for temperature effects, Z_1 and Z_2 are two identical strain gauges (350 Ω at 20°C). Z_1 is the sensing gauge, whereas Z_2 is placed at the same location as Z_1 but is not exposed to stress. This ensures that the two will experience identical temperature changes. The impedances Z_3 and Z_4 are two resistors each 350 Ω. They can be (and sometimes are) identical strain gauges to Z_1 and Z_2. We will assume that they are identical in properties so that these resistors do not, in themselves, introduce temperature-generated errors.

Note: Z_1 and Z_2 must be at the same temperature (the temperature at the location the strain is sensed). Z_3 and Z_4 will normally be at ambient temperature, and again, to minimize errors they should be at identical temperatures, although they do not have to be at the temperature of Z_1 and Z_2. The configuration is shown in **Figure (11.44)**.

b. The best way to see that temperature has no effect is to use the general relation for the resistance of the strain gauge given in **Equation (6.7)** and substitute that into **Equation (11.31)**. The resistance of a strain gauge subject to strain and temperature is (**Equation (6.7)**):

$$R(\varepsilon, T) = R(1 + g\varepsilon)(1 + [T - T_0]) \quad [\Omega].$$

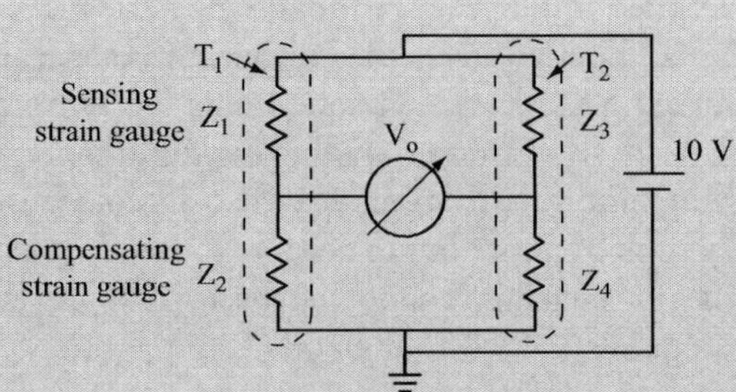

FIGURE 11.44 ■ Temperature compensation in a bridge.

The same strain gauge subject only to temperature is

$$R(\varepsilon, T) = R(1 + [T - T_0]) \quad [\Omega],$$

where ε is the applied strain, g is the gauge factor, T_0 is the reference temperature, T is the sensed temperature, and R is the nominal resistance of the gauge at T_0. By selecting

$$Z_1 = R(1 + g\varepsilon)(1 + [T - T_0]) \quad [\Omega]$$
$$Z_2 = R(1 + [T - T_0]) \quad [\Omega]$$
$$Z_3 = R \quad [\Omega]$$
$$Z_4 = R \quad [\Omega],$$

Equation (11.31) becomes

$$V_o = V_{\text{ref}}\left(\frac{R(1 + g\varepsilon)(1 + [T - T_0])}{R(1 + g\varepsilon)(1 + [T - T_0]) + R(1 + [T - T_0])} - \frac{R}{R + R}\right) = V_{\text{ref}}\left(\frac{1 + g\varepsilon}{2 + g\varepsilon} - \frac{1}{2}\right) \quad [\text{V}].$$

Clearly the dependency on temperature has been eliminated.

c. To calculate the output for the range we write

No strain applied (minimum strain):

At −50°C, maximum strain: $Z_1 = 255.675\ \Omega$, $Z_2 = 255.675\ \Omega$, $Z_3 = 350\ \Omega$, $Z_4 = 350\ \Omega$. The bridge is balanced and the output is zero.

At 800°C, minimum strain: $Z_1 = 1401.05\ \Omega$, $Z_2 = 1401.05\ \Omega$, $Z_3 = 350\ \Omega$, $Z_4 = 350\ \Omega$. The bridge is balanced and the output is zero.

Strain applied (maximum strain):

At −50°C, maximum strain: $Z_1 = 301.185\ \Omega$, $Z_2 = 255.675\ \Omega$, $Z_3 = 350\ \Omega$, $Z_4 = 350\ \Omega$. The bridge is unbalanced and the output is

$$V_o = 10 \times \left(\frac{301.185}{301.185 + 255.675} - \frac{350}{350 + 350}\right) = 10 \times (0.54086 - 0.5) = 0.4086\ \text{V}.$$

At 800°C, maximum strain: $Z_1 = 1650.44\ \Omega$, $Z_2 = 1401.05\ \Omega$, $Z_3 = 350\ \Omega$, $Z_4 = 350\ \Omega$.

The bridge is unbalanced and the output is

$$V_o = 10 \times \left(\frac{1650.44}{1650.44 + 1401.05} - \frac{350}{350 + 350}\right) = 10 \times (0.54086 - 0.5) = 0.4086 \text{ V}.$$

As expected, the output is independent of the temperature—it only depends on the strain. The bridge output varies from 0 at zero strain to 0.4086 V at 2% strain. This may need to be amplified before being used, but it does not depend on temperature.

The methods discussed here are effective in compensating for common-mode effects such as temperature. However, they do not eliminate errors external to the sensors, such as variations of V_{ref} with temperature. As can be seen in **Equation (11.31)** and its use in **Example 11.12**, any change in the bridge voltage V_{ref} will change the output. These have to be compensated for in the construction of the bridge itself. There are many techniques by which this can be accomplished, but this is beyond the scope of this text (see, however, **Example 11.16**). We will normally assume that the bridge itself has been properly compensated.

11.6.2 Bridge Output

The output from the bridge is likely to be relatively small. For example, suppose that the bridge is fed with a 5 V source and a thermistor, $Z_4 = 500\ \Omega$ (at 0°C), is used to sense temperature. Assuming the bridge is balanced at 0°C, the other three resistances are also 500 Ω. This gives a zero output voltage at 0°C. Now suppose for the sake of discussion that at 100°C the resistance of the thermistor reduces to 400 Ω. The output voltage is now

$$V_o = 5\left(\frac{500}{500 + 500} - \frac{400}{500 + 500}\right) = 0.5 \text{ V}.$$

Most sensors will produce a much smaller change in impedance and therefore some sort of amplification will be necessary. The operational amplifier discussed in **Section 11.2** is ideal for this purpose. Two methods through which a bridge may be connected are shown in **Figure 11.45**. In **Figure 11.45a**, the output voltage is connected directly

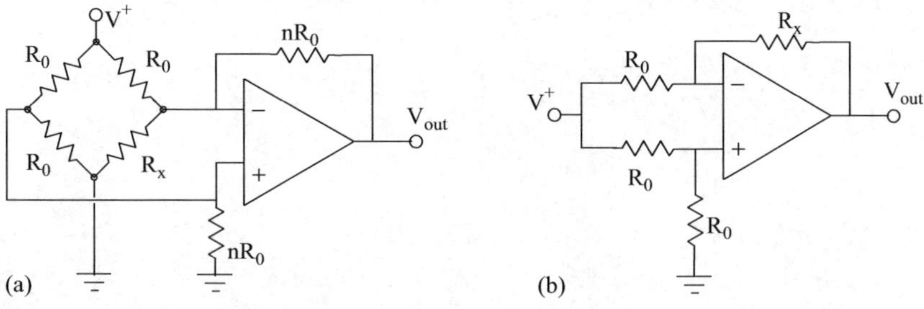

FIGURE 11.45 ■ Amplified bridge circuits. (a) Amplification of the output of the bridge using an operational amplifier. (b) The active bridge. The bridge elements are used as part of the amplifier's feedback.

between the inverting and noninverting inputs. If we assume that the resistance of the sensor changes as $R_x = R_0(1+\alpha)$, the voltage output of the bridge is

$$V_{out} \approx V_{ref} \frac{(1+n)V\alpha}{4} \quad [\mathrm{V}]. \tag{11.38}$$

Note that this circuit provides an amplification of $(1+n)$ but requires that the voltage on the bridge be floating (i.e., the supply of the bridge must be separate from that of the amplifier). The bridge circuit and amplifier in **Figure 11.45b** do not provide amplification, but rather places the sensor in the feedback loop. This is called an active bridge and its output is

$$V_{out} = V^{+} \frac{\alpha}{2} \quad [\mathrm{V}]. \tag{11.39}$$

EXAMPLE 11.13 Bridge circuit to reduce the effect of lead resistance and temperature variations on lead wires on the output

One of the problems associated with the use of resistance temperature detectors (RTDs), and some other sensors, is that the resistance of the connecting wires varies with temperature. In some sensors this may not be a problem, but in RTDs, which have relatively low resistance, this change can introduce a significant error. To eliminate this effect, one common RTD configuration has three wires, as shown in **Figure 3.3b**. This allows connection in a bridge as shown in **Figure 11.46**. The bridge resistances are selected so that $R_1 = R_2 = R_3 = R(0)$, where $R(0)$ is the resistance of the RTD at the reference temperature (usually 0°C). It is also assumed that all three wires have the same length so that their resistances R_l are the same.

a. Show that a bridge measurement can reduce but not eliminate the effect of temperature variations on the lead wires.
b. Show that the three-wire RTD configuration can eliminate the effect of lead wire resistance and temperature variations on the lead wires if the resistance of the RTD wires between points A-A can be measured separately.

Solution:
a. The output of the bridge is calculated as follows:

$$V_a = \frac{V_{ref}}{2} \quad [\mathrm{V}]$$

$$V_b = \frac{V_{ref}}{R(0) + (R(0)+\Delta R) + 2(R_l + \Delta R_l)}(R(0) + \Delta R + R_l + \Delta R_l)$$

$$= \frac{V_{ref}}{2R(0) + 2R_l + 2\Delta R_l + \Delta R}[(R(0) + R_l + \Delta R_l) + \Delta R] \quad [\mathrm{V}]$$

or

$$V_b = \frac{V_{ref}(R(0) + R_l + \Delta R_l)}{2R(0) + 2R_l + 2\Delta R_l + \Delta R} + \frac{V_{ref}\Delta R}{2R(0) + 2R_l + 2\Delta R_l + \Delta R} \quad [\mathrm{V}],$$

where R_l is the resistance of the lead wires (small), ΔR_l is the change in this resistance due to ambient temperature variations, and ΔR is the change of the resistance of the RTD.

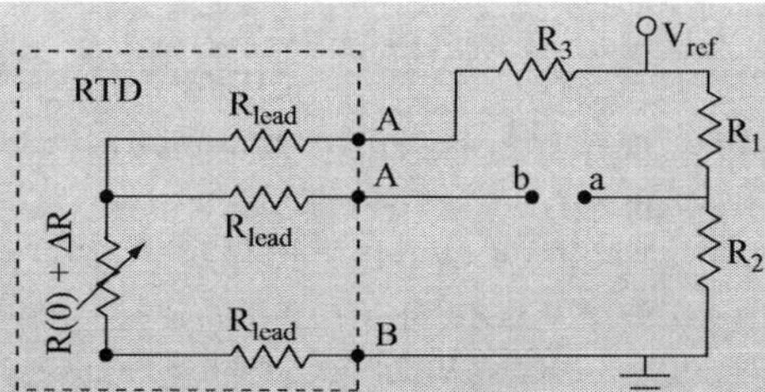

FIGURE 11.46 ■ Three-wire RTD connected in a bridge configuration.

The first term in this relation is approximately $V_{ref}/2$ (it is exactly $V_{ref}/2$ at the reference temperature, where $\Delta R = 0$). Thus V_{ba} is

$$V_{ba} = V_b - V_a = \frac{V_{ref}\Delta R}{2R(0) + 2R_l + 2\Delta R_l + \Delta R} \quad [\text{V}]. \tag{1}$$

Consider a numerical example: suppose the RTD has a nominal resistance $R_0 = 120\ \Omega$ and $\Delta R = 10\ \Omega$. With a 1 Ω resistance for each of the connecting wires and a 0.1 Ω change due to temperature variations, we get for a 10 V reference voltage:

$$V_{ba} = \frac{10 \times 10}{240 + 2 + 0.2 + 10} = 0.39651\ \text{V}.$$

If we assume that the lead wires have zero resistance:

$$V_{ba} = \frac{10 \times 10}{240 + 10} = 0.4\ \text{V}.$$

The error is (0.4–0.39651)/0.4 = 0.008725, or 0.8725%.

If instead we use the two-wire connection rather than the bridge, the total resistance measured is 122.2 Ω and the error is 2.2/120 = 0.01833, or 1.833%. Clearly the three-wire connection is intended to reduce the effects of the lead wire resistance rather than eliminate them. In this case the error is reduced by a factor of 2.1.

b. If the resistance between the two connections marked A-A is measured separately (either through a bridge measurement or through other means), this resistance can be subtracted from the total resistance in the denominator in **Equation (1)**. The resistance measured between points A-A include the effect of temperature on the lead wires and equals $2R_l + 2\Delta R_l$. After subtracting this resistance, the output of the bridge is

$$V_{ba} = V_b - V_a = \frac{V_{ref}\Delta R}{2R(0) + \Delta R} \quad [\text{V}].$$

This is clearly free of any effect due to the lead wires.

Note: Because the lead wire resistances eliminated are the bottom and top resistors, whereas the separately measured lead resistances are the two top resistances, it is essential that all three lead wires be of the same length, otherwise the effect of lead wires will be only partially eliminated. This method of connecting RTDs is the most common method of compensation and allows for long lead wires.

11.7 | DATA TRANSMISSION

The transmission of data from a sensor to a controller or from a controller to an actuator may take many forms. If the sensor is a passive sensor, it already has an output in a usable form, such as voltage or current. It is usually sufficient to simply measure this output directly to obtain a reading. In other cases, such as with capacitive or inductive sensors, this is more complicated since voltage and current are not really an option—more often we will use the sensor as part of an oscillator that produces a frequency that is proportional to the stimulus. Also, and most importantly, the sensor is often likely to be in a remote location. Neither direct measurement of voltage and current nor using the sensor as part of the circuit (in an oscillator) may be an option in such cases. It is often necessary to process the sensor's output locally and transmit the result to the controller. The controller then interprets the data and converts them to suitable forms.

The ideal method of transmission is to convert the sensor's output into digital form locally—at the sensor—and send the digital data to the controller. This method is often employed in "smart sensors" since they have the necessary processing power locally. In most cases a sensor of this type will have a local microprocessor and either be supplied with power from the controller or have its own source of power, perhaps from an onboard battery. The digital data may then be transmitted over regular lines or even through a wireless link. Since digital data are much less prone to corruption, the method is both obvious and very useful.

However, many sensors are analog, and even though their output may eventually be converted into digital form for interpretation and use, it is not always possible to incorporate the electronics locally. This may be because of cost or because of operating conditions such as elevated temperatures. For example, in a car dozens of sensors and actuators are processed by a single central processor, making for a more economical system. It is not practical to supply each sensor with power and electronics to digitize its data when the processor can do that for all of them (in practice some sensors or groups of sensors may have their own processor, e.g., for safety reasons). In other cases, such as, for example, the oxygen sensor in a car, the sensor operates at elevated temperatures, beyond the temperature range of semiconductors, making it impossible to incorporate electronics directly into it. In such cases the analog signal must be transferred to the controller. A number of methods have been developed for this purpose. Three of these methods, suitable for use with resistive sensors and passive sensors, are discussed next.

11.7.1 Four-Wire Transmission

In sensors that change their resistance, such as thermistors and piezoresistive sensors, one must supply an external source and measure the voltage across the sensor. If done remotely, the current may vary with the resistance of the connecting wires and produce an erroneous reading. To avoid this, the method in **Figure 11.47** can be used. The sensor

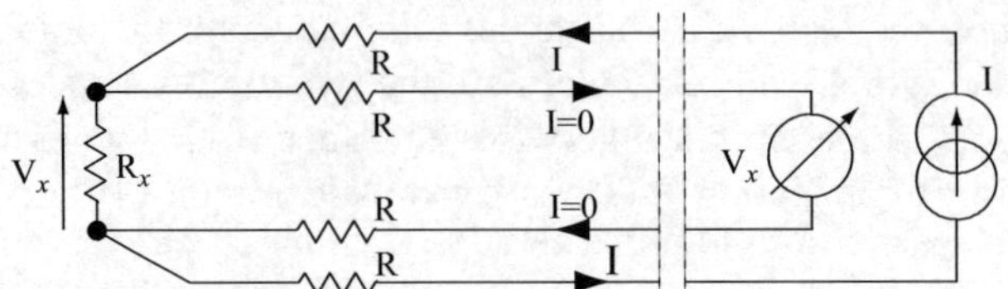

FIGURE 11.47 ■ Four-wire transmission.

is supplied from a current source, I. This current is constant since the internal impedance of a current source is very high. Thus the voltage on the sensor is constant regardless of the length of the wires and their impedance. A second pair of wires measures the voltage across the sensor, and since a voltmeter has very high impedance, there is no current (ideally) in this second pair of wires, producing an accurate reading. This is a common method of data transmission when applicable. A very similar method may be used with Hall element-based sensors in which the current source provides the bias current and the Hall voltage is measured separately.

EXAMPLE 11.14 Elimination of errors in RTDs

RTDs, especially the wire type, have low resistance, typically between 25 and 100 Ω. They come with leads that allow connection to a measuring instrument. Because these leads (usually copper) have their own resistance and their resistance varies with temperature, they introduce errors in sensing. To eliminate this problem, some RTDs come with four wires (see **Figure 3.3c**). The connection in **Figure 11.47** eliminates all errors due to the wires, even if the lead wires are of unequal length. R_x is the resistance of the RTD at the temperature being sensed, whereas R represents the resistance of each of the extension wires. The voltmeter reads the voltage across the RTD, provided its own impedance is essentially infinite. In practical terms it may have a resistance of a few megaohms, which, compared to the low resistance of the RTD, provides virtually error-free reading. The current source must be as constant as possible to eliminate variations in the voltage read, since an x% change in I will show the same as an x% change in R_x and hence will introduce an x% error in temperature.

11.7.2 Two-Wire Transmission for Passive Sensors

In passive sensors, most of which provide output as voltage, it is sometimes possible to measure the voltage remotely using a simple pair of wires since virtually no current is involved in the measurement. This is especially true for DC outputs such as in thermocouples. In most cases a twisted pair line is used because it reduces the noise picked up by the line. In sensors with high impedance it is much more risky to do so because of the inherent noise in the two-wire connection.

11.7.3 Two-Wire Transmission for Active Sensors

A common method of data transmission for sensors (and actuators), and a method that has been standardized is the 4–20 mA current loop. In very simple terms, the output of the sensor is modified to modulate the current in the loop from 4 mA (corresponding to the minimum stimulus value) to 20 mA (maximum stimulus value). The configuration is shown in **Figure 11.48**. Obviously the sensor's output must be modified to conform to this industry standard and this may require additional circuitry. Many sensors are made to conform to this standard so the user only has to connect them to the two-wire line. The power supply depends on the load resistance and the transmitter's resistance, but is between 12 and 48 V. The interface circuitry includes a means of setting the 4 mA and 20 mA values to correspond to minimum and maximum output. These are indicated by the two potentiometers in **Figure 11.48**. The current transmitted on the line is then

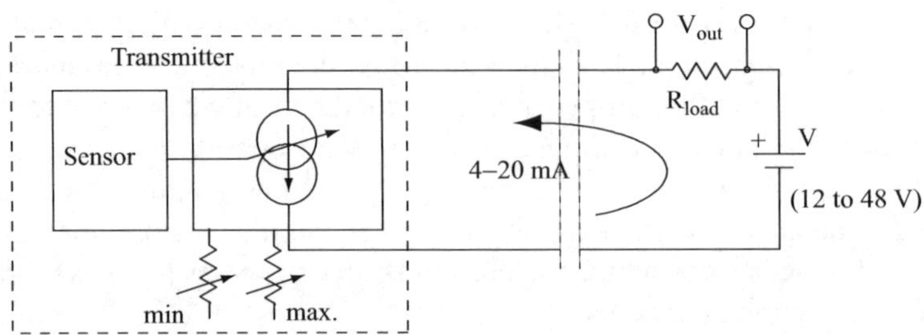

FIGURE 11.48 ■ A 4–20 mA current loop data transmission.

independent of the length of the line and its resistance. The voltage measured across the load resistance is processed at the controller to provide the necessary reading. Most sensors can be fitted with a 4–20 mA loop through appropriate circuitry.

EXAMPLE 11.15 The 4–20 mA loop

A pressure sensor is used in a remote installation to sense air pressure for meteorological purposes. The range required is between 750 mbar (75 kPa) and 1200 mbar (120 kPa). (The lowest air pressure ever recorded on earth is around 850 mbar, recorded during a tornado in the United States, and the highest pressure recorded is 1086 mbar, recorded in Mongolia). The pressure sensor is equipped with a 4–20 mA loop. What is the range of the voltage V_{out} measured on a load resistance of 470 Ω (see **Figure 11.48**) if the loop is properly calibrated.

Solution: Calibration means that the minimum pressure produces a current of 4 mA and the maximum pressure produces a current of 20 mA. Therefore the range is from $V_{min} = 4 \times 10^{-3} \times 470 = 1.88$ V to $V_{max} = 20 \times 10^{-3} \times 470 = 9.4$ V. This is a convenient scale and is typically designed to fit standard measuring instruments.

Note: The actual design of the 4–20 mA loop is rather complex and is sensor specific since it must take into account the sensor's transfer function. Sensors (and actuators) are typically sold with optional 4–20 mA loops, especially for industrial applications.

There are other methods of transmission that may be incorporated. For example, six-wire transmission is used with bridge circuits in which the four-wire method above is supplemented with two additional wires that measure the voltage on the bridge itself (see **Example 11.16**) to allow calibration of the output based on the actual reference voltage at the bridge, and by doing so, eliminating the effects of feed-wire resistances on the output.

EXAMPLE 11.16 Compensation of errors due to source variations in bridge connections

To compensate for errors due to long lead wires for the reference voltage to the bridge, six-wire transmission is indispensable, especially where the impedances of the bridge legs are low and hence the bridge requires relatively large currents. Consider the measurement of force using a load cell. The load cell uses four strain gauges, two sensing compression strains and two sensing tension strain, shown in **Figure 11.49** with opposite direction arrows (see also **Section 6.3** and **Figure 6.11**). Strain gauges are low-impedance devices. We will assume here a common strain gauge resistance of 120 Ω when the gauge is unstrained and a gauge factor of 2.5.

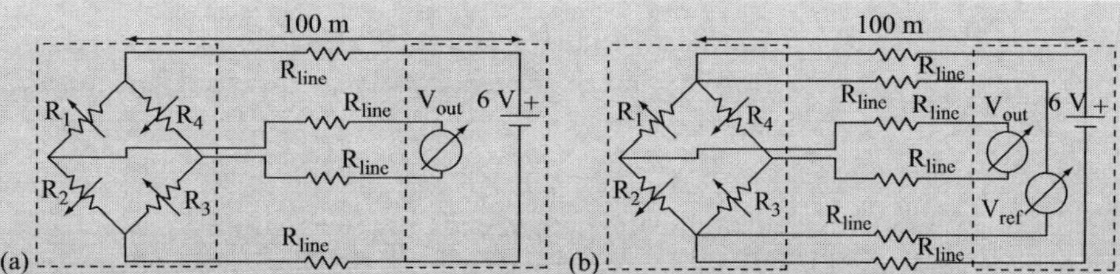

FIGURE 11.49 ■ (a) Four-wire transmission. (b) Six-wire transmission. Arrows up indicate tension, arrows down indicate compression.

a. Calculate the error in output assuming a 6 V reference voltage and 100 m wire lengths using the 4 wire system shown in **Figure 11.49a**. Assume a resistance of 0.25 Ω/m for all wires and that all wires are of the same length. Assume linear strain gauges and a strain of +3% for the tensioned gauges and −3% for the compressed gauges for the measured load. To affect proper measurement, the compressed gauges are prestrained to 3%, that is, when no load is present, the tensioned gauges are at zero strain whereas the compressed gauges are at 3% strain.

b. Show that the bridge output is independent of wire length if the six-wire connection in **Figure 11.49b** is used with the same assumptions as in (a).

Solution: First we need to calculate the resistances in the four legs of the bridge.
For the tensioned gauges the strain is 3%. From **Equation (6.5)**:

$$R_1 = R_4 = 120(1 + 2.5 \times 0.03) = 129\,\Omega.$$

For the compressed gauges the strain is zero:

$$R_1 = R_4 = 120\,\Omega.$$

a. The output of the bridge for the conditions given is

$$V_o = V_{\text{ref}}\left(\frac{R_3}{R_3 + R_4} - \frac{R_1}{R_1 + R_2}\right) = V_{\text{ref}}\left(\frac{129}{129 + 120} - \frac{120}{120 + 129}\right) = V_{\text{ref}}(0.518 - 0.482)$$

$$= 0.036V_{\text{ref}} \quad [\text{V}].$$

Because the only reference voltage that can be measured by the remote instrument is the source voltage (6 V), the expected output is $0.036 \times 6 = 0.216$ V. This is the output the instrument expects (i.e., the calibration output) since the instrument can only measure this voltage.

However, the reference voltage at the bridge is

$$V_{\text{ref}} = \frac{V_s}{R_b + 2\,R_{\text{line}}} \times R_b = \frac{6}{(120 + 129)/2 + 2 \times 0.25 \times 100} \times \frac{(120 + 129)}{2} = 4.2808\text{ V},$$

where R_b is the resistance of the bridge made of the two legs in parallel. Therefore the actual output measured by the voltmeter is

$$V_o = V_{\text{ref}}\left(\frac{R_3}{R_3 + R_4} - \frac{R_1}{R_1 + R_2}\right) = 4.2808 \times \left(\frac{129}{129 + 120} - \frac{120}{120 + 129}\right) = 0.03614 \times 4.2808$$

$$= 0.1541\text{ V}.$$

Since the measured voltage is lower, there is an error of (0.1541 − 0.216)/0.216 = −0.286, or −28.6%.

b. In the six-wire bridge, the measuring instrument measures the reference voltage as 4.28 V. Hence the calibration output is 0.154 V, and this is exactly what the voltmeter measures. The error due to the source wires is eliminated.

When actuators are involved, there are only two ways the power can be transmitted to the actuator. One is to get the actuator close to the source that provides the power, implying that lines must be very short. This is possible in some cases (audio speakers, control motors in a printer, etc.). In other cases it is not practical and the controller and the actuator must be at a considerable distance (robots on the factory floor, etc.). In such cases one of the methods above can be used to transfer data, but the power must then be generated or switched locally at the actuator site. That is, the controller now issues commands as to power levels, timings, etc., and these are then executed locally to deliver the power necessary. Much of this is done digitally through the use of microprocessors at both ends.

11.7.4 Digital Data Transmission Protocols and Buses

When digital data are transmitted between sensors and processors or between processors and actuators, the data are handled by one of the many data protocols available. These protocols are standards that define the interface so that various devices from different sources can communicate with each other. The available protocols include the common serial interface (also known as the RS232 interface), the Universal Serial Bus (USB) interface, the parallel interface (known as the IEEE 1284 interface), and the Controller Area Network (CAN) interface, and many others. These protocols are designed to allow multiple devices to connect and communicate on a single "bus," that is, on a single set of conductors. The number of conductors in a bus varies and depends on the protocol. For example, USB has four conductors, two for power and two for data. A parallel bus may have many more conductors, typically a minimum of about 10. Some busses are specialized, whereas some are general. For example, the CAN interface is a bus interface designed specifically to take into account conditions in vehicles, such as the need for enhanced noise immunity. One of the advantages of these interfaces is that electronic devices designed for their implementation and for translation from one protocol to another are available to the designer. Sensors and actuators with interfaces are available and often offer the quickest and most robust way of interfacing with controllers.

Of particular interest in sensors are the I^2C (two-wire interface) and 1-wire protocols. The I^2C protocol allows connection of multiple devices on two wires. The 1-wire protocol has become very popular for many devices, including sensors. In this protocol both power to the device and data to and from it are passed on a single pair of wires, making this an effective and economical method for sensing over long wires. The term 1-wire is somewhat misleading, since there are in fact two wires. It simply indicates that there is one wire and a ground or that there are two functions (power and data) on two wires.

11.8 | EXCITATION METHODS AND CIRCUITS

Sensors and actuators must often be supplied with power, either from AC or DC sources. First and foremost in excitation is the power supply circuit. In many sensors the power is supplied by batteries, but many others rely on line power through the use of regulated or unregulated power supplies. In addition, the need for current sources (e.g., in Hall elements) and for AC sources (such as in LVDTs) has been discussed in previous chapters. These circuits affect the output of the sensor and its performance (accuracy, sensitivity, noise, etc.) and are an integral part of the overall performance of sensors and actuators.

Aside from the obvious use of batteries to power sensors and actuators, there are two general types of power supplies that may be of use. One is called a linear power supply and the second is called a switching power supply. In addition, there are so called DC-to-DC converters, which are used to convert power from one level to another, sometimes as part of the circuit that uses the power. Inverters for the generation of AC power from DC sources also exist, but are usually restricted to power devices, and hence to some actuators. These will not be discussed here.

11.8.1 Linear Power Supplies

A linear power supply is shown schematically in **Figure 11.50**. It consists of an AC source, usually the line voltage, and a means of reducing this voltage to the required level (a transformer). The transformer is followed by a rectifier that produces DC voltage from the AC source. This voltage is filtered and then regulated to the final required DC level. A final filter is usually provided. This type of regulated power supply is very common in circuits especially where the power requirements are low. Some of the blocks may be eliminated depending on the application. If, for example, the source is a battery (in a car, for example), the transformer and the rectifier are not relevant and filtering may be less important.

More specifically, consider the circuit in **Figure 11.51**. This is a regulated power supply capable of supplying 5 V at up to 1 A. The transformer reduces the input voltage to 9 V RMS. This is rectified through the bridge rectifier and produces 12.6 V ($9\text{ V}\times\sqrt{2}/2$) across C_1 and C_2. These two capacitors serve as filters—the large capacitor reducing low-frequency fluctuations on the line, the smaller capacitor providing high-frequency filtering. The regulator shown is a 5 V regulator, which drops 7.6 V across itself to keep the output constant at 5 V (there is also a small voltage drop on the rectifier, so the voltage across the regulator is slightly lower than the value indicated here). The regulator will do so for any input voltage down to about 8 V. The capacitors at the output are again filters. The current is limited by the capacity of the regulator to dissipate power due to the current through it and the voltage across it. Other regulators

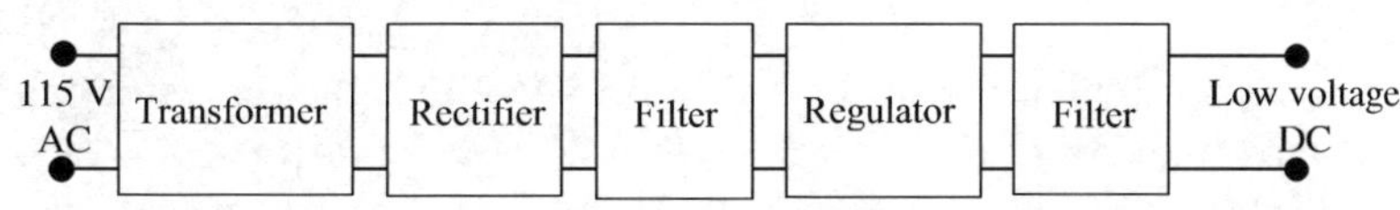

FIGURE 11.50 ■ Structure of a regulated linear power supply.

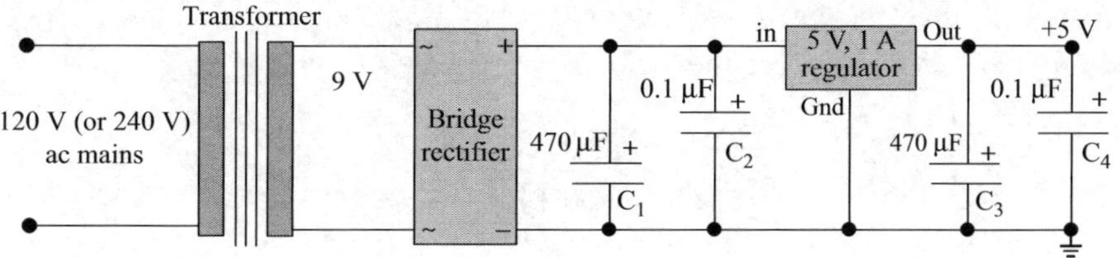

FIGURE 11.51 ■ A 5 V regulated power supply using a fixed voltage regulator.

are available that will dissipate more or less power. These regulators exist at standard voltages, either positive or negative, or as adjustable variable voltage regulators. Discrete component regulators can be built for almost any voltage and current requirements.

This circuit or similar circuits is a common way of providing regulated DC power to many sensor and actuator circuits. The advantage is that they are simple and inexpensive, but they have serious drawbacks. The most obvious is that they are big and heavy, mostly because of the need for a transformer, which must handle the output power. In addition, the power dissipated on the regulator is not only lost but it generates heat, and this heat must be dissipated through heat exchangers, adding to the cost and bulk of linear power supplies.

EXAMPLE 11.17 Linear power supply: efficiency

Consider the linear power supply in **Figure 11.51**. The DC voltage at the "in" pin of the regulator must be at least 3 V above the output—that is, it must be at least 8 V, but can be as high as 35 V (depending on the regulator used). The device is rated at 1 A, 5 V output, and can dissipate at most 3 W.

a. Under the conditions shown in the figure (i.e., the transformer supplies 9 V RMS), calculate the maximum current permissible in the load. Calculate the efficiency of the power supply for that current.

b. Suppose the transformer is replaced with a 24 V RMS transformer. What is the maximum current and the efficiency of the power supply at maximum current?

Notes

1. Efficiency is the useful output power divided by the input power supplied by the transformer.
2. In a bridge rectifier, there are two diodes in series with the regulator and each drops approximately 0.7 V.

Solution: The input voltage to the regulator is the peak voltage the transformer supplies minus the voltage drop in the two diodes of the bridge rectifier. Efficiency is dictated mainly by the voltage drop across the regulator, but the diodes themselves contribute as well.

a. At an RMS voltage of 9 V, the voltage at the input to the regulator is

$$V_{DC} = V_{RMS}\sqrt{2} - 2V_D = 9 \times \sqrt{2} - 1.4 = 11.33 \text{ V}.$$

Thus the voltage drop across the regulator is 11.33 − 5 = 6.33 V. Since the regulator cannot dissipate more than 3 W, the maximum current is

$$I_{max} = \frac{P_{max}}{V_{drop}} = \frac{3}{6.33} = 0.474 \text{ A}.$$

The power supplied to the load is $0.474 \times 5 = 2.37$ W. The input power supplied by the transformer is $0.474 \times 9 = 4.266$ W. Therefore the efficiency is 55.5%.

b. For an input voltage of 24 V RMS, the peak voltage is

$$V_{DC} = V_{RMS}\sqrt{2} - 2V_D = 24 \times \sqrt{2} - 1.4 = 32.54 \text{ V}.$$

Now the voltage drop across the regulator is 32.5 − 5 = 27.54 V. Thus the maximum current is

$$I_{max} = \frac{P_{max}}{V_{drop}} = \frac{3}{27.5} = 0.11 \text{ A}.$$

The regulator cannot supply more than about 110 mA or it will overheat (this type of regulator has a thermal shutdown and will disconnect the load to protect itself from damage).

The efficiency is

$$eff = \frac{I_{max}V_{out}}{V_{in}I_{in}} \times 100 = \frac{0.11 \times 5}{0.11 \times 24} \times 100 = 20.83\%.$$

Notes:

1. Input current and output current in the regulator are taken as identical. In reality there is a small current flowing in the ground leg of the regulator (needed to run the internal circuitry) that further reduces efficiency.
2. Efficiency depends greatly on the voltage drop across the regulator. The highest efficiency is obtained when this voltage is minimized.
3. It is these low efficiencies that favor other types of power supplies, especially the switching power supplies, in spite of their more complex circuitry.
4. Some fixed or variable voltage linear regulators, called "low dropout" (LDO) regulators, require only a fraction of 1 V to regulate and hence they have better efficiencies.

11.8.2 Switching Power Supplies

An alternative method of providing DC power is through the use of a switching power supply. A switching power supply relies on two basic principles to eliminate some of the drawbacks of the linear power supply. These are shown in **Figure 10.52**. First, the power transformer is eliminated and the line voltage is rectified directly. This high-voltage DC is filtered as before. The switching transistor is driven with a square wave, which turns it on for a time t_{on} and off for a time t_{off}. When on, a current flows through the inductor and charges the capacitor to a voltage that depends on t_{on}. The larger t_{on} is, the higher the output voltage. When the switch is off, the current in the inductor L is discharged through the load supplying it with power for the off time. The voltage is stabilized by sampling the output and changing the duty cycle (ratio between t_{on} and t_{off}) to increase or decrease the output to its required value. This change in duty cycle is equivalent to a PWM generator.

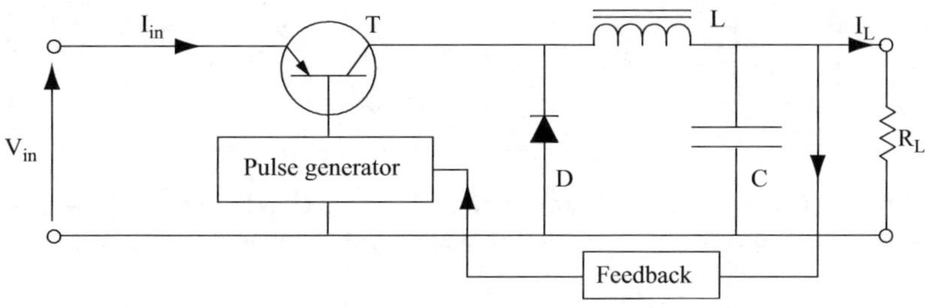

FIGURE 11.52 ■ Switching power supply.

In practical power supplies, additional considerations must apply. First, it is necessary to separate or isolate the input (which is connected to the line) and output. In the linear power supply, this isolation was accomplished by the power transformer. In switching power supplies this can also be done by a transformer, but now the transformer is much smaller, operating at a high frequency. Second, the switching, which must necessarily be done at relatively high frequencies, introduces noise into the system. This noise must be filtered before the power supply can be used. Line-fed switching power supplies include a line input filter made of an inductor in each line and two capacitors—one on the line side of each inductor and one on the power supply side of each inductor to prevent high-frequency noise from getting into the line and possibly affecting other devices connected to the line. As with almost any circuit, there are many variations on the basic circuit shown in **Figure 11.52**, but these variations are not fundamental in understanding its operation. One of the salient features of switching power supplies is that the input voltage can vary over wide ranges with no real effect on the output voltage or efficiency. The common, light-weight dual-voltage power supplies used with many devices are a consequence of this property.

The DC-to-DC converter is a different type of switching power supply. It essentially takes the DC source and converts it into an AC voltage, which can then be converted through a transformer or by virtue of a transient voltage in an inductor or through charging and discharging capacitors to any required level and then rectified back to DC and regulated. In most cases the source will be a battery, but it can also be a rectified AC source. In many instances the purpose is to supply power at a required voltage level when the available voltage level is either lower or higher than that needed. For example, suppose that a consumer product needs to operate on a single 1.5 V cell. Since most electronic components require a higher voltage, say 3 V, a DC-to-DC converter can be used to fill this need. In most cases isolation is not an issue, but if it is, the use of a high-frequency transformer can resolve that as well. DC-to-DC converters are common units in electronic equipment, including sensing circuitry. They come in a variety of sizes and voltage levels and are often used when voltage levels must be changed.

Inductive DC-to-DC converters can be understood from **Figure 11.52**. An alternative to DC-to-DC converters of the type shown in **Figure 11.52** is the charge pump DC-to-DC converter. The idea of a charge pump is to charge a capacitor (or more) and then switch the capacitor and connect it in series with the existing source so that the total voltage across the source and capacitor is higher (approximately twice as high). The process can be repeated and higher voltages can be obtained. For this reason, the device is also called a voltage multiplier. The name charge pump indicates that the charge in the capacitors is being transferred to accomplish voltage conversion. The principle is shown in **Figure 11.53a**. The inverting driver is fed by a square wave and, as a consequence,

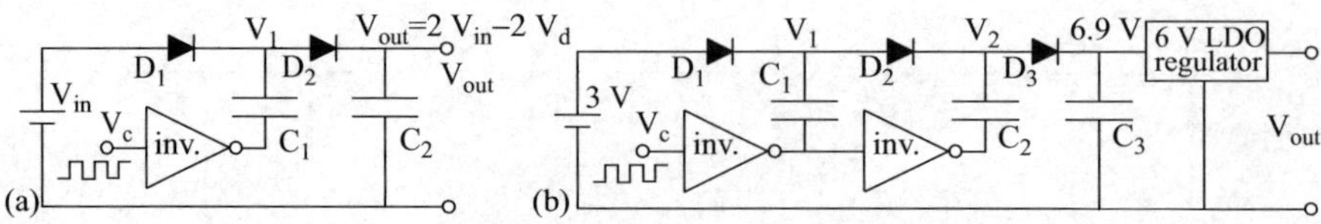

FIGURE 11.53 ■ (a) A two-stage charge pump switching power supply. (b) A three-stage charge pump switching power supply that can supply a regulated 6 V output from a 3 V battery.

whenever the input is zero, the output swings to V_{in}, and vice versa. When the output of the driver is low, C_1 charges to $V_{in} - V_D$, where V_D is the voltage drop across the diode (approximately 0.7 V for silicon diodes and 0.3 V for Schottky diodes). The capacitor charges during a half-cycle. During the next half-cycle, the output of the driver is high and the potential of C_1 now is in series with the output of the driver. The potential V_1 increases to $2V_{in} - V_D$. This causes diode D_1 to be reverse biased, whereas D_2 conducts and charges C_2 to a potential $2V_{in} - 2V_D$. Effectively, the voltage V_{in} has been converted to a higher voltage $V_{out} = 2V_{in} - 2V_D$ by virtue of the transfer of charge. One can add additional stages, each made of a driver, capacitor, and diode, and increase the voltage to the desired level. However, the method has its limits. First, it cannot supply much power since all output current comes from discharging the output capacitor (C_2 in this case). Second, increasing the voltage to higher levels is limited by the operating voltage of the drivers. In addition, voltages that are not multiples of V_{in} (approximately) must be generated using a regulator similar to the fixed-voltage regulator used in **Figure 11.51**. The regulator also serves to regulate the output against variations due to load current. In spite of some serious limitations, the simplicity of the method and economy of components make it a useful method, especially for use with low-power devices. Efficiency is not better than for linear power supplies and usually lower than inductor-based switching power supplies. The method is particularly attractive when the generation of a higher voltage within a circuit is needed, but it can be used for low-power sensors and actuators as well.

EXAMPLE 11.18 Charge pump switching power supply for low-power battery-operated sensors

A sensor requires a 6 V power supply but must operate from a 3 V battery. A regulated charge pump switching power supply appropriate for this purpose is shown in **Figure 11.53b**. Assuming initially that all three capacitors are discharged and that the input to the first driver is high, C_1 charges through D_1 to $V_{in} - V_D$, whereas D_2 is effectively disconnected (it is reverse biased by the output of driver 2, which is high). When the input to the first driver goes low, its output goes high and a potential equal to $V_1 = 2V_{in} - V_D$ is applied across C_2, charging it. D_3 is also conducting and C_2 charges to $2V_{in} - 2V_D$. Next the input goes high again and the output of driver 2 goes high, placing a voltage V_{in} in series with C_2. Now the voltage V_2 increases to $3V_{in} - 2V_D$. This charges C_3 to $3V_{in} - 3V_D$. As long as the capacitors do not discharge, no more charge is transferred, but if a load draws current from C_3, the system replenishes this charge. In this case, the output is 3×3 V $- 3 \times 0.7 = 6.9$ V. The regulator must be a low-dropout (LDO) regulator that can operate at a low voltage difference between the input and output pins. Low-power, low-dropout regulators that require less than 0.1 V difference are readily available, allowing this circuit to operate as the battery discharges and its voltage diminishes. The load current is defined by the load resistance and discharge of C_3 through the load. If the current is too high, the output may lose regulation since C_3 may not be able to supply the higher current and still maintain a charge level sufficiently high to keep the voltage above 6 V. Note also that using Schottky diodes increases the output to 3×3 V $- 3 \times 0.3 = 8.1$ V, allowing a larger margin for the input. For example, with Schottky diodes the battery voltage can decrease to 2.33 V before regulation is lost.

11.8.3 Current Sources

The generation of constant currents is important in many sensors. For example, when using Hall elements, the output is proportional to the magnetic flux density and current. In most cases we must keep the current constant so that the output is only a function of the magnetic flux density. The generation of constant currents can take various levels of complexity. Obviously one can resort to something as simple as a large resistor in series with the relatively low resistance Hall element. In such a configuration the current is not constant, but rather varies a little because the resistance of the sensor is low and its effect on the total resistance is small. More accurate methods of current generation are needed for higher accuracy requirements. A simple constant current source can be built based on the properties of FETs, as shown in **Figure 11.54**. In this circuit, as long as the voltage across the FET is above its pinch-off voltage (V_p) (a fundamental property of FETs), the current is constant and equals $(V_{cc} - V_p)/R$. Of course, the pinch-off voltage is temperature dependent, a problem with all semiconductor devices.

Another simple way of supplying constant current to a load is shown in **Figure 11.55**. In this circuit, the zener diode voltage V_z produces a current in the load equal to $(V_z - 0.7)/R_L$ since the voltage across the base–emitter junction is fixed at 0.7 V and the zener voltage is fixed at V_z. This circuit is more immune to temperature variations since the zener diode and the base–emitter junction are in opposition and hence the variations of the voltage on the zener diode due to temperature are compensated by those of the base–emitter junction (see the discussion on voltage references in the following section).

A more stable circuit is the so-called current mirror circuit shown in **Figure 11.56**. Here a current I_{in} is generated as V_1/R_1 and is kept constant. The collector current in transistor T_3 is virtually equal to I_{in} because the base current in T_1 is very small. The voltage across the base of T_1 keeps the current through the load, I_L, equal to I_{in}, hence the name current mirror. As long as I_{in} is constant, so will the current in the load.

The properties of the voltage follower based on an operational amplifier can be used to generate a constant current, as shown in **Figure 11.57**. The output of the voltage follower is V_i and the current is V_i/R_1. The transistor is necessary to provide currents larger than those possible with an operational amplifier. Many other circuits with

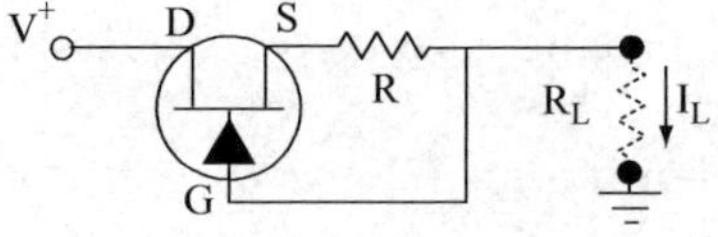

FIGURE 11.54 ■ A constant current generator based on an FET.

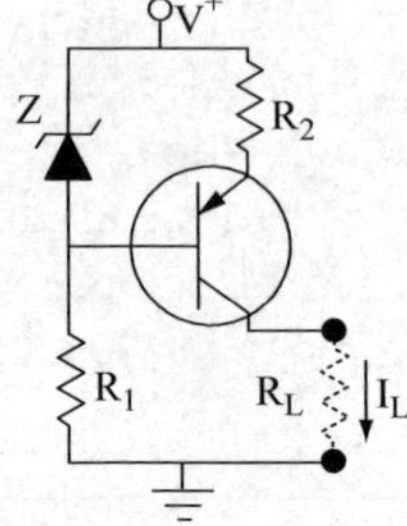

FIGURE 11.55 ■ A zener-controlled constant current generator.

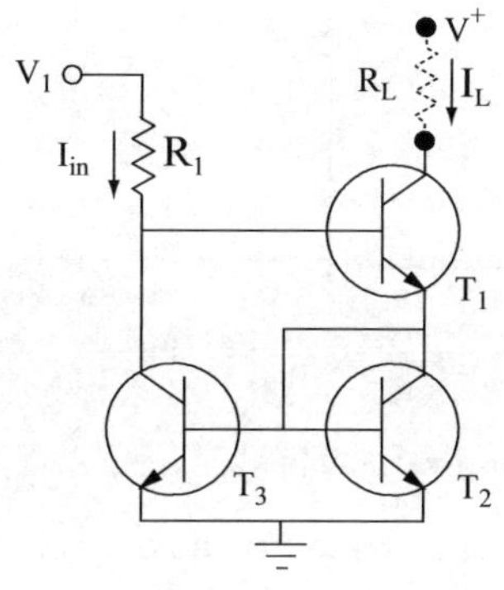

FIGURE 11.56 ■ A current mirror constant current generator.

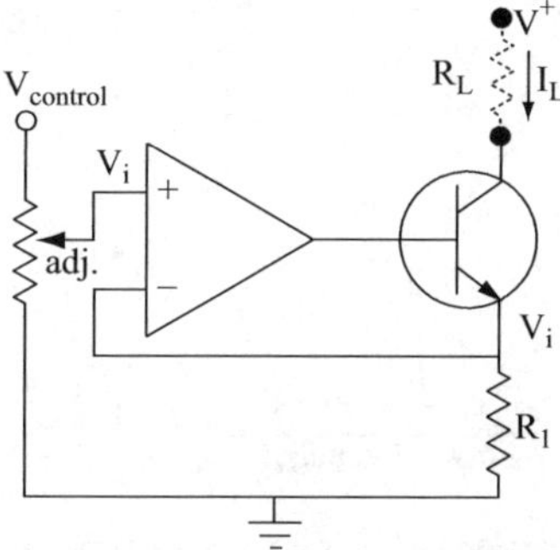

FIGURE 11.57 ■ A voltage follower-based constant current generator.

varying properties may be used, but those shown above represent the basic principles involved.

11.8.4 Voltage References

Some sensor circuits as well as circuits used for interfacing call for a constant voltage reference. Of course, a regulated power supply is a voltage reference, but what is meant here is a constant voltage, usually on the order of 0.5–2 V that supplies very little current, if any, is invariant with respect to source variations, temperature, and external influences, and is used as a reference to other circuits. These reference voltages must be constant under expected fluctuations in power supplies, temperature, etc. The use of voltage references was introduced in conjunction with A/D converters, D/A converters, and bridge circuits.

The simplest voltage reference is the zener diode, shown in **Figure 11.58**. This is a reverse-biased diode, biased at the breakdown voltage for the junction. The resistor limits the current so that the diode does not overheat. As long as the maximum current of the zener diode is not exceeded, the voltage across the diode is kept at the breakdown voltage, and that is constant other than changes due to temperature variations. These diodes are commonly used for voltage regulation and other purposes, such as the constant current generator in **Figure 11.55**. However, there is a special type of zener diode specifically designed for voltage reference applications (called the reference zener diode) in which the breakdown voltage is kept constant and the diode is temperature compensated by using two diodes, one forward and one reverse biased (**Figure 10.59**). In the forward-biased diode, an increase in temperature decreases the forward voltage

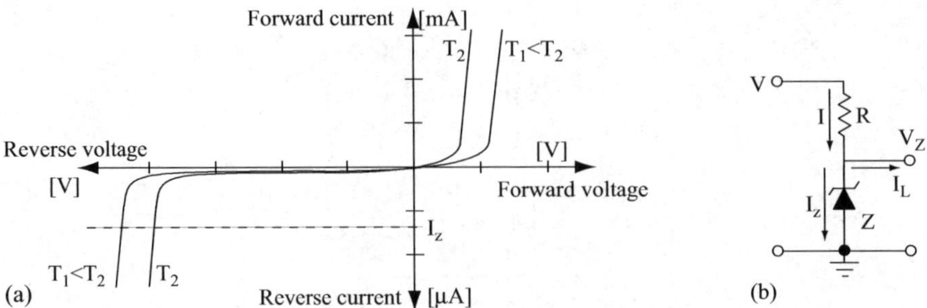

FIGURE 11.58 ■ A voltage reference based on a zener diode. (a) The characteristics of a zener diode showing its temperature dependence. (b) A method of connecting the zener diode.

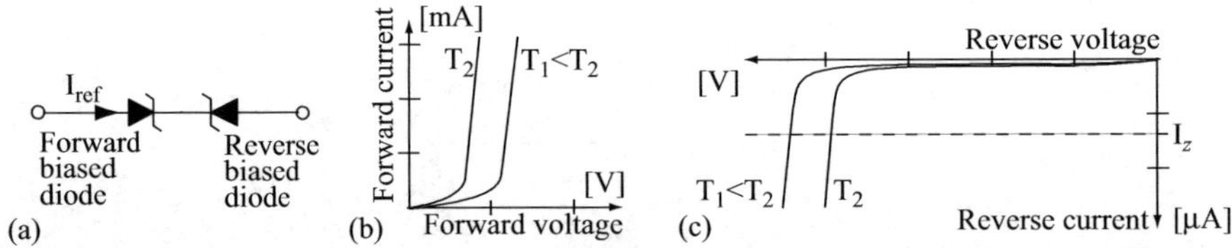

FIGURE 11.59 ■ The zener reference diode incorporating temperature stabilization. (a) Two zener diodes in series. (b) *I–V* characteristics of the forward-biased diode. (c) *I–V* characteristics of the reverse-biased diode.

(by ΔV or about 2 mV/°C), whereas in the reverse-biased diode it increases by roughly the same amount. The total voltage is therefore constant (or nearly so). These diodes are available in fixed voltages down to about 3 V.

Another device that is used for voltage reference is the bandgap reference. It is superior to zener diodes and is available in voltages down to 0.6 V. This device exists as a discrete component but is also commonly integrated within microprocessors and other circuits that require a stable voltage reference.

Voltage reference diodes are available commercially in standard voltages from about 1.2 to more than 100 V.

11.8.5 Oscillators

Many sensors and actuators require voltages or currents that are variable in time. For example, the LVDT requires a sinusoidal source at a frequency of a few kilohertz. Magnetic and eddy current proximity sensors use AC of constant amplitude and frequency to produce an output voltage that is proportional to position. In fact, all transformer-based sensors must use an AC source. Other sensors require special waveforms such as square waves or triangular waves. With the exception of devices that are driven from the 60 Hz (or 50 Hz) sinusoidal line voltage, the source must be generated at the correct frequency and with the required waveform. Often these must be frequency stabilized and amplitude regulated to make them useful sources. There are virtually hundreds of different ways of generating AC signals of any frequency and waveform conceivable, but there are only a few basic principles involved. The most fundamental of these is that an oscillator is an unstable amplifier. That is, starting with an amplifier of some sort, one can provide positive feedback to make it unstable and hence set it into oscillation. The second principle is that this unstable circuit must be forced to oscillate at a specific frequency by means of one of two principles: either an LC tank circuit (or equivalent) is employed or a means of delaying the feedback is used. Further, the circuit must be made to oscillate with a required waveform through use of these or additional components.

It turns out that it is much easier to produce square waveforms, but sinusoidal waveforms can be produced as well. In particular, LC oscillators are usually sinusoidal, whereas those based on delays (typically through RC timing) are usually square waves. A large number of variations of the principles above are employed for specific applications. It is not possible to discuss here all of the types of oscillators, so we will describe a few representative circuits with the understanding that others exist and that those represented here are neither the most important nor the simplest.

11.8.5.1 Crystal Oscillators

The crystal oscillator is based on the natural resonant frequency of quartz crystals or any other piezoelectric material cut and placed between two electrodes, as shown in **Figure 11.60a** (an actual device is shown in **Figure 11.60b**, but there are many sizes, shapes, and variations possible). This device has the equivalent circuit shown in **Figure 11.61** and can oscillate in one of two modes. One is a series oscillation mode, the other a parallel oscillation mode (discussed in **Section 7.7**). When connected in a circuit that can provide the proper positive feedback, it will oscillate at the resonant frequency

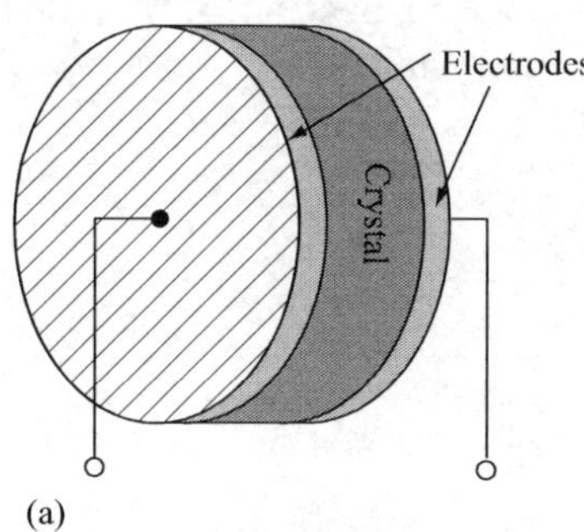

FIGURE 11.60 ■ (a) The basic structure of a crystal used in a crystal oscillator. (b) A 1 MHz quartz crystal. One electrode is shown; the second is partly visible through the semitransparent quartz. The can is sealed against moisture and to protect the device (an older-style crystal is shown to see the details of construction).

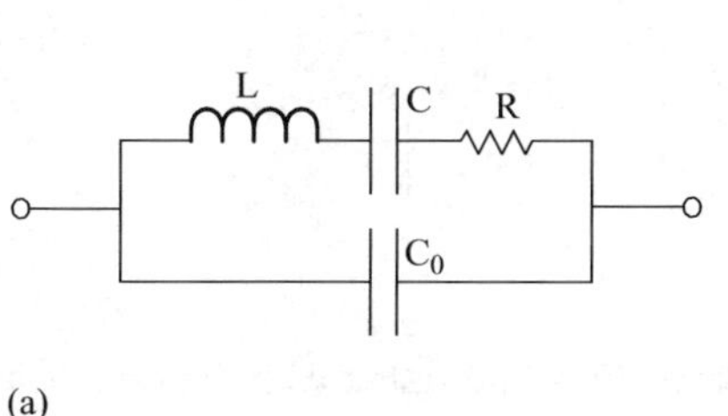

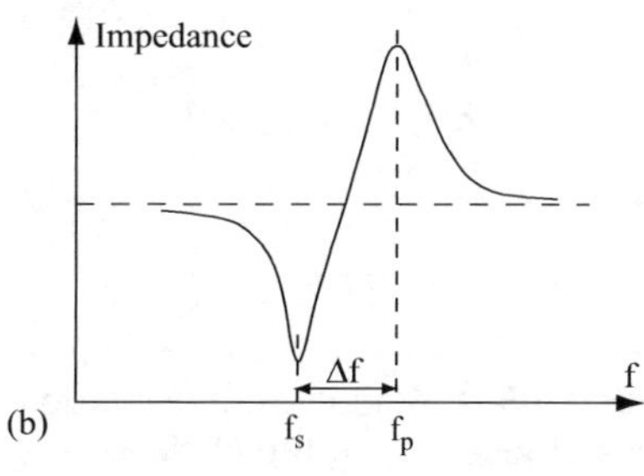

FIGURE 11.61 ■ (a) Equivalent circuit for a crystal and (b) its two basic resonant frequencies.

of the crystal, which depends entirely on the dimensions and composition of the crystal and on the mode of oscillation. In the series mode, the resonant frequency is

$$f_s = \frac{1}{2\pi\sqrt{LC}} \quad [\text{Hz}]. \tag{11.40}$$

In the parallel mode the resonant frequency is

$$f_p = \frac{1}{2\pi\sqrt{LC[C_0/(C+C_0)]}} \quad [\text{Hz}]. \tag{11.41}$$

A simple sinusoidal oscillator is shown in **Figure 11.62**. The details of the circuit are not important here, but it is important to realize that the feedback from output to input (collector to base) is supplied by the crystal. The output frequency is controlled by the crystal and the signal is taken at the collector.

A different approach, one that provides a square wave, is to use two inverting gates, as shown in **Figure 11.63**. Because the gate can only take two states, the output will swing between V_{cc} and ground. The positive feedback is delayed due to the delay of the gate, and again the frequency is controlled by the crystal. These oscillators can be used, for example, in humidity sensors in which the frequency changes with humidity (by changing the mass of the crystal), but they will work equally well to generate the fundamental clock frequency in a microprocessor or the signal from an infrared remote controller. Crystals are available either in parallel or series modes and in a range of frequencies from about 32 kHz (often used in clocks) to about 100 MHz. Higher frequencies can also be generated, and some are available, but at the higher frequencies SAW devices are more effective and more common (see **Section 7.9**). When SAW devices are used they are connected very similarly to crystals.

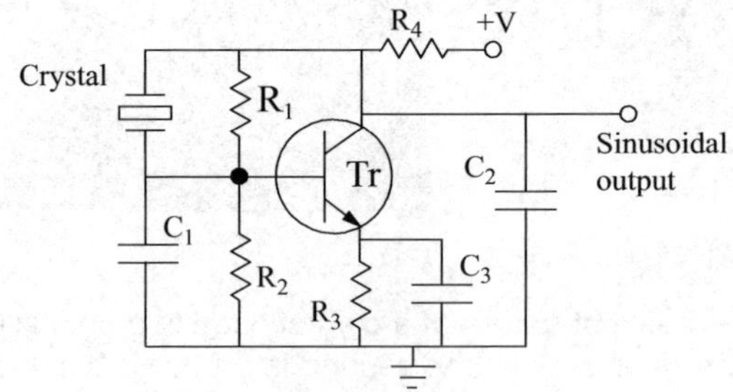

FIGURE 11.62 ■ A sinusoidal crystal oscillator with the crystal in the feedback circuit.

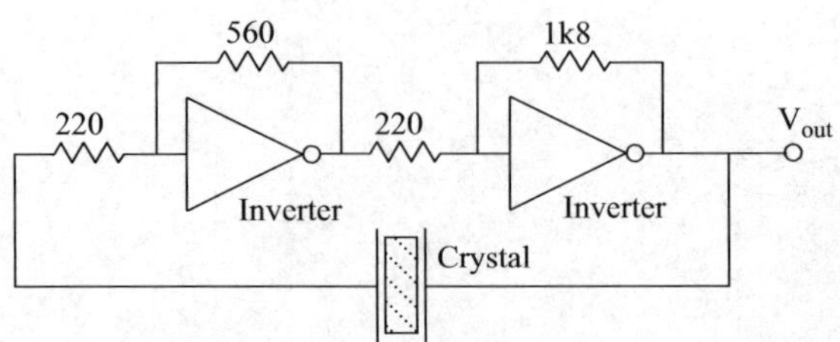

FIGURE 11.63 ■ A square wave oscillator based on inverting gates with the crystal in the feedback loop, ensuring the positive feedback necessary for oscillation.

11.8.5.2 LC and RC Oscillators

Oscillators can easily be built from discrete as well as integrated components without the need for a crystal, although frequency stability is inferior. Four simple square wave oscillators based on the delay of the feedback signal are shown in **Figures 11.64** and **11.65**. **Figure 11.64a** shows a "ring" with an odd number of inverters. Normally, because of the odd number, the output of the last inverter would be in conflict with the input of the first inverter. However, because the inverters exhibit a delay between input and output, the ring operates as a square wave oscillator. Suppose that each inverter has an internal delay of Δt [s] and suppose further that at a given time the input of the first inverter changes state from zero to V_0. After a time Δt, its output changes from V_0 to zero. After an additional time Δt, the second inverter changes state from zero to V_0, and finally, after $3\Delta t$, the third inverter changes state from V_0 to zero. This now forces the first inverter to change state so that after each $3\Delta t$ the output changes state. Given N inverters, each with a delay Δt, the time per half-cycle is $N\Delta t$. Therefore the frequency of oscillation is

$$f = \frac{1}{2N\Delta t} \quad [\text{Hz}]. \tag{11.42}$$

A typical inverter may have a delay of, say, 10 ns. For a three-inverter ring, the frequency is 16.67 MHz. Slower inverters or a larger number of them will produce a lower frequency.

This oscillator is very simple, but it is not controllable other than by selection of the inverters. A better design, one that exists in a multitude of forms, is shown in **Figure 11.64b**. The delay is now controlled through the charge and discharge of the capacitor C through the resistor R_1. The design shown produces a square wave at frequency (see **Example 11.19** for details)

$$f = \frac{1}{2.1972 R_1 C} \quad [\text{Hz}]. \tag{11.43}$$

The inverters are triggered when the input voltage rises above about $V_{cc}/2$. The resistor R_l and capacitor C form a charging circuit. Suppose first that the left gate is on (zero at the input, V_{cc} at the output). The second gate must be off so that its output is zero. Now the capacitor charges with the time constant RC and after a time t_0 will trigger the left gate to change state. Now its output is zero and the capacitor discharges through R_1. The total time constant for charge and discharge defines the frequency.

A third type of square wave oscillator is even simpler and is shown in **Figure 11.64c**. It consists of an inverting Schmitt trigger, a device that flips its output from high to low when the input increases above a voltage V_h and from low to high when the input falls below V_h. Suppose the input to the inverter is low and hence the output is high. The capacitor charges through the resistor to a voltage V_h, at which point the output goes to zero. The time needed to do so is t_c. Now the resistor is connected to ground (output is low)

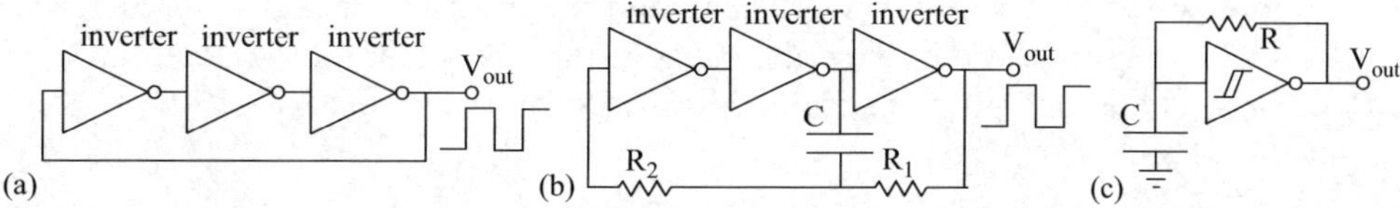

FIGURE 11.64 ■ Square wave oscillator based on the delay in feedback defined by the time constant RC.

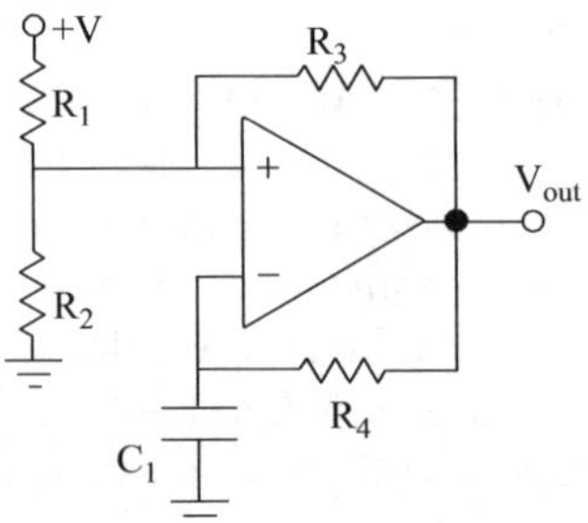

FIGURE 11.65 ■ Square wave oscillator based on charging and discharging a capacitor.

and the capacitor discharges through the resistor until it reaches V_l after a time t_2, at which point the output goes high and the process repeats. The frequency of the square wave is

$$f = \frac{1}{RC(\ln V_1/V_0) + \ln(1 - V_h/V_0)} \quad [\text{Hz}]. \tag{11.44}$$

The circuit in **Figure 11.65** is somewhat similar. The positive feedback through R_3 sets the level at which the amplifier changes state. R_4 and C_1 form the charging/discharging circuit. Suppose that V_{out} is high. Now the positive input will be set at a value that depends on R_3, R_2, and R_1. C_1 charges through R_4. When the voltage at the negative input exceeds that at the positive input the output goes negative and now the capacitor discharges through R_4, repeating the process (see **Problem 11.44**).

EXAMPLE 11.19 Square wave inverter oscillator

Consider the oscillator in **Figure 11.64b**. Assume that the inverters change state at approximately $V_{cc}/2$, where V_{cc} is the supply voltage of the inverters.

a. Based on charging and discharging the capacitor, calculate the frequency of the oscillator.

b. Suppose now that the inverters used change output state as follows:

   1. From low to high when the input is at $V_{cc}/2$.
   2. From high to low when the input is at $2V_{cc}/3$.

c. Discuss the consequences of the results in (a) and (b).

Solution:

a. We start by assuming that the capacitor is fully discharged, the output of the third inverter is at V_{cc}, and its input is at ground level. This means the input to the first inverter is zero. The equivalent circuit is shown in **Figure 11.66a**. The capacitor charges until it reaches $V_{cc}/2$. The resistor R_2 has very little effect on the charging and discharging of the capacitor, as the current into the input of the leftmost inverter is negligible. At this point the inverters change state and the equivalent circuit is that of **Figure 11.66b**. Now the capacitor discharges until the potential on R_1 goes below $V_{cc}/2$ or the potential on the capacitor discharges from $-V_{cc}/2$ to $+V_{cc}/2$:

$$\left(V_{cc} + \frac{V_{cc}}{2}\right)e^{-t_1/R_1C} = \frac{V_{cc}}{2} \rightarrow e^{-t_1/R_1C} = \frac{1}{3}.$$

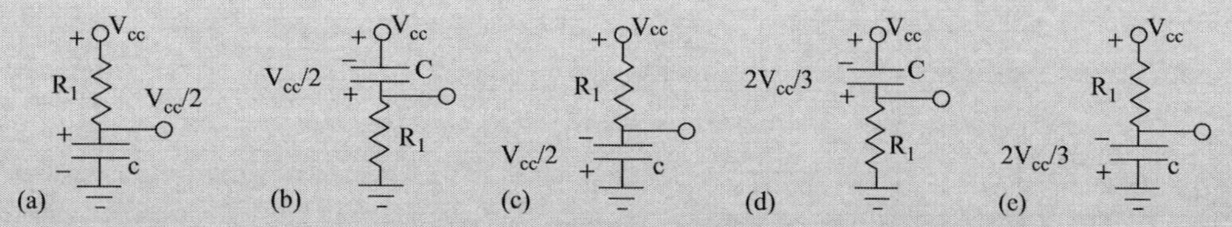

FIGURE 11.66 ■ Equivalent circuits for the ring oscillator in **Figure 11.64b** at various steps in the oscillation process.

Taking the natural log on both sides:

$$-t_1 = R_1 C \ln\left(\frac{1}{3}\right) \rightarrow t_1 = 1.0986 R_1 C.$$

This is the time during which the output of the circuit is low.

Now that the input on the leftmost inverter has gone below $V_{cc}/2$, all three inverters change state again, with the output of the rightmost inverter going high again, as in **Figure 11.66a**, except that now the capacitor changed to $V_{cc}/2$. The equivalent circuit is shown in **Figure 11.66c**. The configuration is exactly the same as in the previous step, so now the timing is

$$t_2 = 1.0986 R_1 C,$$

where t_2 is the time during which the output of the circuit is high. The sequence above repeats indefinitely, alternating between t_1 and t_2. The initial charge of the capacitor to $V_{cc}/2$ only occurs when the circuit is started, because only at that time will the capacitor be fully discharged. The total time is the sum of the two times and hence the frequency is

$$f = \frac{1}{t_1 + t_2} = \frac{1}{2.1972 R_1 C} \quad [\text{Hz}].$$

b. Following exactly the sequence in (a), we start with a totally discharged capacitor and high output. The capacitor charges to $2V_{cc}/3$ when the output goes low. The output stays low until the capacitor discharges from $-2V_{cc}/3$ to $+V_{cc}/2$ (see **Figure 11.66d**):

$$\left(V_{cc} + \frac{2V_{cc}}{3}\right) e^{-t_1/R_1 C} = \frac{V_{cc}}{2} \rightarrow e^{-t_1/R_1 C} = \frac{3}{10}$$

or

$$-t_1 = R_1 C \ln\left(\frac{3}{10}\right) \rightarrow t_1 = 1.204\, R_1 C.$$

Now the output goes high and stays high until the capacitor discharges from $-V_{cc}/2$ to $+V_{cc}/3$ (**Figure 11.66e**):

$$\left(V_{cc} + \frac{V_{cc}}{2}\right) e^{-t_2/R_1 C} = \frac{V_{cc}}{3} \rightarrow e^{-t_2/R_1 C} = \frac{1}{4.5}$$

or

$$-t_2 = R_1 C \ln\left(\frac{1}{4.5}\right) \rightarrow t_2 = 1.504 R_1 C.$$

The frequency now is

$$f = \frac{1}{t_1 + t_2} = \frac{1}{(1.204 + 1.504)R_1C} = \frac{1}{2.708R_1C} \quad [\text{Hz}].$$

Note also that the duty cycle changed from 50% in (a) to $(1.504/2.708) \times 100 = 55.54\%$.

c. Changing the trip voltages at which the inverters change state changes both the frequency and the duty cycle. The longer the capacitor has to charge or discharge, that is, the larger the difference between the high and low voltages at which the state changes, the lower the frequency. The duty cycle, defined as the ratio of time the output is high to the cycle time, depends on exactly what values are used to turn the inverters on and off. By controlling these voltages it is possible to change the duty cycle. Other circuits, such as that in **Figure 11.65**, are better in this respect, as the voltages are fully controllable, whereas in inverters they are typically defined by the internal circuitry of the inverter and vary from one inverter to another.

Examples of sinusoidal oscillators are given in **Figure 11.67a** and **11.67b**. These seem more complex, but again, the important point is that an *LC* circuit is used that oscillates at the required frequency and a feedback is provided from output to input. In **Figure 11.67a**, the feedback is through the lower part of *L*, whereas in **Figure 11.67b** it is through the lower half of the LVDT coil. These circuits will oscillate at

$$f = \frac{1}{2\pi\sqrt{LC}} \quad [\text{Hz}], \tag{11.45}$$

where $C = C_1$ in **Figure 11.67a** and $C = C_1C_2/(C_1 + C_2)$ in **Figure 11.67b**. L is the total inductance of the coil in either figure.

This allows the design of oscillators at almost any required frequency, although at higher frequencies the capacitances and inductances are necessarily low and the effects of stray and parasitic capacitances and inductances must be taken into account, as they can significantly affect the oscillation frequency and the performance of circuits.

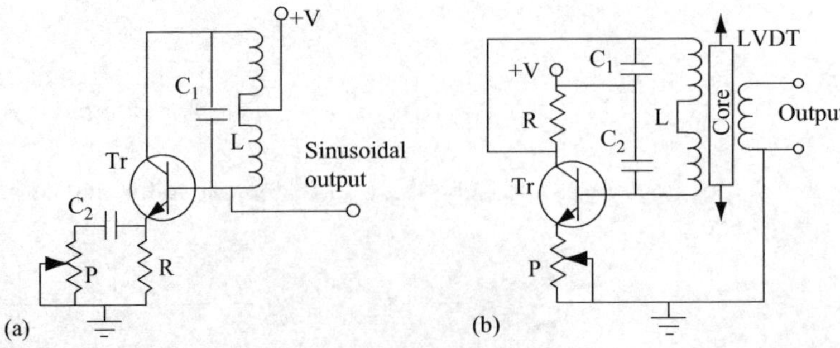

FIGURE 11.67 ■ Sinusoidal *LC* oscillators. (a) Feedback is provided through the lower part of the coil. The frequency is defined by the inductance *L* and capacitance C_1 in the collector circuit. (b) Sinusoidal *LC* oscillator driving an LVDT. Feedback is provided through the lower part of the primary of the LVDT. The capacitance *C* that defines the frequency of oscillation is that of C_1 and C_2 in series.

EXAMPLE 11.20 The metal detector

The metal detector is a common instrument used for diverse applications from "treasure hunting" to detection of nails in a wall to detection of buried mines. Although its main purpose is detection, it can also quantify the detected object by size or type of metal, provided, of course, a calibration curve is available. The common metal detector is based on the change in inductance of a coil in the presence of metal. Ferromagnetic objects increase the inductance of the coil, whereas conducting, nonferromagnetic objects decrease the inductance due to generation of eddy currents in the metal object. Nonconducting, nonferromagnetic materials cannot be detected. To detect and quantify an object, the detector relies on two oscillators. One is made of a relatively large external coil, called the search coil, and the second is an internal oscillator with adjustable frequency. The test starts by adjusting the internal oscillator to the same frequency as the external oscillator. The signals of the two oscillators are subtracted from each other in a mixer, producing a signal with zero frequency (**Figure 11.68**). When a metal object is present in the vicinity of the search coil, its inductance changes, the oscillator frequency changes accordingly, and the frequency at the output of the mixer is entirely due to the change in inductance, and hence due to the detected metal. For small objects in the immediate vicinity of the search coil, the change in inductance is linear with the distance from the coil and the volume of the object. In many metal detectors, the output from the mixer is amplified and fed to a loudspeaker. The tone heard is an indication of the detection, although with some experience one can usually distinguish between ferromagnetic and nonferromagnetic materials and between small or deep buried objects and larger or shallower objects. The output frequency can be measured and correlated with the detected metal, at least to some extent. To see how the instrument operates, consider the following: Assume that the two oscillators are tuned to exactly the same frequency when the search coil is in air and each coil has an inductance L, so that the output frequency is zero. Also assume that as the metal object approaches the center of the coil, the inductance of the coil changes by a value $\Delta L/\text{cm}^3$ for a ferromagnetic object, and by a value $-\Delta L/\text{cm}^3$ for a conducting, nonferromagnetic object.

a. Calculate the sensitivity of the sensor (output frequency per cubic meter of metal) in general terms.

b. Calculate the sensitivity numerically for $L = 500$ nH, $C = 0.001$ μF, and $\Delta L = 0.0001L$.

Solution:

a. The resonant frequency of each oscillator is

$$f = \frac{1}{2\pi\sqrt{LC}} \quad [\text{Hz}].$$

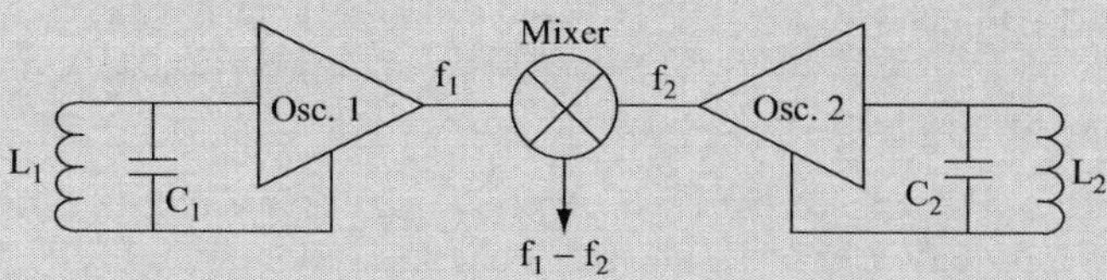

FIGURE 11.68 ■ (a) The principle of metal detection. The frequency out of the mixer is a measure of the metal detected. Oscillator 1 is a reference oscillator, whereas oscillator 2 is the search oscillator.

The frequency of the internal (reference) oscillator remains constant but the external oscillator frequency changes, since the inductance either increases or decreases by $v\Delta L$, where v is the volume of the metal object. We have for the two oscillators:

$$f_s = \frac{1}{2\pi\sqrt{L(1+v\Delta L)C}}, \quad f_r = \frac{1}{2\pi\sqrt{LC}} \quad [\text{Hz}],$$

where f_s is the frequency of the search oscillator and f_r is the internal or reference oscillator. The output frequency is

$$\Delta f = f_r - f_s = \frac{1}{2\pi\sqrt{LC}} - \frac{1}{2\pi\sqrt{(L+v\Delta L)C}} = \frac{1}{2\pi\sqrt{C}}\left[\frac{1}{\sqrt{L}} - \frac{1}{\sqrt{L+v\Delta L}}\right] \quad [\text{Hz}].$$

To calculate the sensitivity we divide the change in frequency by the volume of the object:

$$S = \frac{\Delta f}{v} = \frac{1}{2\pi v\sqrt{C}}\left[\frac{1}{\sqrt{L}} - \frac{1}{\sqrt{L+v\Delta L}}\right] \quad [\text{Hz/cm}^3].$$

b. For the values given:

$$S = \frac{1}{2\pi v\sqrt{1\times 10^{-9}}}\left[1 - \frac{1}{\sqrt{0.001\times 0.5\times 10^{-6}v}}\right] \quad [\text{Hz/cm}^3].$$

Since the sensitivity depends on volume, it is clearly nonlinear. Specifically the sensitivity for $v = 1\ \text{cm}^3$ is

$$S = \frac{1}{2\pi\sqrt{1\times 10^{-9}}}\left[\frac{1}{\sqrt{0.5\times 10^{-6}}} - \frac{1}{\sqrt{0.5\times 10^{-6} + 0.001\times 0.5\times 10^{-6}}}\right] = 355.85\ \text{Hz/cm}^3.$$

Sensitivity increases slightly with volume, but it is approximately 356 Hz/cm^3. This sensitivity should allow detection of reasonably small objects.

Notes:

1. The sensitivity obtained here is very good, primarily because f_0 is high (7.1 MHz). Lower frequencies may be easier to work with but will produce lower sensitivity.
2. Oscillators of this type are not very stable, although, assuming identical components, they should fluctuate more or less equally and any common-mode change will cancel out in the output.
3. A nonferromagnetic object will create the same change in frequency since mixers produce the difference between the high and low frequency, regardless of which is which, that is, one cannot have negative frequency.
4. This example assumed the depth is given. A change in depth reduces the change in inductance in the same way that volume reduces the change in inductance. Therefore the detector cannot distinguish between the two effects (i.e., a deep large object will produce the same output as a smaller shallow object).

11.9 | NOISE AND INTERFERENCE

We had some opportunity to talk about noise in sensors and, in a broad sense, noise is understood as anything that is not part of the required signal, that is, any signal that does not represent the stimulus. It is generally agreed that noise must be reduced as much as possible (elimination is not an option since noise cannot be entirely eliminated). However, more important than eliminating noise is to properly consider it in the design and in the specification of the sensor or actuator. For example, suppose that a temperature sensor generates 10 μV/°C and that a good microvolt meter is capable of reliably measuring 1 μV. This, taken on its own, would imply a resolution of 0.1°C. But if the noise (from all sources) is, say, 2 μV, one can assume that only signals above the noise levels are useful and any signal below 2 μV is useless. Thus the resolution cannot be better than 0.2°C. In many cases, things are worse than this, since the noise can only be estimated. When amplification occurs, noise is also amplified and the amplifier itself can add its own noise. Clearly then, noise cannot be ignored, even when it is small.

There are many sources and many types of noise. We will distinguish between two broad types by separating noise that comes from outside and noise that is inherent to the sensor. These will be termed **interference noise** and **inherent noise**, respectively.

11.9.1 Inherent Noise

Inherent noise is due to many effects within the sensor, some avoidable, some intrinsic. One of the main sources of noise in sensors is the thermal noise or Johnson noise in resistive devices. The noise power density is usually written as

$$e_n^2 = 4kTR\Delta f \quad [\mathrm{V}^2], \tag{11.46}$$

where k is Boltzmann's constant ($k = 1.38 \times 10^{-23}$ J/K), T is the temperature [K], R is the resistance [Ω], and Δf is the bandwidth [Hz]. This noise exists, for example, in simple resistors, and if the resistance is high, the noise can be very high. The Johnson noise is fairly constant over a wide range of frequencies and hence it is often called white noise. Note the units. The quantity e_n is in fact a voltage. In some cases **Equation (11.46)** is divided by Δf to obtain a noise power density given in V^2/Hz. General means of controlling this noise are suggested in **Equation (11.46)**: low temperature, low resistance, and small bandwidth.

A second type of noise is the so-called shot noise produced in semiconductors when a DC current I flows through a semiconductor device. The noise, produced by random collisions of electrons and atoms, is given as

$$i_{sn} = 5.7 \times 10^{-4}\sqrt{I\Delta f} \quad [\mathrm{A}]. \tag{11.47}$$

Although I is DC, the noise depends on a bandwidth Δf over which the noise is considered. Clearly the preference is for lower currents where this noise is concerned.

A third source of inherent noise is the so-called pink noise, which, unlike white noise, has higher energy density at low frequencies. This is a particular problem with sensors that tend to operate at low frequencies (slowly varying signals). The noise spectral density is $1/f$, and at low frequencies it may be larger than all other sources of noise.

Noise levels are very difficult to measure even when the noise is constant. Because it is not generally harmonic in nature, its RMS or even peak values are difficult to ascertain. The noise distribution is not constant (it is usually Gaussian) so that, short of sophisticated measurements, the best we can do is estimate the noise level.

11.9.2 Interference

By far the greatest source of noise in a sensor or actuator comes from outside the sensor and is coupled to it. This type of noise is called interference. The sources of interference can be many. Best known perhaps are the electric sources, including coupling of transients from power supplies, electrostatic discharges, and radio frequency noise from all electromagnetic radiative systems (transmitters, power lines, almost all devices and instruments that carry AC, lightning, and even from extraterrestrial sources). However, interference can be mechanical as well, in the form of vibrations, variations in gravitational forces, acceleration, and others, especially as these refer to mechanical sensors. Other sources are thermal (from temperature variations and the Seebeck effect in conductors), ionization sources, errors due to changes in humidity, and even chemical sources. Some errors are introduced in the layout of the sensors components or in the circuits connected to them through improper circuit design and improper use of materials. In general, electrical sources of noise are called electromagnetic sources (including static discharges and lightning) and are bundled together under the umbrella of electromagnetic interference or electromagnetic compatibility issues.

In some cases a noise is easily identifiable. For example, a common noise in 60 Hz electrical systems, especially those that contain long wires, is a 120 Hz noise (100 Hz in 50 Hz power systems) and is due to power lines. This type of noise is also a good example of a time-periodic noise. Other sources, especially when transient or random, are difficult to identify and hence to correct.

Noise, particularly interference noise, may affect different sensors differently. The simplest is an additive influence (see also **Section 2.2.5**), that is, the noise is added to the signal. The important issue is that the noise is independent of the signal and simply adds to the signal. Therefore additive noise is more critical at low signal levels since it tends to be constant. For example, drift due to temperature variations depends on temperature, but not on the signal level. This type of noise can be minimized by using a differential sensor in which two sensors are used so that one is exposed to the stimulus and both are exposed to the same external noise. Subtracting the two eliminates or at least minimizes the noise.

Another type of noise is multiplicative, that is, it grows with the signal and is due to a modulation effect of the noise on the signal. This noise is usually more pronounced at higher signal levels. The noise may be minimized by using two sensors as previously, but instead of subtracting the reference sensor's output, the sensing sensor's output is divided by the reference sensor's output (see **Section 2.2.5**). Suppose a stimulus is measured (say, pressure) and a noise due to a change in temperature ΔT is present and multiplicative. Assuming that the transfer function is $V = (1 + N)V_s$, where N is the noise function, one sensor senses both the stimulus and the noise and produces an output V_1, which is a function of the stimulus,

$$V_1 = (1 + \alpha\Delta T)V_s \quad [\mathrm{V}], \tag{11.48}$$

and the second sensor senses only the temperature and produces a voltage V_2,

$$V_2 = (1 + \alpha\Delta T)V_0 \quad [\mathrm{V}], \tag{11.49}$$

where V_0 can be assumed constant (i.e., it is only dependent on temperature change), then

$$\frac{V_1}{V_2} = \frac{V_s}{V_0}, \tag{11.50}$$

and since V_0 is independent of the sensed stimulus, the ratio is also independent of the noise. This is called a radiometric method and is most suitable for this type of noise.

Beyond reducing the noise at the sensor, it is prudent and often most effective to reduce the noise before it reaches the sensor. To do so we need to understand the means by which noise can reach a sensor. In terms of electrical noise there are really only four ways this can happen:

1. Noise can get into the sensor through direct resistive coupling, in which the source of the noise and the sensor share a common resistive path. This may be the resistance between the connections of a sensor or through the sensor's body. That is, the sensor is not electrically insulated from the source of the noise. The solution is isolation of the sources of noise (usually current-carrying conductors such as power lines) from the sensor. Often this will require that the sensor is floating electrically.
2. The second type is capacitive coupling. Since capacitance exists between any two conductors, capacitive coupling is very common. Any two wires, any two connectors, any two strips or pads can produce a stray capacitance that can cause coupling. Usually capacitances are small so that their AC impedances are high. This means that capacitive coupling is a problem only at higher frequencies. However, there are sensors, especially capacitive sensors, that use small capacitances to begin with. Any capacitive coupling may be too high for accurate sensing. In such cases the sensor must be electrostatically shielded from the sources that might couple noise. An electrostatic shield is usually a thin conducting sheet, sometimes a conducting mesh, that envelopes the protected area and is grounded (connected to the reference potential). In effect, this shorts the noise source to ground. An example is shown in **Figure 11.69**. The coupling capacitance is shorted, but this also creates a new capacitance between the protected device and ground. Nevertheless, the noise

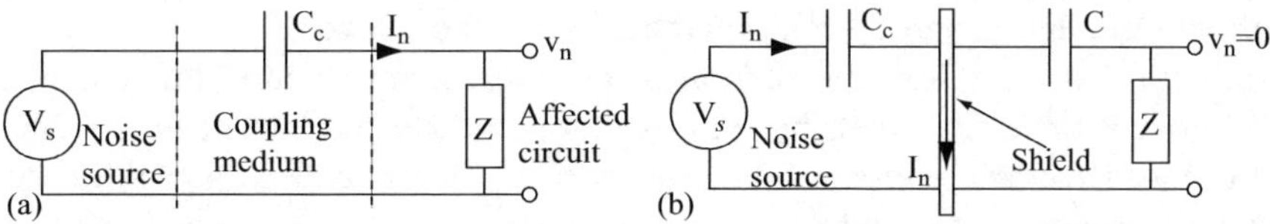

FIGURE 11.69 ■ Electrostatic shielding. (a) Unshielded circuit. (b) Shielding shunts the noise current to ground.

signal is zero (or greatly reduced). Cables leading to the sensor must also be shielded, but the most important point is that the shield must be at a constant potential. For example, shielding a cable and then grounding it at both ends, one to the electrostatic shield at the sensor, the other to the body of the controller, or even at two different locations on the same shield, will immediately produce a loop that may itself generate noise.

3. The third type of coupling is inductive. Inductive coupling is a particular problem between current-carrying conductors, such as between power lines and sensor conductors, and in particular the wires leading to the sensor. For example, the 100 Hz or 120 Hz noise from power lines usually links to sensors through inductive coupling. The solution here is two-fold. For high-frequency sources, a conducting shield just like the electrostatic shield should envelope the conductors leading to and from the sensor. The use of coaxial cables is such an example. This is based on the idea of skin depth (see **Chapter 9**) and simply takes advantage of attenuation of high-frequency fields in conductors. If the noise signal is very low in frequency, a magnetic shield is necessary. This is usually a relatively thick ferromagnetic shield (box) that envelopes the protected device to guide low-frequency (or DC) fields away from the sensor. Proximity sensors often use this type of shield, as was discussed in **Section 5.4.1.1**.
4. The fourth method by which electrical noise can interfere with sensors (and to a lesser extent, actuators) is through radiation or radiated interference. This is based on the fact that any conductor carrying AC is, in effect, a transmitting antenna. Any other conductor becomes a receiving antenna. If that conductor is part of a loop, a current will be induced in the loop. This noise is particularly large from sources of intentional emissions, such as radio and television transmitters, but can occur with any AC source. The reduction of this type of noise relies extensively on reduction in the lengths of wires and on reduction of the size (area) of loops. Shielding is also very effective in reducing radiated interference because many of the interfering sources operate at high frequencies. Other general precautions that must be observed are the use of decoupling capacitors in circuits and power supplies (reducing the AC impedance of the power supply) and twisting of the two wires leading to a device to reduce the area of the loop they form. Coaxial cables, if properly used, can reduce or eliminate most radiated interference. One common cure for many ills is the introduction of a ground plane—a sheet of metal under the circuit (such as a conducting sheet under a printed circuit board or a conducting layer on a multilayer printed circuit board). This helps in reducing the inductance of the circuit and hence will be effective in reducing both inductive coupling and radiated interference.

Mechanical noise, especially from vibrations can often be eliminated or reduced through isolation, but in some sensors, such as piezoelectric sensors, any force (due to acceleration) will produce errors in reading. These errors can be compensated through use of the differential or radiometric methods described above.

In addition to these sources of noise, there are many others. For example, any junction between different metals becomes a thermocouple and introduces a signal in the path. This may affect the reading of the sensor and is called Seebeck noise. It may not be a big problem in most cases, but it is when sensing temperature or when this signal adds to the signal from a stimulus. All in all, the issue of noise is both difficult and ill-defined. Often, finding the source of noise depends on detective work and experimentation.

11.10 | PROBLEMS

Amplifiers

11.1 **Design of amplifiers.** The voltage produced by a thermocouple varies between 0 and 100 μV for the span it senses. The output required for display is between 0 and 5 V. An operational amplifier with an open loop input impedance of 10 MΩ and an open loop gain of 10^6 is used for amplification. Devise a circuit, including the required resistors, to produce the required output.

11.2 **The subtracting amplifier.** Inverting and noninverting amplifiers can be combined into a single unit, as shown in **Figure 11.70**, to obtain a subtractor, that is, to obtain the difference between V_a and V_b.

a. Show that the circuit indeed produces an output $V_{out} = V_a - V_b$ if $R_1 = R_2 = R_3 = R_4$. To do so, first set $V_a = 0$ and calculate the output due to V_b. Then set $V_b = 0$ and calculate the output due to V_a. The superposition of the two outputs gives the required result. Recall that the current into the inputs of operational amplifiers is assumed to be zero.

b. How can the circuit be modified to obtain an output $V_{out} = 5(V_a - V_b)$?

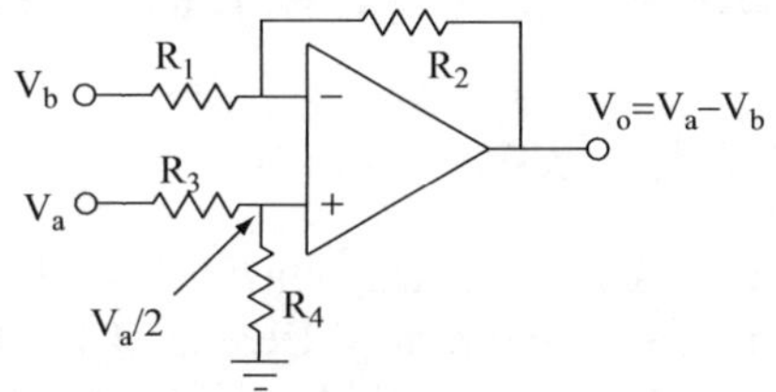

FIGURE 11.70 ■ A subtracting amplifier.

11.3 **Effect of input resistance of operational amplifiers.** A thermistor rated at 1 kΩ at 25°C and having a material constant of 3200 K is used to sense temperature in a vehicle. To reduce the effect of self-heating it is supplied by a 0.2 mA current source (a current source produces a constant current regardless of load and has very high input impedance). The temperature range expected is 0°C–50°C. Because the output must be inverted, an inverting operational amplifier is used to amplify the signal obtained from the thermistor to increase the signal. The circuit is shown in **Figure 11.71**.

a. Calculate the error in temperature introduced by the input resistance of the amplifier.

b. Where in the range is the error expected to be largest and why?

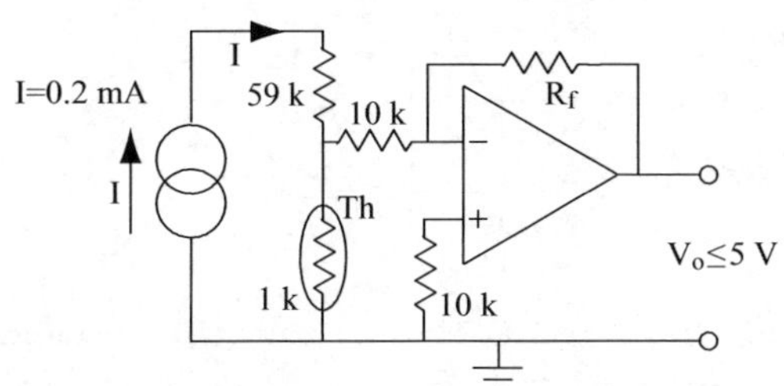

FIGURE 11.71 ■ An operational amplifier used to amplify the voltage on a thermistor.

The voltage follower

11.4 Use of a voltage follower. Consider again **Example 11.1**.

a. Calculate the effect on the input and output impedance if a voltage follower is added at the input (i.e., the microphone is connected to the input of the voltage follower and the output of the voltage follower is connected to the input of the first amplifier in **Figure 11.7**). Use the amplifier data in **Example 11.1**.

b. Calculate the effect on the input and output impedance if a voltage follower is connected at the output of the circuit (i.e., the input of the voltage follower is connected to the output in **Figure 11.7** and the output of the voltage follower becomes the new output of the circuit). Use the amplifier data in **Example 11.1**.

c. Are the changes significant enough to justify either configuration or perhaps both?

11.5 Inverting voltage follower. The voltage follower is a noninverting amplifier. Show how one can build an inverting voltage follower using two amplifiers. Show that with proper selection of components the input and output impedances of the new circuit are comparable to the noninverting voltage follower.

The instrumentation amplifier

11.6 The subtractor. Show how the instrumentation amplifier can be used to subtract two voltages V_a and V_b and obtain the difference $V_a - V_b$. Use the basic circuit in **Figure 11.9** with $R_1 = R_2 = R_3 = 10\ \text{k}\Omega$ as the starting circuit, with $R_G = 1\ \text{k}\Omega$. Show how the input voltages can be connected to an instrumentation amplifier to obtain an output exactly equal to $V_a - V_b$.

11.7 A single operational amplifier subtractor. Consider **Figure 11.72**. The two input voltages are assumed to have negligible internal resistances.

a. Show that the output equals $V_a - V_b$.

b. Discuss the input impedances seen by the sources V_a and V_b and the possible consequences of this circuit.

c. How would you modify the circuit so that both inputs will see high impedances?

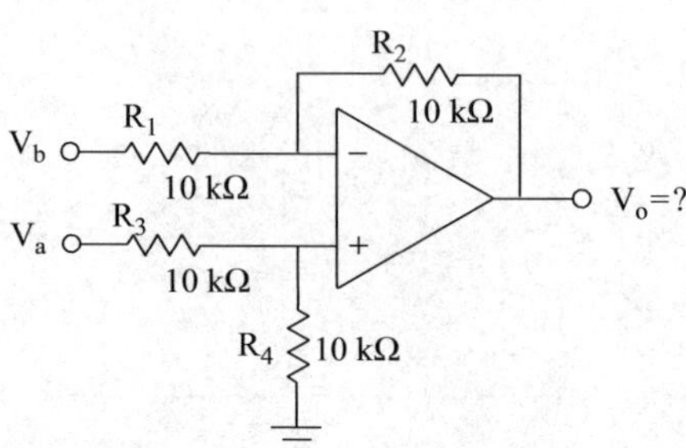

FIGURE 11.72 ■ Subtracting amplifier.

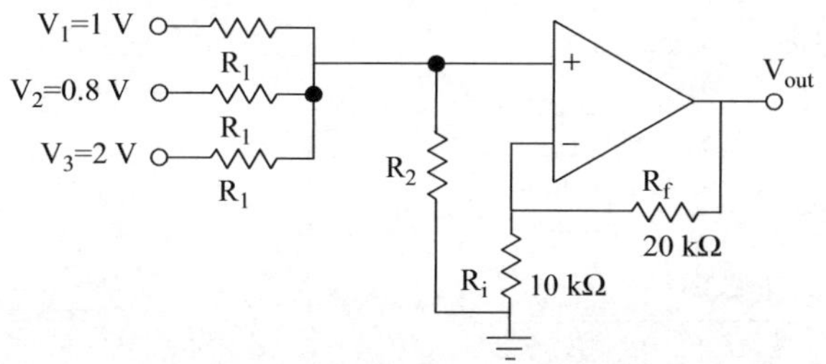

FIGURE 11.73 ■ Summing amplifier.

11.8 **The summing amplifier.** The operational amplifier can be used to add inputs. Consider the circuit shown in **Figure 11.73**, where three voltages must be added.

a. Show that the circuit is in fact an adder.
b. For the circuit given, find an appropriate combination of the resistors R_1 and R_2 that will produce an output exactly equal to $V_{out} = V_1 + V_2 + V_3 = 3.8$ V.

11.9 **The inverting summing amplifier. Figure 11.73** shows a summing amplifier with noninverted output. Show how an inverting summing amplifier can be built to produce an output equal to $V_{out} = -(V_1 + V_2 + V_3)$.

The comparator

11.10 **Digitization of AC signals: control of the duty cycle.** From **Example 11.3**, calculate the pulse width and the duty cycle obtained at the output using $R_1 = 10$ kΩ, $R_2 = 20$ kΩ, and $R_3 = 40$ kΩ. Explain how the pulse width, and hence the duty cycle, can be modified.

11.11 **Incremental display.** The comparator can be used to devise an incremental display as shown in **Figure 11.74**. N comparators are connected as shown. The positive inputs are all connected together and form a single input to which a voltage V_{in} is connected. The negative inputs are connected to a series of resistors, each equal to R, as shown (there are a total of $N+1$ resistors). The output of each comparator drives an LED through a current limiting resistor. Given the voltage $V_{DD} = 12$ V,

a. Explain what happens with the LEDs as V_{in} varies from zero to $V^+ = 12$ V.
b. Calculate the input voltages (V_{in}) at which each of the LEDs turns on or off.
c. What happens if the polarities of the inputs to all comparators are reversed?
d. What happens if the lower P comparators are connected as shown in the figure but the inputs to the rest (N–P comparators) are reversed?

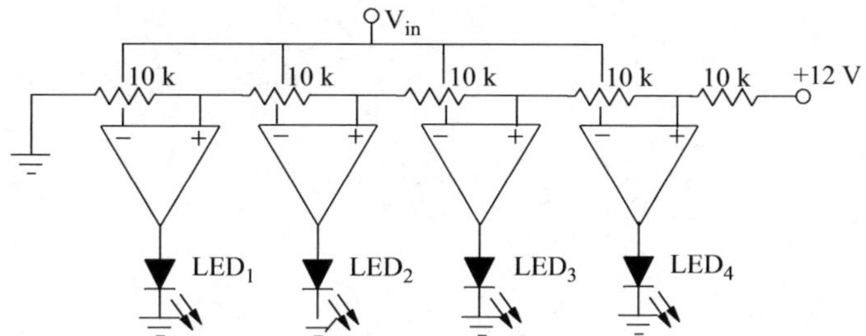

FIGURE 11.74 ■ Incremental display.

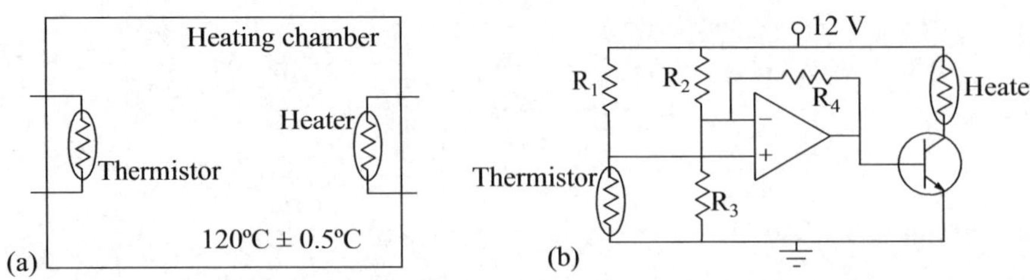

FIGURE 11.75 ■ Control of temperature in a chamber. (a) The chamber showing the heater and the thermistor. (b) The control circuit.

11.12 **Electronic thermostat.** A small chamber must be maintained at a constant temperature of 80°C ± 0.5°C. The circuit in **Figure 11.75** is used to do so. The thermistor serves as the temperature sensor and together with the comparator it forms a thermostat. The transistor is added to allow the large current needed for the heating element. The NTC thermistor has a material constant of 3500 K and a resistance of 10 kΩ at 20°C. Assuming that in the range given the material constant does not vary with temperature, select the resistors R_1, R_2, and R_3 to satisfy the requirements of the thermostat. Assume that the simple model of the thermistor is sufficiently accurate and, for best performance, that the switching on–off occurs with the reference voltage set at 6 V.

11.13 **Use of hysteresis in a comparator.** An electronic ignition in a car uses a Hall element and a rotating cam to generate the ignition pulses (see **Figure 5.39**). The output from the Hall element is sinusoidal with a peak value of 0.8 V. A four-cylinder, four-cycle engine rotates at 3000 rpm. The signal is first amplified to a peak value of 12 V and then digitized using a comparator with hysteresis using the circuit in **Figure 11.14b** (see also **Example 11.3**).

a. Calculate and sketch the signal obtained for $R_1 = R_2 = 10$ kΩ and $R_3 = 100$ kΩ.
b. Calculate and sketch the signal obtained for $R_1 = R_2 = 10$ kΩ and $R_3 = 10$ kΩ.
c. What are the conclusions from the results in (a) and (b)?

Power amplifiers

11.14 **DC motor controller.** A small DC motor is controlled through the use of a simple class A amplifier, as shown in **Figure 11.76**. The speed of the motor is

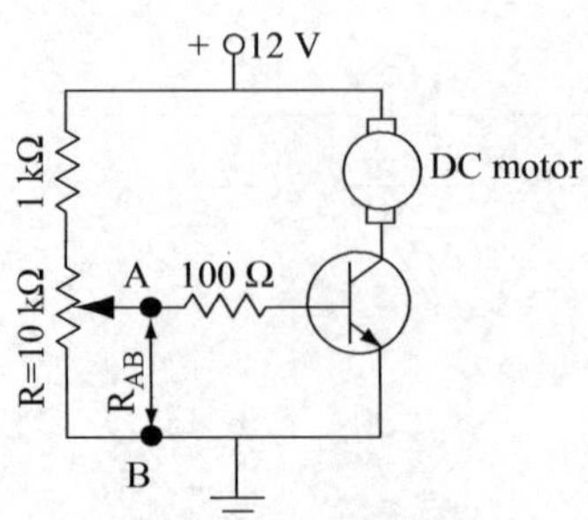

FIGURE 11.76 ■ DC motor speed control.

linearly proportional to the current and it can attain a maximum rated speed of 6000 rpm when the current in the motor is 450 mA. The transistor is an *npn* silicon transistor with a gain of 50 and a saturation current of 500 mA, for a basis current of 10 mA. The voltage drop between the base and emitter is 0.7 V. The speed of the motor is controlled using a linear potentiometer. Assume the transistor gain is constant over the range of currents used here. Calculate and plot the speed of the motor as a function of the position of the potentiometer slider, assuming it varies from 0 to an angle of 300°. What is the range of the potentiometer for speeds between 0 and 6000 rpm?

PWM and PWM amplifiers

11.15 PWM bidirectional motor controller. Using the PWM principle in **Figure 11.18** and the H-bridge in **Figure 11.20**, the speed of a DC motor can be controlled in both directions.

a. Draw a schematic of the pulses and show how the speed can be controlled and how the direction of rotation can be changed using a single PWM generator. The PWM generator is available as a stand-alone unit.

b. Show how the bridge circuit can be modified to affect braking of the motor. To brake the motor it is sufficient to short its connection either to ground or to the $+V$ rail.

11.16 PWM motor controller. Consider the configuration in **Figure 11.21**. The DC motor rotates at 10,000 rpm when connected to 12 V (no load speed). The source voltage is $V^+ = 12$ V and the timing circuit generates a triangular wave of amplitude 8 V and frequency 240 Hz. A 10 kΩ linear potentiometer controls the positive input of the comparator.

a. Calculate and plot the motor speed as a function of the position of the potentiometer. Assume the motor speed is linear with voltage and the "on" resistance of the MOSFET is negligible.

b. Discuss the effect of the frequency of the timing source. How does it influence the performance of the system?

Digital circuits

11.17 Multiple input OR gate. Show how a four-input OR gate can be implemented using

a. Only two-input NAND gates.

b. Only two-input NOR gates.

11.18 Multiple input XOR gate. It is required to build a three-input XOR gate. The XOR function of a two-input XOR gate with inputs A and B may be written as $A \oplus B = A\overline{B} + B\overline{A}$:

a. Verify that the function $A \oplus B = A\overline{B} + B\overline{A}$ indeed produces the truth table of a two-input XOR gate.

b. Find the function of the three-input XOR gate.

c. Show how the three-input XOR gate can be implemented using NAND gates.

11.19 **Delays due to gates and their effects.** Three identical signals are applied to the input of the three-input AND gate in **Figure 11.27c** and, separately, to the three-input AND gate in **Figure 11.28b**. The signals are repetitive pulses with a pulse width of 100 ns and a duty cycle of 50%. The delay at each gate is 20 ns.

a. Calculate and sketch the output signal in comparison with the input signal for the three-input gates in **Figure 11.27c** and **11.28b**.

b. What conclusions can you draw from the result in (a)?

11.20 **Gates as digital switches.** Gates can be used to switch signals. Using NAND gates,

a. Design a switch that can direct the input to one of two outputs based on a command signal C. If $C = 0$, the input appears on line A but not on line B. If $C = 1$, the input appears on line B but not on line A.

b. Design a switch with two inputs A and B, an output O, and a command line C. If $C = 0$, the signal on line A appears at the output O, whereas if $C = 1$, the signal on line B appears at the output O.

11.21 **A 4-bit serial-in, parallel-out (SIPO) shift register.** The shift register in **Figure 11.29** is a serial-in, serial-out (SISO) shift register. Show how one can make it into a SIPO shift register with output available only after data have been entered into the register. That is, how can one get the output in parallel after all the data have been shifted in, but not before that?

11.22 **Electronic timer/counter.** Consider the idea of building an electronic timer to display hours, minutes, and seconds in a 24-hour format. The input is 60 Hz, derived from the mains input. To accomplish this, 4-bit counters are used (these are available commercially as individual integrated circuits) capable of counting to $2^4 = 16$. The circuit must be capable of generating pulses at intervals of 1 s (1 pulse/s), at intervals of 1 min (1 pulse/min), at intervals of 1 h (1 pulse/h), and at intervals of 24 hours (1 pulse/24 h).

a. What is the smallest number of 4-bit counters needed?

b. Show how the 4-bit counters must be modified to produce the necessary signals.

A/D and D/A converters

11.23 **V/F converter.** The V/F converter in **Figure 11.34** is given. It uses a single-polarity comparator operating at 5 V. The Schmitt trigger inverter following it also operates at 5 V and changes output around 2.5 V. That is, when the input rises above 2.6 V, its output goes to zero, and when it dips below 2.4 V its output goes to 5 V (this hysteresis is critical for the operation of this circuit and is indicated by the hysteresis symbol on the inverter). The following variables are given: resistance $R = 100$ kΩ, capacitance $C = 0.001$ μF, and a MOSFET switch with an internal resistance of 250 Ω when its gate is positive and infinite resistance when its gate is zero. Assume ideal components in all other respects.

a. Find the relation between the input voltage V_{in} and the frequency of the output voltage.
b. Characterize the output voltage at an input voltage of 2 V; that is, give the waveform (amplitude, frequency, and duty cycle).
c. What are the components responsible for the width of the "on" and what are the components responsible for the width of the "off" parts of the waveform?
d. Discuss how the frequency range can be changed and what the limits are on this change.

11.24 **V/F converter.** A V/F converter is built as shown in **Figure 11.77**. The operation is as follows: The capacitor C is charged through resistors R_1 and R_2. A reference voltage is supplied to the positive input of CP_2. Initially its output is high since the capacitor is discharged. When the capacitor charges above V_{ref}, its output goes low. CP_1 is high if the capacitor voltage is above V_{in} and low if below. Thus initially the output of CP_3 is low and the transistor does not conduct. As the capacitor voltage rises above V_{in}, CP_1 changes its output to high and CP_3's output goes high. Now the transistor conducts and the capacitor discharges through R_2. This continues until the voltage on the capacitor goes below V_{ref}, when the output of CP_3 resets to zero and the charging process restarts. We will assume here that the comparators have a small internal hysteresis so that they change state to high if $V^+ > V^-$ and to low if $V^- > V^+$, but not when the two are equal. Given the components in the figure, and assuming an ideal transistor (i.e., it acts as a perfect switch) and ideal comparators:

a. Find the waveform across the capacitor as a function of the input voltage V_{in}.
b. Find the output frequency and duty cycle as a function of the input voltage V_{in}.

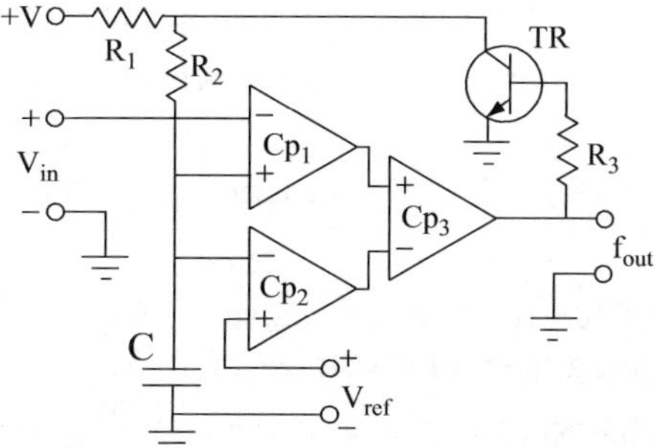

FIGURE 11.77 ■ V/F converter.

11.25 Experimental evaluation of a V/F converter. A V/F converter similar to the one shown in **Figure 11.35b** is evaluated experimentally by varying the input voltage and measuring the output frequency. The results are given in the table below.

a. Plot the data and find the sensitivity of the converter [Hz/V].

b. Find the maximum nonlinearity with respect to a linear best fit of the data. Compare the sensitivity of the linearized transfer function with that found in (a).

V_{in} [V]	2.75	3.0	3.25	3.5	3.75	4.0	4.25	4.5	4.75	5.0	5.25
f_{out} [Hz]	14760	16210	17804	19374	21005	22628	24252	25937	27570	29220	30941
V_{in} [V]	5.5	5.75	6.0	6.25	6.5	6.75	7.0	7.25	7.5	7.75	8.0
f_{out} [Hz]	32602	34278	35993	37680	39357	41077	42788	44458	46131	47858	49466

D/A converter

11.26 A 10-bit A/D converter. A 10-bit successive approximation A/D converter in a microprocessor operates at the clock cycle of the microprocessor, which is 2.5 MHz, and its reference is the 5 V supply of the microprocessor. Assume each operation (setting or testing a bit) takes one cycle. Using the configuration in **Figure 11.37**:

a. Sketch the output sequence for an analog input of 4.35 V and derive the digital output.

b. How long does it take to perform the conversion?

c. Suppose the reference voltage is not stable and reduces to 4.95 V. What is the output now and what is the error incurred due to this change in the reference voltage for the same input voltage (4.35 V)?

11.27 A 14-bit A/D converter using a 12-bit D/A converter. Suppose one wishes to build a 14-bit successive approximation A/D converter, but only a 12-bit D/A converter is available. Assume that the D/A converts 12 bits out of the word and ignores the LSB and the one next to it. Assume a reference voltage of 5 V.

a. Explain why the net effect is a 12-bit A/D converter.

b. Calculate the digital output for an analog input of 4.92 V.

c. What is the additional error incurred with respect to the envisioned 14-bit converter?

11.28 A 4-bit D/A converter. Show the structure of a 4-bit D/A converter based on a resistive network.

a. Select reasonable values for the resistors and show the output for the digital value 1101 for an operating voltage of 10 V.

b. What is the analog voltage step if the converter operates at 5 V?

11.29 A 14-bit D/A converter. In many applications D/A converters need to produce an output that approximates an analog signal with high resolution. In reproducing digital audio or in synthesizing signals it is imperative that the analog levels are very small. Consider a 14-bit D/A converter designed to reproduce digital audio. Show the structure of the D/A converter based on a resistive network.

a. Select reasonable values for the resistors and show the output for the digital value 11010100110110.

b. What is the analog voltage step if the converter operates at 5 V?

c. The step calculated in (b) may be considered as an amplitude distortion in the signal or as a noise. Calculate the noise level introduced by the conversion as a percentage of full scale.

d. Calculate the dynamic range of the digital signal and the dynamic range of the analog signal.

11.30 Errors in D/A conversion. The resistor network D/A converter relies on a resistive network to accomplish conversion. Necessarily the values of the resistors and variations in these values will introduce errors in the conversion. Consider a 4-bit D/A converter (see **Figures 11.38, 11.39** and **Example 11.7** for guidance) with $R = 10$ kΩ.

a. Sketch the structure of the converter and find the analog output for a digital input 1001.

b. Suppose because of production problems, all resistors marked as R are reduced to 8 kΩ and all resistors marked as $2R$ are reduced to 16 kΩ. Calculate the analog output and the error.

c. R is 10 kΩ and $2R$ is 20 kΩ except for the first $2R$ resistor connected to the noninverting input, which is reduced from 20 kΩ to 19.9 kΩ (a 5% reduction). Calculate the output and the error.

d. All resistors marked as R are off by 5% (each is 9950 Ω), whereas all resistors marked as $2R$ are 20 kΩ. Calculate the output and the errors.

e. Discuss the results in (a) through (d).

11.31 A 10-bit PWM D/A converter. A 10-bit PWM is used as a D/A converter, as in **Figure 11.40**. The PWM is set so that when all inputs are "0", the output pulse has zero width (i.e., there is no output pulse). When all inputs are "1" (highest digital input possible), the output pulse width is 100% (i.e., the output is DC). For any other value, the output pulse width is proportional to the input. The frequency of the PWM signal is 1 kHz.

a. Sketch the output of the PWM for digital input 1001101111. What is the duty cycle?

b. Calculate the analog voltage V_{out} for digital input 1100010011 if the pulse height is 5 V.

c. What is the expected error in the analog signal for a PWM pulse height of 5 V?

Bridge circuits

11.32 **Bridge sensitivity.** Obtain the sensitivity of the bridge in **Figure 11.42** if $Z_1 = Z_2 = Z_3 = Z_4 = Z$ and if both Z_2 and Z_3 are sensors whose resistance decrease by dZ due to change in the stimulus. The other two impedances are resistors that do not change with the stimulus.

11.33 **Unbalanced bridge sensitivity.** Consider the bridge in **Figure 11.78**. In this bridge two RTDs are used as sensors, sensing temperature, but the bridge is not balanced. Suppose the two sensors both change (increase) by 2% due to a change in temperature. Calculate the sensitivity of the bridge.

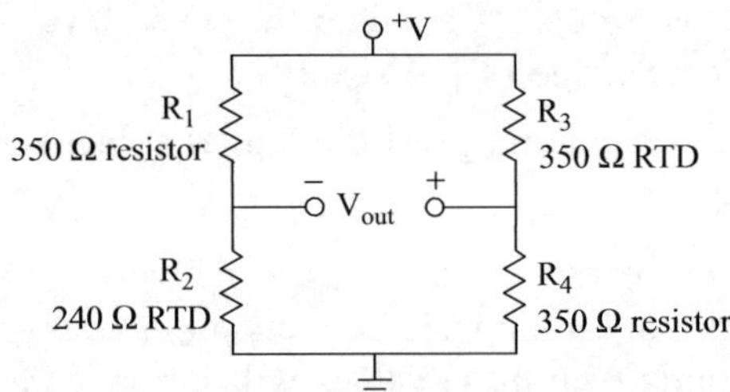

FIGURE 11.78 ■ Unbalanced bridge.

11.34 **A four-sensor bridge.** A bridge can have all four branches act as sensors to increase the output and hence the sensitivity. Consider a force sensor as shown in **Figure 11.79**. Force is measured by sensing the strain in the bending beam using four strain gauges. The upper two strain gauges are under tension, whereas the lower two are under compression. For the lower strain gauges to sense compression strain they must be prestrained, and if initially the four sensors are identical, after prestressing, the bridge is not balanced. This can be taken care of by introducing a bias in the bridge circuit or by using sensors with different nominal resistance for the prestressed sensors so that under no force all four sensors have the same resistance. For the purpose of this problem, assume that the bridge is balanced when no force is applied. Under these conditions, and assuming all four sensors are identical, show that the sensitivity of the bridge is four times as high as that of a bridge with a single sensor.

Note: This configuration also offers temperature compensation since all four sensors are at the same temperature.

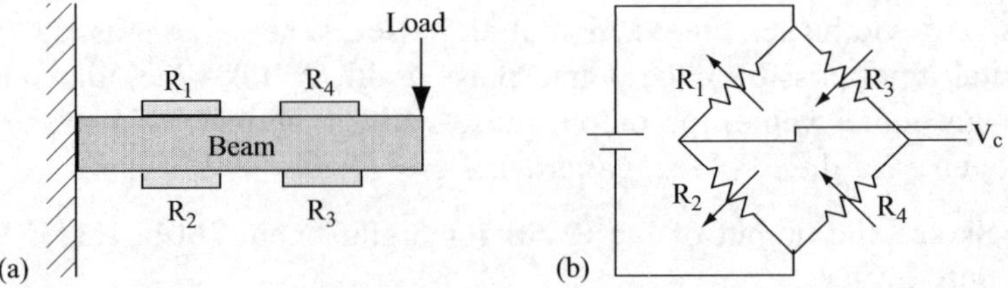

FIGURE 11.79 ■ (a) Force sensor with four strain gauges. (b) The bridge connection of the strain gauges.

11.35 **Temperature and leads compensation in a bridge.** Consider the bridge circuit in **Problem 11.34**.

a. Show that the output is not affected by temperature variations as long as all four sensors are at the same temperature.

b. Calculate the effect of the lead resistance of the sensors on the output. Assume all leads connecting the sensors are of the same length but the resistances of the four sensors are different.

c. Show that if all sensors are identical in resistance and the leads are of the same length, the lead resistance of the sensors has no effect on the output.

d. Show that the output is not affected by temperature or lead resistance as long as R_1 and R_2 are at the same temperature and have the same lead resistances and R_3 and R_4 are at the same temperature (but different than that of R_1 and R_2) and have the same lead resistances (but different than the lengths of the leads of R_1 and R_2).

Linear power supplies

11.36 Maximum efficiency of linear regulators. Consider the regulator in **Figure 11.51**.

a. What is the maximum possible efficiency of the regulator at a maximum current of 1 A, assuming that the mains voltage can vary by ±10%? How can this be achieved?

b. To increase efficiency, the regulator is replaced with a new-generation low-dropout (LDO) regulator that requires only 250 mV potential drop under the same conditions as in (a). Assume the regulator ratings are the same except for the potential drop. What is now the maximum efficiency and how does one need to modify the circuit to achieve it?

11.37 Efficiency of linear regulators. For the design in **Figure 11.51**, obtain the efficiency as a function of load current. Neglect all losses except those in the diodes and the series regulator. Show that efficiency is independent of current.

Switching power supplies

11.38 Charge pump power supply. A five-stage charge pump power supply (see **Figure 11.53**) is built with inverting gates that can operate from a 1.5 V battery. To maximize the output and increase efficiency, the circuit uses Schottky diodes on which the voltage drop is 0.2 V. The switching frequency of the inverters is 100 Hz and the capacitors are 0.1 μF each.

a. Calculate the output voltage the circuit can produce from a 1.5 V battery.

b. If a 3.3 V LDO is connected at the output to regulate the voltage and the LDO has a voltage drop of 100 mV, calculate the maximum current that can be drawn from the regulator and still maintain 3.3 V on the load. Assume that the switching timing of the inverters (10 ms) is sufficiently slow to fully charge the capacitors.

c. What is the answer to (b) if the switching frequency is 1 kHz?

11.39 Voltage control in a switching power supply. A switching power supply is built as in **Figure 11.80**. The PWM generator produces pulses at 10 kHz with pulse widths equal to

$$t_{on} = 10 + 16V_o, \quad t_{off} = 90 - 16V_o \quad [\mu s].$$

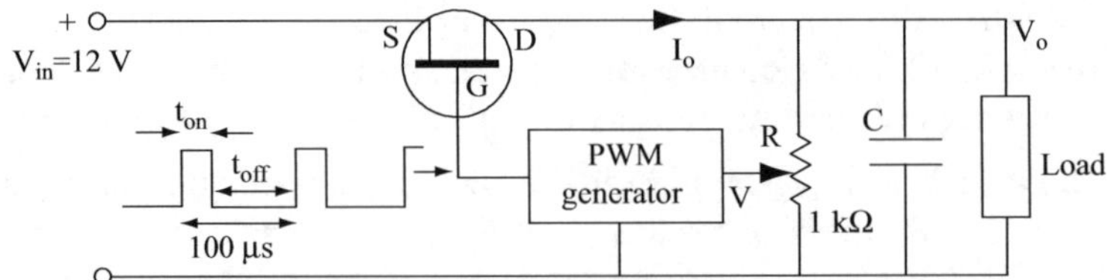

FIGURE 11.80 ■ A switching power supply.

The output voltage is controlled by the potentiometer R.

a. Calculate the range of output voltages that can be obtained by varying the potentiometer from 0 to 1 kΩ.
b. Explain how the circuit operates and how the output is regulated.
c. Suppose that the output is set and regulated at 5 V and the load draws 1 A. Calculate the power dissipated on the MOSFET.

LC and RC oscillators

11.40 LVDT oscillator. The LVDT oscillator shown in **Figure 11.67b** has the following components: $C_1 = C_2 = 0.1$ μF, $R = 10$ kΩ, $V^+ = 6$ V, and a transformer ratio of 1:1 (i.e., the number of turns in the primary equals the number in the secondary). The inductance of the LVDT coil is 150 mH when the moving core is centered with the coil and the current drawn by the transistor is 10 mA. The output frequency is measured as an indication of displacement.

a. Calculate the resonant frequency of the oscillator.
b. The maximum linear range of the LVDT is ±20 mm. If the maximum change in inductance of the coil is ±12% for the maximum displacement, calculate the range of frequencies of the oscillator and the sensitivity of the LVDT.

11.41 Ring inverter oscillator. A CMOS inverter has an input–output delay that depends on the operational voltage linearly. The delay is 8 ns at $V_{cc} = 15$ V and 17 ns at $V_{cc} = 3$ V. A ring oscillator with seven inverters is built.

a. Calculate the oscillator frequency as a function of V_{cc} and the minimum and maximum frequency in the range given for V_{cc}.
b. Suppose V_{cc} varies by ±5% due to poor regulation. What is the change in frequency one can expect as a percentage of the "correct" frequency at the two limits (15 V and 3 V)?

11.42 Schmitt trigger oscillator. Consider the Schmitt trigger oscillator in **Figure 11.64c**. The Schmitt trigger used changes states at input voltages equal to $V_{cc}/3$ and $2V_{cc}/3$. For a resistor $R = 10$ kΩ and capacitor $C = 0.001$ μF:

a. Calculate the frequency of oscillation.
b. Calculate the duty cycle (i.e., the ratio between the time length the output is high and the time length the output is low).

11.43 Schmitt trigger oscillator. Consider the Schmitt trigger oscillator in **Figure 11.64c**. The Schmitt trigger used changes states at input voltages equal to 0.8 V and 1.6 V with an operational voltage $V_{cc} = 5$ V. For a resistor $R = 33$ kΩ and capacitor $C = 4.7$ nF:

a. Calculate the frequency of oscillation.

b. Calculate the duty cycle (i.e., the ratio between the time length the output is high and the time length the output is low).

11.44 Operational amplifier-based square wave oscillator. Given **Figure 11.65**:

a. Calculate the oscillator frequency and its duty cycle.

b. Select components to obtain an oscillator operating at 10 kHz.

Noise and interference

11.45 White noise in operational amplifiers. An operational amplifier is used in an audio system to process audio signals with a bandwidth of 20 kHz. The open loop input resistance of the amplifier is 800 kΩ and its open loop gain equals 200,000. The amplifier is used in two modes: one as an inverting amplifier and one as a noninverting amplifier, both with a gain of 200. Assume that the input to the amplifier behaves as a resistor.

a. Calculate the white noise level at the output of the inverting amplifier at ambient temperature (30°C) if nothing is connected to the input. The feedback resistor (R_f in **Figure 11.6a**) is 200 kΩ and R_i is 1 kΩ.

b. Calculate the white noise level at the output of the noninverting amplifier at ambient temperature (30°C) if nothing is connected to the input. The feedback resistor (R_f in **Figure 11.6b**) is 199 kΩ and R_i is 1 kΩ.

11.46 Effect of noise on Hall element sensors. A Hall element sensor is used to sense low-intensity AC magnetic fields. To do so a semiconductor Hall element with a Hall coefficient of 0.02 and chip thickness of 0.1 mm is connected as in **Figure 5.38**. To accomplish sensing, a current of 5 mA passes through the element and the Hall voltage is measured across the sensor.

a. Calculate the sensitivity of the sensor to magnetic fields and the lowest magnetic flux density that can be reliably sensed for a sinusoidal magnetic flux density at a frequency of 60 Hz, accounting for noise.

b. In an attempt to increase sensitivity, the current in the sensor is increased to 20 mA. What is now the sensitivity and what is the lowest magnetic field that can be reliably sensed, accounting for noise?

CHAPTER 12

Interfacing to Microprocessors

Sensory perception and the brain

The brain is the center of the nervous system and the ultimate processor in the body. Made of some 100 billion neurons, it has many functions ranging from reasoning and thought to memory and self-awareness, and much more. The brain is also the center where much of the sensory data are processed and where motor "commands" originate (some sensory data are processed in ganglia, including the spine). Many of the functions of the brain are only understood from external cues and some are merely assumptions. Some, such as self-awareness, the idea of mind and even emotions, cannot be explained on the basis of neurons. Others, including sensory and motor functions, are well understood and associated with specific structures in the brain.

Most of the neurons are found in the cerebral cortex, a thick layer that covers the brain and is structured in characteristic folds. The cerebral cortex is layered and divided into four lobes—the frontal lobe (in the front-top of the head), the parietal lobe (behind the frontal lobe toward the back of the head), the temporal lobe (on both sides), and the occipital lobe (behind the parietal lobe at the back of the head). A finer division into a large number of cortical areas is based on cell structure and function, with each cortical area associated with particular functions of the body. Lengthwise the brain is divided into two halves, roughly equal, and with replicated structures and functionality. As a whole, the left side of the brain controls the right side of the body, and vice versa. Below the cortex are other structures, including the thalamus, the hypothalamus, the hippocampus, the cerebellum (in the back lower part of the head), the brain stem, and the corpus callosum (a large bundle of nerves connecting various parts of the brain).

Functionally there are three main areas. The first is sensory in nature and involves the thalamus as well as parts of the lobes. The visual area is located in the occipital lobe, the auditory area is in the temporal lobe, and the somatosensory area is in the parietal lobe. Taste and smell are associated with the brain stem and areas above it. The second area is primarily motoric and involves the rear portion of the frontal lobe in conjunction with the brain stem and the spinal cord. The remaining parts of the cortex seem to be involved in the more complex functions of thought, perception, and decision making.

Some sensory functions in humans are more developed than others with corresponding developments in the brain. Other sensory areas seem to be more primitive. The visual area of the brain is perhaps the most developed of the purely sensory areas, whereas the sections responsible for smell seem to be more primitive, as indicated by their association with the brain stem. Motoric functions follow the same pattern, with some functions (like the hand) being more developed, whereas some motoric functions are autonomous and controlled by the spinal cord.

12.1 | INTRODUCTION

In **Chapter 1** we indicated that sensors are input devices and actuators are output devices to controllers. It is now time that we talk about the controller as it relates to sensors and actuators—that is, our discussion will necessarily focus on input to and output from controllers.

What constitutes a "controller" will vary from application to application. In some cases the controller may be no more than a switch, a logic circuit, or an amplifier. In others it may be a complex system that may include computers and other types of processors such as data acquisition and signal processors. More often, however, the choice is on microprocessors. We shall therefore focus the discussion here on microprocessors as a general purpose, flexible, and reconfigurable controller and the ways sensors and actuators relate to these. In fact, microprocessors are often called microcontrollers, but as with sensors and actuators, it is rather difficult to classify them in a simple class. What a microprocessor is, what the difference between a microprocessor and a computer or a microcomputer is, and how a distinguishing set of features is arrived at are all subjective issues. What a microprocessor is to one may be a full-fledged computer to another.

For our purposes, a microprocessor will be viewed as a stand-alone, self-contained single-chip microcomputer. For this to apply, it must have a central processing unit (CPU), nonvolatile and program memory, and input and output capabilities. A structure that has these can be programmed in some convenient programming language and can interact with the outside world through the input/output (I/O) ports. But there are other less obvious requirements. Clearly, for a self-contained system, the microprocessor must be relatively simple, reasonably small, and hence limited in most of its features—memory, processing power and speed, addressing range, and of course, the number of I/O devices it can interact with. Unlike computers, the designer must have access to most features of the microprocessor—the bus, memory, registers, and all I/O ports. In short, the microprocessor is a mere component with flexible features that the engineer can configure and program to perform a task or a series of tasks. The limits on these tasks are only two: the objective limitations of the microprocessor and the imagination (or capabilities) of the designer.

Nevertheless, the basic question of what constitutes a microprocessor has only partially been answered and perhaps cannot be fully and adequately answered. For the purpose of this discussion we will narrow it down to 8-bit microprocessors since these are some of the simplest and are commonly used in sensor/actuator systems, and because they are representatives of all microprocessors (16- and 32-bit microprocessors are also in common use, but the principles involved in interfacing are essentially the same). Even within these there are a number of architectures being used. That is less important to the discussion here, and we will emphasize the Harvard architecture because of its simplicity, flexibility, and popularity. However, this architecture, though common, should be viewed as an example.

12.2 | THE MICROPROCESSOR AS A GENERAL PURPOSE CONTROLLER

In the following sections we will discuss the elements of the microprocessor as a general purpose controller, focusing on those that are important for interfacing. The architecture, addressing, clock and speed, programming, internal devices, memory, I/O, peripherals, and communication will be discussed with a view to the main subject of the chapter, that of interfacing to the microprocessor. The discussion will remain as general as possible, that is, we will try to keep away from specific microprocessors or manufacturers. Nevertheless, especially in examples, it will become necessary to discuss specific terms that may be associated with a specific line of microprocessors from a specific manufacturer. Again, it will be as general as possible without indicating manufacturer or specific part numbers. The reader should view these as representative of other microprocessors and understand that other microprocessors may achieve an identical function or a similar function by different means. Although the specifics vary from one part number to another, and certainly from manufacturer to manufacturer, the issues addressed will be general and apply, with the proper allowance, to all microprocessors.

12.2.1 Architecture

There are about two dozen manufacturers of microprocessors, based on a few architectures. We shall only briefly describe here one popular architecture—the Harvard architecture—used in many microprocessors. The main features of this architecture are separate buses for program memory and operand memory and a small instruction set. This pipelined architecture allows for the retrieval of data while another operation executes. That is, each cycle consists of retrieving the $(n+1)$th instruction while executing the nth instruction. The bus widths vary depending on the manufacturer and on the microprocessor size. **Figure 12.1** shows the bus architecture for a particular device as an example. The data bus is 8 bit, hence its designation as an 8-bit microprocessor. The processor can load 64 kbytes of program into the program memory on the 16-bit instruction bus and can address $2^{15} = 32$k instructions on the 15-bit program address. It can access up to $2^{12} = 4096$ bytes of operand memory on the 12-bit operand address bus, although normally a small portion of the operand memory is reserved for the processor and the user has no access to it. It should be noted that often the limits of the bus are higher than what the actual device supports. For example, the fact that the device shown can access 4096 bytes of operand memory does not imply that it has that amount of memory, only that this is the upper possible limit for that device.

The various bus widths vary from device to device depending on its size and features. A small microprocessor with 512 bytes of memory will only require a 9-bit program address bus ($2^9 = 512$), whereas a large device with, say, 128 kb will require a 17-bit program address bus ($2^{17} = 128$ kb $= 131{,}072$ bytes). The instruction bus must be equal to or wider than the program address bus. The data bus width remains 8 bits for 8-bit microprocessors.

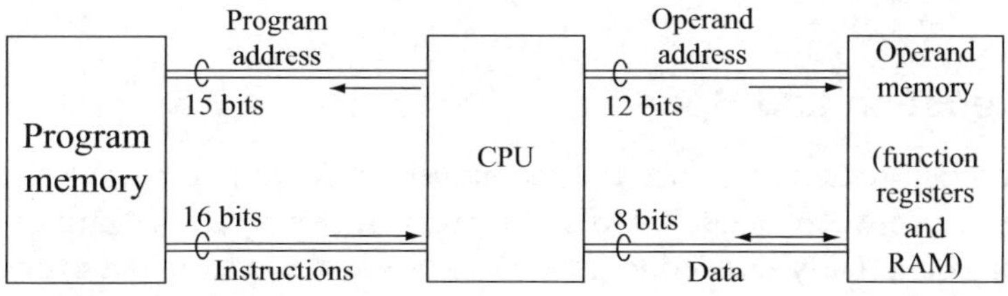

FIGURE 12.1 ■ Bus structure for a sample 8-bit Harvard architecture microprocessor.

The architecture for an 8-bit microprocessor supports direct addressing for the first 8 bits of address space and indirect addressing (variable pointer addressing) for all memory space. The architecture includes a CPU with associated status bits and a set of special functions registers. The latter contain all of the required registers to control the I/O ports, all other peripherals (such as comparators, A/D converters, PWM modules, etc.) as well as timers, status indications, and much more, all available to the user. User writable registers are also provided. Because microprocessors have been designed to respond to specific needs, it is not unusual to find modifications that address these needs, even if this requires deviation from the basic architecture. For this reason, various processors from the same family may have larger or smaller instruction sets to cater to the needs of the processors. The instruction set of most microprocessors varies from about 30 to 150 instructions. Most microprocessors are reduced instruction set computing (RISC) devices.

Memory in microprocessors is also adapted to the specific needs imposed by its use. On the lower end, one can find microprocessors with as little as 256 bytes of memory, whereas there is no specific limit on the high end other than what can physically fit in the device and what is deemed commercially advantageous by manufacturers. In most cases microprocessors have both volatile and nonvolatile memory. The number of peripherals also varies from device to device. Some small devices may include no peripherals at all, whereas larger devices may include dozens of peripherals, including comparators, timers, A/D converters, capture/compare, PWM, communication ports, and other useful functions, and often multiple units of the same peripheral.

The microprocessor communicates and interacts with the outside world through I/O pins and these vary from as few as 4 to 100 or more. Packaging varies from 6 pins to more than 100 pins and the devices come in various configurations (dual in-line packaging [DIP], various surface mount packages, dies, etc.).

12.2.2 Addressing

An 8-bit microprocessor has a word length of 8 bits. That means that integer data from 0 to 255 may be represented directly. To address memory, usually a longer word is needed. Most microprocessors have a 10-bit (1k), 12-bit (4k), 14-bit (16k), or 16-bit (64k) memory address, but longer address words are in use with larger memories. Addressing in microprocessors typically means program addressing. That is, the program memory contains the program to run and it is the instructions of this program that are being addressed. Of course, there is data addressing as well, but data memory is usually very small in comparison to program memory. The size of the program memory defines how long a program can be and, to a large extent, what the microprocessor can accomplish. In essence, program memory defines what can be done and delimits the complexity of the interfacing program. Since microprocessors, by their very nature, are "small" in all respects, including addressing space, efficient programming and efficient use of internal resources is a hallmark of microprocessor applications.

12.2.3 Execution and Speed

Another important issue is the speed of the processor. Most microprocessors operate at oscillator speeds between 1 and 100 MHz. In many microprocessor families (but not all) the oscillator is internally divided to generate a clock, the instruction cycle; that is, the time it takes to execute an instruction is slower than the oscillator speed with typical instruction clocks values between 1 and 20 MHz (1 μs to 50 ns per instruction). But speed is not only a function of the clock or oscillator frequency, how a task is performed can have a significant impact on execution speed. For example, suppose an analog voltage needs to be compared with a reference voltage to detect when the analog voltage is larger or smaller than the reference voltage. One could conceivably convert both voltages to digital representation using A/D converters and perform the comparison as part of the program. But A/D conversion takes a considerable number of instructions. If a comparator is available, the two voltages can be compared directly (within one or two clock cycles) without the need for conversion, speeding up execution of the task.

The basic time unit on a microprocessor is the clock cycle. That has a significant influence on interfacing issues. Nothing, no matter how simple, can be done in less than one cycle. Interfacing requires a program to run, and therefore even the most trivial task will need a few instructions and hence a few cycles to execute. The limitation of execution time can also introduce errors, and these must be taken into account when interfacing sensors and actuators. For example, if a microprocessor operating at 1 MHz requires, say, 30 instructions to activate a switch that turns off a device, and if each instruction is executed in one cycle, it will take at least 30 μs to turn off that device. This delay may be significant and must be taken into account in the design of the system in which the actuator is only one part.

12.2.4 Instruction Set

Microprocessors have a small instruction set—sometimes no more than two to three dozen simple instructions. These are selected to cover the common requirements of programming the device and, in combinations, allows one to perform any task that can be physically performed within the basic limitations of the device. These instructions include logical instructions (AND, OR, XOR, etc.), move and branching instructions (allow one to move data from and to registers and conditional and unconditional branching), bit instructions (allow operation on single bits in an operand such as setting a bit to zero), arithmetic instructions such as add and subtract, subroutine calls, and other instructions that have to do with the performance of the microprocessor, such as reset, sleep, and interrupt. Some instructions are bit oriented, some are byte (register) oriented, and some are literal and control operations. **Table 12.1** lists the various types of instructions with some examples. The limited instruction set means that the user has to do things that are not necessarily intuitive. For example, multiplying a digital number by two is most easily done by shifting the number one position to the left, whereas dividing it by two means moving it to the right one position. On the other hand, multiplying a number by, say, six may require shifting the digits to the left one position to multiply by two and then adding the result to itself three times. Of course, the user is often not aware of how these processes are done unless the instructions are used directly in a low-level language such as the Assembly language. If one chooses this path, each instruction is executed sequentially and the user has direct control of all steps in the most minute details. More often, however, the user will use a high-level language such as C to write more compact and more easily readable and understood programs. These application programs implement the various instructions in machine language and the user does not need to know the exact details. Special purpose

TABLE 12.1 ■ Instructions in microprocessors

Instructions	Examples	Notes
Logic instructions	AND, OR, XOR, 1's, 2's COMPLEMENT	Some generate a carry as well as set other flags for subsequent use.
Integer math instructions	Add, Subtract	Carry and other flags are generated.
Counting and conditional branching	Increment, Decrement, Decrement/Skip, Bit Test/Skip	The main means of creating loops and branching out of loops. Skipping is based on some detectable condition.
Clear and set operations	CLEAR register, CLEAR Watchdog timer, CLEAR bit, SET bit,	Allow manipulation of registers and bits within registers.
Unconditional branch	GOTO, Return, Return From Interrupt	GOTO is a general branch instruction, Return is used to return from a subroutine after its execution.
Move operations	Move to/from register	Allows placing of data for operations.
Other instructions	No operation, Sleep	
Shift operations	Shift left, Shift right	Digits are shifted left or right (through carry).

Notes:

1. Obviously, many more instructions and specialized use instructions exist depending on the microprocessor.
2. A series of flags are set/reset upon completion of an instruction. These may indicate if the operation has resulted in a zero or a negative value or if the 8-bit register has overflowed (carry). These are used internally or by the programmer.
3. Most operations require a single cycle to complete (logic and math instructions, for example). Some may require two or three cycles (branching instructions, for example).
4. The Clear Watchdog timer is issued to prevent the watchdog timer from resetting the program at regular intervals. The timer may also be disabled entirely, but under normal operation its purpose is to ensure the program does not get "stuck" in an unintended instruction or loop.

compilers are then used to "translate" the execution flow into the form required by the processor (machine language).

EXAMPLE 12.1 Programming and execution on microprocessors

As part of a program it is necessary to perform the following operation: $a = 6b + c$, where b and c are integers. Assuming that the numbers are small enough so that all results can be done in 8 bits without overflowing the register, show how this can be done and the time needed for the operation for a 1 MHz clock. Assume each instruction takes one clock cycle. Use $b = 17$ and $c = 59$ as numerical values.

Solution: The result is obviously very simple, and for the numerical values given $a = 161$. But the microprocessor cannot do that directly. Rather, it will perform the following sequence. Somewhere at the beginning of the program the three variables are assigned to three registers that then can be read and written by the CPU.

1. Move the value of b into a work register, w: $w = 00010001$ (17).
2. Shift the contents of w one position to the left: $w = 00100010$ (34).
3. Move contents of w back into register b: $b = 00100010$ (34).

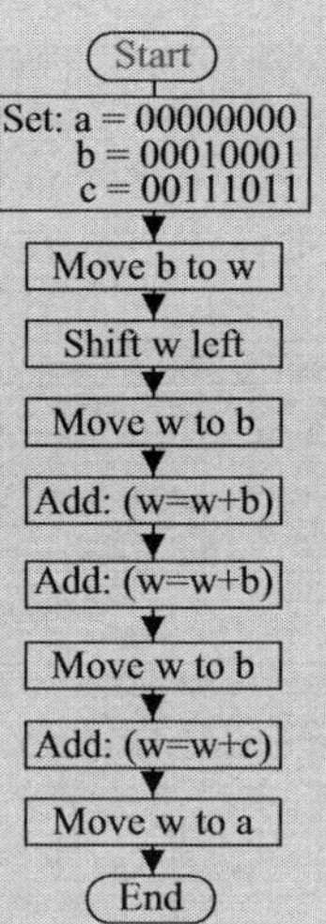

FIGURE 12.2 ■ Flow chart of the program in **Example 12.1**.

4. Add contents of b and w, leave the result in w: $w = 010001000$ (68).
5. Add contents of b and w again, move result in b: $b = 01100110$ (102).
6. Add contents of c and w, leave result in w: $w = 10100001$ (161).
7. Move result from w to register a: $a = 10100001$ (161).

After seven cycles the result is available in register a and can be used for whatever purpose necessary, such as displaying it on a screen or using it to perform additional operations. It takes 7 μs to do so (assuming that no other operations are necessary).

A high-level language such as C will only require one line of code and will look as follows:

$$a = 6 * b + c.$$

Of course, the variables a, b, and c will first have to be declared as integers at the beginning of the program.

Figure 12.2 shows a flow chart of the sequence above. The simple program described here does not require a flow chart, but in most cases a flow chart is a useful step in the programming process. It serves as a "road map" and is invaluable in finding errors in the program, especially since programming is a coding process that is difficult to follow without a guide. The flow chart may be very detailed, it may be just a sketch of the logic employed, or it may be composed of multiple interconnected charts.

Note: This example is very simple and is not written in Assembly, but each line corresponds to one instruction. In reality, some instructions may require more than one cycle to execute and there may be a need to add instructions to branch, to check that the values do not overflow, etc.

12.2.5 Input and Output

Input and output is defined by the availability of pins on the package. Microprocessors are usually limited to about 100 pins (6, 8, 14, 18, 20, 28, 32, 40, 44, 64, and 100 pins are common). Two pins are used to power the device so that, for example, an 18-pin device can have no more than 16 I/O pins. Of these, some may be used for other purposes, such as oscillators or communication, so that usually fewer pins are available for I/O functions. Nevertheless, even a modest microprocessor will have a significant number of pins available for I/O. For example, a 6-pin microprocessor may have as many as 4 I/O pins, while a 64-pin processor can have in excess of 48 I/O pins. I/O pins are grouped into ports, each addressable as an 8-bit word so that each group has up to 8 pins. Different ports may have different properties and may be able to perform different functions. Almost without exception, I/O are tristate, enabling an I/O pin to serve as input, output, or to be disconnected. Most I/O pins are digital, but some may be configured (in software) as analog. I/O pins can supply or sink considerable current, usually in the range of 20–25 mA. This is not sufficient to drive many actuators, but it can drive low-power devices directly or indirectly through switches, relays, and amplifiers. I/O ports are typically assigned to registers and can be addressed, changed, and manipulated in a manner similar to any variable. But ports have additional properties. One of the most important is that they can maintain their status while the processor is in sleep mode. This allows the microprocessor to issue an output on a port and then go into sleep mode, maintaining the output unchanged but consuming little power. Of course, to change the ports the processor must wake up from sleep. I/O ports (although not necessarily all ports or all pins in a port) can also issue interrupts on a change. That is, when the voltage on an I/O pin declared as an input changes, an interrupt to the processor is issued and that wakes it up from sleep to handle the change on the input or interrupts a current task to take on a higher priority task. This capability is again important for power consumption and for functionality. Input pins can also be "pulled up" internally (in software), meaning that the input pin now is connected to V_{DD} through a large resistance. The net effect is that the pin is set at V_{DD} and can detect an input such as a switch connected between that pin and ground when the switch is closed. Although this is a general purpose feature, it is particularly useful in interfacing switches, keypads, and keyboards, but it can also be used with resistive sensors.

EXAMPLE 12.2 Automatic light on upon entry

Many public spaces only require lighting when a person is present, but for safety reasons the light must be on even if a person may not want to turn it on. In such cases an automatic system is used based on a PIR sensor that turns on the light and keeps it on for a predetermined time, after which the light is turned off unless motion is still detected. The PIR sensor detects changes in heat produced by the entry of a person (see **Section 4.8.1** and **Example 4.11**). Upon entry, the PIR sensor detects the infrared radiation emitted by the person entering and produces a small voltage ΔV. This must be amplified to produce a usable voltage. **Figure 12.3a** shows a schematic implementation. The amplifier is selected to ensure that entry forces the output into saturation, that is, the amplifier operates in open loop mode and really only has two states—zero and V_{dd}. Also, because the PIR sensor has high impedance, it is important that the amplifier's input impedance is high. A low-drift FET input amplifier is preferable. This allows the use of a digital input on the pin marked as "1". To minimize power use, the processor is in sleep mode and the pin is set for "interrupt on change." When a person enters, the voltage on pin 1 changes from zero to, say, 5 V. This initiates an interrupt on change, the processor turns on, starting its program, following the interrupt. The lightbulb turns on by setting the output pin "2" high. The transistor conducts and the relay switches the light on. The transistor is necessary since the microprocessor cannot supply enough current to turn the relay on. The timer starts and counts to a time T_0 (5 min).

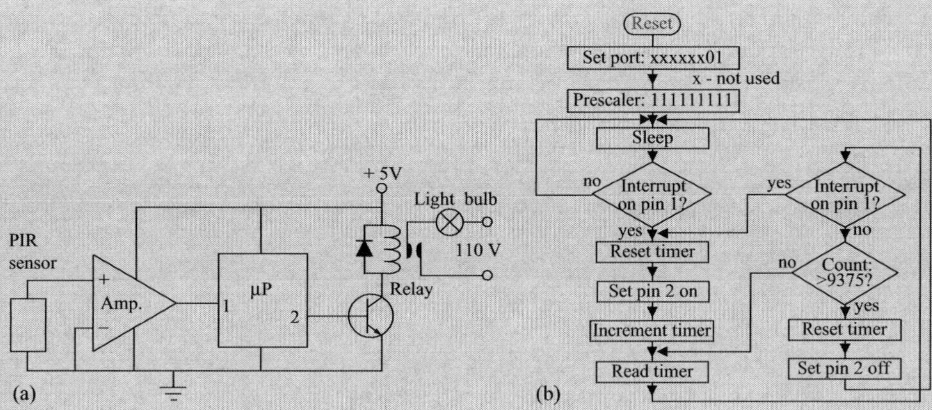

FIGURE 12.3 ■ (a) Schematic of an automatic light-on upon entry. (b) Flow chart of the sequence of operations used to develop a program for the microprocessor.

If no additional motion is detected during the 5 min interval, the light is switched off (by turning off output pin 2). The processor then goes to "sleep" to conserve energy.

Note as well the use of the diode across the relay coil. Its purpose is to protect the transistor from excessive transient currents when the relay is switched off. In many MOSFETs as well as some power transistors, the diode shown is built into the device. The relay also serves to isolate the low voltage circuit of the microprocessor from the high voltage circuit needed to operate the lightbulb.

The following is a simplified sequence of instructions that will allow the processor to react to the sensor and control the light. A flow chart of the sequence is shown in **Figure 12.3b**.

1. Declarations:
   a. Pin 1 set to digital input, pin 2 to digital output.
   b. Oscillator set to minimum frequency possible—we will assume a 32 kHz oscillator (see calculation below), divided internally by 4 for a clock of 8 kHz.
   c. The largest prescaling available is used—divide the clock by 256 (see note 2 below).
2. Start.
3. Set processor to sleep mode. Nothing happens until an interrupt is detected on pin 1.
4. If an interrupt is detected, exit sleep and enter the following loop.
   a. Reset the timer (see note 3 below).
   b. Set output pin on (to 5 V, relay switches light on).
   c. Read timer register (timer counts up and updates continuously as the clock runs).
   d. Has an interrupt been detected?
      (i) Yes: go to 4a to restart the count.
      (ii) No: continue.
   e. Has the count for 5 min (300 s; see below) been reached?
      (i) Yes: reset the timer, turn pin 2 off (light off), go to (3) and wait for next entry.
      (ii) No: go to 4c.

Some data: With a 32 kHz oscillator (a standard frequency available on some microprocessors), after internal division by 4, the clock cycle is 8 kHz. The cycle time is 125 μs. After the prescaler, the time step at the counter/timer is

$$\Delta t = 125 \times 256 = 32,000\ \mu\text{s}.$$

The timer counts in increments of 32 ms. Thus 5 min requires that the counter/timer counts to

$$N = \frac{5 \times 60}{0.032} = 9375.$$

In digital form this is 10 0100 1001 1111 and requires 14 digits to represent. Therefore what is needed is a 16-bit counter/timer (either available in the microprocessor or built-in software). At the end of the count the 16-bit register will read 0010 0100 1001 1111. When this value is reached (detected by the software) the light is switched off and the processor goes to sleep.

Notes:

1. In an application of this type there is an advantage to using a low-frequency oscillator so that the count is relatively low.
2. Most microprocessors have a prescaler that divides the clock frequency by a ratio that can be defined in software. Typically the prescaler can be adjusted from a divide by 2 to a divide by 256 (8 bits). The following section further discusses the prescaler.
3. In an 8-bit processor, even though a 16-bit counter/timer is available, operations are 8 bit. Therefore, detecting if the count has reached is done in two steps. First, the high byte (8 most significant bits) is tested. If this matches, the low byte is tested, and if both match the light is turned off. Alternatively, one can disregard the low byte and turn the light off when the high byte matches or at a count of 9344. This will turn the light off after 299 s, or just 1 s shy of 5 min. Timers are discussed in the following section.

12.2.6 Clock and Timers

The microprocessor must have a timing mechanism that defines the instruction cycle. This is defined by an oscillator that may be internal or external. Usually an RC oscillator is used for internal oscillation, whereas a crystal is the most common way of setting the frequency externally (this requires either dedicated pins or the use of two I/O pins). This frequency is often divided internally to define the basic cycle time. In addition, microprocessors have internal timers that are under the control of the user and are used for various functions that require counting and timing. At least one counter is available, but larger microprocessors can have four or more timers, with some being 8-bit timers (for 8-bit microprocessors) and some 16-bit timers. In addition, a watchdog timer is available for the purpose of resetting the processor should it be "stuck" in an inoperative mode. It counts to a fixed value, and if it reaches the set count, it restarts the program. A timer in a microprocessor is a special purpose register that can be connected to the clock, usually through a prescaler (a divider set by the user in software). The register can be read, can be reset, and can overflow to indicate a full register. Prescaling is software controlled (i.e., the prescaling level can be set, usually from 2 to 256 in increments that depend on the microprocessor). Some microprocessors also have postscaling, or in some cases the user can select either pre- or postscaling. Timers can, of course, be created in software if they are not available, if additional timers are needed, or if longer timers become necessary.

EXAMPLE 12.3 Control of frequency in a power generator

A 120 V AC portable power generator must operate at a fixed 60 Hz frequency. The frequency is regulated by the rotational speed of the generator. To generate the feedback signal needed for the speed control, a small microprocessor measures the frequency and generates a signal proportional to the variation in frequency. Show how this can be done using, as much as possible, internal components of the processor. The given microprocessor operates at a frequency of 10 MHz with an internal division by 4 to generate the internal clock. The microprocessor has an internal comparator and a reference voltage that can be set at 16 equal levels by dividing the power supply voltage. The microprocessor also has a 12-bit A/D converter and an 8-bit timer. An 8-bit I/O port can be set as digital inputs or outputs or as analog inputs or outputs if the comparator or A/D converter is used. Pins can be set individually. The prescaler can be set from 2 to 256, or it may not be used.

The output from the generator is used to power the microprocessor through a transformer and a simple 5 V regulator, as shown in **Figure 12.4a**. A zener diode (the resistor limits the current in the diode to 5 mA) samples the AC voltage and generates a square wave of proper amplitude (5 V) and polarity. The AC voltage sampled is 9 V RMS (12.7 V peak). Note that although the pulse itself is smaller than a half-cycle, the distance between the rise of two consecutive pulses is exactly one cycle (**Figure 12.4b**). Therefore the basic principle here will be to measure the time between the rise of two consecutive pulses. The flow chart in **Figure 12.4c** shows a possible approach to sensing the frequency variation and controlling the speed. The program detects the rising edge of a pulse and counts the clock until a second rising edge is detected. At that point the clock is read and its register compared with the time of a 60 Hz cycle. If the counted time is shorter, the frequency is too high and the engine is slowed down by command from O/p 2. If the count is too long, the frequency is too low and the engine is sped up by command from O/p 1. The process repeats indefinitely to control the frequency continuously.

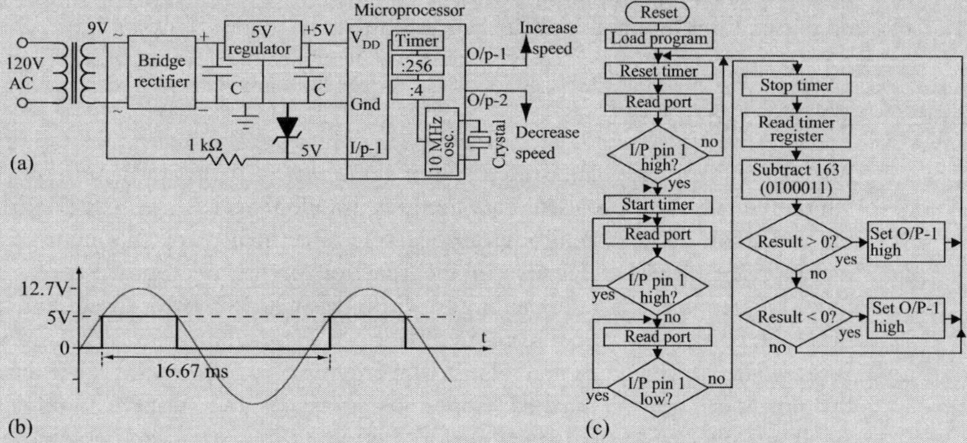

FIGURE 12.4 ■ Frequency control of an engine-driven generator. (a) Circuit showing the power supply and the frequency sampling method. (b) The frequency sampling process. (c) Flow chart of the program.

Internally the clock runs at $10/4 = 2.5$ MHz. The clock cycle is therefore $1/2.5 \times 10^6 = 0.4\,\mu s$. The prescaler divides the clock by 256 so that the frequency at the timer is 9765 Hz, or alternatively, the pulse width is 102.4 μs. That is, every 102.4 μs the timer advances by 1.

The 60 cycle signal has a width of 16.67 ms. That means that the count of the counter after one cycle will be $16.67/0.1024 = 163$, or in digital form, 010100011. This is the comparison value. If the timer value is higher, the engine needs to speed up, if lower, it needs to slow down.

Note: The transformer is needed not only to reduce the voltage to a level that can be easily handled, but also for safety reasons. The flow chart gives the basic steps needed to accomplish the sensing of frequency changes, but the program itself is much more detailed than the chart implies.

12.2.7 Registers

The execution of commands as well as control over the functions of microprocessors, including addressing, is done through registers. There are two types of registers in a microprocessor. The first type is called special purpose, special function, or reserved registers and is dedicated to processor use. These are used to set and control various functions such as the ports, oscillators, flagging, status indication, all peripherals, and some internal functions not available to the user. Most are available to the user and can be modified within given parameters to affect their use. For example, to set a pin in an I/O port as output, the specific bit of that I/O pin must be set to zero. Similarly, to read the status on the I/O pins, the corresponding port register is read just like a data register. The second type of register is called general purpose and it serves as (volatile) memory. It is to these registers that variables used in programming are assigned in a manner similar to the way a variable is defined in a computer program, except that when using registers, the variable is assigned to a specific register by address. Register space is very limited. Some small microprocessors may have no more than a handful of registers available for general use, whereas other, larger devices may have hundreds, but in all cases the number is relatively small. It is not unusual therefore to reuse register space in programs by assigning more than one variable to a specific register, provided that they do not conflict (are not used at the same time).

12.2.8 Memory

Most modern microprocessors, contain three types of memory: program memory, in which the program is loaded, data memory (random access memory [RAM]), and, with the exception of very small microprocessors, electrically erasable permanent read only memory (EEPROM). Because of the function of the microprocessor as a controller, program memory is usually the largest, from less than 256 bytes to more than 256 kbytes depending on the specific device. In most cases the memory is flash memory, meaning it is rewritable at will and is nonvolatile (the program is retained until rewritten or erased) so that power can be disconnected without losing the program. Data memory is usually quite small and may be a small fraction of the program memory (i.e., one-eighth or less) and does not retain data upon removal of power. It can be used for intermediate data retention during execution. EEPROM is nonvolatile rewritable memory used mostly to write data during execution and to retain data when the processor is off. The memory is considered a read only memory (ROM) because it cannot be written externally—only the microprocessor can change the data in it and only through its program. EEPROM is important in a number of situations. It can be used to retain data needed for the program, such as a look-up table, but it can also be written dynamically during the execution and will retain the data for future use. This may be as simple as a count or a code, or it may retain records entered in it, such as the time and date something happened. For example, the EEPROM may retain the time and date a door was opened or, say, the last 20 s of a vehicle's parameters, serving as a "black box" in case of a crash (see also **Example 12.4**).

EXAMPLE 12.4 Speed sensing and odometer in a car—the use of EEPROM

The speed of cars is sensed by counting the number of turns of the drive wheels. In most cases this simply means counting the number of teeth on a toothed wheel on the transmission shaft and relating that number with speed given the wheel diameter. The number of turns per kilometer is also known and that defines the distance traveled. Suppose for the sake of this example that for every turn of the wheel, there are 20 pulses—that is, a toothed gear with 20 teeth is placed on the wheel drive shaft and the wheel has a diameter of 75 cm.

Figure 12.5 shows a schematic of the system. A Hall element backed by a magnet counts the number of teeth as they pass by it. The Hall element output is analog, resembling a sinusoidal signal. This is entered as the positive input to an internal comparator while the negative input of the comparator is connected to a reference voltage internally. The result is that every time the output of the Hall element is higher than the reference voltage, its output goes high, and when the positive input goes below the reference voltage, the comparator's output goes low, thereby digitizing the Hall element output. An internal timer counts these pulses for a convenient period of time to display speed. A possible count is as follows: The wheel has a diameter of 0.75 m and hence a circumference (*WC*) of

$$WC = 2\pi\frac{d}{2} = \pi \times 0.75 = 2.3562 \text{ m}.$$

Since there are 20 pulses/turn, the number of pulses per meter is

$$Pulses/meter = \frac{20}{2.3562} = 8.48825.$$

The speed is usually displayed in kilometers per hour (km/h). Suppose the time reference is 1 s, that is, the counter samples the pulse train for 1 s. A count of 8.48825 pulses/s would indicate the

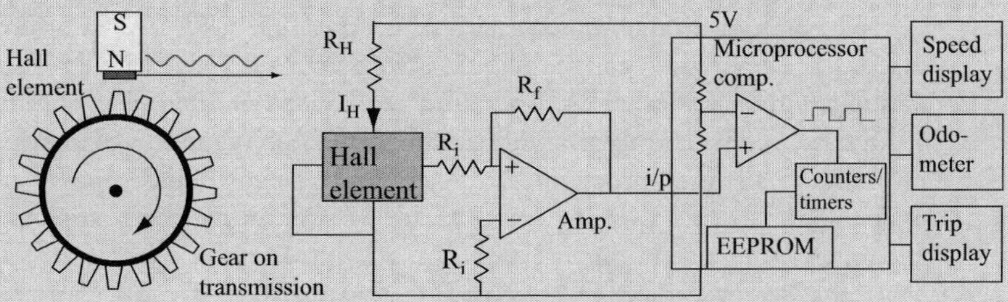

FIGURE 12.5 ■ The components of a digital speedometer/odometer in a car.

car has moved 1 m in 1 s, or that its speed is 1 m/s. More precisely, 1 count/s represents 1/8.48825 = 0.1178 m/s, or 0.424 km/h. The microprocessor can perform these calculations, scale the output as necessary and then update the speed every second (usually faster updates are needed). Of course, the time sampling can be selected as needed to obtain the necessary update rate. After the speed is updated, the timer is reset and a new count starts.

The odometer is typically updated every 0.1 km. That means that a second counter (update timer) counts the pulses and every time the update value (UV) is reached, the odometer value in the EEPROM is read, incremented by 0.1, and saved back to update the EEPROM and the display. The update count value is

$$UV = \frac{100}{2.3562} \times 20 = 849 \text{ pulses.}$$

After the update, the update timer is reset and the count restarts. The displays shown in **Figure 12.5** can be connected through serial or parallel ports depending on needs and the availability of pins on the microprocessor. Typically, serial communication is used to display units that have their own microprocessors embedded in them and capable of serial communication.

12.2.9 Power

The operating voltage of most microprocessors ranges from 1.8 V to about 6 V. Some have a more limited range (2.7–5.5 V, for example), while some can operate at higher voltages. Most microprocessors are based on complementary metal oxide semiconductor (CMOS) technology. This means that power consumption is very modest but frequency and voltage dependent. The higher the clock frequency, the higher the power consumption. Power is also dependent on what the processor does and which modules are functioning at any given time. The user has considerable control over power consumption through choice of frequency, mode of operation, and special functions such as interrupt, wake up, and sleep. These considerations must be reconciled with the requirements of the circuit and, often, a compromise must be reached. Sometimes this is easy, the circuit might need to operate continuously and an interrupt is not an option, meaning that the processor cannot be set to sleep. In other cases, the processor must operate at its maximum frequency. But very often it is up to the designer to decide on the parameters of the processor. If power is an issue, such as when operating from a battery, a reduction in operating frequency is one of the first choices that should be made. Operation at the lowest practical frequency not only reduces power consumption, but also reduces high-frequency emissions and possible issues of interference with other devices. The operating voltage should also be considered, although the limits here are relatively tight. Nevertheless, operation at 3 V rather than 5 V should reduce power consumption considerably, in addition to being a convenient voltage for battery operation. Almost all microprocessors have multiple sleep modes, one of which should be applicable in most cases. In sleep mode a microprocessor's consumption is in the nanoampere range, sometimes below the self-discharge level of a battery in storage.

Thanks to these flexible options, it is possible to design microprocessor interfacing circuits that can operate off a battery for extended periods, sometimes for years. It should be remembered, however, that power consumption is only one parameter in the design of a circuit and it imposes constraints on the circuit. For example, operation at low voltages may also require operation at lower frequencies or may render the oscillator less accurate. It changes the reference voltage for D/A converters and it changes the input and output voltages and currents. All of these must be taken into account in the design.

EXAMPLE 12.5 Battery-operated keypad lock for a safe: power considerations

A keypad lock is a common mechanism for access to restricted space, be it a building or a safe. The system includes a numeric telephone-type keypad that includes the numbers 0 to 9 plus the * and # symbols. The keys are arranged in three columns and four rows and each key is a switch at the intersection of a column with a row. Pressing a key shorts the corresponding column and row. A schematic circuit is shown in **Figure 12.6**. The motor is part of an actuator that physically activates the lock. The circuit operates as follows: The columns are connected to three digital output pins (marked as 1, 2, and 3) that place a 3 V level on each column in sequence. If a switch is pressed, the corresponding digital input pin to the row–column intersection (pins 4, 5, 6, and 7 are set as digital input pins) will be read as high. This allows the microprocessor to identify the column and row, and hence the digit pressed. To program the keypad lock, the user might be instructed to enter a four-digit code followed by *, then repeat to confirm. After that, entering the code followed by # will open the safe if locked or lock it if open. The microprocessor turns on a small, geared DC motor to close or open the lock and also has two limit switches that allow the microprocessor to turn off the motor (through the output pin marked as 8) at the end of the locking and unlocking process. A simple sequential program is required to operate the device. The circuit shown is normally in sleep mode, consuming approximately 8 μA from the battery, including the current in the resistors (R). When entering a code, the current increases to a level that depends on the voltage of the battery and the clock frequency of the microprocessor. The average length of a cycle that includes entering the code followed by the motor opening or closing the lock is 14 s.

a. The keypad lock in **Figure 12.6** has been tested for power consumption by varying the microprocessor's voltage and clock frequency, with the results of the tests shown in the table below. Estimate the current consumption of the circuit (without the motor) at 16 MHz for voltages between 2 and 5 V.

b. If the circuit is operated an average 12 times a day (24 h) and it uses 2 AA batteries capable of 2800 mA/h connected in series (3 V), calculate the length of time the batteries will last at 20 MHz if the motor is fed from a separate source.

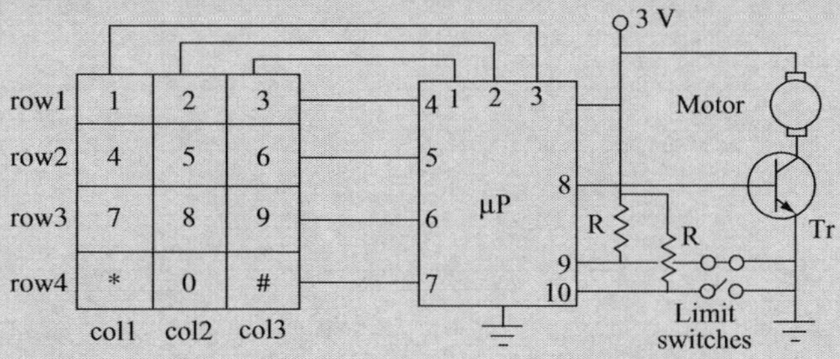

FIGURE 12.6 ■ A keypad-controlled lock.

c. How long will the circuit run under the conditions in (b) if the microprocessor never idles (i.e., it never goes into sleep mode) at 20 MHz and at 80 kHz?

	10 MHz	2 MHz	300 kHz	80 kHz
5 V	2.7 mA	0.84 mA	237 μA	108 μA
4 V	2.1 mA	0.65 mA	186 μA	76 μA
3 V	1.3 mA	365 μA	140 μA	56 μA
2.5 V	1.01 mA	312 μA	122 μA	46 μA
2 V	0.66 mA	205 μA	102 μA	37 μA
1.8 V				34 μA

Solution:

a. Since data for current consumption is available only up to 10 MHz, it is required to extrapolate from the existing data. By plotting the data it becomes apparent that the current increases linearly (more or less) with frequency (see **Figure 12.7**). There are therefore two options: one is to simply extend the curve to 16 MHz using two measurements, and the other is to use a linear least square curve as described in **Appendix A**. We use here direct extrapolation (but see **Problem 12.9**). Using the values at 10 MHz and 2 MHz we calculate the slope of the linear curves in **Figure 12.7** as

$$\Delta = \frac{I_{10} - I_2}{10 - 2},$$

where I_{10} is the current at 10 MHz and I_2 is the current at 2 MHz. Since the slope is assumed to be constant we write

$$\Delta = \frac{I_{16} - I_{10}}{16 - 10} \rightarrow I_{16} = I_{10} + \Delta(16 - 10) = I_{10} + \frac{I_{10} - I_2}{10 - 2}(16 - 10).$$

Now, using the values in the table,

$$\text{At 2 V:} \quad I_{16} = 0.66 + \frac{0.66 - 0.205}{8} \times 6 = 1.001 \text{ mA}$$

$$\text{At 2.5 V:} \quad I_{16} = 1.01 + \frac{1.01 - 0.312}{8} \times 6 = 1.533 \text{ mA}$$

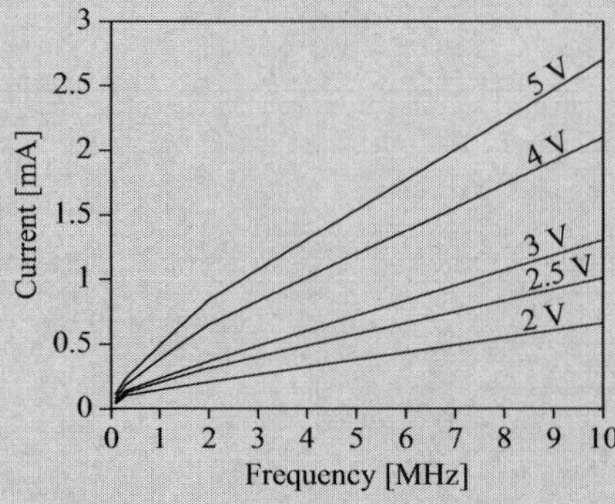

FIGURE 12.7 ■ Current consumption of the circuit in **Figure 12.5** as a function of frequency at various operating voltages.

$$\text{At 3 V:}\quad I_{16} = 1.3 + \frac{1.3 - 0.365}{8} \times 6 = 2.0 \text{ mA}$$

$$\text{At 4 V:}\quad I_{16} = 2.1 + \frac{2.1 - 0.65}{8} \times 6 = 3.187 \text{ mA}$$

$$\text{At 5 V:}\quad I_{16} = 2.7 + \frac{2.7 - 0.84}{8} \times 6 = 4.095 \text{ mA.}$$

Note that the current increases linearly with the operating frequency. **Figure 12.7** shows this clearly, although the measured values at low frequencies vary somewhat.

b. The expected current at 20 MHz and 3 V using the formula in (a) is

$$I_{20} = 1.3 + \frac{1.3 - 0.365}{8} \times 10 = 2.469 \text{ mA.}$$

To evaluate the current consumption, we calculate an average amount of mA-h consumed per day. We then divide the current capacity of the battery by the average consumption in 1 day to obtain the number of days the battery can supply the circuit.

Idle consumption (the circuit is in sleep mode) is 8 μA. Taking 1 day as the basis, this is $8 \times 24 = 192$ μA·h, or 0.192 mA·h/day. The circuit is on for $14 \times 12/3600 = 0.0467$ h and consumes $(14 \times 12/3600) \times 2.469 = 0.11522$ mA·h/day. The total consumption per day is the sum of the two, or 0.30722 mA·h/day. The number of days the battery can power the circuit is

$$N = \frac{2,800}{0.30722} = 9114 \text{ days.}$$

This is almost 25 years. Most batteries only have a shelf life of 6–10 years, but what this calculation demonstrates is that the microprocessor itself is usually not an impediment to low-power circuits, which can operate for years on a relatively small battery. In fact, the limiting factor in the circuit shown here is the motor.

c. Now the microprocessor is on at all times.

At 20 MHz it requires $24 \times 2.469 = 59.256$ mA·h/day. The circuit can run for

$$N = \frac{2800}{59.256} = 47 \text{ days.}$$

At 80 kHz the microprocessor requires $24 \times 0.056 = 1.344$ mA·h/day. The circuit can run for

$$N = \frac{2800}{1.344} = 2083 \text{ days.}$$

This is 5 years and 8.5 months.

12.2.10 Other Peripherals and Functionalities

As indicated above, a microprocessor must have certain modules (CPU, memory, and I/O), but it can have many more. In the development of microprocessors it became apparent that the applications for which they are most suitable require certain common functions. These can be provided externally, but it also became apparent that a more flexible device can be made that includes some of these functions internally. Thus many microprocessors include comparators (for digitization purposes), A/D converters, capture and compare (CCP) modules, PWM generators, and communication interfaces. One or two comparators are provided on many microprocessors and some may have more. Depending on the microprocessor, 8- or 10-bit (sometimes 12-bit) A/D converters are provided, usually in multiple channels (4–16 or more). PWM channels (typically between 1 and 8) are common on some processors. Serial interfaces such as Universal Asynchronous Receiver-Transmitter (UART), Universal Synchronous/Asynchronous Receiver/Transmitter (USART), Serial Peripheral Interface (SPI), two-wire interface (I^2C), Recommended Standard serial (RS-232), and Universal Serial Bus (USB) ports are available on many microprocessors, some supplying multiple interfaces, all under the user's control. Other functions such as analog amplifiers and even radio frequency (RF) transceivers are sometimes incorporated within the chip in specialized devices. The I/O used for these functions is either digital I/O (for communication, for example) or analog I/O (for A/D converters and comparators, for example). All peripherals are available on the data bus.

12.2.11 Programs and Programmability

A microprocessor is only useful if it can be programmed. Programming languages and compilers have been designed for this purpose. The basic method of programming microprocessors is through the Assembly programming language, but programming can be, and very often is done through the use of higher level languages, with the C programming language and its variants leading the way. These are specific compilers adapted for a class of microprocessors. They are based on standard C compiled (such as ANSI C) and modified to produce executable modules that can be loaded onto the microprocessor. With few exceptions, microprocessors can be programmed in circuit, allowing changes to be made or the processors to be programmed or reprogrammed after the circuit has been built. Microprocessors have been designed for integer operations, therefore programming for control, especially sequential control, is simple and logical. However, floating point operations and computation are either not possible or are difficult and tedious. They also tend to require considerable time and memory and should only be attempted if absolutely necessary. The reason for this is that many algorithms that have been built in hardware for the specific optimization of floating point operations on larger computers must be programmed from scratch on microprocessors. In addition, the 8-bit architecture is not very efficient for these types of operations. Floating point operations are the domain of computers in which these facilities are part of the CPU. There are, however, both integer and floating point libraries specifically designed for microprocessors, and these are freely available. These should be used if necessary because they have been optimized. Floating point operations are only practical on larger microprocessors because of their larger memory requirements.

12.3 | GENERAL REQUIREMENTS FOR INTERFACING SENSORS AND ACTUATORS

The general methods of interfacing of sensors and actuators and the circuits needed to do so were discussed in **Chapter 11**. In this section we will review and discuss some of the methods that are unique to interfacing to microprocessors. Many of the methods of **Chapter 11** can be used directly, but in some cases the microprocessor imposes additional requirements or relaxes some of the requirements. Also, the way the sensor and microprocessor interact is quite unique. The microprocessor may be used to monitor the sensor and record/process the data continuously in a dedicated manner. At the same time, it may be performing background functions. The microprocessor may also be part of the sensing algorithm. It may store calibration data, transfer functions, lookup tables, and other information the sensor needs for proper operation. It may initiate modes of sensing, such as range changes, calibration, shutdown, wakeup, preheating, temperature control and compensation, and many others. It may also prevent the sensor data from being available at the output before it has been stabilized. The microprocessor may also be operating in a polling mode in which it reads the sensor's output either continuously, at fixed intervals, or at irregular intervals, according to needs. During periods in which the microprocessor does not read the sensor, other functions may be performed, but the sensor's output is not available. This period may be very short (microseconds) or as long as needed or practical. Poled mode is a common mode of operation because microprocessors are ideally suited for operation in this mode. The speed of most microprocessors means that almost any sensor can be polled at sufficiently short intervals to follow the sensor's output quite well.

Another useful mode of operation is the interrupt mode. Here the sensor or some other event issues an interrupt signal, which then starts the microprocessor's operation. A very common mode of operation is for the microprocessor to either be in sleep mode or engaged in other operations, ignoring the sensor or actuator it monitors. Then, when an interrupt signal is issued, the microprocessor proceeds to read the sensor's output. The interrupt may be initiated at regular intervals (timed) or it may be initiated by the output of the sensor. For example, when the sensor output exceeds a given value (or drops below a certain value), an interrupt may be issued. This may require additional electronics or it may be issued directly, depending on the sensor and microprocessor. The interrupt may even be issued externally by the operator or by a separate sensor (such as a light or temperature sensor) or by feedback from an actuator—the possibilities are many. Of course, the same considerations apply to actuators.

Beyond the mode of operation, interfacing must take into account a number of other issues. The most common of these are discussed below.

12.3.1 Signal Level

The advantage of using a microprocessor to acquire data from sensors and to control actuators is primarily in the flexibility afforded by the microprocessor. Depending on the output of the sensor, the microprocessor may be able to read the data and process it directly or it may require the use of an analog module on board (comparator, A/D converter, etc.), if such a unit is available, or an external circuit must be provided for this purpose. The requirements sensors and actuators impose on the microprocessor stem from the type of signal they generate, the level of the signal, as well as frequency. In most cases the signals are low-voltage, low-power signals, but this is by no means universal. Piezoelectric devices, for example, may generate voltages much above what microprocessors can handle, and electric motors and magnetic actuators almost always require more power than microprocessors can supply. Because of the inherent high impedance inputs of CMOS devices, microprocessors can handle voltage sources very well, but not current sources. Thus sensors in which voltage is the output can be read directly, perhaps with the need of attenuators. In some cases amplifiers are needed to bring the span of the sensor into the input range of microprocessors. Similarly, most sensors whose output is a frequency (square wave or sinusoidal) may be read directly, again with perhaps the need for attenuation or amplification. There are limitations, however, on the frequencies that can be sampled, as will be discussed below. Particular attention must be paid to the polarity of the input signal. Microprocessors cannot tolerate reversed polarities. Therefore any input signal must fluctuate from zero to V_{dd}. If necessary, additional circuitry must be introduced to translate the signal as appropriate.

12.3.2 Impedance

If a sensor or actuator can be connected directly to a microprocessor in terms of voltage levels, we still have to take into account the impedance of the I/O port. When a pin is set as input, it becomes a high-impedance input, on the order of a few megaohms. Typically the current flowing into a pin when in input mode is less than 1 μA. This is ideal when low-output impedance sensors are connected. Thus, for example, many sensors, including resistive, Hall effect, and magnetic sensors, can be connected directly, provided that the voltage level is appropriate. But there are sensors that cannot be connected directly. One example is any sensor that supplies a current as output. These currents must first be converted into a voltage (see **Example 12.6**). More difficult is the problem of connecting high-impedance sensors such as capacitive sensors and pyroelectric sensors. Some of these sensors can have an impedance in the range of 10–50 MΩ, and for these sensors the input of a microprocessor represents an unacceptably high load. In such cases, one of the circuits discussed in the previous chapter (e.g., a voltage follower) may be employed external to the microprocessor. In particular, operational amplifiers with FET input stages are ideal for this purpose.

EXAMPLE 12.6 Interfacing of a current sensor

A microprocessor needs to measure the current supplied by a power supply in a device as part of the energy management system of the device and compute and display the status of the battery powering it. The power supply operates from a car battery and is expected to supply up to 500 mA at a constant voltage of 5 V to a load. The current needs to be monitored and displayed digitally in two ranges using a microprocessor. The microprocessor operates at 5 V, has a 12-bit A/D converter, and we require that the system be capable of sensing in two ranges. In range A it must be capable of measuring in increments of 10 mA. In range B it must be capable of measuring in increments of 0.5 mA.

Solution: As mentioned above, the microprocessor is not equipped to measure current directly. Therefore a small resistance R is inserted in series with the load and the voltage on this resistor is measured by the microprocessor using its A/D converter, as shown in **Figure 12.8**. The converter can resolve down to $5\text{ V}/2^{12} = 1.22$ mV. That means that to sense up to 500 mA in increments of 10 mA we need 50 increments of 1.22 mV. That is, the maximum voltage drop across the sensing resistor is $50 \times 1.22 = 61$ mV. The resistor is $R = 0.061/0.5 = 0.122\ \Omega$. In range B we must be able to measure $500/0.5 = 1000$ increments of 1.22 mV each. That means the voltage across the sensing resistor must be 1.22 V and the resistance must be $R = 2.44\ \Omega$.

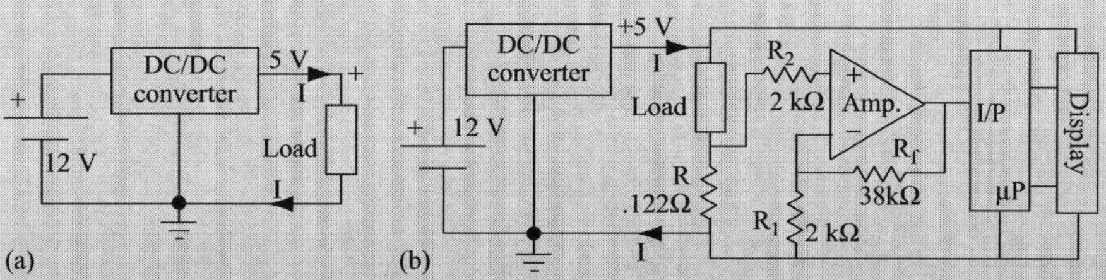

FIGURE 12.8 ■ Interfacing of a current sensor. (a) The power supply before interfacing. (b) The power supply after interfacing. The resistor R was added as a current sensor and the amplifier is added to expand the voltage drop on R.

This looks rather simple, but whereas the solution for range A is acceptable, that for range B is not. In the first case, the series resistor only drops 0.061 V, and hence has little effect on the device being powered. However, in the second case the voltage drop is 1.22 V, and this is a significant percentage of the output voltage. Even from the point of view of power loss in the sensing resistor this is not acceptable—the resistor would have to dissipate 0.61 W. Therefore it is necessary to reduce this resistance to something acceptable, reducing the voltage and then amplifying the voltage to accomplish the resolution needed.

The resistor is selected as $R = 0.122\ \Omega$ and a noninverting operational amplifier with a gain of 20 is added, as shown in **Figure 12.8b**. This boosts the maximum voltage at the microprocessor's input to $0.061 \times 20 = 1.22$ V. Using the configuration in **Figure 12.8b**, the gain is (see **Equation (11.7)**)

$$A_v = 1 + \frac{38}{2} = 20,$$

as required.

Of course, there is no reason not to use a gain of, say, 60 and increase the voltage to 3.66 V. In fact, this would have the advantage of allowing the A/D converter to operate away from its lowest range, improving accuracy. Alternatively, a higher gain will allow a smaller resistor and therefore a smaller influence on the load. The selection between range A and B is done internally by proper programming. The microprocessor may also sense the power supply voltage to calculate power usage and to monitor the battery directly. It can then warn the user or, if so desired, turn off the device when the battery is low to protect the load.

EXAMPLE 12.7 Interfacing an atmospheric charge sensor

An important part of weather prediction and protection of life and property is the monitoring of atmospheric charge or, alternatively, of the atmospheric electric field intensity that gives rise to this charge. Since this electric field changes with weather conditions, it is a good indication of these conditions and, in particular, of the possibility of lightning. The fair weather electric field intensity can be as low as 100 V/m, whereas before and during thunderstorms it can increase to more than 10^6 V/m. A sensor that can detect and quantify these fields is built as shown in **Figure 12.9**. It consists of a conducting plate that, in conjunction with the ground, forms a tiny capacitor. The voltage across this capacitor is a function of the electric field intensity and the

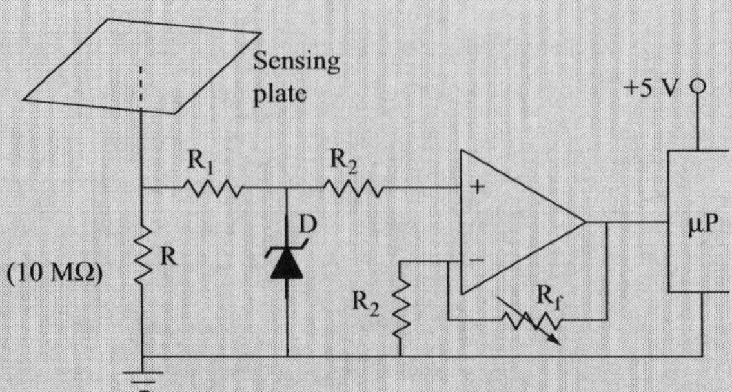

FIGURE 12.9 ■ Interfacing of an atmospheric charge sensor. This implementation assumes the charge on the sensing plate is positive (see **Problem 12.11** for a negative charge on the plate).

charge on the plate is a function of the voltage and capacitance (see **Equation (5.1)**). Although both the capacitance and charge can be calculated, we look here primarily at the interfacing issues.

The first problem we face is that, given a plate, say, at 2 m above ground, its minimum potential (under fair weather conditions) is 200 V and that can increase to thousands of volts per meter when the weather turns stormy. It would seem that the use of a microprocessor is not the best approach. However, if we connect the plate to the input of a microprocessor, we have effectively connected a resistance across the capacitor and the capacitor is essentially "shorted," showing a voltage close to zero. To solve this issue, an operational amplifier with FET inputs is used before reading the input in the microprocessor. This has two effects. First, its input resistance, on the order of 100 MΩ, is sufficiently low to reduce the voltage on the plate (with respect to ground) to a value in the range of the microprocessor. If the voltage is too high, a resistance (R) may be connected to ground to reduce the voltage on the plate. Second, it allows one to adjust the amplification to fit the range of the microprocessor (0–5 V in most cases). The feedback resistance is adjusted to the maximum reading under "stormy" conditions. The zener diode across the input ensures that the voltage cannot rise above 5 V, to protect the circuit.

Consider the following data: The electric field intensity is perpendicular to the ground and the plate, 10 cm × 10 cm, is parallel to the ground and has a capacitance of 330 pF. The charge density on the plate is given as

$$\rho_s = \varepsilon_0 E \quad [\mathrm{C/m^2}],$$

where ε_0 is the permittivity of air (8.854×10^{-12} F/m) and E is the electric field intensity [V/m]. Calculate the voltage at the input of the microprocessor under fair weather conditions ($E = 200$ V/m) with $R_2 = 10$ kΩ and $R_f = 15$ kΩ.

a. When a storm approaches, the electric field intensity increase. What is the voltage at the microprocessor when the electric field reaches 10 kV/m?
b. What is the reading if the electric field intensity increases to 5 kV?
c. Calculate the maximum electric field intensity beyond which the sensor ceases to respond.

Note: The voltage is read in the microprocessor using the internal A/D converter, scaled, and displayed in convenient units such as units of electric field, charge, or even in units such as "fair," "variable," and "stormy" to conform to predicted weather conditions.

The configuration shown assumes the charge on the plate is positive (i.e., the electric field intensity points away from the plate). **Problem 12.11** discusses the same configuration with a negative charge on the plate. In any case, it should be remembered that the potentials anywhere on and within the microprocessor must be positive with respect to ground. One of the challenges in interfacing is to ensure this condition.

Solution:

a. The charge on the plate is the charge density multiplied by the area of the plate (assuming it is uniform on the surface):

$$Q = \rho_s S = \varepsilon_0 ES \quad [\mathrm{C}].$$

Given the capacitance of the plate, the voltage on the plate is found from **Equation (5.1)**:

$$V = \frac{Q}{C} = \frac{\varepsilon_0 ES}{C} \quad [\mathrm{V}].$$

With the values given we have

$$V = \frac{\varepsilon_0 ES}{C} = \frac{8.854 \times 10^{-12} \times 200 \times 0.1 \times 0.1}{330 \times 10^{-12}} = 0.0537\ \mathrm{V}.$$

This voltage is now amplified by the noninverting amplifier. The gain of the amplifier (see **Equation (11.7)**) is

$$A_v = 1 + \frac{R_f}{R_2} = 1 + \frac{15}{10} = 2.5.$$

The voltage at the input pin of the microprocessor is

$$V_{\mathrm{in}} = 0.0537 \times 2.5 = 0.134\ \mathrm{V}.$$

b. When the electric field increases to 5 kV/m we get

$$V = \frac{\varepsilon_0 ES}{C} = \frac{8.854 \times 10^{-12} \times 5000 \times 0.1 \times 0.1}{330 \times 10^{-12}} = 1.341\ \mathrm{V}.$$

And since the amplification remains unchanged, the voltage at the input pin is

$$V_{\mathrm{in}} = 1.341 \times 2.5 = 3.354\ \mathrm{V}.$$

c. As the electric field increases, so does the voltage at the microprocessor. But because of the zener diode, it cannot rise above 5 V. The maximum atmospheric electric field the circuit can indicate is calculated using the same relations, starting with the maximum input voltage, as follows:

$$V_{\max} = A_v \frac{\varepsilon_0 E_{\max} S}{C} \quad [\mathrm{V}],$$

or

$$E_{\max} = \frac{V_{\max} C}{A_v \varepsilon_0 S} = \frac{5 \times 330 \times 10^{-12}}{2.5 \times 8.854 \times 10^{-12} \times 0.1 \times 0.1} = 7454\ \mathrm{V/m}.$$

Above this value the input voltage stays constant. The voltage at the microprocessor increases linearly with the atmospheric electric field from 0 to 7454 V/m. This is sufficient to indicate the approach of a storm.

12.3.3 Response and Frequency

Most sensors and actuators are relatively slow devices and are unlikely to pose a problem for microprocessors in terms of speed or frequency response. But there are many sensors that, while their own response is sufficiently slow, are part of oscillating circuits generating frequencies higher than the capability of a microprocessor. For example, suppose a SAW sensor measuring pressure operates at 10 MHz. This is not an unusually high frequency since the higher the frequency, the higher the sensitivity and resolution of the sensor. But if we use a microprocessor with a cycle time of 0.1 μs, and assuming at least 10 instructions are needed to read and process the input on a pin, clearly this microprocessor cannot measure the frequency of the sensor. In fact, anything higher than perhaps 0.5 MHz is likely to be read erroneously. Of course, one can divide the frequency of the input signal, say, by 100 (must be done externally), but this defeats the purpose by reducing the sensitivity of the sensor. Alternatively a F/V converter can be used and the voltage produced can then be measured through an A/D converter. A process of this type is likely to introduce errors due to both conversions, and these errors come on top of the errors in sensing. Sometimes the interest is in measuring the change in frequency rather than the frequency itself, in which case additional circuitry may be needed to subtract one frequency from another before the microprocessor can be used to measure the frequency difference. When the frequency is high, the only satisfactory solution is to use an external frequency counter capable of measuring the high frequency and then use the digital output of the frequency counter as the input to the microprocessor. In a case like this, the auxiliary circuit is much more complex and more expensive than the microprocessor itself. Nevertheless, some units on microprocessors can operate faster than the example above might imply. For example, capture and compare (CCP) modules can resolve down to less than 10 ns, but these capture the output of internal timers and are therefore most suitable for output operations (such as display). Also, whereas most microprocessors are not particularly fast (normal range up to about 40 MHz clock speed), some can be much faster at a premium in cost. It should be remembered, however, that we have deliberately limited ourselves here to low-level microprocessors. There are processors that are much faster that can be used, as well as single-board computers, and ultimately whole computers or computer systems.

EXAMPLE 12.8 **Quartz crystal microbalance**

A microbalance is a quartz crystal resonating at a relatively high frequency on which an electrode, typically gold, has been plated. Any increase in the mass of the electrode due to an external mass will result in a change in the resonant frequency (see **Section 8.7** and **Example 8.8**). The microbalance measures mass per unit area based on the shift in frequency of the resonator. Based on **Equation (8.17)**, the shift in frequency Δf is

$$\Delta f = \frac{\Delta m}{C_m} \quad [\text{Hz}],$$

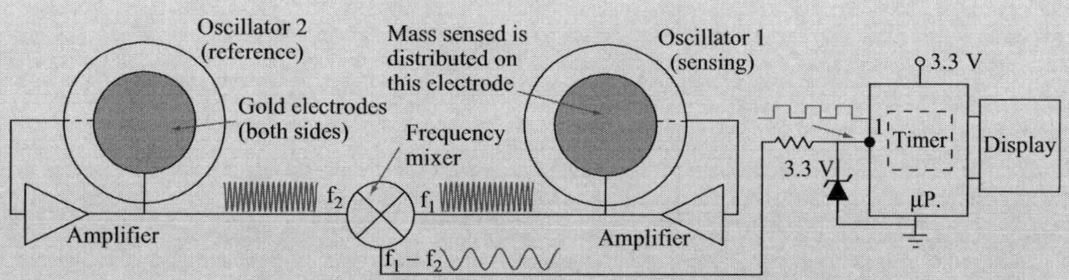

FIGURE 12.10 ■ Schematic of the microbalance.

where Δm is the change in mass of the crystal per unit area [g/cm^2] and C_m is the mass sensitivity factor [ng/Hz/cm^2]. Consider a microbalance resonating at 18 MHz with a sensitivity of 5 ng/Hz/cm^2. The sensor is used to sense the mass of substances coated on the electrode. With an electrode area of 1.5 cm^2, the sensor is used to measure mass up to 100 μg. There are two options that can be pursued. One is to measure the absolute frequency, the other is to measure only the change in frequency.

The first option is not practical with a microprocessor. The base frequency (zero mass) is 18 MHz, which corresponds to a cycle of $1/18 \times 10^6 = 55.55 \times 10^{-9}$ s, or 55.55 ns. To measure this frequency directly would require a clock cycle much shorter than this and such microprocessors do not exist. Further, the change in frequency for the largest mass measured is

$$\Delta f = \frac{100 \times 10^{-6}/1.5}{5 \times 10^{-9}} = 13,333 \text{ Hz}.$$

The change in frequency is significant, but is small compared to the base frequency, meaning that a very accurate determination of the frequency is necessary. On the other hand, the change in frequency is easily measurable and is used here as an alternative to measuring the exact frequency.

This alternative is shown in **Figure 12.10**. Two identical crystals are set into oscillation and the frequency of the second is subtracted from the first. The result is a zero frequency. Now one sensor is used as the sensing sensor, say, the first. The output will change from zero to 13,333 Hz for a change in mass from 0 to 100 μg. The shortest time that needs to be measured is that of one cycle, or 75 μs ($1/13{,}333 = 75 \times 10^{-6}$). A 10 MHz clock cycle (0.1 μs/cycle) would allow accurate determination of the cycle time of the input frequency. The time is measured using the process outlined in **Example 12.3**. Sensitivity is 133.33 Hz/μg and is linear. The microprocessor can display the output in the proper units by scaling the time measured using, for example, a lookup table stored in EEPROM.

12.3.4 Input Signal Conditioning

The power and signal requirements of microprocessors place some restrictions on sensors and actuators. Since microprocessors operate between about 1.8 V and 6 V (5 V and 3.3 V are the most common voltages), it is important that the sensor's signal be in this range as well. To accomplish this, amplification, attenuation, scaling, changes in the offset of the signal, and signal translation are often needed. These are defined here.

12.3.4.1 Offset

Offset is the DC signal on which the sensor's output variations ride. This is a common occurrence in sensors and can be understood from the circuit in **Figure 12.11**. Here a

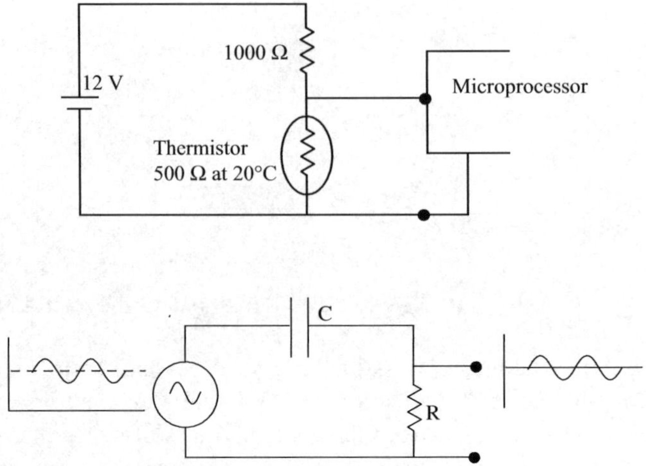

FIGURE 12.11 ■ A simple temperature-sensing circuit connected to a microprocessor.

FIGURE 12.12 ■ Removing the DC offset from a signal.

thermistor is supplied with a current from a 12 V source and the voltage across the thermistor is sensed. Assuming that the resistance of the thermistor is 500 Ω at 20°C and that the change is 5 Ω/°C, the resistance of this sensor will change from 400 Ω to 900 Ω as the temperature changes from 0°C to 100°C. The voltage sensed is $12 \times 500/1500 = 4$ V at 20°C and varies from $12 \times 400/1400 = 3.428$ V at 0°C to $12 \times 900/1900 = 5.684$ V at 100°C. Although the change in voltage is small (only 2.256 V) and perfectly suitable for input to a microprocessor, it rides on a 3.428 V DC signal that raises it to levels above that suitable for a 5 V microprocessor. This can be solved in many ways. One is to remove the DC signal altogether. This can be done, for example, through an instrumentation amplifier by supplying 3.428 V to the inverting input and the signal to the noninverting input and setting the amplification to 1. The output will now be between 0 and 2.256 V, as required. It is also possible to simply reduce the signal to acceptable levels by increasing the value of the fixed resistor, say, to 1500 Ω. With this resistance, the voltage changes from 2.526 V to 4.5 V. Alternatively, the source may be reduced from 12 V to, say, 10 V. In either case, the change in voltage becomes smaller (lower current through the thermistor) but the voltage requirements of the microprocessor are met.

If the sensor's output is an AC signal at a reasonably high frequency, such as that due to an audio microphone or a SAW resonator, any DC offsets can be removed simply by connecting a capacitor in series with the input to the microprocessor. This will remove the DC component, but can cause an even bigger problem, as can be seen in **Figure 12.12**. The AC signal now swings from $-V_p$ to $+V_p$, but microprocessors cannot accept negative voltages. Depending on what the signal represents and what is to be measured, the negative part can be removed through the use of a diode. That is the case when all that is needed is to measure the frequency. The opposite is sometimes done as well; that is, for a dual-polarity signal, a DC signal equal to the mid-level of the signal is added to eliminate the negative polarity.

A useful method of removing the DC bias is to use a bridge, balanced at an appropriate level. An example, using the same thermistor as in **Figure 12.11**, connected in a bridge is shown in **Figure 12.13**. With the resistances given, the output changes from zero at 0°C to 2.3 V at 100°C for the same source voltage (resistance of the thermistor changes from 400 to 900 Ω). This circuit also allows setting a suitable offset,

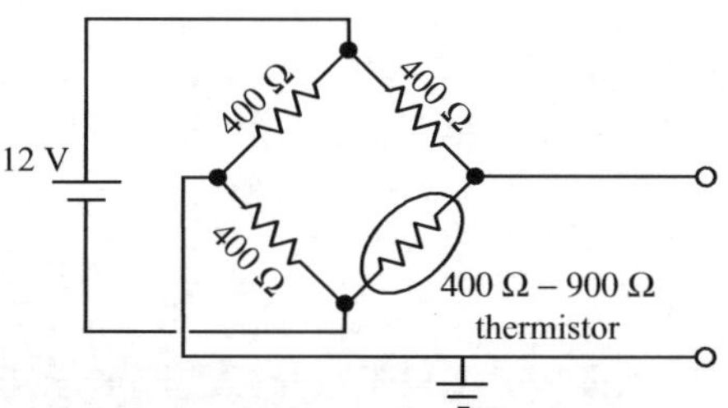

FIGURE 12.13 ■ Bridge connection of the sensor in **Figure 12.11**.

say, 1 V, by decreasing the value of the lower-left resistor to an appropriate value (reducing it to 285.7 Ω will add exactly 1 V offset). Note, however, that the 12 V source must be floating, that is, it cannot be derived from the same source as the circuit to which the bridge output is connected. Usually a simple method such as a reduction of voltage using resistors, as described above, or the bridge configuration is preferred.

EXAMPLE 12.9 Adding DC offset to the signal of a sensor

A high-temperature programmable thermostat is built around a microprocessor using a platinum RTD. The resistance of the RTD is 240 Ω at 20°C and has a temperature coefficient of resistance of 0.003926/°C at 0°C. The thermostat is required to switch off at 350°C and switch on at a temperature lower than the switch-off temperature (in effect, to introduce a hysteresis so that the comparator does not rapidly switch on and off at the set point). The microprocessor operates at 5 V and the idea is to use the internal comparator and internal reference voltage to accomplish the necessary function. Many microprocessors have either a fixed reference voltage or a reference voltage based on the power supply voltage. In most cases a 16-level reference voltage is defined either between 0 and V_{dd} or between some constant value V_0 and V_{dd}. The mode selection is done in software. For simplicity we will use the first option.

A possible implementation is shown in **Figure 12.14a**. A bridge is used, as it allows adjustment of the voltage at the positive comparator pin. However, the use of a bridge requires it to be floating, that is, its 5 V supply must be separate from the 5 V supply of the microprocessor. The negative pin is connected internally to the reference voltage. We need to set the reference

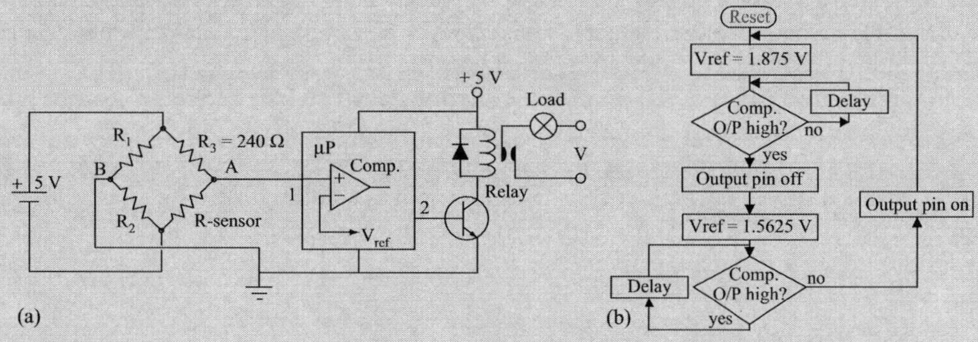

FIGURE 12.14 ■ (a) Implementation of a programmable thermostat. (b) Flow chart of the microprocessor program.

voltage so that the comparator switches from off to on at 350°C. To do so we must first calculate the resistance of the RTD at 350°C. From **Equation (3.4),**

$$R(T) = R_0(1 + \alpha[T - T_0]) \quad [\Omega],$$

where $R_0 = 240\ \Omega$ is the resistance at $T_0 = 20$°C and α is the temperature coefficient of resistance of platinum, given as 0.003926/°C. Since the coefficient is given at 0°C, we must first calculate the resistance of the RTD at 0°C:

$$R(0^\circ C) = 240(1 + 0.003926 \times [0 - 20]) = 221.1552\ \Omega.$$

The resistance at 350°C is therefore

$$R(350^\circ C) = 221.1552(1 + 0.003926 \times [350 - 0]) = 525.045\ \Omega.$$

If we start with a balanced bridge (all resistances are the same, equal to 240 Ω) at 20°C, the voltage at point A of the bridge at 350°C will be

$$V_{in} = \frac{5}{525.045 + 240} \times 525.045 = 3.4315\ \text{V}.$$

The voltage at point B is 2.5 V because R_1 and R_2 are both equal to 240 Ω. The difference, that is, 3.4315−2.5 = 0.9315 V, is the voltage at the positive input pin of the comparator. The internal reference must be set to this value. Now the possible values for reference at the negative input to the comparator can only be set in increments of 5/16 = 0.3125 V. The closest value is 3 × 0.3125 = 0.9575. This may be close, but not close enough—the thermostat will not operate at the required temperature. To solve this problem we can change the resistances R_1 and R_2 to offset the difference. However, since the input to the microprocessor must always be positive, the voltage at point B should be decreased below 2.5 V, otherwise, say, at 20°C, the input would be negative. Therefore we set the reference at 3 × 0.3125 = 0.9575 V and the offset at 0.9575−0.9315 = 0.026 V. This is done by reducing the voltage at point B by the offset value to 2.5−0.026 = 2.474 V. To do so we write

$$\frac{5}{R_1 + R_2} R_2 = 2.474 \rightarrow R_2 = 0.979 R_1.$$

By selecting one resistance, say, $R_1 = 300\ \Omega$, we get $R_2 = 293.7\ \Omega$. A more practical solution is to use a variable resistor (potentiometer), say, a 500 Ω potentiometer, and adjust it so that the voltage at point B (center tap of the potentiometer) is 2.474 V.

If the temperature decreases, we want the thermostat to switch back at a lower temperature. This is done by lowering the reference voltage one step, that is, to 0.9375−0.3125 = 0.625 V (in software) and allowing the comparator to switch to low when that reference voltage is reached. We argue as follows: Since point B is at a voltage of 2.474 V, and switching will occur at 0.625 V, the voltage at point A must be 2.474 + 0.625 = 3.099 V. Therefore the RTD resistance must be

$$3.099 = \frac{5}{R + 240} \times R \rightarrow R = \frac{240 \times 3.099}{5 - 3.099} = 391.247\ \Omega.$$

Now using **Equation (3.4)** with T as unknown:

$$391.247 = 221.1552(1 + 0.003926 \times [T - 0]) \rightarrow T = \left(\frac{391.247}{221.1552} - 1\right)\frac{1}{0.003926} = 195.9°\text{C}.$$

That is, the electronic thermostat will switch off at 350°C and on at 195.9°C. The flow chart in **Figure 12.14b** shows how this can be achieved in software.

Note that in practice the large difference between the on and off temperatures may not be acceptable, in which case a different approach can be adopted, such as setting the second reference voltage on an external pin. The thermostat can be reprogrammed by changing the reference voltages at which it switches on and off.

12.3.4.2 Scaling

The input to a microprocessor can be scaled either by amplifying the signal (scale factor larger than 1) or by dividing it through an appropriate resistive network (attenuation of the signal). Scaling the signal up was discussed in **Chapter 11**. Scaling it down, however, while simple, is fraught with problems unless an amplifier is used. A resistive network is likely to load the sensor unless its internal resistance is low. In the case of the thermistor discussed above, suppose we need to reduce the voltage changes to less than 5 V. One can add a resistance divider, as shown in **Figure 12.15**. Assuming the microprocessor to which the output is connected draws no current from the resistive network, the divider in **Figure 12.15** produces an output between 1.5 V and 2.298 V for the same temperature range. With proper calibration, this method can attenuate the output to a level acceptable as input to the microprocessor. Alternatively, the series resistor in **Figure 12.15** can be raised to a higher value (1500 Ω was shown in **Section 12.3.4.1** to be sufficient). This is a better approach, as well as being simpler. If the sensor has high internal resistance, an isolation unit gain amplifier (voltage follower) can be connected between the sensor and a voltage divider so that the voltage divider does not load the sensor (see **Example 12.10**).

If signals are AC, a transformer may be used as well in which the turn ratio is chosen to reduce the signal as necessary. This method is not usually advisable because transformers are relatively large, require relatively large currents, may be nonlinear, and their frequency response may introduce distortions. With few exceptions this method will be found to be inappropriate in sensors.

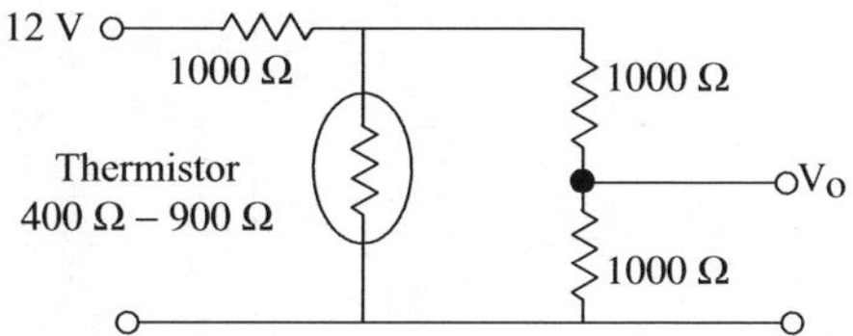

FIGURE 12.15 ■ Reducing the output voltage on the sensor to match the required input of the microprocessor by means of a voltage divider.

EXAMPLE 12.10 Ultrasonic parking "radar"

An ultrasonic parking sensor installed in the rear bumper of a vehicle uses a transmitter to send a beam and a receiver to detect the reflection from objects behind the vehicle to prevent damage to the vehicle. In most cases a single device is used and switched from transmit to receive in a pulse-echo mode (see **Section 7.7**). The distance to the nearest object is determined by the intensity of the reflected beam (hence the colloquial name "radar"). The device can measure distances from about 10 cm to 2 m and warns the driver by a series of beeps whose intensity increases as the vehicle gets closer to the object. The output of the ultrasonic receiver is a signal at 40 kHz whose amplitude varies from 0.1 V (peak to peak) at a distance of 2 m to 12 V (peak to peak) at 10 cm. We will assume here that these voltages are obtained directly from the ultrasonic transducer (usually the output of ultrasonic transducers are fairly low, but that depends on the transmitter signal and the distance). Show how the sensor can be interfaced to a microprocessor that operates at 3.3 V.

Solution: One approach is to detect the peak of the signal and calibrate the distance with that peak. In addition, the signal levels of the sensors and the microprocessors must match. The output impedance of the sensor is high and any loading is likely to reduce the output. A suitable circuit is shown in **Figure 12.16**. A voltage follower is first used at the output of the sensor to ensure it is not loaded by the circuitry (see **Section 11.2.3**). This is followed by a diode and a capacitor to detect the peak value (the diode also ensures that the output is positive only as required by the microprocessor). The capacitor is loaded with a relatively large resistance $R = R_1 + R_2$ so that the capacitor is discharged at a certain rate to follow the peaks of the input signal should these decrease. The time constant RC is selected based on the frequency of the ultrasound signal so that it is longer than one cycle (25 μs). A time constant of, say, 250 μs should be appropriate. Suppose the capacitor is 1 nF. That makes the resistor R

$$R = \frac{RC}{C} = \frac{250 \times 10^{-6}}{1 \times 10^{-9}} = 250,000\ \Omega.$$

By dividing the resistor R into R_1 and R_2 we ensure that, whereas the voltage on R changes between 0 and 12 V, the voltage on R_2 changes between 0 and 3.3 V. The selection of these resistors is as follows:

$$\frac{12}{R_1 + R_2} \times R_2 = 3.3 \rightarrow R_1 = \frac{3.3}{8.7} R_2\ [\Omega].$$

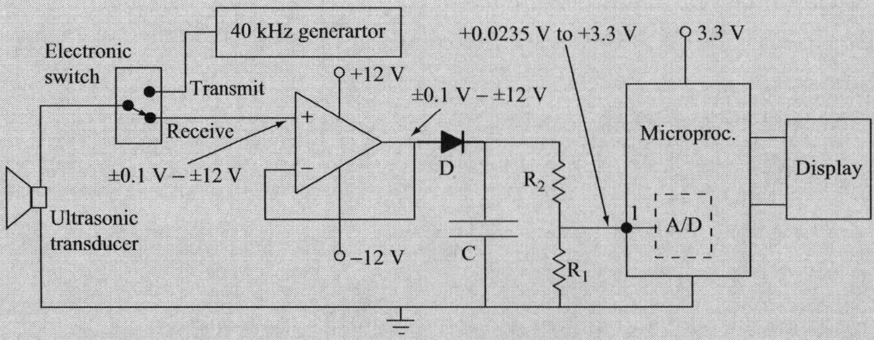

FIGURE 12.16 ■ A pulse-echo parking radar.

and

$$R_1 + R_2 = 250\ \text{k}\Omega.$$

Solving, we get

$R_1 = 68.75\ \text{k}\Omega$ and $R_2 = 181.25\ \text{k}\Omega$.

These are nonstandard resistances. Now the input to the microprocessor will vary from 0.0275 V for a distance of 2 m to 3.3 V for a distance of 10 cm. The analog input is then digitized using the internal A/D converter and the digital representation scaled internally to produce an output. The output of the microprocessor can be displayed on a screen in actual distance, either in addition to or in lieu of the beeps, for a more accurate determination of distance.

Note: The resistors R_1 and R_2 may be difficult to get as standard components. In that case they can be replaced with a 250 kΩ potentiometer with the center top adjusted to produce the correct output or, alternatively, they can be changed to other values since the time constant selected is somewhat arbitrary. We have neglected for the sake of simplicity the voltage drop on the diode (on the order of 0.7 V for a silicon diode or 0.3 V for a Schottky diode). It should also be remembered that alternative methods of implementation are possible.

12.3.4.3 Isolation

Electrical isolation of the signal from a sensor or to an actuator is sometimes needed. This might be the case when the sensor is directly in contact with higher voltages or when it must be floating (i.e., not connected to ground). Actuators may need to operate at much higher voltages than a microprocessor and signal isolation again becomes necessary. In AC systems, such as in an LVDT, it is sometimes possible to use a transformer for isolation purposes. In cases where the signal is digital or has been digitized, an optical coupler may be a useful approach. In an optical coupler, the output from the sensor changes the light intensity of an LED and the LED operates a photodiode or phototransistor to produce a signal proportional to the light intensity (see **Figure 12.17**). Because the LED may require more power than the sensor can deliver, some kind of booster, such as an operational amplifier voltage follower, may be needed. The output from the coupler can then be connected to the microprocessor, subject to matching of voltage levels and impedance. Optical couplers may be used equally well with actuators by driving the LED directly from a microprocessor pin. In this case the phototransistor will be driven from the actuator's power supply. The signal can then be used to operate/control the actuator. Optical couplers are standard components offering

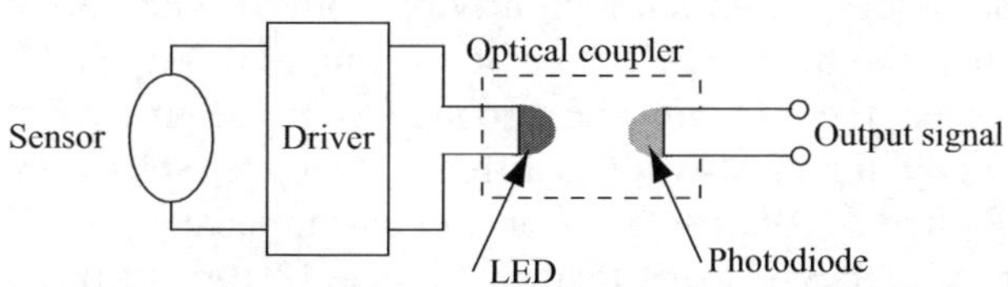

FIGURE 12.17 ■ Coupling through an optical coupler. The driver is needed to boost the current through the LED if the sensor cannot supply the required current.

exceptional isolation up to a few thousand volts. Even so, when additional components are used, their effect on the overall performance (accuracy, noise, sensitivity, etc.) must be taken into account.

In some cases isolation can be achieved through other means. If, for example, a microprocessor is required to turn on and off a high voltage source (a light fixture in a building or a motor in a furnace blower), that can be achieved through a relay (either mechanical or electronic). In these cases the actual high voltage switch is isolated from the low voltage drive of the relay. Of course, depending on the electrical requirements of the relay, it may need to be driven through a transistor or MOSFET because of the limited direct drive capabilities of microprocessor. In still other cases, the means of transferring data provides inherent isolation. An example is transmission through optical fibers, a process similar to the optical isolator except that an optical fiber connects the LED and phototransistor optically. And, of course, one can always opt for an infrared link or a wireless link between the microprocessor and any sensor/actuator that requires isolation.

12.3.4.4 Loading

Anything connected to the sensor represents a load. Microprocessors are no exception, but because their input impedance is high, many sensors can be connected directly without concern for loading by the microprocessor. Of course, there are exceptions to this rule, and we have seen examples of high-impedance capacitive sensors that require additional interfacing circuits such as a charge amplifier or a properly designed FET input voltage follower. In any case, a careful analysis of the loading effects on the sensors and the effects on sensitivity, span, and response must be done whenever anything is connected to a sensor.

12.3.5 Output Signals

The output from an I/O port in a microprocessor is of the same level as the processor's supply voltage (typically 1.8–6 V). The maximum current an I/O pin can drive into a load depends on the way the load is connected, but it is on the order of 20–25 mA per I/O pin. This can be used to drive small loads directly, but for many applications the power must be boosted by an appropriate circuit. Sometimes, when the voltage level is appropriate, only current needs to be boosted, but in other cases both current and voltage will have to be changed. For example, we might want to drive a 12 V DC motor consuming 1 A from the output of a microprocessor. Before we discuss how this can be done, we need to first discuss the issues associated with direct drive power from I/O pins.

The I/O pins of microprocessors are built around complementary MOSFETs. A simplified I/O pin circuit is shown in **Figure 12.18**. In addition to the driving circuitry it includes protection diodes. From the load driving point of view there are two options. One is to connect the load between V_{dd} and the I/O pin, as in **Figure 12.19a**. The other is to connect it between the I/O pin and ground, as in **Figure 12.19b**. The two are equivalent functionally, but the driver has different characteristics in the two modes. The configuration in **Figure 12.19a** has a lower internal impedance (typically about 70 Ω) and can therefore drive larger loads than in **Figure 12.19b**. In the latter, the internal impedance is about 230 Ω and hence can supply a smaller current. Typical values are

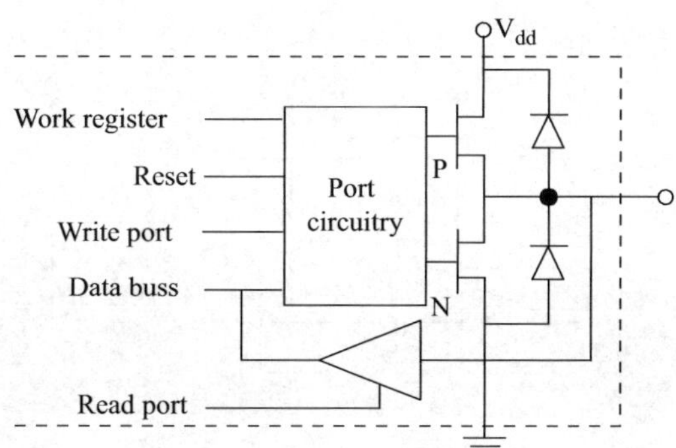

FIGURE 12.18 ■ Simplified I/O circuit for a microprocessor pin.

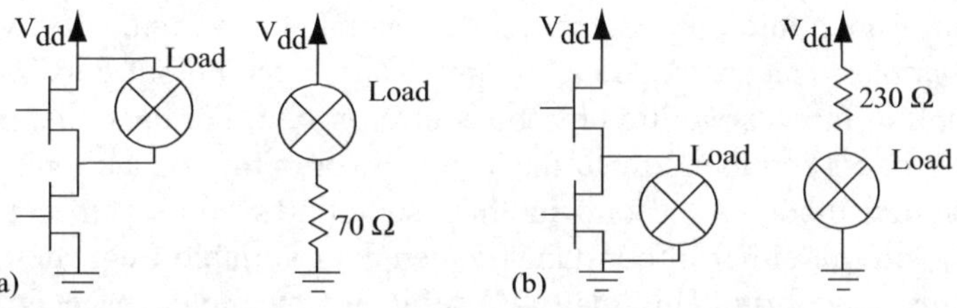

FIGURE 12.19 ■ Connection of a load to the I/O pin of a microprocessor. (a) Sinking current and its equivalent circuit. (b) Sourcing current and its equivalent circuit.

25 mA in sink mode and 20 mA in source mode. The difference is due to the internal resistance of the two types of MOSFETS (the *p*-type MOSFET has higher internal resistance). Even when this difference is not a consideration, the internal resistance is connected in series with the load and must be taken into account in calculations of the current in the load and the voltage across it. These I/O ports can drive a number of devices directly. Small loudspeakers and buzzers, LEDs, small relays, and the like are commonly driven in this mode.

In general, the output pins are not sufficient to drive inductive loads directly, with the exception of small relays. Inductive loads can create relatively large current spikes when switched off. For the small loads that can be driven directly, there is no need for protection against these spikes since the diodes connected across the pin (see **Figure 12.18**) are there for this purpose.

All of the considerations above assume the loads require the same voltage as the microprocessor. Although not a rule, microprocessors include at least some open drain outputs. An example is shown in **Figure 12.20a**. The connection of a load is shown in **Figure 12.20b**. The voltage V_L can be larger than the maximum voltage (V_{dd}) of the microprocessor. In most cases V_L can be up to 14 V ($V_{dd} \leq 6.5$ V), as this is the maximum typical voltage used for programming the microprocessor. The current, however, cannot exceed 25 mA.

If more power is needed, an external circuit must be provided. Some of these circuits were discussed in **Chapter 11** (power amplifiers); others are simple external power transistors or MOSFETs driven by the I/O pin, perhaps from an internal PWM source. Again, it is important to observe all the rules discussed above. The output power of a pin cannot be exceeded, and if isolation is required, appropriate isolation circuits must be used.

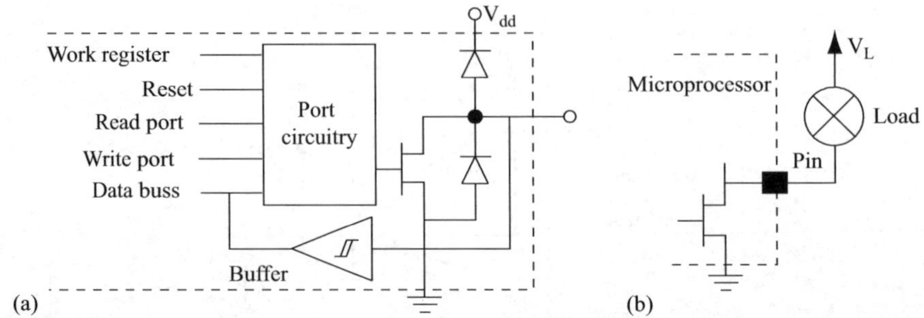

FIGURE 12.20 ■ An open drain output pin. (a) Internal connections. (b) Connection of a load to an open drain output pin. V_L can be larger than V_{DD}.

In the case of outputs, the use of optical couplers is particularly useful since the output pin can easily drive the LED in the coupler directly. Some of the I/O pins are specified as PWM outputs. The considerations above apply to these as well.

The output from a microprocessor can be generated in a number of ways. The most obvious and simplest is a switched DC output. That is, the output port is driven high or low for a period of time as required by the control circuit. For example, an I/O pin may be driven high for 5 s and then low to flash an LED or to turn on a circuit for 5 s. This is the most common mode. A second method sometimes employed is to generate, in software, a waveform, either of constant frequency or variable frequency, and drive the output with this waveform. This may be useful, for example, in generating an alarm pattern or data string or, with proper design, to generate rudimentary speech (speech is usually generated with a much more sophisticated signal processor). The third method is the PWM mode. In the PWM mode, the output is generated through an internal module. This module generates a fixed-frequency square wave whose duty cycle is software controlled. These modules are general purpose and may be used in many applications, but they have been specifically introduced for the purpose of modulating output power in loads, including in switching power supplies, speed control of motors, and the like. They are, in effect, simple D/A modules.

Often the PWM is part of a more general module called a capture/compare/PWM module. In capture/compare mode, it is an input module capable of capturing events and comparing them with internal timings or external events. In the PWM mode, the user defines the period of the PWM (i.e., the frequency). Then the duty cycle can be changed as necessary from zero to the period of the PWM. The resolution can be up to 14 bits but varies based on the processor and on the frequency of the PWM.

EXAMPLE 12.11 Power dissipation considerations

One reason why the output current for an output pin on a microprocessor is limited to 25 mA is to avoid overheating of the microprocessor due to power dissipation in the output drive circuitry. Consider a microprocessor with five I/O ports, each with eight pins. The pins of four ports are set as output pins and each drives an LED at 20 mA. Calculate the power dissipated in the microprocessor if

a. All pins are connected in source mode and all are turned on.

b. All pins are connected in sink mode and all are turned on.

Solution: The power dissipated is due to the internal resistance of the drive MOSFET. In source mode, that resistance is 230 Ω, whereas in sink mode it is 70 Ω (see **Figure 12.19**).

a. Each port has eight pins so that 32 pins are driven. The total current is

$$I = 32 \times 20 = 640 \text{ mA}.$$

The power dissipated in source mode is

$$P = 32 \times I^2 \times R = 32 \times (0.02)^2 \times 230 = 2.944 \text{ W}.$$

b. In sink mode the resistance is only 70 Ω. The current required is the same as in (a):

$$I = 32 \times 20 = 640 \text{ mA}.$$

The power dissipated is

$$P = 32 \times I^2 \times R = 32 \times (0.02)^2 \times 70 = 0.896 \text{ W}.$$

This is much lower. However, microprocessors cannot dissipate that much power. Therefore, even though each pin can carry up to 25 mA, that does not mean all pins can do that simultaneously. For this reason the total current allowable for the microprocessor is limited to less than about 200 mA. If it becomes necessary to drive more current or more pins, external circuitry must be used.

12.4 | ERRORS

The issue of errors has been discussed throughout this text with an emphasis on errors in sensors and actuators. In **Chapter 11** we also mentioned the fact that any circuit used for interfacing necessarily adds its own errors to the overall errors of the system. Microprocessors are no exception. However, because it is a digital device (with the exception of the comparator and A/D modules), the errors are typically those introduced by digital systems—those associated with the device's resolution. Other errors are due to the sampling process on the I/O pins. These are described next.

12.4.1 Resolution Errors

Resolution refers to a number of different issues in microprocessors. In a unit such as an A/D converter it refers to the minimum increment in the input that can be read as a distinct value. For example, a 10-bit A/D converter with a reference voltage of 5 V has a resolution of

$$\frac{5}{1024} = 4.88 \text{ mV}.$$

In converting a sensor's output, the output can only be distinguished in increments of 4.88 mV. This represents an error of $(4.88 \times 10^{-3}/5) \times 100 = 0.1\%$. Perhaps this is acceptable, but it may be higher than the errors of many sensors. A 14-bit A/D converter will resolve down to 0.3 mV, but 14-bit A/D converters are not commonly available in microprocessors.

Then there is the issue of the reference voltage itself. In microprocessors the reference voltage for A/D converters is derived from the power supply. Any error in the reference voltage will add to the total error.

In digital functions, the term resolution refers to the least significant bit (LSB). Clearly, any digital system cannot resolve below the LSB. For example, suppose that the output of an external A/D converter is 10 bits. Its resolution is defined as 1 bit. If we read it in an 8-bit register, the last two bits will be lost, effectively reducing the resolution by 2 bits. Of course, one would not intentionally do so unless there was some good reason for it.

In PWM modules, resolution is defined as

$$PWM_{res} = \frac{\log(f_{osc}/f_{PWM})}{\log(2)}. \tag{12.1}$$

This is a measure of the frequency of the PWM (given a constant f_{osc}, the clock frequency of the microprocessor). The lower the PWM frequency, the higher its resolution.

Another important issue in resolution has to do with calculations on the microprocessor. The fundamental mode is fixed point (integer computations). Every time a register overflows, a carry is generated, but for microprocessors, in which the basic integer is 8 bit, only integers from 0 to 255 are represented directly. More often than not, 16, 24, or even 32 bits are used in subroutines, severely limiting the available memory for other applications. But even at 16 bits, the largest integer represented is 65,535. Any number larger than that is truncated. These errors (sometimes called round-off or truncation errors) are a problem when extensive computations become necessary. When fractional numbers need to be handled, these are typically handled using fixed-point arithmetic, and again truncation beyond the representable values occurs. To overcome these, special math routines, both integer and floating point, have been developed and are freely available. They help with round-off errors by incorporating more accurate algorithms and special programming techniques, but cannot eliminate errors entirely.

EXAMPLE 12.12 Errors due to finite resolution of an A/D converter

A Hall element is set up to produce a Hall voltage of 50 mV/T (see **Section 5.4.2**) and is used to measure magnetic fields that vary between 0.1 T and 1 T. The output of the Hall element is digitized using a 10-bit A/D converter internal to a microprocessor, as shown in **Figure 12.21a**. The microprocessor operates at 3.3 V and the A/D converter uses the 3.3 V source as a reference.

a. Calculate the error in the reading of the microprocessor as a function of the magnetic field.
b. In an attempt to reduce the error it is suggested to first amplify the voltage by a factor of 60, as shown in **Figure 12.21b**, and then digitize the signal. What is the error in the reading of the microprocessor as a function of the magnetic field?

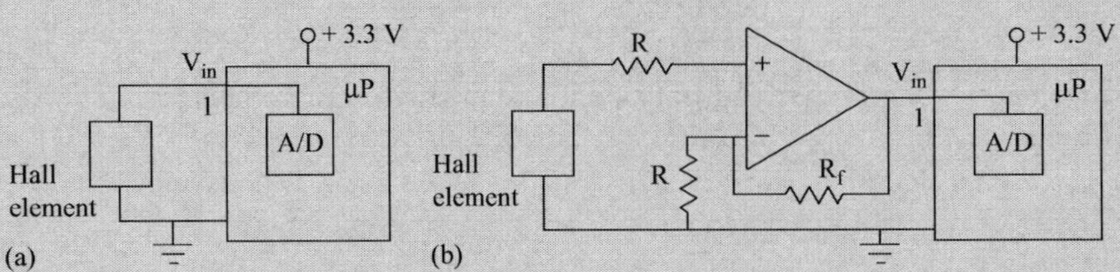

FIGURE 12.21 ▪ (a) A magnetic field sensor connected directly to a microprocessor. (b) The magnetic field sensor's output is amplified before the microprocessor reads the voltage.

Solution:

a. The A/D converter has a resolution ΔV:

$$\Delta V = \frac{3.3}{1023} = 0.0032258 \text{ V}.$$

Because the A/D converter can only read in increments of ΔV, the error will vary from point to point. To calculate the error, we calculate the reading of the A/D converter in the span of the input. This is shown in **Figure 12.22a**. Note the staircase curve showing the digitization in terms of ΔV. The error is calculated as follows:

$$error = \frac{V_{\text{in}} - V_{A/D}}{V_{\text{in}}} \times 100,$$

where V_{in} is the voltage at the input to the A/D converter and $V_{A/D}$ is the equivalent digitized value at the output of the A/D converter. The error is plotted in **Figure 12.22b**. Note that at all values of the input voltage that are integer increments of ΔV, the error is zero. The error increases from one integer increment to the next and is maximum just before the next integer increment of ΔV. Maximum error occurs at the lowest input and decreases gradually as the input increases. In the case here, the error is 50% just before an input of $2\Delta V$ (6.45 mV) and about 6.5% at an input of $15\Delta V$ (48.38 mV). These errors are very large and generally unacceptable.

b. The amplifier must be a noninverting amplifier to ensure the voltage at the input pin is positive. To produce an amplification of 60, and hence a maximum input voltage to the microprocessor of 3 V, we use **Equation (11.7)** and write

$$A = 1 + \frac{R_f}{R} = 60$$

or

$$R_f = 59R.$$

A possible combination of resistances is $R = 10$ kΩ and $R_f = 590$ kΩ. With these the input to the microprocessor varies from $0.005 \times 60 = 0.3$ V for a magnetic field of 0.1 T to $0.05 \times 60 = 3$ V for a magnetic field of 1 T.

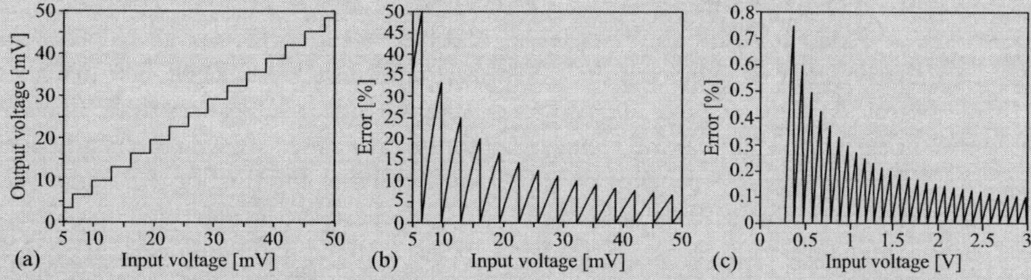

FIGURE 12.22 ■ (a) Output of the 10-bit A/D converter as a function of its input showing the quantization in the digital output (shown here as voltage rather than as a digital representation). (b) Error in input voltage representation for the circuit in **Figure 12.21a**. (c) Error in input voltage representation for the circuit in **Figure 12.21b**.

Using the formula for error in (a), we obtain **Figure 12.22c**. The figure has the same general behavior, but now the error ranges from a maximum of 0.75% to 0.1%, a much more acceptable figure.

As indicated in **Section 11.5**, A/D converters are more accurate in the higher range (see also **Figure 12.26**).

12.4.2 Computation Errors

Microprocessors are designed as general purpose controllers, not as computers. That means that computation is not a major consideration in their operation. Nevertheless, the need to compute values arises, and although not particularly efficient, microprocessors are capable of rudimentary computation. Computation other than simple binary addition and subtraction must be done in software and entails the use of routines, either written by the user or, often, supplied by the manufacturer. Many useful and efficient routines are freely available from various sources. These routines include integer, fixed point integer, and floating point operations. Nevertheless, the user should be careful with noninteger mathematics, especially with floating point calculations—these require more resources than integer operations and entail approximations and truncation of numbers, leading to the introduction of errors, whereas integer operations are faster and are exact. As a rule, one opts for fixed point operations or for floating point operations only when these are absolutely necessary.

Integer computation is exact as long as the variables and the results can be represented with the word lengths available. On an 8-bit microprocessor one can use multiple 8-bit words if necessary so that computations of considerable scope can be carried out with relative ease (see **Appendix C** for details). In practical applications one often needs to use nonintegers. One can easily imagine the division of two integers resulting in a fractional number or the need to scale an integer (or fractional number) by a fractional scaling factor. Or simply suppose the microprocessor is required to use π in an internal calculation. In computers, calculators, and hand computation, this is handled using floating point numbers that use a mantissa and an exponent (e.g., $\pi = 3.14E00$). Because microprocessors do not have sufficient resources to allow floating point computation, fractional numbers are handled using fixed point arithmetic using two integers, one representing the integer part of the number and one representing the fraction. A complete description of the use of binary integers and binary fixed point numbers is given in **Appendix C**.

Aside from issues of resources and the fact that fixed point calculations are slower than integer calculations, one must also take into account the fact that these introduce errors. To understand the problem, suppose one needs to perform the division 10/9. In integer arithmetic the result is 1 (an error of 10%). If we represent the numbers as a fraction, the error depends on how many digits of the fraction we retain. We may write 10/9 as 1.1 or 1.11 or 1.11111111. All of these are inexact and introduce an error in computation. Of course, in a microprocessor the computation is done with binary numbers, but the same principle applies. For example, if the fraction uses 8 bits, the smallest value that can be represented is 1/256, which is more than 3.5% of the fraction 0.1111111. The error in representation can be significant. One can reduce the error by increasing the number of bits allocated to the fraction, but one must be conscious of the limited available resources on the microprocessor and the time needed to complete the calculation.

EXAMPLE 12.13 Errors due to computation

An RTD is connected to a microprocessor as shown in **Figure 12.23**. The input voltage to the microprocessor (pin 1) varies between 0.21 V at 0°C and 1.35 V at 100°C (i.e., the bridge has an offset voltage of 0.21 V at 0°C). The bridge voltage is converted into digital form using an internal 10-bit A/D converter with a reference voltage of 5 V. The output of the microprocessor must send data using a serial port to a display, which converts the data into decimal display.

a. Calculate, using minimum resources, the conversion of the input voltage into the correct value for display.
b. If the display can only display two decimal digits, calculate the error in reading the temperature as a percentage of full scale in (a).

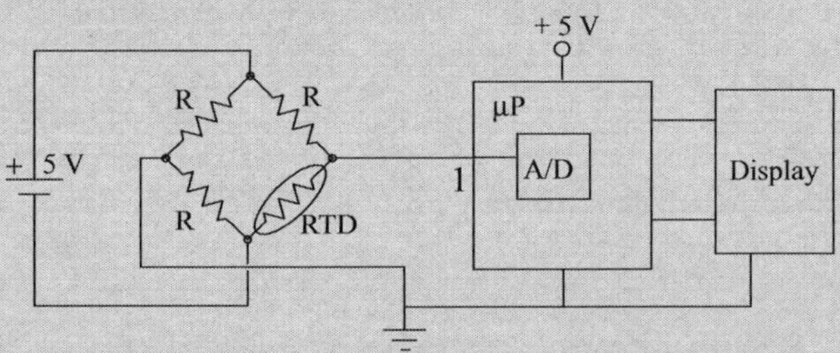

FIGURE 12.23 ■ Bridge circuit used to measure temperature.

Solution: For correct display we must first remove the offset after reading the input voltage through the A/D converter. Then we must scale the range of voltages read so that the output is in the proper range (0°C–100°C).

a. The A/D converter is 10 bit. That is, the resolution of the converter is

$$\Delta V = \frac{5}{1023} = 0.004887585 \text{ V}.$$

That means the reading after the A/D converter is, at 0°C, $r_0 = 1023 \times 0.21/5 = 43$. In digital form this becomes 00000000 00101011. This was written using two 8-bit words, for reasons that will become obvious shortly. This must be multiplied by ΔV to find a representation of the voltage in digital format. In decimal format we would have

$$r = 43\Delta V = 43 \times 0.004887585 = 0.21 \text{ V}.$$

This must be done by the microprocessor using fixed point operations. Since the voltage at the input to the microprocessor is less than 2 V, the integer part of the expression does not need to be larger than 1 bit. Because the fractions we deal with can be small, we will use two 8-bit integers for each value. In this case we need to express ΔV, and to ensure accuracy we use 1 integer bit and 15 bits for the fraction. We write $r_0 = 43$ as an integer and ΔV as a binary fractional number. We write them in binary form as follows:

r: 0 0 0 0 0 0 0 0 0 0 1 0 1 0 1 1 ΔV: 0 • 0 0 0 0 0 0 0 1 0 1 0 0 0 0 0

Each number uses two bytes of data. This is now multiplied by a sequence of shifts and adds as follows (see **Appendix C**):

```
            0000000010100000×
            0000000000101011
            ----------------
            0000000010100000+
           0000000010100000
           -----------------
           00000000111100000 +
         0000000010100000
         -------------------
         0000000011011100000 +
       0000000010100000
       ---------------------
       000000001101011100000
            └┘└─────────────┘
 1 bit integer   15 bits fraction
```

If we translate this back into decimal format we get 0.2099609375 (instead of the exact value 0.21).

At 100°C, $r_0 = 1023 \times 1.35/5 = 276$. In digital form this becomes 00000001 00010100 (requires a minimum of 9 bits, hence the need to use two 8-bit words). This must be multiplied by ΔV to find the digital representation of the voltage:

$$r = 276\Delta V = 276 \times 0.004887585 = 1.34897346 \text{ V}.$$

To calculate r in digital format we multiply ΔV by r_0:

```
                0000000010100000×
                0000000100010100
                ----------------
                0000000000000000+
              0000000010100000
              ------------------
              000000001010000000+
            0000000010100000
            --------------------
            00000000110010000000+
        0000000010100000
        ------------------------
        000000001010110010100000
                └┘└─────────────┘
 1 bit integer      15 bits fraction
```

In decimal format this equals 1.34765625, close to the exact value of 1.35.

Now we subtract the representation for 0.21 V from the representation for 1.35 V. This is done using the 2's complement method: the subtracted number is first 1-complemented (all zeroes become ones and all ones become zeroes). Then we add 1 to the result and that is added with the positive number (see **Appendix C**). The 1's complement of the representation for 0.21 V is

1	1	1	0	0	1	0	1		0	0	0	1	1	1	1	1

Adding 1 to this and disregarding the carry (if any), we get a representation of−0.21:

```
1110010100011111+
0000000000000001
----------------
1110010100100000
```

This is now added to the representation for 1.35 V to remove the offset:

```
1110010100100000+
1010110010000000
----------------
1001000110100000
```

In decimal format this is 1.1376953. That is, the correct voltage varies between 0 V and 1.14 V, although, because of the finite accuracy in representation, the actual values are 0 and 1.1376953.

To scale the input so it represents the temperature, we note that the span of the input is 1.14 V for which the output span must be 100°C. Therefore the input must be scaled by

$s = 100/1.14 = 87.7192982456$. This value must be represented in fixed point format. Since 87 requires at least 7 bits in binary format, we use 8 bits for the integer and 8 bits for the fraction:

0	1	0	1	0	1	1	1	•	1	0	1	1	1	0	0	0

This now multiplies the input voltage after the offset has been removed. To obtain the display reading we multiply the value above by the scaling factor s and obtain

```
                1001000110100000×
                0101011110111000
                ----------------
                0000000000000000+
             1001000110100000
             -------------------
             1001000110100000000+
            1001000110100000
            --------------------
            11011010011100000000+
           1001000110100000
           ---------------------
           111111101101100000000+
         1001000110100000
         -----------------------
         1101000101010110000000000+
        1001000110100000
        ------------------------
        11111010010010110000000000+
       1001000110100000
       -------------------------
       100001110110001011000000000+
      1001000110100000
      --------------------------
      1000110010000000101100000000+
     1001000110100000
     ---------------------------
     11010111111100000101100000000+
    1001000110100000
    ----------------------------
    01100011 11001100 000101100000000
    8 bit integer  8 bits fraction
```

Note that since the two numbers have 15 and 8 bits in their corresponding fraction, there are a total of 23 bits in the fraction after multiplication. We use the first 8 of these for the fraction, truncating the last 15. The representation of 100°C is therefore 0110001111001100. In decimal format this is 99.796875.

The span of the readings of the microprocessor is between 0°C and 99.796875°C.

b. Since the display can only show two decimal points, the display for 100°C is 99.79°C. At 0°C the display shows 0.00 even though the representation of the offset is not exact since it is subtracted from itself. Note that the display does not round off to the nearest integer, it simply truncates the fraction. Of course, one can correct this in software if it is deemed important.

The error is small. In (a) the error at 0°C is zero, and at 100°C the error is $[(99.79-100.0)/100]\times 100 = -0.21\%$. The error is minor and well below the possible error in the sensor. However, one must consider the possibility that this error is added to any other error in the system.

EXAMPLE 12.14 Voltage sensor/monitor: computation errors

A microprocessor in a car is used to monitor the battery and display its voltage digitally on the dashboard (among other functions). To design the circuit, the battery voltage is measured and found to be 12 V when no load is applied (nominal voltage). When charging, the voltage rises to 14.7 V and when fully loaded (lights on and engine off) the voltage is 11.4 V. Design a circuit that will sense the battery voltage and display it using two decimal points. Calculate the errors expected at the three measured points. The microprocessor has an internal 8-bit A/D converter available for use in the design and operates from a regulated 5 V source.

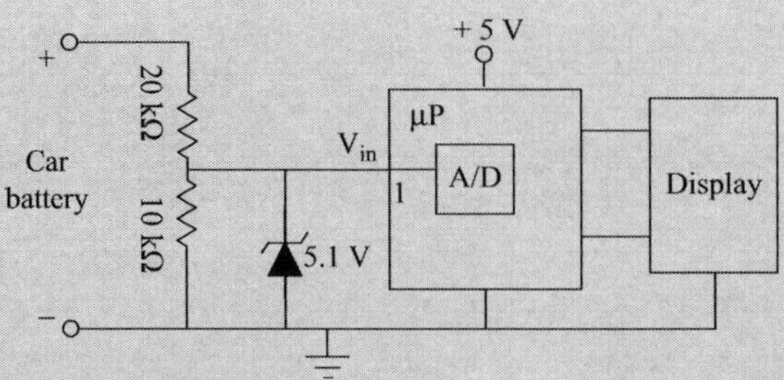

FIGURE 12.24 ■ Voltage sensor and circuit used to monitor a car's battery.

Solution: The design is not unique, but there are certain limits that must be respected. One of these is that the input to the microprocessor cannot exceed 5 V. A possible circuit is shown in **Figure 12.24**. The resistances in the voltage divider are selected to provide a 4 V input on pin 1 at nominal voltage. The resistors are arbitrary, but they should not be too small so that the current consumption is low. They cannot be too high either, since then the minute current that flows into pin 1 (typically less than 1 μA) would affect accuracy. With the values shown, the maximum expected voltage of 14.7 V results in a voltage of 4.9 V on pin 1. The circuit also includes a 5.1 V zener diode across the input to ensure the microprocessor is not damaged if, for some reason, the battery voltage rises above 15 V.

The input voltage is digitized using the A/D converter. The 8-bit A/D converter has 255 steps and hence distinguishes 256 values in increments of $5/255 = 0.019607843$ V. Internally the sensed voltage is stored as digital values ranging from 0 (00000000) to 255 (11111111), with 255 representing 5 V. Note that the output from the A/D converter can be stored in 8 bits.

After reading the input it must be scaled to produce the proper output for display. For example, when the input voltage is 4 V, the A/D converter produces a digital output 11001101 whose decimal value is $(4/5) \times 255 = 205$. We will call this R. The display should show 12.00 V. This means we must scale the reading from $205 - 1 = 204$ to 12.00. This is done using a scaling factor (SF):

$$SF = \frac{12.00}{204} = 0.058823529.$$

An alternative is to send a display value of 400 (and take care of the decimal point in the display):

$$SF = \frac{1200}{204} = 5.8823529.$$

The latter is somewhat more accurate since the scale factor is larger and can be represented digitally more accurately within a limited number of digits, but both will produce acceptable results. Either choice requires a noninteger scale. Selecting the first, we represent within an 8-bit word with 1 digit for the integer and 7 digits for the fraction as follows:

0	•	0	0	0	0	1	1	1

The representation equals 0.0546875, and since it is not exact, it will introduce an error in the displayed voltage.

The display value D is now the product of the A/D converter value R and the scale factor:

$$D = R \times SF.$$

The display values for the low, nominal, and high voltage are as follows:

Low voltage is 10.4 V. The input to the microprocessor is 10.4/3 = 3.4667 V. The A/D converter produces a digital value equivalent to

$$R = \left(\frac{3.4667}{5}\right) \times 255 = 177,$$

or in digital form, 10110001.

The display value is the product of this and the scale factor (0.0000111). Note that the radix point is implicit, but it does not exist in reality. We now multiply the fixed point value scale factor and the integer value R:

```
          10110001×
          00000111
          10110001+
         10110001
        1000010011+
        10110001
    00000010011010111
    9 bit integer   7 bit fraction
```

The decimal value displayed is 9.67 instead of 10.4. This is off by 10.4–9.67 = 0.73 V, or about 7%.

At 12 V the A/D converter reading is 205, or 11001101. Multiplying this by the scale factor, the display value is

```
          11001101×
          00000111
          11001101+
         11001101
        1001100111+
        11001101
    00000010110011011
    9 bit integer   7 bit fraction
```

The decimal value displayed is 11.21 instead of 12.00. The error is about 6.6%. In the high limit,

$$R = \left(\frac{4.9}{5}\right) \times 255 = 250,$$

or in digital format, 11111010. Multiplying this by the scale factor, the display is

```
          11111011×
          00000111
          11111010+
         11111010
        1011101110+
        11111010
    00000011011110110
    9 bit integer   7 bit fraction
```

This is 13.92 instead of 14.7 V, off by 5.6%.

As a rule the errors in this design are too large. They stem partly from the fact that we use an 8-bit A/D converter, but mainly due to the fact that the representation of the scale factor is not sufficiently accurate.

This can be improved by using 16 bits to represent the scale factor, with 15 bits for the fraction and 1 bit for the integer. The digital representation of SF = 0.00588235294118 is

```
[0]•[0][0][0][0][1][1][1] [1][0][0][0][0][1][1][1]
```

Now we repeat the products calculated above with the more accurate scale factor.

For the low value (10.4 V),

```
          0000011110000111×
                  10110001
          ----------------
          0000011110000111+
      0000011110000111
      --------------------
      00000111111111110111+
     0000011110000111
     ---------------------
     000010111000011010111+
   0000011110000111
   -----------------------
   00001010011010001010111
   |_______||____________|
8 bit integer   15 bit fraction
```

This 23-bit result can be (and should be) contained in 16 bits by disregarding the last three digits of the fraction and the first four digits of the integer. The digital value is reduced to 1010.011010000010. The decimal value is 10.40 as required.

For the nominal value, the display will show

```
          0000011110000111×
                  11001101
          ----------------
          0000011110000111+
         0000000000000000
         -----------------
         00000011110000111+
        0000011110000111
        ------------------
        0000100101101000011+
       0000011110000111
       -------------------
       00001100000111011011+
    0000011110000111
    ----------------------
    0000100100000111001101 1+
   0000011110000111
   -----------------------
   00001100000001110001101 1
   |_______||____________|
8 bit integer   15 bit fraction
```

In decimal format this is 12.05, or 0.42% off.

At the high value we have

```
          0000011110000111×
                  11111010
          ----------------
          0000000000000000+
         0000011110000111
         -----------------
         00000111100001110+
        0000011110000111
        ------------------
        0000100101101000110+
       0000011110000111
       -------------------
       00001100000111011011 0+
      0000011110000111
      --------------------
      000001101101001001011 0+
     0000011110000111
     ---------------------
     0000011100101100101011 0+
    0000011110000111
    ----------------------
    00001110101100111010110
    |_______||____________|
8 bit integer   15 bit fraction
```

This is 14.70. The value is exact.

In this case the display values are very accurate. The slight error is primarily due to the 8-bit A/D converter rather than the computation.

12.4.3 Sampling and Quantization Errors

Another source of errors, one that is not as well defined as the resolution errors, is due to sampling. This stems from the fact that any input must be read (or sampled) and this sampling is not continuous or at a fixed rate. Rather, it is defined in the program and depends both on the logic (i.e., what the programmer intended) and on the execution time. Sampling theory says that if a signal is sampled at twice its frequency, the signal can be reconstructed exactly. This of course is the theoretical limit, and in practice faster sampling will be necessary. Also, it assumes a monochromatic signal (no harmonics). In a digital device the sampling itself is digitized, or quantized, that is, we can never sample a signal at its exact level, but rather at increments of, for example, the A/D converter used. Between two samples the signal remains constant and we obtain a staircase representation of the analog signal. In general, analog signals are susceptible to errors both in sampling and quantization. Digital signals are not susceptible to quantization errors since the amplitude is fixed, but they are susceptible to sampling errors, particularly because they are rich in harmonics. For example, suppose that a sensor system is designed to be sampled every 100 μs and it takes 100 instructions to execute the port read instructions (the read itself only takes one or at most two cycles, but the program may involve many other steps of comparison, checking, calculations, etc., before the read is actually performed). For a 0.1 μs processor (40 MHz with internal division by 4 or 10 MHz without division), the total time will be 110 μs ($100 \times 0.1 + 100$). During the 10 μs needed for the instructions the sensor may change back and forth without the processor "seeing" these changes. Granted, this is less of a problem than one may expect since most sensors are rather slow and there are ways of mitigating this issue (see **Example 12.15**). The main point here is that, first, sampling cannot be done at arbitrarily small intervals because of the clock on the microprocessor, and second, programming affects these errors. In the example above, better programming might reduce the number of cycles to, say, 20, reducing the total delay and the errors due to these delays. Quantization of analog signals is further discussed in conjunction with A/D converters.

EXAMPLE 12.15 Errors due to sampling

A microprocessor is used to read an incoming digital signal from a digital capacitive sensor. The frequency of the signal varies in frequency depending on the measurand with maximum frequency of 1 kHz and a 50% duty cycle. The microprocessor determines the frequency of the signal by measuring time. Suppose the incoming signal is sampled at 10 kHz, that is, the time of one cycle is 1 ms and it is sampled every 100 μs. Calculate the error in determining the frequency if

a. Only the half-cycle during which the pulse is high is measured (this may be done so the processor can do something else during the next half-cycle).

b. A full cycle is sampled.

Solution: The main issue to recall is that the microprocessor keeps the sampled input until the next sampling and only changes it if at that point the signal has changed. The maximum error expected is at the highest frequency.

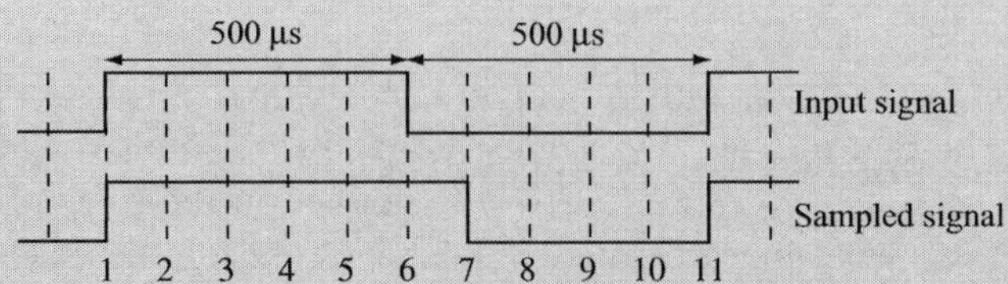

FIGURE 12.25 ■ Sampling the value of a signal on an input pin. Sample numbers are shown at the bottom of the figure.

a. Consider **Figure 12.25**. The first sample (indicated as "1") is read as a logic "1". The signal is retained until the next sample, which happens to be "1" as well. This continues until sample 6. This is "1" as well and is retained until sample 7 is obtained, at which point the sample is "0". Therefore the width of the pulse is read as 6×100 μs, and since it is assumed that the signal has a 50% duty cycle, the microprocessor reads the cycle time as 1200 μs and the frequency is

$$f = \frac{1}{1200 \times 10^{-6}} = 833.33 \text{ Hz}.$$

The error is significant:

$$error = \frac{833.33 - 1000}{1000} \times 100\% = 16.67\%.$$

b. From **Figure 12.25** it is obvious that at sample 11 the signal rises again to "1" and hence the time counted for a whole cycle is 1000 μs. The microprocessor reads the frequency correctly as 1 kHz. This assumes that the sampled signal is equal to "1" when the sampling occurs. If it were read as "0", an additional sample would be required for a total time of 1100 μs. The error in this case is 10%.

Obviously this simple calculation does not take into account delays due to program steps or sampling that may not occur exactly at the determined times in this example. But it emphasizes the need to sample the signal at the highest practical rate. In addition, it may be good practice to sample at least a whole period. Sometimes a number of periods may be sampled either continuously or separately before determining the frequency, thus averaging errors over a longer period of time.

12.4.4 Conversion Errors

The components of the microprocessor are not perfect and all are subject to temperature variations, drift, and manufacturing variations. Any operation, particularly conversion, introduces errors. If one uses the internal comparator with the internal voltage reference, the actual value of the reference voltage will influence the output. For example, if the reference voltage is nominally 0.6 V but can vary between, say, 0.595 V and 0.605 V, comparison will occur at the actual value rather than the presumed reference of 0.6 V. Similarly an A/D converter, apart from the resolution issue discussed above, will introduce its own errors due to internal circuitry, temperature variations, and variations in the reference voltage. Taking again the 10-bit A/D converter discussed above, with a reference of 5 V, a 2.5 V input should produce a digital output 1000000000, or the equivalent decimal value 512. But suppose that instead it produces an output 1000000011, or a decimal equivalent of 515. This would represent an error of 0.29% [$(3/1024) \times 100\%$]. These errors are not constant either—they depend on the value being converted, with lower values incurring larger errors (see **Example 12.16**). Each internal component in the microprocessor is specified in the data sheet and errors expected are indicated in percentages, by specifying minimum and maximum values, or in the case of the A/D converter, in terms of bits. For example, the error in an A/D converter may be specified as ± 1 bit.

EXAMPLE 12.16 Conversion errors

A 12-bit A/D converter in a microprocessor is subjected to a test whereby the input changes from 0 to 5 V and the digital output of the converter is read and compared with the expected digital output for the given input as measured with an accurate voltmeter. The error is calculated as follows:

The digital output full scale of the converter is $2^{12} = 4096$, input full scale is 5 V, reference voltage is 5 V.

The digital output (DO) produced by a given analog input (AI) is read and converted to a decimal equivalent (DE). We calculate

$$error = \left| \frac{\frac{DE}{4096} \times 5V - AI}{AI} \right| \times 100\%.$$

This gives the absolute value of the error as a percentage at any input voltage. The error for the 12-bit A/D converter is plotted in **Figure 12.26**, showing a reduced error with an increase in input voltage.

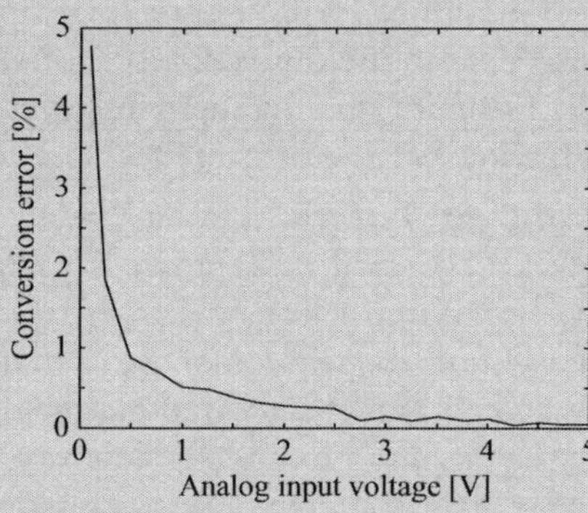

FIGURE 12.26 ■ A/D conversion error as a function of the analog voltage being converted.

This experiment shows as well that it is preferable to operate with higher sensor outputs if A/D conversion is necessary. If the output of the sensor is low, it may be better to amplify its output before digital conversion. Note also that the error is not insignificant, even at higher input voltages, and in some cases may be larger than that of the sensor itself. This is to be expected from a general purpose converter in a microprocessor. Of course, better A/D converters with temperature compensation and stable references are available and can be used external to the microprocessor.

12.5 | PROBLEMS

Instructions

12.1 Use of instructions on a microprocessor. The operation $e = (a \times b + c \times d)/2$ with $a = 2$, $b = 4$, $c = 12$, and $d = 8$ must be performed. Show the sequence the processor must follow to obtain the result.

12.2 Logical operations on a microprocessor. An air conditioning system uses hot and cold air, properly mixed, to maintain a set temperature, measured by a temperature sensor. To perform the operations a microprocessor measures the temperature T and places it in a register we will call t after performing A/D conversion using an 8-bit A/D converter. The user enters the required temperature that we call s as an 8-bit datum. Two valves are controlled by separate outputs that open the hot air (H) or the cold air (C) as necessary. Normally both valves are closed.

a. Show the sequence needed to control the temperature by controlling the two valves.

b. Show the sequence needed to control the temperature so that the H valve is open approximately 1°C above the point in register s and the H valve is opened at the set point in s.

c. If the range of temperatures expected is 0°C–100°C, what is the temperature step that can be set?

Input and output

12.3 Control of a motor. Turn on a small DC motor when a switch has been pressed for at least 3 s, run the motor for 10 s, and turn it off. Show a schematic and a flow chart assuming the microprocessor can count the required times. Pressing the switch while the motor is on has no effect.

12.4 Control of light fixtures. A room has two light fixtures operated by a single switch. When the switch is pressed, fixture 1 turns on. When the switch is pressed again, fixture 1 turns off and fixture 2 turns on. A third press turns both lights on. Continued operation of the switch repeats the sequence. To turn off the lights the switch must be pressed and held for 5 s. Show a schematic with the input and output pins identified and draw a flow chart to accomplish the control. Assume that there is an internal way to count 5 s.

Clock and timers

12.5 Digital ultrasonic distance measurement. A simple and accurate method of sensing distance is to use an ultrasound transmitter (actuator) to send an ultrasonic pulse and an ultrasonic receiver (sensor) to detect the reflected pulse from the target. Since the speed of sound in air is relatively slow ($v = 331$ m/s), the time of flight t to the target and back is a good indication of the distance. The distance d is calculated as $d = vt/2$ and displayed.

a. Draw a schematic to show how this can be done using a microprocessor and explain its operation. Identify the necessary components.

b. Write the basic sequence of operations needed to accomplish the measurement.

c. Draw a flow chart of the measurement sequence. Use a 40 kHz transmitter/receiver and a 16 MHz processor without internal division of frequency.

d. Discuss possible sources of errors in the measurement and means of minimizing them.

12.6 Limitations of timing in microprocessors. Microprocessors operate at a relatively low frequency, typically below 50 MHz. Hence the basic cycle time limits the applications that can be handled or limits the accuracy of sensing. Consider an autofocus camera that uses an infrared beam to automatically focus the lens from 0.5 m to 10 m.

a. What are the requirements of the internal clock if one attempts to measure the time of flight of the beam from the camera to the subject and back? Assume the processor needs a minimum of 10 cycles to measure time. The speed of light in air is 3×10^8 m/s.

b. Based on the response in (a), is this method feasible?

Power consumption and power budget

12.7 Tire pressure sensing: power considerations. The pressure inside a tire is sensed using a pressure sensor and the pressure is transmitted from within the tire to a receiver installed on the vehicle. The sensor, transmitter, microprocessor, and any additional circuits are powered by a 3 V battery (two AA batteries in series) with a capacity of 2800 mA·h. Assume the microprocessor consumes 5 mA when operating and 5 μA when in sleep mode. The pressure sensor consumes 7 mA during operation, whereas the transmitter requires 20 mA during transmission and zero when off. Because batteries cannot easily be replaced, the system is required to operate for the lifetime of the tire without battery replacement (typically 7 years). To resolve this difficulty, it is proposed to add a low-power oscillator, as shown in **Figure 12.27**, that turns the microprocessor on and off so that data are collected every 2 min and transmitted, all in an interval of 5 s. During that process, data collection from the pressure sensor takes 2 s and transmission 3 s. The oscillator operates continuously and consumes 20 μA.

a. Calculate how long the system can operate continuously without any power reduction techniques (i.e., if all components of the system are on at all times).

b. Describe how the proposed oscillator can be used to reduce the overall power requirements of the system.

c. Show and describe how the pressure sensor and transmitter must be connected so that they turn off before the microprocessor goes to sleep and turn on after the microprocessor wakes up.

d. How long will the batteries last with the method proposed?

e. Can you suggest any additional power reduction techniques?

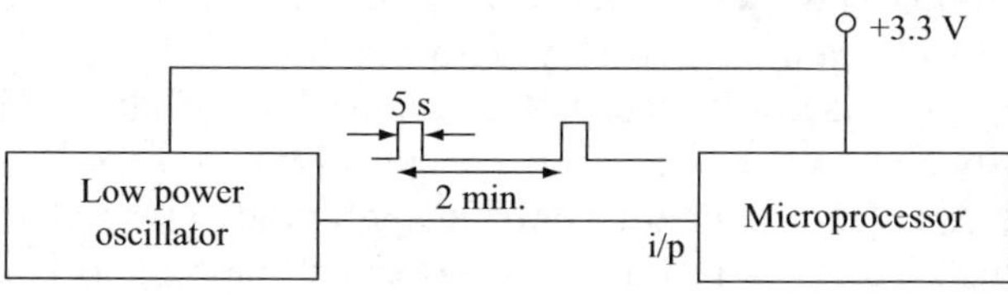

FIGURE 12.27 ■ Method of power reduction in a microprocessor.

12.8 Power budget calculation. A microprocessor in a remote door opener transmitter consumes 5 μA when in sleep mode and an average of 5 mA when in normal operating mode. The transmitter itself consumes 25 mA when on and zero when not in use. The device operates from a button battery with a 650 mA·h capacity. When the transmit switch is pressed the microprocessor and the transmitter follow the following sequence:

1. The microprocessor wakes up and requires 500 μs before it can transmit data. At that time it turns on the transmitter and data are transmitted for 20 ms, after which the transmitter is turned off.
2. If the transmit switch has been released, the microprocessor goes to sleep 300 μs after transmission is complete. The system always completes a transmission that has started.
3. If the switch has not been released, the process in (1) is repeated until the button is released.

   a. Calculate the life of the battery (in days) if an average switch press lasts 0.5 s and the remote opener is used 12 times/day on average. Assume that the sequence in (1) will be completed and followed by (2) even if the button is released earlier.

   b. What is the life of the battery if the microprocessor is programmed to send exactly two transmissions regardless of how long the button is pressed? The 300 μs are still required before the microprocessor can turn off.

12.9 Power consumption in battery-operated actuators. The keypad lock in **Figure 12.6** is modified to serve as a mechanism to unlock a door without a motor. Instead, a small magnetic valve that releases a spring-loaded lever is used. Once the lever is released, the door can be opened by hand. When the door is closed, the lever is armed and locks the door, ready for the next opening.

a. Estimate the current consumption of the circuit [mA] (without the valve) at 1 MHz and at 16 MHz for operating voltages of 2, 3, 4, and 5 V. Use the least squares approximation and compare the results with direct interpolation or extrapolation (see **Example 12.5**). Use the data in the table in **Example 12.5**.

b. The valve requires 350 mA to operate and releases the lever in 450 ms. Calculate the time the battery will last if the microprocessor and the valve operate from a 3 V battery with a capacity of 1900 mA·h. The microprocessor operates at 1 MHz and requires 8 s to dial the code prior to operation of the valve. The door is opened and closed an average of 20 times/day.

12.10 Power reduction techniques. A microprocessor can supply a maximum current of 25 mA on any of its output pins but no more than 200 mA total. It has four ports, each with 8 pins, and of these, 24 pins are used to drive LEDs at the same time, each LED requiring a current of 15 mA for proper visibility. To ensure the maximum current is not exceeded, it is proposed to switch the LEDs on and off at a sufficiently high rate so that to our eyes they look as if they are continuously on and ensure that at any given time only an appropriate number of LEDs are on. An LED turning on and off 16 times/s or faster will be seen as continuously lit.

a. Show how all 24 LEDs can be visible at the same time without exceeding the current limit. For efficiency in design, it is best if all LEDs connected to a port are turned on or off together. Indicate how the LEDs are switched on and off and determine the switching rate.

b. The effective intensity of an LED depends on how long it is on and off. Calculate the effective intensity relative to a continuous 15 mA current in the LED (taking the intensity with 15 mA continuous current as "1").

c. To compensate for the reduction of intensity due to switching, it is proposed to increase the current in the LEDs to the maximum the pin allows. Calculate the effective intensity under these conditions relative to the 15 mA continuous current.

Impedance and interfacing

12.11 Interfacing a charge sensor: negative charges. **Example 12.7** assumed that the charges accumulating on the plate in **Figure 12.9** were positive charges. Show what changes are needed to convert the circuit to sense negative charges. Discuss the effects of these modifications on the performance of the sensor system.

12.12 Energy meter. Design an energy meter to measure and display the energy consumed by the radio in a car. Assume the consumption varies with time, what is being played on the radio, and its volume. You only have access to the power lines feeding the radio. Use a microprocessor to measure the consumption as instantaneous consumption, displaying it at intervals of 1 s [W·h] and as cumulative energy over the lifetime of the car. The microprocessor operates at 5 V, whereas the radio operates at 12 V. The maximum current expected is 2.5 A.

a. Show the necessary sensors and the interfacing circuitry.

b. Discuss and select the clock frequency and the necessary internal components.

c. Write a detailed flow chart of the measuring and display process.

12.13 Interfacing of a pH meter. pH meters have some strict requirements in terms of impedance matching requiring very high impedance interface circuitry. In a microprocessor-assisted pH meter it is required that the impedance seen by the pH sensor be at least 100 MΩ and that the output seen by the microprocessor prior to digitization should vary between 0 and 5 V for pH 1–14. The pH membrane uses a silver/silver chloride (Ag/AgCl) reference membrane, which has a reference voltage of 0.197 V. The device is expected to operate at 25°C. The A/D converter available is 10 bit.

a. Design the necessary circuit and the interfacing components. Show how the output can display pH values directly.

b. Calculate the resolution of the instrument.

c. Discuss the issues involved in this design and the consequences of each.

Frequency and response

12.14 Frequency measurement. In **Example 12.8** the shift in frequency of a microbalance must be measured accurately. The microprocessor operates at 10 MHz (clock frequency) and has all the necessary blocks, including timers.

a. Show how the frequency can be measured using the microprocessor. Discuss possible ways of performing the measurement and their relative advantages.
b. Draw a diagram with the necessary circuitry and write a flow chart for the measurement process.
c. List possible sources of errors in the measurement and estimate these errors based on your design.

12.15 An analog metal detector. A metal detector is proposed as shown in **Figure 12.28**. Two identical *LC* oscillators are built and their frequencies subtracted using a simple analog mixer. The output from the mixer is the frequency difference f_1-f_2 if $f_1 \geq f_2$ or f_2-f_1 if $f_2 \geq f_1$. As well, the raw frequencies f_1 and f_2 are available. The variable capacitor shown is used to balance the oscillators so that in the absence of metal the two frequencies are identical and equal to 400 kHz. When a ferromagnetic metal of a certain size is detected, the inductance of the coil increases by 0.01%, whereas if the metal is nonmagnetic it decreases by 0.01%. The capacitor *C* is 0.001 μF. A microprocessor is used to measure the frequency difference and to indicate three parameters: (1) balanced oscillators before detecting metal, (2) if the detected metal is ferromagnetic, and (3) if the detected metal is nonferromagnetic.

a. Draw the microprocessor and any necessary circuits to accomplish this.
b. Define the clock frequency needed and the frequency measurement process.
c. Write a flow chart showing how the various displays are set and how one can detect which frequency (f_1 or f_2) is higher.

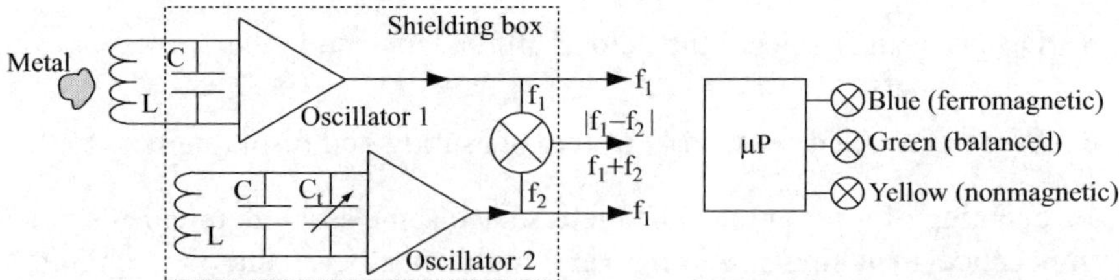

FIGURE 12.28 ■ An analog metal detector. The connections to the microprocessor are schematic, that is, additional circuitry may be necessary.

12.16 A digital metal detector. The metal detector in **Problem 12.15** can be modified to make a digital metal detector. Although the oscillators are still analog, their output is digitized using any of the methods discussed in **Chapter 11**. The schematic is shown in **Figure 12.29**. The oscillators are identical to those in **Figure 12.28**.

a. Select a method of signal digitization that will produce an appropriate signal for the microprocessor.

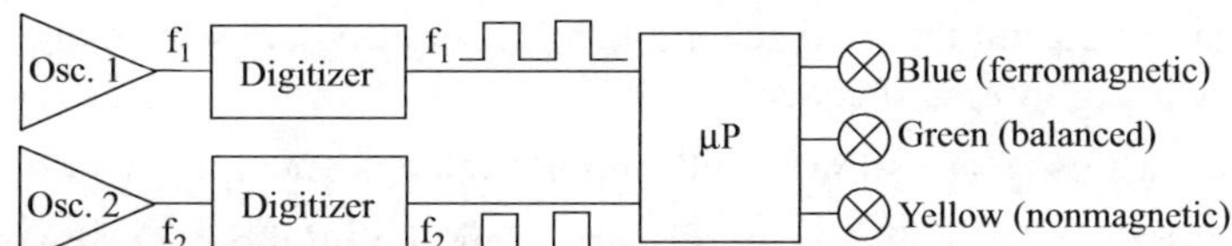

FIGURE 12.29 ■ A digital metal detector.

b. Show how the microprocessor can measure the frequencies and what the limitations are on those frequencies given the clock cycle of the microprocessor.

c. Discuss the sensitivity of this design and its limitations.

Scaling, offsetting, and errors

12.17 Errors due to variations in power supply voltage. In **Example 12.9** the reference voltage is derived from the power supply voltage (V_{dd}) by dividing it into 16 levels. Consider again the data in **Example 12.9**.

a. Suppose now the power supply of the microprocessor voltage changes by $\pm 5\%$. What is the error in the turn-on and turn-off temperatures due to the change in power supply voltage? The bridge is supplied from a separate source and is not affected.

b. Suppose V_{dd} remains constant at 5 V but the bridge voltage supply changes by $\pm 5\%$. What is the error in the turn-on and turn-off temperatures due to the change in the bridge voltage supply?

12.18 Synthesis of a 1 kHz sinusoidal signal. A microprocessor is used to synthesize a 1 kHz sinusoidal signal to drive a loudspeaker as part of an alarm system. To do so an 8-bit external resistive ladder D/A converter is used to generate the waveform. The microprocessor operates at 5 V but the sinusoid must have an amplitude of 15 V (peak-to-peak variation between $+15$ V and -15 V). The D/A converter is set so that when digital data appear on its eight digital inputs the equivalent analog voltage appears on the output pin of the D/A converter. The conversion takes 20 μs.

a. Show the necessary circuitry, including any components required for interfacing.

b. Draw a flow chart showing the main steps needed for conversion. Indicate the sequence of digital values needed to generate the signal and how these may be obtained.

c. Show the waveform obtained on the output of the D/A converter if no filtering is used.

d. What is the maximum ripple on the basic 1 kHz signal.

12.19 Scaling and offsetting data. A sinusoidal signal at 50 Hz with an amplitude (peak value) of 5 V must be digitized using a microprocessor that operates at 3.3 V. The result is displayed on a screen as numerical values for the purpose of monitoring the signal at intervals of 2 ms, starting from the zero crossing point of the signal. A full display will contain the amplitudes for a full cycle. This is updated for each cycle to see how the signal changes.

a. Design a circuit that will accomplish the required objectives together with the interfacing circuits needed.

b. Draw a flow chart showing all important steps and considerations.

c. List the display voltages on the screen, assuming the A/D converter is 10 bit and the display simply displays the numerical (decimal) equivalent of the digital data.

Assume the sinusoidal signal is symmetric about 0 V and the only source available is the 3.3 V supplying the microprocessor. All circuits necessary for interfacing must operate at that voltage.

12.20 Hardware and software scaling. An infrared sensor is designed to operate at 12 V and needs to be interfaced with a microprocessor that operates at 3.3 V to sense infrared light intensity. The sensor has an internal amplifier that produces an output between 0 and 8 V for the infrared span envisaged. The amplifier can supply a maximum current of 2 mA without affecting its performance. The output from the amplifier must be digitized using an internal 10-bit A/D converter for digital display in increments of 5%. The latter displays values from 0 to 100 to indicate relative input infrared light intensity.

a. Show a circuit that will do the following: interface the sensor to the microprocessor and protect the microprocessor from overvoltage at its input.

b. Show a flow chart that will accomplish the requirements for display, that is, will provide a digital signal on an output pin with the proper values for the required span.

12.21 The use of a bridge for signal scaling. One of the functions of bridges is to allow signal scaling. Given the output from an operational amplifier that varies between −15 V and +15 V, design a bridge that will produce a signal between −2.5 V and +2.5 V.

Output signals and levels

12.22 Controlling higher voltage devices. A microprocessor operates at 3.3 V. However, the microprocessor must turn on a low-power buzzer that must operate at 12 V.

a. Show how this can be done using only internal facilities of the microprocessor.

b. Show how this can be done using external components.

c. Discuss the issues associated with the two approaches.

12.23 Control of lights in an office. In an attempt to conserve energy, it is proposed to control the lights in an office using a small microprocessor. The office is divided internally into two sections and has two doors, one leading to each section. The two sections are linked by a narrow opening.

a. Show how one can accomplish the following functions:

1. Turn on lights in the two separate sections of the office as persons enter the space.

2. Turn off lights 30 s after the last person leaves the office.
3. Prevent turning on lights in a section if there is sufficient ambient light in that section.

b. Show the various components needed and how they are interfaced.

12.24 Sink and source modes of output pins. An output pin is connected to a load of 100 Ω. The output of the pin is a square wave at 1 kHz. The microprocessor operates at 5 V. The internal resistance on a pin in sink mode is 230 Ω and in source mode 75 Ω.

a. Calculate the output voltage of the pin in source mode. Compare it to the no-load output.

b. Calculate the output voltage of the pin in sink mode. Compare it to the no-load output.

Errors and resolution

12.25 A/D conversion. In a microprocessor application it is necessary to digitize a 4.6 V analog input then multiply the digital result by 2.7. The reference voltage of the A/D converter is 5 V.

a. Show how digitization may be done using an internal 12-bit A/D converter.

b. Write the sequence needed for multiplication using fixed point arithmetic using 8 bits for the integer and 8 bits for the fraction.

c. Estimate the maximum error expected due to both the conversion and the computation.

12.26 Resolution of digital systems. The output of a microphone varies between 0 and 100 μV. The signal must be recorded digitally. There are two proposed approaches: (1) record the signal directly using an A/D converter, and (2) amplify the signal and then digitize it.

a. In method 1, what must be the A/D resolution for an A/D converter operating at 5 V if the noise associated with errors in digitization cannot exceed 1%?

b. In method 2, what would be an appropriate amplification and what must the resolution of the A/D be to achieve the same error level?

12.27 Errors as an integral part of design. Although errors of all types are to be avoided or minimized as much as possible, one can also use the design to minimize or eliminate their effects. Consider the sensing of force using a platinum strain gauge with a nominal resistance of 350 Ω at 20°C and a gauge factor of 5.1. In an attempt to simplify measurement the voltage across the sensor is measured directly using the circuit in **Figure 12.30**. The resistor R is also 350 Ω. The span of the sensor is 20,000 N (between zero force and a maximum force of 20,000 N), producing a strain between 0 and 3%. An amplifier is used to ensure that the input to the microprocessor is 5 V at the highest strain. Following amplification, the input voltage is digitized using a 10-bit A/D converter internal to the microprocessor.

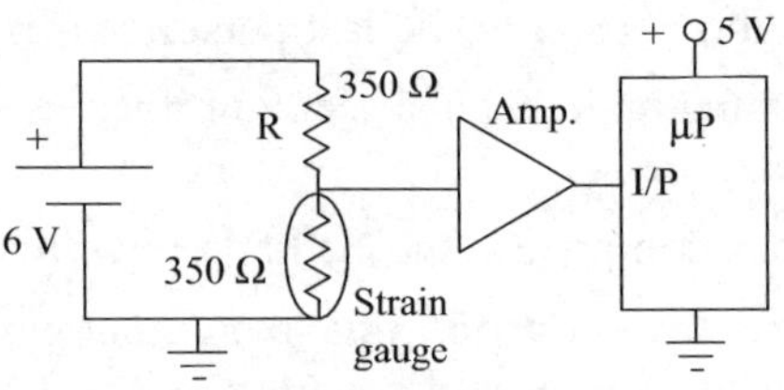

FIGURE 12.30 ■ Connection of a strain gauge to a microprocessor.

a. Calculate the resolution of the sensor.

b. The platinum grade used has a thermal coefficient of resistance equal to 0.00395. Calculate the maximum possible change in temperature that will have no effect on the output.

12.28 Fixed point arithmetic and errors on microprocessors. In an 8-bit microprocessor it becomes necessary to perform the sum $c = a + b$, where a and b vary between zero and one. Calculate the sum for $a = 0.2$ and $b = 0.9$ and the error incurred in the calculation using an 8-bit microprocessor. A single 8-bit word is allocated for each variable.

12.29 Errors due to computation. As part of a computational routine in a microprocessor the product $c = a * b$, where $a = 5.23$ and $b = 17.96$ is needed. Define the computation process and find the actual numerical representation of the result as well as the error involved. Representation of data can only be done in 8-bit words or multiple 8-bit words.

12.30 Errors due to computation. An RTD is connected to a microprocessor as shown in **Figure 12.23**. The RTD senses temperatures ranging from −50°C to 150°C. The input voltage to the microprocessor (pin 1) varies between 0 V at −50°C and 1.85 V at 150°C. The bridge voltage is converted into digital form using an internal 10-bit A/D converter with a reference voltage of 3.3 V. The output of the microprocessor must send data using a serial port to a display, which converts the data into decimal display. The display is on the kelvin scale. Because of the limited resources of the microprocessor, no number larger than 16 bits can be accommodated.

a. Calculate using minimum resources the conversion of the input voltage into the correct value for display in K. Calculate the output for inputs of −50°C, 0°C, and 150°C.

b. If the display can only display two decimal digits, calculate the error in reading the temperature as a percentage of full scale in (a).

12.31 Reduction of sampling errors. A microprocessor is used to measure the frequency from a mass SAW resonator operating at a nominal frequency of 100 MHz. The sensitivity of the sensor is 1800 Hz/μg. The microprocessor has a clock operating at 40 MHz. To be able to measure the frequency, the input frequency from the sensor is first divided by 100 before the signal arrives at the microprocessor. This signal is measured by direct timing. That is accomplished by detecting the transitions of the signal from low to high or from high to low. Assuming that to start the timer requires eight clock cycles and to stop it requires four clock cycles after detection of the appropriate transition and detection of a transition requires two clock cycles:

a. Find the error in frequency measurement if a single cycle of the signal available at the microprocessor is timed.

b. To minimize the error in (a), it is proposed to time 256 cycles of the signal available at the microprocessor. What is the error in the measured frequency?

c. What is the effective sensitivity of the system under the conditions in (a) and (b)?

12.32 Conversion error in an A/D converter. A microprocessor contains a 10-bit A/D converter operating at 5 V. The specifications for the A/D converter indicate that the conversion accuracy is ± 1 bit over the entire range of the converter. Calculate and plot the error as a percentage of the input voltage as it varies from 0 to 5 V.

12.33 Noise-limited conversion. In some applications, such as digital audio, it is necessary to use high-resolution A/D conversion to reduce the quantization error and thus reproduce audio of a better quality. Consider a CMOS A/D converter with an input resistance of 1 MΩ. The audio source connected to it also has an internal resistance of 1 MΩ. The system operates at a temperature of 30°C. Calculate the highest practical resolution of the A/D converter that one can obtain assuming a perfect audio source (no internal noise). Assume an audio bandwidth of 20 kHz and that the A/D converter operates at a voltage of 3.3 V.

12.34 Quantization error in video recording. The video signal from a CCD varies between 0 and 3.3V and is digitized using an 18-bit A/D converter. Assuming noiseless, error-free conversion from analog signal to digital signal:

a. Calculate the quantization error in the signal.

b. If the CCD has a contrast ratio of 4000:1 (i.e., the ratio between the brightest and darkest displayable values), what is the equivalent contrast ratio of the digitized signal when displayed on a screen? Is the contrast ratio limited by the CCD or by the A/D converter?

Answers to Problems

Chapter 1

1.19 $\dfrac{\text{kg}\cdot\text{m}^2}{\text{s}^2}$.

1.21 1.35582 lbf · ft.

1.22 1.9428 mol.

1.23 9.9725×10^{-22} g.

1.24 182.4 g/km.

1.25 (a) 1364 Mb. (b) 1599.5 Gb.

1.26 2048×10^6 bytes (2 GB).

1.27 39.81 μW.

1.28 50, 50, 50, and 8 or 40, 40, 25, and 25, etc.

1.29 (a) 140 dB. (b) 27.96 dB.

1.30 (a) 25. (b) 125.

Chapter 2

2.1 −23.72%.

2.2 (a) $F = 5.6755d + 0.2666$. (b) $F = 7.3272d - 1.5401d^2$.

2.3 $A = 0.00415969$, $B = 8.030935 \times 10^{-7}$, $C = -1.0281 \times 10^{-11}$.

2.4 (b) 16 Ω each.

2.5 (a) 60 V, 60% error. (b) 2235 MΩ.

2.6 0.99 V to 4.95 V.

2.7 $P(\omega) = -\dfrac{0.05}{100}\omega^2 + 0.05\omega \quad [\text{N}\cdot\text{m/s}]$, $\omega =$ speed (rps).

2.8 (a) $P_L = \dfrac{1}{2}\text{Re}\left\{\left(\dfrac{48}{8 + j2 + R_L}\right)^2 R_L\right\} \quad [\text{W}]$.

(b) $P_L = \dfrac{1}{2}\text{Re}\left\{\left(\dfrac{48}{8 + j2 + Z_L}\right)^2 Z_L\right\} \quad [\text{W}]$.

2.9 (a) $P_1(f) = \text{Re}\left\{\dfrac{72}{16 - j2\pi f \times 10^{-3}}\right\} \quad [\text{W}]$,

$P_2(f) = \text{Re}\left\{\dfrac{72}{16 - j0.006f - j2\pi f \times 10^{-3}}\right\} \quad [\text{W}]$,

$P_3(f) = \text{Re}\left\{\dfrac{72}{16 + j0.006f - j2\pi f \times 10^{-3}}\right\} \quad [\text{W}]$.

2.10 130 dB, 130 dB.
2.11 120 dB.
2.12 30 dB.
2.13 160 dB.
2.14 34.77 dB.
2.15 15 bits.
2.16 (a) $\beta = 3578.82$ K, 66.4%, 41.2%.
2.17 (b) 7.74%.
2.19 (b) 1.96%. (c) 2%.
2.21 1.61%.
2.22 At 20°C: $FR = 2.222 \times 10^{-9}$ failures/h, $FIT = 2.22$. At 80°C: $FR = 1.613 \times 10^{-5}$ failures/h, $FIT = 16{,}129$.
2.23 $FR = 7.0588 \times 10^{-6}$ failures/h, MTBF = 141,677 h.

Chapter 3

3.1 −459.67 K.
3.2 1.49×10^{18} eV.
3.3 (a) 54.664 m. (b) 290 mm. (c) 89.58 Ω at −45°C to 166.8 Ω at 120°C.
3.4 0.0063°C at 0°C, 0.00443°C at 100°C.
3.5 (a) 1010°C.
3.6 (a) $a = 3.6595 \times 10^{-3}$, $b = 2.4705 \times 10^{-7}$, $c = -5.8786 \times 10^{-11}$. (b) At −150°C: 15.467 Ω, 1.72% error. At 800°C: 155.257 Ω, −2.79% error.
3.7 (a) 79.5°C. (b) 79.3334°C.
3.8 (a) $T = \frac{1}{\alpha}\left(\frac{V}{R_0 I} - 1\right) + T_0$ [°C]. (b) 1264.88°C.
3.9 (a) $\alpha_{50} = 0.00323$. (b) $\alpha_1 = \frac{\alpha_0}{1 + \alpha_0(T_1 - T_0)}$.
3.10 (a) 33.754 Ω. (b) 0.675 Ω. (c) 2.0252 Ω.
3.11 $a = 8.34188 \times 10^{-3}$, $b = 1.36752 \times 10^{-5}$.
3.12 (a) $R(T) = e^{\left[(y-x/2)^{1/3} - (y+x/2)^{1/3}\right]}$ [Ω], $x = \frac{1.44263 \times 10^{-3} - 1/T}{1.64086 \times 10^{-7}}$,

$y = \sqrt{549.15888 + \frac{x^2}{4}}$. (b) $R(T) = 938e^{3352.34(1/T - 1/298.15)}$ [Ω].

3.13 (a) $R = 133333.33 - 1800T + 6.667T^2$ [Ω]. (b) 133,333.33 Ω.
3.14 (a) $R(T) = 10e^{-\beta/293.15}e^{\beta/T}$ [kΩ]. At 0°C, $\beta = 3505.11$ K. At 60°C, $\beta = 3696.85$ K. At 120°C, $\beta = 3653.58$ K.
(b)

	Resistance at 0°C	Resistance at 60°C	Resistance at 120°C
β calculated at 0°C	23.999 (0%)	2.3798 (8.17%)	0.478 (18.57%)
β calculated at 60°C	25.177 (4.9%)	2.2 (0%)	0.4045 (−3.7%)
β calculated at 120°C	24.906 (3.77%)	2.239 (1.77%)	0.42 (0%)

3.15 $e = 0.0827°C/mW$.
3.16 (a) 4.096 mV. (b) 3.899 mV.
3.17 (a) 4.096 mV. (b) 3.8941 mV.
3.18 (a) 35.365 mV. (b) 36.987 mV (5% difference with respect to (a)).
3.19 (a) 7749.62 μV, 9081.31 μV, 10,730.77 μV, 7229.60 μV, 10,748.61 μV, 8776.40 μV, 9358.53 μV, 9288.1 μV. (b) −16.56%, −2.23%, 15.53%, −22.16%, 15.72%, −5.51%, 0.76%. (c) 241.02°C (20.51%), 175.43°C (−12.28%), 212.59°C (6.29%), 196.47°C (−1.76%), 200.65°C (0.325%), 200.18°C (0.09%).
3.20 (a) 39,111.8 Ω. (b) 0.1425%. (c) 2.803%.
3.21 (a) 658 junction pairs. (b) 13.439 V at 80°C to 12.0085 V at 120°C.
3.22 5296 junctions (2648 pairs).
3.23 (a) 1.073 mV/°C, 0.323 V. (b) +1.516% for +10% change in current, −1.676% for −10% change in current. (c) 2.48%.
3.24 (a) $s = -\frac{b + cT + dT^2}{(a + bT + cT^2 + dT^3)^2}$ [s/K]. (b) Time of flight changes from 0.757 μs at 0°C (273.25 K) to 0.526 μs at 26°C (399.15 K).
3.25 (a) $error = 200d\left(\frac{1}{3.315\sqrt{\frac{T}{273.15}}} - \frac{1}{3.434218}\right)$ [%]. (b) 3.23 m at −20°C and 2.88 m at 45°C.
3.26 345.23 mm^3.
3.27 0.483 mm/°C.
3.28 (a) 0.9346 mm/°C. (b) 0.107°C.
3.29 (a) 0.7723°C/mm. (b) 1900% at high pressure, −2000% at low pressure.
3.30 (a) 2.096 cm. (b) 253°C.
3.31 52.57 cm.
3.32 21,205.75 mm^3, 120 mm long.
3.33 2664.8 mm^3, 15 mm long.

Chapter 4

4.1 1579.14 W.
4.2 0.01 W/m^2.
4.4 3.102 eV to 1.24 MeV.
4.5 (a) 1.0784×10^{15} Hz. (b) 278.2 nm
4.6 5.79285×10^{17} electrons/s.
4.7 (a) 1.513 eV. (b) 1.0718 eV. (c) 6.61 mA.
4.8 Below 0.8862 eV.
4.9 Ge: 38.44 kΩ; Si: 337.4 MΩ; GaAs: 33.9×10^{12} Ω.
4.10 (a) 13.511 kΩ. (b) 6.037 kΩ.
4.11 (a) 3377.75 Ω. (b) 1509.25 Ω.
4.12 (a) $s = \frac{Lhcd\eta T\lambda\tau}{ew(\mu_e + \mu_p)(nhcd + \eta PT\lambda\tau)^2}\left[\frac{\Omega}{W/m^2}\right]$. (b) 519.9 $\Omega/W/m^2$. (c) 1200 nm.
4.13 4 mV (dark), 1.612 V (under illumination).
4.14 (a) 2.94 V. (b) 0.406 mW.

4.15 (a) 644.6 kΩ. (b) 100.0 W/m^2. (c) 611.3 kΩ, 111.5 W/m^2.
4.16 16.3 W.
4.17 (a) 15.04%, (b) 16.29 W.
4.18 $5V_o = 5$ V (laser off) to 0.342 V (laser on).
4.19 Range: zero and 7.933 mW/cm^2; span: 7.933 mW/cm^2.
4.20 8.08 V.
4.21 (a) 0.142 A. (b) 5.96×10^6 m/s.
4.22 9.612 pW/m^2.
4.23 10.54 MHz.
4.24 124.416 MHz.
4.25 $s = 136.48$ $V \cdot m^2/W$.
4.26 (a) 35. (b) 0.4 mW/cm^2.
4.27 (a) $s = 2.965$ V/K.
4.28 (a) 3.542 s. (b) 2.456 s. (c) 90.64 MΩ.
4.29 (a) 5.96 mW. (b) 8.45 mW.
4.30 (a) 945.89 Mbit/s (118.23 Mbyte/s).

Chapter 5

5.1 (a) $C_{min} = 10.628$ pF, $C_{max} = 53.14$ pF. (b) 1.0628×10^{-9} F/m.
5.2 (a) $s = 1.1246$ pF/°C. (b) 0.178°C.
5.3 (a) $F = 1.683 \times 10^{-9} V^2$ [N].
5.4 (a) $R_{empty} = 85{,}454.5$ Ω, $R_{1/4} = 55{,}121.2$ Ω, $R_{1/2} = 37{,}303.0$ Ω, $R_{3/4} = 18{,}181.8$ Ω. (b) $R = -83{,}139t + 80{,}782$ [Ω]. Maximum nonlinearity 5.7%.
5.5 (a) $R_x = \dfrac{2L}{\sigma\pi\left(d - 2 \times 10^{-3}tc_r\right)^2}$ [Ω], t is in years.
5.6 (a) $I_{max} = 1.494$ A, $I_{min} = 1.488$ A. (b) $s \approx \dfrac{3.27}{h}$ [A/m]. (c) 0.667 m.
5.7 (a) $R = \dfrac{9d + 2c - 5b}{6\sigma ab}$ [Ω]. (b) 2.833 Ω and 10.33 Ω. (c) 0.75 Ω/cm.
5.8 (a) $emf = -3.97\cos 2\pi \times 1000x$ [V]. (b) $s = -2.806\cos 2\pi \times 1000t$ [V/m].
5.9 (a) 2.355 m.
5.10 $K_H = 1.06 \times 10^{-10}\left[\dfrac{m^3}{A \cdot s}\right]$, $s = 1.59 \times 10^{-8}$ V/T.
5.11 (a) $K_H = -0.00416\left[\dfrac{m^3}{A \cdot s}\right]$. (b) $s = -0.2082$ V/T.
5.12 $\dfrac{p}{n} = 9$.
5.13 (a) $V_{out} = K_H \dfrac{\mu_0 N}{Rl_g d} P_L$ [V]. (b) $P_{max} = \dfrac{1.4 l_g V}{\mu_0 N} = 668.45$ [W]. (c) 0.315 V. (d) $s = K_H \dfrac{\mu_0 N}{Rl_g d}$ [V/W].
5.14 (b) 8.38×10^{26} carriers/m^3.
5.15 $n = 6.246 \times 10^{24}$ carriers/m^3.
5.16 (a) 10^5 N. (b) 6250 V.
5.17 (a) 32 V. (b) 2560 W.
5.18 (a) $V = avB_0$ [V]. (b) $s = \dfrac{B_0}{a}$ [$V \cdot s/m^3$].

5.19 (a) 7783.64 N. (b) $a = 77{,}836.4\ \text{m/s}^2$, $v_t = 882.25$ m/s.

5.20 (a) $\rho = \rho_0 + \left(\frac{k}{Vg}\right)I^2\quad [\text{kg/m}^3]$. (b) $s = 2\frac{k}{Vg}I\left[\frac{\text{kg/m}^3}{\text{A}}\right]$.

5.21 2.1×10^{-6} T.
5.22 (a) 10.86 N. (b) 43.4 mm. (c) 217.2 m/s^2.
5.23 (b) 0.00604 N·m.
5.25 (a) 30 N.
5.27 (a) 2.8°, $N = 128.571$.
5.28 1.9737°, $N = 182.3985$.

5.29 (a) $\Delta l = \left|\frac{1}{N} - \frac{1}{M}\right|$. (b) 0.25 mm.

5.30 0.185 N.
5.31 (a) 6.98 N. (b) 62.83 N.
5.32 (a) 143,290 turns. (b) 72 turns.

5.33 (a) $emf = \left(N\frac{\mu_0\mu_r}{\sqrt{2}\pi(a+b)}(b-a)c\omega\right)I_1\quad [\text{V}]$. (b) 0.12 V.

5.34 100,000 turns.

Chapter 6

6.1 0.275 Ω/0.001 strain.
6.2 (a) $R(\varepsilon) = 1000\left(1 - 88.833\varepsilon + 11{,}055.5\varepsilon^2\right)\quad [\Omega]$. (b) 922.22 Ω and 1221.89 Ω.
6.3 (a) $R(\varepsilon) = 1000\left(1 + 88.833\varepsilon + 11{,}055.5\varepsilon^2\right)\quad [\Omega]$. (b) 1,221.89 Ω and 922.22 Ω.

6.4 (a) R_2 and R_4 do not change, $R_1 = R_{01}\left(1 + g\frac{F}{acE} + h\left(\frac{F}{acE}\right)^2\right)\quad [\Omega]$,

$$R_3 = R_{03}\left(1 + g\frac{F}{acE} + h\left(\frac{F}{acE}\right)^2\right)\quad [\Omega].$$

(b) $$\frac{dR_1}{dF} = R_{01}\left(\frac{g}{acE} + 2h\left(\frac{F}{a^2c^2E^2}\right)\right)\quad [\Omega/\text{N}],$$

$$\frac{dR_3}{dF} = R_{03}\left(\frac{g}{acE} + 2h\left(\frac{F}{a^2c^2E^2}\right)\right)\quad [\Omega/\text{N}].$$

6.5 (a) $$R_1 = R_{01}\left(1 + g\frac{F_1}{acE} + h\left(\frac{F_1}{acE}\right)^2\right)\quad [\Omega],$$

$$R_3 = R_{03}\left(1 + g\frac{F_1}{acE} + h\left(\frac{F_1}{acE}\right)^2\right)\quad [\Omega],$$

$$R_2 = R_{02}\left(1 + g\frac{F_2}{bcE} + h\left(\frac{F_2}{bcE}\right)^2\right)\quad [\Omega],$$

$$R_4 = R_{04}\left(1 + g\frac{F_4}{bcE} + h\left(\frac{F_4}{bcE}\right)^2\right) \quad [\Omega].$$

(b) $$\frac{dR_1}{dF_1} = R_{01}\left(\frac{g}{acE} + 2h\left(\frac{F_1}{a^2c^2E^2}\right)\right) \quad [\Omega/\text{N}],$$

$$\frac{dR_3}{dF_1} = R_{03}\left(\frac{g}{acE} + 2h\left(\frac{F_1}{a^2c^2E^2}\right)\right) \quad [\Omega/\text{N}],$$

$$\frac{dR_2}{dF_2} = R_{02}\left(\frac{g}{bcE} + 2h\left(\frac{F_2}{b^2c^2E^2}\right)\right) \quad [\Omega/\text{N}],$$

$$\frac{dR_4}{dF_2} = R_{04}\left(\frac{g}{bcE} + 2h\left(\frac{F_2}{b^2c^2E^2}\right)\right) \quad [\Omega/\text{N}].$$

6.6 (a) $M = 49.189 \times 10^6$ kg. (b) 7.92 Ω and 250.56 Ω. (c) 0.02% at 0°C to −0.0305% at 50°C.

6.7 (a) 296.45 Ω for no strain (no force) to 350 Ω, zero to 2.4×10^6 N. (b) −22.3 μΩ/N.

6.8 (a) $V_o = \dfrac{720 + 0.013824F}{526.08 - 0.000885F}$ V. (b) 10,000 N.

6.9 (a) $C(F) = 79.686 - 39.843F$ [pF], $s = -39.843$ pF/N. (b) 2 N.

6.10 350.4635 Ω.

6.11 245.253 Ω.

6.12 8.85 pF.

6.13 0.2125 pF and 0.0797 pF.

6.14 (a) $F = kx = \left(\dfrac{C_F - C_0}{C_F}\right)kd$ [N]. C_0 is the capacitance without applied force, C_F the capacitance with applied force. (c) $P = \left(\dfrac{C_F - C_0}{C_F}\right)\dfrac{kd}{S}$ [N/m²]. (d) $a = \left(\dfrac{C_F - C_0}{C_F}\right)\dfrac{kd}{m}$ [m/s²].

6.15 (a) 4281.25 m/s² (from 2500 m/s² upward to 1781.25 m/s² downward). (b) 0.19 to 8.854 pF (span of 9.044 pF). (c) $s = \dfrac{\varepsilon_0 s 4mc^3}{Ebe^3\left(0.002 + \frac{4mc^3}{Ebe^3}a\right)^2}$ [pF/(m/s²)].

6.16 (a) −7500 to +7500 m/s² (span of 15,000 m/s²). (b) $s = 0.1$ Ω/(m/s²). (c) ±3750 m/s² and 0.2 Ω/(m/s²).

6.17 (a) ±500 m/s² (about ±50 g). (b) 1000 Ω to 3400 Ω. (c) 2.4 Ω/(m/s²).

6.18 (a) 4905 N/m. (b) $s = K_H \dfrac{I\mu_0 N I_c}{d}\dfrac{m/k}{(0.0025 - ma/k)^2}$ [V/(m/s²)]. (c) 1.3962 mV (from −0.1396 mV to 1.2566 mV), $s = -\dfrac{5.124 \times 10^{-13}}{(0.0025 - 0.01a/4905)^2}$ [V/(m/s²)].

6.19 (a) 8 N/m. (b) $s = 2.5$ V/(m/s²). (c) 0.004 m/s².

6.20 (c) $s = -1.185583 \times 10^{-4} \times 101,325e^{-1.185583 \times 10^{-4}h}$ $[\mathrm{N/m^2/m}]$. (d) 70,363 Pa (from 101,325 Pa to 30,962 Pa).
6.21 (a) 1 MPa. (b) 2512 Pa. (c) 2.44%.
6.22 (a) $v = \sqrt{2hg}$ [m/s]. (b) 0.313 m/s. (c) $\sqrt{2hgS}$ $[\mathrm{m^3/s}]$.
6.23 (a) 1.4 m/s. (b) 70 m/s (252 km/h).
6.24 (a) Dynamic pressure 11,454 Pa, static pressure 27,500.16 Pa, total pressure 38,954.16 Pa. (b) −14.93%.
6.25 (a) 11.253 MPa.
6.26 $R = \frac{t_0}{4\pi a^2 \sigma} \frac{(P_0)^2}{(P)^2}$ [Ω].
6.27 (a) $s = 23.873$ s/kg/m²/rad. (b) 4.19×10^{-4} N·m.
6.28 (a) 273.39 Hz/°/s. (b) 1.317°/h.
6.29 31 loops.
6.30 −1075.12 Hz, −1075.12 Hz/°/sec.

Chapter 7

7.1 (a) 795.26 m. (b) 79.53 m.
7.2 $\sigma = 1000$ N/m², $\varepsilon = 5.05$ nm/m.
7.3 Within 0.9%.
7.4 (a) 12.344 MHz.
7.5 (a) 8.334 km. (b) 100 m.
7.6 34.65 μW/m².
7.7 1.486 MHz, 4.95 W.
7.8 (a) $A_3 = Ae^{-\alpha(2L-2h)}$. (b) $h = L - \frac{c_a t}{2}$ [m].
7.9 (a) 0.46 mm. (b) $0.95V_0$. (c) $V = V_0 e^{-2\alpha d}$, where α is the attenuation constant. (d) $t = \frac{2d}{v_c}$ [s].
7.10 (a) 134.
7.11 88.4 mA to 103.2 mA.
7.12 $emf = 0.0133\cos 6283t$ [V].
7.13 (a) $s = -2.655$ V/pF/m. (b) −3.52%.
7.14 (a) $P_{max} = 165,325$ Pa, $s = 0.533$ μV/Pa. (b) $\Delta V = 1.765 \times 10^{-7}\Delta P$ [V], $P_{max} = 104,125$ Pa, $s = 0.176$ μV/Pa.
7.15 9.6 V.
7.16 (a) 2.645 nV to 26.45 mV. (b) 7.56×10^{-4} Pa or 31.5 dB (assuming lowest practical output at 0.1 μV).
7.17 (a) 0.0884 Pa to 2.21 Pa (72.9 dB to 100.9 dB). (b) 2.128 nW to 1.331 μW, 1.25×10^{-5}% to 8.07×10^{-3}%.
7.18 (a) ±40.212 N. (b) ±0.0536 m. (c) 568.88 Pa (150 dB).
7.19 (a) 19.55 N/m. (b) 9.96 Pa (113 dB). (c) 76.44 μW.
7.20 $a = \frac{c_s t_1}{2}$ [m], $b = \frac{c_s(t_3 - t_2)}{2}$ [m], $d = \frac{c_a(t_2 - t_1)}{2}$ [m].
7.21 (a) $s = \frac{\cos\theta/c}{(1 - v_f\cos\theta/c)^2}$ [Hz/m/s]. (b) 5205 Hz. (c) Same except that the shift in frequency is negative.

7.22 (a) $v_f = \dfrac{2h}{T\sin\theta\cos\theta} - \dfrac{c}{\cos\theta}$ [m/s]. (b) $s = \dfrac{2h\sin\theta\cos\theta}{(v_f\sin\theta\cos\theta + c\sin\theta)^2}$ [s/m/s]. (c) 44.4 m/s. (d) $v_f = \dfrac{2h}{T\sin\theta\cos\theta} + \dfrac{c}{\cos\theta}$ [m/s], $s = \dfrac{2h\sin\theta\cos\theta}{(v_f\sin\theta\cos\theta - c\sin\theta)^2}$ [s/m/s].

7.23 (a) $\Delta f = f_0\left(1 - \dfrac{1}{(1 - v_f\cos\theta/c)}\right)$ [Hz]. (b) 9.722 Hz (compared to −5843.07 Hz). (c) 13.33 ms.

7.24 2.547 W/m^2 (124 dB).

7.25 44.88 μm.

7.26 147.5 MHz.

7.27 (a) $\varepsilon = 19.46 \times 10^{-12}$ m/m. (b) $\varepsilon = 5.75 \times 10^{-12}$ m/m.

7.28 (b) $s = -201.04$ Hz/°C.

7.29 (a) 42,254 Hz, $s = 0.417$ Hz/Pa. (b) 149.4975 MHz.

7.30 (a) 410.46 Hz/g. (b) Range: −2.339 to 2.339 kg; span: 4.678 kg. (c) 24.4 mg.

7.31 (a) 16 N.

7.32 (a) 16 N.

Chapter 8

8.1 (a) O_2: 23.145%; N_2: 75.53%; Ar: 1.283%; CO_2: 0.0455%. (b) O_2: 8.6797 mol/m^3; N_2: 32.3545 mol/m^3; Ar: 0.3854 mol/m^3; CO_2: 0.0124 mol/m^3. (c) O_2: 5.227×10^{24} atoms/m^3; N_2: 1.948×10^{25} atoms/m^3; Ar: 2.32×10^{23} atoms/m^3; CO_2: 7.467×10^{21} atoms/m^3.

8.2 (a) 4.762 m^3 air/m^3 methane. (b) 5.59 g air/g methane. (c) 73,923 kJ/m^3.

8.3 (a) 44.0079 g/mol. (b) 24.312 g/mol. (c) 44.0079 g/eq. (d) 12.156 g/eq.

8.4 (a) 990 ppm. (b) 533 ppm.

8.5 0 V (20.9% oxygen) to 31.1 mV (4% oxygen).

8.6 18.372 kΩ, $s = -3187.96P^{-1.11197376}$ [Ω/ppm].

8.7 (a) s(15 ppm) = −69.16 Ω/ppm, s(75 ppm) = −0.46 Ω/ppm. (b) $d\sigma = 12.66 \times 10^{-3}$ S/m/°C.

8.8 (a) 0.2295 V. (b) $s = -\dfrac{3.9276}{P_{\text{steel}}}$ [V/%].

8.9 14.67 mV (on), 8.18 mV (off).

8.10 (a) 0.256 V to 1.0226 V. (b) 0.1% to 1.39%.

8.11 4.046.

8.12 4.53.

8.13 $C = 5.1522 \times 10^{-2}$ g-eq/L.

8.14 (a) 0.363 V. (b) 2.58%.

8.15 (a) From 19.05°C to 30.69°C. (b) 5.2833°C/(mmol/L) to 5.291°C/(mmol/L). (c) Range: 10.488 kΩ to 8.322 kΩ; span: 2.166 kΩ, $s = -1.205 \times 10^{-2}\dfrac{e^{2227/T}}{T^2}$ [kΩ/K].

8.16 (a) 1.45°C/(% sugar). (b) 10.31%.

8.17 $s = 0.0355$ Ω/(% LEL).

8.18 (a) Between 37.3° and 53.3°. (b) 62.08°.
8.19 (a) 57.25°, 39.12°. (b) 66.9°.
8.20 (a) 9,892,570 Hz. (b) 0.18 μm/year.
8.21 (a)
$$C = 17.7 \times 10^{-12}\left(1 + \frac{210.9954 \times e^{20.386-5132/(273.15+t)}}{t+273.15} \times 10^{-6} + \frac{210.9954 \times 10^{0.66077+7.5t/(237+t)}}{t+273.15} \times 10^{-6}RH\right) \text{ [F], where } t \text{ is in °C.}$$
(b)
$$s = 17.7 \times 10^{-12} \times \frac{210.9954 \times 10^{0.66077+7.5t/(237+t)}}{t+273.15} \times 10^{-6} \quad \text{[pF/\% RH]}.$$
8.22 (a) $s = 257.69 \times \dfrac{10^{0.66077+7.5t/(237+t)}}{273.15+t}$ pF/(% RH), where t is in °C. (b) From 133.9125 pF/%RH at 50°C to 124.2965 pF/% RH at 75°C.
8.23 56.5%.
8.24
$$DPT = \frac{237.3\left(0.66077 - \log_{10}\left(10^{1.427815}RH/100\right)\right)}{\log_{10}\left(10^{1.427815}RH/100\right) - 8.16077} \quad [°\text{C}].$$
8.25 (a)

Relative humidity [%]	0	10	20	40	60	80	90
Mass absorbed, 20°C [μg]	0	15.85	41.72	98.64	179.5	295.29	424.99

(b)

Relative humidity [%]	0	10	20	40	60	80	90
Mass absorbed, 60°C [μg]	0	15.09	32.7	73.67	126.5	194.78	237.18

8.26 8.88%.
8.28 (a)
$$s = \frac{30{,}628.52}{P_{ws}} \times 10^{\frac{156.8+8.16077DPT}{237.3+DPT}}\left(\frac{156.8+8.16077DPT}{(237.3+DPT)^2} + \frac{8.16077}{237.3+DPT}\right) \quad [\%\text{RH}/°\text{C}].$$
(b)
$$\Delta RH = \frac{100}{10^{0.66077+7.5T_a/(237+T_a)}}\left(10^{\frac{156.8+8.16077(T_d+\Delta T_d)}{237.3+T_d+\Delta T_d}} - 10^{\frac{156.8+8.16077T_d}{237.3+T_d}}\right) \quad [\%].$$
8.29
$$RH = 100\frac{10^{\frac{156.8+8.16077T_d}{237.3+T_d}}}{10^{0.66077+7.5T_a/(237+T_a)}} \quad [\%].$$
8.30 (a) 2.45 kW. (b) 62.26%, 19.68%.
8.31 (a) 240 g. (b) 111,818 Pa. (c) 217.76 g, 120,011 Pa.
8.32 (a) 7.35 L. (b) 8.09 L. (c) 1.76 MPa and 2.87 MPa.
8.33 (a) 1.66 μΩ. (b) 227 mA.
8.34 (a) 107,279 s (29 h, 48 min). (b) 13.4 GW · h. (c) 778.589 kg. (d) 333.872 kg.

Chapter 9

9.1 1.24×10^{-6} eV at 300 MHz, 1.24×10^{-3} eV at 300 GHz.

9.2 124.071 eV at 30×10^{15} Hz, 124.071 keV at 30×10^{18} Hz, 126.071×10^{16} keV at 30×10^{34} Hz.

9.3 5.4 J/kg.

9.4 (a) 0.924 nA, 2 years, 2 months.

9.5 41.57 nA.

9.6 26.78 mg/m^2.

9.8 (a) 1: 4.185 keV; 2: 3.788 keV; 3: 64.513 keV; 4: 58.403 keV. (b) 1: 1159 pairs; 2: 1049 pairs; 3: 14,562 pairs; 4: 13,183 pairs.

9.9 787.3 nA.

9.10 (a) 2.485 μV. (b) 34.765 μA.

9.11 (a) 5.466 m. (b) 99.113 cm.

9.12 (a) 29.36 nA. (b) 10.5 na/MeV.

9.13 (a) 1.1×10^{-24} A. (b) $s = 4.44 \times 10^{-28}$ A/W/m^2.

9.14 31,722 km/s.

9.15 $f' = \dfrac{10 \times 10^9}{1 - 7.4074 \times 10^{-8}\cos\alpha}$ [Hz],

$$\Delta f = 10 \times 10^9 \left| 1 - \frac{1}{1 - 7.4074 \times 10^{-8}\cos\alpha} \right| \text{ [Hz]}.$$

9.16 (a) $E_r = \dfrac{1 - \sqrt{\frac{2.8+0.336m}{1+0.006m}}}{1 + \sqrt{\frac{2.8+0.336m}{1+0.006m}}} E_0$, $E_t = \dfrac{2}{1 + \sqrt{\frac{2.8+0.336m}{1+0.006m}}} E_0$, where m is the moisture content in percent. (b) $E_r(12\%) = -0.4325E_0$, $E_t(12\%) = 0.5675E_0$.

9.17 (a) 315 m. (b) 72 km/h.

9.18 (a) 7222 Hz. (b) −4.43 km/h (−3.4%).

9.19 (a) $E_r = 0.765E_0$. (b) $E_r = 0.945E_0$. (c) $s = 1508E_0 \times$

$$\left[\left(\frac{0.218\sqrt{\frac{4\pi \times 10^{-7}}{0.218m+2.2}}}{(0.218m + 2.2)^{3/2}\left(\sqrt{\frac{4\pi \times 10^{-7}}{0.218m+2.2}} + 377\right)^3} \right) + \frac{\left(-\frac{0.218\sqrt{4\pi \times 10^{-7}}}{2 \times (0.218m+2.2)^{3/2}} \right)}{\left(\sqrt{\frac{4\pi \times 10^{-7}}{0.218m+2.2}} + 377\right)^2} \right],$$

where m is the moisture content in percent.

9.20 (a) $E_{rec} = \dfrac{\sqrt{\frac{1}{\varepsilon_r(m)}}\cos 30^\circ - \sqrt{1 - \sin^2\theta_i \frac{1}{\varepsilon_r(m)}}}{\sqrt{\frac{1}{\varepsilon_r(m)}}\cos 30^\circ + \sqrt{1 - \sin^2\theta_i \frac{1}{\varepsilon_r(m)}}} E_0$ [V/m]. (c) $E_{rec}(m = 8\%) = -0.3575E_0$ [V/m].

9.21 $s = -\dfrac{2.13 \times 10^{-5}\mu_0\varepsilon_0}{4\pi a\left[\mu_0\varepsilon_0\left(1 + 2.13 \times 10^{-5}m\right)\right]^{3/2}}$ [Hz/% RH], where m is the moisture content in percent.

9.22 (a) $f_{mnp}(t) = \dfrac{1.8715 \times 10^8}{\sqrt{(3.5t + 0.065)}}$ [Hz].

(b) $s = -\dfrac{2.129 \times 10^7}{[3.5t + 0.065]^{3/2}}$ [Hz/m].

(c) 0.235 μm. (d) 0.09%.

9.23 (a) 795.23 MHz to 1.0045 GHz. (b) $s = \dfrac{h}{2\pi(b-h)^2\sqrt{\mu_0\varepsilon_0}}$ [Hz/m]. (c) 530.155 MHz to 795.23 MHz, $s = -\dfrac{2.5\mu_0\varepsilon_0}{4\pi bb\left(\mu_0\varepsilon_0\frac{2.5h+b}{b}\right)^{3/2}} + \dfrac{h}{2\pi b^2\left(\mu_0\varepsilon_0\frac{2.5h+b}{b}\right)^{1/2}}$ [Hz/m].

9.25 (b) N24°5′47″, E56°27′23″.

9.26 32.1 s.

9.27 (a) 1 min, 26 s. (b) 11 min, 11 s.

Chapter 10

10.4 (a) 1.823×10^{-3} μm. (b) 0.25 N.

10.5 (a) $\varepsilon_x = 0.262\cos\theta\cos\phi, \varepsilon_y = 0.262\sin\theta\cos\phi$ [m/m]. (b) $e = (1 - \cos\phi) \times 100$ [%].

10.6 2.86%.

10.7 (a) Range: ±10.185 m/s^2; span: 20.37 m/s^2. (b) 0.183 g. (c) 11 μm.

10.8 (a) $F = 1.195 \times 10^{-8}$ N. (b), b. $T = 1.434 \times 10^{-12}$ N·m.

10.9 (a) 1.328 Ω. (b) 3.852×10^{-11} kg.

10.10 (a) 1.722×10^{-8} μm^3. (b) 17.708 Pa.

10.11 $y_{max} = 1.188 \times 10^{-18} V^2$ [m].

10.12 $I = \dfrac{1}{a(a-r)C}\sqrt{\dfrac{\mu_0 kd}{\pi}}$ [A].

10.13 (a) 1.5 V. (b) 2.9×10^{-14} to 9.38×10^{-14} N. (d) 1.86×10^{-12} and 9.38×10^{-14} N.

10.14 (a) 7.47×10^{-8} N. (b) 2.4×10^{-3} degrees.

10.15 (b) 100 kHz (first five harmonics).

10.16 (a) 4–5 harmonics (80–100 kHz). (b) $F(t) = \dfrac{10}{\pi}\left(\dfrac{1}{2} + \sin 2\pi \times 10^4 t + \dfrac{1}{3}\sin 6\pi \times 10^4 t + \dfrac{1}{5}\sin 10\pi \times 10^4 t + \dfrac{1}{7}\sin 14\pi \times 10^4 t + \dfrac{1}{9}\sin 18\pi \times 10^4 t\right)$.

(c) $S(t) = 12\cos\Big(2\pi \times 10^8 t + 0.5t - 50\cos 2 \times 10^4 t - \dfrac{50}{9}\cos 6\pi \times 10^4 t$ $-2\cos 6\pi \times 10^4 t - \frac{50}{49}\cos 14\pi \times 10^4 t - \dfrac{50}{81}\cos 14\pi \times 10^4 t)$.

10.17 (a) $S(t) = 12\cos\left(1.005 \times 10^7 t + 50\sin\left(6.283 \times 10^4 t\right)\right)$. (b) $k_f = 100$ kHz/V.

10.18 (a) $k_p = 0.625$ rad/V/s. (b) Any signal above 13.333 kHz.

10.19 (b) 0.499 V for f_1, 0.994 V for f_2.

10.25 14ED25B.

Chapter 11

11.3 (a) 4.24%.
11.4 (a) 10^{11} Ω.
11.10 5.195 ms, 25.9%.
11.11 (b) 2.4 V, 4.8 V, 7.2 V, and 9.6 V.
11.12 1315 Ω.
11.21 (a) 8.
11.23 (a) $f = \dfrac{1}{-R_m C \ln\frac{2.4}{2.6} + 0.2\frac{RC}{V_{\text{in}}}}$ [Hz]. (b) Amplitude: 5 V; frequency: 9998 Hz; pulse width: 20 ns on, 10 μs off. (d) 500 Hz to 25 kHz.
11.24 (b) $f = \dfrac{1}{-(R_1+R_2)C\ln\left(1-\frac{V_{in}-V_{ref}}{V^+}\right) - R_2 C\ln\frac{V_{\text{ref}}}{V_{\text{in}}}}$ [Hz],

$$DC = \frac{-R_2 C\ln\frac{V_{\text{ref}}}{V_{\text{in}}}}{-(R_1+R_2)C\ln\left(1-\frac{V_{\text{in}}-V_{\text{ref}}}{V^+}\right) - R_2 C\ln\frac{V_{\text{ref}}}{V_{\text{in}}}} \times 100 \quad [\%].$$

11.25 (b) 2.55%.
11.26 (a) Digital output: 1101111010 (4.345703125 V). (b) 16.4 μs (for serial output). (c) 1110000011 (4.34575195313), 1.01%.
11.27 (b) 111110111100 (analog value 4.9169921875). (c) 0.049%.
11.28 (b) 0.3125 V.
11.29 (b) 3.0518×10^{-4} V. (c) 0.0061%. (d) 84.39 dB.
11.30 (b) 2.81251 V, 0%. (c) 2.81798 V, 0.19%. (d) 2.8098 V, 0.096%.
11.31 (a) 60.899%. (b) 3.4277 V. (c) 0.1%.
11.32 $\dfrac{dV_o}{V_i} = \dfrac{dZ}{4Z}$.
11.33 $\dfrac{dV_o}{V_i} = -0.0098262$.
11.36 (a) 43.6%. (b) 95.2%.
11.37 39.3%.
11.38 (a) 6.5 V. (b) 0.062 mA. (c) 0.62 mA.
11.39 (a) zero and 5.625 V. (c) 6.3 W.
11.40 (a) 7128.5 Hz. (b) 1736 to 1959 Hz, $s = 5.575$ Hz/mm.
11.41 (a)$f = \dfrac{10^9}{269.5 - 10.5V_{cc}}$ [Hz], 8.9 MHz to 4.2 MHz. (b) 0.219% at 3 V, 0.438% at 15 V.
11.42 (a) 721.344 Hz. (b) 50%.
11.43 (a) 7128.5 Hz. (b) 30.5%.
11.44 (a) $f = \dfrac{1}{\left(-\ln\left(\frac{V^+-V_{th}}{V^+-V_{tl}}\right) - \ln\left(\frac{V_{tl}}{V_{th}}\right)\right)R_4 C_1}$ [Hz], $DC = \dfrac{\ln\left(\frac{V^+-V_{th}}{V^+-V_{tl}}\right)}{\ln\left(\frac{V_{tl}}{V_{th}}\right) + \ln\left(\frac{V^+-V_{th}}{V^+-V_{tl}}\right)} \times$ 100%. (b) $C_1 = 0.01$ μF, $R_4 = 4551$ Ω.
11.45 (a) 94.7 μV. (b) 0.12 V.
11.46 (a) $s = 1$ V/T, $V_{sn} = 0.12488B$ [V]. (b) 0.03122 T.

Chapter 12

12.1 (c) 0.392°C.

12.7 (a) 3.64 days. (d) 85 days.

12.8 (a) 13 years 8 months. (b) 14 years 4 months.

12.9 (a)

	16 MHzextrap. least sq.		1 MHzinterp. least sq.	
5 V	4.095	3.195	0.4853	
		0.4378	4 V	3.187
	2.428	0.377	0.338	
3 V	2.0	1.53	0.2326	0.21
2 V	1.001	1.056	0.144	0.126

(b) 5 years 7 months.

12.10 (b) 1/3. (c) 0.555.

12.13 (b) 0.0128 pH.

12.17 (a) Nominal 350°C, varies from 301.68°C to 409.23°C (−13.8% to 16.92%). (b) Nominal: 217.58, varies from 282.1 to 195.37 (−10.2% to 19.4%).

12.18 (d) 19.6 mV.

12.24 (a) 1.515 V. (b) 2.86 V.

12.25 (c) 0.044%.

12.26 (a) 23 bit (minimum). (b) 50,000, 7 bit (minimum).

12.27 (a) 5.88 N. (b) 0.106°C.

12.28 1.09375 (5.68% error).

12.29 93.8515625, 0.084% error.

12.30 (a) 223.14, 273.30, 420.11. (b) −0.005%, 0.075%, −1.52%.

12.31 (a) 11.1%. (b) 0.39%. (c) 250 ns.

12.32 $e = (1/n) \times 100\%$, $n = 1, 2, \ldots, 1023$.

12.33 18 bit (262,144 steps).

12.34 (a) 12.588 μV. (b) 62144 to 1.

Appendix A
Least Squares Polynomials and Data Fitting

Least square polynomials or polynomial regression is a method of fitting a polynomial to a set of data. Suppose we have a set of n points (x_i, y_i) to which we wish to fit a polynomial of the form

$$y(x) = a_0 + a_1 x + a_2 x^2 + \ldots a_m x^m. \tag{A.1}$$

Passing a polynomial through a set of data means selection of the coefficients so as to minimize, in a global sense, the distance between the value of the function $y(x)$ and the values at the points $y(x_i)$. This is done through the least squares method by first writing the "distance" function:

$$S = \sum_{i=1}^{n} (y_i - a_0 - a_1 x_i - a_2 x_i^2 - \ldots a_m x_i^m)^2. \tag{A.2}$$

To minimize this function we calculate the partial derivatives with respect to each unknown coefficient and set it to zero. For the kth coefficient $(k = 0, 1, 2, \ldots, m)$ we write

$$\frac{\partial S}{\partial a^k} = -2 \sum_{i=1}^{n} x_i^k (y_i - a_0 - a_1 x_i - a_2 x_i^2 - \ldots a_m x_i^m) = 0 \tag{A.3}$$

or

$$\sum_{i=1}^{n} x_i^k (y_i - a_0 - a_1 x_i - a_2 x_i^2 - \ldots a_m x_i^m) = 0. \tag{A.4}$$

Repeating this for all m coefficients results in m equations from which the coefficients a_0 through a_m can be evaluated. We show here how to derive the coefficients for a first-order (linear) and second-order (quadratic) polynomial least square fit since these are the most commonly used forms. We assume n data points (x_i, y_i) as above.

A.1 | LINEAR LEAST SQUARE DATA FITTING

The polynomial is first order:

$$y(x) = a_0 + a_1 x. \tag{A.5}$$

The least squares form is

$$S = \sum_{i=1}^{n} (y_i - a_0 - a_1 x_i)^2. \tag{A.6}$$

Taking the partial derivatives with respect to a_0 and a_1 gives

$$\sum_{i=1}^{n} x_i^{\,0} (y_i - a_0 - a_1 x_i) = 0 \tag{A.7}$$

and

$$\sum_{i=1}^{n} x_i^{\,1} (y_i - a_0 - a_1 x_i) = 0. \tag{A.8}$$

It is most convenient to expand these and write

$$n a_0 + a_1 \sum_{i=1}^{n} x_i = \sum_{i=1}^{n} y_i \tag{A.9}$$

and

$$a_0 \sum_{i=1}^{n} x_i + a_1 \sum_{i=1}^{n} x_i^{\,2} = \sum_{i=1}^{n} x_i y_i. \tag{A.10}$$

To calculate the coefficients we denote

$$A = \sum_{i=1}^{n} x_i, \quad B = \sum_{i=1}^{n} y_i, \quad C = \sum_{i=1}^{n} x_i^{\,2}, \quad D = \sum_{i=1}^{n} x_i y_i \tag{A.11}$$

and write

$$n a_0 + a_1 A = B \tag{A.12}$$

and

$$a_0 A + a_1 C = D. \tag{A.13}$$

Solving for a_0 and a_1 in terms of A, B, C, D, and n we get

$$a_0 = \frac{nD - AB}{nC - A^2}, \quad a_1 = \frac{BC - AD}{nC - A^2}. \tag{A.14}$$

Or in terms of the sums:

$$a_0 = \frac{n \sum_{i=1}^{n} x_i y_i - \left\{ \sum_{i=1}^{n} x_i \right\} \left\{ \sum_{i=1}^{n} y_i \right\}}{n \sum_{i=1}^{n} x_i^{\,2} - \left\{ \sum_{i=1}^{n} x_i \right\}^2}, \quad a_1 = \frac{\left\{ \sum_{i=1}^{n} y_i \right\} \left\{ \sum_{i=1}^{n} x_i^{\,2} \right\} - \left\{ \sum_{i=1}^{n} x_i \right\} \left\{ \sum_{i=1}^{n} x_i y_i \right\}}{n \sum_{i=1}^{n} x_i^{\,2} - \left\{ \sum_{i=1}^{n} x_i \right\}^2}. \tag{A.15}$$

This is a linear fit to the data (first-order polynomial fit) and is called a linear best fit or linear least squares fit.

A.2 | PARABOLIC LEAST SQUARES FIT

We start with the second-order polynomial

$$y(x) = a_0 + a_1 x + a_2 x^2. \tag{A.16}$$

The least squares form is

$$S = \sum_{i=1}^{n} (y_i - a_0 - a_1 x_i - a_2 x_i^2)^2. \tag{A.17}$$

Taking the partial derivatives with respect to a_0, a_1, and a_2 gives

$$\sum_{i=1}^{n} x_i^0 (y_i - a_0 - a_1 x_i - a_2 x^2) = 0, \tag{A.18}$$

$$\sum_{i=1}^{n} x_i^1 (y_i - a_0 - a_1 x_i - a_2 x^2) = 0, \tag{A.19}$$

and

$$\sum_{i=1}^{n} x_i^2 (y_i - a_0 - a_1 x_i - a_2 x^2) = 0. \tag{A.20}$$

Again expanding these,

$$n a_0 + a_1 \sum_{i=1}^{n} x_i + a_2 \sum_{i=1}^{n} x_i^2 = \sum_{i=1}^{n} y_i, \tag{A.21}$$

$$a_0 \sum_{i=1}^{n} x_i + a_1 \sum_{i=1}^{n} x_i^2 + a_2 \sum_{i=1}^{n} x_i^3 = \sum_{i=1}^{n} x_i y_i, \tag{A.22}$$

and

$$a_0 \sum_{i=1}^{n} x_i^2 + a_1 \sum_{i=1}^{n} x_i^3 + a_2 \sum_{i=1}^{n} x_i^4 = \sum_{i=1}^{n} x_i^2 y_i. \tag{A.23}$$

Although we could proceed to calculate the coefficients a_0, a_1, and a_2 as for the previous case, the expressions become too complex to handle. It is more practical to write the three equations as a matrix:

$$\begin{bmatrix} n & \sum_{i=1}^{n} x_i & \sum_{i=1}^{n} x_i^2 \\ \sum_{i=1}^{n} x_i & \sum_{i=1}^{n} x_i^2 & \sum_{i=1}^{n} x_i^3 \\ \sum_{i=1}^{n} x_i^2 & \sum_{i=1}^{n} x_i^3 & \sum_{i=1}^{n} x_i^4 \end{bmatrix} \begin{Bmatrix} a_0 \\ a_1 \\ a_2 \end{Bmatrix} = \begin{Bmatrix} \sum_{i=1}^{n} y_i \\ \sum_{i=1}^{n} x_i y_i \\ \sum_{i=1}^{n} x_i^2 y_i \end{Bmatrix}. \tag{A.24}$$

To solve for the coefficients one calculates the various sums first and then proceeds to solve the system of equations.

The extension to higher-order polynomials is obvious and simply adds the next terms to the matrix system in Equation (A.24). Note also that removing the third row and third column from the matrix in Equation (A.24) leads to the linear best fit in Equation (A.15).

Finally, note that the coefficients can only be calculated by hand for a limited number of points, n. In most cases a computation tool such as Matlab will be found useful.

Appendix B Thermoelectric Reference Tables

The thermoelectric reference tables for the most common thermocouples are shown below. For each type of thermocouple we show first the general polynomial, followed by the table of coefficients and by the explicit polynomials for both the direct and inverse use. The output of the direct polynomials is in microvolts (μV). Output of the inverse polynomials is in degrees Celsius (°C). The index 90 indicates the standard used (in this case the International Temperature Scale of 1990 [ITS-90]).

B.1 | TYPE J THERMOCOUPLES (IRON-CONSTANTAN)

$$E = \sum_{i=0}^{n} c_i (t_{90})^i \quad [\mu V]$$

Temperature range [°C]	−210°C to 760°C	760°C to 1200°C
C_0	0.0	2.9645625681×10^5
C_1	5.0381187815×10^1	$-1.4976127786 \times 10^3$
C_2	$3.0475836930 \times 10^{-2}$	3.1787103924
C_3	$-8.5681065720 \times 10^{-5}$	$-3.1847686701 \times 10^{-3}$
C_4	$1.3228195295 \times 10^{-7}$	$1.5720819004 \times 10^{-6}$
C_5	$-1.7052958337 \times 10^{-10}$	$-3.0691369056 \times 10^{-10}$
C_6	$2.0948090697 \times 10^{-13}$	
C_7	$-1.2538395336 \times 10^{-16}$	
C_8	$1.5631725697 \times 10^{-20}$	

−210°C to 760°C

$$E = 5.0381187815 \times 10^1 T^1 + 3.0475836930 \times 10^{-2} T^2 - 8.5681065720 \times 10^{-5} T^3 + 1.3228195295 \times 10^{-7} T^4 - 1.7052958337 \times 10^{-10} T^5 + 2.0948090697 \times 10^{-13} T^6 - 1.2538395336 \times 10^{-16} T^7 + 1.5631725697 \times 10^{-20} T^8 \quad [\mu V]$$

760°C to 1200°C

$$E = 2.9645625681 \times 10^5 - 1.4976127786 \times 10^3 T + 3.1787103924 T^2 - 3.1847686701 \times 10^{-3} T^3 + 1.5720819004 \times 10^{-6} T^4 - 3.0691369056 \times 10^{-10} T^5 \quad [\mu V]$$

Inverse polynomials

$$T_{90} = \sum_{i=0}^{n} c_i E^i \quad [^\circ\mathrm{C}]$$

Temperature range [°C]	−210°C to 0°C	0°C to 760°C	760°C to 1200°C
Voltage range [μV]	−8095 to 0	0 to 42919	42919 to 69553
C_0	0.0	0.0	−3.11358187
C_1	1.9528268×10^{-2}	1.9528268×10^{-2}	$3.00543684 \times 10^{-1}$
C_2	$-1.2286185 \times 10^{-6}$	-2.001204×10^{-7}	$-9.94773230 \times 10^{-6}$
C_3	$-1.0752178 \times 10^{-9}$	1.036969×10^{-11}	$1.70276630 \times 10^{-10}$
C_4	$-5.9086933 \times 10^{-13}$	$-2.549687 \times 10^{-16}$	$-1.43033486 \times 10^{-15}$
C_5	$-1.7256713 \times 10^{-16}$	3.585153×10^{-21}	$4.73886084 \times 10^{-21}$
C_6	$-2.8131513 \times 10^{-20}$	$-5.344285 \times 10^{-26}$	
C_7	$-2.3963370 \times 10^{-24}$	5.099890×10^{-31}	
C_8	$-8.3823321 \times 10^{-29}$		
Error range	0.03°C to −0.05°C	0.04°C to −0.04°C	0.03°C to −0.04°C

−210°C to 0°C

$$T_{90} = 1.9528268 \times 10^{-2}E^1 - 1.2286185 \times 10^{-6}E^2 - 1.0752178 \times 10^{-9}E^3 - 5.9086933 \times 10^{-13}E^4 - 1.7256713 \times 10^{-16}E^5 - 2.8131513 \times 10^{-20}E^6 - 2.3963370 \times 10^{-24}E^7 - 8.3823321 \times 10^{-29}E^8 \quad [^\circ\mathrm{C}]$$

0°C to 760°C

$$T_{90} = 1.9528268 \times 10^{-2}E^1 - 2.001204 \times 10^{-7}E^2 + 1.036969 \times 10^{-11}E^3 - 2.549687 \times 10^{-16}E^4 + 3.585153 \times 10^{-21}E^5 - 5.344285 \times 10^{-26}E^6 + 5.099890 \times 10^{-31}E^7 \quad [^\circ\mathrm{C}]$$

760°C to 1200°C

$$T_{90} = -3.11358187 + 3.00543684 \times 10^{-1}E^1 - 9.94773230 \times 10^{-6}E^2 + 1.70276630 \times 10^{-10}E^3 - 1.43033468 \times 10^{-15}E^4 + 4.73886084 \times 10^{-21}E^5 \quad [^\circ\mathrm{C}]$$

B.2 | TYPE K THERMOCOUPLES (CHROMEL-ALUMEL)

$$E = \sum_{i=0}^{n} c_i (t_{90})^i \quad [\mu\mathrm{V}]$$

Above 0°C the polynomial is of the form: $E = \sum_{i=0}^{n} c_i (t_{90})^i + \alpha_0 e^{\alpha_1 (t_{90} - 126.9686)^2}$. $[\mu\mathrm{V}]$

Temperature range [°C]	−270°C to 0°C	0°C to 1372°C
C_0	0.0	$-1.7600413686 \times 10^{1}$
C_1	$3.945012802 \times 10^{1}$	$3.8921204975 \times 10^{1}$
C_2	$2.3622373598 \times 10^{-2}$	$1.8558770032 \times 10^{-2}$
C_3	$-3.2858906784 \times 10^{-4}$	$-9.9457592874 \times 10^{-5}$
C_4	$-4.9904828777 \times 10^{-6}$	$3.1840945719 \times 10^{-7}$
C_5	$-6.7509059173 \times 10^{-8}$	$-5.6072844889 \times 10^{-10}$
C_6	$-5.7410327428 \times 10^{-10}$	$5.6075059059 \times 10^{-13}$
C_7	$-3.1088872894 \times 10^{-12}$	$-3.2020720003 \times 10^{-16}$
C_8	$-1.0451609365 \times 10^{-14}$	$9.7151147152 \times 10^{-20}$
C_9	$-1.9889266878 \times 10^{-17}$	$-1.2104721275 \times 10^{-23}$
C_{10}	$-1.6322697486 \times 10^{-20}$	
α_0		1.185976×10^{2}
α_1		-1.183432×10^{-4}

−270°C to 0°C

$$E = 3.945012802 \times 10^{1} T^{1} + 2.3622373598 \times 10^{-2} T^{2} - 3.2858906784 \times 10^{-4} T^{3} - 4.9904828777 \times 10^{-6} T^{4} - 6.7509059173 \times 10^{-8} T^{5} - 5.7410327428 \times 10^{-10} T^{6} - 3.1088872894 \times 10^{-12} T^{7} - 1.0451609365 \times 10^{-14} T^{8} - 1.9889266878 \times 10^{-17} T^{9} - 1.6322697486 \times 10^{-20} T^{10} \quad [\mu\text{V}]$$

0°C to 1372°C

$$E = -1.7600413686 \times 10^{1} + 3.8921204975 \times 10^{1} T^{1} + 1.8558770032 \times 10^{-2} T^{2} - 9.9457592874 \times 10^{-5} T^{3} + 3.1840945719 \times 10^{-7} T^{4} - 5.6072844889 \times 10^{-10} T^{5} + 5.6075059059 \times 10^{-13} T^{6} - 3.2020720003 \times 10^{-16} T^{7} + 9.7151147152 \times 10^{-20} T^{8} - 1.2104721275 \times 10^{-23} T^{9} + 1.185976 \times 10^{2} e^{-1.183432 \times 10^{-4} (T - 126.9686)^{2}} \quad [\mu\text{V}]$$

Inverse polynomials

$$T_{90} = \sum_{i=0}^{n} c_i E^i \quad [°\text{C}]$$

Temperature range [°C]	200°C to 0°C	0°C to 500°C	500°C to 1372°C
Voltage range [μV]	−5891 to 0	0 to 20644	20644 to 54886
C_0	0.0	0.0	-1.318058×10^{2}
C_1	2.1573462×10^{-2}	2.508355×10^{-2}	4.830222×10^{-2}
C_2	$-1.1662878 \times 10^{-6}$	7.860106×10^{-8}	-1.646031×10^{-6}
C_3	$-1.0833638 \times 10^{-9}$	$-2.503131 \times 10^{-10}$	5.464731×10^{-11}
C_4	$-8.9773540 \times 10^{-13}$	8.315270×10^{-14}	$-9.650715 \times 10^{-16}$
C_5	$-3.7342377 \times 10^{-16}$	$-1.228034 \times 10^{-17}$	8.802193×10^{-21}
C_6	$-8.6632643 \times 10^{-20}$	9.804036×10^{-22}	$-3.110810 \times 10^{-26}$
C_7	$-1.0450598 \times 10^{-23}$	$-4.413030 \times 10^{-26}$	
C_8	$-5.1920577 \times 10^{-28}$	1.057734×10^{-28}	
C_9		$-1.052755 \times 10^{-35}$	
Error range	0.04°C to −0.02°C	0.04°C to −0.05°C	0.06°C to −0.05°C

−200°C to 0°C

$$T_{90} = 2.5173462 \times 10^{-2}E^1 - 1.1662878 \times 10^{-6}E^2 - 1.0833638 \times 10^{-9}E^3 - 8.9773540 \times 10^{-13}E^4 - 3.7342377 \times 10^{-16}E^5 - 8.6632643 \times 10^{-20}E^6 - 1.0450598 \times 10^{-23}E^7 - 5.1920577 \times 10^{-28}E^8 \quad [°C]$$

0°C to 500°C

$$T_{90} = 2.508355 \times 10^{-2}E^1 + 7.860106 \times 10^{-8}E^2 - 2.503131 \times 10^{-10}E^3 + 8.315270 \times 10^{-14}E^4 - 1.228034 \times 10^{-17}E^5 + 9.804036 \times 10^{-22}E^6 - 4.413030 \times 10^{-26}E^7 + 1.057734 \times 10^{-28}E^8 - 1.052755 \times 10^{-35}E^9 \quad [°C]$$

500°C to 1372°C

$$T_{90} = -1.318058 \times 10^2 + 4.830222 \times 10^{-2}E^1 - 1.646031 \times 10^{-6}E^2 + 5.464731 \times 10^{-11}E^3 - 9.650715 \times 10^{-16}E^4 + 8.802193 \times 10^{-21}E^5 - 3.110810 \times 10^{-26}E^6 \quad [°C]$$

B.3 | TYPE T THERMOCOUPLES (COPPER-CONSTANTAN)

$$E = \sum_{i=0}^{n} c_i (t_{90})^i \quad [\mu V]$$

Temperature range [°C]	−270°C to 0°C	0°C to 400°C
C_0	0.0	0.0
C_1	$3.8748106364 \times 10^{1}$	$3.8748106364 \times 10^{1}$
C_2	$4.4194434347 \times 10^{-2}$	$3.3292227880 \times 10^{-2}$
C_3	$1.1844323105 \times 10^{-4}$	$2.0618243404 \times 10^{-4}$
C_4	$2.0032973554 \times 10^{-5}$	$-2.1882256846 \times 10^{-6}$
C_5	$9.0138019559 \times 10^{-7}$	$1.0996880928 \times 10^{-8}$
C_6	$2.2651156593 \times 10^{-8}$	$-3.0815758772 \times 10^{-11}$
C_7	$3.6071154205 \times 10^{-10}$	$4.5479135290 \times 10^{-14}$
C_8	$3.8493939883 \times 10^{-12}$	$-2.7512901673 \times 10^{-17}$
C_9	$2.8213521925 \times 10^{-17}$	
C_{10}	$1.4251594779 \times 10^{-16}$	
C_{11}	$4.8768662286 \times 10^{-19}$	
C_{12}	$1.0795539270 \times 10^{-21}$	
C_{13}	$1.3945027062 \times 10^{-24}$	
C_{14}	$7.9795153927 \times 10^{-28}$	

−270°C to 0°C

$$E = 3.8748106364 \times 10^{1} T^{1} + 4.4194434347 \times 10^{-2} T^{2} + 1.1844323105 \times 10^{-4} T^{3} + 2.0032973554 \times 10^{-5} T^{4} + 9.0138019559 \times 10^{-7} T^{5} + 2.2651156593 \times 10^{-8} T^{6} + 3.6071154205 \times 10^{-10} T^{7} + 3.8493939883 \times 10^{-12} T^{8} + 2.8213521925 \times 10^{-17} T^{9} + 1.4251594779 \times 10^{-16} T^{10} + 4.8768662286 \times 10^{-19} T^{11} + 1.0795539270 \times 10^{-21} T^{12} + 1.3945027062 \times 10^{-24} T^{13} + 7.9795153927 \times 10^{-28} T^{14} \quad [\mu V]$$

0°C to 400°C

$$E = 3.8748106364 \times 10^{1} T^{1} + 3.3292227880 \times 10^{-2} T^{2} + 2.0618243404 \times 10^{-4} T^{3} - 2.1882256846 \times 10^{-6} T^{4} + 1.0996880928 \times 10^{-8} T^{5} - 3.0815758772 \times 10^{-11} T^{6} + 4.5479135290 \times 10^{-14} T^{7} - 2.7512901673 \times 10^{-17} T^{8} \quad [\mu V]$$

Inverse polynomials

$$T_{90} = \sum c_i E^i \quad [°C]$$

Temperature range [°C]	−200°C to 0°C	0°C to 400°C
Voltage range [μV]	−5603 to 0	0 to 20872
C_0	0.0	0.0
C_1	2.5949192×10^{-2}	2.5949192×10^{-2}
C_2	$-2.1316967 \times 10^{-7}$	-7.602961×10^{-7}
C_3	$7.9018692 \times 10^{-10}$	4.637791×10^{-11}
C_4	$4.2527777 \times 10^{-13}$	$-2.165394 \times 10^{-15}$
C_5	$1.3304473 \times 10^{-16}$	6.048144×10^{-20}
C_6	$2.0241446 \times 10^{-20}$	$-7.293422 \times 10^{-25}$
C_7	$1.2668171 \times 10^{-24}$	
Error range	0.04°C to −0.02°C	0.03°C to −0.03°C

−200°C to 0°C

$$T_{90} = 2.5949192 \times 10^{-2} E^{1} - 2.1316967 \times 10^{-7} E^{2} + 7.9018692 \times 10^{-10} E^{3} + 4.2527777 \times 10^{-13} E^{4} + 1.3304473 \times 10^{-16} E^{5} + 2.0241446 \times 10^{-20} E^{6} + 1.2668171 \times 10^{-24} E^{7} \quad [°C]$$

0°C to 400°C

$$T_{90} = 2.5949192 \times 10^{-2} E^{1} - 7.602961 \times 10^{-7} E^{2} + 4.637791 \times 10^{-11} E^{3} - 2.165394 \times 10^{-15} E^{4} + 6.048144 \times 10^{-20} E^{5} - 7.293422 \times 10^{-25} E^{6} \quad [°C]$$

B.4 | TYPE E THERMOCOUPLES (CHROMEL-CONSTANTAN)

$$E = \sum_{i=0}^{n} c_i (t_{90})^i \quad [\mu\text{V}]$$

Temperature range [°C]	−270°C to 0°C	0°C to 1000°C
C_0	0.0	0.0
C_1	$5.8665508708 \times 10^{1}$	$5.8665508708 \times 10^{1}$
C_2	$4.5410977124 \times 10^{-2}$	$4.5032275582 \times 10^{-2}$
C_3	$-7.7998048686 \times 10^{-4}$	$2.8908407212 \times 10^{-5}$
C_4	$-2.5800160843 \times 10^{-5}$	$-3.3056896652 \times 10^{-7}$
C_5	$-5.9452583057 \times 10^{-7}$	$6.5024403270 \times 10^{-10}$
C_6	$-9.3214058667 \times 10^{-9}$	$-1.9197495504 \times 10^{-13}$
C_7	$-1.0287605534 \times 10^{-10}$	$-1.2536600497 \times 10^{-15}$
C_8	$-8.0370123621 \times 10^{-13}$	$2.1489217569 \times 10^{-18}$
C_9	$-4.3979497391 \times 10^{-15}$	$-1.4388041782 \times 10^{-21}$
C_{10}	$-1.6414776355 \times 10^{-17}$	$3.5960899481 \times 10^{-25}$
C_{11}	$-3.9673619516 \times 10^{-20}$	
C_{12}	$-5.5827328721 \times 10^{-23}$	
C_{13}	$-3.4657842013 \times 10^{-26}$	

−270°C to 0°C

$$\begin{aligned} E = {} & 5.8665508708 \times 10^{1} T + 4.5410977124 \times 10^{-2} T^2 - 7.7998048686 \times 10^{-4} T^3 \\ & - 2.5800160843 \times 10^{-5} T^4 + 5.9452583057 \times 10^{-7} T^5 - 9.3214058667 \times 10^{-9} T^6 \\ & - 1.0287605534 \times 10^{-10} T^7 - 8.0370123621 \times 10^{-13} T^8 \\ & - 4.3979497391 \times 10^{-15} T^9 - 1.6414776355 \times 10^{-17} T^{10} \\ & - 3.9673619516 \times 10^{-20} T^{11} - 5.5827328721 \times 10^{-23} T^{12} \\ & - 3.4657842013 \times 10^{-26} T^{13} \quad [\mu\text{V}] \end{aligned}$$

0°C to 1000°C

$$\begin{aligned} E = {} & 5.8665508708 \times 10^{1} T + 4.5032275582 \times 10^{-2} T^2 + 2.8908407212 \times 10^{-5} T^3 \\ & - 3.3056896652 \times 10^{-7} T^4 + 6.5024403270 \times 10^{-10} T^5 - 1.9197495504 \times 10^{-13} T^6 \\ & - 1.2536600497 \times 10^{-15} T^7 + 2.1489217569 \times 10^{-18} T^8 - 1.4388041782 \times 10^{-21} T^9 \\ & + 3.5960899481 \times 10^{-25} T^{10} \quad [\mu\text{V}] \end{aligned}$$

Inverse polynomials

$$T_{90} = \sum_{i=0}^{n} c_i E^i \quad [\mu\text{V}]$$

Temperature range [°C]	−200°C to 0°C	0°C to 1000°C
Voltage range [μV]	−5603 to 0	0 to 20,872
C_0	0.0	0.0
C_1	1.6977288×10^{-2}	1.7057035×10^{-2}
C_2	$-4.3514970 \times 10^{-7}$	$-2.3301759 \times 10^{-7}$
C_3	$-1.5859697 \times 10^{-10}$	$6.5435585 \times 10^{-12}$
C_4	$-9.2502871 \times 10^{-14}$	$-7.3562749 \times 10^{-17}$
C_5	$-2.6084314 \times 10^{-17}$	$-1.7896001 \times 10^{-21}$
C_6	$-4.1360199 \times 10^{-21}$	$8.4036165 \times 10^{-26}$
C_7	$-3.4034030 \times 10^{-25}$	$-1.3735879 \times 10^{-30}$
C_8	$-1.1564890 \times 10^{-29}$	$1.0629823 \times 10^{-35}$
C_9		$-3.2447087 \times 10^{-41}$
Error range	0.04°C to −0.02°C	0.03°C to −0.03°C

−200°C to 0°C

$$T_{90} = 1.6977288 \times 10^{-2}E^1 - 4.3514970 \times 10^{-7}E^2 - 1.5859697 \times 10^{-10}E^3 - 9.2502871 \times 10^{-14}E^4 - 2.6084314 \times 10^{-17}E^5 - 4.1360199 \times 10^{-21}E^6 - 3.4034030 \times 10^{-25}E^7 - 1.1564890 \times 10^{-29}E^8 \quad [°C]$$

0°C to 1000°C

$$T_{90} = 1.7057035 \times 10^{-2}E^1 - 2.3301759 \times 10^{-7}E^2 + 6.5435585 \times 10^{-12}E^3 - 7.3562749 \times 10^{-17}E^4 - 1.7896001 \times 10^{-21}E^5 + 8.4036165 \times 10^{-26}E^6 - 1.3735879 \times 10^{-30}E^7 + 1.0629823 \times 10^{-35}E^8 - 3.2447087 \times 10^{-41}E^9 \quad [°C]$$

B.5 | TYPE N THERMOCOUPLES (NICKEL-CHROMIUM-SILICON ALLOY)

$$E = \sum_{i=0}^{n} c_i (t_{90})^i \quad [\mu V]$$

Temperature range [°C]	−270°C to 0°C	0°C to 1300°C
C_0	0.0	0.0
C_1	$2.6159105962 \times 10^{1}$	$2.6159105962 \times 10^{1}$
C_2	$1.0957484228 \times 10^{-2}$	$1.5710141880 \times 10^{-2}$
C_3	$-9.3841111554 \times 10^{-5}$	$4.3825627237 \times 10^{-5}$
C_4	$-4.6412039759 \times 10^{-8}$	$-2.5261169794 \times 10^{-7}$
C_5	$-2.6303357716 \times 10^{-9}$	$6.4311819339 \times 10^{-10}$
C_6	$-2.2653438003 \times 10^{-11}$	$-1.0063471519 \times 10^{-12}$
C_7	$-7.6089300791 \times 10^{-14}$	$9.9745338992 \times 10^{-16}$
C_8	$-9.3419667835 \times 10^{-17}$	$-6.0563245607 \times 10^{-19}$
C_9		$2.0849229339 \times 10^{-22}$
C_{10}		$-3.0682196151 \times 10^{-22}$

−210°C to 760°C

$$E = 2.6159105962 \times 10^{1}T^{1} + 1.0957484228 \times 10^{-2}T^{2} - 9.3841111554 \times 10^{-5}T^{3} - 4.6412039759 \times 10^{-8}T^{4} - 2.6303357716 \times 10^{-9}T^{5} - 2.2653438003 \times 10^{-11}T^{6} - 7.6089300791 \times 10^{-14}T^{7} - 9.3419667835 \times 10^{-17}T^{8} \quad [\mu V]$$

0°C to 1300°C

$$E = 2.6159105962 \times 10^{1}T^{1} + 1.5710141880 \times 10^{-2}T^{2} + 4.3825627237 \times 10^{-5}T^{3} - 2.5261169794 \times 10^{-7}T^{4} + 6.4311819339 \times 10^{-10}T^{5} - 1.0063471519 \times 10^{-12}T^{6} + 9.9745338992 \times 10^{-16}T^{7} - 6.0863245607 \times 10^{-19}T^{8} + 2.0849229339 \times 10^{-22}T^{9} - 3.0682196151 \times 10^{-22}T^{10} \quad [\mu V]$$

Inverse polynomials

$$T_{90} = \sum_{i=0}^{n} c_i E^i \quad [°C]$$

Temperature range [°C]	−200°C to 0°C	0°C to 600°C	600°C to 1300°C	0°C to 1300°C
Volt range [μV]	−3990 to 0	0 to 20,613	20,613 to 47,513	0 to 47,513
C_0	0.0	0.0	1.972485×10^{1}	0.0
C_1	3.8436847×10^{-2}	3.86896×10^{-2}	3.300943×10^{-2}	3.8783277×10^{-2}
C_2	1.1010485×10^{-6}	-1.08267×10^{-6}	-3.915159×10^{-7}	$-1.1612344 \times 10^{-6}$
C_3	5.2229312×10^{-9}	4.70205×10^{-11}	9.855391×10^{-12}	$6.9525655 \times 10^{-11}$
C_4	$7.2060525 \times 10^{-12}$	-2.12169×10^{-18}	$-1.274371 \times 10^{-16}$	$-3.0090077 \times 10^{-15}$
C_5	$5.8488586 \times 10^{-15}$	-1.17272×10^{-19}	7.767022×10^{-22}	$8.8311584 \times 10^{-20}$
C_6	$2.7754916 \times 10^{-18}$	5.39280×10^{-24}		$-1.6213839 \times 10^{-24}$
C_7	$7.7075166 \times 10^{-22}$	-7.98156×10^{-29}		$1.6693362 \times 10^{-29}$
C_8	$1.1582665 \times 10^{-25}$			$-7.3117540 \times 10^{-35}$
C_9	$7.3138868 \times 10^{-30}$			
Error range	0.03°C to −0.02°C	0.03°C to −0.01°C	0.02°C to −0.04°C	0.06°C to −0.06°C

−200°C to 0°C

$$T_{90} = 3.8436847 \times 10^{-2}E^{1} + 1.1010485 \times 10^{-6}E^{2} + 5.2229312 \times 10^{-9}E^{3} + 7.2060525 \times 10^{-12}E^{4} + 5.8488586 \times 10^{-15}E^{5} + 2.7754916 \times 10^{-18}E^{6} + 7.7075166 \times 10^{-22}E^{7} + 1.1582665 \times 10^{-25}E^{8} + 7.3138868 \times 10^{-30}E^{9} \quad [°C]$$

0°C to 600°C

$$T_{90} = 3.86896 \times 10^{-2}E^{1} - 1.08267 \times 10^{-6}E^{2} + 4.70205 \times 10^{-11}E^{3} - 2.12169 \times 10^{-18}E^{4} - 1.17272 \times 10^{-19}E^{5} + 5.39280 \times 10^{-24}E^{6} - 7.98156 \times 10^{-29}E^{7} \quad [°C]$$

600°C to 1300°C

$$T_{90} = 1.972485 \times 10^{1} + 3.300943 \times 10^{-2}E^{1} - 3.915159 \times 10^{-7}E^{2} + 9.855391 \times 10^{-12}E^{3} - 1.274371 \times 10^{-16}E^{4} + 7.767022 \times 10^{-22}E^{5} \quad [°C]$$

0°C to 1300°C

$$T_{90} = 3.8783277 \times 10^{-2}E^1 - 1.1612344 \times 10^{-6}E^2 + 6.9525655 \times 10^{-11}E^3 - 3.0090077 \times 10^{-15}E^4 + 8.8311584 \times 10^{-20}E^5 - 1.6213839 \times 10^{-24}E^6 + 1.6693362 \times 10^{-29}E^7 - 7.3117540 \times 10^{-35}E^8 \quad [°C]$$

B.6 | TYPE B THERMOCOUPLES (PLATINUM [30%]-RHODIUM-PLATINUM)

$$E = \sum_{i=0}^{n} c_i (t_{90})^i \quad [\mu V]$$

Temperature range [°C]	0°C to 630.615°C	630.615°C to 1820°C
C_0	0.0	$-3.8938168621 \times 10^{3}$
C_1	$-2.4650818346 \times 10^{1}$	$2.8571747470 \times 10^{1}$
C_2	$5.9040421171 \times 10^{-3}$	$-8.4885104785 \times 10^{-2}$
C_3	$-1.3257931636 \times 10^{-6}$	$1.5785280164 \times 10^{-4}$
C_4	$1.5668291901 \times 10^{-9}$	$-1.6835344864 \times 10^{-7}$
C_5	$-1.6944529240 \times 10^{-12}$	$1.1109794013 \times 10^{-10}$
C_6	$6.2290347094 \times 10^{-16}$	$-4.4515431033 \times 10^{-14}$
C_7		$9.8975640821 \times 10^{-18}$
C_8		$-9.3791330289 \times 10^{-22}$

0°C to 630.615°C

$$E = -2.4650818346 \times 10^1 T^1 + 5.9040421171 \times 10^{-3} T^2 - 1.3257931636 \times 10^{-6} T^3 + 1.5668291901 \times 10^{-9} T^4 - 1.6944529240 \times 10^{-12} T^5 + 6.2290347094 \times 10^{-16} T^6 \quad [\mu V]$$

630.615°C to 1820°C

$$E = -3.8938168621 \times 10^3 + 2.8571747470 \times 10^1 T^1 - 8.4885104785 \times 10^{-2} T^2 + 1.5785280164 \times 10^{-4} T^3 - 1.6835344864 \times 10^{-7} T^4 + 1.1109794013 \times 10^{-10} T^5 - 4.4515431033 \times 10^{-14} T^6 + 9.8975640821 \times 10^{-18} T^7 - 9.3791330289 \times 10^{-22} T^8 \quad [\mu V]$$

Inverse polynomials

$$T_{90} = \sum_{i=0}^{n} c_i E^i \quad [\mu V]$$

Temperature range [°C] Voltage range [μV]	250°C to 700°C 291 to 2431	700°C to 1820°C 2431 to 13,820
C_0	9.4823321×10^{1}	2.1315071×10^{2}
C_1	6.9971500×10^{-1}	2.8510514×10^{-1}
C_2	$-8.4765304 \times 10^{-4}$	$-5.2742887 \times 10^{-5}$
C_3	1.0052644×10^{-6}	9.9160804×10^{-9}
C_4	$-8.3345952 \times 10^{-10}$	$-1.2965303 \times 10^{-12}$
C_5	$4.5508542 \times 10^{-13}$	$1.1195870 \times 10^{-16}$
C_6	$-1.5523037 \times 10^{-16}$	$-6.0625199 \times 10^{-21}$
C_7	$2.9886750 \times 10^{-20}$	$1.8661696 \times 10^{-25}$
C_8	$-2.4742860 \times 10^{-24}$	$-2.4878585 \times 10^{-30}$
Error range	0.03°C to −0.02°C	0.02°C to −0.01°C

250°C to 700°C

$$\begin{aligned} T_{90} = {} & 9.4823321 \times 10^{1} + 6.9971500 \times 10^{-1}E^{1} - 8.4765304 \times 10^{-4}E^{2} \\ & + 1.0052644 \times 10^{-6}E^{3} - 8.3345952 \times 10^{-10}E^{4} + 4.5508542 \times 10^{-13}E^{5} \\ & - 1.5523037 \times 10^{-16}E^{6} + 2.9886750 \times 10^{-20}E^{7} - 2.4742860 \times 10^{-24}E^{8} \quad [^\circ\mathrm{C}] \end{aligned}$$

700°C to 1820°C

$$\begin{aligned} T_{90} = {} & 2.1315071 \times 10^{2} + 2.8510504 \times 10^{-1}E^{1} - 5.2742887 \times 10^{-5}E^{2} \\ & + 9.9160804 \times 10^{-9}E^{3} - 1.2965303 \times 10^{-12}E^{4} + 1.1195870 \times 10^{-16}E^{5} \\ & - 6.0625199 \times 10^{-21}E^{6} + 1.8661696 \times 10^{-25}E^{7} - 2.4878585 \times 10^{-30}E^{8} \quad [^\circ\mathrm{C}] \end{aligned}$$

B.7 | TYPE R THERMOCOUPLES (PLATINUM [13%]/RHODIUM-PLATINUM)

$$E = \sum_{i=0}^{n} c_i (t_{90})^i \quad [\mu\mathrm{V}]$$

Temperature range [°C]	−50°C to 1064.18°C	1064.18°C to 1664.5°C	1664.5°C to 1768.1°C
C_0	0.0	$2.95157925316 \times 10^{3}$	$1.52232118209 \times 10^{5}$
C_1	5.28961729765	−2.52061251332	$-2.68819888545 \times 10^{2}$
C_2	$1.39166589782 \times 10^{-2}$	$1.59564501865 \times 10^{-2}$	$1.71280280453 \times 10^{-1}$
C_3	$-2.38855693017 \times 10^{-5}$	$-7.64085947576 \times 10^{-6}$	$-3.45895706453 \times 10^{-5}$
C_4	$3.56916001063 \times 10^{-8}$	$2.05305291024 \times 10^{-9}$	$-9.34633971046 \times 10^{-12}$
C_5	$-4.62347666298 \times 10^{-11}$	$-2.93359668173 \times 10^{-13}$	
C_6	$5.00777441031 \times 10^{-14}$		
C_7	$-3.73105886191 \times 10^{-17}$		
C_8	$1.57716482367 \times 10^{-20}$		
C_9	$-2.81038625251 \times 10^{-24}$		

−50°C to 1064.18°C

$$E = 5.28961729765T^1 + 1.39166589782 \times 10^{-2}T^2 - 2.38855693017 \times 10^{-5}T^3 + 3.56916001063 \times 10^{-8}T^4 - 4.62347666298 \times 10^{-11}T^5 + 5.00777441034 \times 10^{-14}T^6 - 3.73105886191 \times 10^{-17}T^7 + 1.57716482367 \times 10^{-20}T^8 - 2.81038625251 \times 10^{-24}T^9 \quad [\mu\text{V}]$$

1064.18°C to 1664.5°C

$$E = 2.95157925316 \times 10^3 - 2.52061251332T^1 + 1.59564501865 \times 10^{-2}T^2 - 7.64085947576 \times 10^{-6}T^3 + 2.05305291024 \times 10^{-9}T^4 - 2.93359668173 \times 10^{-13}T^5 \quad [\mu\text{V}]$$

1664.5°C to 1768.1°C

$$E = 1.52232118209 \times 10^5 - 2.68819888545 \times 10^2T^1 + 1.71280280453 \times 10^{-1}T^2 - 3.45895706453 \times 10^{-5}T^3 - 9.34633971046 \times 10^{-12}T^4 \quad [\mu\text{V}]$$

Inverse polynomials

$$T_{90} = \sum_{i=0}^{n} c_i E^i \quad [°\text{C}]$$

Temperature range [°C]	−50°C to 250°C	250°C to 1200°C	1064°C to 1664.5°C	1664.5°C to 1768.1°C
Volt range [μV]	−226 to 1923	1923 to 13,228	11,361 to 19,769	19,769 to 21,103
C_0	0.0	$1.334584505 \times 10^{1}$	$-8.199599416 \times 10^{1}$	$3.406177836 \times 10^{4}$
C_1	1.8891380×10^{-1}	$1.472644573 \times 10^{-1}$	$1.553962042 \times 10^{-1}$	−7.02372/171
C_2	$-9.3835290 \times 10^{-5}$	$-1.844024844 \times 10^{-5}$	$-8.342197663 \times 10^{-6}$	$5.582903813 \times 10^{-4}$
C_3	1.0368619×10^{-7}	$4.031129726 \times 10^{-9}$	$4.279433549 \times 10^{-10}$	$-1.952394635 \times 10^{-8}$
C_4	$-2.2703580 \times 10^{-10}$	$-6.249428360 \times 10^{13}$	$-1.191577910 \times 10^{-14}$	$2.560740231 \times 10^{-13}$
C_5	$3.5145659 \times 10^{-13}$	$6.468412046 \times 10^{-17}$	$1.492290091 \times 10^{-19}$	
C_6	$-3.8953900 \times 10^{-16}$	$-4.458750426 \times 10^{-21}$		
C_7	$2.8239471 \times 10^{-19}$	$1.994710146 \times 10^{-25}$		
C_8	$-1.2607281 \times 10^{-22}$	$-5.313401790 \times 10^{-30}$		
C_9	$3.1353611 \times 10^{-26}$	$6.481976217 \times 10^{-35}$		
C_{10}	$-3.3187769 \times 10^{-30}$			
Error range	0.02°C to −0.02°C	0.005°C to −0.005°C	0.001°C to −0.0005°C	0.002°C to −0.001°C

−50°C to 250°C

$$T_{90} = 1.8891380 \times 10^{-1}E^1 - 9.3835290 \times 10^{-5}E^2 + 1.3068619 \times 10^{-7}E^3 - 2.2703580 \times 10^{-10}E^4 + 3.5145659 \times 10^{-13}E^5 - 3.8953900 \times 10^{-16}E^6 + 2.8239471 \times 10^{-19}E^7 - 1.2607281 \times 10^{-22}E^8 + 3.1353611 \times 10^{-26}E^9 - 3.3187769 \times 10^{-30}E^{10} \quad [°\text{C}]$$

250°C to 1200°C

$$T_{90} = 1.334584505 \times 10^1 + 1.472644573 \times 10^{-1}E^1 - 1.844024844 \times 10^{-5}E^2 + 4.031129726 \times 10^{-9}E^3 - 6.249428360 \times 10^{-13}E^4 + 6.468412046 \times 10^{-17}E^5 - 4.458750426 \times 10^{-21}E^6 + 1.994710146 \times 10^{-25}E^7 - 5.313401790 \times 10^{-30}E^8 + 6.481976217 \times 10^{-35}E^9 \quad [°C]$$

1064°C to 1664.5°C

$$T_{90} = -8.199599416 \times 10^1 + 1.553962042 \times 10^{-1}E^1 - 8.342197663 \times 10^{-6}E^2 + 4.279433549 \times 10^{-10}E^3 - 1.191577910 \times 10^{-14}E^4 + 1.492290091 \times 10^{-19}E^5 \quad [°C]$$

1664.5°C to 1768.1°C

$$T_{90} = 3.406177836 \times 10^4 - 7.023729171E^1 + 5.582903813 \times 10^{-4}E^2 - 1.952394635 \times 10^{-8}E^3 + 2.560740231 \times 10^{-13}E^4 \quad [°C]$$

B.8 | TYPE S THERMOCOUPLES (PLATINUM [10%]/RHODIUM-PLATINUM)

$$E = \sum_{i=0}^{n} c_i (t_{90})^i \quad [\mu V]$$

Temperature range [°C]	−50°C to 1064.18°C	1064.18°C to 1664.5°C	1664.5°C to 1768.1°C
C_0	0.0	$1.32900445085 \times 10^3$	$1.46628232636 \times 10^5$
C_1	5.40313308631	3.34509311344	$-2.58430516752 \times 10^2$
C_2	$1.25934289740 \times 10^{-2}$	$6.54805192818 \times 10^{-3}$	$1.63693574641 \times 10^{-1}$
C_3	$-2.32477968689 \times 10^{-5}$	$-1.64856259209 \times 10^{-6}$	$-3.30439046987 \times 10^{-5}$
C_4	$3.22028823036 \times 10^{-8}$	$1.29989605174 \times 10^{-11}$	$-9.43223690612 \times 10^{-12}$
C_5	$-3.31465196389 \times 10^{-11}$		
C_6	$2.55744251786 \times 10^{-14}$		
C_7	$-1.25068871393 \times 10^{-17}$		
C_8	$2.71443176145 \times 10^{-21}$		

−50°C to 1064.18°C

$$E = 5.40313308631T^1 + 1.25934289740 \times 10^{-2}T^2 - 2.32477968689 \times 10^{-5}T^3 + 3.22028823036 \times 10^{-8}T^4 - 3.31465196389 \times 10^{-11}T^5 + 2.55744251786 \times 10^{-14}T^6 - 1.25068871393 \times 10^{-17}T^7 + 2.71443176145 \times 10^{-21}T^8 \quad [\mu V]$$

1064.18°C to 1664.5°C

$$E = 1.32900445085 \times 10^3 + 3.345093311344T^1 + 6.54805192818 \times 10^{-3}T^2 - 1.64856259209 \times 10^{-6}T^3 + 1.29989605174 \times 10^{-11}T^4 \quad [\mu V]$$

1664.5°C to 1768.1°C

$$E = 1.46628232636 \times 10^5 - 2.58430516752 \times 10^2 T^1 + 1.63693574641 \times 10^{-1} T^2 - 3.30439046987 \times 10^{-5} T^3 - 9.43223690612 \times 10^{-12} T^4 \quad [\mu V]$$

Inverse polynomials

$$T_{90} = \sum_{i=0}^{n} c_i E^i \quad [\mu V]$$

Temperature range [°C]	−50°C to 250°C	250°C to 1200°C	1064°C to 1664.5°C	1664.5°C to 1768.1°C
Volt range [μV]	−235 to 1874	1874 to 11,950	10,332 to 17,536	17,536 to 18,693
C_0	0.0	1.291507177×10^1	-8.087801117×10^1	5.333875126×10^4
C_1	$1.84949460 \times 10^{-1}$	$1.466298863 \times 10^{-1}$	$1.621573104 \times 10^{-1}$	-1.235892298×10^1
C_2	$-8.00504062 \times 10^{-5}$	$-1.534713402 \times 10^{-5}$	$-8.536869453 \times 10^{-6}$	$1.092657613 \times 10^{-3}$
C_3	$1.02237430 \times 10^{-7}$	$3.145945973 \times 10^{-9}$	$4.719686976 \times 10^{-10}$	$-4.265693686 \times 10^{-8}$
C_4	$-1.52248592 \times 10^{-10}$	$-4.163257839 \times 10^{-13}$	$-1.441693666 \times 10^{-14}$	$6.247205420 \times 10^{-13}$
C_5	$1.8821343 \times 10^{-13}$	$3.187963771 \times 10^{-17}$	$2.081618890 \times 10^{-19}$	
C_6	$-1.59085941 \times 10^{-16}$	$-1.291637500 \times 10^{-21}$		
C_7	$8.23027880 \times 10^{-20}$	$2.183475087 \times 10^{-26}$		
C_8	$-2.34181944 \times 10^{-23}$	$-1.447379511 \times 10^{-31}$		
C_9	$2.79786260 \times 10^{-27}$	$8.211272125 \times 10^{-36}$		
Error range	0.02°C to −0.02°C	0.01°C to −0.01°C	0.0002°C to −0.0002°C	0.002°C to −0.002°C

−50°C to 250°C

$$T_{90} = 1.84949460 \times 10^{-1} E^1 - 8.00504062 \times 10^{-5} E^2 + 1.02237430 \times 10^{-7} E^3 - 1.52248592 \times 10^{-10} E^4 + 1.88821343 \times 10^{-13} E^5 - 1.59085941 \times 10^{-16} E^6 + 8.23027880 \times 10^{-20} E^7 - 2.34181944 \times 10^{-23} E^8 + 2.79786260 \times 10^{-27} E^9 \quad [°C]$$

250°C to 1200°C

$$T_{90} = 1.291507177 \times 10^1 + 1.466298863 \times 10^{-1} E^1 - 1.534713402 \times 10^{-5} E^2 + 3.145945973 \times 10^{-9} E^3 - 4.163257839 \times 10^{-13} E^4 + 3.187963771 \times 10^{-17} E^5 - 1.291637500 \times 10^{-21} E^6 + 2.183475087 \times 10^{-26} E^7 - 1.447379511 \times 10^{-31} E^8 + 8.211272125 \times 10^{-36} E^9 \quad [°C]$$

1064°C to 1664.5°C

$$T_{90} = -8.087801117 \times 10^1 + 1.621573104 \times 10^{-1} E^1 - 8.536869453 \times 10^{-6} E^2 + 4.719686976 \times 10^{-10} E^3 - 1.441693666 \times 10^{-14} E^4 + 2.081618890 \times 10^{-19} E^5$$

1664.5°C to 1768.1°C

$$T_{90} = 5.333875126 \times 10^4 - 1.235892298 \times 10^1 E^1 + 1.092657613 \times 10^{-3} E^2 - 4.265693686 \times 10^{-8} E^3 + 6.247205420 \times 10^{-13} E^4 \quad [°C]$$

Appendix C
Computation on Microprocessors

In the following we explore a few issues associated with integer and fixed point mathematics on microprocessors. We do not discuss floating point mathematics since floating point operations are rarely resorted to in interfacing in the context of 8-bit microprocessors.

C.1 | REPRESENTATION OF NUMBERS ON MICROPROCESSORS

C.1.1 Binary Numbers: Unsigned Integers

Internally the microprocessor represents all variables as binary integers, that is, base 2 integers. The integers may be unsigned (i.e., positive numbers) or may be signed (i.e., may be positive or negative). A positive decimal number, say, the four-digit number 3792, makes use of the numbers 0 to 9 and may be represented as

$$3792 = 3 \times 10^3 + 7 \times 10^2 + 9 \times 10^1 + 2 \times 10^0. \tag{C.1}$$

Parallel to the decimal system, binary or base 2 integers make use of the numbers 0 and 1. An 8-bit unsigned integer, say, 10011011, may be represented as follows:

$$\begin{aligned} 10011011 &= 1 \times 2^7 + 0 \times 2^6 + 0 \times 2^5 + 1 \times 2^4 + 1 \times 2^3 + 0 \times 2^2 + 1 \times 2^1 + 1 \times 2^0 \\ &= 128 + 0 + 0 + 16 + 8 + 0 + 2 + 1 = 155. \end{aligned} \tag{C.2}$$

The representation above also shows how numbers can be converted from one system to the other. For the particular case shown here, the decimal equivalent is 155.

To convert a binary number to a decimal number it suffices to sum up the products indicated in **Equation (C.2)**. To convert from decimal to binary we can proceed in two simple ways. A formal method is based on division by 2. The decimal number is divided by 2. If the division is exact, one writes the least significant bit as a "0". If not, it is written as a "1" and the quotient is then divided again by 2 until the quotient is zero. Using again the number 3792, we write

$$
\begin{aligned}
3792/2 &= 1896 \text{---} 0 \\
1892/2 &= 948 \text{-----} 0 \\
948/2 &= 474 \text{------} 0 \\
474/2 &= 237 \text{------} 0 \\
237/2 &= 118 \text{------} 1 \\
118/2 &= 59 \text{-------} 0 \\
59/2 &= 29 \text{---------} 1 \\
29/2 &= 14 \text{---------} 1 \\
14/2 &= 7 \text{----------} 0 \\
7/2 &= 3 \text{------------} 1 \\
3/2 &= 1 \text{------------} 1 \\
1/2 &= 0 \text{------------} 1.
\end{aligned}
\tag{C.3}
$$

The digital representation is 111011010000 and requires 12 bits.

An often simpler method is suggested by **Equation (C.3)**. Find the largest power of 2 that fits in the number and subtract it from the decimal number. Set the digit that corresponds to that power to "1" for the most significant bit (MSB). Find the next largest power of 2 that fits in the remainder and subtract it. Set the corresponding digit to "1" and so on until the remainder is zero. All other digits are zero. In the example above, the largest power of 2 is $2^{11} = 2048$. Digit 11 becomes "1" and the remainder is 1744. Now the largest power that fits in 1744 is $2^{10} = 1024$. Digit 10 is "1" and the remainder is 720. The next largest power of 2 is $2^9 = 512$. Digit 9 is "1" and the remainder is 208. The next power of 2 is $2^7 = 128$. The remainder is 80. Digit 7 is "1", but since $2^8 = 256$ was not used, digit 8 is "0". Continuing the process we obtain the representation 111011010000, as above.

C.1.2 Signed Integers

In the decimal system we use the minus sign to indicate negative numbers (integers or fractions), but in the digital system there is no negative sign. In the common notation, negative integers are treated as positive integers with the MSB serving as a sign digit. If the MSB is "0", the number is considered positive, whereas if it is "1", the number is negative. Consider, for example, the signed integer 01000101. This is a positive number equivalent to 69. The signed integer 11000101 is negative, equivalent to -59. To understand the notation we write the two integers as follows:

0	1	0	0	0	1	0	1
-2^7	2^6	2^5	2^4	2^3	2^2	2^1	2^0

1	1	0	0	0	1	0	1
-2^7	2^6	2^5	2^4	2^3	2^2	2^1	2^0

(C.4)

The first integer is

$$
\begin{aligned}
01000101 &= 0 \times (-2^7) + 1 \times 2^6 + 0 \times 2^5 + 0 \times 2^4 + 0 \times 2^3 + 1 \times 2^2 + 0 \times 2^1 + 1 \times 2^0 \\
&= 0 + 64 + 0 + 0 + 0 + 4 + 0 + 1 = 69.
\end{aligned}
$$

The second integer is

$$
\begin{aligned}
11000101 &= 1 \times (-2^7) + 1 \times 2^6 + 0 \times 2^5 + 0 \times 2^4 + 0 \times 2^3 + 1 \times 2^2 + 0 \times 2^1 + 1 \times 2^0 \\
&= -128 + 64 + 0 + 0 + 0 + 4 + 0 + 1 = -59.
\end{aligned}
$$

Since the sign bit cannot be used as part of the representation of the signed integer, the range of numbers that can be represented is from -128 to $+127$, or in general, from -2^{n-1} to $2^{n-1}-1$, whereas unsigned numbers range from 0 to 255 or from 0 to $2^n - 1$, where n is the number of digits in the representation.

Negative integers are represented using the 2's complement method as follows:

1. Start with the positive value of the negative number. That is, if we need to find the representation of the number $-A$, start with A in binary format.
2. Calculate the 1's complement of A. The 1's complement of a binary integer is found by replacing all zeroes with ones and all ones with zeroes, including the sign bit.
3. Add "1" to the 1's complement to get the 2's complement.

For example, suppose we need to write the integer -59. We first write 59 as 00111011. The 1's complement is 11000100. Adding 1 to this gives 11000101. Note first that this is clearly a negative number. Its value is -59, as was seen above.

In the representation of signed integers we used a "reserved" sign digit, but in the process the range of values that can be represented has been seriously reduced. This is particularly problematic with 8-bit integers. To alleviate this problem, microprocessors have adopted a slightly different strategy. Both signed and unsigned numbers use 8 bits (or 16 bits in 16-bit microprocessors) so that the range of integers is the same for signed and unsigned integers. Two extra bits in a different register are used to indicate carry and borrow. When a carry bit is set, it is an indication that an add function has overflowed the register, whereas when the borrow bit is set, an underflow occurred, indicating a negative number. From the point of view of representation of negative numbers this is the same as using a 9-bit register to represent an 8-bit signed integer.

C.1.3 Hexadecimal Numbers

Binary numbers are particularly useful for computation because the hardware used in microprocessors (and in computers) can represent two states quite easily. The main shortcoming of binary representation outside of computers is that the digital numbers are long. In programming and display it is more convenient to use higher base numbering schemes that are related to base 2 numbers. Two distinct representations that satisfy this requirement are the octal (base 8) and hexadecimal (base 16) numbering schemes. In microprocessors, the hexadecimal representation is most commonly used. The scheme uses the numbers 0 to 9 plus A (= 10), B (= 11), C (= 12), D (= 13), E (= 14), and F (= 15). In this system, the least significant bit (LSB) is multiplied by 16^0, the next by 16^1, then 16^2, and so on. The number 3792 can be written as

$$3792 = 14 \times 16^2 + 13 \times 16^1 + 0 \times 16^0.$$

The representation of 3792 in hexadecimal format is therefore ED0. The subtraction method discussed above for binary numbers applies here as well, with the obvious modifications.

C.2 | INTEGER ARITHMETIC

Since microprocessors were designed for digital control, they can only manipulate binary integers. That means first that integer computation is "natural" in the microprocessor and, second, that computation in any other format must be adapted to the binary integer environment.

Integer computation is exact, that is, there is no loss of accuracy due to round-off errors, as long as it can be done within the number of bits allocated. If, for example, one allocates 8 bits for unsigned integers, the largest value that can be represented is 255. As long as all numbers required for the computation as well as the result are smaller than 255, the result is exact. Allocation of, say, 16 bits allows integers from 0 to $2^{16} - 1 = 65{,}535$.

If negative numbers must be used as well, such as in performing subtraction, one must use signed integers. Since typically the MSB is used for sign indication, a 16-bit signed integer representation will use 15 digits for the integer, allowing representation of integers from -2^{15} to $+2^{15} - 1$, or $-32{,}768$ to $+32{,}767$ (see, however, **Section C.1.2** on how microprocessors handle negative integers). This also shows the main shortcoming of integer computation—its dynamic range is low, that is, the range of numbers it can represent is small. For this reason computers use floating point numbers, characterized by a mantissa and an exponent. By using a short mantissa and an exponent one can describe any number, although numbers are, in general, truncated, hence floating point calculation is not exact.

C.2.1 Addition and Subtraction of Binary Integers

The basic arithmetic operation in a microprocessor is the addition of two binary integers. Almost all other mathematical operations must rely on addition, together with logical operations. Notable exceptions are multiplication and division by powers of 2, since that can be done as logic shifting to the left (shifting the bits one location to the left multiplies the number by 2) or to the right (each shift divides the number by 2).

Addition is performed as with decimal numbers except that $1 + 1$ produces a carry 1, that is, $1 + 1 = 10$ and $1 + 1 + 1 = 11$ (i.e., $1 + 1 + \text{carry} = 11$, whereas $1 + \text{carry} = 10$). For two 8-bit signed integer digital numbers $A = 00110101$ and $B = 00111011$ we have

$$00110101 + 00111011 = 01110000 \quad (53 + 59 = 112).$$

In this case all three values are less than 255 and no carry is generated. But suppose that $B = 01111011$ (123):

$$00110101 + 01111011 = 10110000 \quad (53 + 123 = 176).$$

The result has a "1" where the sign bit should be, but because both integers are positive, the result cannot be negative, hence the resulting integer is properly interpreted as 176.

Suppose now we wish to subtract $B = 01111011$ (123) from $A = 00110101$ (53). First we must write the value $-B$, so we can perform the operation $A + (-B)$. We write

10000100, the 1's complement of B
$10000100 + 1 = 10000101$, the 2's complement of B.

This equals -123 (i.e., $-128 + 4 + 1 = -123$).

Now we add the 2's complement of B and A:

$$00110101 + 10000101 = 10111010 \quad (53 - 123 = -70).$$

This is clearly a negative number, equal to $(-128+32+16+8+2=-70)$, as can be verified using the decimal numbers. Note that since we perform subtraction, the sign bit cannot be interpreted as a carry, it can only be considered as a sign.

C.2.2 Multiplication and Division

The principle of multiplication and division of binary numbers follows the same principles as the multiplication of decimal numbers. The simple long multiplication and long division, as used in hand calculation, can be adapted to binary numbers, and in fact it is much easier to do. However, the term "long" has a particularly clear meaning in binary multiplication and division. Because binary numbers are very long (in comparison with equivalent decimal numbers), the process of long multiplication and division requires a significant number of steps. For this reason microprocessors and computers perform these operations either in special hardware or using optimized algorithms that are much more efficient than long multiplication or division. However, to see the principles involved we will look here at the process by emulating hand multiplication/division.

C.2.2.1 Binary Integer Multiplication

Multiplication is implemented by a sequence of shifts and adds, operations that do not require any special hardware. Consider the multiplication of two 8-bit unsigned integers A = 10110011 (179) and B = 11010001 (209). The result is 37,411, or 1001001000100011. In long multiplication this is done as follows:

```
        10110011×
        11010001
        10110011+
    10110011
    101111100011+
  10110011
  11100010100011+
 10110011        .
1001001000100011
```

The calculation above differs from the common long multiplication in that intermediate sums are computed and instead of multiplying by zero, the multiplicand is shifted one position to the left. The multiplication process requires that the multiplicand be placed in a variable twice as long as itself (16 bit in this case, to allow shifting). The intermediate results are also placed in a 16-bit integer. The multiplier remains unchanged (8 bits). Multiplication is simply a matter of shifting the multiplicand to the left and adding it to the previously calculated intermediate result.

The algorithm is as follows:

1. Place the multiplicand in a 16-bit register (M register).
2. Zero the intermediate result register (I register).
3. If MSB is "1", add the M and I registers. Place result in the I register.
4. If MSB is "0", shift the M register 1 bit to the left.
5. Set STEP = 2 and scan the multiplier register starting from bit STEP.
6. If bit # = STEP is "0", shift the M register 1 bit to the left.
7. If bit # = STEP is "1", shift the M register 1 bit to the left and add it to the I register.
8. Increment STEP.

9. If STEP = 9, go to STOP. If not, got to (6).
10. STOP. The result is in the I register.

It should be noted that the result is twice as long as the original integers. The algorithm itself requires eight shifts and eight adds and longer word lengths, but requires no other type of operation. It is clearly a much slower algorithm than either addition or subtraction. As indicated above, this is a "worst case" algorithm and serves to illustrate the process. Actual multiplication algorithms are much more sophisticated than this and require considerably fewer operations.

We have assumed here that the integers are unsigned. If they are, the multiplication is done by first changing any negative integers into their positive equivalent, multiplying the integers without the sign bits, and then calculating the sign of the product from the signs of the two multiplicands and the multiplier. If the product is negative, it is converted to its negative representation as discussed in **Section C.1.2**.

C.2.2.2 Binary Integer Division

Long division of unsigned integers is similar to long division of decimal numbers and results in a quotient and remainder. As with multiplication, it is done by a series of right shifts and subtractions (which themselves are done through additions). Consider the division of the dividend A = 11101110 (238) by the divisor B = 00001001 (9). The result expected is 26 (the quotient) with a remainder of 4 since we discuss here integer division ($26 \times 9 + 4 = 238$). It is instructive first to look at the division in **Figure C.1a**, as it is done by hand. First the divisor is aligned with the leftmost bit of the dividend. If the divisor is smaller than the number made of digits above it, we subtract the divisor from the (4) digit number above it to obtain a remainder, in this case equal to 101. The MSB of the quotient is set to 1. If this is not to be the case, subtraction is not performed and the MSB stays 0. Next we drop the next digit (fifth digit from the left) of the dividend to the right of the remainder to produce 1011. Now we subtract the divisor from this, producing a remainder equal to 10 and the next digit in the quotient becomes 1 (the quotient now is 11). The sixth digit from the dividend is dropped to the remainder to change it to 101. This is smaller than the divisor. As a consequence, the next digit in the quotient is 0 and no subtraction takes place. The seventh digit is dropped and the remainder becomes 1011. The divisor is subtracted to produce a remainder equal to 10 and the quotient becomes 1101. The last digit is dropped down to the remainder, making it 100. Since this is smaller than the divisor, the quotient becomes 11010 (26) and the remainder is 100 (4), as expected. Now consider **Figure C.1b**, which produces the same result using 8-bit integers. The following algorithm may be used to accomplish this.

1. The dividend (DD register) and the divisor (DR register) are each in an 8-bit register. Create two additional registers: a quotient (Q register) and a remainder (R register).
2. Zero the Q and R registers.
3. Shift the DS register to the left until all leading zeroes clear out and the MSB is "1". Save the number n as the number of shifts that were necessary.
4. Subtract the first n digits of the DS register from the first n digits of the DD register. If the result is negative, set the nth digit of the Q register to "0" and disregard the subtraction.

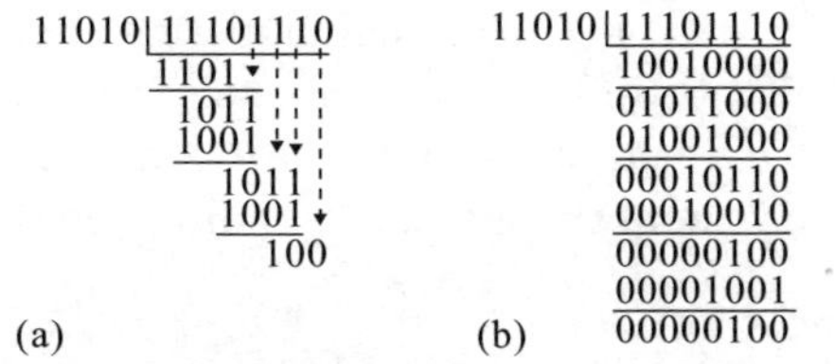

```
11010|11101110
      1101
       1011
       1001
         1011
         1001
          100
(a)
```

```
11010|11101110
      10010000
      01011000
      01001000
      00010110
      00010010
      00000100
      00001001
      00000100
(b)
```

FIGURE C.1 ■ (a) Hand process for long division. (b) Equivalent long division on a microprocessor using 8-bit integers.

5. If the subtraction in (4) results in a positive value, place that in the R register and set the nth digit in the Q register to "1".
6. Shift the DR register one position to the right and increment n ($n = n + 1$).
7. Set the nth digit in the R register equal to the nth digit in the DD register.
8. Subtract the DR register from the R register. If the value is negative, set the nth bit in the Q register to 0. Disregard the subtraction. If $n = 8$, go to (10). Otherwise go to (6).
9. If R $-$DR > 0, set the nth bit in the Q register to 1. If $n = 8$, go to (10). Otherwise go to (6).
10. The quotient is in the Q register and the remainder is in the R register.

This is a long algorithm that can be implemented in software, but it is an inefficient way of performing division. It does exemplify, however, the fact that division can be performed using only shifts and additions, as indicated previously.

C.3 | FIXED POINT ARITHMETIC

In some instances one must use fractional numbers. For example, one may need to find a ratio of two integers or scale an input voltage to a microprocessor by a noninteger value or add a fixed, offset value to a result. In microprocessors this is usually done using fixed point arithmetic, as it requires fewer resources than floating point arithmetic using established routines that are usually available from manufacturers, but from other sources as well. At times it may be useful for the user to write his/her own routines for fixed point computation. Floating point routines are also available.

The basic idea of fixed point arithmetic is to treat the integer and fraction parts of a single integer with the radix point placed implicitly at a fixed location within the integer. By so doing, one has the ability to represent fractional numbers while still operating only on integers. Consider the 8-bit unsigned number shown below with the weights associated with each digit:

1	1	0	0	1	1	0	1
2^7	2^6	2^5	2^4	2^3	2^2	2^1	2^0

The digital number represents the decimal value 205. We could just as easily say that it represents the value 205.0, but by doing so we placed a decimal point (in general we will call this a radix point) at the end of the number. Suppose now we move the decimal point one position to the left. We obtain the value 20.5. In decimal format this can be written as $2 \times 10^1 + 0 \times 10^0 + 5 \times 10^{-1}$. In effect the number has been divided by 10, or more appropriately, scaled by 10^{-1}. Similarly, if we place a radix point, say, between the fourth and fifth digits in the binary number 11001101 we obtain 1100.1101. Following the example for the decimal number, the representation becomes

1	1	0	0		1	1	0	1
2^3	2^2	2^1	2^0	•	2^{-1}	2^{-2}	2^{-3}	2^{-4}

The digital number represents the value 12.8125 and is in fact $205/2^4 = 205 \times 2^{-4} = 12.8125$. For this reason, fixed point numbers are often called scaled integers. Note that the digital number itself has not changed but the weights of the digits have been scaled by 10^{-4}.

The problem with this representation is that the integer can only vary between 0000 and 1111 (or 0 and 15), whereas the fraction can vary between 0000 and 1111 (or 0 and 0.9375 with a resolution of 1/16). If we wish to represent signed numbers as fixed point numbers, the integer in the example above is reduced by 1 bit and will only be able to represent numbers between 0 and 7 (000 and 111). Clearly for the method to be useful, a larger number of bits are needed. In an 8-bit microprocessor, the natural choice is to use 8 bits for the integer and 8 bits for the fraction, allowing unsigned numbers between 0 and 255.99609575 and signed numbers from −127.99609575 and 127.99609575. Of course, one can use, say, 12 bits for the fraction and 4 bits for the integer or any combination that fits the application at hand. However, most routines supplied by manufacturers of microprocessors use 8 (or 16) bits for each part. Any other combination might need to be written by the user, usually as a subroutine to be called when needed.

All operations on fixed point numbers are done as with any integer numbers, with the obvious allowance for the radix point. Addition and subtraction is identical to that of signed or unsigned integers. For example, the sum of the two unsigned fixed point numbers A = 11010010.11010101 (210.83203125) and B = 01000101.00111101 (69.23828125) is

$$\begin{array}{r} 1101001011010101+ \\ \underline{0100010100111101} \\ (1)0001100000010010 \end{array}.$$

The result is 00011000.00010010 with a carry 1. The result is 24.0703125 and a carry equivalent to 256. The decimal result is 280.073125 = 24.0703125 + 256. Obviously the result requires nine digits for the integer rather than eight, but the summation is correct. The main point to notice is that the two numbers are treated as any two integers and that the radix point does not affect the operations performed to obtain the result. Signed fixed point numbers are handled the same as signed integers.

Multiplication and division of fixed point numbers also follows the same process as for integers, but the radix point must be adjusted. For example, multiplication of two 16-bit fixed point numbers, each with 8 bits for the integer and 8 bits for the fraction, results in a 16-bit integer and a 16-bit fraction that must be truncated to a total of 16 bits. Assuming the integer can fit in 8 bits, there will be no loss in accuracy or one would have to use a larger number of bits. As an example, suppose the product of

A = 00000100.11101110 (4.9296875) and B = 00010010.00111010 (18.2265625) is required. The multiplication follows the same rules as decimal multiplication:

$$b = 00010010 \cdot 00111010 = 18.2265625.$$

```
                0000010011101110 ×
                0001001000111010
                0000000000000000+
               0000010011101110
               00000100111011100+
             0000010011101110
             0000011000101001100+
            0000010011101110
            00001000000000101100+
           0000010011101110
        000001000111011111101100+
       0000010011101110
      00000101011111100111101100+
    0000010011101110
00000000010110011101100111101100
|    16 bit integer |  16 bit fraction  |
00000000 01011001 11011001 11101100
         | 8 bit  | 8 bit  |
         integer   fraction
```

The result is

$$c = 0000000001011001 \cdot 1101100111101100.$$

The multiplication results in 16 bits for the fraction and 16 bits for the integer. The 8 LSB must be removed from the fraction and the 8 MSB removed from the integer since only 8 bits are available for each representation even though the internal multiplication is handled with 16 bits for each part. Therefore the result is

$$c = 01011001 \cdot 11011001 = 89.84765625.$$

Because of the truncation of the fraction this is not exact: the correct result should be 89.8512573242 (the error is about 0.8%). Note that truncation of the fraction results in a loss of accuracy but the truncation of the integer does not, as long as the product can fit in 8 bits. If it does not then the result is wrong, since the MSB will be truncated.

Index

T